THE SOCIETY OF NAVAL ARCHITECTS AND MARINE ENGINEERS

TRANSACTIONS

Volume 111
2003

Published in 2004 by

The Society of Naval Architects and Marine Engineers
601 Pavonia Avenue, Jersey City, New Jersey 07306

The volumes of *Transactions* contain professional information of permanent value. Information on the various functions and activities of the Society and a list of members may be found in the 2003-2004 SNAME Membership Directory and Information Book.

ISSN 0081 1661

ISBN 0-939773-44-9 (Hard Cover)
0-939773-45-7 (Soft Cover)
0-939773-45-5 (CD)

Table of Contents

2003 Annual Meeting Activities

Reports

Refereed Papers

The following papers were selected for presentation at the 2003 Annual Meeting by the Society's Papers Committee

Commemoratives

2003 MEMBERS OF COUNCIL

President: Bruce S. Rosenblatt
Treasurer: David H. Klinges

Executive Director: Philip B. Kimball
Assistant Treasurer: Alfred A. Bozzuffi

Robert Axelrod- *Honorary Vice President*
Robert F. Beck- *Chair, Journal of Ship Research Committee*
Suzanne M. Beckstoffer- *Section Chair, Hampton Roads*
Donald M. Birney- *Honorary V. P.; Chair, Pension Committee*
Alfred A. Bozzuffi- *Assistant Treasurer*
Eric G. Bramwell- *Section Chair, Eastern Canadian*
Edward J. Campbell- *Past President*
Carlson S. Chen- *Section Chair, San Diego*
Connie J. Carruthers- *Section Chair, Arctic*
Edward W. H. Clendenning- *Honorary Vice President*
Timothy Colton- *Chair, Advisory Public Service Committee*
Edward N. Comstock- *Vice President (Mid Atlantic 2004)*
Donald P. Courtsal - *Honorary Vice President*
William H. Cowardin- *Vice President, Mid Atlantic*
Joseph J. Cuneo- *Past President*
Charles R. Cushing- *Chair, Fellows Committee*
John C. Daidola- *Chair, Technical & Research Steering Committee*
John M. Dempsey, Jr.- *Chair, Applications Committee*
Richard B. Ervine- *Section Chair, Texas*
Jose Femenia- *Past President*
C. L. French- *Past President*
Robert B. Geary- *Vice President, Northeast; Chair, Papers Committee*
Richard A. Gilmore- *Chair, Investments Committee*
Mahmoud M. R. Haddara-*Section Chair, Canadian Atlantic*
A. Dudley Haff- *Honorary Vice President*
Arthur J. Haskell- *Past President*
John H. Higginbotham- *Chair, Annual Meeting Committee*
Michael C. Hill- *Section Chair, Arctic*
Richard B. Houseman- *Section Chair, Eastern Canadian*
Kenneth F. Humphreys- *Section Chair, Gulf*
Robert P. Hurrell- *Section Chair, Great Lakes/Great Rivers*
Bruce Johnson- *Chair, Electronic Media Committee*
Philip B. Kimball- *Executive Director, SNAME*
Ronald K. Kiss- *Past President; Chair, Awards Committee*
David H. Klinges- *Honorary Vice President; SNAME Treasurer*
Barbara L. Lamb- *Vice President (Pacific 2005)*
Thomas Lamb- *Vice President (Central 2005)*
Kenneth R. Lane- *Vice President (Pacific 2004); Chair, Sections Committee*
Edward M. Lewandowski- *Chair, Marine Technology Committee; Section Chair, Chesapeake*
Walter M. Maclean- *Chair, Professional Engineering Licensure Committee*
Monroe D. Macpherson- *Honorary Vice President*
Michael T. Martin -*Vice President (Northeast 2005)*
Richard M. McCabe- *Section Chair, Los Angeles Metropolitan*
Scott C. McClure- *Chair, Scholarships Committee*
Robert G. Mende- *Honorary Vice President*
Kent Merrill- *Section Chair, New York Metropolitan*
Herron W. Miller, III- *Section Chair, Philadelphia*
Richards T. Miller- *Honorary Vice President*
Michael T. Monahan- *Section Chair, Southeast*
William B. Morgan- *Honorary Vice President; Chair, Publications Committee*
Richard A. Mueller- *Vice President (Central 2004)*
Christopher E. Mulder- *Section Chair, Pacific Northwest*
Perry W. Nelson- *Past President; Chair, Audit Committee*
Peter G. Noble- *Chair, Planning Committee*
Henry A. Olson- *Vice President, Pacific*
David A. O'Neil- *Past President*
Ellsworth L. Peterson- *Honorary Vice President*
W. R. Phillips, Jr.- *Past President*
Donald E. Ridley-*Honorary Vice President*
Paul J. Roden- *Chair, Membership Committee*
Bruce S. Rosenblatt-*President*
Lester Rosenblatt- *Past President*
Susan S. Salancy- *Section Chair, Northern California*
William J. Sembler- *Vice President (Northeast 2004)*
James D. Sharrow- *Vice President Central*
W. M. Simpson, Jr.- *Section Chair, New England*
Robert C. Staiman- *Vice President (Mid-Atlantic 2005)*
Arnold M. Stein- Chair, Member Benefits Committee
Richard L. Storch- *Chair, Journal of Ship Production Committee*
Martin Toyen- *Chair, Finance Committee*

2003 EXECUTIVE COMMITTEE

Tim Colton
Edward N. Comstock
William H. Cowardin
Joseph J. Cuneo
John C. Daidola
Jose Femenia
C. L. French
Edward T. Gaines
Robert B. Geary
Richard Gilmore
John H. Higginbotham
Bruce Johnson
Timothy J. Keyser
Philip B. Kimball
David H. Klinges
Barbara L. Lamb
Thomas Lamb
Kenneth R. Lane
Michael Martin
William B. Morgan
Richard A. Mueller
Perry W. Nelson
Peter G. Noble
Henry A. Olson
Paul J. Roden
Lester Rosenblatt
William J. Sembler
James D. Sharrow
Robert C. Staiman
Martin Toyen

Section Officers

September 1, 2002 to August 31, 2003

ARCTIC
Chair: Chris Hill
Secretary/Treasurer: Colin Revill

CANADIAN ATLANTIC
Chair: Mahmoud Haddara
Vice Chair: Dusty Miller (Nova Scotia)
Vice Chair: Noel Murphy (New Foundland)
Secretary: Dan Walker
Treasurer: Michael Doucet

CHESAPEAKE
Chair: Edward Lewandowski
Vice Chair: Paul Clarke
Secretary: Vincent Skwarek
Treasurer: Jonathan Applequist

EASTERN CANADIAN
Chair: David Morris
Vice Chair: Stephanie White
Secretary: Pierre Demers
Treasurer: Yves Perron

GREAT LAKES/GREAT RIVERS
Chair: Robert Hurrell
Vice Chair: Michael Hawkins (Gt. Rivers)
Vice Chair: Dale Karr (Gt. Lakes)
Secretary/Treasurer: Albert W. Horsmon, Jr.

GULF
Chair: Kenneth Humphreys
Vice Chair: Philip Nuss
Secretary: Kenneth Maloney
Treasurer: Erica Carter

HAMPTON ROADS
Chair: Suzanne Beckstoffer
Vice Chair: Rich Vlaun
Secretary/Treasurer: John Gallahorn

LOS ANGELES MET
Chair: Richard McCabe
Vice Chair: Samuel T. Kelly
Secretary/Treasurer: Richard McCabe

NEW ENGLAND
Chair: William M. Simpson
Vice Chair: Gregory Beers
Secretary/Treasurer: Gerrit J. van Dissel

NEW YORK MET
Chair: Kent Merrill
Vice Chair: Stephen P. Keller
Secretary: Michael Fleahman
Treasurer: Yvonne Traynham

NORTHERN CALIFORNIA
Chair: Susan Salancy
Vice Chair: Don Seaman
Secretary/Treasurer: Paul Snyder

PACIFIC NORTHWEST
Chair: Christopher Mulder
Vice Chairs:
Ben Thompson(British Columbia)
Terrence Dai (Puget Sound)
Godfrey Lee (Columbia River)
Secretary/Treasurers:
Chris Kozikowski (British Columbia)
Elizabeth Reynolds (Puget Sound)

PHILADELPHIA
Chair: Herron W. Miller
Vice Chair: James A. Sanial
Secretary/Treasurer: David Moorhead

SAN DIEGO
Chair: Carlson Chen
Vice Chair: Michelle Pernice
Secretary: Lott Souriyavong
Treasurer: Rich Burns

SOUTHEAST
Chair: Michael Monahan
Vice Chair:
Michael Rosecrans (North)
Drew Hains (South)
Thomas Hagner (Central)
Secretary/Treasurer: Randall R. Fiebrandt

TEXAS
Chair: Richard Ervine
Vice Chair: Paul Griffin
Secretary/Treasurer: Rhonda Fields

SNAME Administrative Staff

Philip B. Kimball, Executive Director

Associate Executive Director (Finance and Administration)	Allan Gray
Information Systems Manager	Joseph Caggiano
Web Site Manager	Connie Cali-Poutre
Events/Office Administrator	Brenda Elmore
Publications Manager	Susan Evans
Member Services Department, Membership	Tommie-Anne Faix
Administrative Asst. (Scholarships, T&R support, PE exam)	Erlinda Faustino
Director of Marketing	Susan Giver
Annual Meeting Manager	Carolyn Marus
Accounts Payable/ Receivable	Anne Mikaelian
Membership/Publication Coordinator	Rich Mouk
Mailroom Coordinator	Raymond Rivera
Member Services Department Manager	Eileen Romanelli
Technical Director	Alan Rowen
Sections, Annual Meeting Papers, T&R Manager	Deborah Sirotiak
Assistant to the Executive Director	Barbara Trentham

2003 Annual Meeting

The 2003 World Maritime Technology Conference was convieved by the Internationalization Committee. The idea was to assemble as many of the world's marine societies as possible under one roof for a three–day conference that would be held every three years, and be hosted on a rotating basis by a different marine society each time.

An initial call for papers was overwhelming with nearly 150 abstracts submitted from SNAME members, the ship production sector and international societies. The major topical areas of "Ship Operations," Design and Technology," Ship Production," and "The Role of Marine Technical Societies in Meeting Industry Challenges in the New Global Economy" initially envisioned 72 time slots, but were later expanded to 108 to accommodate the overwhelming response.

SNAME-related activities including a beginning reception, the closing "On to DC" Reception," Standing and Technical Committee meetings, the Annual Meeting courses, the Council Meeting, the President's Luncheon and Business Session as well as the President's Reception and Banquet were held on Friday and Sunday nights at the elegant and charming Westin St. Francis Hotel, leaving Saturday through Monday for the "World Maritime Technology Conference" at the Moscone Convention Center.

Two major Society awards were presented at the President's Luncheon on Friday afternoon: The Blakely Smith Medal for "outstanding accomplishment in ocean engineering" to Michael M. Bernitsas, and The William H. Webb Medal for "outstanding contributions to education in naval architecture, marine or ocean engineering" to Michael G. Parsons. Other presentations to members included 50-Year Golden Member Certificates, Certificates of Appreciation, the ABS Capt. Joseph H. Linnard Prize, the Vice Admiral E. L. Cochrane Award, the William M. Kennedy Award and the Elmer L. Hann Award.

Later, the student winners of the Lisnyk Ship Design Competition from Virginia Polytechnic Institute, and the two, second place teams from University of Newcastle upon Tyne and Norwegian University of Science and Technology, respectively, received their prizes. In its second year, the first place award in the International Student Offshore Design Competition was made to the University of Michigan team, with second and third places going to teams from Texas A&M University. The luncheon was capped by president Rosenblatt's annual address, which reaffirmed his commitment to continue to strengthen the society's younger member and Section base and international ties while improving overall "value" for all.

The membership, in its Business Meeting, was called upon this year to approve a new governance model, and a new president-elect. On January 1, 2004, the Society will be operating under the new governance rules with Roger Compton starting his year as president-elect before assuming his two-year term as president on January 1, 2005. That evening we dined aboard the "San Francisco Belle" in a cruise around the Bay, under clear moonlit star-filled skies, where 600 guests including our international attendees, were able to meet and socialize with old friends and colleagues.

On Saturday, the World Maritime Technology Conference opened with a flourish, with a traditional Chinese Lion Dance, paying tribute to the large Asian population in San Francisco. The International Luncheon at the Moscone Convention Center was the ideal setting for the presentation of The Elmer A. Sperry Award for 2002 for advancing the art of transportation by past president Perry Nelson to Raymond Pearlson for his invention, development and worldwide implementation of the "Syncrolift," a system for lifting ships out of the water for repair and for launching new ship construction. It was also a fitting occasion for president Rosenblatt to recognize the 50th Anniversary of the watershed paper, "On the Motions of Ships in Confused Seas," by Manley St. Denis and Willard J. Pierson (unrelated to this reaffirmation, sadly both authors passed away in 2003).

Directly following the luncheon was William O'Neil's Keynote Address in which he stressed the importance of the societies gathered at the first WMTC focusing their attention on collaborative solutions to common problems within the global transportation system.

Two events specific to the WMTC were the Intersociety Roundtable and the Principal Sponsors Meeting, each lead by Keith Read, Director General of IMarEST. Both were encouraging because for the first time the international societies were acting with one voice. The idea of an International Federation of marine societies was introduced by president Rosenblatt and endorsed by other society representatives as a future goal. The marine societies all agreed that IMarEST should host the WMTC in March 2006, reportedly to be held in London's Queen Elizabeth Convention Center.

The Westin Hotel was the setting for Sunday evening's International Banquet, at which the Vice Admiral "Jerry" Land Medal for outstanding accomplishment in the marine field was awarded to Richard H. Vortmann, and the David W. Taylor Medal for notable achievement in naval architecture and/or marine engineering was awarded to Robert G. Keane, Jr. Our banquet speaker, C. B. Choo, CEO of Keppel Fels, reminded us with examples that as professionals it is our obligation to take the necessary steps today to promote advancements in the industry that continue to make the use of the sea by mankind sustainable for future generations.

Last year the Executive Committee agreed to Headquarters' plan to partner with The Maritime Group, publishers of Maritime Reporter & Engineering News, Marine News and the Maritime Security Sourcebook, for the two-fold purpose of seeking their help in meeting our goal of 175 booths, more than ever attempted before, and of providing a quantum increase in publicity and promotion of the event.

The Maritime Group's sales and marketing team came through big time by selling out the Expo on July 8, and by saturating the media with notices of the event. In the end we had nearly 1300 registrants and visitors attend the exposition and a crowd of very happy exhibitors because of the additional traffic over past years, resulting in a record number of 98 who signed on the spot for next year

This year we created a separate track for student papers and presentations of the winning teams for the Lisnyk and Offshore Design competitions. Additionally, we held the ever-popular Student Congress, a job fair and a student social event, and offered an opportunity for low cost housing aboard the S/S GOLDEN BEAR, California Maritime's Training Ship. An impressive 127 students from nine schools showed up for the fun, and also elected among their numbers, Steve Olson, Chair, and Matt Unger, Vice-Chair, of the Student Steering Committee

A closing "On to D.C." reception was held at the Moscone Center where all looked forward to a very successful 2004 meeting.

EXECUTIVE DIRECTOR'S REPORT TO COUNCIL, 2003

I am pleased to submit the following report for the fiscal year ending August 31, 2003:

MEMBERSHIP

A comparison of the membership of the Society as of August 31, 2003 and August 31, 2002 is as follows:

	FY 03	FY 02
Members (under age 32)	34	17
Members (over age 32)	3,917	4,077
Affiliates (under age 32)	3	3
Affiliates (aged 32 and older)	77	85
Associate Members (under age 32)	645	503
Associate Members (aged 32 and older)	2,975	3,278
Reciprocal Members	5	5
Student Members	906	1,061
Honorary Members	10	11
Honorary Vice Presidents	4	4
Honorary Fellows	23	25
Fellows	171	157
Total	8,770	9,226

The Society's membership decreased by 5% for fiscal year 2003, however, this is comparable to similar society membership trends.

Membership Retention

Membership retention for fiscal year 2003 reached 87%. This, too, is comparable to similar society retention trends. Of the 8,223 members (number of dues bill sent out, not including students) for fiscal year 2003, 1,094 did not renew their SNAME membership.

Recruitment

There were no funds allocated for a membership recruitment drive during fiscal year 2003.

Applications

The Applications Committee, chaired by John M. Dempsey, Jr., reviewed 356 applications for new membership/upgrade requests during fiscal year 2003. By comparison, applications reviewed/approved in fiscal year 2002 and 2001 were 996 and 1,105, respectively.

The electronic Membership application system is up and running effectively. Marketing Director, Susan Giver, and the Member Services Dept. Manager, Eileen Romanelli, collaborated with Jack M. Dempsey, Jr. in the development of a new leave-behind 'Invitation to SNAME Membership' brochure. This brochure was completed for use at the World Maritime Technology Conference and other events in the fall of 2003.

Susan Giver developed a 'New Member Kit' to be sent to new SNAME members. It is an attractive blue folder with the SNAME emblem and the Society's name imprinted in gold lettering on the front. It presents an image befitting the Society for our new members.

Member Benefits

We were successful in creating and placing a 'Member Benefits' button on the SNAME Web Site home page. This new feature makes is easy for members (and potential members), to view, learn and subscribe to the many benefits of SNAME Membership.

Currently we have the following benefits available to members at a reduced rate: Life and Health Insurance, Long-Term Care insurance, Car and Equipment Leasing, UPS discounts, Avis Car rental and the personalized SNAME MBNA credit card. We are currently working on obtaining discount rates for home and car insurance, plus Dell PC's/accessories for our members

NECROLOGY

The deaths of the following members were reported during fiscal year 2003:

Theodore W. Anderson
Kazuo Hirata
Gunnar C.F. Asker
George O. Hockley, Sr.
Robert C. Ballinger
Arthur Louis Holdeman
Freddryck M. Barfuet
Thomas M. Hopkins
Edgar H. Batcheller
Paul Kaplan
Lawrence J. Bates
Martin C. Kelsey, Jr.
Peter B. Bell
Joseph Francis Kemp
Jack E. Blakeman
Robert Norton Kershaw
Kai H. Boggild
Archibald H. McComb, Jr.
Edward P. Boyle
Joseph A. Melanson
Poul A. Christensen
K. J. Melkvik
Paul S. Crandall
R. Carter Morrell
Herbert Kirschler Daman
Walter Gray Neal, Jr.
Amelio D'Archangelo
Hans A. Ording
Conway D. Davis

Willard James Pierson, Jr.
John G. Davis
Richard H. Roberts
James H. Dickey
Charles S. Robinson
Joseph A. Drelicharz
Lester Rosenblatt
Mathew J. Fairbairn
Robert S. Sanford
William C. Gibson
Rudolph Schoen, III
Jack S. Gilhart
Harvard P. Stewart
William Louis Giordan
Arthur H. Sulzer
Samuel M. Hickson
Carlton E. Tripp
Chester G. Higgins
Vincent R. Van Riper

SNAME/ASNE JOINT STUDENT MEMBERSHIP PROGRAM

The Joint Student Membership Program expired in May of 2003, but continued to remain in force until the two Executive Directors were able to meet in August to discuss the future of the program. It was agreed that the program could continue, with SNAME handling all administrative functions, and both income and expenses shared in proportion to the number of student memberships in each society vs. expenses shared equally as in the first three years of the agreement.

FINANCES

The fiscal year 2003 operating fund financials show that total revenue at the conclusion of the year was about 6.6% below budget. Lower than expected results in sales, educational courses, and advertising income contributed to the shortfall. Actual expenses were less than budget by about 7.9%. Lower than forecasted expenses in publications, Headquarters, Sections, and functions resulted in the savings. Overall, the FY 2003 Operating Fund finished with a deficit of $1,659.

The budgetary process is monitored by the Finance Committee, chaired by Martin Toyen. A break even budget is predicted for Fiscal Year 2004.

A statement of the Society's assets, liabilities, and fund balances are contained in the Report of the Treasurer.

The Audit Committee report for Fiscal Year 2003 will be submitted to Council by January 31, 2004, and the independent Auditor's full detailed report is available for inspection at Society Headquarters.

Finally, the performance of the Society's investments continues to be carefully monitored by the Investments Committee, chaired by Richard Gilmore. The value of the portfolio increased from $7,030,720 on August 31, 2002 to $7,369,852 on August 31, 2003.

SECTIONS

In FY 2003 there were a total of 18 active Sections; 14 in the US, three in Canada, and the Society's newest, the international Section in Greece, which started last year.

President-elect Rosenblatt, representing Headquarters, attended the 50th Anniversary celebration of the Eastern Canadian Section in Ottawa on November 26. Mr. Rosenblatt presented certificates of recognition to the 12 past chairmen, and presented a 50th anniversary rosette for the Section's banner to David Morris, Section Chair.

The 19th Chair-Elect Seminar convened May 30-June 1, 2003, at Saddlebrook Resort in Florida. The two-day seminar was lead by Sections Chairman, Ken Lane. The seminar began on Friday evening at dinner with each Section Chair/Vice Chair describing his/her Section goals for the upcoming year.

Ten chairs-elect and four vice chairs-elect attended the meeting. They were joined by SNAME president, Bruce Rosenblatt; Sections Chair, Ken Lane; Past Sections Chair, Keith Michel; the Executive Director, Phil Kimball; Sections Manager, Deborah Sirotiak; and Member Services Department Manager, Eileen Romanelli.

This year's meeting was of great value to the incoming officers. There was a great deal of discussion on how to improve their individual Sections, and time was allotted for meaningful discussion on how to improve member involvement. Many of the Sections discussed ideas for retaining the younger members of the Society.

The meeting adjourned on Sunday and all agreed that both the format and discussions were beneficial and significant.

PUBLICATIONS

The Publications Committee, chaired by Dr. William Morgan, met at the annual meeting in Boston. The progress of the Ship Design and Construction Revision, edited by Tom Lamb, revision of Principles of Naval Architecture, edited by Randy Paulling, and the revision of Ship Structural Design, authored by Owen Hughes were discussed. It was announced that long time technical editor, Terry Flanagan, was retiring from SNAME. He was thanked for his years of service to the Society. Difficulty in motivating authors, electronic peer review and submission, journal and book marketing and journal backlogs were also discussed.

In June the Committee Chairmanship was passed to Rob Tagg. Dr. Morgan had been chair for eight years and we are all grateful for his dedication. In January Karen Hansen took over the position of Copyeditor from Terry Flanagan. Terry was very helpful in aiding a smooth transition, and Karen is doing a great job. Other projects include:

(a) Membership Directory and Information Book: The 2003-2004 Membership Directory and Information Book was mailed in June 2003 to all paid members of the Society.

(b) Transactions: The 2002 Transactions were mailed in May. Options for hardcover, soft cover and CD ROM continue to be available. Volume 111 will only include SNAME papers, discussions and closures rather than all the papers presented at the World Maritime Technology Conference.

(c) Journals: Online peer review and submission services are being tested to expedite author queries and prompt delinquent reviewers. New author instructions for the web site are also in process. Automated Graphics Systems is now typesetting the Journal of Ship Research. The first issue was produced in September 2003.

(d) Reference Publications: The book TITANIC Ships, TITANIC Disasters: An Analysis of Early Cunard and White Star Superliners was published in time for the Boston. Susan Giver, Director of Marketing, continues to promote this book in various locations.

The revision of Ship Design and Construction , Volume One, was completed and the second Volume should be finished by early 2004.

2002 ANNUAL MEETING

The 110th Annual Meeting was held at the Westin Copley Place Hotel in Boston, Massachusetts. By virtue of a cooperative agreement between SNAME and the National Shipbuilding Research Program (NSRP), a significant increase in attendees was realized due to the overlapping of the Annual Meeting and the Ship Production Symposium. It was a very successful meeting and we look forward to many years of this mutually beneficial effort.

The expanded technical program was composed of 34 sessions of topical and innovative papers, panels, and sessions; as well as a Forum on Maritime Security; courses on "Naval Architecture and Marine Engineering" and "Early Stage Ship Design"; and the Opening SPS Welcome by speaker Michael W. Toner, President of Electric Boat Corporation, sharing his views of today's shipbuilding and repair industry and the collaborative research to advance the industry. Choosing which session to attend presented attendees with a challenge not faced at recent meetings, resulting in one of the best-attended Annual Meetings in years.

The International Maritime Exposition was also contained within the Westin Copley Place Hotel. Although restricted to fewer booths, we were sold out and the exhibitors were pleased with the quality of the visitors, citing more shipyard people in attendance.

Ever mindful of the importance of our student members, this year's Student Program was expanded to include a Student Job Fair, SPS Panel and Student Exchange, a Student Reception at MIT, the Student Congress and a Student Steering Committee Planning Session. On Friday, Professor Boutwell, Chairman of the Student Congress, opened the session with the announcement of the first place winners of the Dr. James A. Lisnyk Design Competition from the Norwegian University of Science and Technology, and the team winners of the First International Student Offshore Design Competition from the Naval and Ocean Engineering Department at the Federal University of Rio de Janeiro.

Our special events program began with the traditional President's Reception. This year it was held at the Museum of Science in Boston where all enjoyed the opportunity to meet old and new friends, sample Boston's cuisine and wander through science and technology exhibits. Student registrants were also able to attend the reception by the generous support of seven marine design firms.

Admiral Robert E. Kramek, USCG (Ret), President of ABS Americas, presented a very thought-provoking Keynote Address on Thursday stressing the need for government support and the important role naval architects play in sustaining a world class maritime industry.

A new event, the Ship Production Symposium Luncheon, took place on Wednesday during this year's program. Distinguished speaker, Captain William G. Schubert, U. S. Maritime Administrator, informed attendees of the ground being gained by the U.S. shipbuilding and repair industry and MarAd's efforts central to that progress. The William M. Kennedy Award for 2002 was presented to Frederick J. Harris during this Luncheon.

President Cuneo presided over a very well attended Annual Banquet on Thursday night. Three prestigious medals were awarded including the Davidson Medal to Nils Salvesen, the Vice Admiral "Jerry" Land Medal to William O. Gray, and the David W. Taylor Medal to R. Keith Michel. Our Banquet speaker, Keith Read, Director General of the Institute of Marine Engineering, Science and Technology (IMarEST), captivated the audience with his humor and yet poignant message regarding the importance of cooperative efforts between mutual member organizations throughout the world.

The President's Luncheon and Business Session, held on Friday, was the forum for President Joseph J. Cuneo's final annual address containing thought-provoking comments necessary for innovation and creativity in shipbuilding.

Many awards were presented at the Friday Luncheon including 24 Fifty-Year Golden member certificates; the ABS Captain Joseph H. Linnard Prize; two distinguished Service Awards; a Certificate of Appreciation; Student Paper Awards including a graduate paper honor prize, a graduate paper award, three undergraduate paper honor prizes, and an undergraduate paper award; a T&R Certificate of Recognition; an Honorary Member certificate; the Vice Admiral E. L. Cochrane Award; and the Dr. James A. Lisnyk Ship Design Competition Awards. The Business Session immediately followed the Luncheon where Amendments to the Bylaws were voted upon and a Report on Elections presented.

Alumni reunions continue to be a popular activity associated with the Annual Meeting. Friday evening was set aside in the program for their individual arrangements.

A new special event was included in our Friday program, bringing the meeting to a close, entitled the "On to San Francisco" Reception. The purpose of the reception was to promote attendance at the first-ever World Maritime Technology Conference and Exposition to be held in San Francisco, California in 2003. A major undertaking for SNAME, originator and host of the Conference, it will assemble 12 international marine societies to provide a forum for discussion and exchange on issues related to the future of the global maritime industry. The Society's annual meeting business requirements will be included at the start of the Conference.

Comparative statistical data of note for the Annual Meeting:

	2000	2001	2002
Total Registration	496	438	605
Banquet Attendance	289	310	345
Closing Reception (new event in '01)		75	175
Dinner Party	125	(No longer held)	
President's Luncheon	427	300	150
President's Reception	550	440	525

2002 International Maritime Exposition (IMExpo)

The IMExpo in Boston was a success. Booths were sold out at 120, with a breakdown of seven bartered booths, two SNAME booths, and 111 paid booths, which generated revenue of $201,779. The space in the Ballroom and Foyer was limited; consequently, 120 booths was the maximum number that could be accommodated.

Attendance at the Exposition was approximately 700 over the two-day period, with Thursday being somewhat more active than Friday. Again, the exhibitors commented on the lack of dedicated Expo time. In fact, three major companies, Caterpillar, Cummins, and Detroit Diesel, insisted on meeting with us to express their concerns about the future of the show. They were looking for more dedicated expo time, better marketing, more concern for exhibitors' needs, and food and beverage in the Expo Hall. They need more attendance to justify continuing to exhibit with SNAME. Following the meeting the concerns were passed on to the WMTC Planning Committee, and the exhibitors' needs have been incorporated into 2003's WMTC schedule.

MARKETING

Exposure for SNAME at various trade shows has continued. Last year, SNAME exhibited at the 6th International Ship Stability Workshop, IBEX, CMA, ASNE Day, OTC, MEETS and the International WorkBoat show. We promoted our publications, conducted membership recruitment, and provided SNAME members and the general public with an increased awareness of the Society, and the many benefits and services available to members and potential members.

OUTREACH

The Office of Naval Research has agreed to fund a national effort to attract young people to the marine industry, and SNAME prepared several proposals to become involved with, and to participate in, this endeavor. The projects will involve reaching out to students at the middle and high school levels with a video and brochures and flyers all produced at Headquarters. Outreach is also continuing through the Society's development of and/or support of various design competitions for college high school and middle school students. The Student Congress at the Annual Meeting is now accompanied by the Dr. James A. Lisnyk Ship Design Competition (LSDC), the International Student Offshore Design Competition (ISODC), and a proposed design competition similar to the Lisnyk, but for naval vessels. Competitions for high school and middle school students include the Nuclear Oceanographic Research Submarine Exploration Program (NORSE) and the High School and Middle School Boat Design and Construction Competition.

TECHNICAL & RESEARCH

Much of the T&R work at HQ this year has been devoted to supporting the Technical Program for WMTC. Nevertheless, two new T&R bulletins and one updated bulletin were released during the past year, and a third may be ready by year-end. Two completed drafts and another that is almost finished are being held back for further work. A CD compendium of over 300 papers on small craft will be released in time for WMTC and the small-craft conferences. In a cooperative effort with Webb Institute, an electronic version of the Proceedings of the 6th International Ship Stability Workshop was released. Twenty-one funded T&R projects are currently in progress. By the end of the calendar year seven projects will have completed successfully.

T&R correspondence is almost totally electronic, and becoming more efficient as more use is made of the Intranet. Use of the web pages provided to the T&R entity chairmen continues to grow, but some of the chairmen have yet to seize the opportunity offered. Reporting and record keeping is much improved, but room remains for further improvement. All T&R entities are now represented on the T&R Steering Committee. Efforts to integrate Section technical interests with the T&R program continue. All Sections now receive announcements and minutes of T&R Steering Committee meetings and the open invitation to participate in meetings is being accepted more often. Missing Section papers are being identified and sought, and a number have already been delivered to HQ to be cataloged and archived.

A new History Committee has authorized by the Executive Committee. The History Committee will provide assistance and guidance for HQ staff and for other SNAME entities on historical matters, will assume the burden of responding to historical inquiries from members and from the general public, will assume responsibilities for caring for the Society's historical artifacts and for writing obituaries, and will improve liaison with marine historical societies, outreach programs, and museums. It is expected that the committee will have an appeal for existing members and may lead to an influx of new members.

The reorganization of the library is perhaps 90% complete. All of the collection has been sorted, cataloged, weeded of duplicates, and reshelved. Numerous volumes, damaged through use, have been repaired. Catalogs are being maintained as new materials are added. Efforts are turning to identifying and replacing missing items including conference proceedings. Restoration of regular book reviews in Marine Technology has enabled an increasing number of new books to be acquired for the library at no cost.

SNAME participated in ICMES 2003, the Ninth International Conference on Marine Engineering Systems, that took place in Helsinki in May. Forty-six papers were presented. This was the second of these triennial conferences in which SNAME has participated through the T&R Program and the NY Metropolitan Section, and in cooperation with IMarEST. The Section provided the core of the Papers Committee, and SNAME members served as vice chairmen of each of the three Technical Committees. The ICMES precedent establishes a model for SNAME participation in the 2006 International Marine Design Conference (IMDC), planned for Ann Arbor, Michigan. SNAME was represented in May at ECMAR in the Netherlands and at IMDC in Greece, sponsored, helped organize, and participated in the Chesapeake Sailing Yacht Symposium in March and MEETS 2003 in August, was wholly responsible for the Propeller/Shafting Symposium in September in Virginia Beach, sponsored and participated in the International Ship Stability Conference in Madrid, Spain in September, and is a sponsor for FAST 2003 in Naples, Italy.

The Events Calendar is updated continuously, with an average of about 18 events being added or updated every week.

General inquiries arrive at the rate of about ten or twelve each week. All are answered.

Professional referrals are initiated at the rate of three per month. Most of these are frivolous, however, with the inquirer backing off once he/she is advised of the service fee, so that only a handful proceed to fruition.

PROFESSIONAL LICENSING

Naval Architecture and Marine Engineering (NAME) P.E. examination development was an ongoing activity. There were 24 authors of the 2003 examination, which was sent to the National Council of Examiners for Engineering and Surveying (NCEES) in January.

The second offering was made of the PERC (Professional Engineering Review Course), developed under the direction of Jose Femenia and formatted by High Caliber Solutions, Inc. There were ten volunteer instructors and 29 registrants for the course. The course used the SNAMeLearning.org web site, during the period January 6 thru March 26 including the extra session requested in 2002. The fourth NAME P.E. examination was given April 11, 2003 to 26 candidates with a realized pass rate of 73%. Based on this year's experience, the third PERC offering is planned for the period January 5 thru March 24, 2004, again being expanded to include eleven Units and additional problem assignments. The 2004 examination will be given on Friday, April 16.

The number of successful 2003 P.E. candidates was about the same percentage level as in 2002. The PERC participants have consistently reported that the PERC instructors contributed much to their success. A NAME P.E. Examination Study Guide is nearing completion to further assist P.E. candidates in their preparations. It is planned that the Guide will be available to all in January 2004 and will include a sample examination with item solutions.

As both the first and second PERC offerings were financially self-sustaining, it is anticipated that the 2004 presentation will be so as well.

Some practitioners in the small craft community remain opposed to P.E. Licensure for small craft designers and the Maine legislature has recently passed legislation exempting those who work only on vessels less than 200 feet in length from provisions of the Maine licensing act. It is not clear what impact this will have on licensed practice in Maine, as commentary in last fall's Maine Board newsletter raised the issue of the industrial exemption, noting that there is rising national concern for the licensed status of those making design decisions within the management structure of industrial concerns.

Continuing education (and professional development) requirements for the renewal of one's P.E. License are continuing to be noted by state boards, the latest one of note being Florida. New York is also in process of revising its continuing education requirements for license renewal, though a final requirement statement has not yet been made public.

AWARDS

The following awards were presented at the Annual Banquet on Thursday, September 26, 2002

The Davidson Medal for "outstanding scientific accomplishment in ship research" was presented to Nils Salvesen by Dick Yue.

The Vice Admiral "Jerry " Land Medal for "outstanding accomplishment in the marine field" was presented to William O. Gray by David A. O'Neil.

The David W. Taylor Medal for "notable achievement in marine engineering" was presented to R. Keith Michel by Bruce S. Rosenblatt

The following awards and presentations were made at the President's Luncheon on Friday, September 27, 2002:

50-YEAR MEMBERSHIP CERTIFICATES

50-Year Membership Certificates representing 50 years of membership were presented by Joseph J. Cuneo to: Raymond O. Anderson (in absentia), Benjamin V.C. Andrews (in absentia), John N. Case (in absentia), Sheldon V. Cass (in absentia), John Irving Coulson, Sr. (in absentia), Donald P. Dickerson (in absentia), Arthur G. Grant (in absentia), William M. Hannan (in absentia), J. Douglas Hearnshaw (in absentia), Martin Ingwersen, Edgar M. Jacobsen (in absentia), Norman Hans Jasper (in absentia), Julius Jermanok (in absentia), Cowan Everette Jones, Jr. (in absentia), Thomas Kennedy (in absentia), D. William Lerch, III (in absentia), John P. McGough (in absentia), Otto Meier, Jr. (in absentia), William B. Morgan, George J. Mortensen (in absentia), Tetsuichiro Nakanishi (in absentia), Howard Andrew Peterson (in absentia), Pao Chi Pien (in absentia), Eugene Schorsch (in absentia).

CERTIFICATE OF APPRECIATION

A Certificate of Appreciation was awarded to William H. Garzke for outstanding leadership and service to the Society as a member of the Education Committee from 1992 to 1994, and Chairman, from 1995 to 2001.

DISTINGUISHED SERVICE AWARDS

Distinguished Service Awards were presented by Ronald K. Kiss for "dedicated personal service and/or technical contributions to the Society and its activities at the national and Section levels," to: Henry A. Olson and Michael L. Powell.

STUDENT PAPER AWARDS

Student Paper Awards for "outstanding student papers presented at a Section during the 2002/2003 program year," were awarded as follows: the Graduate Paper Honor Prize was presented to Yiannis Constantinides; the Graduate Paper Award was presented to Paul H. Miller (in absentia); the Undergraduate Paper Honor Prize was presented to John C. Hootman, Lucas Hurt and Jason P. Updegraph (in absentia); the Undergraduate Paper Award was presented to Melissa D. Harness.

VICE ADMIRAL E. L. COCHRANE AWARD

Edward M. Lewandowski presented the Vice Admiral E. L. Cochrane Award for "the best paper published in Marine Technology," to Michael G. Parsons and Samson C. Stevens.

AMERICAN BUREAU OF SHIPPING-
CAPTAIN JOSEPH H. LINNARD PRIZE

Robert Geary presented the American Bureau of Shipping-Captain Joseph H. Linnard Prize for "the best paper contributed to the proceedings of the Society (2002 Transactions) to Arthur M. Reed and Robert F. Beck.

HONORARY MEMBER CERTIFICATE

Thomas P. Mackey presented an honorary member Certificate to Joseph J. Cuneo.

WILLIAM M. KENNEDY AWARD

Michael L. Powell presented The William M. Kennedy Award for "outstanding service and contribution in the development of systems and planning applying to shipbuilding and ship repair" to Frederick J. Harris at the Ship Production Symposium Luncheon on September 25, 2002.

ELMER L. HANN AWARD

Michael L. Powell presented the Elmer L. Hann Award for the "best paper presented at the 2001 Ship Production Symposium" to Dale G. Karr, K. Peter Beier, Seung Soo Na, and Philippe Rigo, June 13-15, 2001, Ypsilanti, Michigan. The presentation was made at the Ship Production Symposium Luncheon on September 25, 2002.

DR. JAMES A. LISNYK SHIP DESIGN COMPETITION AWARDS

The 2002 Dr. James A. Lisnyk Ship Design Competition Awards were presented by Society President Joseph J. Cuneo and Vice President, ASNE, Glenn Ashe, as follows: The Norwegian University of Science and Technology was awarded First Prize. The winning team headed by Faculty Advisor, Kai Levander (in absentia), were Kjetil Sjolie Strand, Thomas Thompson and Vegard Skar; The Second Prize The Norwegian University of Science and Technology was awarded Second Prize. The winning team headed by Faculty Advisor, Anders Endal (in absentia) were Aage Oscar Langeland, Jr., Charlotte Melsom and Jorn Salthammer (in absentia). The Third Prize was awarded to Virginia

Polytechnic Institute and State University, the winning team headed by Faculty Advisor, Alan J. Brown, were Timothy John Breeton (in absentia), Melissa Sue Gill (in absentia), Moojane Louie, Brian Daniel Wolf (in absentia), and Mario A. Rinza (in absentia).

T&R CERTIFICATE OF RECOGNITION

Robert S. Behr, presented a T&R Certificate of Recognition for "outstanding service in the Society's Technical and Research Program" to Edward M. MacCutcheon (in absentia). Mr. Daidola, Chairman of the T&R Steering Committee, was not in attendance to present the T&R Certificate of Recognition.

GEORG P. WEINBLUM LECTURE

The twenty-fifth Weinblum Memorial Lecture was given by Professor Ronald Wai-Chun Yeung, on November 20, 2002, in Hamburg, Germany.

APPOINTMENTS
(Society Representatives to other organizations and other appointments)

ACCREDITATION BOARD FOR ENGINEERING AND TECHNOLOGY

Board of Directors	Capt. Robert E. Apple
Engineering Accreditation Commission	Dr. Edwin G. Wiggins Dr. Wayne Neu
Engineering Accreditation Alternate	Open at present
Technology Accreditation Commission	Prof. James Harbach
Program Evaluators	Prof. Harold C. Alexander Mr. Robert E. Apple Mr. Edward Barrett Mr. Elwood Baumgart Dr. Rameswar Bhattacharyya Mr. Norman Birrell Prof. Neil Bose CDR Errol M. Brown, USCG Dr. David Burke Prof. Boris S. Butman Dr. Tyne-Hsien Chang Prof. Chrys Chrystomidis Cdr. Kurt J. Colella, USCG Dean Roger Compton Prof. Ralph Davis Mr .Alexander Delli Paoli Dr. Stanley E. Dunn Dr. Jeffrey Falzarano Prof. Jose Femenia Mr. Michael Fleahman Prof. James Harbach Dr. Groves E. Herrick CDR Dwight G. Hutchinson, USCG (Ret.) Prof. Bahadir Inozu Capt. Harry A. Jackson, USN (Ret.) Prof. Paul C. Jackson Dr. Robert G. Latorre Dr. Alaa E. Mansour Prof. Charles J. Munsch Dr. Wayne Neu Dr. Michael G. Parsons Dr. Robert E. Randall CDR Paul Roden, USCG Prof. William J. Sembler Dr. Herman E. Sheets Prof. James R. Sherrard Dr. John Spencer Prof. Lawrence W. Ward Dr. Edwin G. Wiggins
Council of Engineering and Scientific Society Executives (CESSE)	Mr. Joseph Caggiano Mrs. Susan Evans Ms. Susan Giver Mr. Allan Gray Mr. Philip B. Kimball Mrs. Carolyn Marus Mrs. Eileen Romanelli Mrs. Deborah Sirotiak Mrs. Barbara Trentham
Marine Engineering Council Underwriters Laboratories	Open
Maritime Technical Information Facility	Dr. Charles R. Cushing
SPE Liaison Society Executive Directors	Open
National Inventors Hall of Fame Foundation Selection	No representative at present time
Offshore Technology Conference Board of Directors	Mr. Roderick J. Allan
Program Committee 2001/2002	Gregory G. Thory
Program Committee 2002/2003	Gregory G. Thory
Pan American Institute of Naval Engineering	Mr. Bruce S. Rosenblatt

Ship Structures
Subcommittee Dr. Walter M. Maclean

Sperry Board of Award
Capt. Perry W. Nelson, USN (Ret.)
Mr. John M. Dempsey, Jr.
Mr. Philip B. Kimball (Alternate)

U.S. Committee on Theoretical and Applied Mechanics Dr. Dale Karr

SCHOLARSHIPS

Seven scholarship recipients, including two alternates of Graduate Scholarships, started their academic year, bringing the number of Graduate Scholarship awards since 1933 to 375. It should be noted that numbers reported in previous years were sometimes altered by a student's inability to accept the award for a number of reasons. Starting their advanced studies this Fall are: Louis A. Detrisac, Megan A. Hendry-Brogan, Karl M. McLetchie, Benjamin J. Racine, Jon P. Silverberg, Brandon M. Taravella, Andrew D. Wiggins and two alternates; Michael F. DeLorme and Rocky S. Taylor.

The Society's Undergraduate Scholarship Program, inaugurated in 1957, continues to provide assistance. Through academic years 1998-2003, 850 total awards have been made. Students receiving undergraduate Scholarship assistance are attending Florida Atlantic University, Maine Maritime Academy, Massachusetts Institute of Technology, Memorial University of Newfoundland, Texas A&M University, University of British Columbia, University of California/Berkeley, University of Michigan, University of New Orleans, State University of New York Maritime College and Virginia Polytechnic Institute.

In addition, annual contributions to Webb Institute of Naval Architectural have been made to support its program. The Scholarships Committee, under the newly appointed Chairman, Scott C. McClure, continued to give the selection process the dedicated care that has characterized the Scholarships Committee's work through the years. In addition, a thorough review of the guidelines for the graduate and undergraduate selection process was conducted during the year, and will continue on a continuing basis.

OFFSHORE TECHNOLOGY CONFERENCE

OTC attendance this year was 50,655 participants from 110 countries, or 1035 more than last year, and the highest attendance since 1985. The 34th year of the Exposition recorded 2015 booths for a total of 383,350 net square feet of space in Reliant Hall. It was also announced that SNAME member, Rod Allen, was elected to serve as OTC Board Chairman for 2004-2005.
The Society's share of the 2003 Offshore Technology Conference was $443,910.

SNAME WEB SITE

Web site management continues under the direction of the Electronic Media Committee with Connie Cali-Poutré dealing directly with our Web Contractor, EIH. Our web site use continues to be steady totaling between 400,000 and 600,000 hits per month on all pages.

Web Site Enhancements to www.sname.org

There are now eight, password-protected intranets on the SNAME web site. HQ staff and committee chairs are using these intranets to greatly reduce cost of hard-copy mailings.

The 2003 World Maritime Technology Conference web site is actively being used – currently working on technical program page (password protected papers). The on-line registration form is greatly enhanced from previous years and features an automatic e-mail confirmation.

Publications – New format and sub-menus were designed and implemented. Our journals (MT, JSR, and JSP) are all available on-line using Ingenta Select hosting.

2004 Dues renewal – The new form will be completed prior to dues mailing, inserts will be available on-line and automatic e-mail confirmation will be done.

Job Postings on web site – continues to be utilized by both members and non-members. Non-members are charged $125 for a 60-day posting.

Awards – all SNAME awards are listed, with descriptions, list of past recipients, and links to applicable nomination forms.

The SNAMeNewsLetter, is now published semi-monthly, and is e-mailed to all members. It is also available to the general public from the web site (under the "SNAME News" fly-out menu). The newsletter continues to be an effective way of communicating, and we continue to use SNAMeNewsFlashes when necessary to send special and timely messages to our members. The SNAMeNewsFlashes are also accessible on the web site (under the SNAME News fly-out menu). A new fly-out menu on home page was developed for "Member Benefits" on the home page.

New or ongoing projects for our web sub-contractor include the development of an electronic dues notice, an input screen for uploading technical papers, an automation process for sending out the SNAMeNewsLetter, a link for all Section papers to the shopping cart, and a monitoring system to count job postings by members.

MANAGEMENT INFORMATION SERVICES

Headquarters owns 21 PC's, two MAC PC's and five servers running a wide variety of Microsoft products. The five servers consist of two data servers, a mail server, an authentication server and a terminal server.

All PC's have Microsoft 2000 Professional software installed, along with an Internet connection. SNAME is running an Association Management software package from GoMembers.com that consists of Dues Processing, Events Registration, Subscriptions, Order Entry and an Exhibits Management Module.

Over the fall of 2002, Headquarters installed a full T1 connection to the Internet and in the summer of 2003, installed a new file server dedicated solely for our Association Management software applications.

In FY 2003 the computers at Headquarters were available 100% of the time in 9 of 12 months. Overall, our system availability was 99.25% for the Service Center servers and 99.5% for the e-mail server.

SNAME PROPERTIES, INC.

The building was fully rented for the entire year with the exception of 2500 square feet on the third floor, which will be leased as of October 1, 2003.

The mortgage on the building was refinanced on June 26, 2003 at a rate of 4.98% for 9 years.

The building was placed on the market for sale in August 2002, but taken off the market by the Building Committee in December 2002 due to the development of positive cash flow in 2003.

The two second-floor storage rooms were vacated in July, and the mailroom on the fourth floor rearranged to improve storage capacity and efficiency at Headquarters. The second floor rooms were made available for rent to existing tenants in August.

HEADQUARTERS/STAFF

The number of permanent staff at Headquarters was stable throughout the year; however, in the summer we lost two staff members and gained one in the Member Services Department, for a net loss of one, totaling 16 at the end of August.

In order to prepare for the larger registration in 2003 for WMTC we trained two staff members in the registration process. This resulted in a decision to handle registration in house and not outsource in the future representing a significant savings.

As a result of the success of the SNAMeNewLetter we made the decision in November to increase the publication frequency from monthly to semi-monthly, publishing 14 issues in FY 2002, and, projecting up to 22 issues in FY2003.

Dues notices were sent out a month earlier, or in early September, in an attempt to spread out the workload for the Membership Services Department associated with dues renewals. By November 1, we had 1600 members renew their membership and thus qualify for a drawing for a Personal Digital Assistant.

Karen Hanson was hired to replace Terry Flanagan as Editor of the technical journals. Karen started on January 2, and Terry assisted as necessary to effect a smooth transition. Terry was honored for his 35 years of service to the Society during the staff luncheon on December 10.

SNAME executed cooperative agreements with The Society of Naval Architects of Japan and The Ecuador Society of Naval Architects and Marine Engineers (CINAVE) with the electronic exchange of signatures between the presidents of the two societies and president Joseph Cuneo on December 9 and July 22, respectively.

The staff at Headquarters conducted its fourth and most successful annual telethon to date, form February 17 through April 28. Statistics for 2003 vs. 2002 are as follows:

	2003	2002
Length of Telethon, weeks	10 (2/17-4/28)	8 (2/25-4/22)
No. Delinquent Members	1583	1525
Renewed Memberships	664	643
Percent Paid	42%	42%
Amount Collected	$78,000	$75,000

First, second and third place cash prizes were again awarded to staff members based on number of individual renewals recorded. Later we followed up on the 500 dropped members not previously contacted. The results were encouraging as we have confirmed those who have dropped out and why, and cleaned up addresses of an additional 55 members who renewed. At the end of August the retention rate stood at 87%.

Headquarters has assumed the job of processing of checks and credit card transactions previously accomplished as a "lock box" activity through our local bank. Performing this work in house will realize a yearly savings of $7000.

Headquarters initiated an electronic confirmation, which is now automatically sent to registrants for meetings. This replaces staff-generated emails, which will save considerable staff time and effort in the registration process.

For the first time the Annual Meeting brochure material was created for the SNAME web site and downloaded directly by our printer thus eliminating a number of steps and considerable cost in the production process. Future brochures will be electronic only, representing additional savings in production, printing and mailing costs.

The entire staff participated in two brainstorming sessions, called "Discussion Cafes," on February 11 and 25, to develop ideas for increasing revenue and reducing costs, both in general, and, specifically, for the Annual Meeting. This was in response to the Executive Committee's mandate in December for headquarters to "Break Even, or Better" in future annual meetings.

The Office of Naval Research (ONR)-funded Center for Innovation in Design (CISD) sponsored a workshop entitled, The National Naval engineering Education Conference (NNEEC) held in Washington, DC on March 11-13, in which a number of members, including the Executive Director, participated.

The Executive Director, among other members, was invited for a second year to represent SNAME as a judge of "The Nuclear Oceanographic Research Submarine Exploration Program (NORSE)", a middle school science class design competition for teams of eighth graders in Westchester County.

SNAME was informed in May that it would be awarded a sub-contract in "Public Outreach" from the University of Michigan, which was to be receiving a grant from The Office of Naval Research later in 2003 to study innovative ship design as part of the National Naval Responsibility in Naval Engineering initiative.

The Society's second distance learning course, "Professional Sandwich Laminate Design Course - 2003," was developed, tested and will be presented in FY 2003 via SNAMeLearning.org.

Amid the complex planning process surrounding the development of the Society's largest Annual Meeting, the World Maritime Technology Conference, in its 111-year history, the Society managed to successfully create a new governance plan, rewrite its Bylaws and obtain both Executive Committee and Council consensus and approval in 2003 in expectation of final approval by the general membership in the Society's Business Meeting at the Annual Meeting in October, 2003.

I would like to personally thank our president, the Society officers, the Executive Committee members and chairs of standing committees as well as the numerous volunteers who provide the spirit and expertise, which makes the Society stand out from the rest. Also, my personal thanks to our dedicated staff members, all of whom work tirelessly to insure that our members continue to receive value for their membership along with significant benefits and services.

Respectfully submitted,

Philip B. Kimball,
Executive Director

SECONDING SPEECHES

Perry Nelson

It is my honor and privilege to second the nomination of Dr. Roger Compton as president-elect of the Society of Naval Architects and Marine Engineers.

I have known Roger since he was a professor at the Naval Academy during my tour as the head of the Marine Engineering Department. He was an outstanding professor providing support both to the top students as well as those who needed help in their studies.

Dr. Compton created the ABET accredited major program on Naval Architecture, participated on the conceptual design team for the ship hydrodynamics laboratory and directed the Academy's involvement in the design of the 108'yard patrol craft.

Dr. Compton has been an active member of our Society for over 32 years. He has received numerous awards from both SNAME and ASNE. He has served in a number of positions in the local SNAME section including Chairman of the Chesapeake Section.

Dr. Compton is truly dedicated to our Society and it is my honor to nominate him as president-elect.

Bruce Johnson

Mr. President, Fellow Members, It gives me great pleasure to second the nomination of Roger Compton as the next President of this Society.

I began working with Roger at the Isherwood Hall 85-foot Towing Tank in the fall of 1966 when he first joined the Naval Academy faculty. He and Lt Tom Dyer began educating a Purdue fluid mechanics graduate in the field of experimental naval hydrodynamics. Dr Compton's part in this continuing education task took the better part of our 32 year USNA partnership, since I had to learn naval architecture outside the formal classroom. Roger's patience in working with a foot-dragging student of standard tankery methods has been a blessing to me and, I hope, to the Naval Academy.

Roger also introduced me to sailboat racing, as we partnered with Tom Dyer to campaign a Flying Dutchman during the late 60s. During the 70s, we were well matched competitors in the 17' Discoverer Class. Roger occasionally crewed in the 80s on my Catalina 27, a decision he probably regretted the day I collided with the riprap around the Choptank River Light and Roger kept going forward from the main trimmer's position right into the cabin. The things a friendship has to survive.

Dean Compton's contributions to the field of naval architecture are many. Many of you may not realize he was responsible for the concept of double flap wavemakers, now a preferred method of making waves in towing tanks. He also co-developed most of the performance specifications for the analysis software for our Hydromechanics Lab. As Lab Director, he fostered the development of PC-based analysis of towing tank data. This eventually enabled laptops to replace the cumbersome data link system used when the Hydro Lab opened in 1977. Professor Compton directed the Academy's involvement in the design of the 108' Yard Patrol Craft, and was a strong proponent and adviser of midshipman research during his 32 years there.

Dr. Compton served as Chairman of the Department of Naval Architecture, Ocean, and Marine Engineering at the U.S. Naval Academy, and is currently the Dean and a Professor of Engineering at his alma mater, Webb Institute in Glen Cove, NY. While at the Naval Academy he received both the Meritorious and Superior Civilian Service Awards from the Navy. He was named Professor Emeritus in May 2000. He maintains active membership in ASNE, SNAME, and ASEE and served as Chairman of the Chesapeake Section of SNAME in 1993-94. He currently serves as the Western Hemisphere Area Representative on the Executive Committee of the International Towing Tank Conference (ITTC). He was awarded SNAME's Distinguished Service Award and the E.L. Cochrane Prize, and won ASNE's Solberg Award. His is a Life Member of ASNE and is a Fellow of SNAME. The Webb Alumni Association presented him the W. Selkirk Owen Award in September 2000. (Some of these items can be deleted on the fly if covered by other speakers.)

It is a distinct pleasure to second the nomination of Dr. Roger H Compton as President-elect of SNAME.

John Hootman

Mr. President and fellow members, it is both a pleasure and an honor to second the nomination of Dr. Roger H. Compton as the next President of our Society.

Compared to most of the audience here today, and especially my fellow seconders, I have known Dean Compton and this Society for a relatively short amount of time; however, I believe that I have the benefit of a slightly different perspective, that of a student versus a peer.

I have known Dean Compton for five years now, and have had the privilege of learning from him in and outside of the classroom, as have five classes of Webb Institute graduates and 32 classes of Annapolis midshipmen. I believe that it is fair to say that few people in our profession today can claim to have had as large a hand in shaping future professionals or of fostering the aspirations of budding designers as Dean Compton has.

Dean Compton's love of the sea, naval architecture, and sailing is readily apparent. His passion for these interests has aided him in teaching us the art and science of ship design, and in setting the example for a hardworking, honorable, distinguished professor.

He is a challenging professor who demands much from his students, but the experience and lessons learned from his instruction have already paid dividends for myself and others (obviously I might not have felt this way when I was rushing to finish a catamaran ferry design for him at 0530 in the morning on the day it was due).

Nonetheless, Dean Compton's ability to relate to and understand the student perspective has allowed him to foster closer relations between faculty, administration, and students to the benefit of the collective whole. Particularly of note is his firm support of student engagement in activities outside the classroom, ranging from broad support of membership in professional societies to specific instances of inclusion in educational matters of the Webb Board of Trustees and other Institute committees.
I look forward to seeing these interests remain a priority during Dean Compton's term as President. We have made strides in SNAME over the past two years with the Student Steering Committee, but there is much room for continued expansion and improvement. Student involvement in the Society and retention after graduation must continue to improve in order to bolster the foundation of the Society for the future.

Dean Compton's leadership and dedication have benefited the United States Naval Academy and Webb Institute, and I am confident that he will have a lasting and positive impact on our Society as well. Therefore, it is a distinct pleasure for me to second the nomination of Dr. Roger Compton as the next President of our Society.

ACCEPTANCE SPEECH
Roger Compton

President Rosenblatt, fellow members and guests of The Society of Naval Architects and Marine Engineers:

To say that I was totally surprised when Keith called me would be a vast understatement. It wasn't until I received congratulatory calls from President Rosenblatt and Executive Director Kimball – and an office visit from my boss, Ron Kiss – that the full impact of the honor and challenge of Keith's call began to register. To be nominated by my professional colleagues to serve as President of our prestigious society, founded and subsequently guided by giants in our profession's history for the past 110 years, was beyond my wildest expectations. One of those founding giants was William Henry Webb, whose benefaction is directly responsible for any and all professional achievements I have accomplished in my career.

I want to thank Perry, Bruce, and John for their kind remarks and their willingness to speak on my behalf. They represent the best of our society and afford me a segue to highlight what I see right now as areas of concentration for SNAME that I'd like to emphasize. These are in no particular order. First: SNAME needs to continue and increase its emphasis on student involvement in all aspects of its activities. Professional societies' roles in attracting and retaining talented young people are pivotal – from K-12 outreach, to scholarships, to affordable and relevant textbooks, to opportunities for student involvement in T & R activities, to job placement assistance, to faculty development opportunities. John and his contemporaries are the future of SNAME and our profession. Second: SNAME should strive to be inclusive and supportive of the many specialties of naval architects and marine engineers – commercial, military, recreational, and offshore. We should encourage more active involvement by Navy and Coast Guard professionals – both in and out of uniform. Perry – as a career naval officer, as a leader of industry, and as a leader in professional society activities – embodies this inclusivity. Third (and this is related to the last one): We need to continue and become even more actively engaged in exploring collaborative efforts with other technical societies. A prime example is the American Society of Naval Engineers (ASNE). I hope to continue the dialog that has been initiated by President Rosenblatt and ASNE President Sargent. Our overlapping constituencies as well as our operating budgets deserve serious attention. Fourth (and last): Our society should welcome professionals from different backgrounds who can (and history has shown, do) contribute significantly to the work of SNAME. My colleague, co-author, and friend, Bruce, has greatly enriched the technology of our business and has been an inspiration to many future naval architects.

With the mentorship and friendship of President Bruce Rosenblatt and Past President Ron Kiss, and colleagues like Professor Emeritus Bruce Johnson, present and former students from USNA and Webb, and the professional staff at SNAME Headquarters, I will endeavor to strengthen and further the goals of SNAME and to increase the awareness of the American people to the excitement, challenge, and critical national importance of the contributions of naval architects, marine engineers, and offshore engineers to the prosperity, leadership, and security of our country.

Thank you, Keith and members of the nominating committee for your confidence in my abilities to lead this society. President Rosenblatt, thank you for offering me your counsel. Seconders, again, thank you. A special thanks in advance to my wife, Jill, for the time spent at Headquarters and for her unflagging support. My fellow SNAME members, I am truly honored to be nominated as your President-Elect.

TREASURER'S REPORT TO COUNCIL, 2003

Report of the Treasurer

Balance Sheet
at August 31, 2003 and August 31, 2002

	2003	2002
ASSETS		
1. Cash & Cash Equivalents	$464,212	$348,011
2. Accounts Receivable	98,738	27,392
3. Prepaid Expenses		
4. Section Advances	181,000	177,917
5. Other Prepaid Expenses	404,197	338,479
6. Total Prepaid Expenses	585,197	516,396
7. Total Current Assets	1,148,147	891,799
8. Investments	7,369,852	7,030,720
9. Property & Equipment	251,173	227,796
10. Less Depreciation	(230,771)	(198,550)
11. Net Property & Equipment	20,402	29,246
12. Due from SNAME Properties	3,196,329	3,180,844
TOTAL ASSETS	11,734,730	11,132,609
LIABILITIES & NET ASSETS		
1. Accounts Payable	99,178	124,836
2. Deferred Revenue	454,598	371,422
3. Deferred Pension Contributions	60,000	120,230
4. Other Liabilities	190,000	171,174
5. Total - Current Liabilities	803,776	787,662
6. Total Liabilties	803,776	787,662
7. Net Assets	10,930,954	10,344,947
TOTAL LIABILITIES AND NET ASSETS	11,734,730	11,132,609

	8/31/03	8/31/02
REVENUES		
1. Member Dues	$839,192	$872,577
2. Publication Sales	470,907	430,365
3. Contributions	29,977	32,650
4. Functions	938,204	777,130
5. Advertising	25,945	26,427
6. Educational Courses	11,635	31,375
7. Interest & Dividends	158,441	170,047
TOTAL REVENUE	2,474,301	2,340,57
EXPENSES		
1. Program Services		-
2. Publications	421,556	358,979
3. Functions	255,536	286,716
4. Scholarships,Grants	87,774	87,939
5. Total Program Expenses	764,866	733,634
6. Management & General		
7. Salaries and Fees	931,276	933,881
8. Payroll Taxes,Benefits, Insurance	242,308	633,940
9. Subtotal	1,173,584	1,567,821
10. Facility Expenses	150,047	150,188
11. Membership Services	134,423	163,499
12. Professional Fees	27,179	30,939
13. Headquarter & Section Exp.	115,912	105,870
14. T & R Expenses	40,249	56,622
15. Depreciation and Other	48,440	57,073
16. Sundries	54,675	55,008
17. Educational Courses	75,342	87,612
18. Advertising	26,029	29,735
19. Website	48,528	48,402
20. Subtotal (Lines18-27)	720,824	784,948
21.Total MGT & General (Lines17 + 28)	1,894,408	2,352,769
TOTAL EXPENSES (Lines13 + 29)	2,659,274	3,086,403
FROM OPERATIONS INCREASE NET ASSETS	$(184,973)	$(745,832)

SNAME Properties

	2003	2002
ASSETS		
1. Cash & Cash Equivalents	$19,254	$21,519
2. Accounts Receivable	0	0
3. Prepaid Expenses	17,096	21,976
4. Total Current Assets	36,350	43,495
5. Property & Equipment at Cost	3,471,137	3,456,582
6. Less Depreciation	(1,486,747)	(1,394,633)
7. Net Property & Equipment	1,984,390	2,061,949
TOTAL ASSETS	2,020,740	2,105,444
LIABILITIES & NET ASSETS (Net Worth)		
1. Accounts Payable	11,253	14,162
2. Mortgage Current Portion	87,912	76,536
3. Total Current Liabilities	99,165	90,698
4. Mortgage Payable	899,872	992,259
5. Due to SNAME	3,196,329	3,180,844
6. Other Liabilities	19,449	29,848
7. Total Liabilities	4,115,650	4,202,951
8. Net Assets (Net Worth)	(2,094,910)	(2,097,507)
TOTAL LIABILITIES & NET ASSETS	$2,020,740	$2,105,444

	8/31/03	8/31/02
REVENUE		
1. Rental Income	$405,784	$394,533
2. Parking and Other Income	27,711	28,808
TOTAL REVENUE	433,495	423,341
EXPENSES		
1. Cleaning & Rubbish Remova	33,109	27,698
2. Utilities	58,083	56,916
3. Maintenance & Repairs	51,885	42,101
4. Real Estate Taxes	66,638	67,348
5. Property Mtg. Fees	9,600	9,600
6. Mortgage Interest	63,087	86,651
7. Professional Fees	9,580	275
8. Depreciation	92,040	95,159
9. Insurance	4,825	3,934
10. Leasing Commissions	3,885	14,880
11. Security	3,577	3,246
12. Administrative Expenses	70,570	53,450
TOTAL EXPENSES	466,879	461,258
INCREASE NET ASSETS	$(33,384)	$(37,917)

Awards and Citations Presented at the 2003 Annual Meeting

FIFTY-YEAR GOLDEN MEMBERSHIP CERTIFICATES WERE AWARDED TO THE RECIPIENTS LISTED BELOW BY PRESIDENT BRUCE S. ROSENBLATT

Eric L. Barr, Jr. (in absentia)
Frank J. Basile (in absentia)
John O. Bowden (in absentia)
James W. Carpenter (in absentia)
John M. Carras (in absentia)
David G. Davies (in absentia)
Parker C. Emerson (in absentia)
Constantine Foltis (in absentia)
Philippe C. Gaucher (in absentia)
Thomas C. Gillmer (in absentia)
Peter G. Goulandris (in absentia)
Arthur J. Haskell
Thomas M. Hopkins (deceased)
Masasuke Kawasaki (in absentia)
Keatinge Keays (in absentia)
Justin E. Kerwin (in absentia)
Dan Khoushy (in absentia)
Norman J. Mayer (in absentia)
W. Vance Middough (in absentia)
Robert G. Mills (in absentia)
Alfonso M. Morcillo (in absentia)
Henry Muschel (in absentia)
Basil V. Nakonechny (in absentia)
J. J. Nicols (in absentia)
Harry Panagopulos (in absentia)
Milan L. Pittman (in absentia)
Marvin A. Rosenberg (in absentia)
George T. Rundlet (in absentia)
Thomas J. Sartor, Jr.
George C. Stamoulis (in absentia)
Stuart W. Thayer (in absentia)
Elias Venning (in absentia)
Robert E. Williams (in absentia)
Robert B. Zubaly (in absentia)

ABS – CAPTAIN JOSEPH H. LINNARD PRIZE
Presented by: Ronald K. Kiss
Bruce L. Hutchison

CERTIFICATE OF APPRECIATION
Presented by: Bruce S. Rosenblatt
John Angelicoussis (in absentia)
Richard C. Boutwell
Thomas Lamb
William B. Morgan
David R. Zoller

STUDENT PAPER AWARDS
Presented by: Philip B. Kimball

Graduate Paper Award
Richard Camilli

Graduate Paper Honor Prize
Konstantinos Psallidas (in absentia)
John C. Hootman

Undergraduate Paper Award
Karl-Magnus W. McLetchie

Undergraduate Paper Honor Prize
Johanna L. Mathieu (in absentia)

TECHNICAL & RESEARCH PROGRAM CHAIRMAN
Introduced by: John C. Daidola
Robert F. Beck
Robert S. Behr
William T. Bennett, Jr. (in absentia)
Alan A. Gilbert (in absentia)
Henk van Hemmen (in absentia)
Thomas Lamb
Paul B. Mentz
Michael L. Powell
Philip G. Rynn

ELMER L. HANN AWARD
Presented by: Michael L. Powell
Carlos Tomassoni
Ted Huynh (in absentia)
Thomas R. Schiller (in absentia)

WILLIAM M. KENNEDY AWARD
Presented by: Michael L. Powell
Michael W. Butler

VICE ADMIRAL E.L. COCHRANE AWARD
Presented by: Edward M. Lewandowski
William N. France, II
Marc Levadou
Thomas W. Treakle, III
J. Randolph Paulling
R. Keith Michel
Colin Moore

DR. JAMES A. LISNYK SHIP DESIGN COMPETITION AWARDS
Presented by: Bruce S. Rosenblatt and Glenn M. Ashe

Virginia Polytechnic Institute - 1st Prize
Faculty Advisor: Dr. Alan Brown
Students: Jason Albright
Daniel Demko
Matthew Hecht
John Leek
Greg Morrow
Keith Webster

University of Newcastle upon Tyne - 2nd Prize
Faculty Advisor: Prof. Richard Birmingham (in absentia)
Dr. Martin Downie (in absentia)
Students: Alexandros Poutachidis (in absentia)
Michael Anninos (in absentia)
John Monogioudis (in absentia)
Marika Psarros (in absentia)
Andrew D. Smart (in absentia)
Stewart A.K. Baker (in absentia)

Norwegian University of Science and Technology - 2nd Prize
Faculty Advisor: Prof. Kai Levander (in absentia)
Students: Hans Erik Danielsen
Haavard Haaskjold
Erlend B. Sandvik

INTERNATIONAL STUDENT OFFSHORE DESIGN COMPETITION AWARD
Presented by: John J. Filson and Raymond P. Fales

University of Michigan - 1st Place
Faculty Advisor: Professor Thomas Lamb
Students: Louis A. Detrisac
Victor T. D'Souza
Michael Rugnetta
W. Scott Opdyke
Katherine Sultani

Texas A&M University Team #3 - 2nd Place
Faculty Advisor: Dr. Robert E. Randall
Students: Androjan Spence
Enrique Banda
Renee Belton
Stephen W. Faleye
Brandon Holmes
Nikki Ogah

Texas A&M University Team #1 - 3rd Place
Faculty Advisor: Dr. Robert E. Randall
Students: Baron Wilson
Chris T. Chipuk
McAlan Clark
Caroline Hoffman
James R. Peavy
Ramez Sabet

Vice Admiral "Jerry" Land Medal

The 2003 Vice Admiral "Jerry" Land Medal for "outstanding accomplishments in the marine field" was presented to Richard H. Vortmann by Arthur Haskell.

Land Medal Presentation

President Rosenblatt, Honored Guests, Ladies & Gentlemen: It is my pleasure and privilege to present the "Jerry" Land Medal for Outstanding Accomplishment in the Marine field to an innovative manager, a good friend and formidable tennis opponent, "Dick" Vortmann.

After serving in various financial management and strategic planning positions with Kaiser Industries Corporation, he joined the Executive team of NASSCO, which at that time was jointly owned by Kaiser and Morrison-Knudsen. Dick, a quick learner, avid reader and dedicated to achieving his goals, committed himself to becoming a Shipbuilder. He acquired expertise in Engineering, Naval Architecture and Production as he progressed to Vice President of Production, Executive Vice President of Operations and President and Chairman. It was during this period that NASSCO became a wholly owned subsidiary of Morrison-Knudsen.

Subsequently, Dick led the negotiations that resulted in the Employees purchasing NASSCO from Morrison-Knudsen in 1988. He served as President and Chairman of the ESOP until in 1998, when General Dynamics purchased NASSCO from the employees and Dick added the title of Vice President of General Dynamics to that of President of NASSCO.

In addition to his demanding duties at the Shipyard, Dick has participated in many marine oriented activities. He has served as Chairman of the Board of the American Shipbuilding Association, The Shipbuilders Council of America and of the American delegation to the Annual Japanese, European, Korean, Chinese and United States Shipbuilding Conference and as a member of the Transportation Research Board of the National Academies.

The Vice Admiral "Jerry" Land Medal was established by the Executive Committee of SNAME in 1952. Vice Admiral Land was the first recipient and he personally presented the award six times between 1954 and 1960. Based upon an accomplishment of Dick's, in addition to those previously mentioned, I am sure, were he alive today, that Admiral Land would be pleased to welcome Dick to the list of Land Medallists.

President Roosevelt appointed Vice Admiral Land to be the first Chairman of the Maritime Council. This followed his retirement as Chief of the Bureau of Construction and Repair of the U.S. Navy. In his new position he initiated the remarkable merchant shipbuilding effort of World War II. The Liberty Ships, Victory Ships and T-2 Tankers were built during this tour of duty. These programs created shipyards that manufactured ships, a new approach to shipbuilding in which pre-outfitted assemblies were produced and then assembled into completed ships.

Following World War II shipyards in other countries, notably Japan, adopted the manufacture of ships based upon the U.S. wartime model. In fact, National Bulk Carriers, a company I worked for in the 50's and early 60's, hired Elmer Hann from Kaiser's shipyards to introduce the concept to it's Kure Shipyard, which became the model for the Japanese world domination of the large tanker and bulker market in the 60's and 70's. Unfortunately shipbuilders in the United States, for a variety of reasons, some government mandated, abandoned ship manufacturing and returned to ship construction from the keel up, building the hull and then outfitting it.

Dick, as President and Chairman of NASSCO, during the ten years it was an ESOP, led the shipyard in a return to the manufacturing methodology. This required a complete reorganization of Engineering, Planning and Production. It also required extensive facility changes to provide space and flow lines to support pre-outfitting. The program was most significant in the change in management style and employee participation.

While these major changes were going on, the yard continued to build and deliver ships. As a customer during this period I can bear witness that the transition wasn't entirely seamless; but the completion of the Matson containership, *ROBERT J. PFEIFFER*, and of the other contracts undertaken by the yard during that period are testimony to the effectiveness of Dick's leadership and to his determination to successfully complete the transition.

In recognition of his leadership, innovative management and his many other notable achievements, it is my pleasure to present the Vice Admiral "Jerry" Land Medal, for outstanding accomplishments in the Marine field, to a very deserving Richard "Dick" Vortmann.

Land Medal Acceptance

Thank you Art for reading those words of Larry French. And while Larry is obviously not here tonight I want to thank him in absentia for his kind words. As you know Larry is a past president of the Society and a past Land Medal recipient and as such I know he very much wanted to be here tonight but health problems prevented him. He was also my mentor at NASSCO. It was he who gave me every opportunity in the world and led to me, at a very early age, to succeed him upon his retirement from NASSCO.

I very much want to thank the Society and the award committee for this honor of receiving the vice admiral Jerry Land Medal. I can think of no higher honor than to be recognized by your peers.

This award so particularly gratifying to me in that I am only the 3rd person to serve as president of NASSCO since it entered the Shipbuilding world back in 1959 and now I am following in the footsteps of my two predecessors, Larry French and John Banks who both received this award, back in 1975 and 1986.

I have sat up on this dais for many, many years and during these proceedings I have always found myself scanning the list of past years awardees. There is a lot of maritime industry history there on that list. Having been in the shipbuilding industry for some 30 years now, I am proud to say I have had the opportunity to know and work with a great many names on that list. Charlie Cushing, Art Haskell, Perry Nelson, Ed Campbell, Bill Kime, Ron Kiss, Lester, Ed Hood, Elsworth Peterson. The list goes on. I am indeed honored to have my name spoken in the same category of these gentlemen.

On a personal note, it is particularly gratifying to be receiving this award here in San Francisco. San Francisco happens to be my hometown. I grew up here and went to university right across the bay. More relevant it was here 50-55 years ago, where I got my first exposure to Shipyards and Shipbuilding. My father did a 40 year career here in San Francisco with Bethlehem Steel's Shipyard, and he used to take me and my brothers down to the shipyard on weekends and to ship launchings.

Those memories are somewhat bitter sweet in that when I mention Bethlehem Ship it is a stark reminder of the plight of our industry. When I started out back in the early 70's, Bethlehem had just recently closed down it's west coast new building operation. But there were still four full service new building yards on the west coast. Two in Seattle, Lockhead and Todd, a second Todd yard in Long Beach and NASSCO in San Diego. Today only NASSCO exists on the west coast. The other three are gone just like several others on the east coast closed up. Sun Ship, America Ship, Bethlehem Sparrows Point, Quincy, Tampa, Maryland Drydock come readily to mind. This should be a recurring wake up call for all of us in the profession. Without new ship construction in his country there will not be much need for all of us naval architects and marine engineers.

There is no question that the bulk of the commercial sector of the shipbuilding industry has long since moved across the pacific to Japan and Korea, and of late the rapidly growing Chinese shipbuilding industry. We at SNAME along with all other segments of the industry must work diligently to preserve our Jones act-the last bastion of commercial shipbuilding in the U.S. and one of the most effective means of ensuring the Jones Act's continued existence is for all of us, Ship designers and shipbuilders alike, to constantly drive continuous improvement in all our operations to reduce the cost differential, to reduce the productivity differential that exist between out products and those of the far east yards. That has been the driving force for the last 20 years and it will continue to be, but I will stop now before I get too wound up on my soap box.

I want to again thank the Society. I am indeed deeply honored by this award and as I receive this award I would like to thank all the many men and women at NASSCO, now and over the years gone by. It is all those folks who were the ones whose efforts enabled me to receive this award and this recognition tonight. So thank you NASSCO.

Thank you all very much.

David W. Taylor Medal

The 2003 David W. Taylor Medal for "Notable Achievement in Naval Architecture and/or Marine Engineering" was awarded toRobert G. Keane by Edward Comstock.

Taylor Medal Presentation

Good evening, President Rosenblatt, honored guests, ladies and gentlemen. It is my distinct personal pleasure to present the David W. Taylor Medal to Mr. Robert G. Keane, Jr.

Bob Keane has profoundly affected not only the way naval ships are designed but also those who design them. Over a span of five decades, this son of a shipyard pipe fitter has led a sea change in naval ship design as a naval architect, a visionary leader, a teacher, a team coach and a mentor "extraordinaire."

As a member of the Navy's Senior Executive Service (SES) over the past 20 years, Bob Keane rose to the most influential leadership positions of Chief Naval Architect of the Navy, Chief Engineer for Ship Design, and the Central Technical Authority for Total Ship Systems Engineering in the Naval Sea Systems Command. He is, without question, one of the world's leading experts on the design of naval ships.

Bob Keane has built strong technical organizations and has brought robust engineering to the design of naval ships. As Director of NAVSEA's Hull Form, Stability and Hydrodynamics Division, he revolutionized hull form design resulting in the highly acclaimed "seakeeping" hull form and later, as Director of NAVSEA's Ship Survivability Sub-Group, he developed comprehensive ship survivability design principles which were incorporated in every warship since 1984.

However, what distinguishes Bob Keane as one of the most influential leaders of naval ship design in the past 30 years, is his innovative leadership of the Navy's ship design community over the past 17 years -- a time of dramatic change in both design and acquisition. Bob has been responsible for the development and approval of more ship designs, for a longer period, than anyone since World War II. Since 1985, his teams have successfully completed over 35 major warship designs. These Contract Designs became the contractual basis for the acquisition of hundreds of warships including all Flights of DDG 51 following the lead ship, CVN 73-76, LPD 17, LSD 44-49, AOE 6-10, LHD 2-8, MHC 51, T-AGOS 19-23, T-AGS 45, and the Strategic Sealift Ships. The majority of these ships are either serving in the Fleet or are under construction. As older ships continue to be retired, the Navy of the early 21st Century will increasingly reflect the stamp of Bob Keane's technical leadership.

Bob has made major and lasting contributions to the Navy and his profession through his technical leadership, emphasis on teamwork and collaboration, and exceptional abilities in attracting and developing ship design leaders. He is an exemplary role model of day-to-day professionalism and a symbol of the greatness in the Civil Service and our profession at large. While Bob is a superb technical leader, it is his vision of the future and the many innovative ideas that he generates which mark him as a truly exceptional leader. He is never content with the status quo and consistently seeks ways to improve design and acquisition processes including implementing Total Ship Systems Engineering and concurrent engineering principles and establishing the Center for Innovation in Ship Design. His foresight is resulting in revolutionary improvements to the warship design and construction processes throughout the Navy and the shipbuilding industry.

A graduate of Johns Hopkins University, Stevens Institute of Technology, and the University of Michigan, Bob Keane has been very active professionally having served in sixty leadership and deckplate positions in five technical societies. He is a prolific writer having published 35 papers that illustrate his visionary approach, and great desire to effect positive change. He has established and supported NAVSEA Professorial Chairs at four academic institutions. He has received numerous professional awards including the Rosenblatt-Michigan Alumni Award, four Distinguished, Superior and Meritorious Civilian Service Awards, election as a Fellow of SNAME, and receiving the SNAME Distinguished Service Award.

In dramatically re-engineering the warship design process, Bob Keane has led a "sea change" in naval ship design and is "Keeping America's Navy #1 in the World." He is an exceptionally fitting recipient for the David W. Taylor Medal for 2003.

Taylor Medal Acceptance

Ed, thank you for that gracious introduction and thank you for being here tonight and for being such a loyal friend for over 30 years.

Mr. President, members, and friends of the Society. It is because of outstanding professionals like Ed Comstock and the hundreds of other outstanding civilians and military who have worked at the Naval Sea systems command, NAVSEA that I am privileged and honored to stand before you tonight. So, on behalf of these outstanding professionals, I accept this most prestigious award in recognition and honor of Admiral David W. Taylor and I thank the Society for recognizing the contributions that a great ship design organization like NAVSEA and a great shipbuilder like NASSCO have made to the greatest navy in the world, the United States Navy.
Since 1901, the Navy has sustained a cadre of naval officers

and civil servants dedicated to ship design and construction. NAVSEA and its predecessor organizations have been blessed with many great leaders, a number of whom have been previously recognized with the Taylor Medal. I had the honor of working with four of them at NAVSEA: John Nachtsheim, Pete Palermo, Capt. Corky Graham and Pete Gale.
I do feel extremely privileged to be here tonight. I thank all of those who have shown so many kindnesses to me and have helped me over my long career. For the sake of time, I will mention just two who have had a profound influence on my career and my life.

To my Dad who died at 93. He was raised in an orphan home, and started work as a pipe fitter apprentice at Bethlehem Steel Key Highway Shipyard in South Baltimore (BAWMOR) at the height of the depression. He introduced me to my Lord and Savior, he ensured that I got an excellent education, and he introduced me to those sexy curves and lines known as ship hull forms. When I finished graduate school and told him I wanted to design ships, without hesitation, he said "You want to work for Buships, Now NAVSEA. What profound advise! I hope that other generations will have the opportunity to design and build as many ships as we have.

For Ship design is only learned by doing ship design. It is a team port, where people dynamics is equally important as technical dynamics. Our Navy Ship Design Teams have been successful because of those who lead the teams. NAVSEA's Ship Design managers, and the leadership culture that has been passed on from generation to generation.

The man who taught me and many others the art of leadership is Dan Weiler. Dan is the epitome of a great leader. His ambition was first and foremost for the organization, not himself. Those like Dan who build great organizations understand that the ultimate throttle on growth for any great organization is one thing above all others: the ability to get end keep enough of the right people. In my view, the is the greatest challenge for our whole community.

But I am very optimistic about our future. With Admiral Paul Sullivan as NAVSEA's chief engineer and Howard Fireman and his management team as the Navy's new leaders of ship design, all who came up through NAVSEA's ship design ranks. I am confident that they will sustain the greatness of the Navy Ship Design Community. NAVSEA is also bringing to fruition its vision of a collaborative product development environment. NAVSEA is collaborating with our shipbuilders through the national shipbuilding research program (NSRP) and recently established the new center for innovation in ship design (CISD), to create a collaborative enterprise with Shipbuilders, design agents, academia and professional societies like SNAME.

However, not one of our good ideas will become reality until other people embrace it and agree to help move it forward. No matter how clever or gifted we are, nothing in this world happens without teamwork. As teambuilders, we have been given perhaps the most challenging and rewarding work in the world today: teaching people to think and act like a team.

Before closing I must recognize my team leader, who has been my best friend for over 46 years and my companion and wife for over 39 years, Judy, thanks for holding down the home front!
Finally, in closing, I thank God for the day I was hired at NAVSEA because I've been associated with wonderful people. Good people who respected and admired one another. Other than Judy, ship design has been the passionate love affair of my life. No matter how dark the days or how big the task, we had fun. Our experiences went beyond mutual respect to lasting comradeship. I have been a member of a great team and I truly love what we do, largely because I love those I do it with.... And many of you are in the audience tonight. Thank you and God bless you all.

Blakely Smith Medal

The 2003 Blakely Smith Medal for "Outstanding Accomplishment in Ocean Engineering" was awarded to Michael Bernitsas by Robert Sandstrom.

Smith Medal Presentation

Professor Bernitsas, today the Society and your colleagues are taking this opportunity to recognize you for "outstanding accomplishment in ocean engineering". Your achievements span 24 years of technical and educational contributions to the field of offshore mechanics: primarily in the areas of...

- riser mechanics
- structural redesign
- and towing/mooring dynamics and stability.

Professor Bernitsas began his career in Offshore Engineering at the University of Michigan, following completion of his PhD in Offshore Engineering from MIT in 1979.

Mike was the first to demonstrate that global buckling was possible in risers due to internal pressure even when in tension over their entire length. He also proved that the bucking theory for risers, in use since 1941, was incorrect. He found the errors could exceed 80%. And, he developed solutions to correct this deficiency.

Mike's work in perturbation methods has led to significant technical advances for structural redesign, model correlation, structural redundancy and reliability, and most recently topological redesign of structures. These perturbation methods are a numerical technique for determining structural changes to meet performance objectives. His methods provide capabilities to achieve structural changes on the order of 100 to 300 percent without the need for trial and error or repetitive finite element method analysis.

Mike's work on the nonlinear dynamics and stability of towing and mooring has helped to provide a critical and rigorous understanding important in offshore operations.

As part of this work, Mike most recently demonstrated six different dynamical paths leading to those undesirable motions. Some of these nonlinear phenomena result in motions with amplitudes several orders of magnitude larger than motions due to resonance, which classical textbooks and journal papers on mooring dynamics have cited as being the source of the problem.

While at Michigan, Professor Bernitsas has been a visible and prolific contributor to offshore engineering research and education.

- Mike led Michigan's development of their program in offshore engineering.
- Mike has chaired 18 PhD students and advised on 23 additional PhD committees
- Published over 100 peer reviewed articles in leading journals and books
- Initiated and led Mchigan's Sea Grant Industry Consortium in Offshore Engineering
- Participates on multiple committees on the national and international level providing leadership in offshore engineering
- Served as department chair for past nine years

Mike has been unrelenting in defending the importance of Naval Architecture and Marine Engineering at the University of Michigan. The program at Michigan is a national asset. However, major universities like Michigan that compete for research leadership continually reassess their program priorities. Traditional Naval Architecture programs at major universities in the US have been struggling for survival in the face of these pressures.

As Department Chairman, Mike has led the way to ensure that the program at Michigan not only survives, but excels.

Mike, your contributions to the field of ocean engineering as a researcher and educator are a credit to the Society and your colleagues.

I am delighted that you have been awarded this year's Blakely Smith Medal for outstanding accomplishment in ocean engineering.

Smith Medal Acceptance

President Rosenblatt, distinguished guests, ladies and gentlemen, I am honored to accept the Blakely Smith Medal. Bob, thank you very much for your heart-warming citation.
Blakely Smith with his marine background and oil drilling experience became a pioneer and leader in offshore engineering. Following Blakely Smith, eleven more exceptional individuals and pioneers in offshore engineering have been honored by the Society with this Medal. It is no surprise that launching of this engineering discipline and the vast majority of first-developments and inventions took place in this wonderful and progress nurturing country, the United States of America. I am humbled by the accomplishments of Blakely Smith and the first generation of pioneers and I am grateful to the Society for selecting me as the 13th recipient of the Blakely Smith Medal.

I am also pleased to be the first medalist of a younger generation. We were educated by studying the legacy of the first generation – the ocean structures they conceived and built. As we marvel at the creativity, innovative thinking and courage it took to build such complex and large structures – in the frequently harsh ocean environment - we also feel lucky to be the first generation to observe new phenomena and unpredictable dynamics related to those unique and deep-reaching ocean facilities.

Those facilities provide us with numerous challenges to research, phenomena to reveal, and contradictions to explain; ranging from buckling of risers in tension, to vortex induced vibrations, to large amplitude motions of moored platforms due to interaction of excitation with instabilities which may be much more severe than resonance. For engineers with strong background in marine mechanics, analysis and numerical methods, those pioneers made it exciting to work in and contribute to the exploration and utilization of the inner space.
In turn, I find real satisfaction in educating the next generation and observing my students develop new science and technology in industry and academia worldwide.

I would not miss this opportunity to say something about promoting naval architecture and marine engineering to other fields of engineering. I happen to chair the Ocean, Offshore, and Arctic Engineering Division of ASME and act as liaison with SNAME. As such, I was at a recent meeting of the Chairs of the Divisions of the power and energy resources boards. When I introduced our division, one of the other Division Chairs exclaimed "Oh yes, the Naval Architects". He made my day. I only responded with a big acknowledging smile that brightened my face for we are only a minority in that ASME Division.

I have many people to thank for receiving this honor today. First and foremost, my mentor at Michigan Mike Parsons and my colleague Bob Beck; my teachers back at MIT Philip Mandel, Martin Abkowitz, and Chris Chrisssostomidis; and several colleagues in the offshore industry and particularly my friend Bob Sandstrom of ExxonMobil. My gratitude also goes to my family for their support and for being here today to share with me this joyous occasion.

I would like again to thank the Society for this great honor and for providing such a fertile professional forum for development in Ocean Engineering.

Thank you.

William H. Webb Medal

The 2003 William H. Webb Medal for "Outstanding Contributions to Education in Naval Architecture, Marine or Ocean Engineering" was presented to Michael Parsons by Robert Beck.

Webb Medal Presentation

It is my pleasure to present Prof. Michael Parsons with the Webb Medal for outstanding contributions to education in naval architecture, marine or ocean engineering. Prof. Parsons has spent almost his entire career in academics. After graduating from the University of Michigan and a six year tour with the Navy, he enrolled at Stanford University. He received his Ph.D. degree in 1972 and immediately joined the faculty in the Department of Naval Architecture and Marine Engineering at the University of Michigan.

Prof. Parsons is an excellent teacher and has received numerous teaching awards including several presented directly by the students. The courses he has developed in marine engineering and ship design are a fundamental part of our curriculum. The capstone design course, NA 470, has (at last count) a 428 page course pack plus computer codes for preliminary design that were developed by Mike because there was not suitable text or design codes for the course. Schools from around the world have requested copies of the course pack and computer codes. The course has become justly famous for the breadth of ship design knowledge that is presented. As part of the course, and as a general service to the department, Mike developed the Marine Design Laboratory. This state of the art computer laboratory replaced our drawing room and allows students to use the latest engineering design software.

More recently, Prof. Parsons has developed a complementary graduate level design course, NA570. This course builds on NA470 and presents optimization tools specifically for marine design, both ships and offshore structures. The course focuses on multi-criterion optimization and the use of advance decision-making approaches such as genetic algorithms and agents. Mike has also written a chapter entitled "Parametric Design" for the book Ship Design and Construction to be published by SNAME this year.

Prof. Parsons is definitely one of the leaders in naval architecture education. At Michigan he has been chair of the Department of Naval Architecture and Marine Engineering and Associate Dean for Undergraduate Education in the College of Engineering. In 1997, Mike was awarded a prestigious Arthur F. Thurnau Professorship for his outstanding contributions to undergraduate education. Only five or six professors in the entire University are so honored each year. He is Fellow of SNAME and has served as chair of the Great Lakes and Great Rivers Section. More recently, he has been a Vice-President of the Society for the central region.

This is a very short summary of the two pages (in small 10 point font) of his accomplishments that I submitted with the Webb medal nominating form. But what you should really know about Mike is the love he has for teaching and the respect that students have for him. Even when he was dean and was under pressure to stop teaching, he refused and continued to teach NA470. Many students have told me how they often refer back to the NA470 notes throughout their careers in the marine field. NA470 is one of those marvelous courses that make an educational program complete and all students remember fondly. He is a role model for his students. Of the 12 Ph.D. students whose thesis committees he chaired or co-chaired, I believe a third have continued in academics. I can think of no person who is more deserving of receiving the Webb medal for Outstanding Contributions to Education in Naval Architecture, Marine or Ocean Engineering.

Webb Medal Acceptance

Thank you Bob. I would first like to thank the Society for my selection to receive this significant honor. This is important to me and I greatly appreciate the recognition that this represents. When first reflecting on this situation, I re-read the list of past recipients of the Webb Medal and that list certainly adds to its importance. I could not help but think about three of the previous recipients who have been my colleagues over the years.

Francis Ogilvie was my first, full-time chair in the department at Michigan. He tolerated a young, hippie-looking professor who arrived from California with a pony tail halfway down his back. He was always very encouraging and supportive of me and my early work in marine engineering.

Amelio D'Arcangelo was my colleague for about ten years. He was noted for his teaching of our course in preliminary ship design. When he retired, I took over that role and have been teaching "Amelio's course" for the past 18 years. He also played instrumental roles in the creation of the departments at Virginia Tech and New Orleans. Sadly, we all lost Amelio this past year.

Howard Bunch was my colleague for almost twenty years. Howard's major contributions were in introducing more formal consideration of ship production into naval architecture education. He also created the Journal of Ship Production that has elevated the stature of production related research. Howard has always been a wise, funny, and valued colleague.

Occasions like this in one's career also naturally lead to reflection about the mentors and role models who have been a particularly important influence. Two individuals immediately come to mind. First, one of my first bosses and mentors when I was at Naval Reactors while on active duty in the Navy was Al Forssell. Some of you here may have worked with Al. He was one of the best and most effective practicing engineers that I have ever known. He could always tell

Elmer A. Sperry Award

The 2003 Elmer A. Sperry Award for advancing the art of transportation, whether by land, sea or air was awarded to Raymond Pearlson by Perry Nelson.

Sperry Award Presentation

It is my distinct honor to present the Elmer A. Sperry Award for 2002 to Mr. Raymond Pearlson for the invention, development and worldwide implementation of a new system for lifting ships out of the water for repair and launching new ship construction.

The simplicity of this concept has allowed both large and small nations to benefit by increasing the efficiency and reducing the cost of shipyard operations.

To award of this achievement with the Sperry Award truly embraces the awards criterion of advancing the art of transportation, whether by land, sea or air.

Sperry Award Acceptance

Mr. Chairman, Ladies and Gentlemen, I am honored to join the distinguished list of recipients who have received the Elmer A Sperry award.

Inventing a new concept is only the beginning of a process. I would like to acknowledge the important role that others have played in engineering and successfully bringing the Syncrolift® system to the marine industry.

The financial help I received from friends, relatives as well as my bankers should not be overlooked. Their confidence in me made it possible to start my own company in 1959 and to keep my head above water for the many years it took to gain acceptance by shipyard owners throughout the world. There were many of my employees and associates who contributed greatly to the success of Syncrolift but I would particularly like to acknowledge the team of sales engineers who were based in various parts of the world and who assisted me in our sales effort. Three of those naval architects are with us today and I would like to introduce them to you. Howard McKenzie Wilson from London, Trevor Langbridge from Singapore and my son, Douglas Pearlson, from Miami.

Most importantly, I would like to acknowledge the vital role that Marge, my wife of 58 years, played in the Syncrolift story. She not only raised our five children of which two became graduates in naval architecture from the University of Michigan, but she found time to travel with me, assisting with the introduction of Syncrolift to shipyards around the world.

In conclusion, many worthwhile inventions never make it to the marketplace for the lack of financing and encouragement. I hope organizations like those represented at this conference will find ways to support inventors who's innovations will help our industry modernize and grow.

Thank you.

PRESIDENT'S ADDRESS
Bruce S. Rosenblatt

Good afternoon ladies and gentlemen. I always find these luncheons rewarding as they give us the opportunity to pay tribute to so many deserving recipients of our distinguished awards.

It is a sincere pleasure to be with you in San Francisco today, and I am especially honored to be here as your President. I feel this even more so today as we are on the precipice of embarking on this unprecedented meeting. Although today is focused on many SNAME activities, The World Maritime Technology Conference and Exposition, which technically begins tomorrow, will be a meeting for us all to remember for years to come. SNAME along with 11 other naval architecture and/or marine engineering societies have been working for two years to make the World Maritime Technology Conference a special meeting, and I know you will not be disappointed.

Under Phil Kimball's leadership, the SNAME staff has been working "above and beyond" what is expected, but are enthusiastic about its outcome. A special thanks goes out to the steering committee chair, Mr. Tom Mackey. Were it not for Tom's amazing organizational skills, timeless dedication and focused direction throughout these entire two years, as well as a very hard working Steering Committee, made of up of volunteers many of whom are in this room now, this meeting would not have been possible. Last, but surely not least, is the focus of the meeting, the technical content. Bob Kramek and Hank Marcus co-chaired a technical committee that has put together a program whose content is sure to please any and all of us. The attendance for this meeting has way surpassed the numbers that have attended our Annual Meetings in recent years past. I am sure you will all be quite satisfied when you leave in a few days. Our hope is that this meeting will continue, hosted by another society, approximately every three years, and we will be meeting later this afternoon to discuss the initial plans for those future meetings.

You all have the schedules, so I won't dwell on the details, but I hope you get to attend all of the great events planned. From the election of our next President-elect, at the business session just following this luncheon, to the sold-out exposition, which opens tomorrow, or tonight's President's Reception, tomorrow's keynote address and the countless technical papers, panel discussions and T&R sessions, I guarantee you will be quite satisfied.

My father's passing, just this past June, put a damper on my first year as President of SNAME, but I am pleased that my father could be there to see me elected and attend my first Executive Committee Meeting. As I mentioned in the October issue of Marine Technology, my father cared so much about SNAME and the good work it does for our members and the marine industry, and continued to help in any way he could until the very end. This past January first, my first day as SNAME President, he called and said "Mr. President, how may I serve you?" I know much of that phrase contained the words of a proud father, but he meant it, and believed that SNAME could make a difference to our members and our industry. I agree with my father 100% on this matter, and have tried, and will continue to try to help make a difference. But none of us can do this alone. I am committed to be involved, and am in the process of visiting each Section, at least once, but in the end, the success of SNAME will be graded by the value we give to our members. This is where you all come in. Understanding that I am preaching somewhat to the choir, and borrowing some thoughts mentioned by various past Presidents, we need your help!

Our membership is down by about 5% and I am told that percentage is similar to other technical societies. Just about every form of business has been affected by consolidation over the last few years. The marine industry has had more than it's share of this and the recent general economic conditions do nothing to help.

But should we be satisfied to shrink by 5% each year? That question does not even deserve an answer. Even if the various factors leading to this shrinkage are reality, we need to tackle them head on. The one constant in life is change, and, therefore, we need to change as the needs of our members change. I think we have done a decent job of this, but when do you know if it is good enough? I think it is best to never be satisfied.

SNAME has tried to create value for its members in various ways. We have invested heavily in our website and focused on making many of our services more "user friendly." You can now register for Annual Meetings, pay membership dues, buy publications, nominate members for awards and Fellow and many other things, all electronically, on the SNAME web site. These are services our members expect, as they should.

Creating value comes in so many ways, but in an age where more is expected with less, we must honor the time pressures of life and make sure we do give more for less. I am pleased to say that when you all received your dues notices recently; there was no increase in the annual dues, for the third year in a row. That is because we are trying, and I will say succeeding, at least to some extent, in doing more with less.

The time pressures in life I referred to exist not only in our personal lives, but also in our business lives. There we all must continue to do more for less. It is for that reason that we at SNAME are trying to find ways to be more efficient, and use less of your time. We need you all, as volunteers, to keep this society strong, but appreciate and value your time and thus are trying to use it wisely. We have been conducting more meetings by teleconference, or even video teleconference. This practice will increase in the future, and must, if we want to expect your efforts to continue.

How else can we create more value for you, our members? By becoming more relevant! Within minutes from now we will be holding our business session. In addition to the exciting task of electing our next President-elect, someone who I know will do a great job, we will be voting on a change to SNAME's bylaws. This change, perhaps the largest in our history, is a result of listening to you, our members. Many had said SNAME's type of Governance was antiquated. I think the term "dinosaur" was mentioned more than once. Well, we listened, and assuming the changes previously approved by the Executive Committee and the Council are approved today, we will have a more streamlined governance, one that should be more nimble, one that includes both International and student representation, one that will meet more often and one that will tackle critical strategic issues. Why is this important? Because we value your membership, need your help as volunteers, and do all of this to be more relevant to you and to create that necessary value.

Will these changes work? We think so, but if not, let ME know. What I can guarantee you is that we, within our means, will continue to try to create the value needed, increase our membership, and keep us all as satisfied members of this fine society for many years to come.

In closing, thank you all for being here at this meeting. I hope you all enjoy the World Maritime Technology Conference and Exposition, and I look forward to sharing the experience with you as well as seeing you in three years at the Second World Maritime Technology Conference.

BANQUET SPEAKER

Choo Chiau Beng

Mr. Bruce Rosenblatt, President of the Society of Naval Architects and Marine Engineers, Mr. Thomas Mackey, Chairman of World Maritime Technology Conference Distinguished Guests, Ladies & Gentlemen, It is an honour for me to be invited to speak before such an august gathering as we have here tonight. My task is made even more difficult by the fact that this is an after dinner speech. The option to fall asleep is very real. I have been advised to keep the speech light hearted and not to raise topics which might cause indigestion. I hope you will find what I have to say palatable.

In the days when ships were made of wood and man of steel there was pride in craftsmanship. Today when ships are of steel and man a mere mortal the pride is less intense. Advanced technology now shares that pride when ships are built along production line methods. In this regard, the US Navy is one of the strongest navies possessing some of the most advanced naval technology.

The challenges facing Naval Architects and Marine Engineers now are quite different from that of 110 years ago when the Society of Naval Architects & Marine Engineers (SNAME) was founded.

Maritime challenges today cannot all be solved on the drawing boards of naval architects. Pirates with black eye pads wielding knives may not be hogging the news but piracy is still around, especially in the Straits of Malacca. Terrorism is a greater threat, more dangerous than the elements of the sea, and ironically, made more convenient by the increasingly globalised world. Since the sinking of the Titanic in 1912 the International Convention for the Safety of Life at Sea (SOLAS) has made provisions to safeguard life and property. Terrorism has made the contracting governments, on 12 December 2002, to adopt amendments to SOLAS to enhance the security of ships and port facilities. The adoption of the International Code for the Security of Ships and of Port Facilities (ISPS Code), effective from 1 July, 2004, is a grim reminder of the world we now live in.

All of us believe that we should leave the world a better place than when we found it. The environment has not been accorded this philosophy in a serious enough manner although much has been done by various professionals in different disciplines. In the old days ships were designed to protect the cargoes from the sea. Now naval architects and marine engineers spend long hours on how to prevent the cargoes from polluting the seas. Yet countries are refusing refuge to ships in distress. The latest case in point is the Prestige. Marine engines emissions now meet Nox and Sox requirements. In contrast, deforestation goes on unabated. 2,000 single trees are cut down per minute in rainforests. 1,000 acres of trees are cut daily in the US. 65% of Central America has been cleared for pastureland. Some countries such as Ivory Coast, Nigeria, Costa Rice and Sri Lanka are likely to lose all their tropical forests by year 2010 if no conservation steps are taken.

Ballast water exchange restrictions will likely be mandatory next year to prevent the transfer of harmful aquatic organisms and pathogens between ports. IMO guidelines have been in place since 1997. IMO currently recognizes two methods for ballast water exchange at sea – the flow-through and the sequential. A key assumption is that mid-ocean aquatic life forms will not survive in coastal waters and vice versa. It follows from this assumption that any ballast water exchange must take place in open deep waters, giving rise to the problems of longitudinal strength, stability, dynamic loads, slamming and propeller emergence. Here in California the discharge of ballast water containing exotic organisms is prohibited since January this year except by special permit.

The sea never fails to amaze and challenge mankind, rewarding at times, punishing more often than not. Oil and gas is close to my heart because offshore and marine is my business. In Singapore Keppel has endowed the Chair for Offshore and Marine at the National University of Singapore. Professor Torgeir Moan, a Norwegian, has been appointed the first Keppel Professor. Singapore and Keppel believe in R & D and that investments in technology is the way forward. Just as oil discovery moves to deeper waters and engineers come up with all kind of floaters as a solution, jack-ups are moving back into shallow waters in the Gulf of Mexico to drill deeper. Drilling deeper brings about high pressure and high temperature problems which require different solutions. Gas reserves are now estimated to be more than oil reserves and it is forecast that global gas consumption will exceed that of oil by the year 2020. Liquefied Natural Gas or LNG would increasingly be used by Japan, Korea and China. There are now about 180 LNG ships worldwide of either the Moss design or of Membrane tank designs. This will double in the next few years. The CNG (Compressed Natural Gas) ships are about to break into the market, offering a viable solution for stranded gas and short hauls and as an alternative to pipelines and re-injection. The United States wants to use LNG but do not want to have terminals or storage tanks on its soil. Offshore LNG Technical could be a good solution.

Maritime technology has come a long way in the last 50 years. We are now transporting large volumes of ISO containers filled with all kinds of commodities and products, crude oil and product and equivalent natural gas all over the oceans and we are doing it with increasing efficiencies such that today marine transportation continues to be highly competitive with all other means of transportation. Ships are getting bigger, container ships of 2000 TEU were big mainline ships. Now 8000 – 9000 TEU ships are the latest mainline ships. Cruise ships have grown bigger and more luxurious too. However, like the VLCC's and other phenomena, the law of diminishing returns come to play and an optimum size will be found. All these result in cheaper maritime transportation. The high speed ferries have also made commuting more comfortable and fast, in particular in places like Hong Kong, the North Sea and the Baltic.

We are also using the sea for recreation with the huge growth in the fleet of leisure cruise liners, super luxurious cruise ships which bring millions of people to new locations for leisure. We are also producing oil and gas in remote deepwater locations, as deep as some 2 kilometres. These resources are produced at very competitive cost, enabling resources from the deep sea in the Gulf of Mexico, Brazil and West Africa to be exploited for use by man.

As professionals in the marine industry, safety must be paramount in all our undertakings. Although maritime safety in terms of number of ships lost at sea, fatalities or oil spills have improved tremendously over the last 30 years, public perception is still negative. This is because one life lost is one too many. This also explains the bashing of classification societies in recent years because the public look to Class as the custodian of safety standards. This bashing, however, is ironical because we are unable to find an organization with the required knowledge and network to replace Class – not the IMO, the Coastguards, the Flag State or the Port State. All these organizations complement Class. It makes sense therefore for us to support the International Association of Classification Societies (IACS) to accelerate the adoption of unified requirements for the benefit of a unified standard of safety. To be effective, a Class Society must be International. A National Class Society cannot be effective although it may serve a political purpose. When IACS gets its act together safety would be enhanced and competition between Class Societies would be on services provided. Shipyards, Shipowners and society at large stand to gain from this evolution of Class Societies.

Ultimately, it is up to all of us, the professionals in the Maritime and Offshore community to take the initiatives to promote advancements that continue to make the use of the sea by mankind sustainable for future generations.

On this optimistic note I end my speech as we look forward to more restful sleep.

KEYNOTE SPEAKER

William A. O'Neil

Mr. President, Ladies and Gentlemen, Good afternoon, and thank you for giving me the opportunity to address what I feel is going to become one of the most important and productive technical meetings conducted within the shipping industry in this year. The fact that it is a truly international event is certainly a bonus.

Although globalization has become a favourite buzzword of recent years whenever people talk about the changes which are taking place in the supply and demand of materials and services around the world, ships have been operating globally for centuries and, indeed, the ability to move goods from country to country and between continents by sea has always been the basis of international trade.

In some circles globalization is considered to be an economic imperative, whereas in others it is taken as the work of the devil. Whatever your views, there can be no challenge to the fact that high quality, yet cheaper, manufactured goods from Asian countries and particularly China are now overwhelmingly popular in the rest of the world.

The changes in trading patterns and the volumes of bulk and manufactured goods being shifted around the globe have had a dramatic impact on ships and seafarers. The size and power of ships has increased since the middle of the last century to reach proportions previously undreamed of. And many of the cargoes carried today, both in nature and volume, present a potential environmental hazard. At the same time, ships' crews are fewer and the skills needed to perform their tasks are greater and more demanding than ever before. All of this is taking place when there is a major shift in the resource pool of seafarers from the so-called traditional maritime countries to the Asian region where, for instance, the Philippines supplies some 250 to 300 thousand seafarers.

It is therefore fitting and timely that SNAME should have taken the leadership role in bringing together professional naval architects' and marine engineers' organizations from around the globe to meet for the first time at a single conference to discuss and debate the issues facing our industry.

Now well into its second century, SNAME has established a first-class reputation through its involvement in all aspects of research, production, maintenance and operation of ships, yachts, offshore installations and just about every other kind of maritime structure that exists today. It functions in co-operation with government and regulatory agencies, scientific and research laboratories, academic institutions, and the commercial marine industry. A glance at the technical programme before delegates to this event shows that this wide ranging and eclectic scope has been truly reflected in the proceedings of this landmark meeting.

The role of technical organisations and societies within shipping is unusual, in that, thanks to their longevity, and their continued professionalism and dedication, they exert an influence within the industry that is seldom seen in other business sectors. It is important that they do so because, whichever way you look at it, shipping is a fragmented business, made up of so many different professional and technical disciplines which, at first glance, might not appear to have very much in common. Yet whether you are a marine engineer, a deck officer, a naval architect, a dry cargo specialist, a tanker expert, or from any of the other highly specialized branches of this unusually diversified industry, the fact is that most issues, at some point in time, cross the boundary into what is generally thought of as someone else's domain.

That is why an event of this kind, which brings together the widespread technical expertise that exists in every corner of the industry, is so important. It provides a chance for common issues and problems to be discussed from a wider perspective, and for shared solutions to emerge. It offers a great opportunity for synergies to be discovered and developed and I know that the participants will relish the prospect of broadening and enhancing the range and scope of their professional relationships during the course of the conference this week.

This afternoon, a round table of maritime technical societies will look at issues such as how they engage with national governments, with industry, and with IMO, and the topics they should be pursuing in the course of these working relationships. This sort of consideration is timely because, although it is true that the contribution made by the professional societies is greatly valued and can be measured over many years, nevertheless it is also true that nothing remains the same forever. Therefore a review of long-standing arrangements and practices often throws up new and improved ways of working. And the management of these changes is undoubtedly the key to success.

Many of you will doubtless be aware that certain proposals have been made at IMO which, if they came to fruition, would affect the role of classification societies and their relationship with IMO. First of all, and to allay any misapprehensions, let me say from the outset that, despite what you may have read in some sectors of the shipping press, there is no intention that IMO would take over the detailed work of the classification societies, nor that IMO should be responsible for plan approvals, inspections, surveys or matters of that nature. We do not have the mandate, we do not have the resources nor the desire to take on this kind of responsibility and these concepts are not included in the proposals that our Member Governments are now considering.

What is on the table is the notion that IMO would develop "goal-based" standards for ships' construction and equipment; which means that IMO would state what has to be achieved, but would not be involved in the details of precisely how this should be done. In this way, the method of reaching the agreed goals would remain open to differing approaches and, of course, to innovation. It would be clear that, whatever approach was adopted, the overall standard could not be below an internationally agreed level.

This would still leave classification societies, naval architects, engineers and ship builders the freedom to decide on how best to employ their professional skills to meet the required standards. It would also, clearly, be possible to benefit from the wealth of expertise within the technical organizations and classification societies when drawing up the standards. But the key factor is that the standards would be internationally agreed to, transparent and capable of being monitored by national administrations.

Sub-standard ships are the bane of our existence and are the curse of every good ship owner and operator. Unfortunately, they tend to become the measure by which the whole industry is gauged and they provide fodder for those who are seeking to establish an ever more stringent – and costly – regulatory regime. It is our goal and desire to rid the seas of these unwanted ships and their operators and we will continue to do everything within our power to achieve this.

There is no denying that at the outset ships should be built to the highest possible standards and I would welcome any measures which will move us closer to this objective. This becomes particularly significant with the wider employment of double hull ships, where we must ensure that overall hull strength is not compromised by reduced scantlings and that future maintenance issues are not neglected. At present there is no legislation to control or guide these matters so the introduction of a mechanism to ensure harmonised, internationally agreed standards, under the umbrella of IMO, would be a positive step in the right direction.

To push the idea forward a new item has been added to the work programme of IMO's Maritime Safety Committee and interested Governments and international organizations are invited to submit specific proposals on goal-based standards and design philosophies, to help clarify and define their meaning.

The whole question of IMO's involvement in new building standards also throws the spotlight on a growing phenomenon that is becoming increasingly apparent in almost everything IMO does. Despite the huge amount of intensive and successful technical work that is carried out in the various committees and sub-committees, there is no avoiding the fact that the political and economic consequences of IMO's work are gaining in prominence all the time, often to the extent of actually driving the agenda.

This is something that we must acknowledge and deal with. There is always a balance to be struck between the aspirations of politicians, who of course have to act and be seen to act in their constituents' interests, between the commercial interests, and between the technical experts, who traditionally have tried to seek the best solutions to whatever problems have arisen, without always taking into account the political or economic framework that surrounds them. I am somewhat concerned that this balance may have swung too far away from the technical aspects, and that we may be in danger of pursuing solutions which are politically driven and are designed to please electorates rather than being focussed on technical facts. I'm not for one minute suggesting that we should ignore the political and economic dimensions; I acknowledge their importance and recognise that they must be taken into account. However, we cannot sacrifice sound technical advice, with its long-term benefits for safety and protection of the marine environment, to the short-term horizons of politicians. Of course, the technical arguments must be well supported by research and analysis and be put forward in a clear and forceful manner.

I believe that there are four main pillars of IMO, four bedrock qualities that the Organization is able to draw on, and has drawn on time and time again. Its technical expertise is unparalleled and well documented, and there is no doubt that the professional organizations which are coming together at this meeting make a very significant contribution in this regard. I have already mentioned the political and economic dimensions of the Organization's work, which are always present and which will always add their weight to any debate or discussion. But perhaps the single most important quality, one which I think has played a major part in the success of IMO over the years, is an ability to work together for the common good. The Members of IMO have traditionally been able to see that the way forward does not always entail defending their own positions to the hilt, but rather in understanding the views of others and seeking the concessions that eventually lead to solutions that will work for everyone. Some call it compromise or consensus; I prefer to call it common sense.

In IMO's technical committees and sub-committees a huge amount of very detailed, intense work of a highly technical nature is constantly being carried out. This is the real strength of IMO; the unique forum it provides for the world's best experts from within the shipping industry and from within the Member States to come together to thrash out comprehensive solutions to problems and to agree on universal standards that are fair, practicable and, above all, effective. The scrutiny to which any new measure progressing through IMO is subjected to is of the most rigorous nature and, while occasionally some complaints are heard that the processes are too slow, the real proof of the validity of IMO and the measures it has adopted over the years lies in the continued decrease in shipping casualty statistics, whether measured in terms of ships lost, lives lost or damage to the environment.

IMO's track record as a standard-setting body is unparalleled, and that is why the Organization moves with confidence whenever its Members require it to take on new areas of responsibility. The immense amount of work carried out in rapid time to establish a comprehensive international regulatory regime for maritime and port security is among many examples that bear witness to IMO's ability to act quickly and decisively when confronted with a new challenge. I firmly believe that the Organization can and will respond with similar alacrity to whatever new tasks it may be given.

The one proviso, of course, is that the necessary resources are forthcoming, and once again it is worth stressing the importance of effective partnership between the technical experts and their political counterparts. It is, by and large, the technical advisors who devise the work programme and who plot the way forward - very often, I might add, anticipating issues which subsequently may become the politicians' hot potatoes. But there is sometimes a discernible reluctance to supply the necessary funding and manpower resources needed to carry out these programmes to the maximum possible degree of effectiveness. A lean organization may outperform its better fed counterpart in certain areas, but if this is at the expense of limiting its range and scope unduly, it may be a false economy.

IMO is at the pinnacle of the shipping industry, and it has earned its position of leadership through getting the job done, time after time. That is why it is so important, not just for the Organization but for the broad and universally accepted goals of continued improvements in maritime safety and in environmental protection, that regional, political agendas, no matter what part of the world they come from, do not start to undermine the foundations that have been so painstakingly and successfully established and on which the current, very effective, regulatory framework has been built.

The input from the shipping industry's technical bodies has been, and will continue to be, central to the establishment and development of that framework. The level of their participation here in this most prestigious event demonstrates quite clearly that they understand the need for common policies on the issues that affect the maritime industry today.

As I mentioned earlier, the success of IMO has been founded on four essential qualities – technical expertise, political awareness, economic understanding and plain common sense – and these are the qualities that it will draw on as it moves into the future. No-one can predict exactly what that future will hold, but one thing I am certain of is that the participation of the multitude of specialised technical associations that cover all sectors of our industry will continue to play a pivotal role in the Organization's achievements. IMO simply could not function as effectively and successfully as it does without their expertise and assistance.

Ladies and gentlemen, the World Maritime Technology Conference and Exposition this week is providing a forum for discussion and exchange on many key issues related to the future of the global maritime industry, and the programme promises to deliver a wealth of informative technical sessions, lively panel discussions and passionate debate. I wish you a fruitful and positive outcome to your deliberations here in San Francisco, and again offer my commendation to the organisers and their partners for their foresight and vision in bringing us all together.

Thank you.

2003 Annual Meeting Photographs

Following are selected formal photographs from the Annual Meeting.
Some events and awardees are not pictured because formal photographs were not available.

50-YEAR GOLDEN MEMBER CERTIFICATE
Presented by Bruce S. Rosenblatt (r) to Thomas J. Sanotr, Jr. (l)

50-YEAR GOLDEN MEMBER CERTIFICATE
Presented by Bruce S. Rosenblatt (r) to Arthur J. Haskell (l)

CERTIFICATE OF APPRECIATION
Presented by Bruce Rosenblatt (l) to Thomas Lamb (r) for his dedicated service to the Society as Editor of the revised text Ship Design and Construction

CERTIFICATE OF APPRECIATION
Presented by Bruce Rosenblatt (l) to David Zoller(r) for outstanding dedicated service to the Society as chair of the Finance Committee, 1996-2003

CERTIFICATE OF APPRECIATION
Presented by Bruce Rosenblatt (l) to Richard Boutwell (r) for outstanding continuing dedicated service to the students of the Society in his leadership of the Student Congress

CERTIFICATE OF APPRECIATION
Presented by Bruce Rosenblatt (l) to William Morgan (r) for outstanding service to the Society as a member of the Publications Committee 1985-1994 and chair, 1995-mid 2003

CERTIFICATE OF APPRECIATION
Presented by Bruce Rosenblatt (l) to Petros Lalangas (r) who received the award for John Angelicoussis for his dedicated and loyal service to the Greek Section of the Society

GRADUATE PAPER AWARD
Presented by Philip B. Kimball (r) to Richard Camilli (l)

GRADUATE PAPER HONOR PRIZE
Presented by Philip B. Kimball (r) to John Hootman (l) (and Konstantinos Psallidas in absentia)

INTERNATIONAL STUDENT OFFSHORE DESIGN COMPETITION WINNERS:
University of Michigan-1st Place,
Presented by John Filson and Raymond Fales
Faculty Advisor: Thomas Lamb
Students: Louis A. Detrisac, Victor D'Souza, Michael Rugnetta, W. Scott Opdyke, Katherine Sultani

DR. JAMES A. LISNYK SHIP DESIGN COMPETITION
Virginia Polytechnic Institute-1st Prize
Presented by Glenn Ashe
Faculty Advisor: Alan Brown
Students: Jason Albright, Daniel Demko, Mathwew Hecht, John Leek, Greg Morrow, Keith Webster

DR. JAMES A. LISNYK SHIP DESIGN COMPETITION
University of Newcastle upon Tyne 2nd Prize
Presented by Glenn Ashe
Faculty Advisors: Richard Birmingham and Martin Downie
Students: Alexandros Poutachidis, Michael Anninos, John Monogioudis, Marika Psarros, Andrew Smart, Stewart Baker

DR. JAMES A. LISNYK SHIP DESIGN COMPETITION
Norwegian University of Science and Technology-3rd Prize
Presented by Glenn Ashe
Faculty Advisor: Kai Levander
Students: Hans Danielsen,Haavard Haaskjold, Erlend Sandvik, Thorvald Furuseth

VICE ADMIRAL E. L. COCHRANE AWARD
Presented by Edward M. Lewandowski (r) to (from l to r): William France, Randolph Paulling, R. Keith Michel, Colin Moore, Thomas Treakle, Marc Levandou

ABS – CAPTAIN JOSEPH H. LINNARD PRIZE
Presented by Ronald Kiss (r) to Bruce Hutchison (l)

ELMER L. HANN AWARD
Presented by Michael Powell (l) to Carlos Tomassoni (r) (with Ted Huynh and Thomas Schiller in absentia)

WILLIAM M. KENNEDY AWARD
Presented by Michael Powell (r) to Michael Butler (l)

WILLIAM H. WEBB MEDAL
Presented by Robert Beck (l) to Michael

MEDAL PRESENTATIONS

The Vice Admiral "Jerry" Land Medal presented to Richard Vortman (l) by Arthur Haskell (r)

The David W. Taylor Medal prresented to Robert Keane (l) by Edward Comstock (r)

PAST, PRESENT AND FUTURE PRESIDENTS (l to r): Ronald Kiss, Perry Nelson, Bruce Rosenblatt, Roger Compton (president-elect), Joseph Cuneo, Arthur Haskell, Jose Femenia

Papers Presented at the 2003 Annual Meeting

Stern Boat Deployment Systems and Operability

Rubin Sheinberg (FL), USCG Engineering Logistics Center, **Peter V. Minnick** (M), USCG Engineering Logistics Center, **Thomas G. Beukema** (V), USCG Engineering Logistics Center, **Wojciech Kauczynski** (V), Norwegian Marine Technology Research Institute, **Andrew L. Silver** (M), Carderock Division, Naval Surface Warfare Center, **Christopher Cleary** (M), USCG Engineering Logistics Center

ABSTRACT

Ship-deployed boats are essential to the success of many of the U.S. Coast Guard's seagoing missions. In recent years, boat deployment systems have evolved beyond the traditional side-davit system with dual falls, and now include systems that deploy Rigid Hull Inflatable Boats (RHIB) via stern ramps that are integrally designed into the transom of the mother ship. A worldwide assessment of vessels with stern-launch capability undertaken by the Coast Guard confirmed that no established analytical approach was used in the design stage, nor was a standard or criteria set available to the designers of these systems. Consequently, a dedicated effort was undertaken to review existing system evaluation approaches, develop boat deployment criteria and analysis methodologies, conduct stern boat deployment operational tests, and perform percent time operability (PTO) analyses in order to develop a workable design and evaluation methodology for stern-ramp cutters. This paper documents the findings of the worldwide survey and the current state of the art for boat deployment assessment. Specifically, criteria for stern-ramp motions, a ramp availability criterion and stern-ramp deployed boat criteria are covered, along with an example case using these criteria to establish the PTO of stern ramp operations for typical ocean environments.

INTRODUCTION

Ship-based boats are essential to the successful prosecution of many of the U.S. Coast Guard's seagoing missions. The ability to launch and recover boats in a broad range of environmental conditions is necessary to complete these missions. In recent years, boat deployment systems have evolved beyond the traditional side-davit system with dual falls, and now include systems like Rigid Hull Inflatable Boats (RHIB) deployed via stern ramps integrally designed into the transom of the mother ship.

As part of the Coast Guard's need to conduct technical assessments of existing cutters and proposed ship designs to establish the effectiveness of the various boat deployment systems, engineering criteria and methodology must be available. Within the Coast Guard in the mid-1990s, there was a developing interest in stern deployed boats from oceangoing cutters, and the Deepwater Project was considering design ideas for replacement cutters including a new class of Medium Endurance Cutters and a National Security Cutter class. The Naval Architecture Branch of the U.S. Coast Guard Engineering Logistics Center undertook the initiative to develop criteria for design and evaluation of stern launch and recovery systems for small boats from ships up to 400 feet. To this end, a systematic approach was taken to conduct a worldwide assessment of vessels with stern-launch capability, develop boat deployment criteria and analysis methodologies, conduct stern boat deployment operational tests, and perform percent time operability (PTO) analyses.

The purpose of the worldwide search of candidate vessels operating with stern deployment systems was to determine their design criteria and to determine their operating characteristics. The ships identified ranged in length from the U.S. Coast Guard's own 87-ft WPB (Marine Species Class) to the 300-ft Japan Coast Guard Cutter *Erimo* (ex-*Ojika*). The investigation involved meeting with owners, operators and designers to determine the characteristics of their stern deployment systems, the process and the research done by which they selected the stern launch and recovery system and how they optimized it. The survey specifically included questions regarding standards and criteria used in the design of these vessels. The result of this investigation confirmed that no established analytical approach was used in the design stage, nor was a standard or criteria set available to the designers of these vessels.

Consequently, a dedicated effort was undertaken to review existing system evaluation approaches, conduct ship motions studies and generally draw together diverse viewpoints of other investigators into a workable design and evaluation methodology for stern-ramp cutters. The developed approach would have to be applicable to stern deployment of RHIBs and Fast Response Craft (FRC) from various sizes of cutters, would allow comparison to side-boat deployment systems, and would have to be relatively easy to employ given the analysis tools in existence today.

This paper documents the findings of the worldwide survey and the current state of the art for stern boat deployment assessment. Criteria sets for side-davit motion criteria are reviewed, and the recently developed stern-ramp motion criteria, ramp availability criterion and stern-ramp deployed boat criteria are covered. Finally, an example case using these criteria to establish the PTO of boat recovery via a stern ramp for a typical North Atlantic Ocean environment is provided.

NOMENCLATURE

a = Threshold water depth at the ramp sill.
B_{WL} = Breadth on waterline.
Δ = Displacement.
GM_T = Transverse metacentric height.
H_s = Significant wave height.
L_{pp} = Length between perpendiculars.
m_0 = Zeroth moment of the relative motion spectrum.
m_2 = Second moment of the relative motion spectrum.
$\Phi(.)$ = Probability integral.
r_{44} = Roll inertia radius.
σ = Standard deviation of the relative motion process; relative vertical displacement rms.
σ_v = Standard deviation of the first derivative of the relative motion process; relative vertical velocity rms.
T_M = Draft at midship.
T_{RA} = Minimum average ramp availability time interval for the retrieval operation; criterion limit.
T_Z = Mean zero-crossing period.

τ_+ = Time duration that the relative motion exceeds the threshold, "a".

τ_- = Ramp availability duration; time duration that the relative motion is less than the threshold, "a"; i.e., ramp sill (or other required depth) underwater w.r.t. local wave elevation.

$\langle\tau_-\rangle$ = Mean ramp availability duration.

Y(t) = Ramp sill vertical relative motion process; defined to be positive when the selected point on the ship rises above the local water level.

Head Seas are 000° relative wave heading and seas on starboard beam are from 090°.

WORLDWIDE STERN LAUNCH AND RECOVERY CAPABILITY

The process of developing the assessment methodology started with a worldwide search to identify candidate vessels and their designers that presently operate stern deployment systems.

In preparation for the survey, a standard list of questions was developed that would provide uniform responses that could be compared across all the vessels. The questions focused on the operation of the systems as well as any changes that would be desirable. A similar set of questions was developed for the system designers to assess the design criteria used and to determine what, if any, changes would be made for the next generation of stern boat deployment systems.

The following ships were visited during the course of our research:

a. Japan Coast Guard patrol vessel *Erimo* in Tokyo, Japan
b. Mexican Navy ship *Justo Sierra* in Acapulco, Mexico
c. U.S. Navy Patrol Craft *Tornado* in Little Creek, Virginia
d. Canadian Coast Guard Ship *Gordon Reid* in Victoria, British Columbia, Canada
e. Netherlands Antilles and Aruba Coast Guard Cutter *Jaguar* in Curaçao, Netherlands Antilles
f. Finnish Frontier Guard Offshore Patrol Vessel *Telkkä* in Turku, Finland
g. USCG Coastal Patrol Boat *Hammerhead* in Baltimore, Maryland
h. German Sea Rescue Service vessel *Vormann Steffens* in Bremerhaven, Germany
i. Swedish Coast Guard High Endurance Class *KBV 201* in Karlskrona, Sweden

The survey provided valuable information on the design parameters and operation of stern boat launch and recovery systems. In addition to investigating the launching systems, the types of FRCs that were used were also investigated to determine if any special modifications were needed for their use with the stern ramps. It should be noted that all the vessels surveyed for stern deployment capability also had an alternative, traditional boat launch system available.

Other vessels identified, but not visited, with stern boat deployment capability include the Philippines Coast Guard vessels *BRP San Juan* and *BRP Edsa II* and the United Kingdom Customs and Excise ships *HMCC Seeker* and *HMCC Searcher.*

Types of Recovery Systems

The stern launch systems investigated are categorized into four distinct systems with variations on arrangement details. The systems are categorized as follows:

1. Well Dock (as exhibited on the Japanese Coast Guard patrol vessel *Erimo*, Figure 1)
2. Fixed Ramp (as exhibited on the USCG Coastal Patrol Boat *Hammerhead*, Figure 2)
3. Hinged Ramp (as exhibited on the Canadian Coast Guard Ship *Gordon Reid*, Figure 3)
4. Extended Ramp (as exhibited on the Swedish Coast Guard vessel *KBV 201*, Figure 4)

The results of the survey revealed that there were two basic configurations of sloping stern ramps. They are

a. A shaped ramp designed to fit the hull form of the FRC, and
b. A flat ramp with longitudinal tubular rails (or bunks) that provide support and help center the small boat during retrieval.

Well Dock

The stern well dock system is employed on the Japan Coast Guard patrol vessel *Erimo*. This was the largest ship, at 91.4 m (300 ft), identified with a stern launch capability. In the well dock system, the small boat is carried inside the ship in a dry compartment that must be flooded before the boat can be launched.

The *Erimo* is unique in that the 5.5-m FRC is backed out of a stern well under its own power (not launched down a ramp). The arrangement is shown in Figure 1. The boat is kept in a normally dry compartment behind a watertight stern gate. There is a wave-dampening flap installed near the entrance to the stern well that dampens the water motions in the well to maintain the required water depth for the FRC. The wave-damping flap has a roller installed on its upper edge to permit the FRC to pass over it without causing any damage to the hull. The well is lined with fendering material to protect the FRC.

The FRC is a decked-over boat with an operator's cockpit. A foam-filled fender surrounds the boat at the deck level for protection. The fenders prevent damage to the small boat during launch and retrieval. A diesel-driven water jet provides propulsion.

To launch the small boat, the well is flooded, the stern gate opened, the wave-dampening flap is pulled down, and the small boat exits under its own power. The control of the small boat while backing is limited. To

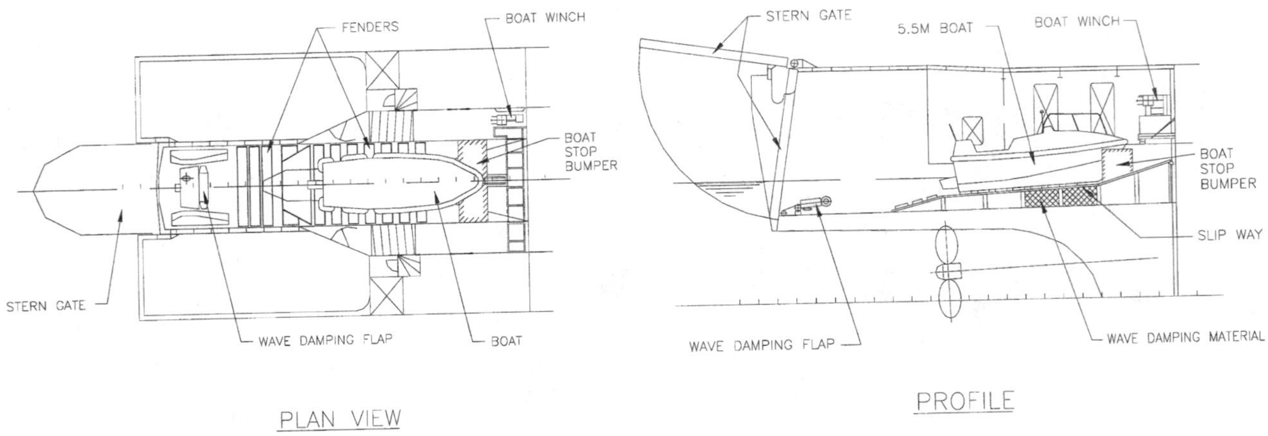

Figure 1. Stern launch well dock arrangement of Japan Coast Guard patrol vessel *Erimo*.

recover the FRC, the boat powers its way into the well dock and is secured. The stern gate is closed and the well is pumped dry.

In addition to the stern well, the ship was equipped with Miranda davits for side launching the rescue craft. The Miranda davits give the ship the ability to launch the small boats up to Sea State 4. The crew prefers to use the side-launched boats because the level of expertise needed is less for the side-launch system. The other ships in the Japan Coast Guard use Miranda davits. The *Erimo* is the only ship that uses a stern well to launch a small boat. Training can take as long as a year before crewmembers become proficient at launch and retrieval operations from the stern well.

Sloped Stern Ramp

Sloped stern ramps can be categorized into two different arrangements; the shaped ramp where the ramp surface is built to suit the shape of the FRC and lined with friction reducing material, and a ramp with a flat surface with tubular rails to guide and support the FRC.

A typical example of a shaped ramp lined with friction reducing material is that of the *Hammerhead* shown in Figure 2. The arrangement of *Jaguar* is very similar.

Shown in Figure 3 is the arrangement of the stern launching ramp on the *CCGS Gordon Reid*. This is an example of a ship with a flat ramp surface with tubular rails. The figure also shows the hinged ramp feature. The arrangement of the *Tornado* is similar without the hinging capability and the stern doors open externally.

The arrangement of the *KBV 201*, shown in Figure 4, provides an example of a downward hinging stern gate. The *Vormann Steffens* uses a similar arrangement of the hinging stern gate.

Summarized in Table 1 are the general characteristics for the ships with stern ramps. The vessels ranged in size from 26.5 m (87 ft) for the U.S. Coast Guard's Coastal Patrol Boat *Hammerhead* to 74.4 m (244 ft) for the Mexican Navy ship *Justo Sierra*. The length of the ship

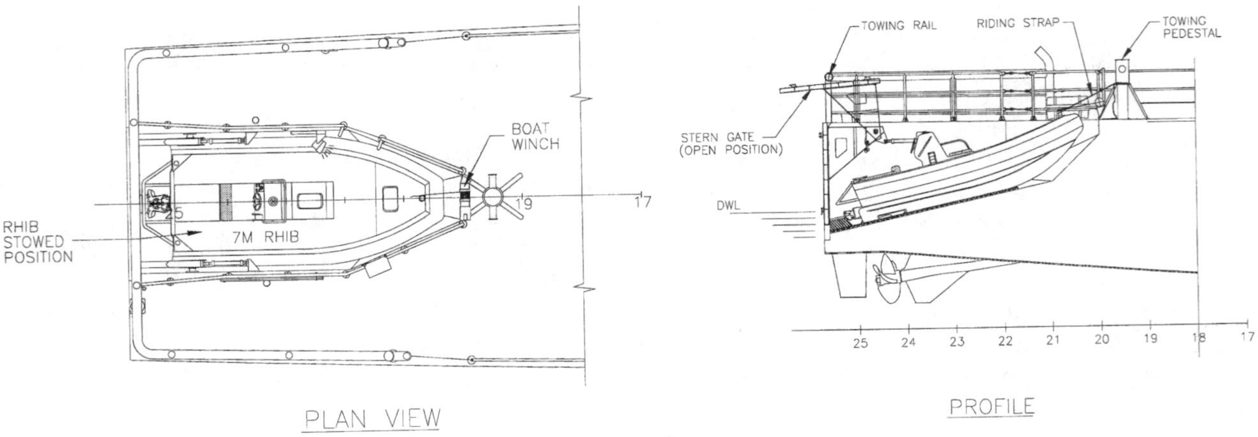

Figure 2. Typical sloped ramp built to the shape of the FRC (*USCGC Hammerhead*).

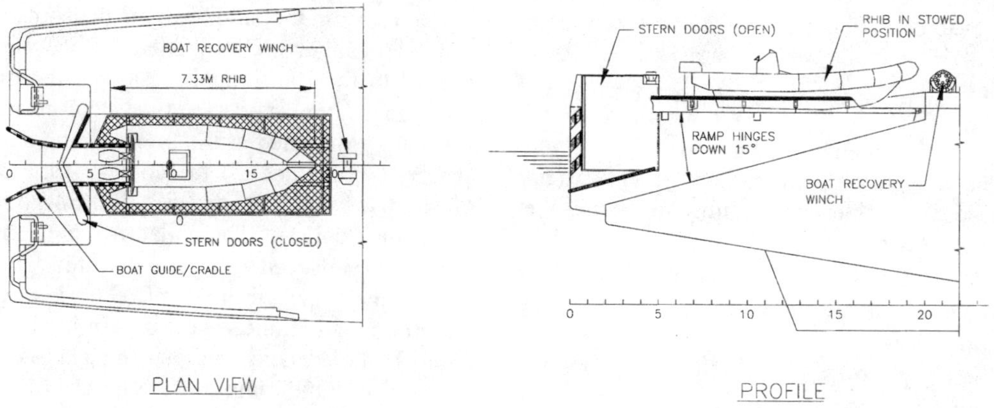

Figure 3. Hinged ramp with tubular rails used on the *CCGS Gordon Reid.*

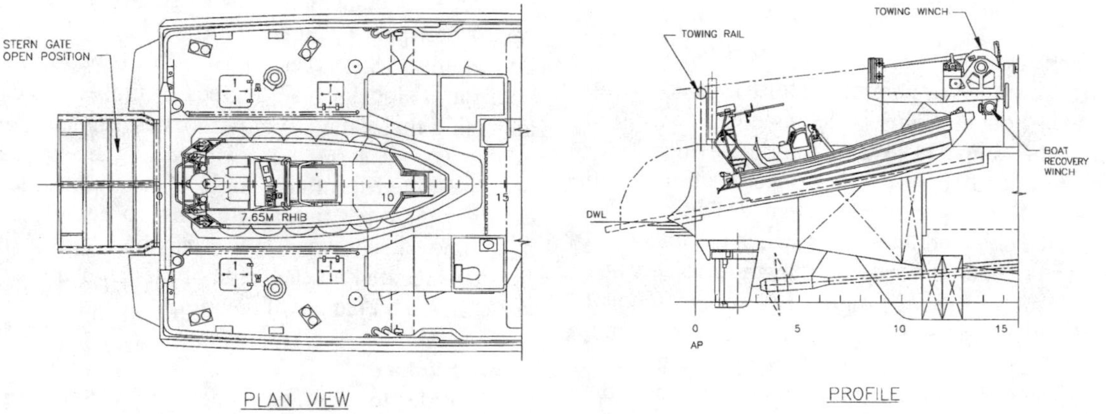

Figure 4. Example of a downward hinging stern gate (*KBV 201*).

Table 1. Ship and Ramp Characteristics

Ship	Year Built	Length of Ship	Type of Stern Ramp	Slope of Ramp	Sill Depth	Operating Sea State
Justo Sierra	1999	244'-0"	Shaped Ramp	8¼°	0"	3
Tornado	2000	179'-0"	Flat Ramp w/Rails	16°-18°	15"	3
Gordon Reid	1990	163'-10"	Hinged Ramp w/Rails	15°	34"	6
Jaguar	1998	140'-5"	Shaped Ramp	14°	12"	4
Telkkä	1999	161'-5"	Ramp w/Cradle	7°	-12"*	2
Hammerhead	1998	87'-0"	Shaped Ramp	12°	14"	4
Vormann Steffens	1989	90'-3"	Ramp w/Self-capture	14°	24"	1
KBV 201	2001	170'-7"	Ramp w/Rollers	12°	12"	4

*Sill on the *Telkkä* is 12 in. above the waterline for ice operations.

affects the motions and accelerations at the stern. These motion responses to operation in a seaway will vary with the ship length, geometry, wave characteristics, ship heading and speed.

Major Elements of Stern Deployment System

The main parameters of the stern deployment system that could significantly influence the ability to recover the boat safely are as follows:

- Ramp sill depth
- Shape of ramp opening
- Ramp slope
- Ramp surface
- Ramp side and clearance
- Overhead clearance
- Stern door or gate configuration
- Water management system
- Capture mechanism
- Stern flaps/wedge

Ramp Sill Depth

One of the most important factors in stern ramp operability is sill depth. This is the submerged depth at the aft end of the ramp. It governs the time available for recovery of the small boat. There is an optimum sill depth for each stern deployment system design and this depends upon the ship and the seas in which the boat must be recovered. In general, greater sill depth translates into a greater ability to operate in higher sea states. Greater sill depth usually ensures more water inside the ramp, thereby providing a longer time of ramp submersion. However, other operational criteria may influence the determination of maximum sill immersion.

Our investigation shows that sill immersion is very important to the ramp design and must be greater than a certain level (0.5 to 0.75 m (1.5 to 2.5 ft) depending on the particular arrangements of the designed system) in order to be effective. Less than this amount of depth and the stern ramp deployment operability and effectiveness could be drastically reduced due to a very short ramp availability time and the high probability of extreme impact accelerations induced during "dry landing".

The vessels investigated had sill depths that varied from 12 inches above to 34 inches below the design waterline. Generally, ships that are operated in colder climates, where ice is a consideration, had ramps with the sill at or above the waterline. This was done to reduce the possibility of ice entering the ramp area. In order to increase the effective sill depth for recovery of the small boat, these ships employed either a stern gate that hinged down forming a ramp extension or a cradle that was extended beyond the stern. Such is the case for the Swedish Coast Guard vessel *KBV 201*, the German Sea Rescue Service vessels and *Telkkä*, a Finish Frontier Guard vessel.

Shape of Ramp Opening

Several of the ships visited had boat well openings that were either well rounded or had stern doors that opened to form a "funnel" to aid in the recovery of the small boat. In every recovery system investigated, where the ramp terminated at the transom, the small boats were observed to bounce off the corner of the ramp entrance. Most of the small boats were RHIBs with inflatable or foam-filled collars. The collars were either covered with specially reinforced materials or fitted with rub strips to protect the collar from damage during recovery. The well-rounded entrance corners permitted the coxswain to fend off the corners, without damage, as the RHIB was powered into the ramp area. The use of square or sharp corners at the intersection of the transom and boat well will increase the wear to the collar of a RHIB.

Ramp Slope

Ramp slope or angle was observed to be more important to launching than to recovery. Ramp slopes varied between 7 degrees on the Finnish Frontier Guard *Telkkä* up to 18 degrees on the *USS Tornado*. All vessels with ramp slopes of 12 degrees or higher are capable of launching their small boat without assistance. In the case of the *Justo Sierra*, the presence of the overhead flight deck restricted the ramp angle. The 8¼-degree slope of the *Justo Sierra's* ramp is too low to permit the FRC to overcome its own static friction and self-launch. Capstans located on each side of the ramp are used to pull *Justo Sierra's* boat down the ramp.

Ramp Surface

In order to facilitate launch and retrieval, most ramp surfaces are lined with high molecular weight plastics, such as Ultra Poly, to reduce friction on the sliding surface. This was true for all ramps except for the ramps on the *Telkkä*, *Vormann Steffens*, and *KBV 201*. These mother ships use a wheeled cradle, rollers, and wheels, respectively, to permit movement of the boat in the ramp. The wheeled cradle of the *Telkkä* permits launching the FRC with a low, 7-degree ramp slope.

The ramp surfaces of the other mother ships were found to be either flat (i.e., no deadrise) with friction reducing tubular bunks or shaped with friction reducing material attached directly to the ramp surface. The friction reducing pads are of composite construction with a cushioning layer, to absorb impacts, below the Ultra Poly surface.

Ramp Side and Clearance

Ramp side clearance was observed on the ships to vary widely from as small as four inches to as much as 18 inches. The clearance must be sufficient to give the coxswain confidence when entering the ramp, but not so much that the FRC will come to rest out of position. Too great a clearance also gives the coxswain increased confidence that translates directly into faster recovery speeds. The faster recovery speeds in turn lead to greater deceleration forces when contacting either the ramp walls or the ramp surface.

One vessel, the *USS Tornado*, was designed to carry two different size RHIBs. To accomplish this, the ship has an adjustable side fendering system and movable longitudinal bunks in order to custom fit each RHIB.

Overhead Clearance

On the mother ships with gates that hinge to the overhead, the clearance between the small boat and the open gate can be of great concern to small boat crews. Of all the vessels visited only thee expressed concern with their overhead clearance during recovery. These were the *Jaguar*, *Hammerhead*, and *KBV 201*. Of the other vessels visited, the *Gordon Reid, Telkkä* and *Tornado* had outward hinging doors with no overhead restriction. The *Justo Sierra* and *Vormann Steffens* would not perform launch operations seas above Sea State 2 thus avoiding high vertical motions. On *Hammerhead* and *Jaguar*, the open gate that is stowed over the ramp area imposes an overhead restriction. On the *KBV 201* there was a tow rail present over the ramp at the transom. With the overhead restrictions, the boat crews had the perception that the overhead obstacle was closer than it actually was and it created apprehension when entering the ramp area.

Stern ramp configurations with overhead constraints, such as hinged stern gates, transom bulwark, overhead deck, etc., could significantly reduce operational effectiveness of the stern ramp system. Therefore, in a stern ramp design, the overhead clearance should be considered as one of the system design and operational constraints, and it should be evaluated with respect to the limiting sea condition.

Stern Door or Gate Configuration

The ramp transom openings are closed either by doors that hinge outward or by gates that hinge up or down. All the stern doors and gates observed were hydraulically powered. The outward hinging doors used on the *USS Tornado* open to 105 degrees to form a funnel shaped entrance to the ramp area. The coxswain would try to get the RHIB into the center of the ramp, but can use this funnel shape to help guide the RHIB if needed. Upward hinging gates must be designed so that there is sufficient overhead clearance for personnel and equipment during launch and recovery. To achieve this clearance the hinge points are placed high on the ramp walls, which also keeps the gate operating equipment out of the way of the small boat during recovery operations.

Two vessels, *Vormann Steffans* and *KBV 201*, use a downward hinging stern gate. The ramps for these vessels end at the waterline at the stern. The surfaces of the gates are covered with wheels and rollers to form a ramp extension when deployed. They provide a sill depth of between 12 and 24 inches at the end of the fully deployed stern gate.

Water Management System

A water management system provides a stern ramp system with the ability to reduce the speed of water washed out of the ramp such that a small boat is not flushed out of the ramp area with the receding water. At the same time, such systems may modify the phase angle of the water motion inside of the ramp delivering more water to the ramp entrance when the mother ship is heaving and pitching up. An effective water management system can contribute to higher operability for stern ramp boat deployment.

The importance of water management was evident on two vessels, the *Gordon Reid* and the *Tornado*. Both of these vessels use tubular rails or bunks to guide and support the RHIB during recovery. With the rails, the RHIBs were supported above the ramp surface that allowed the water to flow out under the small boat's hull.

The ramp surface of the *Gordon Reid* is made of expanded metal that aids in dampening the waves faster and allows the water to dissipate quickly beneath the RHIB. The use of a water management system is especially important when operating in higher sea states where there is considerable wave action in the ramp.

Capture Mechanism

Capture of most small boats during recovery is accomplished by having a deck hand (winch operator) pass the winch line to the bowman in the small boat, who would attach it to the boat. For safety reasons it is not desirable for a crewmember to enter the ramp area and attach the winch line to boat. Use of an automated capture mechanism reduces the number of crew needed to perform the recovery evolution. Three vessels have different methods of capturing the small boat so that the winch line could be attached without the aid of a bowman in the boat.

On the *Telkä*, launch and recovery operations are accomplished with the aid of a longitudinally moving cradle with an integral, hydraulically operated arm that would reach over the RHIB bow and hold the boat in the cradle. With the boat captured in the cradle both the boat and cradle are winched up the ramp as one unit.

The *Vormann Steffens* employs an automatic capture mechanism that would engage the bow of the FRC as it enters the ramp. As the boat crosses the ramp threshold, special pins mounted on the bow of the FRC would be caught by the mechanism and pull the boat up the ramp. At the head of the ramp the boat is put on a riding hook and the mechanism disconnected. The pin feature on the FRC bow and the capture mechanism were specialized designed to mate together. Proper alignment between the bow pins and the capture mechanism is necessary for the system to work, hence the use of the system is limited to lower sea states.

On *KBV 201*, the winch operator attaches the winch line to the small boat with the aid of a long pole. The winch hook is attached to one end of the pole. During recovery, the winch operator uses the pole to reach over to the boat's bow eye and connects the winch line to the boat. After attachment, the pole is pulled free of the winch line and the boat winched up the ramp.

Two sources of power are used on the winches for recovery, electric and hydraulic. One vessel, *Jaguar*, uses an electric capstan for recovery. Recovery winches and capstans have line speeds that range from 100 to 200 feet per minute. This is necessary to pull the FRC quickly up to the stowed position so that the stern doors or gates could be closed.

Stern Flaps/Wedge

One ship visited, *Tornado*, has a stern flap installed to increase the ship's performance. A secondary benefit that the stern flap provides is to effectively increase the sill depth by extending the ramp surface further below the waterline. However, to be effective in increasing the sill depth, the stern flap must have its upper surface at or near the angle of the ramp and should incorporate some method for guiding the small boat into the ramp.

Stern wedges have also been used to increase ship's performance. The use of a stern wedge increases the submergence of the transom, and so, as with a stern flap, it is possible to design a stern wedge that enhances the performance of the stern ramp. In this case, the aft end of the ramp could be lowered into the wedge area increasing the sill depth. Increasing sill depth generally increases the range of sea conditions the boats can safely be recovered in.

LAUNCH AND RECOVERY OPERATIONS

During the survey of vessels with stern deployment capability, it was observed that manning requirements to support boat deployment varied from ship to ship. The Swedish Coast Guard *KBV 201* needed only two people to launch and recover the small boat, one on deck and one in the FRC. The coxswain would pull the remote disconnect from his control station to launch and the winch operator would connect the winch line, via a long pole, to the FRC for recovery. The majority of launches and recoveries could be performed with two personnel in the FRC and one winch operator. A few vessels needed more. These included the *Tornado* and the *Justo Sierra*.

All the ships' operators interviewed responded that they could launch the small boat in any sea condition that the small boat could safely handle, but recovery was limited by the sea state. When the sea state exceeded the maximum for safe recovery, the ship would escort the small boat to calmer water where it could be recovered.

The majority of the ship operators preferred launching with the ship's course set directly into the waves (000°). As an alternative, they would fall off the wave by up to 30 degrees to reduce their pitching motion. Operators on the Finnish vessel *Telkkä* and the Swedish Coast Guard ship *KBV 201* preferred to run with the waves at the same speed as the waves. This gave the *Telkkä* the optimum condition for deploying their cradle and RHIB.

On the *CCGS Gordon Reid*, it was preferred to run in parallel to the crest of the waves (heading of 090° relative to the waves) when they performed boat operations in high sea states. It should be noted that this cutter has dramatically different hull geometry from all the other ships surveyed, and was designed for launch and recovery of the small boat as a primary mission of the ship while operating in a specific area of the world. The necessary space was reserved to install such a system around this mission. Almost all other designs considered stern boat deployment as secondary mission. Additionally, the Canadian Coast Guard employs permanent crew and has an excellent training procedure implemented aboard the ship.

Launch and Recovery Procedure

The procedures for launching and recovery for all the FRCs are very similar. In general, to launch a FRC, the bowman trips the quick release hook and the FRC slides down the ramp and out the transom. On a ramp with a low slope angle, the FRC must be backed out with a winch. The FRC's engines are started when the propellers are in the water.

There are a few exceptions to the launch procedure. Some FRCs are designed to run with the engines dry for a short period of time and do not need to be lowered into the water before starting. In these cases, the engines are started before the quick release line is pulled. Another exception is in the control of the launch. Normally the coxswain is in control of the launch, however, on some ships, it is a deck hand who is responsible for determining when to launch the small boat.

Most of the recovery procedures were nearly identical. The coxswain must time the boat's entrance into the ramp to coincide with the sill's greatest submergence. When the coxswain sees an opportunity, he accelerates the FRC into the transom opening and up the ramp. The winch line is passed to the bowman who attaches it to the FRC, and the FRC is then winched up the ramp to the stowed position.

There are a few differences between the various ships' recovery procedures; most notable would be the method for capturing the RHIB during recovery. On the *Telkkä*, a mechanical arm captures and holds the RHIB in the cradle and then the cradle is winched into the ship.

Table 2. Launch and Recovery Characteristics

Ship	Launch Heading	Ship Speed	Launch Time	Recovery Heading	Ship Speed	Recovery Time
Justo Sierra	0°	1-3 kts	20 sec	0°	0 kts	15-20 sec
Tornado	0°	5 kts	18 sec	0°	5 kts	12-20 sec
Gordon Reid	90°	5-6 kts	10 sec	90°	5-6 kts	8-18 sec
Jaguar	0°	5-8 kts	8 sec	0°	6-10 kts	10-14 sec
Telkkä	180°	6 kts	35 sec	180°	6 kts	12-15 sec
Hammerhead	20°	3-5 kts	7 sec	20°	3-5 kts	9-12 sec
Vormann Steffens	0°	5-6 kts	6 sec	0°	5-6 kts	10-15 sec
KBV 201	180°	5-8 kts	10 sec	180°	5-8 kts	10-12 sec

On the *Hammerhead*, the RHIB is driven all the way up the ramp, captured by a "lasso", and held in place until the winch line is connected to pull the boat to the stowed position. A detailed discussion of launch and recovery procedures is provided by Sheinberg et al. (2001).

Characteristics for Launch and Recovery

Summarized in Table 2 are the launch and recovery characteristics of the ships investigated. The ship speed for most launchings is between 3 and 6 knots. This gives the mother ship enough forward motion to maintain her course but still slow enough for the RHIB to escape the effects of stern wake turbulence immediately after launch.

Launching times vary directly with the launching procedure. The launching time is defined as the time from when the command to launch is given until the boat is clear of the transom. The quickest launches (7 seconds) are experienced on those ships where the small boat's diesel engine can be run dry. On these vessels, after the stern gate was opened all that was involved to launch was to pull the quick release mechanism. Adding a winch to lower the boat into the water before starting the engines increases the launch time to about 10 to 15 seconds. When the cradle or assistance is needed to launch the boat, as was the case for the *Telkkä* or the *Justo Sierra*, the time approaches a maximum of 35 seconds.

During recovery, the majority of the ship operators prefer the same course as they do for launch, head seas to 30° off the waves. The exceptions are the *Gordon Reid*, *Telkkä*, and *KBV 201*. The reasons they prefer their recovery directions are the same as for launch. The mother ship speed for recovery is nearly the same for recovery as for launch. However, *Jaguar's* recovery speed is doubled to nearly 10 knots or twice that of other ships. At the higher recovery speed, the water jet driven RHIB must travel at a higher speed where it has better directional stability, necessary during the recovery operations.

The recovery times are typically quicker than launch times. The recovery time is defined as the time it takes from when the coxswain decides to enter the ramp until the RHIB grounds on the ramp. At that point, the RHIB is attached to the winch and hauled up to the storage position.

Type and Size of Small Boat

The FRCs or small boats observed on the vessels investigated fell into two categories, RHIBs and others. The characteristics of the FRCs are summarized in Table 3.

The majority of the FRCs are RHIBs between 7 m and 7½ m long with one 11-m RHIB. Three ships, the *Erimo*, *Justo Sierra* and *Vormann Steffens*, used small

Table 3. Ship and Boat Characteristics

Ship	Small Boat	Boat Type	Propulsion
Erimo	5.5 m	Fiberglass FRC	Water Jet
Justo Sierra	11 m	Aluminum FRC	Water Jet
Tornado	7 m 11 m	RHIB RHIB	I/O Water Jet
Gordon Reid	7.33 m	RHIB	Outboard
Jaguar	7 m	RHIB	Water Jet
Telkkä	7.4 m	RHIB	Water Jet
Hammerhead	7 m	RHIB	Water Jet
Vormann Steffens	7.5 m	Self-righting FRC	Inboard w/skeg
KBV 201	7.65 m	RHIB	Water Jet

boats other than the RHIB. The *Erimo's* FRC is a fiberglass boat with an operator's cockpit located amidships. The *Justo Sierra's* FRC is an aluminum hulled Interceptor with an operator's cabin and seats aft for a boarding party of four. The *Vormann Steffens'* FRC is a self-contained and self-righting rescue boat.

Power is provided by diesel engines in all but one of the FRCs, the *Gordon Reid's*, which uses gasoline-powered outboards. The outboards provide the RHIB with several advantages. The engines are lighter with a higher power-to-weight ratio and are more directionally stable. They are also very responsive to throttle, very maneuverable, and can be changed out quickly. Their disadvantages include the lower units hanging below the ramp, requiring a cutout or raising before winching up the ramp, a generally shorter life expectancy, and they use the more hazardous gasoline.

Most of the diesel-powered small boats use water jet propulsion with the exception of the U.S. Navy's 7-m RHIB, that uses an I/O drive, and the *Vormann Steffens*, that uses a single propeller protected by a skeg. The larger 11-m boats use twin water jets. Water-jet drives have the advantages of no projection below the hull to interfere with launch and recovery, the diesels can be designed to be run dry before launching to give quicker launch times, and the jet drive has the edge on maneuverability but has a lack of directional stability when operating in the ship's wake. In order to compensate for this they must travel faster in the wake and approach the stern ramp at a higher speed.

Small Boat Design Considerations

When the small boats enter a stern ramp, the impacts due to grounding on the ramp and fending off the ramp corners and walls require special reinforcements on the FRC. The collars on RHIBs must also be reinforced to be more abrasion resistant to the scuffs and scrapes that accompany recovery. The stem and bottom of the FRC must be reinforced to withstand the impacts with the ramp deck caused by recovery and wave motions.

Two RHIBs, those used on the *Gordon Reid* and *Jaguar* use a collar section that wraps completely around the stern. The collar provides additional flotation to prevent the stern from submerging during launch and recovery.

Crew comfort and safety is another concern. As boats travel faster over heavy seas, the more discomfort it transmits to the crew. Long hours of high-speed operation in rough seas can be injurious to the crew. The impacts can cause back, knee, neck, joint, and muscle injury. As a minimum, it will cause fatigue, discomfort and pain. The introduction of ergonomic shock absorbing seats has helped reduce the injury rate, has increased operating times for the boat crews, and should be considered when outfitting the boats.

New steering control stations have been recently introduced to the RHIB industry. As an alternative to the steering wheel and separate engine throttle controls, a handlebar style steering station has been introduced. Along with steering control, the handlebars incorporate controls for speed and water-jet bucket position. This allows the coxswain to control the boat's speed and direction without removing his hands from the handlebars.

HYDRODYNAMIC CONSIDERATIONS ON OPERABILITY AND DESIGN

Stern Wake and Propeller Wash Influence on Recovery

The combined effects of wake of the mother ship, propeller wash and waves on the maneuvering ability of a small boat approaching a stern ramp present a hydrodynamic problem that has not been numerically analyzed due to the complexity of its nature. The factors that influence the wake are the ship's hull form and the propeller wash. They combine to form turbulent eddies that make slow speed transit of the wake difficult. The effects of the wake and propeller wash on a small boat's maneuvering ability when approaching a stern ramp are presently best understood through empirical observations. All the ships, except the *Tornado*, had two propellers. During launch and recovery operations, it was observed that the wake would form a depression between the two propeller washes. This trough would occasionally aid in centering the RHIB during recovery operations. On the *Tornado*, which has four propellers, recovery operations are performed with the two inboard shafts declutched to help reduce the propeller wash on centerline.

Every FRC exhibited difficulty navigating the wake and maintaining a straight-in approach. The stern wake made it difficult to maintain directional control of the small boat. As the sea states increased, the wake effects worsened. The natural tendency of the coxswains was to oversteer when making the approach to the ramp. In all sea states, except flat water, a last minute correction was observed as the RHIB traversed the stern wake and entered the ramp.

The slow speed directional control of the RHIB, when equipped with water jets, is minimal thereby making transit of the wake difficult. The approach speed of the RHIB needs to be approximately twice the speed of the mother ship to maintain good directional stability. On the *Jaguar*, to maintain better directional control on the RHIB, the mother ship recovery speed is increased to between 6 and 10 knots. This permits RHIB recovery speeds of between 12 and 20 knots, providing better directional control during recovery.

On the *Gordon Reid*, the RHIB is equipped with outboard propulsion that exhibited better directional control at the recovery speeds than did the water jet propelled boats.

All stern deployment systems observed are located on the ship's centerline. The recovery course of the RHIB is centered between the more turbulent parts of the wake produced by the propellers. Better maneuverability and directional control of the RHIB can be maintained, increasing the likelihood of a successful recovery.

Hydrodynamic Phenomena at the Stern and Potential Analytical Tools

From the hydrodynamic point of view, the problem of deploying a small boat from the stern of a larger vessel can be divided into three general regimes; wave induced motion of the mother ship and small boat, characteristics of the stern wake and propeller wash, and local effects inside of the stern ramp. The larger hydrodynamic regime concerns the motions induced upon the mother ship and small boat due to the incident ocean waves and the resulting effects of wave sheltering and radiated wave patterns. The local hydrodynamic regime involves those effects generated by the mother ship upon the ocean as the ship is propelled through the sea. This regime includes the stern wake of the mother ship and the propeller wash. Investigations of the third regime, local effects inside the stern ramp, will help with both the ramp design and the operability of using the ramp for launch and recovery.

The first regime is generally solved through seakeeping analysis, while the others require computational fluid dynamics (CFD) analysis for a solution. Unfortunately, stern-ramp deployment operations at sea from a moving ship involve all regimes. If the operation were not near the ship's wake and propeller wash, then seakeeping methods provide the solution. On the other hand, if there were no waves, then a CFD approach to the problem could be employed.

Table 4. Stern Boat Deployment System Hydrodynamic Phenomena and Assumptions

Problem	Description	Assumption
Motion in a seaway	Whole-ship seakeeping motions	Easily solved to a fair degree of accuracy with readily available strip-theory seakeeping programs.
Wave sheltering and reflection	A smaller vessel operating in the lee of a larger vessel will be sheltered to some extent from the incident ocean waves.	Several panel-method seakeeping programs can tackle this problem. The solution, however, concerns the zero-speed case only. For stern ramps, this effect is especially important for the head and following sea directions. Wave sheltering is a relatively small effect at the stern of a frigate as compared to the greater lee provided by the full length of the mother ship for side-davit recovery.
Two-body motions	Radiated waves from a vessel operating in close proximity to another vessel will tend to affect the motions of the other vessel.	Seakeeping programs are beginning to handle this problem. Modeling is tedious and complicated to implement and the solutions may not be accurate. The effect of radiated waves from a larger vessel affecting the motions of the boat is greater for side-deployment systems. This effect is neglected at the stern as wake and propeller wash predominate.
Maneuvering in stern wake	A smaller vessel operating near the stern of a larger vessel will have to pass though the local wave pattern of the larger ship's wake.	This is an extremely difficult and computationally intensive problem to solve. It is known that RHIBs with good acceleration and directional stability can successfully navigate through the wake and onto a stern ramp. Water-jet propelled RHIBs have been observed to be directionally unstable when operating in the wake of a ship, however, utilizing a more experienced coxswain has minimized this problem. Operationally, the ship's speed during launch and recovery should be kept below the level where stern wake becomes a problem.
Maneuvering in propeller wash	A smaller vessel operating near the stern of a larger vessel will have to pass through the currents and eddies generated by the larger ship's propellers.	This is an extremely difficult and computationally intensive problem to solve. It is known that RHIBs with good acceleration and directional stability can successfully navigate through the propeller wash and onto a stern ramp. Water-jet propelled RHIBs have been observed to be directionally unstable when operating in the propeller wash of a ship, however utilizing a more experienced coxswain has minimized this problem.
Interior hydrodynamic effects	This includes a multitude of local hydrodynamic effects impinging on and interior to the stern ramp including beach effects, sloshing, run up, and water drainage, as well as modifications to the undisturbed wave elevation due to the ship's forward motion and attitude.	For the most part, the hydrodynamic effects interior to the stern ramp have not been solved numerically except insofar as they have been inherently captured in the results of stern-ramp model tests. The ramp availability approach assumes an instantaneous wave surface elevation that is undisturbed within the ramp (or at least at the location of the sill). This may be calibrated to a specific ship's ramp configuration and speed based on model test results.

A number of hydrodynamic issues associated with the design of the stern ramp system are described in Table 4 including a listing of the significant hydrodynamic effects and the assumptions taken for each. The first item in the table, relative motions in the seaway and operability assessment of the stern ramp in a seaway, has been predicted using the U.S. Navy's Ship Motion Program (SMP) and Seakeeping Evaluation Program (SEP) suite, and Marintek's Vessel Response (VERES) Program. Each of these programs is based on strip theory and has been well verified. These programs give similar results, and both were used to aid in the investigation of the operability of the stern ramp on various sized cutters in typical ocean environments of the North Pacific and North Atlantic Oceans. However, because each of these programs use strip theory to solve the equations of motion, they are limited to that area and cannot be used to investigate the effects due to other hydromechanics issues such as wave sheltering, two-body motions, stern wake, propeller wash and interior effects to the stern ramp. Not only does stern ramp configuration play a significant role in the severity of impact loads between the mother ship and returning RHIB, but configuration aspects affecting water mitigation within the stern ramp along with the turbulence astern influence the degree of success and safety of the stern ramp system.

To address these other areas of concern, several analytical tools were investigated to determine which would be the best to aid in solving these problems. They were grouped into two main areas; impact loads of small boat on the stern ramp and maneuvering of the small boat across the stern wake and propeller wash near the entrance to the stern ramp. The list of tools that were reviewed was by no means exhaustive, but composed of those readily available to the Coast Guard and Navy.

Four numerical tools investigated to determine if any could aid in determining the impact loads of the small boat on the stern ramp of the mother ship. The first method was based on a time-domain model of an icebreaker impacting an ice sheet (after Popov, 1967). This model would calculate the loads on the small boat and stern ramp of the mother ship. In adapting this model, the mother ship takes the place of the ice sheet and would be assumed to be rigid and stationary with the vertical, transverse and longitudinal motions superimposed on the small boat. A parametric series of approaches would be conducted to determine the impact loads for the different approach velocities and impact locations on the boat. These impact loads, or forces, would then be translated into accelerations acting upon the occupants of the small boat to be compared against established criteria. This approach is simple and would provide a first-cut approximation of the loads and accelerations on the small boat. The mother ship geometry could also be modified to account for different stern ramp designs. However, it is not the most realistic approach because the mother ship would normally be moving with some forward speed, and the mother ship would have wave-induced motions in a seaway.

The second approach investigated to solve the two-body impact loads was to use a program named WAMIT (Wave Analysis MIT) developed by WAMIT Inc. WAMIT is a linear, zero-speed, frequency-domain program that can calculate the motions of two bodies in close proximity to each other. Wave-induced ship motions are calculated independently for each vessel. Although some of WAMIT's shortcomings, lack of determination of impact loads between the vessels and the fact that all calculated results are in the frequency domain, can be overcome through correcting them in a post processor, the fundamental underlying zero-speed assumption of WAMIT makes the complete solution unattainable. Basically, there is no way to solve the two-body hydrodynamic problem of small boats approaching the mother ship with significantly higher speeds. This results in the program failing to account for the influence of the speed difference between the ships on the impact.

The third analytical tool investigated to solve this problem was the Stern Ramp Model (SRM), which uses the Large Amplitude Motion Program (LAMP) developed by Science Applications International Corporation to calculate the ship motions. LAMP is a time-domain ship motion tool that uses non-linear hydrostatics and Froude-Krylov wave forces and linear hydrodynamics. The SRM has a method for tracking the contact between the ship and the boat and determining the reaction forces and kinematics. However, the current configuration of the SRM cannot calculate the hydrodynamic interaction between either the mother ship and the boat, or the planing forces on the small boat.

Because of this limitation, it is difficult to use this model as a stern-ramp design tool. The hydrodynamic interaction between the mother ship with stern ramp and the boat is important in order to determine the veracity of SRM output of the reaction forces on the boat during contact with the mother ship. Additionally, discerning the impact of changes to the ramp design would involve calculating the hydrodynamic interaction between the two vessels.

The last analytical tool reviewed for solving the problem of determining impact loads of one body to another is SWAN 2 developed by Professor Sclavounos at the Massachusetts Institute of Technology (Kim and Sclavounos, 1998). This linear, time-domain code computes calm water sinkage and trim, six-degrees-of-freedom wave-induced motions, and local and global structural loads. Currently there is no provision within the code for calculating two-body motions and impact loads.

Two analytical tools were investigated to solve the second problem of predicting the stern wake and propeller wash in the vicinity of the stern ramp within an ambient seaway. Both involve cutting edge computational fluid dynamics computation methods. The first method is the Reynolds Averaged Navier Stokes code. This code would provide an overall picture of the flow field in the vicinity of the ship. It also can provide propulsor inflow characteristics and the effect of the propulsor in calm water. This method is most commonly used for analyzing different integrated propulsor and hull configurations. Although this methodology can give a good description of the flow field near the ship, the technology has not matured enough to incorporate the ambient wave field into the flow field.

The second method (Lin, 2002) calculates the non-linear interaction of the ambient wave field and the ship-generated wake, as it is moving through calm water. This calculation can be used for all parts of the ship. Although this model shows promise, it is still in the early stage of development. It has been implemented with only double-ended hulls such as the Series 60 hull form and the Wigley Hull. The model still has to be extended to ships with a transom stern. In addition, no ship motions have been calculated for these ships in waves. The program's limitations of not considering the propeller wash aspects in the stern wake, the wave-induced motions of the ship, and the wake behind a transom stern make it very limited in its use currently. However, this approach shows promise because its calculation of the non-linear interaction between the ship generated and ambient waves may be able to provide a more exact definition of the stern wake, which can help in predicting the small boat maneuvering problems behind the ship on its approach to the stern ramp.

Based on the above review, it was concluded that none of the readily available computer programs at present have the capability to calculate all the phenomena considered crucial for simulation of the stern ramp recovery operation and thereby influence the stern ramp design. The parameters considered most critical to influencing stern ramp design are the prediction of water motion inside the ramp, the prognosis of impact acceleration induced on the small boat as it hits the stern ramp, and modeling of the small boat maneuvering in the wake and propeller wash behind the mother ship in the vicinity of the stern ramp. Although some of the available programs may aid in performing calculations of some particular phenomena, none are able to simulate these three effects that might significantly influence the selection of the optimal stern ramp design. Additionally, there was no verified analytical method identified that simulated the motion of water inside the stern ramp.

Model Testing

The review of the available analytical approaches presented earlier indicates that none of the available computer programs can solve the complex hydrodynamic phenomena that occur around the ship stern and inside of the ramp arrangement during an entry of the small boat. At the present time, model tests, especially with large-scale models, are considered the most effective tool for optimizing stern ramp design. Model tests afford the opportunity to model both the complex hydromechanics in and around the stern ramp during the FRC deployment and retrieval and the induced loads on the mother ship and FRC. A series of the model tests with systematic variation of the stern ramp parameters can therefore be the simplest way of selecting the optimal configuration of the stern ramp. Other parameters that can be evaluated during model tests are the limiting sea state, size of the FRC and main particulars of the mother ship. Furthermore, any additional system attached to the stern ramp such as a water-motion control system, boat capture mechanisms, etc., and its effect on the operability of the stern ramp could easily be verified by the model test.

When preparing for a model test, both the model size and wave conditions must be calculated. Each these factors depends on the particulars of available model test facilities and the sea conditions that will be investigated. Generally, it is recommended to use as large a model as possible, preferably with a scale ratio between 1:10 to 1:12.5. There are many benefits of using larger models for testing stern ramp operability. With the large-scale models more measurement equipment will be able to be installed in the small boat. With the extra equipment, it would then be possible to measure not only the boat accelerations but also the boat position in relation to the mother ship, the baseline loads on the boat, and the ability to video the boat's approach to the stern ramp etc. Using large-scale models also reduces the scale effect especially when modeling the time of recovery operation. Additionally, in order to observe and quantify changes in the stern ramp deployment operability as a result of application of different capture mechanisms or any additional design features of the ramp (such as drainage system, damping system, fenders etc.) the relatively large models of such arrangements are necessary.

Most of the recent model tests looking at stern ramp operability and design have been performed at indoor facilities. If larger scale models are required for testing launching or retrieving mechanisms, then outdoor facilities might also be considered as an alternative solution. Such facilities are however not recommended for performing sensitivity or optimization studies on stern ramp parameters. These types of tests require controlled environment and waves of specified heights and periods. Such conditions are difficult to achieve on an open lake or marine bay.

As a part of the USCG's investigations into the availability of using a stern ramp for the launch and recovery of a boat from a frigate or cutter type vessel in higher sea states, a series of model tests were conducted. These tests were performed for a systematic series of stern ramp recoveries under specified wave conditions for a variety of different headings.

The overall objective of these tests was to determine the capability of the stern ramp for small boat deployment and recovery in a variety of different wave conditions. Other beneficial information obtained from the model tests included identification of the key ship responses crucial in the evaluation, comparison of the different boat launch systems, and calibration of the ship motion program used for prediction of the long-term operability.

The model tests were conducted in the deep-water ocean basin at Marintek using frigate- and cutter-type modes as the mother ship and a water-jet driven and radio-controlled planing model as the fast small boat (Werenskiold, April 2001 and March 2003). Two models of the mother ship, one in scale 1:22.88 (the frigate) and the second in the scale 1:12.5 (the cutter) were used in these tests. The full-scale length (L_{pp}) of both mother ships was approximately 119 m (390 ft) and the length of small boat was about 11 m (36 ft). The test matrix for the frigate model (the model of 5.3 m (17.4 ft) in length) was set to measure the mothership and small boat responses at two speeds (5 and 10 knots), five wave headings (0, 45, 90, 135, and 180 degrees off bow), and for three sea states (Sea State 5, Sea State 6 and Sea State 7, respectively). The operability of the stern ramp recovery on the frigate mother ship was also tested for three sill depths. These tested alternative sill depths (0.305 m (1 ft), 0.610 m (2 ft) and 0.914 m (3 ft)) were obtained by trimming the ship model by the stern. Model tests with the frigate-sized cutter mother ship (the larger model of 9.5 m (31.2 ft) length) were aimed at assessing the relative effectiveness of alternative stern ramp widths and different water managements systems. These tests were carried out in one selected sea (Sea State 6) and one wave heading (30 degrees off bow).

During each of the tests, the mother ship model was suspended in a soft spring system that allowed the model to freely heave, roll and pitch. The spring system was designed to provide representative surge, sway and yaw characteristics for a low speed operation in rough water. The mother ship model was self-propelled and was fitted with a propeller whose pitch and speed were representative of full-scale operation. In addition, since the model scale of ship wake and propeller wash affecting the operation of the FRC close to the stern were physically represented, the modeled results closely resembled the full-scale operation. A coxswain sitting on the carriage in a position just astern of the mother ship operated the radio controlled small boat.

The test program included a series of runs with the mother ship alone and the mother ship with the small boat. The tests conducted with the mother ship alone were recorded the water motion inside of the ramp and the relative motion at the ramp entrance. The tests with the mother ship and small boat entering the ramp were performed with different environmental conditions to determine the operability of ramp. The following data were collected during the model tests for both the mother ship and small boat:

- Mother ship motions at six degrees of freedom at the ship center of gravity
- Acceleration in three axes at three locations: ship center of gravity, bridge and stern (operator position).
- Vertical and transverse acceleration at two locations: midship on the main deck and on the leeward side (side-launch operator position).
- Relative vertical wave motions at seven positions: bow, midship on both sides, stern on both sides, and at two locations inside of the ramp.
- Ship speed and propeller revolutions.
- Propeller wake and ship wash at three positions aft of the stern in the FRC approach area.
- Small boat accelerations in three axes close to boat center of gravity.

Results obtained from the model tests with a large model, representing a frigate-sized cutter, have proved that such a procedure can effectively be applied to optimize the stern ramp design. The test results clearly demonstrated that for some stern ramp configurations the number of successful entries were significantly larger than for other types. This was primarily due to the effect of improved water management inside of the ramp and the lower number of collisions with the stern for the fairly formed ramp entrance. Three different ramp configurations were considered within this test series. Ramp no. 1 was a narrow ramp without the water management system. The ramp entrance to the sea was straight and not rounded. Ramp no. 2 had the same ramp entrance form and width as Ramp no. 1, but a water drainage system was added.

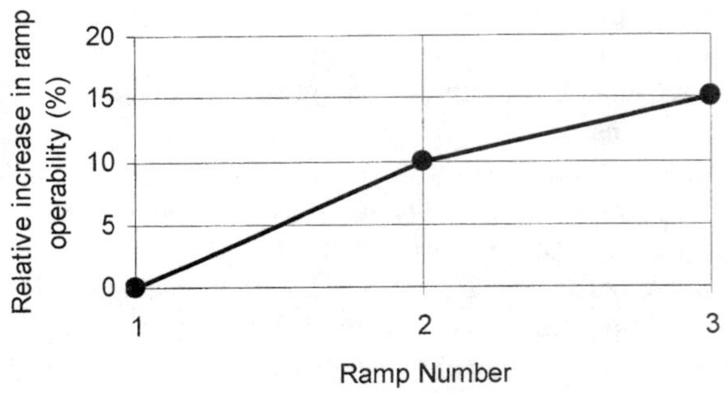

Figure 5. Effect of ramp arrangement on stern ramp operability (from model test results).

Comparing the two configurations (see Figure 5) demonstrated a relative increase in successful recoveries of 10 percent by the addition of an effective water management system. Ramp no. 3 had a wider interior width, its entrance to the sea was additionally widened (with the entrance angle of 30 degrees), and the ramp edges at the transom were rounded. It had the same water drainage system as for Ramp no. 2. In this case, the relative increase in successful recoveries was 15 percent compared to Ramp no. 1.

The model test results were used to verify and calibrate the predicted mother ship response computed from the VERES ship motion program. This program was also used to compute the long-term prognosis of operability of the stern ramp boat deployment. An example of such prognosis is given in the following section.

The second objective achieved by these model tests was the development of the mother ship motion criteria for stern ramp operability. Based on the test results, two mother ship responses have been identified as key parameters for assessing stern ramp system operability. These crucial responses are the relative vertical motion and sway motion at the ramp entrance on the mother ship. Two operability criteria related to these selected mother ship responses have been proposed and the limiting amplitudes of these critical responses have been defined (Kauczynski and Werenskiold, November 2001).

DESIGN CRITERIA FOR STERN-RAMP BOAT DEPLOYMENT

Since 1995, the U.S. Coast Guard has been developing and refining boat-deployment seakeeping criteria. This effort has included compiling analysis techniques, collecting data for analysis through at-sea dedicated trials and model tests, and conducting seakeeping motion, comparison studies. Much of this effort has been directed towards side-davit systems, but recently it was recognized that the Coast Guard would need to develop a consistent set of criteria for stern deployment of small boats to assess stern-boat deployment systems.

For the purposes of performance evaluation of boat deployment from cutters, three criteria sets are considered. The first group is the mother ship motion criteria, which are the same for side-davit and stern-ramp boat deployment. The second criteria set considered takes into account the relative motion of the stern ramp sill in and out of the water, and is embodied in a ramp availability criterion. Combined, these two criteria sets can be used to compare different launch systems on the same cutter or between difference cutter platforms. They can be readily applied using existing strip-theory, frequency-domain seakeeping evaluation programs. The third set of criteria is based on the deployed boat and its capability to affect a successful recovery onto a stern ramp. It imposes a standard for stern-ramp deployed boats to meet so that they can match the level of system performance provided by the mother ship. The deployed boat criteria can be evaluated through model tests or full-scale testing with instrumentation on the small boat, or through the application of impact loads time-domain models. Through a dedicated model test program with a frigate-sized mother ship and FRC, the deployed boat criteria have been related to the motions at the stern of the mother ship. In this way, they too can be evaluated from the point of view of the mother ship motions in a seaway using standard seakeeping analysis approaches.

Mother Ship Motion Criteria

The basic motion-limiting criteria for side-davit boat deployment (both launching and recovery) (Minnick et al., 1999) are given by

Criterion	Limit
Roll	< 8.0° SSA
Pitch	< 2.5° SSA
Vertical Acceleration	< 0.2 g SSA at Boat Station
Lateral Acceleration	< 0.2 g SSA at Boat Station

These criteria are also applied to mother ship motions for stern-ramp, self-propelled boat deployment (recovery as the limiting operation) of a high-performance RHIB or FRC (U.S. Coast Guard, Apr. 2001).

The criteria set was developed as a result of dedicated full-scale trials in the Bering Sea in 1995. A joint Alaskan Patrol (ALPAT) using the *USCGC Boutwell* (WHEC 719) and *USCGC Harriet Lane* (WMEC 903) was conducted for the purpose of quantitatively and qualitatively comparing the operational capabilities of the Coast Guard's 378-ft and 270-ft Class cutters in identical Alaskan sea conditions during a typical ALPAT. During these trials both cutters most often utilized their RHIBs, launched and recovered from single-point davits, for boarding fishing vessels. The single-point davit on the *Boutwell* was fitted with a constant-tension winch. The technical details of this Joint ALPAT are reported in U.S. Coast Guard (February 1996) and by Minnick et al. (June 1999).

The approach used to determine criteria for side-boat deployment was to identify all boat launch or recovery events, correlate the events with the capability assessments ("go" and "no-go" assessment) and the recorded motions data. The data were then rank ordered by degree of difficulty to determine the limiting motions and motion magnitude. The ship motion measurements pertaining to side-boat deployment were pitch, roll, lateral and vertical accelerations at the pilothouse, and the lateral and vertical accelerations at the boat location.

The resulting data set indicated that transverse acceleration was not directly associated with the "no-go" situations. The limiting parameters were found to be pitch, roll and vertical acceleration at the pilothouse or at the boat deployment site. It was assumed that once the decision was made from the pilothouse to launch or retrieve a small boat, actual timing and on-site decision-making transferred to the bosun in charge at the boat station. By comparison of the significant single amplitude (SSA) motions and the minimum "no-go" ship motion measurements it was concluded that a vertical acceleration limit of 0.2 g SSA was the natural break point between the "go" and "no-go" situations and was applicable to both the pilothouse and boat station.

While not a direct validation of this result, an earlier study on the deployment of a towed array from a transom compartment of a naval vessel underway was conducted by Thomas et al. (1992). Array deployment and recovery operations are conducted in head seas or bow seas in order to keep the array cable taut. Other headings can result in surging that cause complications with the winch system and risk fouling the propellers. Array deployment and recovery operations can be hindered in heavy seas due to excessive ship motions at the stern, which interfere with the ability of personnel checking and straightening out the cable as it comes off the winch drum. Using measured motions data along with noted assessments of when a crewmember could successfully deploy the array or not, a similar methodology was used to determine the limiting motion and that motion's magnitude. In this case, the limiting motion for severe degradation of crew performance was also determined to be vertical acceleration with a magnitude of 0.15 g SSA. Both boat deployment and towed array deployment require similar human interactions (reaching, lifting, connecting hooks, etc.) to accomplish the task, so it is reasonable to expect that the magnitude of the vertical acceleration that limits the safety of these tasks would be roughly the same.

Lateral accelerations on the mother ship can also be limiting if the crew on the mother ship cannot perform their duties. This can be referred to as a separate limiting criterion of lateral acceleration. It has been found that if the lateral acceleration on the mother ship exceeds the value of 0.2 g SSA, the crew's performance degrades significantly.

For side-boat deployment using Miranda-type davits, additional constraints, such as a vertical displacement limit, may be justified.

Ramp Availability Criterion

Within the Coast Guard, the problem of evaluating the feasibility and operational effectiveness of recovering a boat via a stern ramp was encountered when developing a series of feasibility studies, using state-of-the-art design synthesis programs, to produce a number of conceptual designs for the National Security Cutter. The completed designs were compared for seakeeping performance using a PTO analysis. This not only identified the need to establish appropriate criteria to evaluate stern-launch systems, but also established that ramp sill emergence in a seaway could easily complicate stern-launched boat recovery from relatively longer cutters, such as a National Security Cutter, in higher sea conditions.

In conjunction with the survey of ships with stern deployment systems, efforts were underway to develop a coherent stern-ramp evaluation methodology. Consequently, a number of parametric studies were conducted that took into account the responses from the operators of stern-ramp vessels, measurements of deployment operations on these vessels, theoretical hypothesis and comparison with dedicated model tests. The seakeeping analytical studies compared side-davit and stern-ramp performance for three sizes of notional cutters: National Security Cutter, Medium Endurance Cutter and Patrol Craft. Limiting responses on all heading and reasonable speeds over a range of sea conditions were investigated and overall PTO evaluations conducted.

This lead to the inclusion of a ramp availability criterion with the mother ship motion criteria into a motion-limiting criteria set for stern-ramp, self-propelled boat deployment (recovery as the limiting operation) of a high-performance RHIB or FRC (U.S. Coast Guard, April 2001). The ramp availability criterion is given by

Criterion		Limit
Mean Ramp Availability	>	Minimum Ramp Availability Duration
$\langle \tau\_ \rangle$	>	T_{RA}

The ramp availability criterion is based on a frequency domain representation of the undisturbed water depth at the entrance to the stern ramp where the threshold distance "a" is set as follows:

a = sill water depth at the ramp sill for RHIBs
a = sill water depth at the ramp sill minus forward draft for displacement boats

Ramp sill depth refers to the vertical distance from the mother ship's still waterline to the lowest point in stern ramp or stern well at the transom. For vessels with a stern flap, the flap's upper surface may allow it to be effective for boat recovery operations. In this case, the effective ramp sill depth may be considered the vertical distance from the still waterline to the lowest point on the upper surface of the stern flap that effectively supports boat recovery. The flap upper surface should have about the same slope as the stern ramp and provide some means of transverse constraint to keep the boat from sliding to one side.

Ramp availability is the period of time during the relative motion process at the ramp sill when the water depth over the sill is deep enough to avoid bottom contact on a boat being recovered. The ramp availability criterion provides a linkage between sill depth, sill emergence frequency, vertical relative displacement on the mother ship and the small boat acceleration performance.

Predicted sill emergence frequency, the relative vertical displacement and ramp availability are readily related in frequency-domain calculations similar to those employed for deck wetness or bow slamming predictions. Different formulations of these interrelated responses may be more readily adaptable to seakeeping response analysis and a PTO evaluation approach over others. A detailed derivation and theoretical discussion of the limits of applicability for this ramp availability formulation are given by Dalzell (2003). A summary of the derivation for the expected average duration for ramp availability and approach implications is presented here.

In the self-propelled retrieval operation onto a stern ramp, the boat approaches and lines up on the ramp and, according to the judgment of the coxswain, accelerates and drives as far up the ramp as possible. At this time, the boat is secured by some mechanism and winched the rest of the way up the ramp. Certainly, it would be unwise for the boat to hit the ramp just as the sill emerges; in fact, there may be a required minimum depth of water over the sill for a reliable landing. It can reasonably be assumed that under severe conditions the coxswain will watch the ship motion relative to the local water surface and start the drive to the ramp when it appears that the ramp is on the downward part of its cycle when the water depth over the sill will be greatest. The coxswain needs time to decide to go, to accelerate the boat, and to move the boat onto the ramp. At the moment of threshold crossing, the water depth over the ramp sill needs to be deep enough to avoid bottom contact on the boat, at least until the bow has "grounded" on the ramp and has been hooked on. Thus, it possible to employ a criterion that involves the statistics of the duration that the depth of water over the ramp sill is greater than some specified depth. Such statistics would define a "ramp availability" time. The associated criterion would be the minimum time required for the retrieval operation.

The overall evaluation and comparison approach involves the conventional linear-random model of the sea and the ship motions. Following the approach by Dalzell (2003), the motion processes are assumed to be Gaussian and zero mean, and these assumptions allow the application of some existing results from threshold crossing theory. First, the relative vertical motion at the location of the ramp sill, Y(t), is assumed to be a random Gaussian zero-mean process. Relative vertical motion is defined as the difference between the absolute ship motion at some point on the static waterline and the local water elevation. (It should be noted that some special ship motion computations beyond the usual strip-theory methods may be required to refine the estimates of the local water elevation just aft of the ship or over the ramp sill.) Relative motion is zero when there are no waves to excite the ship, and under these circumstances, the depth of water over the sill is the nominal still water value.

A fragment of the relative motion process, Y(t), is sketched in Figure 6 along with some definitions that relate the problem to the mathematical threshold crossing problem. The relative motion process is defined to be positive when the selected point on the ship rises above the local water level. Thus, if the relative motion is positive and equal to "a", the water depth over the sill is the original depth less "a". The value "a" is a constant "threshold" that depends upon the geometry of the problem. If "a" is set equal to the depth of water over the sill for no-wave conditions, then during times the relative motion when Y(t) exceeds "a", the sill will be emerged. Alternately, if "a" is set equal to the nominal depth of water over the sill less an allowance for the draft of the boat to be recovered, then during times the relative motion when Y(t) exceeds "a", the water over the sill will be less than the draft of the boat.

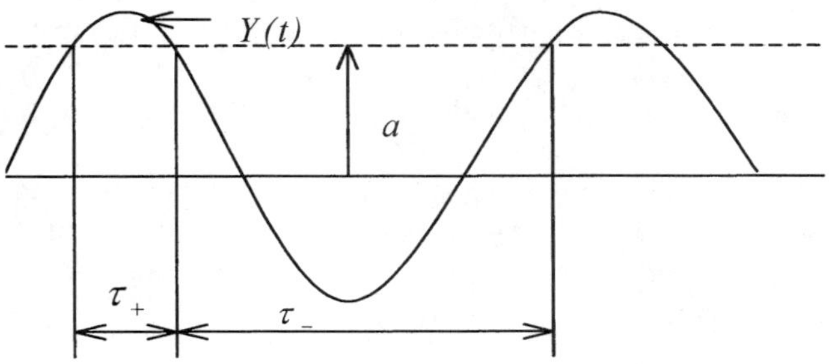

Figure 6. Definition sketch for transom sill relative vertical motion.

As shown in Figure 6, the time duration that the relative motion exceeds the threshold, a, may be denoted τ_+. Similarly, the time duration that the relative motion is less than the threshold may be denoted $\tau\_$. Thus the threshold problem involves the statistics of $\tau\_$ for a specified threshold, a, and random process, Y(t).

The average length of the time intervals during which Y(t) is less than "a" is denoted by $\langle\tau\_\rangle$. The theory yields a solid estimator for this quantity as:

$$\langle\tau\_\rangle = \frac{2\pi\,\sigma}{\sigma_v} \exp[(a/\sigma)^2/2]\ \Phi(a/\sigma)$$

$$\langle\tau\_\rangle = T_Z \exp[(a/\sigma)^2/2]\ \Phi(a/\sigma) \qquad (1)$$

where σ is the standard deviation of the relative motion process, Y(t), σ_v is the standard deviation of the first derivative of the relative motion process, and Φ(.) is the standardized Normal probability integral. Normally, the standard deviations would be computed as the square roots of the zeroth and second moments of the relative motion spectrum, (m_0 and m_2). From a frequency-domain prediction of the motion, they can be obtained as the relative vertical displacement rms and the relative vertical velocity rms.

It should be noted in the equation that functions of (a/σ) modify the mean zero-crossing period, $T_Z = 2\pi\sigma/\sigma_v$, to result in the average time interval. The quantity $T_Z\exp[(a/\sigma)^2/2]$ is an estimate for the mean time between threshold up- or threshold down-crossings. What the probability integral does is to determine the part of the average up-crossing period where the process is expected to be below the threshold. When the threshold is zero, $\Phi(a/\sigma)^2 = 0.5$ and the average time interval during which $Y(t) < 0$ is half the zero crossing period. When the threshold is very large $\Phi(a/\sigma)$ approaches unity and the average time interval during which $Y(t) < a$ approaches the mean time between threshold up-crossings.

The reciprocal of the estimate for the mean time between threshold up- or down-crossings ($T_Z \exp[(a/\sigma)^2/2]$) is the expected number of threshold crossings per unit time, a computation that has been done in support of deck wetness criteria for many years. The only moderately new thing computationally in Eq. 1 is the probability integral, $\Phi(.)$.

The frequency-domain criterion is that ramp operations be considered unsuccessful or more than usually dangerous for a given sea and operational condition if

$$\langle \tau_- \rangle < T_{RA} \qquad (2)$$

where T_{RA} is a minimum average ramp-availability time interval during the retrieval portion of a boat recovery operation.

The survey of ramp-equipped ships reported earlier did ascertain a first-level estimate of appropriate minimum average recovery duration by comparing the measured values of the averages with the stated sea state capability. These are given in Table 2. From this it was determined that a realistic average recovery time for a high-powered RHIB or FRC can be reasonably expected to be near 12 seconds with a minimum physically possible recovery time of 5 seconds.

It should be borne in mind that a recovery time (physical minimum or average) and an average ramp-availability interval are two different measures. The time to complete the boat recovery operation is longer and distinctly different from the actual retrieval interval during recovery (which is related to ramp availability time). Different criterion levels for the minimum average ramp-availability time interval, T_{RA}, have been explored. For high-powered FRCs considered for open-ocean deployment, a minimum average ramp-availability time of about 5 seconds corresponds favorability to the time required for these boats to complete the final segment of their approach and entrance onto a stern ramp.

Finally, it is important to note that ramp-availability time interval does not take into consideration any sway/yaw motion of the mother ship or FRC, impact loads between the FRC onto the ramp, and the propeller wash and other wave/wave turbulences that can occur in the stern wake. Each of these items must be considered in conjunction with the time interval for ramp availability. Additionally, the experience of the coxswain and crew can greatly affect the minimum average time interval for recovery.

Stern-Ramp Deployed Boat Criteria

One of the factors limiting the operability of stern ramp deployment systems is the small boat response in the last phase of the retrieval operation. During an entry into the stern ramp the small boat operability is limited by the boat's structural capability to withstand the impact loads generated on contact with the stern ramp and the surrounding ship structure. Simultaneously, the induced impact accelerations must be kept on a level acceptable for the boat's occupants to assure a safe recovery operation.

Based on data collected from the public transport studies (Hoberock, July 1976) and the operational limits adopted by the IMO Code of Safety for High-Speed Craft, Werenskiold (April 2001) has proposed the following three safety levels related to the acceptable exposure of persons on board a small boat:

- *Safety Level 1*, which represents minor risk of injury to standing people when holding.
- *Safety Level 2*, which represents minor risk of injury to sitting people.
- *Safety Level 3*, which represents minor risk of serious injury to sitting people.

The safety levels are defined in terms of the vertical and horizontal accelerations. The limiting maximum accelerations are based on the safety of personnel on board and include a suitable margin of safety for damage of the small boat (see Table 5). These proposed limits should be regarded as providing a significant safety margin for seated persons in the FRC who are well

Table 5. Limiting Values of the Maximum Accelerations Induced on the Small Boat

Safety level	Criteria: Maximum acceleration not to be exceeded		
	Vertical	Transverse	Longitudinal
1 (Minor risk standing persons)	0.60 g	0.25 g	0.25 g
2 (Minor risk sitting persons)	0.80 g	0.35 g	0.35 g
3 (Risk of injury sitting persons)	1.00 g	0.50 g	0.50 g

prepared for possible impacts when the boat enters the stern ramp. Therefore, Safety Level 3 should be applied for operating crew assumed to be sitting with lap belts, while the more constraining Safety Level 2 is for seated persons on board who are not members of the operating crew. The limiting values correspond to the maximum accelerations and it is expected that such maximum will be induced in a moment when a boat first hits the ramp structure by its bottom or side. Maximum vertical load of 1.0 g at the center of gravity of the FRC is also generally applied as the structural design load for such a high-speed craft.

For practical applications, such as the comparison of the mother ship hull designs, selection of the optimal ramp particulars, evaluation of arrangements of different ramp configurations, or development of an operator guidance system, it is useful to express all the operational criteria of the stern ramp deployment in terms of the mother ship responses.

In order to find a relationship between the mother ship responses and the FRC impact accelerations during FRC entries of the stern ramp, a parametric study was performed with data provided from the model tests. Analysis of the model test results was performed by assessing the number of unsuccessful FRC recovery operations along with consideration of the following factors:

- Vertical impacts between FRC bottom and ship or ramp structure.
- Transverse impacts between FRC sides and ship or dock structure.
- Longitudinal impacts during entry and docking or securing.
- FRC maneuverability and seakeeping performance in open sea or approach area.
- FRC controllability and forward acceleration capability during entry.

An "unsuccessful recovery" has been defined as a case when the boat's vertical or horizontal accelerations exceeded the defined levels. Criteria for assessment of impact loads were related to the accelerations experienced by personnel on board the FRC (as listed in Table 5).

Results of the model tests clearly show that the operability and limitations of stern-ramp recovery systems are dependent on a combination of technical (or physical) factors and human factors. Based on results of this analysis, it was found that the mother ship relative vertical motion at stern is the main factor that contributes to the FRC vertical impact acceleration. The reduction of operability due to FRC longitudinal impact accelerations are mainly caused by the craft hitting a higher part of the ramp structure. This includes the longitudinal component of imposed transverse FRC accelerations, as this is dependent on FRC speed of entry as the mother ship moves to one side. Consequently, transverse stern displacement by the mother ship has been identified to be the fundamental technical factor that causes large FRC longitudinal impact accelerations. Finally, the reduction of operability due to the FRC transverse impact accelerations appears to be mostly dependent on coxswain skill and therefore dependent upon human factors.

Thus, the mother ship's vertical relative motion and sway at the stern are key parameters that influence the magnitude of the impact accelerations on the FRC during ramp entry. Because of this, they have been used as the limiting parameters in two proposed operability criteria for the safe stern ramp operation. The first criterion sets the maximum acceptable amplitude of the relative vertical motion at the stern. The second criterion, the "small boat maneuvering criterion", sets the limit on sway motion at the stern of the mother ship.

Both these criteria can be used to assess risk of the boat recovery operation. They were deduced from dedicated model tests on a frigate-sized mother ship. Different criterion levels may be more appropriate for different sized mother ships, for different entrance shapes or boat-to-ramp well side clearances. These matters are still under investigation.

Stern Vertical Relative Motion Criterion

The limiting magnitude of the relative vertical motion at the stern of mother ship, shown in Table 6, was obtained through analyzing the results of the model tests with a mother ship of frigate size and hullform. The safety levels shown in the table classify the risk of the recovery operation and are related to exposure of persons on board the small boat as shown in Table 5. The percent time the stern ramp is operable (PTO) was defined by the number of successful recovery operations recorded in the model tests. These were computed from the model test results by taken into account number of recoveries with the impact acceleration below the accepted level.

Table 6. Limiting Magnitudes of the Relative Vertical Motion at the Ramp Entrance (rms)

	Ship speed	100% of successful recovery	90% of successful recovery	80% of successful recovery	70% of successful recovery
Safety level 1	5 knots	0.65 m	0.78 m	0.90 m	1.05 m
	10 knots	0.45 m	0.55 m	0.65 m	0.75 m
Safety level 2	5 knots	0.70 m	1.20 m	1.70 m	2.20 m
	10 knots	0.45 m	0.75 m	1.00 m	1.28 m
Safety level 3	5 knots	0.85 m	1.65 m	2.45 m	3.25 m
	10 knots	0.55 m	1.00 m	1.45 m	1.90 m

Referring to Table 6, the amplitudes of vertical relative motion are the lowest for safety level 1, slight risk of injury to personnel standing in the FRC, as this is the most restricted case. Because definitions of safety levels 2 and 3 are more relaxed, the resulting limiting impact load values, and corresponding limiting relative vertical motion values, are higher.

In the model tests, recoveries of the small boat were initiated at arbitrary time points. Such results must therefore be considered as conservative. By assuming that a coxswain with medium experience will steer the small boat during recovery maneuvers, a part of the most "impropriate" recoveries performed in the model tests could be rejected. Therefore, in practical application, it is proposed to use in the limiting ship response corresponding to 90 percent of the successful recoveries.

Small Boat Maneuvering Ability Criterion

During the beam and oblique seas part of the model test matrix, the number of unsuccessful operations was considerable despite the relatively low amplitudes of the vertical motion at the vessel stern. The main reason that so many of the recovery operations were classified as "unsuccessful" was that the magnitudes of the longitudinal impact accelerations measured on the small boat exceeded the established criteria listed in Table 5. Based on model test results, the magnitude of sway at the stern has been selected as a critical ship response, which is directly related to maneuvering problems during stern ramp FRC recovery operations. In beam and oblique seas, the small boat usually maintains more power to keep its position on the centerline behind the ship's stern. On these headings, the mother ship's hull does not shelter the boat maneuvering during approach to the stern ramp, and more power is required to balance the wave and wind effect and to keep the course. This effect will cause higher longitudinal acceleration when boat hits the ramp or ship structure. An analysis of the model test results provides a basis for the settings of the limiting sway amplitudes for a mother ship of frigate size and hullform (see Table 7).

Table 7. Limiting Sway Responses (rms) at the Stern Applied in the Maneuvering Ability Criterion

	Ship speed	100% of successful recovery	90% of successful recovery	80% of successful recovery	70% of successful recovery
Safety level 1	5-10 knots	0.50 m	0.55 m	0.60 m	0.70 m
Safety level 2	5-10 knots	0.50 m	0.75 m	1.50 m	2.25 m
Safety level 3	5-10 knots	0.50 m	2.10 m	4.20 m	6.30 m

Human Factors and Experience of the Coxswain

Another parameter that causes "unsuccessful" recovery operations of the FRC into the stern ramp relates to human factors. In these cases, the transverse accelerations exceeded the limiting values due to the small boat striking the side of the stern ramp at the moment of its entry onto the ramp. The coxswain's ability to keep the FRC in position behind the ship stern and to select the time to start entry are considered to be the main actions that determine the impacts of the FRC on the ramp sides during entry into the stern ramp.

According to model test results, the significance of human factors increases when the higher safety levels are considered. Human factors are strongly dependent on the coxswain skill and experience. For practical application, it is proposed to consider results obtained from the model tests as representative for the level of skill of a coxswain with medium experience. These are presented in Table 8.

Table 8. Loss of operability (in percent) Due To Human Factor (for a medium experienced /skilled coxswain)

	Ship speed	Wave heading off bow				
		0°	45°	90°	135°	180°
Safety level 1	5 knots	5	5	5	5	5
	10 knots	5	10	10	5	5
Safety level 2	5 knot	10	15	15	10	15
	10 knots	15	10	5	15	5
Safety level 3	5 knots	10	15	20	20	20
	10 knots	15	10	20	20	20

The loss of operability due to human factor should be deducted directly from the PTO computed by applying the "technical factors". For a more experienced coxswain, the deduction listed in the table could be decreased by one-third. This effect cannot be completely dismissed since it is based on the induced lateral impact accelerations that can exceed the limit even with a very experienced coxswain.

STERN-RAMP BOAT DEPLOYMENT APPLICATION EXAMPLE

A range of the reduced operability of the stern-ramp deployment caused by the mother ship and boat responses is demonstrated by an example computed for a frigate-type hull. The selected hull has the following particulars:

Length between perpendiculars, L_{pp}	122 m (400 ft)
Breadth on waterline, B_{WL}	16.3 m (53.5 ft)
Draft at midship, T_M	5.0 m (16.4 ft)

Displacement, Δ 5050 tonnes
Transverse metacentric height, GM_T 1.85 m (6.10 ft)
Roll inertia radius, r_{44} 0.346 B_{WL}

It was assumed that the ship has a single stern ramp placed along the ship's centerline. The ship is equipped with bilge keels (of width 1 m (3.3 ft) and a length of 25 percent of L_{pp}) and two rudders. Three alternative sill depths (0.305 m (1 ft), 0.610 m (2 ft) and 0.914 m (3 ft)) have been considered in the computations. The ship speed for the recovery operation was selected to be 5 knots.

The PTO for stern ramp launch and recovery are calculated by comparing motion-limiting criteria with the mother ship motion predictions in representative sea states in one specific geographic region. For these computations the grid point no. 263 in the North Atlantic Ocean, located south of the Gulf of Maine was selected. The joint probability of significant wave height and modal period combination on an annual basis for this grid point are taken from the wave atlas (Naval Oceanography Command Detachment, 1983 and 1985). At this ocean area the statistical data show that the peak of the wave modal period is 9.7 seconds and a probability of occurrence of the wave height greater than 0.305 m (10 ft) is about 28 percent.

The applied set of the operational criteria for stern ramp deployment consists of the seven following criteria that have been discussed with the adopted limiting values as listed:

Roll motion, 8.0 deg SSA
Pitch motion, 2.5 deg SSA
Vertical acceleration at boat station, 0.2 g
Lateral accel. at the boat station, 0.2 g
Ramp availability time, T_{RA} = 5 sec
Relative vertical motion at the ramp entrance,
1.20 m (3.94 ft) rms for Safety Level 2 and 90% successful recoveries
Sway at the ramp entrance,
0.75 m (2.46 ft) rms for Safety Level 2 and 90% successful recoveries

Computations of the mother ship responses were performed by running the VERES ship motion program. This linear-strip theory program calculates 6-degree-of-freedom response amplitude operators. Since amplitudes of the relative vertical motion derived by the program do not include an effect of the altered wave amplitudes in vicinity of the hull as well as disturbances of the water level behind the stern due to the propeller wash and stern wake, the program has been modified by adoption of a set of semi-empirical factors. These factors have been derived by comparison of the ship relative vertical motion at the stern measured during the model tests with the considered vessel and computed by the VERES program. The irregular sea conditions have been modeled by using the Pierson-Moskowitz wave spectrum with the wave spreading factor assumed to be $\cos^2$. All wave directions were considered in the analysis.

Results of the PTO analysis are shown in Figures 7 through 11. The total operability obtained by applying all the listed operational criteria is given first in Figure 7. Each particular criterion is considered separately in the following figures except for vertical and lateral accelerations at the stern and relative vertical motion at the stern, which were not limiting.

In addition, the effect of a range of sill depth on the stern ramp deployment operability computed with the ramp availability criterion is presented in Figure 12. All these results have been computed without including the operability loss caused by possible human factors that for the considered range of the wave height and headings would not be larger than 10 percent.

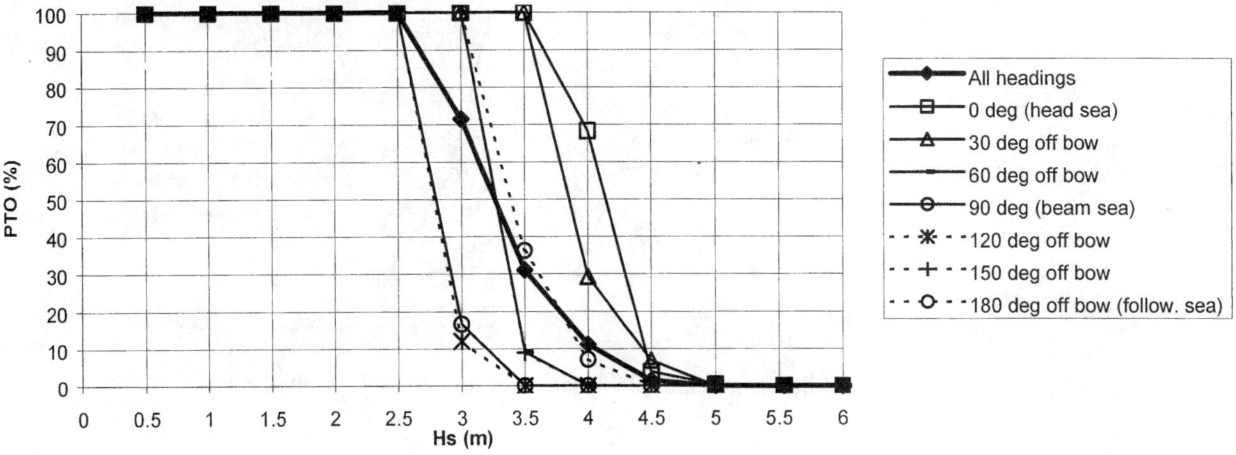

Figure 7. Stern ramp operability from example frigate at 5 kts - All Operability Criteria.

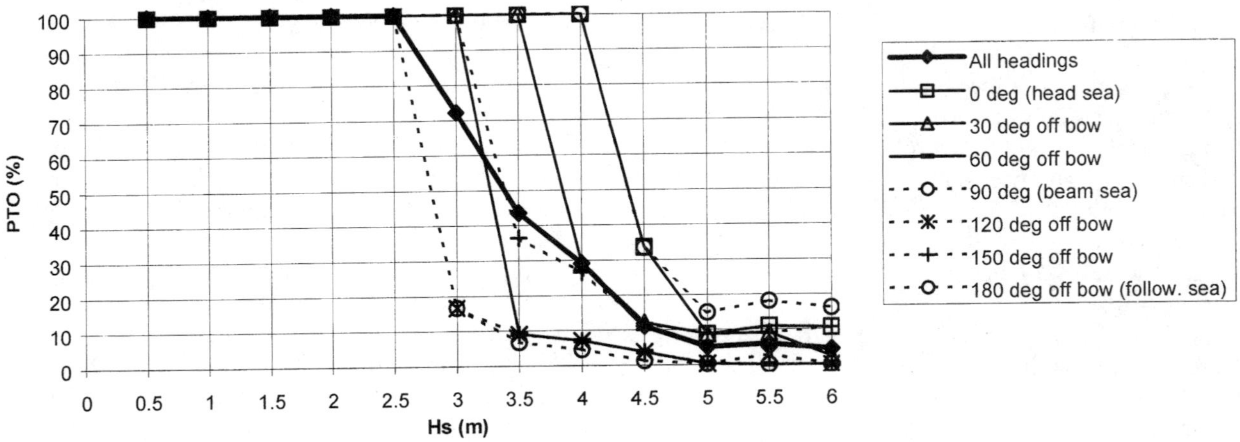

Figure 8. Stern ramp operability from example frigate at 5 kts - Roll Motion Criterion.

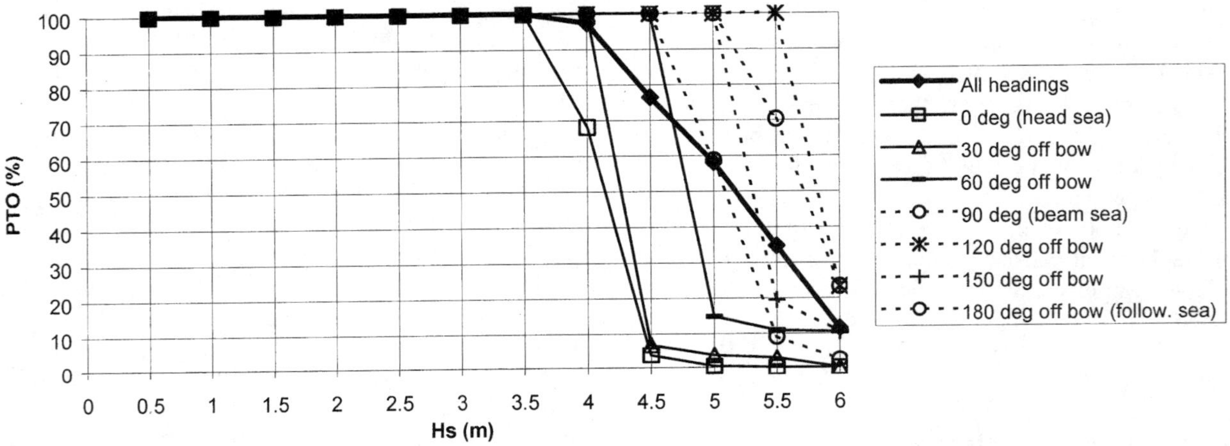

Figure 9. Stern ramp operability from example frigate at 5 kts - Pitch Motion Criterion.

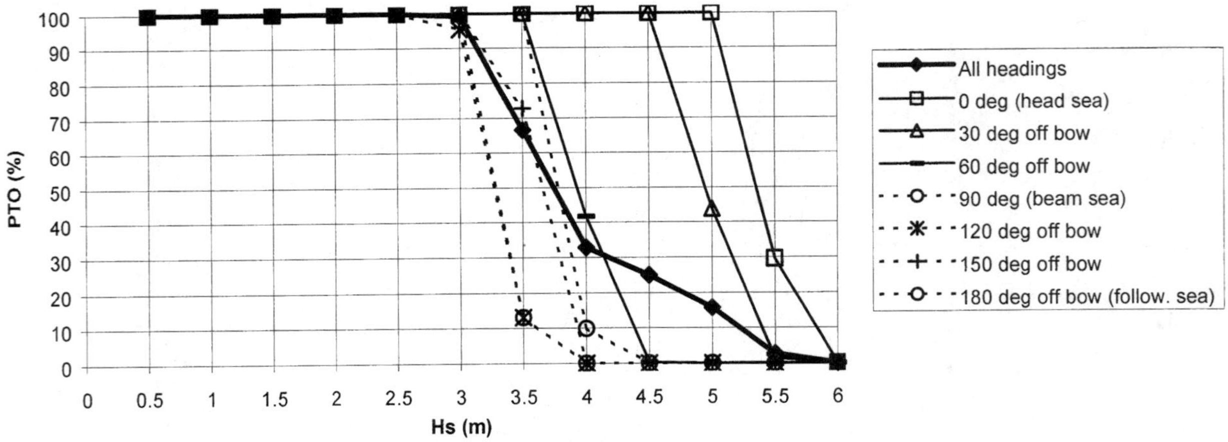

Figure 10. Stern ramp operability from example frigate at 5 kts - Sway Motion at Stern Criterion.

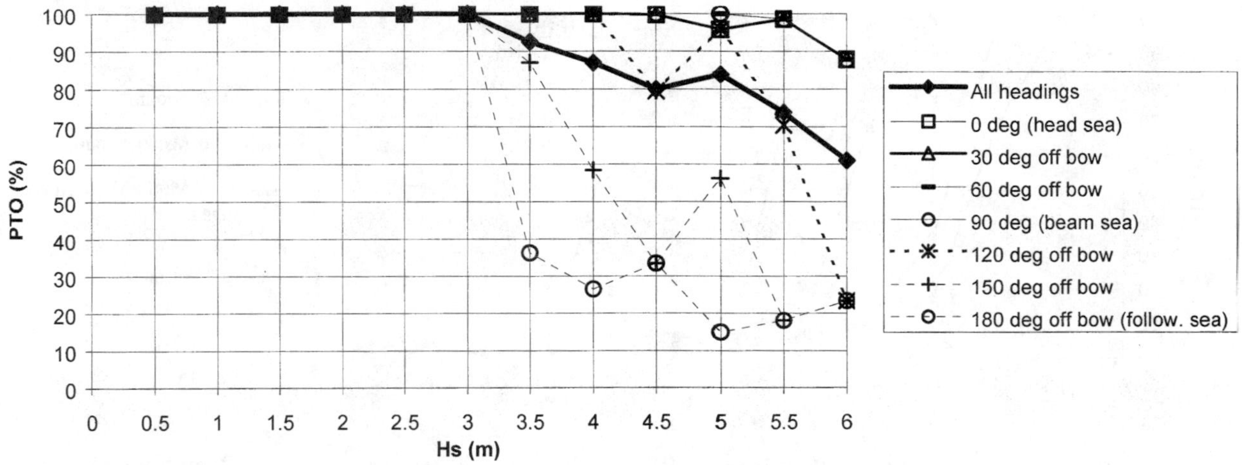

Figure 11. Stern ramp operability from example frigate at 5 kts - Ramp Availability Criterion and 1-ft Sill Depth.

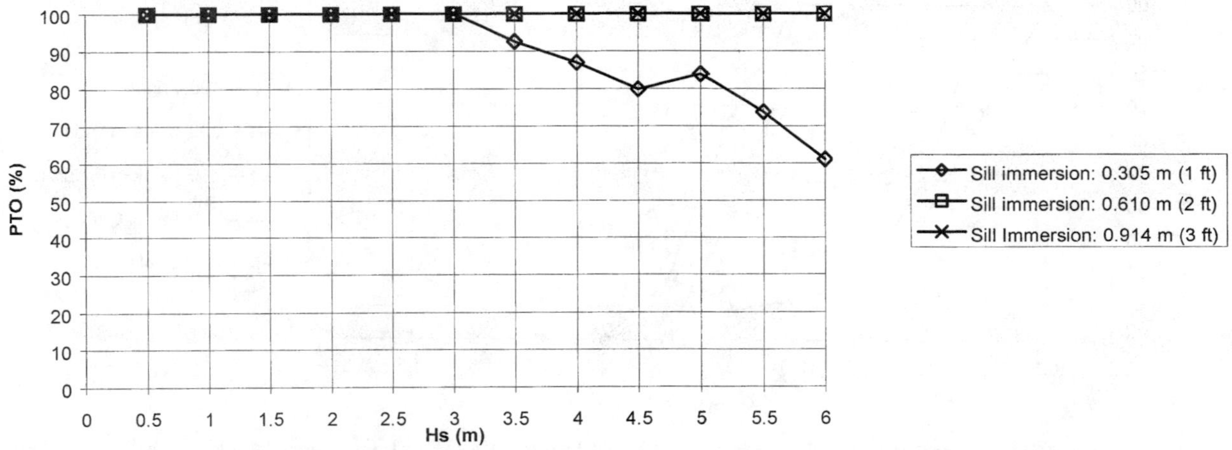

Figure 12. Stern ramp operability from example frigate at 5 kts - Ramp Availability Criterion and Variable Sill Depth.

To investigate an effect of ship size on the operability of stern ramp deployment, similar computations were performed for four additional vessels having an overall length between 26.5 m (87 ft) and 98.5 m (323 ft). The sill depth was set at 0.305 m (1 ft) for all vessels. Comparisons of the PTO for stern ramp deployment for all these ships (including the frigate) are shown in Figures 13 and 14 for the head and beam seas, respectively.

The main conclusions that are important for the stern ramp design and optimization based on results of these computations are as follows:

1. At the considered ocean region, the long-term operability will be 100 percent for seas condition up to a significant wave height up to 2.5 m (8.2 ft) for the frigate-sized vessel.

2. The highest operability will be achieved in head and bow seas (100 percent operability in waves with a significant wave height up to 3.5 m (11.5 ft), or just beyond mid-SS5), and the lowest is in beam seas.

3. The operational criteria that have the largest impact on the limited stern ramp deployment operability for this example are the criteria related to the roll and sway motions. This indicates why the beam direction is not preferable for such an operation.

4. For the frigate-sized ship considered, and when the sill immersion is equal to 0.61 m (2 ft) or more, the ramp availability criterion is not limiting even in seaways with a significant wave height up to 6 m (19.7 ft).

5. Results of computation clearly show a relation between the ship size and a level of the stern ramp operability.

6. Selection of the optimal wave heading for the stern ramp deployment is more important for longer ships. For small ships, the operability of the stern ramp deployment does not differ significantly between the head sea and beam sea directions.

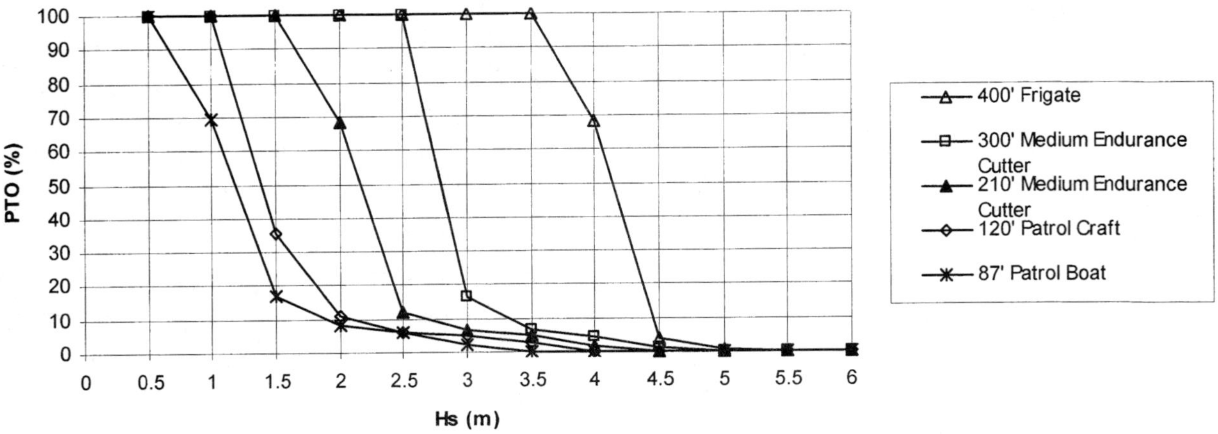

Figure 13. Stern ramp operability for five ships of different length at 5 kts in head seas - All Operability Criteria.

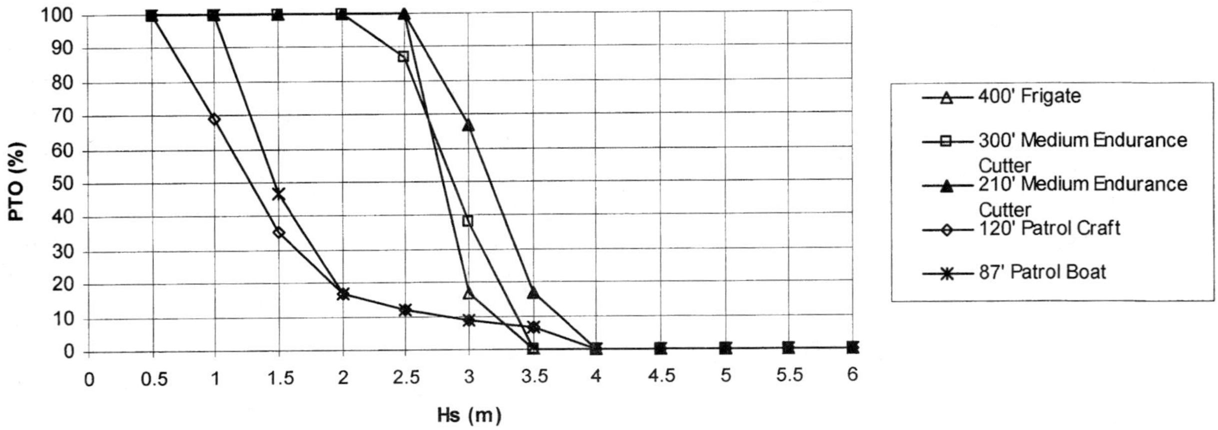

Figure 14. Stern ramp operability for five ships of different length at 5 kts in beam seas - All Operability Criteria.

CONCLUSION

The effort undertaken by the U.S. Coast Guard led to the development of three sets of design criteria for the evaluation of stern-ramp boat deployment. The first addresses mother ship motions for stern-ramp operations and was adapted from similar side-davit motion criteria. The second is a ramp availability criterion that quantifies the effectiveness of sill depth at the stern ramp. Finally, a collection of stern-ramp deployed boat criteria is proposed. The latter have been translated into motion responses at the stern ramp of the mother ship in order to facilitate evaluation and comparison of different mother ship hull designs.

Vertical acceleration at the stern ramp as a mother ship motion criterion was revisited. Here is a situation where the ship's vertical acceleration has a direct bearing on the ability of a crewmember on the ship to perform a task, and requires dexterity, agility and timeliness to secure the returning boat. As long as the securing operation requires a crewmember's ability to toss a line to a boat bowman, lasso a bitt or kingpost, or engage a self-locking device, then the vertical acceleration criterion is needed for crew safety.

For stern-ramp vessels, a ramp availability criterion was defined in terms of fundamental whole ship motions that could be extracted from time-domain seakeeping analysis.

Deployed boat criteria have been introduced that requires the induced impact accelerations on the small boat to be kept on a level acceptable for the boat's occupants to assure a safe recovery operation. Results from the model tests of a frigate-type mother ship show that the operability and limitations of a stern ramp recovery system are dependent upon a combination of technical (or physical) factors and human factors. The technical factors include mother ship vertical relative motion at the stern as contributing to the FRC vertical impact acceleration and transverse stern displacement by the mother ship causing a large longitudinal component of impact acceleration from transverse FRC impacts. Finally, the reduction of operability due to the FRC transverse impact accelerations is assumed to be mostly dependent on coxswain skill and therefore dependent upon human factors.

All in all, the ramp availability criterion and the deployed boat criteria provide the system connection between the mother ship's ability to provide a safe haven and the returning boat's ability to achieve it.

Parametric studies indicate that roll, pitch, and vertical and lateral acceleration at the stern could easily be the limiting factors over the ramp availability criterion for some ships with stern ramps and adequate sill depth. Hence, the criteria set for mother ship motion has been kept with the addition of ramp availability and deployed boat criteria to form coherent criteria for stern-ramp boat deployment.

The stern boat launch and recovery criteria related to the motions of the mother ship were deduced from dedicated model tests on a frigate-sized mother ship and the impact accelerations experienced by a FRC entering the stern ramp. Different criterion levels may be more appropriate for different sized mother ships, for different entrance shapes or boat-to-ramp well side clearances. At present, the model testing approach provides a more direct means to ascertain the acceptability of a particular mother ship and ramp design over present-day analytical means.

With the exception of the ramp sill immersion, the various criteria are not related to other physical ramp particulars such as the ramp breadth, ramp slope angle or entrance form, or the effectiveness of a water management within the ramp. The proposed criteria are at present well suited for feasibility and PTO analyses, but do not provide the means for design and optimization of the stern ramp particulars or configuration. Model testing with large-scale models (mother ship and FRC) have been demonstrated to provide an effective means to optimize the configuration of stern ramp details.

Finally, boat coxswain proficiency with FRC docking into a stern-ramp equipped vessel operating in a seaway depends greatly upon experience. Recent model tests with radio-controlled small boats returning to the stern ramp of a frigate-sized ship have indicated that the effective operability of a coxswain with roughly medium-level experience is about 20 percent less as compared to an experienced coxswain. Simulation-based training or hands-on training in a model basin may provide an effective means to develop coxswain skill prior to facing the challenge in the real ocean environment.

The investigations undertaken have shown that the main particulars and form of a stern ramp can be optimized in order to achieve a higher operability level for a given sea condition. Parameters of the stern ramp designed for operation in a Sea State 5 or higher must be carefully selected and depend upon magnitude of the wave induced motion characteristics of the mother ship and deployed boat. Similarly, the optimal operational conditions, such as wave heading and applied forward speed, might be different for different classes of ships. Since rules of designing an optimal stern ramp have not been established and because a systematic, all-encompassing series of model tests with stern ramp configurations have not yet been performed, selection of the optimal stern ramp may be a process involving some iteration. In this situation, the proposed sets of operational criteria for the stern launch and recovery system should be seen as the first step in the direction of providing guidelines for stern ramp design. More model tests, full-scale measurements and theoretical studies are still required to complete the process of stern ramp design and for evaluation of its operational performance in a seaway. The authors are in the process on developing a detailed design procedure for boat stern launch and recovery system construction.

ACKNOWLEDGEMENTS

The authors wish to thank the Engineering Logistics Center of the United States Coast Guard, the Naval Surface Warfare Center, Carderock Division, and Norwegian Marine Technology Research Institute (MARINTEK) for supporting this effort. In addition, they wish to express their grateful appreciation to the designers, officers and crews of ships visited for their cooperation and valuable assistance. Finally, the authors wish to acknowledge the influential contribution of the late John F. Dalzell (LFL) without whose participation in the early stages of this investigation we may not have achieved all that we had hoped.

REFERENCES

Dalzell, John F., *A Proposed Criterion for Launch Ramp Availability*, Hydromechanics Directorate Technical Report NSWCCD-50-TR-2003/15, 2003.

Hoberock, L. L., "A Survey of Longitudinal Acceleration Comfort Studies in Ground Transportation Vehicles," The University of Texas at Austin, Research Report 40, July 1976.

Kauczynski W., and P. Wereskiold, "Stern Ramp Boat Retrieval Assessment Model", Marintek report no. 601874.00.04, Trondheim, Norway, November 2001. Protected.

Kim, Y., and P. D. Sclavounos, "A Finite-Depth Unified Theory for the Linear and Second-Order Problems of Slender Ships," *Journal of Ship Research*, Vol. 42, No. 4, 1998.

Lin, R.Q., "A Finite Amplitude Steady Ship Motion Model," 24th Symposium on Naval Hydrodynamics,

Fukuoka, Japan, July 2002.

Minnick, Peter V., Christopher Cleary and Rubin Sheinberg, "Operational Comparison of the 270-ft WMEC and 378-ft WHEC Based on Full Scale Seakeeping Trials," Society of Naval Architects and Marine Engineers Chesapeake Section, Baltimore, Maryland, 16 June 1999.

Naval Oceanography Command Detachment, Asheville, NC, U.S. Navy Hindcast Spectral Ocean Wave Model Climatic Atlas: North Atlantic Ocean, Commander, Naval Oceanography Command, NSTL, MS, 1983.

Naval Oceanography Command Detachment, Asheville, NC, U.S. Navy Hindcast Spectral Ocean Wave Model Climatic Atlas: North Pacific Ocean, Commander, Naval Oceanography Command, NSTL, MS, 1985.

Popov, Yu. N., O. V. Faddeyev, D. Ye. Kheisin, and A. A. Yakovlev, Strength of Ships Sailing in Ice, Sudostroyeniye, Leningrad, 1967.

Sheinberg, Rubin, Christopher Cleary and Thomas Beukema, "Worldwide Assessment of Stern Launch Capability", *24th United States Japan Cooperative Program in Natural Resources (U.S./UJNR) Marine Facilities Panel Meeting*, Honolulu, Hawaii, 4 to 9 November, 2001.

Thomas III, William L., Terrence R. Applebee and Alan W. Abbs, *A Method to Define Ship Motion Limiting Criteria*, NSWC Report CARDEROCKDIV, NSWC/SHD-1338-04, October 1992.

U.S. Coast Guard, *Technical Evaluation of 270 WMEC Capabilities in Alaskan Waters*, Technical Branch, USCG Naval Engineering Division, Baltimore, Maryland, February 1996.

U.S. Coast Guard, *Assessment of Stern and Side Boat Deployment Systems: Summary Report*, Report No. ELC-023-01-002, Naval Architecture Branch, Engineering Logistics Center, Baltimore, Maryland, April 2001. Deepwater Project Protected.

Werenskiold, Per, *Summary Report: Stern Ramp Model Tests*, MARINTEK Report No. 601874.00.01, Norwegian Marine Technology Research Institute, Trondheim, Norway, April 2001. Protected.

Werenskiold, Per, *Stern Boat Deployment Tests for DeepWater Notional Security Cutter Design*, MARINTEK Report No. 602011.00.02, Norwegian Marine Technology Research Institute, Trondheim, Norway, March 2003. Protected.

DISCUSSION

Robert G. Allan, Member

The authors are to be congratulated for tackling a very complex subject, and attempting to formulate appropriate design criteria for current and future applications of stern-launch systems for rescue craft. The systematic approach proposed provides useful guidance for all involved in the design process of such systems, from the most practical designer to the sophisticated analyst.

As a practical small craft designer, I have had the privilege of being involved in the design and testing of a number of projects involving stern-launched rescue craft, including the Canadian Coast Guard Type 500 SAR Cutter *Gordon Reid*, one of the vessels evaluated in the authors' study. Therefore, I am very pleased to be asked to comment on the work the authors have done in attempting to bring order both to the design process and more critically to the evaluation of the merits of alternative systems proposed.

There are very few issues raised in this paper with which I would disagree. However, there are some aspects of the design of boat recovery systems on which I would like to elaborate.

Implicit in the design of an effective boat recovery system is the necessity to make the fastest possible transit of what I would call the "transition zone," where the RHIB moves from waterborne support in free water to a fully mechanically supported mode on the ramp or whatever other device is used to stow it on the mother ship. The rest of the process is relatively easy. However, it is in the transition zone where the greatest risks of personnel injury and boat damage exist. This process involves:

- A well-timed entry into the ramp (almost totally operator dependent)
- Safe and accurate landing on and connection to the "shore" (design dependent)
- Removal from, or implementing barriers to, the wave-affected zone (design dependent).

The authors have addressed all these features, and have stressed the critical importance in the entire operation of the operator's skill. The importance of this factor cannot be overstressed, and thus success rates of almost any system could be improved by a proper training regimen. It also perhaps begs the question why well-trained personnel are not maintained and suitably rewarded for their skill in this difficult and often dangerous role.

The authors identify some features that are fairly critical to the proper design of a stern ramp launch/recovery system. Many of these seem to fall under the realm of simple common sense, but it is worthy that they are identified, at the risk of stating the obvious. These features include;

- The use of a "funnel" shape to center the boat in the ramp
- The use of well-rounded transom corners
- A method of reducing the accelerations due to landing on the ramp structure
- Avoiding the use of overhead structures.

In the case of the latter, I would say that the authors have been too gentle in their critique of those designs that incorporated this feature. In the interests of maximizing the safety and speed of the launch and recovery operations, I would not hesitate to state that such a feature should in no case be considered unless the clearance above the maximum height of the RHIB in the worst conceivable operating condition was at least 2 meters. Even at that, the operator of the RHIB is bound to have the perception of having to thread a needle.

Contrary to what is stated in the paper, the landing device in the CCG Type 500 Class cutters comprised a set of simple large diameter PVC pipes (not UHMW), bolted to an adjustable framework to support the RHIB, much as would the "bunks" on a simple boat trailer. PVC was chosen after a series of tests to find a material that not only had a suitably low coefficient of friction to enable self-launching but, more critically, that would deflect appreciably under impact to minimize the accelerations on landing. This material is very inexpensive, and the design permitted easy replacement of the pipes should they wear out or be damaged. Incorporating a simple shock absorber system in the support of these slides could further reduce the impact on the RHIB on landing.

One of the features of the recovery system on the *Gordon Reid* that did not receive much attention in the paper, but that I think is worthy of further comment is the use of the hinged platform within the stern ramp. This ramp comprised a simple lightweight aluminum frame, supporting a deck of noncorrodible, lightweight, reinforced GRP grating (not expanded metal as described in the paper). Above this base were the PVC "runners" discussed above. This hinged platform served two purposes: (1) as described in the paper it had the immediate effect of some wave mitigation from the boat hull interface, and (2) more critically, it enabled the boat to be rapidly removed a large distance from the wave interface, bringing the boat crew and any rescuees onto a horizontal working deck from which they could be far more easily and safely removed from the RHIB.

I would like to reinforce the author's conclusion that at least for the foreseeable future large-scale model testing is the most reliable means of evaluating the suitability of a stern launch/recovery system. In the design of the Type 500 cutters, working in association with Offshore

Research Ltd. of Vancouver, we conducted an extensive series of large-scale (1:10) manned model tests in the open ocean. These tests were conducted to evaluate the same set of data as the authors measured in a laboratory setting, with similar conclusions. I would argue, however, the authors' claim that the use of outside testing is of limited value. By use of a wave buoy in the vicinity to collect sea state data, and bearing in mind the associated influence of wind and irregular waves, it is arguable that outside testing provides a far more realistic test of a system's operability. Indeed, it is less friendly in terms of posttesting analysis, but the results are certainly far more realistic. I am also somewhat surprised by the authors' observations and conclusions regarding best relative heading of the mother ship and rescue boat in recovery operations. All the model (and subsequent full-scale) testing performed for the *Gordon Reid* class proved very conclusively that the best relative attitude was in beam seas. The relative motions between the two vessels were much less in roll than in pitch and yaw, and even though the smaller craft is more exposed to wind and wave influences in this attitude, the reduction in relative pitch appreciably increased the ramp availability and reduced the accelerations of landing in the ramp. In this context, therefore, I would ask the authors, had the operators who preferred a more fore and aft recovery attitude ever even tried a beam-on recovery, and if so, what were the results?

Another observation/comment of the authors, which I would challenge, is that concerning the poor slow-speed maneuverability of a RHIB with water-jet propulsion. In the hands of a well-trained operator, a water-jet-driven boat is actually far more maneuverable than an outboard or stern-drive-propelled boat. However, this capability relies solely on the operator maintaining high engine rpm and controlling the thrust vector with the jet buckets. This is somewhat counterintuitive for most small boat operators who are used to outboards or stern drives, where in close quarters the natural thing to do is reduce engine rpm. In a jet boat the thrust is totally rpm dependent, and hence the operator must enter a confined space with engines close to full rpm for maximum control. Again, this observation of the authors reinforces the need for well-trained personnel in the specific boats used.

In conclusion, I would like to thank all the authors for this excellent paper, which should prove a very valuable first step to the advancement of ever better and safer rescue boat launch and recovery systems. Clearly, there is still much more work to be done in the ability to simulate the relative behavior of two diverse vessel types in close proximity, with so many external influences. However, it is obvious that the authors have charted the first stages of that difficult course very well.

Per Werenskiold, Visitor

Congratulations to the authors on a very complete presentation of an ambitious worldwide assessment of vessels with stern-launch capability, and the effort to develop boat deployment criteria and design analysis methodologies.

The authors provide the most comprehensive state-of-the-art survey and discussion of issues as:

- Vessels with stern ramp capability, vital ship, stern ramp design, and small boat (RHIB) parameters
- Operational experiences, launch and recovery procedures
- Main hydrodynamic effects influencing the operability of stern ramp deployment and alternative numerical modeling methods
- Design process integrating criteria for ship design, design of ramp system, RHIB performance, and human performance
- Use of model tests, especially with large-scale models, as the most effective tool for optimizing stern ramp design.

Having been involved in the establishment of stern ramp operability criteria and project leader for five comprehensive stern ramp model test and detailed design programs, two issues come to mind reading the paper:

1. What types of phenomena shall be considered to optimize the stern ramp system operability?
2. How can assessment criteria precisely reflect these effects to enable a consistent design optimization?

All aspects for answering these questions are considered in the present paper. However, I would like to elaborate this further by adjusting Table 4 in the paper. In particular, the problem areas are given a proposed priority according the importance to achieving a successful recovery operation.

Table 4

Priority/Problem Areas	Phase of Recovery Operation	Type of Aspect	Available Tools
1) Interior ramp hydrodynamic effects	Entry of ramp and boat securing	Ramp design/arrangements	Model tests
		RHIB speed/acc. capability	Ship/RHIB trials
2) RHIB maneuvering in propeller wash	Final approach and decision to start entry	RHIB navigating performance	Ship/RHIB trials
		Coxswain qualities	None
3) RHIB maneuvering in stern wake	Approach at safe distance from ship	RHIB navigating performance	RHIB trials
		Coxswain qualities	None
3) Ship motion in seaway	Influences final approach and entry	Ship stern relative wave motions	Strip-theory ship seakeeping
4) Wave sheltering and reflection	Small effect head sea approach and entry	Ship design	Panel-method 0 kn.
		Masters ship operation	None

Thus, I assume that the most sensitive (limiting) phase of the stern ramp recovery operation is the boat's entrance into the ramp that has to coincide with "sufficient" water level in the sill. In addition, the boat has to be protected inside the ramp until captured and secured.

The authors state that problem areas 1 to 3, listed in the table, are extremely difficult and computationally intensive problems to solve and should be solved numerically by state-of-the-art programs if at all possible. In addition, any design method needs to apply relevant criteria to enable comparison and optimization of designs. The question should be asked if numerical tools can be applied for purposes other than comparison of ship designs based on criteria related to ship stern relative wave motions, no stern ramp effects considered.

The priority areas to be considered should be defined by performance requirements, assessed by quantitative measurements, and performance should be classified according to well-defined criteria. For example:

Problem Areas	Performance Requirement	Quantification	Type of Criteria
1) Interior ramp hydrodynamic effects	The relative wave amplitudes in the ramp entrance (sill) should be as small as possible, and mean value as large as possible.	**Model tests:** Measurement of relative wave in sill	Technical/ consistent
	RHIB should be able to maintain directional control in the stern/ramp wake.	Experiences/ trials	Technical/ qualitative
	RHIB should be able to accelerate onto the ramp with momentum to obtain secure/capture position in the ramp.	Experiences/ trials	Technical/ qualitative
	Coxswain should be skilled and well trained.	Training/ experiences	Human/ qualitative
2&3) RHIB maneuvering in propeller wash and in stern wake	The actual wave amplitudes in the stern wake and propeller wash should be as small as possible.	**Model tests:** Measurement of relative wave in sill	Technical/ consistent
	RHIB should be able to maintain directional control and good seakeeping performance in the ramp approach area.	Experiences/ trials	Technical/ qualitative
	Coxswain should be skilled and well trained.	Training/ experiences	Human/ qualitative
3) Ship motion in seaway	Ship motions should be as small as possible, in particular vertical and transverse displacements of stern.	Strip-theory programs	Technical/ consistent
4) Wave sheltering and reflection	Comment: relatively small effect at the stern of larger ships.		

The recent stern ramp model test programs have all included detailed studies of main parameters that could influence the relative wave motions inside the ramp, in the sill and in the entrance area: water management/damping systems, ramp sill depth, and stern design. For each of the actual projects, major contributions to higher operability for the stern ramp boat deployment are achieved. In addition, parameters that will permit the coxswain to power the RHIB into the ramp without major impacts toward hard structures of ramp entrance and sides are studied in detail.

All the above-mentioned problem areas and main parameters influencing the stern ramp operability are thoroughly discussed by the authors. However, no assessment of the relative importance is indicated. Comments to the above "priority list" and coherence of quantification methods will be of great interest.

One of the most essential ramp design parameters is sill depth. Most projects tested at MARINTEK have sill depth in the magnitude of 1m (3 ft). The PTO analysis (Fig. 12) shows the effects of variable sill depth for an Lpp = 122 m (400 ft) frigate-type mother ship. What would be the design sill depth for the National Homeland Security Cutter?

As the performance of the RHIB and the coxswain's ability to navigate the boat and skill to decide the timing of the entry are most essential for the effectiveness and safety of stern ramp recovery operations, are the authors confident that human factor aspects, such as selection of personnel, simulation-based training, and drills, would be given the highest attention?

While the authors present a state-of-the-art assessment of stern ramp systems, design parameters, model test techniques, analytical tools, and design criteria, they are robust and objective in their approach. In particular, the last sentence of the paper, "The authors (USCG-ELC) are in the process of developing a detailed design procedure for boat stern launch and recovery system construction," should be appreciated by the potential operators and the industry.

Jan O. de Kat, Member

I would like to commend the authors for presenting a thorough study on the rather rare research topic of small boat deployment.The paper is successful in bridging the gap between the practical operational side at sea and the research environment, where model tests and computations are used to simulate reality. It would be appreciated if the authors could expand on some topics addressed below.

First, it would be interesting if the authors could provide more detail on the actual physics of the boat launching and recovery process. For instance, does the coxswain decide to enter the ramp when the sill submergence is at a maximum (i.e., maximum relative motion is anticipated before reentry), or does the coxswain make use of quiescent periods in the sea state (in which case the process would be governed by wave groupiness). Also, how does the water typically behave inside the ramp–sloshing, wave run-up, wave breaking, and so forth.

From the discussion of available computational tools it is clear that such tools are not yet able to predict all of the combined aspects of the relevant physics involved in boat deployment. I was wondering if the authors have considered coupling a VoF (CFD) code to a 6 DOF ship motion code to simulate the internal fluid motion inside the ramp area of the ship in waves.

In the absence of suitable computational tools, the authors resort, rightly so, to model testing techniques. However, I find it puzzling that on the one hand the authors recommend carrying out tests at scale 1:10 to 1:12.5 , whereas on the other hand the majority of tests were carried out at scale 1:22.88 with only a limited number of tests with the large cutter model. Also, could the authors discuss whether there were any differences between the large- and smaller-scale tests as regards boat recovery in sea state 6. One would expect some differences in boat handling ability by the coxswain in the model basin, especially if short duration events have to be judged at smaller model test scale.

Could the authors provide information on how many recovery tests were carried out for each test condition to arrive at the PTO estimates as in Table 6? Presumably the repeat tests were done in different random wave realizations. In view of the human element involved, it would be interesting to hear whether the recovery model tests were carried out always with the same coxswain, and with which skill level.

In the boat deployment application example the authors provide a list of criteria; could they comment on the selected values, such as for limiting roll and pitch? Furthermore, it would be interesting to hear to what extent such criteria might depend on stern ramp configuration. For example, presumably a wide stern ramp would allow a higher level of sway motion at the stern than a narrow ramp entrance.

David W. Byers, Visitor

The science of ship design continues to become more rigorous and truer to a "first principles" approach, which accurately reflects what really happens in the full-scale physical world. These advances are in response to the steady evolution of design tools and methodologies made possible by ever more capable computers. At the same time it is encountering challenges posed by the increasing interest in more radical hull form types, such as high-speed catamarans, trimarans, and so forth for naval or military applications, which require validated design tools and techniques beyond those used in the commercial world, if in fact they exist at all. There is accordingly a critical need to apply a true ship systems engineering methodology to areas of ship design that have historically been based on empirical or "rule-of-thumb" approaches. These historic methods, which frequently combined a large number of uncertainties into safety factors, margins, and so forth, can now be improved as more and more of these previous uncertainties can be more accurately assessed.

The authors of this paper have done a great service to the profession by providing an excellent example of how to apply systems engineering principles to one of these historically "rule-of-thumb" design processes: that of developing a satisfactory means of deploying small boats from the stern of relatively small "mother ships." Their approach to the problem, including breaking it down into its component elements, doing an international survey of existing smaller ships with stern deployment systems, identifying the hydrodynamic phenomena that affect stern boat deployment operations, assessing the state-of-the art in prediction tools, developing criteria for safe operations, and establishing a method for ultimately determining the key metric of percent time operability, are all elements of a rigorous, system-engineering-based approach to problem solving. Particularly noteworthy is the attention paid to the critical human systems component of the total ship system, that is, ensuring that the role of the sailor in small boat deployment operations is defined from the start and reflected throughout the design evolution. As the U.S. Navy moves increasingly to optimally manned ships where functions historically performed by a number of sailors are now performed by just one who relies partially or totally on automated subsystems, it is critical that the remaining human functions and the interfaces with the mechanical components be clearly defined.

Although this paper has focused on the manned small boat operations, the work can be extended to the even more challenging problem of handling an unmanned small boat or other vehicle.

In this respect, the U.S. Navy has already benefited from the work presented in this paper. One of the coauthors, *Mr. Peter Minnick*, is an employee of the U.S. Coast Guard's Engineering Logistics Center (USCG ELC). He was a team member of a ship system concept development project recently completed by the Naval Sea Systems Command Surface Warfare Center Carderock Division's Innovation Center, which I head. The project, conducted between October 2002 and April 2003, was entitled "Surface Combatant Optimized for Unmanned Vehicle Operations" (SCOUVO). One of its challenges was to accommodate a variety of unmanned vehicles (air, surface, and underwater) on a ship platform capable of achieving a speed of 45 knots or more. This resulted in the development of three ship system concepts, each based on a different advanced hull form type, specifically a high-speed monohull, catamaran, and trimaran. All three incorporated a stern ramp for unmanned vehicle handling. The U.S. Coast Guard ELC work described in this paper directly influenced the design of the SCOUVO stern ramp deployment systems. These slides of each concept depict how the stern ramp was incorporated in each case [explain details on slides, particularly variable cradle].

We are pursuing follow-on work to the SCOUVO project leading up to the point of prototype evaluation of some of the subsystems, including the variable cradle and another promising concept, the "homing crane." If we are successful, and as an outgrowth of USCG ELC cooperative involvement with the U.S. Navy's Innovation Center project, the Coast Guard has expressed a willingness to be a partner with the Navy in the evaluation of any such systems of interest to them.

Again, I thank the authors for an excellent piece of work.

Woei-Min Lin, Member

I would like to congratulate the authors of this comprehensive study. This is a very timely paper to address critical design and operation issues of the stern boat deployment system. In particular, the authors have done an excellent job illustrating the key hydrodynamic considerations of operability and design. The discussion presented here comments rather than questions.

I am particularly interested in the possibility of using a physics-based approach to study the design of the stern boat deployment system and to develop guidelines for small boat launch and recovery operations. As the authors correctly pointed out, the hydrodynamic phenomena in the stern region are extremely complicated. None of the advanced computational tools currently available can capture all aspects of the phenomena. However, I believe that the current physics-based tools can be very valuable for certain aspects of the design an operation study as long as we understand the assumptions and limitations of the tools. In particular, the prediction of the local relative motion problem at the stern helps to provide an evaluation of the conditions in which the small boat must operate.

At SAIC, we are continuing the development of the Stern Ramp Model (SRM) that links with the Large Amplitude Motion Program (LAMP). In the past year, we have added the radiation and diffraction waves generated by the mother ship and a local time-domain sloshing model to predict the flow over the stern ramp, and we are in the process of including the propeller effect by simulating rotating propellers with vortex lattice method. In addition, we are developing a more accurate planning boat model as part of the simulation. There are certainly other hydrodynamic phenomena, such as flow turbulence, which are not likely to be modeled by simulation tools in the near future. However, we think that the physics-based simulation tools available today have sufficient fidelity to be useful for design and operation study.

Once again, I would like to congratulate the authors on their significant achievements of this study.

WAVE DAMPING AFTBODY WITH HYBRID PODDED PROPULSORS

Raimo Hämäläinen, **Kvaerner Masa-Yards Turku, Telakkakatu 1**
P.O. Box 666, FIN-20101 Turku, Finland

Jaap van Heerd, **Maritime Research Institute Netherlands (MARIN)**
P.O. Box 28, 6700 AA Wageningen, The Netherlands

ABSTRACT

Nowadays, the shipbuilding industry in Europe is changing from conventional vessels to more advanced ships. Monohull passenger ferries are sailing at higher speeds, ship owners require higher comfort classes, and the seakeeping and maneuvering requirements of the modern ships are much higher. Conventional single and twin-screw propulsion concepts do not always have the capability to fulfil these higher requirements. A rather good solution for improved ship design is the hybrid propulsion concept, where two types of propulsors are mixed.

In this paper, a description is given of some hybrid concepts. The advantages and the drawbacks of the podded propulsion units combined with conventional propeller, the triple screw CPP-POD configuration and the triple podded propulsion concept are reviewed.

The Wave Damping After body (WDA), previously developed for fast monohull ferries decreases dramatically the aft body wave making, which results in a remarkable reduction of the operational cost and can be considered as an environmentally friendly solution, is described.

The Hybrid CRP propulsor developed for a fast Cruise Liner is explained and compared with conventional CPP and FPP designs. The Hybrid propulsion system developed for the World's largest cruise liner is described. A brief review of design of the hull form and the propulsion concept by means of calculations and extensive model experiments is presented.

Finally, the optimization package and the model test results obtained on a modern medium speed Ropax with Controllable Pitch CL-propeller and turnable side PODDED propulsion units are compared to the conventional solution and twin POD case. The excellent wake field at the side and the centerline, the low required power level and the good cavitation pattern and propeller induced pressure pulses are discussed in this paper.

As a concluding remark, the future scenario for hybrid propulsion concepts is given.

Figure 1: WDA hybrid aft body after paint test in MARIN

NOMENCLATURE

Lpp	Length between perpendiculars [m]
B	Breadth moulded [m]
T	Design draught [m]
V	Displacement moulded [m3]
Cb	Block coefficient
C_F^{ITTC}	$=0.075/(10\log Rn-2)^2$
Cm	Midship coefficient
Cp	Prismatic coefficient
CPP	Controllable Pitch Propeller
CRP	Contra Rotating Propeller
CRPA	Contra Rotating Propeller Azipod
Lpp/B	Length/beam ratio
B/T	Breadth/ draught ratio
CFD	Computational Fluid Dynamics
D	Propeller diameter [m]
FPP	Fixed Pitch Propeller
Rn	Reynolds number
Rs	Ship resistance [kN]
PE	Effective power of the ship [kW].
PS	Shaft power [kW]
Ts	Total thrust [kN]
t	Thrust deduction fraction (1-R/T)
w	Effective wake fraction (1-Va/Vs)
Vs	Ship speed [knots]
Va	Speed of advance [knots]
WDA	Wave Damping Aftbody
n	Propeller rotation rate [hz]
N	Propeller rotation rate [RPM]
Fn	Froude number [-], Vs/sqrt(g*Lpp)
h	Head of water, relative to blade tip [m]
d	Shaft line diameter [mm]

INTRODUCTION

The shipbuilding industry in Europe is changing, in part, from simple conventional vessels to more advanced ships. Nowadays, Ropax ferries are sailing at higher speeds, shipowners require higher comfort classes for cruise liners and ferries. The seakeeping and maneuvering requirements of the modern ships are much higher than in the past. More environmentally friendly solutions are required for recently developed ships. That is why fuel economy is playing a more important role. Conventional single and twin-screw propulsion concepts do not always have the capability to fulfil these higher requirements. A rather good solution for an improved ship design is the so-called hybrid propulsion concept combined with Wave Damping Aftbody.

The term hybrid propulsion means that two different propulsor types are combined into one propulsion system.

A general description of the hydrodynamic behavior, efficiencies of different propeller concepts, resistance versus effective wake and the advantages and disadvantages of a CRP propulsion system are described. A number of rules of thumb for propeller loading and optimum POD / propeller locations are given.

A description of the hybrid concept is given for the fast, special CRPA Cruise Liner, for world largest cruise liner and for a medium speed Ropax Cruise Ferry.

The advantages and the drawbacks of the podded propulsion units combined with the conventional propeller concept, the triple POD propulsion system and the triple screw CRPA configuration are reviewed.

In the first part of this paper, the hull form optimization of some ships equipped with a hybrid propulsion concept is shown. The Wave Damping After body (WDA), previously developed for fast monohull ferries dramatically decreases the aft body wave making, which results in a remarkable reduction of the operational cost and can be considered as an environmentally friendly solution. The idea of WDA is explained in recipe form and by examples.

A CRPA propulsion concept designed for a fast large Cruise Liner is described and compared with conventional CPP and FPP propulsion concepts.
The Hybrid propulsion in world's largest cruise liner is explained. A brief review of design of the hull form and the propulsion concept by means of calculations and extensive model experiments is presented.

The optimization package and the model test results obtained on a modern medium speed Ropax with controllable pitch centerline propeller and azimuthing side Podded propulsion units (Figure 2) are compared to a conventional solution and a twin POD case. The excellent wake field at the side and the centerline, the low required power level and the good cavitation pattern and propeller induced pressure pulses are discussed in this paper.
As a concluding remark, the future scenario for hybrid propulsion concepts is given.

Figure 2: Ship model of a Hybrid Ropax Ferry

HYDRODYNAMIC DESIGN OF HYBRID CONCEPTS (GENERAL)

In hybrid projects many critical hydrodynamic aspects have to be taken into account to be sure that the whole ship behaves well.
Some of these items are:

- Low resistance behavior (WDA-effect is required by optimum LCB location and long aft part) in deep and shallow water.
- High propulsive power (propeller usually should be above the baseline).
- Propeller hull excitations (vibration, noise, clearance limitations, etc.).
- Erosion and interactions in CRP-case.
- Cavitation in propeller and appendages.
- Good seakeeping ability (slamming behavior, motions, green water, etc.).
- Good maneuverability (course stability, crabbing, etc.).
- Steering in wind milling case.

The rules of thumb for different hydrodynamic subjects are very useful at the beginning of a project, because they give a rough idea of the critical points and help the designer to concentrate carefully on the correct and most important subjects. They can also help to make good compromises. With hybrid concepts, the most important rules concern the propeller loading, tip speed of the propeller, windmilling operation and erosion.

Some general information and rules of thumb regarding hull lines design (WDA), propeller loading, flow direction and breadth draught ratio (B/T) effects are presented in the next section. (Hämäläinen R., 1998)

Twin screw buttock flow aftbodies

Today we can divide twin screw buttock flow aft bodies into the four different basic types: concave, straight, convex and Wave Damping Aftbody. We are looking at all types in the following section.

Concave form

This type of aftbody form (Figure 3) specially has been used in the older cruise liners. In the modern designs this aftbody type is not used very often because this form does not accelerate the flow in front of the propeller, and there is a risk of flow fluctuation when the vessel has a large block coefficient or in case the LCB is located far aft.

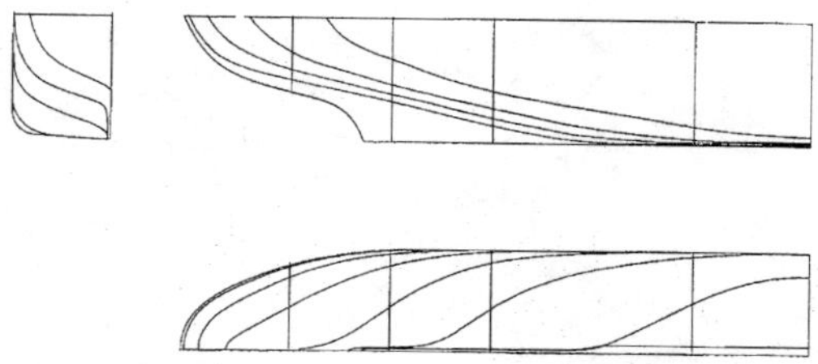

Figure 3: Concave form

Straight form

In the straight form (Figure 4) it is normally possible to draw the buttocks by ruler between the transom and points close to the bottom rounding. In the next example a complex semi tunnel form is shown, which was popular in medium speed ferries some years ago.

This form improves stability by cheeks and makes a bigger propeller diameter possible. The tunnel shape can also be a good alternative to avoid or reduce aftbody slamming. The aftbody slamming is a very important subject and designers have to take this into account when developing aftbody lines.

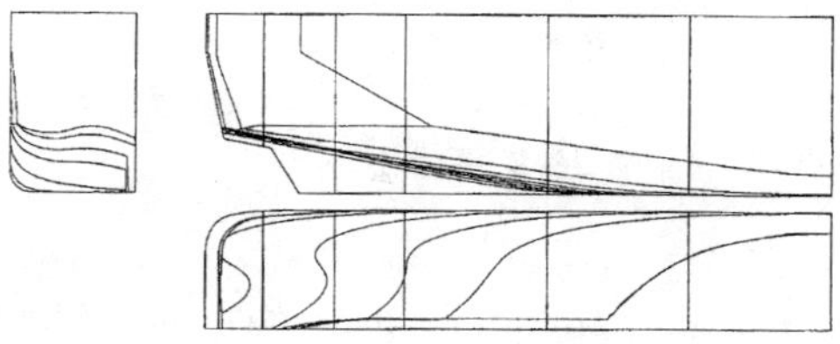

Figure 4: Straight form

Convex (hollow/S-form)

The convex form (Figure 5) is nowadays frequently used when an attempt is made to accelerate the water flow more in front of the propeller. In this form the aft shoulder trough of the wave system is typically visible shortly behind the aft shoulder. The deepness of the trough depends on the fullness (buttock angle) of the aft body, the bilge area form (midship coefficient) and steepness of the aft shoulder in the curve of sectional areas.

Combination.

Usually, a combination of the basic forms is used. Close to the CL location, the verticals can be convex, and close to the side of the vessel they can be straight.

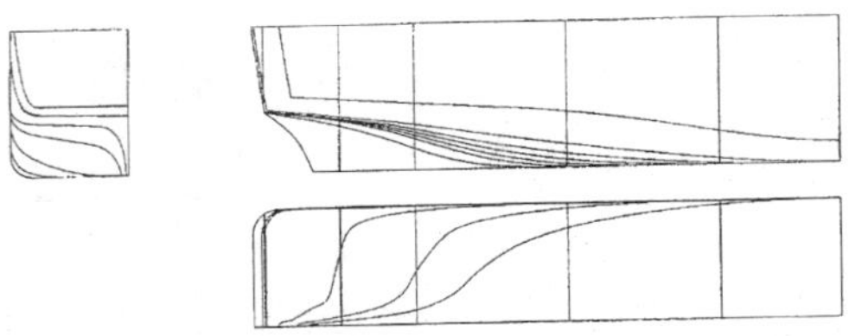

Figure 5: Hollow form

Wave Damping Aftbody

The Shipyard developed a new revolutionary "Wave Damping Aftbody (=WDA)" using the potential flow Shipflow CFD-software and based on experience.

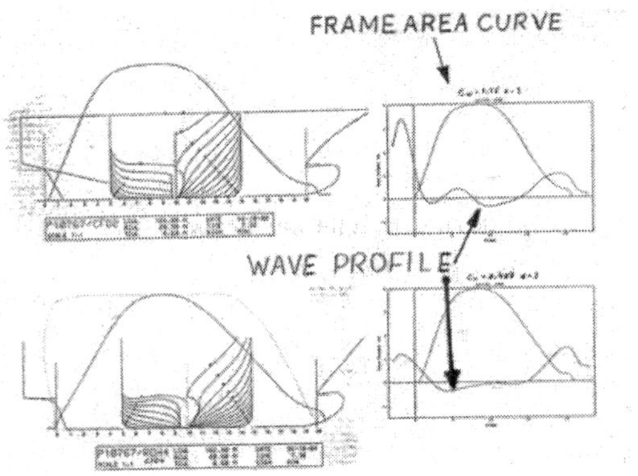

Figure 6: Wave Damping Aftbody

The above figure shows that the wave damping afterbody results in a remarkable difference in wave profile along the hull. Also important differences in sectional (frame) area curve are shown.

The advantage of this aftbody is the absence of the typical aft shoulder crest wave, while the deep wave trough and transom waves are much smaller than in more conventional hull forms. With this aftbody form, the small bilge radius generates more flow in accordance with the buttock directions and prevents the flow turning in the centerline direction and so avoids the typical low-pressure area in the aftbody. This low-pressure area is generated earlier in a typical wave crest location. Verticals in transom location should be inclined so that they direct transom waves for lower amplitudes, which means that they do not give possibilities for higher transom CL waves. Maximum breadth in transom is the most optimum case, the designer should confirm the right transom height by using a wedge form or not, as required.

The waves generated in the aftbody are much smaller than in typical earlier vessels. The wave resistance is much smaller. The WDA was used successfully in the Shipyard's Superfast III, IV and Voyager Cruise Liner Classes. Many WDA studies have been carried out for several vessel types and the following conclusions can be drawn with regard to the power savings:

A big cruise liner	3-5%
A medium speed cruise liner	5-7%
A fast Passenger ferry	7-15%

The same design idea can be used in other projects taking different speeds into account. This type of aftbody works best for the fast, slender vessels as in Superfast III, but is excellent also in a big cruise liner as in The Voyager of the Seas. Today almost everybody tries to obtain the WDA-effect.

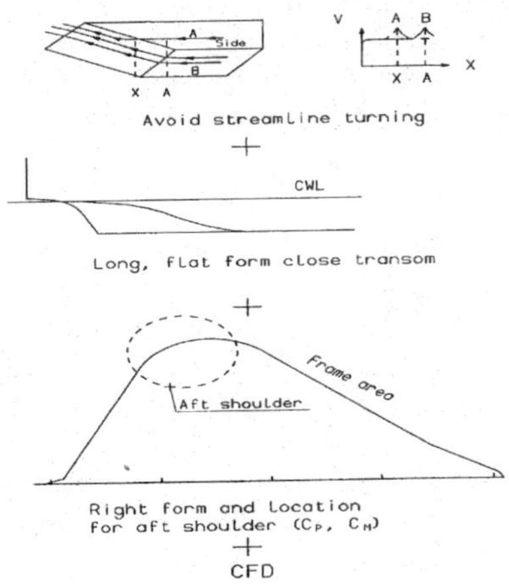

Figure 7: WDA idea in recipe form

The idea of WDA is presented here in a kind of "recipe form" in Figure 7. To ensure a right combination, the CFD tool is needed. The LCB should be located somewhere between -2.5% - -3.5% in order to reach the WDA effect.

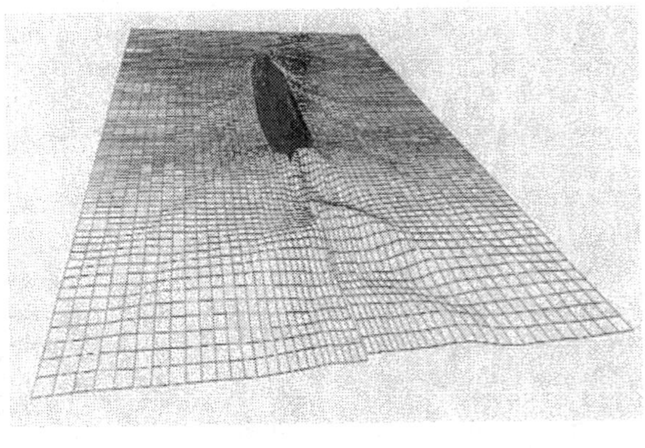

Figure 8: In the left with the WDA effect and in the right without the WDA effect

This new environmentally friendly aftbody decreases generated waves dramatically in the aftbody and also the needed engine power. In Figure 8, the wave pattern of a small cruise liner is presented. In the left hand side the results of a hull design with the WDA are shown and in the right hand side the results of a hull without the WDA effect are visible.

In Figure 9 the wave height at 48 m distance from CL is presented for the Baltic ferries of the 1980's, the Baltic ferries of the 1990's and the recently developed WDA. From this comparison it can be concluded that the difference is remarkable. In the shallow water even bigger differences can be obtained

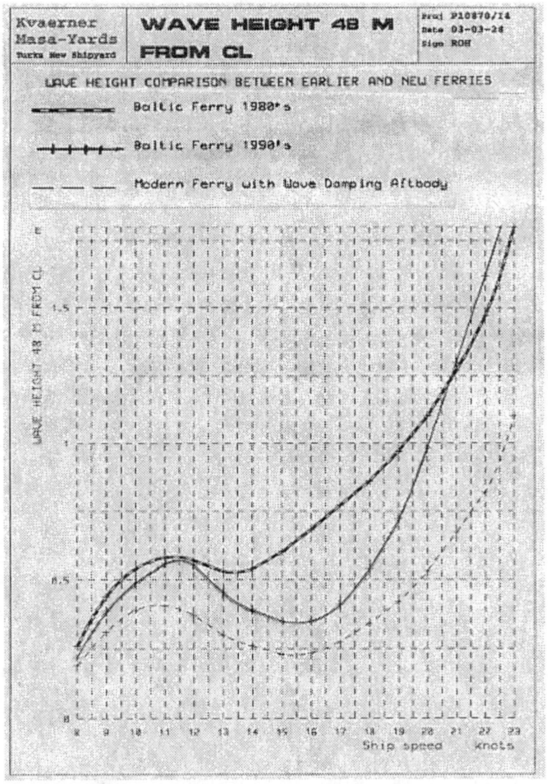

Figure 9: Wave height comparison at 48 m distance from CL

Propeller design

In addition to some general rules given for the development and design of the hull form, this section deals with some rules of thumb with regard to the design of the propeller(s). In order to avoid, as much as possible, cavitation erosion and the risk of vibrations caused by high propeller-induced pressure fluctuations, some simple rules are presented by which the so called "difficulty" of the design can be estimated at an early stage of the project.

Table 1: Difficulty Index

Difficulty Index	Hybrid Cruiser (CPP)	Voyager (side/CL	Hybrid Ropax (side/CL)	Maximum allowable
Tip speed πnD (m/s)	42.5	42.5 / 41.1	45.0 / 45.1	40.45
Propeller loading $(Ps\ N^2)^{0.4}$ (Ps in HP) (N in RPM)	3735	2760 / 2760	2342 / 3360 and 2342 / 3928	3500
Propeller loading $Ps / (1/4\pi D^2)$; (kW / m^2	1190	568 / 611	336 / 824 and 336 / 1218	590

The main source of cavitation and vibration problems is poor wake distribution, with steep gradients and deep peaks of low velocity.

Table 1 gives some so-called "difficulty indices" (DI), their maximum allowed values and the values calculated for CPP Hybrid Cruiser, Voyager and Hybrid Ropax. Note that the maximum allowed values given in Table 1 are only an indication and cannot be considered as fixed values, which may not be exceeded. For each vessel, the calculated "difficulty indices" have to be carefully considered in combination with, for example, propeller tip clearances and the quality of the wake field. Where the calculated "difficulty indices" seem to be critical, a lot of attention should be focussed on the design of the propellers and also on the quality of the construction of the ship, especially in the aftbody.

However, the limitation of power for the conventional twin-screw concept may be derived from Table 1. In case more power is needed to reach the requested speed, the use of non-conventional propulsion systems has to be considered. To reduce propeller loading, the adoption of three or more propellers is a good possibility.

A rather new development with regard to this aspect is the use of Pods in combination with the existing single screw shafting arrangement (Figure 10). This creates a contra-rotating propeller (CRP) concept. The positive effects of this concept are the gain in efficiency caused by the lighter propeller loading and the contra-rotating effect of the propellers. Furthermore, the dimensions of the conventional shafting arrangement can be reduced, and the rudders, headboxes and stern tunnel thruster can be removed. The latter will lead to reduced appendage resistance. The maneuvrability of the vessels using this concept will be improved remarkably.

Figure 10: CRP arrangement

Beam/ draught-ratio and flow direction

The beam/draught-ratio (B/T) plays a very important role in single screw vessels, when we are looking for flow directions, propulsion coefficients and aft body types: horizontal flow, mixed flow or buttock flow aft body.

Thick or thin boundary layer effects have to be taken into account in planning a normal propeller or specially a CRP propeller system. When B/T is low: by reducing the whole wake field, we reduce viscous resistance, by increasing wake we increase hull efficiency. Hull efficiency is much smaller in the buttock flow case because wake is smaller.

When a big B/T-ratio is used, transom width effects body wave making and frictional resistance quite a lot. At higher speed large transom width is recommended. When B/T>4 a buttock flow aftbody is a natural choice. In the big B/T case, we cannot solve the hull efficiency problem by increasing wake locally and reducing the outside propeller. Open water efficiency is somewhat better in a buttock flow afterbody. Rotation efficiency is smaller in buttock flow afterbodies.

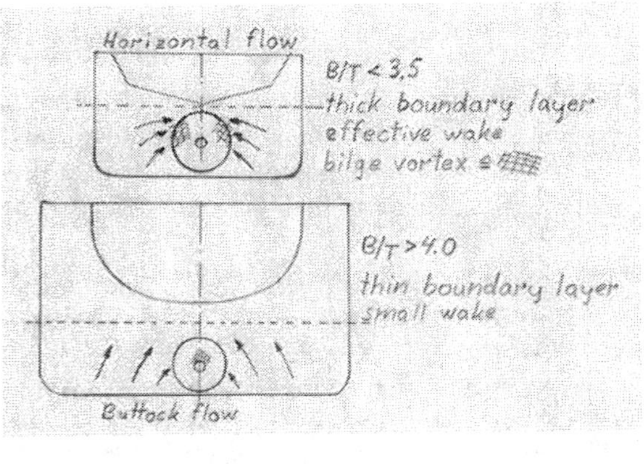

Figure 11: B/T-ratio effect on flow

Pram aft bodies are good at high speed, their resistance behavior is excellent (form factor is small, small boundary layer thickness), but care should be taken with course keeping ability.
Hull efficiency is much higher in a horizontal flow afterbody, because their wake is much higher. Artificial wake increasing is not a good idea in a pram afterbody, because at the same time resistance will increase.
It is most important to realize that rotational energy through a buttock flow aft body is much smaller than in a horizontal flow aft body.

RESISTANCE

Ship resistance can be divided in many ways. A quite common way is the following:

$$R_T = R_F + R_W + R_{VF} + R_{air} + R_{app} + R_{rough}$$

Where,

R_T	Total resistance of ship
R_F	Flat plate frictional resistance
R_W	Wave making resistance
R_{VF}	Hull form induced viscous resistance
R_{air}	Air resistance
R_{app}	Appendage resistance
R_{rough}	Resistance due to hull roughness

The induced resistance in waves and the wave breaking resistance could be increased, but they have not yet been examined in numerical hydrodynamics. In numerical hydrodynamic calculations (CFD) the following division is used in resistance coefficients:

$$C_T = C_F + C_{PV} + C_W \text{ (CFD)}$$

In 3-D model test extrapolation C_T is:

$$C_T = (1+k)C_F^{ITTC} + C_R$$

Where,

C_T	Total resistance coefficient
C_F	Frictional resistance coefficient
C_{PV}	Viscous pressure resistance coefficient
C_W	Wave resistance coefficient
$(1+k)$	Form factor
C_R	Residuary resistance coefficient

The resistance of the fore body is mainly wave resistance and the resistance of the aft body is mainly viscous resistance (friction and pressure resistance). Viscous resistance is made up of two components: frictional and pressure resistance. Frictional resistance is mainly generated due to the setting up of shear stresses in the fluid and the consequent formation of a velocity gradient. In order to reduce frictional resistance for a given speed the product of frictional coefficient (C_F) and wetted surface must be reduced. This is possible only by making the surface smoother or increasing Reynolds number (longer body or higher speed). In order to reduce the pressure resistance (form factor) it is necessary to reduce the loss of static head to a minimum. This in turn implies a reduction in the momentum loses in the boundary layer (=acceleration of flow and separation).

Wave resistance is important when Fn>0.15.

A typical wave resistance curve is shown in Figure 12 with the typical ranges for the various ship types indicated (Holtrop J., 1993). Both the resistance and the non-dimensional resistance curves are shown as a function of the Froude number. The magnitude of the non-dimensional wave resistance depends on the main dimensions, the form coefficients and the shape of the hull form. Most notable is the main hump at a Froude number of Fn=0.5.

At the speeds corresponding to a Froude number of 0.25 the wave resistance is quite important. Splashing, breaking and some spray at the bow are usually present.

Interference of the bow and stern waves becomes important at Froude numbers of about 0.3. At this speed a significant unwanted interference hump on the resistance curve is often found. The transverse waves carry a large part of the energy away.

At Froude numbers greater than 0.4 the wave length becomes in the order of the length of the ship. This results in a sharp increase of the wave resistance beyond this speed towards the main hump.

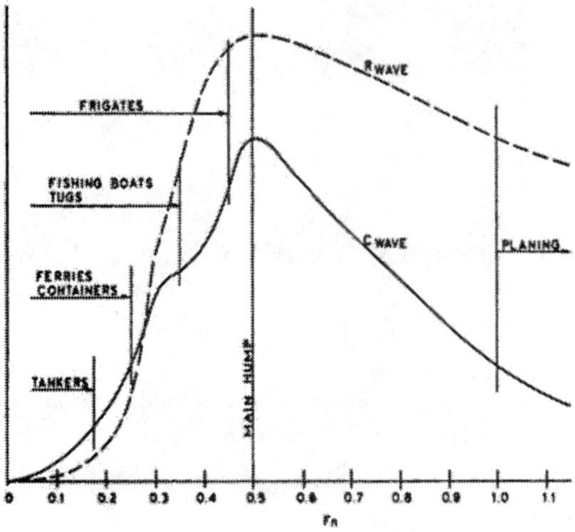

Figure 12: Wave resistance as a function of Fn

In the following Table 2 a typical amount of wave resistances for the different vessel types is shown.

Table 2: Percentage of wave resistance and appendage resistance from total resistance

Type of vessel	Tanker	Cruise Liner	Fast Ropax	Navy Frigate
Wave resistance	5- 15%	15-30%	30-40%	60%
Appendage resistance	1 -10%	7-17%	7-20%	7-15%
Speed	14-20 kn	21-26 kn	26-32 kn	25-45 kn
Fn	0.2	0.25	0.35	0.5

HULL COEFFICIENTS

It is important to carefully select the main dimensions and hull coefficient to reduce the resistance and power demand. The block, prismatic, midship and waterline coefficients used in ships built during the last 10 years is shown in Figure 13 (Levander K.). The selected values vary considerably, but a trend line based on the Froude number can be proposed.

For good hydrodynamic properties the prismatic coefficient *Cp* has to be taken as small as possible. But as a general rule the *Cp*-value should not come below about 0.58; otherwise the curve of sectional areas of the ship will tend to a triangular shape (so-called "diamond" effect). This indicates that the displacement of the ship is heavily concentrated around midships, creating very fine ends with quite strong fore and-aft shoulders, which will lead to a deep wave trough, resulting in an increase of the total wave making resistance. Potential-flow calculations will lead to a more detailed determination of the minimum *Cp* values.

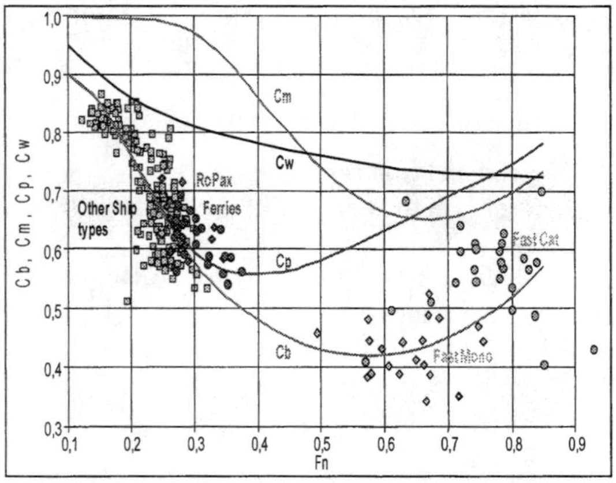

Figure 13: Hull coefficients

FAST HYBRID CRUISE LINER

This project was carried out as a normal commercial project for a client, but at some stage of this project more knowledge was needed to reach all hydrodynamic targets for good results.

The purpose of this R&D project was to confirm the good lines for a conventional propulsion system and a good POD propulsion system at high speed and to investigate CRPA (conventional CPP together with FPP Azipod podded propulsion) in a fast, big Cruise Liner.

In the CRPA case normal propellers are fitted at the forward end of pods placed just downstream of the propellers on the exposed shafts. The pods are azimuthing and the ship is steered by the pods. As a consequence, the pods center line will deviate from the front propeller center due to helm.

In the next table the main dimensions of a fast 3000 passenger cruise liner and the appendages are presented.

Table 3

Item	*Future 2*	
Vessel Type	Cruise Liner	
Lpp	250.00	m
B	32.20	m
T	7.85	m
L/B	7.8	
B/T	4.1	
Cb	0.60	
DWT	6000	ton
Gross ton.	76,000	ton
Passengers	3,000	
Cars/Trailers	-	
Service speed.	27	knots
Max power	64,000	kW

Three different concepts were studied:

- conventional twin screw CPP,
- conventional twin screw FPP and
- CRPA.

In the following pictures 14, 15 and 16 the different aft body appendages are all shown and in Figure 17 the hull lines of a cruise liner are shown.

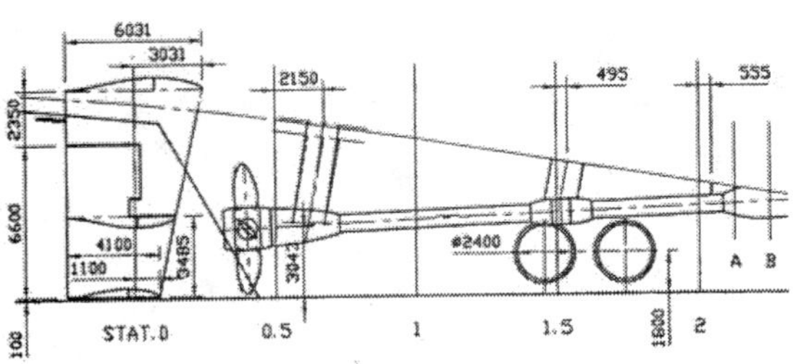

Figure 14: Conventional twin screw CPP aftbody

Figure 15: Conventional twin screw FPP aftbody

Figure 16: Hybrid propulsion: CRPA (=conventional CPP + turnable FPP POD propulsion)

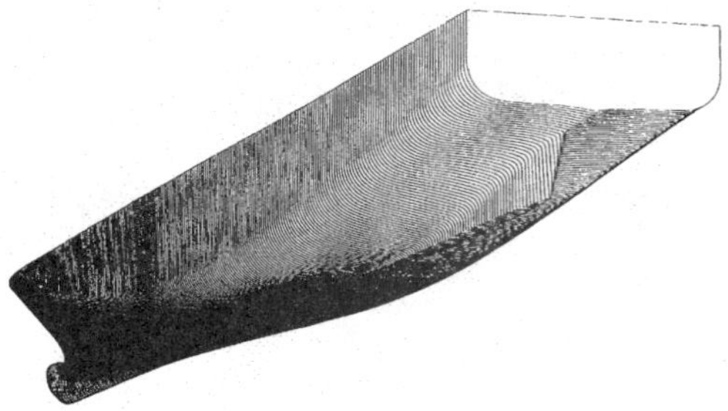

Figure 17: Aft body hull lines of fast big Cruise Liner

FPP and CPP propellers with the diesel electric machinery

From the propeller viewpoint the diesel electric machinery has the advantage of continuous variable speed over the whole rotation speed range from 0 to 100% rpm. Another advantage is that 100% torque can often be utilized within the whole operating range.

An FPP propeller must be designed to reach the MCR point at trial condition. In order to operate at full power in various sea margin conditions, the propulsion system is normally calculated for 10 - 20% over-torque. This mean 10 - 20% over-sizing of the propeller / shaft line / gear / motors / converters etc.

This of course effects the appendage resistance in twin screw ships.

Diesel electric machinery with CPP propellers should ideally have a pitch / rpm combinator between 65 and 100% rpm. CPP arrangement does not need any over-torque capacity, since the MCR point can be reached by pitch adjustment.

Both propeller types are good in diesel electric machinery and their selection depends on the experience of the client.

Propeller hub diameter is bigger in a CPP propeller. This has an effect of about 1.5% for efficiency. Typical values for hub ratio are:

CPP: d/D = 0.30..........0.35

FPP: d/D = 0.17..........0.20

Expanded blade area ratio (A_E/A_O) is limited to about 0.80 for CPP.

CPP pitch angle regulation has effects on pitch distribution and profile configuration. A CPP propeller is more likely to cavitate than an FPP propeller.

The advantages of CPP:

- Good stopping, accelerating and steering properties.
- Possibility for constant thrust at different loading conditions.
- Low fuel consumption with vessels, which operate at different speeds and loadings.
- No need to change rotation direction.
- No need to change rotation speed to change the vessel speed.
- Electric generator can be installed directly to the main engine, because you can keep the rotation speed of the main engine constant.
- Flexibility in different loading conditions.
- No engine overloading or under loading.
- Possibility of bigger propeller diameter with high skew propeller than with FPP.
- Possibility of much smaller propeller hull excitation by high skew propeller use (about 30 - 40% lower) than in FPP.

The disadvantages of CPP:

- Complicated construction.
- High price.
- Need for maintenance.
- Lower efficiency than in FPP (about 2 - 6%).

Model test results:

In Table 4 the following differences between conventional CPP appendages and CRPA appendages are presented:

Table 4

Speed in knots	Resistance PE in %	
	Conventional	CRPA
22	100	101.4
24	100	99.7
26	100	100.9
28	100	101.2
30	100	100.5
32	100	100.5

CRPA seems to be approximately 1% worse in resistance. CRPA case includes two stern thrusters, three bow thrusters, conventional shaft arrangement without rudders, two headboxes and two POD-units.

Appendage resistance comparison between CPP and FPP arrangement showed the following differences in Table 5.

Table 5

Speed in knots	Resistance PE in %		
	Bare hull	CPP	FPP
24	100	116.0	112.0
25	100	115.0	110.7
26	100	114.1	110.6
27	100	113.8	110.0
28	100	112.9	108.0

Appendage resistance of CPP configuration is about 3.5% higher than the resistance of the FPP.

The shaft power comparison between CPP and FPP arrangement gives following differences in Table 6.

Table 6

Speed in knots	Shaft power PS in %	
	CPP	FPP
24	100	94.4
25	100	93.5
26	100	93.7
27	100	94.1
28	100	93.4

FPP arrangement gives about 6% power saving, which is partly caused by a reduction of the resistance of the smaller FPP shaft arrangement (see Table 5) and partly by the higher efficiency of the FPP.

Computer program DESCRP (DESign of Contra-Rotating Propellers), which uses a simple flow induction model and the B- and Gawn series propeller polynomials for the thrust and torque coefficients, was used in preliminary CRP concept planning. In the induction model the actuator disc theory is used in combination with the assumption of a steady rotation of the flow in the slipstream imposed by the front propeller.

The volume flow through the front propeller disc is 10 to 20% greater than through the rear propeller disc. This margin was imposed in the calculations to prevent the rear propeller on the pod being struck by the tip vortices shed from the front propeller.

Two POD sizes were studied:

Type 21 supposes a maximum design power to be delivered of 13 MW at the rotation rate of 116 rpm. Type 23 delivers the same power but at 105 rpm at the preliminary design point of 28 knots. Calculations showed 4.1% for the first and 5.3% gains for the second type compared to the conventional case.

The calculations showed that the optimum diameters are 5.8 and 5.11 m for front and rear propeller. The optimum rotation rates are 72 and 105 rpm. In this kind of concept the pitch for the front propeller should be extremely high (P/D = 2.0). High pitch ratios are good for an optimum CRP when we think of windmilling operation. Feathering the front propeller is even better because it results in an even lower resistance of the in-active propeller.

A more practical solution has to be selected as presented in Table 7.

Table 7

Item	Rear FPP	Fore CPP
Prop diam.	5.09 m	5.76 m
Pitch ratio	1.50	1.82
Volume flow through disc	100%	120%
Blade area ratio	0.63	0.55
Number of blades	4	5
Percentage power	54.3%	45.7%
Prop rotation rate	107 RPM	80 RPM
n-ration	100%	76%

In this case the calculated power gain was 6.3% at a speed of 24 knots and by using POD type 23.

The ratio of the volume flows through the front and the rear propeller discs was at least 1.20. It was decided to conduct the tests with stock propellers. The propellers for which main particulars are presented in Table 8, were used.

Table 8

Item	Rear FPP	Fore CPP
Prop diam.	4.917 m	5.704 m
Pitch ratio	1.52	1.84
Hub ratio	0.335	.335
Blade area ratio	0.616	0.685
Number of blades	4	5
Percentage power	45%	55%
Prop rotation rate	118.1 RPM	90.0 RPM
n-ration	100%	76%

A power comparison between conventional and CRPA is showed in Table 9 and absolute shaft power values are shown in Figure 18.

Table 9

Speed in knots	Shaft power PS in %	
	Conventional	CRPA
24	100	94.6
26	100	94.2
28	100	94.9
30	100	95.4

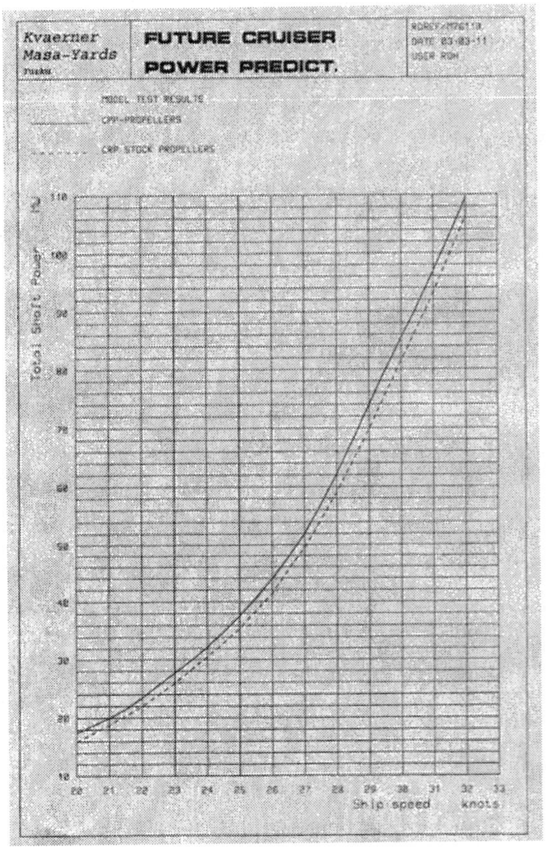

Figure 18: Shaft power comparison between conventional and CRPA

The CRPA arrangement gives about 5 - 6% power saving with stock propellers. New propellers specially designed for CRPA could save even more.

At 27 knots power density for the conventional concept is 1016 kW/m2. At the same speed for CRPA the power density is 532 kW/m2 for the front and 586 kW/m2 for the rear propeller.

Optimum loading was studied for CRPA. See Figure 19

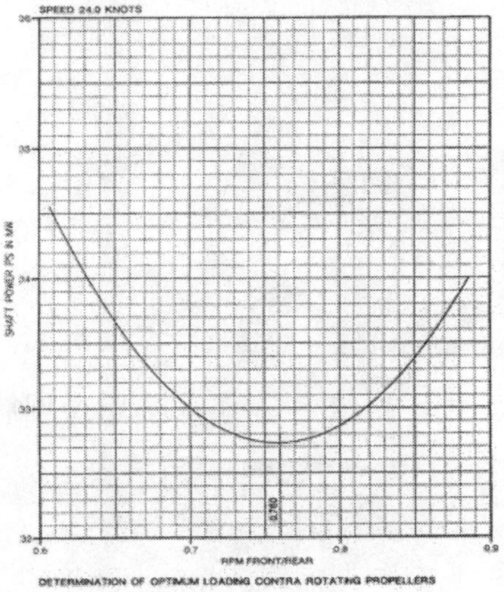

Figure 19: Optimum loading in CRP-case

HYBRID CRUISE LINER

Passenger comfort, operational safety and efficiency were the main targets in the hydrodynamic design of the world's largest cruise liner for Royal Caribbean International. The Shipyard made extensive design studies for several pod alternatives from twin pod to triple pod arrangements before contract. This chapter describes the hydrodynamic advantages and disadvantages among several propulsion alternatives.

Several CL skeg alternatives were studied in order to reach an excellent wake field for the pusher type Fixed Pod configuration.

The revolutionary Wave Damping Afterbody was designed the first time on a large cruise liner by using the Shipyard's Shipflow and the Model Basin's Rapid CFD-programs. The earlier experiences from fast ferries have shown big power level improvements for this new afterbody form.

Figure 20: Hybrid Cruise Liner: pushing type Fixipod and two pulling type Azipods

An extensive design study for a cruise liner was carried out and some of the results are presented below. In the propulsion system comparison several pod alternatives are described, varying from twin pod to triple pod arrangements. In the following example the document describes hydrodynamic advantages and disadvantages between these propulsion arrangements.

Alternative 1:

Two turnable pulling Pods (Ps=14-14.5 MW, Dp=5.6 m), one fixed pushing Pod in gondola (Ps=14-14.5 MW, Dp=5.4 m) and one small course-keeping rudder in CL.

Fixed pod is integrated to gondola. The gondola is close to a "Langenberg-form". See Figure 21.

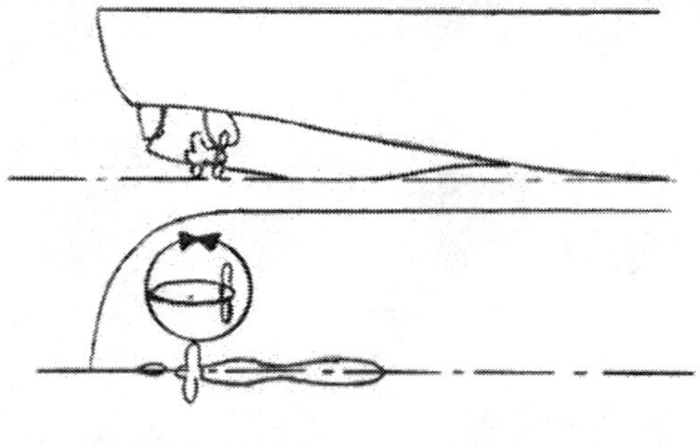

Figure 21: Alternative aftbody 1

Alternative 2:

Two turnable pulling Pods (Ps=14-14.5 MW, Dp=5.6 m), one fixed pushing Pod (Ps=14-14.5 MW, Dp=5.4 m) and one small course-keeping rudder on CL.

Fixed Pod is on CL and a simple CL skeg is ahead of it. See Figure 22.

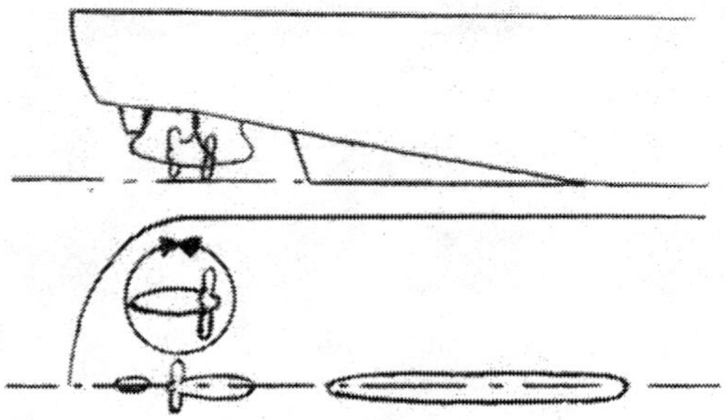

Figure 22: Alternative aftbody 2.

Alternative 3:

Two turnable pulling Pods (Ps=14-14.5 MW, Dp=5.6 m), one fixed pulling Pod (Ps=14-14.5 MW, Dp=5.4 m) and one small course-keeping rudder on CL. Fixed Pod is separately on CL and a simple CL skeg is ahead of it. See Figure 23.

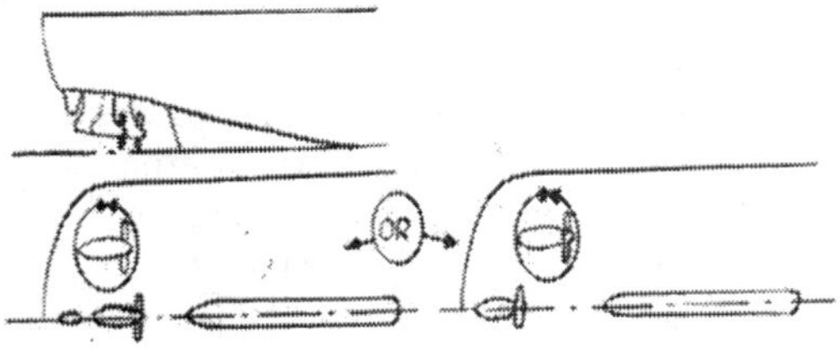

Figure 23: Alternative aftbody 3

Alternative 4:

Two turnable pulling Pods (Ps=21-21.5 MW, Dp=6.1 m) and a normal simple CL skeg. See Figure 24.

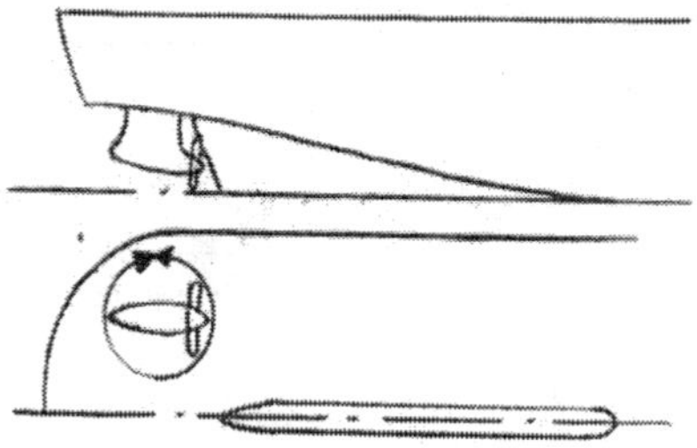

Figure 24: Alternative aftbody 4

Alternative 5:

Two FPP in open shaft arrangement (Ps=21.5 - 22 MW, Dp=6.1 m), two flap type rudders behind propellers and a normal simple CL skeg. See Figure 25.

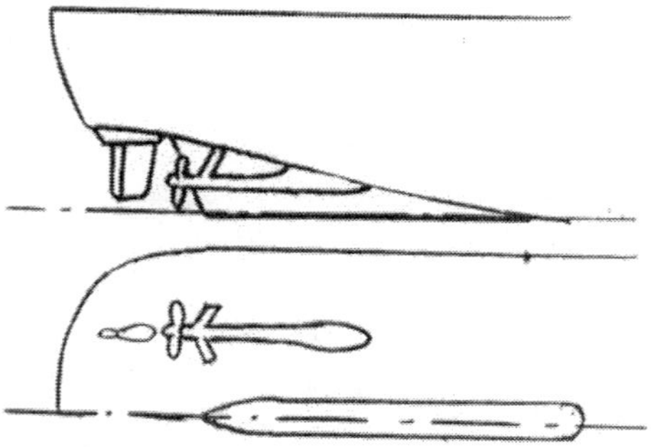

Figure 25: Alternative aftbody 5

Alternatives 1, 2 and 5 were selected for further examinations. See Figure 26, Figure 27 and Figure 28.

Figure 26: Alternative 1 with conventional gondola

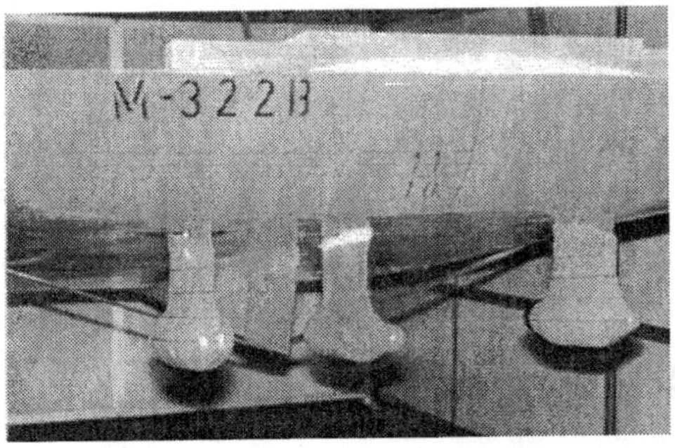

Figure 27: Alternative aftbody 2

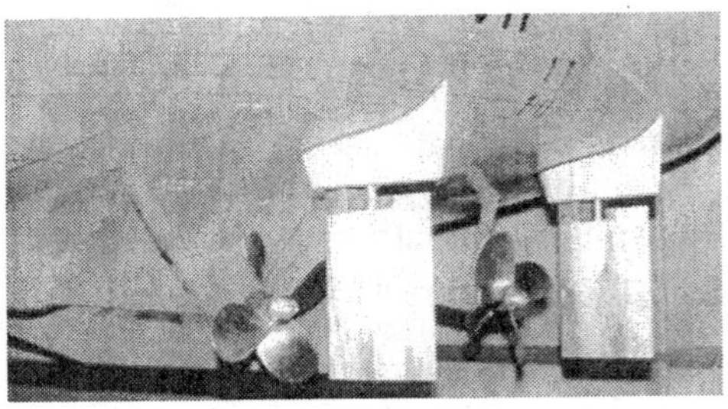

Figure 28: Alternative aftbody 5

Figure 29 shows axial wake fields from alternative 1 (thick gondola M-322 and slim gondola M-322C) and alternative 2 (=M322B).

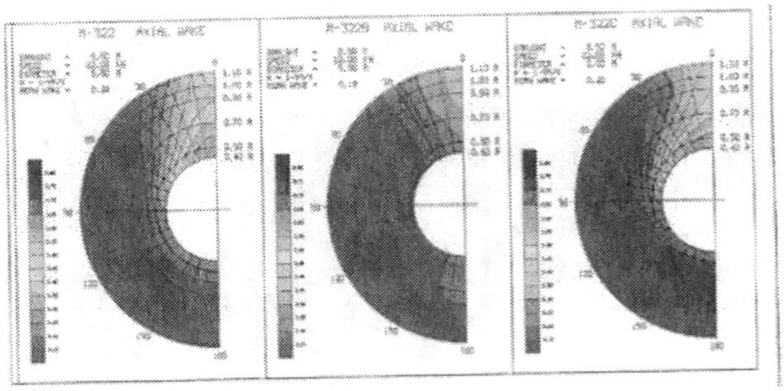

Figure 29: Axial wake fields for thick gondola in alternative 1, alternative 2 and slim gondola in alternative 1

In Table 10 a resistance comparison between the following alternatives is presented: CL-skeg / Thick gondola, Slim / Thick gondola and CL-skeg / Slim gondola.

Table 10

Vs (kn)	CL-skeg / Thick g.	Slim g. / Thick g.	CL-skeg / Slim g.
17	0.96	0.98	0.98
19	0.97	0.98	0.99
21	0.98	0.98	1.00
23	0.99	0.98	1.01
25	0.96	0.97	0.98
27	0.95	0.97	0.98

Alternative 5 shows slightly better results compared with alternative 1. The main reason is in the better aftbody hull form design for wave resistance.

The model tests were carried out at the MARIN model basin in the Netherlands. The excellent wave behavior was observed in the bare hull resistance tests. The WDA idea worked well for this large cruise ship. In Figure 14 the tested aftbody forms are shown. A ducktail was added to the typical cruiser stern to achieve a good WDA effect in the aftbody. A ducktail with a big rounding in the corner significantly improved the aft body wave system by decreasing the wave and pressure resistance at the ship ending.

Cavitation observations and hull pressure measurements were performed on the side propellers (Azipod) and the center propeller (Fixipod) for two loading conditions.

Also the Azipods in turned condition were tested. It can be said that for the cruise ship equipped with three Azipod units, the cavitation behavior is excellent. This means low-pressure pulses and low excitation forces on the hull. Azipod propulsion units are a good choice for vessels with very high demands regarding low noise and vibration levels.

The Pod propulsion concept gives many additional benefits to a classical diesel-electric system. Podded propulsion makes new comfort classes of classification societies easier to reach.

A combination of experience, numerical tool data and good co-operation give good possibilities for success, as was the case in the Voyager-class vessels.

HYBRID ROPAX

The Shipyard started intensive R&D work for the medium speed Ropax in 2001, because there were a lot of unknown subjects and a lot of potential improvements. The earlier Ropax developments were completed 5 years ago (Superfast III and IV, see SNAME 1998) and they were concentrated more on the faster and more slender Ropax vessels.

This R&D work concentrated on the medium speed Ropax (service speed 26 kn) for three different concepts: conventional twin CP propeller, twin FP POD propulsion and hybrid. Figures 30, 31 and 32 show the aftbody of the models.

Figure 30: Conventional twin CP propeller case

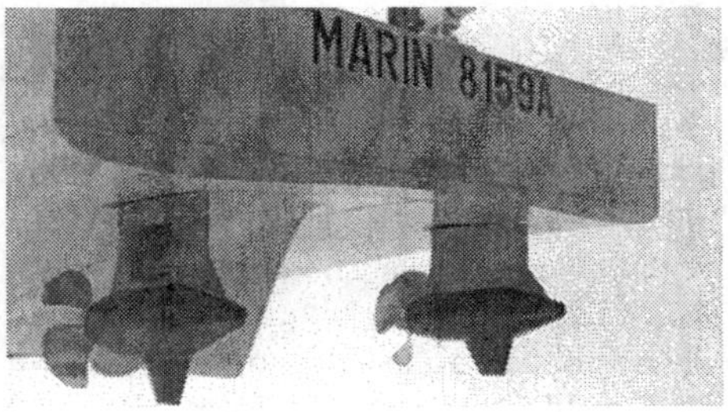

Figure 31: Twin FP POD propulsion

Figure 32: Hybrid case. Two FPP PODs and CP propeller in slender gondola in CL

Main dimensions of the project concept are shown in Table 11.

Table 11

Type of vessel	Convent.	Twin Pod	Hybrid	
Lpp	170.0	170.0	170.0	m
B	28.5	29.0	29.35	m
T	6.6	6.6	6.6	m
L/B	6.0	5.9	5.86.0	
B/T	4.3	4.4	4.4	
Cb	0.62	0.62	0.62	
DWT	4,700	4,700	4,700	
Gross ton	40,500	39,150	40,100	
Passengers	2,250	2,056	1,430	
Trailer lanes	1,585	1,520	1,430	m
Service speed	26	26	26	kn
Prop power	2*20.3	2*17.1	20.3+2*6.6 kW	

The starting power level was obtained from a 1995 tested Ropax, where Lpp was 162.4, B 28.5, T 6.6 m and displacement 19056 m^3.

The WDA-effect was quite well reached in this reference vessel, though the aft part (distance from AP to the transom/ducktail was only 3.8% from Lpp) and was not long enough. This distance should be looked at together with LCB-location: distance can be shorter if LCB is located more forward.

For model test purposes main dimensions were selected according to the conventional concept because there was a similar project going, at the same time where the Owner and the Consulting Office made the model tests for a twin Pods concept. In Table 12 there is a power comparison between the conventional starting level (only the calculated one, selected for 100%), the Owner/Consulting Office twin POD concept, the Shipyard's twin POD concept and the Shipyard's Hybrid concept.

Table 12: Brake power comparison for several concepts

Speed (kn)	Starting Level Conventional	Owner/ Consulting Office Twin POD	Shipyard Twin POD	Shipyard Hybrid
22	100%	93%	75%	71%
24	100%	90%	71%	66%
26	100%	90%	70%	65%
28	100%	88%	73%	67%

The comparison shows large improvements from starting level, at the service speed 35% improvement. The Owner/Consulting Office's results were about 10% better than the Shipyard's calculated starting level.

The Shipyard improved the Consulting Office's power level by 20% in the twin POD concept, where the same propellers and POD-units were used. Resistance difference was about 10 per cent. Clear differences were possible to see in the bulbous bow and in the transom waves. The hull-POD interactions were much better in the Shipyard's design. The Shipyard improved the twin POD case by a further 5% in the Hybrid case. At higher speed it is even more when the right Pod unit size is taken into account.

In Figure 33 a brake power comparison between the four cases is shown.

It is possible to see from these and earlier results that if propeller loading becomes bigger than 700 kW/m^2, then three or more propellers become an attractive alternative.

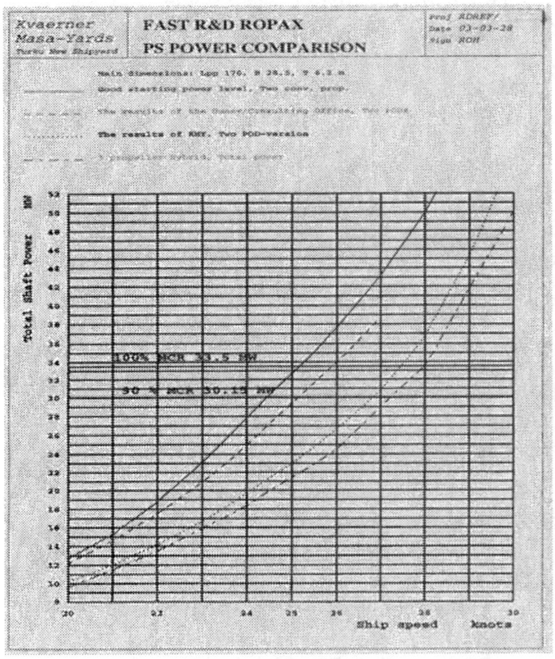

Figure 33: Power comparison between four different alternatives

In the following Figure 34 there are wave profile comparisons between the Consulting Office and the Shipyard.

The bare hull resistance of the Shipyard's design was 10 per cent lower than in the Consulting Office case. This is mainly caused by the improved wave pattern behind the transom.

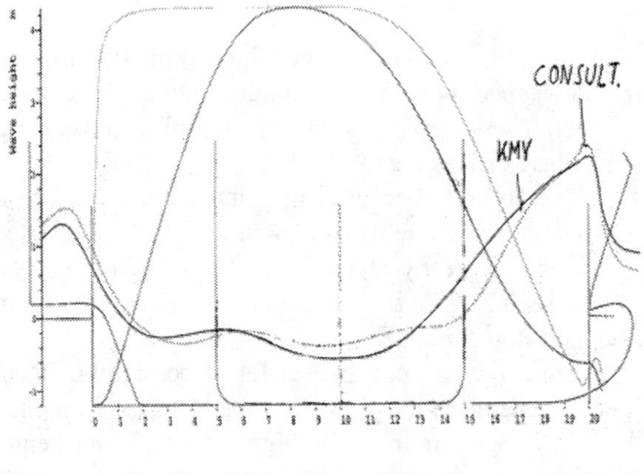

Figure 34: Wave pattern comparison between the Shipyard's and Consulting Office's versions. See specially differences in transom waves

The Hybrid Ropax was better than the twin POD case and it becomes even better when the propeller loading is increased (not possible to see in figure, stock propellers, one pod size). In the model tests the main question of the Hybrid case was: how much power it is possible to put to the CL CP-propeller while keeping excellent vibration and noise levels (Comfort Class) in the aftbody.

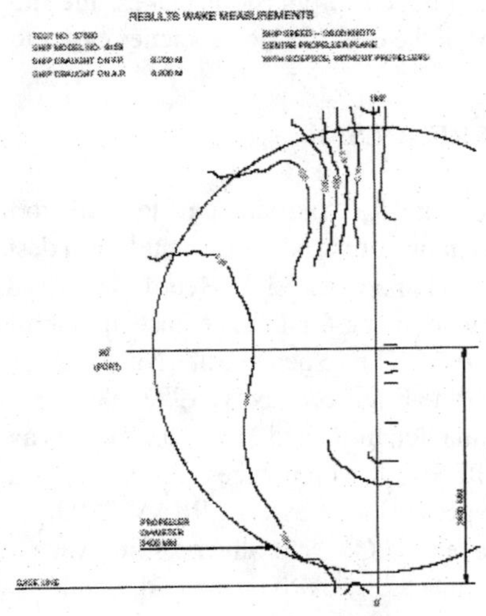

Figure 35: Axial wake of CL propeller

The results of the wake survey on center line propeller location of the hybrid concept at the design draught are presented in Figure 35. The circumferential distribution of the axial wake component is good: the slope of the curves in the upper region is moderate in combination with very small wake peak, having a variation of about 30% of the speed (Figure 35). The tangential and radial wake components observed during tests are only marginally improved.

Compared to wake distributions of comparable ships, the wake distribution is qualified as very good. This wake field is considered an excellent starting point for the design of the propeller.

The cavitation observations were conducted on the 5-bladed propeller model behind the ship model in the Depressurized Towing Tank. of MARIN. The propeller was viewed from both sides and recordings were made of the suction side for 100 per cent MCR. Additionally, observations were done for a higher loaded condition.

The above loading conditions were determined from the open water test results combined with the propulsion coefficient derived from tests with the stock propeller.

The test conditions for the suction side cavitation observations and the pressure side cavitation inception test were chosen such that the propeller loading and the cavitation number were equal to the predicted full scale values. The results show a remarkably good cavitation pattern. Only marginal sheet cavitation at the leading edge of the blade sections is observed at the high load condition.

The pressure side cavitation was observed by means of a video camera, mounted in a submerged streamlined housing behind the ship model. For the same test condition as for the suction side cavitation observations, no pressure side cavitation was found.

By varying the propeller load the margin against pressure side cavitation was established to be 74% of KT for the above trial test condition, which is ample to prevent the occurrence of this type of cavitation.

Summarizing, the cavitation phenomena as observed on the 5-bladed propeller model are very good from an erosion point of view.

The propellers with their dynamic behavior generate a fluctuating pressure distribution over the hull. Although often the pressure fluctuations at a single point on the hull are defined as a criterion for excitation, it is more realistic to determine the total excitation forces as a criterion. However, the level of vibrations also depends upon the structural response.

The propeller induced excitation forces are very low for the 5 bladed center line propeller.

In Figures 36 and 37 different views from the aftbody general arrangement are shown.

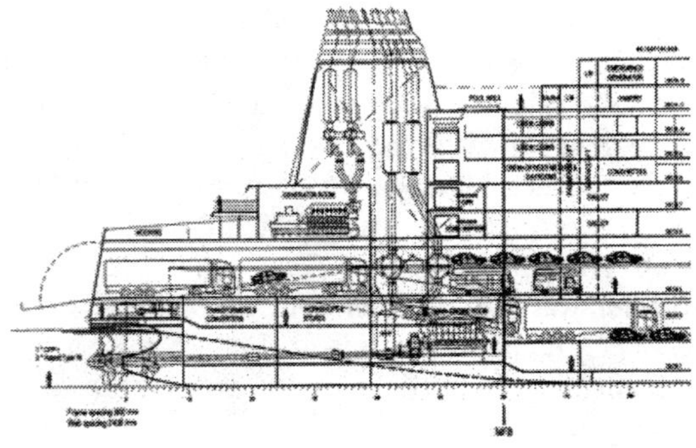

Figure 36: Side view for Hybrid concept

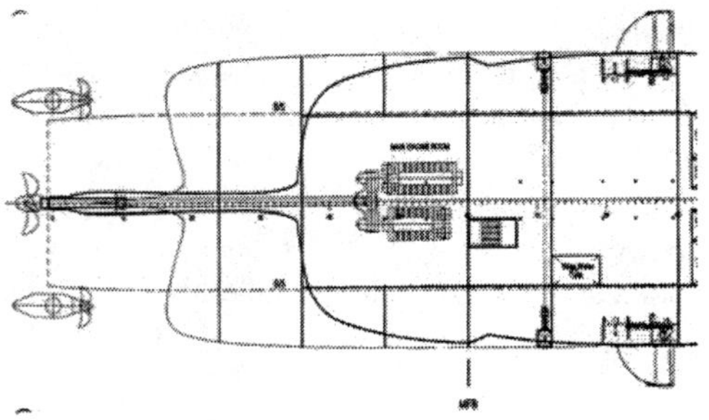

Figure 37: WL view for Hybrid case

The Hybrid concept is very interesting, because it possess a lot of benefits compared to the normally used cases:

- High redundancy.
- Excellent maneuvrability, specially in the harbor at the low speed and in the stopping cases.
- Less propulsion losses than in the twin POD case.
- Lower price than in the twin POD case.
- Good propeller loadings for the medium and high speed Ropax vessels.
- Very flexible propulsion system, which gives many operating possibilities.
- Good possibilities for effective WDA effects with the cheeks and the big tunnel shape.
- Excellent wake field is possible for CL propeller.
- Shorten harbor maneuvering time by 20% compared with the conventional case.
- Increase safety in the harbor steering cases.
- Fuel saving about 1 million USD a year in the typical Ropax routes compared to the conventional mechanical case.
- Gives a lot of possibilities for a good general arrangement: shorter engine rooms, auxiliary engine room in the higher decks, lower height requirements between hull and car deck, many possibilities with the aft ramps, etc.
- Kvaerner Masa-Yards Inc. has patented this propulsion concept.

CONCLUSION

With the aid of CFD tools, new hull form concepts can be developed prior to the model-testing phase of a project. These new tools can be helpful in developing new ideas such as the "Wave Damping Aftbody."

The Hybrid concept is very interesting, because it possess a lot of benefits compared to normal often used cases. It will greatly revise Ropax vessel concepts. The power savings give a lot of abilities to compete better in the hard Ropax market.

This new concept gives a lot of possibilities for a good general arrangement: shorter engine rooms, auxiliary engine room in the higher decks, lower height requirements between hull and car deck, many possibilities with the aft ramps, etc.

In CRPA case normal propellers are fitted at the forward end of pods placed just downstream of the propellers on the exposed shafts. The pods are azimuthing and the ship is steered by the pods. This type of propulsion systems will come more and more to the markets when speeds are increasing.

Passenger comfort, operational safety and efficiency were the main targets in the hydrodynamic design of the world's largest cruise liner. The multi Pod propulsion concept gives many additional benefits over a classical twin propeller system.

Triple Podded propulsion makes the new comfort classes of the classification societies easier to reach.

REFERENCES

- Holtrop, J., "Introduction to hull form design". Training course on CFD in hull form design, 1993.
- Hämäläinen R. and Heerd J., "Hydrodynamic development for a large fast monohull passenger ferry", The Society of Naval Architects and Marine Engineers, Nov. 1998.
- Hämäläinen R. and Heerd J., "The Wave Damping Aftbody Revolutionizes Aftbody Design In All Types Of Fast Vessels", IMAM 2000.
- Raven, H.C., "Non-linear ship wave calculations using the RAPID method," 6th International Conference on Numerical Ship Hydrodynamics, Iowa City, 1993.
- Raven, H.C. and Valkhof, H.H., "Application of non-linear ship wave calculations in design," 6th PRADS Symposium, Seoul, Korea, 1995.
- Levander K. Improving the Ropax concept. Euro conference 2001.

The Nuclear Hazards of the Recovery of the Nuclear Powered Submarine *Kursk*

Peter Davidson *Principal Safety Consultant, NNC Ltd*
Huw Jones *Commander, Royal Navy*
John H Large *Large & Associates, Consulting Engineers*

ABSTRACT

The Russian Federation nuclear powered submarine Kursk sank in August 2000 with the loss of all 118 lives on board. In May 2001 the Russian Federation entered into a contract with the Dutch consortium Mammoet-Smit for the recovery of the Kursk on the condition that it had to be completed within that year. The consortium prepared for this World-first salvage of a nuclear powered and conventionally armed submarine that was very substantially damaged lying at 110m in the icy waters of the Barents Sea. Working at sometimes breathtaking pace, Mammoet-Smit prepared, lifted and transported the wreckage of the Kursk delivering her to a floating dock at Rosljakovo, about 200km south of the foundering site, in just over six months from the contract date. This paper tracks how the nuclear and other hazards of the Kursk, its nuclear reactors and weaponry were assessed and monitored throughout the recovery and salvage program, and it provides an insight into the reasons why the Kursk sank.

THE FOUNDERING OF THE *KURSK*

On Saturday, 12 August 2000 and exactly at 7.29.50 GMT a small and relatively insignificant seismic disturbance was recorded by a Norwegian seismological station. It was followed one hundred and thirty five seconds later with a much more significant event, equivalent to about 3 to 3.5 Richter scale. None of those at the recording stations in Norway, Finland, Scotland, Canada, Alaska and elsewhere realized that this second explosion marked the death knell of an advanced nuclear powered submarine in the Barents Sea.

During the morning of 14 August the rescue centre at Bodø in northern Norway received rumor of an accident on board a then unknown Russian submarine somewhere north of Murmansk. This was the first inkling in the West of a very serious situation, the details of which were to unfurl over the following hours when it became apparent, and was subsequently confirmed by the Russian Federation (RF) Northern Fleet headquarters in Severomorsk, that a submarine had foundered. At about 16.30 that day the Norwegian Moscow embassy was notified by the Russian Federation authorities that there had been an accident to a submarine - that boat was the OSCAR II (RF PLARK class), cruise missile armed and nuclear powered submarine *Kursk*.

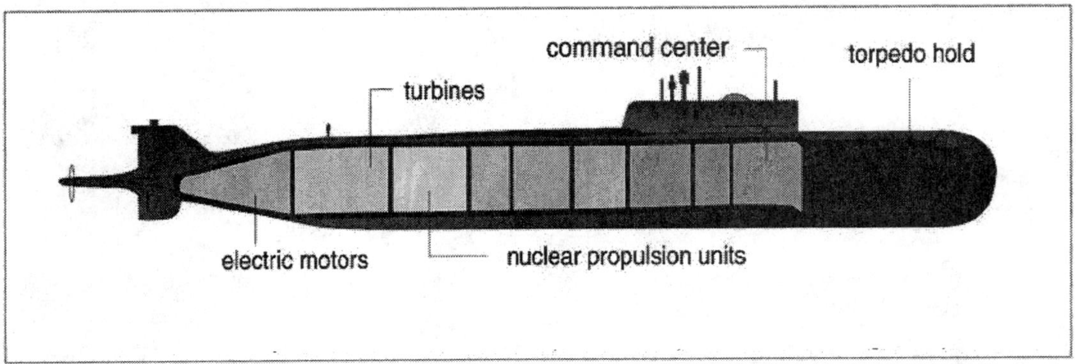

It is now known that *Kursk* left her home base in the Uraguba bay on 10 August, with a total complement of 118 men aboard, two of whom were torpedo designers. She was heading out to participate in sea exercises east of the Rybatschi Peninsula about 200km north of Murmansk. *Kursk* was assigned to an area of the sea cleared as a torpedo range under the supervision of the battle group command cruiser *Pyotr Veliky*, where she was to test fire two unarmed, prototype torpedoes from the forward port 650mm diameter tube that had been specially adapted for the trials.

It was the second firing that went so wrong. Speculation is, and it can only be speculation because the damage to the forward compartments was so great, that the gas generating system of the second prototype torpedo reacted with its main propellant, burning and exploding with an equivalent power to that of 100 to 200kg of TNT.

The prototype torpedoes were of the super cavitating type. This type of deep diving, high-speed torpedo envelops itself in a gas envelope generated at its bow with, essentially, the gas being replenished at the same rate as its progress through the water. The gas generating agent was hydrogen peroxide and, probably, the second prototype torpedo that initiated the sinking was an anti-submarine weapon (ASW) being deep diving and powered by a lithium-fluoride internal propulsion system.

The damage sustained from this first explosion, which alone the *Kursk* could have withstood, can be pieced together from the first of the two seabed debris fields. The debris included plates from the outer (flood) casing and, significantly, components of the port hydroplane hydraulic mechanism and the forward ballast and trim tanks. At the time of the first explosion, *Kursk* was positioned for torpedo firing, at periscope depth in the sub surface layer where to assist with depth control and, to avoid porpoising, she would have been trimmed to negative buoyancy, maintaining her sea depth by driving hydrodynamically against her forward planes at up to about 6 knots.

The centre of the first explosion seems to have been ahead of the foremost section of the pressure hull suggesting that the torpedo was loaded into the firing tube so, if the inner torpedo hatch was and remained closed, the damage to the bow compartment would have been minimal. However, the sonar trace taken by the nearby RF cruiser *Pyotr Veliky* shows continuing activity following the initial explosion spike, which could be interpreted as severe burning and jetting of the torpedo propellant system into the weapon stowage compartment (Compartment N° 1).

It is clear from the sonar records of the very much larger second explosion that this was from five to seven individual events occupying, in all, just over one-fifth of a second. This multi-explosion, equivalent to 2 to 3 tonnes of TNT, is believed to have derived from the detonation of up to 7 fully armed torpedo rounds in the forward port magazine rack. This massive explosion, inside the pressure hull, dealt a catastrophic blow to the *Kursk*, ripping out a very large section of the forward pressure hull (10 x 8m area) and outer casing and, at the same time, sending a reverberating hammer blow through the compartments towards the stern. Structural and flood bulkheads N° 2 and 3 were ripped through, with N° 4 buckling and subsequently collapsing under the hydrostatic flood loading. N° 5, the forward reactor compartment bulkhead, and the remaining bulkheads through to the ninth compartment remained intact.

individuals survived the two explosions and sought refuge in the stern most N° 9 compartment surviving for, it is believed, two to three hours in very cramped conditions on existing oxygen supplies and oxygen

The bow damage showing failure of both flood and pressure hulls- internally the damage extended into the boat, collapsing lateral bulkheads into through to N° 5 Compartment- the red cylinders are gas bottles

breathing apparatus canisters. Whether they perished by hypothermia, nitrogen narcosis or simply lack of oxygen is not known.

What is known is that a number of the crew members subsequently recovered from the N° 9 compartment had sustained quite severe body burns and the water-filled compartment was strewn with dust and ash - the surviving crew had closed the compartment hatch thereby isolating themselves in this final refuge. The source of the fire has not been established, although a survivor trying to recharge an oxygen regenerator plate in the compartment could have sparked it.

The second seabed debris field (at 69°36,99N, 37°34,50E) provides clues to the remaining split seconds of the *Kursk* and for all those crew present in the forward five compartments. The *Kursk* came to rest relatively upright lying on the seabed, with the stem buffered against a sediment bank at an angle of 2° bow down and with the hull pitched to the port side by 1.5°. The major part of the second debris field lay 20 to 30m starboard of the wreck, whereas the pressure hull damage indicates that the major blast direction was upwards and to the port side.

A most telling clue to the dying moments of the *Kursk* was the final position of a 4 by 2m section of forward section casing (the outer flood hull) on the seabed to starboard of the stern, having traveled the 154m length of the hull to its final resting place. This casing plate must have 'swum' from the point of the second explosion through the water down to the seabed; thereafter she drifted down and settled on the seabed at a depth of 110m. Analysis of this gives the *Kursk* at 30-35m above the seabed at the instance of the plate detachment.

When operating submerged, twenty-three crewmembers of the *Kursk* would be positioned aft of the reactor compartment. These crewmembers attended to the steam raising and electricity generating plant generally dispersed about compartments N° 7, 8 and 9. At all times whilst the reactors are operational there are two crew members present in the reactor control room which is located at the higher deck level immediately aft of the reactor compartment. All of these

CONDITION OF THE *KURSK*

Two expeditions to the *Kursk* site were undertaken jointly between the Russian Federation Navy and the Norwegian Radiation Protection Authority. The first of these expeditions was in August, immediately following the sinking, and the second in October 2000 during which twelve of the twenty-three casualties in the stern compartment were recovered.

These two expeditions, particularly the latter, established the radiological regime in and around the *Kursk*. Air, sediment and seawater samples were taken and analyzed, and water samples within the submarine, from compartments N° 3, 4 and 7, were collected and sealed in cans for subsequent gamma spectrometry. Similarly, the remote operated vehicles (ROV) and the diving personnel were rigged to monitor dose rates at various locations about the casing of the submarine

(Amundsen Ingar, 2001).

The preliminary results from these two expeditions did not indicate the presence of radionuclides that may have been released from the submarine reactors or, potentially, from any nuclear weapons carried on board.

The presence of nuclear weapons on board the Kursk at time of the sinking was of particular concern. In 1989, another Russian Northern Fleet submarine, *Komsomolets*, which was lost in the Barents Sea at about 1,700m was leaking from both its single reactor and from two nuclear tipped torpedoes loaded in the bow tubes at the time of the foundering. For the *Kursk,* the Russian Federation Northern Fleet confirmed that at the time of the foundering no nuclear weapons were on board.

At this stage, no attempt was made to sample within the sealed reactor compartment, nor was any significant monitoring undertaken of any thermal gradients in the flood hull in the vicinity of the reactor compartment.

RF NORTHERN FLEET SUBMARINE *KURSK* K141 - TYPE, CONSTRUCTION & WEAPONRY

The *Kursk* is a SSGN (cruise missile armed, nuclear powered) submarine, designated by NATO as an OSCAR II class, commissioned from Sevmash shipyard, Severodvinsk in 1995.

Designed by RUBIN, The Russian State Marine Engineering Design Bureau in St Petersburg, the *Kursk* was 154 m long and 18m beam over the casing or flood hull, with a 11m diameter internal pressure hull, and of submerged displacement 24,000 tons (surface 11,500t). The submarine structure was of double hull construction with nine interconnected watertight compartments, all being normally accessible except for the reactor compartment N° 6 which is passed through via a radiation shield corridor. The outer hull casing comprised 8mm steel plates supported off the pressure hull by webs and struts. The inner pressure hull was an externally ribbed cylindrical form fabricated from 50mm thick high yield steel plate. The void between the casing and pressure hull varied from 1 to 4m within which was located ship's equipment, sonar and the cruise missile silos. The entire outer hull and conning tower was clad with 40 to 80mm thick synthetic rubber tiles serving to both attenuate machinery noise and reduce the reflective echo from incoming sonar signals.

The External Acoustic Layer, showing Noise/Sonar Attenuation Voids revealed by the cutting of the lifting sockets

The power plant comprised two, integrated type pressurized water reactors (OK 650b) each of ~200MW thermal output located in the sealed reactor compartment N° 6. The reactors were arranged in line, in fore-aft fashion, each in its own pressure sealed sub-compartment. Each reactor pressure vessel was housed within a sealed 25m3 capacity water shield tank that was resiliently mounted to absorb shock from the operational submarine when in battle situations. The steam generators were clustered immediately around the RPV with the main circulating pumps above with just over 1m head to assist in natural circulation in the event of pump failure. Fuel comprised annular elements of uranium-aluminum cermet or dispersion type fuel clad in zircaloy, zoned between 20 to 45% (core equivalent 30%) enriched U-235 of 48 assemblies, totaling about 200kg U-235 per reactor core. Gadolinium burnable poison was integrated within the fuel and control was via boron/hafnium absorbers.

Nuclear plant emergency shut down was via control rod injection by spring and pneumatic drive and core cooling was via a relatively conventional ECS with a supplementary bubble tank. As an ultimate safeguard the entire reactor compartment was capable of being flooded with seawater via valves set into the pressure hull.

The *Kursk* submarine had an armament capacity for 24 ship-to-ship cruise missiles (SN-19-GRANIT - NATO Shipwreck) armed with 760kg main charge conventional explosive but nuclear capable for low yield warheads. The missiles were housed in individual pressure sealed silos, pitched forward at 40° arranged in two rows of twelve, each covered by six hatches on each side of the sail (conning tower).

Starboard Missile Bank
Forward Silo Hatches Open

Torpedo munitions comprised 24 torpedoes held in open rack magazines, potentially including torpedoes of nuclear capability, firing from 2x650mm and 4x533mm torpedo tubes in the bow (N° 1) compartment. The armaments could also include ASW Harpoon-type rockets and seabed mines also deployed from the forward torpedo tubes.

Kursk was the latest and most modern attack submarine of the Russian Federation Navy, being assigned to the Northern Fleet operating out of the Northern Kola voyaging into the Barents Sea and beyond. With 49,000 shp through the two 7-blade propellers, she could make 28+ knots when running deep and 15 knots on the surface, being capable of full operations at 600m depth.

MAMMOET-SMIT RECOVERY PLANS

From about January 2001, the Russian Federation Navy and the *Kursk* designers, RUBIN, jointly asked a consortium of companies from the West to tender for the entire recovery of the wreck (with the exception of the totally devastated forward compartment) and, specifically to complete the salvage within the year. This was in order to comply with the promise of President Putin to the relatives of the crew. The first consortium formed, Smit-Heerema-Halliburton, withdrew because Halliburton believed the end of the year recovery deadline could not be safely achieved. In mid May 2001, the Russian Federation and RUBIN, jointly contracted Mammoet-Smit (M-S) to recover the *Kursk* within the year deadline. Although the salvage plan was to be produced by M-S the Russian Federation Navy was to provide a floating dock where the submarine was to be finally berthed.

The M-S strategy was to effect the recovery in three phases, these being:

Phase 1: Preparatory activities, including surveying, radiation monitoring of the submarine, removal of silt around the area of the intended hull cutting operation, and cutting of the hull just forward of the N° 1 bulkhead to sever the most damaged part of the submarine. Then, to give a stable and predictable lift and to mount the rigs, to cut 26 holes through the casing and pressure hull either side of the vertical centerline of the main hull for the subsequent insertion and clamping of the lifting fittings. The positions of these holes were selected by the RF to minimize hull bending during the lift and none were positioned in the reactor compartment. This also included the modification of the *Giant 4* barge by preparing 26 tubes through the barge hull so that the strand jack system, used to lift the submarine, could be fitted.

Phase 2: Installation of the 26 lifting fittings, the lowering through the pre-inserted tubes in the barge hull and connecting of 26 sets of lifting cables, each comprising 54 strands of seven twisted wires each 6mm diameter and the raising of the *Kursk* using Mammoet's strand jack system. The cables would then hold the *Kursk* against a pre-fitted inverted cradle under the barge during transit to a floating dock near Murmansk.

Phase 3: The fitting of two large pontoons, one under each side of the barge, to lift it entirely out of the water to give sufficient clearance of the underslung *Kursk* over the cradles when entering the floating dock, the lowering of the *Kursk* onto the cradles, followed by demobilization and withdrawal of all M-S equipment and personnel.

Severing the remains of N° 1 compartment deployed a heavy cable carrying thick-walled tubular sections coated with a very coarse (~25mm) abrasive. Reciprocating motion was to be provided by two 30 tonne hydraulic rams attached by suction anchors to the seabed.

The cable cutting rig trials on Giant 4 suction anchors (top) acting against an opposite set actual underwater cutting (inset)

The strand jack system relies on two collets on each strand, the upper collets being hydraulically lifted/lowered as a cable group. Additional hydraulics activate the collets under computer control, the timing of the collet activation determining whether the strands are raised or lowered. Each cable lifting system was to be supported by four pneumatic cylinders with 4m strokes and with a large nitrogen gas reservoir, the pressure being matched to the cable load so that large movements due to swell (within the cylinder stroke limits) would have minimal effect on the cable loads.

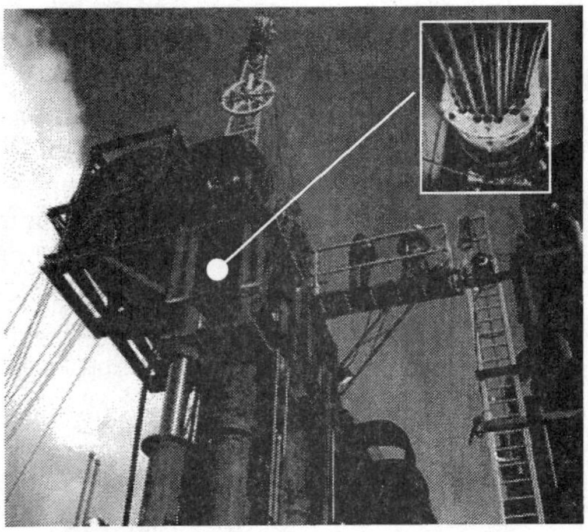

The strand jack (inset) mounted within the cable reel platform with the swell compensator rams underneath

NUCLEAR & RADIOLOGICAL SAFETY

In early June, Large & Associates were engaged by M-S to complete a preliminary assessment of the nuclear hazard and held a number of meetings with RUBIN to discuss and determine the information and data likely to be made available from the RF authorities with respect to radiation and nuclear safety issues. On the basis of this information, Large & Associates was instructed to form and head up the Nuclear Coordinating Group (NCG).

The NCG was headed by John Large of Large & Associates with members Peter Davidson of the UK National Nuclear Corporation (NNC) and Commander Huw Jones of the Royal Navy's Naval Nuclear Regulatory Panel. Later Alan Martin of Alan Marin Associates was seconded onto the NCG to organize the radiological management regime for the recovery spread (the salvage boats). The NCG had direct access to other consultants in the explosive, weapons and salvage fields. John Large also represented and negotiated with the insurers on behalf of Mammoet-Smit for personnel and equipment cover that was ticketed across a number of underwriters at Lloyds, the United States and Russia.

The role of the NCG was to review and evaluate all relevant aspects of nuclear and radiological safety arising from the M-S recovery operations for all stages of the recovery.

The first task was to ascertain what parts of a nuclear safety case were already in place and evaluate them. It quickly became apparent that there was no structured case in existence on which to build.

THE RF APPROACH

The RF approach to safety was essentially deterministic. Any probabilistic treatment was limited to confirming that sequences outside the design basis (which was itself not comprehensively defined) were sufficiently unlikely (e.g. with an annual probability of less than 10-7). There seems to have been no overall integration of the diverse range of technologies covering nuclear propulsion, weapons systems, life support systems and operational systems, to cover the full spectrum of potential interactions between them. Instead, the strategy seemed to consider deliberately each area in isolation with a definition for each area of a worst-case accident that the other areas must withstand. The engineering of the *Kursk* was similarly compartmentalized. This was possibly to minimize the need for detailed interface coverage between the various design bureaus. The flaw in this approach was that there could never have been a full recognition of the wide range of potential challenges, failure modes and consequences (including interactions) arising from internal plant failures and external hazards.

THE NCG STRATEGY

The NCG's overall strategy was framed to suit the RF approach by:

- Establishing the datum condition of the *Kursk* taking into account the effects of the explosions and the degradation over a year of submersion.
- Examining the stability and residual strength of the datum condition, including the degree of defense in depth that might remain available for the essential reactor safety functions.
- Framing limits and conditions for the M-S operations to ensure that the residual strength and stability criteria could not be exceeded, nor the defense in depth totally undermined, together with allowance for unwanted interactions.
- Ensuring that there was an adequate radiological safety management regime in place to protect the M-S employees and contractors.

In light of this, the NCG set out to work with teams of RF specialists to check how each system had been and could be affected by events and thus establish the limits and conditions that had to be maintained during the M-S recovery operations. The actual and potential interactions of the many systems involved warranted a strong probabilistic evaluation but this was not favored nor, indeed, practiced by the RF for its own assessment. Instead, the approach of RF analysts and engineers was, predominantly, underpinned by reliance upon passive safeguards (eg containment, dormancy, etc) for which probabilistic treatment is anyway not usually necessary. However, this reliance required, first, an accurate and reliable assessment of each 'safeguard', particularly the extent to which it may have sustained damage as a result of the original explosions and, then, an account of the degradation that it may have suffered over the year or more that it was submerged in the Barents Sea. Of particular concern to the NCG was the possibility of the M-S operations triggering a further explosion (of a torpedo or missile), and the potential consequences to the reactor plant and safeguards.

On one hand, all that the RF could offer was its assertion and confidence that the M-S salvage of the *Kursk* could be undertaken within the RF's sometimes rather qualitatively defined limits of each of the 'safeguards' but, on the other hand, its engineers and technicians were enthusiastically responsive to any demands placed upon them by the NCG, often responding in detail once trust had been established, and explaining their sometimes brilliantly simple solutions to problems, as they were identified.

In the light of this, the NCG had to conclude that it was not in a position to provide a traditional assessment or review but, instead, had to weigh these RF statements to assess whether, when put together, they provided a sufficiently coherent and persuasive safety demonstration. In doing this, the NCG had to rely largely on its own judgment and experience.

ESTABLISHING DATUM CONDITION OF THE REACTOR PLANT AND SAFEGUARDS

The NCG's strategy required a detailed assessment of the potential damage to the containment, fuel and nuclear shutdown/hold down components of the two nuclear reactor systems.

The determinant of a safe and complete shutdown was, primarily, the ability of each reactor's fuel and shut down systems to function adequately during or immediately following the high levels of impulse loading from the torpedo explosions. Each reactor is supported in a resilient cradle providing shock absorption to cater for an impulse well in excess of 50g, defined by military operations of the boat. To determine the actual loading in the immediate aftermath of the explosions, forensic examination was undertaken on two of the casualties recovered from the stern section, these being identified as the reactor control room operators. These casualties had sustained skeletal damage indicative of body shock loading of just over 50g. This provided a degree of reassurance that the reactor resilient mounts, being below or about the design limit, had not been damaged and that the reactors could have closed down automatically as intended after power supplies had been lost

Other factors relating to the condition of the reactor systems included:

i) The shock level (~50g) would have also temporarily disoriented the reactor control personnel who would not have been expected to recover for several minutes by which time most, if not all, of the power required for operator intervention would have been lost;
ii) the shock would have caused opening of

electrical circuit breakers leading to loss of power, eg to the main coolant pumps;

iii) automatic reactor shutdown sequences would have been initiated probably by the above power loss - this sequence causes the insertion under gravity, with spring and pneumatic assistance, of two diverse means of neutron absorption which then lock into place, and the initiation of decay heat removal which makes use of the large thermal capacity of the water in the shield tank - this sequence cannot easily be interrupted by the operator and it was unlikely that its essentially passive role was impeded by the shock loading;

iv) the reactor shutdown and decay heat removal equipment had a design basis for an inclination of the submarine in excess of 45° for forced cooling during pump spin down, and thereafter a lesser angle of inclination for heat dissipation by natural circulation - the depth of water (108m) in conjunction with the submarine's length (155m) precluded a larger inclination - in view of the perceived accident sequence it is probable that the maximum trim angle was much less than 45°, although this may have been exceeded for a few seconds or more when the hull responded to the expansion of the explosive gas products; and

v) the design provides four barriers to the escape of fission products (or particulates) from the nuclear fuel. These comprise the fuel assembly cladding, the reactor primary circuits, and the reactor shield boundaries.

To confirm the state of the reactor the RF deployed gamma spectroscopy in the range 4 to 8 MeV (characteristic of reactor operation) in the lower regions outside the pressure hull and the thermal gradient in the flood hull space was profiled to detect any thermal input. Negative results suggested that:

i) the reactors remained shutdown;
ii) there was effectively no contamination (eg fuel particulate) in the shield tank, suggesting that the reactor primary circuit containment is complete;
iii) there is no contamination between the shield tank and the pressure hull, suggesting that the shield tank containment is complete; and
iv) the lack of any thermal gradient indicated that no significant heat was being generated in either of the reactor compartments.

On this basis, the NCG's criterion that at least two of the reactor containments be in place was satisfied.

ESTABLISHING DATUM CONDITIONS FOR THE MUNITIONS

Torpedoes: At the time of sailing the *Kursk* was carrying 24 torpedoes, two with dummy warheads, the remainder with conventional explosives, and all stored within N° 1 compartment. Analysis of the acoustic data from the cruiser *Pyotr Veliky* suggested that around seven torpedo rounds were destroyed as a series of explosions in rapid succession. The survey of the second debris field revealed a number of torpedo components but these, collectively, did not account for the remaining 15 or so armed rounds.

Port Torpedo Loading on Kursk

These missing rounds could have been hidden within the hull, particularly in the mangled wreckage of what remained of the bow compartment and some could have been thrown into the wreckage of the second compartment (which was subsequently shown to be the case when the internals of the wreck was dried out and inspected at Rosljakovo). Some or all of the rounds could have burnt during the explosions, some might have fragmented, and others might remain intact and hidden under the submarine hull.

Such was the uncertainty surrounding the presence, state and stability of these missing torpedo rounds that an explosion from this source had to be considered a credible fault condition at any time during the lift operations. Factors in mitigation were:

i) the dispersion of the remaining torpedoes and fragments of torpedoes, made a sympathetic detonation less likely;
ii) detonation would be unconfined and not directed through the hull towards the

reactor compartment (compared to the original explosion that was initially confined by the pressure hull);

iii) the design basis capability of the reactor plant to withstand shock remained available (to an undeclared amount); and

iv) any fragment revealed during silt clearance etc, be removed by the RF, using remotely operated equipment to at least 70m from the submarine - torpedoes and fragments within the cut zone would also be removed.

v) the silt clearing equipment was unlikely to cause detonation of a torpedo or explosive fragment.

The NCG nominated a fault condition whereby the equivalent of two torpedo rounds (~450kg TNT in total) simultaneously detonated during the bow separation operation or the lifting operation. The NCG sought assurance, with explanations, from the RF of each reactor's capability to withstand such an explosion. In addition, an analysis of the effect of the explosion gave the strength requirements of the hull plating of the attending barges and the length limitation for smaller vessels attending the barges, a requirement that these be larger that the sea surface bulk cavitation and gas bubble diameters that would put smaller vessels at risk of sinking. Also, the analysis provided the minimum lashing requirements for the heavy equipment operating on the barge decks, particularly the two 60t crawler cranes working on the *Giant 4* lifting barge, in account that these could topple into the sea and descend onto the *Kursk* in the reactor compartment area or onto the cruise missile silos.

Missiles: At the time of loss, the *Kursk* was armed with 23 SS-N-19 GRANIT cruise missiles with conventional explosives. These missiles were located in forward slanting silo tubes, 12 either side of the submarine, the first being just behind N° 1 compartment and within 3m of the cut line that was to isolate the bow wreckage, and the last two missiles being some 30m ahead of the reactor compartment.

Relevant features of the SS-N-19 missiles are:

i) the propellant fuel is kerosene, with a small (7 Kg TNT equivalent) powder charge for ejection from the silo to the turbojet firing altitude above the sea surface

ii) the launching 'trigger' or arming and firing system (AFS) comprised five independent degrees of protection or latches;

iii) each missile was held within shock mountings within the silo, which itself had the same material characteristics and strength as the submarine pressure hull; and

iv) the missile could be launched only after the silo cap had been opened, which required hydraulic actuation that was no longer available.

Unlike a torpedo round explosion, which was considered to be credible and tolerable, full detonation of a single 760kg missile warhead could not be tolerated at any stage of the lift, conveyance from the wreck site and transfer to the floating dock because this would have imperiled all of those personnel manning the salvage vessels and had the potential to result in a release of radioactivity to the marine environment and hence to the M-S personnel. Thus, it was absolutely essential to determine the most unstable condition for the missile systems and the main fill and ejection charges and if any of the five AFS latches had been enabled by the foundering explosions and the subsequent M-S recovery operations.

This was determined by a series of trials in which fully assembled missiles were subject to a range of conditions simulating the impulse and vibration environments. Particular regard was given to the vibration spectra that was to be generated by the M-S cutting technique deployed to sever the bow section, since there was a possibility that a sympathetic vibration could not only result in the release of the cap of the first starboard side missile silo which had been damaged during the original explosion, but it could also override one of the acceleration/deceleration sensitive latches of the weapon firing system.

ESTABLISHING THE POTENTIAL FAULT CONDITIONS DURING M-S OPERATIONS

Pressure Hull Lifting Sockets: Lifting of the *Kursk* to be secured to the underside of the *Giant 4* lifting barge required the cutting of 26 holes (each ~1m diameter) through the outer hull casing, the removal of any equipment and ship's services in the flood hull space, and cutting through the structure of the pressure hull, thereafter clearing to a depth within the pressure hull to allow for the insertion and fixing of the lifting clamps.

The potential fault scenarios primarily related to cutting through the submarine ship's services occupying the cavity between the casing and pressure hull. Although engineering drawing details had been provided and location trials had been conducted on the sister boat *Orel* (K226), the as-built *Kursk* services installations were found to be markedly differ from the 'design' and/or from the actual installations on the *Orel*.

Saturation Diver Clearing Boat Services Located between Inner and Outer Hulls

Difficulties for the saturation divers undertaking these tasks (surveying the locations and setting up the robotic, high pressure grit cutting equipment) included encountering pockets of explosive gases (three relatively small gas burns/explosions were experienced), and contamination by, particularly, hydraulic gels and asbestos products used in the acoustic tiling bonding system to the outer casing. Procedures had to be introduced for the divers to decontaminate themselves of oils and fibers before entering the saturation chambers on board the diving ship Mayo for shift breaks over each diver's spell of two to four weeks under a full saturation environment.

Lift, Sea, State and Other Factors: Limits on sea state had to be imposed during the lift and transit phases of the recovery operation.

First, lifting operations could not proceed at sea state swell (peak to peak) heights greater than 2.5m because of the limit ram stroke of swell compensation system acting on the strand jacks - this system maintained a uniform cable tension during the lift. The entire 110m lift was scheduled for at a minimum period of 10 hours so a fair weather window of at least this was necessary to ensure safety throughout the lift. If weather conditions deteriorated during the lift then the lift would have to be abandoned and the *Kursk* lowered back to the seabed.

Second, during the transit phase when the *Kursk* was held against the under hull saddles of the *Giant 4* and making way for port to dock with the floating dock, excessive sea state could result in slapping and pounding of the upper casing hull against the saddles and high forces being transmitted into barge frame. In these circumstances, either the *Giant 4* would have to make for sheltered waters or the *Kursk* would have to be lowered to the seabed until clemency resumed. For one particular spell of the open sea transit, over a period of 3 to 4 hours, the distance to the coast and the sea depth precluded both of these options.

Other factors that had to be accounted for included excessive suction binding the *Kursk* to the seabed.

The local seabed at the *Kursk* site comprised silty clays for which M-S had calculated a suction or hold down force of between zero and 11,000 tonnes. To break suction, the plan was to apply a steady but disproportionately higher lift tension to the stern group of lifting cables allowing, over time, this to overcome the suction. This required demonstration that the damaged pressure hull could absorb the bending moment being applied, particularly at discontinuities in the hull form where the forward bulkheads had been blasted through.

In reserve, if the stern lift failed to break suction, a line tethered to two tugs was to be passed under the stern of the *Kursk* with the tugs operating in a seesaw fashion to work the hawser towards the stem. The risks associated with this method included detonation of any torpedo munitions trapped under the hull or, in the event of a hawser failure, whip lashing against the exposed cruise missile silo on the forward starboard side.

In the event, there was no suction, the first movement of the *Kursk* being lateral as the lifting forces allowed her to slip sideways impelled by the tidal stream.

BARGE AND DIVING SUPPORT VESSEL ACTIVITIES - RADIATION RISK

As well as the pre-prepared arrangements for response to a serious mishap to the *Kursk* during recovery (ie torpedo explosion, falling equipment, etc), the barge and support vessel crews had to work under a strict radiological management regime. This regime was administered by a radiation adviser overseeing shifts of health physics monitors surveying and managing contamination, dose receipt and recording, sheltering and other dose mitigation countermeasures.

Giant 4 with the Kursk slung underneath awaits entry into the floating dock at Rosljakovo – two specially designed sinkable pontoons were deployed to raise Giant 4 for the necessary clearance into the floating dock

The NCG cooperated with the RF over analysis of a hypothetical radioactive release from the reactor compartment at the stage when the lifting *Kursk* approached close to the underside of the *Giant 4* barge - this was assumed to be the point at which the barge crew were most at risk of radiation exposure. The conditions assumed for this analysis were:

- expansion of the air/gas bubble drives a discharge of $150m^3$ of water from the reactor compartment via the 6mm diameter instrumentation hole (a known open route into the reactor compartment), taking 36 hours.
- the discharged water contains fission and activation products released from fuel corroded for 14 months by seawater, as determined by a representative test, amounting to some 3×10^{12} Bq (Becquerel). Allowing for dilution in the sea, the total effective dose to a barge crewmember would be less than 1 μSv (at less than 0.1 μSievert/hour). Further development of this model analysis concluded that:
- a larger leak path would not significantly affect the above conclusion.
- if the same amount of fission and activation products were not discharged by the bubble expansion, but remained at the top of the reactor compartment, the 2m of seawater that will fill the space between the pressure hull and the casing would reduce the dose rate to a barge crew member to a few μSv/hour.

To mitigate these risks and those from uncontrolled criticality, discharge of radioactivity or direct radiation resulting in unacceptable levels of exposure, emergency arrangements to protect personnel, including evacuation by the RF Northern Fleet vessels and aircraft, were agreed with the RF Northern Fleet. These actions, triggered by an emergency reference level (ERL) protocol, applied to all personnel present on board M-S vessels.

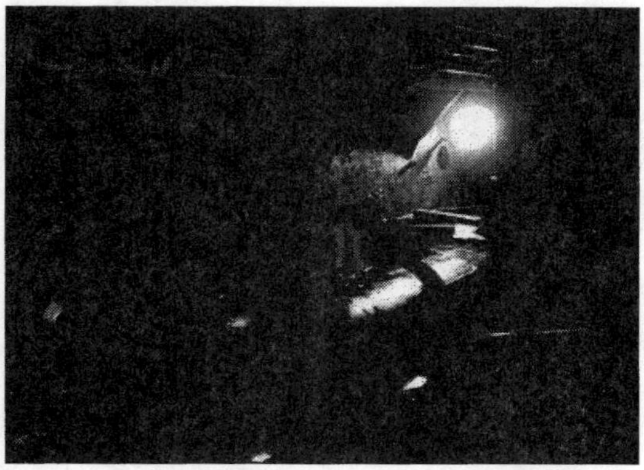

Checking the Sail for Radioactivity
The sail (conning tower) of the Kursk protruding up into the cavity formed in the underside of the Giant 4

OVERVIEW

In completing its task, the Nuclear Coordinating Group adopted the following principles to ensure that the preparatory and recovery operations would not present an unacceptable nuclear or radiation risk to those involved with the recovery and, generally, to the marine environment (NCG 2001):

- **Limits and Conditions:** A clear set of limits and conditions had to be established for all of the operations to ensure that credible hazards would not challenge the capability of the structures, systems and components involved, both on the *Kursk* and or the *Giant 4*.
- **Degradation and Recovery Operations:** The limits and conditions safeguarding the structures, systems and its components would have to account for damage sustained during the sinking, the degradation over the year on the sea bed, and for forces and circumstances introduced by the Mammoet-Smit recovery operations.
- **Defense in Depth:** There should be a number of separate safeguards in place at all times against all of the significant hazards.
- **Tried & Tested Technology:** Since the deployment of novel procedures and

processes introduces additional risk, preferably tried and tested technology should applied.

- **ALARP:** the risk should be reduced to As Low As is Reasonably Practical.
- **Radiological Management Regime:** Radiation doses to those personnel involved in the recovery operation should be controlled below the limits for Class B radiation workers (as defined for the UK).
- **In Contingency:** Contingency plans should exist in the unlikely event of a significant radiological release, particularly for mitigating the impact upon the marine environment.

Mammoet-Smit had contracted to raise the *Kursk* in May 2001 and in just six short months, on 23 October, the *Kursk* was lowered from *Giant 4* onto the cradles of the floating dry dock at Rosljakovo - a quite remarkable and World-first achievement.

A SUCCESSFUL RECOVERY

The recovery of the *Kursk* was a success that derived from a tragedy. The successful and almost trouble free recovery of the sunken nuclear powered submarine *Kursk* was completed by a group of commercial organizations and not by its military operator. This was because the Russian Federation itself did not possess the resources and expertise to do this and, moreover, it had never planned to do so.

The Wreck of the Kursk Rising in the Well of the Floating Dock at Rosljako

In planning and carrying through the entire recovery operation, the Dutch consortium Mammoet-Smit engaged quite remarkable levels of ingenuity of approach to this unique problem. Their strategy of building on their experience of their equipment and of salvage operations in general proved to be sound and ultimately successful.

Because there was insufficient time to generate and evaluate a conventional post-incident nuclear safety case, members of the Nuclear Coordinating Group had to arrive at judgments drawn from their experience in nuclear safety, weaponry and engineering. Moreover, in doing so they had to cross the divide between East and West, accounting not just for the different approaches to nuclear and engineering technologies, but also how the safety reasoning of the original designs could be integrated into the salvage scheme.

This demanding and unique approach was shown to be sound because there was no radiological release or significant radiation hazard to any of the M-S personnel or contactors during any part of the recovery operations.

REFERENCES AND NOTES

AMUNDSEN Ingar, et al, *The Kursk Accident*, Norwegian Radiation Protection Authority, StrålevernRapport 2001:5

NUCLEAR COODINATING GROUP, 1st, 2nd and 3rd Stage Reports of the Nuclear Coordinating Group, Report Ref N° R3065-1B/C/D, throughout 2001

PRINZ C A, *Kursk Submarine Sinking - An Assessment of Potential Hazards*, *Kursk* Foundation, July 2001

Amplification of ship generated wake wash due to coastal effects

B. Allenström (M), H. Andreasson (V), M. Leer-Andersen (V), Da-Qing Li (V)
SSPA Sweden AB

ABSTRACT

This paper discusses the results from extensive wake wash measurements carried out with a model of a ROPAX ship passing a bay. The tests were carried out in SSPA's sea-keeping and maneuvering basin. This model had been subjected to wash wave measurements in open water in previous tests, which gave possibilities for comparisons between restricted and unrestricted waters. Six wave gauges were placed inside and outside the bay, and a significant drawdown was noticed in addition to amplifications of the Kelvin waves. The drawdown and the Kelvin waves that occurred in the bay are difficult to explain, but this paper makes an attempt to analyze the phenomena that occur. The different wave theories that can be used in this case are presented. The methods that are used for analyzing the measured signals are described. One conclusion that can be made is that drawdown caused by the Bernoulli wave can probably be considered a worse problem than ordinary wash waves (Kelvin waves) in restricted and especially shallow waters when looking at large displacement ships.

NOMENCLATURE

A_{bay}	Area of bay (L_{bay} times B)
$A_{bay,cr}$	Cross sectional area of bay (D times B)
B	Breadth of the bay
D	Water Depth
F_h	Depth Froude number
F_n	Froude number
g	Gravity constant
$h_{d,av}$	Mean value of drawdown in bay
k	Wave number
L_{eff}	Effective length of harbour
L_{pp}	Length between perpendiculars
L_{bay}	Length of bay (in y-direction)
L_w	Transverse wave length
L_d	Divergent wave length
T_d	Period of the drawdown at the inner most probe
T_n	The natural resonance period of the bay
T_i	Period of the incident wave at the entrance of the bay.
T_S	Period of seiches at the inner most probe position
U	Ship speed
U_d	Propagation velocity for divergent waves
U_{dd}	Flow velocity due to drawdown
Y	Transverse direction to wave gauge
θ	Divergent wave propagation angle
φ	Cusp angle

1 Introduction

Wake wash is a well-known problem in most coastal areas around the world, and wash waves from high speed crafts have often been pointed out as a high risk for both people and the coastal environment. In a project recently carried out in Sweden (see [Allenström et al., 2003]) wake wash from both high speed crafts and large ferries was studied regarding the effect on the environment in archipelagos. Special interest was directed towards the long periodic waves caused by large ferries, especially effects like drawdown and seiches in bays. To understand this problem better, model tests were carried out in the seakeeping basin at SSPA using a ropax passing a rectangular bay. In the bay, and at the entrance to the bay, six wave probes were placed and the events were video filmed. To be able to conclude what the amplified effects due to the bay are, the test data have been analysed extensively.

2 General theory

Throughout the paper different ship generated wave phenomena and secondary effects are discussed. Therefore, a short description of these (Kelvin, Havelock, Bernoulli and seiche) are presented below.

What is normally associated with ship-generated waves is the Kelvin wave pattern (or Havelock wave pattern for shallow water), as these are the most visible waves. This wave pattern consists of divergent and transverse waves. The point where the divergent and transverse systems meet will have an angle (cusp angle) of 19.28 and will at this point have a propagation angle relative to the ship path of 35.27. When the water is shallow this no longer applies, as the cusp angle increases towards 90.0 for the depth Froude number equal to unity, after which it will again decrease.

The expression Bernoulli wave is used for the formation of the water surface around the vessel hull due to the pressure distribution in the water around it. This pressure, Bernoulli pressure, is a result of the displacement effect of the vessel. The displacement causes a velocity distribution around the vessel, which is balanced by a corresponding pressure distribution that can be estimated by the Bernoulli equation. Approaching the water surface where the pressure is constant (atmospheric pressure) the balance to the velocity distribution takes the form of a rising and lowering of the water surface in a wave-like form, known as the Bernoulli wave. For constant vessel speed the Bernoulli wave does not need any maintaining energy. The energy needed to create the Bernoulli wave is generated with the acceleration of the ship and is recovered at retardation.

The Bernoulli wave depends on the displacement, the vessel speed and the restrictions in the surroundings. In deep and unrestricted water the Bernoulli wave is usually hard to distinguish from other types of waves generated by the vessel (although it is important as it is the starting point of the Kelvin wave system). However, as the water becomes more and more shallow and restricted, the Bernoulli wave can be greatly amplified.

Long periodic waves have been observed in shallow bays in archipelagos or shallow harbours in connection with ship traffic in narrow fairways. These wave motions are characterised by a series of water level rising and lowering of long period, often with the first lowering being the largest in amplitude. The first level lowering results from the direct excitation of the Bernoulli wave and Bernoulli pressure. It is this first lowering of the water level from its still position that is called drawdown.

According to the definition in oceanography, seiches are the long periodic water surface oscillations in closed or half open water bodies such as lakes, harbours and bays. The mechanism of seiche can be described as the free oscillations of water at its eigen frequencies in a restricted area. The seiche needs to be activated. It can be excited by external forces of an impulsive character, for example, an incident wave event or a sudden change in atmospheric pressure. The seiche frequency is determined by the geometry and the water depth of the restricted area. In the present paper, we refer to those long periodic level oscillations that follow the drawdown in the bay as seiches.

Other ship generated wave phenomena exist (such as solitones, wave breaking), but are not pertinent or are difficult to include in this context, and will therefore not be mentioned further.

3 Description of model tests

The tests were carried out in the Maritime Dynamics Laboratory at SSPA. The dimensions of the basin are 88 by 39 meters, and the maximum water depth is 3.5 meters. The ship model used was a 194.4 meter conventional ROPAX constructed to the scale 1:36.9.

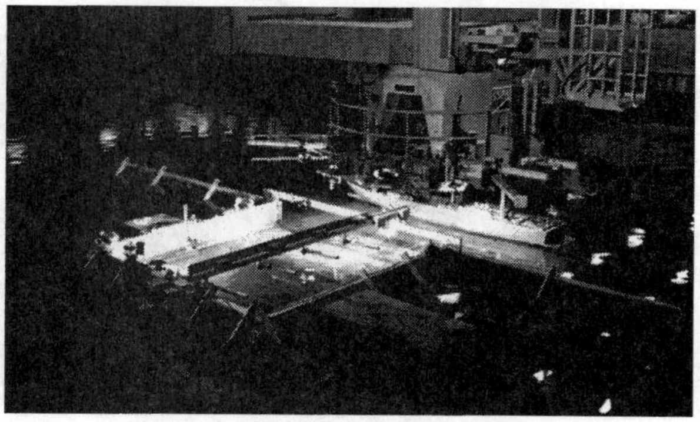

Figure 1:Photograph of the seakeeping basin with the bay

The main parameters for the model are shown below.

	Ship	Model
Length o.a. (m)	194.4	5.268
Length p.p. (m)	172.2	4.667
Beam (m)	28.4	0.770
Draught, fore (m)	6.6	0.179
Draught, aft (m)	6.6	0.179
Block coefficient (-)	0.61	0.61
Displacement (m^3)	19774	0.3946

Table 1: Ship main particulars

The main reason for using this ship model was that extensive wash wave tests had been performed earlier with this model, but in open sea. These measurements had also been compared with wash wave calculations using the SHIPFLOW CFD-code. In other words, the wash waves from this ship were very well known.

The tests with the bay were carried out at two different water depths, 20 and 30 meters corresponding to 0.542 and 0.813 meters in model scale respectively.

A rectangular bay was constructed in the basin according to the figures below. Six wave probes were located in the middle of the bay with the distances to the outer part of the bay (coast line) as shown in Figure 2. One of the sidewalls in the bay was covered with a tape that had a pattern that made the motions of the water surface more visible. This wall was video recorded during the tests. The position of the camera was between wave probe 1 and 2. Another video camera was placed at the end of the bay.

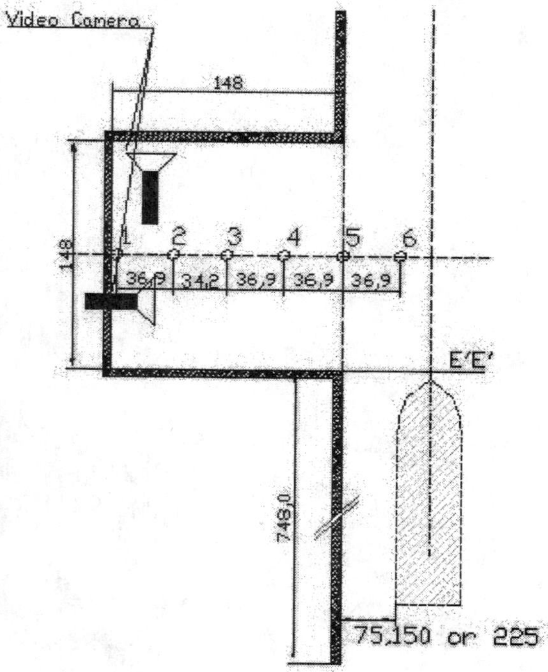

Figure 2: Sketch of the test set-up.

The ship model was self-propelled during the test, and to compensate for the scale effects on the frictional resistance a small towing force was applied. The parameters measured on the ship model were:

- Speed
- Propeller thrust on port and starboard propellers
- Towing force
- Trim fore and aft

A number of tests was performed for different ship speeds and at different distances from the coast line. For each test run measurements were taken with the six wave probes and recordings were made with the two video cameras.

3.1 Test results

A typical result from the tests is presented below. The speed of the ship for this test was 17 knots and it was travelling 75 meters from the coast line.

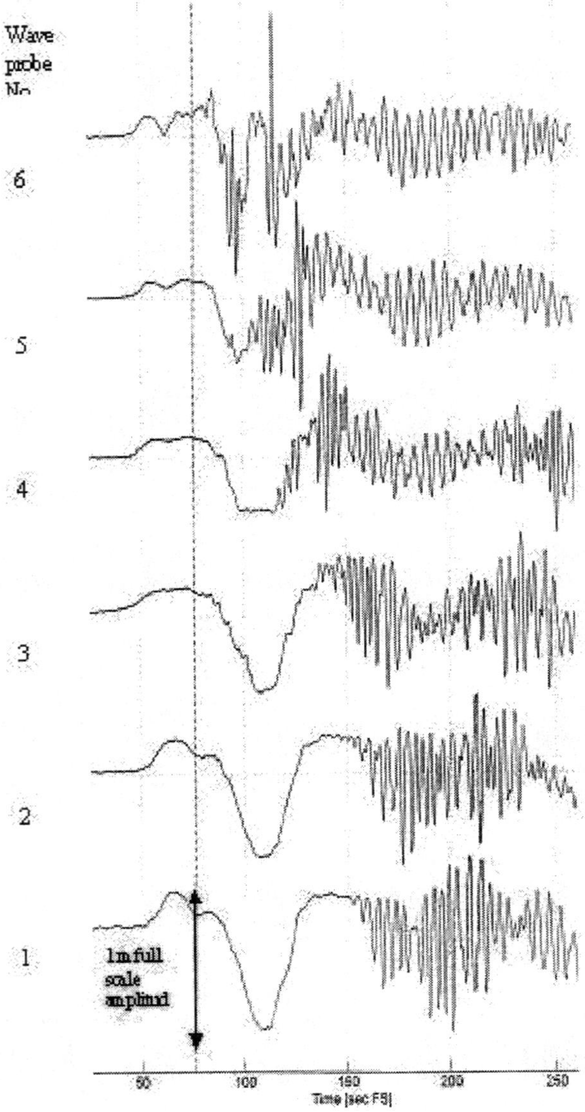

Figure 3: Time histories from wave probes

The vertical dotted line at time = 78 represents the moment when the bow crosses the line E-E shown in Figure 2 (first side wall of the bay). As described above there were two video cameras recording simultaneously. In the pictures below the two views are combined into one picture. The lower part of each still shot shows about a 0.5 m wide section of the water surface. From the videos the following could be observed:

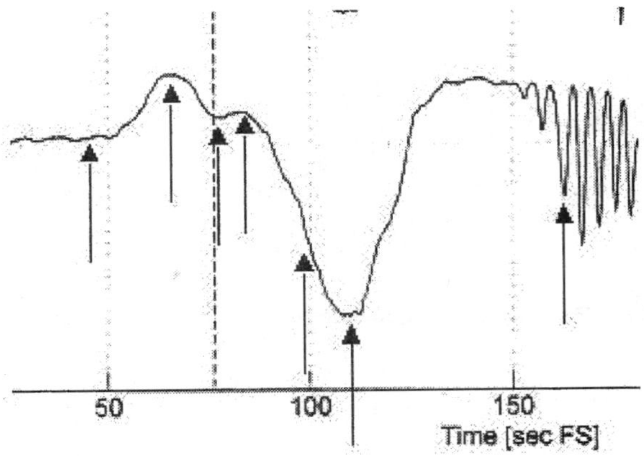

Figure 4: Wave profile at Wave gauge 6. Arrows refer to the still shots shown in the figures below.

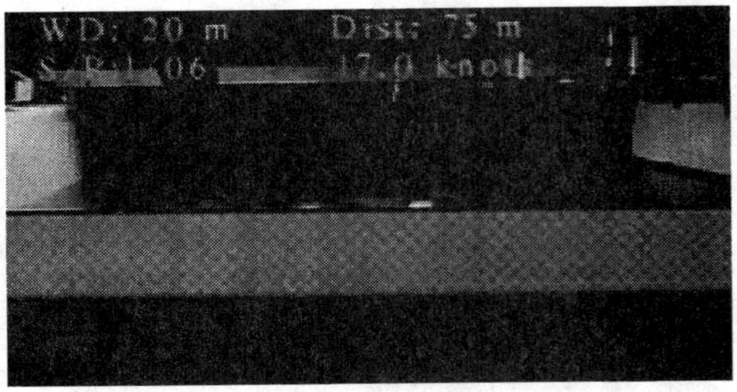

Figure 5: Time t = 48. The ship is approaching the bay. Water surface is calm.

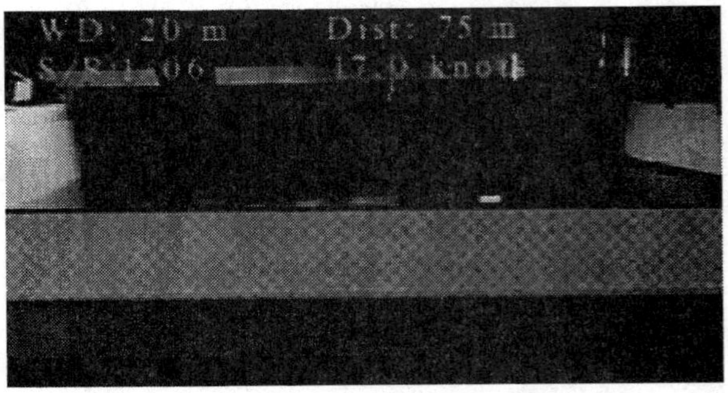

Figure 6 : Time t = 69. Ship bow is 61 m from the bay. Water surface is 0.2 m above calm water level.

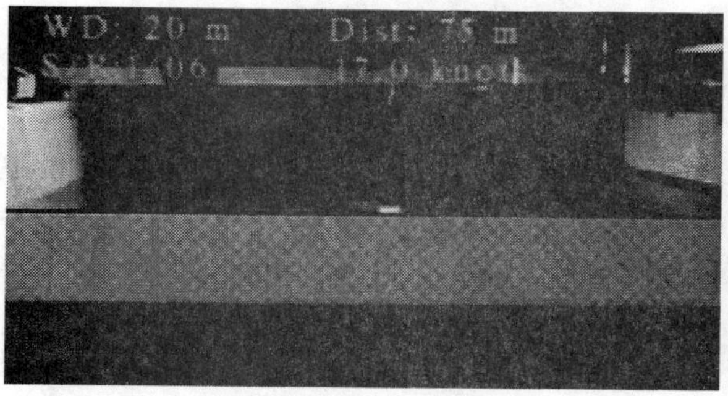

Figure 7: Time t = 74. Ship bow is just passing line E-E. Water surface is oscillating.

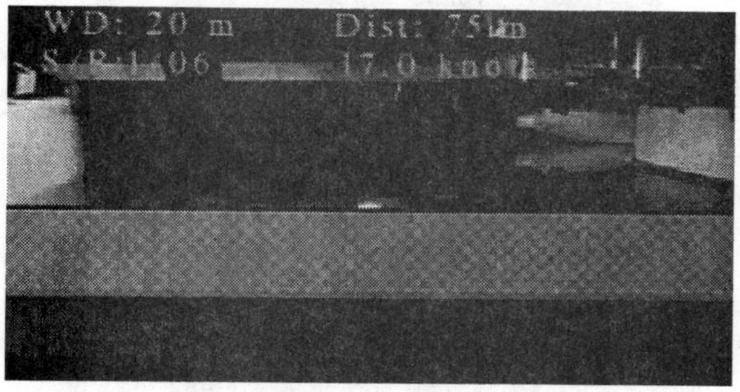

Figure 8 :Time t= 81. Ship bow is about one quarter of a ship length behind E-E.

Figure 9: Time t = 98. Ship is right outside the bay. Water surface is lowering rapidly.

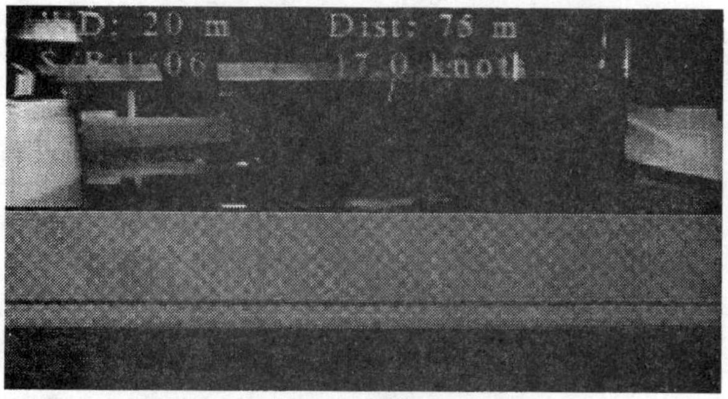

Figure 10: Time t = 110. Ship is leaving the bay. Water surface is at the lowest level, about 0.6 meters below calm water level.

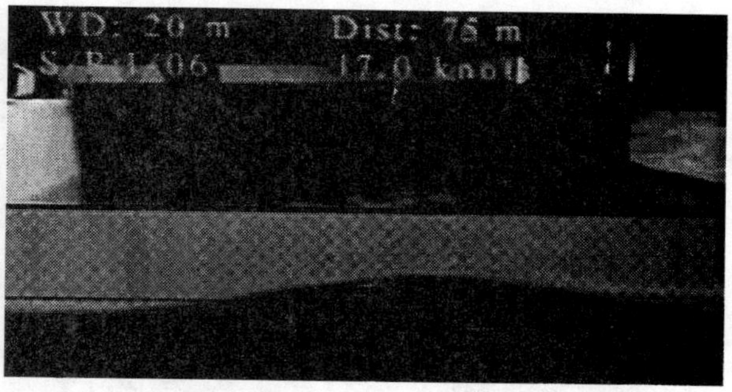

Figure 11 : Time t = 163. Ship has passed the bay. Kelvin waves have reached the inner part of the bay.

The figures above present the situation for the speed of 17 knots and as anticipated the drawdown depends very much on the ship speed. In Figure 12 the effect of ship speed can be seen at the inner part of the bay and at the shore line. It can also be seen that there is a big influence from the water depth. The drawdown at the inner part of the bay is almost three times higher at 20 m water depth compared to 30 m.

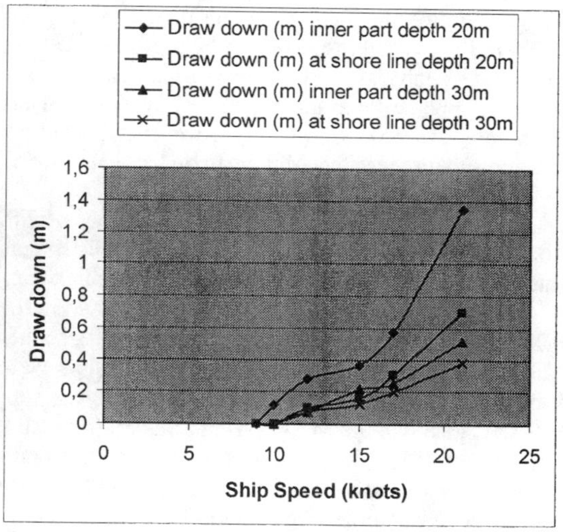

Figure 12 : Speed versus drawdown at distance to shore line 75 m

In addition to the ship speed, the distance between ship and shore also determines the amplitude of the drawdown to a great extent.

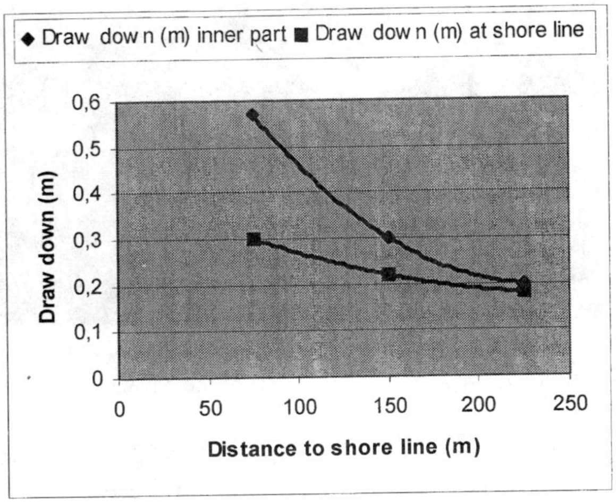

Figure 13: Distance to shore line versus drawdown at speed 17 knots

4 Analysis of model test results

The large quantity of data obtained during the model trials needs more than visual inspection to be useful. Therefore, in addition to analysis of the wave profile measurements themselves, one or more of the following analysis methods are utilized to quantify, evaluate and draw conclusions about the effect of the bay on the wake wash.

- Isolation and identification of different waves with signal filtering is necessary if long periodic events are to be analysed without the short periodic waves and vice versa. Therefore most plots discussed in the following sections have been filtered using either low or high pass filters. Equiripple FIR filters were designed with the Filter Design and Analysis Tool (FDATOOL) in the MATLAB 6.5 Signal Processing Tool Box (SPTB). Both the low-pass and the high-pass filter were of order 204 with filter width 0.1 to 0.2 Hz. The sample frequency is 6.58 Hz (40 Hz in model scale).
- To identify dominating wave period spectrum analysis has been used. The Power Spectral Density of the waves was estimated using the Welch method in MATLAB 6.5 SPTB. The FFT lengths 256 and 1024 were used together with Hanning window and 50% overlap. The lengths of the measured signals varied between 2000 to 3000 samples.
- Potential flow simulation has been used where insufficient data were available. The CFD code SHIPFLOW™ version 2.5 uses a zonal approach to determine the flow near the ship, and is as such especially adapted to calculations involving a ship. The computation domain is divided into three zones. The first zone covers the whole computation domain and in this zone the flow is calculated with potential flow theory. To use potential flow theory, it is necessary to assume inviscid and irrotational flow. The free surface is modelled with non-linear Rankine source panels. For each iteration, the location of the free surface source panels is adjusted to coincide with the location of the free surface, and the panels of the hull are modelled up to the boundary between free surface and the hull. For the simulation performed, where the focus is the waves, only the potential flow calculation was performed. The computation is stationary, which means that the bay cannot be modelled. The effect of the restricted water depth is included in the computation. 11500 panels were used to model the surface and the vessel and the computational domain covers a water surface area of 830×450 m using the centreline of the vessel as symmetry plane.

No further description of the above methods will be given in the paper.

4.1 Long periodic events

These phenomena can not really be described as a travelling wave front, such as the well known Kelvin wave system. The phenomena will be described as long periodic events, i.e. free surface elevations that have a much longer period than the Kelvin waves. These will include the Bernoulli wave, drawdown and seiches.

4.1.1 Bernoulli waves in shallow water

The Bernoulli wave created by any ship is greatly affected by shallow and restricted waters. It will therefore be treated separately in the following discussion.

Bernoulli waves in shallow and open water.

In shallow, open water the fluid flow around the vessel is restrained by the bottom due to the finite water depth. The flow will be more decelerated in the bow and stern region and accelerated amidships in shallow water, compared to the deep-water condition. According to the Bernoulli equation, the higher disturbance velocities will create a higher pressure in the bow and stern region, but a lower pressure amidships in shallow water cases. This in turn will lead to an increased free surface disturbance, which leads to the amplification of the Bernoulli wave in shallow water.

An example of the extent of the Bernoulli wave in shallow water can be seen in Figure 14. It originates from model tests at SSPA in the open basin, where the ship speed was 16 knots (i.e. F_h= 0.76) at a water depth D = 12 m. The wave gauges are marked WG1,...,WG6, where WG6 is the one closest to the vessel. The scale in the y-axis is the wave gauges' distance to the ship centre line, and t= 0 on the time axis corresponds to the moment when the model's F.P. passes the wave gauges. The high frequency Kelvin waves have been filtered out so that only the low frequency waves remain.

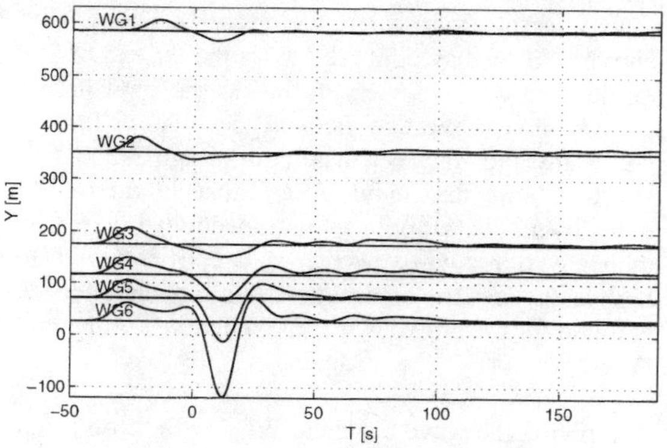

Figure 14: Bernoulli waves (Kelvin waves filtered out) for the Ropax at 16 knots, water depth 12 m

Based on plots like Figure 14, the following observations can be made:

a. The crest-trough-crest pattern, as seen at WG6 in Figure 14 over the time range $-10 < t < 30$, constitutes the Bernoulli wave.
b. The Bernoulli wave is found to exist in the transverse direction all the way from WG6 to WG1 (at y ≈ 3.4 Lpp). The pressure field that causes the Bernoulli wave appears to radiate a long distance in the transverse direction in shallow water.
c. The amplitude of the trough of the Bernoulli wave is very pronounced at probe WG6, but diminishes with the transverse distance to the ship. There is almost no phase shift for the trough at various probe positions, whereas the crests show a little lag.

The model test results described above do not include variations in water depth. Therefore they cannot without additional data support the shallow water influence on the Bernoulli wave. However, [Janson, 2002] has investigated the Bernoulli wave using SHIPFLOW, and a double model solution both with regards to distance from hull and shallow and deep water.

The ratio between the Bernoulli waves amplitude at two distances from the hull calculated using Janson's results (using a different hull and slightly different velocities and water depths), is quite similar to the model tests (3.7 for SHIPFLOW and 2.9 for model tests). Therefore, the model tests and SHIPFLOW calculations seem to be in reasonable agreement. The SHIPFLOW calculations further indicate that very near the hull (almost along the side), the difference between amplitudes in shallow and deep water is not great (about 20%). However, further from the hull it is significant (several hundred percent). Therefore, shallow water effects on the Bernoulli wave increase the amplitudes of the free surface elevation slightly, but greatly increase the distance where it affects the free surface.

Bernoulli waves near a vertical bank

The influence on the Bernoulli wave of a bank in shallow water will be discussed in this section.

One case from [Li, 2000] is used to illustrate the interaction phenomenon between a tanker and a vertical bank in shallow water. Figure 15 contains the free surface elevation in the longitudinal direction. The test conditions in this case are

- U = 12 knots, D = 19 m, L_{pp} = 238 m
- Ship – bank distance y = 65 m, with the bank on the port side.
- Five wave gauges, with one on the inside ($Y_i < 0$, relative to ship centre line) and four on the open side (Y_i = -64 m, 74 m, 97 m, 144 m, 190 m)
- Five pressure probes, three on the bank wall (positions 1-3) and two on the bottom (positions 4-5)

As seen in Figure 15, the wave patterns between the axial position $-0.3 < x/L_{pp} < 0.3$ are recognized as Bernoulli waves whereas the Kelvin waves, appearing after x/L_{pp} = 0.5, are seen to be very small in amplitude at this low speed.

The pressure distribution measured in the longitudinal direction follows the Bernoulli wave quite closely. This pressure (velocity) can be of importance for marine life and erosion, and it is quite apparent that for this low speed and bank/shallow water condition, the Bernoulli wave has a much greater impact than the Kelvin waves.

Comparing the wave profile between wave gauge number 1 (near bank side) and those on the open side, it is seen that waves on the near bank side are amplified due to ship-bank interaction. Wave gauge number 1 and 2 in particular can be closely compared, as they are approximately symmetrical around the ship track, and an amplification on the bank side compared to the open side, shows an amplification with a ratio of approximately 2.

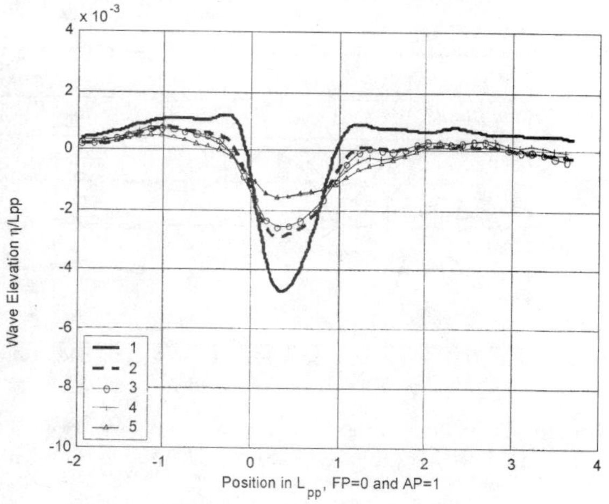

Figure 15: Free surface elevations with bank.

Based on the pressure and wave profiles (pressure profile not shown) such as those shown in Figure 15, the following features have been observed:

- The Bernoulli wave on the near bank side is substantially amplified due to ship-bank interaction and shallow water effect.
- The pressure on the bank is influenced directly by the Bernoulli pressure of the passing vessel. Its distribution has a similar pattern to that of the Bernoulli wave.

While the quantitative data is perhaps less important here, the above example demonstrates the qualitative consequence of interaction between the vessel and the bank. In short, for a displacement ship travelling in shallow water with a bank in its near vicinity, the Bernoulli wave is significantly amplified and by far the most important component of the free surface elevation.

4.1.2 Drawdown in the bay

The previous two sections have dealt solely with shallow water and bank effects on the long periodic waves (Bernoulli wave in this case). The following section will include events that can take place inside a bay if a ship passes the bay entrance.

Isolation of long periodic and short periodic wave events makes it possible to analyse the character of waves of different frequencies. Two long periodic phenomena can exist in a bay. They are drawdown and seiches.

Figure 16 shows wave profiles measured at six wave gauges in full-scale time history. Note that t = 0 is the moment when the vessel's F.P passes the transverse line along the wave gauges, and not as defined in Figure 3 when passing line E-E. WG6 is closest to the vessel, while WG1 is the one farthest from the vessel but innermost in the bay, see Figure 2.

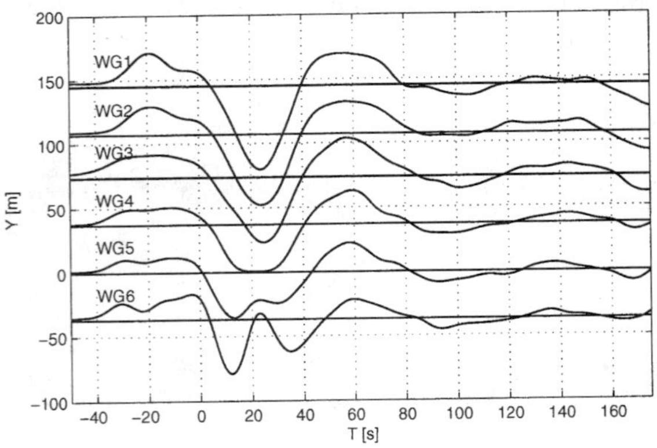

Figure 16. The long periodic waves signals (low pass filtered). U = 17 knots, D = 20 m, distance to shore line 75 m.

Since WG5 is located at the entrance of the bay, the wave motion here (mainly Bernoulli waves) is regarded as incident wave (disturbance) to the water in the bay. The time period -5 < T < 30 [s] in Figure 16 is roughly the time interval from when the vessel's F.P. passes the entrance of the bay until AP passes the far end of the bay. Looking at the wave events inside the bay during this period, one can see that the water level starts to decrease drastically and reaches a maximum trough at about $T \approx 24$ s. It is this maximum trough that is referred to as drawdown. The amplitude of drawdown is much larger than any other wave event. An amplification factor as high as 3 times the amplitude at the bay entrance (WG5) is found for the drawdown at the innermost probe position WG1.

Comparing the wave events at WG5 with those inside the bay in Figure 16, some general features of the long wave events are noted:

- A disturbance (incident wave) appearing at the entrance of the bay causes a water level response all the way into the bay, with only a slight delay in phase.
- The drawdown increases with distance from the bay entrance.
- The drawdown is present in all cases.
- The amplitude of drawdown is much larger than any other wave event.

The reason to separate drawdown from other long periodic events is that the drawdown occurs to a large extent within the period when the Bernoulli wave is passing the entrance of the bay and the water motion at this period should be regarded as a forced (by Bernoulli pressure) excitation and oscillation.

Amplification of drawdown due to bay

To show the amplification of the drawdown and the waves in the bay, the unfiltered wave gauge measurements shown in Figure 16 are compared to the results from a potential flow CFD calculation of the same case but in open water. In other words the difference between the conditions used for the measurements and the CFD calculation is that there is no bank or bay in the calculated model. To validate the results from the SHIPFLOW code with experiments, similar calculations were performed to be compared with the measurements presented in section 3.

The comparison with that experiment shows a relatively good agreement for both period and amplitude of the waves.

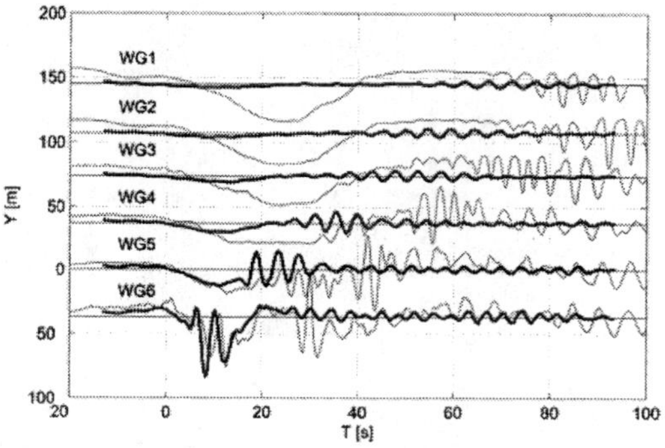

Figure 17: The measured wave signals in the bay compared to open water wave cuts at an equivalent distance as computed with SHIPFLOW (unfiltered signal, thick line). U = 17 knots, D = 20 m, distance to shore line 75 m.

Figure 17 clearly shows the amplification of the drawdown that occurs because the bay is enclosed on three sides. At wave gauge 6 the computation (thick line) hardly shows any drawdown while the measured signal in the bay shows a drawdown similar in size or even greater than that measured at wave gauge 1. The signals at wave gauge 1 show a similar magnitude for both the waves and drawdown.
This small example clearly shows how important the bay enclosure is for amplification of drawdown.

The period of drawdown
An attempt was made to study the variation of the period of drawdown at WG1, denoted by T_d, in relation to other parameters. Because there are too few long periodic signals available, their periods given by the spectrum analysis are not reliable. Thus the drawdown period was directly estimated from the filtered wave profiles, which of course introduces some uncertainties.

The drawdown periods T_d in various test conditions are listed in Table 2. A clear dependence of T_d on the water depth D, vessel speed U, and distance between the vessel to the bay Y can be noticed. T_d decreases with increase of speed U, which should be expected as the Bernoulli wave passes the bay faster. It also decreases with increase of water depth D (fits the observation made previously that the extent of the Bernoulli wave increases with depth Froude number). T_d increases with increase of the bay length L_{bay}, as well the ship-to-bay distance Y. These relations are in agreement with the dependence of drawdown amplitude on these parameters, as discussed in section 3.1.

Estimation of maximum flow speeds due to drawdown
In addition to the safety aspects with drawdown such as mooring problems and motions of ships when loading and unloading, the vertical motions of the water surface is of course connected with a flow into and out of the bay. If the flow velocity is high, the flow itself can be a problem regarding moorings etc., but at a relatively low speed bottom erosion can be a risk.

Due to the long period of the drawdown (in the order of one minute), the water level is assumed to be the same along parallel lines that run across the bay. Then, the water volume moving in and out of the bay can be related to the maximum drawdown of the bay times the free surface area of the bay (A_{bay}). Maximum average drawdown (h_d) is in this context is considered to be the mean value of the maximum drawdown in the inner part of the bay and at the shore line. Then continuity can be used to calculate maximum average velocities moving through the bay entrance as

$$U_{dd} = h_{d,av}\, A_{bay} / A_{bay,cr}\, 0.5T_d$$

where $A_{bay,cross}$ is the cross sectional area of the bay entrance below the free surface, and T_d is the period of the drawdown.

Assuming a period of 50 seconds, even when the ship is travelling at a speed of 21 knots at a distance of 75 m from the bay the flow velocity is not more than 0.15 m/s, see Figure 18. This is to be compared with the value 0.3 m/s that can be taken as a 'rule of thumb' when erosion starts. However, the above method is only a measure of the average flow velocity, and local velocities can be much higher. A smaller water depth will also increase the velocity.

Looking at Figure 12 we can see that the amplitude of the drawdown is highly dependent on water depth. Going from 30 down to 20 meters depth (about a 30% reduction) gives three times higher amplitudes. Another 30% reduction in water depth is still an operational depth for this ship. If we assume that this gives us three times the drawdown amplitude compared to 20 meters depth, and taking into account that the cross section of the bay is also reduced by 30%, we will end up with flow velocities twice the limit for risk of erosion. However, looking at the steep slope of the curve in Figure 18, it can be seen that reducing the ship speed by a few knots eliminates the risk of having too high flow velocities. If the speed is reduced to 8-10 knots no practical problems occur even at distances as short as 75 meters for a ship of almost 200 meters in length.

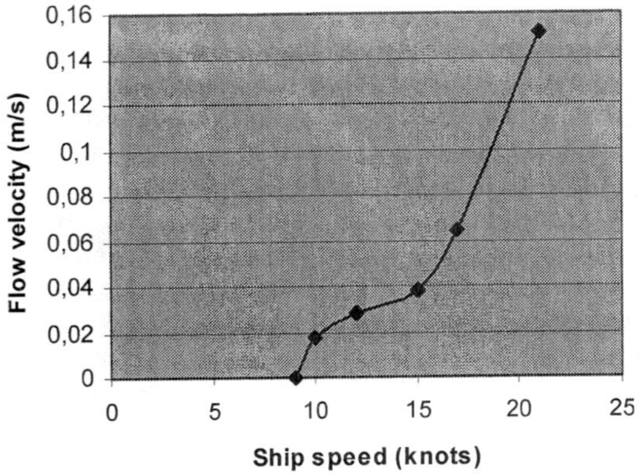

Figure 18: Estimated flow velocity in the bay due to drawdown at 20 meters depth with the ship passing at a distance of 75 meters.

4.1.3 Seiches in the bay

The subsequent long waves that follow the drawdown, starting at $T \approx 50$ s in Figure 16, are characterised as seiches, since these events occur after the vessel has passed the bay. They are free oscillations in the bay.

To see the seiche motion from a different view, the wave cuts are plotted as a function of distance Y, instead of time, using the same data as in Figure 16. The resulting surface elevations are plotted in Figure 19. The x-axis shows the distance in transverse direction with Y = 0 at the entrance of the bay. The y-axis gives the elevation in mm.

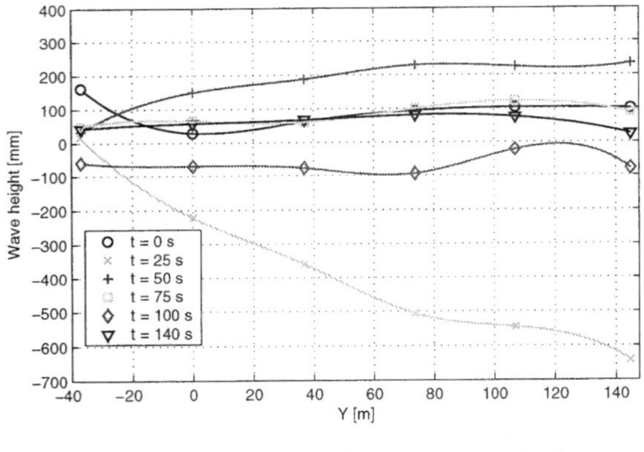

Figure 19: A wave cut through all wave gauges in the transverse direction to show the water level variations for 6 time sequences in the bay. The conditions are the same as in Figure 16.

Some general features of seiches in the bay can be observed in Figure 19.

- The seiches begin at about t > 50s; from the entrance of the bay to the innermost wall of the bay, the water surface rises almost simultaneously, and then falls simultaneously with a similar amplitude.
- No nodal point is found (i.e. the wavelength is at least 4 times longer than the bay), indicating that the oscillation is of lowest order (mode number = 1)
- Seiche amplitudes are rapidly damped. None of them has exceeded the amplitude of the drawdown (comparing the drawdown at t = 25 s with the subsequent wave cuts).

However, there are cases where the drawdown is present, yet no seiche is detectable. Figure 20 shows such an example, it differs from that in Figure 19 only in the vessel speed, here the vessel travels at 12 knots.

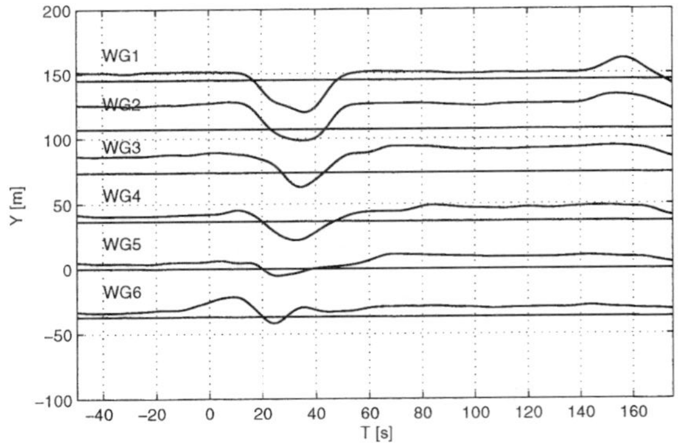

Figure 20: Same conditions as in Figure 16, only U = 12 knots.

The seiche is an eigen-period phenomena, and a reasonable explanation would be that excitation (Bernoulli wave) frequency is too far from the seiche eigen frequency. This will be discussed in the following section.

Comparison of periods of the long wave events with the theoretical natural period.

Three types of periods are compared with the theoretical periods of the bay. They are:

- The incident wave period, denoted by T_i, which is the period of Bernoulli wave at the entrance of the bay,
- The drawdown period, denoted by T_d, which was already mentioned in the previous subsection,
- The actual seiche period, denoted by T_S, which is an average period of the subsequent long waves after the drawdown if it exists.

All periods were estimated directly from the filtered data, instead of using spectrum analysis, because too few long wave signals were available. The theoretical natural periods are T_{n1} and T_{n2}, and can be estimated quite easily for the simple bay geometry used by using the following theory.

The linear shallow water wave theory was used to calculate the natural oscillation period of the bay. Two formulas were adopted for this purpose. The natural seiche period for a rectangular basin with one end (see [Mei, 1983]) open was obtained with

$$T_{n1} = \frac{4L_{bay}}{n\sqrt{gh}}$$

where n is the modal number of the basin, n = 1, 2, ,∞. L_{bay} is a characteristic length of the bay, usually the distance from the open to the closed end. For the 1st order (fundamental) mode, n = 1, the seiche wavelength will be 4 times as large as L_{bay}.

For a rectangular harbour with one end open to sea, Raichlen in [Ippen, 1966] proposed that an effective harbour length L_{eff} should be used in the above formula due to a possible energy radiation from the harbour entrance to the open sea, so that

$$T_{n2} = \frac{4L_{eff}}{n\sqrt{gh}}$$

where

$$L_{eff} = L_{bay} + \frac{B}{\pi}\left[1.051 + \ln(\frac{L_{bay}}{B})\right]$$

These two periods, T_{n1} and T_{n2}, can be used as a lower and upper bound of a possible range of the seiche period, and are compared with those observed in the current experiment to check how close the actual periods of the long waves events to the eigen period of the bay.

The periods T_S, T_d and T_i are compared with the theoretical ones T_{n1} and T_{n2} in Table 2 under various test conditions, to check if there is any correlation between the theoretical seiche periods and the measured ones. In Table 2, the seiche periods T_S are given only for cases where seiches were able to be identified and are marked in bold. For those cases where no seiche was observed, the corresponding cell in Table 2 under column T_S is filled only with a "-". "NA" in the table means data not available. A necessary condition for an oscillation at resonance in a bay is that the period of the incident wave is equal to the natural resonance period of the bay. Therefore, if a seiche occurs, periods T_i, T_d and T_S should be very close to each other, and the actual seiche period T_S should be close to the theoretical seiche period.

The following observations can be made from Table 2:

- Comparing periods T_i, T_d and T_S in Table 2 for each test condition, it is found that they differ from each other somewhat, yet they are of the same order of magnitude.
- Periods T_i and T_d are closer to each other than to T_S; moreover, T_i shows a clear dependence on speed U, depth D, distance Y_i and the bay length L_{bay} in the same way as T_d does, see "the period of drawdown" under section 4.1.2.
- Periods T_i, T_d and T_S are generally greater than theoretical seiche periods T_{n1} and T_{n2}. Further, they are closer to T_{n2} than to T_{n1}. Note that T_{n2} is always larger than T_{n1}. From the formulae where T_{n2} is computed, it is easy to understand that using an effective bay length to account for possible energy radiation from the bay entrance into the open sea is more reasonable. Therefore, only T_{n2} will be used in later discussions.
- The natural seiche period T_{n2} depends only on the bay geometry and water depth. It should be a constant once the bay geometry and water depth are fixed. However, the measured seiche period T_S in Table 2 seems to vary with the vessel speed for the same bay. Furthermore, the seiche occurs only when the vessel speed becomes large enough, i.e. $U \geq 15$ knots; at speeds lower than 15 knots, a seiche is not detectable even though T_i is almost equal to T_{n2} in two low-speed cases.
- The difference between T_S and T_{n2} may be used to explain why the seiche is damped so rapidly in all the cases. If T_S is different from T_{n2}, the seiche will not oscillate at its resonance frequency, hence its amplitude will be decreased by the natural resistance of the system.

D = 20 m	T_{n1} = 42 (s)			T_{n2} = 57 (s)		
L_{bay} = 148 m	U (knot)	Y_i	F_h	T_i	T_d	T_S
	10	75	0.37	26	73	-
	12	75	0.44	68	69	-
	15	75	0.55	70	**64**	**52**
	17	75	0.62	82	**66**	**78**
	21	75	0.77	80	**57**	**92**
	17	150	0.62	90	**82**	**78**
	17	225	0.62	91	**92**	**67**
D = 30 m	T_{n1} = 34 (s)			T_{n2} = 46 (s)		
L_{bay} = 148 m	U (knot)	Y_i	F_h	T_i	T_d	T_S
	12	75	0.36	42	58	-
	15	75	0.45	48	51	-
	17	75	0.51	49	**53**	**48**
	21	75	0.63	63	**49**	**68**
	17	150	0.51	58	61	-
	17	225	0.51	81	84	-
D = 20 m	T_{n1} = 32 (s)			T_{n2} = 43 (s)		
L_{bay} = 111 m	U (knot)	Y_i	F_h	T_i	T_d	T_S
	17	75	0.63	65	**59**	**55**
	21	75	0.77	50	**44**	NA
	21	150	0.77	62	61	NA
	21	225	0.77	63	62	NA

Table 2: Drawdown period versus the 1st order natural resonant period of the bay.

The interaction between the vessel, the restricted water and the bay will induce a drawdown at the innermost bay, and possibly excite seiches in the bay.

4.2 Short periodic events

The following analysis will, unlike the previous analysis, concentrate upon the short periodic waves, i.e. what is better known as the Kelvin wave pattern.

Identification of Kelvin wave system with angle between wave gauges and theoretical propagation angle.

For all tests in the previously described test series, a clear pattern has emerged. High frequency waves emerge first at the outside wave gauge, with a phase shift to the next, and further phase shift to the third and so forth.

When the low frequency waves of the time series are filtered out, plots such as Figure 21 emerge. In this plot the starting point for each high frequency wave train can be identified and the phase shift can be linked to the Kelvin wave pattern, to check whether or not these high frequency waves really are the Kelvin wave system. To do this the propagation angles of the Kelvin waves (and the Havelock waves for shallow water) are necessary.

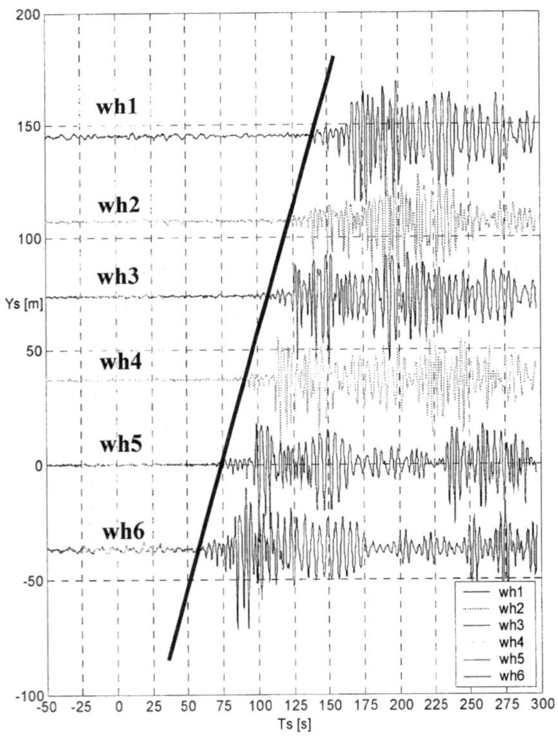

Figure 21: High frequency waves. The line illustrates time lag between gauges.

For deep water the Kelvin wave system can be described as illustrated in Figure 22. The transverse waves will travel at the same velocity as the ship and have a wavelength of

$$L_w = \frac{2\pi U^2}{g}$$

whereas the divergent waves will have a wavelength and velocity as

$$L_d = \frac{2\pi U^2}{g}\cos^2\theta, \qquad U_d = U\cos\theta$$

Transverse and divergent waves meet at the cusp line, which has an angle of φ = 19.27°. At this point the waves will propagate at an angle of θ = 35.16° relative to the transverse direction. Only the waves at the cusp line with this propagation angle will be used in the following analysis, as these have the highest amplitude and will be most visible in the wave profiles. Figure 22 shows the Kelvin wave pattern as an idealized model of the actual wave pattern, as wave systems will be generated at bow, shoulder and so forth of the ship. However, for this analysis the idealized model should be sufficient for deep water.

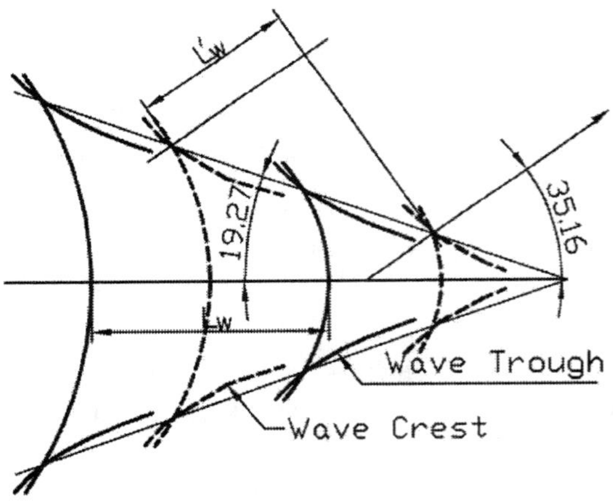

Figure 22: Kelvin wave pattern

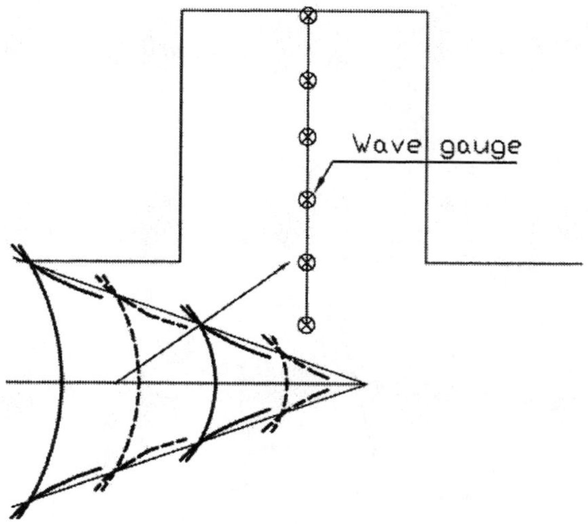

Figure 23: Kelvin wave close to bay

The first divergent wave at the outer wave gauge should be expected when the cusp line (19.27° from the ship centreline) reaches the wave gauge

$$t = \frac{Y_i}{U \tan \varphi}$$

where Y_i is distance from the ship track to the wave gauge and θ is the cusp angle. t = 0 refers to F.P. at the longitudinal position of the wave gauges.

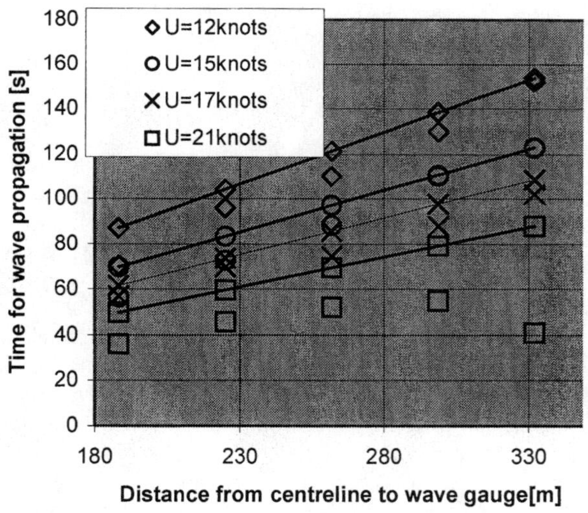

Figure 24: Measured and theoretical wave propagation in the bay:

Figure 24 shows the theoretical and measured time for the propagation of the divergent Kelvin wave, estimated using the equation for t above. Time equal to 0 is when F.P. passes through the transverse position of the wave gauges, and the x-position is related to the distance between the ship centerline and its distance to the bank. Each symbol in Figure 24 relates to a different ship speed, and the points connected to lines are the theoretical values.

As can be seen from Figure 24, the measured propagation times for the divergent waves correspond quite well to the calculated ones. It should not be expected that they match completely, as it can be difficult to determine the exact time when the Kelvin system reaches the wave gauges, even using filtered plots.

The agreement between measured and theoretical wave propagation for the highest velocity is worse than that for the lower velocities. The water depth in the above tests was 20 m in full scale. Therefore the depth Froude numbers are 0.36, 0.55, 0.62 and 0.77. Theoretically, the waves should begin to "feel" the bottom at $F_h = 0.57$, but in reality the effect is not visible before $F_h \cong 0.7$. Therefore the highest velocity (F_h) requires different theory to predict the propagation. For shallow water this is known as Havelock wave formulation, see for instance [Kostyukov, 1959].

$$U = \sqrt{\frac{g}{k}\tanh(kD)}, \quad \cos^2\theta = \frac{8(1-n)}{(3-n)^2}, \quad n = \frac{2kD}{\sinh(2kD)}$$

The cusp angle increases steeply from 19.27° to 90° (at F_h = 1.0), after which it will decrease again. Using the above equation the cusp angle can be calculated (for depth Froude numbers up to 1, where k is the wave number).

The highest Froude number in Figure 24 (U = 21 knots, H = 20 m) corresponds to θ = 28.3°. Once more using the equation above with a cusp angle of θ = 28.3° for U = 21knots, Figure 24 can be corrected to Figure 25.

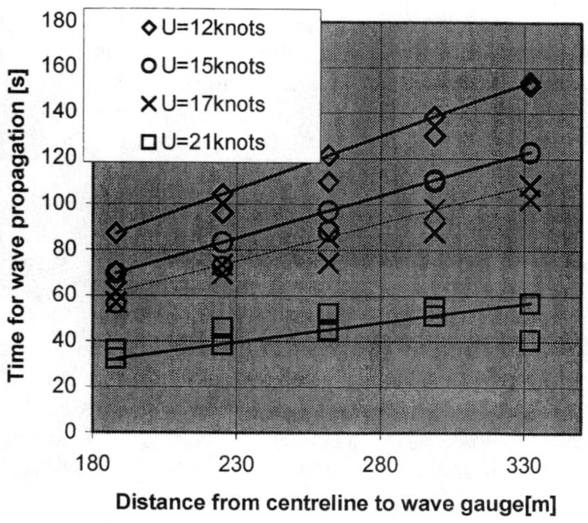

Figure 25: Measured and theoretical Kelvin angles corrected for shallow water.

This figure shows much better agreement for U = 21 knots, than if no shallow water effects were included. However, using the shallow water equations as in the above equation for cases when F_h is closer to unity can give erroneous results, as the crest is not infinitely long. Therefore, as the cusp angle increases, the length of the crest gains in importance.

Several comparisons have been conducted in the same manner as illustrated in Figure 25. For each velocity the average distance to the theoretical value has been calculated, and the results can be seen in Table 3 where H is water depth and Dist is distance from ship track to bank side. U = 21 knots and D = 20 m has been calculated both with correction for shallow water and with no correction.

D [m]	Dist [m]	U = 12	U = 15	U = 17	U = 21
20	75	16,0	5,4	5,0	4,7/14,3
20	150	19,3	16,4	8,9	4,7/21,7
20	225	9,7	4,7	8,5	5,2/52,4
30	75	12,4	8,0	6,8	5,7
30	150	8,2	8,9	5,0	5,2
30	225	10,0	8,2	6,1	4,3

Table 3: Average time difference between predicted and measured wave propagation.

Clearly, the prediction is worst for shallow water with no correction, which is not surprising as it is theoretically wrong to use deep-water Kelvin waves for this condition. It is also apparent from Table 3 that 12 knots seems to produce the poorest comparison. This is probably due to the Kelvin wave system having a very small amplitude making it difficult to identify the exact time at which the Kelvin waves reach the wave gauges. Generally the prediction is close to the measured values. In other words, it is quite possible with this simple technique to predict when the Kelvin waves reach a point in a bay. It should, however, be noted that if the bay geometry and bottom profile is more complex a refinement is necessary.

Another interesting observation is the apparent reflection and amplification of the Kelvin wave system. Figure 26 shows an example of this. In Figure 26 the recordings are numbered as wave events 1 to 11. Event 1 to 6 corresponds to measurements from wave gauge 6 to 1 when the waves are traveling into the bay. Event 7 to 11 corresponds to measurements from wave gauge 1 to 6 when the reflected waves travel from the innermost part of the bay and out. This means that line A shows the propagation of the divergent waves into the bay, as analyzed previously, and line B is the reflection from the back wall of the bay. The angle of the lines representing the phase shift should be expected to be the same moving into the bay as moving out (i.e. the propagation speed in the transverse direction is the same irrespective of the reflections). Positive interference can also be observed at the innermost wave gauge, as a group of divergent waves moves both in and back through this wave gauge (reflected on the back wall). The amplitude here is almost twice as large as at the previous wave gauge, and in fact almost as large as that much closer to the ship.

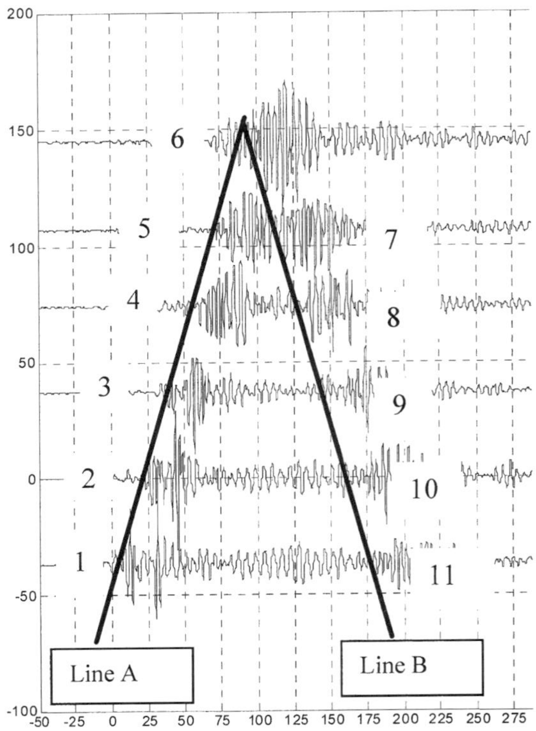

Figure 26: Filtered wave profiles at 17 knots and 20m water depth.

This is illustrated further by Figure 27, where the wave amplitude has been extracted along the lines shown in Figure 26. Wave event number 6 shows the wave amplitude a little after the intersection of the lines in Figure 26 (i.e. closest to the inside wall of the bay), number 7 is the next wave event when the waves are moving out of the channel again.

Except for the innermost wave gauge the amplitude is decreasing, as should be expected for a decaying Kelvin divergent wave. The innermost wave gauge shows a sharp increase in wave height, and no other explanation for this exists except positive interference from waves moving towards and away from the wall. This phenomenon has been found for most other test series. It should, however, be noted that compared to a real bay (sloping banks, more complex shore line) this effect is probably exaggerated. At least it will be more difficult to identify the positive interference for a real bay.

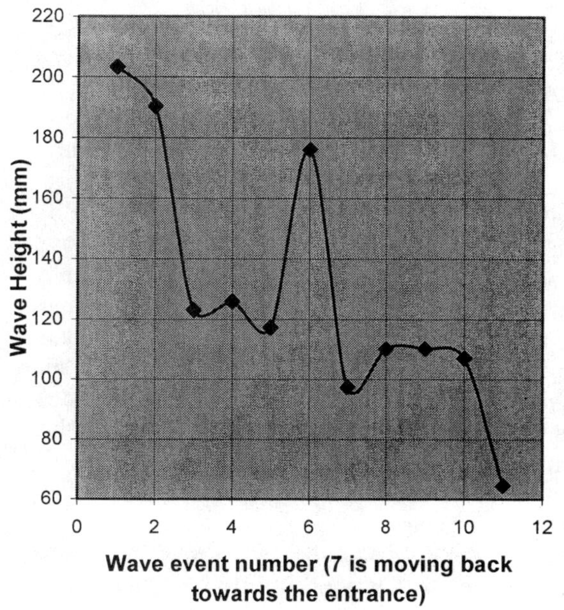

Figure 27: Wave amplitude along Kelvin wave front.

5 Conclusions

When a ship is operating in restricted waters the wake wash is significantly different from the one that occurs in deep and open water. For a relatively large ship like a ropax, drawdown caused by the Bernoulli wave seems to be a worse problem than ordinary wash waves. From the model tests it has been seen that when a ship is passing a bay at a relatively short distance, the drawdown is remarkably larger in the inner part of the bay than at the entrance. Also, Kelvin waves are very high in the innermost part of the bay due to amplification by reflections. When the ship is passing at a longer distance there is still some drawdown, but the difference in the inner and outer part is much smaller than when the ship is passing close to the bay.

The water depth in shallow water plays a very important role for the amplification of Bernoulli waves. Bernoulli waves are already amplified in shallow and open waters. If the vessel is travelling near another construction that forms a further restriction in the water, the Bernoulli waves are even more amplified: in the case of a ship passing near a bank, the Bernoulli wave is more amplified along the bank side than on the outside.

Seiches could also be detected in the bay. While the drawdown is interpreted as a process of forced excitation and oscillation, caused by the Bernoulli pressure of a vessel traveling near the bay, the seiche is the free oscillation of water surface in the bay that follows after the drawdown. The drawdown is always found to be present, but the seiche does not always occur in all the cases tested for this particular bay. Furthermore, the seiche amplitude is found to be smaller than that of the drawdown and rapidly damped.

The period and the amplitude of the bay oscillation caused by a ship passage are complicated and influenced by the topography of the bay, water depth, and ship speed, displacement and distance from the bay. Overall, the phenomena that occurs in a bay when a ship is passing nearby are difficult to explain, but in this paper a serious attempt has been made.

6 References

AT Ippen, "Estuary and coastline hydrodynamics", McGraw-Hill Book Co., 1966.

Mei, C.C. 1983. "*The applied dynamics of ocean surface waves*", Wiley, New York.

Li, D-Q. 2000. "*Experiments on Bank Effects under Extreme Conditions*". SSPA Research Report No. 113, SSPA Sweden AB, Göteborg.

Granath, L. 1992, "*Farledsstränders erosionskänslighet-inventering av strandtyper och skador i Stockholms skärgård (Erosion sensitivity of fairway shores- inventory of types of shores and damages in the Stockholm archipelago)*", Report 1992:10, Länsstyrelsen i Stockholms län. (In Swedish)

Allenström et al, "*The Interaction of Large and High-Speed Vessels with the Environment in Archipelagos-Final Report*" SSPA Research Report No. 122, ISBN-nr.: 91-86532-35-9, ISSN-nr.: 0282-5805, January 2003.

Janson C.E., 2002, "*Bernoulli waves in deep and Shallow Water at Low Speed*, Ship Technology Research, Vol. 49, No. 3

Kostyukov, A. A.; Theory of Ship Waves and Wave Resistance, E.C.I., Iowa, 1959

DISCUSSION

Lars Bergdahl, Visitor

The authors present most interesting results from extensive wake wash measurements carried out with a model of a ROPAX ship in shallow water passing close to a simplified bay. It is concluded that under such conditions the drawdown in the bay cased by the Bernoulli wave can be a worse problem than the more discussed, ordinary wash waves (Kelvin waves). By separating the phenomena by signal filtering into the long-period drawdown and bay resonance and the shorter-period wash waves, the authors are able to carry through clear and enlightening discussions of each of the two phenomena.

By comparing the three cases in shallow water, (1) with open water to both sides, (2) restricted by a quay on one side, and (3) the more complicated case with a quay with a square "bay," the different phenomena and causes are clearly highlighted, and credible conclusions concerning the drawdown, resonance, and their amplification under various circumstances are drawn. This understanding is a first step toward giving the ship operators some tools to assess these important problems beforehand.

The commenter would also like to point out that in the future such well-performed, idealized experiments are of utmost importance for validation of the accuracy and applicability of mathematical procedures being developed at Chalmers (in cooperation with SSPA) and elsewhere (e.g., Danish Hydraulic Institute/Denmark's Technical University, MARIN/Delft Hydraulics, and Duisburg).

Such developments aim mostly at coupling ship resistance, stationary (CFD) codes, such as SHIPFLOW, for the near-field wake to wave propagation models for the far field. Computer-effective wave propagation models are mostly of the Boussinesq type, which is suitable for shallow water. Using a linear slender-ship approximation for estimating the wave generation instead of a Boussinesq-type model (Jiang 2000, Jiang et al 2002), effects of acceleration, deceleration, maneuvering, and changes in the bathymetry of the fairway can be incorporated, but the near field is not as accurately predicted as with a panel code.

The graphs below show an example computation by Claes Eskilsson in the discusser's research group of the water-level changes in the reported experiments for Gauge 1 and Gauge 6. This computation uses a simplified sink source model of the ship combined with a fast spectral element method for the wave propagation. As can be seen, only the long-period waves are well reproduced, but using high-order Boussinesq models, such calculations will include also the Kelvin waves in shallow and not too deep water.

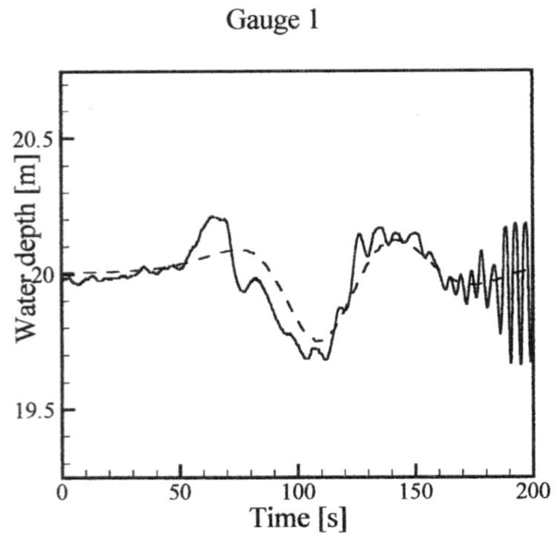

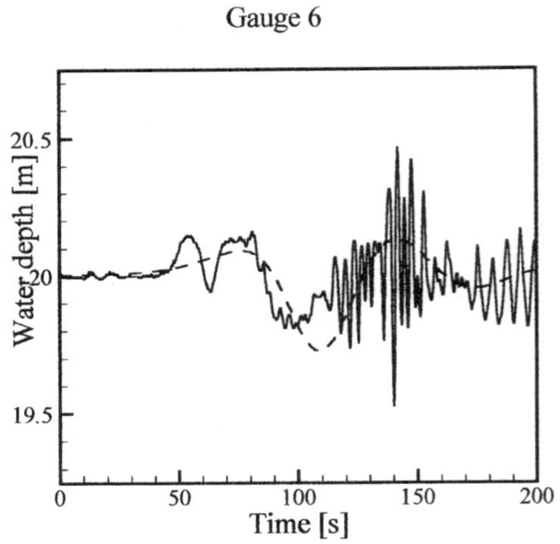

Drawdown in a bay caused by a passing ship. The distance to the coastline is 150 m, and the ship's velocity is 17 knots. The bow of the ship reaches the bay at t = 75.6 seconds. Measured (solid) and computed (dashed) water depth (Fig. 3.16 in Allenström et al 2003).

Additional references

Aage, C., Bell, A., Bergdahl, L., Blume, A. (Chair), et al. 2003 *Guidelines for Managing Wake Wash from High-Speed Vessels*, PIANC Working Group, MARCOM WG 41.

Allenström, B., et al. 2003 The interaction of large and high-speed vessels with the environment in archipelagos—final report, SSPA Research Report No. 122, Göteborg, Sweden.

Jiang, T. 2000 Investigation of waves generated by ships in shallow water, in *Twenty-Second Symposium on Naval Hydrodynamics,* The National Academies Press, Washington, D.C.

Jiang, T., Henn, R., and Sharma, S. D. 2002 Wash waves generated by ships moving on fairways of varying topography, *Proceedings*, 24th Symposium on Naval Hydrodynamics, July 8–13, Fukuoka, Japan.

Stan Stumbo, Member

The authors are to be congratulated on an excellent and important paper. By creating towing tank conditions where drawdown is a worse problem than Kelvin waves, then isolating the phenomenon, the authors have taken the first steps toward methods of analysis of drawdown, seiches, and their amplification when ships pass bays in confined waters. Even though the geometry of the bay is necessarily simplified, the authors shed some light on the subject and give others in the industry some tools to recognize conditions under which these amplified drawdowns and seiches may occur, and thus the ability to warn and train owners and operators to effectively manage these heretofore little-understood aspects of wake wash.

The commenter has observed drawdowns due to the passage of high-speed ferries through restricted channels where there are no well-defined bays and where there is a gradual slope to the shoreline. In these cases, the vessels were traveling at supercritical speeds (depth Froude number greater than 1.0). Although these drawdowns had a much lower amplitude and shorter period than those found in the authors' study, it is hoped that the authors will extend their analysis methods to these and other varying conditions, so to continue to advance the understanding of all aspects of wake wash phenomena.

Tow Simulation

Peter Lacey, LM, ExxonMobil Development Company
Derek Hee, AM, ExxonMobil Development Company
Henry Chen, M, Ocean Systems, Inc.
Vince Cardone, Visitor, OceanWeather Inc.

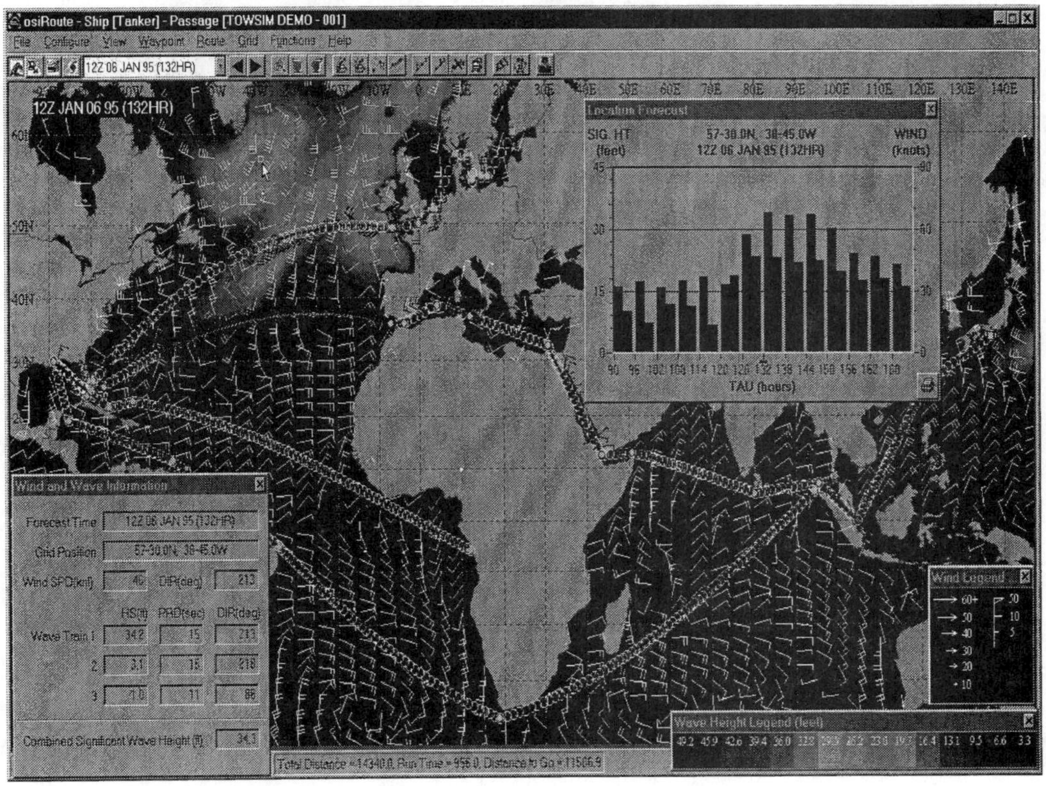

ABSTRACT

Transoceanic tows are undertaken by most deep-water development projects. These tows are necessary since the offshore structures to be towed are often built thousands of miles from the sites of operation. A typical tow configuration uses several ocean-going tugs to move a barge mounted structure at relatively slow speed. A tow from the Far East to the Gulf of Mexico may take three months, and be subjected to a variety of wave heights, wind speeds, and current velocities. Part of the development process is to develop rational tow criteria associated with each tow.

A standard approach to develop tow criteria starts with consideration of metocean route statistics, characterized as probability of encountering a certain wave height or wind speed. While the industry's record of safe tows is good, metocean route descriptions have traditionally relied on sources of environmental data of varying quality and extent. Metocean specialists are essential in deriving a high quality tow criteria. Over the last two decades, advances in hindcasting—the precise description of seastates at discrete times and locations in the past—together with the availability of low cost, high speed computers presents the opportunity to partially automate the tow criteria development process.

ExxonMobil Development Company, in conjunction with Ocean Systems Inc., and OceanWeather Inc., has developed the TowSim computer program, an automated approach that gathers and displays metocean route and response statistics. A TowSim user may easily define and run a thirty-year tow route simulation, literally with just a few mouse clicks. The worldwide metocean hindcast integrated within TowSim is GROW2000, which depicts wind and waves every six hours for a recent thirty year period at a grid spacing of 0.625 by 1.25 degrees. The metocean statistics resulting from a TowSim simulation may be reviewed in a variety of ways using TowSim's post-processing program SimStat. If the user models the tow hull form, then response predictions based on linear ship motion theory may be computed at each step along the route, and similarly compiled and reviewed.

ExxonMobil has used TowSim for the last two years, and has found it a useful tool in developing tow criteria. It provides a solid product for use by the metocean specialists charged with developing the tow criteria, and serves to establish a consistent basis for evaluating tow risk across projects. This paper describes the TowSim program, the underlying hindcast database, its validation with past tows, its support of ongoing projects, its use in solving other problems, and its potential for enhancement.

INTRODUCTION

Floating Production Storage & Offloading vessels (FPSOs), Tension Leg Platforms (TLPs), and Deep Draft Caisson Vessels (DDCVs), are all examples of floating structures currently used by the offshore industry to develop offshore gas and oil reservoirs. As the industry moves into ever increasing water depths, the economics of scale require these concepts to become relatively large. Similarly, large reservoirs are the most attractive prizes, starting at one billion barrels of recoverable oil, and with production rates of 200,000 barrels of oil per day and higher. Such development projects typically require several large structures, together weighing literally hundreds of thousands of tons. The resulting structural designs tend to be a hybrid of ship details and offshore specific details. Examples of offshore specific details are flare towers, production equipment foundations, and riser porches.

To manage project economics and execution risk, the offshore industry trend is to rely on proven ship design and construction methods, augmented as needed by first principles engineering. Deep-water oil and gas fields are located world wide—in most of the world's oceans—while shipbuilding centers tend to congregate in discrete locations. Therefore, blue water ocean tows of large, valuable structures are an increasing and unavoidable factor in offshore project development. In fact, most project execution plans require multiple tows: different fabrication yards may be used for hull construction, production equipment assembly, and integration, with all the associated inter-yard tows and eventual tows out to site for installation.

The offshore industry's record of successful tows is good. Practical mitigation measures can be used to minimize tow risk. For instance, the structure can be ballasted for the tow to minimize primary stresses; weather forecasting can be used during the tow to avoid bad weather; and safe havens can be identified along the intended route for contingencies when bad weather can not be avoided. Also, using a ship design basis for hull scantlings—based on North Atlantic sea states—provides an additional safety margin. Nevertheless, unavoidable tow parameters may expose the tow to environmental risks not typically experienced by ships at sea. For instance, the development execution schedule may require tows during unfavorable months of the year. The structure may require load out of various large pieces on a barge, which require internal shoring for deadweight support, and may expose structural projections to wave damage. A DDCV may require a wet tow on its side, or a multiple section dry tow on several barges. For structures designed to operate in benign seas, the largest loads experienced by the structure may be during the tow.

Therefore, sound environmental tow criteria are required to assess the acceptability of a proposed tow, as well as approve a specific tow at a particular time. A significant amount of skill is required to develop the criteria. ExxonMobil project teams rely on metocean specialists to craft route-specific environmental criteria. The criteria normally include the identification of the maximum wave heights and wind speeds likely to be encountered by the tow sailing on a specified route. To extend the applicability of the criteria, the variability of the maximums are presented as a function of tow departure month, thereby capturing seasonal effects in case the execution schedule accelerates or slips. Different routes are usually considered.

The metocean specialists use the best metocean information available for the routes in question, often a combination of several hindcast databases. Extreme events, such as tropical cyclones, are usually considered separately. A detailed account of the ExxonMobil criteria development process can be found in (Vermersch 1985), which addresses the Trans-Pacific tow of the Harmony and Heritage jacket structures from South Korea to Southern California.

Once the criteria are available, it may be used as input to response based calculations, the results of which can be used to assess tow risk. For the Harmony and Heritage jacket tows, structural fatigue calculations were performed based on predicted barge motions during the tow. As a result, it was necessary to internally brace the jacket structure to reduce the predicted fatigue damage.

Over the past two decades, the accuracy of hindcasts and forecasts has increased (Cox & Cardone, 2000) and (Cardone, Cox, & Swail, 2001). Together with the increase in low-cost computer power, many opportunities arise to use this data which were not available in the past, some of which are described in (Lacey, Chen, & Cardone, 1993) and (Lacey & Chen, 1995). However, the focus of this paper is the opportunity taken to partially automate the tow criteria development process by combining a worldwide, thirty-year hindcast database (GROW2000,) with an existing ship routing program. This synthesis resulted in the TowSim computer program, which has been used by the ExxonMobil Development Company for the last two years.

TowSim allows the user to easily define the tow route between departure and arrivals ports and perform a real time simulation along this route. The thirty-year simulation will typically start at the departure port at midnight on January 1, 1970, and will proceed along the route in six-hour increments. At each time increment, the wind and wave conditions from the hindcast database are noted and saved. After reaching the arrival port, the tow simulation starts again, but six hours later, at 0600 on January 1, 1970. This process is repeated in six-hour increments across thirty years. A five knot tow on a route that goes half way around the world, simulated with departures every six hours for thirty years, takes slightly less than seventy two computer hours to simulate on ExxonMobil's TowSim personal computer.

TowSim output includes wave and wind maximums for the route, six-hour statistics, and mean directional wave spectra. Also embedded in the TowSim code is a six-degree of freedom linear ship motion program adapted from the U.S. Navy's *Ship Motion Program*. Therefore, motions, accelerations, and seakeeping events may also be calculated during the simulation for those hull forms for which linear ship motion theory is applicable. As a future enhancement, for those hull forms for which strip theory does not apply—i.e. most offshore structures—an input subroutine could be added to read response amplitude operators derived from a 3D diffraction program.

ExxonMobil Development Company (EMDC) has found TowSim a useful tool in performing many 'what-if' tow scenarios in support of several ongoing projects. It provides a solid product to the metocean specialists charged with developing tow criteria, and serves to establish a consistent basis for evaluating tow risk across projects. This paper describes the TowSim program, the underlying hindcast database, its validation with past tows, its support of ongoing projects, its use in solving other problems, and its potential for enhancement.

TOWSIM

TowSim is a computer simulation system that recreates voyage conditions encountered during a transoceanic tow. By running multiple simulations of a tow route based on specified vessel characteristics, across thirty years of historical hindcast wind and waves data, statistics on environmental conditions and vessel responses can be accumulated for tows starting in different months. The resulting statistics in terms of joint probability density functions and exceedance probabilities can provide a consistent, comprehensive, and flexible way to develop design criteria. This tool allows the user to quickly screen many 'what if" tow scenarios, and ultimately identify the optimum tow configuration.

TowSim consists of a special version of Ocean Systems' Inc. Shipboard Weather Routing System (osiRoute) and a post processor computer program (SimStat) for statistically compiling the simulation output. The optimum ship routing program was developed for ship captains to plan transoceanic passages and to minimize heavy weather damage and fuel consumption while meeting schedule commitments. The system uses the twice daily long range wind and wave forecast produced by OceanWeather Inc. (OWI) to predict vessel speed and motion responses and to optimize voyages. The benefits of this system have been demonstrated over the last fifteen years by a variety of large ship types sailing on the high seas. Ship types include container ships, crude oil tankers—including VLCCs and ULCCs—as well as U.S. Navy combatants.

In order to model the deep ocean tow, TowSim uses a state-of-the-art high-resolution global historical wind and wave database: Global Re-analysis of Ocean Waves 2000 (GROW2000,) produced by OWI. Archived surface pressure charts, tropical cyclone tracks and intensities, buoy measurements, ship reports and satellite altimeter data over the past thirty years have been used in recreating the global ocean wind and wave environment through a unique hindcast process. The resulting massive database consists of wind and wave data every six hours on a global high-resolution grid: 0.625 degrees of latitude by 1.25 degrees of longitude. One degree of latitude is roughly sixty nautical miles. Uncompressed, this database would occupy over 125 Gigabytes—a data stream more than 25,000 km long.

TowSim's use of the GROW2000 database allows the user to recreate transoceanic tows starting any time after midnight on January 1, 1970. By repeatedly simulating a voyage starting several times a day, for each day across thirty years, tow statistics can be established with minimum sampling error.

The TowSim program, including the GROW2000 database, resides on a powerful PC-based application server, with a Xenon P4 processor and large amounts of memory and disk space. The tow route information is easily entered by mouse clicks on a digital map. The exact waypoint positions and speed profiles can be further refined by zooming in and moving a waypoint with the mouse, or by direct keyboard entry. Great circle or rhumb line sailing can be selectively modeled between each waypoint. The program automatically computes the heading, distance and duration for each leg. Figure 1 shows a typical input screen.

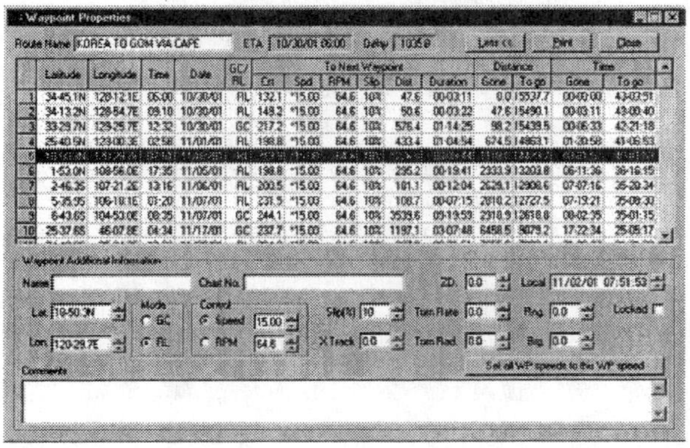

Figure 1. Typical Tow Route Definition Screen

Once the simulation starts, the vessel sails along the prescribed route at the user defined speed, from the departure to arrival port. While enroute, the environmental conditions—wind and waves—are retrieved from the database according to dead-reckoned positions every six hours, as shown in Figure 2.

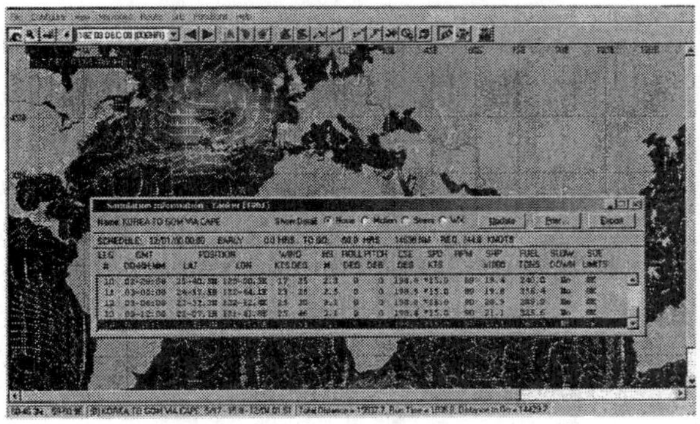

Figure 2. Response Every Six Hours Along Route

For each simulation, all of the environmental conditions experienced enroute are saved for later statistical analyses. Vessel responses, if requested, are also saved. The program also saves other pertinent data including voyage duration, maximum wind and wave height encountered, average speed, and fuel consumption.

The TowSim program can be used to establish preliminary tow criteria based only on environmental conditions, i.e. not considering vessel response effects. However, it does have the capability to estimate heavy weather speed reduction given user prescribed operational response limits, such as maximum roll angle, acceleration, engine limitations, etc... These operational limits together define the Safe Operating Envelope. Figure 3 shows an example of a user selected Safe Operating Envelope.

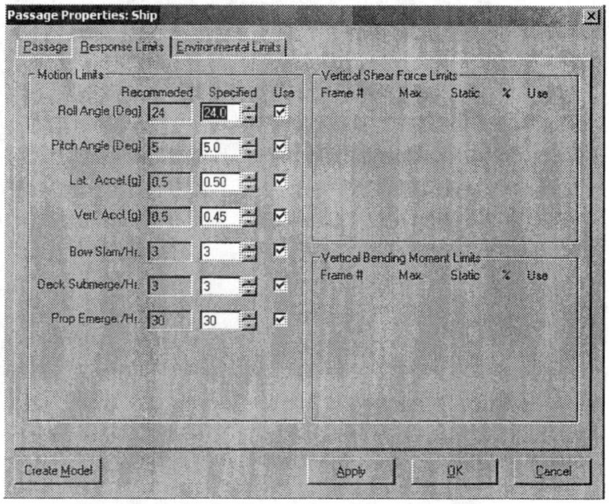

Figure 3. Safe Operating Envelope Screen

To properly model vessel responses, the following technical information for the vessel is required:

1. General Arrangement
2. Hydrostatics
3. Body Plan
4. Bilge Keel Details, if any
5. Engine Specifications
6. Propeller Curve

SimStat: Managing TowSim Output Statistics

The SimStat program compiles the huge volume of output data into meaningful statistics and displays the results in plots and tables. It reads data saved during the simulation, and compiles statistics into three different data sets: (Passage Maximums,) 6-Hourly (enroute conditions), and (Mean Directional) Wave Spectra. The user may select individual months and years to compare the results for monthly and yearly variability.

SimStat Viewer

Figure 4 shows an example of the SimStat opening screen. Here the user can select the project and route name. Radio buttons exist to order the output of Passage Max, 6-Hourly, and Wave Spectra statistics. If simulations have already been performed, a list of existing statistics will appear in the left window, as shown here. To view previously compiled statistics, the user simply double clicks the item name on the list shown on the left side of the window in Figure 4.

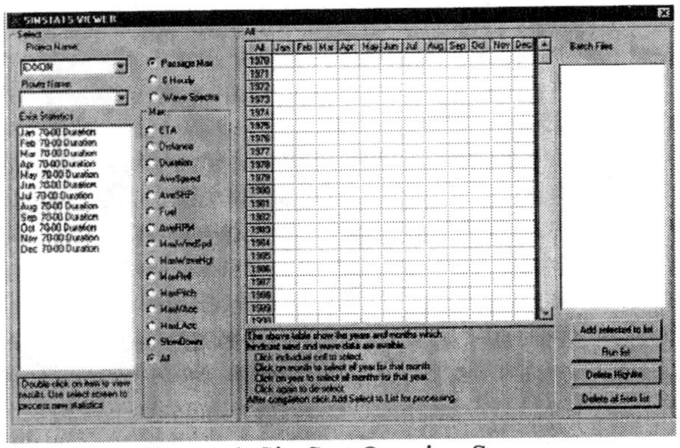

Figure 4. SimStat Opening Screen

To request new statistics, the user selects the type: Passage Max, 6-Hourly, or Wave Spectra, and then selects the month and year by clicking on the cells in the middle window. Clicking the **Run List** button will then start the statistical compilation. As each item requested is completed, it will appear in the left window for later selecting and viewing.

Exceedance Probability of Voyage Maximums

When the **Passage Max** radio button is selected, double clicking the significant wave height or duration existing statistic will show a probability log plot similar to that shown in Figure 5.

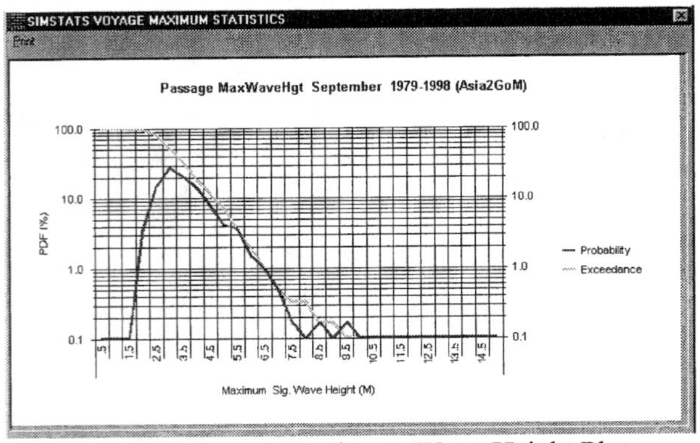

Figure 5. Passage Maximum Wave Height Plot

The Figure 5 plot contains percent probability plotted against significant wave height in a log-linear fashion. Both the probability (% occurrences) and the exceedance (1–cumulative probability) are shown. Figure 6 shows the corresponding wind speed maximum plot with tabulated values. The table in Figure 6 shows the number of observations in each data range, the probability density, cumulative probability and the exceedance probability.

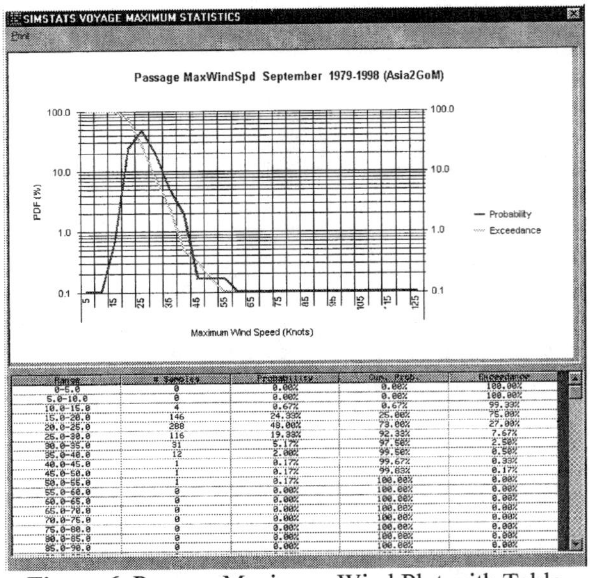

Figure 6. Passage Maximum Wind Plot with Table

Scatter Diagram and Joint Probability Distribution

By selecting the 6-Hourly radio button, scatter diagrams can be generated based on the user chosen duration—month, year, or both.

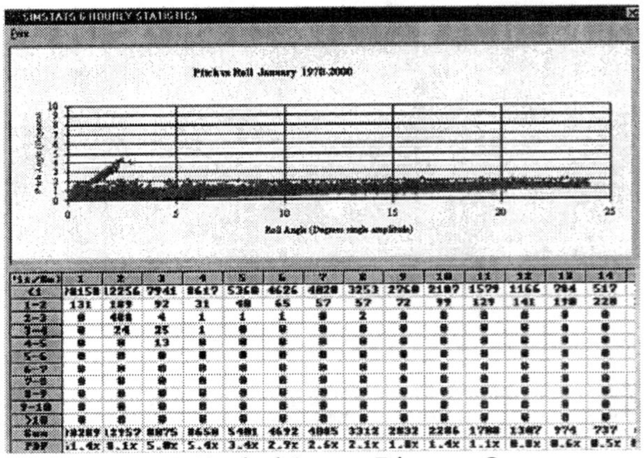

Figure 7. Typical Scatter Diagram Output

Figure 7 shows a typical scatter diagram for the selected pair of parameters: pitch versus roll angle. The table below this plot depicts the joint probability density function and number of observations in each cell range. The row sum and column sum depicts the probability density for each variable.

Note the original directional spectra have been parameterized into up to three wave trains with individual significant height, peak period and direction (Hi, Ti, and Di). Hs is the combined significant wave height and Tp is the highest of the three peak periods. Wind speed is defined as the ten-minute average at the reference height of ten meters above sea level.

Mean Directional Wave Spectra

The thirty-year hindcast uses 46,529 grid points to model the earth's ocean. For each grid point, the raw hindcast data consists of a wave energy matrix composed of wave energy resolved in 24 directions (every 15 degrees) and 23 frequencies per direction. It would require about 72,000 gigabytes of computer storage to save this amount of data. To reduce computer storage requirements to a practical level, the 24x23-wave matrix was reduced to nine parameters—significant wave height, period and direction for the three dominant wave systems—using a 'splitter' program. TowSim then recreates the directional spectra from those nine parameters at six-hourly intervals along the specified route.

For a specified voyage simulation, the directional spectra are sorted into one-meter significant wave height intervals, and the average of the 24x23-wave energy matrix is computed for each wave height group. Figure 8 shows a plot of the spectrum for the 7-8 meter significant wave height group.

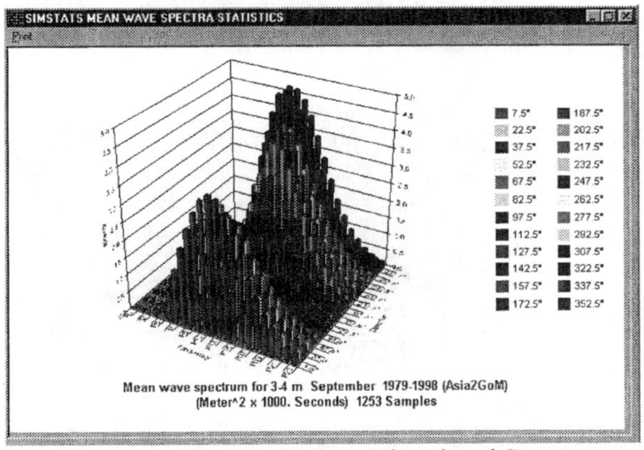

Figure 8. Recreated Mean Directional Spectra

Under the assumptions of linear ship motion theory, the mean square value of the responses is the same whether based on the mean directional spectrum or on the responses individually computed and averaged. The number of samples in each wave height group, divided by the total number of samples, provides the percent occurrence over the many years of simulation. A detailed proof of this assertion is provided in Appendix A.

METOCEAN HINDCAST

Until recently, platform survival type design criteria were typically developed from hindcasts of storm populations drawn from a 20-50 year period of history, and of the type responsible for extreme loads in a given area. However, operational criteria were derived from hindcasts of a few continuous years. Many Joint Industry Projects (JIPs) structured about this approach have been carried out by OceanWeather within the past decade, and ExxonMobil has participated in nearly all of those JIPs (e.g. GUMSHOE, WINX, WAX, SEAMOS, SIMOS, RASMOS, BOMOS.) In recent years, OceanWeather has merged these two approaches and developed long-term hindcast databases for whole basins or the globe itself. The first of the global databases, named Global Re-analysis of Ocean Waves (GROW) spanned 40 years, and could be used to provide at least preliminary extreme and operational criteria at an arbitrary location. Such a data base is also ideal for simulation of a wide range of offshore operations including assessment of offshore tows. GROW is described by (Cox and Swail, 1999). Therefore, it was natural for ExxonMobil to request OceanWeather to provide an updated version of GROW for TowSim.

The GROW data base (predecessor to GROW2000) was generated in early 1998 on a global grid using the well proven ODGP2 wave model. The grid spacing was 1.25 degrees latitude by 2.5 degrees longitude. Winds were specified directly from the new NOAA/NCAR 40-year (1958-1997) Global Reanalysis Project (NRA.) Unfortunately, NRA wind fields do not resolve tropical cyclones and there are biases in extra-tropical storms; nevertheless, GROW has been shown to provide reasonably accurate global wave statistics and estimates of recent trends in global wave climate. OWI has also used GROW successfully to develop initial estimates of site-specific extremes and normal statistics. Where local sea state measurements and satellite altimeter data are used to develop a local unbias factor for GROW time series, the statistics are remarkably close, in many areas, to those developed by detailed hindcast studies.

GROW was not considered suitable for TowSim because of its low spatial resolution, the afore-mentioned tendency of GROW to underspecify sea state peaks in intense extratropical cyclones, and the failure of the NRA input winds to resolve tropical cyclones. These issues were addressed, and corrected, for the North Atlantic Ocean in a 40-year hindcast project of the North Atlantic funded by Atmospheric Environment Service of Canada (AES40.) This work is documented by (Swail and Cox, 2000). In AES40, the wave model resolution was increased to 0.625° latitude by 0.833° longitude, and third generation (3G) wave model physics was used. The wind fields of all major storms were reanalyzed interactively using OWI's IOKA system, and all tropical cyclone winds were modeled using OWI's mesoscale model, but only after individual storm tracks and properties from raw source data were blended into the basin-wide wind fields.

OWI's experience in producing GROW and AES40, and the fruits of several ongoing research programs on marine wind fields accuracy, and accuracy of in-situ and satellite wind data, allowed it to formulate and carry out an update of GROW (henceforth GROW2000) which addressed the deficiencies and made the following enhancements:

1. Decreased the model grid spacing to 0.625 degree latitude by 1.25 degrees longitude for better resolution of tropical cyclones.
2. Removed the NRA winds bias using a set of regressions developed from global evaluation of NRA against adjusted, in-situ data, and corrected satellite scatterometer winds.
3. Specified the time and space evolution of the wind field about each tropical cyclone using OWI's proven mesoscale vortex model.

In GROW2000 the parameters for the tropical cyclones (i.e. the inputs to the mesoscale model) are taken from the best source available. Therefore, for virtually all storms in the greater North Atlantic basin, the inputs were developed (as in AES40) by analysis of raw source data by OWI analysts. The same is true for several hundred Western North Pacific tropical cyclones. In other areas, e.g. Indian Ocean, Australia etc..., OWI mainly used its newly developed algorithm to extract the OWI model parameters from standard historical sources (Global Tropical and Extra-tropical Cyclone CD-ROM) and products of major warning centers (e.g. Joint Typhoon Warning Center, NOAA National Hurricane Center, Australian Bureau of Meteorology, Royal Observatory Hong Kong).

By blending the tropical cyclone wind fields into the corrected time series of NRA winds before the wave hindcast is run, the output of GROW2000 naturally and correctly includes the wind fields and sea states associated with the tropical systems. Most importantly, this improvement includes the patterns of swell radiated into the far field, where the tow route may be affected.

The resulting product used for TowSim is a unified and complete database of global wind and directional spectra on six-hourly intervals for the thirty years presently available. Now that the underlying NRA database has been extended to 55 years (1945-2000), the TowSim 30-year database may be extended to 55 years as well.

Motion Simulation

For motion computations, the wave spectra are parameterized into three wave trains by picking three predominant wave heights, periods and directions while maintaining the total energy. During the simulation, TowSim transforms these nine parameters into directional wave spectra for use in motion prediction estimations. This approach provides the most efficient way to handle the massive amount of data generated by the directional wave spectra model. Experience in real-time ship routing applications shows that the motion predicted using this approach is more accurate than using significant wave height and period from scatter diagrams, which essentially ignores the presence of swells in a sea state.

TOWSIM VALIDATION

As part of the TowSim development process, OWI validated GROW2000 against all available in-situ and satellite data. Furthermore, ExxonMobil carried out extensive validation of the TowSim metocean database, by in-house personnel, and a third party. This validation was required before TowSim was deemed acceptable for use.

Comparisons were made between twelve points in the TowSim hindcast and twelve other points distributed throughout the world's ocean. The data points to which TowSim data was compared were either measurements from ocean weather buoys or points from regional hindcasts that used a tighter grid spacing than GROW2000. The results showed that the TowSim hindcast compared well with 'high confidence' data sources. In many areas, the design data statistics derived from the 30-year time series are unbiased with respect to design data produced by previously completed, intensive, basin-specific hindcast studies.

However, in some areas, especially near shore, a small residual local bias in the hindcast was noted. This bias appeared related to those locations where a tow route came within 40 nautical miles of the shore and where the predominant wave started to 'feel' the bottom. This shallow water bias probably occurs, because the TowSim hindcast does not account for local shoaling or shadowing effects. The bias may be reduced by applying linear regression-based corrections on significant wave heights derived from comparisons of the hindcast and local in-situ measurements or satellite altimeter data. Since TowSim's purpose is to examine deep ocean tows, this correction was not considered essential for the first program version. However, this local unbiasing algorithm may be added as a future enhancement.

Further comparisons between the TowSim metocean database and basin-specific hindcast studies show that the TowSim hindcast has an extreme event bias for some enclosed basins and areas subject to tropical cyclones. For instance, a bias shows up for tropical cyclones in the Gulf of Mexico when compared with the GUMSHOE hindcast. However, little bias is observed for SEAMOS, the South China Sea hindcast study. There are uncertainties in the nature of tropical cyclones and geographical location effects. As a result, ExxonMobil manually applies a bias correction for tows that transit tropical cyclone season areas. Embedding finer grid hindcasts into the TowSim hindcast, or regenerating the TowSim hindcast at a finer grid spacing, would capture tropical cyclone effects more accurately and reduce or eliminate the need to manually apply corrections.

Tow Criteria Comparison

Traditional tow route planners use the Global Wave Statistics (GWS) to determine the waves along a tow route. This database is compiled from decades of shipboard observations and describes large 'blocks' of ocean using scatter diagrams of wave height and period.

While many past studies have been carried out to remove the biases due to observation errors, these traditional wave statistics still contain two fundamental errors; 1. heavy weather bias, since ship captains systematically avoid regions of heavy weather, and 2. temporal and spatial correlation bias associated with extending an ocean area wave characterization over too large an ocean block.

Heavy weather bias is self evident, but an example of the temporal and spatial bias is instructive. One of the most severe areas of ocean is off the South African coast—Block 90 in the GWS. Block 90 extends from the coast to about 400 nautical miles offshore. A comparison using TowSim was run using a route that was 100 nautical miles off the coast and another route that was 250 nautical miles off the coast. The TowSim results showed the waves for the route closer to shore as less severe than the waves further offshore, which corresponds to experience. Running the same routes using the GWS hindcast database showed, incorrectly, that the same seas characterized the inshore and offshore routes.

Harmony and Heritage Jacket Trans-Pacific Tow Metocean Criteria

In the 1970's, ExxonMobil towed two large jackets across the Pacific Ocean, starting in Korea and ending in the waters of Southern California. The metocean tow criteria for this tow was published in (Vermersch 1982). For comparison, TowSim has since been run for the same route. TowSim statistics show lower winds and waves along the route during certain months of the year and slightly larger winds and waves during the summer months.

The tow criteria development procedure for these jackets used two different sources of data: 1. five years of the U.S.Navy's Global Spectral Ocean Wave Model (GSOWM 1985-1990), and 2. the Tropical Cyclone Archive 1945-1970. GSOWM is limited in that it could not adequately resolve tropical cyclones because of its coarse grid size—2.5x2.5 degrees. Also, a simplified parametric wave and wind model (Bretschneider 1972) was used to estimate basic cyclone parameters including wind and wave conditions based on archived central pressure, radius to maximum wind, and maximum wind speed.

For the normal wind and wave conditions estimated by the GSOWM, a given route is broken into several segments, each of which is characterized by an individual GSOWM grid point. The encounter probability of wave height or wind speed thresholds for individual segments are estimated by counting the arrival rate of the respective events during the period of five years. By treating the arrival of such events as a Poisson process, the probability of exceeding a specified threshold is

$$Pe = 1 - \exp(-\lambda \mathrm{T})$$

Where:

λ = the arrival rate of the event per unit time
T = the duration of exposure in each segment

Assuming each segment is independent, the overall encounter probability for n segments is

$$Pe = 1 - (1 - P1)(1 - P2)\Lambda\ (1 - Pn)$$
$$Pe = 1 - \exp\left[-\sum \lambda i \mathrm{T} i\right]n$$

For the tropical cyclone environments during the tow, a "pseudo simulation" approach was taken to estimate the encounter probability. First the departure dates were selected for each month. A search into the tropical cyclone database was performed to get the cyclone parameters that were then used to estimate the wind and significant wave heights. After many repeated simulations, a histogram of tow-maximum significant wave height and wind speed was constructed to estimate the risk of encountering tropical cyclone generated wind and wave thresholds.

The final estimate of the monthly risk of encountering both normal and tropical cyclone generated maximum wind and waves is given by:

$$Pe = 1 - (1 - Pn)(1 - Pt)$$

Where:

> Pn = estimated encounter probability due to normal seastate
> Pt = estimated encounter probability due to tropical cyclone generated seastate

The above approach assumes that the encounter probabilities of each segment are independent, and that the wind and waves generated during normal conditions and generated during tropical cyclones can be treated separately and later combined as two independent consecutive events. In reality, the complex temporal and spatial correlations of environmental conditions between different regions along a route make such quantification suspect or possibly even invalid.

Fortunately, the probability of encountering a specified threshold during a tow is bounded by the estimates based on assumptions of a series of perfectly correlated and a series of mutually independent environmental regions. See detailed discussion in Appendix B. These bounds are tight if the non-encounter probability in any of the regions is dominant, as in the case during tropical cyclone seasons. Furthermore, the independent assumptions yield conservative estimates of encounter probability of environmental threshold during a tow.

TowSim, on the other hand, does not make any of these limiting assumptions. The hindcast grid spacing is tight enough to resolve most of the tropical cyclone generated wind and waves, with future enhancements improving this resolution. Thirty years of 6-hourly synoptic conditions allows a realistic simulation from departure to arrival, anywhere in the globe. This "brute force" approach makes possible thousands of realization of tows from which meaningful statistics can be collected for tow criteria development.

In summary of the validation process, TowSim has shown that its results match well with weather buoy data and fine grid regional hindcasts. There are some areas where biases have been identified and corrections are applied. TowSim, with its fine grid resolution, can reduce some of the conservatism built into most present day metocean tow databases, thereby assisting in the development of more accurate tow criteria.

TOWSIM EXAMPLES

TowSim has been used to create metocean tow criteria for several tows originating from the major offshore fabrication areas—Gulf of Mexico, Korea, Japan, Singapore—to project sites. At least ten different tow criteria have been developed, and TowSim has simulated over fifty individual tows. Many of these runs have modeled "What if?" towing scenarios, needed to answer questions regarding speed or route changes. These type of questions arise especially during the early phases of a project. All of these questions can be addressed by running TowSim and making comparisons between tow runs.

Normal practice is to run a simulation for all months in the year, instead of targeting the projected month when the tow would start based on the project execution schedule. The advantage of this approach highlights a TowSim feature: the tow information can be archived and re-used later, without needing to re-run the simulation. SimStat can always be rerun on the archived tow information, independent of TowSim, should there be schedule slippage, or if other projects are examining the same routes. On the other hand, if a tow cannot occur because of ice, for example, TowSim can be run for the entire thirty years for only the ice-free months. Faster run times will occur with a reduction in the number of start months, or an increase in tow speed.

Since currents are not included in the hindcast, the tow route can be set up so that the tow speed can be manually increased or decreased for those segments of the route that encounter strong, well known, ocean currents, e.g. the Gulf Stream.

INTERESTING APPLICATIONS

TowSim provides a fundamental capability for which many possible applications exist. The attentive reader has already thought of the possibility of generating site specific metocean criteria by simulating a thirty-year circular voyage in one field. However, while quite skillful, the GROW2000 is not of sufficient quality or scope to provide site-specific detailed design criteria (Cardone 2001). Aside from this restriction, TowSim has been used for several interesting applications, not originally foreseen prior to development. Two such applications are described briefly. The benefits gained by each of these unexpected applications more than offset the TowSim development effort.

Example 1: Ship Conversion Fatigue Life

For ship conversion projects, it is very useful to know the seastates the prospective vessel has encountered in its lifetime. From the ship's deck log, or other sources, the departure and arrival locations and times will be known. It is a straightforward task then to model this voyage history in TowSim. As long as the ship is not older than thirty years, each voyage can be discretely modeled and individually simulated. The results may be displayed as a statistical record of the seastates encountered by the vessel, correct in terms of position and time, and limited only by the accuracy of the hindcast. As already mentioned, TowSim can organize and display the historical seastate descriptions in a variety of formats. This flexibility makes it possible to tune the output to the input needs of subsequent analysis programs.

One requirement in converting an existing ship to a moored floating storage vessel is to determine the amount of steel repair required to meet design requirements. An important step in this task is to estimate the amount of fatigue damage the ship has experienced to date, and add to this damage the amount likely to be experienced on site for the remainder of the vessel's design life.

Many different fatigue analysis methods exist, from simplistic to complex. For efficiency, a straightforward, conservative approach can be used with the fallback of reverting to more complex methods if the conservative approach yields unsatisfactory results. That is, an upper bound for fatigue life at hot spot locations can be sought, as opposed to solving for the most accurate, absolute fatigue damage.

One of the simplest (and usually most conservative) fatigue analysis approaches is provided by the Classification Society's hull scantling programs, such as Nauticus, SafeHull, ShipRight, and Veristar. These programs all share the same primary stress loading assumptions, grounded in North Atlantic seastates, and harmonized by the International Association of Classification Societies (IACS). Also, these programs' fatigue methods are calibrated for standard ship structures assuming a fatigue life of about 25 years. So modeling a 20 year plus aged ship in these programs would likely lead to the indication that hotspot details are near the end of their fatigue lives. This indication could be overly conservative given that the ship's trading history did not include North Atlantic, or similarly harsh, seastates.

Such a potentially misleading assessment can be corrected by using TowSim's voyage history seastate descriptions. These descriptions could be input into the Classification scantling program and used in place of the default. In fact, this process was performed for a recent conversion project. The TowSim analysis showed that the vessel actually sailed in seastates considerably more mild than the North Atlantic's. For instance, the maximum significant wave height regime encountered by the vessel during its trading history was 8 meters, while the North Atlantic maximum is 14.5 meters. Fatigue damage is exponentially related to loading, which is similarly related to wave height. So this reduction in wave height means the ship should have significantly less fatigue damage then indicated without correcting for actual voyage history climatology. Making this relatively straightforward adjustment to an existing Classification program, a better assessment of steel replacement amounts and locations was possible.

Also, it's interesting to consider Figure 9, which illustrates the remarkable differences between the IACS North Atlantic and Worldwide wave climates, and the TowSim waves experienced by the tanker in the above study. IACS wave climates were taken from Det Norske Veritas, Note 30.5. The Figure 9 graph is normalized by 100,000 at-sea hours of wave cycles at different encountering wave heights. The number of cycles is computed based on the significant wave height versus zero-crossing period (Hs-Tz) joint probability tables by taking the total number of observations in each cell multiplied by 3600 seconds and divided by the corresponding wave periods. Not only is the TowSim maximum wave height much less than predicted by IACS, there are a lot less cycles in the 4-8 meter range.

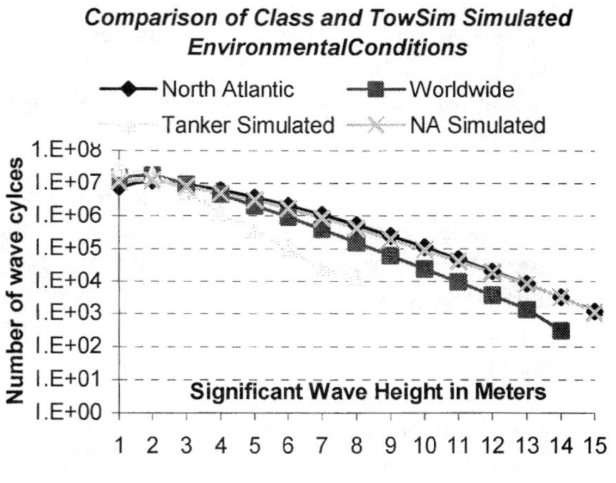

Figure 9. Comparison of Class and TowSim Simulated Environmental Conditions

The IACS North Atlantic wave climate is validated by a TowSim simulation of a typical North Atlantic route: Ambrose Light to Bishop Rock. See Figure 9 'North Atlantic' versus 'NA Simulated' curves, that overlay each other. However, the North Pacific and the Gulf of Alaska route are more severe than the North Atlantic, as indicated by direct measurements in (Lacey and Chen, 1995).

Also, the IACS wave depictions include periods up to, but not exceeding, 17.5 seconds, while the TowSim hindcast database includes wave periods up to 24 seconds. These differences could lead to significant differences in fatigue life estimations for large vessels or structures with longer resonance periods.

There was an additional benefit of using TowSim's underlying GROW2000 database for voyage history assessment. The site metocean environment was also based on the same fundamental technology as GROW2000, but with adjustments such as a tighter grid spacing, and inclusion of shallow water effects. These adjustments, and others done by OWI, increased the hindcast skill to a level necessary to establish the site detail design criteria. This consistency in metocean modeling is a significant analysis benefit.

It is important to note that the fatigue analysis procedure followed above was tuned to a set of variables associated with a particular project: it should not be followed in general. For fatigue methodology for new build projects, see (Kuo and Lacey, et.al., 2001).

Example 2: Storage Optimization

Transporting liquid cargo across the oceans in bulk usually requires a transportation network involving specialized ships, storage vessels, and processing facilities, the cost of which can run into billions of dollars. A challenging aspect of designing such a complex system is how to balance the number and size of ships and storage facilities to ensure a reliable year-round supply of cargo while maximizing the return on investment.

Larger, faster ships can carry more cargo and have better schedule reliability, but require higher initial investment. Smaller and less powerful ships are prone to weather delays, which may lead to larger storage facilities and more ships to achieve the same throughput.

Several sophisticated computer programs exist which can model such a transportation system, and solve for the optimum onshore and floating storage capacities. An input to such a program is the expected ship weather delays. TowSim provides the ability to model a particular ship's parameters, such as hull form and loading condition, and simulate a specific route. Voyage duration statistics, accounting for weather delays associated with a particular hull form on a specific route, can be determined based on thousands of simulated voyages in the hindcasted environment. This process was recently performed for a development project.

Seakeeping and Motion Response Characteristics

An efficient, high speed hull form was digitized according to the input format used by the Standard Ship Motion program. During simulation, the vessel responses were calculated using the hindcast wave data, and compared with the associated criteria values defined by the Safe Operating Envelope. If calculated values exceeded the limits, the ship slowed down in increments of one knot, until either the responses fell within the limits, or the minimum speed of five knots was reached. Five knots is considered the minimum speed to safely ride out the severe weather.

Figure 10 shows TowSim's seakeeping diagram, which depicts the sensitivity of ship motion versus heading relative to seas and swells.

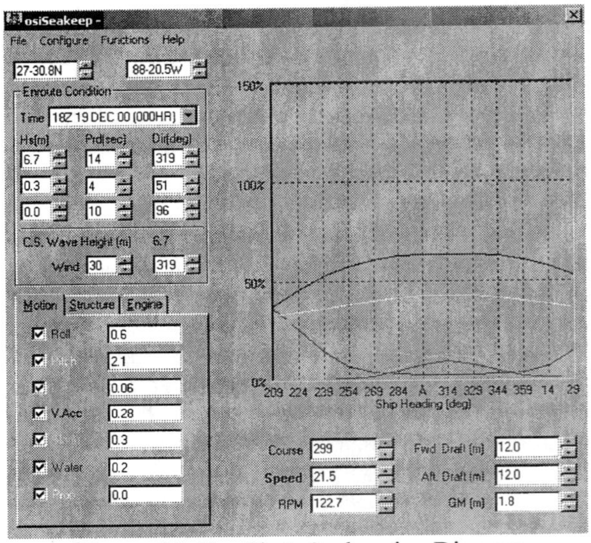

Figure 10. TowSim's Seakeeping Diagram

Speedkeeping Performance Curves

The speed versus power relationships for full and ballast load drafts were fit to a set of power curves. A typical propeller was selected.

The still water resistance curves were used together with added resistance in waves and wind to arrive at the total power requirement for the vessel to maintain the required speed. Figure 11 shows the ship's involuntary speed reduction in waves for different relative headings. Involuntary speed reduction is associated with added resistance from wind and waves as well as hull roughness, while voluntary reductions, not accounted for in Figure 11, are taken by the master to reduce risk of ship damage from bow slamming, green water, excessive motion, propeller racing, and engine racing. An algorithm to account for both involuntary and voluntary speed reductions is described in (Chen, Cardone and Lacey, 1998).

Given its ease of use and accuracy, TowSim simulations are preferred over rules-of-thumb in estimating weather delays to the overall transportation system.

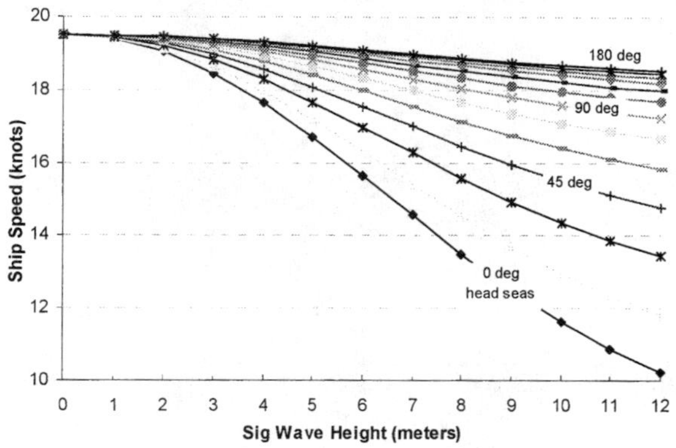

Figure 11. Involuntary Speed Reduction in Waves

FUTURE WORK

Most lasting progress is made in steps, where each step results in a useful product, and a product in which the benefit more than offsets its expense. Revolutionary leaps in progress are possible, but rare, and especially rare in corporate settings. TowSim represents an evolutionary step that can be justified solely by its ability to more easily generate tow criteria.

Nevertheless, two application examples were described above to show how versatile TowSim's fundamental capabilities are. And it's easy to envision upgrades to TowSim that would extend its capabilities.

Future development steps fall within three broad categories: 1. Improvement in the underlying metocean database, 2. Upgrade of computer hardware, and 3. Enhancement of post-processing capabilities.

1. Metocean Database Improvements

As mentioned earlier, merging basin specific hindcast data with the TowSim database can reduce the bias associated with extreme events and shallow water effects, provided, of course, that the basin specific hindcast has a finer grid spacing. Another approach to achieve the same results, would be to regenerate the TowSim hindcast database at a tighter grid spacing.

It's now possible to extend the TowSim hindcast database from 30 years to 55 (1945 to 2000) now that the underlying NRA database has been extended. So regenerating the TowSim hindcast database at a tighter grid spacing should be considered for the 55 years. This change would provide a consistent way to allow two improvements in accuracy with one hindcast update.

Extending the hindcast into the future should be done as well, although this task will take some time. However, extending the hindcast database into the past and the future, will ensure that the TowSim output will become ever more accurate.

2. Computer Hardware Upgrades

TowSim currently runs acceptably on a high end desktop machine. A transoceanic, five knot, thirty year simulation takes all night to complete. Taking advantage of increases in hardware performance is an obvious and easy way to increase performance. However, any improvements in the hindcast database made by decreasing grid spacing will require a significantly more capable computer. Although more powerful computers exist today, they are relatively expensive. It is assumed that if the hardware industry continues to advance at rates comparable to the last twenty years, then reasonably priced computers will be available in the near future—probably in our lifetimes—that can handle a grid point spacing no longer limited by hardware. Limitations in accuracy will more appropriately be related to certainty in the wave model inputs and the physics of the model itself.

Eventually, more capable and reasonable cost computers will allow the direct use of the raw, uncompressed hindcast data. The reduction of the 23x24 element wave energy matrix into nine parameters, and subsequent recreation of the full matrix, introduces uncertainty into the response calculations that would not exist if the full matrix is used. EMDC routinely performs site-based calculations, such as turret moored heading simulations and fatigue assessments, using the full wave energy matrix. While some responses are accurate enough using the recreated directional wave spectra, others, such as some fatigue calculations, are more accurate using the full matrix.

3. Post-Processing Capabilities

The TowSim user interface is already very easy to use. This ease of use can be attributed to the fact that it is a derivative of a product that was honed and fine-tuned over the last fifteen years based on feedback from shipboard operating personnel. The most useful improvement would be to add program features that continue the progress from a wind and waves based tow criteria to one that incorporates typical responses as well.

Knowledge of the metocean route conditions, such as exceedance probabilities of certain wave heights, are meaningful mainly in the context of how certain tow parameters respond in their midst. For instance, a tow with a specific beam may likely experience resonant rolls during part of a voyage, while another tow configuration with a different beam may not. Consideration of the environment alone would not distinguish this difference. Currently, the metocean description and the response calculations are separate processes. However, TowSim provides the ability to analyze many of the most important response parameters. Already, TowSim can calculate the standard seakeeping responses (motions, accelerations, deck wetness, propeller emergence) within the limitations of linear ship motion theory.

Adding an input subroutine to read the motion Response Amplitude Operators (RAOs) from a 3D-diffraction hydrodynamics program would extend this capability to more general shapes. And these are the shapes most often towed by the offshore industry.

One of the more difficult, yet important, responses to calculate is fatigue damage. TowSim has the potential of elegantly solving this fatigue assessment problem. EMDC has performed on-site fatigue assessments for a turret-moored vessel by combining the 3D-wave variance with the structural hotspot 3D RAOs (Lacey & Grant, 2002). The 3D structural RAOs are normally computed during detail design and, for this study, were readily available. Figures 12 through 15 are shown to illustrate the three dimensional nature of the problem, and the importance of orienting the vessel heading properly with respect to the wave regime. A TowSim fatigue subroutine would easily account for these important three-dimensional and directional effects.

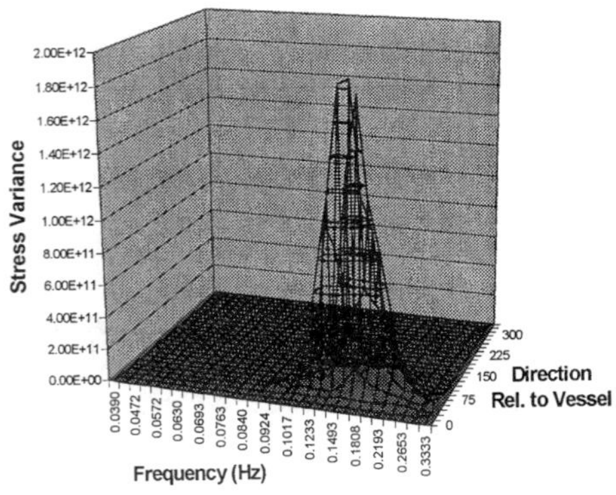

Figure 12. Wave Variance Matrix in ***Absolute*** Coordinates.

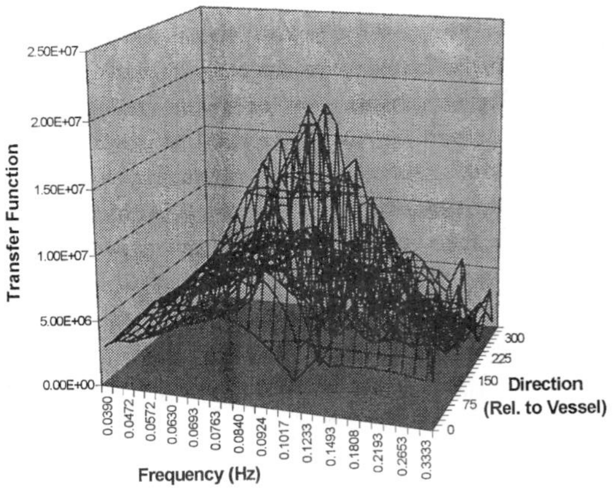

Figure 13. Wave Variance Matrix ***Relative*** to Vessel Heading

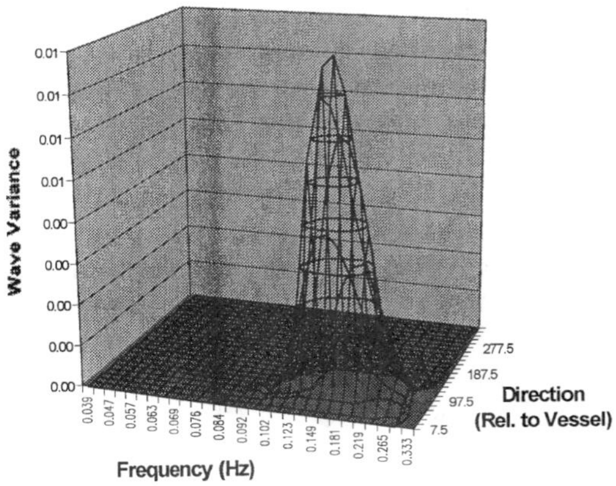

Figure 14. 3D RAO: Stress Transfer Matrix

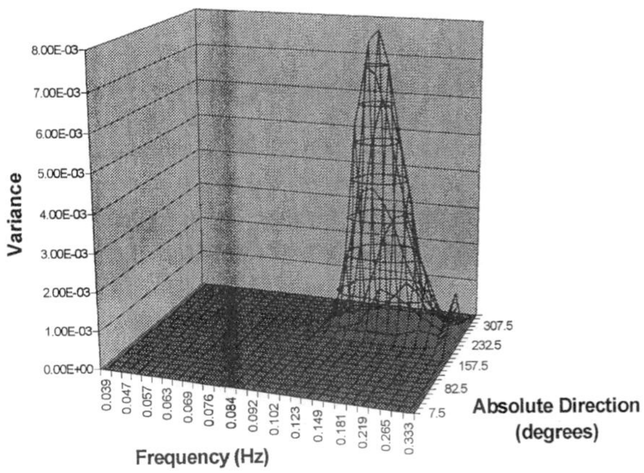

Figure 15. Stress Variance Matrix ***Relative*** to Vessel Heading

These particular calculations used the full wave energy matrix in three-hour intervals; however, the recreated directional spectra could be easily used during TowSim simulation, at six-hour intervals. The fatigue damage could then be calculated as just another response. Fatigue damage could be calculated for any specific voyage in the past, or a complete simulation could be performed to derive a statistical assessment of fatigue damage. This approach using the recreated directional wave spectra would be simple to execute, and would provide an excellent method to compare the relative efficacy of different tow configurations. Or it could be used for comparing relative fatigue damages between different ships with different voyage histories.

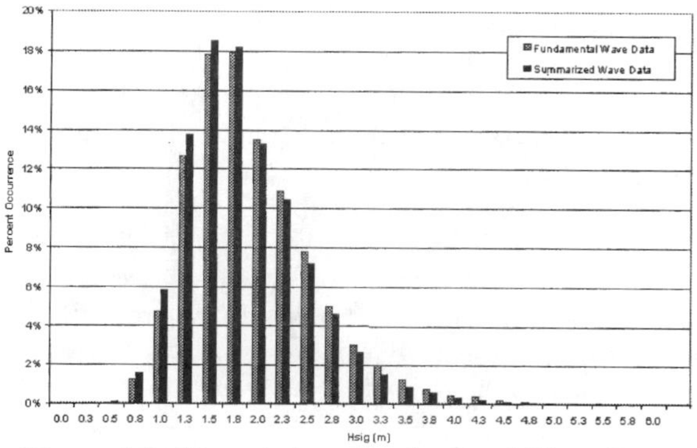

Figure 16. Recreated vs. Fundamental Wave Data

As shown in Figure 16, while the recreated (or 'summarized') data looks promising in comparison to the fundamental hindcast data, the recreation is not accurate enough with techniques used to date, to yield acceptable absolute fatigue damage values. If an estimate of absolute fatigue damage is required, the best approach is to use the full wave energy matrix. In fact, operating directly on the full matrix is becoming increasingly common for a variety of calculations. (Grant & Holboke, 2003). Saving and using the full wave matrix within TowSim will require affordable computers that are vastly more capable then currently available, or are likely to be available in the near future.

CONCLUSIONS

Based on the experience gained during TowSim development, several conclusions seem clear.

TowSim represents a solid first step towards the automation of tow climatology—a necessary input to any tow criteria. The underlying hindcast is very good, but future enhancements will make it better. Over the next several years the hindcast may become as comprehensive and accurate as it is possible to practically become.

Significant progress can be made in the automation of typical responses based on the hindcast data. Such responses are obviously those of classical seakeeping interest: motions, accelerations, deck wetness, propeller emergence—all of importance in transoceanic tows. Other responses can also be added to this list, such as fatigue assessment.

Tow climatology, along with certain responses, are essential in developing a specific tow criteria. However, this information is ultimately used to determine if a particular tow represents an acceptable risk. Developing an objective tow risk assessment methodology, and implementing it in TowSim, would fulfill the program's promise by establishing a rational and consistent tow risk assessment procedure. Ideally, this tow risk assessment would be qualified by comparison against a consistently derived assessment of risk on site. Such risks can be compared using the encounter probability versus the notion of traditionally used Return Period as shown in Appendix B. On-site risk will have already been explicitly accepted, so if the tow risk is less, then the tow must be acceptable

Finally, it would be preferable for the actual tow to use a wave forecasting service that is consistent with the tow criteria development. As noted, the tow criteria is based on GROW2000, which used the same underlying physics on which most of the world's site criteria are based. These programs have been available for use onboard ships at sea for many years (Lacey and Chen, 1995). This use would allow complete consistency in the climatology basis, from design through tow, installation, and to the end of on site operations.

ACKNOWLEDGEMENTS

The authors wish to thank the management of the ExxonMobil Development Company, in particular, Fred Plummer and Roger Leick, for their support and enthusiasm for this work. Without their support, this paper would not have been possible.

APPENDIX A

Response Using Mean Directional Spectra

According to linear ship motion theory, the mean square response, $m0$, is obtained by superimposing the Response Amplitude Operator, RAO, squared and the wave spectral energy in all direction and frequencies:

$$m0 = \sum_i \sum_j RAO(\Theta\Psi)^2 E(\Theta\Psi)$$

Where:

i = number of wave frequencies Θ

j = number of relative headings Ψ

To compute the statistical mean of the mean square response,

$$M0 = \sum_N (m0)n / N$$

Where:

N = the total number of directional spectra in a wave height group, m .

Substituting:

$$M0 = \sum_N \left(\sum_i \sum_j RAO(\Theta\Psi)^2 E(\Theta\Psi) \right) n / N$$

$$\left(= \sum_i \sum_j RAO(\Theta\Psi)^2 \sum_N E(\Theta\Psi) \right) n / N$$

$$= \sum_i \sum_j RAO(\Theta\Psi)^2 Em(\Theta\Psi)$$

Where:

Em = the mean directional spectral energy matrix.

For long term predictions accounting for the spectral shape and directional variability of the wave climate, a covariance matrix for each wave height group is also needed. (Chen, 1982)

APPENDIX B

Return Period Versus Encounter Probability

The offshore industry has traditionally used the notion of return period for specifying design wave heights for a fixed structure. The "100-year return period" wave height has a one percent chance of occurring in any random year. However, if the structure has a 25 year life time, the probability of encountering a 100 year return period wave is actually higher then might be expected, since it is the product of non exceedance considering all 25 years:

$$P_{100\,yearwave} = 1 - (1 - 1\%)^{25} = 22.2\%$$

While the probability of 22% may seem high, it does not necessarily relate to an unsafe design. In fact, this probability of occurrence relates to the level of generally accepted risk for offshore designs.

For transoceanic tows, the voyage may encompass several weather regions. In general, the probability of encountering these 100 year events in each region can be estimated in the following ways:

Return period concept: $P = 1 - (1 - 1/R)^E$

Simple ratio: $P = 1 - E/R$

Poisson process: $P = 1 - \exp(-E/R)$

Where :

E = Exposure duration in the region

R = Return period or number of occurrences per unit time

$(1/R)$ = arrival rate, λ

Notice that when E and R are in years for a fixed location, the above expressions default to the traditional concept of prescribing the design criteria in terms of return period. For large R over E multiples (>8), the probability of encounter estimated by these three methods converges. For example, when the ratio of return period to exposure period is 10, the probability of encounter is around 10%; when the ratio is 20, the probability of encounter is around 5%. In any case, the longer the exposure duration is, the higher the chance of encountering the design seastate.

Endnotes:

CARDONE, V. J., A. T. COX AND V. R. SWAIL. "Specification of the Global Wave Climate: Is This the Final Answer?", 6th International Workshop on Wave Hindcasting and Forecasting, November 6-11, 2000, Monterey, California.

CHEN, H, "Effect of Wave Spectral Shape and Directional Variability on the Design and Analysis of Marine Structures", Behavior of Offshore Structures (BOSS) Conference 1982

CHEN, H., V. CARDONE AND P. LACEY. "Improving Safety and Efficiency with Shipboard Decision Support Systems," International Conference on Maritime Technology: Challenges in Safety and Environmental Protection, Singapore, 17-20 November 1993.

CHEN, H., V. CARDONE AND P. LACEY. "Use of Operation Support Information Technology to Increase Ship Safety and Efficiency." Annual Meeting of Society of Naval Architects and Marine Engineers, San Diego, November, 1998.

CHEN, H, P.LACEY AND R. LOVDAHL. "Advances in Computer Based Onboard Voyage Planning," CyberNautics Symposium, SNAME Joint Sessions Meeting, Los Angeles California, 17-20 April 1995.

COX, A.T. AND V.J. CARDONE. "20 Years of Operational Forecasting at Oceanweather." 7th International Workshop on Wave Hindcasting and Forecasting October 21-25, 2002, Banff, Alberta, Canada

COX, A.T. AND V. R. SWAIL: "A 40-year Global Reanalysis of Ocean Waves 1958-1997 (GROW) ." 1999, Journal of Geophysical. Research, 106, 2313-2329.

GRANT, R, AND M. HOLBOKE. "Optimizing and LNG Marine Terminal Exposed to a Severe Wave Environment." AIChE 2003 Spring National Meeting, April 1, 2003, New Orleans, Louisiana.

KUO, J, P. LACEY, A. MACMILLAN AND N. ZETTLEMOYER. "Fatigue Methodology Specification for New-Build FPSOs," Proceedings of OMAE'01, 20th International Conference on Offshore Mechanics and Arctic

LACEY, P. AND H. CHEN. "Improved Passage Planning using Weather Forecasting, Maneuvering Guidance, and Instrumentation Feedback," *Marine Technology*, January, 1995.

LACEY, P., B. GRANT, C. NORDSTROM AND D. HEE. "Impact of FPSO Heading on Fatigue Design in Non-Collinear Environment," Proceedings of ASME, OMAE 2002, Oslo, Norway, 23-28 June 2002.

SWAIL, V. R. AND A. T. COX. "Evaluation of NCEP/NCAR Reanalysis Project Marine Surface Wind Products for a Long Term Hindcast." Journal of Atmospheric and Oceanic Technology, 2000, 17, 532-545.

DISCUSSION

Bruce L. Hutchison, Member

The authors are to be congratulated on both this paper and the underlying accomplishments on which it reports. As a practitioner who has been engaged in probabilistic analyses of tows and other voyages for the past 30 years, I appreciate the capabilities reported by the authors. Those capabilities address and report on advances in two areas of particular interest.

The first area is that of climatological data. In the past a practitioner embarking on a voyage analysis would find it necessary to assemble climatological data, for the spatial and temporal domains along the route, from a variety of diverse sources. In general, the data of choice were joint probability for such processes as wind speed and direction, wind speed and significant wave height, and significant wave height and some characteristic period (e.g., T_2 or T_P). At best, such data were available on a monthly basis and usually represented different periods of record and frequently different approaches to data collection (e.g., measured data or visual observations from ships). In general, it is not possible to separate sea and swell in such databases and one is usually left to presume that the wave direction is aligned with the wind.

Furthermore, the bivariate joint probabilities that were typically available did not address some relationships, such as that between wave period and wind speed. What would have been desired was a complete multivariate joint probability, $P(U, \chi, H_S, T_P)$. What was available was a subset of bivariate joint probabilities that could be regarded as marginal distributions, $P(U, \chi)$, $P(U, H_S)$, and $P(H_S, T_P)$. In order to develop an estimate of $P(U, \chi, H_S, T_P)$, it was necessary to presume statistical independence between U and T_P.

The GROW database described in this paper and several of its references represents 30 years of wind and wave time series for all major navigable waters of the world. Here is one consistent database that maintains all of the contemporaneous relationships between wind and directional sea spectra. As implemented in TOWSIM, the directional sea spectra are compressed into three separate wave trains, each with its own significant wave height, dominant period, and direction (H_i, T_i, D_i). This allows for representation of the wind-driven local sea and up to two independent swell systems.

The second area of advance is related to the first. A serious shortcoming of voyage analyses in the probability domain (as have been practiced for the better part of the last 30 years) is that there is an averaging process that takes place over the period of record to arrive at the joint probabilities $P(U, \chi)$, $P(U, H_S)$, and $P(H_S, T_P)$. Thus, $P(H_S, T_P)$ may be determined from 30 years of data for a particular spatial domain and month of interest, but $P(H_S, T_P)$ represents only the 30-year *average*. What is lost is all of the variability, good years and bad years. A friend who is both a trained naval architect and a meteorological professional has jested that "normal" is just the average of extremes. There is no assurance that the conditions represented by $P(H_S, T_P)$ or the other usually available joint probabilities, $P(U, \chi)$ and $P(H_S, T_P)$, even exist.

A real problem with voyage or marine operations advice given on the basis of probability domain analyses is that it does not provide guidance that includes consideration of year-to-year variability. A particular marine operation may be judged to be 90% operable in the average climatology, but it may be only 50% operable 1 out of 5 years. If 50% operability is unacceptable, then the operation is in fact, unknowingly, accepting a 20% risk of failure that cannot be detected by probability domain analyses.

Analyses of the operability of marine operations also frequently make use of statistics for mean time between events and duration above or below some threshold (persistence). Again, it must be observed that such measures represent the *average* mean time between events and the *average* persistence, over the period of record (e.g., 30 years).

The advantage of TOWSIM is that, taking place in the time domain, all such year-to-year variability issues may be properly addressed.

In recent years our firm has commissioned several TOWSIM voyage analyses: a semisubmersible, a decommissioned nuclear reactor, and an FPSO. A feature that we have found particularly useful is that it is possible to provide TOWSIM with custom, user-defined subroutines for speed-made-good and for seakeeping responses. In our subroutines for speed-made-good, we have included models for wind resistance based on wind tunnel tests, added resistance, voluntary speed reduction, and even current effects (based on location and season). The subroutines for seakeeping responses allowed us to examine seakeeping responses of particular interest to our programs and to use seakeeping models appropriate to the particular tow.

The Design and Service Experience of the *Polar Endeavour* Class Tankers

Robert A. Levine[1], (M) SNAME, (M) IMarEST, C.Eng., Manager Engineering Integration & Naval Architecture, ConocoPhillips Marine

ABSTRACT

The Polar Endeavour *(Millennium) Class Tankers, represent the first U. S. flag crude oil carriers designed and built for the Alaskan Trade in compliance with the Oil Pollution Act of 1990 (OPA90). These vessels represent the start of a revolution in design for American crude carriers. Not only are they fitted with double hulls, which are deeper than required, but also they are fitted with twin independent engine rooms, and twin rudders. They are the first vessels to be classed R2-S+ and NIBS under the American Bureau of Shipping's (ABS) Guidelines for Redundant Propulsion Systems and Navigation Integrated Bridge Systems. The vessels meet the requirements for Annex VI of MARPOL for air emissions, and are painted with tin free (TBT) anti-foul paints. This paper describes the design development process used and the unique and innovated design, which resulted. Three vessels have entered service in 2001, 2002 and 2003 and two are under construction for delivery in 2004 and 2005.*

[1] The opinions expressed in this paper are those of the author and do not reflect those of Polar Tankers Inc., ConocoPhillips Marine, ConocoPhillips Alaska or the ConocoPhillips Corporation.

ACRONYMS

Acronym	Meaning
+A1E	Compliance with ABS Rules for Self-Propelled Vessels & Equipment Standards
AB	Able Bodied Seaman
ACCU	Compliance with ABS Automatic Centralized Control Unmanned Criteria
ADSSE	Automatic Dependent Surveillance Shipborne Equipment
AIS	Automatic Identification System
AMS	Compliance with ABS Machinery Requirements
ANS	Alaska North Slope
ANTS	Automated Navigation & Tracking System
APS	Compliance with ABS Athwartship Thruster Criteria
ARPA	Automatic Radar Plotting Aid
bbls	Barrel (42 gallons U.S.)
BHP	Brake Horsepower
BHPh	Brake Horsepower-Hour
CAIP	USCG Critical Area Inspection Program
C_b	Block Coefficient
CFD	Computational Fluid Dynamics
COI	USCG Certificate of Inspection
COW	Crude Oil Washing
CP	Controllable Pitch
cSt	Viscosity – Centistokes
DGPS	Differential Global Positioning System
DL	Compliance with ABS Dynamic Load Approach
DLA	Dynamic Load Approach
DNV	Det Norske Veritas
DWT	Deadweight
ECDIS	Electronic Chart Display & Information System
EPA	Environmental Protection Agency
FEA	Finite Element Analysis
g/BHPh	Grams per Brake Horsepower-Hour
g/kWh	Grams per Kilowatt-Hour
GLONASS	Russian Equivalent to GPS
GMDSS	Global Marine Distress & Safety System
GPS	Global Positioning System
H1/3	Significant Wave Height
HFO	Heavy Fuel Oil
HV	High Voltage
HZ	hertz (Frequency in cycles/second)
IG	Inert Gas
IGS	Inert Gas System
IMO	International Maritime Organization
I/O	Input/Output
ISO	International Standards Organization
Kcal/kg	Kilocalorie/kilogram
kJ/kg	Kilojoules/kilogram
KV	Kilovolts
KW	Kilowatts
LAN	Local Area Network (Computer)
LBP	Length Between Perpendiculars
LCB	Longitudinal Center of Buoyancy
LV	Low Voltage
L_{OA}	Length, Overall
L_{PP}	Length, Between Perpendiculars
LSFO	Low Sulfur Fuel Oil
LT	Long Tons
MARPOL	International Convention for the Prevention of Pollution from Ships
MCR	Machinery Control Room
MCR	Maximum Continuous Rating
MCS	Machinery Control Space
MCS	Machinery Control System
MDO	Marine Diesel Oil
MDL	Marine Dynamics Laboratory (SSPA Maritime Consulting)
MDWT	1,000 Deadweight Tons
mil	1/1000 of an inch
MT	Metric Tons
MW	Megawatt
NIBS	Compliance with ABS Navigational Integrated Bridge System Criteria
OPA90	Oil Pollution Act of 1990
OS	Operating Station
PC	Personal Computer
PCU	Power Converter Unit
PMS	Power Management System
PTI	Polar Tankers Inc.
PTO	Power Take Off
psig	Pressure - Pounds per Square Inch Gage
QMED	Qualified Member of the Engine Department
R2-S+	Compliance with ABS Criteria for Redundant Propulsion & Steering Systems with Adverse Weather Endorsement
ROB	Retained on Board
RPS	Redundant Propulsion & Separate
SH	Compliance with ABS SafeHull Criteria
SHP	Shaft Horsepower
SOLAS	International Convention for Safety of Life at Sea
T	Draft
T_f	Draft, Forward
T_a	Draft, Aft
TAPS	Trans-Alaska Pipeline System
TBT	Tri-butyl Tin
UHF	Ultra High Frequency Radio
UWILD	Compliance with ABS Underwater Hull Survey Criteria
VHF	Very High Frequency Radio
VTS	Vessel Traffic System

DESIGN

OPA90 not only created a requirement for new tankers to be fitted with double hulls, it also brought with it cargo owner liability. Cargo owner liability places financial responsibility not only on the Owner/Operator of the vessel, but on the cargo owner as well. For a company like Polar Tankers Inc. (Polar)[2], an Owner/Operator, and through its association with its parent company, a cargo owner, all responsibilities lie at one doorstep. It is recognized that the 80/20 rule applies to tanker accidents, that 80% are caused by human error, and 20% by mechanical failure. Polar had invested large sums of time and money in crew and officer training in order to reduce the human error portion of the 80%. However, there was still an inherent portion of the 80% that resulted from how U.S. tankers were acquired, and a majority portion of the 20% had gone without being addressed.

Polar, in reviewing its existing vessels, and in looking at world class vessels found, in retrospect, not so amazing conclusions. Historically, Polar's tankers were standard shipyard designs, equipped with machinery supplied by the low bidder. Design changes were expensive and served only to correct unacceptable flaws. Chosen equipment, while serviceable, represented a long-term maintenance and repair commitment. Ergonomics, habitability, redundancy, maintainability, and reliability were concepts that were either not known or ignored. Therefore, to reduce the possibility of accidents Polar had to develop a vessel designed to be operated easily and efficiently by the operators with a reduced likelihood of error. The vessels had to first be reliable using proven state of the art equipment. The vessels had to be easily maintained by the crew. And most importantly, to insure the vessel's safety was not compromised in the event of a failure, they had to be fully redundant.

The origin in design of the *Endeavour* Class vessels can be traced to the North Sea shuttle tankers, and in particular, the vessels owned and operated by Knutsen AB, and designed and built by IZAR (formerly Astilleros Espanoles, S.A.). Designed for loading at offshore facilities in heavy weather (Sea State 7 and greater), the twin screw, twin rudder tankers, with high lift rudders and multiple bow and stern thrusters, are capable of dynamic positioning, and operate daily in a highly efficient system. These vessels provided the design basis by which Polar established the design of the *Endeavour* Class vessels.

Knutsen and IZAR also provided insights into equipment and manufacturers. While equipment exhibitions such as SMM™ in Hamburg and Norshipping™ in Oslo provided exposure to a wide variety of equipment, Knutsen and IZAR provided information regarding good, better, and best. They provided information on favorites and preferences, on equipment they were using, and equipment they wished they were using. They opened the door to the world of marine equipment, a world not often visited by American owners and American shipyards.

In addition to the help from Knutsen and IZAR, Polar also sought insight and opinion from within. Polar's predecessors, Atlantic Refining and Arco Marine, Inc., had been operating tankers since 1916. In particular, they had been operating tankers in the Alaskan Trade since the early 1970's and specifically in the Trans-Alaska Pipeline (TAPS) Trade since the pipeline opened in 1978. Polar had at its fingertips data on structural fatigue and problems associated with the severe weather and high cyclical loadings and short runs in heavy weather. Typical tankers may make 8 or 10 voyages in a year, and 250 in a lifetime. Polar's make up to 33 a year and 550-600 in a lifetime.

Long term exposure to the elements provided a wealth of information on machinery reliability and maintainability. We knew what worked, but perhaps more importantly, what didn't. With the voyages and the weather come the sea stories and the anecdotal information. From this information comes the data needed to direct design on ergonomics and habitability. It is our belief that anyone sailing, past or present, in Polar's fleet can proclaim, "I'm responsible for that! I told them it needed to be done." From lighting on deck and in the engine rooms, to athwartship arrangement of bunks, from single messes and lounges to a conference room, from raised catwalks to a cargo control room on "A" deck, Polar employees can claim responsibility.

The *Endeavour* Class tankers represent a transition in design from what the shipyards believed the owner needed in the 1970s, to a ship an owner/operator wants for today and for the future.

Table 1 provides a list of particulars for the *Endeavour* Class tankers. Figures 1 and 2 show the general arrangement for the vessels.

Table 1 - Vessel Particulars

Principal Characteristics	
Length, Overall (L_{OA})	272.96m
Length, Between Perpendiculars (L_{PP})	258.16m

[2]Polar Tankers Inc. and ConocoPhillips Alaska, Inc. are wholly owned subsidiaries of ConocoPhillips. Polar Tankers manages the marine transportation of ConocoPhillips Alaska's North Slope production of Alaska North Slope (ANS) crude oil. Polar Tankers has become part of ConocoPhillips Marine but continues to maintain its Long Beach, California headquarters. ConocoPhillips Marine includes several marine related companies along with Polar Tankers, Conoco Shipping, and Emerald Shipping.

Beam, Molded	46.20m
Depth, Molded (At Side)	25.30m
Depth, Molded (At Centerline)	26.30m
Draft, Design	16.19m
Draft, Scantling	17.52m
Lightship Weight	31,769 MT
Gross Tonnage	85,387 MT
Net Tonnage	41,995 MT
Capacity, Design	125,000 LT
Deadweight, Design Draft	127,005 MT
Displacement, Design Draft	158,774 MT
Capacity, Scantling	139,436 LT
Deadweight, Scantling Draft	141,734 MT
Displacement, Scantling Draft	173,506 MT

Capacities (100%)	
Cargo Tanks (1,014,100 bbls)	$161{,}526m^3$
Ballast Tanks	$62{,}362m^3$
HFO Tanks	$4{,}415m^3$
LSFO Tanks	$615m^3$
DO Tanks	$552m^3$
Fresh Water Tanks	$189m^3$

MODEL TESTING

Background

Historically, Polar's experiences were that model testing was the sole responsibility of the shipyard. As a result, the model test program was limited to resistance and propulsion testing. This testing provided assurance to the shipyard that speed/power requirements would be met. Wind tunnel testing was done on occasion because of smoke problems, and cavitation testing was done only after severe propeller damage was encountered. Review of past Polar vessels showed that shipyards stopped testing when a design point was reached and rarely, if ever, was the optimum design point ever sought. Seakeeping and maneuverability were two concepts that were not studied, and in general were never considered.

In some cases Polar had vessels that performed as designed. In other cases we had vessels that performed to expectations only on those days they encountered calm weather conditions found in model basins. On one class of ships, when we later started studies to improve performance, we received documentation from the model basin showing they had advised the shipyard during the design process that the basin's "Preliminary" lines had been used by the yard as "Final" lines. We believe this information was provided out of fear that the basin's reputation might be soiled because the yard didn't wish to invest in another round of testing. This documentation indicated that the vessel was likely to encounter severe performance degradation in heavy weather, and was likely to experience cavitation and vibration problems. The shipyard, to meet schedule, used their "Final" lines. These vessels, as predicted by the basin, perform poorly in seaways, vibrate, and suffer cavitation problems.

For the *Endeavour* Class vessels, Polar set out to control the hull design. Instead of having the shipyard contract with a model basin, Polar sought out model basins and chose the basin they felt was best suited to work with them in producing an optimum design. Basin selection criteria were facilities, experience designing shuttle type tankers, the ability for the basin, Polar and Polar's design team to work cooperatively, and last of all, cost. Polar's design program included not only development of lines, and resistance and propulsion testing. The program also included computational fluid dynamics (CFD) evaluation of a preliminary set of lines and a set of lines developed by the model basin, computer simulation of maneuvering characteristics for rudder selection, resistance and propulsion testing on the 'initial" lines from the CFD study with a paint flow test and wake survey, development of "final" lines, resistance and propulsion testing with a wake survey using library propeller, propeller design, cavitation tunnel testing with a design propeller, resistance and propulsion testing with a design propeller, seakeeping testing, maneuverability testing, wind tunnel testing, and disabled ship maneuverability. Kamewa, as part of their contract as the provider of the propellers, was required to do resistance and propulsion testing as well as cavitation testing for their final propeller design.

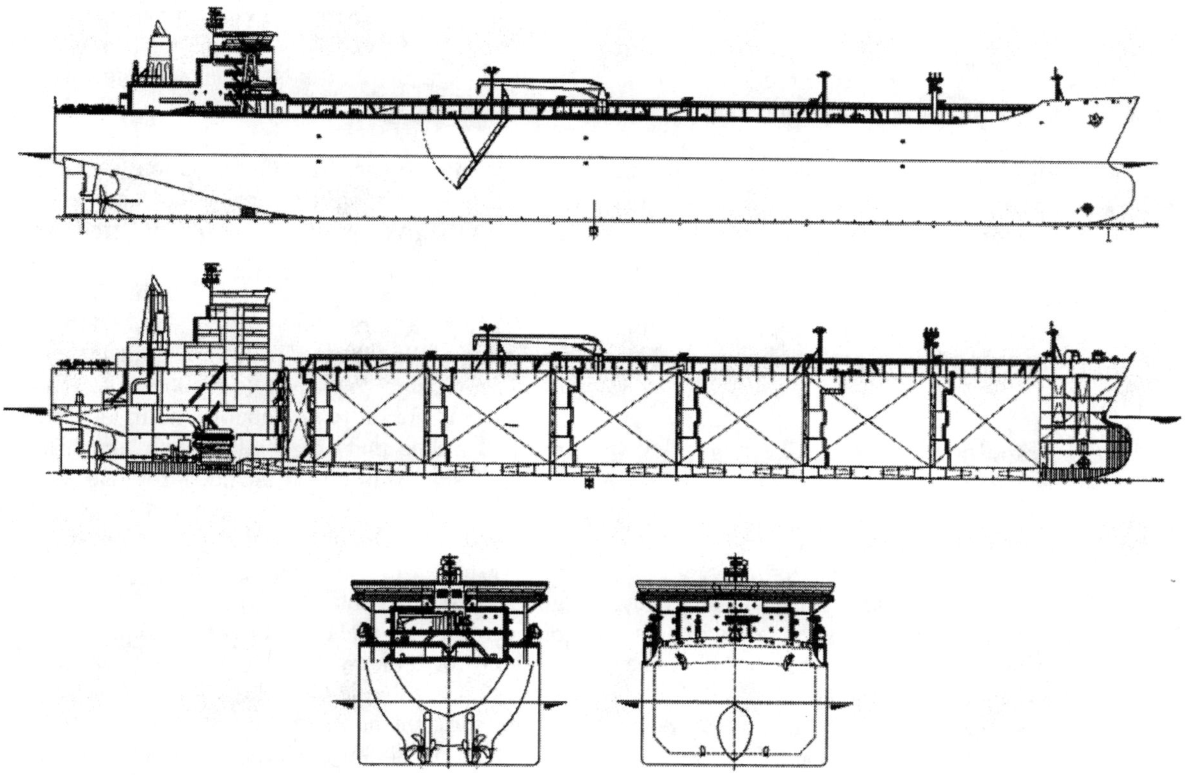

Fig. 1 Inboard Profile, Outboard Profile, Bow and Stern Views

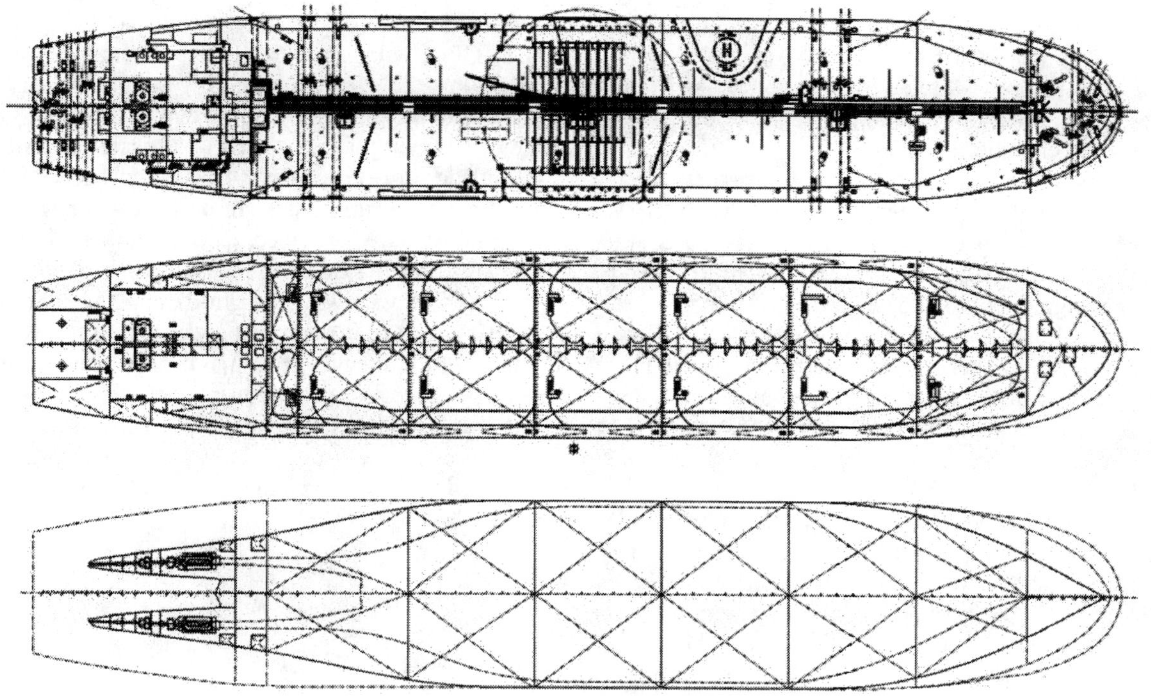

Fig. 2 Upper Deck Plan, In-Tank Plan and Plan on Tank Top

In all Polar committed about $600,000 to the hull design/model test program. This represents less than 0.2% of the cost of the original construction program, and less than 0.06% of the total cost for the five-ship program. If the improvement in efficiency is considered alone, there was a 6% improvement in speed/power from the initial design to the final testing. However, the gains from model testing were far more extensive. Of significance are; the use of a semi-balanced rudders, improved seakeeping, R2-S+ heavy weather compliance, minimization of smoke effects, verification of design models, and data for crew training at bridge simulators.

Resistance & Propulsion Design and Testing

Program

The Resistance & Propulsion portion of the Model Test Program was intended to optimize the overall hull performance of the *Polar Endeavour*. Historically, Polar's tankers were optimized based on still water conditions. Unfortunately, the waters in which they operate are rarely, if ever, "still". Polar believed that a thorough design program could produce a vessel that was not only efficient in "still" conditions, but that would also be efficient in the moderate to rough conditions normally experienced in the Gulf of Alaska. The program developed by Polar in conjunction with SSPA Maritime Consulting (SSPA) involved review of a preliminary (first generation) design, including a Computational Fluid Dynamics (CFD) optimization, and development of a set of second generation design along with testing. Review of the second-generation design and further modification as required produced a final (third generation) design.

As part of the initial design review SSPA questioned Polar's preference for MAN-B&W, 7S50MC-C slow speed diesel engines (11,060 KW @ 127 rpm). SSPA felt consideration needed to given to lowering engine rpm to 106, and using a larger diameter propeller. Polar chose the seven cylinder, 127 rpm engine for a number of reasons. Seven cylinders offered the best answer to vibration problems. Five cylinder engines have inherent problems, which both SSPA and the manufacturer recommended avoiding. Smaller cylinder size meant larger numbers of cylinders (8) and problems with matching cylinder numbers (8) with propeller blades (4). More cylinders also presented problems with engine room length. Larger cylinder size posed similar problems between numbers of cylinders (6) and propeller blades (4). The larger bore engines substituted length problems with breadth problems. It was more difficult to fit the wider engines into the gondolas. Another advantage to using the higher rpm engine was dealing with propeller tip immersion during heavy weather when the vessel is in ballast. The 127-rpm engines resulted in 5.8 m diameter propellers. This gives a 4.2 m immersion in ballast versus the 2.0 to 2.4 m immersion Polar has on other vessels.

Hull Form Design & CFD Optimization

SSPA, the contractor for hull design and model testing, was given an initial set of lines from which to start their design. This set of lines was developed during initial design studies. The initial lines represented a first attempt by hull designers at designing a 125,000 DWT, 16.0 m draft tanker with a total cargo cubic volume of 1,017,000 bbls. The initial design was for a twin gondola stern tanker with a parallel middle body of about 33% of the LBP. The initial design was never intended for use as a final design and it was recognized that further development was required. The only design restriction imposed on SSPA, by Polar, was a result of the arrangement gondolas on the initial design. The gondolas skewed outboard in such a manner that special cradles would be required for building and future dockings. SSPA was instructed to provide gondolas with a flat area that was in the same plane as the horizontal flat keel. This would permit docking with standard keel blocks under each gondola.

SSPA started with the initial lines, generated a new design and then developed a second design for the CFD portion of the study. SSPA's CFD investigation resulted in minor changes to the bow portion of the lines, and virtually no changes to the parallel mid-body. The CFD, along with Polar's requirements for modifying the gondola design, resulted in significant changes to the after body of the tanker. Figures 3 and 4 show the body plans for the initial design and SSPA's second design, respectively [1].[3] The end result was that both the original design and SSPA's design were good. The latter provided smaller static and dynamic wetted surface, smoother pressure distribution in the fore body, lower wave heights along the hull, and a smaller pressure difference between the inside and the outside of the skegs. The initial design had enough free board to keep the deck dry in operational conditions, and the bow flare was suitably shaped to minimize green water on deck. Bow flare slamming was predicted to be low even in storm conditions.

[3] Numbers in brackets refer to references at the end of the paper.

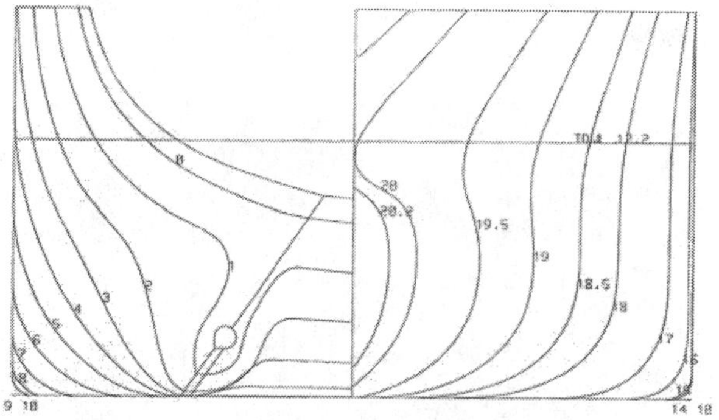

Fig. 3 Initial design - Body plan

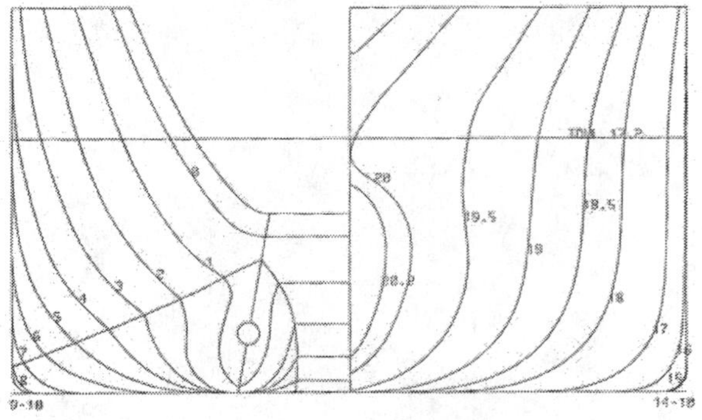

Fig. 4 SSPA second design - Body plan

Figures 5 and 6 show the improvements in flow and pressure distribution between the initial design and the second design from the CFD process [1].

Testing and Development of Hull Form

Resistance and propulsion testing was done first on the design produced from the CFD analysis, and then modifications were made to create the final hull form. SSPA conducted a full set of resistance and propulsion testing on the CFD hull form, Figure 6, including; resistance, propeller rotation (both inboard and outboard), rudder optimization, wake survey, streamline (paint) and self propulsion testing. Self-propulsion testing was done using library propellers. All testing was done using a 1:34.118 scale model. After completion of a first round of testing modifications were made to the hull and a second set of tests, with the exception of the streamline test, were conducted. Testing was conducted at both ballast (9.0 m fore/10.0 m aft) and design (16.0 m fore/aft) drafts.

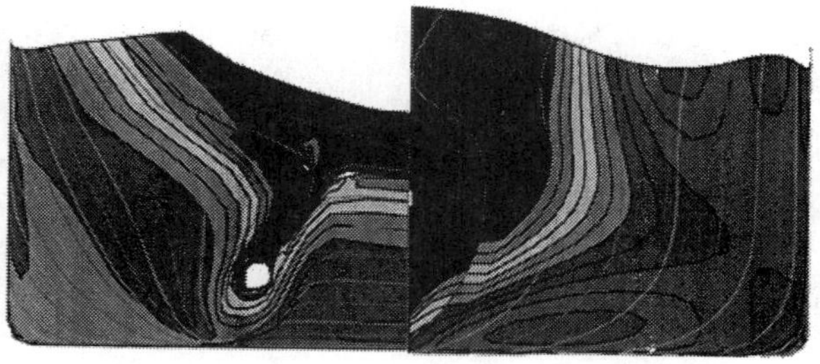

Fig. 5 Initial design - Flow and pressure distribution

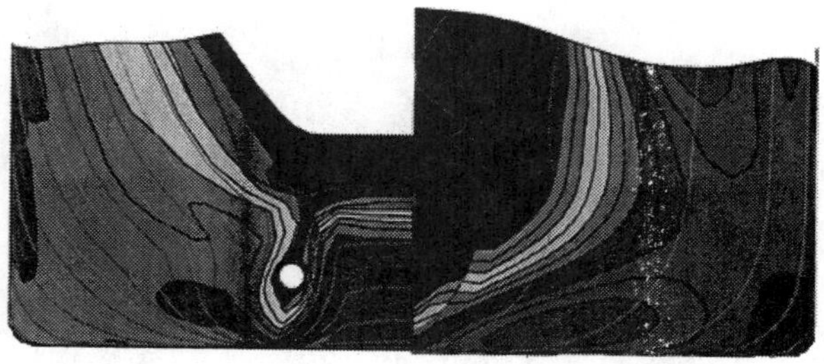

Fig. 6 SSPA second design - Flow and pressure distribution

In the initial testing SSPA found that, as compared to other vessels of similar dimensions and fullness, the resistance and propulsive qualities were "significantly better than average." Due to confidentiality agreements, SSPA compares performance based on a subjective rating system of significantly better than average, better than average, average, and below average. The propeller rotation optimization showed an 8% lower delivered power for inward turning propellers versus outward turning propellers. Optimizing the rudders at a 3° offset (aft edge to centerline) reduced the required delivered power by 300 KW (1.7%).

Two observations were made during the first round of testing. First, the flatness of the hull between the two gondolas aft would, in a following sea, make them susceptible to slamming. And second, there appeared to be a large amount of water being dragged behind the vessel. This was easily illustrated by placing a crumpled piece of paper off the transom. This paper followed the vessel for the length of the towing tank. SSPA felt by experience that this water, which was being dragged, contributed little, if any, to resistance. Wave patterns were found to be favorable and the bulbous bow, particularly in the loaded condition, worked very well.

The flow streamline tests in Figures 7 and 8 showed generally good results, with almost no cross flow over the skegs [3]. There was some tendency for flow separation in the aft body, Figure 8. The test showed that the 3° offset for the rudder placement was correct. The wake distribution, Figure 9, showed a peak at 180° (bottom side of the skeg) [3]. The wake was found to have very high velocities between 60° and 150° at the inner radii. It was noted that improvement was possible by making the skeg fuller in the after part of the skeg.

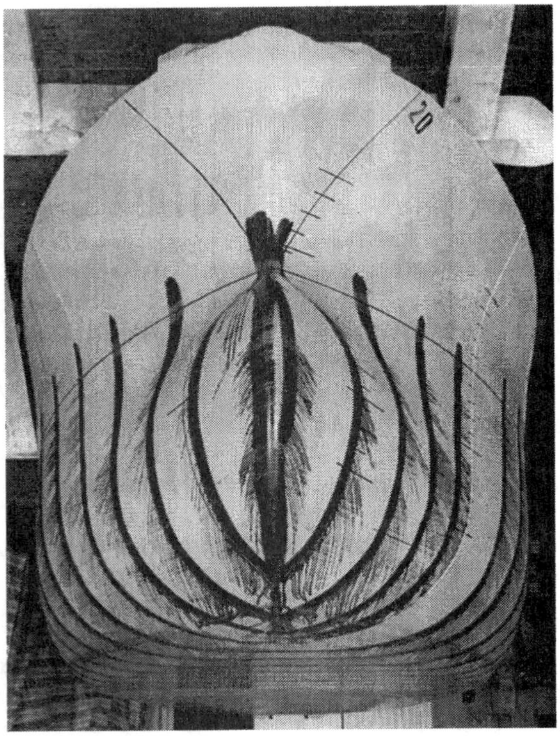

Fig. 7 Forebody streamline test

Work was being done concurrent to this portion of the model test program with regard to loading and compliance with trim and stability regulations. It was found in the full load, scantling draft condition that substantial ballast would have to be carried to compensate for excessive trim by the head due to the full volume of a twin gondola stern. Based on this information it was decided to revise the design and to run a second set of tests. The revision was to include; removing approximately 1150 m^3 of volume from the stern to produce a more favorable LCB, modifying the "flat" area between the gondolas to give it a "vee" shape to reduce effects of slamming from following seas, and to modify the after portion of the gondolas to improve wake distribution. No changes were made to the bow.

Fig. 8 Aft body streamline test

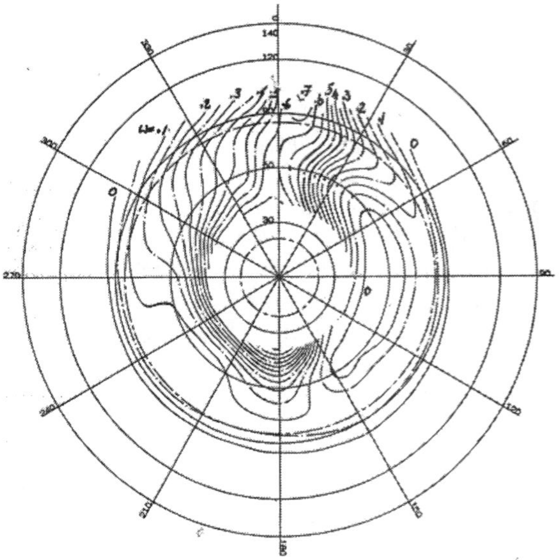

Fig. 9 SSPA second design - Wake distribution

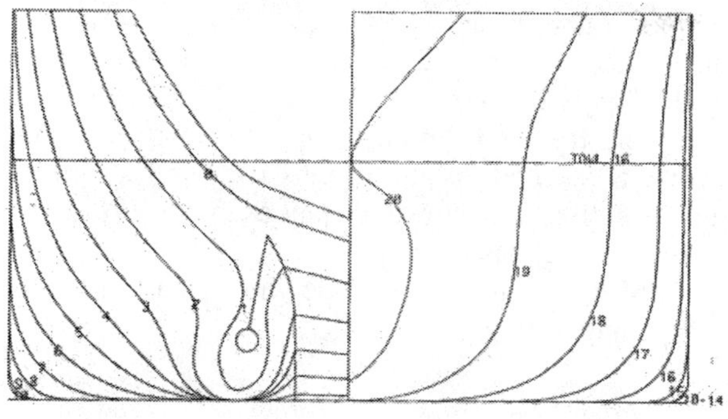

Fig. 10 SSPA final design - Body plan

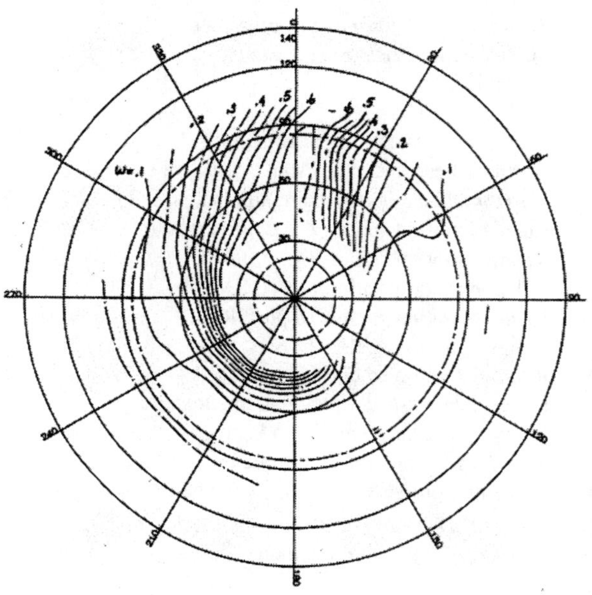

Fig. 11 SSPA final design - Wake distribution

Figure 10 shows the revised body plan for the second round of model testing [4]. This body plan, after confirmation of model testing became the final design plan. For this set of testing the library propellers were again used. Final model testing resulted in inward turning propellers with an 8% lower delivered power than with outward turning propellers. As before, it was found that a 3° offset for the rudders was correct. Figure 11 shows the wake distribution for modified stern [4]. It was found the wake distribution peaked at 180° and that the negative wake between 60° and 150° had been smoothed out. The water being dragged, although not felt to be contributory to resistance, was significantly reduced from the first set of testing. At this point in the test program, Polar and SSPA were satisfied that an optimum hull form had been achieved, and that further development was unwarranted.

Historically, Polar Tankers had less than favorable experiences using library propellers for final designs. While we recognized that design techniques allowed development of suitable designs, our comfort level was not sufficient to accept a design without testing. SSPA, after completion of the second round of testing was tasked with designing a specific propeller for the tankers and to conduct open water and self-propulsion testing. These same propellers were to be tested in SSPA's cavitation tunnel to assure suitable cavitation performance. It was recognized that the propeller designed by SSPA might not be the one recommended by the propeller manufacturer. Provisions were made in the Specification that if "major" changes were made to SSPA's design, they would be considered only if the manufacturer retested and showed equivalence to the performance demonstrated by SSPA.

In testing, cavitation and propeller noise was minimal; primarily do to small, high speed propellers and the accompanying tip clearances and propeller immersion. In all, at the 16.0 m design draft, there was a 0.25-knot increase in speed between SSPA's library and design propellers. The final design propeller tested about 0.05 knot less in speed, when compared to the SSPA design. This small difference between SSPA and Kamewa was attributed primarily to the application of Classification Society rules to the final design.

Hull Form Results

The end result of this portion of the model test program was an improvement in performance of between 7% and 10% from that predicted for the initial design to that of the final design. While it can be argued that it is unfair to make comparisons based on initial lines, from an Owner's perspective, it is not unusual for a shipyard, in their design process, to stop at the initial lines. The cost of the additional time spent in this portion of the design process is more than outweighed by the benefits.

Although it is a case of comparing apples and oranges, there is a tendency to compare the new ships to the ones they ultimately replaced. In comparison to the 120,000 DWT tankers that were in operation, the model testing was predicting an improvement of one knot in speed and a significant reduction in fuel consumption. The new ships are capable of carrying 125,000 DWT vs. the 120,000 DWT of their predecessors, they run a full knot faster, and they consume about 520 bbls/day of fuel as compared to 850 bbls/day. While it is recognized that much of the fuel savings comes from the switch from steam to diesel propulsion, much of the improvement is coming from improved hull efficiency. This improved efficiency is particularly noticeable in the ability to sustain operations in seaways for longer periods of time.

Seakeeping Testing

To Polar's knowledge, little work has been done in the past with respect to tankers and seakeeping. What work has been done has either been related to loading at offshore moorings and dynamic positioning, or was considered proprietary and was not readily shared by owners or model basins. As a result, real life performance, particularly of vessels engaged in the Alaskan trade, differed greatly from the calm water predictions from the shipyards. As a whole Polar's vessels, particularly in winter, suffered from wet to submerged decks, fatigue cracking as a result of cyclical wave loading, and slowing in speed due to seas. It was believed that a modest effort in the area of seakeeping would result in drier decks, reduced motions, and improved at sea performance. An added benefit was data collected would be used to support the structural design effort for the Dynamic Load Approach and the Structural Fatigue Analysis.

Program

SSPA and Polar embarked on a unique effort to both confirm the heavy weather characteristics of the vessel and to support the structural design efforts. The program consisted of two parts; first computer predictions of performance that would be used to develop the lines plans and second model testing to confirm the computer predictions and to provide real time data to the structural design effort.

Computer Predictions of Performance

SSPA used their SCORES computer program to make predictions of the vessels seakeeping characteristics. Using this program, they were able to make recommendations regarding bow flare and to modify bulb design to reduce immergence and to limit slamming. Their predictions indicated that the vessel would be drier than previous vessels, partially due to higher freeboard, and partially to better design. SSPA also encouraged, at this time, the addition of bilge keels to reduce rolling. This suggestion was not accepted because of bad experiences with structural cracking from bilge keels.

After completion of the seakeeping model tests, the computer program was again run to correlate the program to the model. In doing this, SSPA and Polar were able to cooperate in an iterative process, which enhanced data accuracy.

Model Testing in Seakeeping Basin

SSPA, at Polar's insistence, pushed their seakeeping testing in the model basin to its practical limits. First, to obtain data for structural design, Polar asked that pressure sensors be placed not only on the forecastle deck, upper deck, bow flare and forefoot, but that the model be cut amidships so that bending moments and shear forces could be measured. Polar also asked that the side of the vessel be fitted with a grid of pressure sensors so that a CFD model being used for structural design could be calibrated. These demands resulted in one of the most highly, if not the most highly, instrumented tanker models tested in SSPA's model basin.

SSPA recorded the following data during the seakeeping testing:

- Wave height
- Speed, rudder angle
- Thrust starboard propeller
- Thrust port propeller
- Number of revolutions, starboard propeller
- Number of revolutions port propeller
- Surge, sway, heave (re-calculated to the center of gravity)
- Roll, pitch, yaw
- Vertical bending moment amidships, L/2
- Vertical shear force amidships, L/2
- Horizontal bending moment amidships, L/2
- Relative motion station 1 starboard (station 0=AP), RM1
- Relative motion station 1 port, RM2
- Relative motion station 8 port, RM3
- Relative motion station 10 port, RM4
- Relative motion station 12 port, RM5
- Relative motion station bow, RM6
- Longitudinal x-acceleration at one location
- Lateral y-acceleration at two locations
- Vertical z-accelerations at three locations
- Slamming pressure with 5 pressure gauges, P1-P5
- Water pressure with 9 pressure gauges, P6-P14
- Water on deck with three gauges, G1-G3

All testing was done for both the loaded (Tf/Ta=16.0/16.0) and the ballast conditions (Tf/Ta=9.0/10.0). Tests were carried out on a 1:48.333 scale model in six different wave directions from 180 degrees (head waves) to 30 degrees. Tests were carried out for five different wave periods for each of the six headings at a ship speed of 15 knots. Tests were carried out for six headings in irregular waves from two perpendicular directions having a significant wave height of 4 meters and the cross direction having a significant wave height of 3 meters. The ship speed for the irregular waves was 14 knots. In irregular head waves (no cross waves) tests were also carried out in two storm conditions, significant wave heights of 9 meters and 13 meters at a ship speed of 5 knots.

Fig. 12 Model testing ($H_{1/3}$= 9 m irregular head sea)

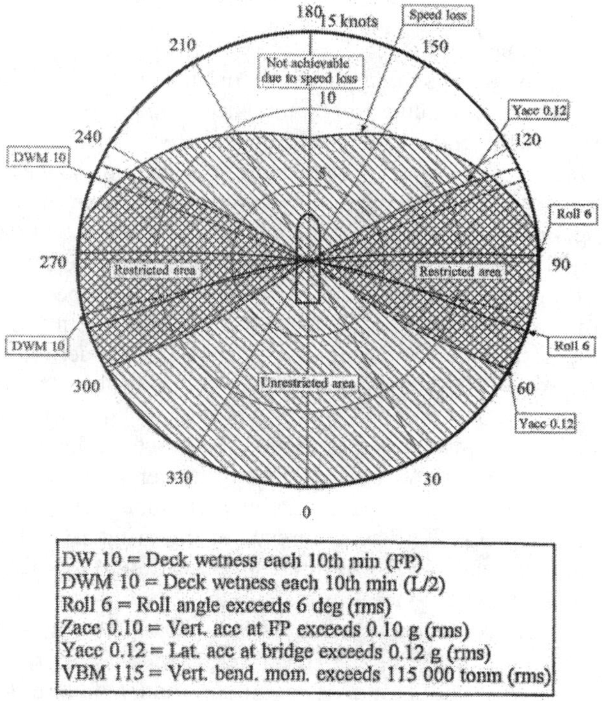

Fig. 13 Operating curves for Full Load, Sea state 7

Figure 12 shows a $H_{1/3}$= 9m irregular head sea [9]. As can be seen from the photo, the deck and the forecastle are both dry. In comparison to our current ships we would expect to have large amounts of green water. In service we have found that the model predictions have been very accurate. The tests also showed that there should be little or no bottom slamming, which has been confirmed in service.

SSPA found that their measured and calculated wave-induced motions, accelerations, vertical bending moment, vertical shear force and added resistance in regular waves showed very good agreement for both loading cases.

The vessel displayed good course keeping in stern seas, wave directions 30 degrees and 60 degrees, for both loaded and ballast conditions in regular and irregular waves. In beam waves, wave direction 90 degrees, the following was observed especially for the full load condition: The tests in regular beam waves demonstrated that for waves with short periods there was a noticeable tendency to yaw into the waves. For long period waves there was a slight tendency to yaw away from the waves. There were some immediate periods with no marked tendency to yaw. The tests in irregular beam seas also revealed a tendency to yaw into the waves though less pronounced than the tendency observed in short regular waves. The ships have confirmed this tendency particularly when finding themselves in the trough of two waves.

The seakeeping testing provided exceptional data in support of the structural design work and in the effects of seas on the vessel. What wasn't clear was, how do we relate the information obtained in the model testing to the operators? Our vessels have long been fitted with a "Hull Monitoring System" provided by MCA Engineers as part of our Critical Area Inspection Program (CAIP) compliance. We use the hull monitoring systems on our vessels to monitor roll, pitch, acceleration, and bow immersion. Our goal is to reduce fatigue cracking by reducing the motions to which the vessel is exposed. To this end we approached SSPA about the possibility of presenting the seakeeping data in a format useable to the operator. What they proposed was an operability analysis, which would transfer the data to a user, friendly environment. Figure 13 shows the operating curves for the full load condition in a Sea State 7 [12]. The operator can look at the predominant wave direction, account for speed loss and determine the most suitable course. MCA has included these plots as part of the Hull Monitoring System so they are readily available to the operator on the bridge.

Results

The seakeeping studies at SSPA provided positive results consistent with Polar's desires. The *Endeavour* Class vessels perform better than previous vessels with regards to deck wetness, slamming, and the ability to maintain speed for longer periods of time, providing weather is not directly on the bow. On the other hand, crew complaints with regards to rolling are somewhat higher than expected, although not unanticipated.

The ability to use the model test data to support structural design was very significant. The data was used to confirm the CFD analysis used for the Dynamic Load Approach and in the Spectral Fatigue Analysis. Data was also used to confirm shear force and bending moment predictions.

Unanticipated was the useful data obtained in way of the operability analysis. While naval architects have long understood the effects of cyclic loadings on ships, and ship's crews have long understood the effects of sea state and direction on comfort, finding a common discussion point was problematic. The operability analysis has provided the naval architect a means for illustrating information on vessel motions. This information, which is intuitively obvious to the senior officer based on their experiences and training, can then be conveyed to junior officers as part of their training. To make the information readily available, the operability plots are included as screens on the vessel's Hull Monitoring System, which measures roll, pitch, acceleration and bow immersion. Junior officers have quick access to information on direction to alter course to minimize vessel motions.

It should be noted that while providing an operator with additional information on seakeeping of the vessel is important, it does not substitute for their experience. These plots are a useful training aid for experienced Masters to teach junior Mates. We find them to be only a reference to the senior Masters.

Maneuvering Computer Design & Model Testing

Background

Having made the decision to build twin-screw, twin rudder tankers, improved maneuverability was an inherent benefit when compared to the existing single screw vessels in Polar's fleet. However, decisions still had to be made with regards to enhanced maneuverability and just how much enhancement was required. It was determined quickly that there was no need, short or long term, to fit the *Endeavour* Class vessels for dynamic positioning. Therefore there was no need for multiple thrusters fore and aft. However, decisions still needed to be made as to whether a bow thruster would be fitted, and whether high lift rudders would be utilized.

Polar's senior officers wanted the ability to be able to dock and undock a vessel without tug assistance. Although it was never intended to eliminate docking tugs, the officers felt that it was advantageous to be able to fully control the movement of the vessel on and off the docks from the vessel's bridge. They wanted the new vessels to have the ability to move sideways and at angles, as desired. It was determined that tugs would always be used when winds reached 30 knots, so there was no need to supply thrusters for conditions greater than 25 knots. The officers also wanted the ability to have greater rudder control for maneuvering in port and during transits through Prince William and Puget Sounds.

Because these vessels were new to the Polar fleet and to U.S. shipping in general, Polar also felt a need to be able to acquire enough maneuvering data through computer simulation and model testing to allow the construction of accurate bridge simulation programs. This would allow officers to be thoroughly trained in ship handling prior to having ever set foot on a new vessel. This portion of the model test program provided necessary data to allow Polar's officers and port pilots to be trained in advance of the arrival of the *Polar Endeavour* on the West Coast. In general, since putting the vessel in service the officers have found it to be slightly more responsive than the initial simulator models predicted. These models have since been adjusted and now portray accurate representations for ship handling. A scale model of the ship has also been built for training at Warsash, based on the modeling information, to aid in ship handling training.

Computer Simulation

Computer simulations were used to assist Polar in selection of rudder type and size, range of rudder movement, and in sizing the bow thruster. To ensure an objective point of comparison in selection between alternatives, maneuverability projections were made against single screw tankers and IMO Resolution A.751. Prior to the start of testing Polar had interest in fitting conventional Mariner rudders, primarily because of simplicity and past experience. There was a high comfort level increasing the size of the rudders and allowing rudder movement to increase from the normal 35° to 35°, to 45° to 45°, providing that rudder movement was limited to 35° at speeds over 12 knots.

Table 2 shows a comparison of rudder sizes used in the study [2]. The total rudder area ultimately used was 3.66% of LppT. The difference in the original 3.03% and the final 3.66% was in the modification of the stern from the initial to final lines.

The computer simulations focused on turning and zigzag tests. Comparisons were made in normal conditions and assuming a single rudder failure to determine ability to meet IMO criteria in a disabled condition. All simulations were done using SSPA's PORTSIM simulation program. Tables 3 and 4 show the results of the simulations for turning circles and zigzags in normal operations [2].

Table 2 – Comparison of rudder sizes used in study

Ship	1	2	3
Stern Type	Twin Screw	Twin Screw	Single Screw
Rudder Type	Conventional Mariner	High Lift	Conventional Mariner
Rudder area, moveable [m^2/rudder]	60.8	44.9	89.8

Rudder area, total [m²/rudder]	78.6	55.8	111.6
Height [m]	10.5	9.0	12.8
Total Area [% of LppT]	3.03	2.24*	2.24*

**Based on SSPA formula for minimum recommended area (% of LppT) = CbB/T*

Table 3 - Turning tests, Study ships vs. IMO requirements
Approach speed = 16 knots
35° rudder

	IMO	Ship 1	Ship 2	Ship 3
Advance/Lpp	4.5	2.18	2.00	2.22
Tactical Diameter/Lpp	5.0	2.47	2.20	2.20

Table 4 - Zigzag tests
Approach speed = 16 knots

	IMO	Ship 1	Ship 2	Ship 3
10°/10° Zigzag				
Initial Turning Ability / Lpp	2.5	0.99	0.91	1.08
1st Overshoot (°)	20	6.2	6.6	8.7
2nd Overshoot (°)	35	8.2	8.7	16.6
20°/20° Zigzag				
1st Overshoot (°)	25.0	13.0	14.3	17.9
2nd Overshoot (°)	n/a	13.5	14.3	21.5

As indicated, both the conventional Mariner rudder and the high lift rudder produce results that are significantly better than the IMO requirements and that are better than a comparable single screw vessel. Based on the computer predictions, Polar's knowledge of Mariner rudders, and no requirement for dynamic positioning, it was determined best to proceed using oversized Mariner rudders.

As part of the performance evaluation simulations were made for disabled conditions. Predictions were made assuming one of the rudders failed. As can be seen from the data, the *Endeavour* Class tankers should meet IMO maneuverability requirements even with a single failed rudder and a failed propeller. Tables 5 and 6 show the results of the simulations for turning circles and zigzags, under starboard single screw operation [2]. These results were later updated when the R2-S+ testing was done. However, rather than upgrade the tables, operability plots were developed.

For the purpose of the computer simulations it was assumed that a single 2800 KW tunnel thruster of 3.0 m diameter and an anti-suction tunnel would be installed forward to produce 360 kN maximum thrust. This provided an ability to crab laterally at 0.5 knots against a 25-knot wind, with a 2-knot current running along the length of the vessel. In order to turn the ship 90° at ballast draft in about 500 seconds a 2207 KW (3000 HP) thruster was determined to be required. Ultimately a 2207 KW thruster of 2.4 m diameter with a 0.8 m diameter anti-suction tunnel were fitted.

Model Testing

Free sailing model tests were carried out in order to study the maneuvering characteristics in general, and with respect to IMO Resolution A.751 in particular. The 1:48.333 scale model used for the seakeeping testing was also used for the maneuvering tests. Turning tests to port and starboard at 35°, zigzag tests at 10°/10° and 20°/20° with 1st execute to port and starboard, and reverse spiral tests for both full load and ballast conditions were conducted. Table 7 shows the maneuvering characteristics for the *Endeavour* Class tankers in their final design configuration [8].

IMO stopping ability was not tested during the maneuvering tests.

Results

Because there was no intent to dynamic position the *Endeavour* Class vessels it was decided that oversized Mariner rudders were a better solution than high lift rudders. The final rudders were approximately 35% larger than what is considered normal for rudders on a vessel of this size. As a means of limiting loading on the rudders an electronic limit was installed which limits rudder movements to 35° at speeds 12 knots and above and allows the rudders to go to 45° at speeds below 12 knots.

It was determined between the wind tunnel testing and the maneuvering testing that a 2207 KW (3000 HP) bow thruster would be used. This thruster would be capable of turning the vessel in a 25 knot wind, and would have the capabilities to work in unison with the twin screws and twin rudders to allow sidewise motion of the vessel at low and zero speeds.

Table 5 - Turning tests in a disabled condition
Approach speed = 11.5 knots
35° rudder

	IMO	Ship 1 Port/Stbd	Ship 2 Port/Stbd
Advance/Lpp	4.5	2.59/3.14	2.37/2.87
Tactical Diameter/Lpp	5.0	3.04/3.78	2.75/3.33

Table 6 - Zigzag tests in a disabled condition
Approach speed = 11.5 knots

	IMO	Ship 1 Port/Stbd	Ship 2 Port/Stbd
10°/10° Zigzag			
Initial Turning Ability / Lpp	2.5	1.31/1.34	1.16/1.24
1st Overshoot (°)	20	5.5/5.1	5.35/5.5
2nd Overshoot (°)	35	6.4/9.9	6.9/9.0
20°/20° Zigzag			
1st Overshoot (°)	25.0	11.3/8.8	11.4/9.4
2nd Overshoot (°)	n/a	9.5/13.6	10.3/13.2

Table 7 – Final maneuvering characteristics

	IMO	*Endeavour* Class
Turning Test 35°		
Advance/Lpp	4.5	3.19
Tactical diameter/Lpp	5.0	3.21
Zigzag Test 10/10°		
ITA/Lpp	2.5	1.55
1st overshoot [deg]	20	5.9
2nd overshoot [deg]	35	8.6
Zigzag Test 20/20°		
1st overshoot [deg]	25	11.8
2nd overshoot [deg]	n/a	9.8

Performance

The ultimate goal was for the vessels to exceed the IMO requirements for maneuverability, to be able to crab on and off docks under their own power, and to be able to turn at zero speed in their own length. These goals were based on the concept that the vessels would be highly maneuverable at sea, when maneuvering, and when docking. This goal was achieved based on computer predictions, model basin testing, full scale testing, and most importantly in service. The *Endeavour* Class vessels maneuvering capabilities closely match those predicted in computer and model testing. Since their introduction, the vessels continue to have tugs secured, but they dock and undock under their own power and control. They can turn, at zero speed, in their own length.

Propeller Design & Cavitation Testing

Background

Polar relied historically on the shipyards for hull and, subsequently, for propeller design. After some unexpected propeller problems on one class of vessels, it was discovered that the shipyard and the model basin had decided, together, to forego any specific propeller design and instead to use a library propeller as a "good enough" fit. Later in the life of the vessel this caused both performance and cavitation problems, both of which could have been avoided. Polar required that propeller design and cavitation studies be done to ensure that the *Endeavour* Class vessels had propellers that were "optimized" within the design limits of the engines and hulls. SSPA started with a "best" selection from their propeller library, tested with the model, then designed a specific propeller for the *Endeavour*. The designed propeller was tested with the final hull design and was cavitation tunnel tested to ensure performance. Propeller manufacturers were allowed to either accept the SSPA design, or modify the design, under the provision that any significant change would require a return to SSPA to repeat performance and cavitation tests.

Program

The initial test program consisted of open water testing to confirm the performance of the library propeller and then resistance and propulsion testing using the library propeller. Based on these results, along with the wake surveys, a specific design propeller was provided by SSPA. SSPA was not constrained in design, other than an owner's constraint limiting back skew to 35°, and a requirement to meet ABS Rules.

Library Propeller

Figure 14 shows the open water curve for the library propeller [3]. The library propellers alone provided satisfactory results. However, both SSPA and Polar were confident that improvements in efficiency and noise could be achieved by using a specifically designed propeller.

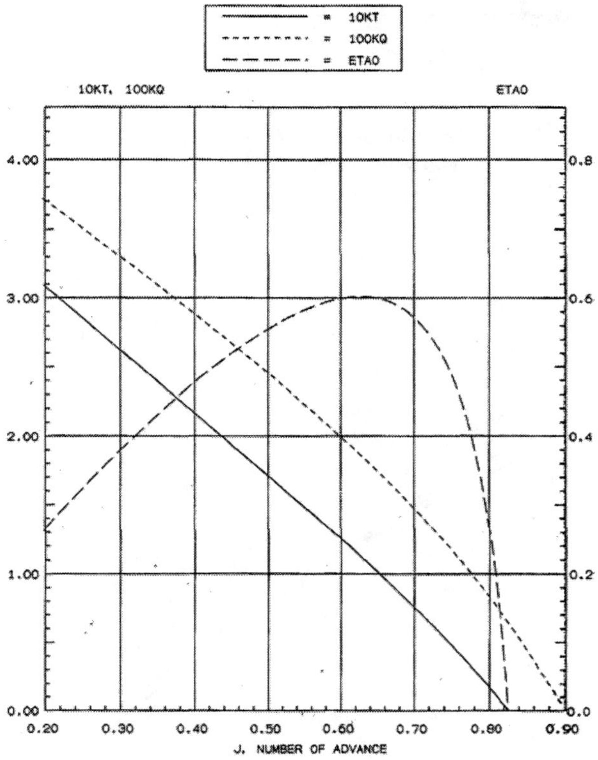

Fig. 14 Open water curve for library propeller

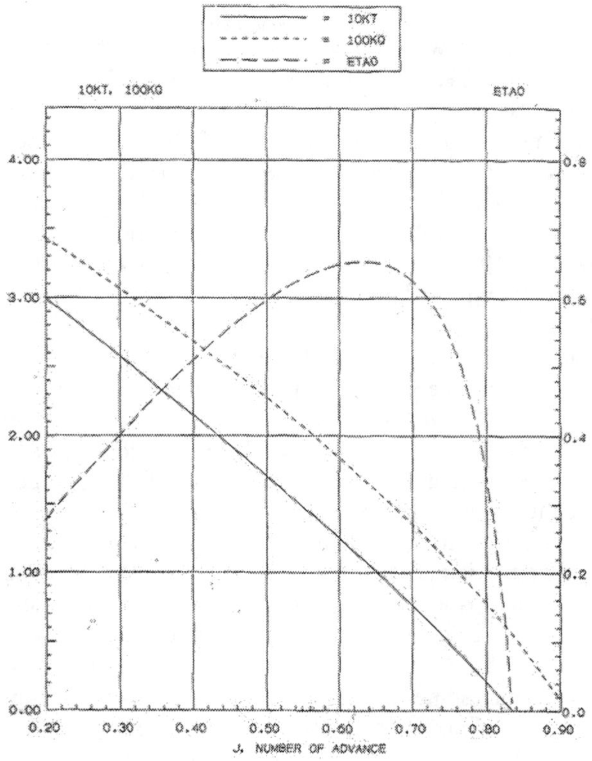

Fig. 16 Open Water Curve for SSPA Propeller

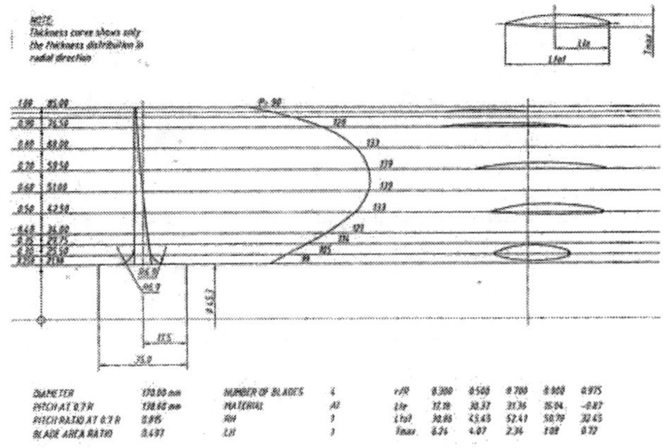

Fig. 15 SSPA propeller design

SSPA Propeller Design & Testing

In designing a specific propeller for the *Endeavour* Class vessels, SSPA again approached Polar about their constraints. Polar again reiterated their desire to retain propeller speed at 127 rpm. In discussing Polar's somewhat arbitrary decision to limit back skew to 35°, SSPA indicated that they felt no need to use higher skews. Polar's reluctance to increase skew angle was based primarily on inexperience. Unlike passenger or naval vessels where noise is of significance, tanker owners worry more about reliability and maintainability then comfort.

Open Water

Figure 15 shows the SSPA designed propeller [5]. The back skew angle selected by SSPA was about 25°. Figure 16 shows the open water curve for the SSPA design propeller [5]. The new propellers showed a 2.5% improvement in efficiency over the library propellers.

Cavitation Tunnel

Cavitation tunnel testing was done using SSPA's large cavitation tunnel. Because of the higher propeller speed (127-rpm), the smaller diameter (5.8 m), and the high tip clearance (4.2 m in ballast) very good performance was expected. Not surprisingly, excellent performance resulted. Cavitation observation tests were done at two loading conditions, equivalent to 90% MCR at 14 knots at full load draft and 15 knots at ballast draft. This is equivalent to a sea margin of 70%. The predicted vibration velocity at blade frequency using SSPA's empirical method was about 0.75 mm/s at full load and about 2.1 mm/s at ballast draft. According to SSPA this is a very conservative estimate. The ISO6954 norm regards vibration velocities of less than 2.2 mm/s as fully acceptable.

Kamewa Propeller Design & Testing

Having been awarded the contract to supply propellers for the *Endeavour* Class vessels, Kamewa was given an option to accept the SSPA design, or to offer an alternative design. On review of the provided design Kamewa discovered that adjustments had to be made for compliance with ABS Rules to meet specific design particulars. Most of these adjustments were associated with requirements for blade root strength and thickness. Kamewa also initially approached Polar with a proposal to increase back skew above 35° in hopes of recovering some performance losses resulting from changes to blade root thickness. Ultimately, Kamewa found that a 33° back skew was consistent with their optimum.

Figure 17 shows the final Kamewa designed propeller, with a back skew angle of about 33° [19]. Figure 18 shows the open water characteristics for the Kamewa propeller in comparison to the SSPA propeller [19]. As can be seen from the two curves the two propellers are extremely close in performance, but both are better than the library propeller.

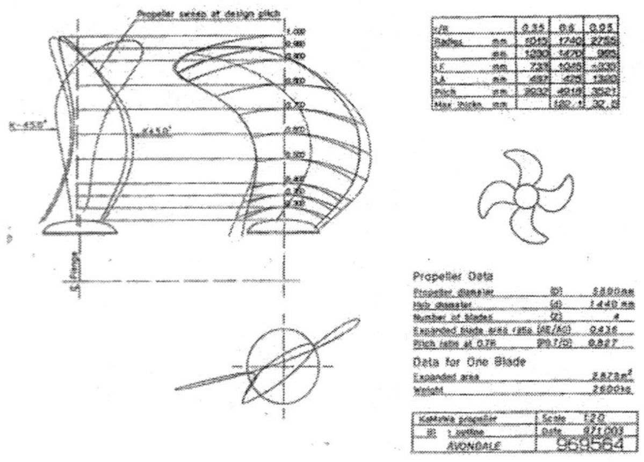

Fig. 17 Kamewa design propeller

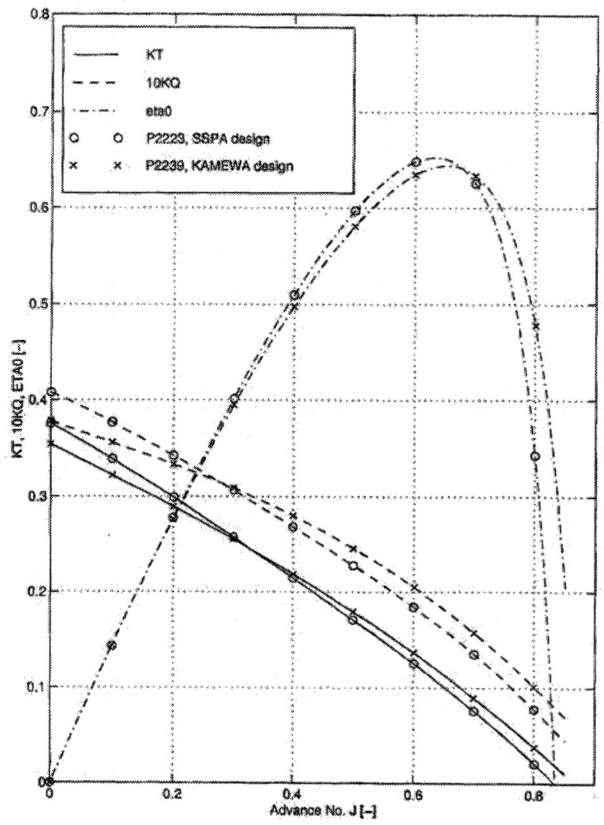

Fig. 18 Propeller comparison - Kamewa vs. SSPA

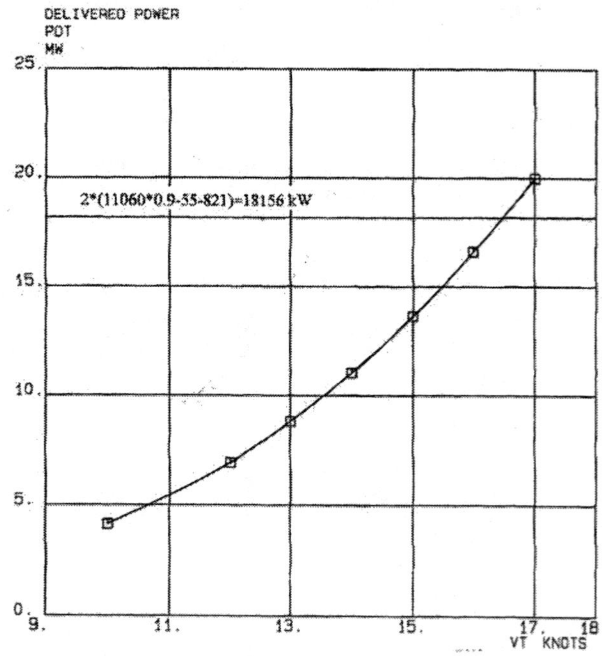

Fig. 19 Final speed-power curve for *Endeavour* Class tankers

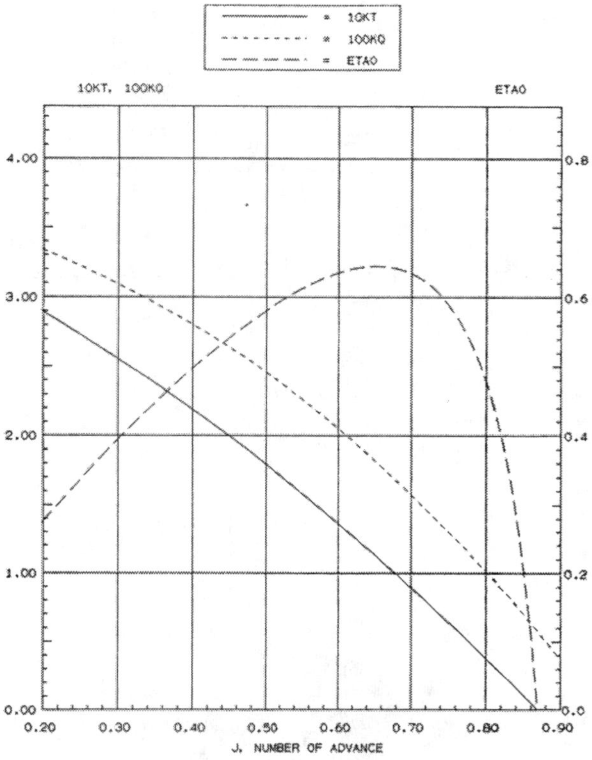

Fig. 20 Final Kamewa propeller curves

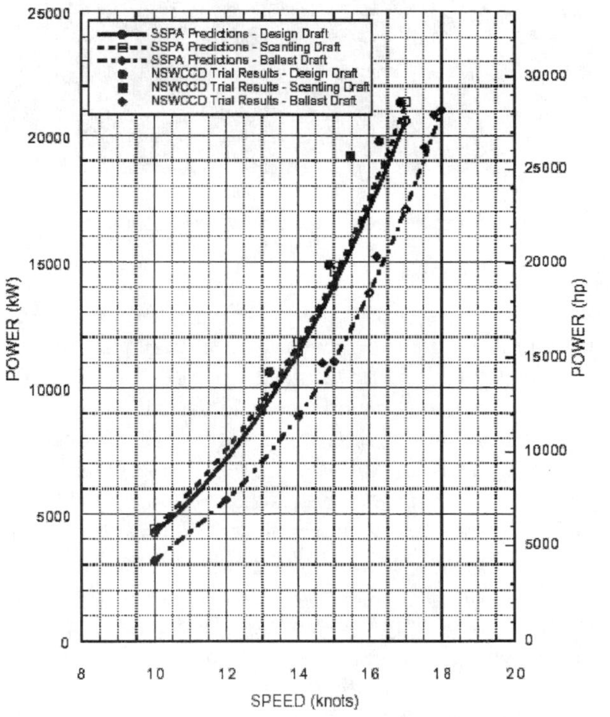

Fig. 21 Sea trials results – SSPA predictions vs. sea trials data

Fig. 22 Cavitation Test Results

Open Water

Figure 19 shows the final speed power curve for the *Endeavour* Class tankers [19]. Figure 20 shows the open water curve for the selected propellers [19]. Although there was a slight decrease in performance between SSPA's propellers and Kamewa's due to blade root thickness, the difference was within measurement accuracies. Figure 21 shows the results from Sea Trials [26]. It should be noted that the Sea Trial testing was somewhat clouded by currents and poor weather.

Cavitation Tunnel

Like the SSPA designed propellers, the Kamewa propellers performed superbly in the cavitation tunnel. Cavitation observation tests were done at two loading conditions, equivalent to 90% MCR at 14 knots at full load draft and 15 knots at ballast draft. This is equivalent to a sea margin of 70%. The predicted vibration velocity at blade frequency according to SSPA's empirical method was about 0.72 mm/s at full load and about 2.1 mm/s at ballast draft. This according to SSPA is a very conservative estimate. No vibration problems were anticipated due to values measured pressure pulses. Figure 22 shows the impact of cavitation on the propeller [20]. The loss of paint near the blade root was from handling versus cavitation.

Results

The final propeller design resulted in propellers that met Polar's expectations and that will provide excellent performance. Although having a moderate back skew of only 33°, it is still a very large amount for most tankers. Cavitation tunnel results predicted negligible cavitation and propeller-induced noise. To date there have only been diver inspections of the propellers but there have been no reports of any damage of any kind. We have had no reports of any cavitation noise or propeller-induced vibrations from the crew.

Wind Tunnel Testing

Program

Unusual to tanker design are wind tunnel tests. In general, the design of tankers does not require information provided by wind tunnel studies. Stack heights are generally high enough that smoke problems rarely exist. Information on wind loads is useful, but there are other suitable ways to predict loading.

The Carquinez Bridge in the San Francisco Bay area limits the air draft of large tankers. The air draft in effect limits the height of the radar scanners. This in turn limits stack height to ensure there are no blind spots on the radar screens. Based on experiences with the design of Polar's 190,000 DWT vessels we knew this process resulted in smoke problems on the stern of the vessel. Similar hull depths, and requirements for bridge visibility that results in higher bridge heights, led us to believe we would encounter similar problems on the *Endeavour* Class vessels. These factors led to a decision to do testing, recognizing we could only reduce, but not eliminate smoke problems. It was also decided that if we were to do testing it would be a good opportunity to validate wind force data.

Another issue associated with current tanker design is the affect of bunker tank vent locations on the intake of fumes and odors into the accommodation block ventilation system. This problem has recently been exacerbated by recent SOLAS rules, which require raising bunker tank vents above the upper deck level to prevent accidental down flooding by boarding seas.

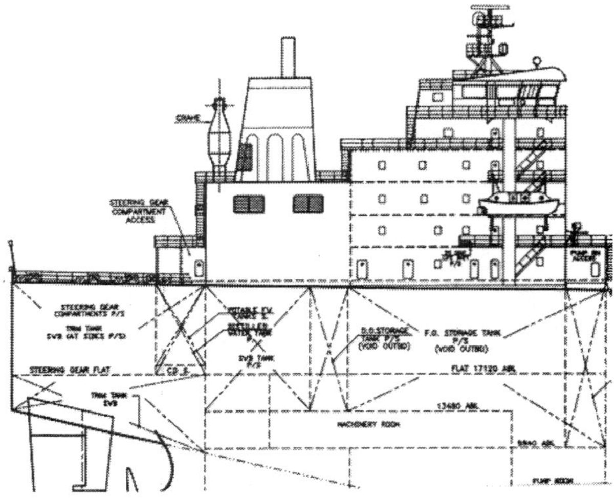

Fig. 23 Original house and stack configuration

Fig. 24 Scale model used in wind tunnel testing

Stack Gas Trajectories

As expected, the early design configuration shown in Figure 23 provided results that were less than satisfactory. Smoke tests indicated that a large amount of smoke would be drawn into the dead area immediately aft of the house and forward of the stack and onto the after deck area of the stern. Using a 1:250 scale model for the testing, shown in Figure 24, the first efforts were made to modify the configuration of the exhaust pipes coming from the stack [10]. Angled pipes replaced straight pipes. Various angles from vertical were tried, including angling the stacks off centerline. Angled pipes worked much better than straight pipes, but the farther off centerline the pipes were placed the worse the performance became.

While this process provided improvement, on a relative scale improvement was from very poor to poor. Further improvement was needed. At this point SSPA, based on previous experience, and the author, who was involved in similar tests done years earlier, concluded that some form of foil was needed on the house top to modify the airflow to the stack. Polar's 190 MDWT tankers were fitted with a radiused leading edge on the wheelhouse, and a foil about 1 m above the house top. In the case of the *Endeavour* Class vessels the wheelhouse is fully enclosed from the centerline to the bridge wings. SSPA, based on previous experience, recommended not putting a radius on the leading edge of the house front, but instead extending the wings for the full width of the enclosed bridge. These modifications were made with an improvement in the amount of smoke being kept between the stack and house and on the after deck. While not perfect, the improvement could be rated on a relative basis from poor to fair/good. Short of raising the stack 6 m, good to excellent performance was not possible.

Wind Load Testing

Testing for wind loads was done in both ballast and loaded conditions. The tests provided no new information over that found in numerous publications. The tests served only to provide measured versus calculated values for wind loading. This data was used to aid in mooring studies and in the sizing of the bow thruster. The data did confirm existing data as being accurate.

Ventilation Intakes

Eliminating fumes and odors from bunkers and lubricating oil vents will likely be a problem for vessels under the new SOLAS rules for vent locations. As a rule of thumb the accommodation block air intakes should be above and away from vents. This, of course, is easier said then done. Rules push the vents to the "B" (third) deck level, and intakes higher. Because the SOLAS rules had not yet been promulgated at the time the initial design was being done for the *Endeavour* Class, no effort was made to raise the air intakes from their original positions immediately below the "A" deck level (galley and wheelhouse intake), and on the "B" deck (accommodation block intake) between the house and the stack. The accommodation ventilation intake was far enough away to ensure that no fume or vapors would be drawn into the house. The same could not be said for the galley and wheelhouse intakes. It was virtually impossible to locate the intakes far enough away from any vent to ensure that no fumes would be drawn into the house. Rather than redesigning the accommodation block, it was decided best to place intakes port and starboard to allow one side to be secured when fuel or lubes were being transferred on that side.

Results

The wind tunnel testing proved to be very accurate in its predictions. Air foil angle was set to an "optimum" based on subjective nighttime sea trial observations. A determination was made as to when the least amount of smoke was present over the after deck and between the house and stack. In service, smoke is drawn into the area between the house and stack and affects the afterdeck. However, smoke is no worse than expected. In the case of the air intakes, the starboard side air intakes were eliminated as part of the yard's design process. Unfortunately, this also eliminated the ability to select the side air was being drawn into the galley and the wheelhouse. This has proven to be a design error. Bunker fumes have plagued the *Polar Endeavour* since she entered service. This problem has been resolved on later ships by raising the wheelhouse and galley vent intakes through the "A"-Deck platform to provide a distinct isolation. Since this correction has been made the complaints have disappeared.

Disabled Ship Testing

Program

The Disabled Ship Testing program provided perhaps the most unique and creative portion of the model test program. Up until that point in time, analysis of disabled ships was limited to drifting studies and some computer simulations to predict performance of twin screw/twin rudder ships in disabled conditions. Being the first ship to seek compliance with ABS R2-S+ requirements, computer simulations were required for the first time to prove vessel capabilities. ABS was concerned how accurate computer simulations were, however, and if they were truly indicative of the vessel performance. After long discussions between Polar and ABS, and Polar and SSPA it was determined feasible to return to the model basin using the maneuvering model to verify the computer model. In addition, a unique opportunity occurred during sea trials to test disabled performance capabilities in Sea State 6 conditions. This furthered knowledge of disabled ship performance.

Computer Simulations

Compliance for the ABS R2-S certification requires:

"A vessel fitted with multiple propulsors (hence multiple propulsion systems) which has the propulsion machines and propulsors, and associated steering systems arranged in separate spaces (propulsion machinery space and steering gear flat) such that a fire or flood in one space would not affect the propulsion machine(s) and propulsor(s), and associated steering systems in the other space(s) will be assigned the class notation R2-S [27]. *"*

For the plus (+) notation;

"Upon a single failure, the propulsion and steering system is to be continuously maintained or immediately restored within 2 minutes, as in the case when an alternate standby type of propulsion is provided (e.g., electric motor, diesel engine, water-jet propulsion, etc.) such that the vessel is capable of maneuvering into an orientation of least resistance to the weather, and once in that maintaining position such that the vessel will not drift for at least 36 hours. Using all available propulsion and steering systems including thrusters, if provided may achieve this. This is to be possible in all weather conditions up to a wind speed of 17 m/s (33 knots) and significant wave height of 4.5 m (15 ft.) with 7.3 seconds mean period, both of which are acting

concurrently in the same direction. The severest loading condition for vessel's maneuverability is also to be considered for compliance with this weather condition [27]. *"*

These weather conditions are approximately equivalent to a Sea State 6 or Beaufort 8 sea.

SSPA, using their Portsim simulation program, was able to provide computer predictions for single failures required under R2-S+, and also multiple failures, i.e., rudder and propeller, and two rudders. Portsim predicted compliance, however ABS had no basis to determine the accuracy of the computer predictions. Because we could not provide correlation data for disabled conditions, we reached agreement with ABS to do disabled model testing. Once this model testing was complete, it was agreed that the Portsim model would be updated to reflect the model data. Of particular interest in the Portsim predictions was that it was easier for the vessel to proceed en route than it was to hold station per ABS R2-S+ requirement. When asked, ABS agreed that proceeding under control to the vessel's destination satisfied their requirement for holding station.

Model Test Confirmation

A full series of model tests were performed to validate the Portsim model. These tests were designed to study the behavior of the vessel in various failed conditions. For single failures, model tests were carried out either with one propeller free-wheeling (wind-milling) at design pitch or with one rudder locked amidships. For multiple failures, model tests were carried out with either one propeller free-wheeling at design pitch and a rudder locked amidships, or with a propeller free-wheeling at zero pitch and the rudder locked hard over at 45°. The tests were carried out at two sea states with long-crested irregular waves and steady wind. Table 8 lists the maneuvering test criteria used [21].

Course-keeping and steering tests were run to determine minimum speed to maintain heading and maximum attainable speed. Drifting tests were also conducted to supplement the **Disabled Tankers Report of Studies on Ship Drift and Towage,** published by the Oil Companies International Marine Forum [28].

Table 8 - Maneuvering Test Criteria for R2-S+

Sea State	Sine wave height (m)	Zero Crossing Period (sec.)	Mean Wind Velocity (m/s)
4	1.9	6.3	9.8
6	5.0	8.8	19.3

Sea Trials

Polar intended from the very beginning to verify the results of disabled vessel testing during sea trials. It was intended, as with any sea trial, to perform a series of tests in calm weather conditions that could be extrapolated to heavy weather conditions. During trials, rather than having calm conditions at the time disabled vessel testing was to occur, there were Sea State 6 conditions. To facilitate testing, and to obtain the most useable data, it was decided the best way to acquire that data was to have the vessel perform 10°/10° and 20°/20° zigzags at increasing 30° heading increments until the compass rose was boxed. One propeller was set to zero pitch and was allowed to freewheel, and the corresponding rudder was locked hard over at 45°. Except for one 10°/10° maneuver, the vessel was able to maintain steerage and headway. In the case of the one maneuver, the vessel went into "irons" (i.e., lacked sufficient speed and stopped) and it had to wait until it naturally came back into the wind before control could be regained. This result was consistent with predictions from Portsim.

Results

The data acquired during the Disabled Ship Testing portion of the design provided some interesting problems for SSPA and Polar. The ship outperformed either SSPA's or Polar's expectations. In fact, the performance was so good that a device for transferring the information to ship's crews had to be developed.

The *Endeavour* Class vessels satisfied the ABS R2-S+ requirements for single failures in Sea State 6. Fig 25 and 26 show the polar plots for single failures (i.e., propeller or rudder) in a Sea State 6 as required for compliance for R2-S+ [22]. As can be seen from the figures it is more difficult for the vessels to maintain position than it is for them to proceed to their destinations. In neither the loaded, nor the ballast case, do the vessels, after a single failure, encounter any difficulties proceeding. The inner ring of the "donut" is the minimum speed for the vessel to maintain steerage. The outer ring is the maximum speed attainable. As can be seen, single failure compliance is attainable.

Further to the single failures was the question of multiple failures. Fig 27 and 28 show the polar plots for multiple failures (i.e., propeller and rudder) in a Sea State 6 [21]. As can be seen, it is still possible to control the vessel, however, the ship's crew must be aware of limitations and must avoid certain "dead" areas where the ship could go into "irons". The "donut" reduces to a "C" in the area from 30° to 90°. It was in this region that during sea trials steering control of the vessel was lost during the 10°/10° maneuver. During the 20°/20° maneuver for the same failure steering control was maintained.

To provide the ship's crews with this data in a useable and understandable format it was decided to develop a supplement to well known publication **Peril at Sea and Salvage, A Guide for Masters** [29]. For this document to be useful it had to present the work done by naval architects at SSPA in a form known and useable to ship's personnel. Working with SSPA a stand-alone supplement was written. This supplement provides polar plots along with written explanation. Using "Peril at Sea" and the Supplement the ship's officers have enough data at hand to respond to single and multiple failures.

HULL STRUCTURAL DESIGN

Structurally the *Endeavour* Class vessels were designed and built for a 30-year fatigue life. The vessels are built primarily with mild steel, with little or no reduction in scantlings. Where high tensile steel is used (upper deck in way of the cargo block, sheer strakes, and highly stress details) margins above regulatory allowable minimums are applied. Critical structural areas were studied by finite element analysis, dynamic load approach and spectral fatigue analysis.

Criteria

Based on Polar's Alaskan operating experiences a number of design criteria were established for the structure. First, it was determined that mild steel (primarily ABS Grade A, with some B and D) would be used for structure. If high tensile steel were to be used, a reduction in scantling was allowed, providing there was a minimum 10 percent allowance above the ABS requirements, including corrosion allowance. Structural steel in cargo and ballast tanks was to be a minimum of 12 mm in thickness and was to be no more than 25 mm. For the upper deck, because the double hull construction results in mild steel plate thickness well above 25 mm, the deck is constructed of ABS Grade DH plate. However, by using Grade DH plate, ABS rules allow for a plate thickness of 19.5 mm, including corrosion allowance, but 22 mm plate was used. Sheer and deck stringer strakes are Grade EH plate.

Structural design was developed based on careful engineering. Polar Tankers required that the ships be designed for a 30-year life. Compliance required not only compliance with ABS' SafeHull A and B programs, adjusted to 30 years, but finite element analysis (FEA), structural fatigue analysis and compliance with ABS' Dynamic Load Approach (DLA). Structural details were studied for areas known to experience structural fatigue cracking, based both on ABS and classification society publications, and for areas found to be prone to cracking based on Polar's experiences. During seakeeping model testing, models were instrumented to aid designers in correlating the motions model to actual measurements. Structure was modified based on results of these analyses.

Engineering

Structural engineering for the *Endeavour* Class vessels was a cooperative effort. John J. McMullen Associates led the design efforts in producing the midship section shown in Figure 29. Avondale (Northrop Grumman Ship Systems - Avondale Operations) assisted in the development of the midship section and was responsible for the scantling plans. Avondale was also responsible for running ABS SafeHull Phase A and B. MCA Engineers was responsible for the Finite Element Analysis, Dynamic Load Approach, and the Spectral Fatigue Analysis. A detailed description of the vessels structural design is found in **Millennium Class Tanker Structural Design – From Owner Experience to Shipyard Launching Ways** by James Read, et al. [30].

	Port side	Starboard side
Propeller	Operable	Operable
Rudder	Operable	Locked at 0°

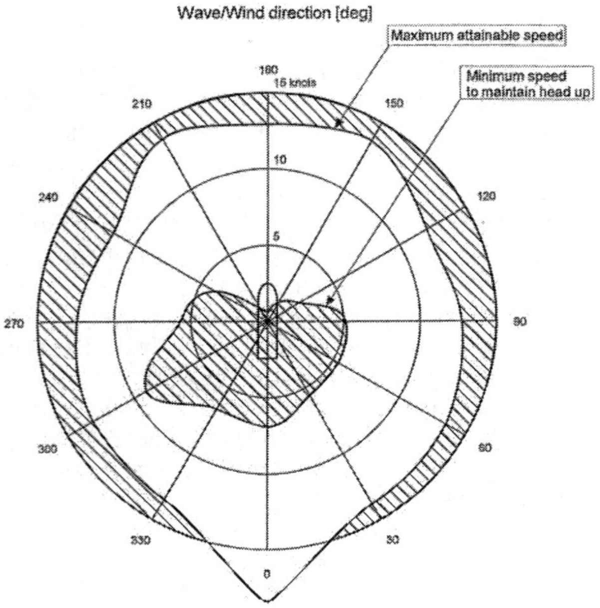

Fig. 25 Operability curves for single rudder failure - Full Load condition

	Port side	Starboard side
Propeller	Operable	Operable
Rudder	Operable	Locked at 0°

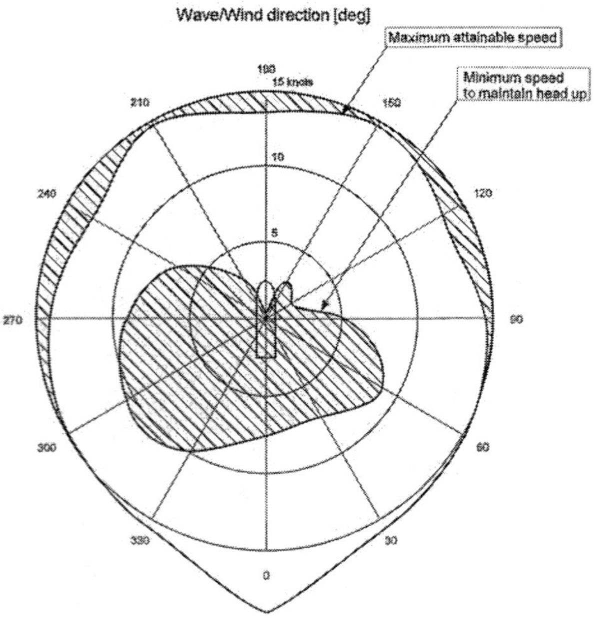

Fig. 26 Operability curve for single rudder failure - Ballast condition

	Port side	Starboard side
Propeller	Operable	Free-wheeling at zero pitch
Rudder	Operable	Locked at hard Stbd (45°)

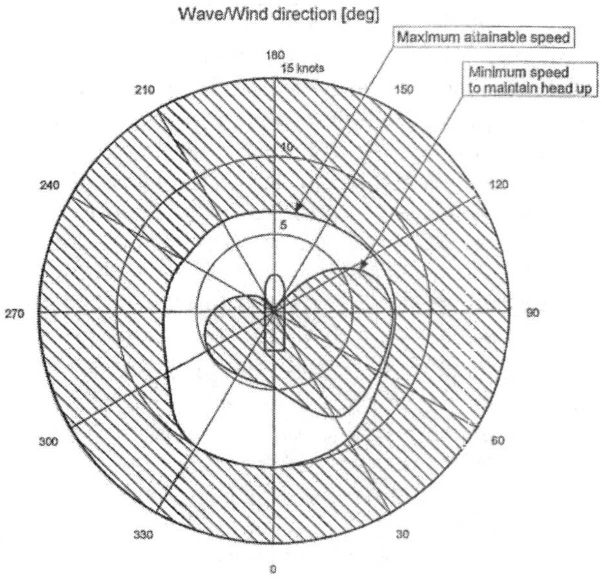

Fig. 27 Operability curves for rudder and propeller failure - Full Load condition

	Port side	Starboard side
Propeller	Operable	Free-wheeling at zero pitch
Rudder	Operable	Locked at hard Stbd (45°)

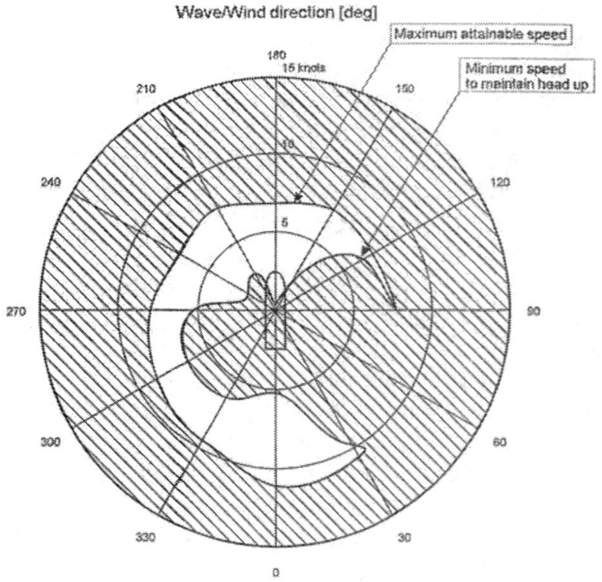

Fig. 28 Operability curves for rudder and propeller failure, Ballast condition

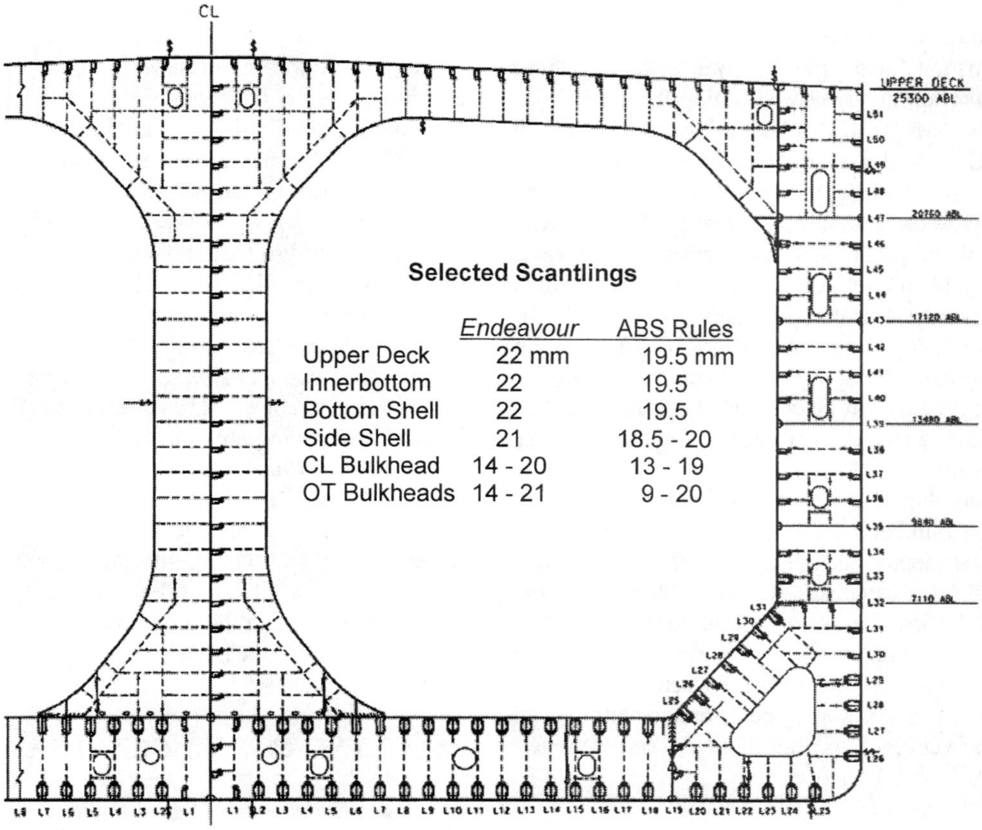

Fig. 29 Midship Section and selected scantlings compared to ABS Rule requirements

ABS SafeHull A & B

The *Endeavour* Class vessels comply with the requirements for ABS' SafeHull Phase A and B certification. As naval architects it is still our position that SafeHull is a good program for checking compliance with the Rules, but it should not be relied on as the primary design tool.

Prior to involving JJMA in the design process Polar found that the naval architects we were working with were using SafeHull A as their primary design tool. SafeHull provided a very efficient tool for making multiple runs to find minimum scantling sizes, which would satisfy Rule requirements. Unfortunately, minimum scantling size and a 30-year fatigue life are not necessarily compatible.

Using conventional methods for designing the midship section and SafeHull Phase A and B a suitable midship section was developed. Through the FEA, DLA and Spectral Fatigue Process some errors were found and changes had to be made. However, in general SafeHull Phase A and B provided an adequate resource for checking Rule compliance during the design process. Some additional structural modifications were found necessary as a result of running SafeHull. Some plate scantlings required increasing and some panel breakers were required.

Finite Element Analysis

MCA Engineers was originally tasked to create one Global, three General Local and nine Detail finite element models for the *Endeavour* Class vessels. The details originally selected were done so based on historical experience operating vessels in the Alaskan trade and information published by the Tankers Structures Cooperative Forum [31].

MCA and JJMA used the information generated in the FEA work to refine the development of the midship section. As the work continued, further FEA studies were done both because of questions generated in the DLA and Spectral Fatigue portions of the structural design and also because of the need to support shipyard production. Because of difficulties in producing mushroom cuts, Figure 30, the shipyard requested we investigate allowable tolerances. The availability of the

FEA let us study the problem and find a workable solution.

Dynamic Load Approach

The Dynamic Load Approach was a new structural analysis tool when it came to designing the *Endeavour* Class tankers. MCA, JJMA, Avondale and Polar had never used DLA and we were unsure as to application or value. Before running the DLA analysis MCA had to first verify wave induced pressure profiles. MCA had historical operating data and projections for our vessels, plus the model test data from SSPA, which they used to verify their computer generated *SPLASH®* model. Using *SPLASH®* and *SCORES-II®*, a strip theory ship motions program, dominant wave patterns and equivalent loads can be calculated. In turn various finite element analyses are performed to analyze particular details.

Ultimately the DLA analysis resulted in increasing the centerline bulkhead thickness at several locations, increasing the depth and thickness of the forepeak frames and one forepeak stringer, increasing the thickness of the lower transverse bulkhead stringers at the connection of the toes to the longitudinal wing bulkhead, adding two brackets between the vertical bulkhead stiffeners and the sloped hopper plating, and modifying the vertical bulkhead stiffeners above the top stringer.

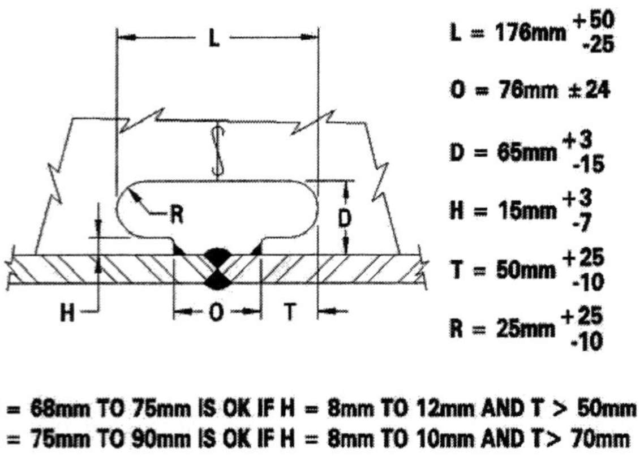

Fig. 30 Construction tolerance for "mushroom" cutouts

Spectral Fatigue Analysis

The fatigue analysis did not produce any surprising results. As would be expected, most of the changes generated in the fatigue studies were of minor significance and involved softening of bracket terminations, changing the shapes of cutouts, modifying the location of erection joint cutouts, changing plate thickness and using full collars at a number of locations. More extensive modifications were required in the hopper area (i.e., the triangular area where the double bottom meets the double side) where some structural modification was required to improve fatigue life. Even after this was done, peening was used on welds to ensure the desired life was achieved. A significant stress riser was found in the connection of the deck longitudinals to the vertical bulkhead stiffeners. Modifications were made to this design to achieve the desired fatigue life.

Structural Design Conclusion

Through the first two years of operations we have not seen any structural cracking in the cargo block of the *Endeavour* Class vessels. This is not unusual in that the cyclical loadings should be nowhere near the point of reaching a crack. However, there is a difference between not seeing any and finding a few. We believe that the *Endeavour* Class vessels have been as carefully designed from a structural point of view as was possible. I do not believe if we were to approach the design again we would alter the approach that was taken. I am confident that the design is the best that we could achieve given the tools that were available at the time the vessels were designed.

MACHINERY & ELECTRICAL

The *Endeavour* Class tanker machinery and electrical systems are designed to fully comply with the latest redundant propulsion requirements. The vessels have two fully independent engine rooms separated by a watertight bulkhead that extends from the keel to the top of the stack. Normal operating practice is to run with a split plant, although cross connection of switchboards and service systems are possible. While both plants are routinely controlled from the starboard machinery control room, full redundancy within the control system ensures a single control failure does not result in the loss of both plants. In case of a catastrophic failure in the starboard machinery control room, the port engine room can be operated from the port machinery control room.

Criteria

The vessels have been certified with ABS as +A1E, Oil Carrier, +AMS, +ACCU, SH, DLA, R2-S+, UWILD, +APS, NIBS. ABS' *Guide for Propulsion Redundancy* was first published in June 1997 and the *Endeavour* Class tankers were contracted for on July 1, 1997 [27]. Because the ABS Guide was new and the *Polar Endeavour* was the first application there were some growing pains with regards to rule interpretations. Most of these issues were minor however, and were resolved easily without serious consequence. To our knowledge, the *Endeavour* Class tankers are the first vessels certified by ABS with a Classification Notation for redundancy.

Design

At the time the *Endeavour* Class design and specifications were being developed only Det Norske Veritas' (DNV) redundant propulsion guidelines were available. Consequently, DNV's RPS (Redundant Propulsion and Separate) guidelines were used as a design standard [32]. As the design developed and ABS' Guide for Propulsion Redundancy became available, ABS' guidelines were used. When there was a difference in the two, the more stringent was used.

Operation

The *Endeavour* Class vessels are designed for operation in one of three default modes, e.g., In Port (Hotel), Cargo or Sea. These modes represent prescribed design alignments of the power management system, as determined during the design stage. The power management system controls one 8.6MW/6.6KV PTO generator, one 6.6KV/480V transformer, one 1000 KW/480V power converter unit (PCU), and one 1000KW diesel generator in each engine room. The power management system also has control of the emergency generator, and the high voltage (6.6KV) and the low voltage (480V and 120V) switchboards.

While the engineers can choose alternative arrangements in case of an upset condition or for maintenance, the power management system will align the plant in accordance with the default settings. Should the designated alignment not be achieved, the power management system will seek alternative arrangements until it is able to stabilize the plant. Initially, this alignment represented a 'best' guess of how the plant was to be operated. Consideration was given to the fact we were making a shift from single screw steam plants to a twin-screw, redundant, diesel plant with PTO generators and a Power Management System (PMS) that could make decisions and take actions faster than any combination of crewmembers we had on any of our tankers. In service, with a small amount of tweaking, the PMS system has proven to be an incredible asset and a key component to the success of the ships.

In Port

The In Port mode, Figure 31, is essentially the hotel condition, where the vessel has only those loads necessary to maintain the accommodations [33]. In this mode, one of the two diesel generators will typically be placed on line, powering both engine rooms, and the second will be placed in standby. The In Port mode may also be used for loading and discharging segregated ballast by powering the high voltage switchboard from the low voltage (LV) switchboard through the transformer. Alternately, all electric power may be taken from one of the main engine PTOs. In service it has been found more often than not that the hotel load exceeds 1000 kW and that both diesel generators must be placed on line.

Cargo

The Cargo mode, Figure 32, is used for discharging cargo [33]. In this mode one of the main engines and its PTO generator is selected to power all hotel load needs and the cargo pumps in a non-split plant configuration. The bus-tie breakers connect the two HV switchboards and the LV switchboards remain split. Ship service power is provided by the PCUs. While pumping cargo, the working engine's clutch is disengaged and the shaft brake is engaged preventing the propeller from turning. The second engine is secured with the clutch engaged.

Sea

The Sea mode, Figure 33, includes operations for both maneuvering and free running conditions [33]. In normal maneuvering mode, the ships will be operated in a split plant configuration, with both PTO generators and both PCU generators on line. For maneuvering, the plant will be operated at 106-rpm (50HZ). From about 12-15 knots, the plant will be operated on a combinator curve where both pitch and rpm will be varied to provide maximum propeller efficiency. Above 15 knots, the vessels will be operated at 127-rpm (60HZ). This provides for constant frequency allowing the PCU generators to be secured and the transformers to be placed on line, conditions permitting. The plants will be kept split under normal at sea conditions. Depending on maintenance needs, both engine rooms can be powered off a single PTO generator.

Split Plant Operation

The value of the split plant operation has been proven to us when the vessel was operating both with and without the plants split. In one case while operating with the plants split on the maiden voyage, a fuel hose to the main engine failed causing a black out of the starboard engine room. The ship's crew worked feverishly to restore essentials before tackling the failed fuel hose problem. In the mean time, the ship slowed to a comfortable 12.5 knots on the remaining shaft. After repairing the failed hose and returning to the control room one of the engineers crossed over to the port control room where an ABB technician had been working through the entire ordeal, oblivious to the failure on the other side!

In the second case the ship was being operated without the plants split. Through a combination of adding loads (i.e., IG blower, engine room ballast pump, fire pump, and deck machinery) while shifting docks, the PCU generators became overloaded. Before the diesel generators came on line the plant blacked out. Power was restored in less than a minute, but the lesson was clear, vessel transits at sea and between docks need to be done with a split plant.

Even with the lesson of split plant operation learned, the Power Management System is being revised. The revision will prevent adding loads that would overload the system because of intermittent startup currents and will refine the load shedding configuration to assure non-essential equipment is taken off line rather than risk tripping the plant.

Transition Modes

A transition period exists each time the operating mode is changed. To get underway (i.e., transition from Cargo mode to Sea mode), the second engine is started and all ship service power is transferred to its associated PCU generator. The first engine's shaft brake is released and the engine is slowed, allowing the clutch to be engaged (at 35 to 50 rpm), after which the engine is returned to maneuvering speed (106 rpm), and the PTO generator is placed back on line. The two plants are then split ensuring a single failure will not result in a total loss of propulsion or steering.

Machinery Design

The machinery plant for the *Endeavour* Class vessels reflects an evolution in the design of the plants that were in common use on the North Sea shuttle tankers. Much of the design is a result of the creative engineering of the marine engineers at IZAR, Renk and VULKAN, along with our ability to approach their ideas with an open mind. None of the proposed changes would have been possible had it not been for the reality checks that the engineers at JJMA provided, ensuring that we were progressing on a sound engineering basis.

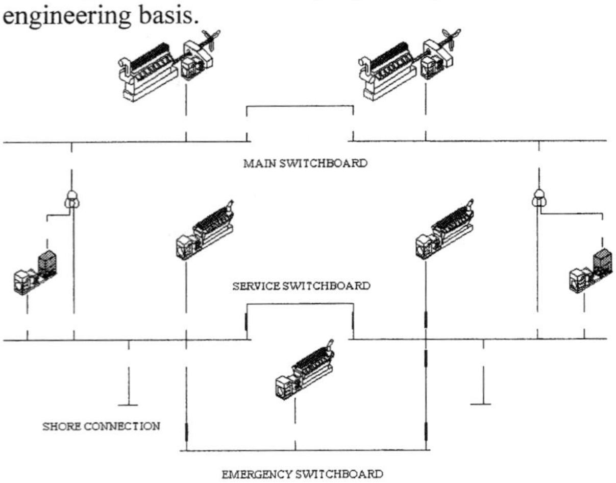

Fig. 31 Plant configuration - In Port (Hotel)

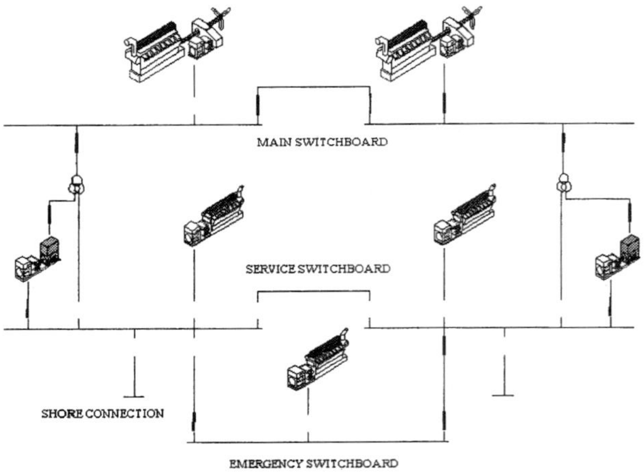

Fig. 32 Plant configuration - Cargo

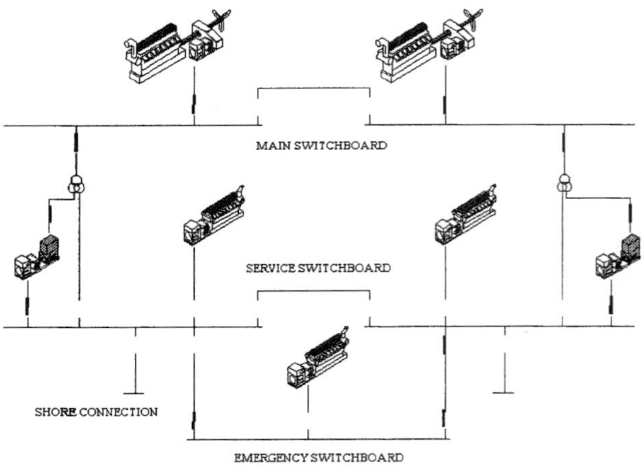

Fig. 33 Plant configuration - Sea Mode

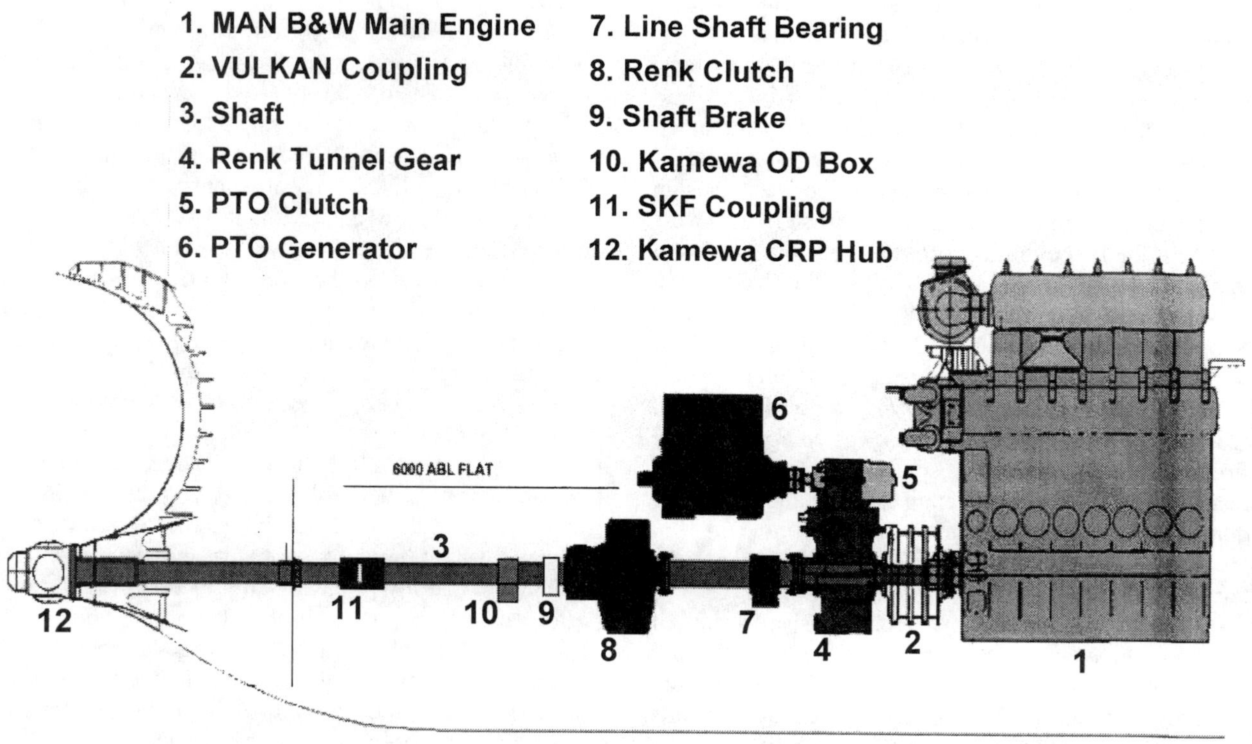

Fig. 34 Drive train components

Drive Train

The drive train, Figure 34, on initial inspection, appears typical of slow speed diesel, PTO generator and CP wheel installations. However, closer examination yields a unique solution utilizing a tunnel gear (Renk), flexible rubber coupling (VULKAN) and a hydraulic friction clutch/thrust bearing combination (Renk) designed with fully redundant hydraulics. Although representing a number of innovations, the drive train was developed based on the proven technologies of a number of leading manufacturers.

Engines

The *Endeavour* Class tankers are supplied with two MAN-B&W 7S50MC-C slow speed diesels rated at 11,060 kW (15,015 BHP) at 127-rpm. The engines are rated for a specific fuel consumption of 171 g/kWh (126 g/BHPh) when burning fuel with a lower calorific value of 42,700 kJ/kg (10,200 kcal/kg) at ISO conditions. In simpler terms, the vessels are expected to burn 500 bbls/day of 700 cSt fuel with the engines at 90% MCR and the PTO generators providing ship service power. This equates to 16.5 knots loaded (125,000 DWT) and 17.3 in ballast at a delivered power of 18,156 kW (24,347 BHP). In comparison, our 120,000 DWT steam vessels burned about 850 bbls/day at 15.5 knots loaded and 16.2 knots in ballast at 17,600 KW (23,600 SHP).

Polar, in their desire to build an environmentally friendly vessel, elected to upgrade the engines on the *Endeavour* Class tankers to make them MARPOL Annex VI compliant. This decision was made in the summer of 1999 and testing occurred in the fall of 1999. At the time MAN-B&W predicted a loss in fuel economy of about 3%. Test bed results showed that at 11,060 kW (15,015 BHP) at 127-rpm the engines had a specific fuel consumption of 178 g/kWh (131g/BHPh). At lower powers this difference would be more pronounced. Still after entering service with the Annex VI Low NOx configuration the average daily consumption was still only 520 bbls/day.

In October 2002 MAN-B&W and IZAR approached us with a belief that they could improve fuel economy. They felt that in the original testing they had sought "a solution" rather than the "optimum solution". In a series of tests where fuel valves, turbocharger diffusers, and turbocharger nozzles were varied an optimum solution was sought to find a compliant engine. After completion of the modifications and retesting, the engines showed consumption of 170 g/kWh (125.3 g/BHPh). Overall, with minor modifications to the turbochargers, i.e., changing the inlet nozzles and changing the fuel valves to conventional valves, it is possible to comply with Annex VI and to achieve better fuel economies.

During the first year of service the engines have suffered from an undue amount of cylinder head erosion in way of the nozzles. At first this was blamed on the low NOx nozzles. However, MAN-B&W quickly discounted the possibility that the nozzle design had anything to do with the erosion. Erosion rates were such that the first cylinder heads showed a loss of 3 to 4 mm of material in the first 60 days of full operation between the shipyard and the West Coast of the United States. Working with MAN-B&W and Goltens, Polar began removing the cylinder heads on a rotational basis to have them clad with Inconel® in way of the areas of erosion. By the time this process was complete on the first engine the depth of erosion on the last heads had reached 10 mm. On future engines, the cladding will be done long before erosion depths reach 10 mm. In fact, the last two engines were clad before leaving the factory.

It is believed that the erosion is a result of operating the engine with a controllable pitch propeller and with the PTO on the generator curve, rather than the propeller curve. MAN-B&W, based on previous experience, believes that as load is reduced there is some fuel, which is not burned and it forms carbon deposits on the cylinder heads near the nozzles. These carbon deposits later burn at higher temperatures resulting in the erosion. The Inconel® cladding offers a workable solution for both Polar as an operator and MAN-B&W as the engine designer. The cladding appears to have fully solved the problem.

Tunnel Gear

The Renk type BSL 225 tunnel gear provides power to the PTO generator directly from the main engine by way of the VULKAN coupling. The drive for the propeller shaft connects directly to the main engine and passes through the tunnel gear, without making physical contact. The tunnel gear steps the input speed from 127-rpm to a 1200-rpm output speed to the generator. It is fitted with a multi-disk clutch that allows the generator to be connected or disconnected, as well as with three gear-driven oil pumps. The pumps provide hydraulic power for the clutch after the gear has been brought up to speed, and for lubrication of the gear.

The tunnel gear also offers another capability we hope never to use. In the event the generator, VULKAN coupling or gear become disabled, the VULKAN coupling can be removed so the gearbox and generator are essentially removed from the drive train. This enables the ship to continue operation on both shafts, at full power. Similar failures on ships without tunnel gears could render the shafts inoperative.

VULKAN Coupling

The VULKAN RATO S 7330 coupling provides the connection between the tunnel gear and the main engine. It consists of 18 rubber wedges, each weighing 500 kg, that provide for misalignment and vibration isolation. Because of the unique design of the VULKAN couplings, requirements for torsional vibrations could be met without adverse effects on design or cost.

After sea trials on the *Polar Endeavour* there was some question as to the performance of the VULKAN coupling and it's predicted dampening with regard to torsional vibrations. After much discussion it was decided to change one of the rows of the coupling from rubber to silicone rubber. The silicone provided greater flexibility and would not stiffen as it heated. The new silicone couplings were placed on the second ship for trials. It was at this time that it was discovered that an internal bushing used to assemble the coupling was binding and that it was having an affect on the coupling performance. After trials the coupling was disassembled the bushing was bored, lubricated and reassembled. Now with the silicone rubber elements in place, because of their additional flexibility, in certain conditions generator instabilities are encountered. We believe the ultimate solution to the instabilities lies in the clutch in/clutch out speed, which is easily changed through software modification. We are currently working with IZAR, Norcontrol, VULKAN, and JJMA to define the problem and to find a permanent solution.

Clutch/Thrust Bearing

The VULKAN coupling and tunnel gear options were presented early in the detail design process. They were accepted as viable alternatives to mechanical couplings and conventional gears, after careful study, soon after detail design started. Acceptance of the hydraulic disk clutch was not as quick. All of the clutches we examined were conventional gear clutches that were manually operated. Clutching/de-clutching with gear clutches requires a crewman to go to the clutch, open the cover and jack the clutch in or out, making sure required alignments are maintained. Although inconvenient, the mechanical clutching arrangement was found acceptable, although access was limited.

The gear clutch was ultimately found to be unacceptable due to concerns of inherent hammering of the clutch resulting from axial vibrations, generator harmonics, and unbalances in the main engine (i.e., misfiring). To deal with these problems, Renk proposed a combination thrust bearing and friction clutch, type TB/FC 500, Figure 35 that had no gears that could hammer. The clutch could also be engaged/disengaged remotely. To satisfy our redundancy requirements, Renk designed the clutch with two independent hydraulic circuits, each capable of transmitting 100% of the rated torque. Under normal operations both circuits are engaged, however the ship could operate on a single circuit.

In service the Renk clutch has performed admirably. If there is a complaint, it is that we did not have foresight in locating the support equipment needed for the hydraulic systems. While the pumps and filters must be located in the bilge area to assure suction, the control equipment does not. The controls would have been more serviceable had they been placed a deck above the bilge where there was better access.

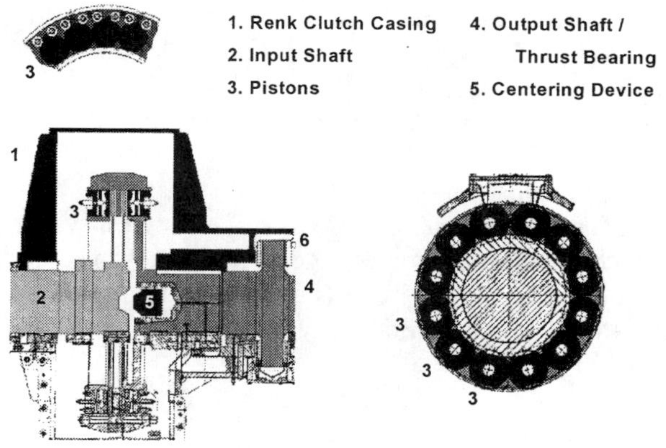

Fig. 35 RENK hydraulic clutch, Mechanical arrangement

Propeller

The ships are fitted with two Kamewa controllable pitch propellers. The propellers are four bladed, 5.8m diameter, Nickel Aluminum Bronze, with a 33° skew. The propellers are designed with a 30-year fatigue life for the blades, hub and internals and are finished in accordance with ISO R484 Class I requirements. To ensure performance of the Kamewa designed propellers, all self-propulsion and cavitation model testing was re-done using the final propeller design.

Polar did consider acquiring one set of stainless steel blades polished to an ISO R484 Class S finish. This change was never brought to fruition. Should these propellers suffer from unusual cavitation wear, the stainless steel blades would be one way to improve efficiency and prevent wear simultaneously. At this time the propellers on the Endeavour have been examined three times by divers, at six, 12 and 18 months. The propellers show no cavitation or other damage. In fact, the propellers have surprisingly retained their original level of polish. While stainless steel blades will always be a possibility, based on current knowledge, they will most likely never be needed.

Steering

Steering is accomplished through Porsgrunn "IMO" type rotary vane steering gears. The Porsgrunn units meet or exceed all IMO standards for tanker steering gears. Although the steering gears are of the "IMO" configuration, they are installed in compliance with the USCG Code of Federal Regulations (CFR). The primary difference between the two is that IMO allows the two steering gear pumps to operate simultaneously providing higher operating speeds. Whereas, the CFRs require that the two pumps be operated independently. From an operator's standpoint we could find arguments for both, but strong preferences for neither.

The steering gears are capable of operating from 35° to 35° at vessel speeds greater than 12 knots, and 45° to 45° at speeds below 12 knots. The Raytheon Integrated Bridge controls switch from 35° to 45° electronically.

To date the greatest issues encountered with the steering system have been due to a lack of experience. Historically, all of Polar's vessels were single screw fitted with ram steering engines. Twin screw, rotary vane systems offer new challenges and experiences. We have discovered differences in response rates between the rotary vane and the ram systems and cycling due to the inherent leakage of the vanes. This in turn has led to the premature failure of some amplifiers. We have also found some anomalies in where the "correct" position of the rudders should be when at sea. One vessel tends to have its rudders stay in one position while another has them skew to a different position. There is some thought that this makes the steering gears work harder, but in checking with other operators the amount they work is consistent with their experience. We are investigating installing "staged" pumps rather than the single screw pumps provided. We believe this will cushion the slamming of the rudder experienced when small corrections are needed. Our ram type steering gear are generally fitted with three pumps. Depending on the degree of change required, one, two and three pumps come on line in series. We have talked with one company that has considerable experience with this type of retrofit on rotary vanes and they have indicated we would be pleased with the change.

Electrical

Development of the electric system represents the one area where there was a great amount of design change. Based on Avondale's recommendations the system evolved from earlier designs to the twin PTO generators, twin transformers, twin PCU generators (historically Power Converter Units have been called motor generator sets in the United States) and twin diesels now installed. The development of the design centered on providing maximum operability along with maximum redundancy for a vessel built with two completely separate engine rooms. Design was dictated not only by compliance with the R2-S+ requirements, but with providing a plant that was truly operable.

A sole source vendor in accordance with the Ship's Specifications is providing the electric system. ABB will provide all motors, controllers, switchboards, generators and automatic controls for the electric system.

Arrangement

Figure 36 shows the one-line diagram for the electric system. As can be seen each engine room's design is equivalent to a single screw power plant. A PTO generator and transformer/PCU generator combination provide one source of electric power, while the diesel generator set provides the second. Bus-tie breakers, providing both redundancy and flexibility in operation, interconnect each engine room's switchboard.

PTO Generators

The Power Take Off (PTO) generators are 8.6MW/6.6kV, 1200-rpm machines operating off the PTO gearbox. The PTO generators are designed to generate power between 50 and 60Hz. They may be clutched in at main engine speeds up to 80-rpm or de-clutched at any speed. They are designed to operate in parallel with each other for closed transition power transfers only.

PCU Generators

The PCU generators are powered from an AC/DC inverter that receives 690V input and provides a 660V DC output. The DC output is used to power a DC motor, which in turn drives a 1000 kW, 480V AC generator. The AC generator is identical to the diesel generators. The PCU generators are designed to deliver constant 60Hz ship service power at any PTO generator output frequency between 50 and 60Hz.

Diesel Generators

The diesel generators are conventional sets powered by Bergen 6-cylinder diesels. The diesels are capable of running on heavy fuel but have been provided with MDO as the fuel source. The diesel generators are intended for use in the In Port mode and during some transitional periods. They are mainly intended as backups to the PCU generators.

Transformers

The 6.6KV/480V transformers are proving to be far more useful than first expected, particularly when weather conditions permit their use. The transformers eliminate the use of the inverters on the PCU generators and the DC motors, saving approximately 500 kW per shaft in energy. Because the transformers are used at periods where there is little variation of propeller pitch and governor movement, fuel rate tends to be steadier, giving better overall fuel economies. Also, the transformers are virtually maintenance free in comparison to the PCU generators.

To date, we have found that normal variation of frequency has the greatest effect on the purifiers. None of the other equipment has demonstrated problems associated with minor frequency variations. Because of the behavior of the purifiers, the transformers are secured and the PCU generators are placed on line. We are currently experimenting with variable speed motor controllers for the purifiers in hopes we can isolate them from changes in frequency. This will enable us to keep the transformer on line for longer periods of time.

Switchboards

The *Endeavour* Class vessels electrical systems have both high and low voltage switchboards to support the high load requirements (i.e., cargo pumps and bow thruster) and the normal daily ship's loads (i.e., 480V and 120V)

High Voltage (HV)

The vessels are fitted with two high voltage (HV) switchboards, one in each machinery control room. The HV switchboards are designed to operate at 6.6KV from 50 to 60Hz. The starboard HV switchboard services two cargo pumps, the cargo stripping pump and a segregated ballast pump. The port HV switchboard services two cargo pumps, the crude oil washing pump, a segregated ballast pump and the bow thruster. The port HV switchboard is also fitted for the future installation of an 8.6MW shore power breaker, which will allow for powering the vessel from a shore power source for cargo discharge. Each HV switchboard connects to a step down transformer, which can power a PCU generator or the 480V switchboards directly. Two bus-tie breakers interconnect the two HV switchboards, one in each machinery control room.

Low Voltage (LV)

The vessels are fitted with two low voltage (LV) switchboards, one in each machinery control room. The LV switchboards are designed to operate at 480V/60Hz constant frequency. Each switchboard can be powered from a 6.6kV/480V transformer, a 1000kW PCU generator or a 1000kW diesel generator set. Each switchboard powers a 120V, 60Hz service switchboard. Two bus-tie breakers interconnect the two LV switchboards, one in each machinery control room. The two 120V switchboards are not interconnected and must be powered from their respective switchboards. A unique feature of the LV switchboards is their division into two sides with one vital service feed (i.e., fuel pump) on each side of the switchboard. Each side of the switchboards may be isolated from the other by a disconnect breaker. This feature allows maintenance of the switchboard without securing the plant. A 1000kW shore power connection is provided for each switchboard.

Power Management System

The key to the electrical system is the Power Management System (PMS). To operate and control the electric system ABB, in conjunction with Avondale Shipyard and Polar, developed a system specifically for the *Endeavour* Class tankers. The PMS incorporates such features as load shedding and sequential restart, along with full control of the electrical system. The operators can choose any allowable configuration of generators and switchboards, and can designate standby scenarios. The PMS, on sensing upset conditions, will determine the proper course of action to maintain the plant in its current operating condition. This course of action could be closing bus-tie breakers or bringing standby generators on line. In each case, available power will be sensed and the time needed to bring the plant back will be determined and the fastest choice will be taken. The PMS can start and stop generators, parallel generators, open or close tiebreakers, and shed load or restart essentials, as required to keep the plant on line.

The PMS always starts from default conditions that have been determined in advance. However, the ship's crew has the ability to alter configuration according to the conditions that they are currently working. The value of the PMS to date has been its flexibility. As we've gained knowledge on the operation of the vessels, added equipment, learned about impacts of load shedding and sequential restart, we've been able to address these issues without redesigning the system, or replacing it.

The other value has been the acceptance and trust the operators have shown in the equipment. To date problems encountered have mainly been associated with loose connections. In some cases we are improving the strength of housings, in other cases it will require periodic inspection and tightening of fittings by ship's personnel.

Automation

The automatic controls for the *Endeavour* Class vessels will play a major role in their success in service. Besides certification for +AMS and ACCU by ABS, the ability to operate the plant as designed is fully dependent on the design, construction, operability and maintainability of the automation. Prior to deciding on the Specifications for the automation we consulted with the automation experts at what was then our Cherry Point Refinery for input as to state-of-the-art automation used in refineries. We found automation systems should be designed with redundant busses to assure operation in the event of a single failure, should rely on personal computer based technologies using current microprocessors, should have sufficient capacity to handle any conceivable future expansion, should not rely on distributed processors, and should be "backwards compatible". "Backward compatibility" is referring to the ability of the automation system to accept and utilize advancements in technology without becoming obsolete. For instance, current pressure sensors collect signals and send the signals to Input/Output (I/O) processors, which in turn send the signals to the master processor. The next generation of pressure sensors will have the I/O processors built-in.

Automation for the power plant, Figure 37 includes not only the two engine rooms but also covers the power management system (electrical), main engine safety systems, main engine governors, power plant operation, and propeller control [33]. ABB Industri AS of Norway is providing the main automation, power management and cargo control systems, and is integrating the Kamewa CANMAN® controls for the propeller and bow thruster, Saab Tank gauging system and the Norcontrol engine governor and safety systems.

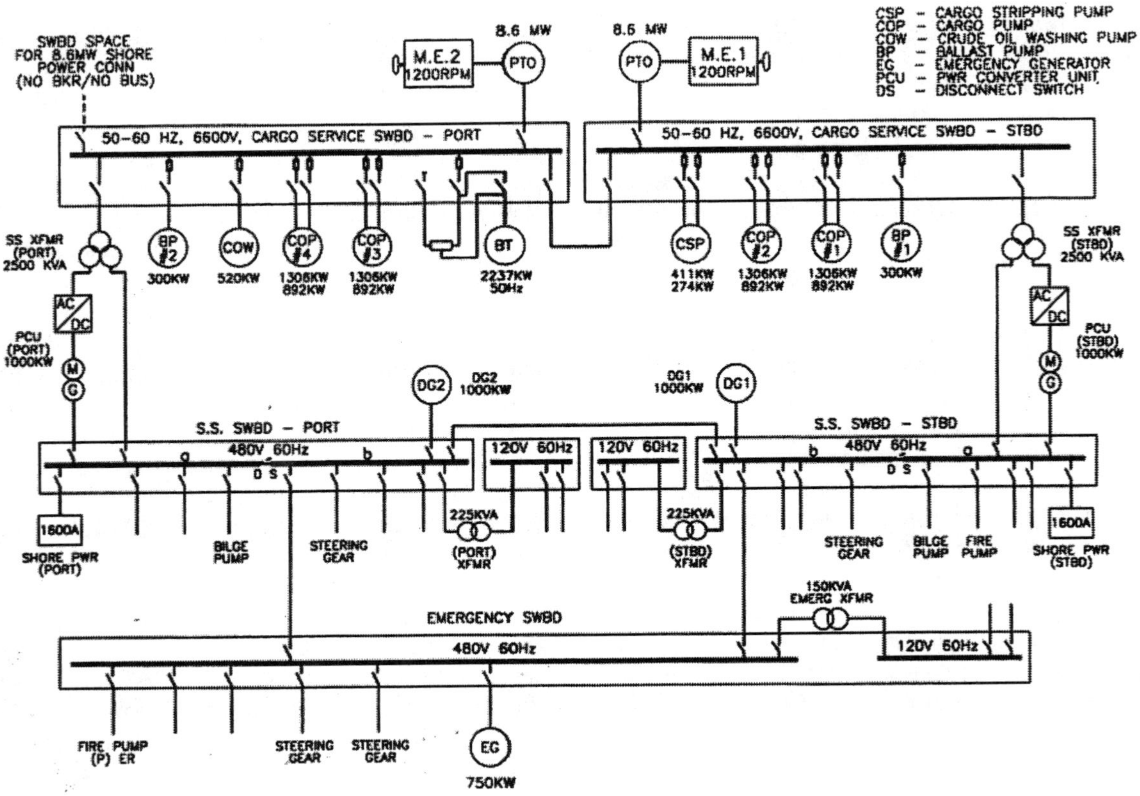

Fig. 36 One-line diagram

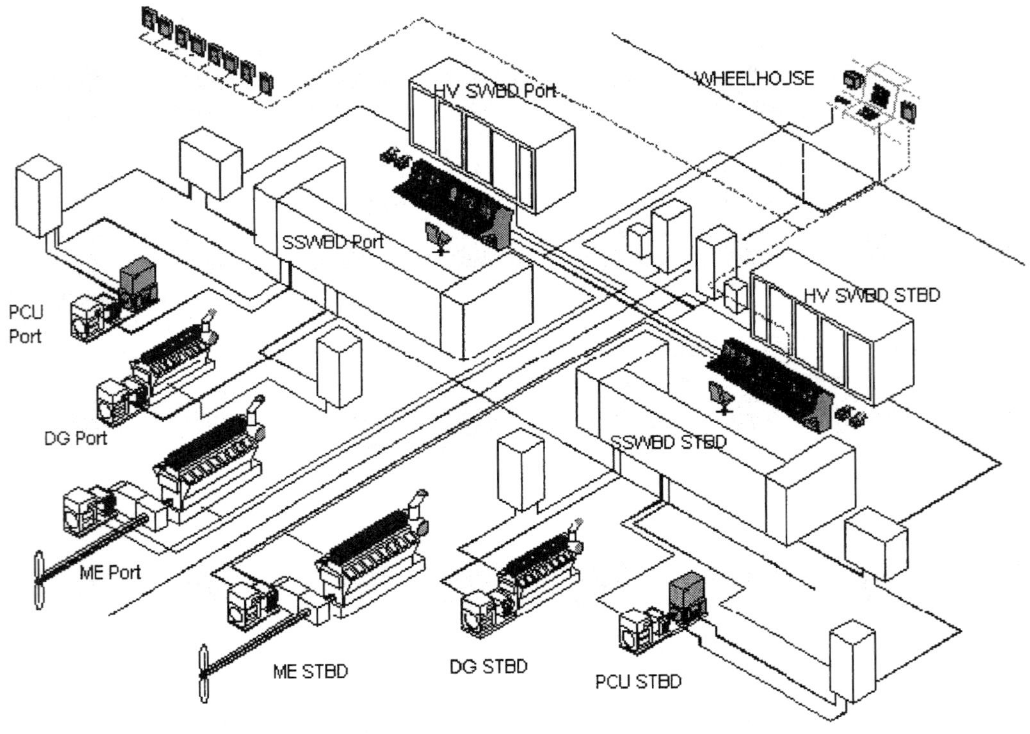

Fig. 37 Automation control system arrangement

ABB

The ABB automation system is based on two redundant PCs (Advant Controller AC450) located in each engine room. These PCs handle all engine room, power management and cargo control requirements. Each PC is capable of controlling about 5,000 input/outputs individually. The *Endeavour* Class tankers have a total of about 4,000 points that are being monitored or controlled. As installed, one PC is capable of operating the vessel, and four are fitted. Two PCs, one in each engine room, are kept on line at all times, with the second set of PCs in standby. The ABB system is based on commercially available Pentium II® processors and a UNIX® operating system. The system utilizes a redundant data bus, and efforts have been made to isolate vital components so that no single failure will result in the loss of the plant. All equipment being used is off-the-shelf ABB supply.

The ABB automation system may be operated from any one of the seven control stations located throughout the ship (two in each machinery control room (MCR), two in the cargo control room, and one on the bridge). Each control station consists of a 21" color monitor and a dedicated keyboard with track ball. Certain functions may be locked out of specific control stations (e.g., starting of main engines from the cargo control room) based on normal operations. However, access to all operating functions at any control station is possible by use of keys and password protections. Control mimics are fully developed. However, as the vessels have entered service and the crew has found need for change we have found it fairly easy to have ABB make modifications, and to have the modifications transferred to the sister vessels. The mimics are a combination of system mimics (i.e., lube oil service, fuel oil service, compressed air, etc.) and function mimics (i.e., starting the main engine, bringing a propeller shaft on line, etc.). The function mimics include all those items needed for the function (e.g., start air, lube oil, fuel oil, etc. for starting the main engine).

Of particular interest in the ABB system has been the ability to upgrade the system as we've found need. In particular is the inclusion of additional alarm and monitoring points that were not present in original scope. We've also added the ability to connect the system directly to Oslo, where a qualified service technician is available 24 hours a day. This technician, upon being called by the Chief, and being connected by the Chief can access the system to analyze problems and to correct them. Software can be changed while the vessel is at sea operating, instead of waiting for it to return to port, when the problem no longer exists. Another upgrade that is expected to be made in the not to distant future is to change from paper alarm, bell and data loggers to electronic loggers, making data recovery quicker, easier and more efficient. Instead of wafting through thousands of points trying to find the first point of failure in the tree, the chief will be able to scroll through the list. ABS and the USCG have given approval to our proposed data logging system. Again, we believe this will be an industry first.

An unexpected benefit of the ABB system has been the increased awareness of the engine room and engine performance that has come with placing an Operator's Station (OS) on the bridge. The Deck Officers no longer call the engine room when getting ready to leave the dock to check when engines will be ready. They now monitor the OS station mimic screens to seen when shafts are clutched in and when the Power Management System is lined up properly. They know in advance if the engine room is having problems because they can see the same monitors and alarms on the bridge. At sea they now monitor propeller speed, power and torque. When sea states are such that the loading of propellers is too high they have ready access to data to use in conjunction with the Hull Monitoring System to change course or to slow. Because of the new data available, at recent Bridge Team Management training courses there has been a greater call for information on the engine room to improve engineering understanding.

Norcontrol

Norcontrol is providing the governor and the engine safety shutdown system for both main engines. A Norcontrol Model 8800E electronic governor is being provided, along with a Norcontrol Model 8810 ship safety system. Both of these systems are furnished as part of the main engine package. Of particular note has been Norcontrol's full cooperation in interfacing its controls with the ABB system.

To date there have been no significant issues with the Norcontrol system. We have experienced some problems with governor instabilities that are associated with the flexibility of the VULKAN couplings. This problem appears at this time to be somewhat inherent in the design, but certainly is neither incurable nor something that we cannot live with.

CARGO SYSTEM

The cargo system on a tanker is the ship's life/blood system. No one ever takes notice of a successful transfer, however, if there is a problem everyone notices. Simple delays or slowness in loading or discharge results in claims and complaints from the terminal. Cargo retained on board (ROB) results in claims and lost tonnage on following voyages. Then there is every tanker owner's and crewmember's nightmare, the error or failure that result in a spill, large, small or miniscule, which must be avoid at all costs.

Polar decided on a number of radical departures from what had been the norm for their cargo transfer systems. First, because of the machinery plant design, conventional steam turbine cargo pumps were to be replaced by electric pumps powered from the PTO generators. Another radical departure was the shift in control of bunker loading and the transfer of engine room ballast from the engineering to the deck department, effectively breaking an 85-year tradition. Reasoning for this change is simple, control. If the deck department was to control hull trim and stress during cargo loading and discharge they needed full control of all operations. The last change, which required the most adjustment, was the elimination of a conventional cargo control console in favor of a modern electronic system featuring two CRT control consoles, a dedicated tank level console and the loading computer, Figure 38.

Fig. 38 Cargo control console

Criteria

The design criterion for the cargo system was simple. The vessels had to be able to load fully in 12 hours, and had to discharge ballast using gravity and pumping in 10 hours. Cargo pumps were, by Specification, to be capable of averaging a discharge rate of 50,000 bbls/hr. over the entire discharge, including stripping and tank washing operations, or to deliver 60,000 bbls/hr when the tanks are half full, whichever resulted in the higher rating, with a rail pressure of 150 psig. Cargo, ballast and fuel loading control are from a cargo control room located on the "A" (second) deck level.

The cargo control system was to consist of a stand-alone package that would be capable of operating all cargo, ballast, and valves associated with the loading of bunkers. Valve control was to be hydraulic for cargo tank valves, deck and pump room valves, and for engine room valves. Tank gauging was to be by radar (SAAB) for cargo tanks and pressure (Autronica) for ballast and service tanks. CargoMax® was to be used for the loading computer.

Initially, it was decided that the cargo watch officer's control was to be limited control of the starting and stopping of the cargo pumps, and in selecting high and low speeds. Specifically, it was decided that the engineering watch officer was to have sole control for the variation of main engine speeds between 106 (50Hz) and 127 (60Hz) rpm.

Design

The cargo systems on the *Endeavour* Class vessels are simple. It is this simplicity that makes them easy to load and discharge. The vessels are fitted with 12 cargo and two slop tanks, arranged two across with a centerline bulkhead. The ballast tanks are arranged into a forepeak, six pairs of "J" tanks in the cargo block, two after ballast, and two after trim tanks. Tank arrangements were shown previously in Figures 2 and 3.

The cargo system consists of two bottom mains, each serving alternate pairs of cargo tanks. Each main splits at the pump room to supply the cargo pumps. The four electric two-speed motor driven cargo pumps deliver 60,000 bbls./hr. when the tanks are half full, with a delivery pressure at the ships rail of 150 psig. The maximum loading rate is 110,000 bbls./hr. The cargo system is designed for loading through a four-header manifold that combines and drops into the two-cargo mains, one port and one starboard. The cargo pumps discharge into two discharge headers that run through the cargo tanks with risers at the cargo manifold, thereby reducing chance of deck spills.

The cargo system is also fitted with a dedicated electric motor-driven crude oil washing pump that powers the crude washing machines. Stripping is accomplished by a dedicated motor-driven positive displacement (rotary) stripping pump, which is connected to a dedicated stripping main. The stripping pump discharges to the cargo pump discharge headers, downstream of the cargo pumps. A stripping eductor, powered by the COW pump, is fitted as a backup to the stripping pump. Schematics of the pump room and tanks are shown in Figures 39 and 40, respectively. Please note that Figures 39 and 40 are mimic displays from the ABB cargo control/automation.

Segregated ballast is loaded into the cargo area double-hull ballast tanks by gravity and by two motor-driven ballast pumps. The segregated ballast mains are run on both sides of the centerline bulkhead, with connections to both the port and starboard headers in each tank. This installation provides redundancy should a ballast pump or tank valve fail. Stripping is accomplished by the use of an eductor powered from the fire pumps.

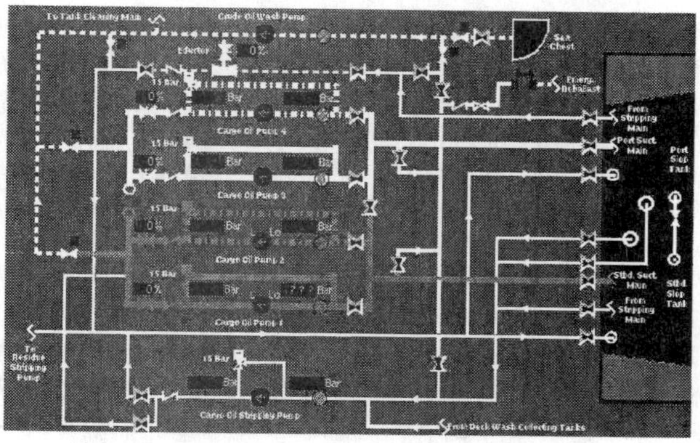

Fig. 39 ABB display mimic of Pump Room cargo piping

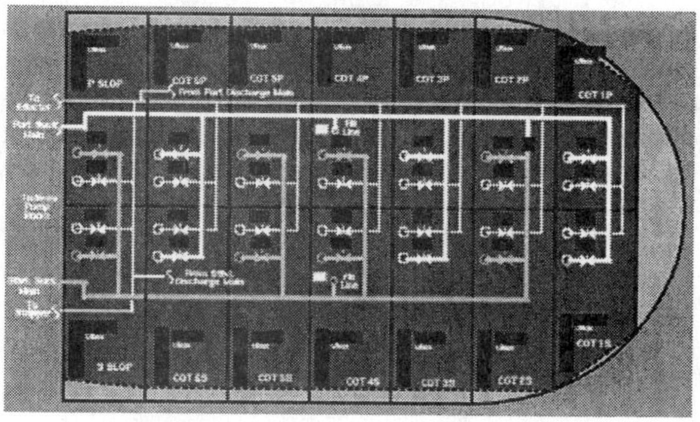

Fig. 40 ABB display mimic of in-tank cargo piping

Unique Features

As the design of the Machinery Control System (MCS) developed, Polar decided it was in their best interest to utilize the MCS for both machinery space automation and cargo control. Despite recommendations from the shipyard for standalone systems, Polar felt that they would have the greatest amount of control from a single system. This decision has proved to be far wiser that ever anticipated. Originally, the *Endeavour* Class vessels were to discharge primarily at two docks where full discharge rates are the norm. However, as a result of corporate mergers and resulting changes in trading patterns, the vessels discharge at a large number of docks and at various rates.

Using a single MCS/Cargo Control System for control of engine speed allowed the flexibility to give the cargo control officer not only the ability to start and stop motors and to vary motor speeds between fast and slow, but also to control engine speed to meet dock needs. For ballast transfer operations the cargo control officer controls all loading and discharge in the engine room, including the starting and stopping of pumps and the opening and closing of valves. All this has enabled the machinery control room to remain unattended during loading and discharges, allowing the engineers to focus on maintenance needs.

Also unique to the *Endeavour* Class vessels is the valve actuation arrangement in the ballast tanks. Polar, finding the placement of hydraulic actuators inside ballast tanks to be somewhat of an anomaly in logic, insisted that ballast tank valves be actuated by reach rod. Unfortunately, because of the lengths and the numbers of bends and elbows needed, it was not practical to use reach rods. Rather, than abandon their concerns and accept normal actuators, Polar designed a system for enclosing the hydraulic actuators (U.S. Patent No. US 6,176,248 B1). This system provides a secondary barrier that catches any leakage from the actuator and its associated fittings. It also provides a means for sensing the presence of hydrocarbons in the void space between the actuator and the secondary barrier. To date there has been one instance of a leaking actuator. The secondary containment would have been completely effective, but it was found that modifications had been made from the patent design and there were problems with installation, which provided a means of escape for a very small amount of oil. This problem is being addressed.

The use of having the cargo pump discharges return to the manifold through the cargo tanks, rather than rise in the pump room and run down the upper deck is catching on in the tanker industry. This simple change reduces the pipes on deck, limits the numbers of flexible expansion joints, and thereby reduces the risk of leaks or spills. The *Endeavour* Class vessels not only took this opportunity, but also eliminated the flexible expansion joints in the crude oil wash (COW) and inert gas (IGS) lines on the upper deck, further reducing the chance of leakage.

One of the easiest tasks to accomplish and one of the most appreciated by the deck officers was raising the Cargo Control Room from its normal position on the upper deck (first) to the "A" (second) deck level. Historically, cargo control rooms had limited visibility, either due to the low numbers of windows or to the various obstructions found on the upper deck. The *Endeavour* Class Cargo Control Rooms are above these obstructions and two 2000 x 800 mm windows provide more than adequate visibility. Add to this a closed circuit TV system that monitors the pump room and the upper deck and watch officers have far better control of transfer operations.

NAVIGATION & BRIDGE EQUIPMENT

The wheelhouse was one of the more interesting design problems. It brought together the traditions of years of seagoing experience on ARCO and Polar's vessels, new state of the art electronics, at the time unseen by anyone but the naval architects, and European insight into how "modern" ship bridges needed to be arranged for better comfort and efficiency.

Wheelhouse

The first argument encountered was one of closed versus open design. Should the bridge wings be fully enclosed? Or, should they be left open, as they had always been? Or, should some compromise position be found. European operators, particularly in the North Sea areas had long gone away from open bridges, finding both improved operator comfort and better performance of electronics. The improved comfort of operators translated to better attentiveness and better awareness during watches. The better performance of electronics translated to reduce maintenance due to exposure to salt air. Arguments for the open bridges were that you couldn't hear as well and that you couldn't sense the wind, particularly during docking situations.

It was finally decided to proceed with enclosed bridges on the *Endeavour* Class vessels. There was a caveat however, that they either be fitted with an opening window or a door on each of the bridge wings so that the Master could get a feel for the wind. As it turned out both of these options had to be abandoned because of the slope (25°) of the windows.

Since entering service we've heard nothing except extremely positive feedback from our crews regarding enclosed bridges. We've also heard one anecdote regarding a pilot who needed to move from the center console to the bridge wing for the docking operation in Valdez. It seems the pilot proceeded to put on his winter coat, gloves, and when he got to his hat, was asked by the ship's Master where he was going. At that point the pilot remembered they docked the *Polar Endeavour* from the inside.

Integrated Bridge

Navigation and Bridge Control

The integrated bridge, Figures 41 and 42, was designed and built to meet the requirements of ABS' *Guide for Bridge Design and Navigational Equipment/Systems (NIBS)* and to meet Polar Tanker's requirements for operating in the North Pacific/Gulf of Alaska environment [36]. The vessels are being built with a fully enclosed bridge that extends the full beam to one meter past the maximum beam of the vessel on each bridge wing. The decision to enclose the bridge wings was based on operability concerns resulting from the ergonomics of "open" bridges and the need to protect the bridge electronics. Operation of the vessel is centered at the bridge console where all electronics are fully integrated. The integrated bridge, built by Raytheon Marine, includes a fully redundant bus design to ensure availability of all navigation equipment in case of any single bus or component failure.

The integrated bridge is equipped with an Automatic Navigation and Tracking System (ANTS) designed in accordance with DNV Watch One requirements. The ANTS system will steer automatically according to a programmed course when the autopilot is engaged. It will also monitor the course and will alert the operator should the vessel deviate from the prescribed course more than the allotted amount. The ANTS system is equipped with a fully adaptive autopilot.

Important to the bridge design is the center console, the two wing consoles, the GMDSS station and the navigation planning station. The center console is setup similar to an aircraft cockpit. The navigation officer sits immediately to the left of the center section of the console, immediately in front of his control location is an ARPA radar, and controls

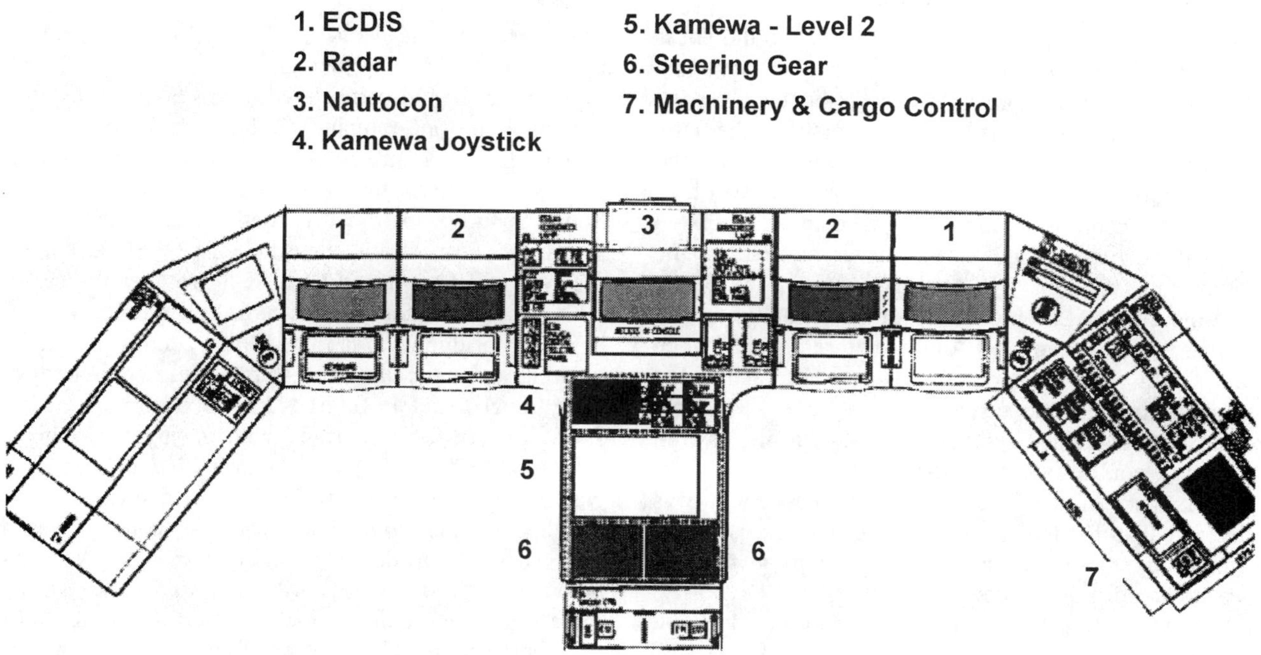

Fig. 41 Layout of integrated bridge equipment

Fig. 42 *Endeavour* Class integrated bridge

for the ANTS autopilot and its overrides. To the left is the ECDIS master console, and to the right the Nautoconning navigation information display. On the center section to his left are controls for the Hull Monitoring System, navigation lights, radios, and echo depth sounder. In the center console are the Level I, II, III and IV controls for the Kamewa controllable pitch propellers, rudders and bow thruster, as well as the helmsman's station. On the right side of the center console the navigation controls are duplicated in a near mirror image with an ARPA radar before the chair, then an ECDIS slave immediately to the right. The right wing portion of the center console houses GMDSS radio controls, clock controls, CCTV controls, as well as engine controls and the ABB operator's station.

As part of the bridge enclosure process it was decided that for the able-bodied seaman (AB) watch stander to remain an effective part of the bridge team, at sea, their position needed to be upgraded. Polar ABs had long been part of all bridge team management training; they had not however, qualified for ARPA training. It was decided that with the new Integrated Bridge and what was envisioned for further advancements in our Bridge Team Management program, that the AB watch standers all needed to be ARPA qualified. The first group of ABs was trained with the help of a senior Master from the *Polar Endeavour*. This group completed the course successfully, and gratefully. The program is viewed as a success both on the part of management and on the part of personnel.

Behind the center console are the GMDSS radio station and the navigation plotting station. The navigation plotting station is fitted with an independent ECDIS that can be used for route planning. This ECDIS can serve as a spare to the main ECDIS in the event of a failure except for replacing the ANTS. Normal course tracking must be used when on the backup ECDIS. Both ECDIS are fitted with special prediction software from SSPA that combines wind, propeller, GPS, bow thruster, etc. data and predicts where the vessel will be. This software is considerably more accurate then dead reckoning programs previously used.

The navigation plotting station was originally fitted with one DGPS and two ADSSE transceivers, which were capable of receiving both GPS and GLONASS signals. The ADSSE transceivers provide positioning data at sea and positioning and movement data to the Valdez, Alaska Vessel Traffic System (VTS) in Prince William Sound. ADSSE is a requirement of 33 CFR §165.1703(c)(6) as legislated by OPA90. With the promulgation of the SOLAS requirement for ships to be provided with Automatic Identification System (AIS) equipment, it was decided that the DGPS would be upgraded to an AIS system, and that one of the ADSSE transceivers would be changed out in favor of an AIS system. The one remaining ADSSE transceiver is to remain, first to assure compliance with the USCG VTS until such time as the AIS system in Prince William Sound is fully functional. Second, even after the AIS system is in operation, the ADSSE's GPS/GLONASS receiver will be retained because it provides the vessel with its second means of electronic positioning.

Each bridge wing is fitted with a control station for docking and undocking the ship. These stations have both Level I and Level II Kamewa controls, and the *Polar Endeavour* was fitted with a single 27" high-resolution monitor capable of switching between the Nautoconning display or the ECDIS display. As installed on the *Polar Endeavour* the console was a total success for docking and undocking. A much higher degree of control was achievable, and at the same time much quicker docking and undocking times were being achieved. What was found was the consoles were not particularly user friendly. The Level I controls were too far away from the window and that when forced into the choice between Nautoconning and ECDIS displays, the Nautoconning display was most often used. This meant the predictor software wasn't being used.

It was found that the having radar displays on the wing consoles would enhance safety when embarking and disembarking pilots. Between the *Polar Endeavour* and the *Polar Resolution* the console was redesigned, the console was shortened, moving the Level I control forward against the window. The 27" high resolution monitor was replaced with two 17" flat screens. Each flat screen is now able to display not only the Nautoconning and ECDIS displays but also both 3 and 10 cm radar displays. This change has resolved all issues with bridge wing console use.

Joystick

In developing the bridge design, careful consideration was given to facilitating operations. Close attention was paid to both ease of operation and redundancy in control. As a result of design efforts, the bridge is fitted with Kamewa CANMAN® controls. The controls feature a fully integrated joystick, Figure 43, that is capable of controlling both rudders, both controllable pitch propellers, and the bow thruster from a single lever. The single joystick may be operated in an at sea mode where it operates the propellers as a throttle and the rudders as a helm lever. It may be operated in a docking mode where the rudders, propellers and bow thruster are activated in a "point and shoot" methodology. The joystick is also capable of adjusting the center-of-pivot from the bow to midship to the stern. This first level of controls is commonly referred to us as Level I controls.

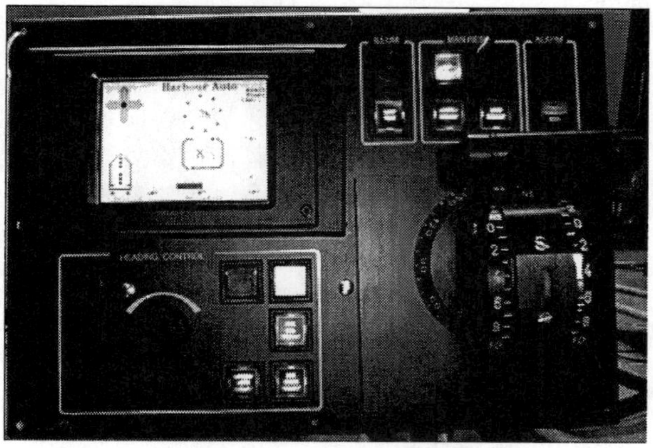

Fig. 43 Kamewa joystick control (Level I)

A second level of control, utilizing individual controls for the two propellers and rudders, as well as the bow thruster is also installed. The second level consists of a port and starboard joystick with throttle and steering control, and a bow thruster lever control. We commonly refer to this as Level II control.

In addition, for steering, a fully adaptive autopilot, follow-up helm lever (Level III), and non-follow-up controls (Level IV) are installed. When in ANTS control, follow-up override tillers are available at the port ECDIS and next to the port radar. For throttle control, pitch can be changed through the CANMAN® system (Level III) and there is an engine order telegraph installed (Level IV).

One common complaint our crews have had about the joysticks, since the beginning, is that they are not electroshaft (servo-motor) controlled. When the pitch and rpm is varied the controls in the engine room automatically move to indicate the variation. The same is not true on the bridge. While control position can be followed when someone is working in Level II at the center console, it is very difficult to follow when someone is working in Level I. It is almost impossible to follow if the vessel is in a docking mode with control on the bridge wing. A mate must follow the indicators for the bow thruster, the two propellers, and the two rudders simultaneously in order to tell what is happening. There is no way to predict what was ordered.

Based on our requests, Kamewa has been working with us for the past two years to develop an electroshaft control to replace the existing units. Kamewa had electroshafts from other services, which our operators felt were not as ergonomic as the original designs. We expect to see the first prototype by the end of 2003.

ACCOMMODATIONS

Any discussion of the *Endeavour* Class vessels must include the accommodations. From the exterior only the wheelhouse, which has been discussed in detail earlier, indicates any difference from any other vessel. Once inside, the accommodations offer insights into changes to an ever evolving culture. Changes that Polar made with full concurrence of Company management, the Officer's Association and the Employee's Union.

History

The segregation of Officer's and crew on ships has probably existed, as long as there have been ships. The segregation of deck and engine officers has probably existed ever since they started fitting engines in ships. For Atlantic Refining (ARCO Marine's predecessor company) the segregation of the deck and engine departments started with the two island ships with the deck officers and the able-bodied and ordinary seaman forward, and the engineers, firemen, oilers, and everyone else aft. When the forward house moved aft the segregation continued. The first ships had a deck with the Master and the Radio Officer, a second with the deck officers, a third with the engineering officers and the rest of the crew resided on the remaining decks. In the next generation of ships the segregation still existed; it was much the same, only slightly different. This time the deck department occupied one deck, the engineering department another, and everyone else the remaining decks.

Lounges, laundries, and messes were segregated by officers and crew. It was not uncommon to have tables in the officers' messes with four place settings (i.e., Master, Chief Mate, Second Mate, Third Mate). You ate at your assigned seat, without deviation. On ships with long tables again there was assigned seating according to position. Shore staff and visitors were directed to empty seats. Seating in the crews' mess for other than unlicensed personnel was by invitation, often with raised eyebrows. Lounges provided one of the few inner sanctums, other than a private cabin, where crew could socialize away from the public. But even then, the crew and officers generally socialized in separate lounges.

Philosophy

In the mid to late 1980's ARCO Marine began its bridge and engineering team management training programs. This program focused on a philosophy of equalizing the roles and impact each person had in emergency situations. As this program matured management also realized that there were some anomalies in the philosophies they were touting in that they were looking for an equalizing in roles and impact, while at the same time openly segregating by working group (deck and engine) or by class (licensed and unlicensed). To this end the naval architects were tasked to make the following changes to the accommodations for the *Endeavour* Class vessels:

- Equalize rooms – same rooms for officers and crew
- Separate crew spaces from public spaces
- Eliminate Officer/Unlicensed segregations
- Institute No Smoking Policy
- Provide onboard computer based training facilities
- Improve sound and noise isolation
- Improve lighting

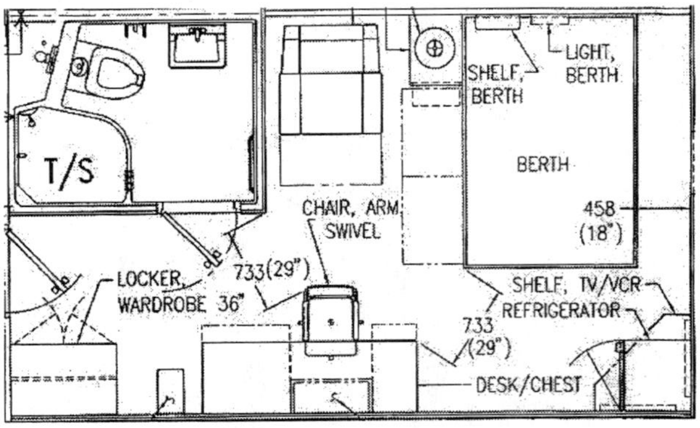

Fig. 44 Typical cabin

Rooms

It was Polar's intent to make all rooms identical, junior officers and crew. This was done as much as possible. There are variances between rooms as a result of structural differences in the steel structure, as opposed to joiner work. Figure 44 shows a typical cabin. Each cabin has its own private toilet/shower, a double bed, recliner, wardrobe, dresser, desk, desk chair, radio shelf with radio, TV/VCR, refrigerator, and nightstand. The officer's cabins are dedicated because of watch call equipment installations, and the Steward's cabin is assigned because of a LAN installation. Other than that all crew cabins are identical. The Master and Chief Engineer do have offices, day rooms and bedrooms. However, they have identical toilet/showers and beds.

To encourage the removal of some of the long, standing barriers between the Engineering and Deck departments it was decided that some innovated ideas would be tried. The first was to have the Chief Mate and the First Engineer share an office. The Mate and First's cabins are identical to the crews except they don't have desks; instead there are doors, which lead to the shared office. One advantage that has been found to this arrangement is that the Mate and First Engineer are in much closer contact, facilitating discussions on work that needs to be done.

The second experiment was to put the Master and the Chief Engineer on the same deck, by themselves, and to place a door between their offices. To date I've heard mixed reactions. I've heard some Chiefs act no differently then when their cabins were two decks below. I've heard that others keep the door open and the conversations run freely. This is something that is most likely a function of individual personalities. The presence of the proximity of the rooms and the presence of the door will help, but it can't cure the problem if there is a lack of communication.

Crew/Public Decks

The *Endeavour* Class accommodations are laid out so that the when in Port the Public (i.e., contractors, port personnel, shore personnel, etc.) need only have access to the Upper and "A" decks. "B", "C", "D", and the Wheelhouse remain off limits. The exception would be, those having business to do with the Chief or Master on "D" deck or having to do maintenance in the Wheelhouse.

The Upper Deck on the *Endeavour* Class vessels has men's and women's change rooms, hospital and a riding crew room as well as number of utility spaces. "A" deck has the Cargo Control Room, Mess, Conference Room, Training Room, Library, Gymnasium, Smoking Lounge, and Galley. These spaces are considered "public".

Of interest, all crew, licensed and unlicensed, participate in the cleaning of the vessels and in the polishing of the decks. Access to the "private" decks on the *Endeavour* Class vessels may not be done in work clothes. If a crewmember has to make a quick sojourn to his cabin and they do not want to change, surgical booties are required. The same is true of shore visitors.

Public Spaces

Riding Crew

The *Endeavour* Class tankers were equipped with a single riding crew room capable of holding six in three bunk beds. The room is fitted with two toilets. At the time of design it was determined that because of the proximity of the change rooms the riding crew could use the showers in the change rooms rather than duplicate facilities. At those times the riding crew rooms are full, this proves to be a burden. In hindsight, the riding crew rooms should have been fitted with individual showers. This would allow for better segregation for a riding crew and would ease the strain on the change rooms at times when the vessels are carrying maximum crews (i.e., when cleaning for shipyard).

Change Rooms

Providing change rooms was somewhat of an experiment based on the European method of operating tankers. The European ships all have change rooms and the crew never carries engine room or tank dirt with them into the accommodation block. On the other hand, U.S. ships almost never had change rooms, and crews were expected to change in their cabins. As part of the isolation of crew spaces from the working and public spaces change rooms were provided. Each change room was provided with a locker room area and a shower area.

The concept of the change room is very good and has been very well received. It has been found that it is not practical to limit dirty clothes to the Upper Deck level, that "A" deck level is more practical. Dirty clothes are not tolerated above "A". It was found that the men's locker room is too small and the women's is too large. This of course is subject to the crew mix at any given time. If we were designing the locker rooms again I would look at the riding crew room to see if it was possible to compact it into a "minimum" area in order to expand the men's locker room as much as possible. In other comments we received was a recommendation to have a dedicated "oily" clothes washer and dryer in the locker rooms so as not to bring dirty clothes back into the house, when they need washing.

Conference Room

Historically none of our ships have been fitted with conference rooms. When meetings needed to be held lounges were commandeered. If someone was watching a movie or reading, it was too bad. When we went to shipyards no matter how much protection was placed on the furniture, invariably someone would have a screwdriver or other tool in their pocket, which would end up tearing the upholstery.

In designing the *Endeavour* Class tankers it was decided there would be a dedicated conference room for meetings. This room would serve for meetings; safety, maintenance, human relations, etc. The room would be equipped with white boards and a TV/VCR for training videos. During shipyard periods, shipyard meetings would be held in the conference room rather than in the lounges. At worst a chair would be damaged rather than a sofa or a recliner. Carpeting was no longer at risk, since the deck in the conference room is linoleum.

The conference room has proven to be a better benefit than anyone ever expected. The room is used for all the anticipated meetings. It was the unanticipated meetings that surprised us. Starting within the first days out of the shipyard the conference room began being used daily. Each day the maintenance crew and the officers gather first thing in the morning and list the daily maintenance work on the white board. Assignments are made, work is divided, and as work is completed people return to find out what is next on the list. This daily use has made it a favorite location with the crew.

Mess

The introduction of a single mess for officers and crew initially brought comments ranging from "it's about time" to "if they do that it's time for me to retire." After a year and a half of service the single mess is alive and well, and is one of the better advancements that have been made in our fleet.

Initially the mess was to have had seating for 32 at eight tables each seating four. This arrangement was rejected by the officers as being no better than having two separate messes. They felt that ultimately the crew would migrate to tables as far apart as possible from the officers, and that in some cases the deck and engine officers would separate. Ultimately two long tables of 16 were fitted. Although not capable of serving all crew in a single serving it is possible most of the time to get everyone to one table. One comment that did return after the *Polar Endeavour* left was that the 2 m space that was left between the tables was too large and that a 1 m space was more than sufficient.

Lounge

Like the mess a single lounge was thought to be an issue of concern, but not as great as with the mess. With the advent of small TVs and personal VCRs, over time the crew congregated less and less in the lounges and spent more and more of their free time in their rooms. The single lounge in effect brought no real comments one way or the other.

In service the single lounge seems to have had the exact opposite effect as what might have been expected. Each cabin on the *Endeavour* Class vessels is fitted with a TV/VCR which is connected to the ship's AM/FM/Antenna system which allows it to receive to satellite stations, one as selected by the Master and one as selected in the lounge. Yet, it is reported that the lounge is once again being used as a meeting place for watching movies as a group. Where for many years, everyone went their separate ways, they are starting to come back together again.

It cannot go without mentioning that the *Endeavour* Class vessels are smoke free with the exception of a small smoking lounge located on the "A" deck. This concept was presented early in the design stages and was accepted readily, without comment. Since implementation there have been no adverse comments. In general, the non-smokers have been very happy, and the smokers have been able to get along.

Sound and Noise

At the time the Specifications were written for the *Endeavour* Class vessels our knowledge of sound and noise standards was non-existent. Historically, we relied on the shipyard and the joiner contractor to select materials that would provide accommodations that would be suitable for the crew. Recognizing that additional effort had to be made to limit noise exposure we did include IMO Resolution A.468 (XII) "Code on Nose Levels On-Board Ships" as a Specification requirement. We also included this somewhat cryptic phrase in the Specifications: "The acoustic properties of joiner bulkheads between adjoining living spaces shall be such as to prevent transmission of noise associated with normal levels of conversation, radio and television reception and personal activity in adjacent spaces."

For future projects three changes would be made to the approach to accommodation sound and noise. First, the accommodation block would be treated in the same manner as passenger quarters on cruise vessels. Second, it would be ensured that the space between the deck and the false ceilings would be considered with regards to sound insulation. On the *Endeavour* Class vessels many joiner bulkheads stopped at the ceiling, leaving an open void between rooms. Third, the Machinery Control Rooms would be treated as joiner spaces. Although the rooms are quiet, a false ceiling and joiner bulkheads would have further improved sound insulation.

Lighting

Historically, lighting has been one area where Polar has had the greatest number of crew complaints, particularly with respect to deck lighting. Luckily, lighting quality is easily fixed by including the right requirements in the Specifications. The required average illumination levels, not including a 0.70 maintenance factor, are shown in Table 9.

We have had no complaints and only positive feedback about the lighting levels on the *Endeavour* class vessels. The lighting levels are consistent with our desire to never hear another complaint regarding lighting.

Table 9 - Lighting Requirements

AREA	Average Illumination (Foot Candles)
• Wheelhouse • Machinery Control Room • Cargo Control Room • Machinery Spaces • Inert Gas Generator Room • Cargo Pump Room • Electronic Equipment Rooms • Training Library • Dayrooms • Serving Area • Hospital • Change Room • Toilets • Staterooms • Baths	20
• Cargo Manifold • Mooring Stations	15
• Upper Deck (Cargo Block Area) • Incinerator Room • Battery Room • CO_2 Room • Foam Room • Hydraulic Room • Elevator Machinery Room • LAN/UPS Room • Garbage Room • Medical Locker • Riding Crew Berthing • Slop Chest	10
• Unassigned	5

MANNING

The immediate reaction to having a ship with two engine rooms and with the high degree of sophistication that the *Endeavour* Class vessels have is that crew size would increase over what had been our normal manning levels. From the onset our goal was to choose equipment and to design the vessel so as to minimize manning requirements. As we developed manning requirements we worked closely with the U.S. Coast Guard in New Orleans to establish a manning scheme that was not only suitable for our needs, but was acceptable for meeting regulatory requirements.

The *Endeavour* Class vessels have accommodations for 32, six of which are in a riding crew room. So, actual crew size is limited to 26. Because we were operating with crews of 22 on existing ships, we were confident that room for a 26-person crew was more than adequate.

In determining manning requirements we sought the Coast Guard's concurrence in providing a maintenance gang rather than a conventional arrangement of having 3 Ordinary Seaman and 3

Qualified Members of the Engineering Department (QMED).

Ultimately the Certificate of Inspection (CoI) calls for a crew of 20; eight officers (Master, Chief Mate, Second Mate, Third Mate, Chief Engineer, First Assistant Engineer, Second Assistant Engineer, Third Assistant Engineer), three Able Bodied Seaman, six Maintenance personnel, plus 3 other persons in the crew (Steward, Cook/Baker and Messperson). The Maintenance personnel must consist of at least three individuals holding QMED ratings with endorsements as Junior Engineers or Oilers and at least three individuals holding Ordinary Seaman endorsements.

Polar took it upon itself, because of the sophistication of the navigation equipment to require that the Able Bodied Seaman have RADAR/ARPA endorsements.

The Endeavour Class is currently carrying 21 crewmembers and occasionally carries 22 in the crew. In addition to the above listing of crewmembers, the new vessels generally carry an additional Third Mate to ensure compliance with OPA90 requirements for work time. The need for the additional Third Mate is a result of the ship's trading patterns, as opposed to unexpected work as a result of faulty design. As needed, depending on special maintenance projects, an additional Third Assistant Engineer may also be added.

COATINGS

Concluding a paper on the *Endeavour* Class tankers without mentioning coatings would be somewhat of an injustice, particularly in light of the rubbish that has been spread in the press by purported experts with little or no background or knowledge of tankers or tank coatings. The *Endeavour* Class tankers were coated in accordance to a very careful plan based on 20 years of knowledge, at the time, of operating in the Alaskan trade.

General

Coatings, particularly ballast and cargo tank, are significant contributors to the life, health and longevity of a vessel. They play a major role in the maintenance costs and because of this they were a major focus of attention during the development of the Specifications for the *Endeavour* Class vessels. In developing the Specifications, Polar was careful to limit paint suppliers and products to those, which it had tested in ARCO's laboratories and in field-testing. Polar also made sure that the Specifications were consistent with recommendations of the Society for Protective Coatings (SSPC) and their own experiences. Polar's Specifications with regards to paint were very specific and offered little or no latitude for change. However, Polar was willing to listen to proposals from both Avondale and International Paint Company, the ultimate supplier, with regards to betterments.

Tank Coatings

Tank coating application is a relatively simple process. If you follow the manufacturer's recommendations the coatings will be good. If you don't you will ultimately have problems. The manufacturer's recommendations are well written, generally at a ninth grade or lower level so that they are easy to understand. In our case, we have our own full time paint inspector and two International Paint inspectors that report directly to us on site. Still, it is often times hard to get people to follow directions.

Prior to signing the construction Contract with Avondale, we met with Avondale senior production officials to discuss the painting of ship's tanks, specifically cargo and ballast tanks. At issue was how to paint tanks, which were to be built of 150-ton units over a period that would span close to 12 months (note: the time span has actually run about 19 months). After a reasonable discussion it was determined that the only practical way was to paint the tanks when they were complete, rather than in the blast house as each individual unit was completed. The cure time of epoxy paints and the physical damage of assembly made painting in the blast house and localized repair impractical.

Our Specification requires that for tanks all edges be rounded to a 2 mm radius, that all weld spatters be removed and that they be blasted to a near white metal (SSPC SP-10). The 2 mm radius is usually achieved by edge grinding. However, care must be given to assure that it is indeed round and that sharper edges are not introduced.

Now, with that said, over the last four years we've seen numerous failures in the painting process. When directions are followed we've had no problems, when they are not we've seen many. Here are some of the causes of problems:

- Unclean surfaces prior to painting (i.e., dust/rust blooms)
- Securing dehydration units at night or on weekends or between paint coats
- Inadequate ventilation (i.e., air changes)
- Securing ventilation over weekends
- Too thick an application of paint in a single coat (i.e., up to 24 mil. wet film)
- Too long between coats (i.e., 9 day window expires)
- Full coat used rather than tie coat (i.e., to make up for 9 day window full spray, results in over millage, poor curing)

- Too thick a total application of paint for all coats (i.e., up to 85 mil. dry film vs. specified 16 mil.)
- Poor mixing
- Improper use of solvents/thinners
- Improper use of nozzles
- Poor spray techniques
- Too sharp (<2mm) radii
- Weld spatter
- Pin holes

Cargo Tanks

With over 24 years of operation in the Alaskan trade we've found that corrosion in the cargo tanks is limited to the tank bottoms. On two ships that entered service in 1978 and 1979 there is virtually no corrosion in the overheads, the same is true of a third, which entered service in 1982. None of these three ships had their cargo tanks coated when they were built. The first two carry ballast water regularly in the cargo tanks, whereas the third does not. We attribute this lack of corrosion in the overheads to the fact that ANS crude is waxy, and the structure gains some protection from the wax. Moisture is not able to stay in contact long enough to corrode the structure, so loss of steel is not known. Because of this there is no reason to coat the overheads of the cargo tanks. It should be noted that coating tank overheads was discussed, but the burden of maintaining a coating system to assure failures do not result in structural cracks (i.e., corrosion concentrates in areas where coatings fail) was not warranted. In the end, ANS trade tankers do not have the same problems that other tankers have, and do not require the same treatments.

Bottom corrosion is a different story. In recent years bottom corrosion has been attributed to microbial blooms. But, as an operator we've been well aware of the problem since our 190,000 DWT tankers first entered service in the winter of 1978. At that time we discovered pits, ranging from 6 mm to 12 mm in diameter and up to 6 mm deep. The pits were generally filled with a grayish colored water, which when rubbed would polish the steel in the bottom of the pit to a shinny silver color. At the time there were theories of microbial action, hydrogen sulfides, and sulfuric acids, but being good ship operators we took the easy way out and painted the bottoms of the tanks with two coats of epoxy paint. For the last twenty years, two coats epoxy on the bottom surface and up 400 mm on the vertical surfaces has been the proven solution for us.

The original Specification called for International Interline 604 for coating the cargo and slop tanks. This is conventional epoxy paint. Before starting the cargo tank coating on the third ship, the shipyard and International Paint came to us proposing that we upgrade from Interline 604 (80% solids) to Interline 925 a 100% solids paint. We had used 925 on three tanks on the *Polar Resolution* after experiencing some paint failures. The 925, when applied according to directions provides an excellent coating. Interline 925 has been accepted as a substitute for Interline 604 for the cargo tanks on the last three hulls.

Ballast Tanks

The ballast tanks on the *Endeavour* Class tankers hold approximately 62,362 m^3 of ballast (100% capacity). The tanks vary from the openness of the 3 m wing walls and double bottoms, to the mazes of the after trim tanks. The structure is filled with brackets, headers, rat holes, cut outs, collars, etc. that make painting a nightmare. The Specification was written and the paint was selected based on painting the ballast tanks once at mid-life of the vessel, or 15 years after first entering service. International Paint has provided a 5-year warranty on the coatings. Our belief being that if the paint is good for 5 years it will be good for 15. After the first year of service on the *Polar Endeavour* we've heard no reports of paint failures.

As mentioned earlier the ballast tanks are prepared with a 2 mm radius on all structure and all weld spatter is removed. The tanks are then blasted to a near white blast (SP-10) and cleaned for inspection. After passing inspection the tanks receive their first coat of International Intergard 403. The tank is inspected at this time to see the first coat is proper. This coat of paint is followed by two stripe coats, one hand and one spray, and then a final spray coat is applied. At this point a final inspection is held. This process has proven to be very successful.

Final Comments on Tank Coatings

One of the biggest problems we've had with coatings has come not from the painters, but from the damage done after the painters have finished their work. The Far Eastern and European yards all build around a schedule that calls for delivery within days of completion of the tanks. For our vessels the time between the "final" paint inspection and the delivery can be 6 to 8 months. During this time the tanks are not locked and the tanks are not isolated from hot work and outside damage. This damage has had a great impact on the paint. To date we've not seen any failures in service as a result. But it would not be a surprise if there were.

Anti-foul

Polar made a decision that the *Endeavour* Class vessels would not be painted with TBT containing paints, although the original Specifications did call for them. At the time we were somewhat limited in alternatives, as the EPA had not yet approved International Paint's Ecoloflex for domestic use. As a result we painted the *Polar Endeavour with* International's Interspeed BRA 640 CDP and received permission from the EPA to put a test patch of Ecoloflex on the *Polar Endeavour*. International Paint, in fairness to them, did advise us that the CDP was a lesser quality paint than what we were used to and that they fully expected that we would have less than satisfactory results. After less than a year of service the CDP had a full growth of grass, which had to be scrubbed, which was not surprising. What was surprising was that the test patch of Ecoloflex faired no better. While part of the growth was attributed to the paint we also discovered that we were getting cracks similar in appearance to dried earth in a drought. The cracks in turn provide a place where microorganisms can grow. These microorganisms provide nutrients for grass growth. It has been determined that the cracking is a result of the prolonged periods of time the anti-foul paints are left in fresh water.

We are currently working with International regarding the possibility of painting the third ship, *Polar Discovery*, and the remaining vessels with either International Paint's Intersleek or with Ecoloflex using a different application scheme. We are currently costing a change order to upgrade the bottom paint.

As an aside, for the last 20 years we have not painted the flat bottoms of our vessels with anti-foul paint. We've had little or no growth that has warranted the application and maintenance of bottom paint on the flats, other than anti-corrosives. In the past two years there has been an increase in the appearance of small barnacles, which has prompted us to rethink our position.

CONCLUSION

If this paper reads somewhat like a primer for a naval architecture/marine engineering design class, it should. We were given what for me will most likely be a once in a career opportunity to design a tanker as I was taught in college. We were allowed to follow the design spiral as it took us through the revolutions. For the first time in my professional career engineering decisions for ship construction were being made by Owner/Operator and not by the Shipyard or worse yet the bean counters.

Some will argue that the cost of the ships was excessive. Jones Act ships or any ship built in the United States is expensive when compared on a world basis. However, much of the cost of these ships lies in the detail of the structure, which is not included in foreign construction, and in the redundant propulsion plants. Granted, the cost will never equal the cost in the Far East yards, but the differential is probably not as excessive as some would like to believe.

The point here is not the cost of ships. The point is that ships, be they built in the Far East, Europe, or the United States are expensive. They represent significant capital investments by the Owners. Our total engineering investment at the start of this project was approximately $2.5 million, roughly 0.6% of the initial two ships construction cost. Model tests results produced savings of over 6%, improvements in plant design, and other changes adds an additional 4% to the improvement total. Selection of equipment to reduce maintenance is already showing considerable paybacks. The ability to improve ergonomics and to satisfy crew wishes is priceless. The results of these efforts are shown in the fact that in the first 29 months of operation the *Polar Endeavour* and the *Polar Resolution*, together, have only experienced 12 hours of unscheduled down time.

I hope that both students and working engineers that have taken the time to read this paper have reflected back and realized, that the *Endeavour* Class tankers are living proof that the ideas laid out in *Principles of Naval Architecture*, *Ship Design and Construction* and *Marine Engineering* and in your college design courses do exist, and they do work!

ACKNOWLEDGEMENTS

I'd like to acknowledge the people of ARCO Marine and Polar Tankers who have taken the time to talk to me, or someone involved with this project to pass on their thoughts and ideas. It is these thoughts and ideas that have been incorporated in the design of these exquisite ships. Each of you should take pride when you look at the ship and claim proudly you'd passed on your ideas to the designers, and they were heard.

I'd like to thank all the people that have worked in our offices in Avondale, particularly those who have been there on a permanent basis. It has been a long and grueling project that has tried our mettle. We can be proud of our accomplishment. Without all of us the *Endeavour* Class tankers don't exist, as we know them. In particular I'd like to recognize James Read, Francis Iwancio, and William Croke who have been together with me since the start of this project, and William Young and Dave Jones our two senior steel inspectors.

To the many consultants and contractors who have given considerable efforts, these efforts have not gone unnoticed. We have come to rely on your expertise to help us with our shortcomings and to depend on you for both engineering and moral support.

I'd also like to acknowledge the working men and women of Northrop Grumman Ship Systems Avondale Operations and its sister operations, who toil daily in the heat and humidity of the Gulf Coast. Building ships under these conditions is no easy task. Building fine ships is an indication of their personal dedication.

No acknowledgement is complete without recognizing John L. Sullivan, the father and inspiration of the Millennium (*Endeavour*) Class Tankers. I had an arrangement with John Sullivan, before he retired, he handled the business end of the project, and the project team handled the engineering. John often complained that we got the fun end of the deal. I never argued. Even so, without John there are no ships, period. It was John who sold the ships to the Board of Directors, each and every ship. It was John, who prevented their cancellation when money got tight. It was John who brought twin-screw, twin rudder ships to the Alaskan trade; when he could have sold single screw, single rudder ships. It was John who changed the face of tanker transportation forever. John also helped with the machinery section of this paper, portions of which have appeared in earlier papers. We are all in his debt.

Finally, I must thank my wife Michele. If you ask her she'll tell you that for the last eight years our marriage has been on hold while I've had a love affair with a series of big blue boats. This is a tough point to argue, if I try I'm not likely to win. I thank her for sticking by me and giving me all the support I need, through thick and thin.

REFERENCES

1. GUSTAFSSON, Lars T. "Hull form design and CFD optimisation of the 165 000 DWT ARCO tanker." *125 000 / 165 000 DWT Tanker Design and Model Testing,* Report 7975-1, SSPA Maritime Consulting, July 10, 1997.

2. TRÄGÅRDH, Peter. "Initial Manoeuvring Simulation Study for the 165 000 DWT ARCO Tanker." *125 000 / 165 000 DWT Tanker Design and Model Testing,* Report 7975-2, SSPA Maritime Consulting, June 23, 1997.

3. GUSTAFSSON, Lars T. "Resistance, self propulsion, streamline and wake measurements for the 125 000 DWT ARCO tanker. Ship model 2834-A with stock propellers." *125 000 / 165 000 DWT Tanker Design and Model Testing,* Report 7975-3, SSPA Maritime Consulting, June 15, 1997.

4. GUSTAFSSON, Lars T. "Resistance, self propulsion, streamline and wake measurements for the 125 000 DWT ARCO tanker. Ship model 2834-B with stock propellers." *125 000 / 165 000 DWT Tanker Design and Model Testing,* Report 7975-4, SSPA Maritime Consulting, July 10, 1997.

5. GUSTAFSSON, Lars T. "Open water and self propulsion tests for the 125 000 DWT ARCO tanker. Single screw and "PTO" self-propulsion tests. Ship model 2834-B with SSPA design propellers." *125 000 / 165 000 DWT Tanker Design and Model Testing,* Report 7975-5, SSPA Maritime Consulting, July 10, 1997.

6. LINDELL, Per and ROSENDAHL, Ted. "Propeller design and analysis for the 125 000 DWT ARCO tanker" *125 000 / 165 000 DWT Tanker Design and Model Testing,* Report 7975-6, SSPA Maritime Consulting, November 3, 1997.

7. LINDELL, Per. "Cavitation tunnel tests for the 125 000 DWT ARCO Tanker." *125 000 / 165 000 DWT Tanker Design and Model Testing,* Report 7975-7, SSPA Maritime Consulting, June 16, 1997.

8. TRÄGÅRDH, Peter. "Free sailing Manoeuvring Tests in MDL for the 125 000 DWT ARCO Tanker." *125 000 / 165 000 DWT Tanker Design and Model Testing,* Report 7975-8, SSPA Maritime Consulting, July 9, 1997.

9. LUNDGREN, Jan. "Seakeeping calculations and model tests in regular and irregular waves for the 125 000 DWT ARCO tanker." *125 000 / 165 000 DWT Tanker Design and Model Testing,* Report 7975-9, SSPA Maritime Consulting, July 10, 1997.

10. ROSENDAHL, Ted. "Wind load characteristics and studies of exhaust gas dispersal and flow visualisation for the 125 000 DWT ARCO tanker." *125 000 / 165 000 DWT Tanker Design and Model Testing,* Report 7975-10, SSPA Maritime Consulting, August 29, 1997.

11. TRÄGÅRDH, Peter. "Captive tests and final manoeuvring simulations for the 125 000 DWT ARCO tanker." *125 000 / 165 000 DWT Tanker Design and Model Testing,* Report 7975-11, SSPA Maritime Consulting, November 4, 1997.

12. LUNDGREN, Jan. "Operability analysis for the 125 000 DWT ARCO tanker." *125 000 / 165 000 DWT Tanker Design and Model Testing,* Report 7975-12, SSPA Maritime Consulting, November 7, 1997.

13. GUSTAFSSON, Lars T. "Resistance and self propulsion tests for the 165 000 DWT ARCO tanker. Ship model 2834-C with SSPA design propellers." *125 000 / 165 000 DWT Tanker Design and Model Testing,* Report 7975-13, SSPA Maritime Consulting, July 10, 1997.

14. TRÄGÅRDH, Peter. "Free sailing Manoeuvring Tests in MDL for the 165 000 DWT ARCO tanker." *125 000 / 165 000 DWT Tanker Design and Model Testing,* Report 7975-14, SSPA Maritime Consulting, August 27, 1997.

15. TRÄGÅRDH, Peter. "Final manoeuvring simulations for the 165 000 DWT ARCO tanker." *125 000 / 165 000 DWT Tanker Design and Model Testing,* Report 7975-15, SSPA Maritime Consulting, November 4, 1997.

16. LUNDGREN, Jan. "Seakeeping calculations and model tests in regular waves for the 165 000 DWT ARCO tanker." *125 000 / 165 000 DWT Tanker Design and Model Testing,* Report 7975-16, SSPA Maritime Consulting, August 29, 1997.

17. LUNDGREN, Jan. "Operability analysis for the 165 000 DWT ARCO tanker." *125 000 / 165 000 DWT Tanker Design and Model Testing,* Report 7975-9, SSPA Maritime Consulting, November 19, 1997.

18. LUNDGREN, Jan. "Operability analysis and guidelines for vessel crews for operation of the 125 000 DWT ARCO tanker." *125 000 / 165 000 DWT Tanker Design and Model Testing,* Report 7975-18, SSPA Maritime Consulting, December 22, 1997.

19. GUSTAFSSON, Lars T. "Open water and self propulsion tests for the 125 000 DWT ARCO tanker. Ship model 2834-B with Kamewa design propellers." *125 000 / 165 000 DWT Tanker Design and Model Testing,* Report 974343-1, SSPA Maritime Consulting, January 1, 1998.

20. LINDELL, Per. "Cavitation tunnel tests with Kamewa design propellers for the 125 000 DWT ARCO Tanker." *125 000 / 165 000 DWT Tanker Design and Model Testing,* Report 974343-2, SSPA Maritime Consulting, November 4, 1997.

21. OTTOSSON, Peter and TRÄGÅRDH, Peter. "Disabled Ship Studies." *MILLENNIUM CLASS TANKER,* Report 2000 0208-1, SSPA Sweden AB, June 22, 2000.

22. OTTOSSON, Peter and TRÄGÅRDH, Peter. "ABS R2S+ Compliance." *MILLENNIUM CLASS TANKER,* Report 2000 0208-2, SSPA Sweden AB, June 29, 2000.

23. TRÄGÅRDH, Peter. "Supplement to Peril at Sea and Salvage, A guide for Masters." *POLAR ENDEAVOUR CLASS TANKER,* Report 2000 0208-3, SSPA Sweden AB, February 15, 2002.

24. TRÄGÅRDH, Peter. "MANOEUVRING OF A TWIN-SCREW TANKER including engine and rudder failure modes." *International Symposium and Workshop on Forces Acting on a Manoeuvring Vessel,* Val de Reuil, France, September 1998.

25. LUNDGREN, J., CHEUNG, M.C., and HUTCHISON, B.L. "Wave-Induced Motions and Loads for a Tanker. Calculations and Model Tests." *PRADS 98: Proceedings of the Seventh International Symposium on Protocol Design of Ships and Mobile Units*, The Hague, The Netherlands, September 1998.

26. GRIGGS, Douglas, HILL, Brian and INTOLUBBE, Steven. "POLAR ENDEAVOUR Builder's Sea Trial Results." *Carderock Division, Naval Surface Warfare Center*, West Bethesda, Maryland, 2001.

27. AMERICAN BUREAU of SHIPPING. "Guide for Propulsion Redundancy". New York, NY, June 1997.

28. OIL COMPANIES INTERNATIONAL MARINE FORUM. "DISABLED TANKERS Report of Studies on Ship Drift and Towage." Witherby & Co. Ltd., London, 1981.

29. INTERNTIONAL CHAMBER of SHIPPING and OIL COMPANIES INTRNATIONAL MARINE FORUM. "Peril at Sea and Salvage, A Guide for Masters." Witherby & Co. Ltd, London, 1998.

30. READ, James, STENSENG, Arne, HULLA, Rod and POULIN, Darold. "Millennium Class Tanker Structural Design – From Owner Experience to Shipyard Launching Ways." *Ship Structure Symposium 2000 Meeting*, Washington, D.C., 13 June 2000.

31. TANKER STRUCTURE CO-OPERATIVE FORUM. "Guidance Manual for Tanker Structures." Witherby & Co. Ltd., 1997.

32. DETNORSKE VERITAS. "Redundant Propulsion." *Rules for Classification of Ships.* Det Norske Veritas, Høvik, Norway, January 1996.

33. ABB INDUSTRI – MARINE DIVISION. "Functional Design Specification Power Management System." ABB Industri, Oslo, Norway, July 1998.

34. LEVINE, Robert A. and SULLIVAN, John L. "The Millennium Class Tanker Design." *The 1999 MotorShip Conference*, Athens, Greece, 1999.

35. MERK, Karl-Heinz and HULTCH, Harmut. "RENK AG Propulsion Plat PTO – System 11/8,6 for Millennium Class 125,000 DWT Tanker." Augsburg, Germany, January 1999.

36. AMERICAN BUREAU of SHIPPING. "Guide for Bridge Design and Navigational Equipment/Systems". New York, NY, January 2000.

DISCUSSION

Alfred Bozzuffi, Member, **Rod Hulla,** Visitor, and **Ivan Zgaljic,** Member

First, we would like to congratulate the author and Polar Tankers, Inc., on the excellent new vessels, which in many respects perform at levels much better than expected and predicted. We would also like to thank the author for sharing in this fine paper not only his knowledge and experience in designing and building innovative ships, but also the in-service results, both the good and the bad. We are sure that many naval architects and marine engineers will look for the paper when searching for information and new ideas.

As the structural design was dealt with in Jim Read's paper, cited by the author, there is not much to say in this regard. However, we agree that in today's marketplace, the prospective ship owner needs to understand that the yard's "standard designs" have in many cases been optimized, using classification society software in an iterative process, to develop minimum scantlings. Down the road, acceptance of such vessels may hold more than a few surprises, especially when renewals are required at much earlier stages in the vessel's life.

Even though we are sure that the readers obtained a good picture about the propulsion system installed in these vessels, we would like to mention a few additional details that contribute to the vessels' overall maneuverability, habitability, and durability. By taking advantage of the CP propeller, the propulsion system is designed to be free from excessive torsional vibrations in all normal operating modes. Other than during the propulsion clutch engagement process, the engine is not required to pass through the major torsional resonance during maneuvering or normal operation in any operating mode. This not only reduces fatigue stress on all propulsion components, but also eliminates the associated annoying hull and accommodation vibrations normally present on similar vessels when passing through barred speed ranges during maneuvering operations.

Another feature contributing to the reliability and durability of the plant includes the high-precision, case hardened and ground gears for the PTO drive. The gears are suitable to operate continuously with vibratory torque of ±15% of the rated torque in the no-load condition and ±40% in the full-load condition. In this way, the gears can operate without limitations in all normal operating modes and can accept higher than normal firing unbalance in the main engine cylinders.

Yet another aspect of the innovative design and quality suppliers includes the propulsion shafting system. Because of the disconnect clutch in the propulsion train, independent thrust bearing, and longer than usual shafting system, great attention was given to the shaft alignment procedure. The obstacles of how to properly incorporate the propulsion clutch in the shaft alignment calculation model and how to avoid excessive axial loads resulting from the shafting longitudinally restrained at two points have been successfully resolved with extensive help from Renk, MAN B&W, and IZAR. Cooperative efforts and great precision of the shipyard when assembling the shafts resulted in a trouble-free and reliable propulsion shafting system.

To conclude, the designing and building process of Polar tankers has been an exciting and extremely fulfilling process for us at JJMA, and we are thoroughly thankful for the opportunity to be included in such efforts.

The Articulated Tug-Barge – A Case Study

Douglas M. Wolff, P.E.
Member, Vice President, Elliott Bay Design Group, Ltd.

ABSTRACT

This paper reviews the regulatory and operational advantages and disadvantages of the articulated tug-barge unit, in which an independent tug, acting as a detachable power module, is connected into the stern notch of the cargo-carrying barge by means of a pair of retractable pins. In addition, this paper describes the contract design process of a 160,000 barrel capacity ocean service petroleum products barge and its mate, a 6260 kW (8400 BHP) twin screw tug. Extensive tank testing was performed to predict accelerations in the pilothouse as well as the dynamic forces at the pin connections; the testing regime is described and results are discussed.

NOMENCLATURE

ABS – American Bureau of Shipping
ATB – articulated tug-barge
DMI – Danish Maritime Institute
EBDG – Elliott Bay Design Group
SBOT – Sause Bros. Ocean Towing
USCG – United States Coast Guard

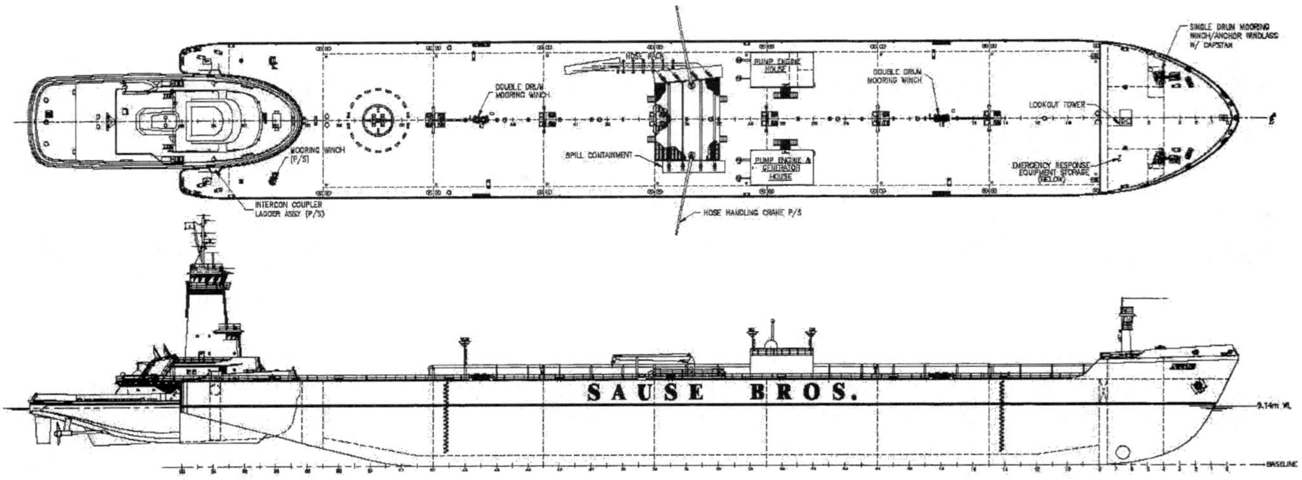

Figure 1 Tug and Barge Profile and Deck Plan

INTRODUCTION

At least 43 articulated tug-barge units are currently in service in the Western hemisphere, serving all three North American coasts as well as the Great Lakes, Hawaii and the Caribbean. Interest in ATBs is rapidly growing, due in large measure to regulatory influences and the expected growth in floating processing, storage and offloading facilities.

In late 1999, Sause Bros. Ocean Towing teamed with Elliott Bay Design Group to design an ATB for all-weather service on the west coast of the United States, a route which is well known for stormy winter weather conditions, few ports of refuge, and an unforgiving leeward shore.

Selection of the ATB configuration was driven by USCG regulatory considerations, while the size and functional arrangements were driven by the capabilities of the oil terminals to be served.

REGULATORY REGIME

USCG regulations dictate a higher standard of safety, and larger crew, for a tankship or cargo ship than for a tug towing an unmanned barge. Additionally, the USCG ruled in 1981 that an ATB would, for enforcement purposes, be considered separately as a tug and a barge. This statutory differential forms the basis for the prevalence of coastal transportation tugs and barges in the United States. A further benefit of the ATB concept is that, although tank ships and oil barges are required to be USCG-inspected, a tug measuring less than 300 U.S. regulatory gross tons is an uninspected vessel.

Regulations pertaining to tank ships, tugs and oil barges are found in the Code of Federal Regulations (Titles 33 and 46), USCG Navigation and Vessel Inspection circulars, and the USCG Marine Safety Manual, as well as SOLAS, which is applied to vessels on international voyages. To enhance marketing and operational flexibility, SBOT made the decision early in the design process that the subject vessels would be designed and constructed in compliance with relevant SOLAS regulations.

ATB Advantages

By far the largest advantage of the ATB as compared with a U.S.-flag tank ship or cargo ship lies in the USCG manning regulations. Depending on the level of automation, the regulations mandate a ship crew size of between 15 and 24, whereas an ATB tug will require a crew of no more than seven. The cost savings resulting from this crew size differential has been the primary economic driver for the expansion of the ATB fleet.

A safety argument favoring the ATB (as compared to a ship) lies in the ability to disengage the tug and shift to a towing mode in the event of a major fire or collision. In addition, the tug can be used independently of the barge for oil boom deployment or other purposes.

Compared to a towed barge configuration, the ATB offers several significant advantages, including increased speed, reduced fuel consumption, ability to transit in higher sea states, access to the barge at all times, and elimination of the vulnerable tow wire connections. Another benefit is that of reduced accelerations on the tug, resulting in greater crew comfort and reduced fatigue.

A scheduling advantage, particularly for short runs, lies in the ability for one tug to service two or more barges. An example of this would be the case wherein an empty barge is delivered to a terminal for loading, and a previously loaded barge is taken away immediately, avoiding down time for the tug and crew.

This efficiency must be carefully balanced with the issue of crew rest, but can result in near-full utilization for the expensive propulsion unit.

Some study has also been given to the use of ATB units where the voyage includes both river and coastwise legs. A smaller, less powerful tug could be used for the river service segment, and a more powerful tug could take over for the coastwise segment. This scheme would require careful scheduling, but might create significant economic advantages on a heavily traveled route.

ATB Disadvantages

Disadvantages of an ATB as compared with a ship include speed, maneuverability, and the inherent "weak link" connecting the tug and barge.

Significant hydrodynamic inefficiencies are unavoidable when the tug operates in a notch. Severe discontinuities in hull shape at the notch introduce turbulence, resulting in increased resistance and reduced propeller efficiency. Additionally, the propeller diameter is limited by the light draft of the tug. These issues combine to reduce total hull and propulsive efficiency so that the ATB can be expected to sacrifice one or more knots at full speed.

ATB maneuverability may be marginally compromised because the beam of the tug limits the transverse distance between rudders. Rudders on a ship can be further apart transversely and so can provided greater turning moment, particularly when one engine is backing down. For this project, no effort was made to document any differential in maneuverability.

Use of a connection system, no matter how reliable, will always introduce the risk of unintentional disconnection. Although a number of operators report no such events, others have experienced connector failure resulting in damage to both the tug and barge, and forcing a change to the towing mode. These events, however, have been very rare and have not slowed the pace of introduction of new ATB units.

Compared to a traditional towed barge configuration, ATB disadvantages include increased acquisition cost and reduced barge interchangeability.

CONCEPT DESIGN PROCESS

Certain fundamental requirements were established at the start of the design process – these requirements were the "drivers" in establishing length, beam and depth.

SBOT had determined a market requirement for a clean products transport with a capacity of about 160,000 barrels. Individual tank size was established at about 12,000 barrels, the typical size of the tanks at the terminals to be served.

Schedule reliability was deemed to be very important. To achieve this requirement, it was necessary that the ATB have the ability to at predictable speed in expected sea conditions, as well as possessing the ability to cross the notorious Columbia River bar in all but the most exceptional conditions. A comprehensive testing program, discussed below, was conducted to ensure that performance goals would be achieved.

The owner's construction philosophy is to specify heavier-than-required scantlings and to perform regular maintenance for longevity. This philosophy dictated that hull materials at the sheer strake, hull and forebody be increased above those required by the American Bureau of Shipping. SBOT decided to class the vessel with ABS and to apply for the SafeHull® designation. SafeHull® is a global finite element analysis of all critical hull structure, with special emphasis on hull girder bending loads and cargo tank sloshing loads.

PERFORMANCE PREDICTION PROGRAM

To ensure safe and efficient performance when subjected to the weather conditions of the North American west coast, SBOT decided to invest in both computational fluid dynamics (CFD) analysis and a rigorous model testing regime. The goals of this prediction program were, in order of priority:

- Accurately estimating the forces at the connection pins
- Determining accelerations at the upper pilothouse
- Estimating speed in waves and still water
- Optimizing bow shape for excellent seakeeping and minimum resistance
- Optimizing skegs for effectiveness and minimum resistance
- Optimizing barge aft hull form for minimum resistance.

Reliable program results are obviously dependent on the quality of the program input, quality of the calculations, and the quality of testing models and methods. Key among the various inputs is the data used to create the waves for the model testing.

Wind and Wave Data

Wind and wave data used in determining the model wave spectrum were collected by National Oceanic and Atmospheric Administration weather buoys moored off the U.S. west coast between 1983 and 1993. Data was recorded every ten minutes and then processed by NOAA to create climactic data summary tables and plots.

Throughout the selected geographic range, annual mean wind speed shows two peaks, averaging 14 knots (peaking at 16.6 knots) at 320° true and 12.5 knots (peaking at 15 knots) at 170° true.

The annual trend in significant wave heights showed the majority of waves were approximately two and a half meters. When the monthly trends were examined they nicely followed that of the wind trends with lower significant wave heights in the summer months of one and a half to two meters. The larger waves of the winter months ranged from two and a half to three meters in height.

Wave periods tended to be longer in the winter months than in the summer months. On an annual basis, the average wave periods ranged between six and eight seconds.

Analysis of the sea spectrum buoy data resulted in the following Ochi wave spectrum parameters:

- Wind driven wave height: 2.70 m
- Wind driven wave frequency: 8.4 seconds
- Wind driven shape parameter: 3.0
- Swell driven wave height: 4.30 m
- Swell driven wave frequency: 14.3 seconds
- Swell driven shape parameter: 2.4

CFD Analysis

DMI performed analysis for the tug and barge in both light and loaded conditions. Water velocity distribution, wave patterns and tuft diagrams were created; see Figure 2 for a sample tuft diagram. Smooth water speeds were estimated at 13.90 knots and 12.90 knots, respectively. The following recommendations arose from the CFD analysis:

- Making the barge aft waterlines finer would reduce a predicted wave trough aft. This recommendation was incorporated into the lines plan.
- Reducing the slope of the barge aft buttocks would reduce the likelihood of flow separation. This recommendation was also incorporated into the lines plan.
- Modifying the barge stem to a vertical profile would reduce wave-making resistance somewhat. This recommendation was not incorporated because of concerns that a vertical stem profile would have reduced seakeeping capability.
- Changing the angle of attack of the skegs would result in improved flow patterns and reduced resistance. This change was incorporated into the line plan.

Still Water Resistance Tests

Using the lines plans as revised after inclusion of CFD recommendations; Danish Maritime Institute fabricated wood models of the tug and barge at a scale of 23.45:1. The scale was selected to suit the model basin dimensions and to permit use of existing propellers for the tug. Propeller characteristics were as follows:

Diameter:	0.1559 m
Pitch ratio:	0.7550
Length @ 0.7R:	0.0654 m
Blade area ratio:	0.6320

4 blades, no rake.

Barge appendages included the transverse bow thruster tunnel and two skegs. Tug appendages included three rudders, two shaft struts and two propeller bossings. The tug model was self-propelled.

Paint streamline tests indicated that the tug lines were as efficient as could be expected, given the fact that the tug has bilge chines for simplicity of construction. Flow lines on the barge were very reasonable except that some separation was evident at the radiused intersection of the flat of bottom with the stern rake bottom slope. DMI recommendations to reduce the angle of the aft rake bottom slope were incorporated into the lines plan. Flow line results were also used to fine-tune the angle of attack of the skegs.

Still water resistance tests were performed at ballasted and full load barge drafts (9.14 m and 3.66 m, respectively) and with no trim. For both conditions the tug was ballasted to the design draft of 6.40 m. Both drag resistance and free running resistance testing were performed. Results showed that the calculated powering predictions were fairly accurate and that light draft condition required less power than the loaded draft

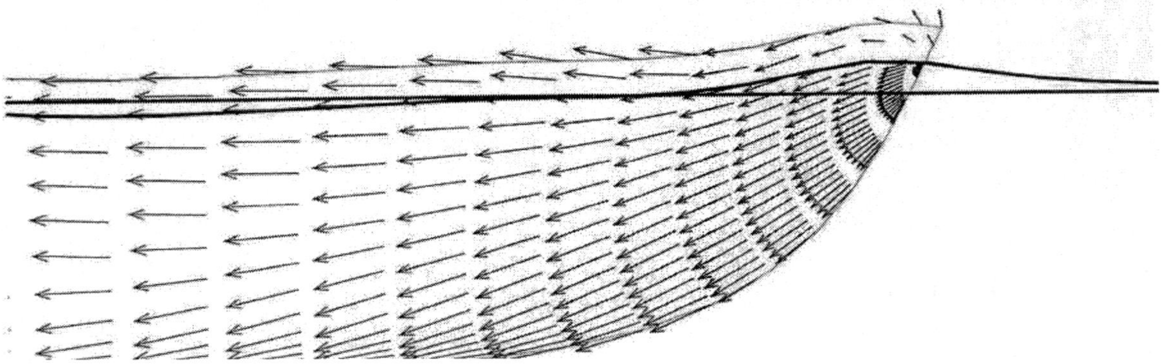

Figure 2 Tuft diagram

condition. Top speeds for the designed engine power in still water for light and loaded conditions were determined to be 13.9 and 12.9 knots, respectively. See Figure 3 for a plot of the results.

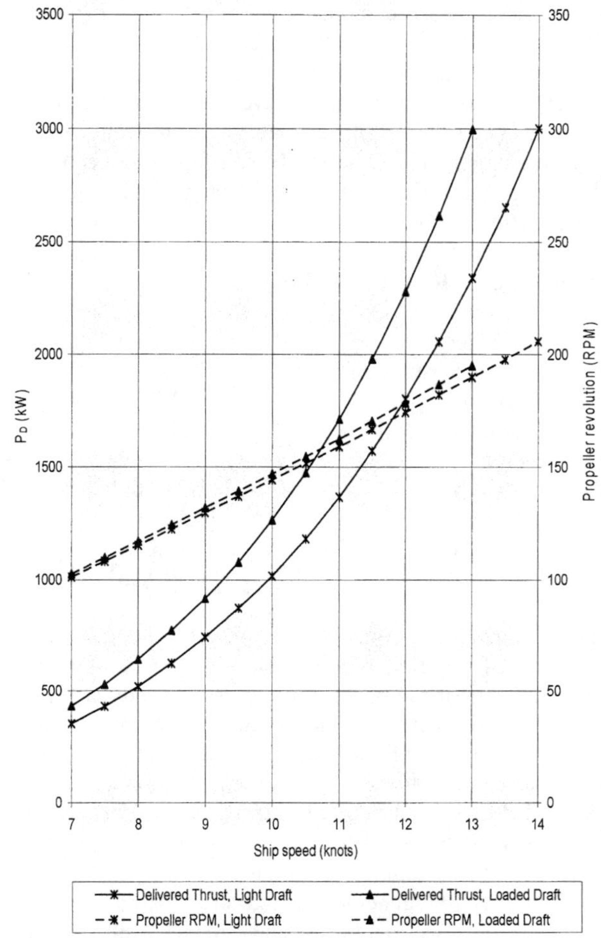

Figure 3 Resistance curves

Seakeeping Tests at DMI

Initial seakeeping tests were performed at DMI. Free running self-propulsion tests were conducted using the designed irregular seas for both head and following directions. Seakeeping responses were also measured at zero speed for quarter head seas, beam seas, and quarter following seas. The following parameters were measured and recorded:

- Wave height and period
- Tug heave and pitch
- Propeller thrust, torque and RPM
- Connection pin loading (3 directions)
- Accelerations at the upper pilot station (3 directions)
- Speed

Duration of each test was about ten minutes, corresponding to 50 minutes in full scale. The condition of the tug and barge for the seakeeping tests are shown in Table 1.

Parameter	Tug	Barge
Draft, m	6.40	9.14
VCG, m	6.14	6.71
GM_T, m	2.01	10.06
K_{xx}	40% LOA	40% LOA
K_{yy}	25% LOA	25% LOA
K_{zz}	25% LOA	25% LOA

Table 1

Loading from a 12m/s wind speed was imposed and the maximum speed of the ATB in the light condition, in both head and following seas, was measured at 9.0 and 12.5 knots, respectively.

Forces at the pin connections were measured at a variety of vessel speeds and directions relative to the wave train. As expected, the forces peaked when the vessel wallowed at zero speed with the waves approaching from an angle of 45° off the bow. Forces were compared with the design loading of the pins and found to be within the range of safe operation.

Seakeeping Tests at SSPA

Additional seakeeping tests were performed at the Marine Dynamics Laboratory of SSPA in Göteburg, Sweden, using the irregular wave spectrum described previously. Free running self-propulsion tests at 11 knots were conducted at the following headings: 0° (head seas), 30°, 60°, 90°, 120°, 150° and 180° (following seas). The autopilot was allowed to steer the model in a soft mode, without forcing it to remain exactly on course.

Table 2 gives the condition of the tug and barge for the seakeeping tests.

Parameter	Symbol	Tug	Barge
Draft forward, m	T_F	6.40	9.14
Draft aft, m	T_A	6.40	9.14
Displacement, MT	Δ	1756	28102
Vertical center of gravity, m	VCG	6.14	6.71
Metacentric height, m	GM_T	2.01	4.00
Roll radius of gyration, m	I_{xx}	5.59	9.51
Pitch radius of gyration, m	I_{yy}	10.29	40.75
Yaw radius of gyration, m	I_{zz}	10.29	40.75

Table 2

Motions in the pilothouse were also measured and recorded at a variety of vessel speeds and directions (relative to the wave train). NORDFORSK establishes the following motion limits to ensure maximum crew effectiveness:

- Vertical acceleration (RMS): 0.15 g
- Lateral acceleration (RMS): 0.12 g
- Roll (RMS) 4°

Measured values at the pilothouse fell below these limits with the exception of the zero speed condition in a beam seaway. It should be noted that bilge keels were not fitted on the model for any of the testing.

Maneuvering tests demonstrated excellent maneuverability at all tested speeds and relative directions, although a constant rudder angle of about 20° was required to maintain heading in a beam sea.

Survival State Seakeeping Tests at SSPA

Survival state seakeeping tests were performed primarily to measure the loads at the tug-barge pin connections and were based on the "hundred-year storm" with the following Ochi wave spectrum parameters:

- Wind driven wave height: 6.34 m
- Wind driven wave frequency: 8.4 seconds
- Wind driven shape parameter: 3.0
- Swell driven wave height: 7.41 m
- Swell driven wave frequency: 12.5 seconds
- Swell driven shape parameter: 2.4

In comparing these values with those defining the normal sea state, one can see a significant increase in wave height. In practical terms, this spectrum indicates the expectation of encountering a 14 meter wave every 3½ hours.

Measured loads were considered when selecting the connection pin size.

CONNECTION SYSTEMS

The great majority of connection systems currently in service in North America were designed and manufactured either by Bludworth Cook or Intercontinental Manufacturing.

The Bludworth Cook system is based upon a single hinge point at the forward end of the notch. The tug is fitted with retractable skid pads at amidships port and starboard, and a hydraulically-actuated clamp at the stem. The tug enters the notch with one pad retracted (for side clearance) and advances until the clamp engages the hinge mechanism. Once the clamp is secured, the retracted pad is extended. The two pads are lubricated and allow the tug to pitch about the stem hinge in an arc against the notch side plating. The tug is coupled in roll and heave to the barge.

Similarly, the Intercon system allows the tug to pitch independently of the barge, but couples the ATB in heave and roll. The tug is fitted with retractable, hydraulically-actuated pins port and starboard, forward of amidships. The pins are retracted as the tug enters the notch, and are extended once aligned with ladder castings that are built into the notch sides.

SBOT performed their own research on connection systems and selected the Intercon system for this project. The selected pin diameter was 1626mm (64"), one size larger than indicated by the forces determined from the seakeeping tests, but in keeping with SBOT construction philosophy.

BARGE CHARACTERISTICS

Length, molded	162.96 m
Length, design waterline	159.48 m
Beam, molded	23.77 m
Depth, molded amidships	13.11 m
Design draft	9.14 m
Cargo capacity	161,940 bbl
Ballast capacity	70,500 bbl
Potable water capacity	11.0 m^3
Fuel oil capacity	60.1 m^3

Barge Arrangements

One of the first required decisions regarding the arrangements of the barge was whether to have a flush deck or a raised trunk. The raised trunk configuration results in less total weight of steel. The Owner, however, favored the flush deck configuration based on simplified maintenance and enhanced crew safety while working near the vessel side. Accordingly, the flush deck design was adopted, see Figure 1.

Pump engine/generator houses are located forward of the amidships manifolds and are set well off of vessel centerline for visibility of the full length of the deck from the tug pilothouse.

The foc'sle provides space for the cargo control room, workshop, day room, spill response equipment room, potable water tank, and chain lockers. The bow thruster room below the foc'sle is accessed by an inclined ladder and houses the thruster, thruster engine, air compressor, oily water pump, bilge pump and bilge manifold. Arrangements are provided in Figure 4.

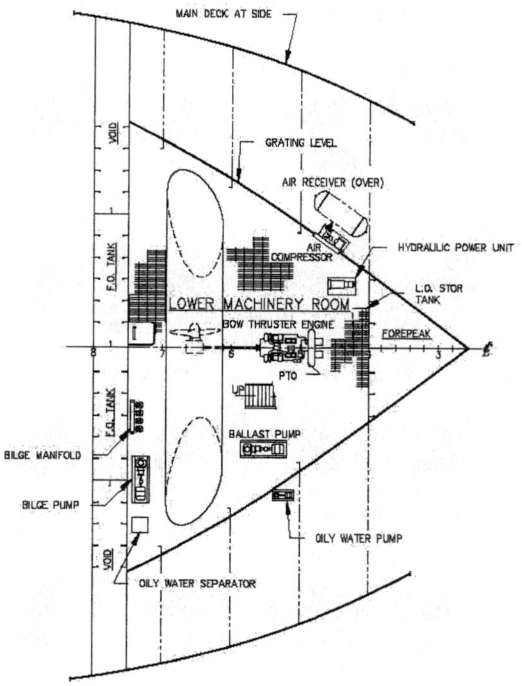

Figure 4 Bow thruster machinery

Tank Configuration

Seven pairs of tanks form the cargo block. At 99% full, ten tanks have an approximate capacity of 11,850 barrels each and four have an approximate capacity of 10,500 barrels each. Tank capacities were standardized to suit the shoreside tank capacities of the terminal facilities. In accordance with SOLAS regulations, wing and double bottom tanks are fitted, and can be used for ballast. An additional pair of ballast tanks for trim control is located in the stern rake aft of the cargo block. Total ballast capacity is 70,500 barrels, served by two Byron Jackson 2-stage deep well pumps, each with a capacity of 5000 barrels per hour at 21.3 m (70') head.

A diesel oil storage tank is located in the cofferdam space between the cargo block and the bow thruster machinery space. The forward part of the port side pump engine house contains two slop (stripping) tanks with 26.5 cubic meter (7000 gallon) capacity each, a fuel oil day tank, a dirty oil tank and a hydraulic oil storage tank.

Structural Features

The entire barge is constructed of mild steel. Scantlings were originally determined from the ABS Rules for Barges; however, the Owner elected, late in the design, to have the barge SafeHull® certified, thus applying ABS tank ship rules. The effect of the application of the ship rules was an increase in scantlings at the transverse bulkheads and stools. The final midship section is shown in Figure 5 and includes some specific scantling increases that reflect SBOT preferences.

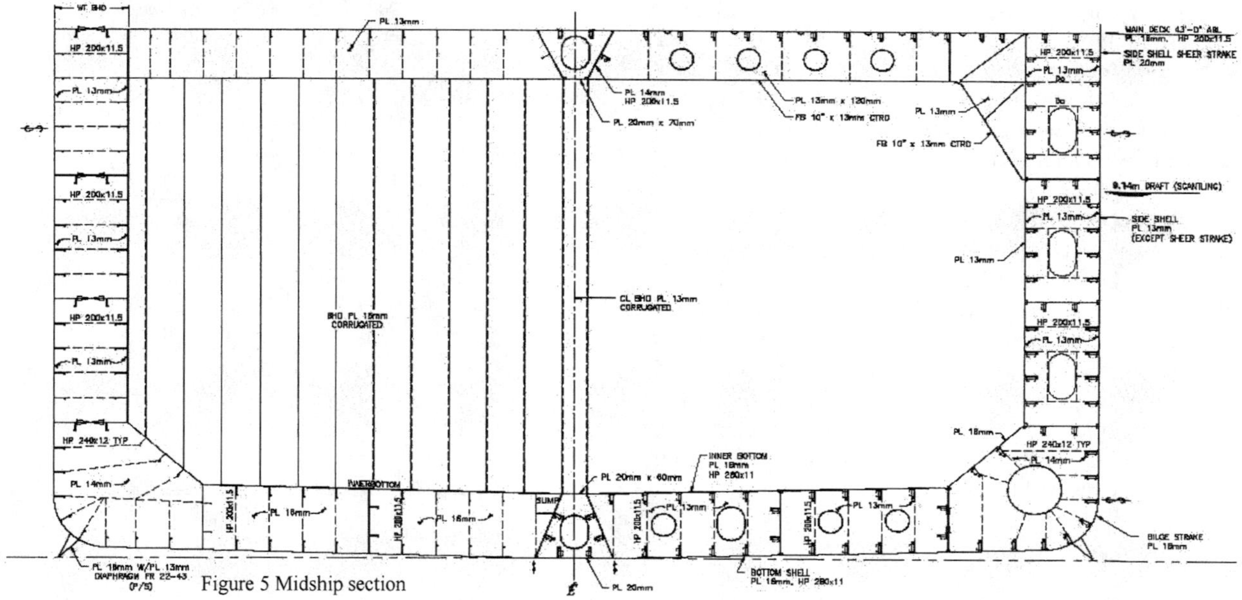
Figure 5 Midship section

Cargo System

The four main cargo pumps are Byron Jackson 4-stage, diesel engine-driven, rated 5000 barrels per hour at 114 meter head (3500 GPM at 375'). These pumps are located in the #4 cargo tanks. The Detroit Diesel Series 60 engines (rated 395 HP at 1800 RPM) that power the pumps are installed in the pump engine houses port and starboard. Pump engines are compressed air start and air cooled.

Stripping pumps are located on the main deck forward of the port side pump engine house and are powered by hydraulic motors.

The manifold, shown in Figure 6, is located amidships and consists of four headers, each fitted with 10" discharge valves port and starboard. Additionally, each header pipe is fitted with a 10" hose connection at the hose manifold aft at the aft side of the load/discharge manifold.

Cargo stripping and vapor recovery systems are installed; discharge headers are provided at the manifold port and starboard.

Machinery Installation

A diesel engine-driven transverse tunnel thruster is located forward. Based on windage studies, the thruster is rated at 559 kW (750 horsepower) to provide control of the forward end of the barge during docking maneuvers.

Deck machinery consists of two fixed boom cranes for hose handling, two windlasses with single mooring drums, two double drum mooring winches, and two single drum mooring winches. All deck machinery is powered by a common medium pressure hydraulic system An electro-hydraulic power unit as well as engine-driven hydraulic pumps are located in the port side pump engine house.

Air compressors are located in the bow thruster room and in each pump engine house. Compressed air is used for starting the thruster engine and cargo pump engines (generator engines are electric start), for control air, and for general services.

Electrical Installation

Two 95 kW diesel generator sets provide ample power for barge electrical requirements. The rating of the generators was selected to avoid the USCG-inspection requirements for generator sets rated at 100 kW or more. Generators are located in the starboard pump engine house.

A shore power connection is provided, but the Owner chose not to provide an umbilical connection between the tug and barge.

TUG CHARACTERISTICS

Length, molded	41.27 m
Length, design waterline	41.10 m
Beam, molded	14.02 m
Depth, molded amidships	6.48 m
Design draft	6.40 m
Propulsion power	6260 kW
Potable water capacity	68.1 m^3
Fuel oil capacity	870.5 m^3

Arrangements

The ATB configuration imposes specific and challenging tug design requirements. Most obviously, a pilothouse must be located high enough so as to provide visibility when the barge is light and the tug has a full fuel load. Equally obvious is the need for the tug to fit into the notch. Not so apparent are the tight tolerances required to guarantee proper operation of the connection system, coupled with the need for adequate clearances to prevent binding when the tug pitches within the notch. Because of these issues, fendering selection must be made early in the design process. Fendering details are shown in Figure 7.

Below the main deck, the tug is designed to take advantage of the archaic (and nonsensical) United States admeasurement rules. Careful placement of

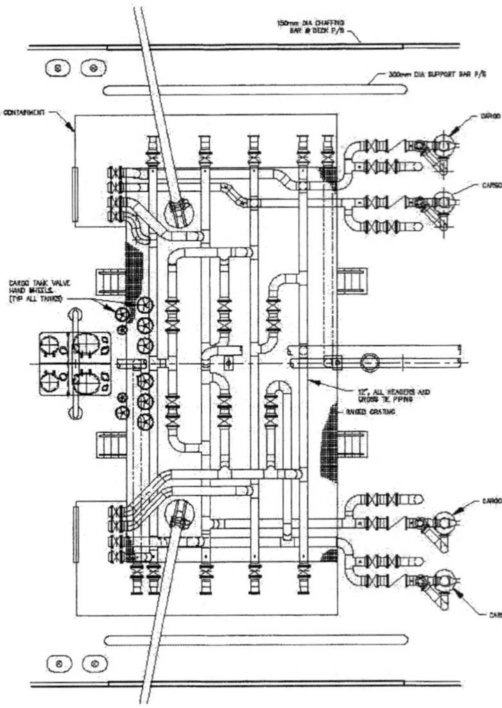

Figure 6 Cargo manifold

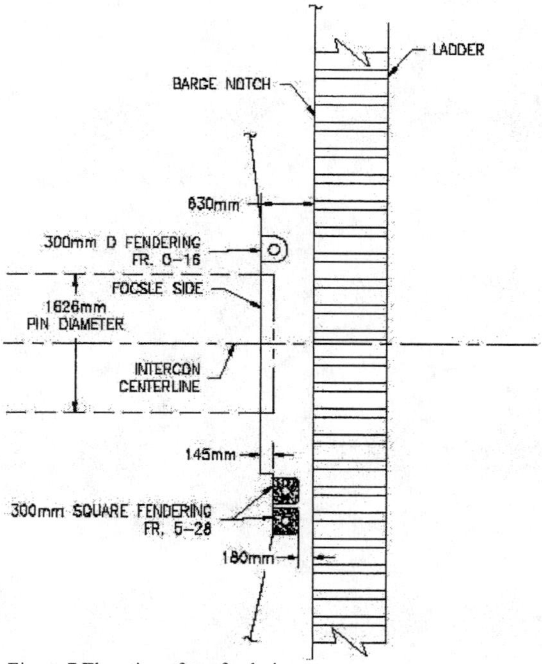

Figure 7 Elevation of tug fendering

ballast tanks and transverse non-tight bulkheads results in gross tonnage less than 200, the point at which additional manning requirements significantly increase operational costs.

On the main deck, arrangements must be designed around the coupling rams and machinery, which occupy prime real estate forward of amidships. Arrangements are shown in Figure 9.

Two complete pilothouses are provided; see Figure 8. It is expected that the lower pilothouse (at the 03 level) will be used whenever possible for reduced motions. The upper pilothouse (at the 06 level) will be used only when the barge is light or when improved visibility is required.

Structural Features

Because the vessel will be classed, the structure and machinery were designed in compliance with the ABS rules for small ships.

Several structural features are noteworthy. Most apparent is the forest of transverse bulkheads below the main deck. These bulkheads serve only to reduce the admeasured regulatory tonnage and would appear only on a U.S flag vessel. Elimination of these redundant bulkheads would reduce construction cost, lightship weight and maintenance cost.

SOLAS regulations mandate a full double bottom in the engine room. For the tug, however, a waiver from this requirement was sought. The basis for appeal was that a double bottom under the main engines would raise the main engines, also raising their exhausts and consequently the main deck, increasing hull depth and raising the center of gravity, reducing stability. The USCG approved the waiver request, but stipulated that a double bottom must be installed on vessel centerline inboard of the main engine foundations, and that the shell plating thickness outboard of the bottom must be increased. These requirements were incorporated into the design.

The Intercon coupler unit is furnished within a structural box assembly that includes the transverse bulkheads immediately forward and aft (located at frames 17 and 23) as well as the main deck below, the 01 Deck above, and the side shell structure. This entire assembly is fabricated to match the lofted lines plan and will be dropped in place as a unit before the deckhouse and tower are set on the vessel.

To reduce both topside weight and maintenance, the tower structure above the 03 level is aluminum and is welded to the steel structure below using explosion-bonded bimetallic strips.

Deck Machinery

For deck machinery, the tug is equipped with an electric bow winch/capstan, a 610mm (24") diameter electric stern capstan, a towline reel with 180m of 305mm (12") circumference emergency line, and a hydraulic crane for handling the rescue boat. Perhaps surprisingly, no tow winch is provided since towing is not part of routine operations. Should towing become necessary, the emergency tow wire can be attached to the tri-bitt on the main deck aft.

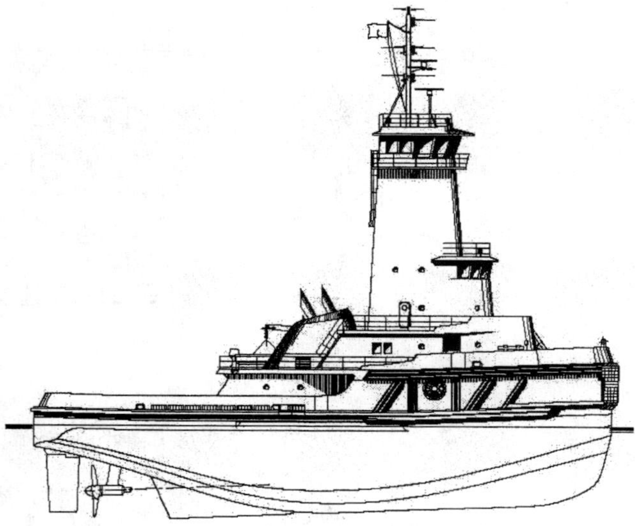
Figure 8 Tug outboard profile

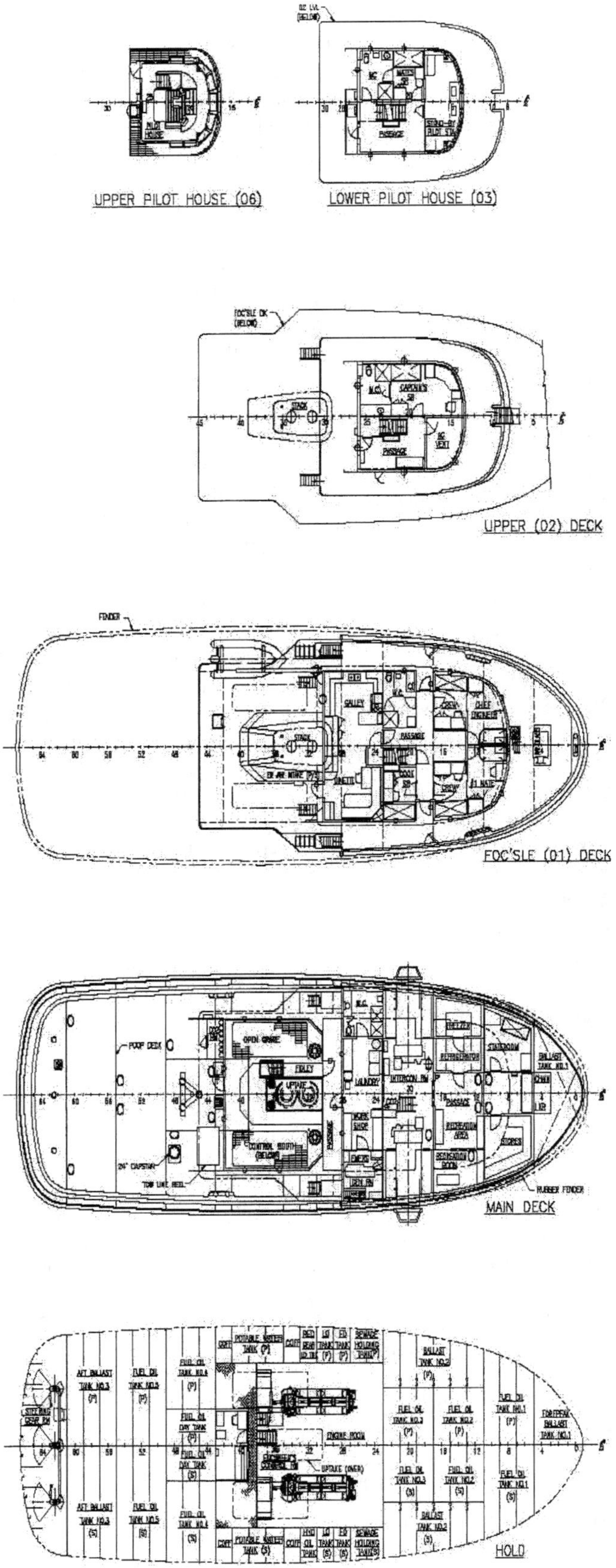

Figure 9 Deck Arrangements

Propulsion Machinery

In keeping with fleet standards, the owner elected to have two simple drive lines, each consisting of a medium speed diesel engine, reduction gear, steel shafting and open propeller. The machinery arrangement is shown in Figure 10.

Because the ATB is expected to operate regularly within California state waters, the main engines must be certified for compliance with California air quality emissions standards. The EMD 16-710G7B engine, rated 2983 kW (4000 BHP) at 900 RPM, is so certified and was chosen; a significant factor in the selection was the familiarity of the fleet engineers with the EMD line of engines.

The selected reverse reduction gears are Falk 3790 MRH with a 5.57:1 ratio, resulting in a shaft speed of 161.6 RPM. A shaft brake is fitted onto each forward shaft coupling.

Propeller shafting is ABS Grade 2 steel, 12.45 meters long, covered with 5 mm of glass reinforced plastic where exposed to sea water. Bearing liners are shrink-fit stainless steel, 19 mm thick. Propellers are fixed pitch, open, five blade stainless steel, 3660 mm diameter.

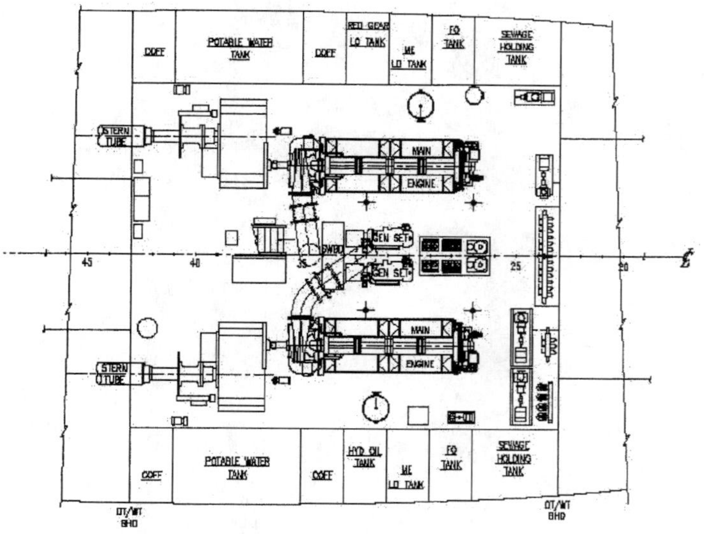

Figure 10 Tug machinery space

For enhanced maneuverability, rudders are Tenfjord high lift flap type with independent port and starboard controls.

Electrical System

Power to the paralleling main switchboard is provided by two 150 kW diesel generator sets located in the engine room. A 200 amp shore power receptacle is provided, as is a 100 amp connector for power to the barge.

An 85 kW diesel generator set provides power to the emergency switchboard, which is connected to the main switchboard.

Stability

As part of an ATB combination, the tug is required by USCG regulations to meet stability criteria as an independent towing vessel. The combination of high propulsion powering requirements, beam restrictions (to fit in the barge notch), and high vertical center of gravity (due to the tower structure) causes a need for careful analysis, and compromise, to ensure compliance. Required compromises include the use of aluminum for the tower structure and an increase in hull depth for greater freeboard.

CONCLUSION

SBOT is currently negotiating with several shipyards for the construction of the tug and barge. Other ATB's have recently come into service on the North American west coast. It is expected that market conditions will very soon justify acquisition of the subject vessels, and that they will be in service in 2005.

REFERENCES

Danish Maritime Institute Report No. 1 "Still Water and Seakeeping Tests", DMI 2000114, dated 7/19/2000

Danish Maritime Institute Report No. 2 "Seakeeping Tests", DMI 2000114, dated 9/14/2000

Danish Maritime Institute Report No. 3 "Seakeeping Tests in Survival Sea State", DMI 2000114, dated 11/6/2000

American Bureau of Shipping "Rules for Building and Classing Steel Vessels"

American Bureau of Shipping "Rules for Steel Vessels Under 90 Meters (295 Feet)"

OCHI, M.K., "Wave Statistics for the Design of Ships and Ocean Structures," SNAME *Transactions*, Vol. 86, 1978, pp.47 – 76

NORDFORSK (VTT, Marintek, DMI, SSPA), "Assessment of Ship Performance in a Seaway" 1987

Time-Dependent Risk Assessment of Aging Ships Accounting for General / Pit Corrosion, Fatigue Cracking and Local Denting Damage

Jeom Kee Paik (M), Pusan National University, **Ge Wang** (M), American Bureau of Shipping, **Anil Kumar Thayamballi** (LM), Chevron Shipping Company LLC, **Jae Myung Lee** (AM), and **Young Il Park** (AM), Pusan National University

ABSTRACT

The present paper is a summary of recent research and developments in the areas related to time-dependent risk assessment of aging ship structures, jointly undertaken by the Pusan National University and the American Bureau of Shipping. The age-related structural degradation (i.e., corrosion and fatigue crack) and mechanical damage (i.e., local dent) are considered as types of damage.

The mathematical models for predicting corrosion and fatigue cracking damage are developed as a function of ship age. Two sets of time-dependent corrosion wastage models for 34 longitudinal member groups of tankers / FPSOs, and for 23 longitudinal member groups of bulk carriers are developed by statistical analysis of corrosion measurement data. While initial crack characteristics can be analyzed by the fatigue analysis using the S-N curve approach or detected by survey, crack growth behavior can be assessed by the fracture mechanics approach. The crack size prediction model as a function of ship age is presented in the present paper.

Extensive theoretical, numerical and experimental studies are in this work carried out to investigate the ultimate strength reduction characteristics of ship panels that are wasted due to general or pit corrosion, or have fatigue cracking and local denting damage. The prime concern is with derivation of relevant design formulations for the ultimate strength of structural members with pitting, cracking or local denting damage.

Time-dependent variations of ultimate longitudinal strength reliability/risk of a bulk carrier, a double hull tanker and a ship-type FPSO under vertical bending moments are studied accounting for the effects of general / pit corrosion, fatigue cracking and local denting damage, in both deterministic and probabilistic form. It is aimed at developing more rational repair and maintenance scheme that explicitly takes into account the ultimate longitudinal strength. It is concluded that the insights and methodologies developed from the present study will be very useful for assessing time-dependent risk of aging ships and also establishing damage tolerant design procedures for new ships.

INTRODUCTION

Failures in ship structures, including total losses, continue to occur worldwide, in spite of ongoing continuous efforts to prevent them. Such failures can have enormous costs associated with them, including lost lives in some cases. One of the possible causes of marine casualties is the inability of aging ships to withstand rough seas and weather, because the ship's structural safety becomes reduced during later life although it is quite adequate at the design stage and perhaps some 15 years beyond.

Ship structures suffer various types of damage while in service. Some types of damage such as corrosion and fatigue cracks are related to ship age, while others such as local dent, collision or grounding damage are mechanical damage caused by accidental loading or impact. Structural damage reduces the load-carrying capacity of the structure, and may cause leakages, resulting in pollutions, cargo mixing or gas accumulating in enclosed spaces. In severe cases, such structural damage may conceivably lead to catastrophic failure or total loss of ships.

In the design and operation of ship structures, there are a number of uncertainties that must be dealt with. Wherever there are uncertainties, a risk of failure exists. For a structure, the risk of failure will for our purposes be defined as the probability that the load-carrying capacity is smaller than the extreme or accidental load that the structure is subjected to.

To minimizing and/or preventing loss of life and financial exposure caused by ship structural failure, it is of vital importance to keep the safety and reliability at an acceptable level even later in life.

As measures of the structural integrity, three terminologies are typically used, namely hazard, reliability and risk. A hazard is meant to be a physical situation with a potential for human injury, damage to property / environment or both. The systematic identification of the hazards with a given system, including potential causes, effects and possible corrective / preventive measures is called the *hazard identification.* Reliability is a measure of the probability of a system or component which ensures to perform the intended function for a specified period of time and under specified operating conditions. The analysis of reliability for a system or component under applied loads in normal or extreme conditions is called the *reliability assessment.* Risk is a probability associated with both the frequency (event/year) and potential consequences (effects/event) of an accident or a combination of accidents. The analysis of risk which predicts the behavior of a system in terms of accidents, frequencies and consequences in a systematic manner is called the *risk assessment.*

The present paper is a summary of recent research and developments, jointly undertaken by the Pusan National University and the American Bureau of Shipping. The following seven subjects were studied theoretically, numerically and experimentally (Paik 2002):

- Time-dependent corrosion wastage models for the structures of bulk carriers, oil tankers and ship-type FPSOs,
- Time-dependent fatigue cracking damage models,
- Ultimate strength of structural members with general/pit corrosion,
- Ultimate strength of structural members with fatigue cracking damage,
- Ultimate strength of structural members with local denting damage,
- Time-dependent risk assessment of ship structures taking account of corrosion, fatigue cracking and local denting damage, and
- Repair/maintenance scheme for aging ships.

PROCEDURE FOR RISK ASSESSMENT

The procedure for risk assessment of ship structures is similar to that of other types of steel structures. The associated risk may be written, for our purposes, as follows:

$$\text{Risk} = P_f = \text{Prob}(C \le D) \qquad (1)$$

where P_f = probability of failure, C = load-carrying capacity, D = demand.

The safety and reliability of a structure is the converse of the risk, i.e., the probability that it will not fail, namely

$$\text{Reliability} = \text{Prob}(C>D) = 1 - P_f \qquad (2)$$

The result of a standard reliability calculation, normally carried out after transformation of design variables into a standardized normal space, is a reliability index β which is related to the probability of failure P_f by

$$P_f = \phi(-\beta) \qquad (3)$$

where ϕ is the standard normal distribution function.

Broadly speaking, risk assessment involves consideration of not only the probability of failure, but also the consequences of failure, ideally in quantitative terms, see for example, Ditlevsen & Madsen (1996) and Ayyub (2001). In our study, the reliability assessment is termed as the converse of the risk assessment, however. This practice has been common in the field of structural reliability for some time now.

A major aim of structural risk assessment for merchant cargo vessels is normally to determine the level of risk in terms of probability as related to total loss from structural causes, and increasingly, the possibility of environmental pollution. An reliability assessment in a "design by analysis" approach usually includes the following tasks:

(a) Specify the required target value of the reliability index (or a target level of risk),
(b) Identify all unfavorable failure modes of the structure or fatal events,
(c) Formulate the limit state function for each failure mode identified in the item (b) above,
(d) Identify the probabilistic characteristics (mean, variance, distribution) of the random variables in the limit state function,
(e) Calculate the reliability against the limit state with respect to each failure mode of the structure,
(f) Assess if the predicted reliability is greater than or equal to the target reliability, and
(g) Repeat the above steps otherwise, as required.

TARGET RELIABILITY OF SHIPS AGAINST HULL GIRDER COLLAPSE

The required level of structural safety and reliability may vary from one industry to another depending on various factors such as the type of failure, the seriousness of its consequence or perhaps even the cost of adverse publicity and other intangible losses. Appropriate values of target safety and reliability are not readily available, but may be determined by examining statistics of failures. In doing so, the fundamental difference between a risk assessment and a reliability analysis needs to be acknowledged when interpreting such results.

The methods to select the target level of safeties and reliabilities may be categorized into the following three groups:

- "Guesstimation": A "reasonable" value as recommended by a regulatory body or professionals on the basis of prior experience. This method may be employed for innovations for which a statistical database on past failures does not exist.
- Analysis of existing design rules: The level of risk that one has traditionally lived with is estimated by calculating the reliability that is implicitly embedded in existing design rules which have been successful in the past. This method is often used for revisions of existing design rules such as reformatting a traditional experience based standard to a reliability based standard.
- Economic value analysis: The target level of safety and reliability is selected to minimize total expected costs during the service life of the structure. This is perhaps the most attractive approach, although it is often difficult to undertake in practice.

Figure 1 shows the reliability indices of some types of ships, which have been previously obtained by different investigators and calculation methods, as a function of the year of the publication (Paik & Frieze 2001). The categories of vessels, governing failure modes and service life are identified. For ships with 20 years of service life, the effect of corrosion damage was accounted for in calculating the β values. It is seen from Figure 1 that the calculated reliability index decreases as the calculation is made in more recent years. This trend does not mean that vessels themselves are becoming less reliable as well, at least not to the extent implied by the trends shown. It certainly is of course true that ship structures have become more efficient over time. But more importantly, some of the calculation results shown are perhaps more notional than the others, and the failure modes considered have become more sophisticated as well. Also, most calculations tend to use the design wave environment as the notional basis, although the actual experience may be relatively more benign, but hard to pin down particularly in trading vessels.

For instance, the calculations in the year of 1974 were made by Mansour (1974). The early results of Mansour demonstrate high notional β values. This appears to be mainly a consequence of obtaining the probability of failure under any wave load cycle rather than under an extreme load. Over time, Mansour has vastly refined and further developed the early calculation methodology from some 25 years ago, and has even successfully unified calculation procedures under the two different wave load criteria (Mansour et al. 1997).

Also, many early pioneering calculations typically used "first yield" as the failure criterion. The first-yield criterion ignores loss of plate effectiveness due to any propensity for buckling and so the location of the neutral axis of a ship's hull girder during the actual ultimate failure process will not be correctly simulated. This can result in somewhat lower levels of stress being determined for the compression region of the hull girder in addition to the basic panel strength being too high in some cases, implying a higher predicted hull girder bending strength and similarly higher reliability when compared to a reliability calculation based on a more refined prediction of ultimate hull girder bending strength.

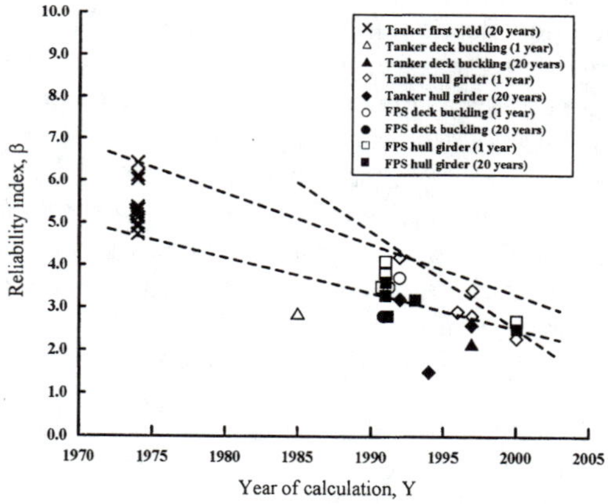

Figure 1 Variation of the calculated notional reliability indices over the passage of the years for ships during 1974 to 2000 (FPS = floating, production and storage unit) (Paik & Frieze 2001)

Whereas the β values determined in 1991 average around 3.5, these calculated in 2000 average 2.5, and all of these ignore age-related degradation effects that will cause a decrease in the β values further in comparative terms. However, there is to date no escaping the fact that even today's calculations result in reliability indices that are other than notional and comparative, mostly because of the uncertainties in the loads involved. This situation is expected to continue in the foreseeable future. We can, however, improve the value of comparative and notional reliability measures further by appropriately using the results of continuing advances in load prediction and ultimate strength assessment procedures.

Based on the above varied results, and for purposes of use with the more recent (advanced) methodologies for ultimate hull girder strength calculations, it is considered that $\beta = 2.5$ may be a speculative but good target reliability index to aim for in respect of ultimate hull girder strength.

LIMIT STATE EQUATIONS

For hull girder ultimate strength reliability assessment, four levels of failure modes may be considered, namely the primary, secondary, tertiary and quaternary failure modes.

The *primary failure mode* represents a condition which hull girder collapse takes place by involving buckling collapse of compression flange (i.e., deck panel in sagging or bottom panel in hogging) and yielding of tension flange (i.e., bottom panel in sagging or deck panel in hogging). The *secondary failure mode* indicates a limit state when the flange in compression reaches the ultimate strength. The *tertiary failure mode* is typically considered when support members in the compression flange fail by lateral-torsional buckling or tripping. The *quaternary failure mode* is shown when plating between support members in the compression flange reaches the ultimate strength.

The last three failure modes are not necessarily meant to be total loss of a ship, while a ship's function will be totally lost if the primary failure mode occurs. The limit state equation of each failure mode noted above can be expressible, as follows:

Primary failure mode:

$$F_I = x_{1u}M_{1u} - (x_{sw}k_{sw}M_{sw} + x_w k_w M_w) \le 0 \quad (4a)$$

Secondary failure mode:

$$F_{II} = x_{2u}M_{2u} - (x_{sw}k_{sw}M_{sw} + x_w k_w M_w) \le 0 \quad (4b)$$

Tertiary failure mode:

$$F_{III} = x_{3u}M_{3u} - (x_{sw}k_{sw}M_{sw} + x_w k_w M_w) \le 0 \quad (4c)$$

Quaternary failure mode:

$$F_{IV} = x_{4u}M_{4u} - (x_{sw}k_{sw}M_{sw} + x_w k_w M_w) \le 0 \quad (4d)$$

where M_{iu} = hull girder capacity for the *i*th failure mode, M_{sw} = still-water bending moment, M_w = wave-induced bending moment, k_{sw}, k_w = load combination factors related to still-water bending moment and wave-induced bending moment, respectively, x_{iu}, x_{sw}, x_w = factors taking account of modeling uncertainties associated with hull girder capacity, still-water bending moment and wave-induced bending moment, respectively.

Table 1 Samples of mean and COV of variables related to the modeling uncertainties and load combinations

Variable	Distribution	Mean	COV
x_u	Normal	1.0	0.10
x_{sw}	Normal	1.0	0.05
x_w	Normal	1.0	0.15
k_{sw}	Fixed	1.0	-
k_w	Fixed	0.9	-

Table 1 represents samples of mean and COV (coefficient of variation) of variables related to modeling uncertainties and load combinations in Equation (4), which will be adopted for reliability assessment of ships later to be illustrated in the present paper.

HULL GIRDER STRENGTH MODELS

To predict the ultimate hull girder strength, two types of structural idealizations are relevant, namely plate-stiffener combination model or plate-stiffener separation model, as shown in Figure 2.

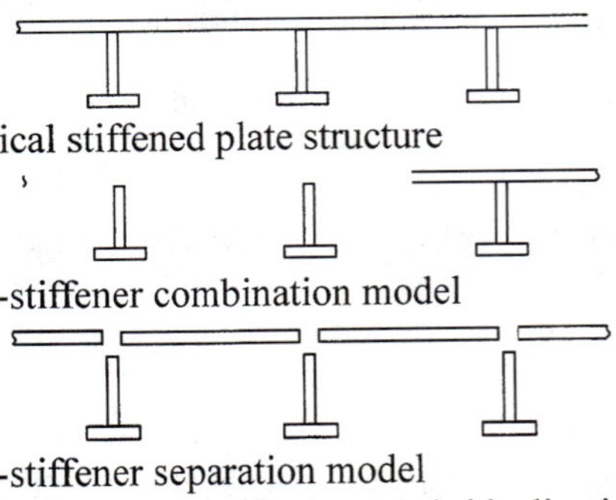

(a) A typical stiffened plate structure

(b) Plate-stiffener combination model

(c) Plate-stiffener separation model

Figure 2 Two types of structural idealization for a stiffened panel (Paik & Thayamballi 2003)

The formulations for predicting the hull girder capacity are now presented in conjunction with the four types of failure modes described in the previous section.

Primary Failure Mode

The ultimate bending moment of a ship hull with positive sign for hogging and negative sign for sagging can be calculated by (Paik et al. 2002, Paik & Thayamballi 2003)

$$M_{1u} = \sum_C \sigma_i A_{ei}(z_i - g_u) + \sum_T \sigma_j A_j (z_j - g_u) \quad (5)$$

where $g_u = \frac{\sum_C \sigma_i A_{ei} z_i + \sum_T \sigma_j A_j z_j}{\sum_C \sigma_i A_{ei} + \sum_T \sigma_i A_j}$, σ_i = longitudinal bending stresses of the ith structural member with negative sign in the compressed part and positive sign in the tensioned part, which are given by $\sigma_i = \frac{z_i - g}{D - g}\sigma_{Yeqd}$ for hogging and $\sigma_i = \frac{g - z_i}{g}\sigma_{Yeqb}$ for sagging, z_i = coordinate of the ith element measured from the base line to the deck with $z_i = 0$ at the base line, g = neutral axis, which is given as $g = \frac{\sum_C A_{ei} z_i + \sum_T A_j z_j}{\sum_C A_{ei} + \sum_T A_j}$, $\sum_C(\,)$, $\sum_T(\,)$ = summation for the part in compression or tension, respectively, A_{ei} = effective cross sectional area of the ith element in compression, A_j = cross sectional area of the jth element in tension, σ_{Yeqd}, σ_{Yeqb} = average equivalent yield stresses at upper deck or outer bottom panels, D = depth of the ship.

In calculating the longitudinal bending stress value defined in Equation (5), the following criteria should be satisfied, namely

$$\sigma \le \sigma_{Yeq} \quad \text{for tensioned elements} \quad (6a)$$

$$|\sigma| \le |\sigma_u| \quad \text{for compressed elements} \quad (6b)$$

where σ_{Yeq} = equivalent yield stress, σ_u = ultimate compressive stress of the structural element.

When a continuous stiffened plate structure is idealized as an assembly of the plate-stiffener combination units as shown in Figure 2(b), for instance, the ultimate compressive stress of each unit can be predicted, while the ultimate strength of tensioned member is considered to equal the yield stress, as follows (Paik & Thayamballi 2003)

$$\sigma_u = \frac{\sigma_{Yeq}}{\sqrt{0.995 + 0.936\lambda^2 + 0.170\beta_x^2 + 0.188\lambda^2\beta_x^2 - 0.067\lambda^4}}$$

$$\text{and } |\sigma_u| \le \frac{\sigma_{Yeq}}{\lambda^2} \quad (7)$$

where $\beta_x = \frac{b}{t}\sqrt{\frac{\sigma_Y}{E}}$, $\lambda = \frac{a}{\pi r}\sqrt{\frac{\sigma_{Yeq}}{E}}$, a = plate length (in the ship's longitudinal direction) between two adjacent transverse frames, b = plate breadth (in the ship's transverse direction) between longitudinals, t = plate thickness, E = Young's modulus, σ_Y = yield stress of plating, σ_{Yeq} = equivalent yield stress of both plating and stiffener.

On the other hand, when the ship hull structure is modeled as an assembly of plate-stiffener separation models as shown in Figure 2(c), the ultimate hull girder strength can also be calculated from Equation (5), but as a function of the ultimate strengths of plating and longitudinals (support members). In this case, the ultimate compressive strength of plating can be predicted from Equation (11), while that of support members without attached plating may be calculated as the minimum value of the limit loads obtained for the three failure modes, namely lateral-torsional buckling (tripping), beam-column type collapse or Euler column buckling adjusted by plasticity correction (Paik & Thayamballi 2003). Under axial tension, the ultimate strength of plating or support members may be considered to equal the yield stress.

Secondary Failure Mode

The hull girder strength formula based on the secondary failure mode can be given by

$$M_{2u} = \sigma_u Z_e \quad (8)$$

where σ_u, Z_e = ultimate compressive stress and elastic section modulus, respectively, at the compression flange of a ship.

Again, the ultimate compressive stress of the compression flange may be predicted from Equation (7) as long as the deck or bottom panel is idealized as an assembly of plate-stiffened combination units as shown in Figure 2(b), or as an assembly of plate-stiffener separation elements as shown in Figure 2(c).

Tertiary Failure Mode

The hull girder strength formula based on the tertiary failure mode can be given by

$$M_{3u} = \sigma_u Z_e \quad (9)$$

where σ_u = ultimate compressive stress of stiffeners (support members) at the compression flange, Z_e = as defined in Equation (8).

The ultimate compressive stress σ_u in the tertiary failure mode is meant to be the minimum value of the three strengths of longitudinals at the compression flange, namely elastic-plastic tripping strength, beam-column type collapse strength and Euler column buckling strength adjusted by plasticity correction. Closed-form formulae for predicting the ultimate strength of support members considering types of stiffener profiles (e.g., flat-bar, angle and T-bar) are relevant in textbooks (Paik & Thayamballi 2003).

Quaternary Failure Mode

The hull girder strength formula based on the quaternary failure mode can be given by

$$M_{4u} = \sigma_u Z_e \tag{10}$$

where σ_u = ultimate compressive stress of plating between stiffeners at the compression flange, Z_e = as defined in Equation (8).

The ultimate compressive stress of plating between stiffeners can be predicted as follows (Paik et al. 2003a) (For symbols not specified below, Equation (7) is referred to):

For $\frac{a}{b} \geq 1$:

$$\frac{\sigma_u}{\sigma_Y} = \begin{cases} -0.032\beta_x^4 + 0.002\beta_x^2 + 1.0 & \text{for} \quad \beta_x \leq 1.5 \\ 1.274/\beta_x & \text{for} \quad 1.5 < \beta_x \leq 3.0 \\ 1.248/\beta_x^2 + 0.283 & \text{for} \quad \beta_x > 3.0 \end{cases} \tag{11a}$$

For $\frac{a}{b} < 1$:

$$\frac{\sigma_u}{\sigma_Y} = \frac{a}{b}\frac{\sigma_{xu}}{\sigma_Y} + \frac{0.475}{\beta_y^2}\left(1 - \frac{a}{b}\right) \tag{11b}$$

where σ_Y = yield stress, $\beta_x = \frac{b}{t}\sqrt{\frac{\sigma_Y}{E}}$, $\beta_y = \frac{a}{t}\sqrt{\frac{\sigma_Y}{E}}$.

σ_{xu} in Equation (11b) is taken as $\sigma_{xu} = \sigma_u$ defined in Equation (11a), but replacing β_x by β_y.

Effect of General/Pit Corrosion on Plate Ultimate Strength

Corrosion wastage in ship plates can reduce their ultimate strength. Two types of corrosion damage are usually considered, namely general (or uniform) corrosion and localized corrosion. General corrosion reduces the plate thickness uniformly, while localized corrosion such as pitting appears non-uniformly in selected regions, e.g., the vessel bottom in cargo tanks of crude oil carriers.

The ultimate strength of a steel member with general corrosion can be easily predicted, i.e., by excluding the plate thickness loss due to corrosion.

Based on a series of experimental and numerical studies on steel plated structures (Paik 2002), however, it is realized that the plate ultimate strength reduction characteristics due to general corrosion are quite different from those due to pit corrosion. The so-called equivalent plate thickness reduction approach, which represents a pitted plate with a plate of an "equivalent" thickness, is not sufficient for accurately predicting the plate's ultimate strength.

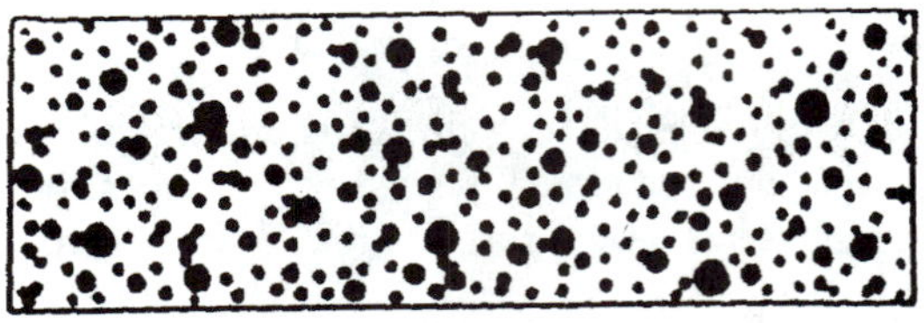

Figure 3 A sample of the pitting intensity diagram (Degree of pit intensity = 20%)

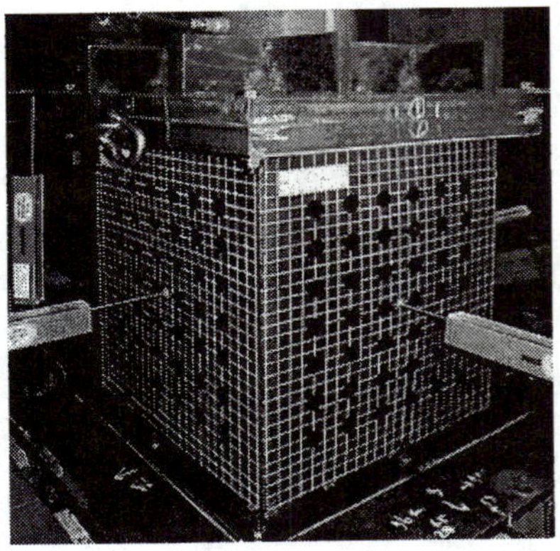

Figure 4(a) Collapse test set-up on the box column type of a plated structure with idealized pits

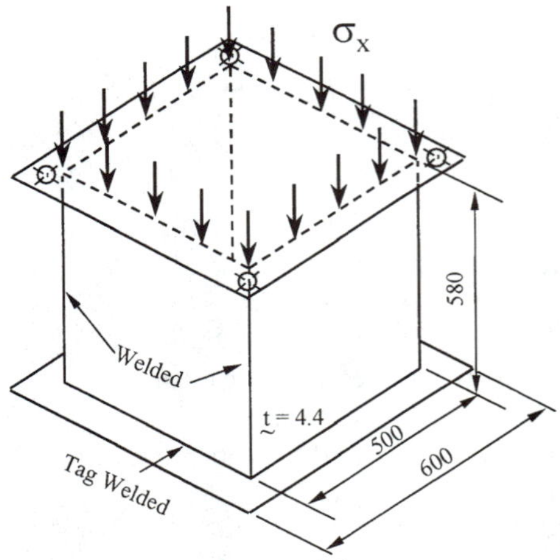

Figure 4(b) A schematic view of the test structure

Figure 3 shows a sample of pitted plates. Figures 4 and 5 represent examples of the structural models used for the experiment and numerical computations undertaken in the present study.

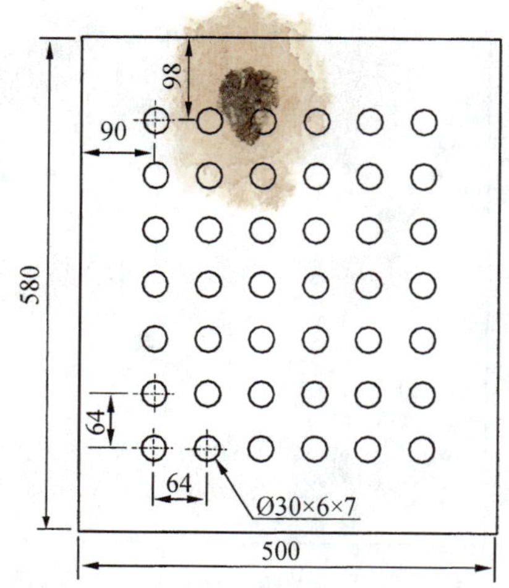

Figure 5(a) Idealization of pit size and location

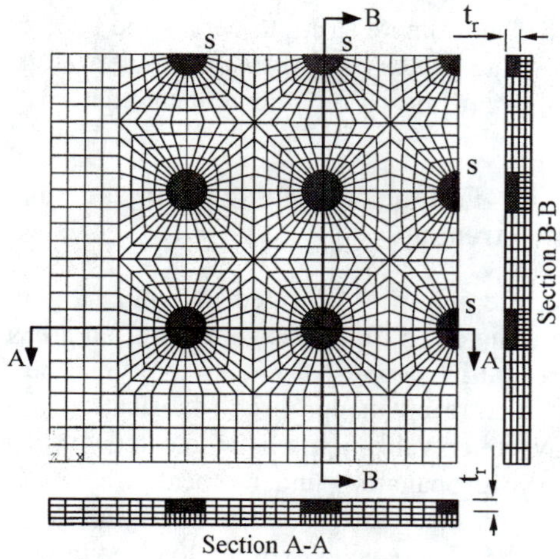

Figure 5(b) Finite element modeling of a plate with regular pit corrosion (s: symmetric boundary condition)

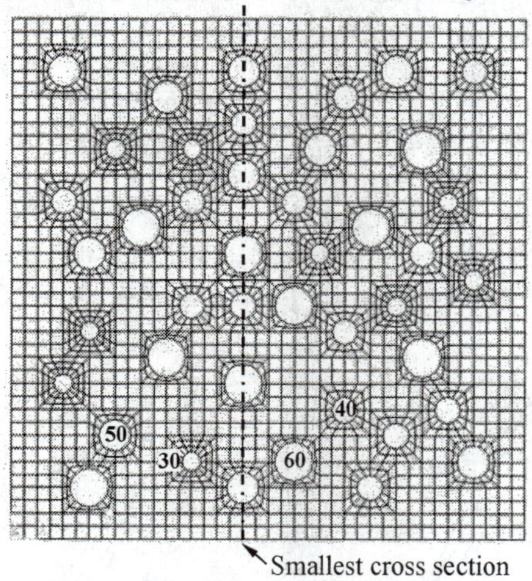

Figure 5(c) Finite element modeling of a plate with random pits $t_r = t$ (through thickness pit), $a \times b \times t = 800 \times 800 \times 15$mm, DOP=11.3% with $(A_o - A_r)/A_o = 0.56$ (The digits inside the figure represent the diameters of pits in mm)

To assess scale for breakdown due to pit corrosion, a parameter denoted by DOP (degree of pit corrosion intensity) is often used. DOP is defined as the ratio percentage of the corroded surface area to the original plate surface area:

$$\mathrm{DOP} = \alpha = \frac{1}{ab}\sum_{i=1}^{n} A_{pi} \times 100\ (\%) \tag{12}$$

where n = number of pits, A_{pi} = surface area of the ith pit, a = plate length, b = plate breadth. For a circular type of pit corrosion, $A_{pi} = \pi d_{ri}^2/4$ with d_{ri} = diameter of the *i*th pit.

A series of experimental and numerical studies for steel plated structures with pits and under axial compressive loads or edge shear were performed by varying the DOP, the depth of pit, the regularity of pit, the plate thickness and the plate aspect ratio (Paik 2002).

It is found from the experimental and numerical studies that the ultimate strength of a plate with pit corrosion can be estimated using a strength knock-down factor that can be calculated using the following formulations:

For axial compressive loading:

$$R_{xr} = \frac{\sigma_{xu}}{\sigma_{xuo}} = \left(\frac{A_o - A_r}{A_o}\right)^{0.73} \tag{13}$$

where R_{xr} = a factor of ultimate compressive strength reduction due to pit corrosion, σ_{xu} = ultimate compressive strength for a member with pit corrosion, σ_{xuo} = ultimate compressive strength for an intact (un-corroded) member, which can be given by Equation (11), A_o = original cross sectional area of the intact member, A_r = cross sectional area involved by pit corrosion at the smallest cross section, see Figure 6.

For edge shear:

$$R_{\tau r} = \frac{\tau_u}{\tau_{uo}} = \begin{cases} 1.0 \text{ for } \alpha \le 1.0 \\ -0.18\ln\alpha + 1.0 \text{ for } \alpha > 1.0 \end{cases} \tag{14}$$

where $R_{\tau r}$ = a factor of ultimate shear strength reduction due to pit corrosion, τ_u = ultimate shear strength for a pitted plate, α =DOP as defined in Equation (12), τ_{uo} = ultimate shear strength for an intact plate which can be given by (Paik & Thayamballi 2003)

$$\frac{\tau_{uo}}{\tau_Y} = \begin{cases} 1.324(\tau_E/\tau_Y) & \text{for } 0 < \tau_E \le 0.5 \\ 0.039(\tau_E/\tau_Y)^3 - 0.274(\tau_E/\tau_Y)^2 + 0.676(\tau_E/\tau_Y) + 0.388 \\ \quad \text{for } 0.5 < \tau_E/\tau_Y \le 2.0 \\ 0.956 \text{ for } \tau_E/\tau_Y > 2.0 \end{cases} \tag{15}$$

with $\tau_E = k_\tau \frac{\pi^2 E}{12(1-\nu^2)}\left(\frac{t}{b}\right)^2$, $k_\tau = 4\left(\frac{b}{a}\right)^2 + 5.34$ for $\frac{a}{b} \ge 1$ or $k_\tau = 5.34\left(\frac{b}{a}\right)^2 + 4.0$ for $\frac{a}{b} < 1$, $\tau_Y = \sigma_Y/\sqrt{3}$, σ_Y = yield stress of material.

Figure 7 compares Equation (13) with the numerical and experimental results. Figure 8 compares Equation (14) with the non-linear finite element computations. It is evident that the strength knock-down factor approach proposed in the present study is useful for predicting the ultimate compressive or shear strength of pitted plates.

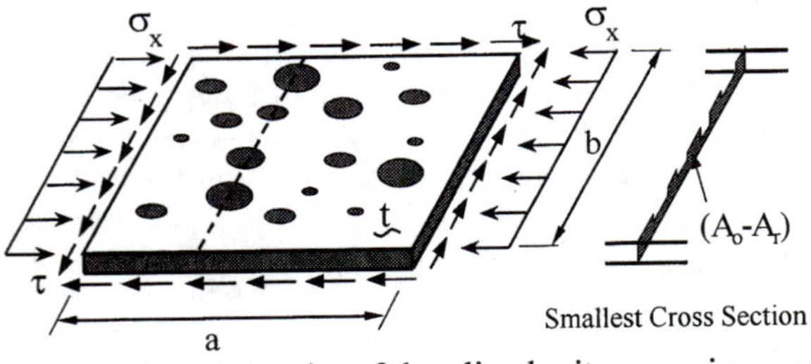

Figure 6 A schematic of localized pit corrosion and definition of the smallest cross sectional area

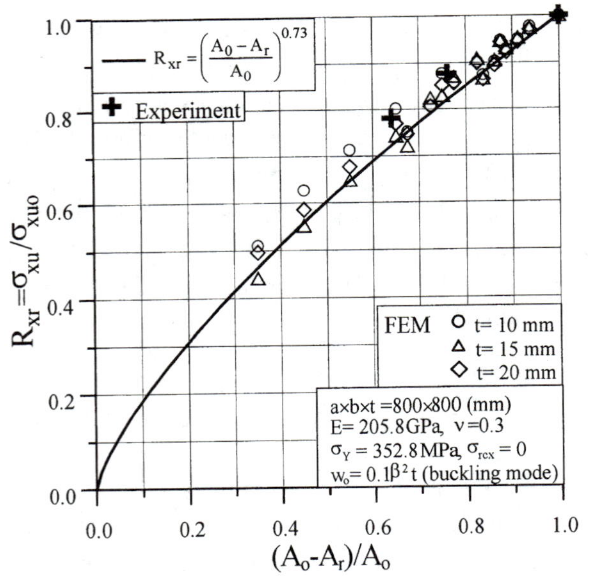

Figure 7 The ultimate compressive strength reduction factor as a function of the smallest cross sectional area for a plate with pit corrosion

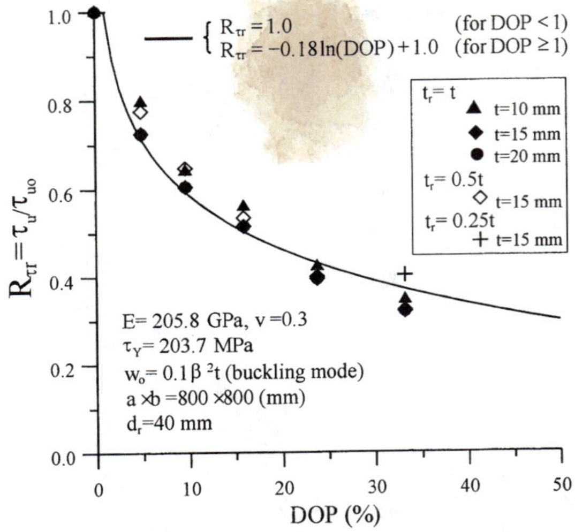

Figure 8 The ultimate strength versus the DOP ratio for a steel plate with pit corrosion under edge shear (Symbols: non-linear finite element calculations)

Effect of Fatigue Cracking Damage on Plate Ultimate Strength

Under the action of repeated loading, fatigue cracks may be initiated in the stress concentration areas of the structure. Initial defects or cracks may also be formed in the structure by inappropriate fabrication procedure and may conceivably remain undetected over time. In addition to propagation under repeated cyclic loading, cracks may also grow in an unstable way under monotonically increasing extreme loads, a circumstance which eventually can lead to catastrophic failure of the structure. This possibility is usually tempered by the ductility of the material, and also by the presence of reduced stress intensity regions in a complex structure that may serve as crack arresters even in an otherwise monolithic structure.

For residual strength assessment of aging steel structures under extreme loads as well as under fluctuating loads, it is thus often necessary to take into account a known or premised crack as a parameter of influence.

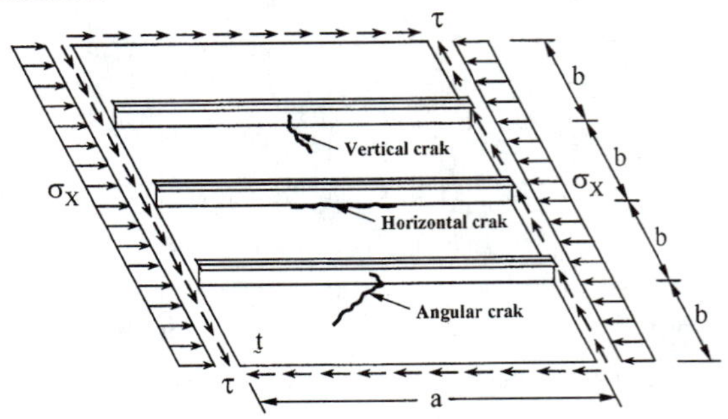

Figure 9 A schematic of a stiffened steel panel component with three types of crack orientations and under axial compression or edge shear

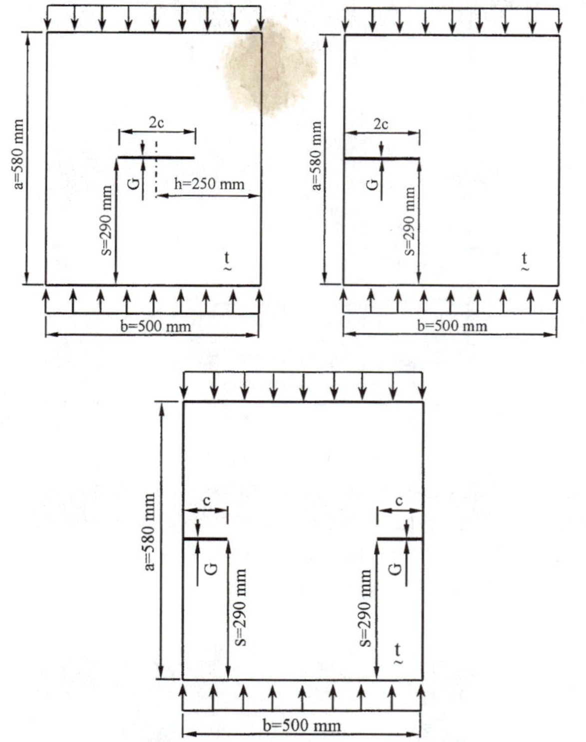

Figure 10 Various crack locations in the test structure considered in the present study

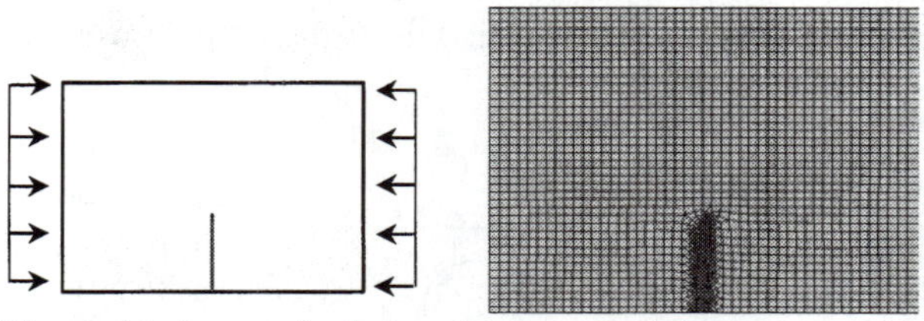

Figure 11 A sample finite element mesh in a plate with one edge crack and under axial compression

Figure 9 shows a schematic of a stiffened steel panel component with three types of crack orientations and under axial compression or edge shear. Strictly speaking, the ultimate strength behavior of panels depends on the types of crack orientaions, among other factors. While pending further studies, the present study considers the cross sectional area as a primary parameter which governs the ultimate strength reduction characteristics, regardless of the crack orientations.

A series of theoretical, numerical and experimental studies for steel plated structures with premised cracks and under axial tensile/compressive loads or edge shear were carried out by varying the location and size of cracks and the plate thickness. Figure 10 shows various crack locations in a plate considered in the present study. Figure 11 shows a sample finite element modelling for a plate with edge crack at one side and under axial compressive loads.

Based on the results of the experiment and non-linear finite element computations obtained from the present study, the strength knock-down factor approach is again suggested for predicting the ultimate strength of a plate with premised cracks, as follows:

For axial tensile / compressive loading:

$$R_{xc} = \frac{\sigma_{xu}}{\sigma_{xuo}} = \frac{A_o - A_c}{A_o} \tag{16}$$

where R_{xc} = a factor of the ultimate tensile or compressive strength reduction due to cracking damage, σ_{xu} = ultimate axial strength of cracked plating, σ_{xuo} = ultimate axial strength of uncracked plating which may be taken as $\sigma_{xuo} = \sigma_Y$ for axial tensile loading and $\sigma_{xuo} = \sigma_u$ with σ_u as defined in Equation (11), A_o = cross sectional area of uncracked (original) plating, A_c = cross sectional area involved by cracking damage.

For edge shear:

$$R_{\tau c} = \frac{\tau_u}{\tau_{uo}} = \frac{A_o - A_c}{A_o} \tag{17}$$

where $R_{\tau c}$ = a factor the ultimate shear strength reduction due to cracking damage, τ_u = ultimate shear strength for a plate with premised cracks, τ_{uo} = ultimate shear strength for an intact plate, as defined in Equation (15).

Figure 12 compares Equation (16) with the experimental and numerical results. Figure 13 compares Equation (17) with the numerical computations. It is again apparent that the proposed formulae reasonably predict the plate ultimate strength at somewhat pessimistic side.

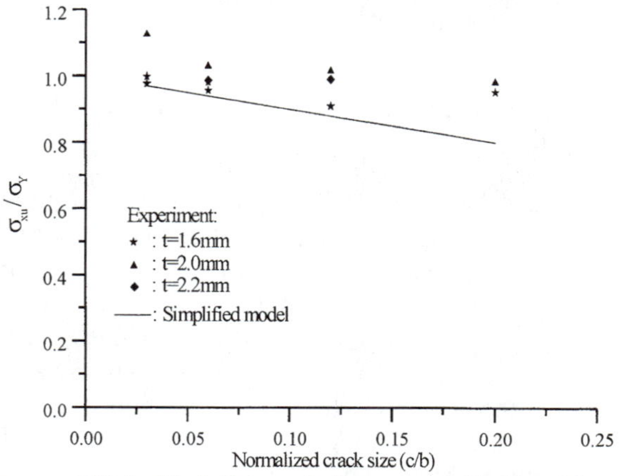

Figure 12(a) Variation of the normalized ultimate tensile strength of a steel plate with a single center crack as a function of the crack size

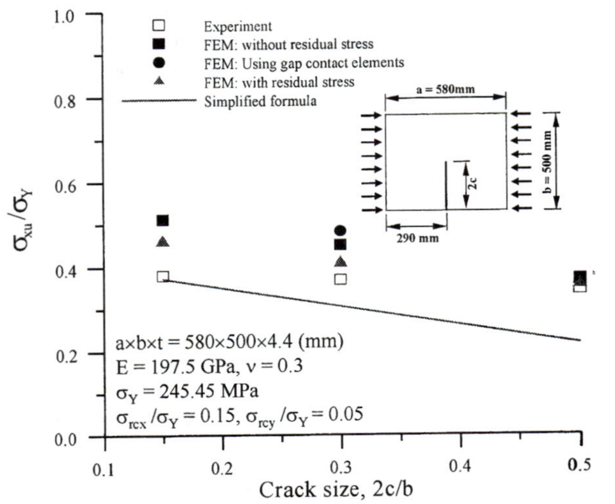

Figure 12(b) Variation of the normalized ultimate compressive strength of a steel plate with a single edge crack as a function of the crack size

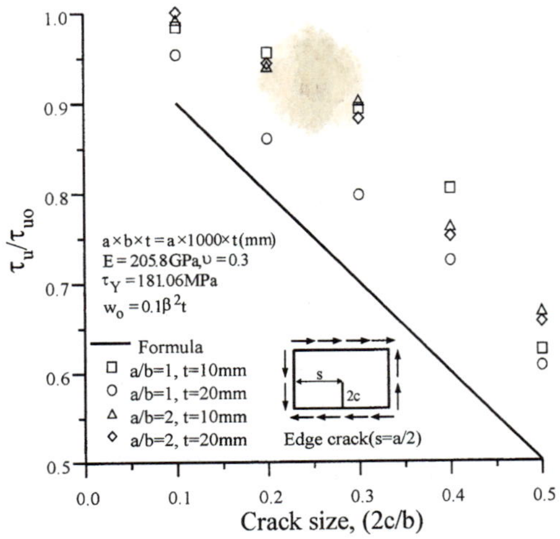

Figure 13 Variation of the ultimate shear strength of a steel plate as a function of the crack size, varying the plate thickness and aspect ratio (Symbols: non-linear finite element computations)

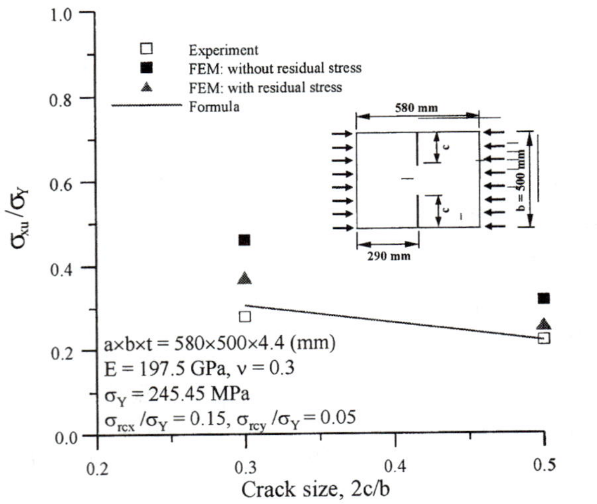

Figure 12(c) Variation of the normalized ultimate compressive strength of a steel plate (with multiple cracks) as a function of the crack size

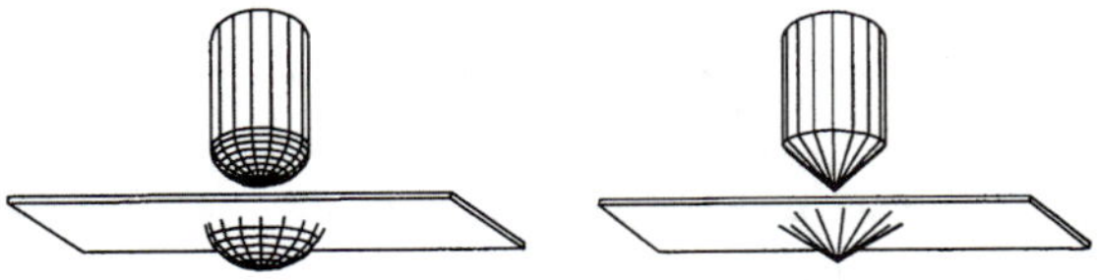

Figure 14 Two types of local denting considered in the present study (left: spherical shape, right: conical shape)

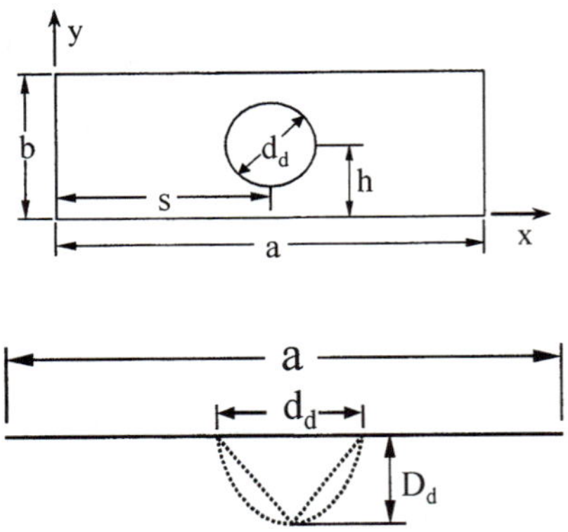

Figure 15 Geometric parameters of the local dent

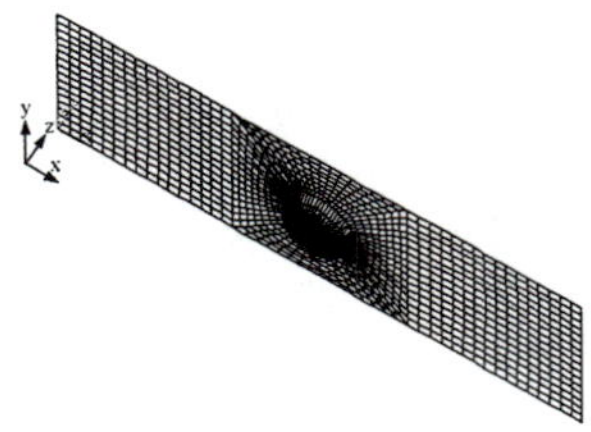

Figure 16 A sample of finite element modeling for a steel plate with local denting and under axial compressive loads

Effect of Local Denting Damage on Plate Ultimate Strength

Plate panels in ships and offshore platforms can have mechanical damage. The inner bottom plates of cargo holds of bulk carriers may be damaged during loading of iron ore when iron ore cargo strikes the plates. In unloading of bulk cargoes such as iron ore or coal, excavators can result in impacts to the inner bottom plates mechanically. Deck plates of offshore platforms may be subjected to impacts due to dropped objects from cranes. Mechanical damages can be denting, cracking, residual stresses or strains due to plastic deformation, and coating damage.

The ultimate strength of plates with dent damage under axial compression or edge shear was studied using non-linear finite element method, with varying the location and size of dent, the plate thickness and the plate aspect ratio.

Two types of denting were considered, as shown in Figure 14. Figure 15 defines the geometrical dimensions of the local dent. Figure 16 shows examples of the finite element modeling for a plate with local dent at the plate center. It is concluded that the influence of spherical dent on the plate ultimate strength is similar to that of conical dent, while the former is slightly worse than the latter in terms of load-carrying capacity. Therefore, the spherical dent may be taken as a representative of local dent shape for the purpose of the plate ultimate strength prediction, regardless of the actual shape of denting (Paik et al. 2003b).

Based on the results and insights developed from the present study, empirical formulae for predicting the ultimate compressive strength of dented plates are derived, when $d_d / b < 1$, as follows:

For axial compression:

$$R_{xd} = \frac{\sigma_{xu}}{\sigma_{xuo}} = C_3\left[C_1 \ln\left(\frac{D_d}{t}\right) + C_2\right] \quad (18)$$

where R_{xd} = a factor of the ultimate compressive strength reduction due to local denting, σ_{xu}, σ_{xuo} = ultimate compressive strengths of dented or intact plates, respectively. The coefficients C_1, C_2 and C_3 are empirically determined by regression analysis of the computed results as follows:

$$C_1 = -0.042\left(\frac{d_d}{b}\right)^2 - 0.105\left(\frac{d_d}{b}\right) + 0.015 \quad (19a)$$

$$C_2 = -0.138\left(\frac{d_d}{b}\right)^2 - 0.302\left(\frac{d_d}{b}\right) + 1.042 \quad (19b)$$

$$C_3 = -1.44\left(\frac{H}{b}\right)^2 + 1.74\left(\frac{H}{b}\right) + 0.49 \quad (19c)$$

where $H = h$ for $h \le 0.5b$ and $H = b - h$ for $h > 0.5b$. For axial tension, it is considered $R_{xd} = 1.0$.

For edge shear:

$$R_{\tau d} = \frac{\tau_u}{\tau_{uo}} = \begin{cases} C_1\left(\frac{D_d}{t}\right)^2 - C_2\left(\frac{D_d}{t}\right) + 1 & \text{for } 1 < \frac{D_d}{t} < 10 \\ 100C_1 - 10C_2 + 1 & \text{for } 10 \le \frac{D_d}{t} \end{cases} \quad (20)$$

where $R_{\tau d}$ = a factor of the ultimate shear strength reduction due to local denting, τ_u = ultimate compressive strengths of dented plates, τ_{uo} = as defined in Equation (15). The coefficients C_1 and C_2 are empirically determined by regression analysis of the computed results as follows:

$$C_1 = 0.0129\left(\frac{d_d}{b}\right)^{0.26} - 0.0076 \quad (21a)$$

$$C_2 = 0.1888\left(\frac{d_d}{b}\right)^{0.49} - 0.07 \quad (21b)$$

Figures 17 and 18 compare Equations (18) and (20) with non-linear finite element computations for dented plates under axial compressive loads and edge shear, respectively.

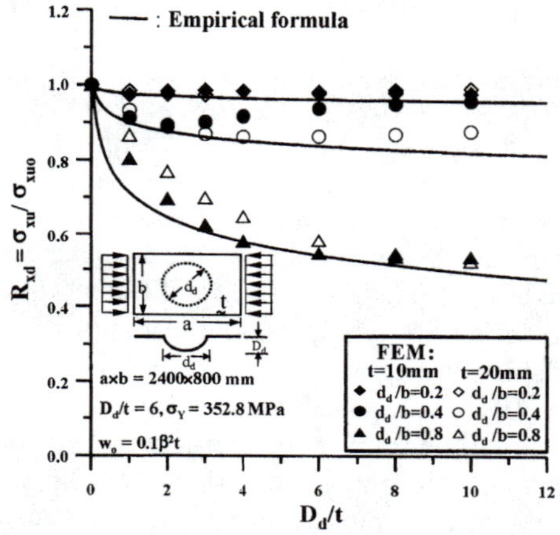

Figure 17 Accuracy of Equation (18) with Equation (19) for predicting the ultimate compressive strength of plates with dent damage

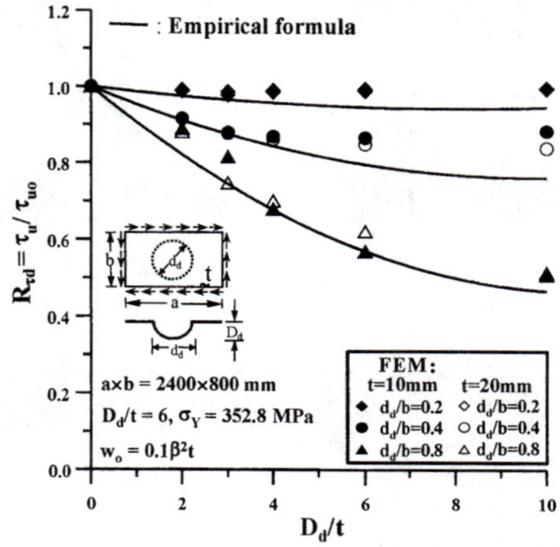

Figure 18 Accuracy of Equation (20) with Equation (21) for predicting the ultimate shear strength of plates with dent damage

Effect of Combined Corrosion, Fatigue Cracking and Local Denting Damage

When pit corrosion, fatigue cracking and local denting damage exist simultaneously, the ultimate strength of a member under axial loads or edge shear may simply be calculated by virtue of a multiplicative model, as follows:

$$\sigma_{xu} = R_{xr} R_{xc} R_{xd} \sigma_{xuo} \tag{22}$$

$$\tau_u = R_{\tau r} R_{\tau c} R_{\tau d} \tau_{uo} \tag{23}$$

where R_{xr}, R_{xc}, R_{xd}, $R_{\tau r}$, $R_{\tau c}$, $R_{\tau d}$ = as defined in Equations (13), (16), (18), (14), (17) and (20), respectively, σ_{xuo} = ultimate axial strength of an intact member, τ_{uo} = ultimate shear strength of an intact member.

TIME-DEPENDENT CORROSION MODELS

Corrosion Mechanism

The corrosion characteristics in a ship structure are influenced by many factors such as type of corrosion protection, type of cargo, temperature, humidity and so on. This means that it should be able to make estimates of corrosion depth for various different structural members grouped by type and location for different types of ships or cargoes.

Figure 19 is a proposed corrosion process model for a coated area in a marine steel structure. The corrosion behavior is in this model categorized into three phases, on account of a) durability of coating, b) transition to visibly obvious corrosion, and c) progress of such corrosion (e.g., Paik & Thayamballi 2003).

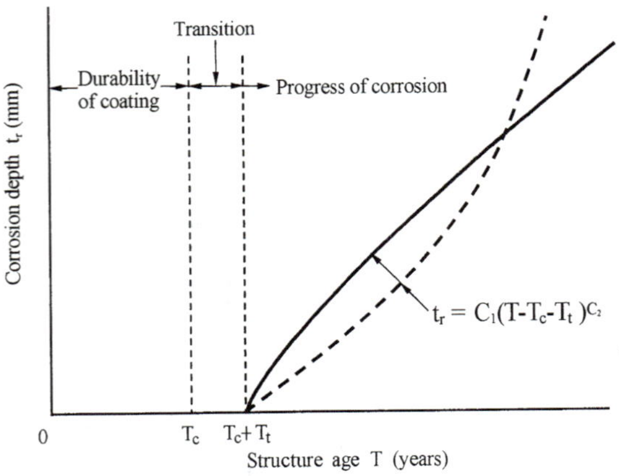

Figure 19 A schematic of a proposed corrosion process model for marine structures

The curve showing corrosion progression as indicated by the solid line in Figure 19 is convex, but it may in some cases be a concave (dotted line). The convex curve indicates that the corrosion rate (i.e., the curve gradient) increases in the beginning but decreases as the corrosion progress proceeds. This type of corrosion progression may be typical of a static immersion environment in sea-water, since the relatively static corrosion scale at the steel surface can eventually disturb the corrosion progression. On the other hand, the concave curve represents a case where the corrosion rate accelerates as the corrosion progress proceeds. This type of corrosion progression may be likely to happen in changing immersion conditions at sea, particularly in dynamically loaded structures where flexing continually exposes additional fresh surface to the corrosion effects. It is considered critical if the 'domino effect' of coating breakdown, leading to accelerated corrosion, which reduces scantlings and can ultimately lead to structural failure, is to be minimized or avoided (Contraros 2003).

The life (or durability) of a coating corresponds to the time when a pre-defined and measurable extent of corrosion starts after either a) the time when a newly built ship enters service, b) the application of coating in a previously bare case, or c) repair of a failed coating area in an existing structure to a good intact standard. The life of coating typically depends on the type of coating systems used, details of its application (e.g., surface preparation, stripe coats, film thickness, humidity and salt control during application, etc.), and relevant maintenance, among other factors. While the coating life to a pre-defined state of breakdown is a random variable, it is often treated as a constant parameter.

After the effectiveness of coating is lost, some transition time, i.e., duration between the time of coating effectiveness loss and the time of corrosion initiation, may be considered to exist before the corrosion 'initiates' over a large enough and measurable area. The transition time is often considered to be an exponentially distributed random variable. Three types of corrosion models have been suggested as follows:

Paik & Thayamballi (2002):

$$t_r = C_1 (T - T_c - T_t)^{C_2} \tag{24}$$

where t_r = depth of corrosion (thickness loss due to corrosion), T_c = coating life, T_t = transition time between coating durability and corrosion initiation, T = structure age, C_1, C_2 = coefficients taking account of the characteristics of corrosion progress.

Guedes Soares & Garbatov (1999):

$$t_r = t_{r\infty} \left[1 - \exp\left(-\frac{T - T_c}{T_t} \right) \right] \tag{25}$$

where $t_{r\omega}$ = depth of corrosion when the corrosion progress stops, with other symbols defined in Equation (24).

Qin & Cui (2002):

$$t_r = t_{r\omega}\left\{1 - \exp\left[-\left(\frac{T - T_{st}}{\eta}\right)^{\alpha}\right]\right\} \tag{26}$$

where T_{st} = time when accelerating of corrosion stops, α , η = coefficients to handle the corrosion decelerating, with other symbols defined in Equations (24), (25).

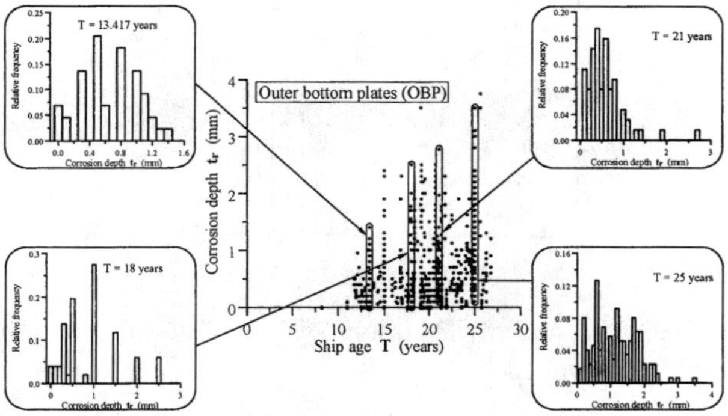

Figure 20 The corrosion depth versus the ship age obtained from the thickness measurements for outer bottom plates of aging bulk carriers

Figure 20 shows an example of the corrosion depth for outer bottom plates of existing bulk carriers as a function of age. The relative frequency of the corrosion depth tends to follow the normal function for younger ships, while it follows a log-normal or Weibull function for older ships.

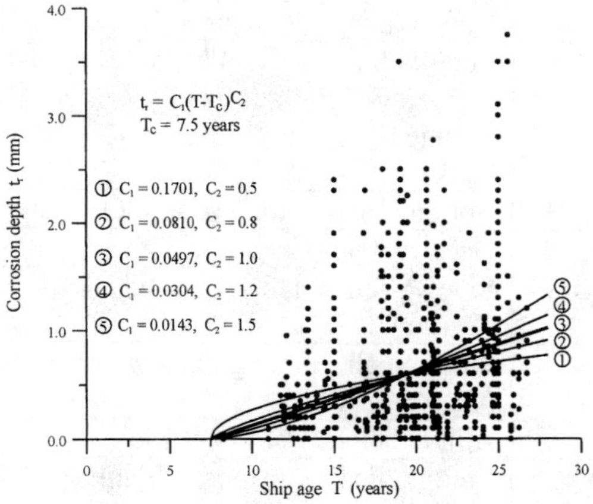

Figure 21 Sample formulations of the corrosion depth measurements on outer bottom plating of aging bulk carrier structures as a function of ship age, varying the coefficient C_2

Figure 21 plots Equation (24) which shows curve fits to a sample of statistical corrosion data for outer bottom shells of bulk carriers, varying the coefficient C_2 , provided that T_c = 7.5 years and T_t = 0. The trend of the corrosion progress slightly varies with the coefficient C_2 as would be expected, but its effect may be ignored as the ship gets older. For practical design purposes, therefore, it is assumed to be constant regardless of time, i.e., with $C_2 = 1.0$.

Provided that the corrosion initiates without a transition time after the effectiveness of coating is lost, i.e., $T_t = 0$, Equation (24) is simplified to

$$t_r = C_1(T - T_c) \tag{27}$$

where the coefficient C_1 corresponds to the annualized corrosion rate.

Figure 22 represents a schematic of the probability density distribution for the coefficient C_1 obtained by statistical analysis of corrosion measurement data for outer bottom plates of bulk carriers. The figure indicates that a Weibull density can be an adequate fit in this case.

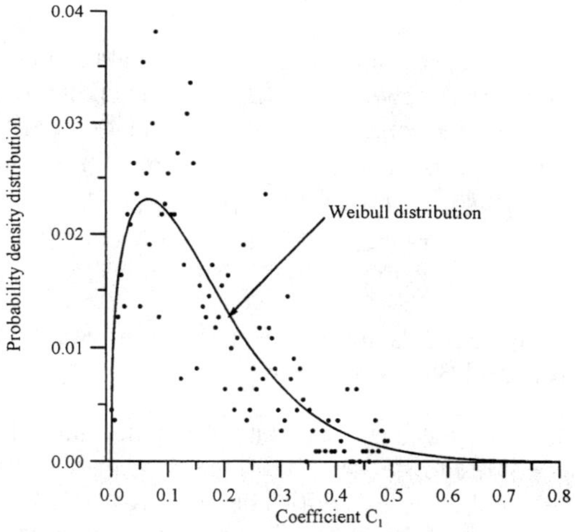

Figure 22 The Weibull type probability density function fitted to the coefficient C_1 with $C_2 = 1.0$; data for the outer bottom corrosion of bulk carriers

The probability density function of the coefficient C_1 may then be assumed to follow the Weibull distribution:

$$f_{C_1}(x) = \frac{\lambda}{\alpha}\left(\frac{x}{\alpha}\right)^{\lambda-1} \exp\left[-\left(\frac{x}{\alpha}\right)^{\lambda}\right] \tag{28}$$

where α and λ are scale and shape parameters, respectively, which will be determined through a probability density fit using the method of moments, the maximum likelihood method or other appropriate method. The choice of a Weibull density function offers some flexibility as it is capable of representing a range of types of exponential behavior.

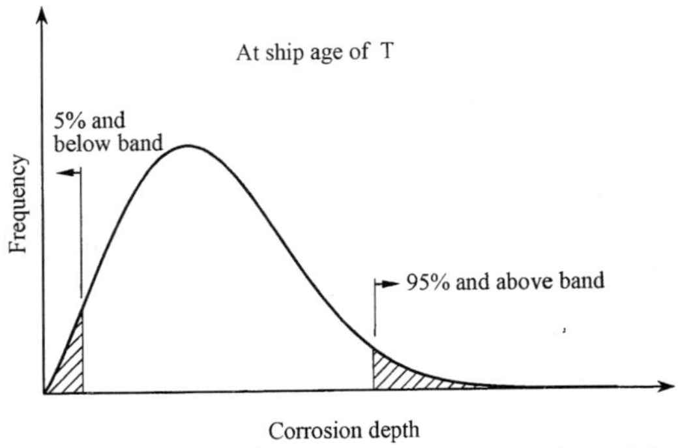

Figure 23 The 95% and above band or 5% and below band for developing the upper or lower bound corrosion rate models

Mean and COV (coefficient of variation) of the coefficient C_1 can then be calculated by statistical analysis once the corrosion measurement data are available. The most probable (or average) level of corrosion damage will be predicted if all gathered corrosion measurement data are used for the statistical analysis. Since the statistical corrosion data are usually very scattered, however, it may also be of interest to investigate the upper or lower corrosion statistical characteristics of the coefficient C_1, the former being related to 95% and above band or severe level, while the latter being related to 5% and below band or slight level, as shown in Figure 23.

Corrosion Models for the Structures of Single Hull Tankers and Ship-type FPSOs

Corrosion measurement data for a number of 230 aging single hull tankers carrying crude or product oil has been collected by entities based in Korea and made available to this study. Figure 24 represents the distribution of the ship age that had been surveyed for this purpose.

A total of 33,820 measurements for 34 different member groups (defined by locations, categories and corrosion environments of members) which include 14 categories of plate parts, 11 categories of stiffener webs, and 9 categories of stiffener flanges were obtained and available for the present study.

Table 2 defines the identification of member location/category groups. Figures 25 and 26 represent the most probable (average) or severe level of mean and COV of the coefficient C_1 (annualized corrosion rate) for the 34 different member groups, respectively. It is noted that these results were obtained assuming T_c =7.5 years and T_t = 0. The corrosion depth of individual member categories can then be predicted from Equation (27) using mean and COV of the coefficient C_1 defined in Figures 25 or 26 (Paik et al. 2003c).

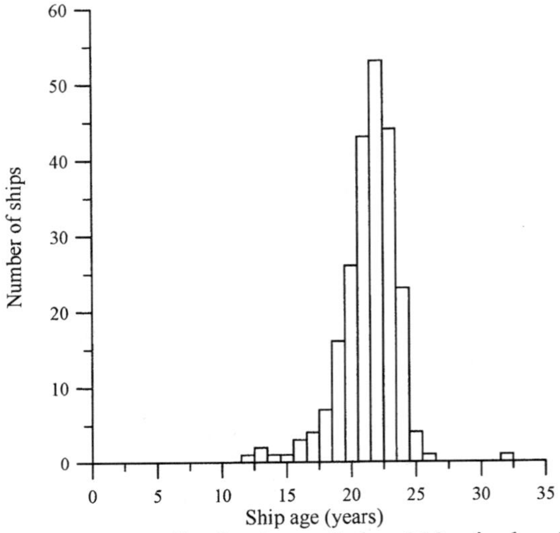

Figure 24 Age distribution of the 230 single skin tankers surveyed

Table 2.a Identification of member location/category groups for plating of tankers or FPSOs

ID	Member type
B/S-H	Bottom Shell Plating (Segregated Ballast Tank)
A/B-H	Deck Plating (Segregated Ballast Tank)
A/B-V	Side Shell Plating above draft line (Segregated Ballast Tank)
B/S-V	Side Shell Plating below draft line (Segregated Ballast Tank)
BLGB	Bilge Plating (Segregated Ballast Tank)
O/B-V	Longitudinal Bulkhead Plating (Segregated Ballast Tank)
B/B-H	Stringer Plating (Segregated Ballast Tank)
O/S-H	Bottom Shell Plating (Cargo Oil Tank)
A/O-H	Deck Plating (Cargo Oil Tank)
A/O-V	Side Shell Plating above draft line (Cargo Oil Tank)
O/S-V	Side Shell Plating below draft line (Cargo Oil Tank)
BLGC	Bilge Plating (Cargo Oil Tank)
O/O-V	Longitudinal Bulkhead Plating (Cargo Oil Tank)
O/O-H	Stringer Plating (Cargo Oil Tank)

Table 2.b Identification of member location/category groups for stiffener web and flange of tankers or FPSOs

ID (stiffener web)	Member types	ID (stiffener flange)	Member types
BSLBW	Bottom Shell Longitudinals in Ballast tank – Web	BSLBF	Bottom Shell Longitudinals in Ballast tank – Flange
SSLBW	Side Shell Longitudinals in Ballast tank – Web	SSLBF	Side Shell Longitudinals in Ballast tank – Flange
LBLBW	Longitudinal Bulkhead Longitudinals in Ballast tank – Web	LBLBF	Longitudinal Bulkhead Longitudinals in Ballast tank – Flange
BSLCW	Bottom Shell Longitudinals in Cargo oil tank – Web	BSLCF	Bottom Shell Longitudinals in Cargo oil tank – Flange
DLCW	Deck Longitudinals in Cargo oil tank – Web	DLCF	Deck Longitudinals in Cargo oil tank – Flange
SSLCW	Side Shell Longitudinals in Cargo oil tank – Web	SSLCF	Side Shell Longitudinals in Cargo oil tank – Flange
LBLCW	Longitudinal Bulkhead Longitudinals in Cargo oil tank – Web	LBLCF	Longitudinal Bulkhead Longitudinals in Cargo oil tank – Flange
BGLCW	Bottom Girder Longitudinals in Cargo oil tank – Web	BGLCF	Bottom Girder Longitudinals in Cargo oil tank – Flange
DGLCW	Deck Girder Longitudinals in Cargo oil tank – Web	DGLCF	Deck Girder Longitudinals in Cargo oil tank – Flange
DLBW	Deck Longitudinals in Ballast tank – Web		
SSTLCW	Side Stringer Longitudinals in Cargo oil tank – Web		

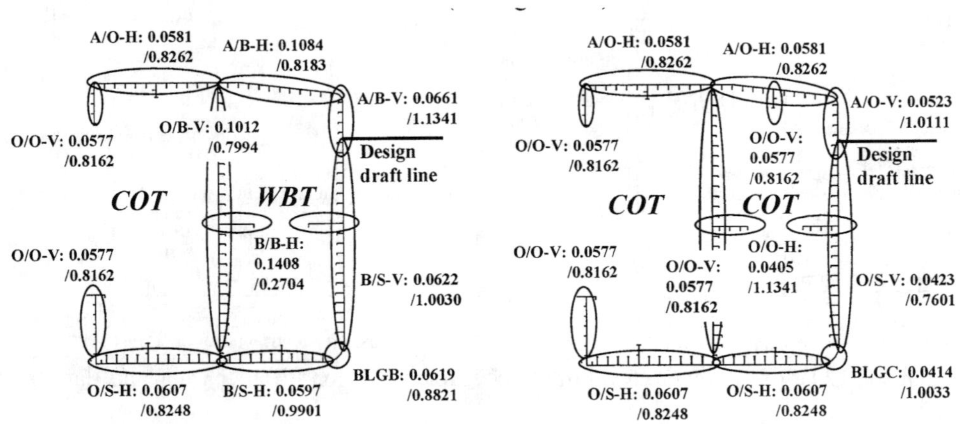

Figure 25(a) Mean and COV of the most probable (average) level of the coefficient C_1 for plating in a single skin tanker or FPSO

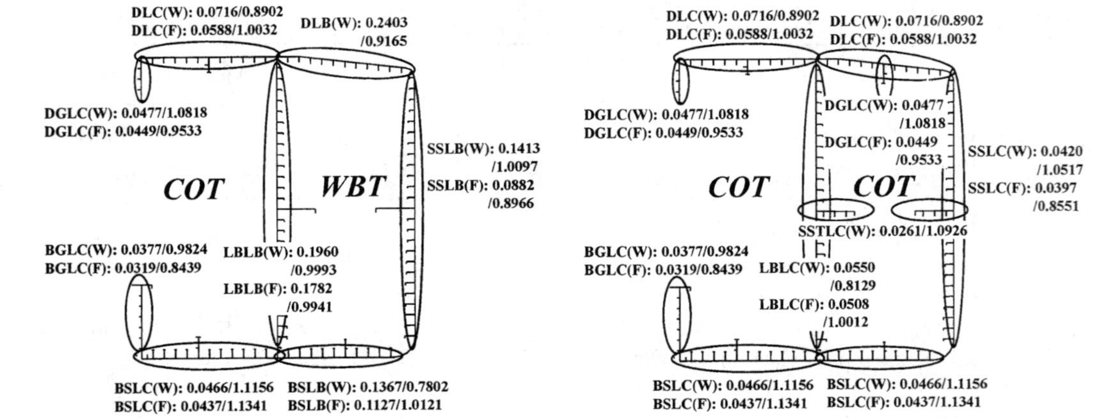

Figure 25(b) Mean and COV of the most probable (average) level of the coefficient C_1 for stiffener webs and flanges in a single skin tanker or FPSO

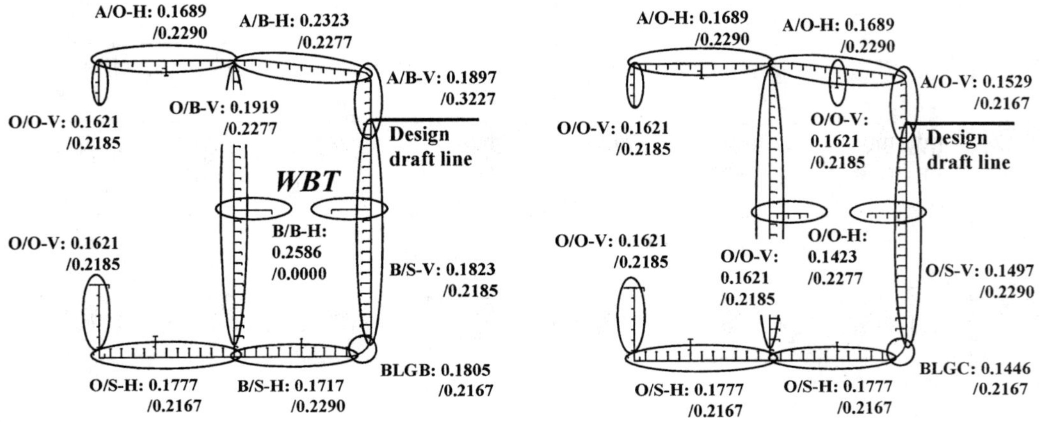

Figure 26(a) Mean and COV of the severe level of the coefficient C_1 for plating in a single skin tanker or FPSO

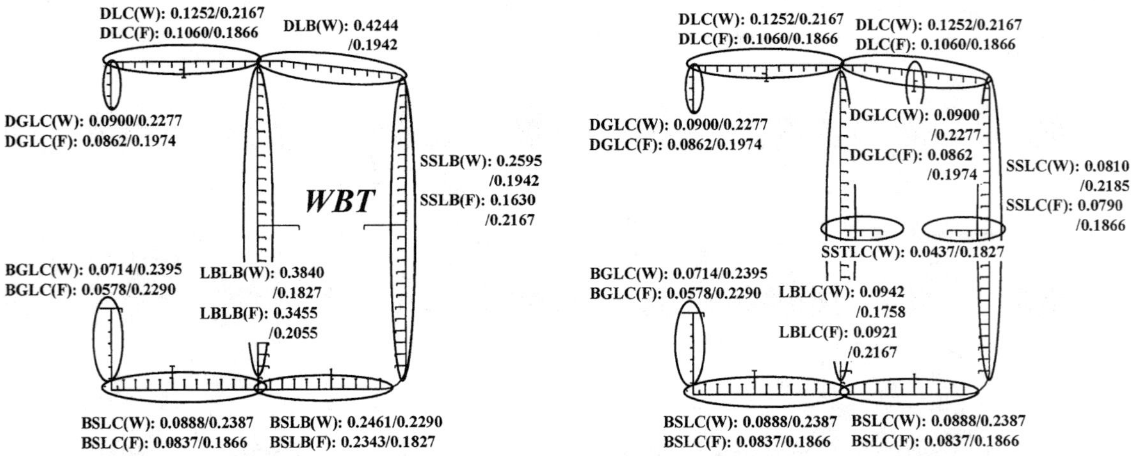

Figure 26(b) Mean and COV of the severe level of the coefficient C_1 for stiffener webs and flanges in a single skin tanker or FPSO

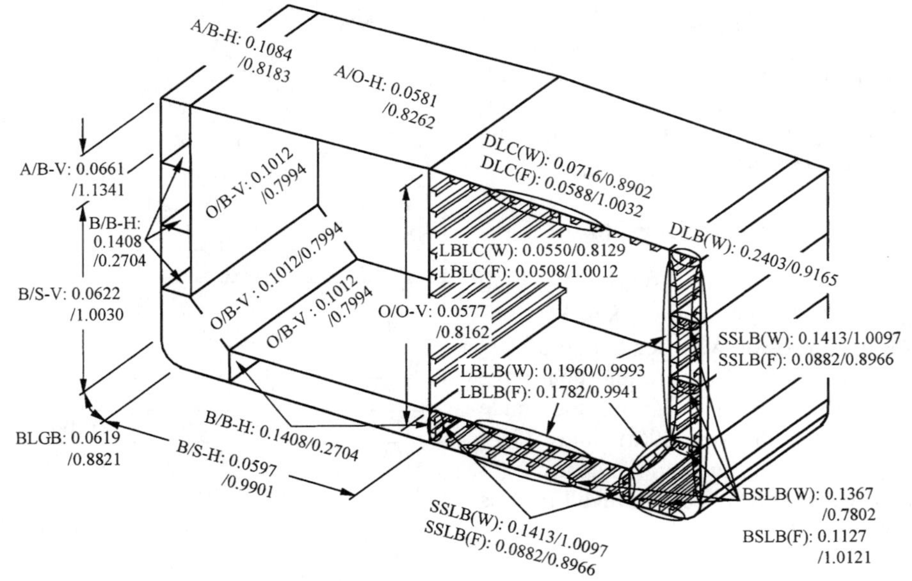

Figure 27 Mean and COV of the most probable (average) level of the coefficient C_1 for the 34 member location/category groups of a double skin tanker or FPSO

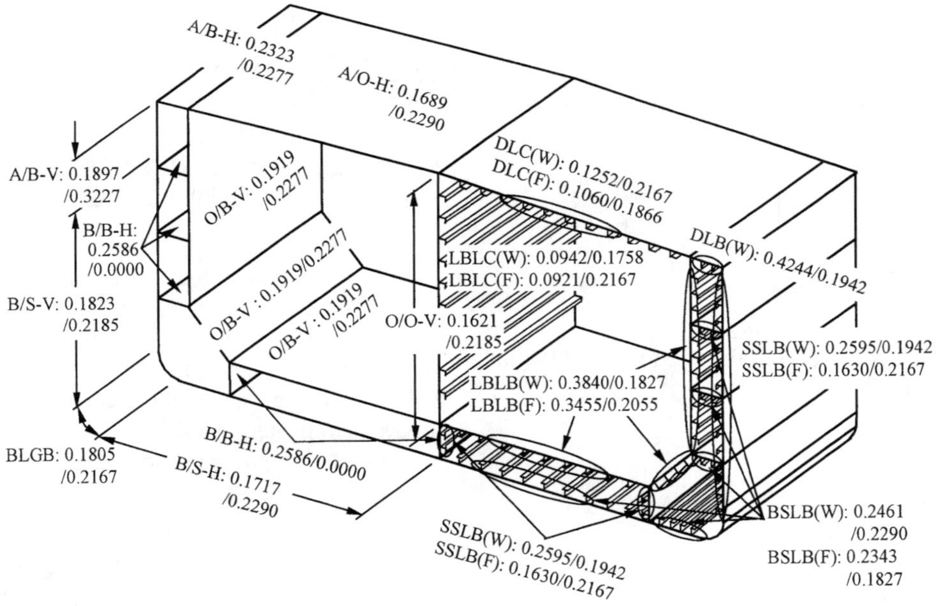

Figure 28 Mean and COV of the severe level of the coefficient C_1 for the 34 member location/category groups of a double skin tanker or FPSO

Corrosion Models for the Structures of Double Hull Tankers and Ship-type FPSOs

Compared to single hull tankers, very few double hull tankers have reached 15 years old, and only a couple of FPSOs have been in operation more than 10 years (Wang and Spong 2003). Corrosion wastage measurements from double hull tankers and FPSOs are limited and a statistical analysis as mentioned in the previous section does not seem to be available for some years. In view of the many similarities between single hull tankers and double hull tankers, the mean and COV values of the coefficient C_1 for the single skin tanker structures as defined in Figures 25 and 26 may be used to those for double skin tankers (Paik et al. 2003c). To supplement limited service performance data, trading tanker experience has provided much of the basis and framework for FPSO design, maintenance and inspection planning.

Figures 27 and 28 represent the most probable (average) or severe level of the annualized corrosion rate for the 34 different member groups of double hull tankers or FPSOs, respectively. Therefore, the corrosion depth of individual member categories can then be predicted from Equation (32) with T_c =7.5 years.

Corrosion Models for Bulk Carrier Structures

The number of ships surveyed was 109 and their size was in the range of 6,095 DWT to 138,655 DWT. Measured data for the corrosion wear for a total of 109 bulk carrier structures carrying iron ore and coal have been collected by entities based in Korea and made available to this study. Figures 29(a) and 29(b) represent the distribution of the age and the deadweight for the 109 bulk carriers that had been surveyed for this purpose.

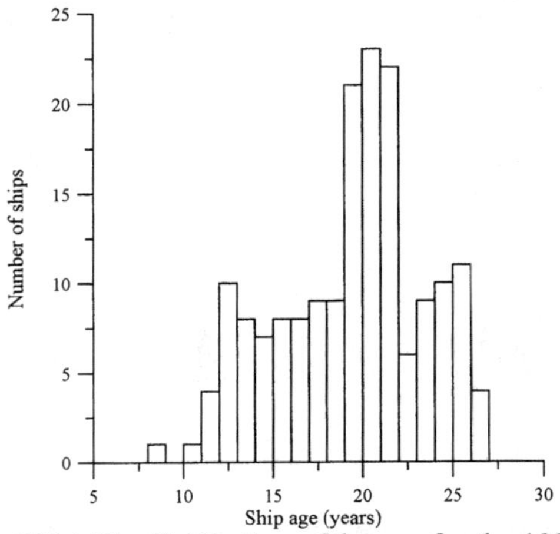

Figure 29(a) The distribution of the age for the 109 bulk carriers surveyed

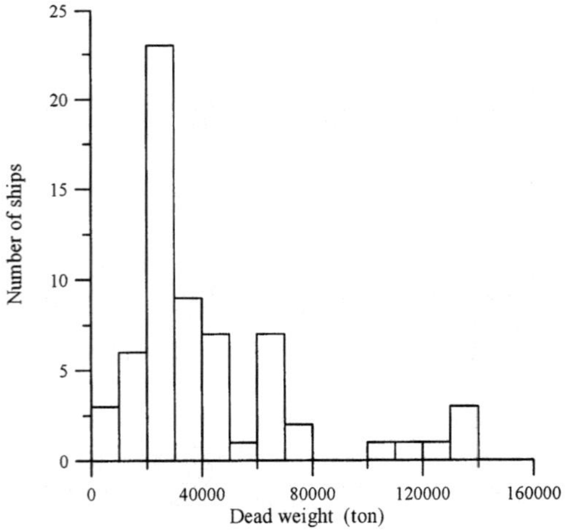

Figure 29(b) The distribution of the dead weight tonnage for the 109 bulk carriers surveyed

Table 3 Identification of the 23 member groups for a bulk carrier structure

ID	Member type
OBP	Outer bottom plates
IBP	Inner bottom plates
LSP	Lower sloping plates
LWTSS	Lower wing tank side shells
SS	Side shells
UWTSS	Upper wing tank side shells
USP	Upper sloping plates
UDP	Upper deck plates
BG	Bilge girders
OBLW	Outer bottom longitudinals – web
OBLF	Outer bottom longitudinals – flange
IBLW	Inner bottom longitudinals – web
IBLF	Inner bottom longitudinals – flange
UWTSLW	Upper wing tank side longitudinals – web
UWTSLF	Upper wing tank side longitudinals – flange
USLW	Upper sloping longitudinals – web
USLF	Upper sloping longitudinals – flange
UDLW	Upper deck longitudinals – web
UDLF	Upper deck longitudinals – flange
LWTSLW	Lower wing tank side longitudinals – web
LWTSLF	Lower wing tank side longitudinals – flange
LSLW	Lower sloping longitudinals – web
LSLF	Lower sloping longitudinals – flange

A total of 12,446 measurements for 23 longitudinal members (defined by locations and categories of members) which include 9 categories of plate parts, 7 categories of stiffener webs, and 7 categories of stiffener flanges were obtained and available for this study, as indicated in Table 3.

Figures 30 and 31 show mean and COV of the average or severe level of the coefficient C_1 for the 23 member location/category groups of a bulk carrier structure, respectively. As previously described, the average corrosion rate is based on all gathered corrosion data, while the severe corrosion rate is based on 95% and above band of the corrosion measurements, see Figure 23. Therefore, the corrosion depth of individual member categories can then be predicted from Equation (27) with T_c =7.5 years, except for inner bottom plates(IBP) and lower slope plates(LSP) which may take a shorter coating life, i.e., T_c =5 years.

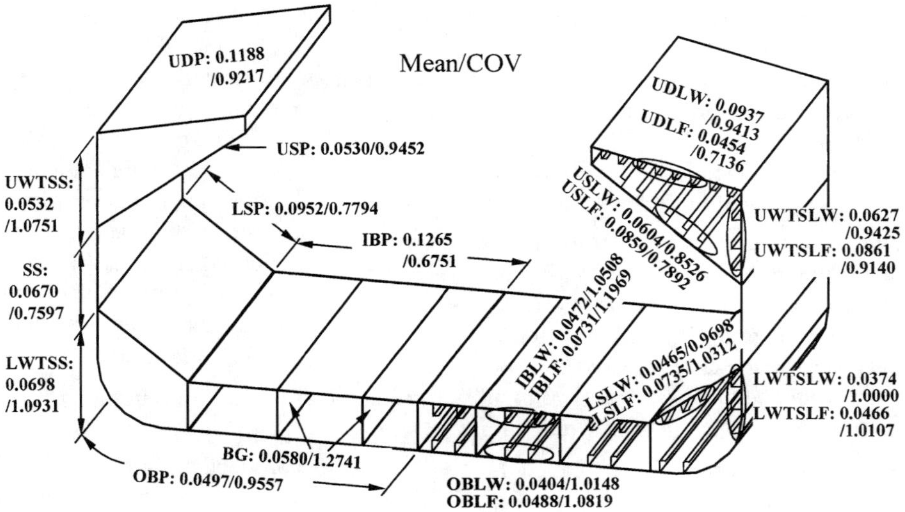

Figure 30 Mean and COV of the average level of the coefficient C_1 for the 23 longitudinal members of a bulk carrier

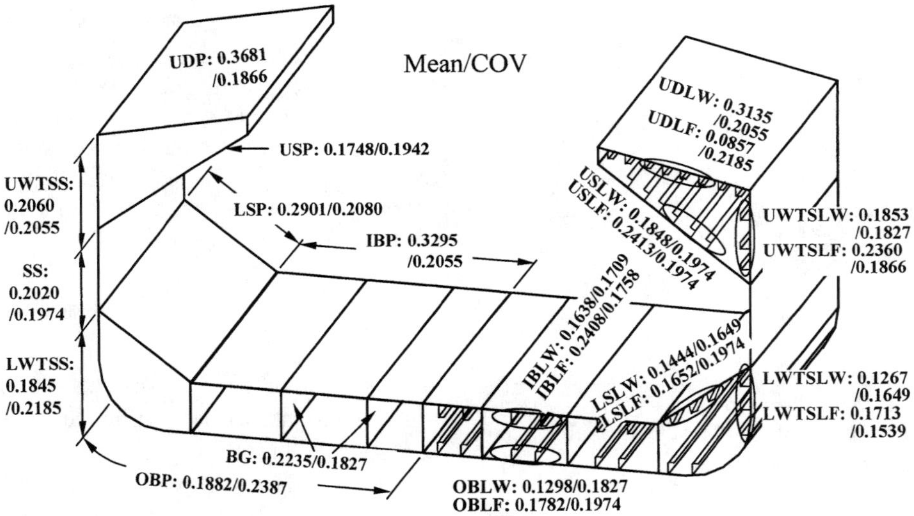

Figure 31 Mean and COV of the severe level of the coefficient C_1 for the 23 longitudinal members of a bulk carrier

TIME-DEPENDENT FATIGUE CRACKING DAMAGE MODELS

Fatigue Cracking Mechanism

Fatigue cracking damage has been a primary source of costly repair work of aging ships. Cracking damage has been found in welded joints and local areas of stress concentrations, e.g., at the weld intersections of longitudinals, frames and girders. Initial defects may also be formed in the structure by fabrication procedure and may conceivably remain undetected over time. Under a cyclic loading or even monotonic extreme loading, cracking may propagate and become larger with time.

Since cracks can conceivably lead to catastrophic failure of the structure, it is essential to properly consider and establish relevant crack tolerant design procedures for ship structures, in addition to implementation of close-up survey strategy. For reliability assessment of aging ship structures under extreme loads, it is often necessary to take into account a known (existing or premised or anticipated) crack on the ultimate limit state analysis as a parameter of influence. To make this possible, it is required to develop time-dependent fatigue cracking model which can predict the cracking damage in location and size as the ship gets older.

Figure 32 shows a schematic of fatigue cracking progress as a function of time (age) in steel structures. The fatigue cracking progress can be separated into three stages: initiation (stage I), propagation (stage II) and failure (fracture) stage (stage III) (ISO 2394 1998).

It may be assumed that no initial defects exist so that there is no cracking damage until time T_I. While the fatigue cracking damage is affected by many factors such as the stress ranges experienced during the load cycles, local stress concentration characteristics, and the number of stress range cycles, the initiation of cracking can be evaluated by the fatigue analysis.

On the other hand, when any crack is detected in an existing structure at time T_0, it has normally a certain amount of crack size (length), denoted by a_0 called the initial crack size, which must be detectable. For the assessment of time-dependent risks of a structure, however, it is often assumed that the initial crack size is small, say 1.0mm.

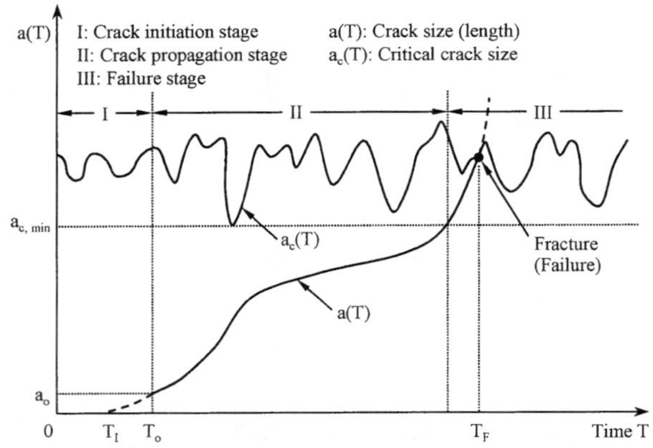

Figure 32 A schematic of the crack initiation and growth for a steel structure with time

Fatigue cracks propagate with time progressively in ductile material. They may however become quite unstable in brittle material. Crack propagation is affected by many parameters such as initial crack size, history of local nominal stresses, load sequence, crack retardation, crack closure, crack growth threshold and stress intensity range in addition to stress intensity factor at the crack tip which depends on material properties and geometry. The fracture mechanics approach is often used to analyze the behavior of crack propagation.

The time-dependent cracking damage model may also be composed of the three separate models:

- A model for crack initiation assessment or detection
- A model for crack growth assessment
- A model for failure assessment

Crack initiation at a critical structural detail can be theoretically predicted using the S-N curve approach (S = fluctuating stress, N = associated number of stress cycles). The Palmgren-Miner cumulative damage rule is applied together with the relevant S-N curve.

This normally follows three steps: (a) define the histogram of cyclic stress ranges, (b) select the relevant S-N curve, and (c) calculate the cumulative fatigue damage and judge the initiation of crack.

Cracks at critical joints and details can be detected during inspection when the crack size is larger enough, usually about 15-30mm. Since the present study focuses on the integrity of existing aged ship structures, it is assumed that the crack with length of a_o at a critical joint or detail has initiated at T_0 years.

Crack growth can be assessed by the fracture mechanics approach which considers that one or more premised cracks of a small dimension exist in the structure, and predicts the fatigue damage during the process of their crack propagation including any coalescence and break through the thickness, and subsequent fracture. In this approach, a major task is to pre-establish the relevant crack growth equations or 'laws' as a function of time (year).

The crack growth rate is expressed as a function of the stress intensity factor at the crack tip, on the assumption that the yielded area around the crack tip is relatively small. The so-called Paris-Erdogan law is often used for this purpose, and being expressed as follows:

$$\frac{da}{dN} = C(\Delta K)^m \tag{29}$$

where ΔK = stress intensity factor at the crack tip, C, m = constants to be determined based on tests, a = crack length, N = number of stress cycles.

For steel structures with typical types of cracks, the stress intensity factor formulae are given in textbooks (Broek 1986, Paik & Thayamballi 2003). In ship stiffened panels, cracks are often observed along the weld intersections between plating and stiffeners. For a plate with cracking, ΔK may be given, when stiffening effect is neglected, as follows:

$$\Delta K = F\Delta\sigma\sqrt{\pi a} \tag{30}$$

where $\Delta\sigma$ = stress range (or double amplitude of applied fatigue stress), a = crack size (length), F = geometric parameter depending on the loading and configuration of cracked body.

Fatigue Cracking Damage Model

In the time-dependent risk assessment considering the growth of fatigue cracking, it is more efficient to express the crack growth behavior in a closed-form.

The crack length, $a(T)$, as a function of time, T, can then be calculated by integrating Equation (29) with regard to the stress cycle, N. In the integration of Equation (29), it is often assumed that the geometric parameter, F, is constant, i.e., assuming that the geometric parameter, F, is unchanged with the crack propagation. This assumption is reasonable as long as the initial crack size, a_o, is small.

In this case, the integration of Equation (29) after substituting Equation (30) into Equation (29) results in

$$a(T)=\begin{cases}\left[a_o^{1-m/2}+\left(1-\frac{m}{2}\right)C\left(\Delta\sigma F\sqrt{\pi}\right)^m (T-T_o)\omega\right]^{\frac{1}{1-m/2}} & \text{for } m\neq 2\\ a_o \exp\left[C\Delta\sigma^2F^2\pi(T-T_o)\omega\right] & \text{for } m=2\end{cases} \quad (31)$$

where a_o = initial crack size, a = total crack size.

It is roughly considered that in ships the expected number of wave load cycles occurs once in every 6-10 seconds, and hence

$$N \approx (T-T_o)\times 365\times 24\times 60\times 60/10=\omega\times(T-T_o)$$
$$\text{or } \omega\approx 365\times 24\times 60\times 60/10$$

where T = ship age, T_o = age at the initiation of the cracking with $a=a_o$.

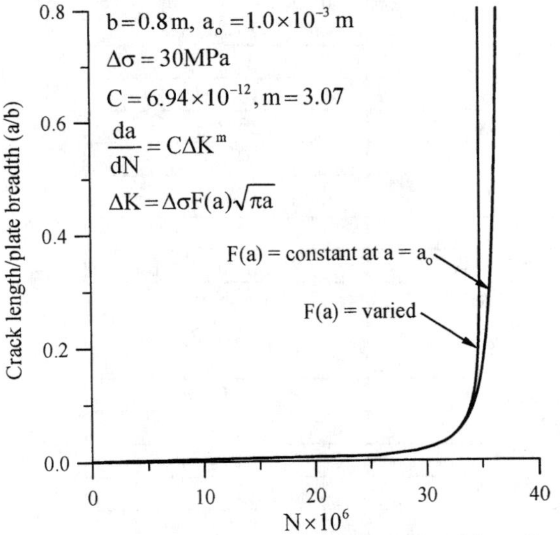

Figure 33 A comparison of Equation (31) with a direct (numerical) integration of Equation (29) for a small initial crack size

Figure 33 shows a sample application of Equation (31) by comparing with a direct integration of Equation (29) which accounts for the effect of crack growth on the geometric parameter, F, i.e., being varied with time. It is seen from Figure 33 that Equation (31) slightly overestimates the fracture life as the crack propagates. This is expected because Equation (31) was derived under the assumption that the geometric parameter, F, remains unchanged with time. However, it is considered that their difference for a small initial crack size is negligible, and Equation (31) gives a reasonable tool for the crack growth assessment.

TIME-DEPENDENT RISK ASSESSMENT OF SHIPS: ILLUSTRATIVE EXAMPLES

The methodology developed in the present study has been automated within the GUI-based computer program TRAAS which stands for Time-dependent Risk Assessment of Aging Ships. Some illustrative examples of the methodology are now shown.

The Object Ships

The time-dependent risk assessment for three vessels, namely a 105k dwt double hull tanker, a Capesize bulk carrier and a 254k dwt single hull tanker-type FPSO, is now carried out with focus on the primary failure mode, i.e., hull girder collapse, taking into account the influence of age related structural degradation (corrosion and fatigue cracking) and local denting.

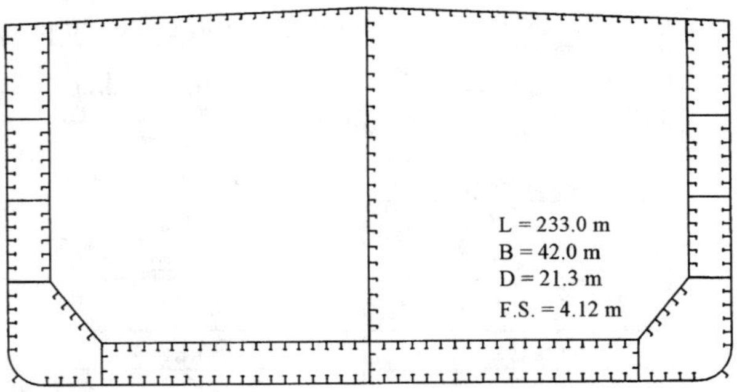

Figure 34(a) Mid-ship section of a 105k dwt double hull tanker

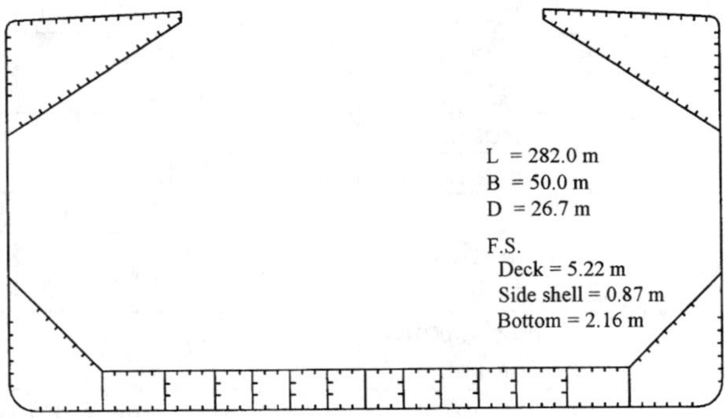

Figure 34(b) Mid-ship section of a 170k dwt single side-skin bulk carrier

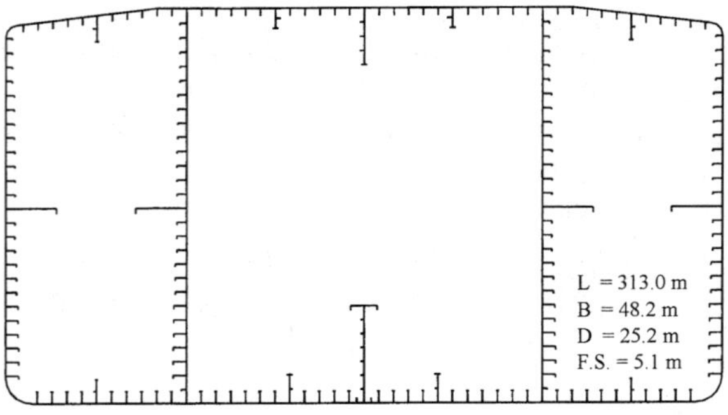

Figure 34(c) Mid-ship section of a 254k dwt single hull tanker–type FPSO (floating, production, storage and offloading unit)

Figure 34 shows mid-ship sections of the three ships considered. Table 4 indicates the principal dimensions and the sectional properties of the ships.

Progressive Collapse Analysis and Safety Measure Calculations

The progressive collapse behavior of the three ship hulls under vertical bending moment when no structural damage exists was analyzed by ALPS/HULL. An elaborate description of these results are found in the 2002 SNAME annual meeting paper by Paik et al. (2002).

The safety measure calculations for the three ships were then made as a ratio of the ultimate hull girder strength to the IACS total design moment, where the ultimate hull girder strength is referred to for an average level of initial imperfections (initial deflection and residual stresses).

Generally, the three vessels have enough safety margin as long as structural damage does not exist. Nevertheless, a variety of structural failure modes including local plate buckling, yielding and lateral-torsional buckling of stiffeners or longitudinals take place before the ship hulls reach the ultimate limit state.

Traditionally, the longitudinal strength safety measure is defined as a ratio of the section modulus Z to the IACS required section modulus Z_{min} . An alternative to measure the safety of a ship's longitudinal strength is the ratio of ultimate strength M_u over the maximum total bending moment M_t . Figure 35 compares these two safety measures.

Table 4 Hull sectional properties of the three object vessels

Property		Double hull tanker	Bulk carrier	Single hull tanker-type FPSO
LBP (L)		233.0 m	282.0 m	313.0 m
Breadth (B)		42.0 m	50.0 m	48.2 m
Depth (D)		21.3 m	26.7 m	25.2 m
Draft (d)		12.2 m	19.3 m	19.0 m
Block coeff. (C_b)		0.833	0.826	0.833
Design speed		16.25 knots	15.15 knots	-
DWT		105,000 DWT	170,000 DWT	254,000 DWT
Cross-sectional area		5.318 m^2	5.652 m^2	7.858 m^2
Height to neutral axis from base line		9.188 m	11.188 m	12.173 m
I	Vertical	359.480 m^4	694.307 m^4	863.693 m^4
	Horizontal	1152.515 m^4	1787.590 m^4	2050.443 m^4
Z	Deck	29.679 m^3	44.354 m^3	66.301 m^3
	Bottom	39.126 m^3	62.058 m^3	70.950 m^3
σ_Y	Deck	HT32	HT40	HT32
	Bottom	HT32	HT32	HT32
M_p	Vertical moment	11.930 GNm	20.650 GNm	22.615 GNm
	Horizontal moment	19.138 GNm	31.867 GNm	31.202 GNm

Notes: FPSO = floating, production, storage and offloading unit, I = moment of inertia, Z = section modulus, σ_Y = yield stress, M_p = fully plastic bending moment.

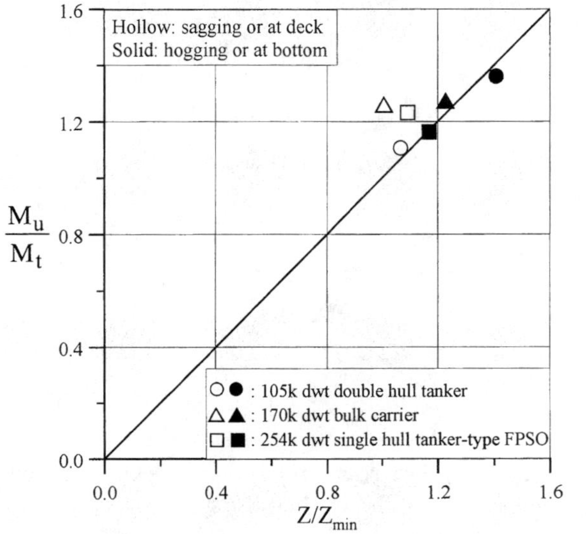

Figure 35 A comparison between the section modulus based safety measure and the ultimate hull girder based safety measure for the three object ship hulls with average level of initial imperfections but without any structural damage

Table 5 Scenarios for operating condition and sea states of the three object vessels

Parameter	Double hull tanker	Bulk carrier	FPSO
Operating speed	0.6 × design speed = 9.75 knots	0.6 × design speed = 9.09 knots	10 knots*
Significant wave height (H_s)	$1.1\sqrt{L}$, L in feet = 9.27 m	$1.1\sqrt{L}$, L in feet = 10.198 m	12.2m at GOM, 18m at NS
Storm duration	3 hours	3 hours	3 hours

Notes: * indicates an equivalent design speed for load prediction purposes, while a FPSO remains at site once installed, GOM = Gulf of Mexico, NS = North Sea.

Scenarios for Operating Conditions and Sea States

It is important to realize that hull girder loads depend on operational condition and sea states. In this regard, it is necessary to establish relevant scenarios for operational condition and sea states in predicting the wave-induced bending moments.

Table 5 represents significant wave height, operational speed and storm duration which the three object vessels would encounter. While the double hull tanker and the bulk carrier considered are ocean-going vessels, two service locations, namely Gulf of Mexico and North Sea, are considered for the ship-type FPSO.

It is noted that for wave load prediction purposes, the ship-type FPSO is assumed to have an equivalent operational speed of 10 knots in waves, while it usually remains at a specific location once installed. The significant wave heights of the FPSO considered are representative values obtained from the wave statistics in the corresponding installation sites (e.g., Wang et al. 2002).

Based on the steepest boundaries of global wave scatter diagrams, the 1979 ISSC Environmental Conditions Committee I.1 defines extreme climatic waves of limiting steepness which are close to Hogben's contemporary work (Hogben et al. 1986, Hogben 1990), giving the zero-up-crossing wave period T_z, as follows:

$$T_z = \sqrt{13H_s} \tag{32a}$$

where H_s = significant wave height (m). Therefore, the number N of wave peaks which a vessel encounters during a storm duration can approximately be given by

$$N = T_s / T_z \tag{32b}$$

where T_s = storm duration in seconds. For 3 hours storm duration with $H_s = 12.2$ m, for instance, the number of wave peaks then becomes $N = 3 \times 60 \times 60 / \sqrt{13 \times 12.2} = 857$.

Scenarios for Structural Damage

Ship structures typically suffer age related structural degradation such as corrosion and fatigue cracking. Inner bottom plates of bulk carriers may also have local denting damage during loading and unloading of cargo. The age related structural damage needs to be dealt with as a function of ship age, while local dent can be treated as time-independent damage.

Corrosion Damage

Annualized corrosion rate models for different structural member groups by type and location, considering plating, and webs and flanges of stiffening were in the present study developed by statistical analysis of corrosion wear measurements. These models can be used to predict the general/pit corrosion depth (diminution) of primary members as the ships get older. Since the structural characteristics and corrosion environments of tankers are similar to those of ship-type FPSOs, the corrosion models developed for tanker structures may be employed for FPSO structures (Paik et al. 2003c, Wang et al. 2003a, 2003b).

In the present risk assessment, two levels of corrosion rates, namely the average and severe level are considered, the former being based on all collected data of corrosion measurements and the latter being based on 95% and above band of the measurements. While it is assumed that corrosion starts immediately after the breakdown of coating, the coating life of all structural members in the double hull tankers, bulk carriers and FPSOs is considered to be 7.5 years except for inner bottom plates and lower sloping plates in the bulk carrier which have 5 years of the coating life (Paik et al. 2003d).

Figure 36 shows the average level progress of corrosion depth for some selected members as the vessels get older.

There are several types of corrosion possible for mild and low alloy steels in marine applications. The so-called general (or uniform) corrosion refers to uniform wastage of thickness. Localized corrosion such as pitting is more likely to occur in bottom plates.

The ultimate strength behavior of ship structures with general corrosion is somewhat different from that with pitting corrosion (Paik & Thayamballi 2003). More realistic assessment of structural integrity of aging ships should take into account both pit and general corrosion.

For any structural member with general corrosion, it is considered that the member thickness is uniformly reduced following the corrosion rate with time. The ultimate strength of corroded plating is predicted by using the design formulae excluding the plate thickness loss. The ultimate strength of a structural member with pit corrosion is calculated by that of un-corroded member multiplied with the ultimate strength reduction factor derived in the previous sections.

In the present risk assessment, it is assumed that the most heavily pitted cross section of any structural member expands over the plate breadth. This may provide somewhat pessimistic evaluation of residual strength, but should be sufficient for illustrating the risk assessment approach.

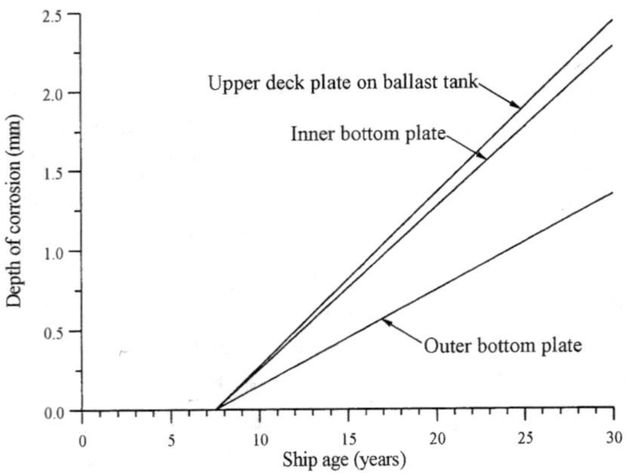

Figure 36(a) Progress of corrosion depth for some members in the 105k dwt double hull tanker

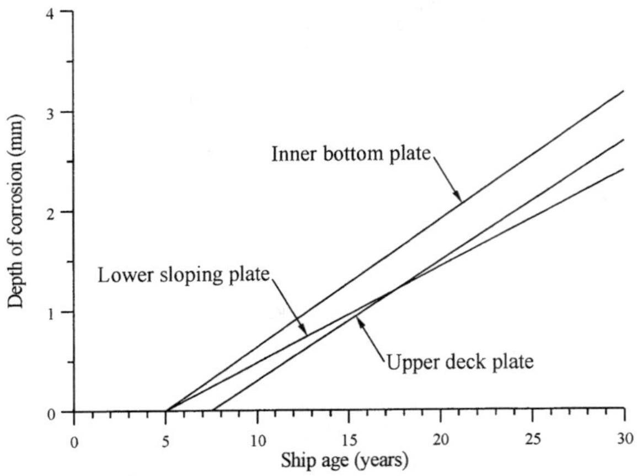

Figure 36(b) Progress of corrosion depth for some members in the 170k dwt bulk carrier

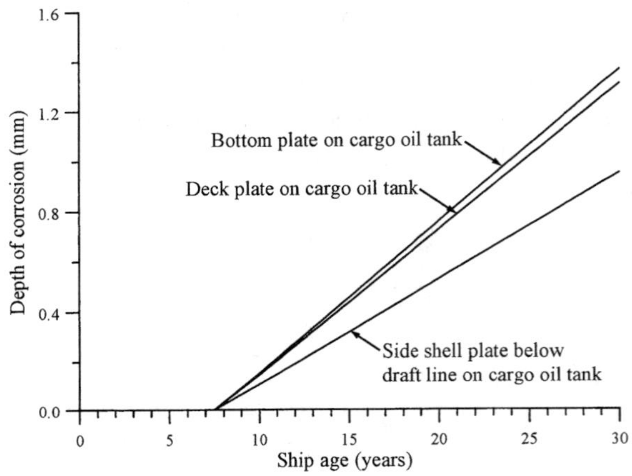

Figure 36(c) Progress of corrosion depth for some members in the 254k dwt single hull tanker-type FPSO

Fatigue Cracking

The time-dependent fatigue cracking model has been established in the present study. The crack length of any critical area is then predicted by a closed-form formula as a function of ship age. It is assumed that cracking initiates at all stiffeners and plating after 5 years of the ship age. The initial crack size is considered to be 1.0mm.

While the constants of the Paris-Erdogan equation are considered to be the same at all joints, the crack growth characteristics may be different because the stress ranges affecting the stress intensity factors at individual joints are different from each other.

Fatigue loading is random in nature. The long-term distribution of the fatigue loading is at most known. Therefore, some refined methodologies are used to generate random loading sequence (Broek 1986). With fatigue loading sequence and amplitude known, the dynamic stress range, $\Delta\sigma_i$, at the ith joint may then be given by

$$\Delta\sigma_i = 2 \times \sigma_{xi} \times SCF_i \times k_f = 2 \times \sigma_{xi}^* \times SCF_i \qquad (33)$$

where σ_x = cyclic "peak" stress amplitude which may be given by $M_w z / I$, $\sigma_x^* = k_f \times \sigma_x$, M_w = wave-induced component of bending moment, I = time-dependent moment of inertia, z = distance from the time-dependent neutral axis to the point of stress calculation, k_f = the knock-down factor accounting for the dynamic stress cycles.

SCF in Equation (33) is the stress concentration factor at any critical joint, which takes into account geometric effects. For the present illustrative example purposes, it is assumed that the SCF at joints of plating and stiffeners (or support members) is 2.1. However, hold frame connections to upper or lower wing tanks and side shell of bulk carrier have SCF=3.75, and hopper knuckles have a much larger SCF, i.e., 10.5. It should be noted that SCF's for different joints can be determined using more refined calculation approaches, and those for typical joints in ship structures may be obtained from some established tables (ABS 2002). The knock-down factor k_f accounting for the dynamic stress cycles is taken as 0.25 for the sake of simplicity.

Figure 37 shows time-dependent hull girder bending stress ranges calculated from Equation (33) for some selected critical joints of the three object vessels. Figure 38 shows the resulting crack growth characteristics which are evaluated by Equation (31).

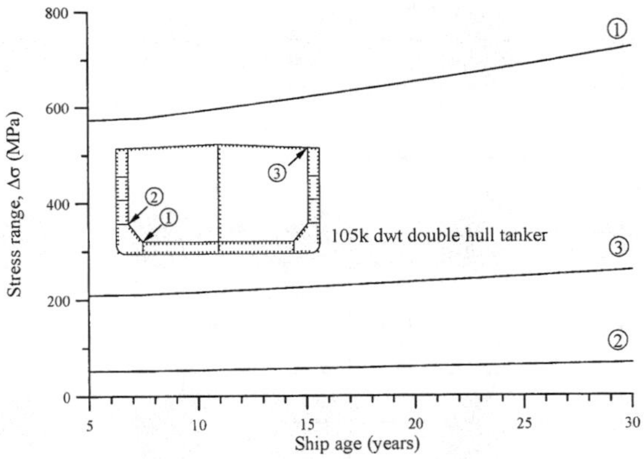

Figure 37(a) Time-dependent hull girder bending stress range at some critical joints in the 105k dwt double hull tanker

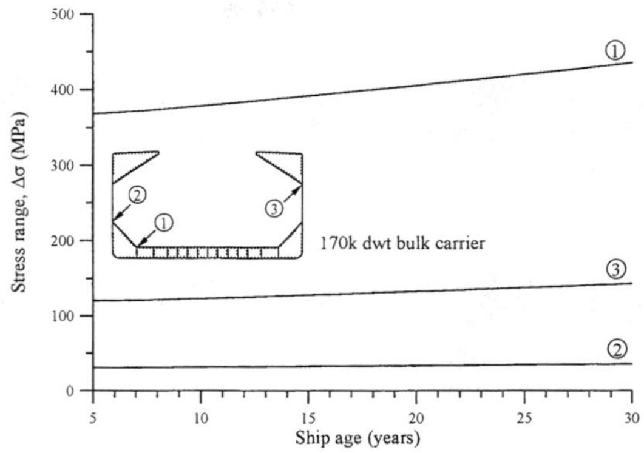

Figure 37(b) Time-dependent hull girder bending stress range at some critical joints in the 170k dwt bulk carrier

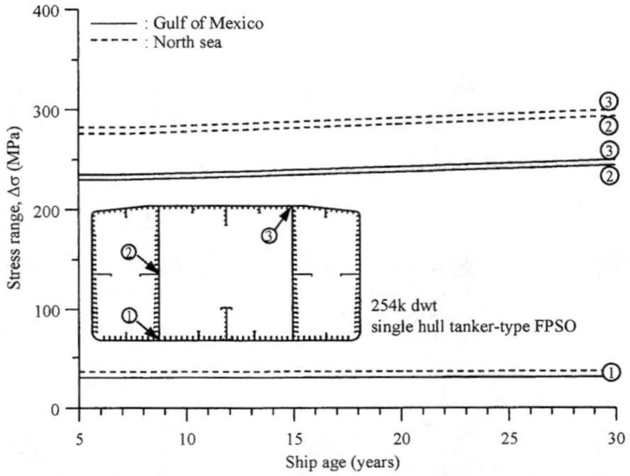

Figure 37(c) Time-dependent hull girder bending stress range at some critical joints in the 254k dwt single hull tanker-type FPSO

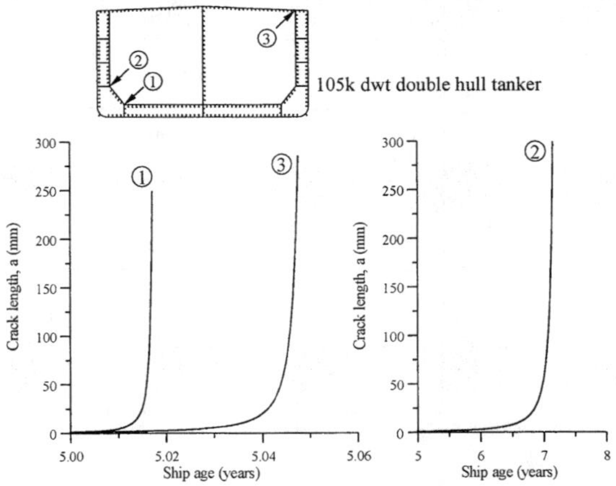

Figure 38(a) Crack growth characteristics of some critical joints in the 105k dwt double hull tanker

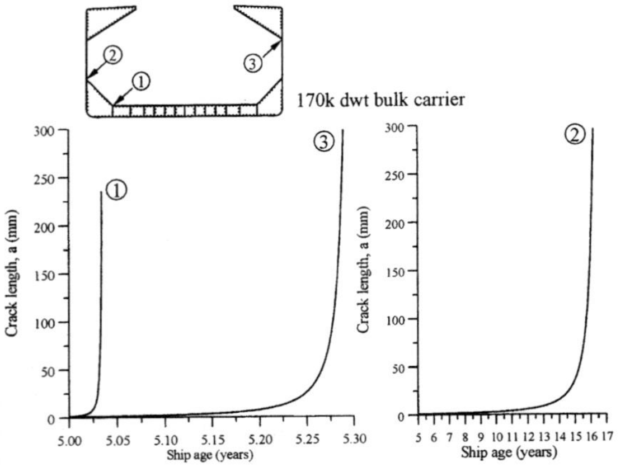

Figure 38(b) Crack growth characteristics of some critical joints in the 170k dwt bulk carrier

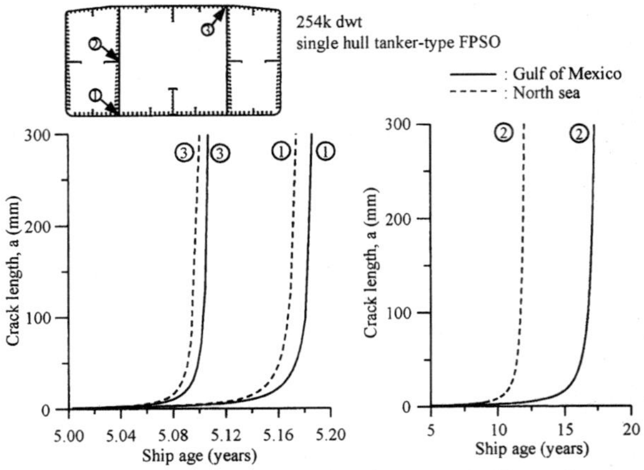

Figure 38(c) Crack growth characteristics of some critical joints in the 254k dwt single hull tanker-type FPSO

Table 6 The probabilistic characteristics for random variables used for the present risk assessment

Parameter	Definition	Distribution Function	Mean	COV
E	Elastic modulus	Normal	205.8GPa	0.03
σ_Y	Yield stress	Log-Normal	As for each member	0.10
t_p	Thickness of plating	Fixed	As for each member	-
t_w	Thickness of stiffener web	Fixed	As for each member	-
t_f	Thickness of stiffener flange	Fixed	As for each member	-
T	Ship age	Fixed	As for each age	-
T_c	Coating life	Normal	5.0 years	0.40
			7.5 years	0.40
C_1	Corrosion rate	Weibull	As for each member	As for each member
a_o	Initial crack size	Normal	1.0 mm	0.20
C	$\frac{da}{dN} = C(\Delta K)^m$	Log-Normal	6.94E-12	0.20
m	$\frac{da}{dN} = C(\Delta K)^m$	Fixed	3.07	-
d_d	Diameter of local dent	Normal	0.3b	0.10
			0.5b	0.10
			0.8b	0.10
D_d	Depth of local dent	Normal	1.0 t_p	0.10
			2.5 t_p	0.10
			5.0 t_p	0.10

Notes: a = crack size, b = plate breadth, N = number of stress cycles, ΔK = stress intensity factor

At a given age, the ultimate strength of structural members with known (premised) fatigue cracking damage is then predicted by the simple design formula developed in the present study. It is considered that fracture takes place if the crack size (length) of the member reaches the critical crack size, assumed to be either the plate breadth or the stiffener web height. Table 7 lists the probalistic characteristics for random variables used for the risk assessment associated with fatigue cracking.

Local Dent

Inner bottom plates of bulk carriers can have local dent damage caused by mishandled loading and unloading of dense cargo such as iron ore.

It is assumed that inner bottom plating of the bulk carrier have local dent damage after 5 years. For the present illustrative purposes, the size of local dent is assumed to be the same at all inner bottom plating, namely $D_d / t = 2.5$ and $d_d / b = 0.5$, where D_d = dent depth, d_d = dent diameter, b = plate breadth between bottom longitudinals, t = plate thickness.

The ultimate strength of plating with local dent damage is then predicted by design formulae developed in the present study. The COV associated with the local dent related parameters is assumed as defined in Table 6.

Still-water Bending Moment Calculations

M_{sw} in Equation (4) is taken as the maximum value of the still-water bending moment resulting from the worst load condition for the ship, considering both hogging and sagging. The related detailed distribution of the still-water bending moment along the ship's length can be calculated by a double integration of the difference between the weight force and the buoyancy force, using the simple beam theory.

For convenience, the mean value of M_{sw} is often taken from an empirical formula that has been suggested for a first cut estimation of the maximum allowable still-water bending moment by some Classification Societies in the past. That approximate formula is given by

$$M_{sw} = \begin{cases} +0.015CL^2B(8.167 - C_b)(\text{kNm}) & \text{for hogging} \\ -0.065CL^2B(C_b + 0.7)(\text{kNm}) & \text{for sagging} \end{cases} \quad (34)$$

where
$$C = \begin{cases} 0.0792L & \text{for } L \le 90 \\ 10.75 - \left(\frac{300 - L}{100}\right)^{1.5} & \text{for } 90 < L \le 300 \\ 10.75 & \text{for } 300 < L \le 350 \\ 10.75 - \left(\frac{L - 350}{150}\right)^{1.5} & \text{for } 350 < L \le 500 \end{cases}$$

with L =ship length (m), B = ship breadth (m), C_b =block coefficient at summer load waterline.

Table 7 indicates the calculated still-water bending moments for the three vessels computed at the operational conditions and sea states assumed. The COV associated with still-water bending moment of a merchant cargo vessel is normally large, perhaps as high as 0.4. The variation in still-water bending moment is usually assumed to follow the normal distribution.

Wave-induced Bending Moment Calculations

For reliability assessment of newly built ships, M_w in Equation (4) is normally taken as the mean value of the extreme wave-induced bending moment which the ship is likely to encounter during its life time, which is given for unrestricted worldwide service by the IACS, as follows

$$M_w = \begin{cases} +0.19CL^2BC_b \,(\text{kNm}) \text{ for hogging} \\ -0.11CL^2B(C_b + 0.7)\,(\text{kNm}) \text{ for sagging} \end{cases} \quad (35)$$

where C, L, B, C_b = as defined in Equation (34).

For the safety and reliability assessment of damaged ship structures in particular cases, short-term based response analysis may be used to determine M_w when the ship encounters a storm of specific duration (usually 3 hours) and with certain small encounter probability. The MIT sea-keeping tables developed by Loukakis & Chryssostomidis (1975) are useful for predicting the short-term based wave-induced bending moment of merchant cargo vessels. The MIT sea-keeping tables are designed to efficiently determine the root-mean-square value of the wave-induced bending moment given the values of effective wave height, B/d ratio, L/B ratio, ship operating speed, the block coefficient, and sea state persistence time. USAS-L program which automates the MIT sea-keeping tables can be downloaded from the internet web site (Paik & Thayamballi 2003).

The most probable extreme value of the wave-induced loads, M_w, i.e., mode, which we may refer to as a mean for convenience and its standard deviation, σ_w, can then be computed based on up-crossing analysis as follows

$$M_w = \sqrt{2\lambda_o \ln N} + \frac{0.5772}{\sqrt{2\lambda_o \ln N}} \quad (36a)$$

$$\sigma_w = \frac{\pi}{\sqrt{6}}\sqrt{\frac{\lambda_o}{2\ln N}} \quad (36b)$$

where $\sqrt{\lambda_o}$ is the root-mean-square value of the short-term wave-induced bending moment process at amidships. N is the expected number of wave bending peaks. For example, if a peak normally occurs once in every T_z seconds obtained from Equation (32a), and hence in a storm of 3 hour persistence time, $N = 3 \times 60 \times 60 / T_z$.

Table 7 indicates the wave induced bending

moments for the three vessels computed at the operational conditions and sea states assumed. The COV associated with the wave-induced bending moment can be defined as $COV = \sigma_W / M_W$ from Equation (36) on which the short-term response is based upon, while it is sometimes assumed to be 0.1 when M_W is predicted from Equation (36a).

Table 7 Mean and COV of still water and wave-induced bending moments for the three object vessels

Parameter			Double hull tanker	Bulk carrier	FPSO	
					GOM	NS
M_{sw}	Mean	Sag	-2.318 GNm	-4.210 GNm	-5.058 GNm	-5.058 GNm
		Hog	2.559 GNm	4.673 GNm	5.584 GNm	5.584 GNm
	COV		0.40	0.40	0.40	0.40
M_w	Mean		3.35 GNm	5.93 GNm	8.87 GNm	11.42 GNm
	COV		0.09	0.09	0.09	0.10

Notes: GOM = Gulf of Mexico, NS = North Sea.

Figure 39 shows variations of the wave-induced bending moments as a function of significant wave height, obtained by using the MIT sea-keeping tables, for the three object vessels in a storm during 3 hour persistence time. The IACS formula values by Equation (35) are also compared. It is seen from Figure 39 that the wave-induced bending moment significantly increases as the significant wave height increases. Also, for the particular scenarios of operational condition and sea states presumed in the present study, it is observed that the short-term based wave-induced bending moments are smaller than the IACS formula by 10 to 20%.

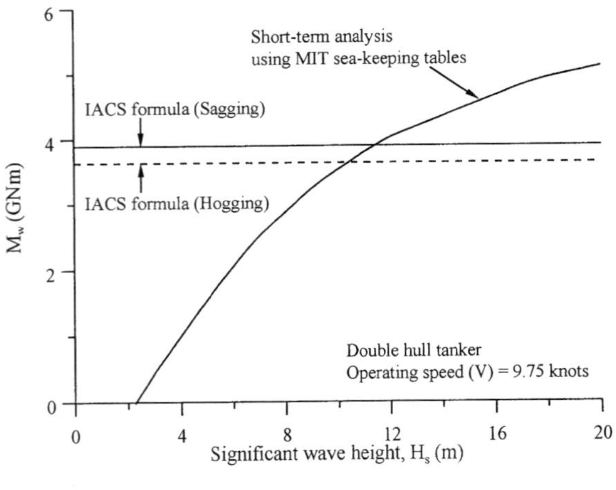

Figure 39(a) Variation of wave-induced bending moment of the 105k dwt double hull tanker as a function of significant wave height

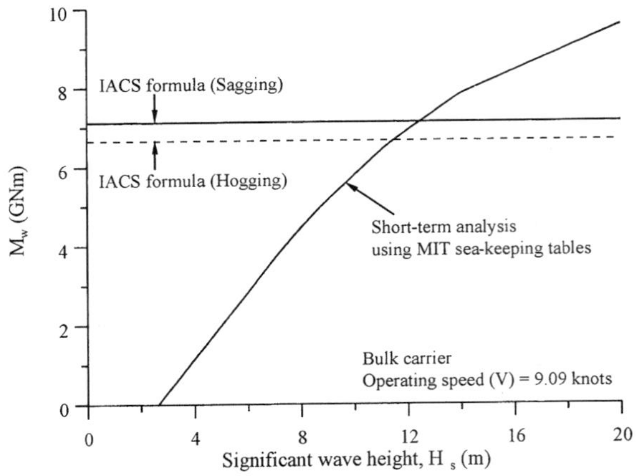

Figure 39(b) Variation of wave-induced bending moment of the 170k dwt bulk carrier as a function of significant wave height

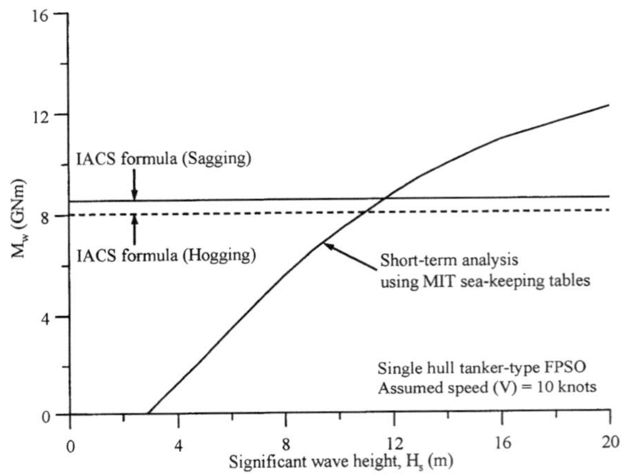

Figure 39(c) Variation of wave-induced bending moment of the 254k dwt single hull tanker-type FPSO as a function of significant wave height

Time-Dependent Risk Assessment

The formula for hull girder ultimate strength is eventually expressed as a function of design parameters related to geometric, structural scantlings and material's properties. When time-dependent structural degradation (e.g., corrosion, fatigue cracking damage) and local dent is considered, the value of member thickness at any particular time is a function of such damage. This results in

$$M_{us} = M_{us}\left(E_i, \sigma_{Yi}, t_{pi}, t_{wi}, t_{fi}, T, T_{ci}, C_{li}, a_{oi}, C_i, m_i, d_{di}, D_{di}\right) \quad (37a)$$

$$M_{uh} = M_{uh}\left(E_i, \sigma_{Yi}, t_{pi}, t_{wi}, t_{fi}, T, T_{ci}, C_{li}, a_{oi}, C_i, m_i, d_{di}, D_{di}\right) \quad (37b)$$

where M_{us}, M_{uh} = ultimate hull girder strength in sagging or hogging. The subscript *i* indicates the *i*th structural member.

In reliability assessment, all the parameters noted in Equation (37) are treated as random variables, with the probabilistic characteristics (i.e., mean, COV and distribution function) as defined in Table 6. The structural damage scenarios are divided into the following five groups:

- Intact (undamaged)
- Corrosion damage alone
- Corrosion + local dent damage (only for the bulk carrier in hogging)
- Corrosion + fatigue cracking damage
- Corrosion + fatigue cracking and local dent damage (only for the bulk carrier in hogging)

Figures 40 to 42 show the time-dependent characteristics of ultimate hull girder strength, reliability or risk of the three object vessels varying the damage scenarios when no repair or renewal is made. As the vessels get older, the corrosion depth and cracking size (length) grow and thus the ultimate hull girder strength or the reliability index decreases, while the level of the risk against hull girder collapse increases.

The reliability indices of the three object vessels against hull girder collapse are about 2.5 which is considered to be adequate considering the target value of the reliability index for merchant cargo vessels as previously noted. At the age of around 15 years the safety and reliability of the three vessels tends to become smaller than 90% of the newly built states. If repair and maintenance is not properly undertaken, the level of risks increases very rapidly.

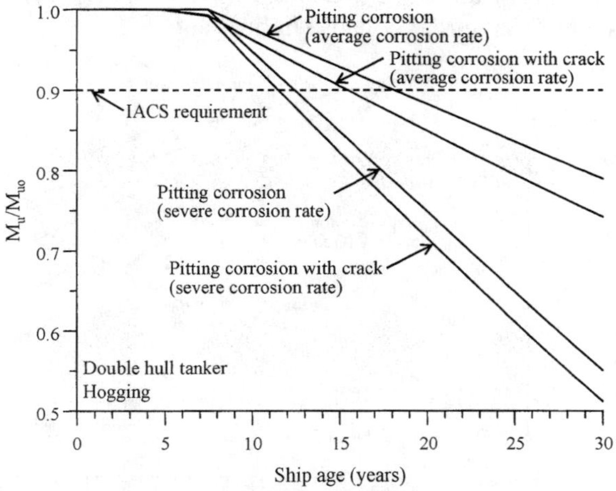

Figure 40(a) Time-dependent ultimate hull girder strength of the 105k dwt double hull tanker against hull girder collapse in hogging

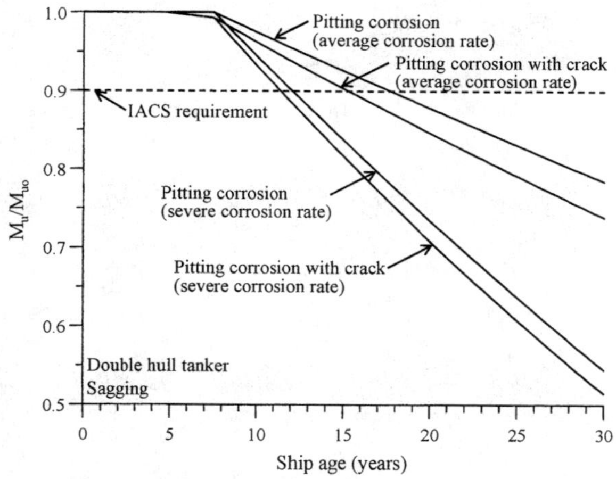

Figure 40(b) Time-dependent ultimate hull girder strength of the 105k dwt double hull tanker against hull girder collapse in sagging

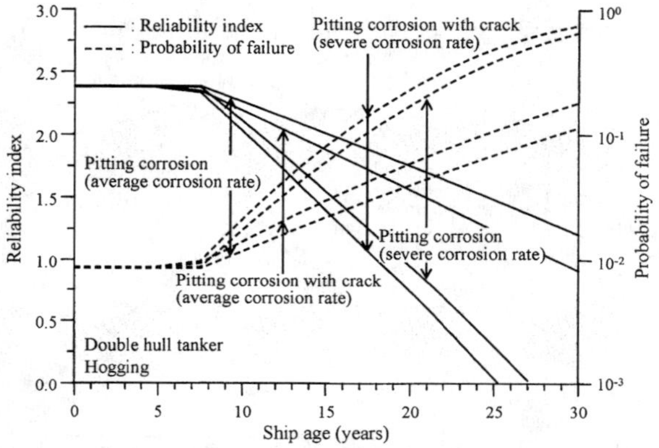

Figure 40(c) Time-dependent reliability or risk (probability of failure) of the 105k dwt double hull tanker against hull girder collapse in hogging

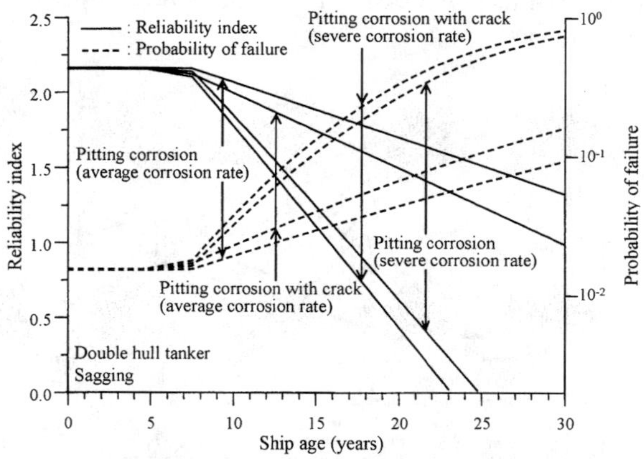

Figure 40(d) Time-dependent reliability or risk (probability of failure) of the 105k dwt double hull tanker against hull girder collapse in sagging

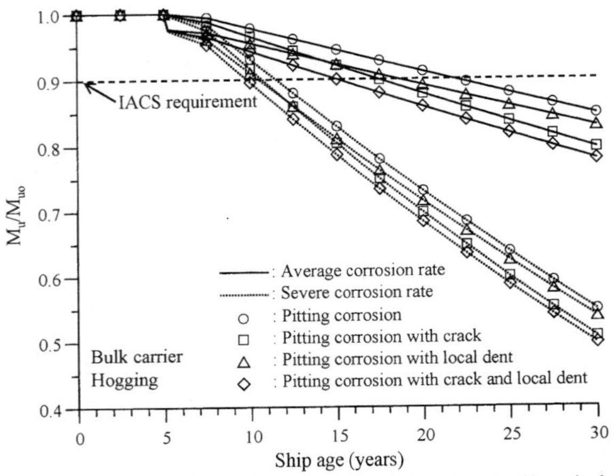

Figure 41(a) Time-dependent ultimate hull girder strength of the 170k dwt bulk carrier against hull girder collapse in hogging

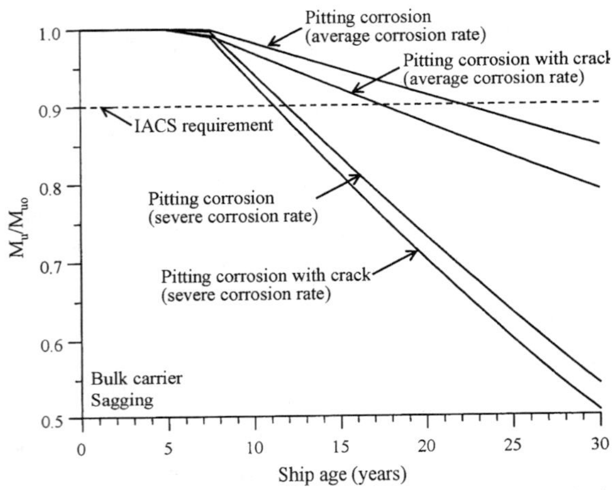

Figure 41(b) Time-dependent ultimate hull girder strength of the 170k dwt bulk carrier against hull girder collapse in sagging

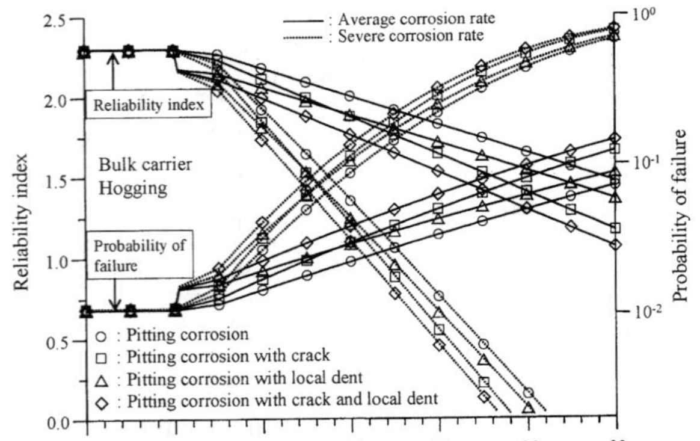

Figure 41(c) Time-dependent reliability or risk (probability of failure) of the 170k dwt bulk carrier against hull girder collapse in hogging

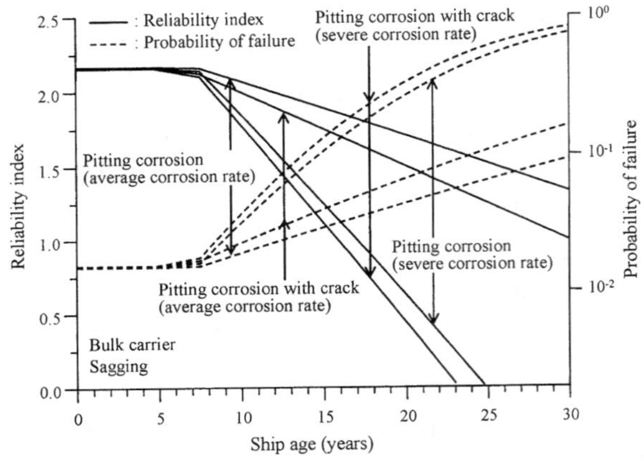

Figure 41(d) Time-dependent reliability or risk (probability of failure) of the 170k dwt bulk carrier against hull girder collapse in sagging

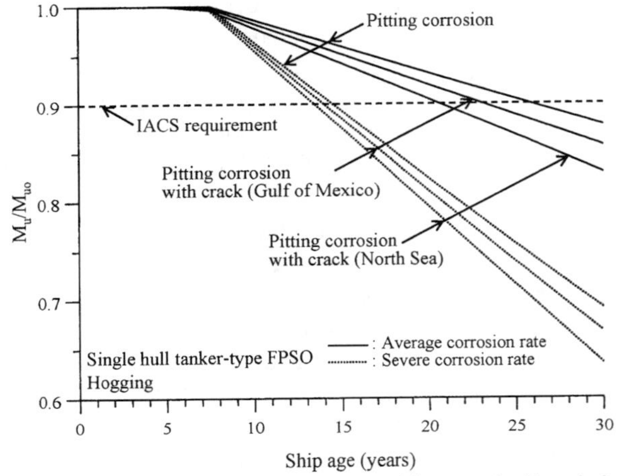

Figure 42(a) Time-dependent ultimate hull girder strength of the 254k dwt single hull tanker-type FPSO against hull girder collapse in hogging

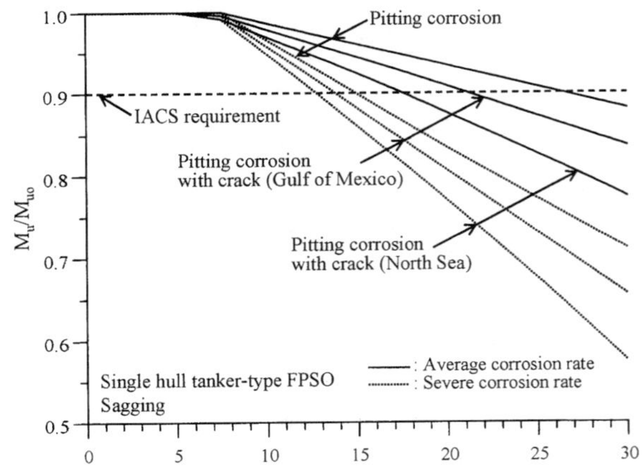

Figure 42(b) Time-dependent ultimate hull girder strength of the 254k dwt single hull tanker-type FPSO against hull girder collapse in sagging

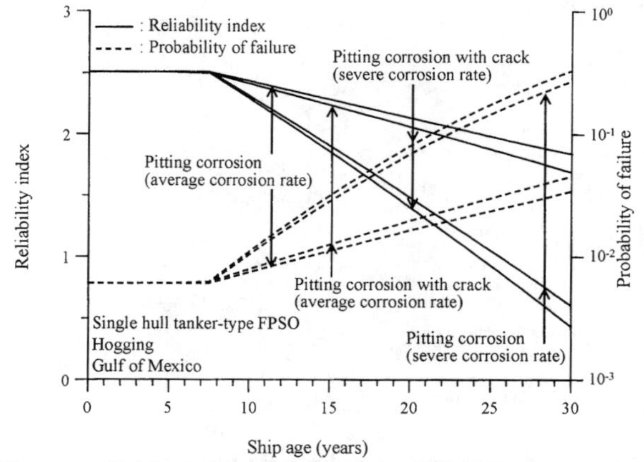

Figure 42(c) Time-dependent reliability or risk (probability of failure) of the 254k dwt single hull tanker-type FPSO against hull girder collapse in hogging (Gulf of Mexico)

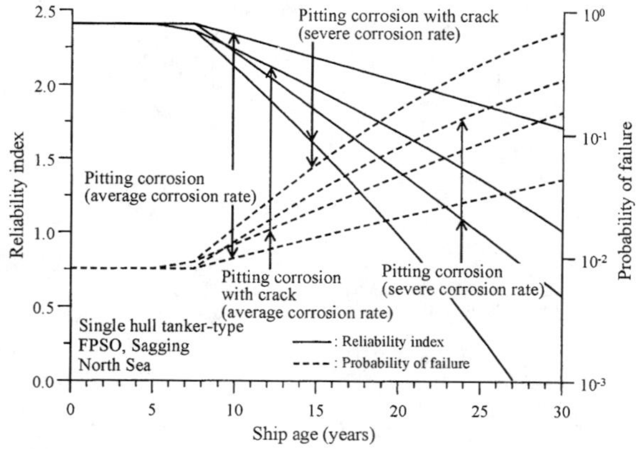

Figure 42(f) Time-dependent reliability or risk (probability of failure) of the 254k dwt single hull tanker-type FPSO against hull girder collapse in sagging (North Sea)

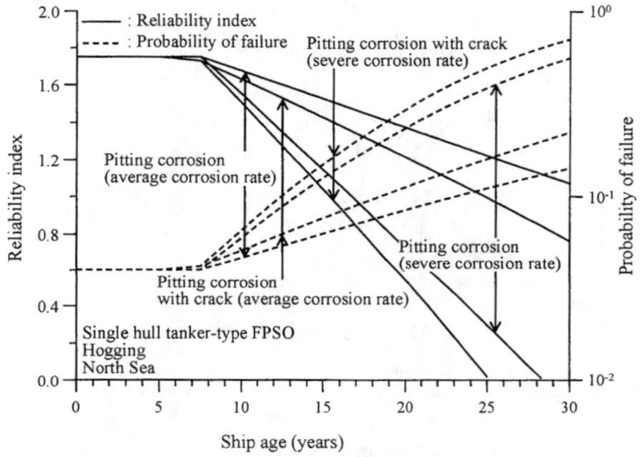

Figure 42(d) Time-dependent reliability or risk (probability of failure) of the 254k dwt single hull tanker-type FPSO against hull girder collapse in hogging (North Sea)

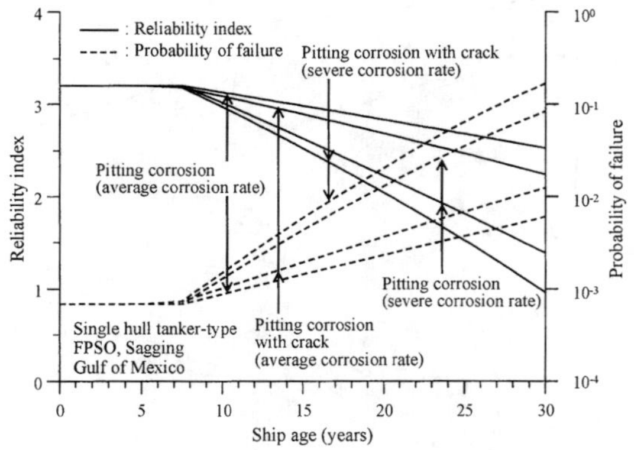

Figure 42(e) Time-dependent reliability or risk (probability of failure) of the 254k dwt single hull tanker-type FPSO against hull girder collapse in sagging (Gulf of Mexico)

CONSIDERATIONS FOR REPAIR SCHEME

To efficiently keep the ship's safety and reliability higher than a critical level, a proper and cost-effective scheme of repair and maintenance must be established (TSCF 1997). In this regard, some considerations for repair strategies of heavily damaged members are now made.

The IACS requires to keep the longitudinal strength of an aging ship at the level of higher than 90% of the initial state of new building. While the IACS requirement is in fact based on the ship's section modulus, it is in the present study applied for establishing repair schemes so that the ultimate hull girder strength of an aging ship must be greater than 90% of the original ship even later in life.

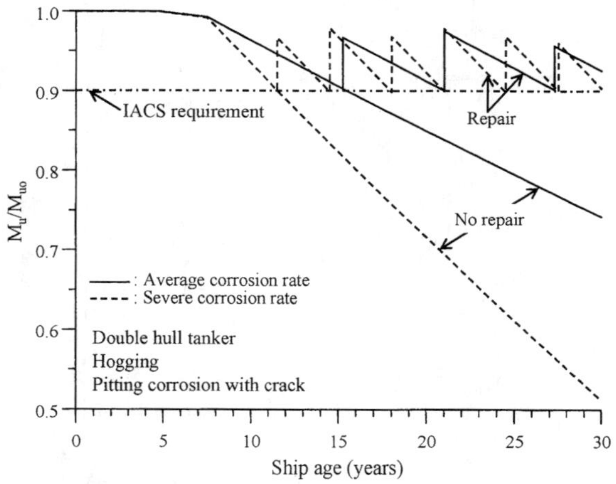

Figure 43(a) Repair and the resulting time-dependent ultimate hull girder strength of the 105k dwt double hull tanker in hogging

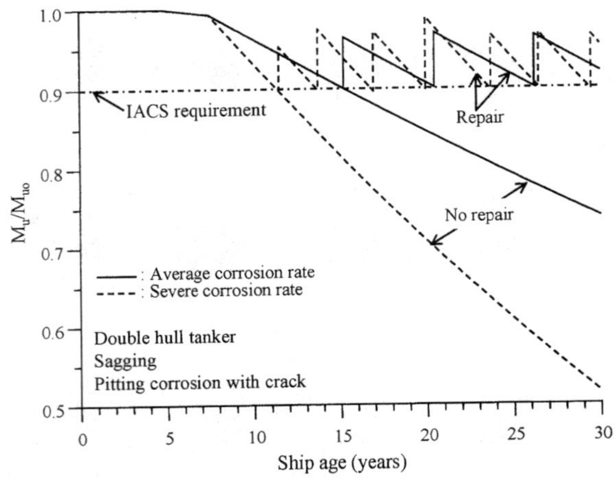

Figure 43(b) Repair and the resulting time-dependent ultimate hull girder strength of the 105k dwt double hull tanker in sagging

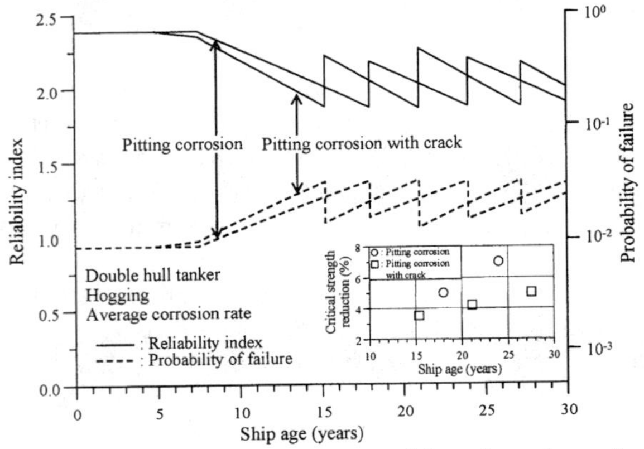

Figure 43(c) Repair and the resulting time-dependent reliability index of the 105k dwt double hull tanker in hogging and the critical ultimate compressive strength reduction percentage of heavily damaged members (average corrosion rate)

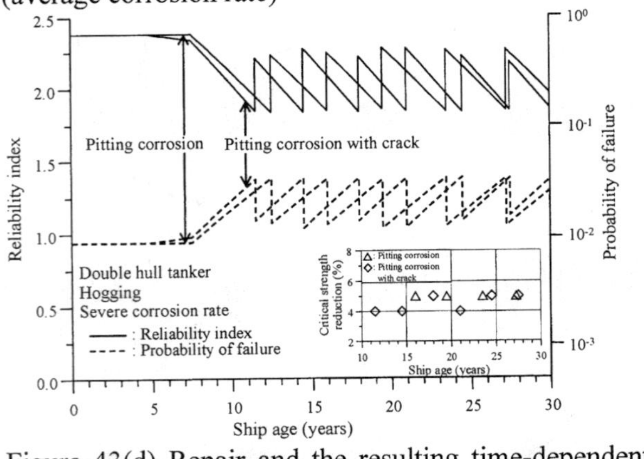

Figure 43(d) Repair and the resulting time-dependent reliability index of the 105k dwt double hull tanker in hogging and the critical ultimate compressive strength reduction percentage of heavily damaged members (severe corrosion rate)

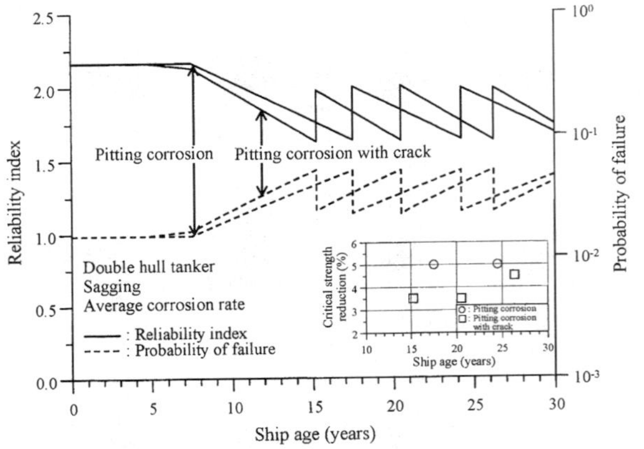

Figure 43(e) Repair and the resulting time-dependent reliability index of the 105k dwt double hull tanker in sagging and the critical ultimate compressive strength reduction percentage of heavily damaged members (average corrosion rate)

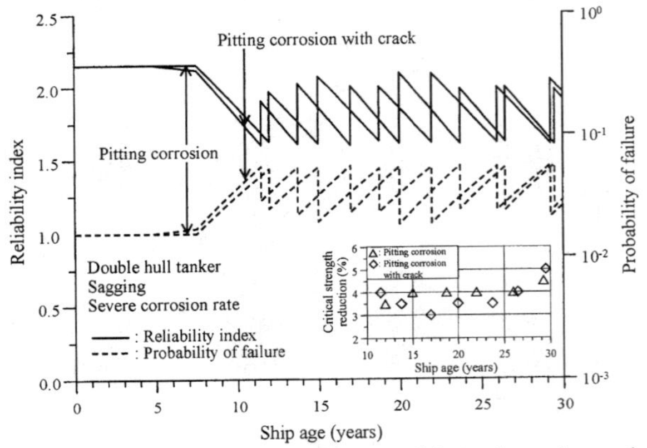

Figure 43(f) Repair and the resulting time-dependent reliability index of the 105k dwt double hull tanker in sagging and the critical ultimate compressive strength reduction percentage of heavily damaged members (severe corrosion rate)

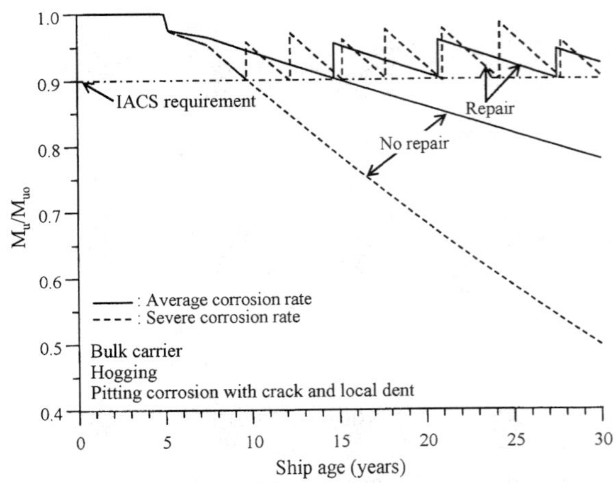

Figure 44(a) Repair and the resulting time-dependent ultimate hull girder strength of the 170k dwt bulk carrier in hogging

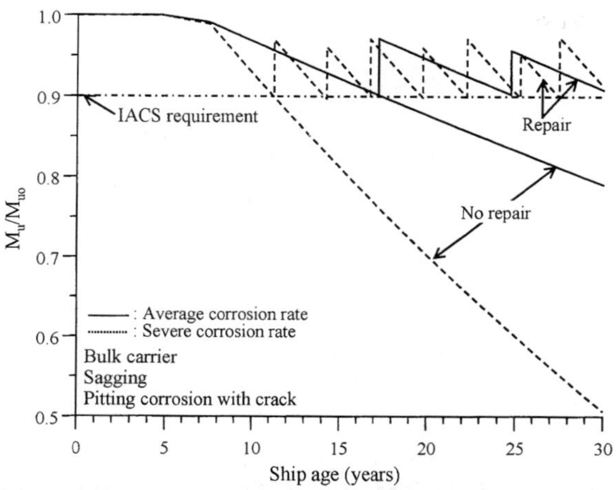

Figure 44(b) Repair and the resulting time-dependent ultimate hull girder strength of the 170k dwt bulk carrier in sagging

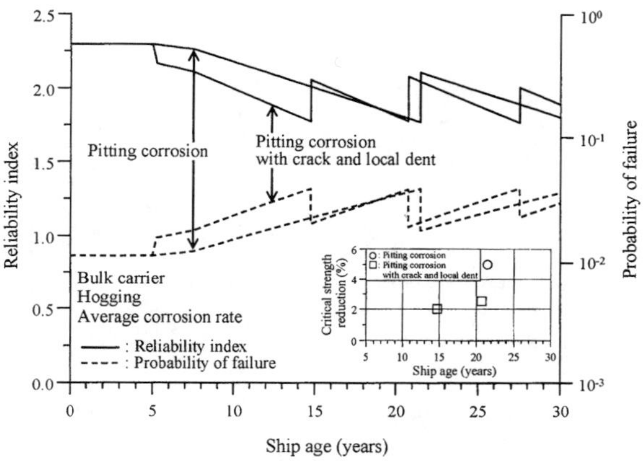

Figure 44(c) Repair and the resulting time-dependent reliability index of the 170k dwt bulk carrier in hogging and the critical ultimate compressive strength reduction percentage of heavily damaged members (average corrosion rate)

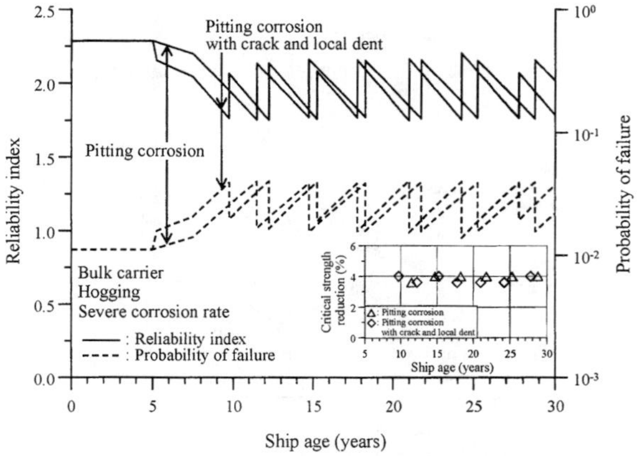

Figure 44(d) Repair and the resulting time-dependent reliability index of the 170k dwt bulk carrier in hogging and the critical ultimate compressive strength reduction percentage of heavily damaged members (severe corrosion rate)

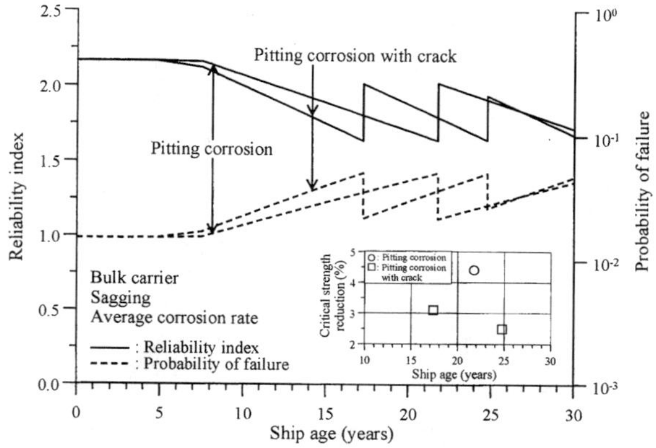

Figure 44(e) Repair and the resulting time-dependent reliability index of the 170k dwt bulk carrier in sagging and the critical ultimate compressive strength reduction percentage of heavily damaged members (average corrosion rate)

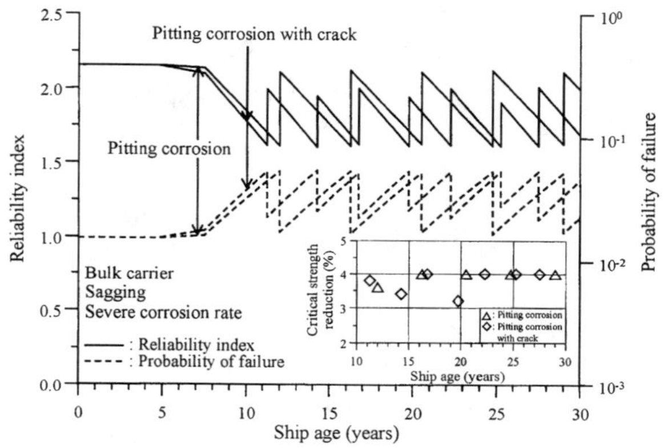

Figure 44(f) Repair and the resulting time-dependent reliability index of the 170k dwt bulk carrier in sagging and the critical ultimate compressive strength reduction percentage of heavily damaged members (severe corrosion rate)

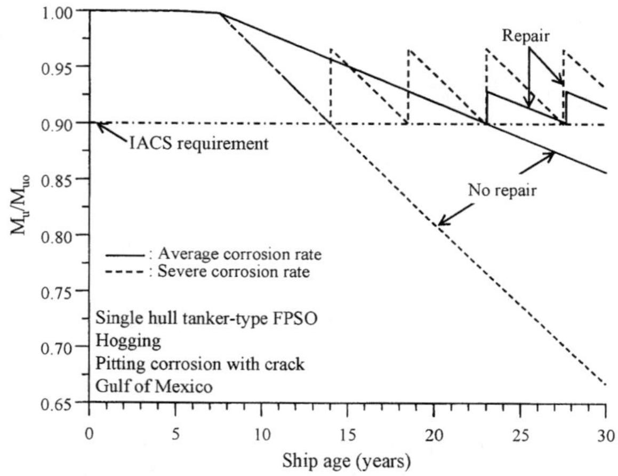

Figure 45(a) Repair and the resulting time-dependent ultimate hull girder strength of the 254k dwt single hull tanker-type FPSO in hogging (Gulf of Mexico)

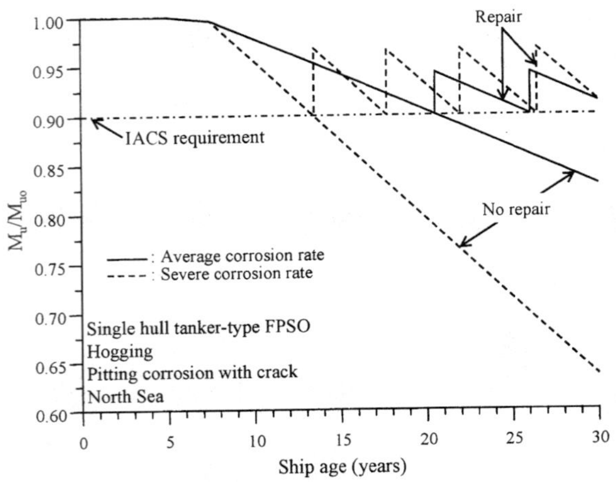

Figure 45(b) Repair and the resulting time-dependent ultimate hull girder strength of the 254k dwt single hull tanker-type FPSO in hogging (North Sea)

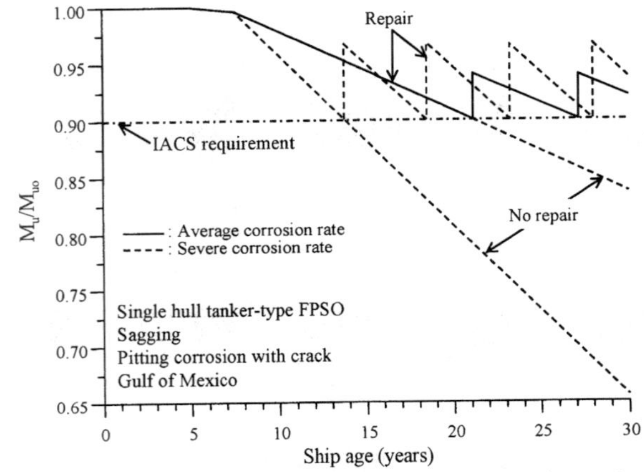

Figure 45(c) Repair and the resulting time-dependent ultimate hull girder strength of the 254k dwt single hull tanker-type FPSO in sagging (Gulf of Mexico)

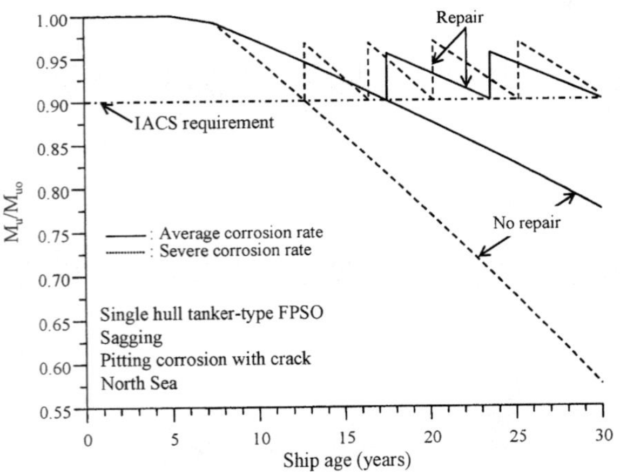

Figure 45(d) Repair and the resulting time-dependent ultimate hull girder strength of the 254k dwt single hull tanker-type FPSO in sagging (North Sea)

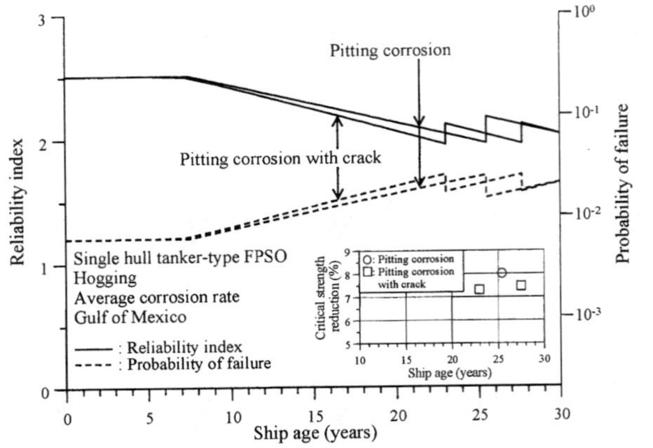

Figure 45(e) Repair and the resulting time-dependent reliability index of the 254k dwt single hull tanker-type FPSO in hogging and the critical ultimate compressive strength reduction percentage of heavily damaged members (average corrosion rate, Gulf of Mexico)

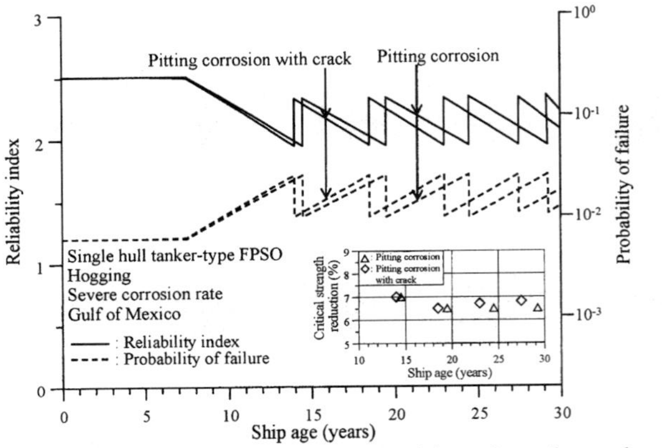

Figure 45(f) Repair and the resulting time-dependent reliability index of the 254k dwt single hull tanker-type FPSO in hogging and the critical ultimate compressive strength reduction percentage of heavily damaged members (severe corrosion rate, Gulf of Mexico)

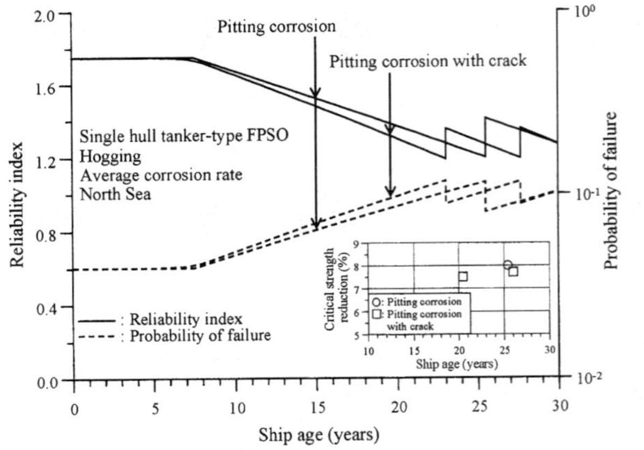

Figure 45(g) Repair and the resulting time-dependent reliability index of the 254k dwt single hull tanker-type FPSO in hogging and the critical ultimate compressive strength reduction percentage of heavily damaged members (average corrosion rate, North Sea)

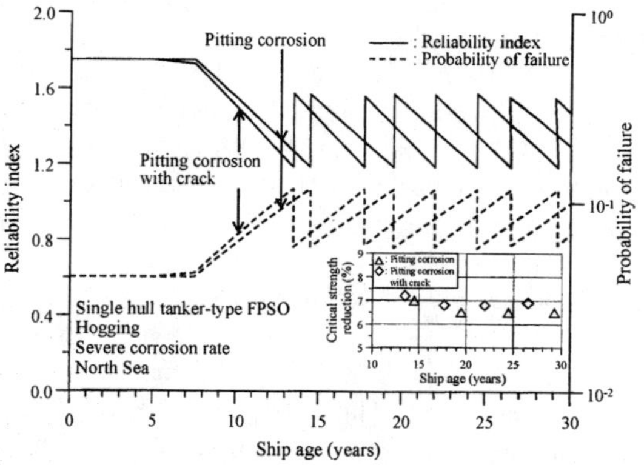

Figure 45(h) Repair and the resulting time-dependent reliability index of the 254k dwt single hull tanker-type FPSO in hogging and the critical ultimate compressive strength reduction percentage of heavily damaged members (severe corrosion rate, North Sea)

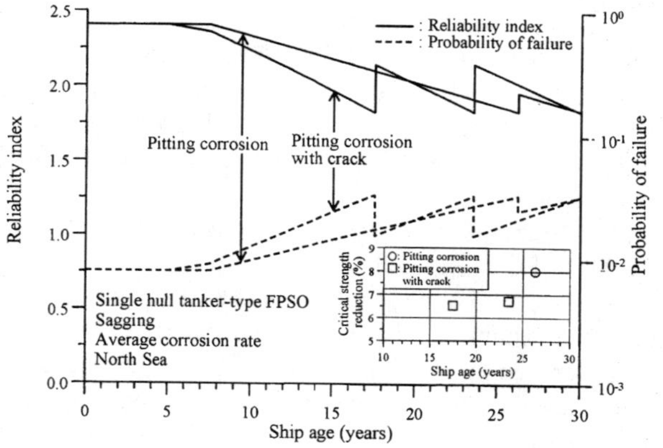

Figure 45(k) Repair and the resulting time-dependent reliability index of the 254k dwt single hull tanker-type FPSO in sagging and the critical ultimate compressive strength reduction percentage of heavily damaged members (average corrosion rate, North Sea)

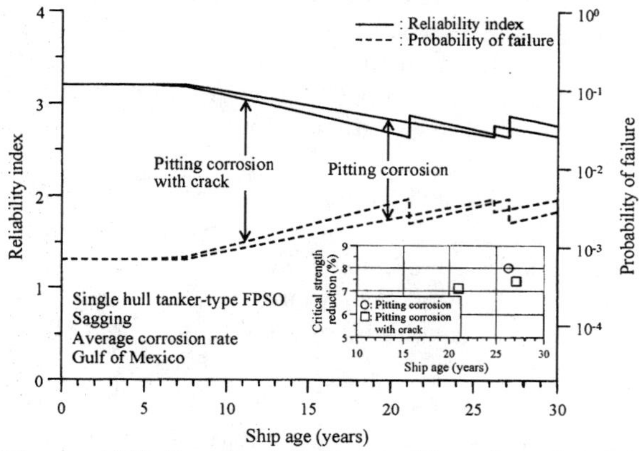

Figure 45(i) Repair and the resulting time-dependent reliability index of the 254k dwt single hull tanker-type FPSO in sagging and the critical ultimate compressive strength reduction percentage of heavily damaged members (average corrosion rate, Gulf of Mexico)

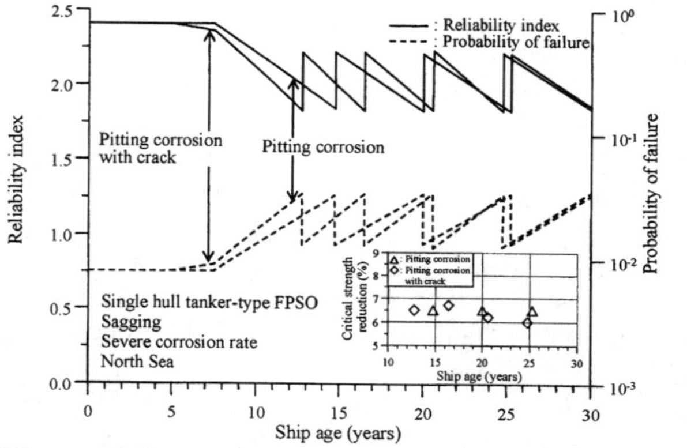

Figure 45(l) Repair and the resulting time-dependent reliability index of the 254k dwt single hull tanker-type FPSO in sagging and the critical ultimate compressive strength reduction percentage of heavily damaged members (severe corrosion rate, North Sea)

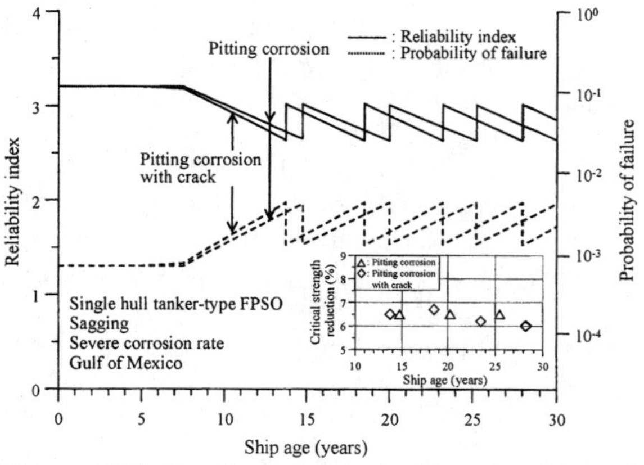

Figure 45(j) Repair and the resulting time-dependent reliability index of the 254k dwt single hull tanker-type FPSO in sagging and the critical ultimate compressive strength reduction percentage of heavily damaged members (severe corrosion rate, Gulf of Mexico)

The renewal criterion of any structural member is based on the member's ultimate strength rather than the plate thickness. This is because the latter approach, i.e., based on the percentage of member thickness loss, can not reveal the effects of fatigue cracking or local dent damage and even pit corrosion, while it may properly handle the thickness reduction due to uniform corrosion.

On the other hand, the former approach, i.e., based on member's ultimate strength, is adequate to measure the reduced strength of damaged structures. In the former approach, therefore, any structural member category will be repaired if the percentage of its ultimate strength loss due to age related degradation and/or mechanical denting exceeds a critical value.

Therefore, the present work considers that heavily degraded (or damaged) members are renewed (or repaired) to their original state, immediately before the ultimate longitudinal strength of an aged ship becomes a value smaller than 90% of the original ship.

Figures 43 to 45 show time-dependent ultimate hull girder strength, reliability index and probability of failure for the three object vessels calculated after repair of heavily damaged structural members so that the ultimate hull girder strength must always be greater than 90% of the original state.

It is evident from Figures 43 to 45 that the structural safety and reliability of aging vessels can be controlled by proper repair and maintenance strategies. Repair criteria based on the member's ultimate strength can be better controlled. It is also seen in the illustrations that the critical ultimate strength reduction percentage of structural members varies in the range of 2 to 5 % of the original state. For double hull tanker and FPSO, the critical strength reduction percentage for repair increases with time, but it tends to decrease for bulk carrier.

CONCLUDING REMARKS

In the present study, a procedure for the time-dependent risk assessment of aging vessels taking account of general/pit corrosion, fatigue cracking and local denting damage has been established with focus on total loss by hull girder collapse. The procedure was then applied to the time-dependent risk assessment of a 103k dwt double hull tanker, a 170k dwt bulk carrier and a 254k dwt single hull tanker-type FPSO. Some considerations for developing repair schemes were made to keep the vessels' ultimate longitudinal strength at an acceptable level.

It is concluded that the procedure developed in the present work will be useful for the damage tolerant design of new ship structures and also for establishing cost-effective repair schemes of existing aging ships.

ACKNOWLEDGEMENTS

The present study was undertaken with support from the American Bureau of Shipping, the Korea Ministry of Maritime Affairs & Fisheries and the Korea Science and Engineering Foundation. The authors are pleased to acknowledge their support. Thanks are also given to B.J. Kim, M.J. Ko and D.H. Lee who are graduate students at the Department of Naval Architecture and Ocean Engineering, Pusan National University, for their contribution to the experiments and numerical computations.

REFERENCES

ABS (2002). Steel vessel rules, American Bureau of Shipping, Houston.

Ayyub, B.M. (2001). Elicitation of expert opinions for uncertainty and risks, CRC Press, New York.

Broek, D. (1986). Elementary engineering fracture mechanics, Martinus Nijhoff Publishers.

Contraros, P.D. (2003). The domino effect – coating breakdown – corrosion – structural failure – leading to possible design ramifications, Presented at Lloyds List Events – Marine Corrosion Club, Prevention and Management of Marine Corrosion.

Ditlevsen, O. and Madsen, H.O. (1996). Structural reliability methods, John Wiley & Sons.

Guedes Soares, C. and Garbatov, Y. (1999). Reliability of maintained, corrosion protected plates subjected to non-linear corrosion and compressive loads, *Marine Structures*, Vol.12, pp. 425-445.

Hogben, N, Dacunha, N.M.C. and Olliver, G.F. (1986). Global wave statistics, Unwin Brothers.

Hogben, N. (1990). Long term wave statistics, Ocean Engineering Science, Vol.9, Part A, Section 8, John Wiley & Sons.

ISO 2394 (1998). General principles on reliability for structures, International Standard, International Organization for Standardization, Second Edition, June.

Loukakis, T.A. and Chryssostomidis, C. (1975). Seakeeping series for cruiser stern ships, *SNAME Transactions*, Vol.83, pp.67-127.

Mansour, A.E. (1974). Approximate probabilistic method of calculating ship longitudinal strength, *J. of Ship Research*, Vol.18, No.3, pp.201-213.

Mansour, A.E., Wirsching, P., Luckett, M., Plumpton, A. (1997). Assessment of reliability of existing ship structures, Ship Structure Committee, SSC-398.

Paik, J.K. (2002). Ultimate strength of aging ships – Time-dependent risk assessment taking account of general/pit corrosion, fatigue cracking and local denting, Report submitted to the American Bureau of Shipping, No. RD 2002-11, December.

Paik, J.K. and Frieze, P.A. (2001). Ship structural safety and reliability, *Progress in Structural Engineering and Materials*, John Wiley & Sons, Vol.3, No.2, pp.198-210.

Paik, J.K., Lee, J.M. and Lee, D.H. (2003b). Ultimate strength of dented steel plates under axial compressive loads, *International Journal of Mechanical Sciences*, Vol.45, No.3, pp.433-448.

Paik, J.K., Lee, J.M., Hwang, J.S. and Park, Y.I. (2003c). A time-dependent corrosion wastage model for the structures of single- and double- hull tankers and FSOs and FPSOs, *Marine Technology*, Vol.40, No.3.

Paik, J.K., Lee, J.M., Park, Y.I. and Hwang, J.S. (2003d). A time-dependent corrosion wastage model for bulk carrier structures, *International Journal of Maritime Engineering* (in press).

Paik, J.K. and Thayamballi, A.K. (2002). Ultimate

strength of ageing ships, *Journal of Engineering for the Maritime Environment*, Vol.216, No.M1, pp.57-77.

Paik, J.K. and Thayamballi, A.K. (2003). Ultimate limit state design of steel-plated structures, John Wiley & Sons.

Paik, J.K., Thayamballi, A.K. and Lee, J.M. (2003a). Effect of initial deflection shape on the ultimate strength behavior of welded steel plates under biaxial compressive loads, *Journal of Ship Research* (in press).

Paik, J.K., Wang, G., Kim, B.J. and Thayamballi, A.K. (2002). Ultimate limit state design of ship hulls, *SNAME Transactions*, Vol.110.

Qin, S.P. and Cui, W.C. (2003). Effect of corrosion models on the time-dependent reliability of steel plated elements, *Marine Structures*, Vol.16, pp. 15-34.

TSCF (1997). Guidance manual for tanker structures, Tanker Structure Co-operative Forum, Witherby & Co. Ltd., London.

Wang, G., Spencer, J. and Elsayed, T. (2003a). Estimation of corrosion rates of oil tankers, Proceedings of the 22nd International Conference on Offshore Mechanics and Artic Engineering (OMAE'03), 8-13 June, Cancun, Mexico.

Wang, G., Spencer, J. and Sun, H.H. (2003b). Assessment of corrosion risks to aging ships using an experience database, Proceedings of the 22nd International Conference on Offshore Mechanics and Artic Engineering (OMAE'03), 8-13 June, Cancun, Mexico.

Wang, G.. and Spong, R. (2003). Experience based data for FPSO's structural design, Proceedings of the Offshore Technology Conference (OTC'03), 5-8 May, Houston, TX.

Wang, G.., Tang, S. and Shin, Y. (2002). A direct calculation approach for designing a ship-shaped FPSO's bow against wave slamming load, Proceedings of the 12th International Offshore and Polar Engineering Conference (ISOPE'02), May 26-31, Kitakyushu, Japan.

DISCUSSION

Joško Parunov, Visitor

I would like to congratulate the authors on this very interesting paper that describes a number of the most important items concerning the safety of aging ships. In the future, this study will be an unavoidable reference for researches in the fields of corrosion progression and influences of corrosion, fatigue, and dent damage on ship load-carrying capacity.

The authors suggest that corrosion progress on oil tankers and FPSO ships may be similar. However, there are some substantial functional differences between these two types of structures that put this suggestion in question. Typical oil tankers are normally operated with cargo tanks either completely full or empty, and the liquid cargo motion is thus restricted. On the other hand, the FPSO vessel is always in partial loading condition due to a different loading and offloading procedure. Namely, cargo tanks are filled with produced crude oil until the shuttle tanker arrives (loading cycle). The shuttle tanker for discharging of the FPSO normally arrives before all cargo tanks are completely full to avoid the risk of overfilling. When the crude oil is discharged (offloading cycle) from the FPSO vessel, the tanks will never be completely empty because production continues even during the offloading phase. Therefore, some of the tanks will always be partially filled. Sloshing represents the dynamic magnification of internal pressure acting on cargo tank boundaries and internal structure due to liquid motion in partially filled tanks. Because of the mentioned reasons, sloshing is much more important for a FPSO vessel than for an ordinary oil tanker. The sloshing pressure could appreciably exceed its static value and cause both structural overstress and fatigue life reduction. After reading the section of the paper where the corrosion mechanism is related to changing the immersion conditions and dynamic loads, the question that may be asked is, does sloshing also accelerate the corrosion process? Could this effect be of such importance to produce a “domino effect” and to result in the concave curve of corrosion progression?

I think that the authors should be very careful when applying extensively the MIT seakeeping tables, especially to FPSO vessels. These seakeeping tables were calculated for merchant ships some 30 years ago. Even today, they can give a fairly accurate prediction of wave bending moments of oil tankers in full load condition. However, the mass distribution of FPSOs is quite different than for ordinary oil tankers due to the topside equipment and mooring system. Also, the hull form of new FPSOs may significantly deviate from the tanker form. The intention is to gain more cargo capacity while the seakeeping performances are less important. Thus, FPSO vessels with a block coefficient larger than 0.9 are not rare today, and this is outside the range of MIT tables. Because the FPSO vessels have no forward speed, extrapolation of seakeeping data outside their range of validity should be performed, and this introduces additional uncertainty. The seakeeping tables are originally applied for the one-parameter Pierson-Moskowitz spectrum, and the correction procedure should be applied when analyzing the two-parameter ITTC spectrum. For the case of the JONSWAP spectrum, which is quite often the design requirement for FPSOs, the tables are not applicable at all. It should also be noted that all these differences could have larger impact on the accuracy of calculated wave bending moments than on the ship motions.

I would also comment on the authors' choice of design sea states for seagoing ships that are specified in Table 5. The formula applied in the paper for the selection of significant wave height, $H_s = 1.1\sqrt{L}$, was used in the past as the wave height of the static trochoidal wave to give the same bending moment at midship as in the rules of classification societies. Nowadays, the design significant wave height is usually the result of the statistical analysis of different sea states occurring with various probabilities on an assumed shipping route. If reference sea states in Table 5 are just for illustrative purposes, I think that the authors should avoid such statements as “the short term based wave-induced bending moments are less than the IACS formula by 10-20%” or “the reliability indices of the three object vessels against hull girder collapse are about 2.5 which is considered to be adequate.” These statements bring a clear message to the reader that the authors are convinced of their correct choice of loads and statistical models used in their analysis. If so, I would kindly ask the authors to explain their choice of the design sea states for the double-hull tanker and bulk carrier. How can it be supported by arguments that the design sea states for ships are much lower than for FPSOs located in the Gulf of Mexico (about H_s = 10 m for ships against H_s = 12.2 m for FPSOs in the Gulf of Mexico). Would that imply that ships are not safe enough when sailing in the Gulf of Mexico? It is also very difficult to understand the meaning of the design speed of 10 knots for FPSOs. This design speed could lead to large overestimation of actual wave loads. This is not in line with the intention of the reliability methods, where we are interested in realistic loads that are neither overestimated nor underestimated. I would also like to discuss equation (32a). The recommended limiting wave steepness of extreme sea states for analysis of ship structures reads 1/15 (Bittner-Gregersen et al 1995). This wave steepness leads to mean zero-up-crossing period $T_Z = \sqrt{9.6H_s}$, which is a much lower T_Z than given by equation (32a). The limiting wave steepness could have great impact on extreme wave loads, especially in head seas. Generally, by increasing the wave steepness, that is, by reducing the wave periods, the vertical wave bending moment at midship increases.

My last comments are related to the safety indexes that are discussed and calculated in the paper. It may be noted that results of Professor Faulkner from the late 1970s are missing in Fig. 1. His recommended safety indexes of 3.0 for merchant ships and 1.5 for naval ships are placed well below the proposed trend line in Fig. 1 (Faulkner & Sadden 1978, Faulkner 1981). Perhaps this would mean that his research was well ahead of his time. In fact, if the extreme value distribution is used to describe the wave loads and if the model uncertainty is taken into account in the limit state equation, the target safety indexes would certainly be around these values, irrespective of the time period when the research has been performed. I think that the assumed target reliability of 2.5 against hull girder collapse proposed by the authors should be related to the return period for calculation of the wave and still-water loads. The assumed target reliability would not be the same for return periods of 1 year and 20 years. I would also like to turn the authors' attention to the fact that FPSOs normally have a significantly higher reliability than seagoing ships. The annual reliability index of a new FPSO is higher than 3.71, which is not the case in Figs. 42 (c–e) (Det Norske Veritas 1992). The fairly high reliability of FPSOs is mainly the consequence of the offshore regulation applied in design, which is more severe than the rules for the construction of seagoing ships. For example, the extreme wave loads in the design of FPSOs are calculated for the return period of at least 100 years, whereas the extreme loads in ship rules refer to the return period of about 20 years. Modeling uncertainty of calculated wave loads is for several reasons lower for structures located on a fixed location, as FPSOs, than for oceangoing ships. This is also a contributing factor for increasing the calculated reliability of FPSOs compared to ship hulls (Parunov & Senjanović 2003).

Additional references

Bittner-Gregersen, E. M., Cramer, E. H., and Løseth, R. 1995 Uncertainties of Load Characteristics and Fatigue Damage of Ship Structures, *Marine Structures,* **8**, 97–117.

Det Norske Veritas. 1992 Classification Notes No. 30.6, *Structural Reliability Analysis of Marine Structures*, July.

Faulkner, D., and Sadden, J. D. 1978 Toward unified approach to ship structural safety, *Transactions of the Royal Institution of Naval Architects*, paper No. 3, Spring Meeting.

Faulkner, D. 1981 Semi-probabilistic approach to the design of marine structures, *Transactions of the Society of Naval Architects and Marine Engineers*, 213–230.

Parunov, J., and Senjanović, I. 2003 Incorporating model uncertainty in ship reliability analysis, *Proceedings*, Society of Naval Architects and Marine Engineers Annual Meeting and World Maritime Technology Conference, October, San Francisco, CA.

Design Of An Articulated Tug and Barge (AT/B) To Shuttle Crude Oil On The Bohai Gulf, China.

Joseph P. Fischer, P.E. [1]
Fellow

ABSTRACT

The paper describes the design of an articulated tug and barge (AT/B) to shuttle crude oil from a platform in the Bohai Gulf of China to a receiving terminal at the Port of Tanggu, China, a distance of approximately 33 nautical miles. An Articouple system of connecting the tug and barge was selected to provide a large part of the operation in the push mode. A 21,000 BBL tank barge making a daily round trip from the platform to the terminal along with a 3500 HP, twin screw tug is described. The vessels are of shallow draft and ice classed.

1. GENERAL

1.1 Introducion

An articulated tug and barge (AT/B) was designed to shuttle crude oil on the Bohai Gulf, China for Apache China Corporation, LDC.

Apache China Corporation LDC is a wholly owned subsidiary of Apache Corporation of Houston, Texas. Apache China Corporation has an interest in a petroleum contract with PetroChina Company Limited, a subsidiary of state-owned China National Petroleum Corporation. PetroChina produces two-thirds of China's gas and oil and is listed on the NYSE.

The platforms are designed by Worley International, Inc. of Houston, Texas and overall project management is by Hallibuton, Kellogg, Brown and Root Services Corp. of Houston, Texas. Contract Plans and Specifications for both the tug and barge were prepared by Bay Engineering, Inc. of Sturgeon Bay, Wisconsin.

1.2 Location

The Bohai Gulf is located at the northwest end of the Yellow Sea. A large part of the Gulf has been plotted by the PRC (People's Republic of China) for oil exploration and production by various oil interests. PetroChina Corporation has a plot in the Zhao Dong Block. There are two (2) platforms planned, one drilling platform (ODA) and one production platform (OPA). The platforms are located about 5 nautical miles offshore and about 33 nautical miles from the Port of Tanggu, near the city of Tianjin, China. The main water depth at the production platform site is about 1.8 meters with an average tide of 2.4 meters with a two tidal cycle time of 24 hours and 50 minutes.

The platform location is subject to rough weather with wind from the northeast. The average sea state has a significant wave height of 1.0 meters with wind speeds of 13 meters/second (sea state 3). A single year storm has a significant wave height of 3.38 meters (sea state 5) and a wind speed of 24.1 meters/second. The platform is far enough north to experience ice in cold years. The port of Tanggu is normally ice free.

In April of 1999, a meeting was held at the Dagong Oil Field to discuss the pros and cons of a pipeline versus water transportation. The economic merits of a tug and barge eventually prevailed.

1.3 Concept Design

Discussions were held early on with Apache Corporation, PetroChina Ltd, Worley International, Kellogg, Brown and Root, and Bay Engineering, Inc. on various forms of marine transportation for the crude oil.

A production curve of estimated barrels of oil per day (bopd) showed on initial production at start up of about 5000 bopd and climbing to 25,000 bopd after

[1] President, Bay Engineering, Inc.
Sturgeon Bay, WI, USA

several years and then gradually reducing to about 3000 bopd after 15 years.

If a single tug and barge (AT/B) could be built to satisfy the first few years, a second barge could be added at a later date to satisfy the peak demand in a drop and swap operation. This full utilization of a single power plant was estimated to have a lower capital cost than a single large tanker or two (2) smaller tankers to meet the peak demand. If the production curve were truncated by reducing the peak production and holding this production for a longer period of time, the production and transportation equipment could be smaller and better utilized.

A list of pros and cons for a tug and tank barge versus a self-propelled tank ship pushed the pros in favor of the tug and tank barge as long as the unit could minimize loss of speed and/or operational down time because of weather and ice.

A review of various means of connecting the tug and barge with a simple and fast disconnect system that maximizes the time pushing led to the Articouple as the system of choice. This system utilizes two transverse rams on the tug with a toothed head, each extending outward into a vertical rack on the side wall of the barge notch. The tug has a single degree of freedom, pitch, relative to the barge and the connector can be designed to meet the anticipated sea state. The initial Articouple design was done by T. Yamaguchi of Japan [1] [2] [3]. This system uses hydraulic cylinders to extend and retract the rams. A similar design using electric driven gears and a power screw to move each ram was developed by R. Hill and L. Kuhlman under the trade name INTERCON [4]. Similar connection designs using hydraulic cylinders have been done by the author after the expiration of the original patent [5].

1.4 Vessel Requirements

A tug-barge system of transporting crude oil from the Zhao Dong Platform needed to consider the following factors:

- The platform berthing area should have sufficient dredged area to berth on either side of the loading dock with a turn-around basin.
- The vessels operating draft shall be about 3.0 meters.
- The dredged distance from the platform to open water is about 2.3 nautical miles.
- In periods of low tide, it may be necessary for the tug and barge to ground on the mud at the loading dock. The bottom structure shall be strengthened to allow static grounding in 2.0 meters of water.
- Both tug and barge should be capable of operating in ice of 30 cm at 100% coverage. Each vessel shall be built to ABS ice class AO.
- Each vessel shall carry its own equipment and power and shall be independent of shoreside equipment except for loading of cargo, fuel, water and stores.
- The barge shall be unmanned. There will be no persons accommodated on board while at sea. The barge may be permissively manned while in port. Qualified tankermen shall be berthed on the tug.

1.5 Design Standards

As the vessels would probably be built in the Far East and be registered under Chinese Flag, the question of Classification was discussed with Apache China Corporation, LDC. Some shipyards prefer working with certain classification societies with whom they are most familiar. Considered were American Bureau of Shipping (ABS), China Classification Society (CCS) and Lloyd's Register (LR). All are members of International Association of Classification Socities (IACS).

The design of both vessels, tug and barge, are to meet ABS Rules for Ice Class AO. It was decided to follow ABS Rules for the contract design and the classification for detail design and construction to be negotiated with the Builder. The tug was designed according to ABS Rules for Building and Classing Steel Vessels Under 90 Meters (295 Feet) In Length, 2001 and the barge was designed according to ABS Rules for Building and Classing Steel Barges, 1991.

It was also decided to meet applicable U.S. Coast Guard requirements in 33CFR and 46CFR as a standard of safety.

The requirements for Safety of Life at Sea (SOLAS) are only applicable to vessels engaged on international voyages and are not required but are used as a standard of safety.

In addition, the vessel designs were to meet MARPOL regulations for pollution prevention.

The design also used ASTM Standards, Volume 01-07, Standards Relating to Ships and Marine Technology, Section F, as a standard of quality.

1.6 Shuttle Schedule, Capacity and Speed

A single dedicated tug and barge of 21,000 barrels capacity can make a round trip from the platform to the receiving terminal in Tanggu Harbor and back in a 24 hour period. Table 1 shows a trip schedule used in the design process.

The design loading rate at the platform is 900 cubic meters per hour. With the crude oil heated to a specific gravity of 0.93, this equates to 837 metric tons per hour and 5660 barrels per hour.

The design unloading rate at the Tanggu terminal is also 900 cubic meters per hour. Average rates are estimated at 85% of the design values.

The speed of the AT/B in the push mode is estimated at 8.00 knots in the platform channel, 9.20 knots in open water in the loaded condition, and 9.66 knots in open water in the ballast condition. There is no speed reduction in the Tanggu channel.

With One Tug & One Barge

ITEM	UNITS	
Conversion	Bbls/m^3	6.29
Capacity	Bbls	21049
Capacity	m3	3346
Cycle Time	Hrs.	24
Specific Gravity	S.G.	0.93
Delay	%	15%
Delay Factor	100-%	0.85
Platform Loading Rate, Design	m3/Hr	900
Platform Loading Rate, Design	MT/Hr	837
Platform Loading Rate, Design	Bbls/Hr	5265
Platform Loading Rate, Average	m3/hr	765
Tanggu Unloading Rate, Design	m3/Hr	900
Tanggu Unloading Rate, Design	MT/Hr	837
Tanggu Unloading Rate, Design	Bbls/Hr	5265
Tanggu Unloading Rate, Av.	m3/Hr	765
Platform Channel Distance	N. Miles	2.3
Open Water Distance	N. Miles	23
Tanggu Channel Distance	N. Miles	7.0
Platform Channel Speed	Knots	8.0
Open Water, Loaded	Knots	9.2
Open Water, Light	Knots	9.66
Platform Turning & Mooring	Hrs.	0.50
Hose Connection & Test	Hrs.	0.50
Platform Loading Time	Hrs.	4.38
Platform Hose Clear	Hrs.	0.50
Platform Mooring	Hrs.	0.50
Margin Time at Platform	Hrs.	1.00
Platform Channel Travel	Hrs.	0.35
Open Water Travel, Loaded	Hrs.	2.94
Margin Time at Anchorage	Hrs.	1.00
Channel Time, Tanggu	Hrs.	0.90
Maneuver & Mooring at Tanggu	Hrs.	0.50
Tanggu Hose Connection & Test	Hrs.	0.50
Unloading Time at Tanggu	Hrs.	4.38
Tanggu Hose Clearing	Hrs.	0.50
Mooring Time Tanggu	Hrs.	0.50
Margin Time at Tanggu	Hrs.	1.00
Channel Time, Tanggu	Hrs.	0.90
Open Water Travel, Light	Hrs.	2.80
Platform Channel Travel	Hrs.	0.35
Total Trip Time	Hrs.	24.00
Time at Platform	Hrs.	7.38
Time at Tanggu	Hrs.	7.38
Time at Travel	Hrs.	9.24

Table 1: Schedule

Figure 1 shows an estimated speed/power curve in sea state 3. The average speed is estimated at 85% of the trial speed in sea state 3.

A further constraint on the operation is imposed by the Port Authority of Tanggu. Because of the relatively narrow channel between the Port and the open sea, only single direction traffic is permitted in the channel. Every two (2) hours the direction is reversed. This means if you miss the window for travel in the desired direction, a vessel must wait, either at the entrance anchorage or at the terminal, until the required travel direction is permitted.

With time allocation for maneuvering in port, and cargo hose connection and testing, a very tight schedule is arrived at. Spare time of 1.00 hour is allowed for queing at the Tanggu channel entrance for the inbound, loaded AT/B. Spare time of 1.00 hour is allowed for queing at the unloading terminal at Tanggu to meet the directional constraints and 1.00 hours is allowed at the loading platform. This schedule determines the tug speed and horsepower and barge pumping capacity.

2. BARGE DESIGN

2.1 Barge Hull Form

The general arrangement of the barge is shown in figure 2.

The principal characteristics of the barge are shown in Table 2.

• Length, overall	99.50 m
• Length, loadline	95.52 m
• Beam, molded	18.00 m
• Depth at side, molded	5.40 m
• Draft, DWL	3.10 m
• Camber	0.18 m
• Sheer, Forward	0.60 m
• Sheer, Aft	0.00 m
• Light Ship Weight	1636 MT
• Displacement at DWL	4770 MT
• Deadweight	3134 MT
• Tonnes per Centimeter	16.47 MT
• Crude Oil Cargo @ 95%	21049 BBL
• Diesel Oil Cargo @ 95%	46.2 m^3
• Block Coefficient	0.873
• Prismatic Coefficient	0.881
• Waterplane Coefficient	0.935

Coefficients are based on L of 95.52 meters

Table 2: Barge Characteristics

The barge has multiple chines for ease of construction. It is intended that the bow lines in the ice belt be faired as conical and cylindrical development,

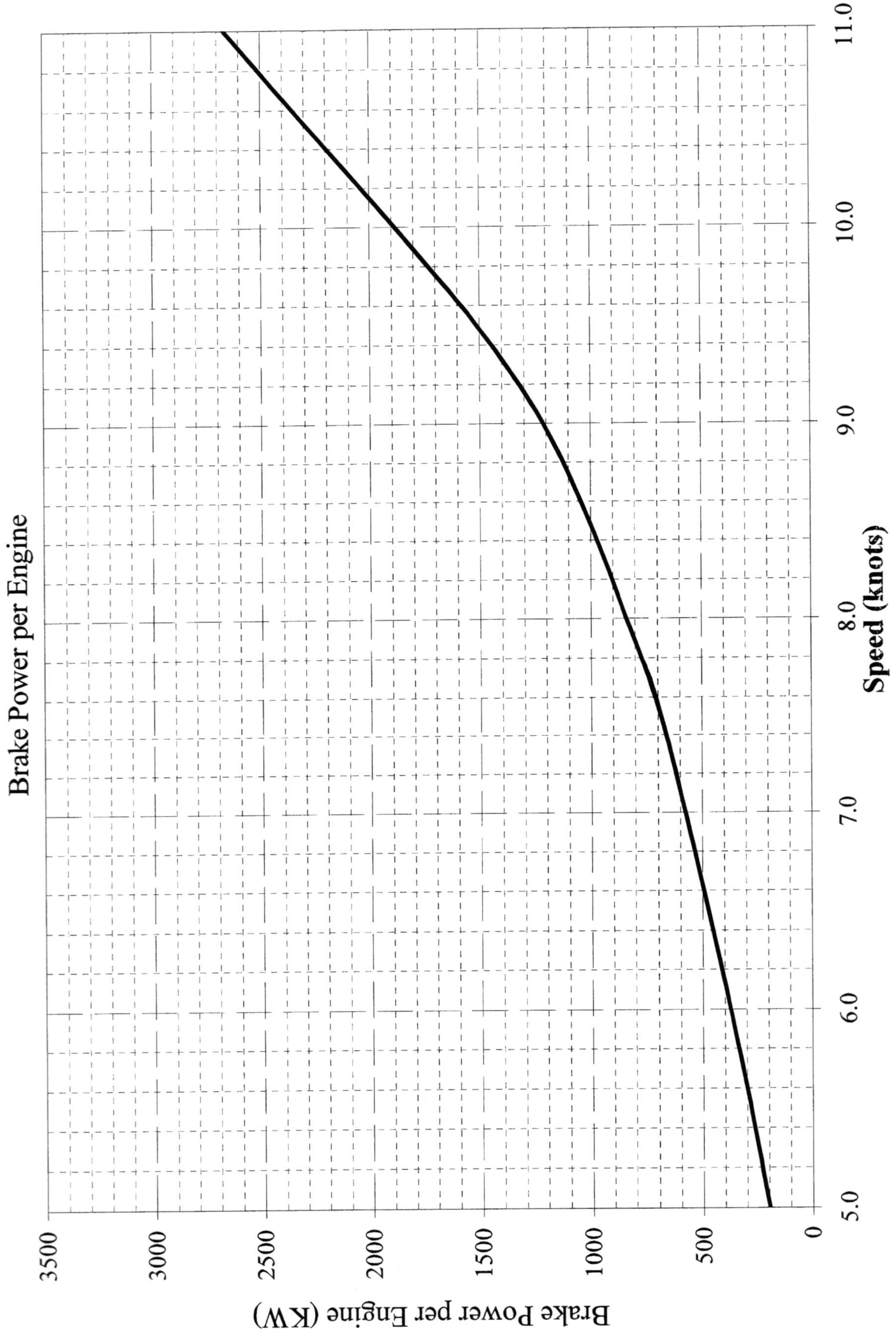

Figure 1 - Speed and Power, Sea State 3

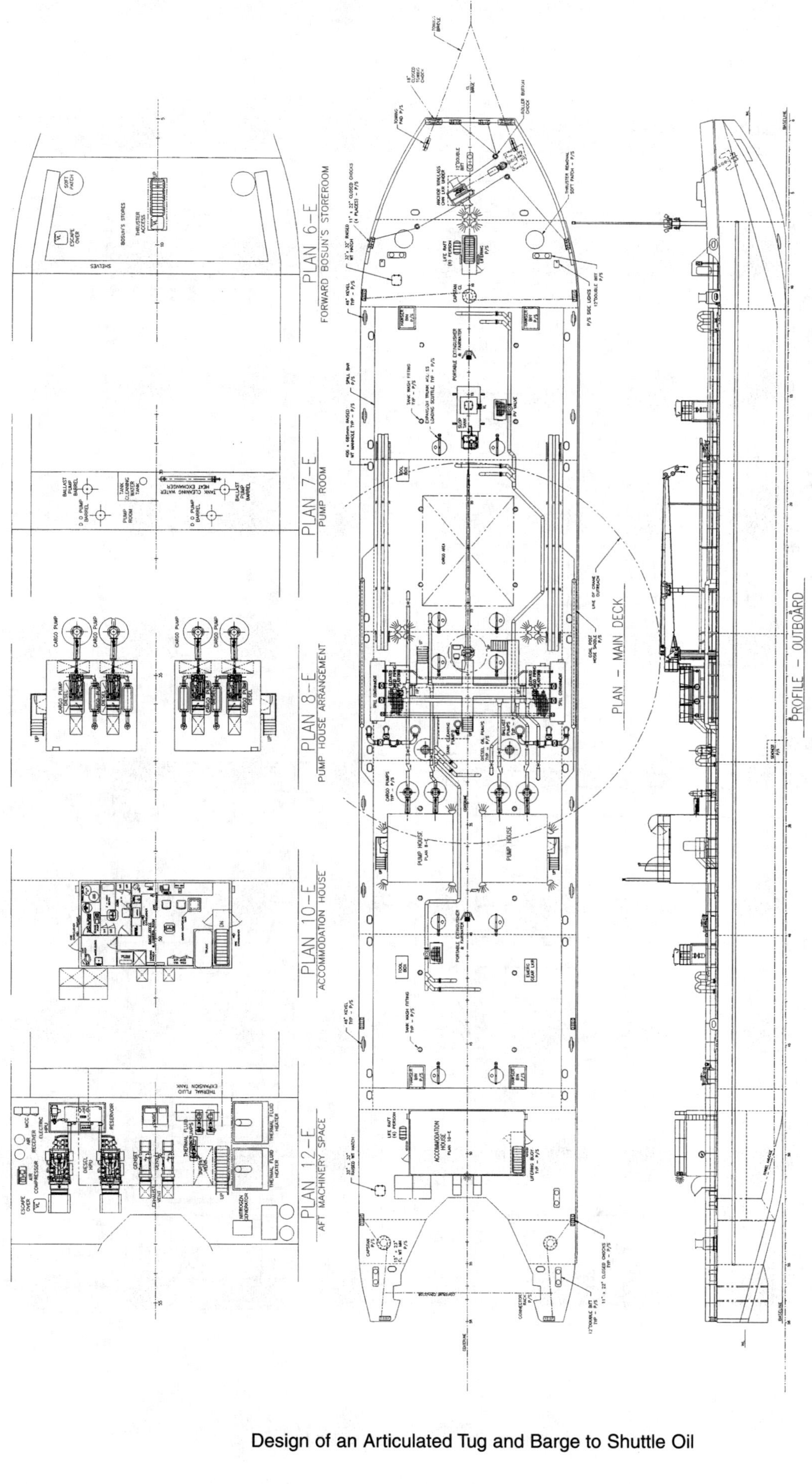

Figure 2 - Barge Arrangement

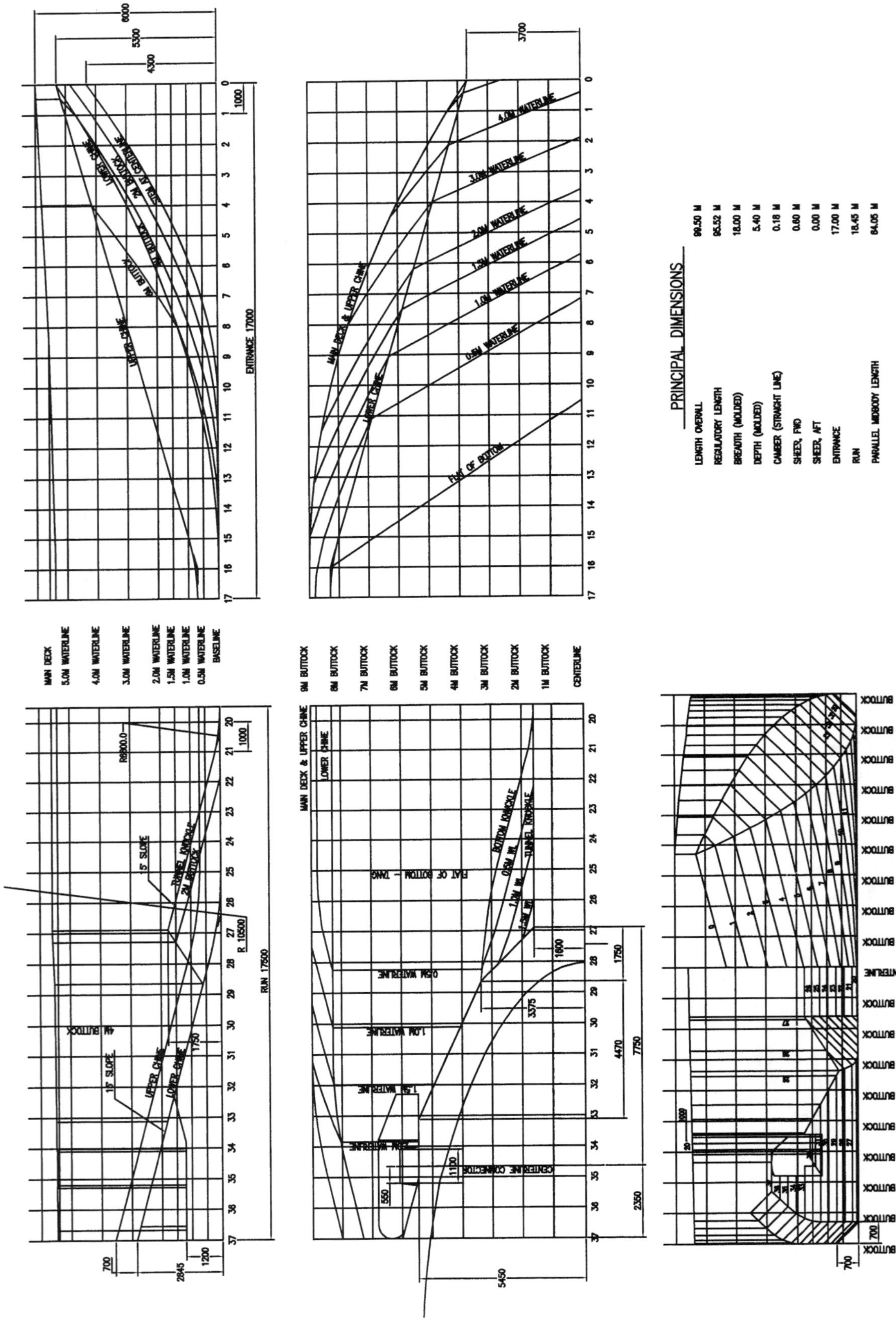

PRINCIPAL DIMENSIONS

LENGTH OVERALL	99.50 M
REGULATORY LENGTH	95.52 M
BREADTH (MOLDED)	18.00 M
DEPTH (MOLDED)	5.40 M
CAMBER (STRAIGHT LINE)	0.18 M
SHEER, FWD	0.60 M
SHEER, AFT	0.00 M
ENTRANCE	17.00 M
RUN	18.45 M
PARALLEL MIDBODY LENGTH	64.05 M

Figure 3 - Barge Lines

which will cause a slight curvature to the shell plate. This will stiffen the plating that is subjected to ice loads.

The bow is of "spoon" shape to reduce resistance and allow operation in ice.

The stern has a tunnel at centerline to reduce the pressure drag normally experienced by a deep notch for the tug. A skeg is fitted on each side of the notch to provide support to the Articouple rack when the barge is at light draft.

Sea worthiness is enhanced by adding a small amount of sheer to the bow. An unmanned barge does not need to meet the loadline requirements for minimum bow height.

The barge lines are shown in figure 3.

2.2 Barge Capacity

The barge has four (4) main cargo tanks, port and starboard for a total of eight (8) tanks for crude oil. The total crude oil capacity at 95% full is 21,049 barrels.

The barge is also intended to carry No. 2 diesel oil (MDO) back to the platform as cargo. The number 3 diesel oil cargo tanks, port and starboard, near midships have a total capacity of 446.2 cubic meters at 95% full. The barge has separate tanks for fuel oil for the diesel pumps, cargo heating, generator, a slop tank and a wash water tank. The slop tank is used to drain the hoses and is discharged back to the platform. The salt water ballast capacity is 2,324 metric tonnes at 100% full. The tank arrangement and capacities are shown in figure 4.

The arrangement shows the pump room and diesel oil cargo tanks located near midship. This reduces the hull sagging moment when carrying crude oil and also reduces the hogging moment on the return trip with ballast and diesel oil cargo.

2.3 Pollution Prevention

The barge has a double hull, see Figure 5, to meet the pollution prevention requirements of MARPOL 73/78 [7]. These requirements are also invoked in U.S. Coast Guard regulations [8] and are further explained by ref. [9].

The barge is designed to meet these regulations as follows:

The minimum width of side tank for vessels with a deadweight less than 5000 tonnes from ref. [8] is w=(0.5+DWT/20,000) meters. For a deadweight of 3134 tonnes, the minimum side tank width is 0.776 meters. Actual side tank width is 1.400 meters. This results in a b_i/B ratio of 0.078 where b_i is the actual side tank width (1.400m) and B is the molded beam (18.000m). The added width is used to obtain the shallow draft in the loaded condition. It also increases the vessels stability, increases the allowable cargo tank length, and provides easier access for construction and inspection.

The minimum height of double bottom from ref. [8] is B/15 which is 1.200 meters. The actual tank top height is sloped 2/100 for drainage from 1.200 meters at centerline to 1.356 meters at the side tank bulkhead.

The maximum allowable length of cargo tank from ref. [8] for vessels having a centerline longitudinal bulkhead inside the cargo tank is L_a=[0.25(b_i/B)+0.15]L. This results in a maximum tank length of 19.10 meters for normal tank vessels. Section 12 of ref. [9] allows an equivalent length of cargo tank for tank barges of 150 meters or less in length. For center tanks with b_i/B less than 0.20, the maximum tank length for the barge is L_a=[0.25(b_i/B)+0.20]L which is 20.96 meters. Actual cargo tank lengths are 12.60 meters for the end tanks and 14.40 meters for the center tanks.

The allowable volume of a center cargo tank from ref. [8] is 50,000 cubic meters. This is far in excess of the largest cargo tank volume of 470 cubic meters.

The maximum hypothetical oil outflow from [8] is determined from an assumed extent of damage. For side damage, the assumed length of damage, l_c, is 1/3 $L^{2/3}$ which is 7.0 meters. The assumed transverse extent of damage, t_c, is B/5 which is 3.60 meters. The assumed vertical extent is from baseline upwards without limit. As the shortest length of cargo tank is 12.60 meters, only a single bulkhead can be damaged. The transverse extent, t_c, is greater that the side tank width of 1.400 meters. The factor $K= 1-b_i/t_c$ is 0.611 for each tank. For damage to cargo tanks 1 and 2, which is the same for cargo tanks 4 and 5, the outflow of oil, O_c, is $K_1C_1+K_2C_2$ equal to 511.3 cubic meters where C_1 is the #1 tank capacity of 390.5 cubic meters and C_2 is the #2 tank capacity of 446.3 cubic meters. The allowable outflow, O_A, is 30,000 cubic meters.

For bottom damage, the assumed vertical damage from [8] is B/15 which is 1.200 meters. As the double bottom is a minimum of 1.200 meters high, there is no assumed bottom damage.

2.4 Barge Machinery

An unmanned tank barge is not required to have a fixed fire fighting system according to IMO, ABS and U.S. Coast Guard. None is furnished.

A tank barge having a deadweight less than 20,000 tonnes does not require an inert gas system nor a crude oil washing system (COW) according to IMO, ABS and U.S. Coast Guard. As an option, a nitrogen generator is used to provide a blanket of inert gas to the cargo tanks as they are emptied. With nitrogen purged to the atmosphere when loading at the platform, there is no need for a vapor recovery system.

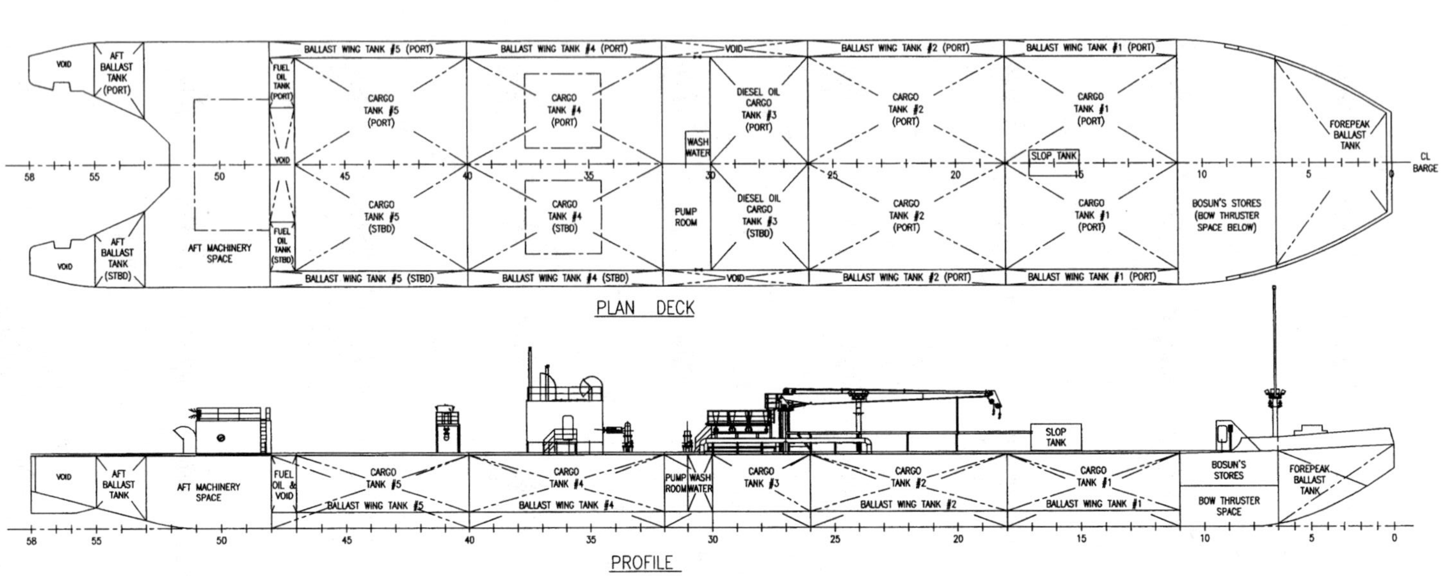

CRUDE OIL CARGO TANKS (95% FULL)						
TANK	FRAMES	VOLUME (BBL)	WEIGHT (MT)	VCG (m ABL)	LCG (m aft Fr 0)	TCG (m off CL)
CARGO #1 (port)	11-18	2,456	363.11	3.284	21.900	3.875 P
CARGO #1 (stbd)	11-18	2,456	363.11	3.284	21.900	3.875 S
CARGO #2 (port)	18-26	2,807	414.96	3.284	35.400	3.875 P
CARGO #2 (stbd)	18-26	2,807	414.96	3.284	35.400	3.875 S
CARGO #4 (port)	32-40	2,807	414.96	3.284	60.600	3.875 P
CARGO #4 (stbd)	32-40	2,807	414.96	3.284	60.600	3.875 S
CARGO #5 (port)	40-47	2,456	363.09	3.284	74.100	3.875 P
CARGO #5 (stbd)	40-47	2,456	363.09	3.284	74.100	3.875 S
TOTAL		21,049	3,112.24	3.284	48.000	0.000

DIESEL OIL CARGO TANKS (95% FULL)						
TANK	FRAMES	VOLUME (m^3)	WEIGHT (MT)	VCG (m ABL)	LCG (m aft Fr 0)	TCG (m off CL)
CARGO #3 (port)	26-30	223.1	194.09	3.284	46.200	3.875 P
CARGO #3 (stbd)	26-30	223.1	194.09	3.284	46.200	3.875 S
TOTAL		446.2	388.18	3.284	46.200	0.000

BALLAST TANKS (100% FULL)						
TANK	FRAMES	VOLUME (m^3)	WEIGHT (MT)	VCG (m ABL)	LCG (m aft Fr 0)	TCG (m off CL)
FOREPEAK BALLAST	0-7	338.9	347.40	3.912	5.330	0.000
BALLAST WING #1 (port)	11-18	201.2	206.26	1.474	21.902	5.676 P
BALLAST WING #1 (stbd)	11-18	201.2	206.26	1.474	21.902	5.676 S
BALLAST WING #2 (port)	18-26	230.1	235.82	1.474	35.400	5.677 P
BALLAST WING #2 (stbd)	18-26	230.1	235.82	1.474	35.400	5.677 S
BALLAST WING #4 (port)	32-40	230.1	235.82	1.474	60.600	5.677 P
BALLAST WING #4 (stbd)	32-40	230.1	235.82	1.474	60.600	5.677 S
BALLAST WING #5 (port)	40-48	230.1	235.82	1.474	75.000	5.677 P
BALLAST WING #5 (stbd)	40-48	230.1	235.82	1.474	75.000	5.677 S
AFT BALLAST (port)	53-55	73.0	74.82	3.275	92.825	6.538 P
AFT BALLAST (stbd)	53-55	73.0	74.82	3.275	92.825	6.538 S
TOTAL		2,267.9	2,324.48	1.954	45.355	0.000

MISCELLANEOUS TANKS (95% FULL)						
TANK	FRAMES	VOLUME (m^3)	WEIGHT (MT)	VCG (m ABL)	LCG (m aft Fr 0)	TCG (m off CL)
FUEL OIL (port)	47-48	25.1	21.84	3.286	81.300	5.995 P
FUEL OIL (stbd)	47-48	25.1	21.84	3.286	81.300	5.995 S
SLOP TANK	15-17	9.9	9.39	6.600	24.600	0.000
WASH WATER	30-31	18.5	18.99	3.390	50.700	1.196

SPECIFIC GRAVITY	
CRUDE OIL	0.93
DIESEL OIL	0.87
BALLAST WATER	1.025
SLOP	0.95

Figure 4 - Barge Capacity Plan

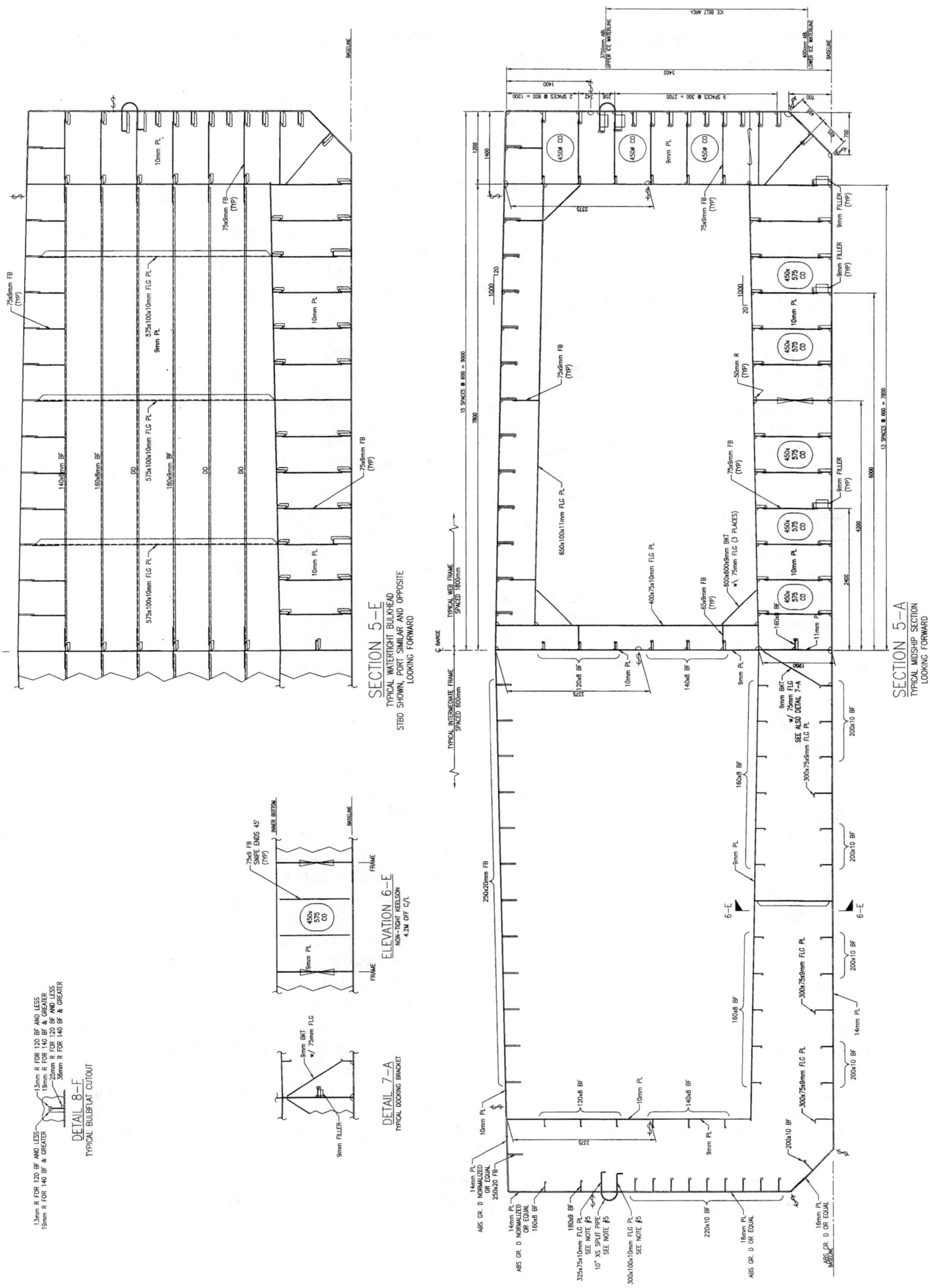

Figure 5 - Barge Midship Section

In lieu of a crude oil washing system (COW), the tanks may be washed by hot water utilizing the thermal oil heaters and a fixed line with hose connecting to a portable cleaning gear.

The barge has four (4) barrel type cargo oil pumps, each driven by a 340 KW diesel engine located in houses on the main deck.

Cargo oil is heated to 60°c by hot oil utilizing one of the two (2) thermal oil heaters located below deck, aft.

The thermal oil heaters are also used to heat the fresh water to 60°c used to clean the cargo oil tanks. It is not intended to wash the cargo tanks after each trip, but rather to selectively wash the cargo tanks for inspection and reduce build-up of wax. The oily wash water is retained in the cargo tank and discharged back to the platform while loading.

Two (2) Schottel Pump Jet bow thrusters, model SPJ57T, are located forward and outboard along the deadrise so as to be above the bottom for operation in shallow water. Each thruster is powered by a 185 KW hydraulic motor. The thrusters are controlled locally from the deck of the barge and remotely from the pilot house of the tug by means of an umbilical cord.

Two (2) 85 KW diesel generators are located in the aft auxiliary machinery space.

Two (2) auxiliary diesel engines are located in the aft auxiliary machinery space, each driving multiple hydraulic pumps mounted on a gear. The hydraulic system also powers the following equipment:

- Anchor Windlass
- Deck Crane
- Mooring Capstans
- Diesel Oil Cargo Pumps
- Ballast Pumps
- Tank Cleaning Pump
- Nitrogen Generators

An electric hydraulic power unit is also used to furnish limited hydraulic power to the smaller horsepower units in lieu of operating the larger diesels at smaller loads.

2.5 Barge Outfit

The barge has accommodations for two (2) persons while moored at the platform. The barge is "unmanned" at sea.

The barge has a single bower anchor at 1 ½ times the weight of each bower anchor required for equipment numeral class. The optional circular E notation for class is not obtained.

Three (3) powered Mooring capstans are provided for ease in mooring the vessel.

A towline is rigged from the bow leading along the vessel's side to the aft end. The towline is held by break-away clips. The towline can be connected to the tug towline before the tug disengages from the barge.

3. TUG DESIGN

3.1 Tug Characteristics

The tug shall be capable of a dual-mode operation, ie., both pushing and towing; as defined by U.S. Coast Guard [6]. The tug arrangement is shown in Figure 6.

The principal characteristics are shown in Table 3.

Characteristic	Value
• Length, overall	30.60 m
• Length, loadline	29.40 m
• Length, class	28.862 m
• Beam, molded	10.80 m
• Depth, at side, molded	4.50 m
• Draft, DWL	3.10 m
• Camber, Main & Focs'le	0.10 m
• Horsepower, Continuous	3500 BHP
• Crew	12
• Accomodations, total	14
• Light Ship Weight	441.9 MT
• Displacement at DWL	575.7 MT
• Tonnes per Centimeter	2.90
• Fuel Capacity @ 95%	52.83 MT
• Fresh Water Capacity	28.08 MT
• Block Coefficient	0.612
• Prismatic Coefficient	0.730
• Waterplane Coefficient	0.912

Coefficients based on L of 29.4 meters

Table 3: Tug Characteristics

3.2 Tug Hull Form

Like the barge, a multiple chine hull is selected for ease of construction. It is intended that the bow lines in the ice belt be faired as a conical development, which will cause a slight curative to the transverse frames, which will stiffen the plating and reduce the "starved horse" appearance common with straight frames.

The stem is cut away at approximately 39 degrees from horizontal for breaking ice.

The sides are sloped about 14 degrees from the vertical for ease of operation in ice.

A structural bolster is located in way of the Articouple connector. The bow profile above the main deck is raked to insure adequate clearance in the barge notch with the tug pitched at extreme angles.

Twin rudders are used behind each Kort nozzle in order to obtain the rudder area necessary to turn the AT/B unit and still maintain a high aspect ratio foil for high efficiency. This also lessens the chord length when compared to a single rudder for each screw allowing the propeller to be located further aft and decreasing the angle of the aft buttocks, which is favorable for good

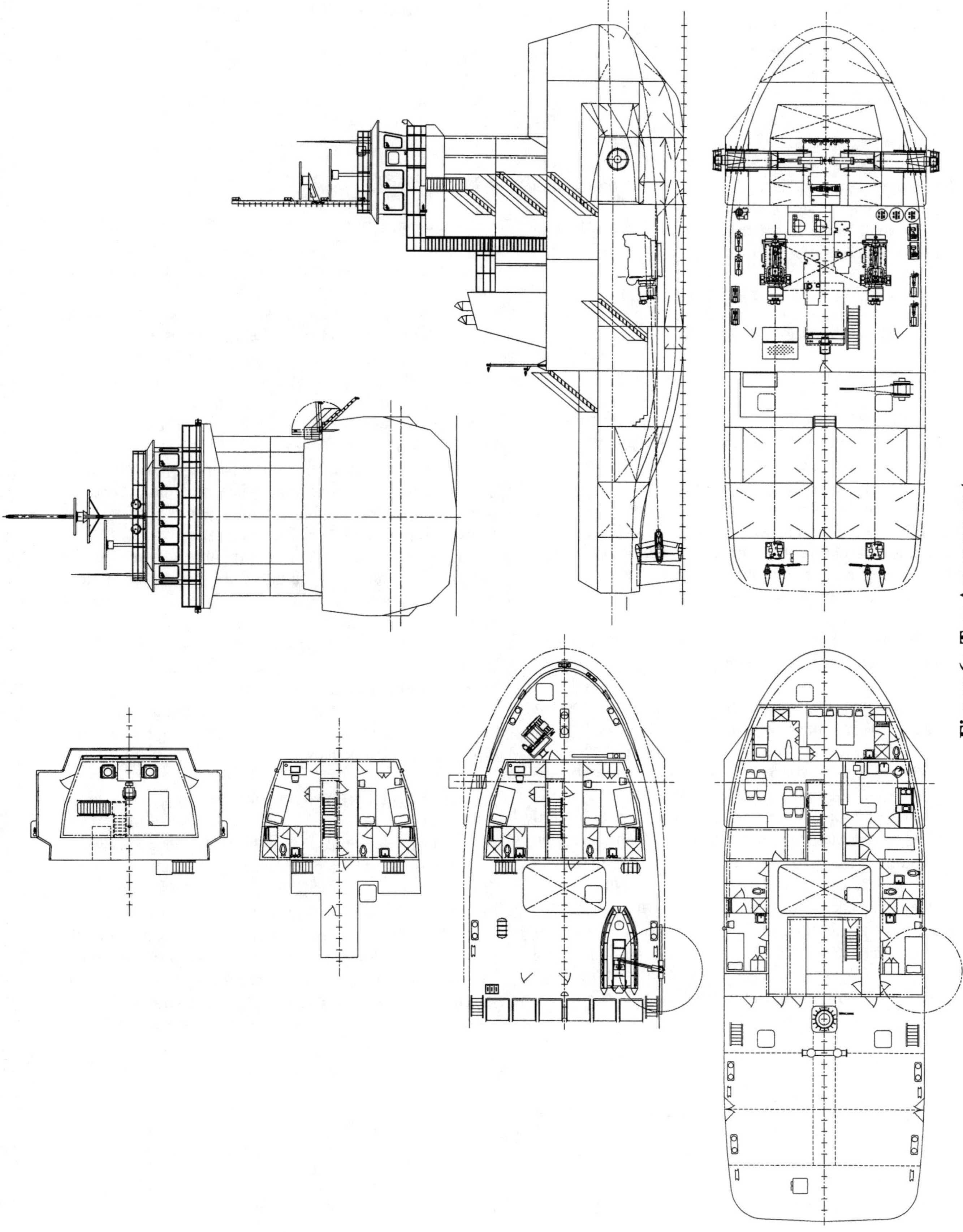

Figure 6 - Tug Arrangement

water flow to the propellers. The total rudder area is about 2.30 percent of the total underwater lateral plane in the push mode.

The Tug Lines are shown on Figure 7.

The estimated geometric freeboard assigned to the tug is about 0.30 meters. The tug will be limited to an operating draft of 3.2 meters by scantling draft.

The tug pilot house is raised to satisfy the navigation bridge visibility requirements of ABS while in the push mode with the barge in ballast condition. These rules are the same as IMO and U.S. Coast Guard requirements for bridge visibility.

The intact stability of the tug meets the following criteria in all conditions of loading.

- Weather Criteria (46 CFR 170.160 & 170.170)
- Dynamic Criteria (46 CFR 170.173)
- Towing Criteria (46 CFR 173.095)

3.3 Tug Machinery

With a shallow draft tug, twin propulsion units offer a larger propeller diameter per horsepower than single screw. Propulsion arrangements considered open propellers, propellers in Kort nozzles and Z drives.

Kort nozzles were selected as offering the greatest protection to the system when grounded. Z drives were felt to be vulnerable to damage in shallow water and when resting on the bottom. A geared diesel of 1750 HP driving a fixed pitch propeller is felt to be adequate and cost effective.

The tug has two (2) ship's service generators of 240 KW each and a single emergency generator of 120 KW.

Engine cooling arrangement presented another challenge. When operating in ice, sea chests tend to become plugged with ice. ABS rules for ice class A1 through A5 requires the raw water cooling to be re-circulated. While not required for ice class A0, the engine room double bottom is fitted for re-circulation of cooling water when operating in ice. When sitting on the bottom at the platform, the problem may not be ice but mud. Re-circulating the cooling water in the summer may cause the water to be overheated. In this case, water to air heat exchanges mounted on the aft Focs'le deck are utilized for engine cooling. This redundancy offers three means of cooling for the main propulsion diesels and diesel generators.

The tug has a sea chest on each side between the upper and lower chine leading to a common sea bay. The sea bay supplies raw water to the following:

- Fire Pump
- Fire and General Service Pump
- Sanitary Water Pump
- Ballast System
- A/C Cooling
- Engine Cooling (alternate)

3.4 Tug Outfit

An electric powered capstan is located on the main deck aft along with a double bitt for towing. The towline is stowed on a reel below deck and is lead to the capstan by fairleads.

A single bow anchor is stored on the Fosc'le deck and deployed by a davit. The anchor is rigged to a winch on the Fosc'le deck at time of deployment.

A similar anchor is stored on the main deck aft of the Fosc'le and is deployed by the davit on the 01 deck. The anchor can be rigged to the towline and deployed by the capstan.

The tug meets all SOLAS requirements with inflatable life rafts to accommodate the total number of persons on each side plus a rigid inflatable rescue boat (RIB) equipped to SOLAS requirements.

Fresh (potable) water is available at the Port of Tanggu. The tug was originally designed with a flow through sewage treatment system using a class II marine sanitation device. Discussions with the Chinese informed us that the vessels' route was in a "zero discharge" zone and the tug was designed with holding tanks for both black and grey water.

Taking on fuel and water and discharging sewage and garbage occurs at the Port of Tanggu while the crude oil is being discharged.

4. ACKNOWLEDGEMENTS

I wish to thank Mr. Hugh L. Pryce, Project Director of Apache China Corporation for his support of our design team at Bay Engineering, Inc. I also wish to thank Mr. Neville Jones of Apache Energy, Ltd, Australia, for his valued input into the design details and Mr. Alan Wang of Kellogg, Brown & Root for his enthusiasm and support, especially in International relations with the Chinese participants on the project.

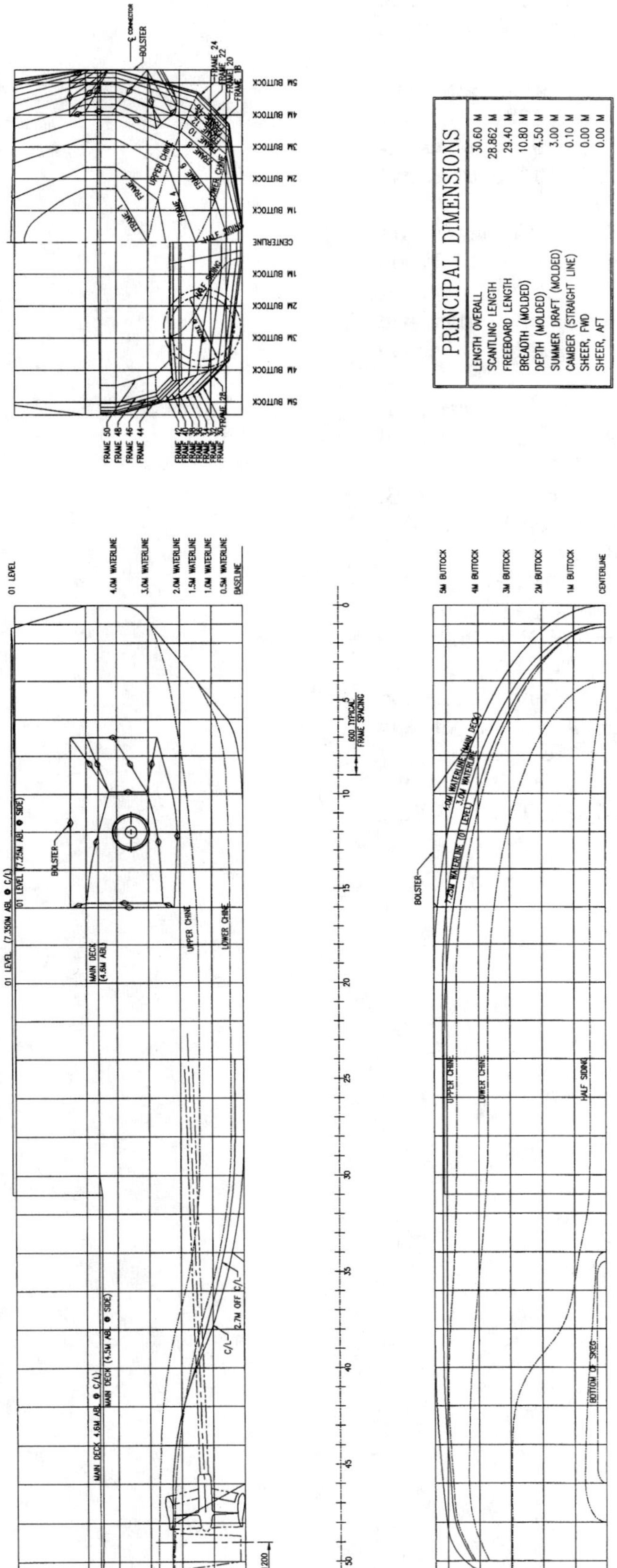

PRINCIPAL DIMENSIONS	
LENGTH OVERALL	30.60 M
SCANTLING LENGTH	28.862 M
FREEBOARD LENGTH	29.40 M
BREADTH (MOLDED)	10.80 M
DEPTH (MOLDED)	4.50 M
SUMMER DRAFT (MOLDED)	3.00 M
CAMBER (STRAIGHT LINE)	0.10 M
SHEER, FWD	0.00 M
SHEER, AFT	0.00 M

Figure 7 - Tug Lines

5. REFERENCES

[1] Yamaguchi, T. "Means of Connecting Pusher Boat and Barge" U.S. Patent 3,844,245. Oct 29, 1974, assigned to Taisei Engineering Consultants, Inc.

[2] Yamaguchi, T. "Twelve Years Experience with Articouple Pusher-Barge System", 1984.

[3] Yamaguchi, T. "Utilization of Articouple-Pushers at Marine Construction Sites" 10th International Tug Convention, Sydney 1988.

[4] Hill, R.P. and Kuhlman, C.J. "Apparatus for Coupling Tugboats to Barges", U.S. Patent 4,688,507, Aug 25, 1987.

[5] Fischer, J.P. "Design Notes on Articulated Tug/Barge Connections", GL&GR Section SNAME, Oct 29, 1999

[6] U.S. Coast Guard, NVIC 2-81, Change 1, "Inspection Guidance Regarding Integrated Tug/Barge Combinations".

[7] MARPOL 73/78 Convention, Regulations 22,23, and 24 of Annex 1 of the International Convention for the Prevention of the Pollution from ships.

[8] US Code of Federal Regulations, 33 CFR Part 157, Rules for the Protection of the Marine Environment Relating to Tank Vessels Carrying Oil in Bulk.

[9] U.S. Coast Guard, NVIC 6/94, Ch 1, "Guidance for Issuing International Oil Pollution Prevention (IOPP) Certificates Under Annex 1 of the International Convention for the Prevention of Pollution from Ships, 1973, as Modified by the Protocal of 1978 Relating There To (MARPOL 7/78)

Aftbody Slamming and Whipping Loads

G.K. Kapsenberg (V)[1], A.P. van 't Veer (V)[1], J.P. Hackett (LM)[2], and M.M.D. Levadou (V)[1]

[1] Maritime Research Institute Netherlands,
[2] Northrop Grumman Ship Systems

ABSTRACT

This paper describes a method of measuring aftbody slamming loads on a ship model in the seakeeping basin. The method uses a large number of small pressure transducers mounted on the flat counter of the stern. The impact force is obtained by integrating the pressures. Experiments have been carried out on a scaled model of a modern cruise ship. The model was split into two segments at midships to simulate a simplified 2-node vibration mode. A structural mathematical model was made of the model as used for the tests. The analysis shows that the measured pressures can be used to predict the whipping loads on the segmented model. A fatigue analysis was made for the vertical bending moment midships. It appears that for a following sea case the potential for fatigue damage is dominated by the whipping contribution. The extreme vertical hull bending moment is determined by sailing in high head seas. The results of the experiments show that the whipping component only contributes a small percent increase to the extreme vertical bending moment. It is concluded that the measured loads can be used with a finite element model of the ship's structure to predict whipping stresses in the actual ship structure.

SYMBOLS

A_i	[m^2]	Area
AZ	[m/s^2]	Vertical acceleration
B_H	[Nm]	Damping of spring
c_{ij}		Hydrostatic restoring coefficient
C_H	[Nm]	Spring stiffness
FE	[Pa^3]	Fatigue effect
FZ_{SLAM}	[kN]	Vertical slam force
HF		High frequent part (due to whipping) on a signal
H_S	[m]	Significant Wave Height
I_{ij}	[m^4]	Moment of Inertia
k_{yy}	[m]	Radius of Gyration
M	[kg]	Mass
MY	[kNm]	Vertical Bending Moment
N	[-]	Number of classes of the rainflow count
n_i	[-]	The number of load cycles in class i
P	[kPa]	Pressure
T_P	[s]	Peak period of wave spectrum
V	[m/s]	Velocity
WF		Wave frequent part of a signal
x_{SLAM}	[m]	Longitudinal distance of center of effort of slam force to midships
z_i	[m]	Heave motion
θ_i	[deg]	Pitch motion
κ	[-]	Tuned non-dimensional damping coefficient
σ_i	[Pa]	The mean stress of class i
ω_n	[rad/s]	Frequency of encounter

INTRODUCTION

The goal of every commercial ship owner is to maximize the revenue producing portion of the vessel and minimize the non-revenue producing portion such as the engine room. A "shoe box" with no propulsion plant, ballast tanks, etc. would be considered the ideal solution. As a result, over the years an increasing demand was placed on design naval architects to produce ship designs with full beam main decks along almost the entire length of the ship. For containerships this allows more containers to be stowed on the main deck. Likewise, for cruise ships, a full width main deck allows a full width superstructure that contains more passenger cabins. Competition is now so intense that this trend has expanded to include filling out the aft portions of the hull between the full load waterline and main deck. This added hull volume means more container stowage capacity for container ships, more space for passengers on cruise ships, and a full breadth vehicle deck for more vehicle stowage and/or a vehicle ramp on ferries. Owners rationalize that as long as the filling out of the hull occurs above the full load waterline, there should be no negative impact on calm water powering and fuel economy.

This design philosophy has produced hull forms with progressively flatter and flatter sterns. Some ships have been built with zero or near zero deadrise and buttock angles near the transom. Such shapes have hydrodynamic benefits in calm water as demonstrated by Hämäläinen and van Heerd (1998); however, they are a far cry from the traditional cruiser sterns seen on Mariner hull forms and liners like the *SS United States*.

At the same time that this hull form development was occurring, a new propulsion system was being developed. The idea was to move the electric propulsion motor from inside the hull to an azimuthing pod located external to the hull. This allows the elimination of the propeller shaft and shaft struts, the rudder, and stern tunnel thrusters. This concept provides greater flexibility for the design engineer to locate elements of the propulsion system; hence, an ability to provide additional internal volume to the revenue producing portion of the vessel. It also produces 10 to 15% improvement in calm water powering and significant improvements in vessel harbor maneuverability. The azimuthing pods encouraged design naval architects to create stern designs that were even flatter than their conventional screw propelled sister ships to simplify the pod installation. One vessel's counter approached a flat horizontal surface in way of the pods.

The first ship equipped with podded propulsion was the *Seili*, an ice going supply vessel built by Kværner Masa in 1991. The first application to a cruise ship was in 1997, the *Elation*. Since then, podded propulsion has become the industry standard on cruise ships. The French Navy has now ordered a large LHD type amphibious assault ship outfitted with twin pods, the first naval application for podded propulsion.

SLAMMING AND VIBRATION PROBLEM

It was not long before a serious drawback of the flatter stern design surfaced. Some of the new cruise ship designs began to report slamming and subsequent vibration problems in following seas at zero and low ship speeds. Yet ferries did not report these problems. While the hull forms of cruise ships and ferries are much the same, the operating profiles for the ships are quite different. Ferries sail at low speeds inside the harbor, and then come up to cruising speed and remain there until they reach their destination. Cruise ships spend large amounts of time operating at low speeds, even in the open seas. The reported slamming and vibration problem occurred in very mild sea states when the waves had a length on the order of the length of the ship. In the process of investigating the zero speed following sea stern slamming and vibration problem, questions began to arise with respect to possible hull girder fatigue life. Later issues of ultimate hull girder longitudinal strength surfaced. These issues brought together the sciences of hydrodynamics and structural engineering.

SUBJECT VESSEL

The cruise ship reported on herein is a typical state-of-the-art design with a twin podded propulsion system. The particulars of this 1,900 passenger cruise ship are in Table 1. Its hull form is a variation of a proven design. Figure 1 shows the body plan with pod. Although the stern of the subject vessel is somewhat flat, it has considerably more shape aft than other vessels that have reported stern slamming problems. Note that the body plan shows a very reasonable buttock rise moving aft, and section curvature near the skeg. The minimum deadrise angle exceeds 3 degrees.

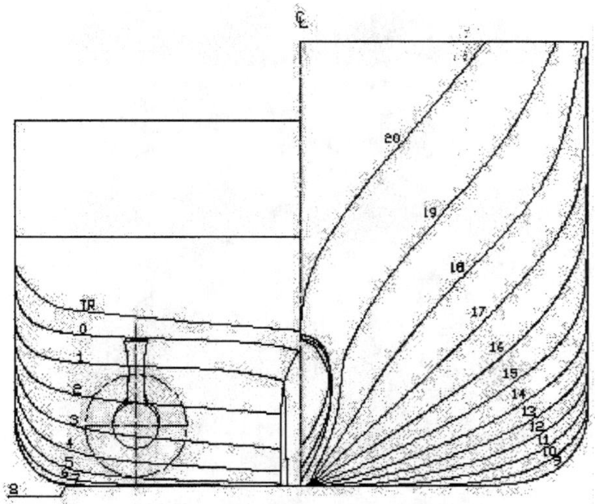

Figure 1 Profile and body plan of the cruise ship.

Prior to performing the special stern slamming test program reported here, a conventional seakeeping model test program was conducted. This program included measuring motions, accelerations, deck wetness, bow and stern emergence, and slamming. In order to investigate slamming, the model was instrumented with three pressure panels on the bow flare, one on the stem and one on the flatter portion of stern near the transom. The size of each panel was 50 mm x 50 mm. The forces recorded from these panels were representative of the local structural loads imposed on the ship. All of these test results were within acceptable limits. The purpose of the special tests was to determine the global structural loading and response. Specifically, it was to confirm if sufficient aft shape was present and if sufficient structural rigidity and strength existed to avoid the vibration and structural problems present in those ships that encountered severe slamming.

Length between perp.	233.00	m
Beam	32.20	m
Draft	8.00	m
Displacement	39000	ton
Design speed	21	kts
GM	2.10	m

Table 1 Main dimensions of the cruise ship.

EXPERIMENTAL METHOD

In general, impacts due to slamming contain energy in high frequency bands (1-20 Hz). At very low deadrise angles (below 5 deg), the high frequency content increases. Also, hydro-elastic effects might be important for these cases, as reported by Faltinsen (1996) and Bereznitski (2002).

If the high frequency component of the slamming pressure is important, it needs to be accurately measured. This means that the experimental set-up needs to have a high resonance frequency, at least twice as high as the frequency content of the slamming signal. In addition to this, underwater video recordings of the aftbody of the model slamming in waves, Figure 2, revealed a large number of air bubbles underneath the stern during the impact. This indicates that there are strong local effects.

The conventional test methodology is to isolate the aftbody of the model from the rest of the model. The aftbody is then connected to the main portion of the model in such a way as to allow a system of strain gauges to be installed along the joint. Such a system will have a relatively low resonance frequency and will hence be insensitive to high frequency contents in the pressure signal. The measured force will be the result of the dynamics of the aftbody connection to the hull. Deriving the impact forces on the stern is not straightforward. Therefore, this testing method was disregarded.

A large array of 33 pressure sensors was used to allow adequate definition of the impact, capturing both the high frequency content of the input and the local effects. The array used is illustrated in Figures 3 and 4. This set-up allowed direct measurement of pressures including the high frequency content in the signal, without the dynamics of a strain gauge system.

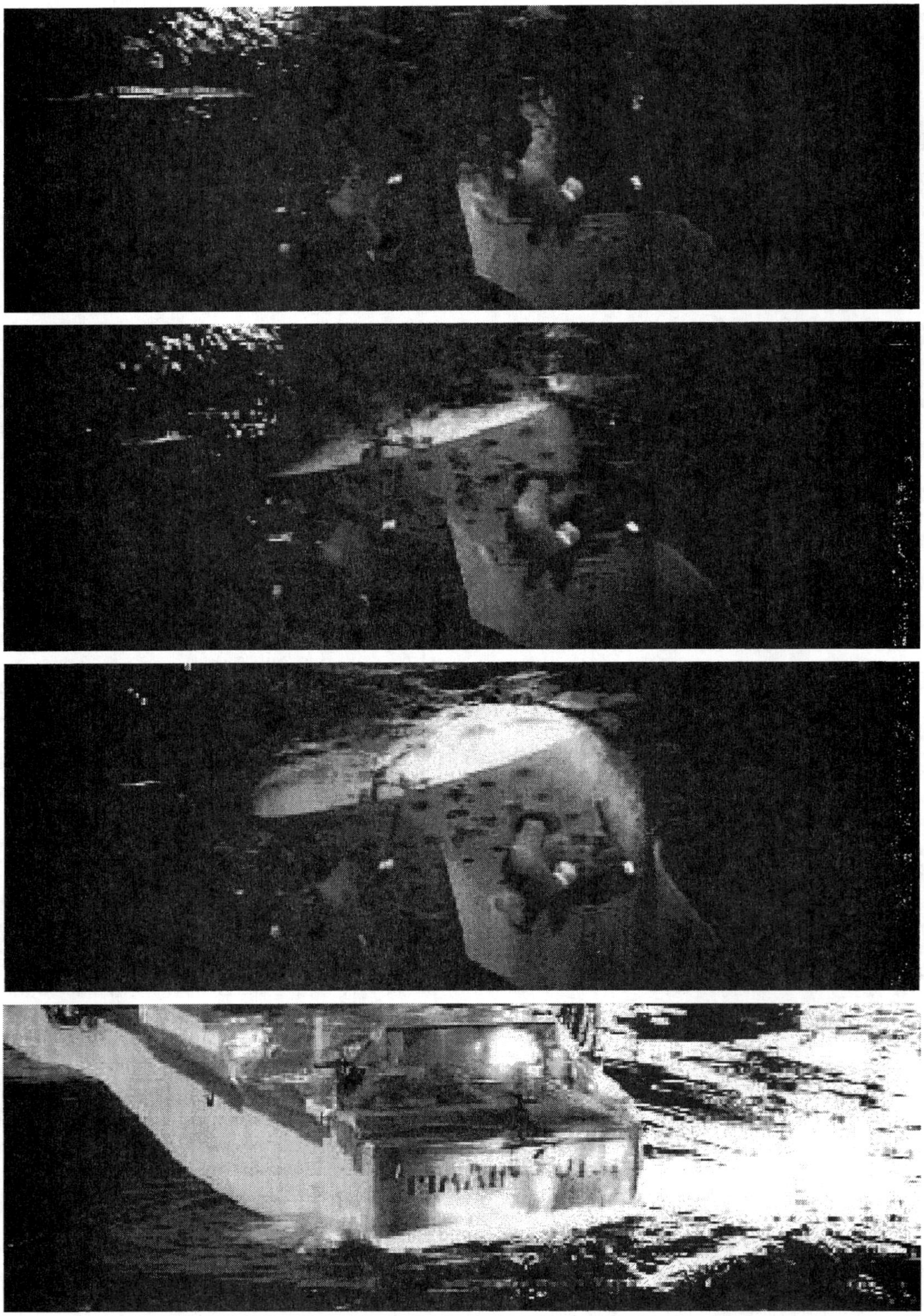

Figure 2 Sequence of pictures from a video showing one slamming event in following waves.

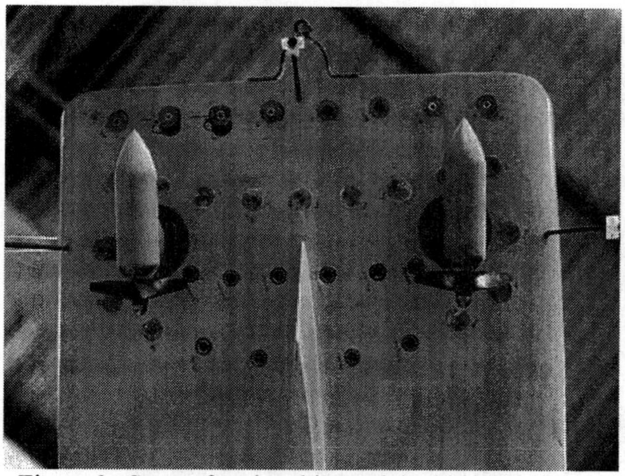

Figure 3 Stern of cruise ship model with podded propulsion and pressure gauges.

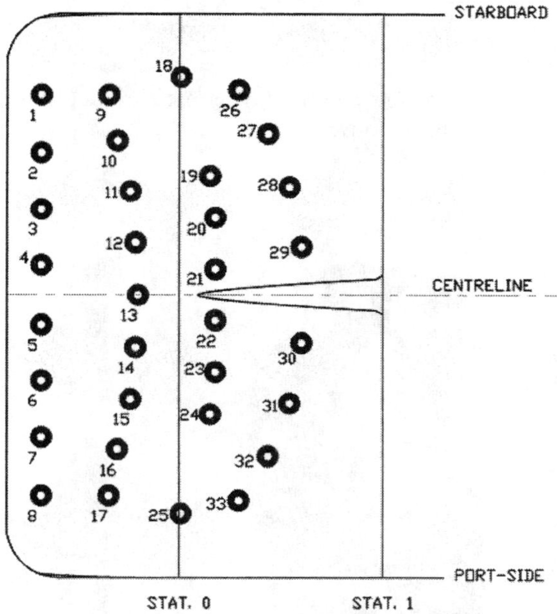

Figure 4 Location and identification numbers for the pressure transducers model.

It is important with this set-up that all components in the model are as rigid as possible. This essentially rules out measuring the longitudinal stresses at midships by means of a strain gauge system, just as a segmented stern model measuring the total impact force is not considered acceptable, as discussed before. The whipping component of the vertical bending moment is one of the important consequences of aftbody slamming, so a way to measure it was developed. A two step approach was taken:

1. Measure the impact loads (pressures) on a very rigid model.
2. Measure the whipping loads on a model cut in half while the frequency of the 2-node deformation is scaled from model to ship.

The pressures at the stern were measured for both the flexible and rigid models. The assumption was that the total pressure would be the sum of the impact pressure and the pressure due to the deformation. If this was the case, based on the results of the model with different natural frequencies, it would mean that the hydro-elastic effects could be ignored, at least for practical purposes.

The objective of step 2 was to demonstrate that the physics of the problem were understood by being able to calculate the response of the segmented model. Using the measured pressures on the model allows the calculation of the impact forces on the stern. The impact forces are then applied to a mathematical whipping model of the tow tank model in the water.

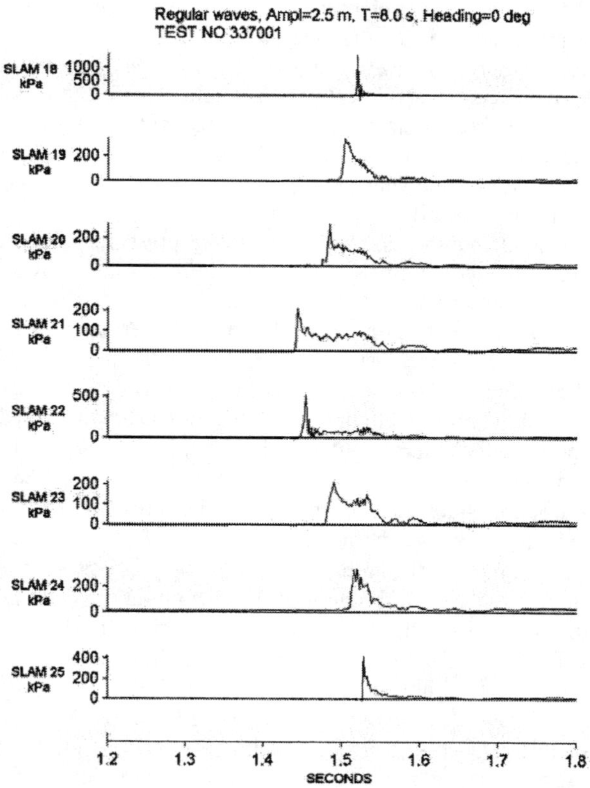

Figure 5 Recordings of pressure gauges 18-25 during a slam.

The validation then consists of the comparison of the calculated whipping moment to the measured values. Positive validation would yield a reliable tool, when used in conjunction with a detailed structural model of the ship, for determining the full scale whipping response of the ship.

ANALYSIS OF THE IMPACT PRESSURE

When a slam occurs during the model tests in waves, short duration impulses are measured on each of the pressure gauges. Figure 5 shows a time history for eight of the pressure transducers. This figure shows that there are important differences in the peak of the impact pressure (from 200 to 1000 kPa) and in the timing of the peak as a function of the different locations of the gauges. Also, the shape of the pressure pulses as a function of time can be quite different for the various transducers.

There appears to be a high-pressure ridge that travels with a rather high velocity over the aftbody of the model. The velocity of the pressure ridge was determined by using the time difference associated with the pressure peak passing the different gauges. Figure 6 shows the analysis of a slamming event. This event was selected from tests in a sea state characterized by a JONSWAP spectrum, H_S = 4 m, T_P = 8 s and a peakedness parameter $\gamma = 3.3$. The figure shows the 100 kPa isobars at different time steps. The time is shown as the small number in each isobar. The pressure gauges are indicated as small dots. The figure further shows that the impact started on starboard side, close to the aft end of the skeg, 2 m forward of the aft perpendicular (AP). The distance between the isobars is used to derive the velocity of the pressure ridge. These velocities are indicated on the figure with the associated arrow showing the direction of travel. The initial velocity of the pressure ridge is 13-14 m/s forward and to the starboard side, and 21 m/s aft. The pressure ridge velocity moving aft increases to 30 m/s after 0.3 s; athwartship it decreases to 6 m/s. The high pressure ridge covers the full length of the aftbody in 0.5 s after the first impact, then it expands mainly athwartship at a velocity of 15-20 m/s to starboard and about 45 m/s to port. The total impact takes about 1 s.

Figure 7 shows a pressure map of the second stage of this impact, 0.55 s after the initial impact, when the pressure ridge moves athwartship to both port and starboard. The peak pressure is now on the order of 100 kPa. Figure 8 shows the second impact occurring at two locations simultaneously, one on the centerline about 2 m aft of the skeg and the second close to the centerline on the starboard side. These two spots join to one large area only 0.15 s after the impact, which covers the full length of the aftbody. The high pressure area further expands to athwartship at a speed of 20 – 40 m/s. The total impact lasts about 0.5 s.

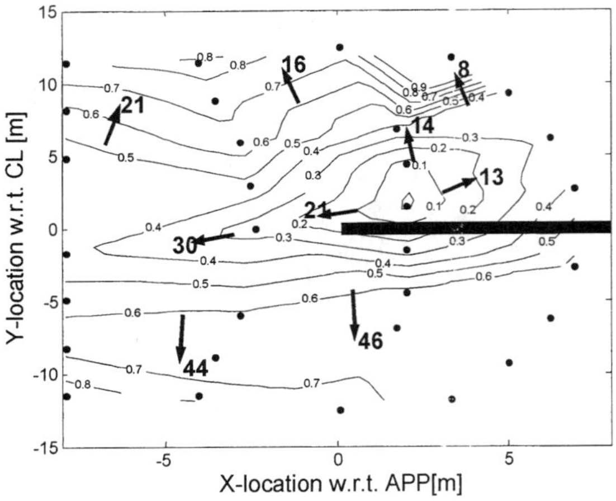

Figure 6 Contour plot of moving pressure fields over the aftbody.

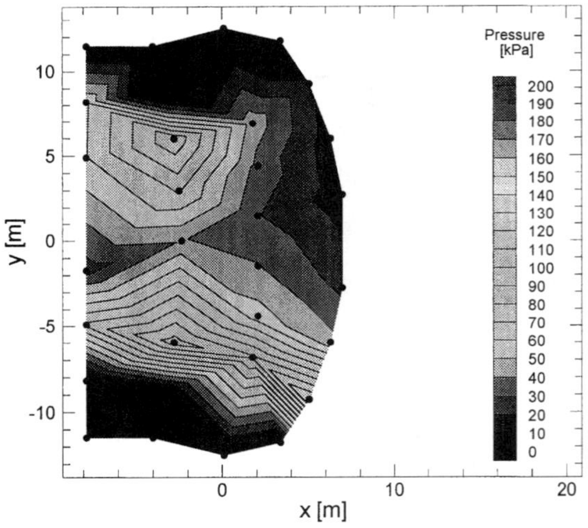

Figure 7 Pressure map, at t = 0.55 s, of the impact shown in Figure 6.

Analysis of a large number of impacts in different wave conditions shows that most impacts have a duration between 0.4 to 0.75 s. This is illustrated in Figure 9.

The typical impact was found to be a moving pressure front with peak pressures between 300 to 500 kPa in the initial stage of the impact, diminishing to 100 to 200 kPa in the second stage. The width of the high pressure ridge is about 2 – 3 m. The high pressure area grows very quickly longitudinally (velocity 30 - 40 m/s) until it covers the full length of the stern. After this it travels athwartships at a velocity of 20 – 30 m/s. This makes the total duration of the impact on the order of 0.5 s for the subject vessel.

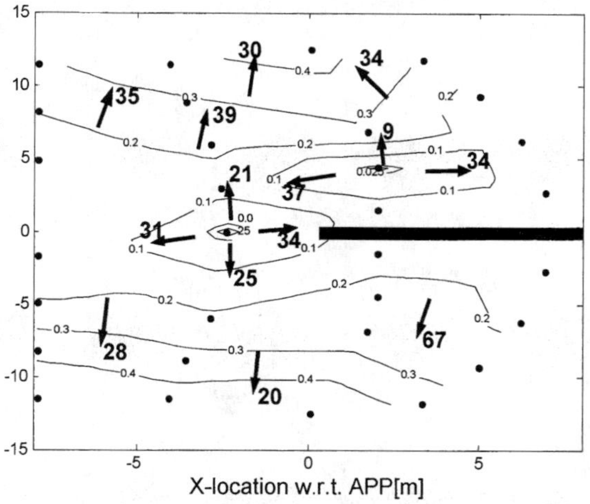

Figure 8 Contour plot of pressure field showing two locations of the initial impact.

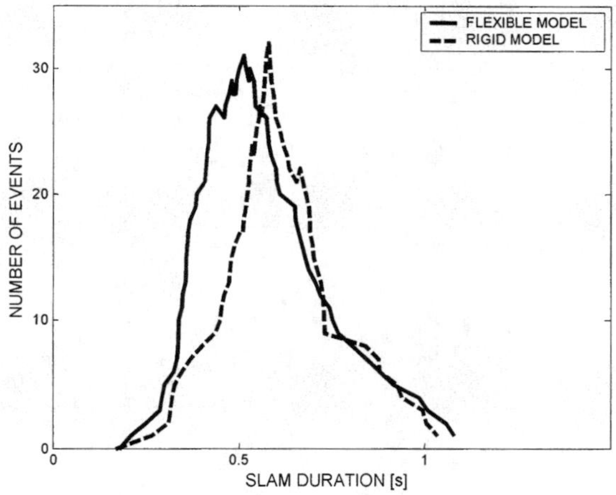

Figure 9 Statistics of the duration of the impact in a 4 m Sea State.

PRESSURE INTEGRATION

Knowing the details of the pressure distribution on the aftbody, the impact force on the aftbody as a function of time can be determined. This requires pressure integration in space. The width of the pressure ridge requires a density of pressure gauges which is far greater than practical. With the current experimental set-up it is possible that the high pressure ridge is at some instant "lost" because it is in-between the pressure transducers. Therefore, it is important to carefully select a spatial integration method of the pressure signals. A simple and robust pressure integration technique, denoted (S), was tested first. The method makes use of a so-called Delaunay triangulation as illustrated in Figure 10. This triangulation is unique; no other data points fall within the circle drawn through the corner points of each triangle. The triangles are used in the pressure integration.

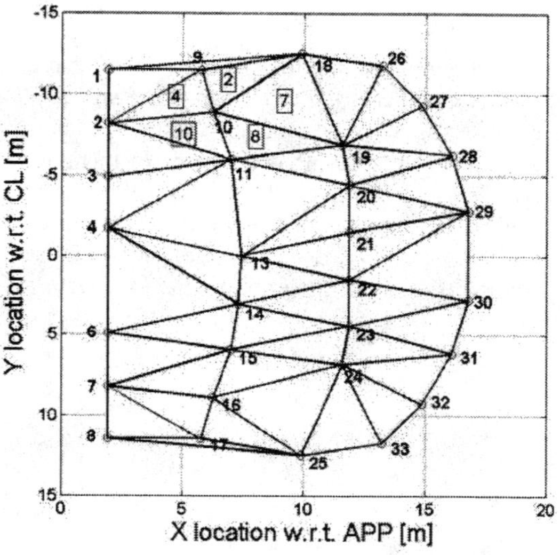

Figure 10 Delaunay triangulation of the pressure gauges.

The measured time trace of the pressure at a point is multiplied by 1/3 of the area of all surrounding panels. In formula:

$$FZ_{SLAM}(t) = \sum_{\substack{\text{pressure}\\ \text{pick-ups}}} P(t) \cdot \sum_{\substack{\text{adjacent}\\ \text{panels } i}} A_i / 3 \qquad (1)$$

This procedure uses the complete area enclosed by all the grid points (329.9 m^2). A similar procedure was used to calculate the center of effort of the impact force as a function of time. For example, the pressure measured in P10, see Figure 10, is multiplied by $^1/_3$ of the area of triangles 2-4-7-8 and 10.

The proposed integration method is robust and simple, but it requires a density of pressure gauges that is high relative to the width of the high pressure ridge. Since this is not the case with the current set-up, the resulting force, $FZ_{SLAM}(t)$, will be very spiky due to an over-estimation of the force at the moment the ridge travels over one of the pressure gauges, and an under-estimation when the ridge is just between the gauges.

To overcome the expected inaccuracy noted in the simple integration method, a more advanced pressure integration is applied. This technique, denoted as method (C), regards the pressure as a moving pressure ridge over a panel. The shape of the pressure ridge p(t) is considered constant for each panel. Since the velocity V of the pressure ridge can be determined, the time signal at each point can be transformed from the time domain to the spatial domain:

$$p(s) = p(t) \cdot V \qquad (2)$$

The direction and velocity of the moving pressure ridge is determined by the pressure-time signals at the three corner points. The velocity vector at each corner point is known, but normally only one of them actually crosses the subject panel. The pressure signal at this point is used to calculate the force on the panel in a strip-wise manner for each time step. Figure 11 illustrates this spatial integration process over a single panel.

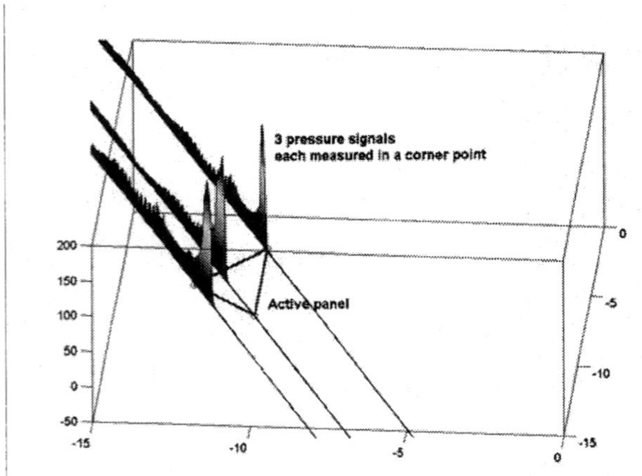

Figure 11 Impression of moving pressure front over a panel.

The results of both pressure integration methods are illustrated in Figure 12. The time trace of the signal is different, the peak of the signal is very different, but the total impulse appears to be almost identical. The impulse $\int F\,dt$ is 7278 Ns for the simplified integration method (S) and 7445 Ns for the advanced method (C). The analysis of other slams with both methods showed that the difference in impulse between the two methods is no greater than 3%.

This result was then applied to a mathematical model of the structure of the model hull as it was in the towing tank. The objective was to show that the response of the mathematical model was similar to the response of the real model in the tank. Verification of this demonstrates an "understanding" of the physics of the problem.

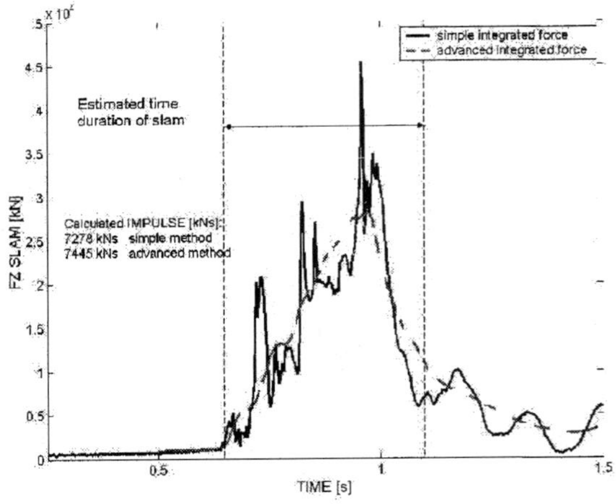

Figure 12 Result of the Simple (S) and Advanced (C) pressure integration methods for one slam.

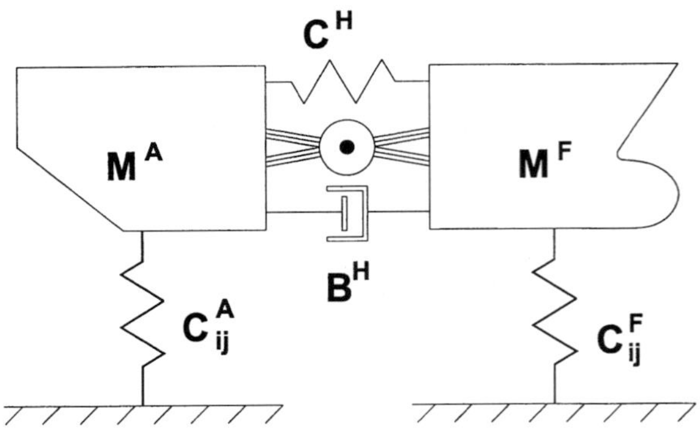

Figure 13 Dynamic model of the cruise ship in the towing tank.

MATHEMATICAL MODEL TO CALCULATE THE WHIPPING LOADS

A schematic of the model in the tank is illustrated in Figure 13. The model is represented by two masses (forward and aft hull), connected by a hinge, a spring and a damper. Both masses (hull halves) are supported by a hydrostatic spring. This spring c_{ij} consists of the coefficients c_{33}, c_{35}, c_{53} and c_{55} to take into account heave and pitch restoring forces as well as the cross coupling. Hydrodynamic damping is ignored because of the high frequencies of interest.

The equations of motions of this two-mass system are built using:

a) The pitch motion equation of the forebody (index 1) around the CG of this part of the model, and of the aftbody (index 2) around its CG. The distance from the CG of part 2 to the impact location is x_{slam}:

$$I_{yy}^F \ddot{\theta}_1 + C_{53}^F z_1 + C_{55}^F \theta_1 = -C_H(\theta_1 - \theta_2) - B_H(\dot{\theta}_1 - \dot{\theta}_2)$$
$$I_{yy}^A \ddot{\theta}_2 + C_{53}^A z_2 + C_{55}^A \theta_2 = C_H(\theta_1 - \theta_2) + B_H(\dot{\theta}_1 - \dot{\theta}_2) + FZ_{slam} \cdot x_{slam} \quad (3)$$

b) The heave motion equation of the forebody and aftbody is:

$$M^F \ddot{z}_1 + C_{33}^F z_1 + C_{35}^F \theta_1 = 0$$
$$M^A \ddot{z}_2 + C_{33}^A z_2 + C_{35}^A \theta_2 + FZ_{slam} = 0 \quad (4)$$

c) The hinge keeps the two parts connected. The distance from the local CG to the hinge is defined by x_1 and x_2:

$$z_2 - x_2\theta_2 = z_1 - x_1\theta_1 \quad (5)$$

Equation (5) is used in equation (4), after which the two equations of (3) are summed and written as one single equation. Similarly, equation (4) can be substituted in equation (3), after which the system can be written as a differential equation:

$$\mathbf{M}\frac{d\,y(t)}{dt} = F(y,t) \quad (6)$$

in which **M**, the 6x6 mass matrix, can be derived from the equations (3) through (6). The y(t) vector contains the unknown displacements and rotations and their time derivatives:

$$y(t) = (z_1, \dot{z}_1, \theta_1, \dot{\theta}_1, \theta_2, \dot{\theta}_2)^T \quad (7)$$

The right hand side vector force F(y, t) contains the restoring forces and the slamming force. The differential equations are solved using an explicit Runge-Kutta 4th order scheme.

DYNAMIC CALIBRATION

The model system is calibrated by hitting the model on the extreme aft end and calculating the response. The response in this respect is either the moment in the connecting spring or the angular deformation. The angular deformation is the deformation mode with the lowest eigen frequency. The response of this system on a triangular load, force as a function of time, was calculated, see Figure 14. The duration of the load was varied while the impulse, $\int F\,dt$, was kept constant for the different cases.

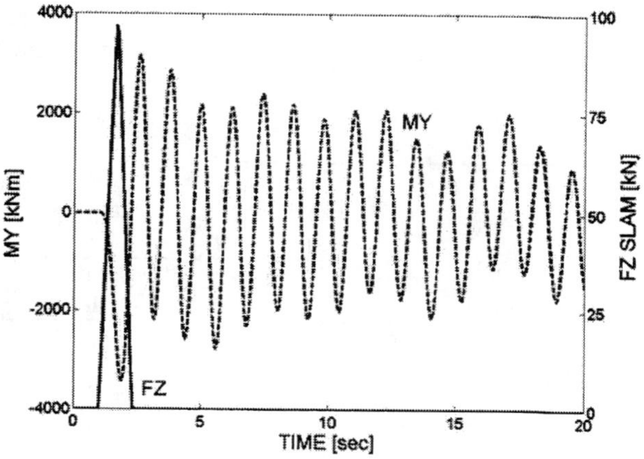

Figure 14 Whipping response of the model on a triangular impulsive load.

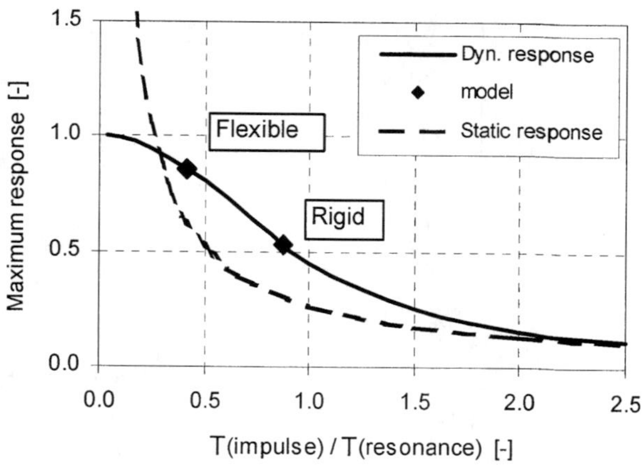

Figure 15 Response of the schematized model to an impulse at the AP.

The results of the calculations are shown in Figure 15. The duration of the impulse is normalized by the period of the angular deformation mode of the system.

This figure shows that for very short impulses it is not the peak value of the force that is important, but the impulse. When the duration of the impulse is very long, the response of the system goes asymptotically to the static case. The response of the system to a constant force is denoted as the static response in Figure 15.

The model of the cruise ship was subjected to an impulsive load when lying at zero speed in calm water. This test was performed on both the flexible model and the rigid model to determine the effect of model stiffness on response frequency. The frequency of the wet eigen mode of the model and the damping determined from the tests, are presented in Table 2. The low damping of the flexible model was expected, see Betts (1977) and Bishop (1979). The high damping of the rigid model was quite surprising. The most likely explanation is associated with the beams added to the segmented model to make it whole again. Internal friction of the screw connections of the beams to the wood model was creating frictional damping.

	Natural Frequency [Hz]	Damping [-]
Rigid model	1.75	0.049
Flexible model	0.83	0.007

Table 2. Results of model hammer tests.

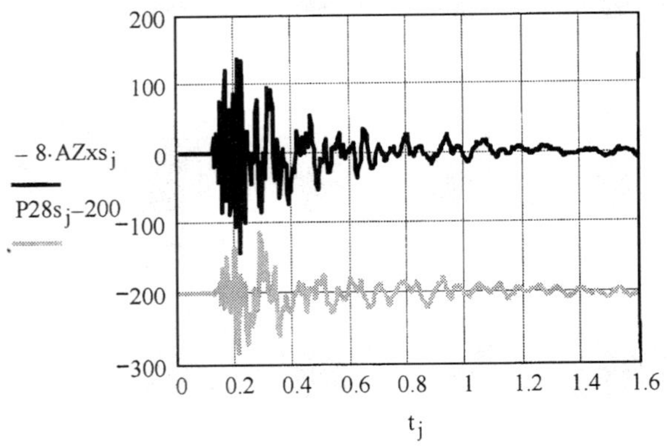

Figure 16 Response of flexible model on a hammer impact.

As was illustrated in Figure 9, the duration of the slamming impact is on the order of 0.5 s. This allows the points for the rigid and flexible models to be plotted in Figure 15. The figure shows that the response of the rigid model is not quasi-static; in fact, the rigid model was not rigid enough. The response of the flexible model was closer to a quasi-static response than that of the rigid model.

The response of the flexible model on the impulse is shown in Figure 16. This figure shows the pressure measured by gauge P28 and the local vertical acceleration AZ. The signals are shifted in the figure for clarity. High frequency local vibrations are dominant in the initial stage of the impact; these are damped in about 2 s. After this time the simplified two-node deformation mode is dominant. About 0.5 s after the initial impact, the pressure at P28 is reasonably in phase with the local vertical acceleration (times –8). This means that the measured pressure is only due to the added mass effect. A similar relation between the local vertical acceleration and the pressure was found for the rigid model.

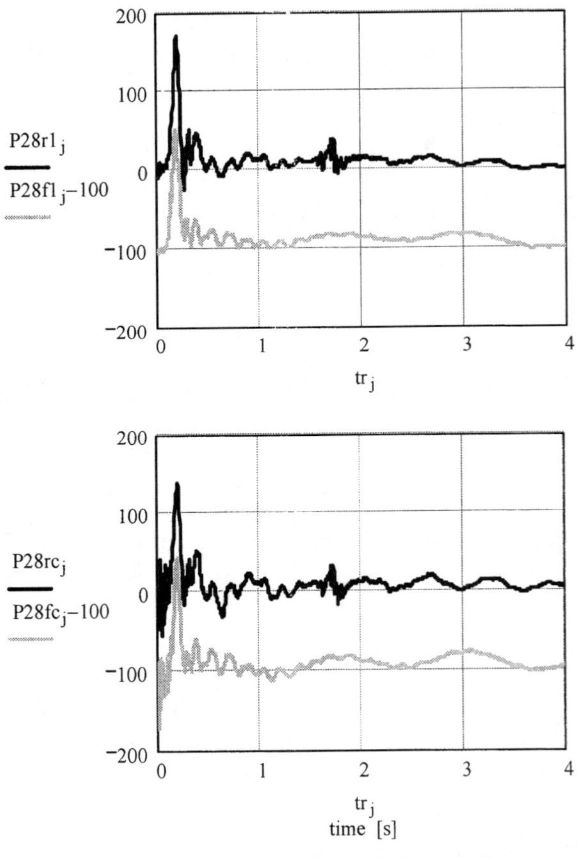

Figure 17 Pressure at gauge 28 during an impact on the flexible (P28f) and on the rigid (P28r) model. Top graph: as measured; bottom: corrected for local deformation.

This result is used in the analysis of the impacts in regular waves. The measured load is split into parts: the first due to the actual wave impact and the second due to the local deformation. The pressure due to the actual wave impact is defined:

$$P_{IMPACT} = P_{MEASURED} - P_{LOCAL\ DEFORMATION} \quad (8)$$

For P28 the pressure due to the local deformation is:

$$P_{LOCAL\ DEFORMATION} = -8AZ_{X=P28} \quad (9)$$

The results of this analysis are shown in Figure 17. The difference in the peak values is quite low, and it is reduced by correcting them for the local deformations as illustrated in the lower of the two plots. Note that the local vibrations are dominant over the two-node bending mode at the time of the impact. A comparison of the integrated pressure for the rigid and the flexible model is shown in Figure 18 (regular following waves, amplitude 2.6 m, period 8 s.)

From these results it is concluded that the hydro-structural interaction is low, and the pressure due to the whipping of the model is an order of magnitude lower than the pressure resulting from the impact. Hydro-elastic effects are again lower, so it is concluded that, for practical purposes, they can be ignored for the problem of aftbody slamming.

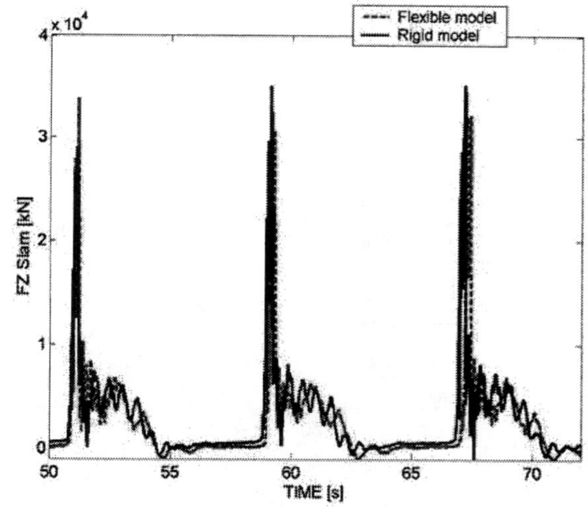

Figure 18 Slamming force for the flexible and rigid model.

TUNING OF THE WHIPPING MODEL

The mathematical model presented before requires as input the properties of the tow tank model (at prototype scale) and the stiffness C_H and damping B_H of the connecting spring. The value of C_H follows from the natural period of the two-node vibration and by assuming uncoupled motions between heave and pitch, thereby neglecting the z-displacements in equation (3):

$$\ddot{\theta}_1 + 2\kappa\omega_n\dot{\theta}_1 + \omega_n^2\theta_1 = 0$$

$$\kappa = \frac{2B_H}{\omega_n \cdot I_{yy}^{F+A}} \qquad \omega_n^2 = \frac{4C_H + C_{55}^{F+A}}{I_{yy}^{F+A}} \quad (10)$$

The damping BH follows from the whipping moment decay curve, which was obtained by hitting the model with a hammer. This damping curve contains both hydrodynamic and structural damping. The first component is normally considered negligible at this frequency. The data for the proposed mathematical model are listed in Table 3.

		Aft	Forward
Mass	[ton]	22237	16636
k_{YY}	[m]	40.87	43.81
LCG	[m]	-53.53	44.30
C33	[kN/m]	36521	27521
C35 = C53	[kN]	104575	-7743
C55	[MNm]	22040	38620
C_H	[MNm]	4247	
κ	[-]	0.0077	

Table 3. Data of whipping mathematical model. The data are relative to the CG of the segment. LCG is relative to the midships cut.

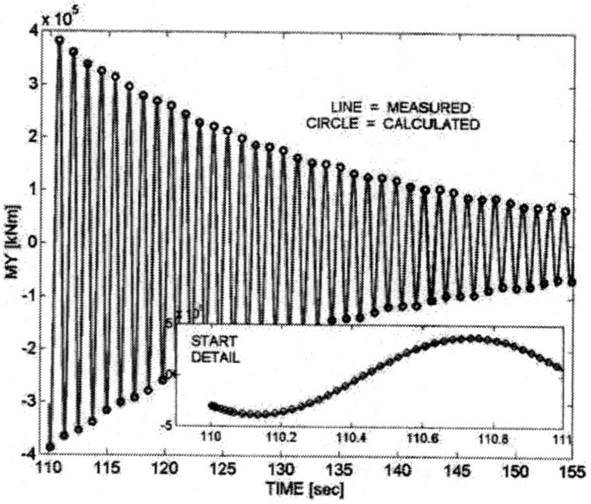

Figure 19 Decay of the response to an impulse of the physical model and the mathematical model.

Figure 19 presents the correlation between the mathematical model and the decay curve of the physical model. The plot shows the response to an impulse. The initial conditions for starting the curve are obtained using the measured vertical bending moment (MY) at that time. The initial part of the simulation is shown in the detailed graph in the figure. Obviously, in the decay curve, the vertical slam force (FZ) is zero. The tuned non-dimensional damping coefficient was $\kappa = 0.0077$. This is a low value in comparison to full-scale ships, which have damping on the order of 2 to 3% of the critical value.

USE OF THE WHIPPING MODEL TO ANALYZE SLAMMING EVENTS

The objective of the whipping model was to analyze slamming events to check if the physical process was understood. If the slamming force, as derived from the pressure gauge measurements, is applied to the whipping model, the response must correspond to the response of the model in the tank. When that is the case, the loads can be applied to a Finite Element (FE) structural model of the ship

A first check was carried out to compare the response of the model to the simple integration method (S) and the advanced integration method (C). This comparison is shown in Figure 20. The calculations start at T = 381 s with initial conditions zero, thus with no displacement, rotations or whipping moment present in the hull.

The slamming force shows some spikes when the simple integration method is used, but the model does not respond to such short duration, low impulse spikes. The result of the simulation shows an identical response of the model with respect to the vertical bending moment.

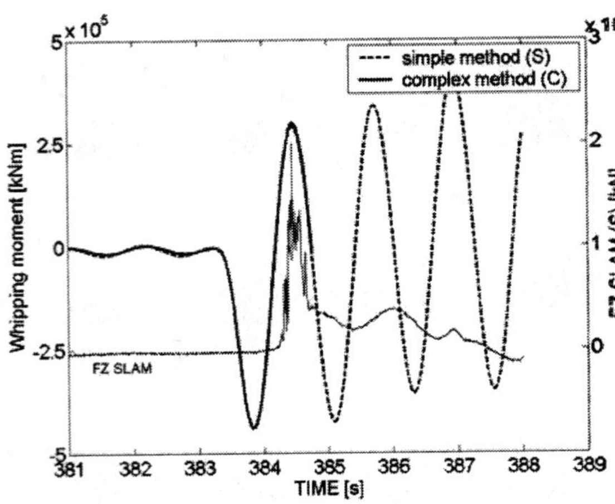

Figure 20 Calculated whipping response due to a single slam.

RESULTS OF EXPERIMENTS IN WAVES

Tests were carried out in following sea conditions at zero speed. The wave conditions consisted of long crested irregular waves with significant wave heights (2 and 4 m), using the JONSWAP spectrum with a peakedness parameter 3.3, and a peak period of 8 s. The energy spectrum of the vertical bending moment midships is shown in Figure 21. The figure clearly shows two peaks, the first at a frequency of 0.8 rad/s due to the waves and the second at 5.2 rad/s due to whipping.

The measured MY (in the 4 m Sea State) is plotted in Figure 22. The total signal (top time trace) is split into a whipping component (using a high-pass filter with cut-off frequency 3 rad/s), and a wave component. Both components are also shown in Figure 22. The vertical bending moment due to whipping is clearly dominant in this condition. The results of the whipping model will be compared to the whipping component of the vertical bending moment midships. The comparison will mainly be made for the 4 m Sea State as specified above.

Earlier discussion showed the whipping response of the vessel due to a hammer impact. The validation was made using zero initial conditions in the calculations. In waves the situation is different. Figure 23 shows that whipping due to previous slamming events continues when the vessel is hit by the next wave.

The timing between the existing whipping deformation and the next slam impact appears to be crucial for the resulting response. For example, a relatively low impact force is able to minimize the whipping moment at t = 26 s. A similar event is found around t = 110 s. On the other hand, the impacts at

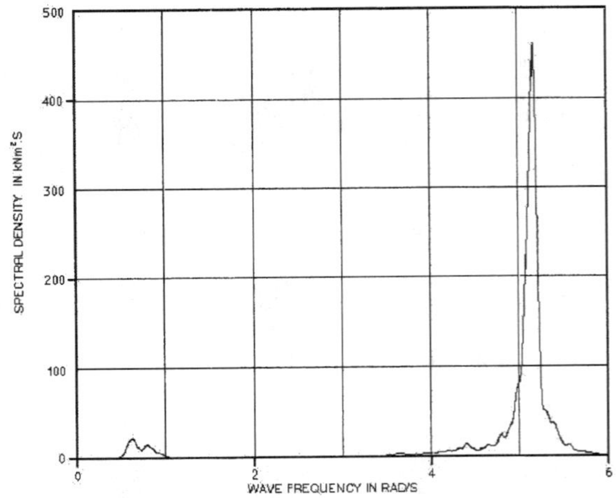

Figure 21 Spectrum of vertical bending moment showing wave frequency and whipping part.

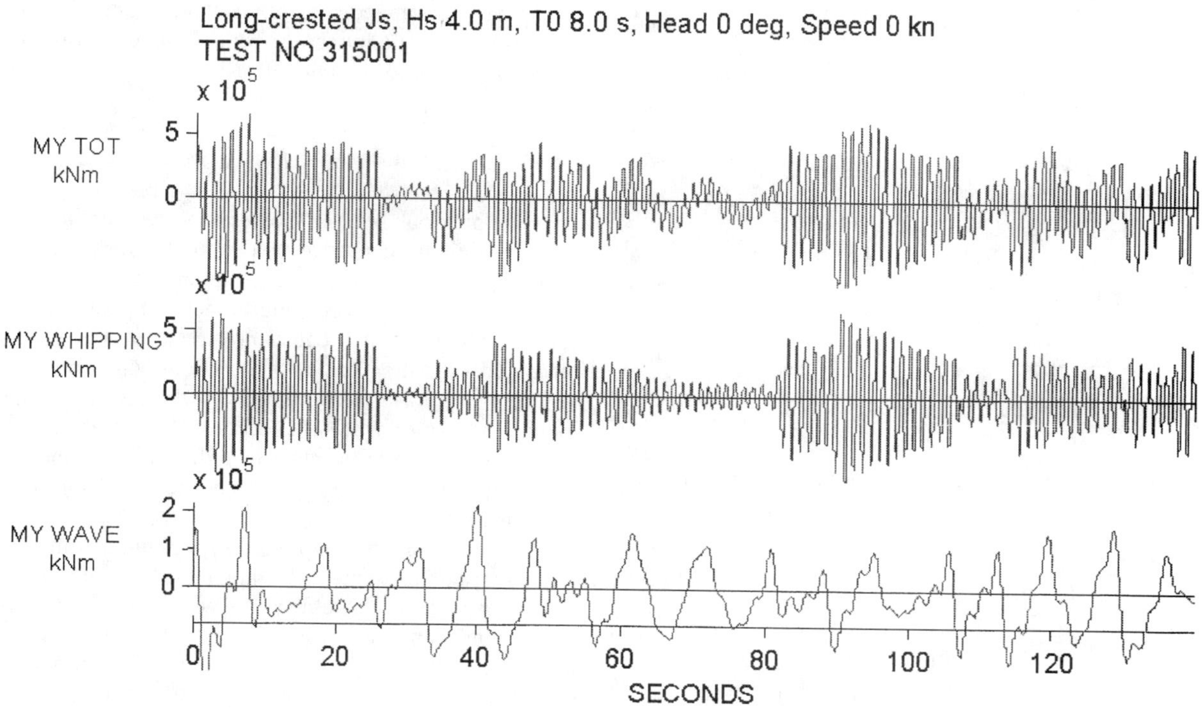

Figure 22 Time traces of the vertical bending moment midships. Top: the total signal; middle: the whipping component; bottom: wave frequent component.

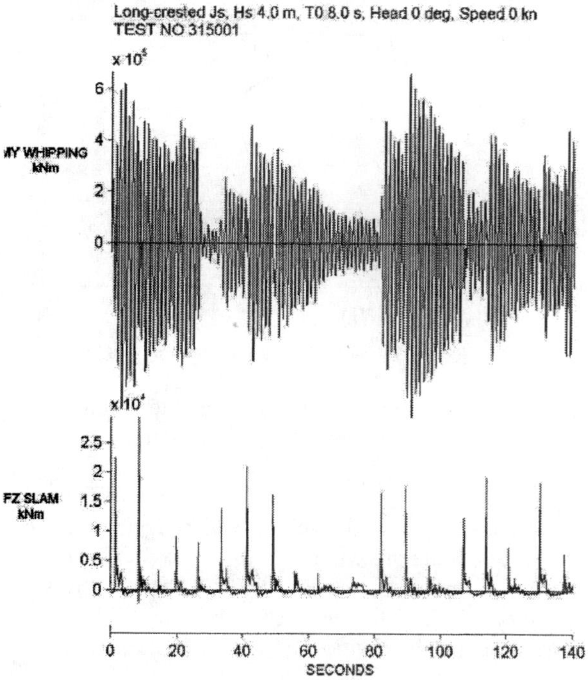

Figure 23 Whipping bending moment and slam force from tank tests.

t = 34 and t = 41 s increase the whipping moment more than could be expected due to a "favorable" timing.

The whipping load is calculated using the initial conditions (angle of the fore and aft segments, and angular velocities) from the model tests. The amplitude of the whipping load is used to obtain the difference in the angle of the fore and aft segments. The angular velocities are obtained from two angular rotations at

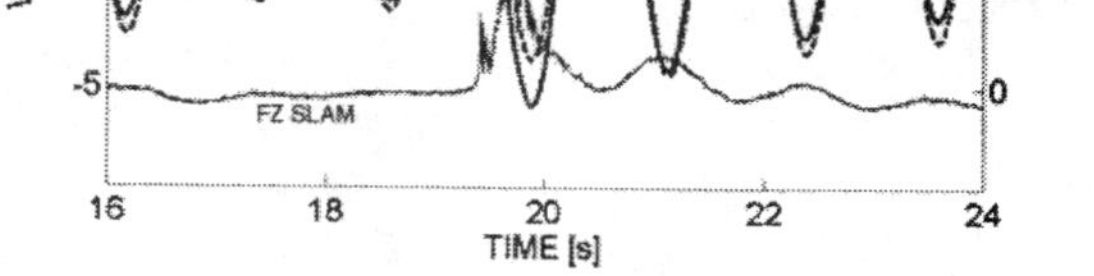

Figure 24 Measured and calculated whipping response due to an impact force.

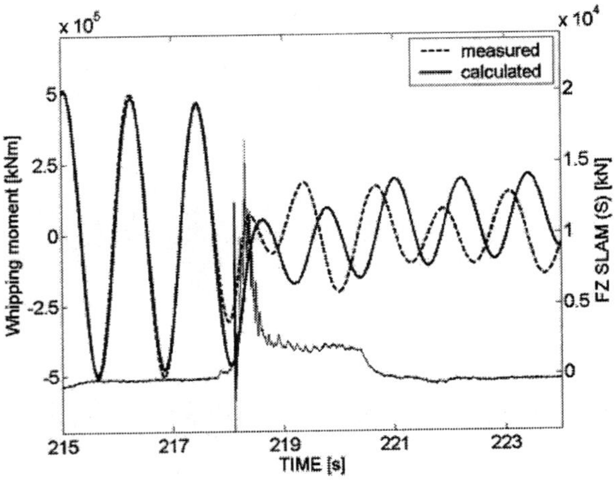

Figure 25 Example of a slamming event that causes a decrease of the whipping moment.

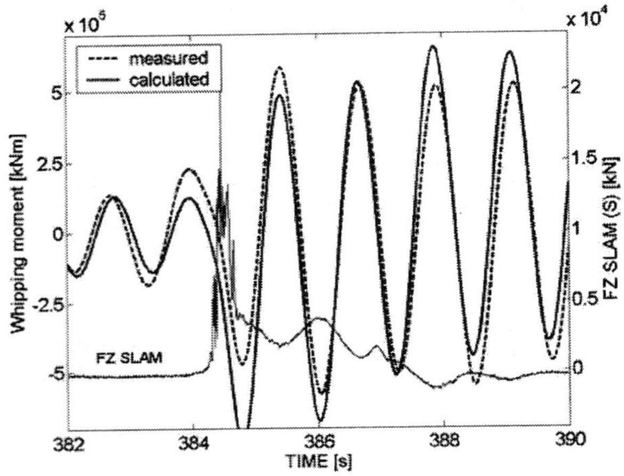

Figure 26 Example of a slamming event that causes an increase of the whipping moment.

consecutive time steps. Figure 24 through Figure 26 present the results of the simulations with the whipping model compared to the measured vertical bending moment for three cases. Figure 24 shows how a relatively small impact increases the whipping moment some 20%.

Figure 25 shows how a slam that occurs out of phase with the whipping motion reduces the moment by a factor of four. The calculations predict the trend correctly, but the phasing of the whipping moment after the impact is incorrect.

Figure 26 gives an example of how a favorable tuning increases the whipping moment by a factor of four. The prediction of the whipping model agrees very well with the actual measured moment.

The general good agreement between the calculations using the whipping model and the actual measured vertical bending moment shows that the whipping phenomenon is "understood", and that the response of the model can be reliably predicted by the whipping model for individual cases.

STATISTICS OF WHIPPING LOADS

The previous section showed that the amplitude of the whipping moment is dependent on the timing of the slam relative to the existing whipping motion. When simulations with the whipping model were carried out, each slamming event could be predicted (using the measured slamming force) if the initial conditions just before the impact were tuned. If this is not done, inevitably some drift will occur. This in turn will change the tuning of the slam with the whipping motion and will have a large effect on the resulting peak whipping moment.

Now consider a long record of measured whipping moments and compare the statistics of the measured whipping peaks to those from the whipping model. The initial conditions for the whipping model are not tuned to match the experiments; therefore, individual slamming events can be very different between experiment and simulation.

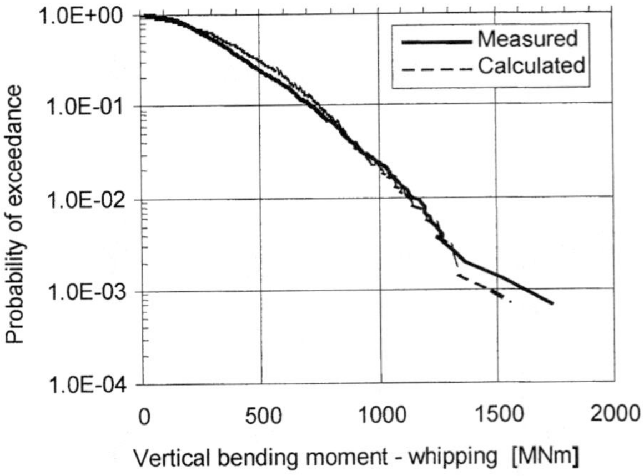

Figure 27 Extreme values in a 4 m Sea State. Results of the whipping model and test data.

The measured peaks of the whipping moment in the 4 m Sea State are compared to the peaks calculated with the whipping model, Figure 27. The vessel was at zero speed in a following sea condition. This result shows that the distribution of the peaks is similar, so that the differences, which develop due to "drift" in the whipping model, are inconsequential to the extreme whipping moment.

Figure 28 shows a similar plot for a 2 m Sea State. The differences between actual measurements and the whipping model at low probabilities of exceedance are important, but the tails of the distributions are again similar. These results mean that a simulation with the whipping model as presented in this paper and the measured slamming force cannot be used for fatigue assessments. However, the results of the whipping model can be used for the prediction of extreme values.

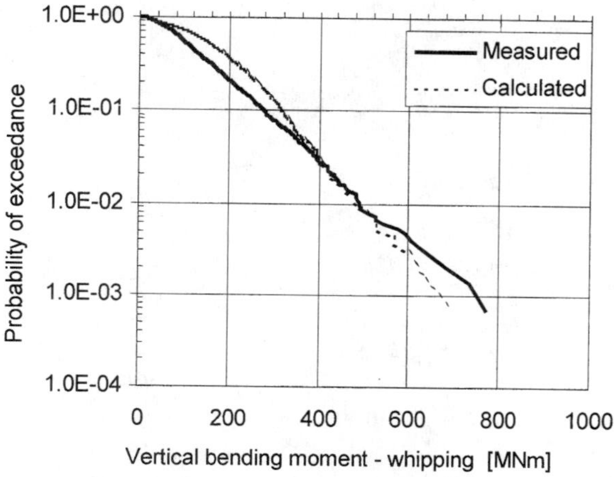

Figure 28 Extreme values in a 2 m Sea State. Results of the whipping model and test data.

EFFECT OF WHIPPING ON FATIGUE

In addition to the tests in following seas at low ship speeds, a series of experiments in high head and bow quartering seas at low ship speeds was carried out. A fatigue analysis (rainflow count) has been made of the measured vertical bending moment and of the moment from which the whipping part has been removed by using a low-pass filter. This filtered moment is indicated as the wave vertical bending moment.

The fatigue effect (FE) is in general defined by:

$$FE = \sum_{i=1}^{N} \sigma_i^3 \cdot n_i \tag{11}$$

Since, in the elastic regime, the stresses are proportional to the bending moment and because the structure is not being considered, equation (12) is a basis for comparing the FE in different wave conditions.

$$FE = \sum_{i=1}^{N} MY_i^3 \cdot n_i \tag{12}$$

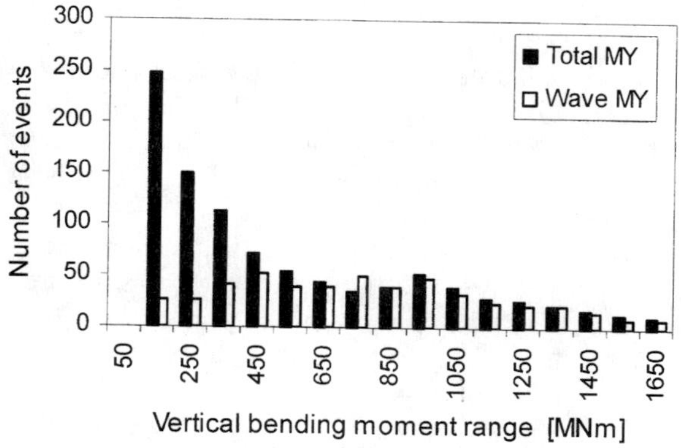

Figure 29 Fatigue in head seas, 9 m Sea State at 9 kts speed.

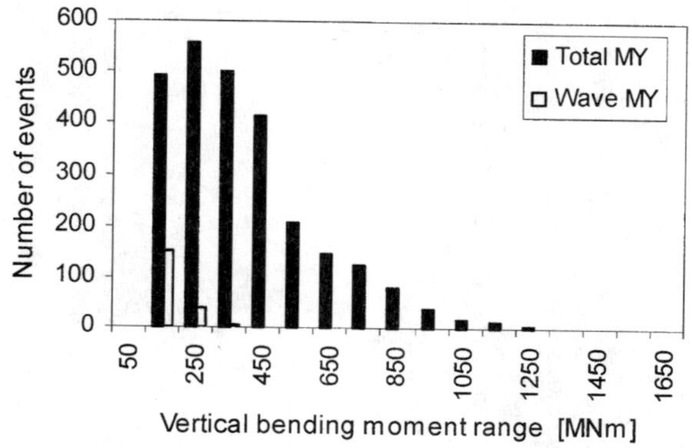

Figure 30 Fatigue in following waves, 4 m Sea State (Tp = 8 s) at zero speed.

	Head seas, H_S = 9 m; V = 6 kts	Following seas, H_S = 4 m; V = 0
Total MY	2.60 E+11	1.73 E+11
WF part of MY	2.07 E+11	8.62 E+08

Table 4. Fatigue effect, $(MNm)^3$, in one hour on the MY midships for the model in the tank.

The result of an experiment in a high (9 m) Sea State is shown in Figure 29. This figure shows the total and the wave MY. The whipping part has an effect mainly on the lower MY ranges. This means that the fatigue portion in one hour is 80% dominated by the first order wave forces and 20% by the whipping component, see Table 4.

Figure 30 shows a similar plot of a test in a 4 m Sea State in following waves conditions (the same test as shown in Figure 22). This figure shows that the results are dominated by the whipping response of the ship. There is a very large number of load cycles at medium ranges of the MY due to the whipping response of the vessel. The fatigue effects due to the total signal is a factor 200 times higher than those due to the first order wave loads only. The fatigue effect in this condition is of the same order as for the 9 m head seas condition mentioned before.

The extreme slamming events that are reported in this paper only exist for a narrow range of sea states, which means a narrow range of the peak period of the wave spectrum. The worst case is in a sea state with a peak period of 8 s.

Figure 31 and Figure 32 show test results in a 2 m Sea State with a peak period of 8 s and 14 s, respectively. Figure 31 shows that for the Tp = 8 s case, the fatigue effect is dominated by the whipping contribution. Figure 32 shows that for the Tp = 14 s case, the wave contribution is dominant. The number of load cycles is also an order of magnitude lower in this case, which makes the fatigue effect a factor 10 lower than the Tp = 8 s case.

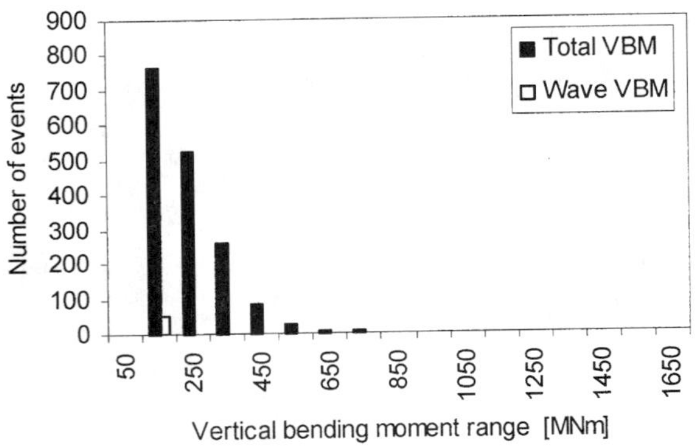

Figure 31 Fatigue in following waves, 2 m Sea State (Tp = 8 s) at zero speed.

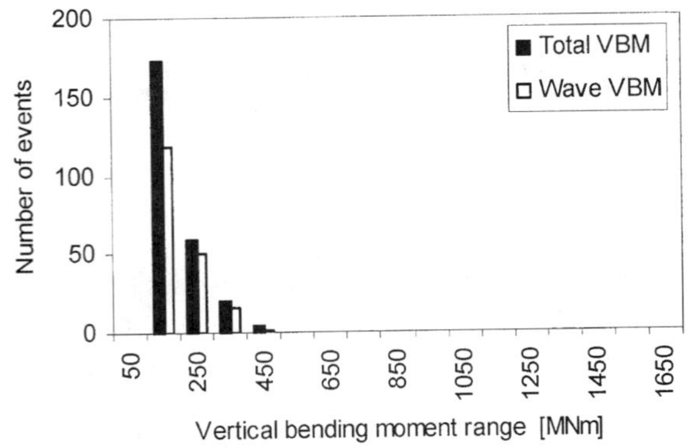

Figure 32 Fatigue in following waves, 2 m Sea State (Tp = 14 s) at zero speed.

EFFECT OF WHIPPING ON EXTREME LOADS

The extreme values of the sagging and hogging component of the MY have been collected for the mentioned tests in the 9 m head seas and the 4 m following seas conditions. This analysis has also been done on the total MY signal and the filtered signals (the wave frequency and the whipping part).

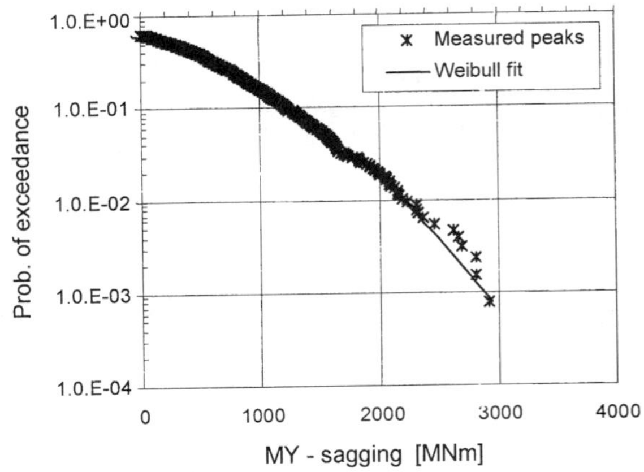

Figure 33 Extreme values (sag) in head waves, 9 m Sea State.

Figure 33 shows the extreme values of the test data in head seas and a 3-parameter Weibull distribution fitted through the data. This fit describes the shape of the distribution of the maximum values quite well. From here on use has been made of the fitted Weibull curve for comparisons.

Figure 34 shows the results for the sagging component of the MY in the same 9 m Sea State. The plot shows the extremes of the wave component (WF) and the whipping component (HF) separately. These two signals were used to create the line "WF + HF". This line is constructed using the rules:

$$MY_{SAGGING} = MY_{WAVE\ FREQUENCY} + MY_{WHIPPING} \tag{13}$$

$$\Pr\{MY_{SAGGING}\} = \Pr\{MY_{WAVE\ FREQUENCY}\}$$

where Pr is the probability.

This means that the probability of an extreme sagging moment is equal to the joint probability of an extreme moment due to the first order wave forces and an extreme whipping moment. The occurrence of extreme events of these two phenomena is therefore considered independent. The timing of the whipping event is always such that the first peak, which is the extreme, occurs at the same time as a sagging peak.

Because of this last aspect, the same rule does not apply to extreme hogging moments as is illustrated in Figure 35. For this particular ship, the rule appears to be that the whipping contribution is reduced to 50% at the time of peak of the hogging moment.

$$MY_{HOGGING} = MY_{WAVE\ FREQUENCY} + \tfrac{1}{2}\, MY_{WHIPPING}$$

$$Pr\{MY_{HOGGING}\} = Pr\{\ MY_{WAVE\ FREQUENCY}\} * Pr\{MY_{WHIPPING}\} \qquad (14)$$

Another interesting feature from these measurements is the sag/hog ratio. This ratio is for most ships larger than 1 due to non-linearities in the hull form (flare at the bow and/or the stern). It is understandable that this ratio increases for increasing wave height as is illustrated in Figure 36. This figure shows that the sag/hog ratio is quite large, about 1.8, for large vertical bending moments.

Figure 37 and Figure 38 show similar plots for the 4 m following seas case. The whipping component is now very dominant. The peak of the whipping moment occurs at the same time as the peak of the hogging moment (unlike the head seas case). Therefore, the two added together give a distribution of extreme values which is comparable to the distribution of the total MY. Adding the two components together for the sagging peaks gives an over-prediction of the actual MY, as is illustrated in Figure 37. In fact, the extreme MY due to the whipping contribution is always larger than the extremes of the total signal.

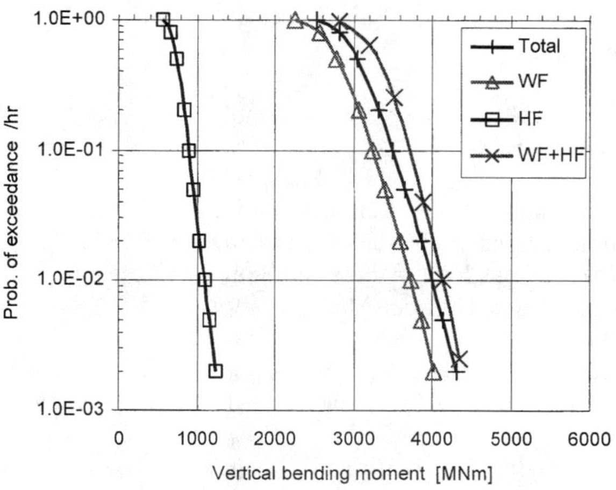

Figure 34 Extreme values (sag) in head waves, 9 m Sea State.

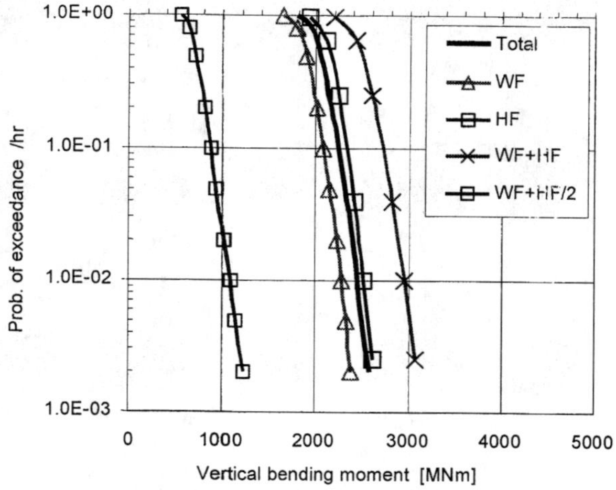

Figure 35 Extreme values (hog) in head waves, 9 m Sea State.

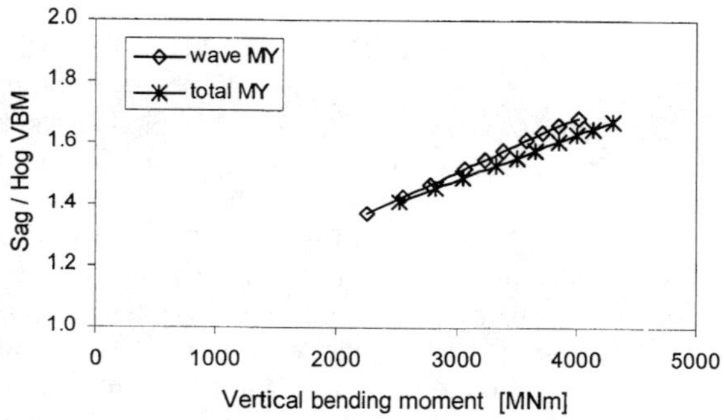

Figure 36 Sag/hog ratio as a function of the magnitude of the MY in head waves, 9 m Sea State.

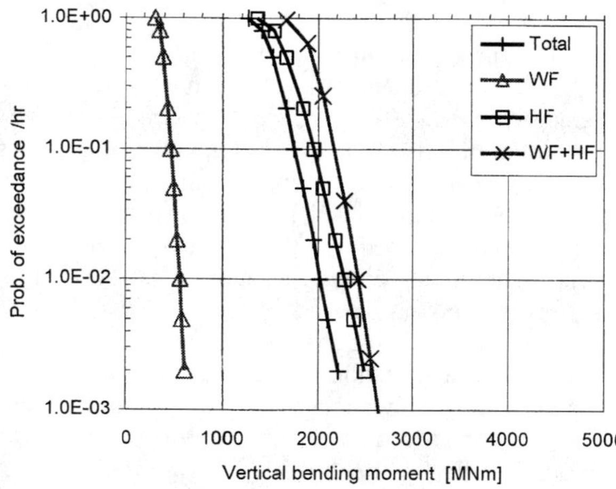

Figure 37 Extreme values (sag), 4 m following Sea State.

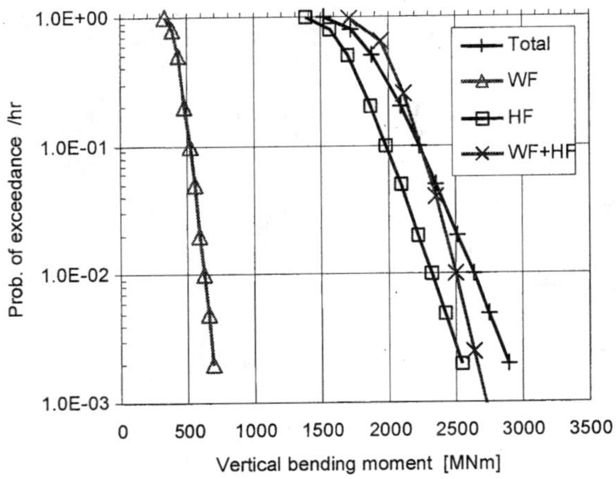

Figure 38 Extreme values (hog) component, 4 m following Sea State.

PREDICTING FULL SCALE WHIPPING LOADS

The foundation has now been laid to predict wave loads for the prototype ship. This paper demonstrates that the impact load can be derived from the measured pressures. Applying this load to a structural model of the segmented model, it is possible to reproduce the whipping bending moment at midships. The next step is to apply the measured impulse to a full 3D structural model of the ship and do a time domain simulation to determine the whipping response and peak stresses in the ships structure.

CONCLUSIONS

This paper has introduced a reliable method for measuring the excitation of a ship's hull due to aftbody slamming impacts. The measured impact force has been derived from a large array of pressure gauges. Tests with two models with varying stiffness indicated that the pressure due to the whipping response of the model can be superimposed on the impact pressure. This means that there is hydro-structural interaction, but no important hydro-elastic effects.

The impact force has been used in a simple structural schematization of the segmented model that was tested in the towing tank. The predictions with this whipping model compared very well to the measured vertical bending moment. It is therefore concluded that the impact force can be applied to a structural model of the ship to determine the whipping moment and peak stresses for the full scale vessel.

Further conclusions are:

- A vessel having a modern stern shape lying at zero or low speeds in following wave conditions can experience heavy stern slamming. This slamming starts occurring in mild sea conditions if the wave length is on the order of the ship length.
- Stern slamming can be considered as an impact that starts at a certain point and then expands first longitudinally and later athwartship over the stern. The high pressure ridge passes a single point on the stern in about 0.05 s. Due to the size of the stern the total impact takes about 0.5 s.
- Instrumenting the aftbody of the model with a large array of pressure gauges allows an accurate measurement of the impulsive force.
- The measured impulsive force can be used on a dynamic structural model of the ship to predict the whipping loads.
- The fatigue effects due to stern slamming in a 4 m following Sea State at zero ship speed is on the same order as the effects in a 9 m head seas Sea State at a ship speed of 6 kts.
- The peak of the whipping vertical bending moment (MY) in head seas occurs together with the peak of the sagging MY. In following seas the timing is different: then the peak of the whipping MY occurs at the same time as the peak of the hogging MY.
- The design value of the extreme MY should account for a larger than normally used sag-hog ratio.

ACKNOWLEDGEMENT

The authors acknowledge the permission of Northrop Grumman Ship Systems to publish the results in this paper.

REFERENCES

Bereznitski, A. and Kaminski, M.L., "Practical Implications of Hydroelasticity in Ship Design", *Proceedings* Int. Offshore and Polar Engineering Conference, Kyushu, 2002.

Betts, C.V, Bishop, R.E.D. and Price, W.G., "A Survey of Hull Damping", Transactions RINA, Vol. 119, 1977.

Bishop, R.E.D and Price, W.G., "Hydro-elasticity of Ships", *Cambridge University Press*, 1979.

Hämäläinen, R. and Heerd, J. van, "Hydro-dynamic development for a large fast monohull passenger vessel", Transactions SNAME Vol. 106, 1998.

Haugen, E.M, Faltinsen, O.M. and Aarsnes, J.V., "Application of Theoretical and Experimental Studies of Wave Impact to Wet Deck Slamming", *Proceedings FAST-97 Conference*, Sydney, 1999.

Faltinsen, O.M., "Slamming", *Colloquium for Ship and Offshore Hydrodynamics*, Hamburg, 1996.

Faltinsen, O.M, "Water Entry of a Wedge by Hydroelastic Orthotropic Plate Theory", *Journal of Ship Research*, Vol. 43, No 2, 1999.

Faltinsen, O.M., "Hydroelastic Slamming", *Journal of Marine Science and Technology*, Vol. 5, No. 2, 2000.

Kurimo R., "Sea Trial Experience of the First Passenger Cruiser with Podded Propulsion", *Proceedings of the 7TH Int. Symposium on Practical Design of Ships and Mobile units (PRADS)*, The Hague, 1998.

Kvålsvold, J. and Faltinsen, O.M, "Slamming Loads on Wet Decks of Multihull Vessels", *Proceedings Hydroelasticity in Marine Technology Conference*, Trondheim, 1994.

Comparative Naval Architecture of Passenger Ships

Philip Sims (M), Naval Sea Systems Command

Frontispiece - Same scale models of *Britannic,* the *Titanic's* sister ship, and the *Majestic,* which was the largest in-service ship in the world only 22 years earlier

Abstract

Numerous physical descriptions of Titanic *and dramatic histories of her sinking have been published. What have not been well explored are the fundamental questions in assessing the quality of any machine: "Compared to what?" and "Why did the designers do what they did?" This paper starts by comparing the* Titanic *to ships before and after her. The ship features that seem frivolous to us (a huge fake fourth stack) or bad engineering practice (shallow watertight bulkheading forward) are explained. The operational background for owner/operators to make decisions like providing an inadequate number of lifeboat seats and proceeding full speed through a scattered ice field are described. Authors of popular texts have assumed Safety of Life at Sea (SOLAS) conferences issued design rules to insure that if a ship, designed to modern rules, had the same damage as the* Titanic, *it would not sink but this is not so. The post-SOLAS 1929 ships are compared to* Titanic *and modern ships. In the larger framework, the paper describes the continuing evolution of passenger ships, the sources of danger to them and if there still is the possibility of "iceberg-like" surprises for modern ships.*

Introduction

Comparative Naval Architecture was an analysis philosophy used in the late 1970s by the US Navy to determine why the Russian Navy was building different, and apparently more heavily armed, ship types. It involved reverse engineering that assumed the other designer made a rational decision about how to use the technology and information he had available to him. This paper will use a Comparative Naval Architecture approach (described in greater detail in Appendix I) to discuss in four sections the unusual features of the *Titanic,* the post-*Titanic* ships of the 1920s and 30s, modern passenger ships and risk of unanticipated levels of damage to future ships.

Section I: Features of the Titanic

The Fourth Fake Smokestack

The reason most often given for the *Titanic*'s false fourth smokestack is that a series of Blue Ribbon winning four stack German Liners, plus the *Lusitannia* and her sisters, had established four stacks as the symbol of a modern high speed ship, so customers preferred them. [Maxton-Graham, 1972] However, the Germans had already commenced work on the large liners *Imperator* and *Bismarck* which were three stack ships (including one fake). The publicity value of having less- than-four stacks did not deter them and the Germans, having fitted a fourteen foot long gilded eagle on the bow of the *Imperator* to insure that she was the longest ship in the world, were not averse to frivolous appendages for their own ships to achieve advertising goals. There must have been other reasons than fashion for fitting a ship with very large active and false stacks.

The 1870 through early-1890s steam passenger ships, although intended to operate 100% of the time under steam, were single screw so they retained a sail rig for back-up power in case the steam machinery failed. One now thinks of shafts and propellers as very reliable, but that was not the case for the built-up propellers of the early years, since the single screw *Ethiopia* (in 1880) *Aurania* (1898), *Umbria* (twice 1892, 1893), and *Etruria* (1900) all had prop/shaft failures. [Gibbs, 1957] The steamship *City of Brussels* had to finish voyage to Ireland under sail in two times in 1870.

Figure 1 shows the single screw *City of Peking* of 1874 with her full sail rig and modest smokestacks. The fitting of the next generation of passenger ships with twin screw machinery meant that the sail rig could be dispensed with since the steam plant had its own redundancy.

The masts and spars of an auxiliary sail passenger ship had a favorable effect on the motions of the ship. They raised the center of gravity so the transverse metacentric height (GM) of the ship (which governs the force that brings a ship upright after a roll) was not too large. The mast, spars and supporting wires created air drag as the ship rolled back and forth acting as a damper on the motions (acting as a kind an aerial bilge keel).

Figure 1 ***City of Peking*****, a single screw steamship with emergency sails**

Figure 2 ***Campania*****, a twin screw steamship**

As early as 1873, naval architects were doing roll experiments on real ships that showed that rolling speeds were not only a function of metacentric height but radius of gyration. [White, 1900] Although calculating an actual total ship gyradius was beyond their mathematical means, the engineers of the time knew the fundamentals of how to compensate for the loss of the sail rig. The first generation of pure steam ships, Figure 2 shows the *Campania* of 1893, had huge, elaborate smokestacks.

Since righting arm goes up as a function of beam cubed times length, the large dimensioned *Great Eastern* of 1859 had a great deal of stability. She also had a round midships section, no bilge keels, no superstructure plus spindly little masts and stacks, hence, a low gyradius. It is not surprising that she was a wicked roller, to point of self destruction of her paddle wheels, in a sea state high enough to excite motions in a large ship.

The large increase in ship sizes between 1870-1914 meant the ships would have similar severe motion problems unless the ships were fitted with large stacks (both active and fake) which had high weight and added to the radius of gyration. Their structure and bracing wires also replaced the air drag of the deleted mast and their shrouds during rolling, further dampening the motions. Thus, in addition to their primary function of lifting the unpleasant coal cinders well away from the ship, stacks had a secondary utility as a means of giving the ships pleasant roll motions.

The German naval architects regressed to three stacks, because with their love of marble and thick wood paneling, they were having no problems with excessive stability. In fact, the *Imperator* had nine feet cut off the top her as-built stacks as one of several measures to regain stability. [Miller, 1987] A contrary example of stability tuning was the *Kaiser Fredrick* that had 14.5 feet added to her already substantial funnels when rebuilt in 1899. [Kludas, 1975]

The Counter Stern

Sailing ships had sterns of various shapes. The ropes from the helm wheel went down to an actual tiller forward of the rudderstock. Having the wheel far aft, while not providing the best view of the sea ahead, allowed the captain to view how his sails were set. The coming of the smokestacks and paddle wheels further blocked the view ahead and made seeing the masts irrelevant. The place to steer the ship was ahead of the stacks and paddle wheel boxes; this required running a mechanical link aft. For a ship still relying on manual maneuvering, this meant running ropes or chains aft as shown in Figure 3. The way to get leverage was a quadrant behind the rudderpost, the swing of which encouraged making the stern shape to be semi-circular. The need to fit an aft watertight bulkhead forced the chains to be located high in the ship, since an attempt to put them low in the ship would require watertight penetrations. Making the chain penetrations watertight would greatly increase the system complexity and add friction, along with the risk of a jamming.

Figure 3 Chains-linked-to-quadrate steering

The first steam steering engines, directly over the rudder aft, were controlled either by chains or by long rotating control rods. For the same reasons as for mechanical chains, that motivated a high location with a straight run aft from the bridge clear of the bulkheads. The reciprocating steam steering gear put a great deal of heat and humidity into the compartment air, which would condense inside the space. The steam machines also leak water and oil directly into a compartment. Fitting the steam steering engines in a high aft location had the advantage of easy ventilation and drainage, so the counter stern configuration continued to be the preferred shape as shown in Figure 4. It also assured a short mechanical connection to an aft emergency steering wheel

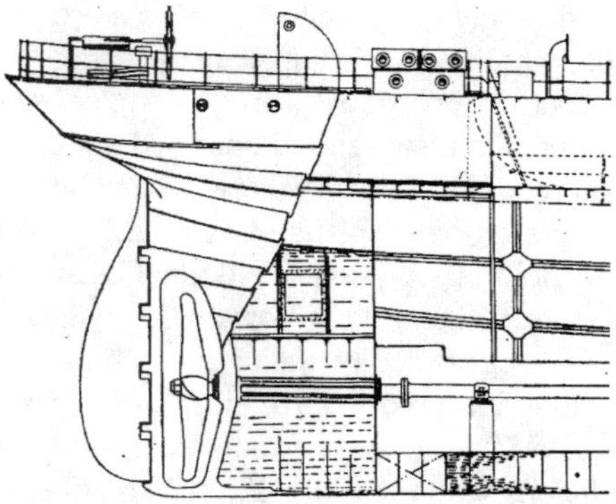

Figure 4: Counter stern allow high steering gear location.

The *Titanic* used a transitional system with "telemotor" control (hydraulic lines, which could make watertight penetrations,) from forward but a direct mechanical link from auxiliary steering station just above. [Shipbuilder, 1911] An exception, due to her armed merchant cruiser wartime role, was the *Lusitania* which had her steering machinery located below the waterline in a cramped compartment to protect it from shell hits. Her counter stern was empty and fitted to be fashionable or, possibly, a decoy target for raider shells.

Designing a vehicle to survive an accidents

Designing a vehicle to survive an accident requires that the accident be foreseen, probable, subject-to-affordable-solution and agreed upon. "foreseen" is self-explanatory. A ship being caught by a Tsunami wave and thrown ashore, such as the cruiser *Memphis*, is a foreseen event but ships are not designed to resist that because it is not a probable event. It is feasible to design commercial airliners to provide individual ejection seats for every passenger; that equipment would save many lives. However, the expense of planes designed that way would make commercial passenger flight economically unaffordable.

"Agreed upon" is the most difficult feature to achieve in a competitive world. If one liner company has lavish safety spending and their competitor does not, the safety conscious line is punished twice. Once for having a higher cost structure and, secondly, if the shoddy liner sinks, tarring the entire industry as unsafe including the good ships. The "agreeing upon" can be informal as prior to the *Titanic*, national rules, or international rules.

Without it being imposed by statute, the maritime community had reached consensus on the foreseen, probable, subject-to-affordable-solution and agreed upon accident in the late 1800. The expected accident was the "T-bone" ramming of one ship by another. Claims of being "practically unsinkable" were made with this accident in mind. The expected hole in the side of the ship was a V, as seen in Figure 5, made by the straight vertical stem typical of ships of that time.

Figure 5: The expected damage

Number of Lifeboats

Sailing ships had never carried enough lifeboats. All of the horror stories of the age of sail, such as the 1816 Raft of the *Medusa*, or the 1629 abandonment of the survivors of the *Batavia,* can be traced back to lack of lifeboats. Sail equipped steamers continued the pattern. The previous great North Atlantic disaster was the *Arctic* of 1854 that rammed another ship, the *Vesta,* but also put holes in her own bow. The uncompartmented wooden ship tried to head for Newfoundland but continued to sink. Since there were not enough lifeboats for everyone, the crew took the ones that were launched leaving the passengers behind. [Professional Mariner, 2002]

The reason sail-equipped ships did not carry enough lifeboats was there was no place to put them on deck without fouling the standing and running rigging. What few boats were carried were placed on the centerline for that reason and, hence, difficult to launch in a hurry. As ships became more reliant on steam and were fitted with screw propulsion, there was naturally created a place to put lifeboats on the deck edge: outboard of the stacks. Figures 1 and 2 show the correlation of the reduction in sail rigging with expansion of the number of deck edge lifeboats.

In the expected accident, a T-bone ramming of large metal-hulled steamships, there would be two ships on scene. The rammed ship, which might sink, and the ramming ship, protected by the modern invention of the forepeak bulkhead, which would survive. The lifeboat's sole role was to shuttle people from the severely damaged ship to the survivor; since both ships had lifeboat outfits, there certainly was no need for an individual ship to carry a 100% outfit.

Parliament chartered a "Bulkhead Committee" to investigate the claims that the transverse watertight bulkheads could greatly enhance the safety of ships. In 1891, they concluded that bulkheads spaced to allow a ship to float with two adjacent compartments flooded was highly recommend for ocean going ships greater than 425 feet long. As a sweetener, to try to induce industry co-operation, they recommended that the 1888 rules on number of lifeboats could be reduced on a ship so fitted. [Board of Trade, 1891] An extensive lifeboat outfit was disliked by the operators due to first cost, maintenance costs and the ruining of the upper decks, for passenger pleasure, with their clutter as in Figure 6.

In the 1909 *Florida* and *Republic* collision, the whole system worked to perfection with the ramming ship, protected by its collision bulkhead, becoming the rescue ship for the fatally injured ship. The lifeboats adequately served their role as shuttlecraft.

Figure 6: Post-*Titanic* lifeboats fitted to the *Olympic*

Other possible causes of sinking

It was hard, in the early 1900s, to imagine a reason a modern ship would sink in deep water. The Hague Treaty of 1905 required even submarines to stop and search ships in wartime, plus insure safety of passengers and crew. If the accident were grounding, the proximity of land would require a shuttle of boats and not a full outfit.

It is well documented that the *Titanic*'s crew was not afraid of icebergs. Icebergs were a proven killer of wooden hulled, non compartmented ships and greatly feared by those vessels. However, modern compartmented metal-hulled ones had a much different record. In 1856, the wooden Collin lines steamship *Pacific* disappeared without survivors in a large ice field. In the same field, the Cunard iron hulled *Persia* smashed her bow and damaged her paddles but, protected by the collision bulkhead, continued to New York. [Brinnin, 1971] By 1912, there was a 30 year proven track record of surviving ice collisions with big (over 325 ft long, over 2,500 Gross Register Tons) modern (metal hull, steam powered) ships as discussed in Appendix II. There had been at least 53 ice incidents survived by the bigger ships. The biggest modern ship sinking, with loss of life, due to ice damage was the diminutive 2,484 Gross Ton, 327 ft long *SS Vicksburg* way back in 1875. Two larger ships, the *Iowa* (4,329 GRT) in 1891 and the *Notting Hill* (4,021 GRT) in 1884, were abandoned without loss of life after hitting ice. Their slow flooding allowed ships that happened to be in the vicinity to take on board the crew and passengers with the ships abandoned to sink slowly. For modern ships smaller than 2,500 Gross Tons, only 1 in 8 were lost after an iceberg collision during the same period. Given this splendid record, there was no reason to be overly concerned with icebergs. They were a nuisance to collide with but there was no reason to expect that they could sink a large modern ship fitted with extensive watertight bulkheading.

The Titanic's Structure

The *Titanic*'s Second Officer Charles Herbert Lightoller expressed the theory, in his account of hitting the iceberg, that the sides of the *Titanic* were weaker than previous ships he had served on. He stated afterwards [Lightoller, 1960]:

> "... *Each blow stove in a plate, below the water line, as the ship had not the inherent strength to resist.*
>
> *Had it been, for instance, the old* Majestic *or even the* Oceanic *the chances are that either of them would have been strong enough to take the blow and be bodily thrown off without serious damage. For instance, coming alongside with the old* Majestic, *it was no uncommon thing for her to hit a knuckle of the wharf a good healthy bump, but beyond perhaps, scraping off the paint, no damage was ever done. The same, to a lesser extent, with the* Oceanic."

The 19th century started off with ships built of wood to rules-of-thumb and, after great leaps in materials science, ended with large steel ships designed with a good understanding of sea loads. The first classification rules for metal ships' structures were issued in 1855 in England. The first iron ships had conservatively sized structure due to the designers knowing that they had a limited ability to estimate loads and strength. At the end of the century, the engineers felt that they had a good approach to both predicting loads and structural strength.

Loads on ship structure are tertiary (point loads such as keel blocks, pier piling impacts, tug bow pushes, equipment foundations), secondary (panel loads like water pushing up on the bottom of the hull) and primary (entire hull bends like a girder). For very much smaller ships, the tertiary loads predominate. For larger ships, the need to resist primary loads dominates. The damage that Lightoller complains about was due to tertiary loads that are primarily resisted by the side plates.

The increase in dimensions and weights of passenger ships from 1841 to 1940 are shown in Table I. Corlett [1971] has a midships section of the first iron hulled passenger ship, the *Great Britain.* John [1887] has midships sections of three steel 1880s Atlantic passenger ships. Hanscom [1903] has the midships section of the large passenger-cargo ship *Minnesota* of 1904. While it was a US built ship, Lloyds classed it. Lure [1907] shows that the *Lusitania* was a relatively strong ship with a High Tensile Strength (HTS) sheer strake and doublers, with the doublers being three plating strakes high at midships. The *Titanic*'s plating is from Garzke [1981]. The plating thicknesses of the ships are compared in Table II. It can be seen that the *Titanic's* hull plate was not especially thin. The *Titanic's* hull plates are similar to the larger *Queen Mary*. [Pigott, 1937]. The 1931 Italian liner *Rex* is very similar to the *Titanic* in dimensions, displacement and gross tonnage. It had noticeably lighter scantlings, although that was compensated for by the use of HTS steel; it did not have a sheerline doubler plate but one was fitted to the deck stringer plate [Bates, 1947]. While using stronger steel allows equivalent strength in longitudinal bending for less weight, the thinner HTS plates are not as resistant to side loads because the plate thickness helps provide section modulus to resist that kind of loading.

Strength of Seams

Relying on closely spaced transverse frames to hold the hull in shape, large wooden ships were built with longitudinal seams which had no strength at all since they were caulked to keep out the water but not physically connected. The *Great Eastern*, which was an exceptionally large ship for its time, introduced longitudinal framing plus a double hull but the brilliant structural engineer Brunel was satisfied with the external longitudinal seams being joined by a single row of rivets. The British shipbuilders generally used a double row of rivets as standard for longitudinal seams since that construction gave a good tight joint. The naval architectural books at the time had little design guidance on either expected loads or required strength of longitudinal seams.

Energy in a High Mass Hull

The sudden spurt in ship size at the turn of the century caused a large increase in mass. The *Titanic* had four times the mass of the 1880 generation of passenger ships so, when it bumped the pier, four

Table I: Passenger Ships Dimensions 1841-1938

Name	Date	Length, over all	Beam	Gross Tonnage	Displacement
Great Britain	1841	289 ft	50.5 ft	5,780	3,675 tons
City of Rome	1881	560 ft	52 ft	8,144	13,500 tons
Servia	1881	515 ft	52 ft	7,392	12,300 tons
Oregon	1883	500 ft	54 ft	7,375	12,500 tons
Campania	1893	601 ft	65 ft	12,950	21,630 tons
Kaiser Wilheim der Grosse	1897	627 ft	66 ft	14,349	20,550 tons
Kronprinzessin Cecile	1903	685 ft	72 ft	19,503	32,130 tons
George Washington	1909	699 ft	78 ft	25,569	39,435 tons
Minnesota	1904	630 ft	73.5 ft	21,000	33,000 tons
Lusitania	1907	760 ft	88 ft	31,550	37,960 tons
Titanic	1912	883 ft	92.5 ft	46,328	52,000 tons
Vaterland	1914	907 ft	100 ft	52,820	69,000 tons

Europa	1930	890 ft	102 ft	49,746	55,500 tons
Manhattan	1932	668 ft	86 ft	24,289	33,250 tons
Conti Di Savoia	1932	815 ft	96 ft	48,502	40,000 tons
Rex	1932	880 ft	97 ft	51,062	45,800 tons
Normandie	1935	981 ft	118 ft	83,423	67,500 tons
Queen Mary	1938	1020 ft	118 ft	80,770	77,400 tons
America	1940	723 ft	93 ft	33,961	35,440 tons

Table II: Passenger Ships Plating 1841-1938

Name	Sheer Strake	Waterline Plate	Turn of Bilge	Bottom Plate
Great Britain	.5 in	.63 in	.63 in	.69 in
City of Rome	.75 in	.69 in	.75 in	.69 in
Servia	.81 in	.69 in	.75 in	.75 in
Oregon	1 in	.81 in	.94 in	.875 in
Minnesota	1 in +1 in	.85 in	.90 in	.85 in
Lusitania	1 in + 1.05 in	1.05 in	1.05 in	1.05 in
Titanic	1 in +1 in	1.0 in	1.0 in	1.0 in
Rex	.75 in	.75 in	.81 in	.81 in

times the energy had to be absorbed as point loads by side plates only slightly thicker (1.0 in vice .69 in). The acceptable speeds which the officers were accustomed to using, when coming alongside a pier, were learned by maneuvering the earlier generation of smaller ships. They learned that coming up and gently bumping a pier at a certain speed would do no harm. They tried coming alongside at the same speed with the larger ships and had problems. Thus to the operators, not used to handling the mass of huge ships, the big ships seemed more likely to receive damage, hence weaker. Officer Lightoller came close to diagnosing the problem in the paragraph following the one quoted above:

"Then ships grew in size, out of all proportion to their strength, till one would see a modern liner brought with all the skill and care possible, fall slowly, and ever so gently on a knuckle, to bend and dent a plate like a piece of tin"

It was not physical size but mass that made the difference in ship response to bumping. The big ships did not have weaker structure but it was their greater mass (hence energy) that made them seem to be more easily damaged by bumping into things at what had before been safe speeds.

The even more massive *Queen Mary* was not seen as an especially weak ship, even though her plates were not much thicker than those of the *Titanic,* because her officers had become accustomed to handling a very heavy ship. One also suspects that another aid to resolving the pier bumping problem was better fendering technology that provided the large rubber shock absorbers along the pier.

The largest ship to collide with an iceberg prior to 1912 was the *Kronprinz Wilhelm* in 1907 (14,000 tons weight and 14 knots) and it scraped down her side without causing much damage. [Schaum, 1981] In comparison to the energy in the 1907 collision, the *Titanic* (56,000 tons travelling at 21 knots) had nine times as much energy stored in her moving hull as any previous ship that had hit an iceberg. However, the actual collision did not transfer much of that energy to the iceberg – there would have been a noticeable loss of speed and people plus crockery being thrown about. The impact was felt but it was not violent, unlike previous head-on impacts such as the *Arizona* of 1879.

Structural Response to Loads

The three fundamentals of structural strength are materials, fasteners and shape. There have already been widespread discussions about whether material or fasteners were the problem. [Hooper, 2003; Woytowich, 2003]

The steel and rivets or that era do not meet modern standards but ships built at the same time as the *Titanic* in the same yard had successful 40-year careers. [Hackett, 1996] There are three specific arguments against their being anything excessively wrong with the steel or rivets of *Olympic* class ships. The first two are the collisions where the *Olympic* sank other steel ships with little damage to herself.

In May of 1918, the *Olympic* took the opportunity to run down the attacking U-*103*. The submarine was 234 ft long, 21 ft beam with a surfaced displacement of 838 tons. The captain reported the violence of the impact "…jumped us off our feet on the bridge." The *Olympic's* bow was bent sideways eight feet to port but only dented and not torn open. [Eaton, 1990]

The *Olympic* also ran down the Nantucket lightship on May 15, 1934. The victim, Light Vessel *LV 117*, had a length of 135 ft, a beam 30 ft, and a draft of 13 ft. The ship displaced of 630 tons. The *Olympic* had "very little damage". A ship that destroys other ships, with little damage to herself, is not built of brittle materials or with defective fasteners.

The third specific argument is evidence from the *Titanic* on the bottom. The photographs of the wreck, compiled by artists into underwater wreck paintings, show "stress wrinkles" in bow due to the ship hitting the bottom. [Ballard, 1987] In places where the stress was not enough to initiate tearing, the wrinkled side shell plate still has its rivets in place. One can not get that kind of plastic deformation with excessively brittle steel or weak rivets.

A Ships Sensitivity to Damage

For a typical set of ship lines, the floodable length curve has a "camel" shape. The bow and stern can accept a large length of flooding because, the hull being narrow, there is less flooding water to pull the ship down and cause trim. In the midships area, the curve rises because, while a large amount of flooding water pulls the ship down, it does so without creating much trim. The forward and aft quarter length points in the curve are lower because the flooding causes a combination of increased draft and strong trim. Thus, ships are at greater risk from a fixed amount of damage at the quarter points.

Shape of the hull

Figure 7 is a schematic of three major bridge types: an arch, a suspension bridge and flat bridge. A force applied on an arch creates compression throughout the material. A force applied to a suspension bridge creates a membrane tension load along the structure. A side force on a girder bridge produces a strong shearing load and does not use the compressive/tensile strength of the structural plate.

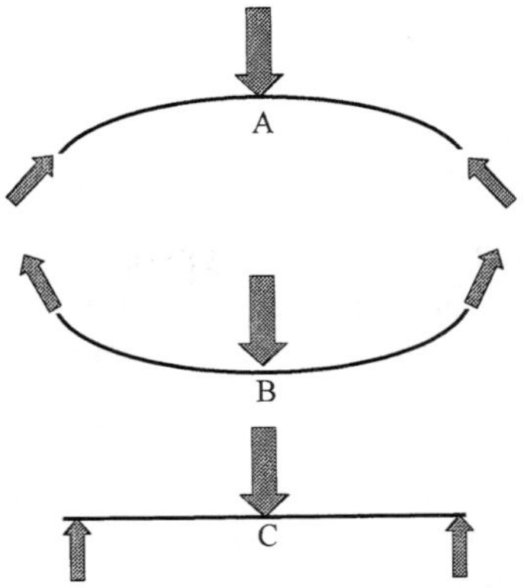

Figure 7 Response of three bridge shapes (arch, suspension and flat) to a load

When a highly curved ship's side is impacted with a side load, the side shell responds similarly to a loaded bridge acting as either an arch in compression or membrane in tension. In both cases, the hull plate contributes to resisting the load. At the joint, it is not just the cross sectional strength of the rivet carrying the load but, because of the compression of the two plates together due to rivet shrinkage, there is additionally the friction between plates transferring the load away. Experiments showed that the plate friction was 40% of the strength of the joint. However, if a side load pushes on flat side plates, they respond like a girder bridge. The plates are not in either compression or tension and pass the predominately shearing load to the joint. At the joint, only the longitudinal strength of the rivet shank and its head carries the load.

The frontispiece photograph shows the smaller ship, the *Majestic*, has a more highly shaped hull with much of the surface concave in or rounded outward. The most highly shaped part of the hull is the sensitive quarter points. The flat plate parts of the hull (right near the bow and midships) are parallel to the direction of the ships direction of movement. The sudden increase in sectional area in the quarter points, needed to get low hull width for cargo and machinery, creates a shoulder wave in that region. The *Majestic* needed a highly shaped hull, once the length was set for speed plus carrying capacity and the beam for stability, the hull needed enough draft to keep the screws submerged. It had been ships with highly shaped hull that had been having success with icebergs the previous decades.

As a ship gets larger, excessive draft becomes a problem. Large ships naturally evolve into a U shaped midship section that is carried forward. A long, almost vertical, narrowing toward the bow is easier to construct and easier to arrange the spaces within, than a hull with shape. The weight of a large ship is going up as a cubic (length, beam, depth) while the amount of drydock load bearing surface is increasing only as length times beam. Making the hull as U shaped as possible increases the available bottom support for a large ship.

The second ship in the frontispiece, the *Britannic*, shows the extensive flat side hull surfaces inherent in a U shaped large ship. A higher portion of the underwater hull is almost flat and the flat plate area covers the sensitive quarter points. Instead of being in areas parallel to the ships motion, the flat plate area leading away from the bow forms a wedge shape that, once contact is made, continues to push on the object over a long length.

The fluid flow around the bow has changed with the change in hull shape. The gradual widening as the ship goes aft eliminates the sharp sectional area change at the quarter points (that created a pushing away wave there). The large extent of flat plate provides a surface that creates strong Bernoulli suction along the entire bow area. When the suction of the Titanic pulled the liner *New York* free from her moorings during the Southampton departure, this demonstrated that, due to size and hull shape, the *Olympic* class ships were capable of producing a substantial Bernoulli force suction on a nearby object. Once in close proximity to a floating object, the large flat surface would tend to resist being pulled away by the rudder.

The Low Bulkheads of the *Titanic*

Because of their side armor, warships coaled through their upper decks. A series of canvas chutes were rigged from the weather deck to the top-of-bunker deck via loading holes (that normally had portable cover plates installed). The coal was dumped on the weather deck from shore or barge mounted cranes. The crew shoveled the coal into the top of the chutes and then, after finishing, de-rigged the installation. Chute rigging, coal shoveling and disassembly were manpower intensive but the worst feature was the inevitable leakage of coal dust into the ship. This resulted in even more labor consumed by cleaning. For a highly manned warship, the labor involved was considered free, since the crew was carried to man the guns, and, one suspects, the officers felt the exercise was good for crew and its discipline.

Passenger ships could not afford to waste any manpower, had to meet strict port turn-around times and, above all, could not afford to get coal inside the ship to bother the passengers. Their high freeboard would require use of very large cranes if the coal were to be landed on the top deck and the owners would be locked into buying coal from the person who owned the crane on the dock. Thus passenger ships needed an automated system that could get coal from a variety of small barges that could be brought along the outboard side of the ship. Figure 8 shows a schematic of such a system. [Reisenberg, 1922]

Figure 9, also from Riesenberg, shows a photo of such a system in action.

An obvious question is how were the side doors sealed. One authority says, contrasting the Royal Yacht to normal warship practice:

"In Royal Yachts coaling shoots between decks are objectionable, and in such cases coaling is not carried out overall as in warships, but through the side as in Merchant ships. Square ports are cut in the side plating just above the deck forming the crown of the bunkers. A port lid, hinged at the bottom and provided with flanged side plates, to form guides for the coal, are fitted to the ports. The port is cased-in dust tight, with plating on the inside of the ship and a hole being cut in the crown of the bunker inside the casing. Coal can be thrown direct into the bunker. Water tightness of the ports is obtained by red lead joints and they are closed home by hooks and bottle screws which are manipulated through hand holes in the steel casing. Wooden stages abreast each port are slung from the side of the ship.

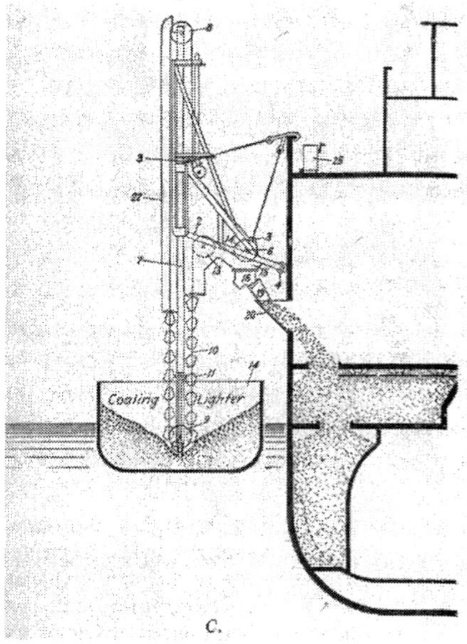

Figure 8 Merchant Ship Coaling from a Barge via Side Ports

Figure 9 - Side port coal loading of S.S. *George Washington*. The machinery stands out because the hull is painted in temporary wartime gray.

The coal bags are landed on these stages, and men stand on them for working the bags, as shown in Fig. 86. The arrangement is termed a cheese port." [Admiralty, 1909]

The Fig 86 referred to in the original text is reproduced here as Figure 10.

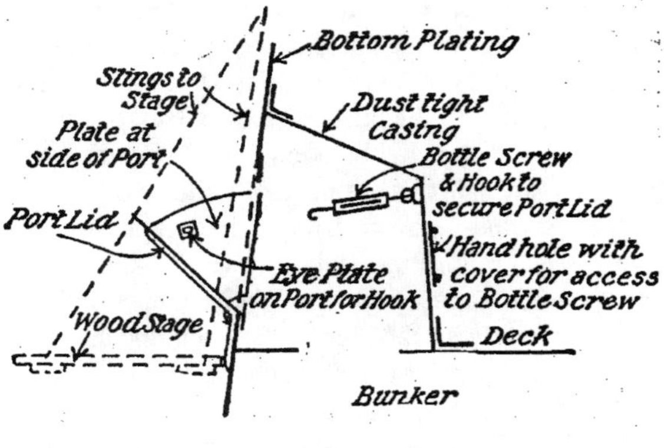

Figure 10: Configuration of coaling port and door attachment.

The use of side coaling ports explains why the almost universal hull color of 1900 vintage ships was black. Since the sides would be inevitably streaked with coal dust, black was the best color. Photos of the WWI hospital ships, such as the *Britannic*, show smudges along their white sides. Modern white hulls are possible only with oil fuel.

The water tightness of side coaling ports after years of service would be questionable. Coaling chutes would continually bang the port lid. The red lead joints would have lumps of coal crushed into the soft material every time that the lid was closed. The crew had to reach into the dark and dirty trunk through a small hole to attach hooks and screw them tight and had no means of inspecting or testing the tightness of the closure.

The *Titanic* has been widely criticized for having low (although closely spaced) transverse bulkheads forward. Figure 11 shows the location of the foremost coal bunker and the associated side port loading door on the profile of the *Titanic*. Note that the loading doors are just below "E Deck" at the top of "F Deck". Figure 12 shows a flooding line that would start to reach the possibly leaky coal ports; the *Titanic*'s formula of low-height but closely spaced bulkheads was in order to protect the coal ports. In a bit of irony, probably because she was such a new ship, the *Titanic*'s coal ports appeared to have been essentially watertight during the ships actual sinking since there was not a rapid increase in flooding rate when the waterline reached them. The existence of known low-in-hull flooding routes, that unexpectedly actually held tight, may be an explanation of Thomas Andrew's pessimistic ("two hours at most") projection of ship's survival time after his inspection.

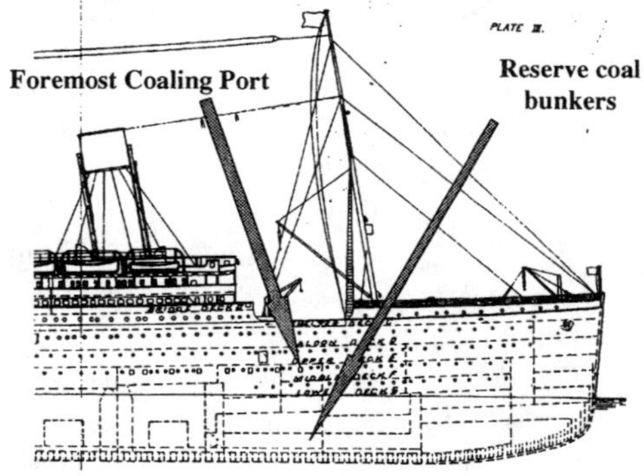

Figure 11: Location of forward coaling ports on the *Titanic*

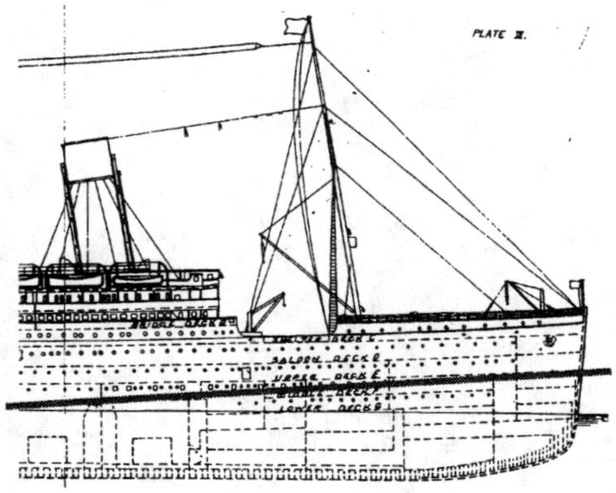

Figure 12: *Titanic*'s bow flooding waterline that begins to reach the coal ports.

The post-*Titanic* improvements to the *Britannic* are raised bulkheads forward, double hull in the center part so the ship and "increased pumping capacity". One can not expect a ship to pump out the water of a direct hole, but fitting increased pumping might counter leaking through battered coal ports. The widely discussed "fix' to the *Titanic* of fitting high bulkheads forward is really only valid for oil fueled ships.

Bow Shape

The Titanic had a nearly vertical stem and only slight flare in the bow. Figure 13 shows a modern cruise ship bow that is a combination of radical rake forward, flare to the sides and bulbous bow. Bows of this configuration are almost universal on modern passenger ships. A modern bow is a multi-purpose device. The rake and flare push bow spray away from the ship and allow mooring access above the bulb. It also provides intact buoyancy far forward and above the water. It, along with the modern transom stern placing weight well aft, insures that that the primary bending moment of a modern ship is hogging.

If the Titanic were fitted with a modern bow would it had made any difference? If one assumes the trim inducing extra weight is compensated by transom stern, could the intact buoyancy of the bow and bulb, forward of the actual damage keep the ship up? Rough calculation on a hypothetical modern bow fitted to the *Titanic* shows that it is just not enough floatation there to do the job (even assuming an intact bulb forward of the forward perpendicular). The hull flare further aft gives additional hull volume that the slow leak would have to flood but that additional volume gives only a few more minutes afloat at the flooding rated postulated by Hackett.

Figure 13: Modern Bow on the *Royal Princess*

Other Flooding Routes

The *Titanic*'s boiler rooms had ash ejectors discharging in between the middle and upper deck [Shipbuilder, 1911] that would also be hard to keep reliably sealed. A careful check of survivor testimony in a timeline prepared by William Garzke shows flooding water in the ship that was not coming from the initial damage [Maryland Science Center]. One possible source is plumbing drainage, normally well above the waterline, which had become a route of inflooding.

An ambiguous safety device is the size of portholes fitted to ships. Three German liners (the *Bremen*, *Salle* and *Main*), next to a Hoboken pier in 1900, caught fire due to a fire started on the pier. People aboard the ships died because they could not go above through the fire or escape out the too-small portholes. The safety lesson-learned was that portholes should be large enough to squirm out off. [Miller, 1984] However, open large portholes helped rapidly flood the *Lusitania* and *Britannic.*

Self Closing Watertight Doors

Wooden ships are almost impossible to transversely compartment since their flexible working, in a seaway, would open up the dried-out bulkhead seams. The invention of the iron hull was soon followed by the invention of the transverse watertight bulkhead. The crews did not like this obstruction of their previously unrestricted fore and aft passage within the hull, especially lack of one-level access to the coal bunkers. This became one of the many struggles where naval architects try to make things tight and marine engineers like to put holes in them. The compromise solution was bulkhead doors but open doors are a problem. The steamship *Oregon* was rammed by a schooner in 1886 and the crew could not get her doors shut due to coal dust in the tracks. [Gibbs, 1957] After the *Cromartyshire* rammed the *Bourgogne*, the engine room crew fled leaving a bulkhead door open so the ship sank. [Talbot ,1912] Seeking to make sure doors would be closed in an accident, the Norddeutscher Lloyd company engaged the British engineering firm of Stone and Company to develop the first remotely controlled self-closing watertight door. They were first fitted to Kaiser *Wilhem der Grosse* in 1897 causing concern by the crew that their escape routes being blocked. They were not accepted by mariners until the doors were fitted with local controls to allow the below-decks crew to open them to escape but they were also fitted with a timer so the door would automatically self-close again and not depend on crew actions.

In spite of hull improvements based on the *Titanic* sinking, the sister ship Britannic sank because floating debris jammed a major door open.
By contrast, the *Titanic*'s doors were successful as safety devices allowing crew to assess damage and attempt pumping, via hoses through the door, and then allowing the sealing up of the bulkhead.

The sinking of the *Empress of Ireland*, Figure 14, by the ramming of the collier *Storstad* was the naval architect's nightmare. It was the foreseen, agreed upon, economically preventable accident and the system failed, resulting in a quickly capsized ship and more passengers drowned than in the *Titanic* sinking. A RINA paper [Reid, 1916], based on a careful inspection of the bow of the surviving ship, shows that the collision was a single point penetration of a classic "T-bone" ramming and not a raking blow. By examining imprint of the rammed ship's rivets and side doors left on collier's bow, the exact location of the hit was established.

Figure 14: The ship that failed the test, the *Empress of Ireland*

The formal investigation concluded that the *Empress of Ireland*'s self-closing watertight doors didn't close. Massive flooding ran down the length of the ship, followed by capsizing due to sheer loss of waterplane.

By contrast, the *Kaiser Wilhelm II* had taken the precaution of shutting her watertight doors while travelling in a fog on June 17, 1914 so when she was rammed by the *Incemore,* the crew could control the flooding. [Shaum, 1981]

Operators

Could the *Titanic*'s crew have done anything differently? There were lights of a stopped ship visible in the distance shortly after the collision. Observers on board guessed a distance between 4 and 20 miles. Later calculations, based on the geometry of the lights, placed the *California* between 7 and 16 miles from the Titanic. [Kuntz, 1998] It is intuitively obvious that pushing a ship through the water with damage forward makes the situation worse, as it creates additional pressure forcing water into the ship. In fact, as early as 1878, model tests were conducted on what would happen to the battleship *Inflexible*, after her unarmored ends were holed, as she was driven forward.

The *Inflexible* has a small heavily armored central box upon which the ship would float in still water. The model tests were sponsored by Mr. R. E. Froude for the Committee of Designs, which sat in 1877-1878, to find out what would happen if the damaged ship moved ahead. The results were, as shown in Figure 15, increasing trim by the bow and in some of the tests the bow was driven under. [Admiralty, 1909]

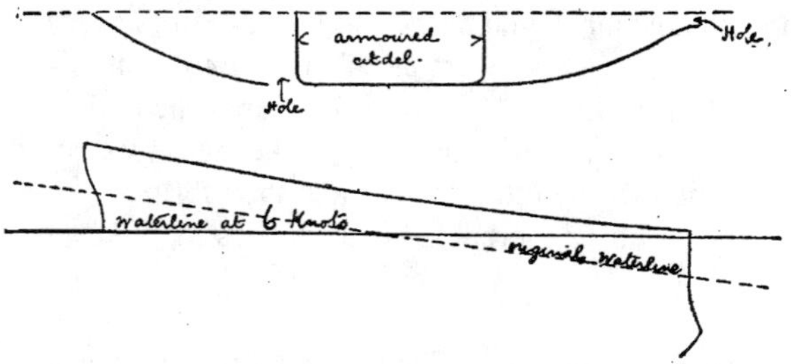

Figure 15: HMS Invincible Damaged Ship Model Tests

If it is obvious that going forward makes things worse, what about going backwards? The back part of a hull moving through water, in a region of suction, is generally exposed to less flooding pressure not more. [Saunders, 1956] Backing may not improve flooding in a damaged bow but it doesn't make it worse. In 1916 the crew knew enough to back the badly-bow-damaged battlecruiser *Seydlitz* the final leg of the trip back to Germany after Jutland.

The *Titanic* could have backed toward the ship in the distance for an hour at 15 knots, pulled along side and still had time to get people into the boats. Either the crew didn't think of backing or Thomas Andrews pessimistic time estimates convinced them they didn't have enough time to try it, since they did not know how long it would take to lower the lifeboats. A contributing factor was the builder not providing the crew with damage scenarios and responses options that a crew could use.

The Ships that did not sink

With the famous wartime sinkings of the *Lusitania* and *Britannic* coupled with the peacetime losses of *Titanic* and *Empress of Ireland*, there is a public impression that ships of the time were unable tolerate damage. In fact, the training and attitude of the crew could determine the fate of the ship with that technology. Appendix III recounts how tough passenger ships of that era could be if the crew was determined to keep them afloat.

Section II: Post-*Titanic* Ships

Short Stacks and Lifeboats

By 1929, the streamlined *Bremen,* shown in Figure 16, set a new fashion of short stacks. Why did the *Bremen* not need large stacks? The *Bremen* had short stacks because, being a product of the "streamline era" of design, tall stacks would be seen as totally old fashioned. It was hoped that, as an oil burner, the cinder problem no longer existed so the hot exhaust air would carry itself away. In fact, the ship's stacks shown were too short, even for an oil-fueled ship, since soot and drops of unburned oil landed on deck. [Brinnin, 1971] The stacks were first extended by the Germans and then further heightened by the French when she was the *Liberte*.

Figure 16: The new 1920s fashion – short stacks

What took the place of the tall stacks in motion control? The post-*Titanic* requirement for a full compliment of lifeboats plus the invention of the gravity davit (which lifted the boats above the promenade deck) put a large mass of boats at the corners of the ship. The lifeboats replaced stacks on the 1930s ships as part of the means of providing a high gyradius, hence, nice sea performance. The gravity davit, lifting boats up as in Figure 17, also converted the unwanted deck clutter of lifeboats to a welcome supplier of shade or light rain protection to deck chairs. Since the boats could be lowered and passengers put aboard by a limited number of non-technical crew, there was no longer a large number of seamen around the boat tempted to take it for themselves, as in the *Arctic* incident

After WWI, rudders were powered with compact electro-mechanical machines (not requiring so much ventilation and drainage) controlled electrically (so no straight-line high connections required). Steering gear could be at any level in the stern and so, except for a few deliberate fashion statements like the *Independence* class of 1950, sterns could be any shape desired.

Changes in Design Practice

By 1916, naval architectural texts contained, for the first time, guidance on designing longitudinal seams. [Holms, 1916]. For ships over 480 ft in length, and having plated more than 0.86 in thick, the new Lloyds' Rules called for all longitudinal seams to be triple riveted over 80 percent of the ship's length.

The later passenger liners raised the bulkheads but no references have been found of them having their overboard drainage fitted with check valves to prevent flooding. A book on design, published in 1955, says that drains from fixtures on decks 5 ft or more above the deepest load waterline can be discharged overboard via scupper valves at or near the load waterline. [Ericson, 1955] This brings into

Figure 17: The installation of Gravity Davits solved the lifeboat clutter problem

question the whole concept of "margin line" in the floodable length calculation for ships with overboard drains.

The introduction of mechanical ventilation followed by a limited amount air-conditioning in pre-WWII ship meant that some portholes began to be replaced with portlights that reduced the possible porthole flooding path.

1929 SOLAS Requirements

The *Titanic*'s collision with the iceberg opened up holes in 6 compartments (fore peak, holds 1 through 3, Boiler Rooms numbers five and six). Even with sonar imaging below the mud line, the exact extent of damage to the *Titanic*'s shell is not known. {Garzke, 1996] However, the distance between the forward bulkhead of Hold 1 and the aft bulkhead of Boiler Room #6 is 201 ft, so that is the minimum extent of the damage. Placing the fore peak and boiler room #5 damage in their centers defines the actual damage length as approximately 250 ft on an 883 ft waterline hull or 28% of the hull length..

An extensive paper on the possible sinking modes of the *Titanic* was prepared by Hackett and Bedford [1996]. It provides the *Titanic*'s floodable length curve which shows that the *Titanic* could survive four compartment flooding forward, so the designed survivable damage is the just under the length of three shortest compartments forward (Holds 1 though 3), thus a total 147 feet.

A 1947 SNAME paper contains the floodable length curves and compartmentation of many mid-20th Century Ocean liners [Bates, 1947]. By looking at the floodable length curves in Bates, it is possible to compare the post-1929 SOLAS ships damage resistance forward to the actual *Titanic* damage and the proportional length of damage.

The *Titanic* of 1912 had compartmentation that was equivalent to that of the later passenger liners of the 1920s and 1930s. The length of damage suffered by the *Titanic* (either the absolute length or a similar percentage of length between perpendiculars) would have sunk every post WWI ship that has had its floodable length curve published: the *Europa, Manhattan, Conti Di Savoia, Rex, Normandie* and *America.* The1933 Italian Liner *Rex*, figure 17, was very close in gross tonnage and dimension to the *Titanic*. Figure 18 is the floodable length curve of the *Rex*.

The fact that the Safety of Life at Sea (SOLAS) 1929 conference ships were compartmented no better than the *Titanic* reflects an international consensus that the *Titanic's* damage was both statistically unlikely (made more so by involutions like the Ice Patrol) and compartmentation to resist it was unaffordable to implement. The probable, agreed upon, affordable maritime threat remained the T-bone ramming.

One observer notes that the *Normandie*, due to having wing oil tanks in the forward engine room, would have survived shallow side penetrations such as the iceberg inflicted on the *Titanic*. [Garzke, 2002] This improvement was a fall out of introducing oil fuel, and not from improved international standards.

Figure 17 - The 1933 Italian liner *Rex* had same dimensions as the Titanic but that was no longer of remarkable size

Section III Modern Ships

Stacks on Modern Ships

The post WWII *Rotterdam, Canberra* and the Moore-McCormack twins, *Argentina* and *Brazil*, illustrated how small smoke pipes could be on modern diesel ships – they look like kingposts. However, in order to be fast, the *SS United States* had

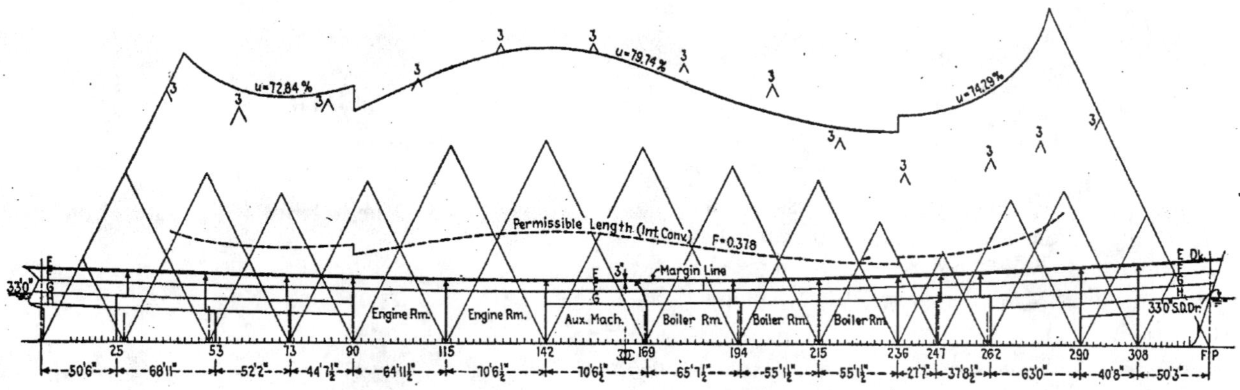

Figure 18 - The Floodable Length Curve of the *Rex*

a long hull and was fitted with a weight saving aluminum superstructure. A secondary effect of aluminum superstructure is to reduce the radius of gyration, making the ship more sensitive to sea induced roll motions. The ship's beam was set by both the need to fit the high-power steam machinery and troop transport stability needs. The long length, wide beam and low gyradius could have resulted in severe motions. Unlike her contemporaries, the *United States* had huge stacks that not only supplied an increase in gyradius but had an obvious above water stabilizing fin shape - streamlined airfoil shape facing ahead but presenting large flat high-drag plate to the side.

It turns out that another used of false smokestacks is as weight and moment reserves. If a ship is modernized, the imitation stack can be removed to compensate for growth elsewhere. The *America* lost her forward false stack to become the Chandris liner *Italis* and the *Kungsholm* of 1966 lost her forward fake funnel during modernization to become P&O ship *Sea Princess*.

Today's ships are getting bigger (more dimension-driven GM to absorb) but the luxury suites with balconies and swimming pools on top of a tall superstructure provide the raised center of gravity, air drag in roll and high radius of gyration that the huge stacks used to provide. The in-the-water stabilizing fins provide active roll control rather than relying on the less efficient passive anti-roll air drag of stacks and their supporting wires.

Modern SOLAS Requirements

The current Safety of Life at Sea rules use the length of ship, number of passengers carried, passenger space below the margin line, whole ship volume below the margin line, and machinery space volume to determine a factor of subdivision. The *Titanic*'s compartmentation would be required to meet the two compartment SOLAS standards. Hackett conducted the calculation and found the *Titanic*'s compartment lengths meet the modern criteria even with the low bulkheads forwards. They found the *Titanic* also meets the additional modern SOLAS rules on residual righting arm (lessons the *Andrea Doria* accident) except for not having enough residual stability in one damage case. That case is well away from the iceberg's actual damage. If fact, over much of her length, including the bow region, the *Titanic* greatly exceeds current SOLAS requirements. This is because the probable, agreed up and affordable threat contained in the modern standards is still the monohull-on-monohull T-bone ramming.

Lifeboats and liferafts

One very visible change on modern cruise liners from ships like the 1950s*United States* and 1930s *Queen Elizabeth* is that the row of lifeboats no longer completely lines the length of the ship. This is due to fitting of inflatable liferafts to insure the survival of the passengers. Since liferafts have no propulsion, in a rammed ship sinking, ramming ship remains afloat scenario, the solid lifeboats will return to their original shuttle function, even if it means towing rafts to the surviving ship.

Flooding Paths

One of the improvements in modern ships is the internal drainage system required by pollution control standards that prevent overboard discharge. It also prevents inadvertent flooding of a severely damaged ship as it sinks/trims to the margin line. However, there is some risk in deep penetration collisions of the internal pipes becoming flooding pathways themselves. Rescan [1955] mentions the need for check valves when a drain goes through a watertight transverse bulkhead and the pipe should be placed not less than one-fifth of the maximum beam of the ship inboard of the shell.

Modern air conditioning, hence the fitting of permanently closed windows in living spaces in the hull, has also eliminated the possibility of portholes becoming unexpected flooding route.

Growth in Ship Size

As during the decades prior to the Titanic, the last decade of the 20th century saw a remarkable spurt in ship size shown in Table III, a continuation of Table I. [Payne 1981; Hansa, 2001, P&O, 1996, Seaward, 2001] Figure 19 is a profile Comparison of the *Titanic* of 1912 to the *Voyager of the Seas* of 2000. The white ship in the foreground is the *Titanic* of 1912 (46,329 tons GRT, Length overall of 882.5 feet) while shaded in the background is the *Voyager of Seas* of 2000 (137,399 GRT, Length overall of 1020 feet). A person poolside on the *Voyager* is at the same height above the sea as a person perched on the top of the stack of the *Titanic*.

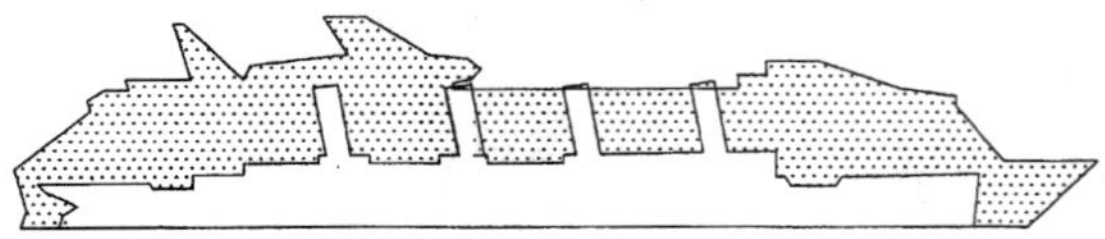

Figure 19 – Titanic in front of Voyager of the Seas

The owners of the even larger at 153,000 gross tons *Queen Mary II*, which is under construction, have take pains to point out that their ship is designed as an ocean liner (suitable for all oceans all the time) rather than a cruise ship (some oceans, some of the time)

Design to Rule

When preparing international standards, the rule writer tries to anticipate how it will be applied but fail to foresee novel interpretations. The 1948 SOLAS rules did not take into account that oil prices would be much cheaper in the United States than Italy so the post-war Italian line ships would have abnormally large fuel tanks in order to fill up for a round trip in the United States. Abnormally large tanks lead to unexpected listing after damage both because of their size and the crew trying to save loading time by not ballasting them. The *Andrea Doria* also had a classic "design to rule" error, where the ship designers assumed, since the 1948 SOLAS definition of the ramming penetration would not breach the side of a centerline longitudinal passageway, there was no need for a watertight door at its generator room end. The hypothetical ramming hole did not factor into account an ice strengthened bow like that of the *Stockholm.* The *Andria Doria* lost generating power because that passageway door was not fitted in a major transverse bulkhead. [Moscow, 1959] One suspects that original rule writers never thought that their rule would allow door-less access through major transverse bulkheads. The Italian Line's near sister ship *Cristofo Colombos* was quickly fitted with a watertight door at the end of her similar longitudinal passageway.

Engineers working with design standard become familiar with the phenomenon of "minimum-maximum" inversion. If it is a feature that the customer will not notice, the minimum acceptable standard will become the maximum that will be used. Hoping that, in a competitive environment, engineers will on their own do better than the minimum agreed upon standard has not shown to be a realistic outcome.

Density of Modern Ships

A look at Tables I and Table III shows the changing ratio of the weights and volumes of passenger ships. The *Titanic* had displacement-to-gross tonnage ratio of 1.1, which was typical for her era. Technology improvements, such as replacing low heating value coal with higher-energy content oil burned in improved steam plants (replaced, in turn, by diesel propulsion), reduced machinery weights. The introduction of welding took an additional 4% off of the hull structural weights so the 1950s ocean liners, such as the *Queen Elizabeth 2*, had a weight to volume ratio of about 0.8. The latest generation of cruise ships has a weight to volume ratio of .5 to .55.

One would expect that as modern ships picked up desirable machinery such as air conditioning, pollution control and elevators, there would be forces making them denser. Passengers, however, like large open atriums and shopping areas that have a low density. In older ships, the pillars in the theaters were always a nuisance to the passengers. Recently, the industry has provided the customers with what they desire by fitting the ships with large theaters without pillars. The weakness introduced into ship by these large unsupported spaces are certainly adequately covered by modern analysis of the known loads. The risk is the unknown, unanticipated loads on a ship such as would occur after major flooding.

In addition to large spaces, improved structural analysis has allowed a more sophisticated, hence lightweight, main hull structure. This has to be the source of a major portion of the density reduction since structure is always the major component of a ship's weight. However, there is something to be said for semi-sophisticated structure. The unsophisticated structure of the DC-3 has no known fatigue limit, and a few are still flying, while the improved lightweight structure of modern planes has actual fatigue limits. The fact that marine structural knowledge is not yet perfect is evidenced by the fitting of external box girders to the UK Type 42 Batch III and the French *George Leygues* class destroyers to overcome hull weakness.

Table III: Post WWII Passenger Ship Dimensions

Name	Date	Length, oa	Beam	Gross Tonnage	Displacement, mt
United States	1952	302 m	30.5 m	52,072	45,400
France	1961	316 m	33.7 m	66,438	51,815
Raffaello	1965	276 m	33.4 m	45,900	50,678
Queen Elizabeth 2	1969	293 m	32.0 m	67,140	49,657
Sky Princess	1984	240 m	27.8 m	43,692	29,629
Royal Princess	1984	231 m	29.2 m	44,588	24,841
Sovereign of the Seas	1988	268 m	32.2 m	73,192	35,000
Norwegian Sun	2001	258 m	32.0 m	78,809	39,000
Voyager of the Seas	2000	311 m	47.8 m	142,000	64,000

Section IV: Unusual Sources of Damage

Traditional Risks

From experience, it is easy to gauge the risk of travelling on a modern ship. The highest risk factor is anything with rubber tires on a public street, either getting to ship or visiting tourist sites. The next highest risk factor is anything with wings; the July 2000 Concorde crash killing 113 people, the majority headed to New York to join a cruise ship, was the highest modern loss of passengers related to an intended sea trip. However, in spite of modern equipment, classic seafaring accidents have continued to occur. Although the cruise ship was the aggressor, the 1999 ramming by the *Norwegian Dream* of the container ship *Ever Decent* in the English channel, shows that collisions of monohull-on-monohull continue to be possible, in spite of radar and collision avoidance aids.

However, the coming of the large bulbous bow has certainly changed the way rammed ships are penetrated from the classic vertical V tear in the side plate with a wedge shaped penetration into the hull. The first news pictures of the destroyer *Radford*'s collision showed only a tilted 5-inch gun and minor above water damage due to the flare of the cargo ship's bow. As figure 20 shows, the underwater damage was much more severe than could be deduced from the above water pictures. A modern ship's bow, with flare and a bulb, can cause superficial topside damage, a mid-level of undamaged plate and severe penetration into the hull low in the ship.

There is the possibility of an even worse form of bulbous bow penetration if the proposed very high-speed monohulls or trimarans with wave piercing bows join the world maritime community in any great numbers. Figure 21 shows the concept of one such ship. With a conventionally shaped bow, the flared above water portion tends to lock the collision to a side ramming. With a wave-piercing bow, there is a risk of under-running the victim, cutting off the bottom of its hull girder. The broken hull girder combined with the unusual loads from flooding water could result in structural breaking of the rammed ship's hull.

Risk of "Iceberg" Damage to Modern Ships.

The number one safety device for dealing with icebergs is radar - icebergs always have the right of way. We do not expect modern ships to absorb glancing blows from icebergs. In spite of modern global positioning systems, reefs can still inflict long raking damage. The *Monarch of Seas* grounded off St. Martin flooding 3 of her 18 compartments. It was severe enough damage to have the captain deliberately place the bow aground on softer bottom as the damage was sorted out. It was reported that

Figure 20- Underwater Bulbous Bow damage to the USS *Radford*

Figure 21- High-speed trimaran concept with wave piercing bow.

460 tons of plating had to be replaced. [Marine Log, 1999] By contrast, if the *Titanic* had survived, her repairs (consisting of plate 10 feet above and below the estimated 250 feet of damage) would have required the replacement of about 100 tons of steel. Photographs of *Monarch of the Seas* in drydock show, in addition to the flooding holes, a large area of the bottom wrinkled by the collision but still intact. A major advantage of modern ships is that the use of welded steel construction which is less likely to open up a seam when loaded from the side by a glancing collision. A communications link back to the owner's naval architectural staff provided the operator with resources not available in previous major flooding incidents.

Just as bulbous bows change the dynamics of monohull rammings, the existence of a multi-hull ship expands the possible shapes of collision damage. The SWATH type vessel, as shown in Figure 22, can cause two widely spaced bulbous bow rammings. They are probably not a statistically significant threat due to their limited numbers, slow speeds and their assigned area of operations. Their rough water oceanographic or towed array surveillance missions normally have them assigned to the high latitudes which are generally away from the main shipping lanes. The one exception is the largest SWATH, the *Raddison Diamond*, of 11,740 tons displacement (20,400 GRT) with molded breath of 32 meters, which operates in areas frequented by other cruise ships. However, that is just one ship and no sister ships have followed.

Figure 22: The configuration of a Small Waterplane Twin Hull (SWATH) vessel

A naval architect tends to think that the large powerful *Stockholm,* with its ice breaking bow, hitting the *Andrea Doria* as the worst case collision. The potentially most damaging modern collision could be with small ships/barges or large catamarans. The DD 990 (USS *Ingersoll*) collided in the Straits of Malacca with a small tanker, *Matsumi Maru #7*, on 20 June 1992 and sustained four compartment flooding as the small but dense ship bounced down her side.

Figure 23 shows two ships. The steel hulled monohull on the left is the agreed upon threat and any passenger ship meeting SOLAS rules would survive being rammed by it. The aluminum hulled catamaran on the right, one of many possible hull shapes for that high speed vessel type, can only make shallow penetrations into a steel ship they hit. However, the multiple collision holes will have a considerable longitudinal extent.

A large high-speed catamaran can have a beam of 25-30 meters which means, in a collision, the twin hulls can place twin holes 20-25 meters apart in the ship they hit. In Figure 18, the bulkheads at frames 236 and 262 are 19.9 meters (65.3 feet) apart. A collision with a large catamaran at that sensitive point would have caused four compartment flooding and loss of the *Rex.* If hit in the most sensitive part of their hull, that damage would be fatal to every one of the ships whose floodable length curves have been published. Published profile plans that include bulkhead locations show a large catamaran would open up three compartments on ships like the *Royal Princess*, *Oriana* and *Norwegian Sun*. [Naval Architect, 1985, The Motor Ship 1995 and Hansa, 2001] The ship with the best compartmentation to resist such damage is the bow region of the *Titanic* with its closely spaced bulkheads.

Figure 23: Two ships, a monohull and a catamaran

Unlike SWATHs, the catamaran ferries exist in considerable numbers and, in their ferry business, constantly cross main shipping lanes perpendicular to the flow of long distance traffic.

Another exotic new threat is giant Wing-in-Ground Effect (WIG) proposed for high-speed transport, shown in Figure 24. Named the "Pelican", the vehicle is reported in the press as having a wingspan of 500 feet and a speed of 200 to 225 mph at a cruising altitude of 20 feet above sea. [Cole, 2002] The worst part of a collision between a ship

Figure 24: A Wing-In-Ground Effect WIG Concept

and a WIG would be a World-Trade-Center-type of fuel fed fire but the reported possible cargo of 17 Army tanks (being of steel and dense) would become projectiles capable of doing immense structural damage.

The Way Forward

A designer of a future ship must be expected to create the best design for his vessel that he can, within physics and international maritime rules, regardless of what that vessel might do to another ship in collision. The problem of possible adverse interaction between various vessel types must be addressed by the maritime community at large. Using modeling and simulation, the maritime community should create a catalogue of unexpected, but possible, accidents between various ship types. The first priority for the results of such investigations is to provide the crews of existing ships information on what might possibly happen, and what their options are, after such an collision. The next, and most difficult, stage is for the maritime community to agree that the designs of future ramming and future rammed ships must be changed in order to reduce the damage inflicted.

It may be necessary to put something resembling a bowsprit on commercial wave piercing hulls to prevent diving under. Forcing more damage to the side of a victim is better than loss of the lower flange of the hull girder. Another option is to insure the bow structure, while strong enough for routine sea loads, is weak enough that it will rapidly collapse under ramming loads. For wide-but-shallow damage from multi-hulls, it may require a return to Brunel's original (and *Britannic's* backfit) formula of a narrow double hull up past the waterline or dramatically lowering the size of ships that should be designed to a three compartment standard. A modern double hull would require adequate cross flooding to prevent list from affecting use of the lifeboats and rafts.

Passenger airliners used to be navigated like ships. After clearing the local airport vicinity, the captain set his own chosen flight height, course and speed. He depended on the crew lookout and maneuvering to avoid any collision. That system did not work that well, cumulating in a two-airliner collision over the Grand Canyon in June 30, 1956. The resulting loss of 128 lives led to implementation of nationwide positive air traffic control by a third party for the entire flight. [Job, 2001] If large numbers of WIGs begin crossing the oceans, they would be so fast and dangerous that depending on own vessel lookouts and sensors and their own maneuvering carries the same risk as with airplanes. The implication of their speed is the need for equivalent of worldwide air traffic control by a third party, using transponder and satellite tracking, for both ships and WIGs.

Conclusions

Often noted but unexplained features of the *Titanic*, such as the fake stack and low bulkheads forward, can be explained by the problems facing the designers and through comparison to other ships. In discussing the failure mode of that ship's shell plates, hull shape has to be considered, along with plating and fasteners. What has not currently changed since the *Titanic's* era is the definition of the foreseen, probable, agreed-upon accident. It is still a monohull-on-monohull ramming; the level of compartmentation of modern ships has not changed very much and, in some respects, is less rigorous. The *Titanic*'s raking iceberg damage would have brought down all the post-1929 SOLAS ships of the 1930s because the agree-to damage had not been changed by the international community.

The foremost safety improvements of modern ships over older ships are electronic: radar, navigation and communications linking back to the owner's naval architectural staff. A full set of lifeboats and rafts plus the tear resistance of a welded hull are major improvements.

Modern ships have other features, not normally thought of as safety improvements, such as oil fuel, air conditioning and pollution control. They serve to eliminate side scuttles, portholes and outboard drains that could become in-flooding routes.

In spite of the marvelous safety record of modern ships, there are still concerns about marine safety. As in *Titanic*'s era, ships have undergone a major escalation in size while, simultaneously, their density is being decreased. There are increasing numbers of vessels that are capable of creating, not the expected *Stockholm*-into-*Andrea Doria* deep-V ramming hole, but *Titanic*-like damage (very shallow but wide) on the ships they collide with. Additional analysis and research into the interaction of the different hull types, being introduced in significant numbers to the world's oceans, should be considered by the maritime community. The goal would be to determine if new design standards should be placed on both possible ramming victims and the ship doing the damage.

Acknowledgements: This paper is compiled from a series of short papers presented to the Marine Forensics Analysis Panel (SD-7) in response to ideas raised during the panel meetings. The author would like to thank Mark Bebar for his reviewing and comments on the early draft.

Photo Credits. Older ship photos used by permission of the Steam Ship Historical Association collection at the University of Baltimore. Other sources are Figure 22 (Australian Navy), Figures 21 and 23 (USN) and Figure 23 (Boeing). Figures 3, 4, 5, 6,10, 11, 12 are from the quoted references. Remaining figures are by the author.

References

Admiralty, *Manual Of Seamanship, Vol. II,* By Authority Of The Lords Admiralty, His Majesty's Stationery Office, London, 1909, pg 479 and pg 558-560

Ballard, Robert D., Dr., *The Discovery of the Titanic,* Warner Books, New York, 1989, pg 121-122, 218

Bates, James L., and Wanless, Ivan J., "Aspects of Large Passenger Liner Design", *Transactions of the*

Society of Naval Architects and Marine Engineers, Vol 55, Society of Naval Architects and Marine Engineers, New York, 1947. This reference also has midships sections of the *Europa*, *Manhattan*, *Conte De Savoia*, *Normandie*, and *America*.

Board of Trade, "Report of the Committee appointed by the President of the Board of Trade to Consider and Report Up the Spacing and Construction of Watertight Bulkheads in the Ships of the Merchant Marine", Her Majesty's Stationary Office, London, 1891, pg 7.

Brinnin, John Malcom, *The Sway of the Grand Saloon*, Delacorte Press, New York, 1971, pg 194, pg 452.

Cole, William, http://www.boeing.com/news/ frontiers/ archive/2002/september/i_pw.html Information is repeated by Wilson, Cynthia, "The Next Biggest Thing?" *St. Louis Post-Dispatch*, October 9, 2002

Cornet, E.C.B., "The Steamship "Great Britain"", *Transactions of the Royal Institution of Naval Architects, Vol 113*, Royal Institution of Naval Architects, London, 1971.

Eaton, John P., and Haas, Charles A., *Falling Star: Misadventures of White Star Line Ships*, (W.W. Norton & Company, New York) 1990 pg 142.

Rescan, CE, "Hull Piping Systems", *Design and Construction of Steel Merchant Ships*, Society of Naval Architects and Marine Engineers, New York, 1955, pg 375.

Garzke, William, Jr., Brown, David K., Sandiford, Arthur, Woodward, and Hsu, Peter K., "The Titanic and Lusitania: A Final Forensic Analysis", *Marine Technology and SNAME News*, October 1996, Vol 33, Number 4, Society of Naval Architects and Marine Engineers, New York.

Garzke, William H., Jr and Woodward, John, B, *Titanic Ships, Titanic Disasters*, The Society of Naval Architects and Marine Engineers, Jersey City, 2002,

Gibbs, C.R. Vernon, *Passenger Liners of the Western Ocean*, Staples Press, London, 1957, pg 70, 71, 118, 187, 207.

Gleams, Albert, Vice Admiral, USN, *A History of the Transport Service*, George H. Doreen Co, new York, 1921, Quote, pg 85. Ship Weights Tables, pg 68 & 264.

Hackett, C., and Bedford, J.G., "The Sinking of S.S. Titanic; Investigated by Modern Techniques", Joint Meeting of The Royal Institution of Naval Architect and The Institution of Engineers and Shipbuilders in Scotland, 10 December 1996

Hansa, "Schiffbau Kreuzfahrtchiffe, *Norwegian Sun*" *Hansa International Maritime Journal*, October 2001, pg 34.

Hanscom, Charles R., "Description of the Design and Building of the 21,000 Ton Steamships *Minnesota* and *Dakota*, *Transaction of the Society of Naval Architects and Marine Engineers, Vol XI*, Society of Naval Architects and Marine Engineers, New York, 1903

Holms, A. Campbell, *Practical Shipbuilding – A Treatise on the Structural Design and Building of Modern Steel Vessels*, Longmans, Green & Co., 1916.

Hooper, Jennifer J., Foecke, Tim, Graham, Lori and Weihs, Timothy, "Metallurgical Analysis of Wrought Iron From the RMS *Titanic*", *Marine Technology and SNAME News*, Volume 40, Number 2, April 2003, Society of Naval Architects and Marine Engineers, New York.

Job, Macarthur, *Air Disaster, Volume 4*, Aerospace Publication Pty Ltd, Fyswick, Australia, 2001, pg 69-79.

John, W., "Atlantic Steamers", *Transactions of the Institution of Naval Architects, Vol XXVIII*, Institution of Naval Architects, London, 1887.

Kludas, Arnold, *Great Passenger Ships of the World, Volume 1: 1858-1912*, Patrick Stephens, Wellingborough, 1975, pg 54.

Kuntz, Tom, ed., *The Titanic Disaster Hearing: The Official Transcripts of the 1912 Senate Investigation*, Pocket books, New York, 1998, pg 509.

Lightoller, Commander, "Titanic" in *The Story of the Titanic as Told by Its Survivors*, ed. by Jack Winocour, Dover Publications, Inc, New York, 1960, pg 286.

Lure, W.J., "On Some Points of Interest in Connection with the Design, Building and Launching of the *Lusitania*, *Transactions of the Institution of*

Naval Architects, Vol XLIX , Institution of Naval Architects, London ,1907

Marine Log, Vol 104, No 4, *Royal Repair*, Simmons-Boardman, New York April 1999, pg 38.

Maryland Science Center, *Science of the Titanic Exhibit*, "Titanic Timeline", 2002.

Maxtone-Graham, John, *The Only Way to Cross*, Macmillan, New York, 1972, pg 31.

Miller, William H., *Famous Ocean Liners*, Patrick Stephens, Wellingborough, 1987, pg. 31.

Miller, William H. Jr., *The First Great Ocean Liners in Photographs,* Dover Publication, New York, 1984, pg 33.

Moscow, Alvin, *Collision Course*, Grosset & Dunlap, New York 1959 pg 300.

Motor Ship, "Ship Report-Oriana: Cruise Ship cater to British Market", The Motor Ship, June 1995. Pg 77.

Naval Architect, "Fit for a Princess", *The Naval Architect*, Royal Institution of Naval Architects, January, 1985, pg E40.

P&0, *Royal Princess Facts Sheet*, hand out to passengers, 1996.

Payne, S.M., "The Evolution of the Modern Cruise Liner", Royal Institution of Naval Architects, London, 1989. pg 172.

Pigott, S.J., "Some Special Features of the S.S. *Queen Mary*", *Transactions of the Institution of Naval Architects, Vol LXXIX*, Institution of Naval Architects, London, 1937.

Professional Mariner, "The Loss of the Arctic, 1854" *Professional Mariner #59*, pg 61.

Reid, John, "Notes from a Collision Investigation", *Transactions of the Institution of Naval Architects, Vol LVII* , Institution of Naval Architects, London, 1916, pg 211 and Plate XXI

Riesenberg, Felix, *Standard Seamanship for the Merchant Service,* New York, D. Van Norstrand Company, Inc., 1922.

Russell, James C. and Moore, William E., *The United States Navy in the World War,* Pictorial Bureau, Washington, 1921, p 142.

Schaum, John H, Jr. and Flayhart, William H., III, *Majesty at Sea: The Four Stackers*, W.W. Norton and Company, New York, 1981, pg 31.

Seaward, "Looking Seaward", Spring 2001, Seaward International, Inc., Clearbrook Virginia, pg 1. Gross and displacement of *Voyager of the Seas*

Shipbuilder, The, *Ocean Liners of the Past: White Star Triple Screw Atlantic Liners Olympic and Titanic*, Amereon House, 1995, Reprint of 1911 *Shipbuilder* Article, pg 121-124 and Fig 44, pg 45.

Talbot, Frederick A., *Steamship Conquest of the World*, J.B. Lippincott Company, Philadelphia, 1912, pg 89, 167.

White, Sir W.H., *A Manual of Naval Architecture,* London, John Murray, 1900, pp 154-155.

Woytowich, Richard, "Riveted Hull Joint Design in RMS *Titanic* and Other Pre-World War I Ships", *Marine Technology and SNAME News*, Volume 40, Number 2, April 2003, Society of Naval Architects and Marine Engineers, New York.

Appendix I: Comparative Naval Architecture

During the early 1970s, the US Navy's ship designs were routinely compared unfavorably to the new ships of the expanding Soviet Navy. A pair of sample criticisms were "A comparison with the Soviet *Kresta II* class of approximately the same tonnage shows the *Spruance* to be outclassed in all aspects except for anti-submarine operations and the possession of gas-turbines" and "a *Krivak*, although lacking her own helicopter, is more than a match for any Western destroyer and, in any event, could outrun anything opposed to her." [Moore, 1974]. The USN started a formal process of reverse engineering called Comparative Naval Architecture (C-N-A). [Kehoe, 1980] In the C-N-A process, the unusual rules of evidence are suspended as one set of designers try to work backwards from dimensions and photographs of a foreign ship to the requirements and technology that lead the competitor team to produce such a ship.

Open literature search revealed that the Italians had taught the Soviet designers to design large warships in the 1930s. [Meister, 1972] Italian designers have a long tradition of producing fast, heavily armed ships. They achieved this by using light structure, tight machinery boxes, and small store plus fuel loads predicated on Mediterranean sea states and distances from homeport.

One of the things that made the Soviet ships appear to be well armed was the presence of large missile tubes on deck, whereas the DD 963 and FF 1052 classes has only an unimpressive ASROC box As a product of this work, the American designers produced alternative studies of existing USN ships whose topsides were covered with large missile tubes that the commentators found so impressive. How was this increase in firepower achieved? It turned out that the Harpoon missile designers had put a major effort into fitting the missile with tightly folded wings and the solid fuel booster behind the missile, so it would fit into existing submarine torpedo tubes and aircraft bomb bays. The original US designers had put the folded missiles in magazines inside the ship.

The C-N-A ship designers just took the 16 Harpoon and ASROC missiles, already carried in the internal magazine behind the ASROC box of the FF 1052, and replaced them with on-deck Harpoon tubes of an existing design. Thus the "Sovietized" FF 1052 was suitably cluttered with missile tubes. [Kehoe, 1976].

In addition the Russian missile designers only partially folded their wings (increasing width), did not fold the tails and had the booster slung under the missile (both greatly increasing its depth). If an alternative Harpoon, with partially unfolded wings and booster underneath, were assumed then the ship would need big impressive deck tubes. However, the fitting of numerous big missile tubes have certain drawbacks: they increased the radar cross section, raised the center of gravity, interfered with underway replenishment plus boat handling operations, and provide more targets for enemy weapons to detonate than missiles hidden inside the ship.

The intellectual case made by C-N-A supported the American Navy proceeding further away from big tubes on deck to Vertical Launching Systems that lack any visual impact at all. No one today believes big missile tubes on deck are a sign of clever engineering.

Appendix I Additional References

Kehoe, James, W., Jr, Capt, Graham, Clark, Cdr, Brower, Kenneth S. and Meier, Herbert, "Comparative Naval Architecture Analysis of NATO and Soviet Frigates", *Naval Engineers Journal*, October 1980 (Part I) and December 1980 (Part II).

Kehoe, James, W., Jr, Capt "Warship Design – Ours and Theirs", *Naval Engineers Journal*, February 1976, pg 99.

Meister, Jurg, *The Soviet Navy I,* Part of "Navies of the Second World War" series, Doubleday & Company, Inc., Garden city, 1972, pg 35.

Moore, John E., Capt, "Forward", *Jane's Fighting Ships 1974-75*, Franklin Watts, Inc., New York 1974. pg 73, 75.

Appendix II: Passenger Ships Hitting Icebergs

Whenever a wooden ship came into heavy contact with ice, the results were invariably bad - splintered planks and split seams leading to sinking of the uncompartmented ship. However, early in the metal ship era, the ship operators discovered that iron was stronger than ice. By 1865 a Royal Navy Rear Admiral could write: " I have two years' experience in ramming the ice...... That was not like an isolated case of a vessel falling foul, but we ran the vessel's nose dead on to the ice, and did the ice more injury than the vessel, for the vessel never was injured in several years such service."[Belcher, 1865] By 1899, a technical paper on a distinctive type of vessel called an "Ice-Breaker" was presented. [Swann, 1899] The specialized type, with a riveted hull, had been in service since 1871. The Russian icebreaker *Yermak*, built by Armstrong in 1898, served until the 1960s. The was no fundamental incompatibility of riveted construction ships and hitting ice.

The Canadian National Research Council produced a list of every ship that they could discover which had ever hit an iceberg [Hill, 2000]. Data from 1874 through April 1912 from the web site has been cross-indexed by matching the ships' names, where possible, with ship dimensions and construction material taken from contemporary Lloyd's Register of Ships. Wooden hulled steam ships, sailing ships and sealing/fishing ships, the traditional fatal victims of icebergs, are discarded due to their inherent natural vulnerability. Then the iron/steel ships are sorted into those that sank and those that survived. By sorting the iron hull ocean-going ships data into 1) ships larger and 2) ships smaller than the biggest ship ever sunk by an iceberg with loss of life, a pattern emerges. It turns out that big (over 325 ft long, over 2,500 Gross Register Tons) modern (metal hull, steam powered) ships had been hitting and surviving iceberg collisions for about 30 years prior to the *Titanic* with 53 incidents survived by the bigger ships. The two exceptions to this grouping, the *Iowa* (4,329 GRT) in 1891 and the *Notting Hill* (4,021 GRT) in 1884, were abandoned afloat without loss of life. The *Iowa* personnel were rescued by the *SS Chester* which left the sinking ship, down by the bow, behind still on the surface as she proceeded on. The *Notting Hill* hit the iceberg on Feb 2nd and the ship,s crew and passengers were taken off three days later by the *SS State of Nebraska* and ship, with 18 inches of water in the hold, was left to slowly sink. To this day, such undramatic sinkings do not attract much attention since Hocking's compendium of disasters at sea doesn't even list the incidents [Hocking, 1990]. Hill included in his list the disappearance of the 6,694 ton cattle carrier *Naronic* in 1893 as ice incident although the board of inquiry and modern investigations conclude the iceberg damage note-in-a-bottle was a Hoax. [Eaton, 1990] Even the under 2,500 gross ton metal-hulled ships survived iceberg encounters by a 7 to 1 ratio.

What was not being accounted for in the record of successful iceberg impacts, was how close to failure some of the incidents were. Figure AII-1 is the famous *Arizona* (5,147 GRT) collision of 1879 that showed that a modern passenger ship could survive a direct, full speed collision with an iceberg. [Talbot, 1912] It was widely taken a proof of the strength of modern ships. However, the ramming tore the hull plates to within 2 feet of the collision bulkhead. [Maw 1884] Another surviving ship that almost had fatal damage was the *Nessmore* which, during one of her two (1883, 1890) iceberg collisions, had the following happen: "The British steamship *Nessmore* was passing through the southern fringe of the great icefield and hit a berg head-on, stoving in her bows. The ship made port safely and docked. Then it was found that she had scraped along a spur for some distance just above the keel, for there was the corroborative mark on her hull, while four frames were found broken, and many plates had been cut almost through." [Talbot, 1912]

This background of the late 19th and early 20th century large ships having a proven record of success in surviving collision with icebergs, while no one ashore was assigned to consider on how incidents might have been worse, explains the willingness to push the *Titanic* at top speed through known ice area – the crew knew they were taking a risk but the damage risk that they expected was, at worst, a few buckled bow plates and minor flooding that could be controlled by the pumps.

Figure AII-1: The iceberg damaged bow of the SS *Arizona*

Appendix II Additional References

Belcher, Rear-Admiral Sir, E., "On the Construction of Armour-Clad Ships of War", *Transactions of the Institution of Naval Architects, Volume VI* (1865, London).

Eaton, John P., and Haas, Charles A., *Falling Star; Misadventures of White Star Line Ships*, W.W. Norton and Company, (1990, New York) pgs 9-10, pg 50-61.

Hill, Brian T., *Database of Ship Collisions with Icebergs*, Institute for Marine Dynamics, National Research Council Canada, 2000 http://www.nrc.ca/imd/ice/bergs2_01e.html

Hocking, Charles, *Dictionary of Disasters at Sea during the Age of Steam 1824-1962*, London Stamp Exchange, London 1990)

Maw, William H., *Recent Practice In Marine Engineering, Vol. I., Text* (London, Offices Of "Engineering,", 1884), pg 145-148

Swan, H.F., "Ice-Breakers", *Transactions of the Institution of Naval Architects, Volume XLI* (1899, London).

Talbot, Frederick A., *Steamship Conquest of the World*, J.B. Lippincott Company (Philadelphia, 1912) pg 158, 167.

Appendix III: The Tough Survivors

The seized German liner *Kronprinzessin Cecile* (19,503 GRT; 32,130 tons displacement) was serving as the US troopship *Mount Vernon* when she was hit by one torpedo from the *U82* on September 5, 1918. Captain Dismukes had trained his crew that "Men, remember that one torpedo hit cannot sink your ship, but keep your water-tight doors shut". [Gleaves, 1921] The hit was on a bulkhead between two fire rooms and resulted in one hundred and fifty feet of flooding in a 685 foot long ship. The 7,000 tons of flooding water increased the undamaged draft from 30 feet to 40 feet. The ship slowed to 6 knots and took on a 15-degree list. [Gibbs, 1952] Thirty-five of the crew were killed on impact. Sailor Thomas F. Buckley, knocked down by explosion, jumped to his feet and yelled out "Remember boys, we are all Americans and it's only one hit." [Russell, 1921]. The crew turned to for damage control and got the speed up to fifteen knots as they returned to Brest.

Figure AIII-1; the *Justicia.*

The best example of how tough a WWI era ocean liner could be, if the crew kept her tight, is the *Justicia* shown in Figure AIII-1. The ship had been laid down as the *Statendam* for Holland American Lines but taken over by White Star as a troopship. On July 19th 1918, the 32,234 Gross Ton, 776 foot long ship was hit by at least one torpedo at 2 pm followed by another hit two hours later. As the wounded ship was taken under tow, it became the center of a multi-submarine vs destroyers battle. [Hurd, 1929] Mid-day on the 20th, a third torpedo attack was launched against the *Justicia* and it sank with the loss of 16 lives out of a crew of over 600.[Gibson, 1931] However, it took longer to finally sink after the third attack than the *Titanic* took after the iceberg collision.

Both these ships show the importance in keeping a ship, of a given design technology, afloat as long as possible of having a crew which expects damage and keeps the ship tight as possible prior to the flooding.

Appendix III Additional References

Hurd, Sir Archibald, *History of the Great War: he Merchant Navy, Vol III,* (Longmans, Green & Co, New York, 1929), pg 291-2

Gibson, R.H., and Prendergast, Maurice, *The German Submarine War 1914-1918,* (Constable and Co, London, 1931) pg 313.

DISCUSSION

William H. Garzke, Jr., Member

Mr. Sims is a valued member of the Marine Forensics Panel and has contributed much to the success of this Panel's activities. In 1996 he coauthored a paper on tumblehome warships to give an idea of the depth of his interest in ship design and operation.

Under the section, "*Titanic*'s Structure," Mr. Sims has quoted some structural theory by Second Officer Charles Lightoller of the *Titanic*. It should be realized that Lightoller's comments were used by Joseph Conrad in his blistering essay on the *Titanic*. Officer Lightoller was instrumental in suppressing the structural failure in his testimony before the U.S. and British inquiries into the loss of this vessel. By insisting that the ship sank intact, he provided crucial evidence that was false and misleading. If the British inquiry had concluded that the ship sustained a structural failure, there would have been further investigation to determine the cause. We now know that the ship did fail at the surface, and it was due to the massive flooding in the bow portion of the ship.

Naval Architect Edward Wilding used the standard longitudinal bending moment approach of the L/20 wave to prove that the ship did not have a structural failure. That calculation ignored the flooding in the bow and should have been done using the still water bending moment approach. However, the L/20 wave calculation is not a scientific approach to determine a ship's bending moment; rather, it is a comparison measure to determine structural adequacy between ships. The cargo ship *Wolverine State* was instrumented to record stresses in a seaway after World War II. There were no stresses recorded of the magnitude predicted by the L/20 wave. More sophisticated methods are now available to determine stresses that ships experience at sea.

On page 7 of the paper, Mr. Sims makes an important point about ships and icebergs. The potential energy is not 9, but 10, in the *Titanic* because she was making 22.5 knots at the time of the collision. As I have said in my presentations on the *Titanic*, she sank because she was going too fast in an area known to be dangerous with ice. In other words, "Speed kills!"

On page 8, the author discusses the shape of the hull and its effect during a collision. The major damage in the *Titanic* was in cargo hold No. 3 and boiler room No. 6. This is where the heaviest flooding took place after the first 30 minutes of the casualty. It was in the quarter point of the vessel where the hull began to flare out to its fullest width. This verifies the point that once contact is made, a force is exerted over a long length.

On page 11, Mr. Sims notes that flooding was coming in from areas above the margin line. When I examined the "Big Piece" shortly after its recovery in August 1998, I noticed that the plumbing drains above D Deck had no check valves because the discharges were located above the margin line. Once the *Titanic* sank below these openings, water was forced up into sinks and toilets. Examination of survivor testimony verified this in one area of the ship.

The author's comments on the size of stacks are most interesting. This might explain why the Italian liner *Rex* that had small stacks was outfitted with gyrostabilizer, one of the first passenger ships so equipped.

It should also be noted that the *Titanic* would meet many present-day SOLAS requirements, as shown by John Bedford and Chris Hackett in their 1995 RINA paper referenced by the author.

Author's Closure

The author would like to thank *Mr. Garzke* for his additional thoughts on the evolution of passenger ships. It is interesting that the Italian ship *Rex* was pioneering mechanical stability devices (very large gyroscopes) given the ship's New York–Gibraltar–Italy route through the relatively calm southern part of the North Atlantic. During the same period, the Germans experimented with antiroll tanks on the *Albert Ballin* class due to both the smaller size of their ships and their true North Atlantic route.

Mr. Garzke's comments were the only written input, and there were no questions from the floor. However, the paper later resulted in a fair amount of not-for-attribution hallway conversations with the author. The reticence of individuals to submit public discussion material may be due to the extremely competitive nature of the cruise ship design and construction industry. Comparative naval architecture is to regular ship design as art criticism is to art, and the artists are not always appreciative of being subject to it.

Just as the 1850s will always be remembered as the era of radical experimentation to perfect the clipper ships, the decade and half since 1990 will be remembered as the great era of innovation in cruise ships, especially with respect to lightweight structure and large size. These ships are the great Gothic cathedrals of our time.

The Impact of Evolutionary Acquisition On Naval Ship Design

Jack Abbott, Life Fellow, AOC, Inc., **Richard DeVries,** Member, AOC, Inc., **William Schoenster,** Visitor, NAVSEA/PEO Ships, **John Vasilakos,** Member, CSC, Inc.

ABSTRACT

New Department of Defense strategies have led to an approach of "Evolutionary Acquisition" for the design and construction of Navy warships. This strategy requires that future ships have an inherent flexibility in mission re-configuration, be capable of accommodating new technologies over the life of the ship and keep pace with the shorter life cycles associated with higher usage of commercial-off-the-shelf (COTS) products. Traditional ship design has typically focused on only the initial configuration of the combat system suite and its supporting systems. How can a ship be designed in the future so as to minimize the impacts to the basic hull and support systems caused by changing requirements either during acquisition while in production of different flights or later during modernization or conversion? This paper will discuss a ship design process that can address this new requirement of "adaptability" without making the ship significantly bigger or the removal/replacement cost of the original equipment prohibitive. Important aspects of the process include increased use of Technology Management and a modular Open Systems Approach.

INTRODUCTION

The Office of the Secretary of Defense (OSD) and the Navy have begun to implement a new strategy for the acquisition of weapon systems; a strategy that seeks to address both the uncertainties of current political events and the rapid pace of change in technology. In particular, there is a recognition that fielded systems must not only satisfy initial operational requirements, they must also possess inherent flexibility to meet new threats and missions, be capable of accommodating new technologies over the life of the weapon system and keep pace with the shorter life cycles associated with higher usage of commercial-off-the-shelf (COTS) products. In the case of the Navy, this means getting both advanced and appropriate technologies into the hands of the warfighter at minimum time and cost to allow a typical warship to remain a cost effective military system over a 35-50 year life cycle. Secretary John J. Young (ASN RDA) in a memorandum on the Surface Combatant Family of Ships, dated March 26, 2002, stated that an "approach is being used for developing the DD(X) / CG(X) / LCS family of ships, where technology development and maturation lead to evolutionary requirements addressing mission need. This approach calls for construction of multiple flights of multi-mission ships with progressive capability improvements. A focus on modular and scalable architectures and systems will allow the Navy to field innovative technologies, as they become available, across multiple classes of ships."

Traditional ship design in the Navy involved a lengthy process that started with requirements definition as captured in the Mission Needs Statement (MNS) and the Operational Requirements Document (ORD). Design and engineering of the ship were then conducted for both the hull and the set of specific systems installed within it. The architecture of the ship typically focused on only the initial combat system suite comprised of Sensors for surveillance (air, surface and underwater), Ordnance/aviation for power projection and the Command, Control, Computers, Communications and Intelligence (C4I) needed to direct overall ship combat operations. If the technologies in these areas changed dramatically with time or there was a need for a different mix of components due to changing mission requirements, changes to the original configuration would be required. How can the design of future naval ships achieve successful integration of such systems while minimizing the impacts to the basic hull and support systems caused by alternate configurations – either during acquisition while in production of different flights or later during modernization or conversion?

The approach proposed in this paper includes use of Modular Open Systems enabled by a robust continuous Technology Management process as the way to provide cost effective life cycle upgrades and mission reconfigurations. The focus of the paper will be on presenting a new ship design process to accommodate Evolutionary Acquisition in Navy ships and on providing some examples of its impact on ship construction.

BACKGROUND

Over a quarter of a century ago, a paper was presented to the Society of Naval Architects and Marine Engineers at the 1977 annual meeting in New York called "Modular Payload Ships in the U.S. Navy" [Reference 1]. Its message was to rethink how surface combatants were designed and constructed. The paper included the results from a Navy 6.2 Research & Development program called SEAMOD, which showed that:

1. The time and cost of ship construction can be reduced through the procurement of equipment to a standard interface, modular packaging of equipment to ease outfitting and testing of equipment prior to ship installation.
2. The time/cost of mission reconfiguration can be reduced through the use of modular mission packages.
3. The time/cost of technology upgrades can be reduced through the use of pre-defined interface standards for a specified functional area.

Not only did the paper cover the naval architecture aspects of incorporating physical modularity into the ship's hull for weapons and electronics, it recommended redesign of the Combat Direction System (CDS) – going from a highly centralized architecture (e. g. Naval Tactical Data System) to a distributed CDS using projected high powered micro processors. At the time, this was thought impractical, of course – as they say – the "rest is history." Not only are the goals and design techniques presented in that paper still applicable today, the main question posed at the end is also still valid: "the [concept] can be done, the question is, will it be done?"

The Ship Systems Engineering Standards (SSES) program followed the SEAMOD program. SSES (1980-1985) began development of modular interface standards for the next destroyer program, the DDG51. References [2] and [3] describe the approach and progress of the SSES program as of 1982. The SSES program succeeded in developing zone and module/module station interface standards for the main weapon stations on the DDG51 called the "A" and "B" module stations. These module stations were designed to accommodate the VLS, MK 41 with a full tomahawk load or an assortment of alternate gun systems.

In 1985, further efforts to develop modular payload ships within the U.S. Navy ceased, while at the German shipyard of Blohm & Voss modular frigates and corvettes were being constructed for third world navies using a concept called MEKO. Reference [4] describes this concept that included SSES standards for its modular weapons. The Royal Danish Navy began implementation of modularity concepts (1986) for a new class of patrol craft. This concept (called STAN FLEX) uses modularity to quickly change missions of the craft. In a matter of days, the mission could be changed to one of four capabilities: surveillance, mine laying, missile launch or mine countermeasures. Reference [5] includes a discussion of STAN FLEX.

With "Acquisition Reform," leaders within the Department of Defense (DoD) were starting to rethink the approach to acquisition of major weapon systems such as ships. In 1995, DoD established a special office at the Secretary of Defense level called the Open System Joint Task Force (OSJTF). OSJTF was formed to promulgate the policy of DoD Instruction 5000.2 requiring use of modularity and interface standards as a part of an Open System Architecture (OSA) for all defense weapon systems (Army, Air Force, Navy and Marines). Within the Navy, a program called Affordability Through Commonality (ATC), described in reference [6], was funded to investigate the potential benefits of modularity and standardization. To implement the concept of OSA within the Navy, the ATC program expanded its efforts by establishing an Integrated Product Team (IPT) called "Total ship Open System Architecture" (TOSA) in 1998. This paper reflects recent efforts of the TOSA team to develop "modular and scalable architectures and systems allowing the Navy to field innovative technologies, as they become available, across multiple classes of ships." The combination of programmatic, technical and business procurement strategy that will implement such an approach is called "Evolutionary Acquisition" – see reference [7].

CONCEPT DEFINITION

The approach to achieving successful "Evolutionary Acquisition" involves three major integrated activities, (1) Acquisition Schedule (2) Technology Management and (3) Modular Open Systems approach.

Acquisition Schedule

Today's Acquisition Schedule emphasizes providing "just in time" technology to the warfighter. This translates to a schedule of acquisition that plans for upgrades over the life of the weapon system procurement as a way to minimize risk of premature technology adoption. In the case of ships this means that the configuration of the initially built ships would include a combat system that is tailored to the mission as required at the time and fielding the appropriate systems and technologies that are available and proven. As the life cycle of the ship (and production run for the class of ships) proceeds forward in time, planned upgrades (called flights) will be established that permit both technology upgrade of new and improved systems for the various combat system functions and/or modification of the combat system equipment mix to provide different mission capabilities than originally required. This strategy of "flexibility" in acquisition timing is based upon the realization that the old concept of procuring new "leap frog" technologies that have not eliminated performance and cost risks prior to integration into weapons system platforms are no longer acceptable. Such prior "coupling" of new maturing technology to platform design and construction caused time delays and cost overruns. Where "modernizations" used to be infrequent over the life of a ship (2-3 times over 30 years) now changes in both technology and operational scenarios happen at such a fast pace that the goal is to make the platform "transformational" in that it will allow quick incremental upgrades or mission change when required. As a result, requirements for new major weapon systems include not only the initial performance capabilities but also the ability to change to future capabilities with minimum cost and time needed for the reconfiguration. This is the challenge of "Evolutionary Acquisition" to the Naval Engineering community. Executing this policy will require a robust "Technology Management" process that supports a "Modular Open Systems" approach.

Technology Management

Technology Management Problem – The technology Management problem comes from the fact that a Navy ship class typically takes 10 to 15 years to develop and field "concept to first delivered ship" and is followed by another 10 years of production; upon delivery, each ship is expected to have 30 to 35 years of useful operational life. However, it is difficult to predict future changes to threats, ship missions, and requirements and the underlying technologies to meet these changes across such a long acquisition and operational life cycle. To offset the challenge of change, the Department of Defense (DoD) typically has conducted high risk, advanced technology development concurrently with weapon systems development with the intent to develop breakthrough "leap-ahead" technologies that are far enough out to preclude fielding obsolete systems and meet current and projected requirements in one step. However, to further compound the long life cycle issue, new factors have emerged in the last decade that challenge this leap-ahead approach. First, changes in technology and threats are accelerating rapidly, such that certain technologies selected early in development risk being eclipsed; creating an obsolescence issue (e.g., supportability or effectiveness) later in development or shortly after a weapon system is fielded. Second, DoD and commercial market relationships have changed, whereby DoD no longer leads in the development of critical technologies and is more dependent on commercial technologies and their shorter product life cycles. The transition rate (e.g., amount of innovation) and the speed (e.g., process) of inserting newer, innovative technologies into critical ship systems across this time spectrum have been difficult and costly for many reasons. Figure 1 depicts a typical ship life cycle

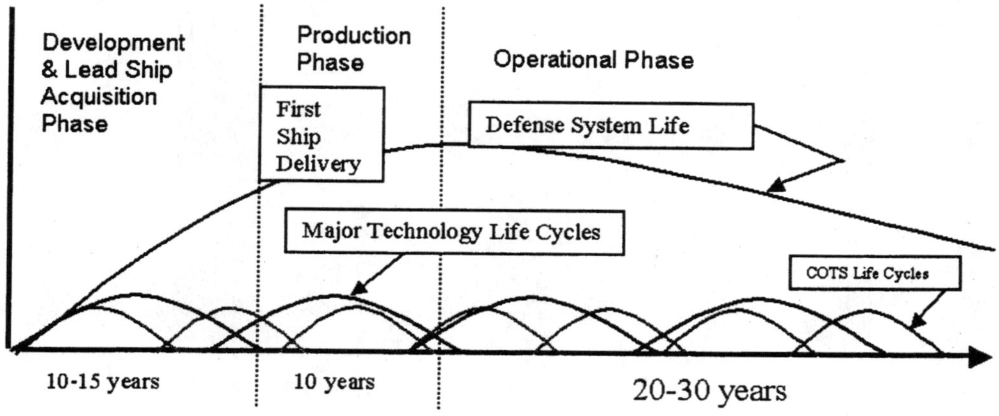

Fig. 1 Ship, Technology, and COTS Life Cycles

against notional technology and commercial off-the-shelf (COTS) life cycles.

Technology Management Process Issues – The low transition rate and the speed of inserting newer, innovative technologies is caused by the following:

- There is no overarching or strategic technology management process connecting the technology development and the product development communities in the surface warship arena. This is particularly critical in light of the shorter COTS technology life cycle issues depicted in Figure 1; though more recently, there have been efforts made to create elements of this function within the Navy program offices.
- There is no transition path to insert technology on the platform, caused in part by lack of adaptability. Either the platform cannot be modified (e.g., stability, structural) or it is too complex and costly.
- The technology development effort does not develop beyond an early prototype stage, or if development does continue to maturation, it winds up being late – well after the start of construction of a new ship class. At this point, even if the new technology is inserted on later ships of the class, the cost to back-fit the change on earlier ships is always a budget challenge, and ends up competing with other system upgrades seen as equally or more critical. The technology transition either stops or slows down, as depicted in Figure 2. This can leave the warfighter to "make do" with outdated equipment.

The Navy's technology development is performed initially by the Science and Technology (S&T) community (e.g., Navy government labs, universities, industry, etc) that are managed by the Office of Naval Research (ONR) or it can come directly from the industrial R&D technology base. Technologies can be picked up by the actual weapon systems platform for immediate implementation (if fully matured) or funded for further development. The "valley of death" is a gap in the development process that may occur for the following reasons:

- Lack of a handoff strategy between S&T and product development organizations.
- Incompatible technology-to-platform interfaces.
- No perceived value by end users.
- Lack of transition funding needed to continue the development.
- S&T community interest in working on technologies of scientific interest but not necessarily to end users (e.g., product developers).
- Technology cannot be made to work in development.
- Business or contract obstacles.

Processes and Attributes to Improve Management of Technology Transition and Insertion – As part of an attempt to improve Technology Management (TM) within the Navy, research was undertaken [Reference 8] to conduct an assessment of current Navy processes (Virginia Class Technology Insertion Program) as well as commercial processes (e.g. Boeing Phantom Works) of technology transition and insertion. Detailed

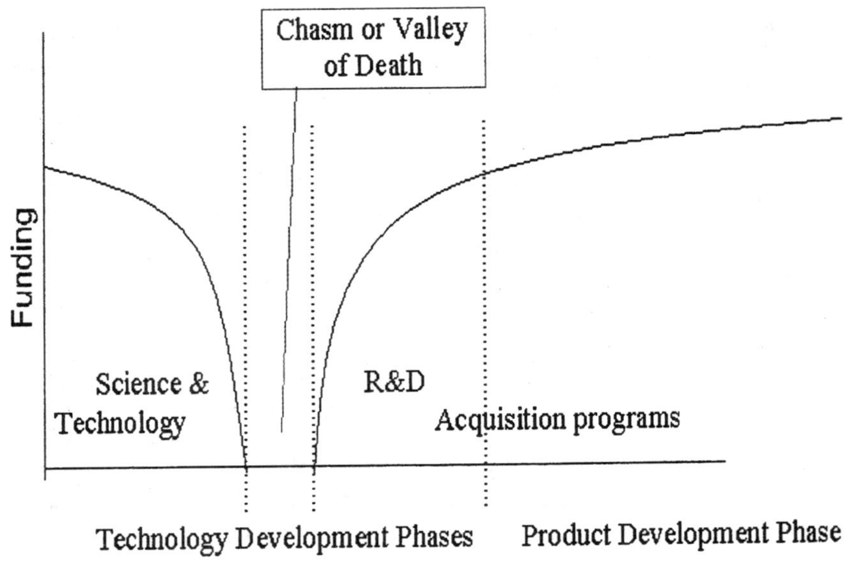

Fig. 2 Technology Development Timeline and "Valley of Death"

findings are contained in this reference from which key conclusions are abstracted for this paper. After evaluating the above two organizations, it became clear that successful Technology Management processes are comprised of the major components shown in Figure 3. As stated earlier, there are two closely linked factors that can directly influence each other, but which impact cost effective life cycle adaptability of ships.

occurring with ONR Future Naval Capability (FNC) programs).

- Better define technology-to-platform interfaces and integration issues.
- Articulate stated value to end-users and promote stronger advocacy by stakeholders.
- Identify clear funding needs to continue development to a higher technology readiness level

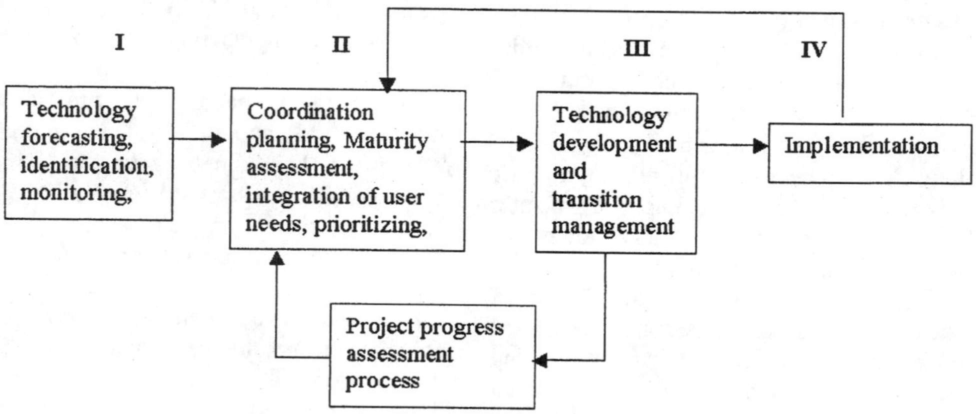

Fig. 3 Functional Phases of the Technology Management Process

with lower attendant risk.

- Make timely decisions to stop failing technology efforts.
- Promote good communications.
- Identify early items of benefit to the acquisition program.

These factors are (1) effective Technology Management processes that can provide for more planned, timely, and orderly transition and insertion of newer technology into critical ship systems; and (2) use of these TM processes to influence technical design approaches, which later determine the ease of life cycle adaptability of the ship. This latter factor is the main focus of this paper.

The TM research study set about to demonstrate the validity of these factors and investigated the following:

- Existing technology management processes
- Examples of past Navy technology transition and insertion efforts to identify "lessons learned"
- Examples of design approaches that were selected as an outcome of technology management processes and what effect these design approaches had on platform adaptability.

A key takeaway of the TM process is that it provides a basis for developing robust future capabilities as part of the interfaces developed during the ship design process. These interfaces can accommodate change both during acquisition and across the life cycle, while minimizing Non-Recurring Engineering (NRE) costs and ship modification periods.

Regarding attributes, the following actions are needed to improve technology transition and insertion as part of an overall Technology Management program:

- Force agreements on handoff strategies in the technology development process (such as is

Efforts to date to implement such an approach have included development of improved Technology Management Plans for application to surface combatants. The purpose of the TM Plans is to systematically assess, balance and match existing and emerging technologies to current and projected requirements and capabilities for future ships, with applicability to in-service ships. This new process will allow current and projected requirements (capabilities) to be achieved via technology management across various flights of ships. It will allow visibility to current and emerging transformational technologies across mission areas to influence the initial design and to accommodate future technology insertion opportunities. These long-term TM Plans include (1) Continuous monitoring of "forces of change" that might result in a need for technology insertion, (2) Technology development and insertion to achieve increased capabilities in response to forces of change, and (3) Support for the development and management of architectures and interfaces to facilitate technology insertion (to be developed using the Modular Open Systems approach for design).

1. The continuous monitoring of "forces of change" includes the following:
   - New technologies or "Technology Opportunities," representing current and future technologies that are or may be available for use and which may influence requirements/ capabilities.
   - New operational architectures including new missions, new threats, etc.
   - New markets, as increased use of commercial systems has increased market exposure and resulted in the need for technology refresh.
   - New regulations (environmental, safety, etc.)
2. Technology development and insertion is a time-phased development and fielding of specific technologies based on requirements, forces of change and technology opportunities.
3. Configuration Management of defined architectures and interfaces is needed to implement and manage technology insertion without creating costly total system impacts, which also creates a defined path for upgrades. Development of the Architectures and Interfaces is discussed as part of the Modular Open Systems Approach.

An illustration of the Technology Management process is shown in Figure 4.

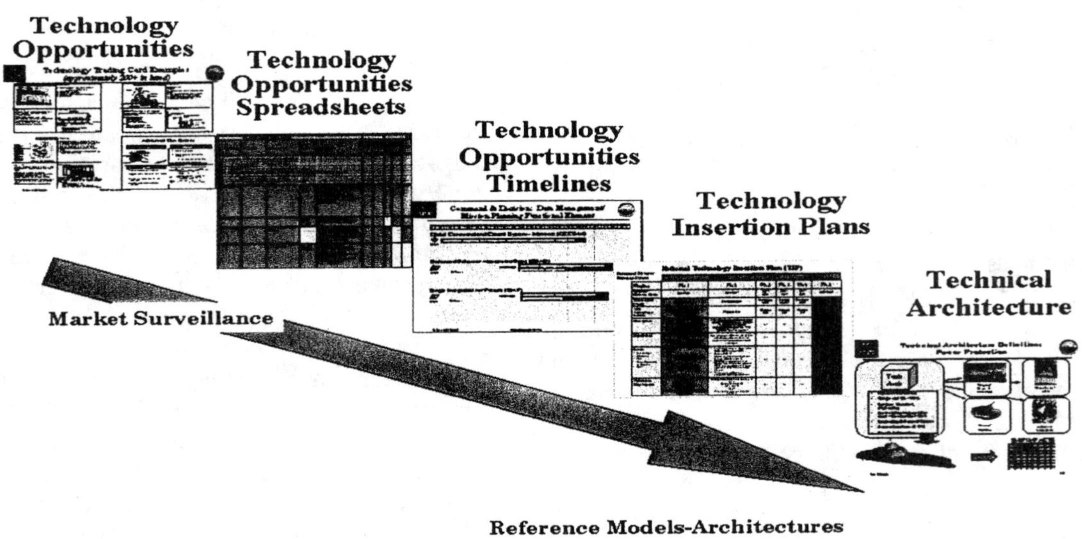

Fig. 4 Technology Management Process

To capture Technology Opportunities, the TOSA Team formed an ad-hoc, cross-functional Technology Management Integrated Product Team (IPT) with representatives from the broader Naval and DoD R&D communities to capture information on current and future technologies, their projected capabilities, technology readiness levels (See Appendix) and timelines for availability/maturity. TOSA developed and is populating a database of these Technology Opportunities.

Modular Open Systems Approach

As defined by the Open Systems Joint Task Force (OSJTF) in the Office of the Secretary of Defense (OSD), a Modular "Open Systems approach is both a business and engineering strategy for developing new systems and modernizing existing ones. As a business strategy, this approach enables program teams to build, upgrade and support systems more quickly and affordably. This can be achieved through the use of commercial products from multiple sources and leveraging the commercial sector investment in new technology and products. The technical portion of the Open Systems approach is focused on a system design that is modular, has well defined interfaces, is designed for change and, to the extent possible, makes use of commonly used industry standards for key interfaces. This system design is best accomplished by using a sound systems engineering processes."

The successful application of the OSJTF approach results in a design that emphasizes modularity based on functional partitioning. This means modules are based upon grouping of components and parts with similar functions. As the system evolves and technology changes, only those modules that are directly affected need to be changed. As a result, the overall system can be easily modified or upgraded. An example of functional partitioning following a functional breakdown structure for a ship is shown in Figure 5.

To insure that system modules will work together when integrated, their module-to-module station interfaces need to be defined and controlled by the use ship testing and "Just in Time" installation. To achieve the benefits of a modular open system, the OSJTF has established a set of both business and technical indicators for DoD managers of major acquisition programs (A Category > $50M) to help them evaluate

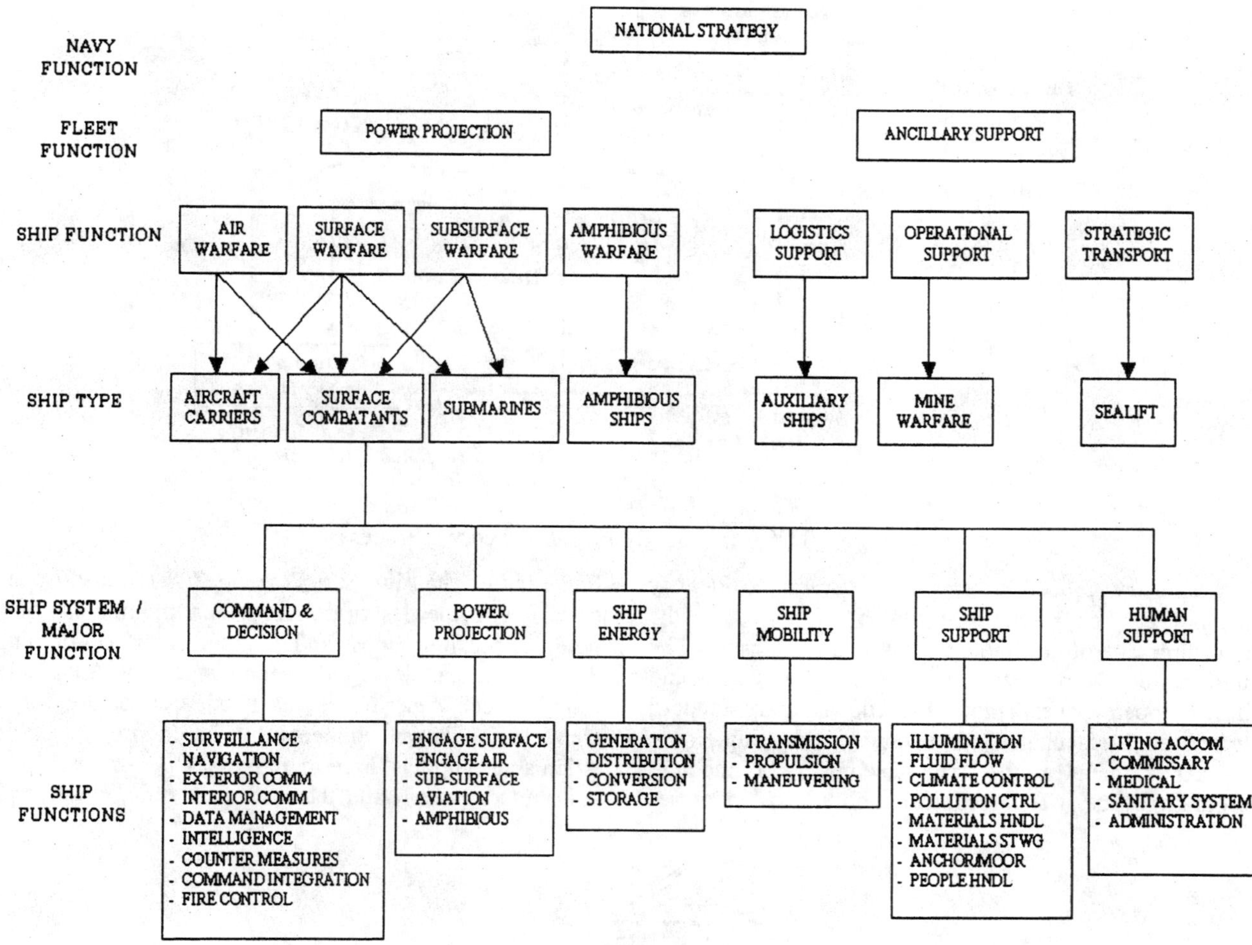

Fig. 5 Ship Functional Breakdown

of standards. These key interfaces need to be defined by widely used, non-proprietary, "open standards". By calling out open standards, a wide range of products can be used in the system's design.

The net benefits of this approach are systems that are flexible, easily upgradeable and able to take advantage of competition by using a wide range of commercially available products. This facilitates exploitation of commercial technologies and market forces, accommodates future technology upgrades and results in reduced total ownership costs. A ship design that utilizes the Modular Open Systems approach will benefit from competitive purchasing, de-coupled equipment selection from the construction process, off-whether their program is following this approach.

THE NEW SHIP DESIGN PROCESS

While each engineer-architect may have a unique twist on how they approach the ship design process, some key steps are always performed. These include the analysis of requirements, the decomposition of the requirements into functions and equipment, and the synthesis of the design. These are iterated through several requirements and design loops until the design converges on a reasonable answer.

When Technology Management and the Modular Open Systems approach are applied to ship design, the process used by TOSA for systems engineering is not much different. The biggest changes include a greater emphasis on functional analysis, the definition and control of key interfaces and increased efforts in market research-technology projection as shown in Figure 6.

the services, interfaces, standards, and their relationships. It provides the technical guidelines for implementation of systems upon which engineering specifications are based, common building blocks are built, and product lines are developed." For a ship, this translates to the description of zones, module stations, module configurations and critical interfaces. The TOSA team has developed a template, "The Adaptable Ship," which describes a technical architecture for a ship. It is shown conceptually in Figure 7. The Adaptable Ship has three main parts, Open zones, Open distributed systems and the rest of the ship, closed and

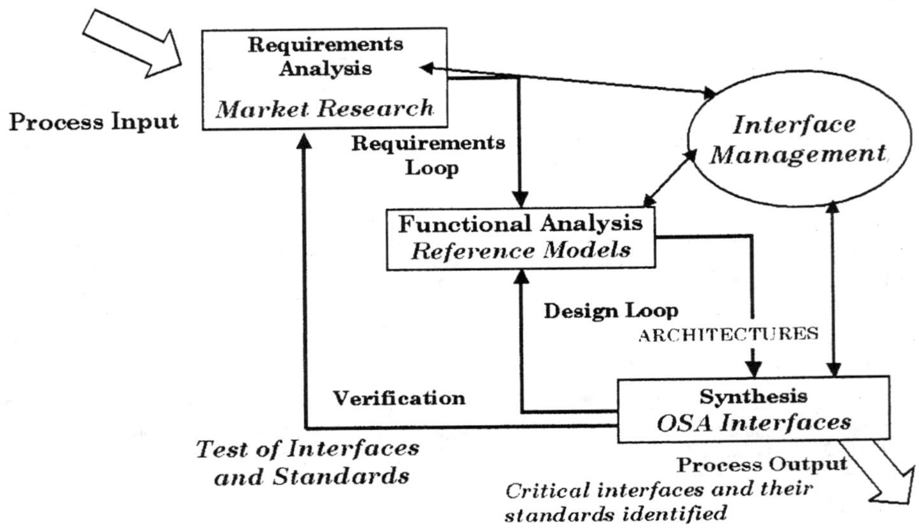

Fig. 6 TOSA and the Systems Engineering Process

The new process uses market research to identify technology opportunities, which leads to the development of technology insertion plans. The information from these steps is used to define and design open zones, module stations, and standard interfaces, the "technical architecture" of the ship.

As defined by the Joint Technical Architecture program office, "The technical architecture identifies

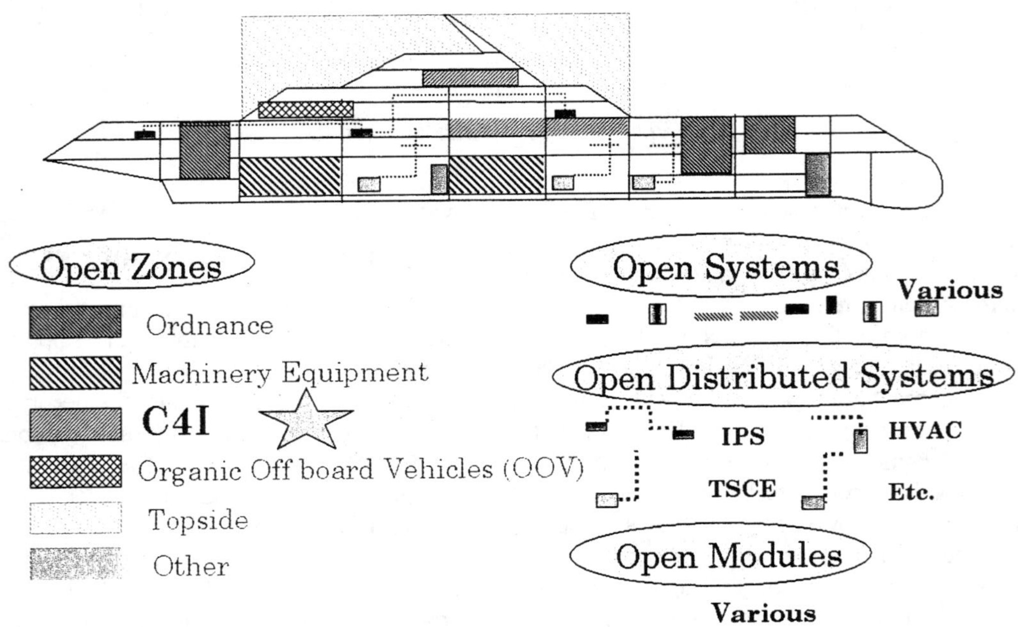

Fig. 7 The Adaptable Ship

custom designed. Open Modules can be designed to interact with either the Open Zones or the Open distributed systems.

The Open Zone is a volume of the ship that provides space, structural support and services for a group of similar functions located therein. It can be a single compartment or multiple adjacent compartments. This can be one major function, such as exterior communications, weapons mounts and magazines, or a collection of similar functions, such as the numerous battle management functions performed in the Combat Information Center (CIC). The Open Zone includes one or more mission module stations as well as the spaces that support the module equipment. A mission module station is the specific volume, structure and support service connections (e.g. electrical power, data, air, water, and other fluids) built into the ship to receive a corresponding module. The actual definition of Open Zones will depend on the functional analysis of the ship's systems and the selection of potential modules. Interfaces between the Open Zone and the ship are defined. Some interfaces between equipment inside the Open Zone and its internal boundaries can be defined and controlled as required. In other words, the Open Zone has been opened at the space/compartment level and it can contain one or more open systems, including modules and module stations.

The A sized (32 missile cells) and the B sized (64 missile cell) VLS modules, installed in the DDG51 Class are examples of an Open Zone (in this case a weapons zone). The "Turnkey" modular open approach used for the DDG radio communications system is a similar concept. In Turnkey, the Communications Center and RadioTransmitter Room is the controlled Open Zone, and the Navy's Shipboard Exterior Communications Integration Branch does the arrangements and equipment selection within the Open Zone. The Turnkey process results in the "just in time" insertion of the latest technology during the construction of each ship. Figure 8 presents a detailed view of Open Zones, modules and module stations.

The Open Zone concept is not well suited to ship systems with components distributed throughout the ship. These systems benefit from being opened as individual systems. They still take advantage of OSA and modularity concepts, but the resulting open architecture is not based on the ship's compartments. Their open architecture is limited to the components and related open modules. These open distributed systems have open standard interfaces for their modules and associated distributed equipment, ducting, piping, cables, etc.

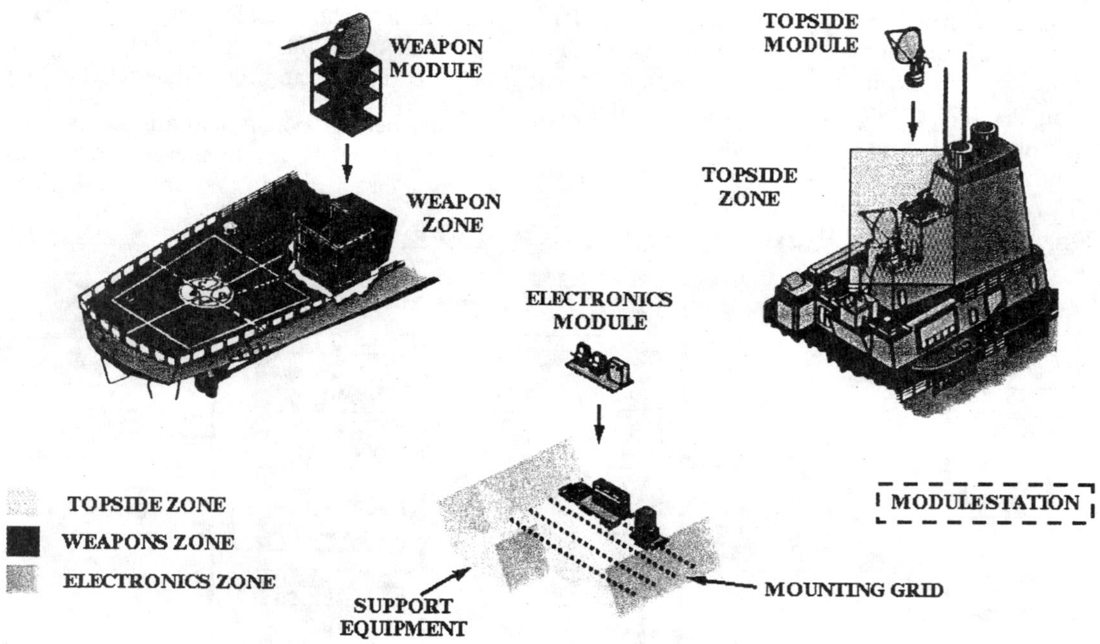

Fig. 8 Open Zones and Modules

Finally, many systems or components should not be designed as an open system. They may be unique to the Navy and have no commercial equivalents, there may be no expected upgrades needed during the life of the system (e.g. propeller shaft), or the system may be in such a critical configuration that it allows no flexibility. Whatever the reason, the OSA concept does not force itself on every system or subsystem. The candidate selection activity is an important part of the

TOSA process and will select only those systems that are technically and economically feasible to be opened.

To summarize, technical architecture development incorporates a systems engineering process as illustrated in Figure 6. It is developed based on: requirements analysis, market research, functional analysis, technology insertion plans and design synthesis. These key steps will be discussed in the following sections of the paper.

New Design Team Roles

As delineated by the OSJTF, the preferred strategy for implementing a modular open systems approach is to employ an Integrated Product Team (IPT) comprised of government and industry representatives. This IPT must include all of the stakeholders involved in the acquisition and deployment of the product. As a minimum, the IPT should include those who design, specify, build, test, operate and maintain the system. For ships, this would include Navy Operations specialists, naval architects, combat systems engineers, technologists, systems developers, production engineers, cost analysts, test engineers, ILS engineers and Configuration Management (CM) engineers. The responsibilities of the IPT include gathering and analyzing data by conducting continuous market research and analysis, documentation of performance requirements in both operational and engineering terms and development of consensus candidate interface standards at both the zone and module levels.

In the case of the Navy, an industry-government IPT was established in 1998 at NSWC Carderock, MD called the Total ship Open Systems Architecture (TOSA) IPT. TOSA has been implementing the policies of the OSJTF by applying OSA to the traditional ship design and acquisition process. The resulting process addresses the whole ship at all stages of the ship's life cycle and results in a modular Open Systems Architecture for the whole ship. An initial overview of the TOSA concept was presented in reference [9]. The ship's modular architecture addresses multiple levels of both external and internal interfaces. The external interfaces address system architecture requirements from platform to platform; and the internal interfaces address system to system (and equipment) interface requirements. The hierarchy of this family of standards is illustrated in Figure 9.

Requirements Development and Design criteria

Requirements development is an *iterative process* and is focused on defining and developing functional and performance requirements, and clarifying constraints that limit design flexibility at each level of development. Evolutionary Acquisition and related changes in Defense Department policies have had a significant impact on requirements development and thus on design criteria. This is one of the factors driving the need to revise the traditional ship design process. Some of these changes include:

- Technology driven requirements definition

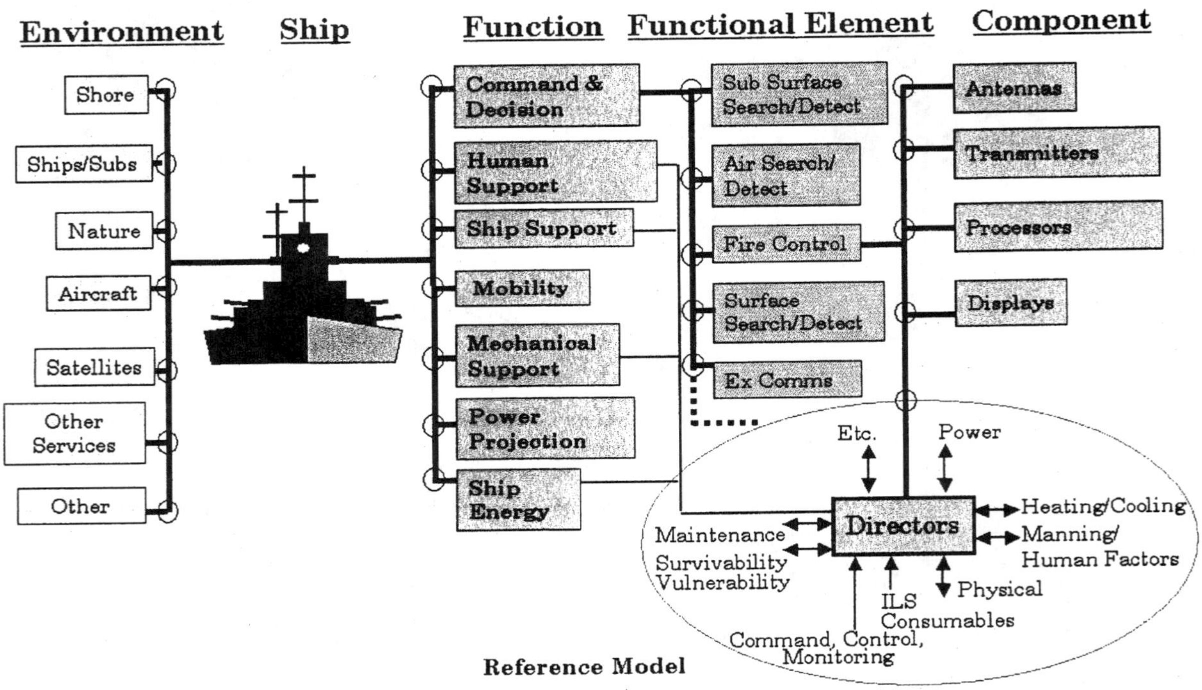

Fig. 9 Interfaces at Multiple Levels

- New national strategy to address evolving threats
- Emphasis on the use of commercial products
- Cost as an independent variable
- Reduced manning

Technology driven requirements definition basically means that the requirements are based on what can feasibly be achieved. *The ships performance criteria and mission profiles are not fixed but are dependant on what can be delivered by the technology.* This can affect criteria such as speed, endurance and weapons payload. This approach requires an extensive market research effort to identify the technology opportunities.

The Cold War is over and new enemies and threats drive the development of the national strategy. With changes in strategy come changes in missions. These new missions are sometimes ill defined. Often the mission that a ship is initially designed for will not be the same mission that the ship will perform when it becomes operational.

There is a greater emphasis on using commercially available products, commercial off the shelf (COTS). In some cases, this is because the commercial products are technologically more advanced than the military items. In the other cases, it is because the larger civilian market keeps costs under control and can lead to a more affordable system.

The emphasis on COTS is also related to the requirement of addressing Cost As an Independent Variable (CAIV) in many new weapons programs. CAIV means that the overall weapon system cost becomes a requirement equal in weight to the system's performance criteria. The designer must trade-off an option's ability to achieve a greater range of performance against its cost. Often a weapon system cannot achieve a desired level of performance and still meet the cost goals with the current status of technology. That system will need to accept new components when the technology matures and becomes affordable.

Finally, reduced manning can lead to the increased use of automation, interchangeable modules, built in test and analysis capabilities and off-ship training and maintenance. These factors in turn lead to the need for the latest technology.

Combined, these factors lead to a ship design that must be flexible and accommodate change. Life cycle upgradeability and reconfigurability will have a higher priority than before. This means that the design must support a "range" of possible equipments not just a "point design" for one configuration. Also such maintenance approaches as real time on-line monitoring, Condition Based Maintenance and direct Supply Support to achieve improved ship readiness with reduced crews will become very important in future Navy ships. Design criteria will include an assessment of "openness" to use COTS and future technologies as key to the design's acceptability.

As part of this effort, the design team should identify internal (e.g. program, service, DoD, etc.) and external constraints. Laws and regulations calling for use of commercial standards, the compliance requirements contained in open standards, capabilities of allied or coalition forces that will interface with the system, and the present and future technology base are among external constraints related to open systems. DoD guidelines that call for use of a Modular Open Systems approach, domain technologies that necessitate commonality and reuse across platforms, and the system of systems interoperability requirements are among the internal open system related constraints. The goal of establishing and refining operational and design requirements that result in the proper balance between performance and cost within affordability constraints has not changed. What has changed is a new expectation that 'requirements' of new ships may come as much from 'Technology Push" as from operational "Requirements Pull." The question from Navy leadership is "what can technology do for me" to create improved ship operational capabilities in the future.

Technology Management – Market Research

Market research is an essential step in the systems engineering process. It is even more important to the open systems approach. It involves the identification of relevant weapon system programs, the commercial market and its capabilities, and the forecast of future trends in technology. This entails the surveillance of the market and the subsequent analysis of specific products. Market research is guided and supported by the functional analysis efforts. The main phases of market research are described below. An excellent tutorial on how to perform market research is located on the web at: http://www.acq-ref.navy.mil/marketresearch/index.html

Market Surveillance

Market surveillance begins early in the design process and continues throughout the ship's life cycle. Market surveillance will establish the framework for further technology management efforts by identifying available technologies, manufacturers, products, and product developments. The initial steps identify potential sources, especially new sources and technologies, to ensure that systems are truly open and to reach suppliers that have not participated in defense or government acquisitions. Market surveillance also forecasts future trends in technology.

Market surveillance includes the surveys of program offices, navy research facilities, manufacturers, commercial ships, and existing naval systems to identify the sources of products and technologies. The World Wide Web is an excellent resource at this stage. In addition, several directories are helpful, such as the Thomas Register. The surveys will provide the design team with points of contact for further research and will determine the size of the market and potential new technologies.

Market surveillance will identify key ship systems and their corresponding technologies, manufacturers, and equipment, (present and projected). Sufficient general product information is required at this time to make informed decisions during the early stages of ship design.

Market Analysis

The products and technologies identified by market surveillance will be the focus of a more detailed analysis. Market analysis will help identify systems that are suited for an open system and to design and develop open zones, modules, and individual open systems. Market analysis will also be used at different times throughout the process to identify potential products, services, and technologies to meet specific requirements as they begin to take shape later in the process.

Not all ship systems can or should be opened. One factor in selecting the system to open is the size of the market, availability of alternative commercial components that meet the Navy's requirements, and the cost (including logistic support) of the commercial items. A naval warship design also needs to consider the number of alternative weapons systems that are available for a given function and whether the ship will be required to be reconfigurable for multiple missions.

As the ship design matures, additional and more focused market analyses will be performed to determine the actual equipment that satisfies the requirements and fits in the proposed reference model. Characteristics, values for the equipment's interfaces, prices, and maintenance needs will be collected and tabulated. The objective of the focused market analysis is not to select just one or two items that meet the requirement, but to collect enough data (through phone calls, visits, and/or written matter) to synthesize the open architecture interfaces. This is accomplished through surveys of weapons programs, suppliers, users, and related industry organizations. Potential suppliers and programs were identified during the market surveillance step. In a focused market analysis, the aim is to examine all potential suppliers of products, services, and technologies to meet more specific requirements.

Evaluation of Market Data

The quality of the data collected by the market research efforts needs to be evaluated. The manufacturer's data may be incomplete, or not detailed enough to determine whether the product meets the needs of the applicable system. Additional information regarding interfaces, packaging and what the "lowest repairable unit (LRU)" are often the hardest to obtain, but are important to the design of the modular and open architecture. For example, if certain components were always sold together in one package, the critical interfaces would define the boundary of that package. The critical interfaces are also set to match components that are replaced as part of periodic maintenance. It may be necessary to contact alternate sources of information regarding the product, including applications, field service, or design engineers.

The evaluation may find few or no equipment/components to satisfy the requirements (in spite of the efforts in the candidate selection step). This may be due to the severity of requirements that are unique to that ship's requirements. The design team can develop enablers, adapters, modules, etc, to meet these unique requirements.

Functional Analysis

Functional analysis is a part of most system engineering approaches, though many times it is quickly passed over when the engineer jumps to equipment-component selection. In the Modular Open System approach, functional analysis is a key step in insuring a flexible, modular design. Functional analysis identifies the key functions of a ship's system as shown in Figure 5 and partitions the system's requirements into functional components and parts. These components can then be grouped into modules with similar functions – isolating changes in one component's technology from the overall system.

Functional analysis begins by decomposing the overall ship system's mission into a smaller number of fundamental building blocks (e.g., weapons entity, power plant entity, processing entity, etc.). Each functional building block can be further decomposed during the engineering processes. The selection of these blocks represents the initial design decisions for a system. At each step, the performance requirements for the system are allocated to each decomposed functional block – the functional elements.

The functional elements are then partitioned and grouped together into possible module packages. With an idea of the equipment that may be allocated to the functions, we know which functional elements have a high rate of change due to evolving technology, which need to be replaced often because of maintenance or equipment failure and which will change due to changes in mission. Much of this knowledge base on technology turnover comes directly from the TM process introduced earlier. All of these are reasons to group those items together onto modules so that changes in one module do not impact the rest of the system. In addition functions can be partitioned among modules for productivity and market reasons.

A description of the functional elements and their interrelationships is basically a functional architecture and must be coordinated with the software open architecture of the ship. As an aid in this step, the OSJTF uses the graphical tool called the reference model to describe it. A reference model provides a "high level, generalized view of the weapons system family." Its purpose is to provide a common conceptual framework and define the functional architecture for the different ship design teams and resources. A reference model provides a framework for how to apply standards – particularly how to identify interfaces critical to the Open System Architecture as illustrated in Figure 9. Reference models are developed for different aspects of the system. For a ship design, this means the development of several reference models, a general overall ship level model and major ship system models. These models will guide the development of the Open Zones, module designs and related module stations in the ship.

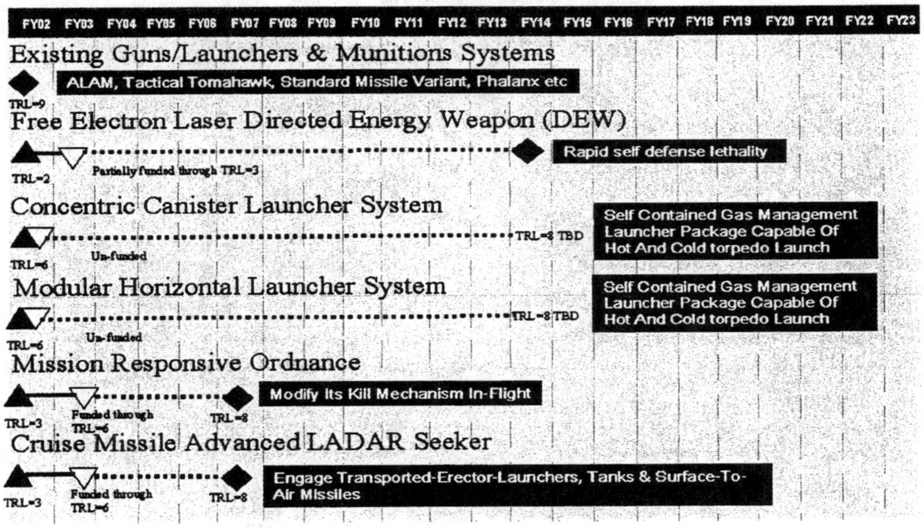

Fig. 10 Technology Opportunity Timelines

Technology Management – Technology Insertion Plans

The various products and technologies that were discovered by Market Research can be installed in the ship at different times in its lifetime. These technologies have different capabilities, levels of maturity, risks and cost associated with them. In addition, the ship has its own schedule and operating profile that may conflict with planned upgrades or modernization. We propose the development of a new tool, the Technology Insertion Plan, to help control the impact of new technology.

The Technology Insertion Plan will facilitate the timely and orderly insertion of potential new technologies while addressing these and related issues. Technology Insertion Plans lay out options for technology introduction into the ship. As part of the proposed ship design process, Technology Insertion Plans benefit the development of modular architectures. They guide the development of a technical architecture that will enable the insertion of new technology to achieve increased capabilities in response to forces of change.

Development of Technology Insertion Plans will begin with the analysis of the technology opportunities uncovered by market research. Technology opportunities represent current and future technologies that are or may be available for use by the ship and can perform some of the ship's functions. Technology opportunities may also influence requirements. The analysis will identify the status of the technology, how far from production or achieving operational status the technology may be and any funding required for programs that are still in the research and development phase. Timelines will be developed that show technologies available within a ship's lifecycle for each of the ship's major functions. An example of a timeline is shown in Figure 10.

Technologies that will become part of the mission system will also need to be integrated into coherent, functioning mission suites. The various mission systems in a ship must work together; disparate technologies are not useful if they are not integrated. Several candidate mission suites will be developed to address a range of potential capabilities. Extensive

Systems Engineering tradeoffs involving performance, cost and risk will be conducted on the candidate mission suites. These Systems Engineering tradeoffs will result in Technology Insertion Plans (TIP) that will identify the desired "Evolutionary Acquisition" of improved combat systems during the series production of a class of ships. Mission capabilities will be documented for each candidate suite.

A well formed TIP will identify shortfalls in the ship's ability to meet requirements; map technology opportunities to capabilities and requirements and identify funding and timelines; define class and flight configurations; and provide a schedule for ship installation to overcome the shortfalls. The development of the technical architecture will be based on the Technology Insertion Plan. The zone, interface and module characteristics and configuration will be developed to suit the candidate systems and their planned installation.

The management of architectures and interfaces to facilitate technology insertion is a continuous process for a system's entire lifecycle with continuous life cycle systems over the platform life cycle. Such details as structural approach for modules, HVAC interfaces, electrical and data interfaces will be addressed with system developers. Developers of future, yet to be identified systems, will be able to start with this set of "design to" interfaces and zone architectures so that early in the R & D process the "install-ability" of a new technology can be assessed before arbitrary decisions are made on the system design. In synthesis, the functional architecture will be transformed into the technical architecture, which includes the design and definition of critical interfaces (external and internal), zone, modules and module stations as well as the configuration of the systems architecture and system elements. Synthesis will also define alternative system concepts and define alternative product and process solutions. The optimum synthesis approach is to use an Integrated Process Team with government and industry to develop the zones, module stations and define the critical interfaces.

The synthesis process for zone and module stations is summarized in Figure 11.

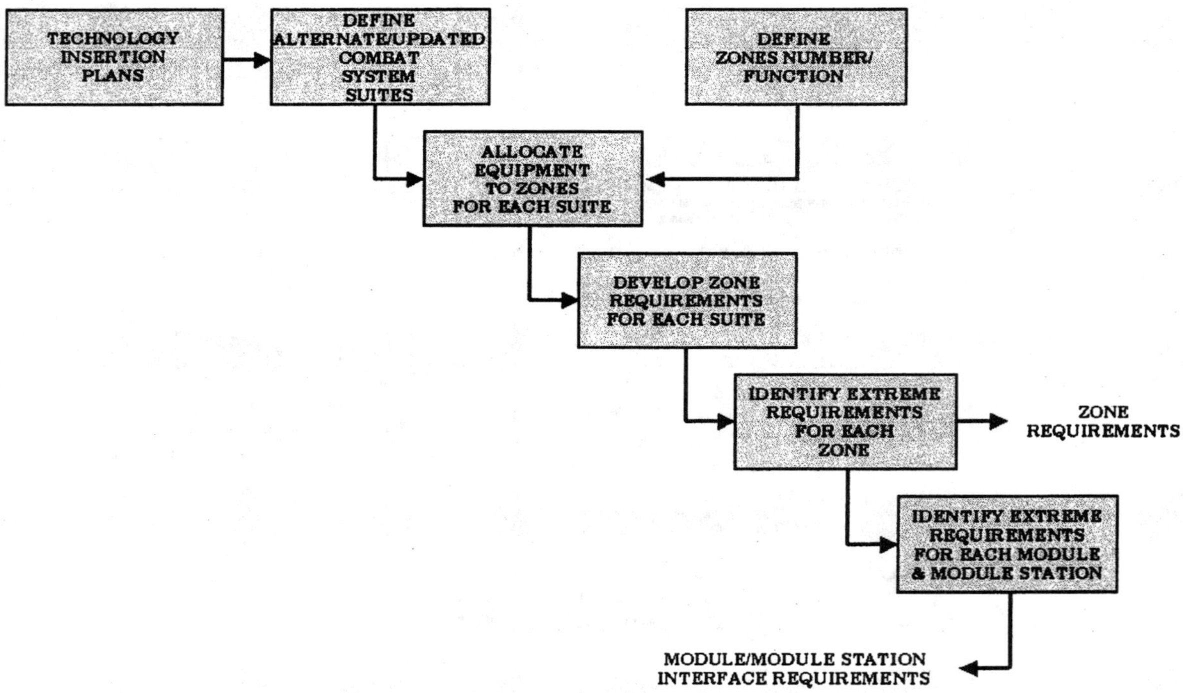

Fig. 11 Synthesis Process

benefits. A technology insertion plan should be maintained and revised to reflect new developments in both technology and mission requirements.

Technical Architecture Synthesis

Ship design efforts will require far more detailed design earlier in the design process for "open zones" to validate that the zone/ship can accommodate alternate

Synthesis will begin with the list of candidate systems including the equipment identified by the Technology Insertion Plans, and from the market research results. Mission related equipment would be part of integrated mission suites. Each mission suite will require the integration of its components into a coherent total mission system that will work with the ship's core command and control architecture. To facilitate reconfigurability, the command and control architecture must be robust and flexible enough to integrate a range of possible mission systems.

Concurrently, the number and types of zones are selected and designated volumes of the ship are identified as the potential locations for the zones. The types and actual arrangement of the zones will depend on the ships mission requirements, the ship's arrangements and the proposed module designs. Typical zones include Weapons, Command and Control, Topside, and Off-board vehicles (i.e. helicopters and unmanned vehicles) Zones. The candidate equipment is then allocated to the zones, and zone characteristics are developed to accommodate them. The results from the functional analysis will also guide the selection of zones and the allocation of equipment. A detailed system characterization, which will identify essential minimal characteristics for the mission systems, will be performed for those items allocated to the zones and modules. These include:

- Performance, cost, other.
- System components (e.g. detachable items)
- Interface requirements (Data & Control I/O, Mechanical & Electrical)
- Component physical and environmental requirements

The data collected in the market research step will be a key source of information for this step. The extreme requirements for each zone are then identified. Trade off studies may be performed at this stage to refine the zone dimensions and characteristics as well as to investigate the impact on the overall ship design. Key steps in the design of the zones include:

- Development of alternative zone arrangements with the different modules for each potential mission
- Assessment of space, weight and cost impacts of the different configurations
- Integrate all zones with the core Command and Control Architecture
- Performance of trade off and integration efforts to select the most feasible module configurations

As each zone design matures, module stations will be arranged and interfaces for the mission modules will be developed. The types and characteristics of the modules that will be installed will drive the module station configuration.

While the zones architectures are developed, the engineers will also define the modular packages that will be installed in the zones. There are different types of modules – ISO based containers, custom fitted independent enclosures and open-unenclosed. The selection of a module type is driven by several factors. Some systems are part of reconfigurable mission suites and will need to be changed often, quickly and at forward operating bases. These types are candidates for self-contained modular packages (custom or ISO container). Other systems are part of the ship's core system; those that support functions for all missions and operations, and will probably only need to be changed to insert upgrades and new technologies. These types can be installed at a shipyard during an overhaul period and can be manufactured without independent structural enclosures – an open module. Other factors to consider include:

- Technology and market forces
- Threat, Operational and Market Forces of Change
- Mission and operational needs including the speed of installation (hour, day, week, etc) and location (tender, forward basing, etc) of reconfiguration
- Transportability (DoD and Commercial infrastructure)
- Specific System Needs (Alignment, shock etc)

In designing the module packages, input from the functional analyses will be used to group components with similar functions and similar maintenance profiles together into modules. The modules should be designed to allow the incremental improvement of part of the system through the upgrades of individual hardware or software modules with newer modular components without redesigning entire systems or large portions of the modular package.

Other factors that need to be considered in the design of the module, module stations and zones are described in Figure 12.

The factors inside the inner circle need to be addressed by the specific zone or module designers, and will be defined and controlled by the technical architecture. The factors outside the inner circle are general ship design requirements and will be referenced by the technical architecture.

The final efforts in the synthesis of the technical architecture is to calculate the configuration, size, weight limitations, services etc. of each zone and to define the critical interfaces (and select the standards for them). Interfaces are the functional and physical characteristics required to exist at a common boundary or connection between systems or items, also the common boundary where physical interaction, support services, and data/information exchange takes place. The technical architecture defines and controls those critical, key interfaces that make the mission system and platform mutually responsive or dependent. These include the interfaces between the Open Zone and the ship and those between the module station and the module. Some interfaces between other equipment inside the mission zone and its internal boundaries can be defined and controlled as required.

As listed in Figure 12, interfaces include physical features such as dimensions, foundations, and weights; power; information such as data flow and signal definitions; and environmental concerns such as shock, EMI, and temperature.

Standards will then be identified and evaluated for the critical interfaces of the module/module station. The evaluation will consider the following important attributes:

- Standards should be open, available to the public
- Standards should be widely used
- Standards should support the future evolution of technology

Often the existing standards are not adequate or not available for the selected critical interface. In that case, new standard interfaces or OSA specifications will need to be developed. The OSA designer (or an Interface working group) develops these standards based on a statistical analysis of the market. The standard interface development will address the form, fit and function of the critical interface.

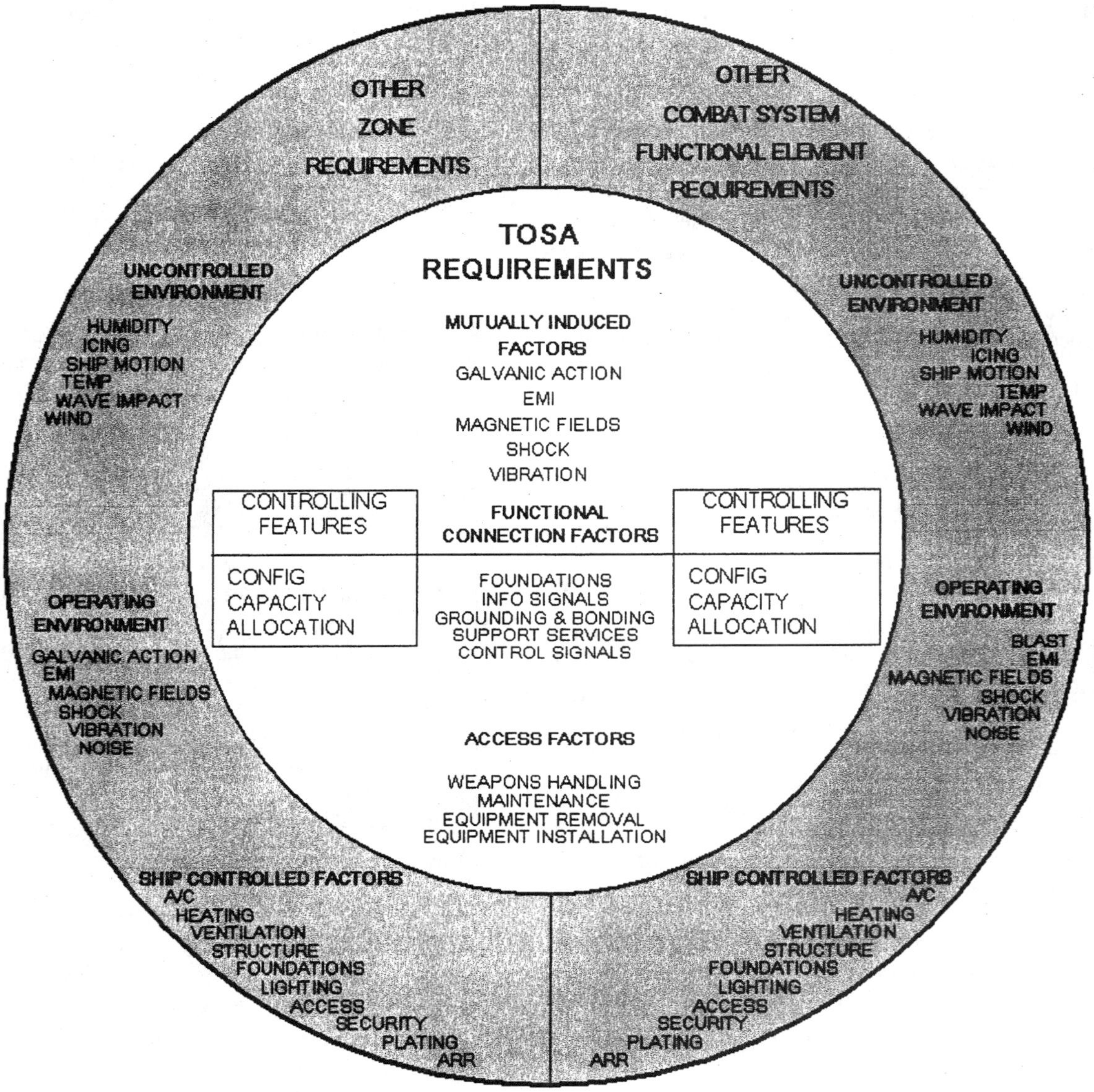

Fig. 12 Design Factors

For further details regarding the development of critical interfaces and their standards see www.acq.osd.mil/osjtf/). These activities result in a Total ship Opens Systems Architecture for Navy ships as illustrated in Figure 13.

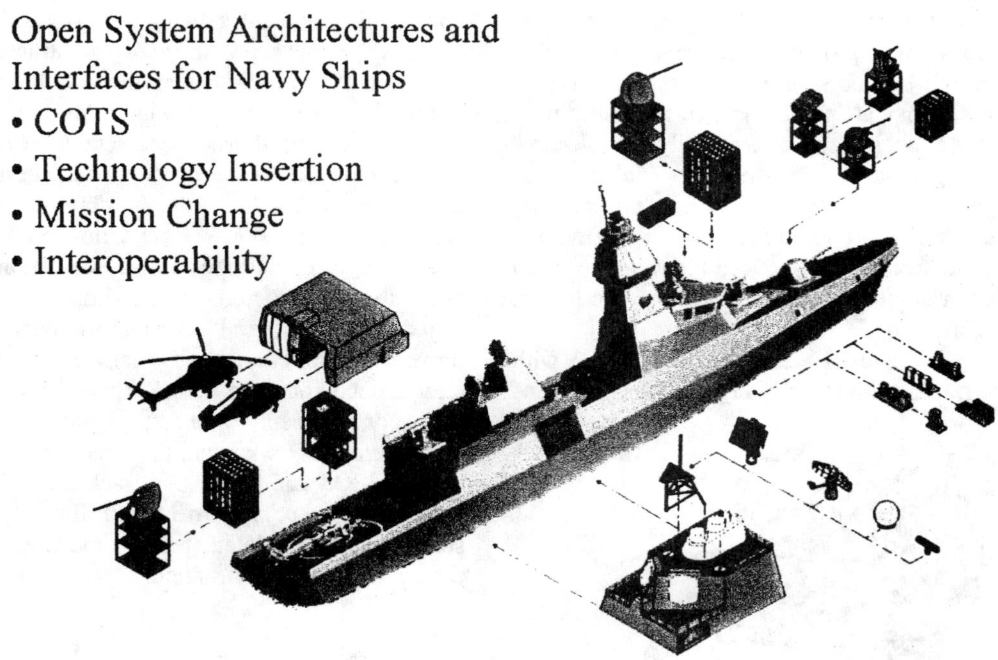

Fig. 13 A TOSA Ship

IMPACT ON SHIP CONSTRUCTION

The inclusion of the open systems approach in the design process for naval ships will result in many benefits during the ship construction. The primary impact will be a schedule reduction in production timelines due to parallel sequence fabrication, assembly and testing. Open structural, HVAC, electrical power, lighting, data distribution, and piping interfaces within a functional element zone will allow the ship platform to reach a higher percentage of completion more quickly unencumbered by the impact of the numerous workers required to install equipment in a piecemeal fashion. With the majority of the interfaces being mechanical connections, the erection units of the ship platform can be final painted with very little pick-up required to achieve a final coating, resulting in a cleaner environment within the ship platform. In parallel with the ship platform construction, sub-system modules can be completed and tested as a completed system in the shipyard shop (or suppliers shop). This will not only result in overall schedule compression but will also result in man-hour reductions to complete the ship. Typical shipyard work cost factors for the various construction areas using the fabrication shop as a factor of 1 are generally grouped as follows:

- Fabrication shops – 1-3
- Assembly shop prior to paint – 3-6
- Post paint outfitting shop– 7-12
- On launching platform prior to launch – 15-20
- On ship post launch – 25-35

By accomplishing the majority of the testing on the modules in the post paint outfitting shop, a cost reduction is projected of approximately 50% over what would normally be accomplished on the launching platform and 75% of the cost during post launch where the majority of testing currently takes place. Similar savings in construction costs will be experienced due to the removal of construction work that is currently accomplished on the launching platform or on the ship post launch. Depending on the degree of implementation, modular payload ships cost less to build than conventionally designed ships, although a small weight penalty may occur in some instances.

Examples

The following examples of a machinery module and a C4ISR space installation illustrate the impact of open interfaces during ship construction. Currently the majority of equipment is installed during the 'Post Paint' Outfitting Shop process within the machinery spaces. Machinery spaces are generally a high work content area and the cost factor for the work area is heavily skewed to the 10-12 factor range due to the large size of the equipment and the congestion of material and personnel in the work area. Each piece of equipment is mounted on a unique foundation that is integrated into the primary structure and individually designed shock mounts are incorporated when required. Interconnecting equipment requires dogleg flexible piping that must also be supported by shock mounts. These interconnecting piping components require that the equipment be located significantly apart in order to be able to accommodate the associated flexible piping requirements (this impact is especially troublesome when the piping is large). These factors increase the work required to install and in addition increase the weight of the ship due to entrained fluids. The module described below would have reduced the weight of each of the chiller installations by approximately one ton on the DDG51 class ships.

As shown in Figure 14, the use of a module with standard interfaces eliminates many of these problems

Fig. 14 200-ton Chiller/FW Circulating Pump/Salt Water Treatment/Power Panel/Speed Control Module

and consequently reduces the work content. The modules can be assembled and tested in the fabrication shops with a work factor of about 3. The foundations for attaching to the primary structure are all a standard design differing only in location. All these factors further reduce the install cost during the 'Assembly Shop Prior to Paint' process. Work content cost can achieve a 75% reduction depending on the location where the work is accomplished. Further, the work in the 'Outfitting Shop Post Paint' process is reduced, thereby shifting the work cost factor to the 7-9 range due to the lower work content. The time duration during the 'Outfitting Shop Post Paint' process is also significantly reduced because only one module needs to be loaded and mechanically attached to the interfaces in lieu of multiple pieces of equipment – each with multiple piping/hoses, etc.

In addition, the module can be tested in the fabrication shop for all the parameters to ensure that the equipment will meet the system requirements, prior to being placed on the unit. Once installed on the erection unit, which then becomes part of the ship, the only testing to be done is the complete system test (stage 8).

In C4ISR spaces, the development of the Shipboard Modular Arrangement Reconfiguration Technology (SMART) System is another open interface technology that can have a significant impact on ship construction schedules and costs. The SMART System currently consists primarily of the structural interface using the ISO 7166 foundation track and the electrical interface consisting of "plug and play" power and lighting configurations. Considerable development has been completed on the open HVAC interface and to a lesser degree on the data distribution within C4ISR spaces. The ISO 7166 foundation track used in the open structural interface is an adaptation of the seat track found on commercial airlines. The ISO 7166 foundation track and its support system provide a universal interconnection between the ship and equipment within the C4ISR space. Once the size of the compartment and the projected maximum equipment load is known, the structural support system and the ISO 7166 foundation track can be designed and installed in the compartment, with final grooming of the space being completed without knowing the details of the equipment to be installed. As seen in Figure 15, the

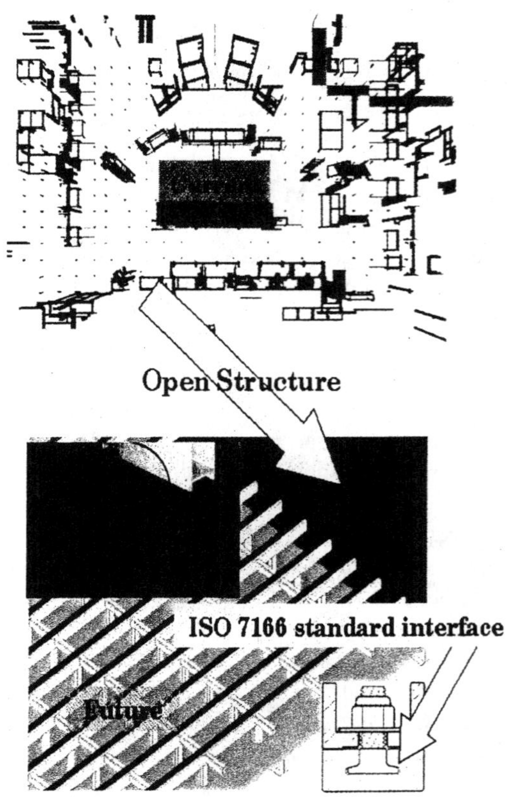

Fig.15 Open Structural Interfaces

system is a repeating pattern that lends itself to the use of jigs and fixtures. The repeating nature of the installation results in a rapid learning curve. As long as the equipment can fit through normal shipboard access doors or can be disassembled to do so, the space can be completed before the equipment is finalized and can even be groomed in a turnkey site prior to installation, further removing the cost and schedule impact of the late arrival or changes to such equipment. Once the equipment is delivered, interfacing sub-bases can be constructed in the shop using the equipment as the template. The repeating nature of the ISO 7166 foundation track allows the mating holes to the track to also be completed prior to moving the equipment to the ship. Thus, there is no hot work or drilling that needs to take place within the space. On Arliegh Burke, the lead ship of the DDG51 class, nearly 80% of the foundations had to be rebuilt or reworked due to the design information not matching the later delivered equipment, at significant schedule slip and cost to the government. Such shipboard impacts are all but eliminated when the SMART System ISO 7166 foundation track is used as the open interface between the ship and the equipment. The equipment can be designed, fabricated, and installed in a "just-in-time" scenario. As a result, evolutionary changes in the equipment are easily accommodated in the ship with the SMART system ISO 7166 foundation track concept.

All the deck tiles are standard 10 ½"x24" panels that are simply dropped in between the ISO 7166 foundation track and held in place by a snap-in plastic strip that has undergone and passed extensive shock tests. The installation of the deck tiles for an average sized C4ISR space can be accomplished in less than ½ man-day.

A preferred variation of the open SMART System ISO 7166 foundation track concept is to construct the system during the low cost "fabrication shop" or "installation prior to paint" processes as a complete platform with shock mounts connecting the platform to the ship's structure as shown in Figure 16.

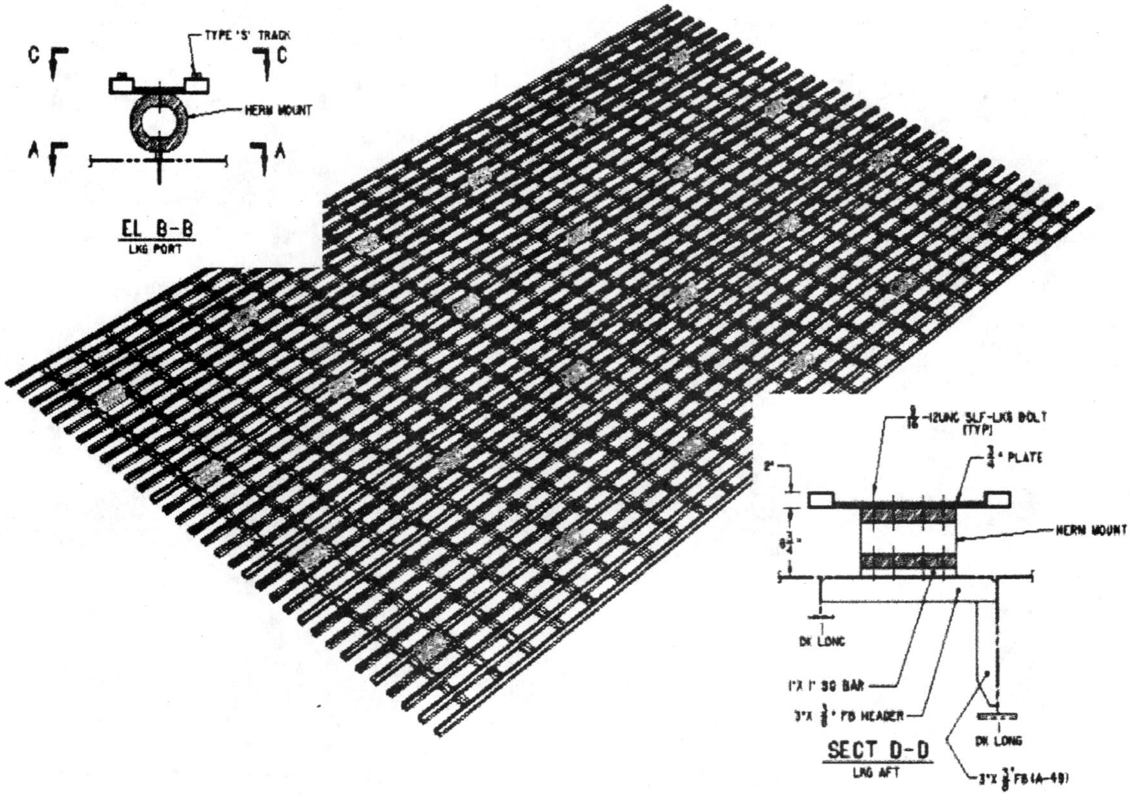

Fig. 16 Open SMART System ISO 7166 foundation track built as a platform

The space shown is the 30'x50' CIC on DDG51 where locations for 24 shock mounts have been provided. The foundations for the shock mounts are incorporated into the primary structure of the deck below and the installation is accomplished using mechanical fasteners prior to erecting the unit above, and the final painting of the deck can be completed prior to erecting the platform. The concept not only reduces installation man-hours but also decreases weight due to the lower forces transmitted to the structure during a shock event as documented in Figure 17 with a related decrease in assembly and erection man-hours. The results shown do not have the benefit

Equipment ID	Equipment	Fixed Support	Rafted Support
14	Console, Combat Display	46.4	11.8
7	Console, Combat Display	66.0	-8.9
21	Console, Combat Display	-59.9	-9.8
13	Console, Combat Display	-36.0	-14.2
19	Console, Combat Display	-24.4	-12.9
358	Console, Combat Display	-43.2	-9.7
18	Console, Combat Display	-48.1	-8.9
17	Console, Combat Display	-57.7	9.9
216	Advanced Tomahawk Display Center	-44.7	-8.9
215	Advanced Tomahawk Display Center	-41.8	-8.7
370	Projection Plotting Unit	17.8	9.6
371	Projection Plotting Unit	19.2	12.5
384/385	Control-Display Optoelectronic	21.7	8.8
9	Console, Combat Display	25.4	-12.7
388	Console, Combat Display	24.6	13.7
213	Advanced Tomahawk Display Center	-40.2	-10.6
242	Console, Combat Display	-33.9	-11.2
12	Console, Combat Display	-29.5	7.1
102	Control-Indicator	-23.9	7.3
8	Console, Combat Display	-13.3	8.9
344	Monitor, Monochrome	12.7	-10.7
232	Panel, Remote Control	13.1	6.4
20	Console, Combat Display	13.3	4.7

Fig. 17 Shock loading in "g's" transmitted to the equipment based on a fixed support vs. a shock mounted rafted support on DDG51

of "tuning" the raft by removal of shock mounts. The 4x6 arrangement of the shock mounts has a 30+% weight capacity margin.

The open SMART System HVAC concept shown in Figure 18 uses the area under the false deck as a supply or return plenum eliminating all overhead ductwork. The deck tiles under the equipment are cut

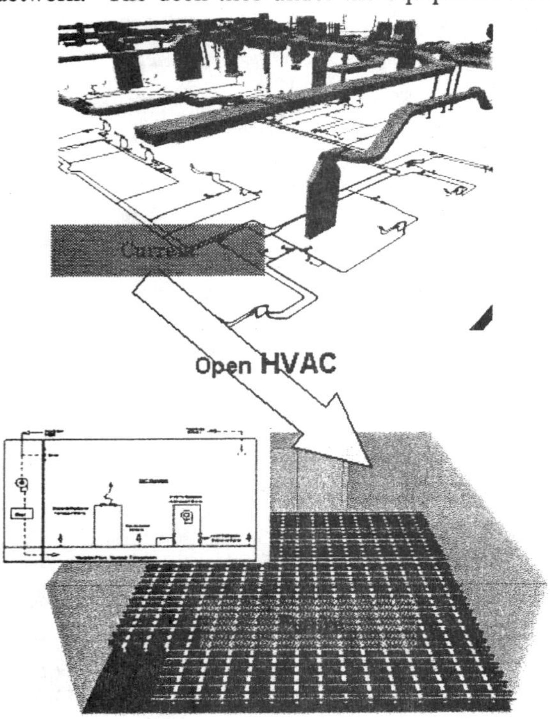

Fig. 18 Open SMART System HVAC

with a standard opening to supply cooled air to or extract exhaust air from the equipment. The supply or exhaust ducting can be built integral with the boundary bulkhead of the space as shown in Figure 19, thus

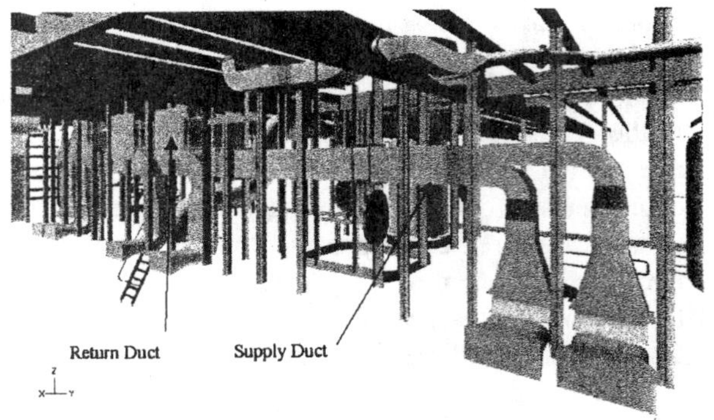

Fig. 19 Supply and return ducting built integral with the boundary bulkhead (Bulkhead plating removed for clarity)

accomplishing all the HVAC ducting installation during the low cost "fabrication shop" or "installation prior to paint" processes. In addition, the amount of ducting is significantly reduced and the overhead is left clear to run other support services. Considerable computational fluid dynamics and land based testing have proven the viability of this concept.

The open SMART System power and lighting concept shown in Figure 20 has been developed with

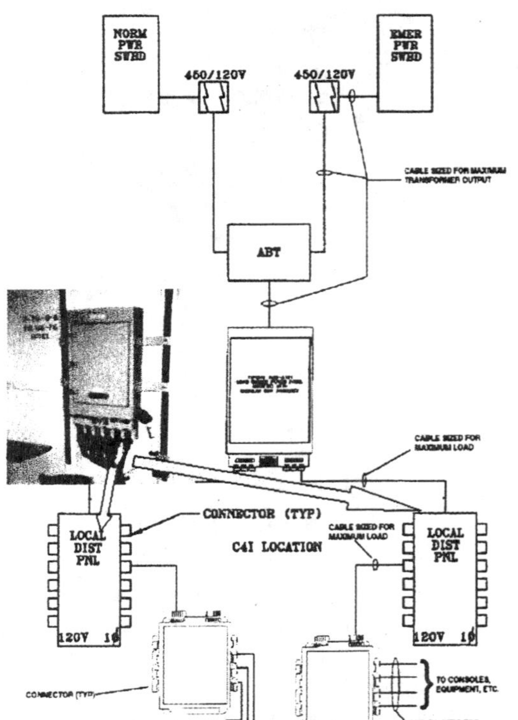

Fig. 20 SMART System power distribution concepts

components that have passed shock qualification tests. With redundant connections included in the design for future evolutionary developments, the "plug and play" open SMART System power system can also be designed and installed prior to the development of the details of the equipment. Within the turnkey land based site, the cables can be cut and connectors installed, thereby reducing the onboard installation efforts to pulling the cables and attaching the connectors on the equipment and the workstation connection boxes. The open SMART System lighting system, shown in Figure 21, can also be installed with the final location of the

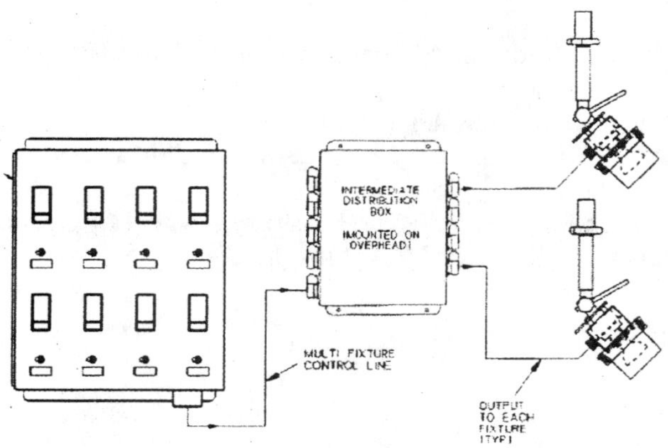

Fig. 21 SMART System lighting concept

clamp-on lights determined once the compartment arrangement has been finalized.

The open SMART System data distribution concept shown in Figure 22 is a similar concept again

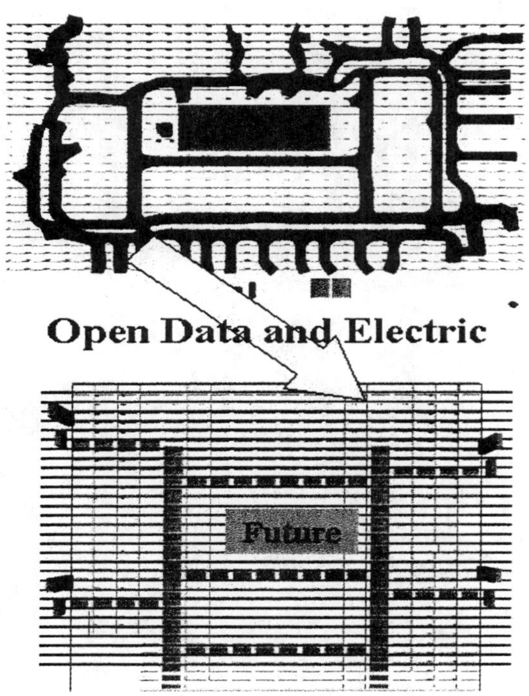

Fig. 22 SMART System data and electric distribution

with the interconnection cabling or fiber optics distribution tubes cut to length at the turnkey land based site. This allows parallel construction scheduling with a commensurate reduction in schedule duration.

CONCLUSION

The new DoD strategy of Evolutionary Acquisition wherein weapon platforms must be designed to accommodate new technologies and systems over time brings with it new challenges to the naval ship designer. This paper presented a new ship design process that embraces the principles of Technology Management and Modular Open Systems to enable a ship acquisition program to bring current and relevant combat capabilities to the warfighter at minimum time and cost. This new process places greater emphasis on functional analysis and partitioning, market research and technology projection. It requires the definition and control of key interfaces for zones, modules and module stations on adaptable (modular payload) ships. Modular Open Systems with associated architectural interfaces must be developed to provide the framework for not only existing/projected system installation requirements but as a set of "design to" requirements for 'next generation' systems to develop against. Finally, the development of the interface standards must include the affected system developers to insure that their knowledge and experience is incorporated in the final standards.

As was noted in the paper, not all systems are amenable to such an approach. Some will not change often and others will be so far outside the 'main stream' set of interface characteristics that they will require a 'custom installation.' For these systems the design approach will be "business as usual." However, the authors contend that the advantages of using Modular Open Systems with well thought out interface standards for ship installation of most payload equipment will allow the Navy to:

a) Simplify the acquisition, construction, and modernization of ship platforms and payloads.
b) Hasten the introduction of new-technology weapon systems (payloads) into the fleet.
c) Quickly convert the type and mix of combat system elements to counter new and changing threats.

Sound familiar?

References
[1] J.W. Abbott, "Modular Payload Ships in the U.S. Navy," Transactions, Volume 85, pp. 350-395, 1977, The Society of Naval Architects and Marine Engineers
[2] G.W. Broome Jr., D.W. Nelson, and W. D. Tootle, "The Design of Variable Payload Ships," Naval Engineers Journal, Volume 94, No. 2, pp. 147-178, April 1982, The American Society of Naval Engineers
[3] D. H. Thompson Jr., and L. M. Thorell, "The Construction of Variable Payload Ships," Naval Engineers Journal, Volume 94, No. 2, pp. 179-199, April 1982, The American Society of Naval Engineers
[4] H. D. Ehrenberg, W. Schmidt, "MEKO Technology: Modular Naval Surface Ship Design – The Flexible Answer to Changing Naval Requirements," INEC 92, September 1992
[5] W. A. Webster, "Is it SEAMOD Time?" ASE Annual Technical Symposium, April 1991, Association of Scientists and Engineers
[6] M. L. Bosworth and J. J. Hough, "Improvements in Ship Affordability," Transactions, Volume 101, pp. 705-726, 1993, The Society of Naval Architects and Marine Engineers
[7] Secretary of Defense Memorandum, *The Defense Acquisition System*, August 29, 2002
[8] W. Schoenster, "Technology Transition and Insertion in Navy Ship Acquisition Programs," Masters Thesis, University of Maryland, Summer 2001
[9] R. DeVries, K. T. Tompkins, and J. Vasilakos, "Total Ship Open Systems Architecture," Naval Engineers Journal, Volume 112, No. 4, pp. 59-77, July 2000, The American Society of Naval Engineers

APPENDIX

Technology Readiness Levels (TRL) and Definitions

Technology Readiness Level	Description
1. Basic principles observed and reported.	Lowest level of technology readiness. Scientific research begins to be translated into applied research and development. Examples might include paper studies of a technology's basic properties.
2. Technology concept and/or application formulated.	Invention begins. Once basic principles are observed, practical applications can be invented. The application is speculative and there is no proof or detailed analysis to support the assumption. Examples are still limited to paper studies.
3. Analytical and experimental critical function and/or characteristic proof of concept.	Active research and development is initiated. This includes analytical studies and laboratory studies to physically validate analytical predictions of separate elements of the technology. Examples include components that are not yet integrated or representative.
4. Component and/or breadboard validation in laboratory environment.	Basic technological components are integrated to establish that the pieces will work together. This is relatively "low fidelity" compared to the eventual system. Examples include integration of "ad hoc" hardware in a laboratory.
5. Component and/or breadboard validation in relevant environment.	Fidelity of breadboard technology increases significantly. The basic technological components are integrated with reasonably realistic supporting elements so that the technology can be tested in a simulated environment. Examples include "high fidelity" laboratory integration of components.
6. System/subsystem model or prototype demonstration in a relevant environment.	Representative model or prototype system, which is well beyond the breadboard tested for TRL 5, is tested in a relevant environment. Represents a major step up in a technology's demonstrated readiness. Examples include testing a prototype in a high fidelity laboratory environment or in simulated operational environment.
7. System prototype demonstration in an operational environment.	Prototype near or at planned operational system. Represents a major step up from TRL 6, requiring the demonstration of an actual system prototype in and operational environment, such as in an aircraft, vehicle or space. Examples include testing the prototype in a test bed aircraft.
8. Actual system completed and "flight qualified" through test and demonstration.	Technology has been proven to work in its final form and under expected conditions. In almost all cases, this TRL represents the end of true system development. Examples include developmental test and evaluation of the system in its intended weapon system to determine if it meets design specifications.
9. Actual system "flight proven" through successful mission operations.	Actual application of the technology in its final form and under mission conditions, such as those encountered in operational test and evaluation. In almost all cases, this is the end of the last "bug fixing" aspects of true system development. Examples include using the system under operational mission conditions.

DISCUSSION

Millard Firebaugh, Member, **Alfred Malchiodi,** Visitor, and **Tom Plante,** Visitor

In their paper, the authors explore changes in the design process that can facilitate modularity in the architecture of warships. Modularity can facilitate technology insertion during service life, and modularity can also reduce construction costs and cycle time. Rather modestly, Jack and his fellow authors recounted the efforts in the Naval Sea Systems Command some 25 years ago in SEAMOD and the generation of Ship System Engineering Standards that were to have been foundational steps in the direction of modularity for surface combatants. The program lost traction, but not before it pointed the way toward the MK 41 VLS launcher interfaces, which have served the Navy well. Jack Abbott was the advocate and technical manager for those programs at that far off time. It must be with pride and some sense of opportunity delayed that he sees his ideas born again in current thinking. In celebration of Jack's new currency, let's look at "modularity" across a broad spectrum of naval engineering activity.

Modularity to facilitate construction has a long history. Modular concepts were tried out in World War II here in the United States and in Germany. With respect to submarines, in the late 1960s, the construction of the experimental submarine *Dolphin* at the Portsmouth Naval Shipyard was based on an early version of end-loading with some modularity. Starting with the *Trident* in the mid-1970s, Electric Boat pioneered in modular construction of submarines. New techniques and processes were developed by Electric Boat to facilitate the ability of the shipyard to build larger, more complete modules in a shop environment, as opposed to the traditional on-hull outfitting. Key innovations included a land level facility, special fixtures for building hull cylinders using automated welding processes, fixtures of manufacturing and outfitting assemblies and modules, dimensional and accuracy control processes, and lifting and handling of very heavy weights. Large packages of machinery and deck modules were end-loaded into open cylinders rather than using the traditional concept of completing the hull first and then cutting accesses in which to feed components, foundations, pipe, and cable to assemble the ship systems within the confines of the pressure hull. In the United Kingdom, then Vickers, now BAE, created the SMITE in which the submarine propulsion plant was assembled and tested off-hull in a shipyard facility and loaded into the hull as a single unit. As the advantages of the end-loading and packaging of major areas of the ship became evident, Electric Boat re-engineered the SSN 688 Class submarines it was building so they, too, could be built with modular techniques. Electric Boat and Newport News Shipbuilding designed the SSN21 Class with more emphasis on end-loading and large packages than the Ohio Class or the re-engineered 688s had been able to incorporate. The Virginia Class submarines, designed by Electric Boat and being built in a teaming arrangement by Electric Boat and Newport News Shipbuilding, have achieved a very high state of modularity. Essentially, the entire ship is an assembly of 21 modules, including seven large deck packages. It was this level of modularity that enabled the EB/NNS team to share the work on each ship without a large penalty for inefficiency. Very complete modules, for example, the entire forward compartment command and control spaces, are built and tested off-hull in a shipyard facility. In the Virginia Class Command and Control Systems Module (CCSM), there is modularity within the modules. The CCSM contains structurally integrated enclosures that are designed for modular electronics. The modular electronics facilitate ship construction (e.g., the entire combat system of 26 subsystems is now installed within 4 weeks) and technology insertion through the life of the class. And, by virtue of the fact that the CCSM can be inserted fully tested into the ship as a complete unit, late in construction, due to its modularity, it can benefit initially from up-to-date electronics, including a refresh prior to delivery.

In commercial ships, LNG tanks, built with the Moss-Rosenberg spherical tanks, technology dating from the late 1960s, are an example of modules created for ease of construction and reduction of cycle time. The LNG tanks come to the hull as completed units allowing tank fabrication and hull construction and outfitting to proceed in parallel. Marinex is now in the marketplace with a more geometrically effective tank shape than the Moss-Rosenberg sphere, which will have the same benefit in terms of modular cargo tanks.

Modularity to facilitate technology insertion has had a good run in submarines. The SSBN Classes were designed to accommodate a succession of missiles increasing in size. The conversion of the first four Ohio Class submarines to SSGNs was facilitated by the ability of a *Trident* missile tube to accept the *Tomahawk* cruise missile module, which contains seven missiles. Twenty-two of the 24 missile tubes per ship will be able to accept the cruise missile modules. The DARPA submarine payload and sensors program developed modular approaches for future submarine payloads, some of which are funded for further development. The modular approach in SSGN to utilizing the inherently simple interface of the large *Trident* missile tube is a harbinger of other exploitations of Ohio Class submarines for new modular payload and sensor capabilities in the future.

Modularity in HM&E is the next challenge for naval architects. Ideas are now on the table that can introduce real modularity in HM&E in what have typically been characterized as distributed systems. Propulsion pods hung under the stem of a surface combatant are feasible. These pods would be a self-contained electric drive motor and propeller unit. They could be built and tested off-hull for ease of construction and could be removed waterborne for maintenance or for upgrade. These units eliminate the need for propulsion shafts. Future surface combatants could have electronic spaces like the Virginia Class CCSM in which the entire package is built and tested off-hull and loaded as a unit. The off-hull fabricated LNG tank designs suggest the possibility of very large modules with different kinds of nearly self-contained functions, all accommodated by a basic hull architecture with electrical and cooling water connectivity in the next generation of naval auxiliaries. A key concept in facilitating these advanced modular architectures is to convert as much ship functionality as possible to electrical components and systems. Electric ships will give new meaning to the idea of "plug and play."

In the future, we need to look at the potential for warships in which the hull is a hydrodynamic form with the necessary inherent strength to support functional modules in propulsion, auxiliaries, electronics, habitability, sensors, weapons, and control. The hull/hydro module will support all the other modules. Such an approach holds the promise to reduce the costs and cycle time for design and construction, reduce the costs of maintenance, enhance the availability of the ship, and facilitate in-service technology insertion. Modules! Modules! And more modules!

Paris Genalis, Member

The authors have written a very timely paper reflecting a new approach to solving a centuries-old problem. I was delighted to see their opening remarks that reminded us of earlier attempts, such as the SEAMOD, SSES, and others. They also mentioned the Opens Systems Architecture task force that the Office of the Secretary of Defense established to address this problem, not only for ships but also for other defense systems. Lest we think this is a modern day problem, let me mention that a cursory scan of seafaring history will quickly reveal that the Phoenicians and the ancient Greeks had created solutions to facilitate trade of wet and dry cargo, which varied with the seasons, by use of different containers that fit the same (nearly "standard") shape of the holds of their ships.

There has always been a tension between the designer and the operator. The designer's wish to provide for future flexibility to adopt new technology and adapt to new missions carries with it a cost in space, weight, and margins that the operator does not need now. These options provide a cost-avoidance in future expenditures for updates and upgrades. However, the fulfillment of military requirements is focused on capabilities now and not on possible options for the future. Similarly, the commercial operator will quickly point out that profits now are much more important than possible cost avoidance in the future. It takes a far-seeing owner/operator to subscribe to, and maintain, a long- term view.

During the lifetime of the ship, additional costs need to be accounted for. Different flights of the same ships have different configurations, possibly different components, and different combat systems. To provide for long-term support of various flights of ships, we need to develop a much more sophisticated configuration control system and perhaps revamp the logistics system that supports such an architecture of ship systems. That cost is often significant and may require doctrinal and cultural changes that may be very difficult.

We must also reconcile our own policies that reflect our conflicting desires. Responsible acquisition executives demand that each program manager have a complete picture of the project's costs—the life cycle cost or total ownership cost. They also expect that the design provide flexibility to adjust to future technology and requirements. While it is possible to provide costs for flexibility built-in, it is of course impossible to cost the unknown technology for the same mission, let alone the future technology and hardware for a new mission. We have attempted to resolve this untenable situation for the program manager by insisting that clear total costs are available for each stage of the design. This may be the best we can do, but we must recognize that this is another form of suboptimization, because we don't have the ability now to see the future options that will be available to the designer of the modifications. Possibly this is good enough, but it is a suboptimization nonetheless.

Thus, our attempts to save money by providing for future cost avoidance results in costs that must be born now and perhaps over the life of the ship. This clearly results in a quandary of how to balance the current, and known, cost against the future and unknown cost avoidance. This perhaps explains the many previous attempts to address the problem, which always seem to be not quite satisfactory in balancing current and future desires. Constrained resources have been the stimulus that keeps bringing the problem to the fore, and it is that day's technology that has attempted another solution.

Today's approaches reflect capabilities that are now at our disposal: better analysis tools that can expand our grasp of the "big picture," better estimating processes based on the availability of databases for design procedures and material and labor costs, the dogmatic belief that tomorrow's computing devices will be more powerful. It is, however, far easier to imagine and demand new performance than it is to develop the capability to deliver it. We must be aggressive in our attempts for far-reaching solutions, as we are vigilant not to overreach and expect today to solve tomorrow's problem.

Cyrus Azani, Visitor

Generally speaking, I found the paper to be well written, easy to follow, and reflective of OSJTF thinking. As the main author of the OSJTF guide on the modular open systems approach (MOSA), I can say that the paper is consistent with the OSJTF guidance provided therein. The final version of this guide, *A Modular Open Systems Approach (MOSA) To Weapon System Acquisition*, Version 1.0, April 2003, is now posted at http://www.acq.osd.mil/osjtf/implement/implement_tools.html and should be considered as an additional reference in the paper.

There is need for key interface definition so that the reader understands the distinction between such interfaces and other types of interfaces within a system or system of systems. Here is the definition used in the guide: Key Interface: An interface for which the preferred implementation uses an open standard to design the system for affordable change, ease of integration, interoperability, commonality, reuse, and other essential considerations such as criticality of function.

Some emphasis is needed on test and evaluation challenges of using open systems. A large number of vendors claim that their systems/products are open, and without some sort of confirmation or verification tests, we will face lots of problems in interoperability among systems and in integrating subsystems into a whole greater than the sum of constituent parts. (Please see my paper published in the ITEA Journal and posted at the OSJTF website titled "The Test and Evaluation Challenges of

Following an Open System Strategy" for a discussion of some of these challenges.)

Captain Chuck Goddard, Visitor

The authors have done a superb job of covering an important topic that will change the Navy's approach to design and acquisition. This paper embodies what the DDX program spiral design and acquisition approach is striving to achieve. As many in the professional naval design community know, the DDX program technology development efforts include developing and testing an advanced gun system, dual band radar suite, advanced vertical launch system, integrated topside design, total ship computing environment, integrated power system, optimized manning, and a wave piercing, tumble home hull form. The evolution of these technologies is expected to continue across the life cycle of the program.

To spiral DDX capability over its life cycle, we are implementing both the management processes and design attributes identified in this paper to forecast, assess, plan, and respond to present and future needs. From the technology development efforts cited above, expected threat and mission changes influenced our spiral development thinking, which in turn impacted our design approach. For example, the advanced vertical launch system is being designed to accommodate the current family of missiles as well as expected size growth of future missiles. To accommodate the expected future changes, the design will incorporate larger diameter and longer missile cells. Another area is electronics, which have traditionally been a challenge during initial installation and life cycle management. DDX will incorporate electronic modular enclosures or a box-in-a-box concept. This approach will allow for separate but concurrent testing and ease of installation during new construction, as well as ease of repair and upgrade across the ship's service life. Access routes and ways are being provided to fully enable this concept. Of note, the DDX design team is also evaluating the reconfigurable foundation (e.g., smart track concepts outlined in this paper) for the CIC spaces. These are a few examples of the extensive DDX spiral design approach. As the authors have clearly pointed out, these types of design approaches provide the Navy the ability to easily and affordably install transformational technologies aboard not only the DDX, but also all future combatant ships.

Authors' Closure

The authors would like to thank all the discussers for the time they took to read our paper and provide excellent and insightful comments. It is truly an honor to present a paper to the SNAME members. The remarks by all are in the truest tradition of providing critique and assessment of a paper thesis by one's peers at the annual meeting.

With regard to the comments by *Millard Firebaugh and his colleagues* at Electric Boat, the kind words regarding past efforts on this concept are most welcome. It is always interesting to see how well things are done in the submarine community. The discussion of use of modular open systems is certainly an indication that this approach is alive and well in the design and construction of submarines. It is obvious that this concept has resulted in efficiencies in submarine construction in both time and cost by installing the entire combat system within 4 weeks.

Dr. Paris Genalis has hit on a key issue in that "It takes a far-seeing owner/operator to subscribe to and maintain a long-term view" to achieve the tenets of the paper. One only has to look at the LPD 17 program, where easily reconfigurable combat spaces have been reduced from 14 to 11 and now to 3. However, with the formation of our consortia, design approaches are being developed by the shipbuilding industry to remove current construction weight and cost penalties.

It is comforting to note that the TOSA efforts have remained in sync with the policies of the OSJTF. Furthermore, OSJTF has expressed full support of TOSA efforts within the Navy to implement the modular open systems approach. They once called us their poster child. In addition, they sought our input in the development of the MOSA guide referred to by *Mr. Azani.*

Going forward, the challenge is to apply this concept to real ships. We are delighted to see that *Captain Goddard* expects that the DDX will include this approach for technology insertion and reconfigurability. However, before the DDX is fielded, we have our hands full with work on the LCS. Like everything new, the application of these concepts must be done using good systems engineering principles.

Marine Systems In Refuse Hauling Infrastructure

By Dr. John C. Daidola, P.E.[1] (F), Richard C. Rodi, P.E.[1] (M), Benidetto Stazzone[1] (V), Harvey A. Brodsky, P.E.[2] (V), Russell L. Filtz[3] (V)

ABSTRACT

The collection and disposal of refuse is considered from the standpoint of delivery to its final disposal site. A review is provided of historical aspects of two major urban areas as well as current features of those systems. Details are provided of a marine-based transfer system utilized to deliver material from local collection sites to disposal or transfer sites. The potential profitability of transporting material generated in the U.S. to disposal sites worldwide is examined as well.

INTRODUCTION

The collection and disposal of refuse is a massive industry spanning the entire globe. Major urban areas such as the New York City (NYC) may generate as much as 20,000 tons per day. Across the United States over a million tons are generated per day, around the world literally millions of tons per day.

Collection of refuse begins at the source, usually by truck. Further complication has occurred with the recent separation of recyclable components. Although in many cases the collected material is transported directly by these trucks to local disposal sites, it is often brought to staging areas and then transported to its final destination. This latter approach has increased as local disposal sites have been eliminated due to depletion, environmental concerns, or public outcry.

NYC had historically developed a marine-based system wherein collection trucks would carry their load to Marine Transfer Stations where it was loaded into specially constructed hopper barges for transport to a local landfill where the barges were unloaded. The closing of this landfill several years ago led to the end of the marine transfer systems and the beginning of a massive trucking operation to sites in New Jersey. From there the refuse may be transferred to larger, long-haul trucks or containerized and transported by trucks to final destinations in other states or to rail heads for similar transport. This system has increased truck emissions and proven costly and difficult to operate and maintain. Thus the City is planning to revert to a marine transportation system.

Another example of an urban area is the City of Hong Kong (HK), which had in the more recent past adopted incineration technology to dispose of its refuse. However, as experience indicated incineration was adversely affecting the air quality HK has migrated to marine based transfer stations for the containerization of waste for shipping to remote disposal sites.

A modern marine transfer system will likely incorporate utilization of a containerized system for transport of both refuse and recyclable materials. Facilities are required for the containerization of the materials and loading onto barges or ships. Components of this system include the facilities where the containerization occurs, vehicles, compactors, conveyors, containers, handling systems and vessels for transport.

This paper presents the historical evolution of the marine hauling of refuse in NYC evolving into the current conceptualized system utilizing containerization. M. Rosenblatt & Son has supported all aspects of the barging in this Marine Transport System since 1960. A similar system is being utilized in Hong Kong and is discussed as well. The characteristics of this system provide a model for other locations.

[1] AMSEC LLC/ M. Rosenblatt & Son
[2] Greeley and Hansen LLC
[3] Shaw EMCON/OWT, Inc.

Consideration is also given to the potential market for marine transportation of the containerized waste material to major centers of distribution or directly to the disposal sites they serve. This aspect has not yet been developed for delivery of material over long distances.

THE REFUSE INDUSTRY
U.S. Production

The generation of garbage, trash or refuse has been a part of our sociological infrastructure through the centuries. Figure 1 shows the production per person in the City of New York. Interestingly, per capita, the production at the turn of the 19th and 20th centuries is approximately the same.

The production throughout the United States is depicted in Figure 2, showing over one million tons per day. Approximately 40 percent of the nation's waste is generated by five states [3]. California leads at 45 million tons a year, followed by New York, Florida, Texas and Michigan. The national data indicates that each U.S. resident produces 4.3 pounds a day of trash, twice as much as a generation ago.

Disposal Options

Sixty-one percent of U.S. trash is consigned to landfills, while 30 percent is recycled and nine percent incinerated [3]. At an average of $80 per ton for dumping or "tipping," Alaska has the nation's highest fees. It is followed by New Jersey, Delaware, Vermont, New Hampshire, and Minnesota.

As a result of high local tipping costs, some states have gone to and continue to increase their exports of trash to other states for disposal in landfills. Waste disposal has now grown to a $36 billion a year industry and the decision to transport waste interstate is usually a question of economics. The average fee shown in Figure 2 and available in a state such as Virginia is in the order of $35 a ton. Pennsylvania, Virginia, Indiana, Michigan and Illinois are the nation's leading trash importing states.

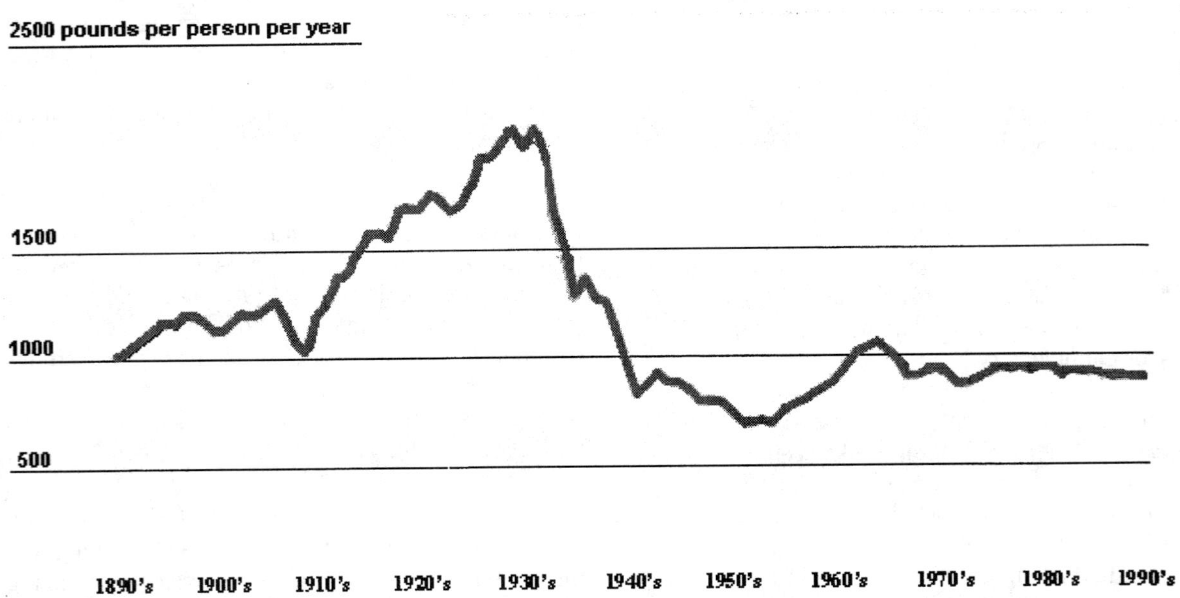

Figure 1: New York Refuse Rate [1]

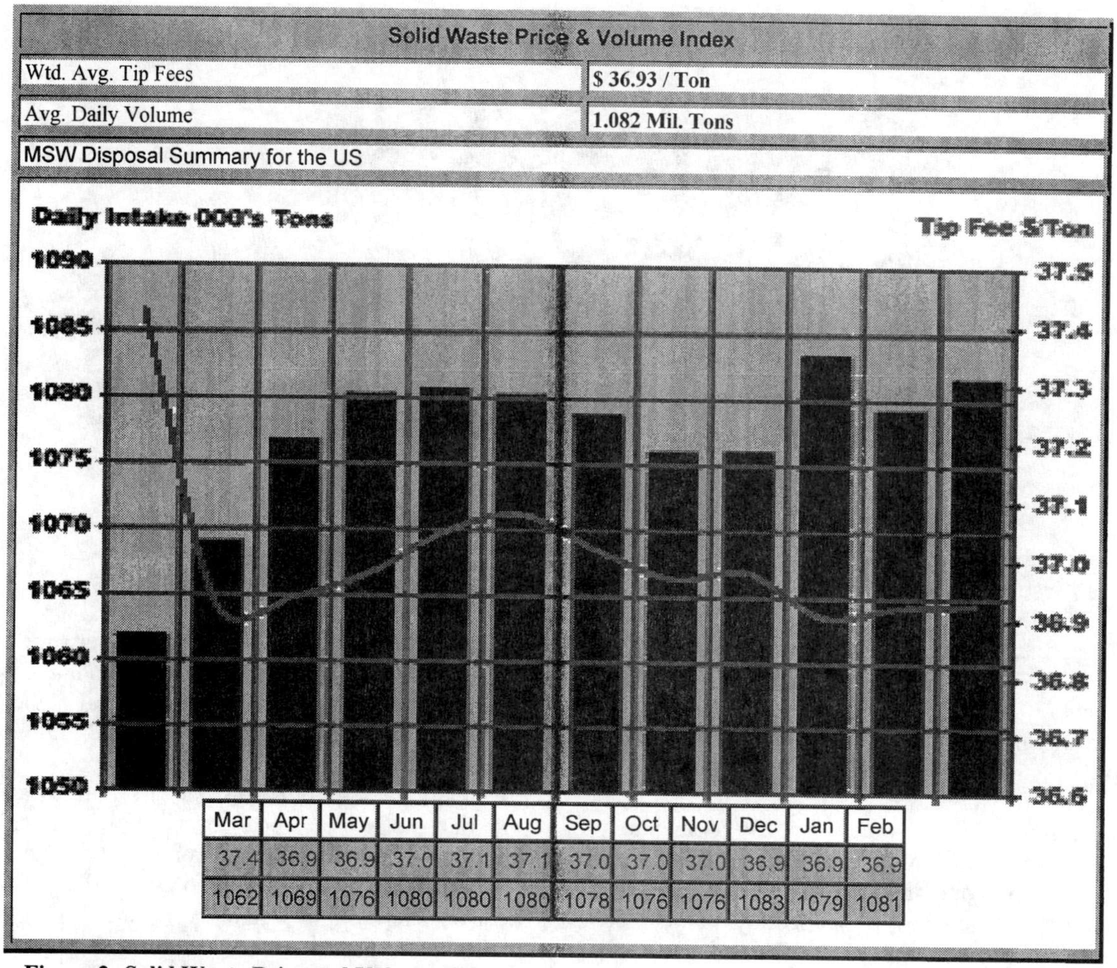

Figure 2: Solid Waste Price and Volume [2]

Transfer Options

Refuse is typically collected at the source by truck, which transport it to a local disposal site or to an intermediate transfer station. The latter has increased with the interstate transport wherein local collection trucks are not appropriate for long haul transportation. Methods utilized for transportation from these transfer stations have included over the road trucks, containerization for trucking or rail and marine transportation by barge.

HISTORICAL ASPECTS OF A MARINE TRANSFER SYSTEM

The problem with what to do with what they throw away has concerned residents of the nation's largest city, NYC, almost since its inception. More recently, the mayor suspended part of the city's recycling effort, citing the lack of a profitable recycling market. He decreed that the Department of Sanitation (DSNY) would no longer make separate pickups of plastic and glass. Instead, these items are again being mixed in with the bulk of the 928 pounds of garbage each resident throws away in a year.

In 1896, a new sanitation commissioner reorganized the department and instituted a systemized plan. For the first time, New Yorkers were required to separate their garbage at curbside. Much of it was sold to private contractors, who boiled it down for grease and fertilizer, or used it for landfill and other purposes.

Figure 3: Dumping off Coney Island [1]

But by the 1920's, this earlier recycling effort had also been defeated by a falling market in trash. The city had some 80 open dumps, including the monumental ash heap in Flushing Meadows that F. Scott Fitzgerald described memorably in "The Great Gatsby" – " a fantastic farm, where ashes grow like wheat into ridges and hills and grotesque gardens."

Garbage lined much of the Manhattan waterfront along the Lower East Side. From there it was transported to one of New York's 22 incinerators or – until the practice was banned by the United States Supreme Court in 1934 – towed out and dumped into the sea, Figure 3. The sea often threw it back onto local beaches, along with the city's sewage.

Over the years, improved public policy and real estate pressures consolidated the NYC dumps into one at the Staten Island Fresh Kills landfill, and neighborhood objections put the incinerators out of business. Now most of the city's refuse is simply shipped out of town, to landfills in Pennsylvania, Virginia and elsewhere.

For most of the last century the refuse in NYC was transported to its final destination by barge. For approximately half of that time the system revolved around collection from eight distribution centers, Marine Transfer Stations, and barging to the Fresh Kills landfill.

Marine Transfer Stations

DSNY historically used a network of eight marine transfer stations (MTSs) as destinations where collection vehicles deliver municipal solid waste. These MTSs, locations shown in Figure 4, are unique to NYC and were developed to provide efficient waterborne transport of solid waste. The stations provided local destination for collection vehicles serving the boroughs of the Bronx, Manhattan, Queens and Brooklyn and contributed to the efficiency and reliability of the DSNY's waste management operations.

Collectively the MTSs are capable of transferring approximately 14,000 tons per day of municipal solid waste from land-based collection vehicles to barges for waterborne transport. Each MTS barge eliminated the need for 30 truck transfer trailers that would otherwise move over the city's roadways.

Figures 5 and 6 depict schematic views of the most common MTS design, which is simple, robust and reliable. City collection vehicles enter the second level of the MTS building (truck-receiving level) and weigh-in over the scale. Vehicles proceed onto the truck receiving floor and back into position at one of 12 unloading bays. There are six bays on each side of the truck receiving floor. Each unloading bay has an opening in the floor, which is clear to the barge slip level below the truck receiving floor. A hopper barge is positioned in the slip under the openings. The truck receiving floor is approximately 20 feet above high tide level. Each barge slip is approximately 40 feet wide by 140 feet long. Each DSNY collection vehicle carries, on average, approximately 10 tons of waste. MTS hopper barges carry approximately 620 tons, the equivalent of loads from 60 to 70 collection vehicles. (In contrast, a typical transfer trailer contains the loads of two to three collection vehicles.)

The Plan view of the first level in Figure 5 shows barges positioned in two slips. The section view shows a collection vehicle in a tipping bay dumping into a barge. Waste receiving operations are very simple. There is no handling of waste: it is unloaded by the force of gravity directly from trucks into the barge. There are no mechanical parts that can fail or become inoperable (except trucks) to affect the reliability of unloading operations. As shown in the plan view in Figure 5, the existing truck receiving floor is approximately 80 feet wide by 100 feet long and can accommodate up to six vehicles simultaneously

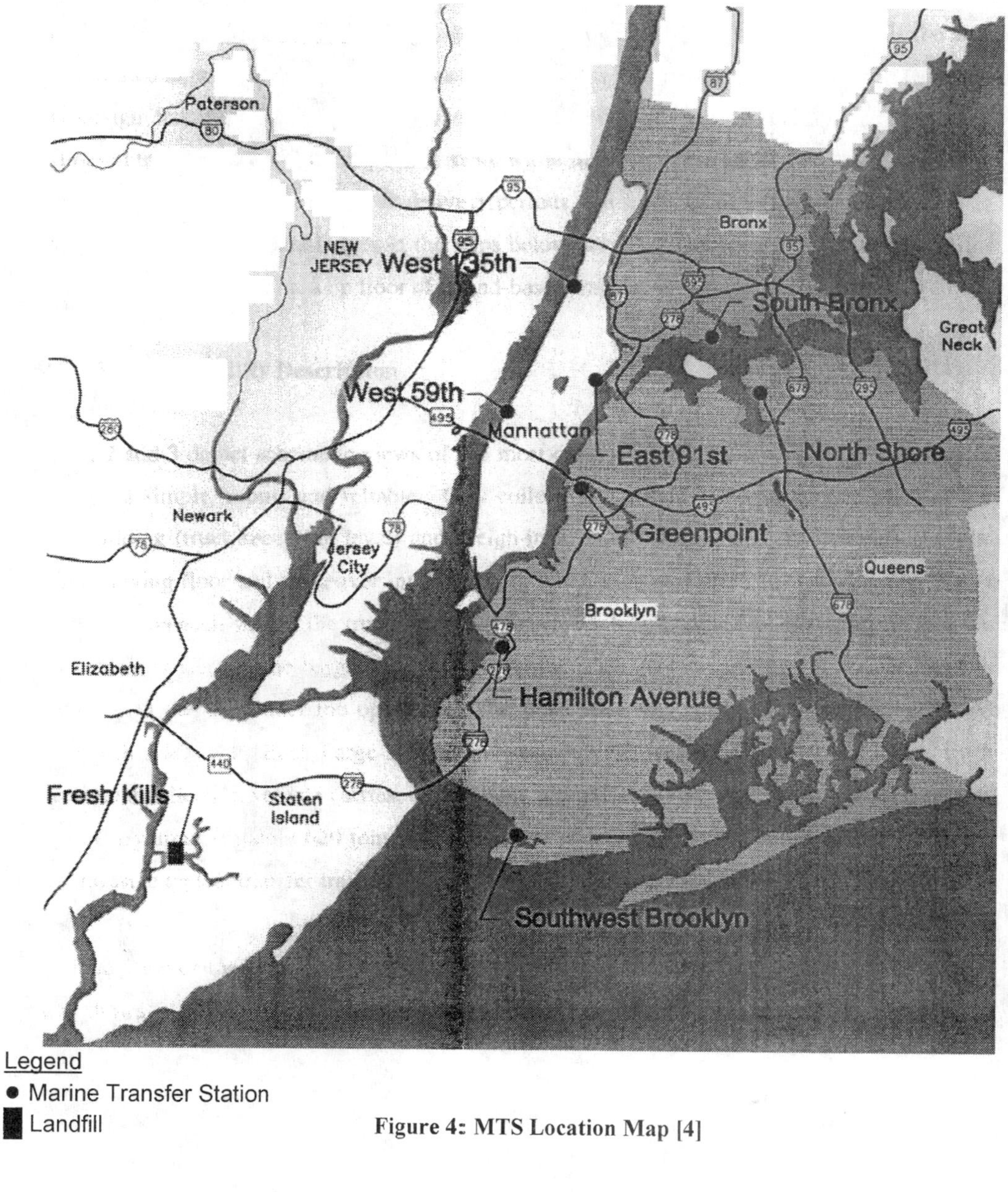

Figure 4: MTS Location Map [4]

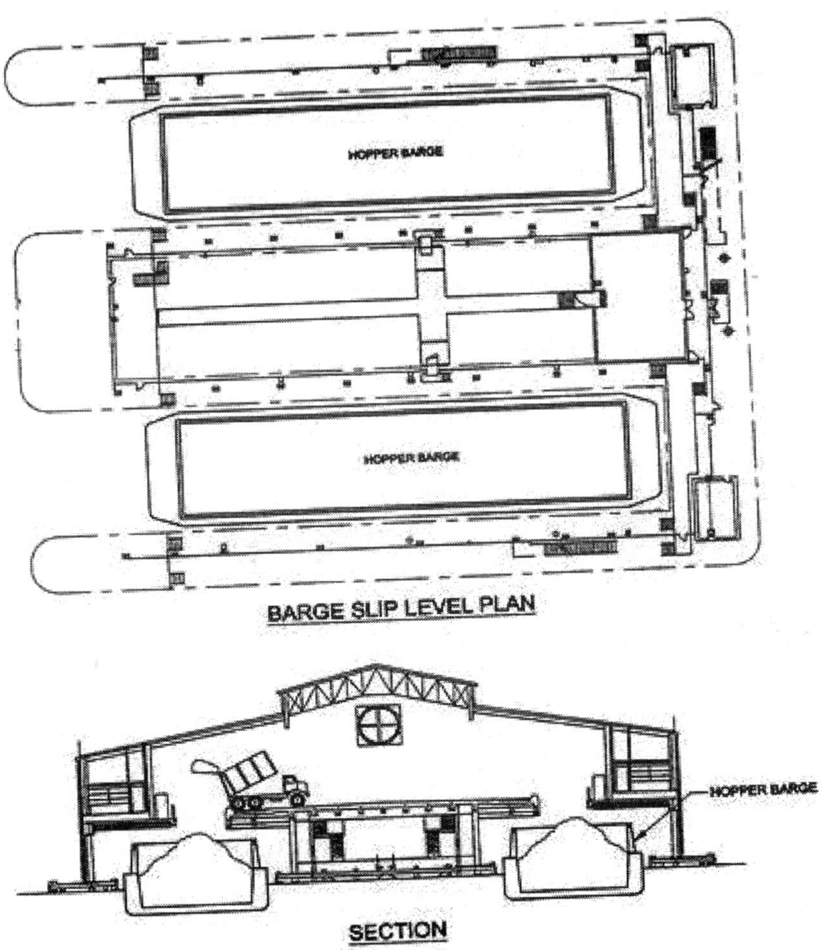

Figure 5: Marine Transfer Station - Slip Level [4]

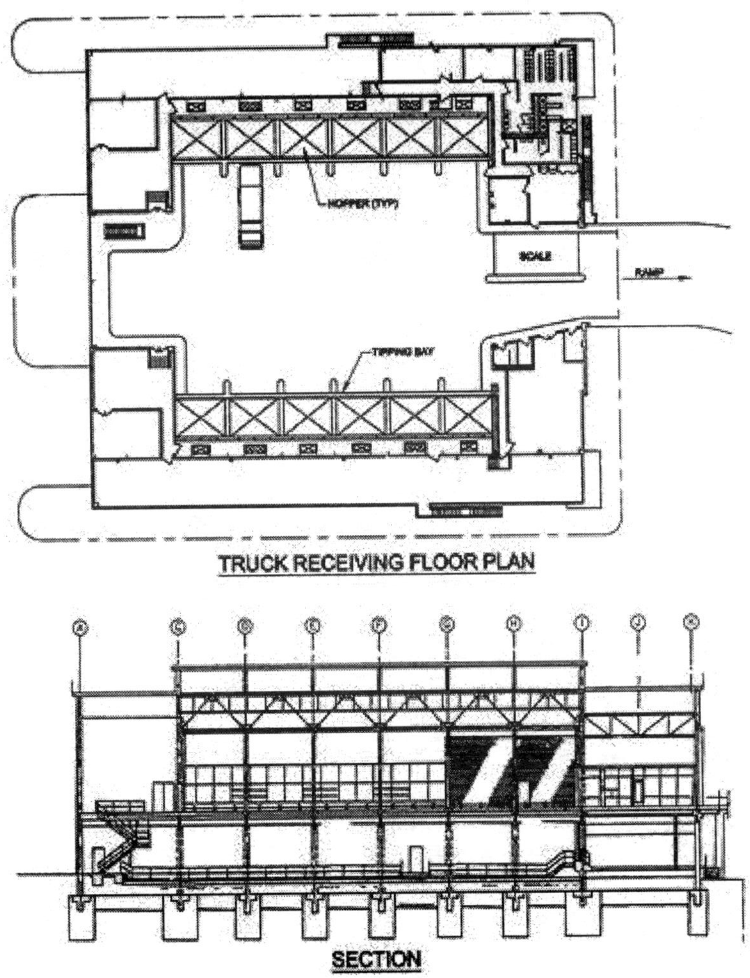

Figure 6: Marine Transfer Station - Truck Level [4]

Barges

The number of barges utilized in this operation reached a peak of approximately 110. The design of the last series was accomplished by M. Rosenblatt & Son, Inc., and built by NABRICO is shown in Figures 7 and 8.

Figure 8 indicates some of the prominent features, heavy side and steel structure and innerbottom covered with asphalt. These features, as well as others, were developed as a result of the rough treatment of this equipment, especially from the clamshell bucket unloaders at the Fresh Kills Landfill. These were later replaced by hydraulic diggers, Figure 9.

Recent Changes

Since the closure of the Fresh Kills landfill, previously the ultimate destination for NYC municipal solid waste, the DSNY has utilized a series of land-based transfer stations for overland export of DSNY managed waste to remote disposal facilities. This mode of disposal has added to air and noise pollution within the City limits, has increased maintenance and operating costs of collection vehicles and has contributed to increased deterioration of City streets and highways. As a consequence and as a result of a decree by the Mayor, currently NYC is developing a design using the existing MTS network upgraded and modified to provide a system where the waste could be received at the MTSs, placed into sealed containers, loaded onto barges and transported to an inner-harbor port facility where the containers could be off-loaded onto rail, ship or truck for transport to remote out-of-city disposal facilities.

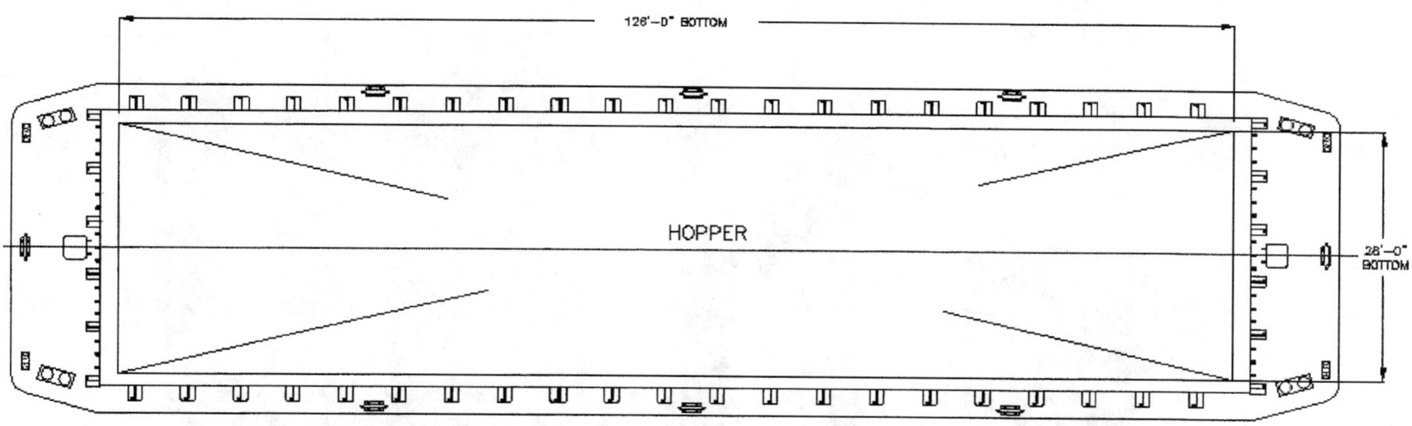

Figure 7: DSNY Barge – Plan View

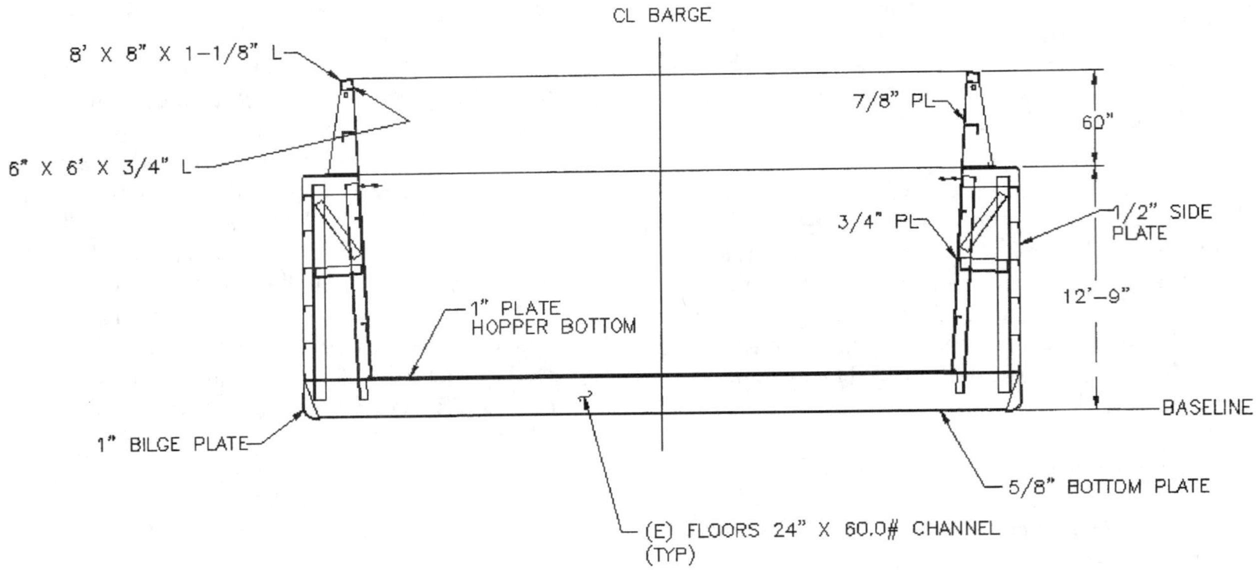

Figure 8: DSNY Barge – Midship Section

Figure 9: Hydraulic Diggers [6]

AN OVERSEAS EXAMPLE

Hong Kong (HK) has been operating marine based transfer stations for over ten years as well. These stations were developed to take the place of older incineration technology, which made the already poor air-quality of the region even worse. The difference in HK is that they went for compaction into ISO containers right away, instead of using loose-fill garbage "scows" as NYC initially operated.

A prime example of HK's marine transfer stations is their newest, the West Kowloon Transfer Station (WKTS) in Cheung Sha Wan along the western coast of the mainland part of HK, Figures 10 and 11. This station is also the largest, with a design tonnage of 3,500 metric tons per day, and a peak of 4,000 metric tons during emergencies and Chinese New Year.

Figure 10: Rendering of West Kowloon Transfer Station

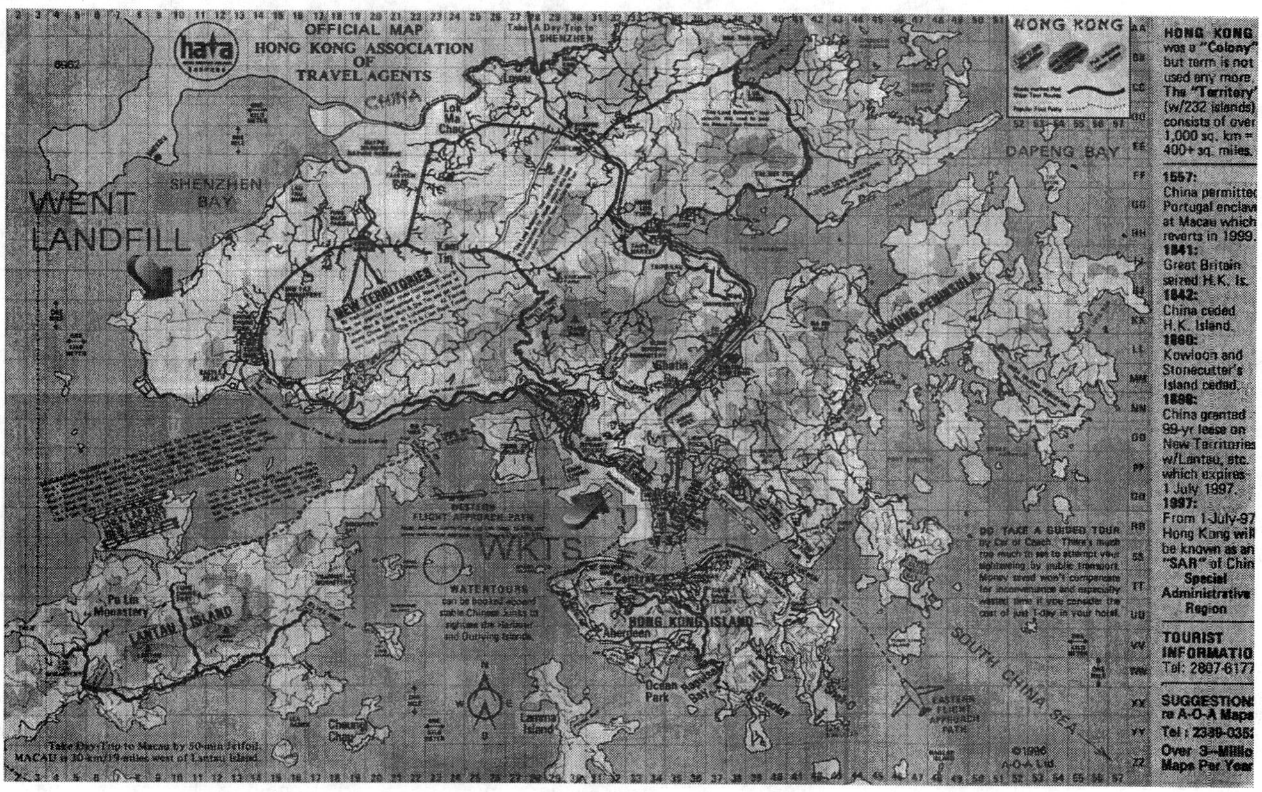

Figure 11: locations of WKTS and Associated Landfill

The facility was constructed on reclaimed land from the ocean, which also presented challenges for the project team. During the bidding process, it was assumed that the soil (mostly sand) could withstand normal bearing pressure. However after winning the project and obtaining samples, it was found the whole site was largely unconsolidated. A specially designed machine was brought in to compact the sand, by high-pressure water spray and a vibratory head, to a depth of over 18 meters, Figure 12 and 13.

This operation cost the project timeline almost 2 months, which the Environmental Protection Department (EPD) would not allow as a change to the start-up date! The surface of the site dropped by 3-4 feet due to the deep compaction.

Figure 12: Vibratory Compaction Rig

The facility was given a site area of 1.9 hectares (about 4.7 acres). This area sounds adequate, but the facility had to support the transfer building, full administration and maintenance buildings, on-site ramp for second floor tipping, enough room on the quay to store 210 containers plus spares, and still allow room for a full-size fire engine to circle the building.

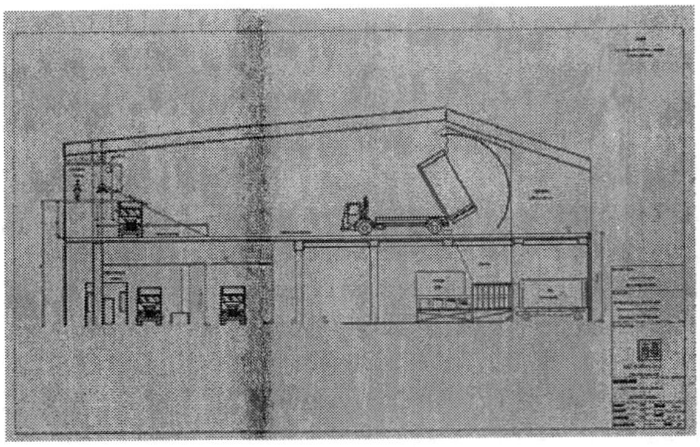

Figure 14: Section of Transfer Building

Figure 13: Vibratory Head of Compaction Unit

Construction of the facility was scheduled for 17 months, but this included design and did not anticipate the deep-compaction. Actual construction started in May of 1996 and facility was commissioned on May 26, 1997, a little less than 13 months! The construction contract was unusual as there were no progress payments during construction. The operations and construction contractors partnered and had to fund the entire construction period. Once operational testing was complete, a single payment was made to cover construction. Needless to say, the team was a bit nervous during commissioning, as there was a payment of over $80 million hanging on the outcome.

The transfer building, Figure 14, was designed as 2-levels, with the upper floor housing the vehicle tipping and waste hoppers, and the lower floor comprising the waste compactors, container handling dollies, leachate treatment system, laboratory, and employee facilities. Also enclosed within the first floor are the outbound vehicle traffic lanes, outbound scales, and a three-lane vehicle wash.

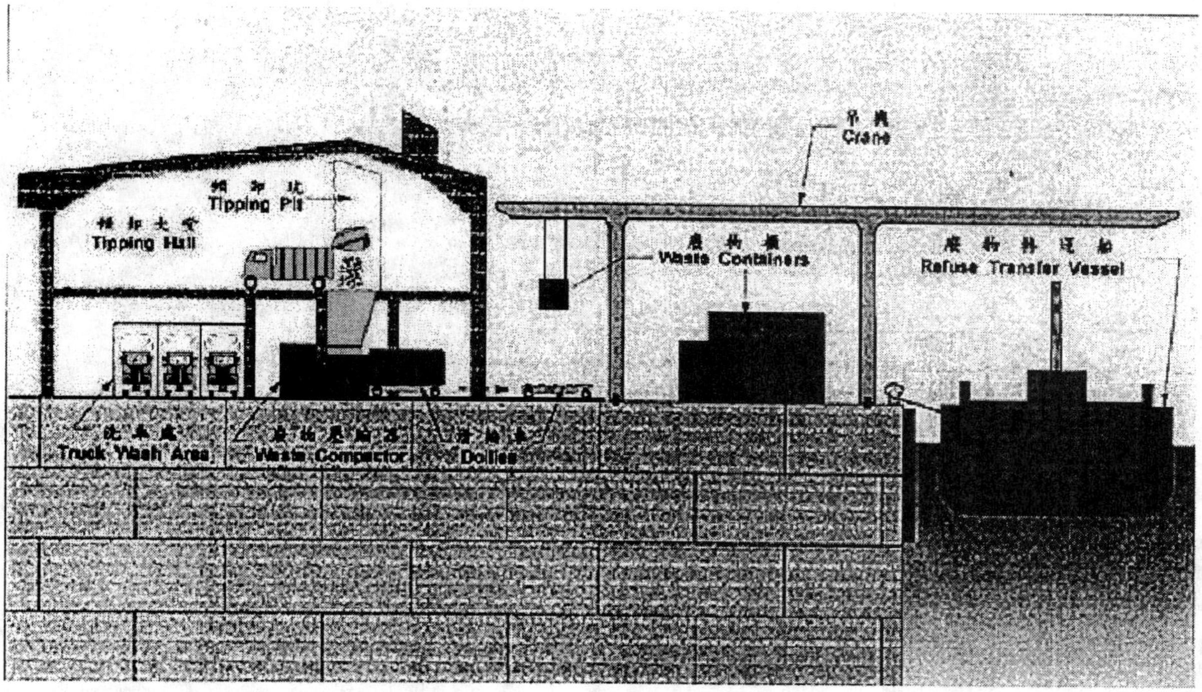

Figure 15: Schematic of Transfer Concept

The main concept of the transfer station was to compact waste into specially designed ISO containers, which are then moved out of the building on rails, picked up by a custom gantry crane and loaded into custom designed ocean-going container ships, Figure 15.

Figure 16: Hydraulic Compactors with Pre-compaction Doors

Loading of the waste into the ISO containers is accomplished by ten, large hydraulic, compactors and six moving floor hoppers feeding eight of the compactors, Figure 16. Two of the compactors are fed directly and are primarily used for bulky waste and have a pre-compaction door for double compacting. The normal compactors have a full cycle time of 16.5 minutes, which includes loading and unloading by the crane, for an average payload of 15.1 tons in a 20-foot container.

The compaction system was designed to have extreme redundancy, as was the entire facility. Even though the design tonnage is 3,500 metric tons, the facility can actually process over 6,000 metric tons in 16 hours with one compactor out for servicing.

The ISO containers used are specially designed for compaction stresses and are ISO certified at nominal dimensions of 20' L by 8' W by 8'6" H. Tare weight of the container is 9,260 pounds with a maximum gross of 44,780 pounds and a payload volume of 37.3 cubic yards. Figure 17 shows the ISO container used at WKTS and the extra stiffeners required for compaction are obvious. Other specific design features for waste handling are: a tapered interior for ease of unloading; vent holes at the rear to prevent vacuum during discharge; fire hose port in case of a burning load; dual horizontal doors hinged at one side; special door gasket materials to withstand leachate attack; and special door cinching cams to prevent leakage. Lloyd's Register Industrial Services certified the containers produced in Yangzhou, China.

Figure 17: Compaction ISO Container

Figure 17 shows the custom designed ISO container being dropped onto a rail-mounted trolley. The extra compaction stiffeners are plainly seen. The trolley will then move the container into the building to be latched onto the compactors as shown in Figure 16. The trolleys are moved electrically by means of a cable reel with a cable tray mounted between the rails. The spreader shown is a standard 20-foot fixed-length rotating Bromma. The reeving is angled outwards to provide some anti-sway effect. Active non-sway reeving was not used.

Figure 18: Landside Leg of Gantry Crane

The ISO containers were handled primarily by means of two custom designed gantry cranes, Figures 18 and 19. A total of four cranes were required, two at the transfer station and two at the landfill reception area. A single crane could handle each site, but redundancy was required. The cranes are rated with a duty cycle of 24 per hour, which defined as latching onto a full container at the trolley, loading it onto the container ship, and placing an empty back onto the trolley.

Figure 19: Quayside Leg of Gantry Crane

The gantry cranes operate on the same set of rails and can be parked at either end of the site without impacting operations. The cranes are capable of stacking containers 3-high on the quayside and lifting a fourth over-top to the ship. Figure 19 also shows that the cranes were designed with luffing booms so that the ships could berth easily.

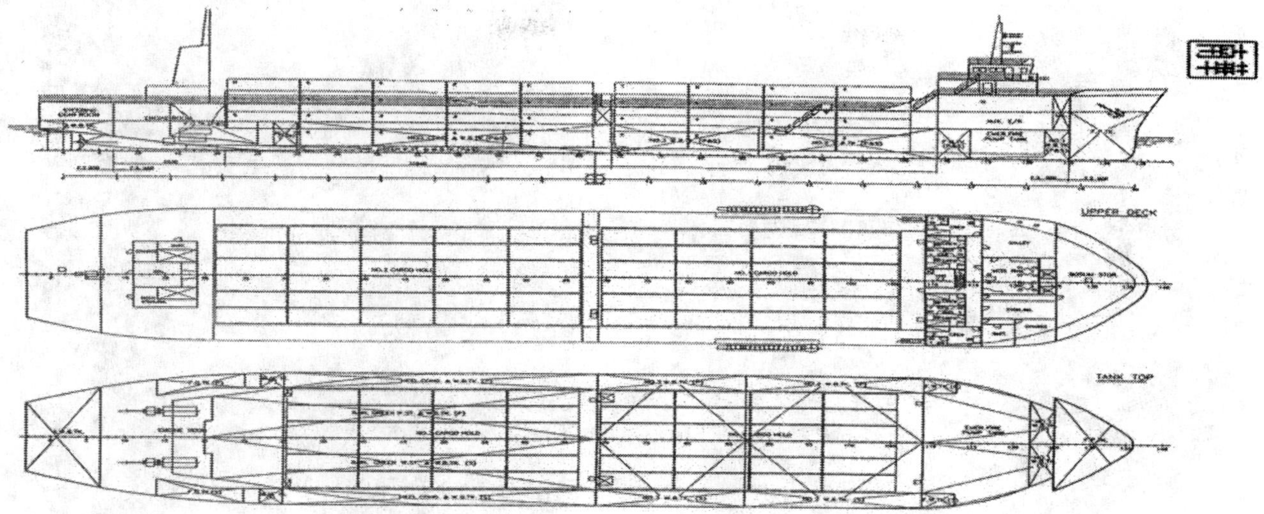

Figure 20: Drawing WKTS Container Ship

The cranes are powered via cable reel, also seen above, with the cable laid directly onto the concrete, outside of the rail. A concrete curb protected the cable from being washed over the side during storms.

The cranes operate at the following speeds: spreader hoist – 131 fpm; trolley travel – 394 fpm; crane travel – 262 fpm; spreader rotate – 1 rpm; boom luffing – 5 min. The cranes are certified to operate in winds up to 45 mph. Maximum design wind speed for an anchored crane is 110 mph.

The ships used at WKTS were specially designed, oceangoing, 210 TEU container ships, Figure 20 and Table 1. Two ships were built and swap places every day between the transfer station and the landfill. A full-time crew lives and works onboard each vessel.

Figure 21: Container Ship Being Unloaded

The hold is arranged in two sections and containers are arranged 6 rows wide, 9 rows long, and 4 deep, Figure 21.

At the time the ships were being commissioned, the team was unaware that they were actually going to be the largest ships ever operated under their own power in Hong Kong Harbor. All other large ships, such as cruise liners, are tugged through the harbor. As such, it was difficult to find captains for the ships, as most were not comfortable with the challenge.

Name(s):	Ngon Sheun, Lai Wan
Number of ships used:	2
Length overall:	94.5m
LBP:	90.0m
Beam (width):	20.8m (hold 15.8m)
Draft:	3.8m
Deadweight:	4,000 tonne
Container capacity:	210 TEU @ 25 tonne each
Speed:	10 knots
Propulsion:	2238kw – 2 Caterpillar 3512B
Propellers:	2
Crew:	6

Table 1: Container Ship Principal Characteristics

Figure 22: Cell Guides on the Ship

The ships were fitted with cell guides to make handling the containers faster, Figure 22.

Figure 23 shows the "luffing" feature of the crane booms in the background. Any water collected in the holds is tested and either pumped to the leachate treatment plant at the transfer station, or to the sanitary sewer as required.

Figure 24: Loading Empties at the Landfill

Figure 23: Container Ship Being Loaded

Figure 24 shows the operation at the landfill side. The cranes operate in a similar fashion as the transfer station except instead of loading a trolley on rails they place containers on vehicles. The vehicles shown are specialized off-road Volvo tractors with self-tipping frames for the containers. The vehicles are 6-wheel drive and articulated to operate on rough terrain and in bad weather.

In summary, the WKTS has turned out to be a shining example of how marine transport of waste can be done efficiently and with environmental sensitivity. The facility is one of the largest, and possibly the largest, waste transfer facilities in the world.

A MODERN MARINE TRANSFER SYSTEM

Introduction

Hauling of refuse begins with local trucks and ends in the same manner when the source of disposal is local. Otherwise, it is necessary to transfer the material to more suitable long distance transportation. The re-packaging of the material is likely to containers, top or end loading, for hauling by truck, train or waterborne vessel. Some type of compaction of the material is also desirable to increase density and reduce volume.

Particularly in urban areas, the transport of refuse by local trucks to re-packaging sites and the subsequent pick up and hauling by long haul means can result in an operation contributing to air and noise pollution within the city limits, and cause increased maintenance and operating costs of the local collection vehicle if they are to travel a distance to the re-packaging area. Furthermore, the increased road use by heavy vehicles will contribute to deterioration of the roadways.

An opportunity to mitigate or reduce these unwanted results is by utilizing a marine transportation system. The sections that follow this introduction describe the principal components of this system.

Marine Transfer Stations

The epicenter of a modern marine refuse transportation system is the point at which the collection trucks dispose of the load and it is transferred to waterborne transportation. In concert with current methods of shipping, it should be a segment of an intermodal transportation system. It is a reasonable consequence that accordingly the transfer of refuse should be to intermodal containers standardized for this purpose and allow movement by highway, rail and vessel.

Figures 25-28 depict what could be considered a typical MTS consisting of three levels housing three main process areas, personnel areas and an access ramp. A plan view and section of the typical MTS are shown in Figures 25 and 26. An administration and operations building will house two of the process areas, a portion of the pier level where container handling and barge loading will take place are uncovered. The physical layout of the facility and the elevations of each floor are determined by the requirements for receiving, handling and processing the incoming stream of municipal solid waste. Three-dimensional renderings in Figures 27 and 28 show the typical MTS without roof structure and provide a view of the three levels of the facility.

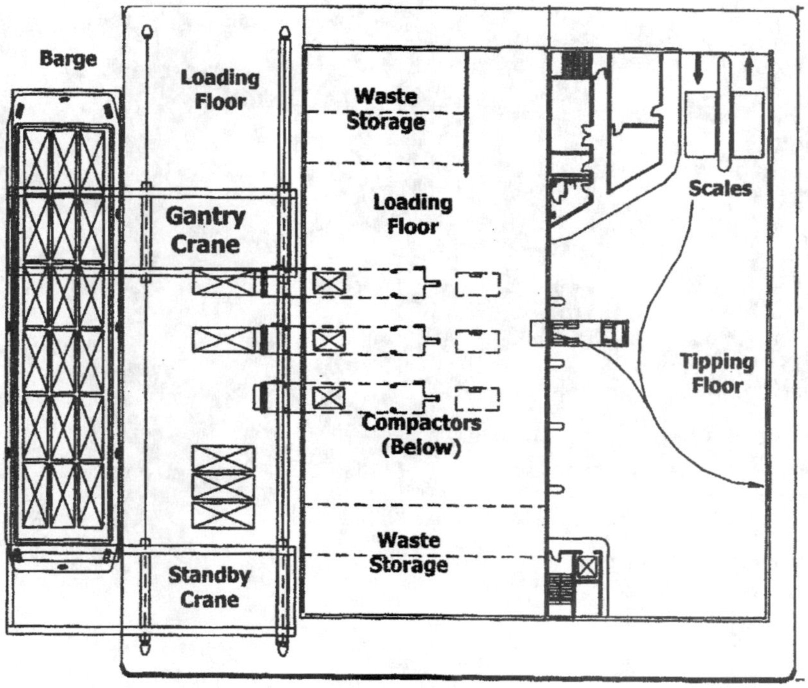

Figure 25: Typical MTS Floor Plan [5]

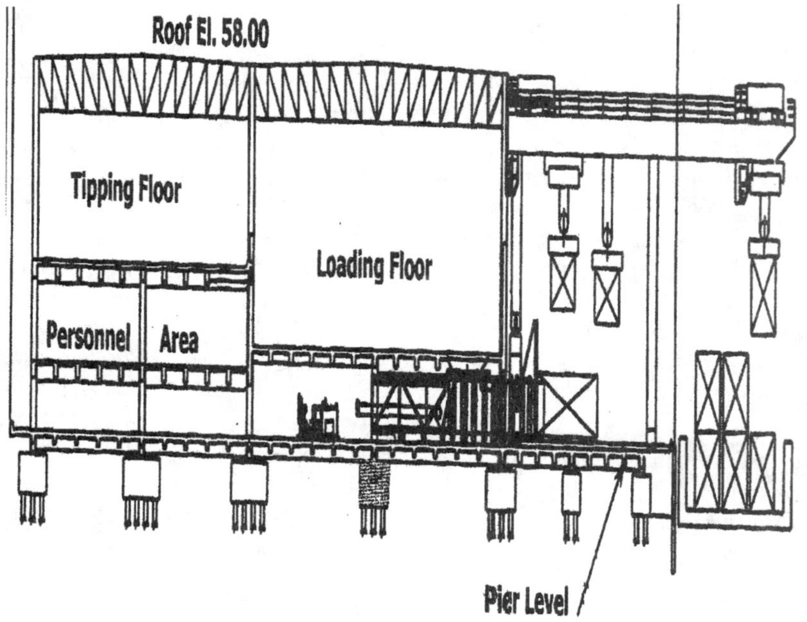

Figure 26: Typical MTS Section [5]

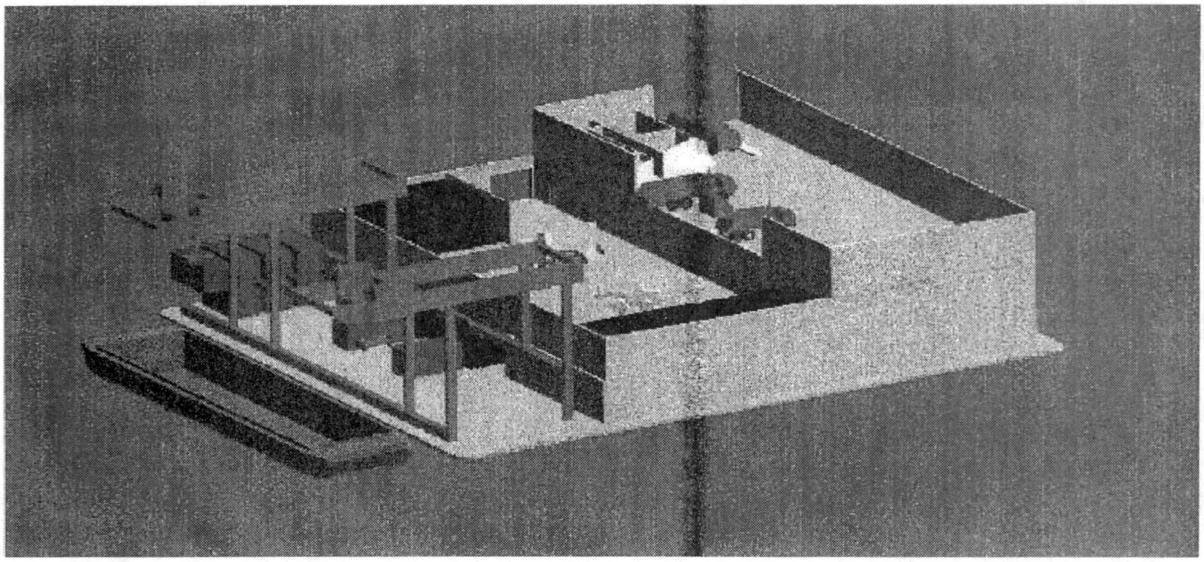

Figure 27: MTS 3-D Rendering – View 1 [5]

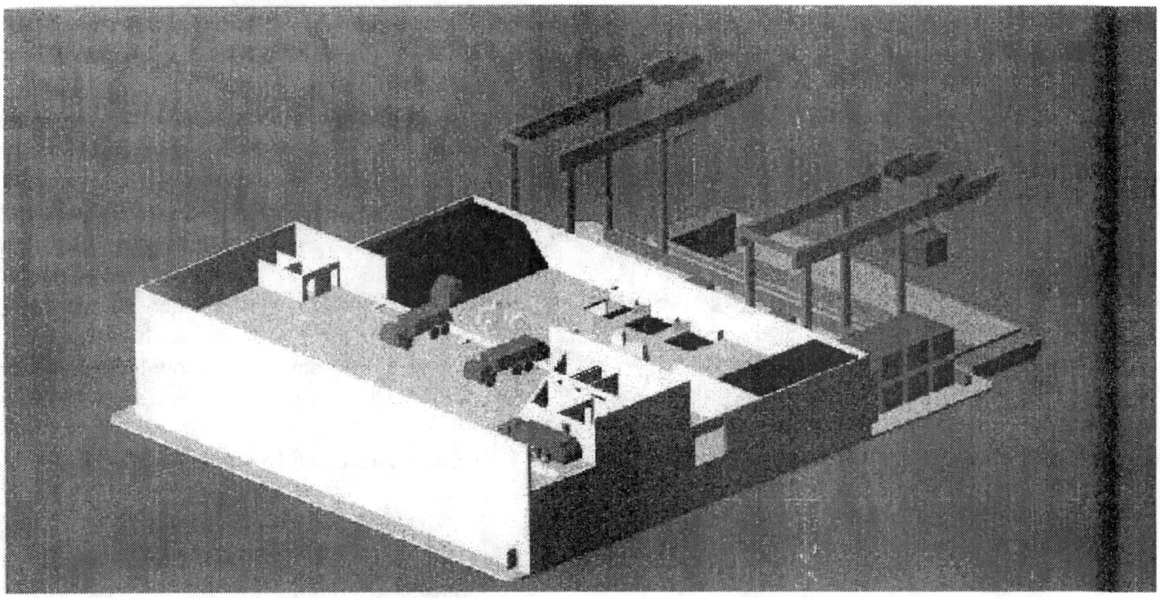

Figure 28: MTS 3-D Rendering – View 2 [5]

Entrance

The entrance to the typical MTS is located at the tipping floor level and provides access for collection vehicles to the tipping floor. A weigh station, consisting of incoming and outgoing truck scales along with a weighing office are located at the immediate entrance to the building. Radiation detection equipment to monitor incoming vehicles for radioactive material is located ahead of the incoming scales.

Tipping Floor

The major portion of the tipping floor consists of the collection vehicle maneuvering area and the truck bays. Enough open space has been provided to allow a collection vehicle to enter the tipping floor, maneuver into position and back into one of the truck bays while two other bays are occupied. Along with the weighing office, a portion of the personnel areas are located near the entrance to the tipping floor. Also located at the tipping floor level are stairway and elevator access to the personnel areas immediately below.

Loading Floor

The loading floor on the typical MTS is located below the tipping floor to provide sufficient height to comfortably perform the tipping operation. The loading floor consists primarily of areas for material handling, with areas segregated for four operations. The tipping area directly below the tipping bays will receive the material brought in by the collection vehicles. Immediately beyond the tipping area are smaller areas designated for the incoming waste to be sorted for the removal of objectionable materials not amenable to compaction. To either side of the main floor are areas for temporary storage of the waste material. Piles of the incoming solid waste will be created at these locations to handle diurnal variations in the rate of incoming vehicles. At the end of the floor opposite the tipping area are the floor openings that allow loading of the compactor hoppers located in the lower level. The floor openings are designed to allow wheeled scraping machines to push the solid waste across the floor and directly into the hoppers.

Pier Level

The Pier level contains the compaction equipment and the equipment for maneuvering containers and loading them onto barges. The main body of the compactors is located directly beneath the Loading floor within the enclosed structure. The containers, maneuvering equipment and primary and standby gantry cranes are located on the open area of the pier level. A portion of the open pier is allocated for maneuvering and staging of containers during the compaction and loading operations.

Compactors

Pre-load compaction equipment is provided to compact the waste. Waste enters the compactor from a hole in the loading floor. The waste falls directly into a compactor's charger which remains open while it is filled with waste. When it is full, a hydraulic ram pushes the load to the front of the compaction chamber and compresses it against a guillotine gate in the down position at the end of the chamber. The compaction unit sits on load cells that accumulate the weight of waste fed to the unit. This compaction cycle is repeated until the compressed load meets preset limits for weight or until six charges of the waste has occurred. At this point, the guillotine gate is raised and the ram is activated, moving forward to push the compressed waste into a steel container. The container then pulls away from the compactor and the rear door is closed to provide a tight, leak-proof seal.

The volume of the waste is reduced by approximately 50 percent in the compactor. The density of the waste at the loading floor is about 450 pounds per cubic yard. The density of the compacted waste is between 800 and 900 pounds per cubic yard. Each compactor can fill 4 19-ton containers per hour. This will result in a rate of about 75 tons per hour for each compactor. Table 2 provides the components of a compactor cycle time.

An alternate to compaction is the utilization of tamping of waste tipped into open top containers below where the compactors would be located. This type of an operation, which could be accomplished by the same vehicle performing the scraping, eliminates the use of the more costly compacting equipment. The increase in density achievable is however not as great as with the compactor. Therefore, more containers throughput is required or the utilization of larger containers.

Container Handling Systems

Overhead Bridge Crane/s or Container Gantry Crane/s (CGC) can be utilized as options for container handling for vessel/compactor loading/unloading operations. The CGC for this application is similar to the cranes used at the marine terminals for moving containers to and from container ships. The cranes must also be configured to rotate the containers for loading/unloading at the compactors. Table 3 provides the components of a CGC cycle time.

COMPACTOR CYCLE TIME

Load per Container (Bale)	=	19 tons
Volume per Bale (before compaction)	=	100 cubic yards
Number of Charges per Bale	=	6 charges
Volume per Ram Charge (Bale)	= 100/6 =	16.7 cubic yards
Time for Loading Charge	=	.33 minutes
Average Time to Wait for Charge	=	.25 minutes
Total Time per Charge	=	.58 minutes
Total Compactor Time per Bale	= 6*.58 =	3.5 minutes
Time to Eject Bale	=	3.5 minutes
Time to fill container	=	7 minutes
Time for Operator to close doors	=	0.5 minutes
Total Time to Fill Container	=	7.5 minutes

Table 2: Compacter Cycle Time [5]

Maximum Gantry Crane Cycle Time

Gantry Crane Travel Speed	=	120 feet per minute
Crane Bridge Trolley Speed	=	100 feet per minute
		75 feet per minute
Crane Hoisting Speed	=	25 feet per minute (movement 3 feet or less)
Furthest Barge Position from Compactor	=	84 feet
Gantry lowers hook to empty container*	=	0.41 minutes
Hook up and lift	=	0.41 minutes
Rotate container	=	0.25 minutes
Gantry to compactor	=	0.28 minutes
Trolley to compactor	=	0.56 minutes
Lower Container	=	0.12 minutes
Disconnect and lift hook	=	0.12 minutes
Total time to unload empty container and place in front of compactor	=	**2.15 minutes**
Gantry to second compactor (full container)	=	0.32 minutes
Lower hook	=	0.12 minutes
Hook up and lift	=	0.12 minutes
Trolley to barge	=	0.56 minutes
Gantry to barge	=	0.28 minutes
Rotate container	=	0.25 minutes
Lower container	=	0.41 minutes
Disconnect and lift hook	=	0.41 minutes
Total time to load full container onto barge	=	**2.47 minutes**
Total cycle time	=	**4.62 minutes**
Extra time remaining	=	**2.88 minutes**

***Begin above furthest empty container on barge**

***Summary of AMSEC Compactor Service by Crane, Time Study**

Table 3: Crane Cycle Time [5]

Barges

Type	Deck	Deck	Deck
Size, LOA x B x D	280'x75.5'x18.25'	200'x48'x13.5'	150'x32'x12.5'
Lightship Weight, (s.Tons)	1630	735	387
Container capacity, TEU	288	80	36
Container deadweight	18.54 ST	18.54 ST	18.54 ST
No. of Barges reqd. per day	0.61	2.19	4.86
Total Number required	3	7	16
Unit Cost, M$	$2.31 M	$1.04 M	$0.55 M
Fleet Cost, M$	$6.93 M	$7.29 M	$8.78 M
No. of Barges per Tow	1	2	4

Table 4: Barges for 3250 Tons Per Day

Barge Size & Type	300' Deck	200' Deck	150' Deck
No. of Cranes reqd. incl. spares	2	2	2
No. of Container Handling Trucks required, incl. spares	2	2	2
No. of add'l Tugs required	0.15	0.27	0.30
No. of Shifts Worked	1	1	1
No. of Personnel required per shift:			
Sanitation Supervisor	1	1	1
Sanitation Men	2	3	5
Crane Operators	1	1	1
Oilers	1	1	1
Maintainer (1 shift only)	2	2	2
Capital Costs, Annualized, (MS)			
Barges	$1.01 M	$1.06 M	$1.28 M
Cranes	$0.18 M	$0.18 M	$0.18 M
Container Handling Trucks	$0.17 M	$0.17 M	$0.17 M
Facility	$0.06 M	$0.05 M	$0.05 M
Total Capital Costs:	$1.41 M	$1.46 M	$1.68 M
Operations & Maintenance Costs, Annualized, (MS):			
Maintenance & Repair	$1.06 M	$1.09 M	$1.24 M
Personnel	$0.39 M	$0.44 M	$0.55 M
Tug Operations	$0.20 M	$0.36 M	$0.40 M
Unloading Costs	$3.12 M	$3.12 M	$3.12 M
Total O & M Costs:	$4.77 M	$5.01 M	$5.31 M
Total Annualized Costs, including 10% Contingency:	$6.79 M	$7.12 M	$7.68 M
Average Cost, ($/s. Ton):	$9.37 /s. Ton	$9.83 /s. Ton	$10.60 /s. Ton

Table 5: Average Cost Per S. Ton

Destination	Shipping per TEU – New York (20 Tons)	Allowable Tipping per Average ($35) U.S. $/ton	Allowable Tipping for Max ($105) U.S. $/ton
Western Europe (ship)	$950.	$25.	$95.
Far East (ship)	$1100.	$17.50	$87.50
Africa w/transfer (ship)	$2500.	-----	$17.50
Midwest, U.S. (truck)	$750.	$35.	$105.
South, U.S. (truck)	$1500.	-----	$67.50
West, U.S. (truck)	$2000.	-----	$42.50

Table 6: Residual Tipping Fees – From New York (Compared to equal MidWest Transport/Tipping Cost)

Transport Vessels

Vessels utilized for transporting containerized refuse can be barges or ships. As a result of the dimensional constraints of an MTS, a barge or smaller ship will be required for direct service. These will also be suitable for local or moderate distanced centers for disposal or inland distribution. If international transport is intended, then a transfer to an ocean going vessel will be required as well.

As an example of comparative vessel utilization and cost, consider the barges shown in Table 4 intended as options to carry 3250 S tons per day. An application of this system in NYC for trans-harbor movements reflecting costs in 1994 [6] is provided in Table 4. The results indicate a significant savings from the equivalent trucking costs of $12.36/S. ton at that time.

OPPORTUNITIES FOR SEABORNE TRANSPORT

As previously mentioned, the average tipping fee in the U.S. is $35/ton. Some have reached $105/ton [7]. The case for seaborne transport is then rooted at tipping fees at overseas destinations and the difference in shipping costs internationally. Table 6 indicates average container shipping costs to overseas destinations and the residual remaining for tipping fees when compared to current shipping costs from NYC to local states

As can be seen as the tipping fees in the U.S. continue to rise the allowable tipping costs for overseas destinations could be as high as $87.50 a ton when compared to $105 in the U.S., although only $17.50 when compared to the U.S. average of $35/ton.

The system might incur a requirement for additional containers if the time required to ship to the destination and back is more than that for U.S. destinations.

CONCLUSIONS

- A Marine-based refuse hauling system provides significant opportunity to address a complex and costly transportation system requirement.

- There is economic incentive for a seaborne transportation system of refuse if overseas destinations are identified with interest in receiving the material.

- Historical data available on per capita production of refuse suggests that a continual source of refuse material will be available for some time in the future.

REFERENCES

1. New York Times, November 22, 2002
2. Chartwell Information, Wastinfo.com, 1/13/2003
3. Trabay, Mark, "More States are Taking Out Their Trash," USA Today, October 13, 1998.
4. HDR Engineering, Inc., "Feasibility Report in MTS Conversion," for the City of New York Department of Sanitation, April, 1999
5. Greeley and Hansen, "MTS Conversion Conceptual Design Development, For the New York City Department of Sanitation," September, 2002
6. M. Rosenblatt & Son, Inc., "Barge Fleet Sizing and Barge Design, For the New York City Department of Sanitation," October, 1994.
7. New York Times, May 4, 2003, "Some Green in That Garbage," Page 45.

Exploiting Technology to Optimize the Design of Large Composite Vessels

Albert W. Horsmon, Jr.[1] (M) and **Bryant Bernhard**[2] (AM)

ABSTRACT

Advances in polymer technology have allowed designers to exploit the advantages offered by composite construction in marine structures up to 72m LOA (Visbey 2003). Large, self sufficient, ocean going modern yachts and commercial vessels of over 34m LOA are routinely built of fiberglass and offer the same design challenges faced in metal ships. Under the sculpted shapes and glitzy facades reside intricate composite structures and engineering details to satisfy the most diehard of "big ship" architects. These vessels require the full expertise of Naval Architects and Marine Engineers to optimize the designs for shipyards to construct large and complex but beautiful structures. All the advantages of composite construction, but in ship size doses, are brought to the benefit of the composite "ship" customer.

This paper briefly examines the evolution of composite design, explores the use of modern techniques to optimize ship structures and reviews current manufacturing methods that allow modern composite structures to be constructed in an efficient, cost effective manner. This paper explores details of the structural design and construction of an ABS classed 46m sandwich composite trideck motoryacht and a 41m sandwich composite ferry design as examples of the application of modern composites technology in the building of a "fiberglass ship". Class rules are discussed along with detailed analysis of laminate schedules and laminate requirements for various loaded areas of the structure. Mechanical fastening and the use of high strength adhesives for providing foundations for machinery and equipment and vibration considerations are addressed for Marine Engineering applications. Finite element analysis methods are exercised to exploit further enhancements in vessel design. This paper also provides examples of

[1] Professional Engineer, Naval Architect, ATC Chemical Corp.
[2] Professional Engineer, Naval Architect

fabrication details with practical considerations to assure designs are compatible with normal ship production techniques.

NOMENCLATURE

ABS American Bureau of Shipping
ASTM American Society for Testing and Materials
FRP Fiber Reinforced Plastic
HAP Hazardous Air Pollutants
IMO International Maritime Organization
NDT Non-destructive Test
NSRP National Shipbuilding Research Program
PVC Poly Vinyl Chloride
PVB Poly Vinyl-buteral
SAN Styrene Acrylo-Nitrile
VOC Volatile Organic Compounds

INTRODUCTION

In the past several decades, advances in the material sciences introduce several options for the selection of advanced technology materials suited for the marine environment. These evolutionary developments allow marine craft designers and builders the discretion of choosing the best structural configurations for the discrete components of marine vessels.

In the not so distant past, when ships were limited in size and carrying capacity by the material characteristics of the wood of which they were built, and the joining methods of the construction techniques, designers and shipbuilders turned to steel as the primary build medium, enhancing the capabilities of the vessels. It is almost difficult to remember that the first 100 years of steel ships were riveted. Advances in metallurgy and welding technology of merely a generation ago allowed these changes. With the introduction of quality marine grade aluminum and advances in aluminum welding technology, aluminum was introduced successfully as an alternative to steel construction in components of ships and the primary structure in boat construction. Now aluminum ships of 100m are not uncommon in the speed sensitive offshore industry.

As polymer technology advances, composite structures are gaining in popularity among designers and owners because they offer distinct advantages over the limits of traditional wood, steel or aluminum construction.

High technology composite materials, when properly applied, afford the end user of a marine vessel enhanced capabilities for a given design. These positive attributes are most remarkable in high speed and ergonomically sensitive craft applications where strength, damage tolerance, low weight and appearance are paramount to the desires of the owners and operators of the vessels. Fiberglass (FRP) composite materials are applied to fully utilize the strengths of the intrinsic components while minimizing any compromise resulting from inherent properties of any one of the elements. By utilizing properly designed and arranged composites, the resulting structure is better suited for its application, by default, than selecting a single build material, such as aluminum or steel. The selection of a single material does not afford the designer an opportunity to minimize the effects of the drawbacks of the material selected.

EVOLUTION

Evolution of Composite Design.

Original fiberglass boat scantlings were basically set the same as wood boats. That is, they used the same thickness as wood planking, but left out some of the framing because the need to attach all the planks was eliminated. As builders gained experience with the continuity and strength of a continuous skin, and lamination techniques became a bit better, laminates became a bit thinner.

Then core materials came along, so the basic theory was to make the sandwich have at least the same moment of inertia as the single skin laminate that it replaced. But, how to size the skins? A very thin skinned thick core laminate could match the moment calculation, but would be impractical. Enter various "rules of thumb" one of which is outlined below.

"Rule of Thumb" from the 70s

This "rule," (Johannsen 2001) developed by Swiss foam engineer Karl Brandl, looks somewhat simplistic now, but served us well in the 70s and early 80s. It produced many successful boats, especially when a large sandwich vessel was a boat in the 16m range. For the sake of history, and to document experience, we would like to summarize it here once more.

This rule assumes, as was the case in the earlier years of fiberglass boatbuilding, that the laminate thicknesses of a single-skin fiberglass boat are over-dimensioned by 1/4 to 1/3 of their wall thickness to compensate for the relatively low stiffness of FRP in comparison to its other outstanding mechanical properties. If this extra thickness is now taken out, and the remaining thickness split in half, and a core is inserted, a sandwich panel resulting in generally better overall mechanical properties is obtained.

Based on practical experience, a simple formula was developed. Assuming a fiberglass laminate consisting of woven roving and mat, the average hull thickness for single skin construction can be calculated by dividing LOA by 1000.

Using a 12m cruising sailboat as an example, the average single skin thickness would be 12m / 1000, giving an average hull thickness of 12mm. The means the topsides were probably 9.5mm, increasing at the waterline to about approximately 12mm and increasing gradually towards the keel area to 15mm. To calculate the core and skin thickness of the sandwich, one would use the single skin thickness of 12mm as 100%. The core would then be 150% of the single skin thickness (20mm), the outside skin would be 40%, or 5mm, and the inside skin 30%, or 4mm.This would save 30% in the fiberglass lay-up, and would result in a stiffer hull shell, which is insulated, and has a higher impact strength. For powerboats, with flatter surfaces, the formula would use 200% for the core, and the same thickness for the skins.

It must be borne in mind that the basis of the thickness criteria came from a period before the multi-axial stitched reinforcements were on the market. They are generally higher in strength, but not as thick, and therefore not quite as stiff. These type of reinforcements, however, are ideally suited for sandwich structures, as the skins provide the strength, and the core provides the stiffness.

There are limiting factors to this simple rule. As boats get larger, the skin thickness Laminate skin thickness does not increase proportionally to the real size and performance of the vessel. More hull stiffness is carried by the transverse and longitudinal stringers and bulkheads. However, the core thickness does increase to provide more panel and overall shell stiffness. Also, higher strength skins using more advanced resins and reinforcements are thinner and lighter. Such structures rely more on the core to provide the required stiffness. There is no consideration for the frame spacing, the weight or the speed of the vessel in this rule. Laminates form these basic rules applied to a 21m motor yacht are summarized in Table 1.

Table 1. Single Skin and Sandwich Laminate Thickness Comparisons of Various Early Design Rules

		Rule of Thumb		ABS 78 Rule		ABS 90 Yacht Guide		ABS 97 High Speed Guide	
	Units	Single Skin	Cored	24 kts	35 kts	24 kts	35 kts	24 kts	35 kts
Core Thickness	mm		42.7	29.5	58.7	32.3	46.5	34.3	61.2
Outer Skin	mm	21.3	8.5	8.5	8.5	4.5	4.5	4.5	4.5
Inner Skin	mm		6.4	6.4	6.4	3.1	3.1	3.2	3.2
Laminate Weight	kgs/m^2	32.8	26.9	25.7	28.4	15.5	16.8	15.7	18.2

1978 ABS Rules for Building and Classing Fiber Reinforced Plastic Vessels

The 1978 ABS Rules (ABS 1978) have a fairly simplistic treatment of sandwich panels. For power vessels, the overall sandwich thickness must be:

< 31 kts - Sandwich Thickness =
(0.586 * k2 * V * s)/Shear Strength of the Core

>31 kts - Sandwich Thickness =
($0.0195 * k2 * V^2 * s$)/Shear Strength of the Core
where: k2 = a factor based on panel aspect ratio taken from a table
V = Vessel speed in kts
S = the panel span in inches

The "basic" laminate was considered mat and woven roving with given minimum physical properties, but allowances are made for laminates with better or worse properties. This part of the '78 Rules is a bit simple, but the whole rule book was fairly comprehensive. It included requirements (criteria) for the stiffness of frames and girders, details of construction for most parts of a vessel, quality control, and for attachment of machinery, among others. Using the same skin thickness and applying this rule to the sample 21 m (70 ft) yacht gives the laminates the table.

One must realize that the only way these rules and the other ABS rules and guides become requirements is if some authority requires that a vessel meet the rules. ABS requires the vessel to meet the rules or guides if the vessel is to be classed with ABS. Full classification requires plan review, survey while under construction, and issuance of ABS Certificates. To remain in class, periodic reinspections are also required. At one time, ABS conducted plan review only, but too many vessels were having problems because they were not actually built to the approved plans. Now ABS only performs full classification. ABS will Class commercial vessels of all lengths, but pleasure craft must be over 24 m (79 ft.) in length in order for ABS to Class the vessel. The U.S. Coast Guard and other government authorities can require that a vessel structure meet a certain rule, as can the owner or designer of a private vessel

1990 ABS Guide for Building and Classing Motor Pleasure Yachts

The 1990 ABS Yacht Guide (ABS, 1990),(and at the time the 1990 High Speed Craft Guide) is similar to most of the other classification society guides in that it develops pressures for various areas of the hull, topsides and superstructure, then uses that pressure for the load to determine the required sandwich panel properties.

The 1990 Yacht Guide has criteria for the sandwich to meet a minimum section modulus and moment of inertia. There is also a criterion that relates the core shear and compressive modulus times the skin compressive modulus against the compressive and tensile strength of the skins to assure that the skins will not buckle on the compressive side. With the 80 to 130 kg/m^3 cores used in most marine hulls, once the shear strength criterion is met, the other two requirements are usually met. When the 0.55 shear strength factor from the 1997 ABS High Speed Craft Guide is applied, this increase in the amount of the allowable ultimate shear strength used in design is only for linear foams that exceed 40% shear elongation.

The 1990 ABS Yacht Guide also has a general criteria for minimum skin thickness that are a good reference point, but are considered a bit excessive for the lightest current construction. The 1990 ABS Guide for Building and Classing High Speed Craft was very similar to the Yacht Guide except that it had higher service factors for slamming loads. This because the high speed craft guide was intended for application to different classifications of commercial vessels such as pilot boats, rescue craft, and patrol boats which see a much rougher service life than a pleasure yacht. However, considering the service life of a regularly used sport fishing boat - the service factors in the yacht guide probably do not account for some of the loads, so the higher factors of a high speed guide may apply.

1997 ABS Guide for Building and Classing High Speed Craft

The 1997 High Speed Craft Guide (ABS , 1997) is similar to the 1990 Yacht Guide but a bit more involved in developing pressures for various parts of the hull, deck, interior and superstructure. However, it is the latest of the classification society guidelines and has a number of refinements. It has a more involved method for calculating pressures in the forebody area where different combinations of flair and rake can concentrate pressures forward. It also has a modified method for minimum skin thickness. Most significant for sandwich construction is that ABS recognized that different cores behave differently in an overload situation and assigned different safety factors on shear strength based on each core material's ability to absorb damage. This ability to absorb damage is related to the core's shear elongation at failure.

Taking the 21 m (70 ft) yacht through the 1997 ABS Guide is not a reasonable direct comparison. However, it is not uncommon for hull forms to be used for a variety of vessels types. Therefore, for this part of the comparison, the same hull is considered as used for an unrestricted service patrol craft and laminates are developed.

Other classification society rules have similar developments of laminates with different criteria for safety factors. The International Organization for Standardization has developed a new standard for small craft up to 24 m in hull length, ISO 12215. This standard was developed mainly at the request of the European Union to regulate the quality of small boats manufactured and sold in EU countries.

DESIGN

Local Loads

Local loads on a boat are basically the pressure loads from the vessel pushing through or bouncing over the waves. These can be static head pressures, impact slamming pressures or hydrodynamic pressure. The vessel must be designed to handle the worst case of these loads. For a planing vessel that has a speed to length ratio greater than 9, the full impact pressure should be applied to the whole bottom area because the boat can become airborne and slam down on any part of the bottom. The speed/length ratio is the speed in knots divided by the square root of the waterline length in meters (ABS 1997). For pleasure yachts and large commercial vessels, this is rarely the case and impact pressures need only be applied to the forward parts of the vessel. Scott (Scott 1996) has a full explanation of all the likely local load cases.

Concentrated Loads

Some concentrated loads are also considered in the local load set. Machinery foundations, davit bases, mast mounts, stays, chain plates, etc. are all special local loads that must be accounted for. For thin-skinned sandwich panels, local compression inserts and additional laminate are required to distribute these loads. Propulsors and other large machinery should be mounted with isolators on skids or foundations and fastened to the primary boat structure. This is analogues to steel or aluminum designs only through bolts and/or adhesives replace welding.

Methacrylate adhesives are increasingly popular in high stress applications due to their ability to produce durable, high-strength bonds to polyester and vinyl ester resins and gel coats. Their chemical bonds typically are stronger than the local substrates and exhibit high elongation. This combination is ideal for equipment mounting which either is sensitive to structure borne vibrations or would drive vibrations into the structure.

Global Loads

The global loads on a composite hull must be considered a bit differently because of the low modulus of the commonly used E-glass material. Generally, for a normally proportioned vessel under 30m LWL, hull bending is not critical and the local pressure loads are the most critical loading case.

For vessels over 30m, longitudinal bending in waves must be considered. Even with low modulus E-glass laminates, this is usually not much of a problem up to 50m hulls.

As hull length approaches 50m, hull girder stiffness must be carefully studied. Discontinuities in main deck structure due to stair towers, steps in decks, routing the stiffening structure around other obstacles, etc. can combine to reduce hull girder stiffness.

The ABS Guides (ABS 1997) have reasonable design criteria for the various aspects of composite vessels, as do most of the other classification societies. Variations in safety factors and loading criteria make for the differences between the rules and guides.

Deckhouses are typically not included in the hull girder calculation because of their inherent lack of stiffness due to their light construction. A typical hull girder arrangement for a 41m composite vessel is shown below in Fig. 1. The geometric properties for hull bending calculations are shown in Table 2 and Table 3, and the resulting stresses from those calculations are given in Table 4.

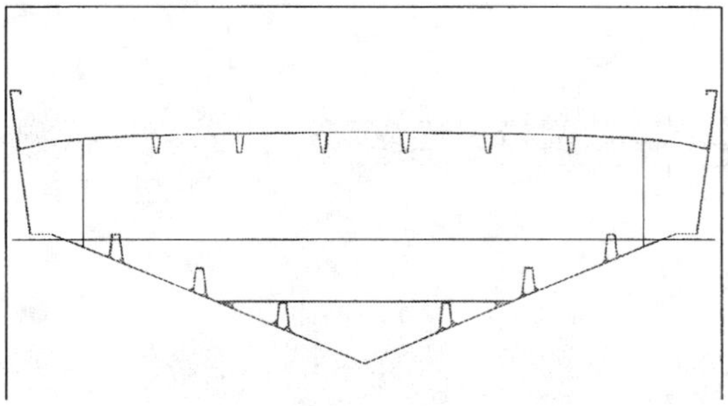

Fig. 1. Typical Midship Section for High Speed Craft Primary Bending Calculations

Table 2. Typical Geometric Properties for High Speed Craft Primary Bending Calculations

Area:		0.3857 m^2
Perimeter:		80.80 m
Centroid:	X:	0.00 m
	Y:	1.80 m
Moments of inertia:	X:	1.72 m^4
	Y:	3.46 m^4

Table 3. Hull Section Modulus Calculations

Sagging Moment (from ABS)	12,646 kN-m
Hogging Moment (from ABS)	9,208 kN-m
Moment of Inertia =	1.72 m^4
c bottom =	1.80 m
c top =	2.16 m
c deck =	1.55 m
Section Modulus (Bottom) =	0.96 m^3
Section Modulus (Top) =	0.80 m^3

Racking loads must also be carefully considered, especially in low modulus composite vessels. Usually bulkheads, and well placed web frames or ring frames, will handle these loads. Transferring this structure through continuous decks to hold the deckhouse in place requires special details that are in keeping with the manufacturing process and sequencing.

Table 4. Typical Primary Stresses

Stress (keel) =	13.23 N/mm^2	Tension - Sagging
Stress (keel) =	9.63 N/mm^2	Compression - Hogging
Stress (Bulwarks) =	15.84 N/mm^2	Compression - Sagging
Stress (Bulwarks) =	11.53 N/mm^2	Tension - Hogging
Stress (Deck) =	11.36 N/mm^2	Compression - Sagging
Stress (Deck) =	8.28 N/mm^2	Tension - Hogging
Lower Hull Panel	1.54 m	below Neutral Axis
Primary Stress =	11.27 N/mm^2	Tension - Sagging
Primary Stress =	8.21 N/mm^2	Compression - Hogging

ANALYSIS TECHNIQUES

Finite Element Analysis

The natural frequency of a component is a measure of it's tendency to oscillate (vibrate) naturally. Frequency the measure of what rate a mass vibrates and is typically expressed in Hertz (Hz) or cycles/second. The greater the magnitude of frequency, the greater the stiffness of the product. Every structure has its own unique tendencies to vibrate, which a function of the material properties and geometry. The mode refers to the easiest deformed shape a component will move to while excited to it's natural frequency. The primary deformation mode most structures take in the design of marine structures of this type are bending, torsion and swaying.

Understanding a component's primary deformation mode will allow the designer to modify the materials used or geometry to stiffen the structure as required. It is not always intuitively obvious what the primary mode of deformation a part will take, therefore using FEA early on in a design to calculate critical areas can be very useful to avoid designing in problems which may be more difficult to fix later. The strength and stiffness of each component make up the stiffness matrix of the entire structure, therefore, the more closely the parts are controlled to contribute to the primary loading of the final structure, more efficient structures are created. Also, creating models of the components of the complex vessel will allow for combining all of the components and analyzing global bending and stiffness. Local stresses caused by global loading can be mitigated through careful modeling and analysis of the individual parts.

Finite element analysis (FEA) has been used widely for global analysis of metal and composite hulls for over 25 years. However, the detail to which a sandwich composite hull can be modeled is limited by the ability of most standard FEA programs unless they are modified to accurately handle the differences between the thin, high modulus and strength skins, and the thick, low modulus and strength core material. Following typical classification society rules will not yield an optimized design as the rules are written with overlapping factors of safety. However, owners of 46m yachts, commercial vessels and now large patrol frequently request the approval of a classification society, so ignoring the class rules is usually not an option.

The class rules use simplified methods to determine the response of panels and built up sections to loads. However, the class societies will all allow the use of alternative loading and/or analysis methods if they are shown to be accurate. The method used here is to use ABS loading criteria as specified, but perform much of the analysis with FEA.

43 Meter Yacht Case Study

The following case study is an example of utilizing FEA to analyze and develop laminate requirements for a recently built 43m yacht. The FEA program FEANICS (Horsmon 2000) is used here to develop optimized panels. Skin stresses, core stresses, skin buckling loads and limit stresses, and adequate panel stiffness are all analyzed for a number of critical panels for the structure for the 43m yacht. An example of the analysis is presented below. The midship section of the yacht is presented in Fig. 2.

Fig. 2. Midship Section of 43m Motor Yacht

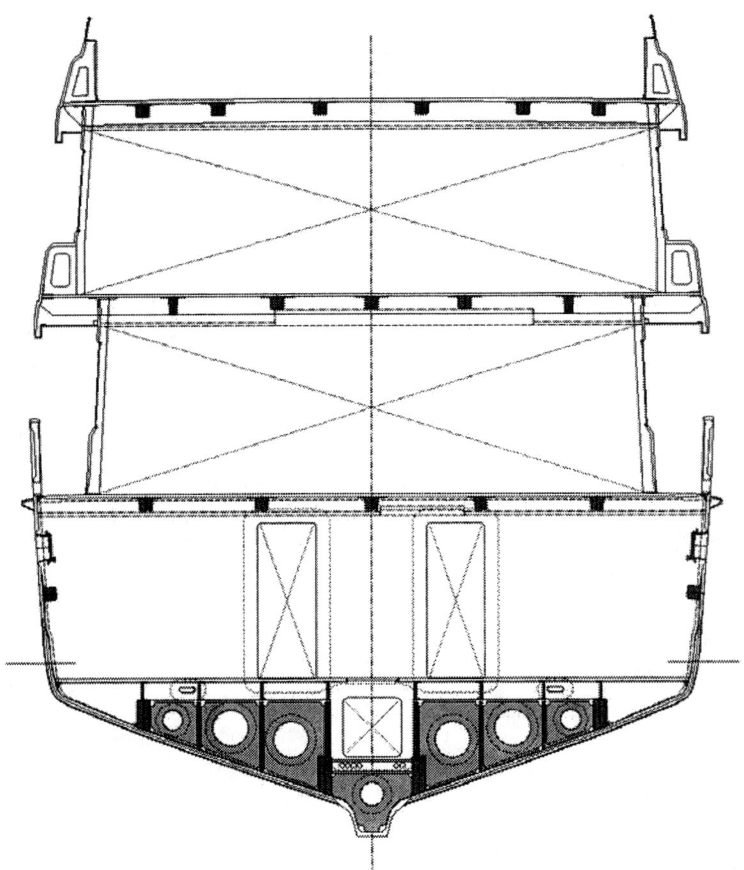

Main Deck Analysis

Calculations according to class rules show the main deck panels would require 100mm core to allow 2.1m spans between main deck stringers. The laminate below was analyzed by FEANICS to achieve a stiffness of L/200 (span divided by deflection at center is less than 200).

The deflection with a 64mm core shown in the FEANICS results is 9.55mm (0.376 in.) giving a deflection ratio of 2100/9.55 = 220 so there is actually capacity for increased span. The through thickness deflection is shown in Fig. 3. The laminate schedule is given in Table 5.

Table 5. Main Deck Laminate Schedule

Ply	Weight (gm/m^2)	Thickness (mm)
Gel Coat	450	0.5
1.5 OZ CS Mat	1500	1.1
XM 2408	1900	1.4
CM 2408	1900	1.4
Core-Bond	450	0.6
Priming Resin	450	0.0
A500 Core-Cell	5567	64.0
2 X CM 2408	3800	2.8

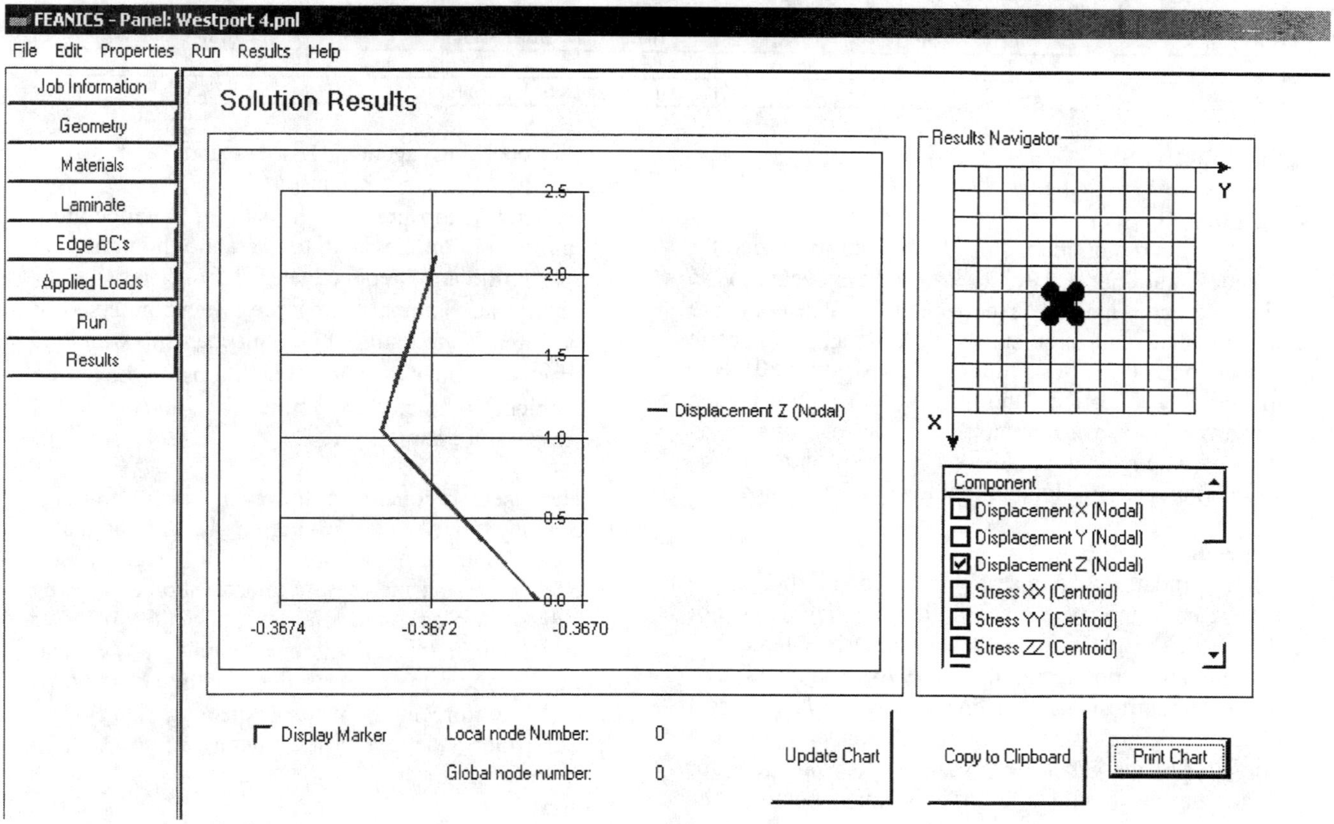

Fig. 3. Typical Solution Results from FEA to Calculate Panel Stresses

Large Deck Span Analysis

Another area of concern in this vessel is the pilot deck and flybridge deck spans. These are free standing decks with spans of 7.3m from side to side of the deckhouse and over 9m between supporting bulkheads. Deck beam height is limited by the desire for large unobstructed areas, the need to keep deck to deck height to a minimum to achieve a sleek look, and the tendency for large salons to have punched up center areas.

With low modulus E-glass in the deck beams not stiff enough to get the desired stiffness, carbon fiber is the typical solution. Some builders are using aluminum I-beams in their designs, but it is difficult to optimize the shape to get the best stiffness, especially if there is camber in the deck. A spreadsheet showing the development of the carbon beam is given in Table 6.

Table 6. Carbon Fiber Beam Development for Flybridge support for 43m Yacht

		Flybridge longs. 46-60		Outboard figured at 200mm deep by x 100mm wide			
	width		Area	dist. center			item inertia
Item	base b	height h	= b X h	d	A X d	A X d^2	I o
Outer Skin	610	2.54	1548.4	1.3	2.E+03	2.E+03	8.32E+02
Core t	0	36	0.0	20.6	0.E+00	0.E+00	0.00E+00
Inner Skin	610	2.54	1548.4	39.9	6.E+04	2.E+06	8.32E+02
0	0.00	0	0.0	41.1	0.E+00	0.E+00	0.00E+00
Tabbing A	508	2.95	1496.8	42.6	6.E+04	3.E+06	1.08E+03
Carbon BX 1200 shear webs B	5.9	210.82	1242.3	149.5	2.E+05	3.E+07	4.60E+06
Carbon C-LA 2012 Uni cap	1143	3.43	3919.3	253.2	1.E+06	3.E+08	3.84E+03
Carbon C-BX 1200 wraps	203.2	2.95	598.7	256.4	2.E+05	4.E+07	4.33E+02
		261.29	10353.9		1.46E+06	3.24E+08	4.61E+06
Frame Height	220.1						
ABS Yacht '90 Required	Units	Dna	14.1		FRP Stiffener	Limits	
1175	cm^4	Ina	12256	Height	not >	30 X sidewall thickness	
		SMo	870	210.8	not >	88.4	
171	cm^3	SMi	1018	Crown width	not >	20 X crown thickness	
8.1	cm^2	Shear A	12.4	101.6	not >	127.5	
			Bending Deflection, mm		3.19	Defl. Ratio	1955

MATERIALS

Laminate Properties

Steel and aluminum ships have been in service for decades, and their reaction to long term exposure to the elements and loading experienced in a seaway are documented and predictable. Although composite vessels have been in continuous use for over 50 years in marine applications, the history with highly loaded large vessels is not commonly known. Claims of no loss of strength or stiffness, or more importantly, elongation, are anecdotal, but not well documented.

Degradation

Degradation of glass reinforced plastics in a marine environment can effect all three components of the GRP; the glass fiber, the fiber/matrix interactions and the laminating resin. (Juska 1996) Degradation of the resin manifests itself as blister growth. This is most common in orthophthalic laminates made with ortho polyester resins. These resins are not recommended for larger marine structures for other reasons, including the lack of secondary bond strength, therefore will not be included in this analysis. Isophthalic and vinyl ester laminates generally are not subject to blistering. E glass fiber is generally considered inert and is not attacked by seawater, although it can be damaged by large out of plane deflections. Structural considerations are discussed elsewhere in this paper.

Findings by Tom Juska (Juska 1993), while working in support of to support the Organic Composite Ship Structures Project sponsored by the Office of Naval Research, validate the claims that vinyl ester laminates will not significantly degrade during extended service in a marine environment, if the glass fiber sizing or finish composition includes a vinyl-specific coupling agent. His research and testing indicate that this is true because:

- realistic exposures such as low temperature immersion or elevated temperature humidity do not significantly reduce material properties
- vinyl ester laminates degrade when immersed at elevated temperature by a mechanism which is inactive under service conditions, which are ambient temperature immersion and elevated temperature humidity

His research concludes, however that the findings in his study, that woven roving/vinyl ester laminates cured at room temperature will not significantly degrade in a marine environment, should not be generalized to predict the long term durability of other classes of laminates for this environment. With that caveat, designers must be prudent in the selection of resin & reinforcement when expecting long term exposure, such as large marine structures.

Orientation

Scott (Scott 1996) describes in great detail the orthotropic nature of mat and woven roving laminates. Alternate layers of woven roving at +/- 45 degrees could be used to even out the humps in the directional strength chart in Scott's Fig. 10, to try to get a more isotropic laminate, but there may be problems with large lumps when the laminate overlaps bunch up.

Keep in mind these orthotropic properties are only in-plane properties. Out of plane properties are quite different and somewhat less. There are special 3D knitted fabrics that have a large number of strands sewn through the plane of another 2D fabric to achieve the desired strength. It is very difficult to get composites to exhibit the quasi-isotropic behavior of metals, but that's a big advantage of composites. The fibers (most of them) can be oriented in the direction of stress. Symmetry is not usually needed, and is rarely achieved in standard marine laminates.

The charts of different properties in various directions in *Marine Composites* (Greene 1999) illustrate the variability seen in most all areas of composites. Even though this data is all based on the mat/woven roving laminates, the same variability is exhibited in knitted biaxial materials, unidirectional materials, and combination materials. Those considerations also apply to the properties in Appendix A in *Marine Composites*. Also consider that Appendix A is mostly based on a 50% glass content. The glass content achieved by different manufacturing methods varies from 30% to 70% as listed in Table 7.

Table 7. Glass Contents to be Expected Using Various Lamination Methods

Lamination Method	Glass Content by Weight
Low Quality Hand Lay-up	30%
Good Quality Hand Lay-up	40%
Lower Quality Impregnator, Wet Bag, SCRIMP, VARTM	50%
Higher Quality Impregnator and Wet Bag, SCRIMP, VARTM	60%
Autoclave	70%

Secondary Bonding

Given the size of typical marine structures it is usually necessary to join the separately laminated parts with secondary bonds. In particular, stiffeners are laminated onto a cured skin and bulkheads and decks are joined to the hull by tabbing. It is also standard practice to fabricate large substructures in multiple stages of a few plys in each stage.

Resins used in boatbuilding typically exotherm in about 30-45 minutes after initiation, so lamination over "cured" laminate is usually necessary. Bonds which form between the successive laminations are called "secondary" bonds in the boatbuilding industry. Secondary bonds can be classified into two types:

- Type 1. Cross-linking reactions occur across the bondline, so the thermoset network is continuous, in spite of the delay.
- Type 2. Cross-linking reactions do not occur across the bondline, so the thermoset network is discontinuous and the bond is adhesive.

Test Methods

Marine Composites (Greene 1999) gives a good explanation of the various test methods and what mechanical laminate property is actually being determined. Table 8 (Horsmon 1994) gives examples of problems that are related to test methods.

With sandwich composites, the laminates and cores are tested separately and together. Core testing brings it's own set of problems. For example, making a compressive test sample square instead of round can give lower results for some cores. The rate of testing can be varied within the parameters of the test methods, but some cores are strain rate sensitive and show different results at different rates. These problems and others are being addressed by the SNAME HS-9 Panel.

Table 8. Test Method Problems and their Associated Solutions

PROBLEM	SOLUTION
Samples are fairly small, boats are big.	Use larger samples, adapt the test method, and consider the hydromat panel test.
Samples are often made up in a lab on a table. Boats are built out in a shop in hot, dusty conditions.	Take the sample from a cutout on a boat, or make the sample in the shop alongside the boat.
A sample cut 5% off axis from a panel will show up to 15% reduced strength for some tests.	Cut carefully, larger samples reduce consider source of errors.
Testing usually returns an average value; most design criteria call for some minimum value.	Know the statistical spread on the test results. Adjust the safety factor to account for the unknowns.

The hydromat test jig was developed separately to test marine panels. The hydromat test has been developed into ASTM D6416. In some cases, it returns results similar to 3-point bend tests, but it gives more reliable results for a wider range of large marine panels subject to lateral loads.

Concise, repeatable material properties that show up consistently in all of the tests should not be expected. However, as stated in the references, and in ABS and other classification society guides, testing is necessary to support the design, prove the capability of the builder and confirm assumptions.

Testing is also a form of quality control, but better if used as process verification. If testing is used as quality control, it has been done too long after the lamination. Then it is too late to change a laminate or laminating process when it is shown to have low physical properties. If the process is verified by testing under conditions similar to those in the shop before lamination begins, then small sample tests can be used to verify that the larger part will actually be built as expected.

The NAVY requires a Lamination Process Description (LPD) be written prior to approving a lamination process. (Bernhard 1994) This LPD includes a description of all facets of the lamination process used while producing the composite parts. The crew and materials require qualification prior to commencing any construction. The qualification procedure consists of producing panels in the production environment using production materials, techniques and tools and testing the panels. These test results will qualify the process, materials and personnel and assure repeatability of the laminate quality.

Sandwich Composite Structure

Strength, stiffness and weight are the major considerations when designing large waterborne structures of FRP. Sandwich composite structures are used throughout the design to offer strong, stiff structures without a large weight or cost penalty. Recent studies have confirmed the applicability of sandwich panels for use in ships decks and bulkheads (Roberts 2002)

The sandwich composite design is an iterative process, to direct the weight and load spiral in the other direction, towards a lighter vessel. This design process involved balancing:

- the stiffness of the panel;
- the strength of the panel;
- panel span;
- engine placement;
- arrangement of internal structure
- bottom pressure;
- use of commonly available fabrics and resins, and
- use of commonly available core density and thickness.

A case study of designing a 41m high speed ferry in both composite and aluminum construction and comparing the two resulted in a side by side analysis of framing scenarios using best shipyard practices. Balancing the aforementioned alternative choices during the composite ship design process resulted in a final sandwich composite laminate, most of which is shown in Table 9. (Horsmon and Bernhard 1999). One can see that the fabric and core variations have been kept to a minimum, but that the number of fabric layers and directionality, and the density and thickness of core material used, are carefully chosen to give a reasonably ordinary but light sandwich laminate.

An example of how the composite laminate can be easily optimized can be seen by comparing the differences in the laminates for the engine girders and the other frames. The engine room for this concept vessel is on the order of 8 m (25 ft.) long and must support the weight and torque of large diesel engines. High density foam (200 kg/m^3 A1200) is used to help support the weight of the engines and to prevent compression of the core from bolted connections. The girders are a minimum of 500 mm (20 in.) high and are capped with multiple layers of unidirectional fiberglass (A260) to raise the section modulus. A transverse deep web is used to break up the span of the engine room and allow a change in girder height to set the generators aft.

Comparably, 4.6 m (15 ft.) compartment and is cored with lower density foam, only needs to be 300 mm (12 in.) high and has only 3 A260 tapes on the top to meet section modulus requirements.

Beside the ability to create very large panels that are also light, there are many other reasons why a sandwich with a tough core material is used in place of single skin fiberglass, or aluminum. These reasons include, but are not limited to:

- impact resistance
- stiffness
- light weight
- thermal insulation
- sound insulation
- vibration dampening
- corrosion resistance
- lower maintenance

These topics are discussed in detail in the following section.

Table 9. Sandwich Composite Laminates (Weights typical for Infused Process)

HULL BOTTOM		Weight	Thickness	
US fabrics	Metric Fabrics	gms/m^2	mm	
GELCOAT		450	0.51	
1.5 oz Mat	450gm Mat	1500	1.27	
+/- 45 XM 2408	EBX 800/225	1900	1.42	
0/90 CM 2408	EB 800/225	1900	1.42	Outer Skin
CM 2408	EB 800/225	1900	1.42	5.53
Core-Bond		450	0.64	
Priming Resin		450	0	
A600 Core-Cell		3520	32.0	
1.5 oz Mat	450gm Mat	1500	1.27	
Core-Bond		450	0.64	
Priming Resin		450	0	Core t
A600 Core-Cell		3520	32.0	66.55
CM 2408	EB 800/225	1900	1.42	Inner skin
CM 2408	EB 800/225	1900	1.42	
XM 2408	EBX 800/225	1900	1.42	4.26
		23690	76.85	
Weight per meter		=	**23.69**	kg/m^2

TOPSIDES		Weight	Thickness	
US fabrics	Metric Fabrics	gm/m^2	mm	
GEL		450	0.51	
1.5 OZ MAT	450gm Mat	1500	1.27	
XM 2408	EBX 800/225	1900	1.42	Outer Skin
CM 2408	EB 800/225	1900	1.42	4.11
Core-Bond		450	0.64	
Priming Resin		450	0	Core t
A500 Core-Cell		3340	38.0	38.640
CM 2408	EB 800/225	1900	1.42	Inner skin
XM 2408	EBX 800/225	1900	1.42	2.84
		13790	46.1	
Weight per meter		=	**13.79**	kg/m^2

FOREDECK		Weight	Thickness	
US fabrics	Metric Fabrics	gm/m^2	mm	
GEL		450	0.510	
1.5 OZ MAT	450gm Mat	1500	1.27	
XM 2408	EBX 800/225	1900	1.42	Outer Skin
CM 2408	EB 800/225	1900	1.42	4.110
Core-Bond		450	0.64	
Priming Resin		450	0	
A500 Core-Cell		3340	38.0	Inner skin
2 X CM 2408	EB 800/225	3800	2.820	2.820
		13790	1.814	
Weight per meter		=	**13.79**	kg/m^2

KEEL		Weight	Thickness	
US fabrics	Metric Fabrics	gm/m^2	mm	
GEL		450	0.51	
1.5 oz Mat	450gm Mat	1500	1.27	
2 XM 2408	2xEBX 800/225	3800	2.84	Outer Skin
4 X CM 2408	4xEB 800/225	7600	5.68	9.79
Core-Bond		450	0.640	
2xPriming Resin		900	0	Core t
A1200 Core-Cell		11180	50	2.345
4 X CM 2408	4xEB 800/225	7600	5.68	Inner skin
2 XM 2408	2xEBX 800/225	3800	2.84	8.520
		37280	69.46	
		=	**37.3**	kg/m^2

MAIN DECK		Weight	Thickness	
US fabrics	Metric Fabrics	gm/m^2	mm	
GEL		450	0.51	
1.5 OZ MAT	450gm Mat	1500	1.27	
XM 2408	EBX 800/225	1900	1.42	Outer Skin
CM 2408	EB 800/225	1900	1.42	4.110
Core-Bond+prime		900	0.64	
A500 Core-Cell		5630	64.00	
2 X CM 2408		3800	2.84	2.84
		16080	72.100	
		=	**16.08**	kg/m^2

BULKHEADS		gm/m^2	mm	
XM 2408	EBX 800/225	1900	1.42	
CM 2408	EB 800/225	1900	1.42	
Core-Bond+priming		1800	0.64	
A500 Core-Cell		6680	75.0	
CM 2408	EB 800/225	1900	1.42	
XM 2408	EBX 800/225	1900	1.42	0.112
		16080	3.137	
		=	**16.08**	kg/m^2

LOWER TRANSOM		gm/m^2	mm	
GEL		450	0.51	
2 XM 2408	2xEBX 800/225	3800	2.84	
4 X CM 2408	4xEB 800/225	7600	5.68	0.336
Core-Bond+priming		1800	0.64	
2xA1200 Core-Cell		11180	50.0	
Priming resin		300	0.00	
4 X CM 2408	4xEB 800/225	7600	5.68	
2 XM 2408	2xEBX 800/225	3800	2.84	
		36530	2.419	
		=	**36.53**	kg/m^2

ENGINE GIRDERS		gm/m^2	mm	
4 XM 2408	4xEBX 800/225	7600	5.68	
3 CM 2408	3xEB 800/225	5700	4.26	
10 X 6 A260	10xUD900	14440	9.91	
Priming Resin		1200	0	
5x25mm A1200		27950	125	
3 CM 2408	3xEB 800/225	5700	4.26	
4 XM 2408	4xEBX 800/225	7600	5.68	
Sidewall thickness			9.94	
Cap thickness			19.85	
Weight per meter			**27.6**	kg/m

LONG. DEEP WEBS		gm/m^2	mm	
XM 2408		1900	1.42	
CM 2408		1900	1.42	
Prime Resin		900	0	
2x38mm A400		5000	76.0	
CM 2408		1900	1.42	
XM 2408		1900	1.42	
		13500	81.68	
		=	**13.50**	kg/m^2

LONG. FRAMES		gm/m^2	mm	
6 E-TLX 2400	6xTRI 800	5700	4.00	
5 X A260	5xUD900	7220	4.95	
Priming Resin		600	0	
3x25mm A400		5040	75.0	
Sidewall thickness			4.953	
Cap thickness			8.953	

Impact Strength

A sandwich using a foam core with a high degree of toughness is more impact resistant than a single skin laminate of equal or even higher weight. This has been experienced since the Dutch built the 23 m (77 ft.) sandwich composite pilot boats for the Port of Rotterdam in the mid-60s. These vessels are still in service in India. A common test to validate the impact strength of competing designs is to drop a 45 kg (100 lb.) impacter onto a 600 x 600 mm (2 ft X 2 ft) from a height of 3 m (10 ft.). The effect this impact has on several different designs is shown below. Fig. 4 shows a single skin FRP laminate which is holed due to the impact. Fig. 5 shows a cross linked PVC core. Note the shear failure on both sides of the impact location and delamination. Fig. 6 is a balsa cored panel which failed in shear and delaminated from the inner and outer skins. Fig. 7 shows a damage tolerant SAN core which contains the impact locally and does not exhibit any delamination or shear failure, only local impact damage.

Fig. 4. Single Skin Impact Damage

Fig. 5. Cross Linked PVC Core Impact Damage

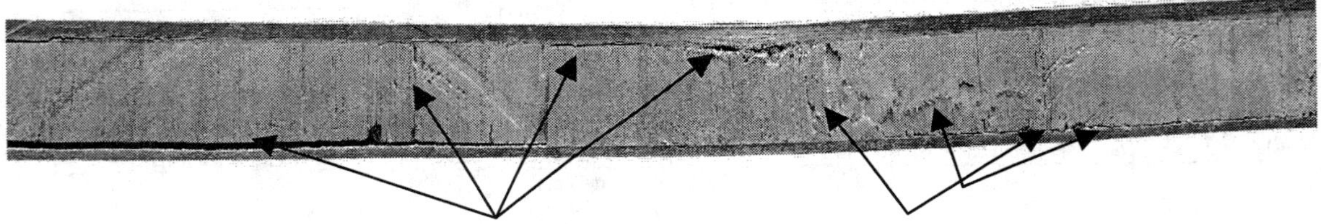

Fig. 6. Balsa Core Impact Damage

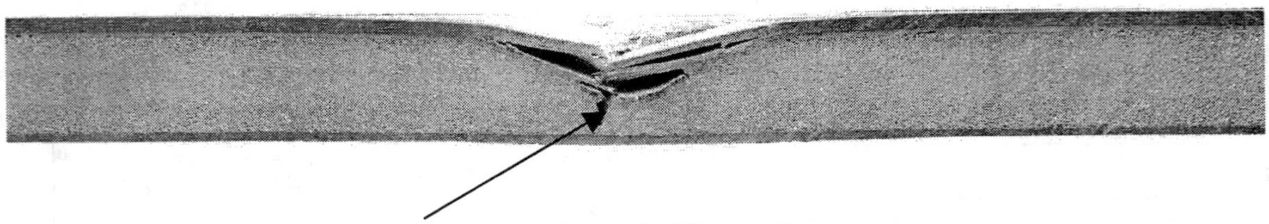

Fig. 7. SAN Core Local Impact Damage

The reason for the higher impact strength can also be explained by the principle of design used in car windshields, where a thin layer of polyvinyl-buteral film is laminated between two sheets of glass. The resilient PVB film in the center absorbs the shock of an impact and thus prevents penetration. The film also adheres to any glass splinters.

This principle also holds true for the sandwich. The foam absorbs the impact energy, and holds the skins together. It protects the inner skin from impact damage. However, not all core materials can do this as well. The more brittle PVC foams, end grain balsa and honeycomb materials used in various other areas of marine construction do not fare well under severe impact, and should only be used in non-impact areas. It is not just grounding or the hull bottom that sees high impact loads. Wave slamming loads also occur on the sides, foredeck and the house front. Shock loads are carried to other parts of the structure. Damage tolerant design and construction must be considered throughout all parts of a high performance marine structure.

Survivability of sandwich composite marine structures is well accepted in other areas. Sweden has committed to building their latest 70 m (230 ft) patrol craft in composite (Lonno, 1998). The Swedish have also built a number of patrol craft and mine hunters in sandwich composite. The US Navy is investigating expanded use of composites in a number of areas, one of which is in an advanced material transporter (Nguyen et al, 1993).

Stiffness

The sandwich cross section selected for the vessel in this design is nearly 75 mm (3.0 in.) thick and weighs in at 22.4 kg/m^2 (4.6 lbs/ft^2). A single skin fiberglass laminate would need to be over 25 mm (1.0 in.) to meet the ABS Rules and nearly 50 mm (2 in.) at 58 kg/m^2 (12 lbs/ft^2) to achieve the same stiffness. The difference in stiffness is dramatic, and stringers and other stiffeners cannot possibly be designed to achieve the same overall stiffness of a sandwich. The aluminum bottom shell plate for the design is 11 mm (7/16 or 0.4375 in.), weighs 30.3 kg/m^2 (6.2 lbs/ft^2), and needs more stiffeners than the composite shell. The FRP vessel ends up being stiffer than a single skin fiberglass, and just as stiff but much lighter than an aluminum vessel of the same hull design.

Weight

The other potential pitfall in building a heavier boat is that the general philosophy about designing, building and outfitting a heavier vessel, often encountered with yards as well as owners, is that as they are already making a heavier structure, a little more weight elsewhere will not hurt. Heading down that road can easily add 6,800 – 9,000 kg (15-20,000 lbs.) to a boat in the size range being considered. Careful attention to details can head the whole project in the other direction. Part of the penalty paid by overweight performance craft – the heavier the vessel, the higher the loads are for a given speed, more fuel must be carried to cover a specified endurance, so the vessel ends up being much heavier than intended, or some performance is given up. A weight comparison of a fiberglass sandwich construction design versus a aluminum design 41m ferry is shown in Table 10. (Horsmon and Bernhard 1999) Note the structural weight of the aluminum design approaches half of the full load weight, whereas the composite design yields approximately only one third of the full load.

Table 10. Weight Comparison of FRP & Aluminum 41m Ferry

	FRP Sandwich	Aluminum
Structure	65,454	102,762
Propulsion	34,090	34,090
Electrical & Electronic	9,090	9,090
Systems	15,909	15,909
Furnishings	17,045	17,045
Glass	3,818	3,818
Load	34,500	34,500
Fuel	12,727	12,727
Water	3,786	3,786
Full Load Estimate (kg)	**196,423**	**233,730**

Thermal Insulation

Thermal insulation provided by the foam in all areas of the boat is an important consideration. Frequently the water the boat sits in is colder than the outside temperature. With metal boats, condensation forms, and becomes apparent as bilge-water. The insulating effect of the foam nearly eliminates condensation and bilge-water formation from condensation. Machinery and electronic equipment lasts longer in a low humidity environment. The thermal insulation in the house is "free" and need not be added as part of outfitting.

Sound Insulation/Vibration Damping

Noise and vibration travel well through an aluminum boat. These are not just noises from the main engines and from hitting waves while under way, but also the subtle noises from generators and wave lapping while moored. The tougher foams are excellent for damping these noises and vibrations. Boats with cored structure are simply quieter.

The ride quality of a composite vessel will be superior to an aluminum vessel while they share identical hull forms, because the composite vessel is compliant. The damping effect of the composite panels absorbs many of the short duration accelerations. These same accelerations are transmitted into the primary structure of an aluminum vessel. By the same token, machinery hull borne vibrations are decreased in composite vessels over stiff metallic construction.

Both shocks from hydrodynamic loading, and machinery initiated vibrations should be minimized in marine vessels to provide a comfortable ride to the passengers and prolong the service life of the vessel, as is evidenced in common fatigue studies.

Esthetics

Composite vessels generally look better than aluminum vessels in commercial form. A roughly done composite commercial vessel has a fair shape but may show a bit of fabric print through. Commercial aluminum vessels usually have the "hungry horse" look where the structural framing can be measured from outside the vessel. Yacht quality finish is built into a composite vessel – aluminum yachts take days and tons of fairing compound.

Fire Protection

Fire protection and composites seem to be mutually exclusive terms. However, and as with weight gain, proper attention to details can produce a safe structure in either aluminum or fiberglass. Both materials have a reduced heat deflection temperature compared to the steel fire protection standard, so fire proof thermal insulation is a must for the highest fire hazard areas.

Test done at Westport Shipyard years ago showed similar survivability to fire for both an aluminum and a balsa cored sandwich composite deck panel with a load. Westport built a scaled ASTM (American Society for Testing and Materials) E-119 oven (1/16 the volume but capable of producing the same time/temperature curve. Both the aluminum and FRP deck panels survived 10 minutes with typical passenger deck loading but the aluminum had 260°C (500°F) temperature rise while the FRP had only 10°C (50°F) temperature rise. Passengers could have walked on that deck without being burned.

A Norwegian design (Hoyning and Taby 1998) is already capable of meeting the tough IMO High Speed Craft Code, ostensibly without much of a weight penalty over a similar performance aluminum craft.

The U.S. Coast Guard has been working with the Passenger Vessel Association, through a "Natural Working group," to develop a guideline to fairly judge a fire protection equivalency of fiberglass passenger vessels compared to a "standard" steel vessel. The result of the two year effort is the draft document titled "Performance Based Fire Engineering Guidelines for Equivalency" which is now being tested against a large composite passenger vessel. After this exercise, the document is likely to be converted into a USCG Navigation and Inspection Circular (NVIC) for general use

DETAILS

The key to making composite structures work is in the details. It is common knowledge that composites do not corrode, have high strength for low weight, have low panel stiffness that can be solved with sandwich structures, etc. Despite these advantages, composite construction has some features that require special details. Following is a partial list of details from a sandwich core manufacturer's design guide (Johannsen, et.al 2001)):

- bulkhead detail with putty fillets,
- bulkhead detail on foam pad,
- solid centerline detail,
- cored centerline detail,
- stringer with putty fillets - separate tabbing,

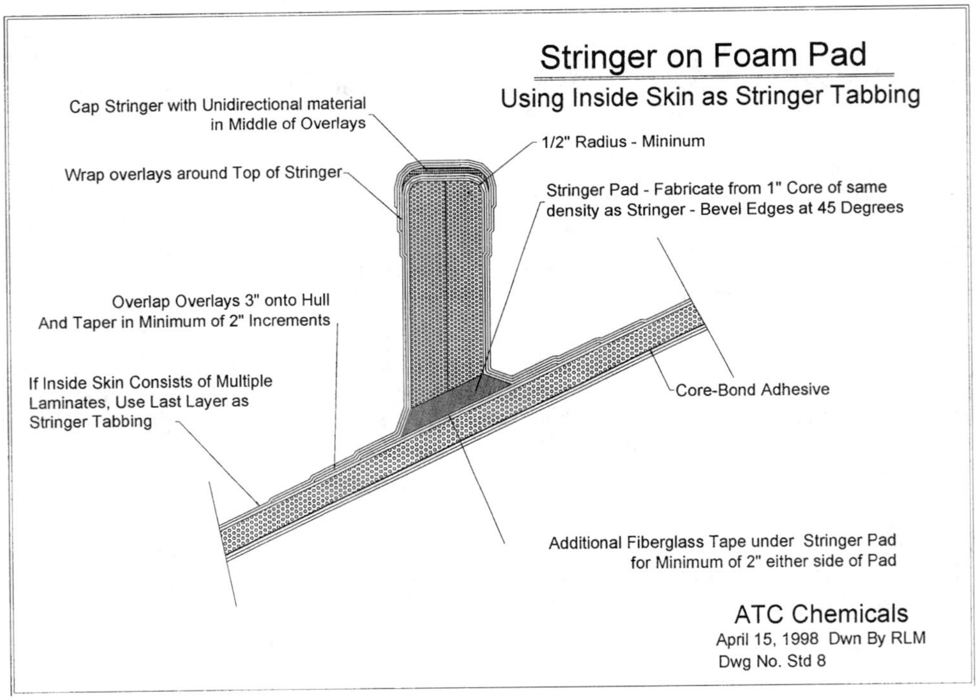

Fig. 8. Composite Stringer Detail

- stringer on foam pad - separate stringer tabbing,
- stringer with putty fillets, inside skin tabbing,
- stringer on foam pad, inside skin tabbing,
- stepped chine, high density insert,
- stepped chine, foam wedge,
- stepped chine detail - putty radius,
- hard chine with laminate overlaps,
- hard chine with high density insert,
- hull to deck joint detail w/flush deck,
- hull to deck joint detail w/bulwarks,
- through hull fittings,
- winch pads,
- mast step and keel bolts
- ballast keel attachment, and
- chain plates.

Many of these details are extensions of details required by classification society guides and rules expanded to capture the experience of various builders and manufacturing processes.

Fig. 8 shows a typical method of laminating a stringer to the hull (or other panel) shell. A foam pad is used at the base to avoid hard laminating acute angles. The tabbing laminates are wrapped from each side of the stringer, over the top and stopped. This sequence makes for easier laminating and doubles the thickness of the layers at the top, where the strength is needed.

Fig. 9 depicts a flat chine detail where the bottom outside skin is carried though the chine and up the side a bit, the side skin is carried down through the chine and onto the bottom a bit. Then the bottom and side core is installed with an extra wedge inside to stiffen the chine. Then the inside skins are similarly carried through the chine.

Fig. 10 shows two methods of attaching through-hull fittings to a sandwich structure. On the left, the position of the fitting is preplanned and a section of high-density core material inserted in place of the fitting. Then the hole for the fitting is drilled and the fitting inserted with appropriate bedding compound. On the right, the fitting is inserted into an existing low density cored sandwich laminate by drilling the location, routing out the core, filling the hole with a high density putty, then re-drilling the hole and installing the fitting

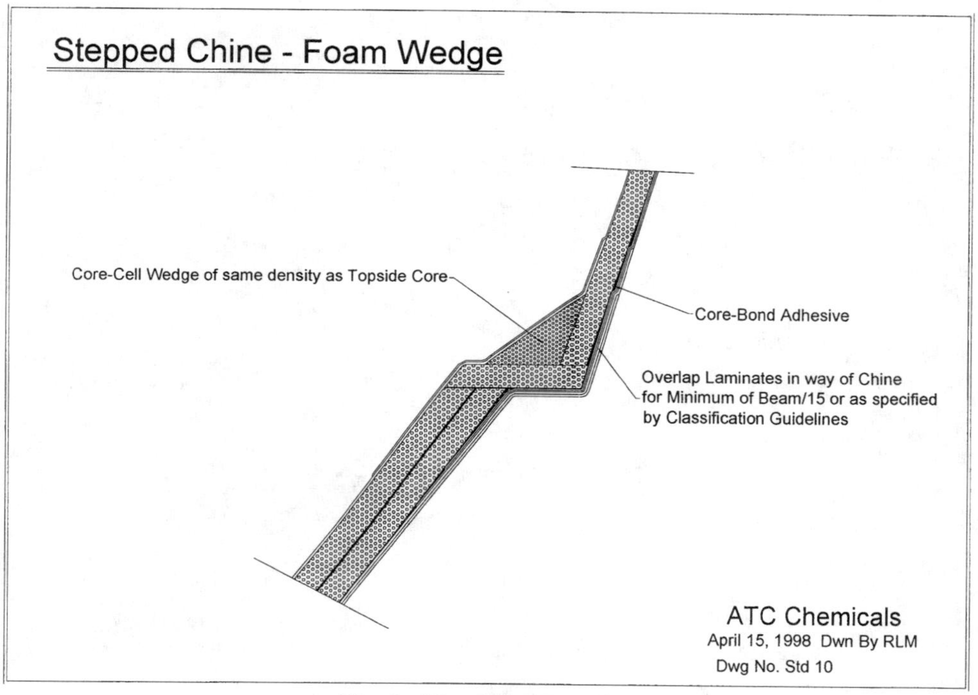

Fig. 9. Flat Chine Detail

Concentrated Load Attachment

Fig.11 shows how a winch can be attached to a sandwich deck. High-density core is installed and the skin laminates locally doubled to distribute the loads from the winch. . These local reinforcements are faired out into the rest of a panel or to major stiffening members. With a large backing plate, the winch can be bolted on. If the backup to the bolts is only washers, a tubular insert is recommended to take up the compression loads from the bolts.

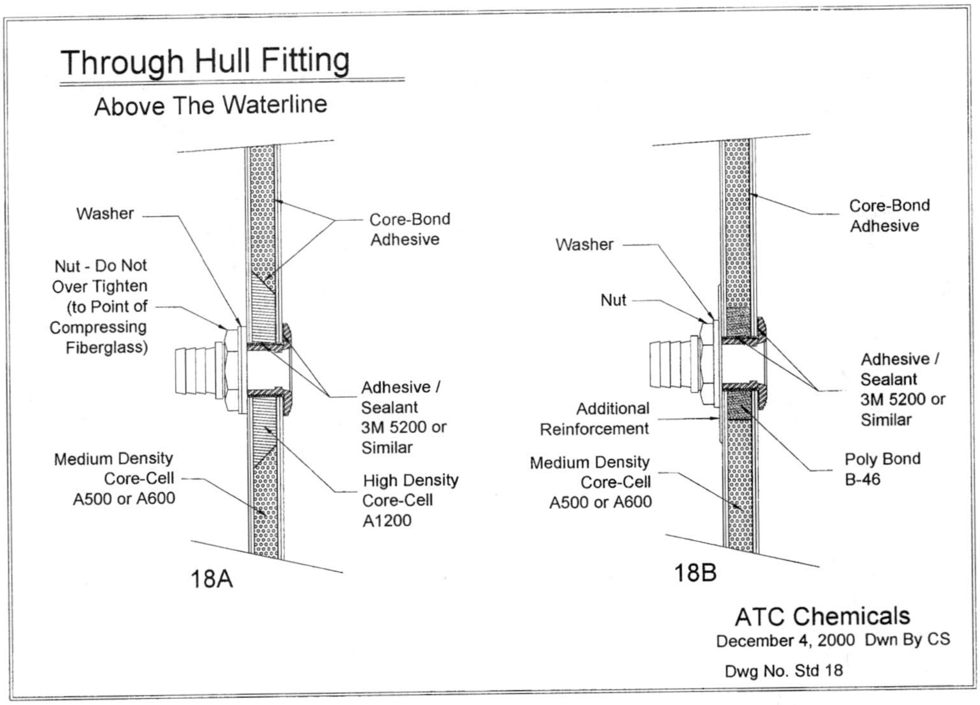

Fig. 10. Through Hull Fitting Attachment

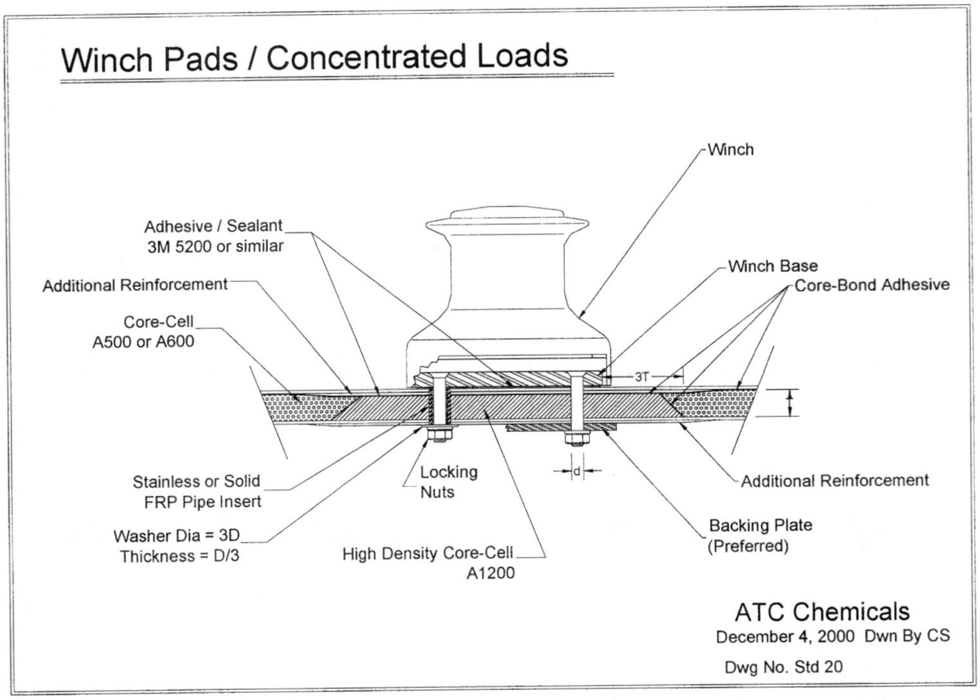

Fig. 11. Concentrated Load Attachment

MANUFACTURING METHODS

Forming the Shape

There are many different ways to form composites. A complete structure may use a number of these methods. The basics are covered here. Full coverage of all systems is beyond the scope of this paper but can be found in the many references.

Molds – Female, Male

Female molds are the most common method of establishing the shape for composite parts. They are usually made by applying fiberglass with special tooling resins over another part or a temporary male shape. Female molds can also be made directly from combinations of wood frames and sheets covered by fairing and mold compounds.

To allow easier worker access, some female molds for boat hulls are made with a split seam down the middle. Each side of the hull can be laminated more down hand, leaving some of the laminate layers out of the keel and stem area. Then the molds are bolted back together and the rest of the laminates are used to join the hull sides together.

Matched female molds are used for resin transfer and resin injection molding. These are heavily formed tools to take the pressure of the laminating methods. The extra tooling can be justified by multiple runs of smaller parts roughly 1m square.

Male molds are, simply enough, just the opposite of female molds. However, as it is the outside surface that is usually the exposed "show" surface, the mold does not need to be finished near as nicely, so it is easier to make the male mold. However, the outside surface must then be faired and painted. Male molds seem to be popular for mid size sailboat construction as many of these are custom projects, and working over a 10-20m hull is easier than making a female mold for one project and working inside.

One-Off and Custom Methods

A number of methods are available for making one-off and custom composite parts. Generally, if three or less parts are to be made, it is more cost effective to use a one-off method.

An early one-off method was described by Johannsen (Johannsen 1973). Sections at regular stations are cut out (usually wood) and placed on a sturdy frame. Battens are nailed over the stations to form a male shape. Then structural foam sheets or planks can be temporarily formed over and nailed to the battens. When the planks are all fitted, with the butts and seems bonded, they are attached to the battens from the inside with wood screws. The temporary outside nails or screws are removed and the holes plugged. The outside of the foam can be faired as necessary and is now a foam "plug." The required fiberglass skins are laminated to the outside and faired. Then the whole assembly of wood, foam and fiberglass is turned over and placed into and secured to a sturdy cradle. The wood forms and wood screws are removed, and the inside skin is laminated.

With the one-off method, the hull shell (or other part) is formed without going through the time and expense of making a male plug and female mold that may only be used once or twice.

Fig. 12. Bead and Cove Planks

Fig. 13. Bead and Cove Construction

A natural extension of this method is the bead and cove method perfected by Bilodeau (Bilodeau 1998). With a new kind of SAN (Styrene Acrylo-Nitrile) linear structural foam, Bilodeau discovered that the foam could be easily and reliably machined into different width bead and cove planks, shown in Fig. 12. The sections are made into the male frame based on the thickness and density of foam used. No battens are necessary. Each bead and cove plank is bonded to the previous plank line with specially made putty that is strong enough to bond the foam but soft enough to fair the foam shape. See Fig. 13.

As each line of planks is bonded together, the foam shell becomes quite stiff, eliminating the need for battens. Some temporary fasteners are used, but many fewer than with the previously described method. Wider planks are used in areas of less curvature. Plain flat sheets are used in flat or mostly flat areas. The foam is faired and the laminates applied as before. One-off construction requires extra fairing and painting, but is still efficient for custom work.

Build-Up From Panels

Large parts can also be made by joining a number of flat or nearly flat panels. Westport Shipyard in Westport, Washington used to make 3m x 9m gelcoated FRP panels on a large flat glass table. They could pick this panel off the table and place it into a basic female slat frame. Some laminates were left out of the butts, which were joined later with additional layers, similar to joining hull halves. Then they had a finished outer skin in a female frame to which they could apply core material, the inner skin, and the rest of the structure. They had variable beam bow molds, hinged at the stem, to allow the building of a number of different sized hulls from the same tools.

ATL Composites in Queensland, Australia have developed the Duflex® system to build [mainly boat] structures out of joined, pre-laminated, flat sandwich panels. Standard panels are 1.2m x 2.2m with a "Z-Joint" machined into the edges. This joint has a 15-1 scarf in the top laminate, a square butt for the core, and another scarf for the bottom laminate. They can be NC cut at the factory into numbered pieces that are joined at the job site using epoxy adhesives.

There are many variations of the methods mentioned. Easily shapeable composite materials are an enabling technology.

MODULAR CONSTRUCTION

The same kind of advanced and reliable adhesive technology that allows one-off methods such as Duflex also allows modular construction in composites. The VISBY corvette (Kockums) is being built with flat sandwich panels with some material held back at the butts and seems. Then additional material was placed into the joints and they were infused with vinyl ester resins. A similar method is used to join major hull sections.

Some key optimization decisions need to be made to determine if the structure is to be built as large parts or modular units later joined together. Continuous laminating FRP allows large parts with an uninterrupted smooth surface. Steel parts must be built up from limited plate sizes that can be butt welded efficiently and quickly. Composite parts usually require additional reinforcements at large joints.

A compromise solution is for large hull parts to be manufactured in a continuous fashion with smaller parts such as bulkheads, deck sections and superstructure sections planned into separate modules. Modular construction is a relatively recent feature for most large yacht builders.

One type of modular construction used by many pleasure boat builders and fabricated (multiple) parts suppliers is to use molded grid sections for the framing system. A separate mold is made to include the longitudinal and transverse framing system for a boat. It is generally easier to laminate on this mold than down inside a female mold. Then the liner or grid is removed from the mold and bonded to the shell to form the stiffened part. Some of these grid molds are male molds that also contain a partial liner for the hull. When installed, there is a smooth gelcoated surface on the inside.

Most of the advantages of steel or aluminum modules, such as zone outfitting, inside shop construction, and zone painting; can be realized with composites. The main consideration for composite modules is that they are likely to be less stiff than similar sized metal modules, so heavy outfitting attachments and rigging for movement must be specially considered.

Accuracy control can be a problem as the various thermoset resins used for most polymer composites will shrink some after lamination. However, the edges are easy to trim and the butt and seem joining adhesives and laminations can fill moderate gaps of a well-designed scarf joint.

Lay up Techniques

Larger vessel hulls are typically laid up with a transverse shingle lay up. Shingle lay ups are employed to maximize the coverage at a single work area and maintain primary bond status of the part. The shingle lay up is employed by laying the new ply with a minimum of two inch overlap of the preceeding ply, and three layers wet on wet. The overlap on subsequent layers should be no less than six inches away from the overlap underneath.

Three layers wet on wet is typically the maximum limit due to high exotherm in the areas of overlap. Tied to the exotherm and number of wet on wet pairs variable is the promotior and catalyst levels of the resin. The gel time of the resin used in the impregnator ideally should be long to avoid resin from gelling in the impregnator bath. Both the gel results and promotional system should be long and cold for the introduction of multiple layers of laminate on the work surface to avoid excessive exotherm. The gel time for lamination on vertical surfaces should be short to prevent the resin from running out of the material as it stands on the hull sides. Alternatively, a thixotropic agent can be added to the resin to retard run out, but it has adverse effects on the workability of the resin, which may again cause voids.

A primary bond should be achieved in the hull shell structure as the large vessel will be subject to high primary bending stresses. Using general purpose polyester resin systems the maximum amount of time that should be allowed between layers is 24 hours in order to achieve a primary bond. The reason behind the primary/secondary bond definition is the supposition that after a laminate cure of 24 hours, a subsequent laminate polymer bond will not be initiated with the previous laminate. This causes a lower interlaminar shear strength value in the final matrix. This phenomenon is due primarily to the lack of continuation in the polymer chain reaction exhibited most frequently in general purpose polyester resin systems.

In vinyl ester systems, the polymer chains are typically less affected by the cure time from laminate to laminate and will in fact bond very well between laminates laid upon each other after the initial layer has cured and exhibits barcol hardness of over 50. Barcol hardness is generally considered an indication of the level of resin cure. Testing has shown that there is no loss in interlaminar shear strength in vinyl ester laminates made with up to seven days elapsed between laminations. (Bernhard 1994) (Juska 1996)

MATERIAL APPLICATION

High volume lamination applications where large surface areas are to be covered, such as the hull, decks, superstructure, bulkheads and platforms, do not lend themselves to the use of classic production boat building lamination techniques. Small part lamination is typically done by hand lay-up, open faced molding, with the material brought dry to the work area and wet out with catalyzed resin using a chopper gun. The material is either wet out in situ or next to the part and carried to the part and applied. This technique suffers many disadvantages while building larger parts, including poor quality control, little glass to resin ratio control, logistical problems transporting the reinforcing material and a high rate of VOC emissions, because the resin is atomized and subject to high velocities at the gun head.

One of the nice advantages of composites is that they can be easily formed into complex shapes. The materials are applied with a liquid resin that, with the right fabric for the job, can be easily sprayed or applied to a mold surface. Special spray fillers and putties can fill or build up more difficult places. A brief look at the different material application methods follows. Scott (1996) and Greene (1999) have more in depth descriptions with multiple photographs and diagrams.

Resin Mixing

Polyester and vinyl ester resin mixing can be done a number of ways. The simplest and most common is to measure the proper volume of resin into a bucket, then measure the right amount of catalyst out and stir it into the bucket. On a nice 21°C day, the laminator may get 1/2 to 3/4 hour working time before the resin starts to gel. That is IF the proper amounts were measured, the resin is really at room temperature, the catalyst chart was read properly, and the catalyst is fresh, all among other variables. This is often where the laminating process that seems simple from the outside becomes a bit complicated.

Catalyst ratios for poly and vinyl ester resins are usually on the order of 1.5% to 2.5% of Methyl Ethyl Ketone Peroxide (MEKP), sometimes Benzoyl PerOxide (BPO), adjusted for the shop temperature and laminating method. Too much catalyst, and the resin can kick too quickly, lose some of its physical properties and cause too much exotherm. Too little and the resin will not cure right. The resins can react differently with different makes of catalyst. Mixing by weighing the liquids instead of judging volumes is better, but there are still many variables to keep track of. Thus the use of internal mix laminating equipment is becoming more popular.

Epoxy resins are easier to mix, as they are more nearly equal parts of resin and hardener. Epoxies give off fewer HAPs, but exposure can sensitize workers to the point they can no longer be around the resin.

Sprayup

The most basic of lamination methods is to spray the resin from a mixing gun into (or onto) the mold. Some guns chop and blow short glass fibers into the stream, chopper spray gun (see below). Some are just used to get resin onto the part to wet out dry fabrics. Older guns mix the catalyst in the air outside the nozzle. Newer technology mixes the catalyst in the gun near the nozzle to give better mixing. This provides for quick easy coverage of the mold or part. However, all the tiny droplets also present a lot of surface area to the atmosphere and produces higher Hazardous Air Pollutants (HAPs). Volatile Organic Compounds (VOCs) was the previous defining negative term for the airborne byproducts of laminating with organic resins. Either way, reduction of airborne pollutants is quickly changing the lamination methods of the composites industry.

Chopper-Spray Gun

A chopper gun is a special spray gun that has rotating knives that chop continuous glass strand rovings into short glass fibers. It is easy to form these wetted fibers into various shapes. Typical of products produced solely from chopper guns are fiberglass bathroom fixtures and hot tubs. Because there are no binders with chopped glass as there is with fiberglass mat, there are advantages of having a chopped skin coat on a boat for blister protection. A chopped laminate generally has a low glass and low strength, but is quite economical, apart from controlling the HAPs.

Hand Layup

Hand layup is characterized by manually wetting dry fabrics and consolidating this wetted fabric into the mold. Fabrics can be back wetted, where resin is first applied to the part, then the dry fabric laid in, then a serrated roller used to force the fabric down into the resin so that the resin flows up through the fabric and forces the air out. Resin drainage before fabric application must be watched to make sure there is enough resin to properly wet the upper portion of a vertical surface. Resins can be formulated with a thixotrope to keep them from draining too fast, another variable to be accounted for. Glass content can be anywhere from 30% to 50 % depending on the type of fabrics used and the skill of the laminators.

With the messiness involved in hand laminating methods, workers are usually clothed in throw-away protective suits (Tyvek coveralls are popular) including booties, most wear some kind of respirator.

Impregnating Systems

The most common tool used to overcome the disadvantages of open faced molding is the impregnator. Resin impregnators are sets of rollers that hold a pool of catalyzed resin. As the dry fabric passes through the resin pool, it gets wetted and is passed down through the machine ready to be placed into or on the part. Fabrics can be cut to strips and dropped down on top of stringers too. HAPs are reduced as there is no spraying and less exposed resin. Glass content is higher than hand layup and can range from 45% to 65% glass by weight, depending on the type of machine, the type of fabric, and how the machine is set. Worker exposure is reduced, but protective measures are still usually needed.

The impregnator is an indispensable tool to lay-up large vessels with large panels. The impregnator system is in use throughout the United States by builders such as Westport Shipyards for hard chine planing craft with few non-developable surfaces. It also proves to be quite successful in the lamination of round bilged trawler type hulls. The impregnator is used to wet out the reinforcement, usually suspended on overhead cranes with rotating turrets from a bridge above the work area so wetted fabric can be placed at any angle in a part where the material is dropped slowly, and continuously to the working surface.

Seven roll impregnators are commonly used feed the material and assure quality laminate, which possesses the desired glass to resin ratio and is free of air voids. The seven roll impregnator wets out the material in the bath then presses the material through a series of rollers to insure proper bundle impregnation. The final set of rollers, with a micrometer clearance adjustment, squeezes the excess resin from the laminate to achieve the proper glass/resin ratio. See Fig. 14.

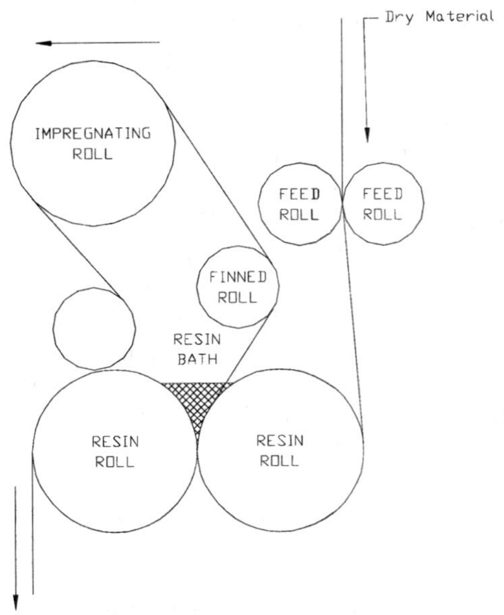

Fig. 14. Seven Roll Impregnator

Limitations of the impregnator include the lamination of hard to reach and work areas such as the stern and keel. Areas which historically involve high labor intensity to eliminate bridging and resin pools are not good candidates for the impregnator because the time involved to work the areas increase the likelihood or the resin "kicking off" in the bath. This causes a loss or the resin volume in the bath (about five gallons or 20 liters) and the necessity to unload the reinforcement from the rollers. (Bernhard 1994).

Resin Infusion

With infusion, dry fabrics and/or cores are situated in the mold then enclosed in a vacuum bag. The vacuum pulls catalyzed resin into the fabrics and core, from many ports for large parts, until the whole part is totally wetted out. This method gives a high glass content (60-70%), a generally void free laminate and a reliable bond to the core in a sandwich laminate. However, infusion is not for the unskilled and the weight savings in the skins is partially offset by the weight gain in the core.

The most common of the infusion methods is the Seemann Composites Resin Infusion Molding Process (SCRIMP). This is a patented process that uses a number of vacuum and resin ports along with a distribution medium to pull resin into the part. There are a number of variations on this theme, but the basic principle is the same. Infusion reduces HAPs and makes for a worker friendly environment.

Prepregs

With pre-impregnated fabrics, the catalyzed resin is already in or on the fabric. Some forms of prepregs must be refrigerated until they are placed in the mold, but then they cure very slowly at room temperature giving a lot of working time. They are usually supplied with epoxy resins and carbon fiber reinforcements, and used in more advanced projects such as aircraft and racing yachts.

The basic approach with prepregs is to apply the fabrics to the part in multiple layers to suit the design. Then a release film, bleeder cloth and vacuum bag (or sealing bag for an autoclave) is placed over the part, vacuum is pulled (or pressure is applied) and heat is applied. Cure temperatures can range from 80- 135oC for anywhere from 1 to 10 hours, depending on time, temperature and pressure.

Prepregs yield high reliable glass contents. They are the most expensive form of material, but there is little waste. Additional adhesive films must be used to provide the extra resin to bond to core materials. As a closed molding system with very little extra solvent for the resin, there are no HAPs.

Quality Control

Quality Control in composite construction is more challenging than for steel or aluminum mainly because non-destructive test (NDT) methods for metals are fully developed and cost effective. The outward appearance of a weld is a fairly good indication of its quality. The outward appearance of a multi-layered composite E-glass laminate reveals that the surface layer has a reasonable glass content and that there are an acceptable number of air inclusions in the surface and first subsurface laminates. Past that, there are no economical NDT means of determining the quality and strength of the laminate. Hammer testing is a crude way of testing for voids and lack of bonds. Moisture meters can check for water content of the laminate. Thermography can see through a laminate to the core but not past it. X-rays and certain ultrasound techniques can show some additional defects. Special fibers can be bonded into a laminate to show a degradation in strength beyond a norm.

However, the better methods are not cost effective for the majority of builders. Basically, the marine industry can not afford the level of QC common in the aerospace industry. The best QC is to ensure quality in the building process. The latest guide from the American Bureau of Shipping (ABS 1997) has an extensive section on needed QC measures to achieve quality construction.

SUMMARY

The use of composites in the marine industry has a long and varied history. Few doubt the advantages of composites for small craft. The hanging question is how large can a plastic vessel be and still be a small craft. The 57m US Navy MHCs are surely not small craft. Sweden has extended the composite size range up to 73m with the VISBY Corvette.

Over 30 years ago a GRP freighter was researched (Scott 1971). There were a number of problems identified, the main ones of which were fire protection and global stiffness. Fire protection concerns have been addressed in modern passenger and military vessels (Høyning 2002) where the need for high speed encourages the use of light weight composites, albeit with additional safety measures. Global stiffness can be addressed by a number of means, including rearranging the structural shape to be stiffer (Horsmon 1994) or through extensive use of carbon fiber (Kockums).

Composite vessels can be lighter and better performing than aluminum (Horsmon and Bernhard 1999). There are obviously certain advantages and disadvantages to every material. The evolution of composite materials, design, and products continues.

This paper is only introductory. The interested reader is encouraged to investigate additional information in the references and related materials.

REFERENCES

American Bureau of Shipping Rules for Building and Classing Fiber Reinforced Plastic Vessels, 1978, ABS Americas, New York, NY.

American Bureau of Shipping Guide for Building and Classing High Speed Craft, 1997, ABS Americas, New York, NY.

American Bureau of Shipping Guide for Building and Classing Motor Pleasure Yachts, 1990, ABS Americas, New York, NY.

Bernhard, B., “First of a Class: Production of Large Military FRP Displacement Hulls,” *MACM Proceedings*, April 1994

Bilodeau, A., “Building Your Hull with the Bead & Cove System Using Core-Cell Foam,” Published by the author, St. James City, FL, 1998.

Bohlmann, R. E. and Fogarty, J. H., “Demonstration of a Composite to Steel Deck Joint on a Navy Destroyer,” *Marine Applications of Composite Materials (MACM)* 2002, Melbourne, Fl, March 2002.

Cahill, P., "Composite Materials on Naval Surface Combatants: The Integrated Technology Deckhouse Project," *Journal of Ship Production*, February 1992.

De Fever, A., “Restoration of the Riveted Steel Hull of the Steam Yacht MEDEA,” presented to the San Diego SNAME Section April 3, 1991.

Fox, John, “Selecting Structural Cores,” *Professional Boatbuilder*, No. 52, April/May 1998.

Greene, E., "Marine Composites," Ship Structures Committee Report SSC -360, June 1990, revised to a 2nd edition by Eric Greene and Associates, Annapolis, MD, 1999.

Horsmon Albert W. Jr , David Eby and Ronald C. Averill “FEA Sandwich Panel Analysis Using Zig-Zag Theory – Compact, Accurate, Verified, Demonstrated,” *Fifth International Conference on Sandwich Construction*, Zurich, September, 2000.

Horsmon Albert W. Jr., and Bernhard, B., “The Advantages of Advanced Sandwich Composites over Traditional Aluminum Construction,” *FAST 99 Conference,* Seattle, WA, August 1999.

Horsmon, A. W. Jr., "Permanent Composite Cladding of Deteriorating Steel Hulls," *Journal of Ship Production*, May 1992.

Horsmon, A. W. Jr., “Composites for Large Ships,” *Journal of Ship Production*, November 1994.

Høyning, Bjørn, “Meeting Commercial and Military Requirements for Passive Fire Protection of Composite High Speed Craft,” *Marine Applications of Composite Materials (MACM)* 2002, Melbourne, Fl, March 2002

Johannsen, T. J, “One-Off Airex Fiberglass Sandwich Construction,” *Chemacryl Plastics*, Toronto, Ontario, 1973.

Johannsen, T. J, A. W. Horsmon Jr., R. Mazza, and J.R.D. Benkelman, *ATC Core-Cell Design Manual*, ATC Chemicals Inc., Oakville, Ontario, February, 2001.

Juska, T., "Effect of Water Immersion on Fiber/Matrix Adhesion", CARDIV-SME-92/38, Jan 1993

Juska, T., "Properties of Marine Laminates", Advanced Composites for Future Surface Ships Program, Organic Composite Ship Structures Project, 1996.

Kockums Karlskronavarvet, .Karlskrona “VISBY Class Corvette, The True Stealth Vessel,”, Sweden. http://www.kockums.se, 2003

Mazza, Robert and Lou Codega, “The Cored Bottoms Controversy.” *Professional Boatbuilder*, No. 51, February/March 1998.

Nguyen, L. B., Kuo, J. C., Critchfield, M. O. and Offutt, J. D., "Design and Fabrication of a High-Quality GRP Advanced Material Transporter," *ASNE Small Boats Symposium*, May 26-27, 1993.

"Of Material Interest," *Advanced Materials & Processes*, Vol. 142, No. 1, p4, July, 1992.

“Opening the Gates of Opportunity for Composites on the Waterfront,” *Composites Technology*, Vol. 5, No. 2, March/April 1999, pp 27-31.

Roberts, Boyle, Wienhold, Ward & White, “Strains and Deflections of GFRP Sandwich Panels Due to Uniform Out-of-Plane Pressure,” *Marine Technology*, October 2002

Scott, R. J. and Sommella, J. H., "Feasibility Study of Glass Reinforced Plastic Cargo Ship," *Ship Structures Committee Report SSC -224*, 1971.

Scott, R. J., "Fiberglass Boat Design and Construction," SNAME, 1996.

USCG, “Navigation and Inspection Circular 8-87 – Notes on Design, Construction, Inspection and Repair of Fiber Reinforced Plastic (FRP) Vessels, USCG, 1987.

Quality Management of Formal Safety Assessment (FSA) Process

J. Dasgupta
Indian Register of Shipping

ABSTRACT

*The **Formal Safety Assessment (FSA)** process, proposed during MSC 66 at the IMO and subsequently released as Interim Guidelines (MSC/Circ.829-MCPC/Circ.335) has generated a number of applications over the past few years, particularly in MSC 73, MSC 74, MSC 75 and MSC 76. These applications cover a wide area of interests in the marine field, most important one being the FSA Studies on Bulk Carrier Safety by various Administrations and IACS. It becomes imperative that the FSA studies being conducted now towards the development or upgrading of future Rules are based on systematic quality management principles, to avoid a wide variation in the quality and conclusions from the FSA process. The five-step FSA process is considered suitable for application of the "process approach" as per ISO 9001:2000 requirements. This application of the Quality Management Standard would ensure that the quality objectives of each of the Steps are identified, and attained through use of appropriate techniques and data by the FSA team. The "process flow" during each of these steps are required to be worked out, and the pertinent issues viz. necessary inputs, activities, sub-processes, critical parameters, controls and measurements are to be identified to arrive at the desired outputs. Such a 'process approach' would assure transparency of the activities behind the FSA as well as reliability of the concluded results.*

The paper presents a proposal for the application of the above quality management methods to the FSA procedures using a 'process approach'. Typical 'Process Flow Diagrams' for the various steps of FSA process along with the quality parameters are included in the paper. A proposal for internal validation of an FSA application prior to submission to the decision makers is also covered in the paper.

1. INTRODUCTION

Formal Safety Assessment (FSA) is a rational and systematic process for the proactive management of safety based on principles of hazard identification, risk analysis and cost-effectiveness evaluation of the efforts in controlling the risks. FSA can be used as a tool to help in the development of new safety regulations or in analysing an existing set of regulations, and thus achieve a balance between various technical and operational issues, including human element and costs.

The basic methods of the FSA are the various safety management tools developed in the 70's'and 80's' These have been adopted in a number of countries mainly by the industries like nuclear power, chemical process plants, offshore petroleum and some aspects of aerospace engineering. Traditionally, the maritime industry has been reactive in its development of rules and standards for ship safety. Table 1 lists the major maritime casualties that have resulted in IMO Rule Developments over the past few years.. Two notable contributors to the change in industry's attitude were the Lord Cullen Report on the Piper Alpha (1990) and the Lord Donaldson Report on the grounding of the tanker Braer (1994). As a result of the recommendations of the Lord Cullen Report, offshore platforms operating under the UK administration are required to perform safety cases for each individual platform. The Lord Donaldson Report that followed recommended that the maritime industry consider a similar approach for ship operation (Lord Donaldson, 1994).

The essential aspects of FSA lie in the proactive control of risks [1]*. The '*Interim Guidelines for the Application of Formal Safety Assessment (FSA) to the IMO Rule-Making Process*' was adopted at the 68th. Session of MSC held in May 1997 and was circulated as MSC/Circ.829 (MEPC/Circ.335) in December 1997 [2]. These guidelines were intended to facilitate trial applications of the FSA process. The trial applications covered a wide area of interests e.g. High Speed Crafts, Helicopter Landing Facilities on Large passenger Ships, Hazard Identification of Ballast Water Exchange, Safety of Lifeboat Launching Devices and most importantly the FSA Studies on Bulk Carrier Safety by various Administrations and IACS. The Interim Guidelines were revised by IMO during MSC 74, 2001, and released as the revised IMO FSA Guidelines MSC/Circ.1023-MEPC/Circ/392 published in 2002 [3]. It is likely that many more applications of FSA will now be carried out and submitted to the regulatory authorities for acceptance.

2. FSA PROCESS [2], [5]

The modalities of FSA application to shipping have now been established. This generic approach is designed to comprise the following steps (Fig. 1):

a) Hazard identification;
b) Risk Analysis;
c) Risk control option
d) Cost benefit assessment; and
e) Decision making recommendation.

The characterisation of hazards and risks should be both qualitative and quantitative, consistent with the available data, and should be broad enough to include the range of options for reduction of risks.

A typical FSA exercise in a ship type according to the IMO Guideline would proceed as follows:

Problem definition:

The problem under analysis and its boundaries should be carefully defined. While defining the problem, the following parameters may be considered relevant;

a. Ship category (e.g. type, length or gross tonnage, new or existing etc.,)
b. Ship systems (e.g. type layout, subdivision, propulsion, etc.)
c. Ship operation (e.g. in ports and/or during navigation, etc.)
d. Accident category (e.g. collision, explosion, fire, etc.)
e. Risk category (e.g. injuries and/or fatalities to passengers and crew, environmental impact, damage to ship or port, etc)

Table 1: Major Maritime Casualties which Influenced Rule Development

March 1967	Torrey Canyon	West Coast of England
March 1978	Amoco Cadiz	Northern Coast of France
September, 1980	Derbyshire	North Pacific
March 1987	Herald of Free Enterprise	Zeebrugge
March 1989	Exxon Valdez	West Coast of Alaska
April 1990	Scandinavian Star	Skagerak
January 1993	Braer	The Shetland Isles
September 1994	Estonia	South of Uto (Finland)
February 1995	Sea Empress	Milford Haven
January, 1998	Flare	Nova Scotia
December 1999	Erika	West Coast of France

* [ ] denotes References at the end of the paper.

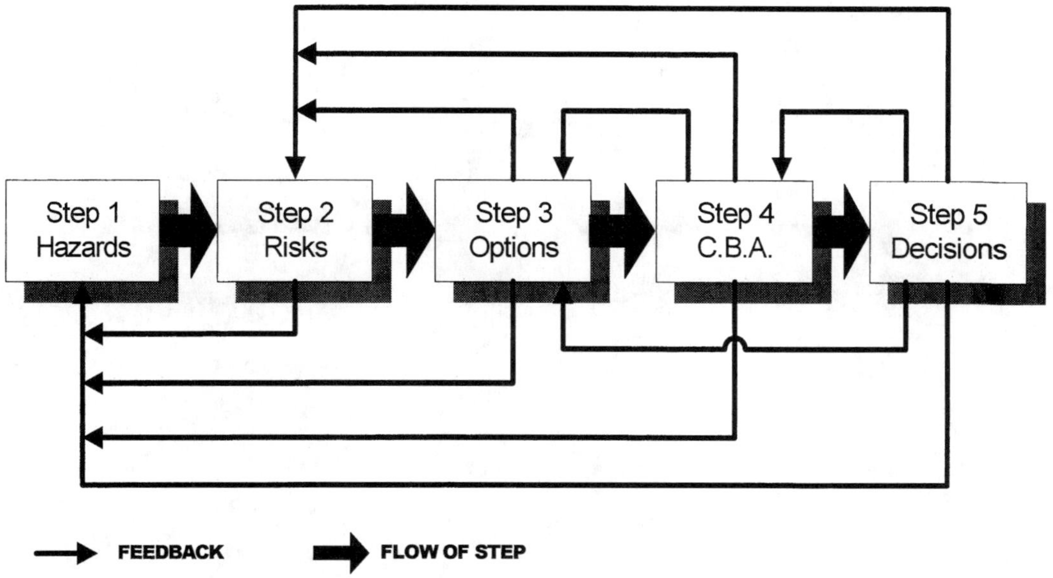

Fig. 1 Information flow in FSA Process [4]

Prior to application of the FSA steps additional information would require to be compiled on the following:

- Identification of existing design concepts and review of existing rules/regulations
- Identification of existing operational procedures/concepts
- Compilation of materials under consideration and their properties.
- Identification of involved parties responsible/liable for safety.

In general, the problem under consideration should be characterised by a number of functions. Where the problem refers to a type of ship, these functions include carriage of payload, emergency response, communication, manoeuvrability etc. Where the problem relates to a type of hazard, for instance Fire, the functions include prevention, detection, alarm, containment, escape, suppression etc. It is imperative that a comprehensive view is taken of the ship 'hardware' (i.e. technical & engineering system) dynamically integrated to the 'software' (i.e. human behaviour governed by organisation & management infrastructure).

Identification of Hazards (FSA Step 1):

The purpose of Step 1 is to help identify the causes of accidents and to generate a prioritised list of accident categories, specific to the problem under review. There are several methods available to document this identification exercise e.g. Brainstorming, HAZOP, FMEA followed by qualitative Fault Tree Analysis (FTA) and Event Tree Analysis (ETA) for specific cases. The accident categories are screened and prioritised discarding scenarios judged to be of minor significance using a risk matrix or similar other technique. The screened and prioritised list of accident categories provide the input for the risk contribution tree in Step. 2.

Risk Analysis (FSA Step 2):

Risk is the combination of frequency and consequence. In this FSA Step the risk associated with an accident data is assessed by constructing and quantifying a diagram called '*the risk contribution tree*', based on accident data and expert judgement to display the distribution of risk. The fault trees and event trees from Step 1 are quantified in developing the risk contribution tree (Fig. 2).

The purpose of Step 2 is to identify the distribution of risk, thus allowing attention to be focussed upon high-risk areas, and to identify and evaluate the factors, which will influence the level of risk. Thus Step 2 aims to establish the relationship between the regulatory provisions and the occurrence of accidents, enabling appropriate regulatory changes to be introduced/amended to reduce risk.

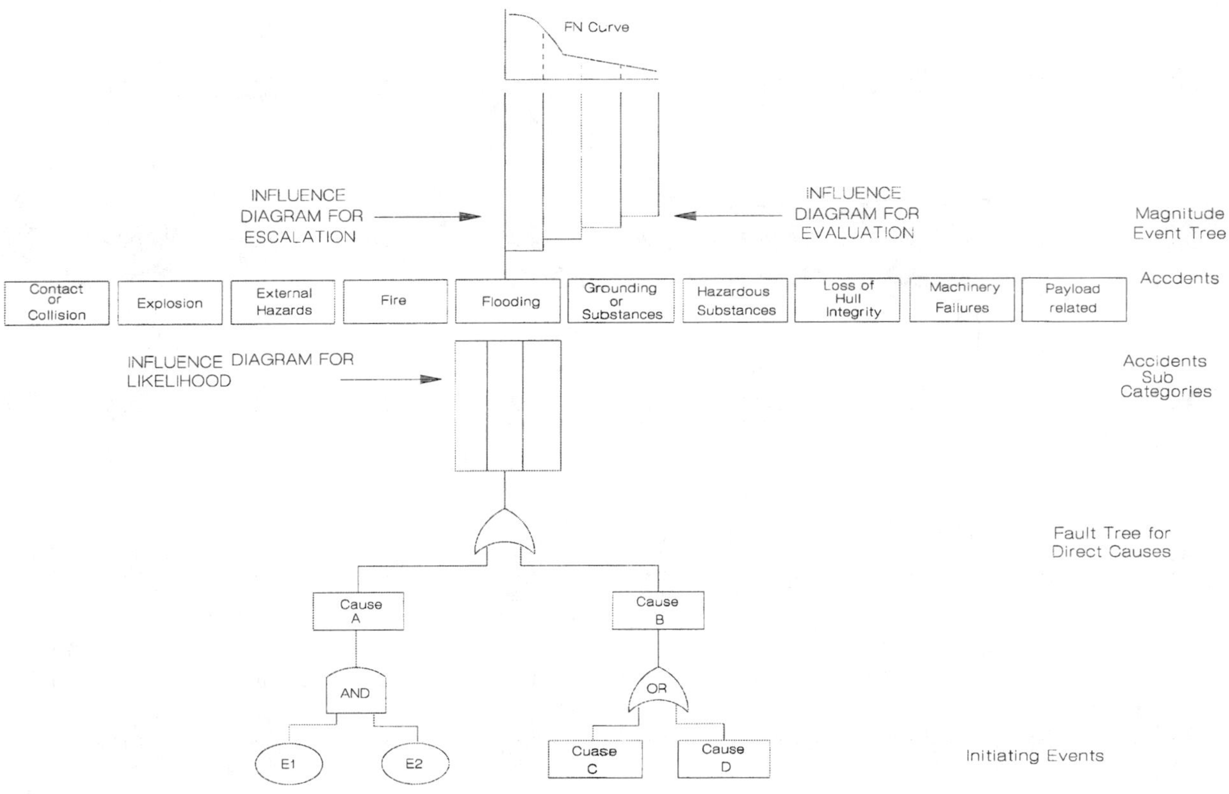

Fig. 2 The general structure of the risk contribution tree[2]

Risk Control Options (FSA Step 3):

A risk control measure is a method of controlling single element of risk and the risk control option is appropriate combination of risk control measures. FSA Step 3 comprises of 3 stages:

- Attention on areas needing control by identifying areas of the risk contribution trees representing risk, high consequence, high frequency and low confidence.
- Identification of new risk control measures not controlled sufficiently by existing measure. The techniques used include use of Causal Chains & Risk Control Attributes to encourage development of appropriate measure at a selected control point in sequence of: **{Causal factors >> failure >> circumstance >> accident >> consequence}**
- Derivation of the risk control options i.e. to first decide whether preventive or curative measures are to be adopted; and then group the suitable risk control measures as per the chosen approach to produce a number of possible and practical risk control options.

The results from Step 3 comprises a range of risk control options, which are again assessed for their effectiveness as per Step 2, and then passed on to Step 4 for assessment of costs and benefits.

Cost Benefit (Effectiveness) Analysis (FSA Step 4):

When the basic risk is estimated to be within an "as low as reasonably practicable"(ALARP) region, Risk Control Options (RCOs) should be developed and Cost Benefit Assessment (CBA) or Cost Effectiveness Assessment (CEA) should be carried out. CBA or CEA may be used to select reasonably practicable risk reduction measures. The purpose of the assessment is to provide a basis for decision-making about RCOs. In a conventional CBA, the acceptance criterion is simply that the benefits outweigh the costs. In CBA the analyst converts all risks to monetary units. CEA presents a ratio of costs to benefits, and avoids putting a value to the benefit (e.g. life saved). The value judgement is left to the decision-maker when deciding which RCOs to implement.

Decision Making (FSA Step 5):

The purpose of this Step is to compare and rank the risk control options on the basis of the principle of ALARP (As Low As Reasonable Practicable). The comparison should account for all those entities that are significantly influenced in the area under concern and ensure that they are equally affected by the proposed new regulation as far as practicable. The scope of this Step is also to provide a feed back to review the information generated in the previous steps so as to re-iterate the process if so required for decision making.

The long-term prospects of FSA implementation provide for various options for applications. These options, apart from Rule making at IMO also provide for individual Administrations or Shipowners to apply FSA as a management tool in controlling risks as a part of the Safety Management System (SMS).

The five-step FSA procedure is a combination of a number of sub-processes which are enumerated in the IACS Training Manual, and are given in Fig. 3.

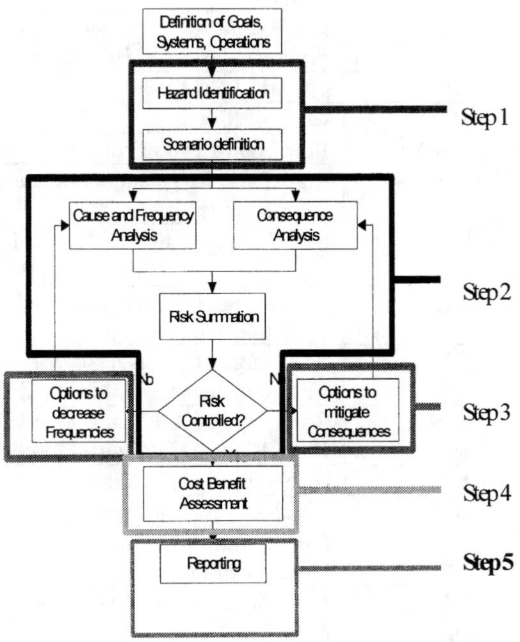

Fig. 3 Flow Chart of FSA Process [6]

In 1999 the International Association of Classification Societies (IACS) submitted a paper to IMO providing guidance on how to incorporate the human element through the application of Human Reliability Analysis (HRA). The proposal by IACS was accepted in principle at the MSC 72 Session in May 2000 and it was decided to annex Guidance on the application of HRA to the Interim Guidelines [7]. The MSC 72 also approved the following work plan on FSA matters:

- finalization of the Guidelines on application of FSA to the IMO rule-making process;
- further consideration of the integration of the human element and FSA into the IMO rule-making process;
- development of risk evaluation criteria with regard to maritime safety and the protection of the marine environment; and
- development of a training package for the establishment of a basic understanding of the FSA process.

In addition to the work being carried out on bulk carrier safety, in which FSA studies play an important role. The first two issues have been completed through the publication of the revised Guidelines (MSC Circ. 1023), and IACS has already prepared and commenced the FSA Training Programmes through the Classification societies. The third issue however, remains to be finalized.

3. APPLICATIONS

Since the publication of the IMO 'Interim Guidelines', various organizations and authorities have conducted a number of studies. These FSA related studies have endeavoured to substantiate the regulatory initiatives and include the following:

a. *FSA Studies Reported to IMO:*

- Formal Safety Assessment of Helicopter Landing Area by DNV & International Council of Cruise Lines (ICCL).
- Formal Safety Assessment of Helicopter Landing Area by Italy.
- FSA Illustrative Case Study regarding accidental pollution from crude oil tankers.
- Bulk Carrier FSA co-ordinated by UK MCA
- Bulk Carrier FSA conducted by Japan
- Bulk Carrier FSA conducted by Korea
- Bulk Carrier FSA conducted by Norway / ICFTU
- Redundant propulsion study submitted by Germany
- Joint Nordic Project on Safety Assessment of HSC Operations (Sweden)
- High Speed Catamaran Ferries FSA submitted by UK
- Trial application of FSA to the transportation of dangerous goods on passenger / ro-ro cargo vessels (Finland)
- Research project on risk assessment (Japan)
- Risk analysis of large passenger ships by the United States

b. *IACS studies*:

- Ballast Water Exchange HAZID
- Bulk Carrier FSA Work Package 5 (watertight integrity of fore end)
- Tanker Pump-room study

c. *Individual Class Societies studies:*

- DNV Cost Benefit Analysis of Safety Measures for Existing Bulk Carriers, to support IACS decision to Strengthen bulkheads between Hold 1 & 2.
- Formal Safety Assessment of Life Saving Appliances by DNV
- Joint North-West European Research Project on Safety Assessment of Passenger/RoRo Vessels
- Risk Analysis in connection with Solo Watch-keeping by DNV
- Concerted Action on FSEA by BV, DNV & GL on behalf of EEIG Unitas
- Paper on the application of FSA methodology to No.1 cargo hold flooding of Bulk Carriers, by Korean Register
- Loading and Unloading of Bulk Carriers by Lloyds Register
- Collaborative FSA Studies by CCS, IRS, KR & NK.

Many of these studies have actually resulted in the development / amendment to the Regulations. The IMO, based on the aforementioned FSA studies, decided that SOLAS regulation III/28.2 should only apply to ro-ro passenger ships and not to all Passenger ships above 130 metres as originally stipulated.

The major concerted application of FSA has been on improving the safety of Bulk Carriers(BCs). The MSC 75 & 76 have seen a number of submissions on Bulk Carrier safety and have accepted regulatory changes to SOLAS Chapter XII based on the FSA reports. These Bulk Carrier FSA studies sought to address the existing safety regime of Bulk Carriers given in Table 2.

The FSA study coordinated by UK-MCA (UK/INT) has been the most exhaustive and applied the FSA methodology to dry bulk shipping. The types of operations covered the complete dry bulk-shipping activities from loading to discharge terminals through passage at sea, including structural loading, stability & sea motions, ballast water exchange at sea, life saving, main machinery configurations etc. The study was initiated by considering an exhaustive range of accident & incident categories for which the data was available in commercial databases e.g. LMIS (LR-Fairplay); a total operating life of 145,582 vessel years were analysed to collect the historic data on Loss-of-life. (It is interesting to note that in keeping with IMO's safety objectives, all the BC-FSA studies have only addressed the risk of loss of life, and not the risk to property, where the FSA can be equally applied). The initiating events of recorded incidents leading to any form of loss of life are indicated in Figure 4.

The proximate initiating event for fatalities involving bulk carriers are indicated in Figure 5, and the derived *potential loss of life (PLL)* for each event are indicated in Fig. 6. The data, when compared to the individual risk of fatality to crew on different ship-types (Fig. 7) and the frequency of fatality per accident type for all ships (Fig. 8), indicated that fatality risk of crew on bulk carriers remained in the ALARP (*As_Low_As_Reasonably_Practicable*) region.

Table 2: Regulations on BC Safety

SOLAS	BC	Ch. XII – Safety Measures of BCs
SOLAS	ISM Code	Ch. IX – ISM Code
SOLAS	ESP	Ch. XI – ESP & IMO Res. A 744(18)
SOLAS	BC Code	Ch. II-2, VI & VII Code of Safe Practices for Solid Bulk Cargoes
SOLAS	LSA	Ch. III , LSA Code
SOLAS	IMDG	Ch. II-2 & VII
ILLC 1966		ILLC'66 & 1988 Protocol
MARPOL 73/78		MARPOL 73/78 and associated amendments
STCW '95		STCW Convention & Protocol
IACS URs	S12	Side structure of single skin BCs
IACS URs	S17	Longitudinal Strength in flooded condition
IACS URs	S18	Scantlings of Transverse WT Bhds considering Hold Flooding
IACS URs	S19	Scantlings of Aft Transverse WT Bhd of No. 1 Hold with Flooding
IACS URs	S20	Allowable hold loading of BCs
IACS URs	S21	Hatch cover scantlings
IACS URs	S22	DB Strength for No. 1/ No.2 Holds with flooding
IACS URs	S23	Implementation of IACS UR S19 & S22
IACS URs	S24	Detection of water ingress in cargo holds

The availability of the large data meant that there was little opportunity to synthesize an accident scenario, and the risk analysis was carried out based on the data analysis. The UK/International study identified following groups of risk control options (RCOs):

a. Closing devices e.g. hatch covers, pipes, valves etc. these were a combination of structural modifications and enhancements to reduce criticality of human factors;
b. Hull envelop i.e. side structure to ensure integrity of primary & secondary barriers;
c. Ship operations i.e. overall management of the ship including its interfaces with the shore-based stake-holders; and
d. Evacuation of the crew.

INCIDENTS

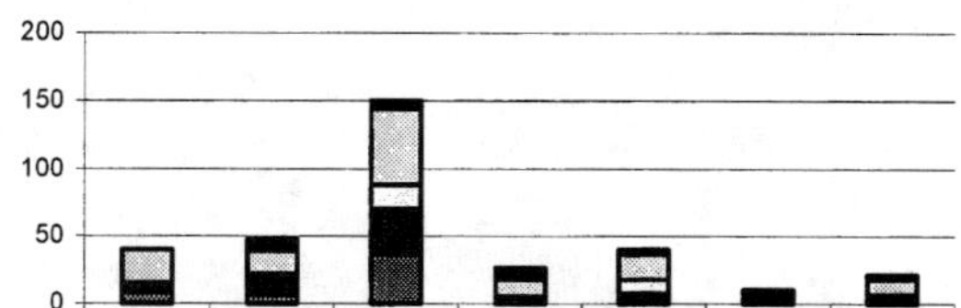

	Hatch Cover	Pipes & valves	Side shell	Side shell(colli	Bottom shell	Deck	Side shell Penetrati
Mini	0	8	6	8	4	1	3
Hsize	25	17	56	13	18	4	11
Hmax	4	3	18	2	10	0	3
Pmax	3	13	34	2	4	3	2
Csize	8	6	36	1	4	2	2

Fig. 4: Initiating events of recorded incidents [9]

FATALITIES

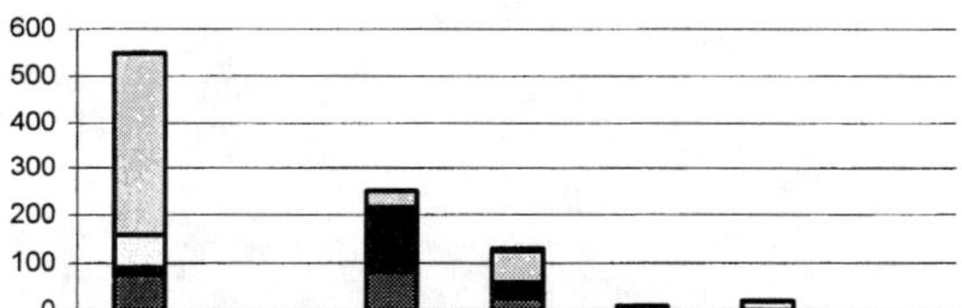

	Hatch Cover	Pipes & valves	Side shell	Side shell(colli	Bottom shell	Deck	Side shell
Mini	0	0	0	6	0	0	0
Hsize	387	0	33	66	6	21	0
Hmax	69	0	3	2	0	0	0
Pmax	17	0	131	29	4	0	0
Csize	74	0	82	27	0	0	0

Fig. 5: Proximate Initiating events for Fatalities[9]

Fig. 6: PLL Per Ship-Year For Each Event [9]

In view of the risk being in the ALARP region, the Cost effectiveness of the RCOs was given the sole governing criteria for selection and decision-making. The UK/INT study used the NCAF (Net Cost of Averting a Fatality) criteria, allowing for the benefits accruable to all stakeholders associated with bulk carrier shipping.

The Japanese FSA study on Bulk-carrier safety had also used the same casualty database, but arrived at a different set of recommendations for decision-making. It used GCAF (Gross Cost of Averting a Fatality) to arrive at the decision taking criteria.

The International Association of Classification Societies (IACS) had carried out a comprehensive study on the watertight integrity of the fore-end of bulk carriers earlier. The results from this study had been used in both UK/INT. & Japanese studies. IACS also took-up major initiatives and had introduced FSA Studies on the side shell integrity of bulk carriers.

A comprehensive list of the various risk control options are presented in Table 2. The number of recommendations made in all these studies, and the supplementary work carried out by different agencies make it quite difficult to analyse the recommendations. The IMO MSC/76 also decided to refer a few of these recommendations for further consideration by the DE Sub-committee prior to taking a re-look during MSC 77 or MSC 78.

4. REASONS FOR QUALITY MANAGEMENT

The enormous complicity of the FSA process, especially when applied to problems encompassing a number of stakeholders, engineering systems and cost variations, makes it imperative that the FSA is carried out in a transparent and systematic manner. It is also the basic premise that FSA be a tool for supporting the decision making process at IMO. It should make decision-making procedures more rational and provide a proactive and holistic approach, thus reducing the number of ad-hoc proposals and implementations.

FSA should also provide reliable information on hazards, risks, risk control options, their costs and benefits, in a rational, structured and auditable manner, in order to improve the decisions regarding the management of the risks identified.

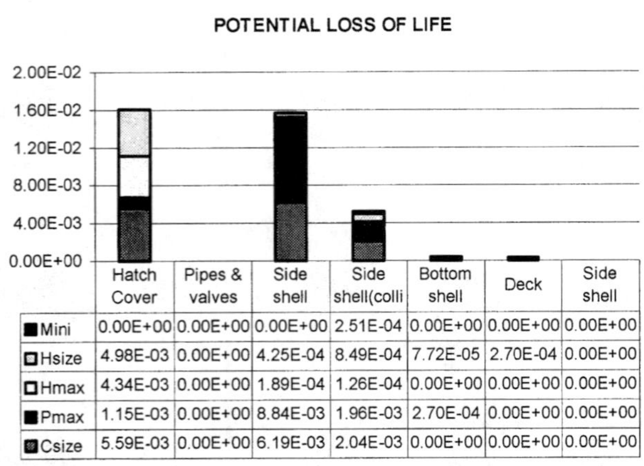

	Hatch Cover	Pipes & valves	Side shell	Side shell(colli	Bottom shell	Deck	Side shell
■Mini	0.00E+00	0.00E+00	0.00E+00	2.51E-04	0.00E+00	0.00E+00	0.00E+00
□Hsize	4.98E-03	0.00E+00	4.25E-04	8.49E-04	7.72E-05	2.70E-04	0.00E+00
□Hmax	4.34E-03	0.00E+00	1.89E-04	1.26E-04	0.00E+00	0.00E+00	0.00E+00
■Pmax	1.15E-03	0.00E+00	8.84E-03	1.96E-03	2.70E-04	0.00E+00	0.00E+00
■Csize	5.59E-03	0.00E+00	6.19E-03	2.04E-03	0.00E+00	0.00E+00	0.00E+00

Table 2: Risk control options in the Bulk Carrier FSA studies & their acceptance at IMO MSC 76

Sr. No	List of Recommendations	New Constructions					Existing Ships				
		UK/INT	IACS	ICFTU	Japan	MSC 76	UK/INT	IACS	ICFTU	Japan	MSC 76
1	Double hull - all holds	Y	Y		Y (HS)	Accepted	Y (CS)				Not Accepted
2	Improved hold coating	Y	Y			Not Accepted	Y	Y			Not Accepted
3	Tighter steel repair standards	Y				Accepted	Y				Accepted
4	Additional hold frame corrosion margins				Y	Not Accepted				Y	Not Accepted
5	Installation of Forecastle		Y			Not Accepted		Y (CS)			Not Accepted
6	Bulwark / breakwater	Y	Y			Not Accepted		Y			Not Accepted
7	Redesign of the Ballast system	Y				Not Accepted					Not Accepted
8	Strengthening the hold frames to UR S12					Accepted					Accepted
9	Better corrosion control of side shell					Accepted				Y	Accepted
10	Strengthening fore deck fittings	Y	Y			Accepted	Y				Accepted
11	Redesign hatch covers for both vertical & horizontal loads	Y				Not Accepted	Y				Not Accepted
12	Positive indicators for closed hatch covers / access hatches	Y				Not Accepted	Y				Not Accepted
13	Hatch covers to comply with UR S21					Accepted					Not Accepted
14	Free-fall lifeboats	Y		Y		Accepted	Y (CS)				Not Accepted
15	Water ingress alarms	Y	Y	Y		Accepted	Y	Y	Y		Accepted
16	Personnel immersion suits			Y		Accepted			Y		Accepted
17	Improving terminal interfaces	Y				Accepted	Y				Accepted
18	Risk-based ESP targeting	Y				Accepted	Y				Accepted
19	Training of PSC Officers	Y				Accepted	Y				Accepted
20	Weather routing & passage	Y				Not Accepted	Y				Not Accepted
21	Improved Loading/Stability Info	Y				Accepted	Y				Accepted
22	Mandatory BC Code	Y				Accepted	Y				Accepted

Notes: - HS : Handy Size, CS : Cape Size

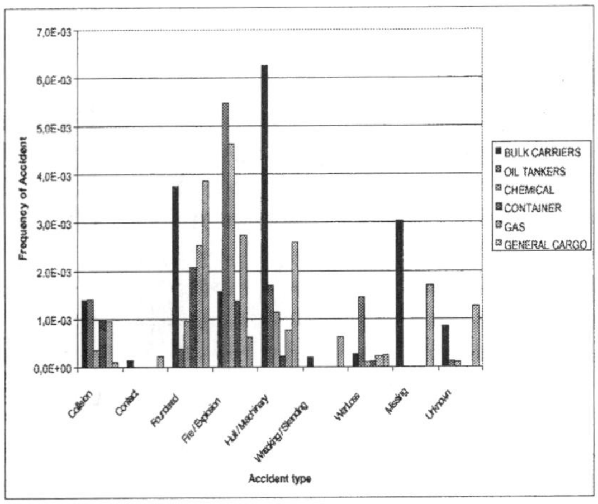

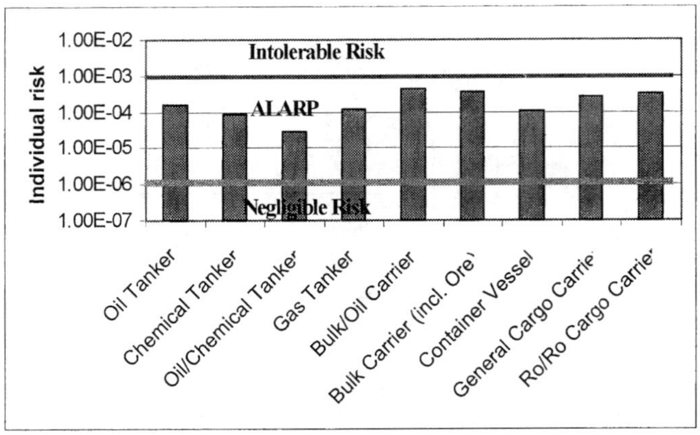

Fig. 7: Data analysis for HAZID[10]

Presently there exist the following options to evaluate the quality of a FSA study;

a) To carry out a complete parallel study on the same problem, at least a part of it towards validation
b) To compare the study to descriptions of accidents/incidents based on available data
c) Examining the process behind the study.

The first approach can be used for assessing the differences between methods and teams performing the FSA study, but would involve considerable resources. The second approach concerns evaluation of synthesized accident scenarios; all information on incidents, accidents and actual working procedures can be used as feedback to the recommended RCOs. The third approach is based on critical evaluation of the process – planning, decisions, expertise, resources etc. behind the FSA recommendations.

The MSC Circ. 1022 has provided a practical guideline for application of FSA. This document augments the MSC Circ. 1023 "Guidelines for Formal Safety Assessment (FSA) For Use in IMO Rule Making Process". Based on these two documents it is possible that the FSA applications are planned, executed and monitored under a Quality Management System. This would be an indirect evaluation as indicated in c) above, and should help both in establishing a faith in the entire FSA process, as well as in ensuring that the proposals put forward to the Authorities have received rigorous internal scrutiny. The application of the Quality Management Principles would imply that following aspects are taken care of:

1. *Customer Focus*: The specific requirement of the industry vis-à-vis a perceived risk is duly addressed
2. *Leadership*: The leadership in the FSA process establish an unity of purpose, create and maintain the internal environment to facilitate complete involvement of the team-members, especially during the brain-storming sessions
3. *Involvement of People*: The capabilities of the team-members are fully utilized
4. *Process Approach*: Desired results are achieved efficiently
5. *System Approach to Management*: Manage the interrelated Steps of the FSA Process effectively
6. *Continual Improvement*: Establish bench-marks of the achievement and improve upon them with each application of FSA
7. *Factual Approach to Decision Making*: The data and information are appropriately presented to facilitate decision making
8. *Mutually Beneficial Supplier Relationships*: Sub-contracted modules are duly evaluated and integrated.

5. ISO 9001:2000 QUALITY MANAGEMENT SYSTEM & PROCESS APPROACH [11]

The ISO 9000 series of standards are a powerful tool for the application of the Quality Management System. It defines a quality management as a set of interrelated or interacting processes that achieve the quality policy and quality objectives. The ANSI/ISO/ASQ Q9001-2000 clause 1.2 states:

> '*All the requirements of this International Standard are generic and are intended to be applicable to all organizations, regardless of type, size, and product provided.*'

An approach to the development and implementation of the quality management system consists of several steps including the following:

a. Determination of the needs and expectations of clients
b. Establishment of the quality policy and quality objectives
c. Determination of the processes and the necessary responsibilities

d. Determination of necessary resources
e. Establishment of methods to measure the effectiveness and the efficiency of each process
f. Application of these measures
g. Determining means of preventing non-conformities & eliminating their causes
h. Striving for continual improvement

Any activity that uses resources to transform inputs to outputs can be considered a process. Often, the output from one activity/process goes into another activity/process as an input. The systematic identification and management of the processes employed within an organization, and particularly the interactions between such processes, are referred to as '*process approach*'. The FSA related activities are structured into six different stages with the Problem Definition and subsequent five Steps in a manner that the above *process approach* can be conveniently applied. It would involve identification of each FSA Step as a sub-process involving a number of interrelated activities, and establishment of the means to facilitate, monitor and control these activities to achieve the desired objectives. The fact that a complete 5-Step FSA is not always necessary to propose a Risk Control Measure/Option, such adoption of structured Step-wise approach would facilitate assurance of the quality of the FSA process.

6. APPLICATION OF QMS TO FSA[6]

Problem definition:

When beginning the assessment, it would be necessary to organize a meeting, at least with the Team Leaders/ Coordinators of the various subsequent FSA Steps (Sub-processes). MSC Circ.1022/MEPC Circ. 391 provides some guidelines for the practical application of FSA and use of experts. Based on these and the end requirements, it is worthwhile to develop a check-list, while deciding on the issues to be referred prior the commencement of the FSA process; it is important to identify:

- **Why** is the FSA exercise being undertaken?
- **What** knowledge & skills are required for the assessment?
- **Whether** human reliability issues are to be covered in the study?
- **Who** will be in the FSA team?
- **What** will be their roles during the various stages?
- **Who** will be responsible & accountable for what stage?
- **Which** of the ship categories, systems, operational aspects, influences, accident categories, data sources & consequences will be used in the assessment?
- **What** supporting Rules, Regulations, Codes & Standards of Construction are to be considered?
- **Whether** any specific feature requires a detailed analysis e.g. use of FMEA in case of specific critical equipment, or HAZOP in case of hazardous cargo.
- **How** many steps of the FSA sub-processes are to be completed as a minimum?
- **When** will the assessment be completed?
- **Who** will receive the conclusions drawn from the report?

One of the key issues remaining to be resolved is the establishment of risk tolerance criteria for safety of people, safety of property and safety of the environment within the FSA framework. Norway submitted a paper in MSC 72 (MSC 72/16) on the '*Decision parameters including risk acceptance criteria*'[12], which is yet to be formally ratified at the IMO. Present position at the IMO does not permit assignment of a monetary value to human life. Accordingly, all FSA studies on safety of personnel have focussed on the '*Potential Loss of Life (PLL)*'. It is therefore, not possible to have a combined criteria catering to safety of personnel, property & environment. Work by Nathwani et. al. [13] and Rackwitz[14] have endeavoured to address the issue of risk to personnel based on economic activity of a specific population. Health & Safety Executives, (HSE), UK have established risk acceptance criteria for general and offshore industries. It would be necessary to establish an overall objective of the FSA process based on these studies.

On the basis of the above identified requirements and the overall objective (criteria) subsequent Steps (sub-processes) can be commenced.

Step 1 (Hazard Identification)

The objectives of the HAZID process in Step 1 are:

a. To identify all potential hazardous scenarios which could lead to significant safety or operability consequences.
b. To decide whether the current design, construction & operations ensure that the risk from each identified scenario is at a suitably low level.
c. If not, to recommend further continuation of the FSA process which will decide on how to reduce the risk to a suitably low level.

This implies that the HAZID process quality objectives should include the following:

a. All potential scenarios leading to significant safety or operability issues are to be identified.
b. Decisions are to be made on the acceptability of the risk level from each of these scenarios.

c. Where the risk is intolerable, clear recommendations are to be made for further progress on the FSA Steps.

It is then necessary to decide which factors could influence these quality objectives, and how the quality system could ensure their achievement. In addressing these quality objectives following factors are considered important:

- Team composition, and qualifications
- Methodology followed (e.g brainstorming method)
- Use of appropriate technique in adopted Methodology
- Completeness of coverage
- Judgment criteria

A Typical Process Flow Diagram and the Controls & Measurements for use in each activity are presented in Appendix 1 towards establishment of Quality Management during execution of Step 1. The critical issues are to be based on the specific requirements for the FSA exercise and may be any / all of the factors mentioned above.

Step 2 (Risk Analysis)

The objectives of the Risk Analysis process in Step 2 are:

a. To identify accident scenarios that are significant contributors to risk
b. To obtain a realistic quantitative measure of the probability of occurrence of these risk contributors
c. To obtain a realistic evaluation of the potential consequences associated with those accident scenarios

Towards these objectives following critical issues of the sub-process may need to be addressed:

Modelling of accidents

- The most significant accident possibilities and accident type in the object to be modeled are to be carefully assessed. We need to think what all information the modeling should give and where they are to be used.
- It is necessary to consider whether it is worth continuing the analysis with the modeling or more detailed examination of hazards with identification methods are necessary.
- How the controlling systems for those hazards and disturbances not modeled are studied.
- The modeling methods used are to be applicable in the scenario and for the objectives. In case of modeling limitations, the limitations of the methods need to be covered.

Estimation of the accident frequencies

- Whether the numerical estimation of the frequency of the accidents are to be considered. It is often adequate to rank hazards, for example, into the following groups: probable, improbable, very improbable and insignificant.
- What kinds of assumptions; simplifications and restrictions can be made when estimating the frequencies?
- What kind of the data are needed when estimating the frequencies and where from the data can be obtained. The references of the numerical values used in the estimations.
- If subjective judgments are necessary, estimates are to be obtained from well-versed experts. Subjective judgments, whenever used should always be indicated.
- Accuracy and reliability of the data used in the estimations should be assessed. Sensitivity analysis should be carried out for the effect of the accuracy of the data (at least for the most important systems and components).
- Calculation formulas, principles and computer programs are to be noted; their starting points and scope of applications checked and their reliability, validity and limitations should be assessed.
- Information on real incidents are to be collected along with their root-causes and compared with the estimated frequency of accidents and critical events.
- Significance of the uncertainties in frequency estimations is to be evaluated.

Estimation of consequences

- Need to consider what kind of accident types to be examined and what kind of accident types cause so negligible risk that they can be excluded from the consequences estimations.
- Need to evaluate what kinds of assumptions and simplifications can be made when estimating the consequences.
- To obtain the methods and models available for consequence estimation, check their applicability to the scenario and phenomena to be studied. Available consequence models and programs should cover all the significant hazards and the possible operating conditions. Any limitations, defects and restrictions of the methods are to be evaluated, and ways how they can be overcome. Similarly, the reliability and validity of the calculation programs and their applicability to the scenario are also to be estimated.
- The kind of data needed when estimating the consequences and where the data can be obtained from are to be decided. The references of the numerical values used in the estimations should be available.

- Assure that the damage criteria to be used are appropriate and that possible factors e.g. effects of heat radiation, effects of wind & waves, pressure effects of explosions as well as issues pertaining to evacuation & rescue are included in the consequence estimations.
- If the domino effect is possible it is necessary to estimate its consequences.
- Significant uncertainties in consequence estimations should be evaluated, also the significance of types of accidents not covered by the models used.
- Lastly, it is necessary to collect information on real accidents and relate it with the estimated consequences.

Estimation of risk

- It is necessary to consider whether the estimation of individual or societal (group) risk is needed to obtain the goals set for analysis.
- Necessary simplifications and assumptions that can be made in risk estimation should be decided.
- Factors that may cause the most uncertainty in risk estimations are to be evaluated and estimate their influence on the results.

The Process Flow and the Controls & Measurements for use in each activity are presented in Appendix 2 towards establishment of Quality Management during execution of Step 2.

Step 3 (Risk Control Options)

The objectives of the Risk Control Options process in Step 3 are:

a. To specify risk reduction measures for scenarios identified in Step 2
b. Group these measures into possible Risk Control Options

Following issues are to be considered during this process:

Firstly, prior to proposing new measures the reasons for lack of effective risk control by existing measures must be investigated. (It is possible that the existing measures inherently cannot control the risk, or that they can but are not being implemented or complied with.)

Secondly, the process of generation of Risk Control Measures should be as comprehensive as possible and should use as much risk and hazard information as is available. The methodology user should refer back to Step 1 and develop new Risk Control Measures for hazards identified in Step 1 that are not sufficiently controlled by existing measures.

Thirdly, the HAZID team should be invited for a new meeting. The team should be strengthened with people with broad design experience & creativity. Factors listed for Step 1 as contributing to the effectiveness of such sessions are to be re-considered. The result of the meeting should be a list of RCOs that seems promising and should be evaluated. Often, the RCOs are implemented with the best intentions, but turn out to reduce safety.

Fourthly, the Risk Control Measures identified are combined into a Risk Control Measures Log. The Risk Control Measures Log is then reviewed with particular reference to the influence of and individual Risk Control Measure on other risks identified in Step 2. This influence (dependency attribute) can be beneficial, reducing the risk or can be detrimental, increasing the risk. The analysis of the RCOs tends to be static (assuming everything except the RCO is unchanged). Often the RCO introduce other effects.

Finally, a draft of the Risk Control Measures Log should be reviewed and comments incorporated prior to generating the Risk Control Options Log.

At any stage of this process it may become necessary to refer back to Step 2, or even Step 1, for more information or further analysis.

The Process Flow of Step 3 and the Controls & Measurements for use in each activity are presented in Appendix 4 towards establishment of Quality Management during execution of Step 3.

Step 4 (Cost Benefit Assessment)

The objective of the assessment is to provide a basis for decision-making about the RCOs identified in Step 3 through evaluation of costs & benefits associated with their implementation.

Costs:

Costs for implementing an RCO that should be included are Marginal Costs as compared to the base case. Costs should be as comprehensive as possible and include

- Cost of equipment
- Cost of redesign and construction
- Cost of documentation
- Cost of training
- Cost of inspection, maintenance and drills
- Cost of auditing
- Cost of regulations
- Reduced commercial use (e.g. reduced deck space with commercial use)
- Operational limitations (e.g. reduced loads, speed)

(It may be difficult to agree on realistic cost estimates because of the large differences between countries.) There will usually be large costs differences for the same RCO if implemented for a new ship as compared to an existing one. However, by applying the same acceptance criteria on cost effectiveness this may show that while an RCO for new building should be implemented, the same RCO should not be implemented for existing ships. This way the Grandfather clause will appear as rational.

Benefits:

All benefits are to be the marginal benefits as compared to a base case established in Step 2, and be comprehensive including:

- Reduced probability of fatalities or number of fatalities
- Reduced number of injuries and severity of injuries
- Reduced negative effects on health
- Reduced probability of severity of pollution
- Reduced economic losses

There are benefits e.g. to the owner of the ship that is not relevant in FSA. For example reduced insurance premiums as result of RCO should **not** be included, because the FSA includes the corresponding risk reduction explicitly. To include an eventually reduced insurance premium would count the same effect twice.

Quantification Of Cost Effectiveness:

The criteria of Cost Effectiveness should be clearly defined to be either *Gross Cost of Averting a Fatality (GrossCAF)*, or *Net Cost of Averting a Fatality (NetCAF) as follows:*

$$GrossCAF = \frac{\Delta C}{\Delta R};$$

$$NetCAF = \frac{\Delta C - \Delta B}{\Delta R}.$$

Some authorities favour prioritisation of *GrossCAF*, followed by evaluation of NetCAF for each stakeholder separately. Others prefer to prioritise *NetCAF* alone.

Discount Rates:

Discount rates used in safety investments serve a different purpose. The purpose is to ensure some fairness between investment in saving current and future lives. The discount rate will change priorities between lives saved now and in the future (between generations) and the cost of capital is not relevant in this context. How the investment in the ship and its safety is financed is not a concern of the safety regulator. It should also be obvious that a risk premium on discount rates is irrelevant in a risk analysis, since the risks are modelled explicitly.

In the literature two alternatives may be found. Many engineers treat investment cost as a loan at a discount rate similar to "risk free investment" and use no depreciation on lives saved. Both are converted to annual costs and annual benefit (e.g. lives saved). It should be clear from the arguments above that this cannot be correct. It can also be proved that the result of this is that the same decision on investing in safety delayed one year is better than investing now.

The Step 4 activities may not necessarily follow a process; accordingly, no specific process diagram with control mechanisms have been presented. *The possible way to measure, review and control the effectiveness of this Step could be through auditing the data used.*

Step 5 (Decision Making)

Purpose of Step 5 of the FSA procedures address the following:

- Note the hazards & their associated risks
- Review the Risk Control Options (RCOs) which keep Risk 'As Low As Reasonably Practicable (ALARP)'
- Compare & rank RCOs based on the associated costs & benefits
- Specify recommendations for presentation to decision makers in a simple & concise manner

The decision process in Step 5 should examine the risks and their inherent uncertainties associated with the procedures/terminal details in the FSA process. It is to be followed by a judgement on the tolerability of the risks to individuals or groups e.g. (crew/passengers/others) for their activities.

Based on the assessment of the risk and the purpose of the FSA study it is to be decided as to how far can the risk be reduced taking cost aspects into account. It is to be ensured that all parties involved are equitably affected with respect to the final recommendations.

On completion of the decision process the results are presented in order to recommend areas requiring new rules or areas where existing rules are to be reviersed.

The report should be internally reviewed for assurance of its quality & process. Internal review may be carried out as per the diagram given in Appendix 5. In addition it should be examined that the Controls & Measurements have been exercised for each of the FSA Steps as per the Process Control measures. IMO MSC Circ.1022 stipulates following terms of reference in the Guidelines for the review process prior to decision-making:

- evaluation of the methodology applied and verification that it is in accordance with the FSA Interim Guidelines;
- evaluation of the appropriateness of the scenarios applied, assumptions made and limitations of the FSA study with regard to the significance of the findings;
- evaluation of whether the risks and risk control options have been evaluated in an appropriate manner; and

- presentation of the results of the review, in clear and comprehensive terms, including any recommendations for the IMO rule-making process.

As in any exercise towards idealisation and quantification of physical outcomes, the techniques in FSA are also subject to a degree of uncertainty. These uncertainties are divided into two categories:

a) Those which arise out of variability in the observations, and therefore represent randomness in samples (called aleatory uncertainties); and
b) Those which come from ambiguity of knowledge about the physical phenomena (called epistemic uncertainty)

Most risk analysis problems involve both known statistical samples, and unknown or partially known mechanisms. FSA techniques predominately use a frequency approach and therefore require examination of the uncertainty in the analysis. In the FSA framework the objective is to analyse the system by functions and to compute the probability of system failure as a function of the performance of the components. This also accounts for the possibility of external events such as storms or behaviour of other ships. The appropriate level of treatment depends on the scope of the FSA study, the magnitude of risk and the cost-effectiveness of the identified RCOs. Normally best estimates / central value treatment is adequate for most shipping regulations.

7. CONCLUSIONS

FSA results represent a Risk-based optimisation study in the development of the Rules, and need to be monitored after implementation of the RCOs. Once the QMS is implemented, the process variations in the FSA would be fairly well understood, and the success of the Rules (or their effectiveness) after implementations can then be analysed in terms of the process variations during the FSA study behind these rules. Suitable remedial action can then be initiated without much difficulty.

8. ACKNOWLEDGEMENT

The paper is a reflection of the author's concern over the correctness of the FSA applications & interpretations towards Rule developments & Rule changes. Indian Register of Shipping has supported the work, and the paper in part produces some results of a collaborative study of FSA carried out with the China Classification Society, Korean Register of Shipping and ClassNK.

REFERENCES

1. IMO MSC 62/24/3, 2 March 1993
2. IMO MSC/Circ.829 (MEPC/Circ.335),17 November 1997
3. IMO MSC/Circ. 1023(MEPC Circ. 392)
4. Interim Report of Collaborative FSA Studies by CCS, IRS, KR & NK – Internal communications
5. DASGUPTA, J. "Application Of Formal Safety Assessment To Rule Making Process", INMARCO 1998
6. International Association Of Classification Societies (IACS), FSA Training Modules, 2002
7. IMO MSC 72/23; REPORT OF THE MSC 72, PARA16.2.4; MAY 2000
8. IMO MSC 76/5/5; "BULK CARRIER SAFETY International Collaborative FSA Study: Final Report"; UK Submission, March 2002
9. IMO MSC 75/5/1; "BULK CARRIER SAFETY International Collaborative FSA Study: Progress Report"; UK Submission, September 2002
10. IMO MSC 76/5/5; "BULK CARRIER SAFETY International Collaborative FSA Study: Final Report"; UK Submission, March 2002, ANNEX 2
11. HOYLE, D. ISO 9000 Quality System Handbook, 4th. Ed. 2002
12. IMO MSC 72/16 "Decision Parameters Including Risk Acceptance Criteria"
13. NATHWANI, J.S.; LIND, N.C.; PANDEY, M.D.; "Affordable Safety By Choice: The Life Quality Index Method", Institute Of Risk Research, University Of Waterloo, 1997
14. RACKWITZ, R.; "How Safe Is Safe Enough? An Approach By Optimisation And Life Quality Index", Proceedings Of ASRANET Conference, 2002

APPENDIX 1: A TYPICAL PROCESS FLOW CHART & DETAILS FOR FSA STEP 1

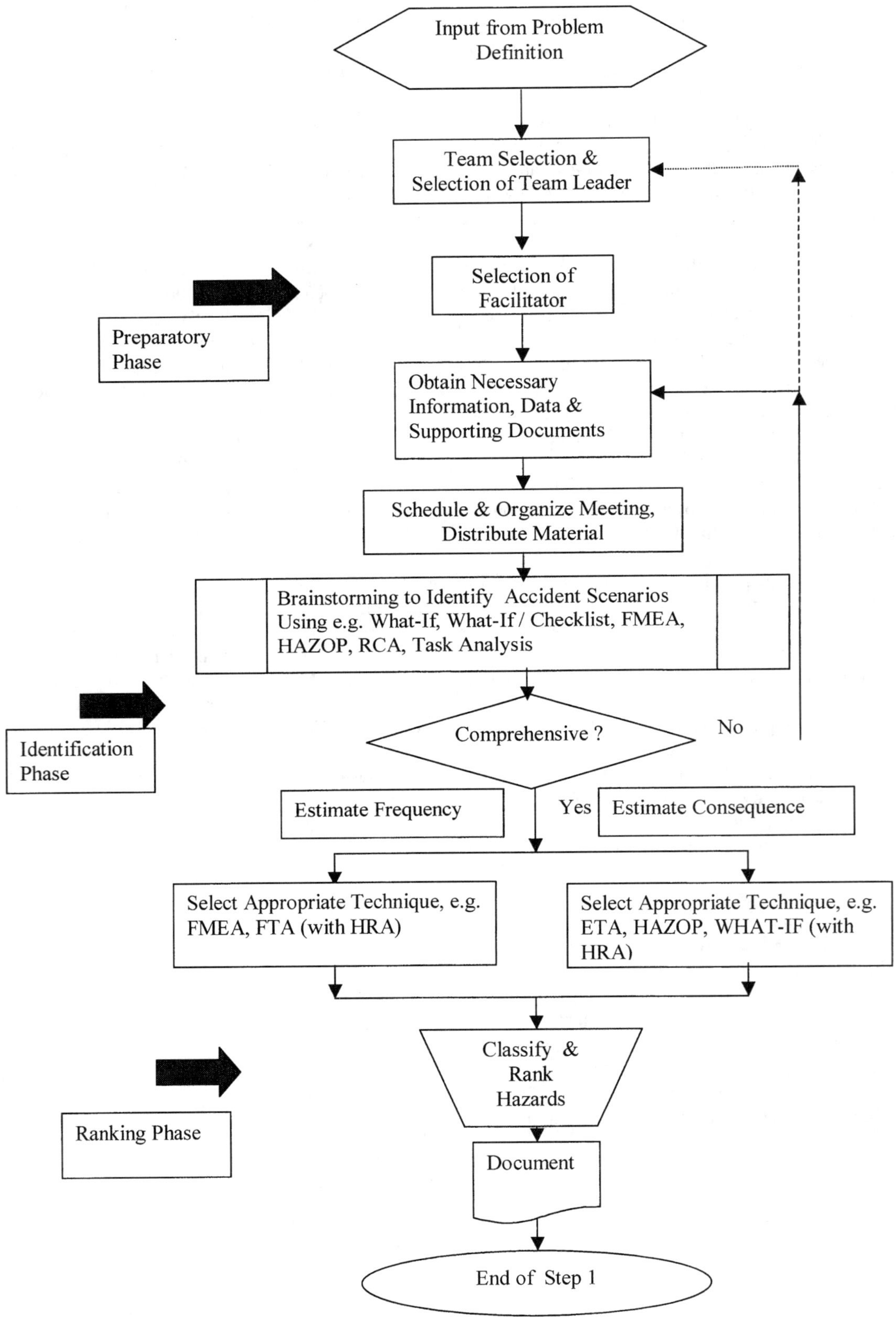

Table A 1: Process Parameters for FSA Step 1

Sr. No	INPUT	ACTIVITY	BY WHOM?	INTERACTING / SUB-PROCESS	CRITICAL ISSUES	CONTROLS	CONTROLLING MEASUREMENT	OUTPUT
1	Scope & Details from Checklist	Input from Problem Definition	FSA Coordinator	Making Check-lists	Scope of Project & Relevance w.r.t. Rules & Regulations	-	-	Scope of HAZARD IDENTIFICATION Process
2	Bio-Data of Team Members	Team Selection & Selection of Team Facilitator (Leader)	FSA Coordinator & Team	-	Contribution of Members, Availability of Members	Qualification & Experience of Members, Planning & Scheduling of Meeting Dates	Size of Team, Rejection of Proposed Dates	HAZID Team & Facilitator
3	Bio-Data of Team Members	Selection of Recorder	Facilitator & HAZID Team	-	Ability to Capture Relevant Inputs	Use of Software, Tape-recorder	Monitoring of Records by Facilitator	Selected Recorder
4	Experience of Members, Casualty Statistics, Data	Obtain Necessary Information, Data & Supporting Documents	Team Members	Root-Cause-Analysis of Accidents/Incidents	Validity of the Input Data	Input from Reputed Data-bases, Relevant Experience of Members	Convergence characteristics of Inputs	Assimilated data
5	Members' Engagements	Schedule & Organize Meeting, Distribute Material	Facilitator (Leader)	-	Time for completion of HAZID Process, Cost of Meetings	Selection of stand-by Alternative Members	Time to decide the Meeting Date	Scheduled Date of Meeting
6	Experience of Members, Casualty Statistics, Data	Brainstorming to Identify Accident Scenarios Using e.g. DELPHI, What-If, What-If / Checklist, FMEA, HAZOP, RCA, Task Analysis	Facilitator & HAZID Team	What-If, What-If / Checklist, FMEA, HAZOP, RCA, Task Analysis	Environment & Members' Attitude	Evaluation of Inputs	Divergence of Inputs	Identified Accident Scenarios
7	Accident Scenarios	Comprehensive ?	Team Leader (Facilitator)	-	Divergence & Relevance of Inputs	-	Classification of Inputs	Decision on Continuity of Process
8	Each Accident Scenario	Estimate Frequency : Select Appropriate Technique, e.g. FMEA, FTA (with HRA)	Team	FMEA, FTA (with HRA)	Validity of Result	Validation with Reported Data / Experience	Variation from Reported Data	Frequency Estimates, F
9	Each Accident Scenario	Estimate Consequence: Select Appropriate Technique, e.g. ETA, HAZOP, WHAT-IF (with HRA)	Team	ETA, HAZOP, WHAT-IF (with HRA)	Validity of Result	Validation with Reported Data / Experience	Variation from Reported Data	Consequence Estimates, C
10	Accident Scenarios, F & C	Classify & Rank Hazards	Team	-	Establishing Risk Matrix, Elimination of Irrelevant Scenarios	Experience of Members & Reported Data	% of Generated Scenarios	Ranked Scenarios
11	Accident Scenarios, F & C	Document Results	Team	-	Reasoning on Ranking, Size of report, Clarity	Documentation Standard (IMO /IACS)	Step 1 Document, Contents & Presentation	Output From Step 1.

17

APPENDIX 2 : A TYPICAL PROCESS FLOW CHART & DETAILS FOR FSA STEP 2

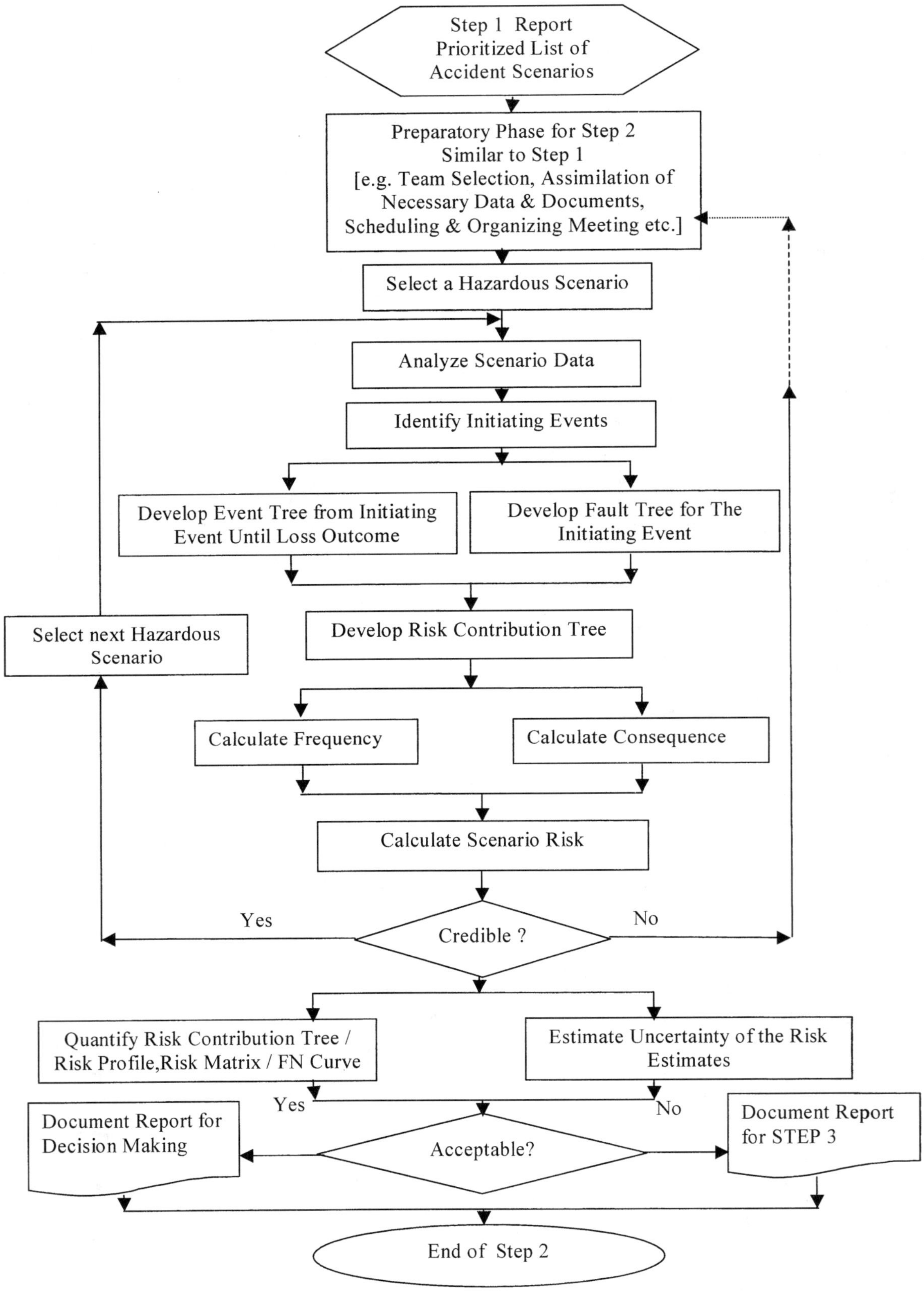

Table A 2: Process Parameters for FSA Step 2

Sr. No.	INPUT	ACTIVITY	BY WHOM?	INTERACTING / SUB-PROCESS	CRITICAL ISSUES	CONTROLS	CONTROLLING MEASUREMENT	OUTPUT
1	Prioritized List of Hazardous Scenarios	Input from Hazard Identification	FSA Coordinator		Assessment of most significant accident scenarios identified	-	-	-
2	Prioritized List of Hazardous Scenarios	Preparatory Phase for Step 2 Activities	FSA Coordinator & Team	Making Check-list for Step 2 Assessment (Refer Indicative Checklist)	Contribution of Members, Availability of Members	Qualification & Experience of Members, Planning & Scheduling of Meeting Dates	Size of Team, Rejection of Proposed Dates	Step 2 Team & Facilitator; Planning Document
3	Identified Hazard List of scenarios	Select Hazard Scenario	Facilitator & Risk Assessment Team		Ability to Capture Relevant Inputs	Mapped Results from Step 1		Selected Scenario
4	Scenario Data	Analysis of Scenario Data of Incidents / Accidents	Team Members	Root-Cause-Analysis (RCA)of Accidents/Incidents	Validity of the Input Data	Input from Reputed Data-bases, Relevant Experience of Members	Convergence characteristics of Inputs	Assimilated data
5	Scenario Data, Experience of Members, Step 1 Details, RCA	Identifying Initiating Events	Team Members	Grouping of Initiators	Correct Sequence of Events	Drawing-up Preliminary Event Tree	Avoidance of Double Counting, Number of Events	Identified Initiating Events
6	Scenario Data, Experience of Members, Initiating Events	Developing Event Tree	Facilitator & Risk Assessment Team	ETA, What-If, What-If / Checklist, FMEA, HAZOP, RCA, Task Analysis	Common Cause Failures, Correctness of Models, Domino Effects	Evaluation Against HAZID Report	Structure of Event Tree, Time to Construct	Event Tree
7	Scenario Data, Experience of Members, Initiating Events	Developing Fault Trees	Facilitator & Risk Assessment Team	FTA, FMEA	Correctness of Model, Assumptions in System Definition	Evaluation of Minimal Cut Sets	Structure of Fault Tree, Time to Construct	Fault Trees with Cut-Sets
8	Event Tree & Fault Trees	Developing Risk Contribution Tree	Team	ETA & FTA	Construction of FTAs for Initiating & Critical Events	Validation with Reported Data / Experience	-	Risk Contribution Tree (RCT)
9	Historical Data	Calculation of Frequency of Events	Team	-	Validity of Result	Validation with Reported Data / Experience	Variation from Reported Data	Frequency Estimates, F
10	Historical Data	Calculation of Magnitude of Events	Team	-	Validity of Result	Validation with Reported Data / Experience	Variation from Reported Data	Consequence Estimates, C
11	Scenario, F & C	Credibility Check	Team	-	Validity of Result	Validation with Reported Data / Experience	Variation from Reported Data	Credible Scenario
12	RCT, Freq. & Consequences	Quantifying RCA, Risk Matrix, FN Curve	Team	Risk Matrix, FN-Calculation	Validity of Result	Validation with Reported Data / Experience	Variation from Reported Data	Risk Assessment
13	RCT, Freq. & Consequences	Evaluation of Uncertainty	Team	-	Reliability of Estimates	Validation with Reported Data / Experience	Variation from Reported Data	Uncertainty Estimates
14	Risk Assessment, Uncertainties	Acceptance of Risk Results	Team	Reference to Acceptance Standard, Step 3 & Step 4	Applicability of The Acceptance Standard	Validation with Published Data	Risk Evaluation	Output from Step 2
15	Step 2 Output	Documentation	Team	-	Coverage, Clarity	Documentation Standard (IMO/IACS)	Contents & Presentation	Step 2 Document

APPENDIX 3 : PROCESS FLOW CHART & DETAILS FOR FSA STEP 3

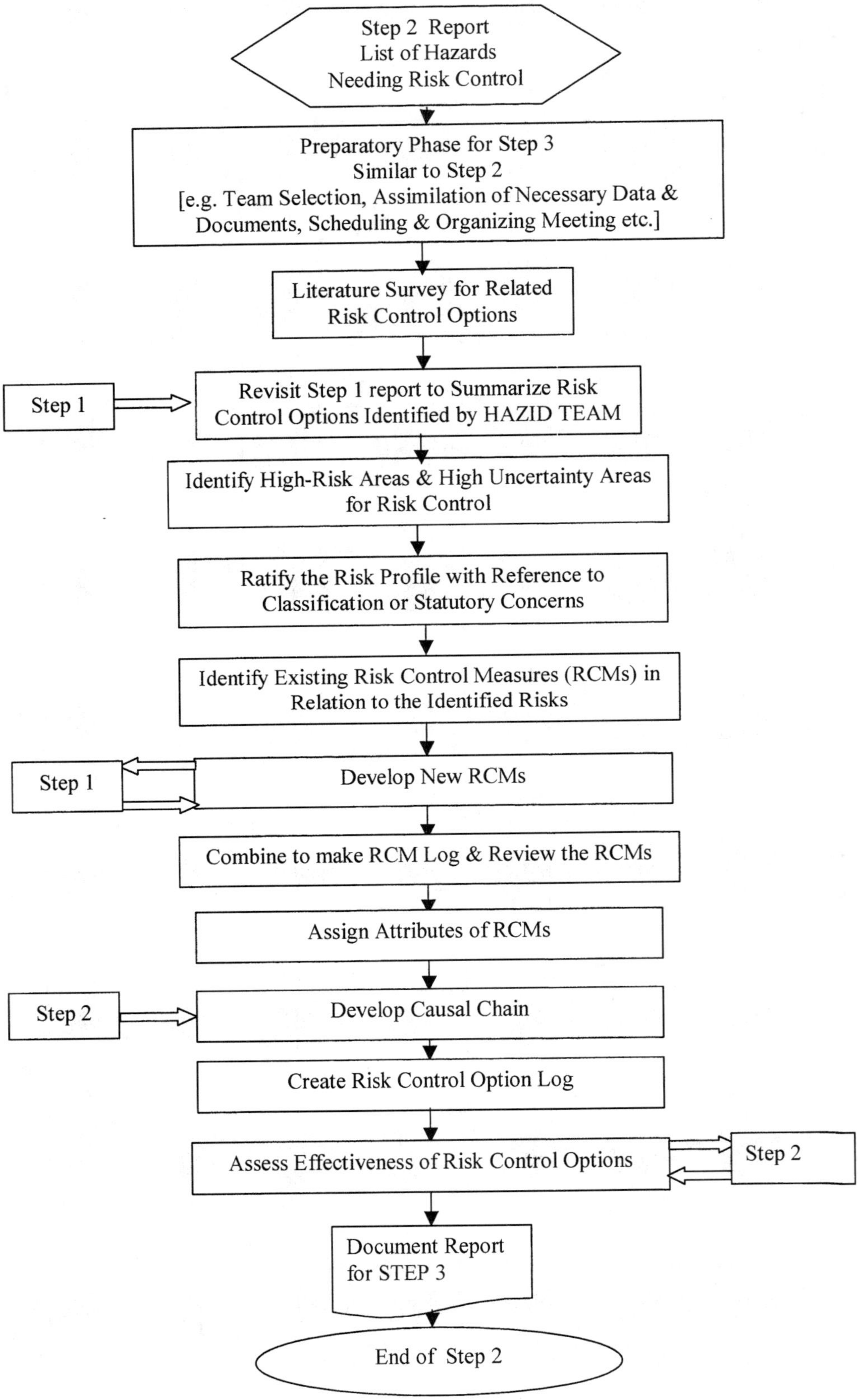

Table A 3: Process Parameters for FSA Step 3

Sr. No.	INPUT	ACTIVITY	BY WHOM?	INTERACTING / SUB-PROCESS	CRITICAL ISSUES	CONTROLS	CONTROLLING MEASUREMENT	OUTPUT
1	List of Hazards Needing Risk Control	Preparatory Phase for Step 3	FSA Coordinator & Team	Making Check-list for Step 2 Assessment	Contribution of Members, Availability of Members	Qualification & Experience of Members, Planning & Scheduling of Meeting Dates	Size of Team, Rejection of Proposed Dates	FSA Coordinator & Team
2	Rules, Regulations, Industry Practices	Literature Survey for Related Hazards	FSA Team	Data Acquisition, Theories on Risk Controls	Establishment of Base-value of Risks, Validity of Data	Qualification & Experience of Members, Published Literature & Reports	-	Background Information
3	Step 1 Report	Summarize Risk Control Options Identified by HAZID TEAM	FSA Team	-	Validation of Risk Levels in Step 1 & Step 2	Comparison of Reported Risks	Risk Comparability	Preliminary RCOs
4	Step 2 & Step 1 Reports	Identify High-Risk Areas & High Uncertainty Areas for Risk Control	FSA Team	-	High Probability High Severity & Low Confidence Areas	Prioritizing Areas Needing Control	Nodes in Event Trees & Failures in Cut-sets of FTs	Areas Needing ontrol
5	Hazards & Rules / Regulations	Ratify Risk Profile w.r.t Class/Statutory Concerns	FSA Team	-	High Probability High Severity & Low Confidence Areas	-	-	Ratified Risk Profiles
6	Step 1, Rules, Rules & Reg., Procedures	Identify Existing Risk Control Measures (RCMs)	FSA Team	Analysis of Rules & regulations	Effectiveness of Existing RCMs (Theory & Implemenation)	Analysis of Incidents / accidents	Incident/Accident Data	List of existing RCMs
7	Step 1	Develop New RCMs, By brainstorming	FSA Team	Step 1, Brain-storming	Flexibility/Practicality ; Performance Based / Prescriptive in Nature	Acceptance by Team	Divergence of opinion	List of New RCMs
8	-	Make RCM Log & Review	FSA Team	-	Comprehensive Coverage	-	-	RCM Log
9	-	Assign Attributes of RCMs	FSA Team	Brainstorming & Sorting	Interaction between Attributes	Avoiding Over-reliance on Single Category	-	RCMs with Attributes
10	Step 2	Develop Causal Chain	FSA Team	Step2	Recognition of Underlying Influences	Validation with Event Trees	-	Pivotal Events for RCMs
11	-	Create Risk Control Option Log	FSA Team	Selection of Approach	Practicability of Regulatory Option	Preventive / Distributed Approach	Prevention / Mitigation	RCO Log
12	Step 2	Assess Effectiveness	FSA Team	Step 2	Reduction of Risks/ No New Risk	Risk assessment	FN Curve / Risk Profile	Reduction in Risk
13	-	Report for STEP 3	FSA Team	-	Clear & Concise Report	Reporting Standard	-	Step 3 Report

APPENDIX 4: PROCESS FLOW CHART FOR FSA REVIEW

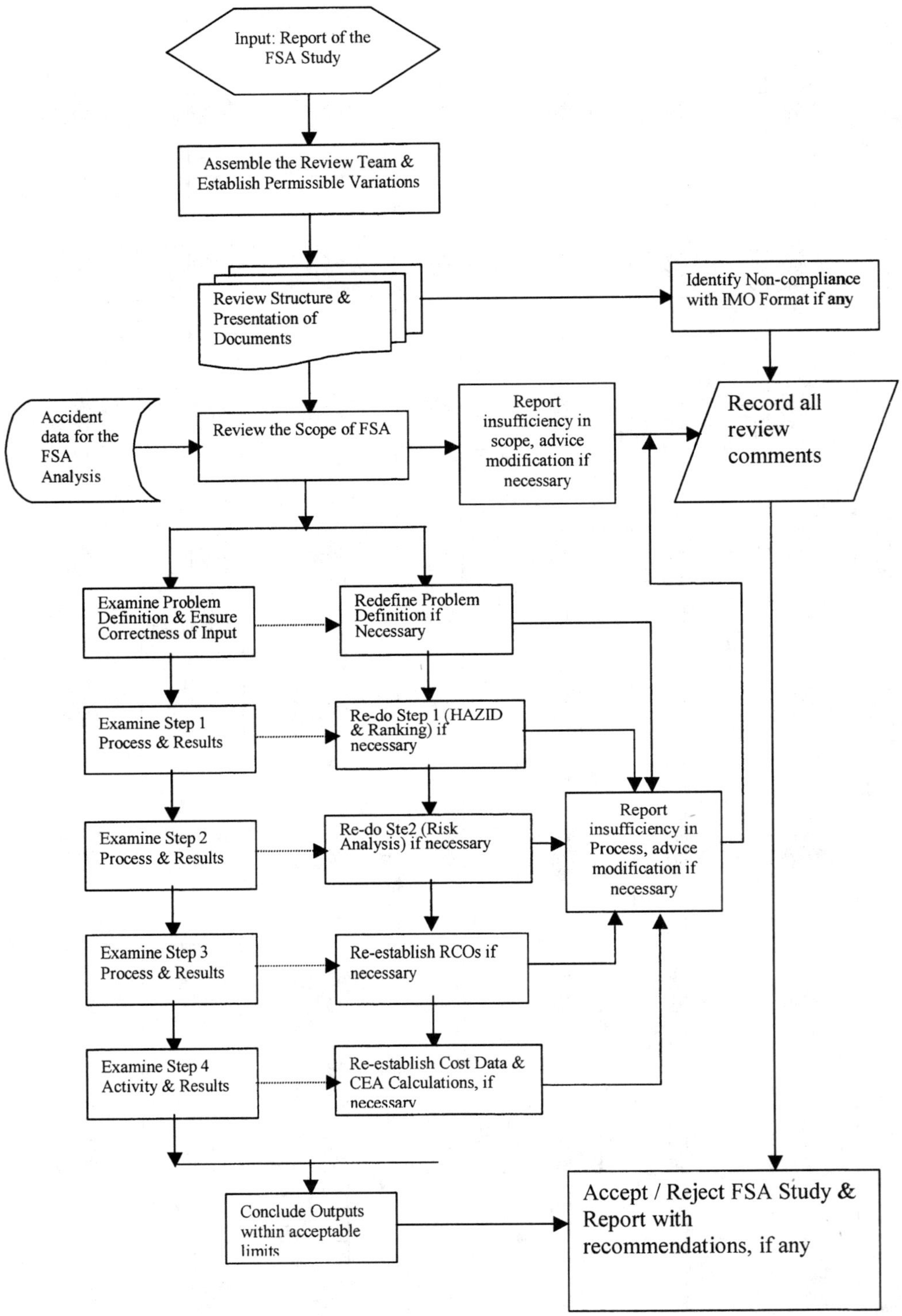

Thermo-Mechanical Analysis of Plates undergoing Line Heating using Pseudolinear Equivalent Constant Rigidity System

Malabika Adak, Reseach Scholar, Department of Ocean Engineering and Naval Architecture, IIT Kharagpur.
N.R. Mandal, Associate Professor, Department of Ocean Engineering and Naval Architecture, IIT Kharagpur

ABSTRACT

Line heating technique involves application of controlled thermal cycles. In developing the technique of line heating, which is nonlinear and involves plastic deformation of the medium at high temperature varying in both time and space, analytical solutions turned out to be inadequate. At the same time conventional numerical techniques proved to be highly time consuming and thereby prohibitively expensive in real life situations. In this investigation, the initial non linear problem with modulus varying with temperature, was transformed into a pseudolinear equivalent system of constant rigidity, that was solved by applying linear analysis. The analysis procedure presented in this work and the results thus obtained holds a great promise for analyzing the complicated thermoelastic-plastic problems as encountered in line heating process.

1. INTRODUCTION

A substantial part of fine form ships' hull are composed of curved plates having double curvature. Shaping of these plates is done either by mechanical means using hydraulic press or by line heating carried out by skilled workers. Automation of the mechanical method of plate forming operation was tried out in Japan using multiple piston press way back in 1976. The results of which were not very encouraging and soon it was realized that the solution lies in the line heating technique of plate forming.

In developing the technique of line heating, an analytical or numerical solution of the thermoelastic-plastic problem is of great importance. Due to the complexity of the problem, which is nonlinear and involves plastic deformation of the medium at high temperature varying in both time and space, analytical solutions turned out to be inadequate. At the same time conventional numerical techniques proved to be highly time consuming and thereby prohibitively expensive in real life situations.

Several investigators have tried to solve this thermo-elasto-plastic problem using various schemes.

Line heating technique involves application of controlled thermal cycles. The heating and cooling cycles produce thermal strains and these strains are used to generate the desired plate shape. This process has been gainfully used in several shipyards by skilled workers. However with time every shipyard started facing a grave problem of rising labour cost and at the same time lack of such skilled workers. This prompted researchers to look into this problem to work out a methodology to automate the process of plate forming by line heating.

Watanabe and Satoh [1] used the analytical solutions resulting from the theory of elasticity for prediction of thermal deformations due to welding and line heating. However, since elastic solutions are limited, application of the method is also limited.

Later Masubuchi [2] and Fujita and Nomoto [3] have calculated stresses in butt welded plates assuming temperature independent material properties and uniform temperature and stress fields through the thickness of the plate.

With increasing computational power of present day computers, numerical methods based on finite element and finite difference schemes have been extensively used. Masubuchi [4], Kristina and George [5], Argyris et al [6], Rybicki et al [7], Papazoglou and Masubuchi [8], Ueda et al [9] worked out thermo-elasto-plastic problems using high speed workstations but the resulting deformations showed poor accuracy compared to the experimental results.

Several investigators Nomoto et al [10], Ueda and Murakawa [11, 12], Rybicki and Stonesifer [13] developed the simulation method for line heating. They obtained shrinkage moments which cause bending deformations of the plate through experiments, and developed a computer program which could analyze thermal deformations and display deformed shapes. Thus rigorous thermal elasto-plastic analysis was replaced by elastic analysis in conjunction with experimental data. This resulted in improved computing efficiency. However, the developed method could be applied to the line heating processes where the heat input was known. Also, as longitudinal bending moments were not accounted for, in some cases deformed shapes did not agree with the experimental results due to longitudinal deformation.

Commercial software such as ABAQUS [14], and ANSYS [15] have also been used to simulate line heating process. However this process has several drawbacks, such as irregular mesh size, extraordinary mesh shape, and repetition of mesh generation for each line heating task. Recently, using neural networks, the limitations of slow response could be overcome by Kuo et al [16]. Good results were obtained, but this required vast data samples to train the neurons.

With this background, work was initiated in the Department of Ocean Engineering and Naval architecture at IIT Kharagpur, to work out a possible solution of this highly nonlinear complex phenomenon of thermoelastic-plastic problem. In the earlier works of the authors, a computationally efficient technique based on the concept of pseudolinear equivalent constant stiffness system has been developed [17]. The method was rigorously tested and successfully implemented on beam model for estimating angular deformations due to thermal cycles. In the present investigation the same concept has been further extended to solve plate problems.

2. PROBLEM FORMULATION

The problem of thermal deformation of plates was handled in a decoupled manner. First the heat flow problem, considering temperature dependent material properties, was solved. The result of this solution i.e. the time dependent 3-D temperature profile over the plate was used as input to the solution of the problem of mechanical response of the plate undergoing heating by a moving heat source.

Therefore as a first step towards the development of the technique, a reliable and efficient method for determining the temperature history due to a moving heat source has been developed based on a centre difference implicit scheme using Jacobi iteration method. Here it was assumed that the heat conduction process is uncoupled from the stress strain state of the plate.

In the present investigation, the line heating situation was simulated by bead-on-plate welding, and accordingly experiments were carried out using submerged arc welding to verify the theoretical results. This was done since the heat input could be easily estimated in case of welding. Here the purpose of the work was to establish the method of analyzing thermal deformation of plates using Pseudolinear Equivalent Constant Rigidity System.

2.1 Heat Flow

The temperature T (t, x, y, z) as a function of time (t) and spatial coordinates (x, y, z) satisfies the following diffusion equation at every point within a finite domain,

$$\frac{\partial T}{\partial t} = \nabla\left(\frac{\lambda \nabla T}{\rho c}\right) + \frac{Q}{\rho c} \tag{1}$$

The axis system and the heat source model are shown in Fig.1 [18]. Heating direction being y-axis, finer grid of 1mm was taken along both x-axis and z-axis. After the initial transient state, near the starting point of the heating line, the temperature field repeats itself leading to a quasi-stationary state. Hence a coarse grid, numerically equal to the speed of the moving torch, was chosen along the heating axis. Considering the heating line to be on the line of symmetry only one half of the plate about y-axis was taken as calculation domain as shown in Fig.2.

The heat loss, due to radiation, conduction through electrode and heat consumed towards burning of flux and melting of electrode, were accounted for by the arc efficiency parameter η.

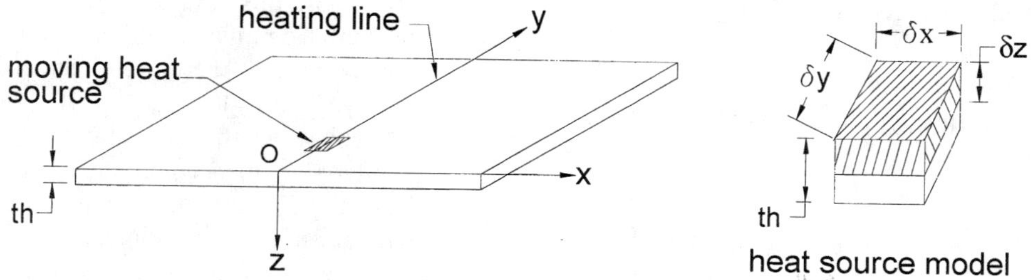

Fig.1 Axis system and heat source model

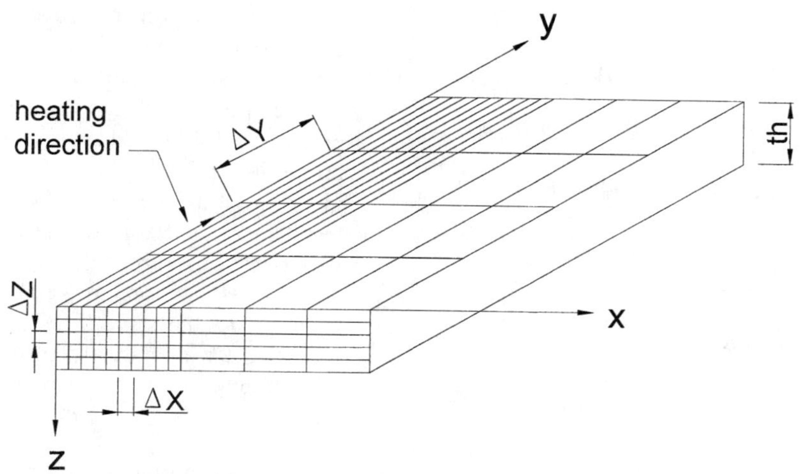

Fig.2 Grid system and mesh dimensions

In the present study a value of 0.8 was taken for η. Therefore, the net heat available to the plate was ηVI. This was considered to be uniformly distributed over the shaded part of the heat source model as shown in Fig.1. Therefore, the heat input per unit volume was given by,

$$Q = \frac{\eta V I}{\delta x \, \delta y \, \delta z} \tag{2}$$

Boundary Conditions

Along the line of symmetry, i.e. the y-axis, the following boundary condition was defined:

at x = 0;

$$\frac{\partial T^n_{i,j,k}}{\partial x} = 0 \Rightarrow T^n_{-1,j,k} = T^n_{1,j,k} \tag{3}$$

Along the other three boundaries and on the top and bottom surfaces of the plate, heat dissipation will occur through convection. Therefore, the following boundary conditions were defined for various surfaces:

At the plate edge along x-axis, let **x = xx** ;

$$\lambda \frac{\partial T_{xx,y,z}}{\partial x} + h\left(T_{xx,y,z} - T_0\right) = 0$$

$$T_{xx+1,y,z} = T_{xx-1,y,z} - C_x\left(T_{xx,y,z} - T_0\right) \tag{4}$$

where $C_x = \frac{2.\Delta x.h}{\lambda}$.

at y = 0;

$$-\lambda \frac{\partial T^n_{i,0,k}}{\partial y} + h\left(T^n_{i,0,k} - T_0\right) = 0$$

$$\Rightarrow T^n_{i,-1,k} = T^n_{i,1,k} - C_y\left(T^n_{i,0,k} - T_0\right) \tag{5}$$

At the plate edge along y-axis, let **y = yy** ;

$$\lambda \frac{\partial T^n_{i,yy,k}}{\partial y} + h\left(T^n_{i,yy,k} - T_0\right) = 0$$

$$\Rightarrow T^n_{i,yy+1,k} = T^n_{i,yy-1,k} - C_y\left(T^n_{i,yy,k} - T_0\right) \quad (6)$$

where $C_y = \frac{2.\,h..\,\Delta y}{\lambda}$.

At the plate top surface, **z = 0;**

$$-\lambda \frac{\partial T^n_{i,j,0}}{\partial z} + h_t\left(T^n_{i,j,0} - T_0\right) = 0$$

$$\Rightarrow T^n_{i,j,-1} = T^n_{i,j,1} - C_{zt}\left(T^n_{i,j,0} - T_0\right) \quad (7)$$

where $C_{zt} = \frac{2.\,.\,\Delta z.\,h_t}{\lambda}$,

h_t = top surface convection coefficient.

At the plate bottom surface, **z = zz** ;

$$\lambda \frac{\partial T^n_{i,j,zz}}{\partial z} + h_b\left(T^n_{i,j,zz} - T_0\right) = 0$$

$$\Rightarrow T^n_{i,j,zz+1} = T^n_{i,j,zz-1} - C_{zb}\left(T^n_{i,j,zz} - T_0\right) \quad (8)$$

where $C_{zb} = \frac{2.\,.\,\Delta z.\,h_b}{\lambda}$,

h_b = bottom surface convection coefficient.

In addition to these boundary conditions, the initial condition over the entire plate was defined as,

$$T^0_{i,j,k} = T_0 \quad (9)$$

where T_0 = ambient temperature.

Finite Difference Scheme

The governing equations were represented in a finite difference form following centre difference implicit scheme. The numerical solution of equation (1) with appropriate boundary conditions (3) - (9), required fine grids along x and z axes and a small time step to ensure conservation of energy during each time step. The fine grids in the 3-D problem led to a very large sparse matrix resulting in a limitation to the calculation domain size. This limitation was overcome by using Jacobi iteration scheme to solve the system of difference equations. The coefficient matrix of this system of equations not being diagonally dominant, exhibited the problem of non-convergence. Therefore, under-relaxation was employed in the Jacobi iteration scheme for dampening out the oscillations. The Jacobi iteration and iteration to accommodate variation of thermo-physical properties with temperature ran simultaneously. The entire code was written in VC++ environment. The model used 71 × 16 × 8 grid system for 6 mm plate for the calculation of temperature history over 1500 time steps, each time step of 0.2 second.

The variation of specific heat and thermal conductivity with temperature have been incorporated in the calculation of temperature history. The temperature dependence of these parameters is given in Table 1 [19]. The various mechanical properties, as have been used in this analysis are given in Table 2 [19]. Solution of this thermo-mechanical problem assumes complexity because of variation of all physical properties with temperature. This leads to the material non-linearity of the problem. At the same time, because of moving heat source, the temperature profile continuously changes with time, keeping the problem from start to end in a transient state.

2.2 Thermal Strains and Bending Moment

Due to the temperature differential along the x, y and z axes, thermal strains will develop varying with time. These in turn will give rise to bending moments causing the plate to deform. The bending moments were calculated separately about the x and y axes. Since the temperature profile, due to the moving heat source, attains a quasi stationary state, therefore a strip of plate perpendicular in the direction of heating was considered for calculation of moment M_x about y-axis, the heating direction. Whereas the moment M_y about x-axis was calculated considering plate strips parallel to heating direction, the y-axis. The average of the moments thus calculated about x-axis was considered. These moments were calculated over a time domain covering the heating and cooling phase.

Considering the plate along x-axis, due to the temperature distribution, at a given time instant, there will be a thermal strain distribution along a plate strip as given below,

$$\varepsilon(n, x, y, z) = \alpha(n, x, y, z)\Delta T(n, x, y, z) \quad (10)$$

where

$$\Delta T(n, x, y, z) = T(n, x, y, z) - T(n\text{-}1, x, y, z) \quad (11)$$

Table 1 Temperature dependent thermal properties

T	293	373	573	773	973	1073	1273	1473	1673
λ	0.054	0.052	0.045	0.039	0.032	0.025	0.027	0.029	0.031
c	397	435	546	686	894	928	679	739	799

Table 2 Temperature dependent mechanical properties of C-Mn steel

T	293	373	573	773	973	1073	1273	1473	1673
α	12.03	12.15	12.45	12.75	13.05	13.20	13.50	13.80	14.10
E_1	209	204	191	155	97	67	9	6	3
σ_1	398	379	305	192	41	36	28	20	12
σ_2	562	542	466	356	210	124	29	21	13

The forces due to these strains acting on an element b_x dz as shown in Fig. 3 are given by,

$$F_x(n, x, y, z) = E_1(n, x, y, z)\, \varepsilon(n, x, y, z) b_y\, dz \qquad (12)$$

when $\varepsilon(n, x, y, z) \le \varepsilon_y(n, x, y, z)$

or

$$F_x(n,x,y,z) = [E_1(n,x,y,z)\ \varepsilon_y(n,x,y,z) + E_2(n,x,y,z) \{\varepsilon_y(n,x,y,z) - \varepsilon(n,x,y,z)\}]\ b_y\, dz \qquad (13)$$

when $\varepsilon(n, x, y, z) > \varepsilon_y(n, x, y, z)$

Therefore moment due to the thermal strains at different nodes in x-direction at time step n, will be given by,

$$M(n,x)=\sum\left[\frac{F_x(n,x,y,z)+F_x(n,x,y,z+1)}{2}\right] \times[z-th/2] \qquad (14)$$

Similarly, the forces acting on the strips along the y-axis and the moment due the thermal strains in y-direction will be given by,

$$F_y(n, x, y, z) = E_1(n, x, y, z)\, \varepsilon(n, x, y, z)\, b_x dz$$

when $\varepsilon(n, x, y, z) \le \varepsilon_y(n, x, y, z)$

or

$$F_y(n,x,y,z) = [E_1(n,x,y,z)\ \varepsilon_y(n,x,y,z) + E_2(n,x,y,z) \{\varepsilon_y(n,x,y,z) - \varepsilon(n,x,y,z)\}]\ b_x\, dz$$

when $\varepsilon(n, x, y, z) > \varepsilon_y(n, x, y, z)$

$$M(n,y)=\sum\left[\frac{F_y(n,x,y,z)+F_y(n,x,y,z+1)}{2}\right] \times[z-th/2]$$

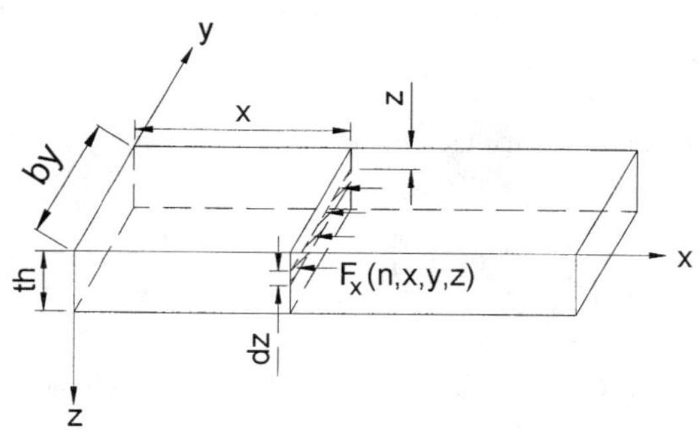

Fig.3 A portion of plate strip considered in the analysis

2.3 Constant Stiffness System

When heating over a line is carried out on a plate, like in the case of bead-on-plate welding, a time dependent temperature profile is created as discussed above in section 2.1. This causes variation of modulus leading to variation of rigidity of plate in a time domain. The solution of such problems is very complex. The purpose of this work was to develop a method and concept that was used to derive accurate and convenient solutions to these complicated nonlinear plate problems. The methodology developed in [17] regarding computation of reduced (or equivalent) modulus for beams, will be extended here to apply for the inelastic analysis of plates.

By using the reduced rigidity, the variable stiffness problem was simplified to that of a constant stiffness system.

For the evaluation of the reduced rigidity, it was essential to know the stress strain characteristics of the material undergoing thermal cycles. The inelastic analysis was carried out, using bilinear approximation of the stress-strain curve of the mild steel as shown in Fig.4.

Let the curvature of the neutral surface produced by the bending moment M_y be r_y. Therefore at a distance z from the neutral surface, the unit elongation ε is given as,

$$\varepsilon = z / r_y .$$

Therefore $dz = r_y \, d\varepsilon$.

Let h_1 and h_2 be the distances from the neutral surface to the lower and upper surfaces of the plate respectively, we have,

$\varepsilon_1 = h_1 / r_y$ and $\varepsilon_2 = h_2 / r_y$

$$\text{Therefore} \quad \Delta_y = \varepsilon_1 + \varepsilon_2 = th / r_y . \tag{15}$$

$$\text{Similarly} \quad \Delta_x = th / r_x$$

Now the total strains and stresses are given by,

$$\varepsilon'_x = \frac{1}{1-\nu^2}(\varepsilon_x + \nu\varepsilon_y) \tag{16}$$

$$\varepsilon'_y = \frac{1}{1-\nu^2}(\varepsilon_y + \nu\varepsilon_x) \tag{17}$$

$$\sigma_x = \frac{E_x}{1-\nu^2}(\varepsilon_x + \nu\varepsilon_y) \tag{18}$$

$$\sigma_y = \frac{E_y}{1-\nu^2}(\varepsilon_y + \nu\varepsilon_x) \tag{19}$$

Therefore Eq.(18) and (19) could be written as,

$$\sigma_x = E_x \varepsilon'_x \tag{20}$$

$$\sigma_y = E_y \varepsilon'_y \tag{21}$$

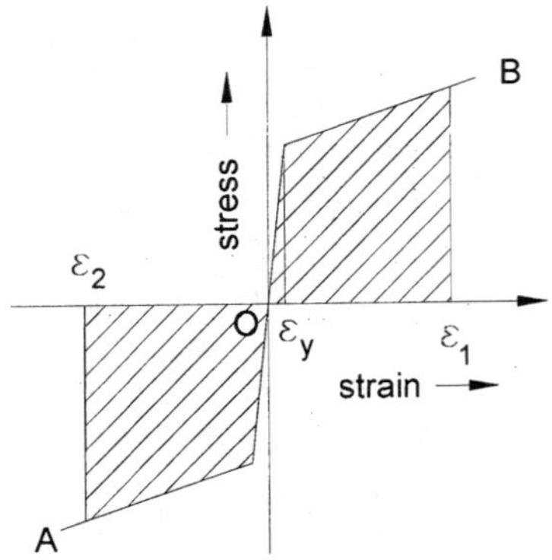

Fig.4 Bilinear approximation of stress strain curve

At a given cross section, the bending moment M_y of the plate with unit sides will be given by,

$$M_y = \int \sigma_y z \, dA = \int_{h2}^{h1} \sigma_y z \, dz \tag{22}$$

Substituting for z and dz , one obtains,

$$M_y = r_y^2 \int_{\varepsilon 2}^{\varepsilon 1} \sigma_y \varepsilon_y \, d\varepsilon_y \tag{23}$$

From Eq.(15) and (23), one obtains,

$$M_y = \frac{th^3}{\Delta_y^3} \frac{1}{r_y} \int_{\varepsilon 2}^{\varepsilon 1} \sigma_y \varepsilon_y \, d\varepsilon_y \tag{24}$$

Therefore the reduced rigidity D_{ry} along the y-coordinate can be written as,

$$D_{ry} = \frac{th^3}{\Delta_y^3} \frac{1}{1-\nu^2} \int_{\varepsilon 2}^{\varepsilon 1} \sigma_y \varepsilon_y \, d\varepsilon_y \tag{25}$$

Similarly, the bending moment M_x and the reduced rigidity D_{rx} along the x-corodinate of the plate will be,

$$M_x = \frac{th^3}{\Delta_x^3} \frac{1}{r_x} \int_{\varepsilon 2}^{\varepsilon 1} \sigma_x \varepsilon_x \, d\varepsilon_x \tag{26}$$

$$D_{rx} = \frac{th^3}{\Delta_x^3} \frac{1}{1-\nu^2} \int_{\varepsilon 2}^{\varepsilon 1} \sigma_x \varepsilon_x \, d\varepsilon_x \tag{27}$$

The integrals in Eq.(25) and (27) represent the moment with respect to the vertical axis through the origin O of the shaded area of the stress strain curve as shown in Fig.4.

An iterative procedure was used to determine D_{rx} and D_{ry}. Since Δ_x and Δ_y were unknown, the iteration was initiated by assuming a value for Δ_x and Δ_y. With the assumed values, Eq.(24) and (26) were used to determine M_{yreq} and M_{xreq}. The procedure was repeated by assuming new values for Δ_x and Δ_y, until M_{yreq} and M_{xreq} became equal to the actual moments M_y and M_x respectively, that were obtained from the thermal load. With these values of Δ_x and Δ_y, the reduced rigidities D_{rx} and D_{ry} were calculated from Eq.(27) and (25) respectively.

2.4 Deflection

The curvature-moment relationship for rectangular plates can be written as,

$$\frac{\partial^2 w}{\partial x^2}=\frac{1}{1-\nu^2}\left(\frac{M_x}{D_{rx}}-\nu\frac{M_y}{D_{ry}}\right) \tag{28}$$

$$\frac{\partial^2 w}{\partial y^2}=\frac{1}{1-\nu^2}\left(\frac{M_y}{D_{ry}}-\nu\frac{M_x}{D_{rx}}\right) \tag{29}$$

With known values of D_{rx} and D_{ry} the deflection 'w' due to the thermal moments M_x and M_y was calculated by numerically integrating Eq. (28) and Eq. (29) respectively.

3. EXPERIMENTS AND RESULTS

The experimental setup comprised of instrumentation for precise measurement of deflection due to a thermal load. To simulate the case of thermal loading, bead-on-plate welding was carried out on test specimens of 6mm thick C-Mn steel plates. The dimensions and control points of measurement of deflection of the plate are shown in Fig.5. The composition of the C-Mn steel used are given in Table 3.

The measurements were taken using a LVDT as shown in Fig.6, prior to welding and after welding once the specimen cooled down to room temperature at the control points as shown in Fig.5. The difference of these two measurements indicated the resultant deflections of the plates due to the thermal load caused by the welding arc. The welding parameters and the results, both calculated and measured for the control points 31, 32 and 33 as shown in Fig.5, are given in Table 4. The deviation, between the calculated and measured deflections at the plate edges were, found to be within 7.5%.

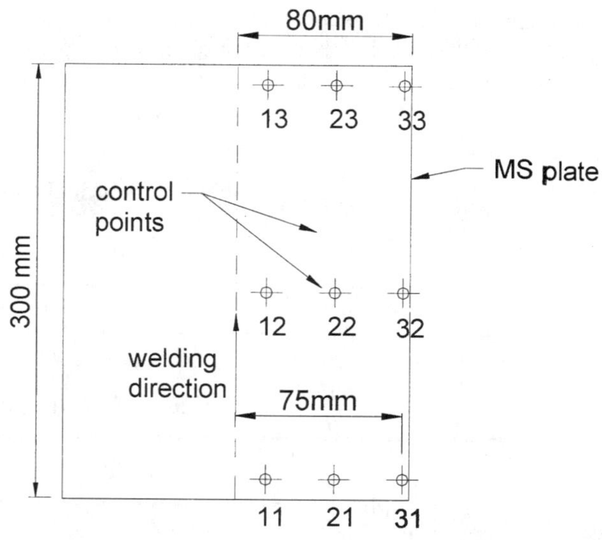

Fig. 5 Specimen dimensions and control point positions

Fig. 6 Deflection measurement using LVDT

4. CONCLUSION

The concept of pseudolinear equivalent constant rigidity system has been developed in this investigation in case of a plate undergoing variation of rigidity due to a continuously changing temperature profile. The method has been effectively implemented in this investigation to solve nonlinear thermo-elasto-plastic problem of angular deformation of rectangular plates. The case of line heating has been simulated here through bead-on-plate welding carried out by submerged arc welding. The theoretical results compared well with those of the experimental ones. The results are encouraging for further extending this concept to evaluate the deformation pattern due to multi-line heating. The present method was found to be computationally more efficient and simpler to model compared to FEM solving similar thermo-elasto-plastic nonlinear problem. The analysis procedure presented in this work and the results thus obtained holds a great promise for analyzing the complicated thermoelastic-plastic problems as encountered in line heating process.

Table-3 Chemical Composition of C-M_n Steel in %

C	Si	Mn	P	S	Cr	Ni	Mo	Cu	Al
0.19	0.37	1.57	0.023	0.027	0.06	0.03	0.01	0.04	0.046

Table - 4 Welding Parameters and Deflections

Sl. No.	Plate thickness (mm)	Welding			Deflection (mm)					
		Current (A)	Voltage (V)	Speed (mm/s)	Calculated at location			Measured at location		
					31	32	33	31	32	33
1	6	480	24	12	1.45	2.18	2.48	1.37	2.03	2.46

References:

1. Watanabe, M. and Satoh, K., *Effect of Welding Conditions on the Shrinkage Distortion in Welded Structures*, Welding journal, Vol. 40, 1961.

2. Muraki, T., Bryan,J.J., and Masubuchi, K., *Analysis of Thermal Stresses and Metal Movement During Welding*, ASME Journal Engineering Materials and Technology, Vol.97, No.1, Jan. 1975, pp.81-91.

3. Fujita, Y., and Nomoto,T., *Studies of Thermal Stresses in Welding with Special Reference to Weld Cracking*, First International Symposium on Cracking and Fracture in welds, Tokyo, Nov. 1971, 12 pp.

4. Masubuchi, K., Analysis of Welded Structures, Pergamon Press, Oxford, U.K., 1980.

5. Kristina, A. et al, *Artificial Neural Network Applied to Arc Welding Process Modeling and Control*, IEEE Transaction on Industry Applications, Vol.26, No.5, sept./oct.1990, pp.824-830.

6. Argyris, J.H., Szimmat, J., and Willam, k.J., *Computational Aspects of Welding Stress Analysis*, Computer Methods in Applied Mechanics and Engineering, Vol.33, 1982, pp.635-666.

7. Rybicki, E. F., Schmueser, D. W., Stonesifer, R. B., Groom, J.J., and Mishler, H. W. , *A Finite Element Model for Residual Stresses and Deections in Girth-Butt Welded Pipes*, Journal of Pressure Vessel Technology ,Vol.100, 1978, pp.256-262.

8. Papazoglou, V.J. and Masubuchi, K., *Numerical Analysis of Thermal Stresses During Welding Including Phase Transformation Effects*, Journal of Pressure Vessel Technology, Vol.104, 1982, pp.198-203.

9. Murakawa, H., Ueda, Y., Zhong, X. M. , *Buckling Behavior of Plates Under idealized Inherent Strain*, Transactions of JWRI, Vol.24, No.2, 1995, pp.87-91.

10. Nomoto, T. et al., *Development of simulator for plate bending by line heating,* Journal of the Society of Naval Architects of Japan, Vol. 168, 1990.

11. Yokio Ueda, Hidekazu Murakawa, Ahmed Mohamed Rashan, Yasuhisa Okumoto, and Ryoichi Kamichika, *Development of Computer-Aided Process Planning System for Plate Bending by Line Heating(Report 1)- Relation Between Final Form of Plate and Inherent Strain*, Journal of Ship Production, Vol.10, No.1, Feb. 1994, pp.59-67.

12. Yokio Ueda, Hidekazu Murakawa, Ahmed Mohamed Rashan, Yasuhisa Okumoto, and Ryoichi Kamichika, *Development of Computer-Aided Process Planning System for Plate Bending by Line Heating(Report 2)- Practice for Plate Bending in Shipyard Viewed Aspect of Inherent Strain*, Journal of Ship Production, Vol.10, No.4, Nov. 1994, pp.239-247.

13. Rybicki, E. F., Stonesifer, R. B., *Computation of Residual Stresses Due to Multipass Welds in Piping Systems*, Journal of Pressure Vessel Technology ,Vol.101, 1979, pp.149-154.

14. ABAQUS User's Manual, Hibbit, karlsson and Sorensen Inc., Farmington Hills, Michigan, 1993. 26. ANSYS User's Manual, Swanson Analysis System Inc., Houston, Pennsylvania, 1992.

15. ANSYS User's Manual, Swanson Analysis System Inc., Houston, Pennsylvania, 1992.

16. Kuo, H.C., Wu, L.J., and Kuo, Y.H., *A Study on Temperature and Distortion Measurement of Ship Plate for Heat-Bending Process*, Measurement Technology and Intelligent Instruments, 1998, pp.44-51.

17. Adak Malabika and Mandal N.R., *Thermo-mechanical analysis through a Pseudolinear Equivalent Constant stiffness System*, communicated to Journal of Engineering for the Maritime Environment .

18. Mandal N.R and Adak Malabika, *Fusion Zone and HAZ prediction through 3-D simulation of welding thermal cycle*, Journal of the Mechanical Behaviour of Materials, Vol.12, No.6, 2001, pp401-414.

19. Gundersen_; et al, *Modeling of residual stresses in weld simulated restrained C-Mn steel specimen,* Procedings of Ninth (1999) International O_shore and Polar Engineering Conference, Brest, France, May30 - June04, pp187-194.

Nomenclature

E = elastic modulus at room temperature
E1= elastic modulus at different temperatures
Dr= reduced rigidity
M = moment
T = temperature
V = arc voltage
I = arc current
F = force
h = convection coefficient
b_x = strip width along x-axis
b_y = strip width along y-axis
h_1, h_2 = distance from neutral surface
c = specific heat
th = plate thickness
r = radius of curvature
t = time
n = index (time step)
λ = thermal conductivity
ρ = density
σ = stress
$\varepsilon_1, \varepsilon_2$ = elastic strain
ε_y = yield point strain
ν = Poisson's ratio
η = arc efficiency

On the safety and training of High Speed Craft navigators along the coast of Norway.

Norvald Kjerstad. Associate Professor in Nautical Science at Aalesund University College, Norway.

ABSTRACT

This article will present the results of a comprehensive survey done among Norwegian High Speed Craft (HSC) navigators and includes details related to operation of vessels along the demanding inshore waterway of Norway, as well as the navigator's educational background and competence. The survey was part of the preparations related to the construction of a State of the Art HSC-simulator and training regime for HSC navigators at the Aalesund University College. As presented in the article there is an alarming gap between the reported use of electronic navigation aids and the knowledge of the limitations of these navigational aids.

1. INTRODUCTION

After many accidents and incidents with High Speed Crafts (HSCs), where as much as 80% is caused by human error, there has been a strong focus on reducing this category. IMO launched the HSC-Code in 1994, with a major update in 2000 (IMO 2001a), where requirement for technical and operational improvements are described. IMO has also updated the STCW-convention (IMO 2001b) and introduced the ISM-code (IMO 2002) for a general improvement on maritime education and management aspects. In addition to this, the maritime authorities in some countries including Norway have introduced special requirements (NMD 2002), qualifying for special endorsement in the navigator licence. In spite of this accidents still happen, and accident reports often conclude with elements of human error or operational weaknesses. Through a lot of informal contact with HSC navigators we also claim that many navigators feel that much of the education and training required from IMO is not very relevant for navigating modern HSCs along the demanding coast of Norway.

So, what is a HSC ? In the HSC-code this is defined as: a craft capable of maximum speed, in meters per second (m/s), equal to or exceeding:

$$3.7\nabla^{0.1667}$$

Where, ∇ = Volume of displacement corresponding to the design waterline (m^3). Further is the <u>design waterline</u> defined as the waterline corresponding to the maximum operational weight of the craft with no lift or propulsion machinery active. Further is a <u>passenger craft</u> defined as a craft carrying more than 12 passengers.

In the Norwegian regulations there are different regulations for ships longer or shorter than

24 meter, and the "high speed limit" is set to 20 knots (NMD 2002).

Since many ships in use today is not build and operated according to the HSC-code the article will here have a broader approach, and include all non-displacement vessels above 15 meter which on a daily year round and commercial basis is carrying passengers in the coastal waters. At present there are more than 100 ships of this type operating along the coast of Norway. All taxi-, ambulance- and rescue vessels are excluded in this paper. One should be aware of the fact that many of the relatively small HSC, sailing in national waters, are not defined to be within the international SOLAS-convention (IMO 2001c).

Based on this regulatory framework and informal indications from sailing navigators we decided to do a comprehensive survey of different element of HSC navigation, based on the navigator's own views. The main goal with the work was to map the relevant element related to operational safety and training on the HSCs, as well as correlations between different aspects. Secondary a more general approach to coastal navigation for all vessels will be investigated and discussed. The result and the conclusion we intend to use as a tool for developing new and improved training programs, as hopefully will improve the safety for coastal navigation in the long term. The survey was launched in the spring of 2002, and was complementary to the work Aalesund University College (AUC) was doing on the construction of a State of the Art HSC simulator, and related course syllabus (Kjerstad 2002). Gathered information from the survey will also be important in future R&D with the simulator, e.g. in studies of accident (Kristiansen 2002) as well as development of new and better operational systems.

The definition of coastal navigation in this paper will be the internal waters along the coast of Norway. Probably no other places on earth are year round navigation routes more complex and challenging than along this coast. The distance from the Swedish border in the south to the border of Russia in the north is approximately 1500 nm (2800km) (excluding Arctic archipelagos and long fjords). Most of this area is sheltered and navigable for most ships, even if many straits are narrow and shallow. One should also bear in mind that most of this area is affected by almost 24 hour darkness and frequent snowstorms during the winter season.

2. METHOD

The intensions have been to collect and analyse data for three main purposes. First we would like to find different operational aspects related to the human element, and second we would like to do a survey of technical elements relevant for the design of the new simulator concept at AUC. Third we would like to collect a variety of data as background for future research, as described in the discussion. This includes:

- Competence and background of the navigators.
- Demand and attitude to training and retraining.
- Safety culture and routines during navigation.
- Opinion on navigation system performance.
- Opinion on bridge design and ergonomics.
- Support from shipowner / operational manager.
- The job situation for the navigators.
- Attitude to rules and regulations.
- Available instruments and performance.

To accomplish this goal we found that a comprehensive survey based on a questionnaire was the best approach. This would also help us to gain a better understanding of the reality and reason for all the rather informal information that pops up in different occasions, e.g. during observations onboard different vessels. Since little work has been done in this field we also found this explorative method would help us to develop theories for further work and longitudinal studies in the future. The scales and layout of the questionnaire were designed in accordance to relevant methods and literature in this field (Converse & Presser 1986, Håvold 2001, Ringdal 2001), and reviewed by four experienced navigators and two experienced scientists. In addition to data fields for basic information and comments, a six-division *Likert* scale was constructed to provoke an attitude rather than invite to easy average answer in the middle of the scale. All questions were formed as assertions, summing up to a total of 127 data fields. To avoid bias of the "yes / agree" effect a mixture of positive and negative assertions were made.

A number of HSC and coastal shipping companies, representing the different regions of Norway were selected, and the questionnaire was distributed to the ships via the company mail. Because of difficulties to call for response for improving feedback from the ships, prices were promised to three of the responding navigators. A total of 9 companies, representing 51 HSCs were contacted. From this group 26 HSCs responded. For analyse of certain issues we also decided to send questionnaire to a reference group. The reference group was based on displacement ships sailing with speed of 14 – 20 knots, and with the major activity along the same coastline as for the HSCs. From this group 49 of 58 contacted ships responded, of these 11 were coastal steamers (ref. fig. 1). The questionnaire analyse will consequently be based on a total of 75 responding navigators, which give a mixed feedback on 69%. The selection of the ships was chosen to represent a balanced geographical distribution. The fundamentals of the responding ships are given in table 1.

Table 1
Average of the fundamentals of the ships responding on the questionnaire.

Group	Number of ships, n	Ship length (meter)	Max. speed (knots)	Age (year)
HSCs	26	30.1	31.8	9
Other coastal vessels	49	98.5	15.9	10

All statistical analyses presented in the next chapter have been done by the SPSS software (SPSS 1999), and with good guidance by Green & Salkind 2002 and Ringdal 2001. Missing data in the questionnaire will be excluded from the calculations in SPSS.

3. RESULTS

In this chapter I will present the most of the results from the survey. Different areas of interest will be arranged in sub-chapter. Even if not all elements of information will be analysed in the discussion, I have decided to publish them for potential background for other research in the field of HSC and coastal navigation. Reliability for the questions and scales used has been tested, and the Cronbach Alpha coefficient lies typically between 0.4 and 0.9. The questionnaire was in its origin in Norwegian, but for those interested, an English version can be provided from the author.

3.1 Ship characteristics and instrument performance

Table 2 shows comments on available instruments above minimum requirement. Since no ships have electronic chart complying with full ECDIS standard, the commonly used term ECS (Electronic Chart System) is used. ENC is used as term for the Electronic Nautical Chart (chart database). Results regarding use of Radar and ECS will be presented in chapter 3.4.

The navigators were also asked to answer on certain assertions in the questionnaire. This was based on a mix between positive or negative statements, but to make it easier to interpret and to avoid ambiguity, negative are converted to positive in the tables below (tab. 3, 7, 8, 10 and 13). A six level *Likert* scale, as indicated below, was used for measuring response:

1. Strongly disagree
2. Moderately disagree
3. Mildly disagree
4. Mildly agree
5. Moderately agree
6. Strongly agree

The column indicating, "% agree" is the total response within scale 4 – 6. The score is the mean value of the answers on the scale from 1 – 6. The Standard Deviation (SD) indicates the distribution of the answers. The score level based on this type of ordinal level scales has in it self no meaning else to indicate the mean value in the distribution of answers. The "percent agree" is the sum of all answers in category 4 – 6. By this presentation the highest value of "score" and "% agree" can be interpreted as a "good" or satisfactory result. Low scores will consequently be studied closer.

Table 2
Navigation equipment above minimum requirements, and comments on the use.

HSCs	
ECS	76% had ECS, but only 6% of the used ENC was authorized.
GPS	28% had 2 or more receivers, 92% had DGPS.
Auto Pilot and waypoints	33% had option for track steering mode, 63 % used waypoint database, 52% had edited the database on board.
Night Camera	4% had night camera.
Other ships	
ECS	98% had ECS, but only 13% of the used ENC was authorized.
GPS	90% had 2 or more receivers, 100% had DGPS.
Auto Pilot and waypoints	66% had option for track steering mode, 92 % used waypoint database, 88% had edited the database on board.
Night Camera	10% had night camera.

Table 3
Navigators response on the assertions in the questionnaire, related to system performance. Type 1 is HSCs, and 2 is other vessels.

Assertions	Type	Score	SD	% agree
The GPS has never failed on his ship	1	3.3	1.8	37.5
	2	3.1	1.7	38
Accuracy of our GPS is always better than 10m, while sailing at full speed	1	2.6	1.3	24
	2	3.2	1.5	43.7
The DGPS corrections never fail along the coast	1	3.6	1.4	54.2
	2	2.7	1.1	49
The electronic chart database has no errors in the coastal waters	1	2.0	0.8	0
	2	2.2	1.1	8.8
Information on ECS is easy to read and interpret	1	3.4	1.4	50
	2	3.5	1.3	80.9
Radar performance is good in rain and rough seas	1	1.9	1.7	28
	2	2.5	1.2	42.9
Night camera contributes to significantly better safety	1	4.6	1.4	90
	2	3.8	1.4	72.7
The instruments on the bridge are perfectly located	1	3.8	1.7	52
	2	3.9	1.6	58
Light on the bridge don't make problem for night watch keeping	1	2.4	1.7	44
	2	3.5	1.3	65.3
The company provide needed equipment for safe operation	1	3.4	1.3	48
	2	4.6	1.4	79
Control leavers etc. are perfectly located on the bridge	1	4.3	1.5	64
	2	4.0	1.4	65.4

On an invitation to do a general comment on the instrument performance, the only significant response was to the radar. As many as 44% of the HSC navigators, and 50% of the navigators on the coastal steamers, meant that the radar was too sensitive to rain and sea clutter. Since only eight of the vessels had Automatic Identification System (AIS) installed, the navigators were asked if they believed AIS would contribute to reduce risk of collisions along the coast. To this question 54.2% from HSC and 71.8% of the other ships answered positive.

3.2 Navigator's background and training needs

In Norway there are different ways to obtain navigator education. The different systems have also had a relatively rapidly changing structure, from short courses to more comprehensive University College studies. Now (2003) there are 4 University Colleges, 1 Naval academy, and approximately 20 upper secondary schools (including Technical Community Colleges) giving nautical education according to the highest level (management level) in STCW. In addition, a limited certificate can be obtained from other institutions and correspondence schools as well. It is therefore interesting to see which navigator background recruit to the fleet sailing in the very demanding coastal waters of Norway. In this chapter the educational background of the navigators are presented, as well as their experience and attitude to the education they have been given at the different institutions. For the HSC group 84% of the answers came from Captains and 16% from Chief Officers. For the other ships, 38% were Captains, 38% Chief Officers, 19% First Mates and 5% Second Mates. In table 4 it is indicated what kind of licences the different groups were holding, and in table 5 the education level and experience is presented. In table 6 and 7 the included elements in, and attitude to the education is presented.

Table 4
International navigator licences hold by the navigators in the survey.

Licence / Group	Number of ships, n	Class 1 [%]	Class 2 [%]	Class 3 [%]	Class 4 [%]	Class 5 or other [%]
HSCs	26	48	16	4	24	8
Other	49	74	18	4	4	

Table 5
Educational background and average seafaring experience for the navigators in the survey.

	HSC (n=26)	Other (n=49)
UniversityCollege level (2-3years after upper secondary) [%]	12	18
Upper secondary school (2 – 4years) [%]	68	64
Other education (typical 3 years upper secondary, before 1981) [%]	20	18
Experience as navigator [years]	16.1	14.3
Experience as HSC navigator [years]	9.4	0.6
Experience from coastal navigation [years]	14.3	11.1
Special regional permit (check-out or certificate) [%]	92.3	69.4
Average age [years]	41.2	40.0

Table 6
Included elements in the nautical education.

Subjects and issues which was included in the education	HSC (n=26) [%]	Other (n=49) [%]
Visual simulator training	32	42
Coastal navigation on training ship	84.6	74
Navigation with GPS	36	56
Navigation with ECS / ECDIS	12	30
Navigation with Radar	96	100
Coastal navigation	84	70
HSC navigation	4	2
Navigation with automatic track steering	4	10
Stability and floating characteristics for catamarans	8	2

Table 7
Attitude to quality and relevance of the elements of the nautical education. Type 1 is HSC navigators and 2 is from the other ships (1) ref. tab. 3 and chapt.3.1).

Assertions	Type	Score 1)	SD	% agree 1)
My education was good regarding coastal navigation	1	2.7	1.5	24
	2	2.6	1.4	26
Training ship was very helpful for coastal nav. training	1	5.2	1.0	91.7
	2	5.0	1.1	91.8
Simulator was very helpful for coastal nav. training	1	3.8	1.5	71.4
	2	4.1	1.4	68.3
Education in electronic navigation was very good	1	1.2	1.2	12
	2	1.7	1.3	30.6
A dedicated HSC would be important for coastal training	1	4.8	1.1	92
	2	5.0	1.0	95.5
Good education of damage stability for this ship type	1	1.9	1.4	20.8
	2	2.8	1.5	31.9
Good education in the limitation of the GPS	1	2.7	1.7	33.3
	2	3.2	1.7	48.9
Navigation textbooks were very good regarding present job	1	2.3	1.4	32
	2	2.9	1.4	38.7
The school was well equipped for modern education	1	3.4	1.2	40
	2	3.6	1.5	47.9
Celestial navigation training has been important for me	1	2.1	1.6	20
	2	3.2	1.7	48
STCW-95 upgrade has been important for safety	1	2.4	1.8	48
	2	2.8	1.5	60
The company has given me opportunity to have training on the instrument on board this vessels	1	3.1	1.6	40
	2	3.8	1.5	55.1
I was a better first-time navigator than today's graduates	1	3.4	1.3	58.3
	2	2.9	1.3	35

In addition to the assertions in table 7 all navigators were asked to comment on insufficient and superfluous subjects in their education. Most significant results of the comments are:

Insufficient:

- 40% found electronic navigation system limited
- 17% found coastal navigation limited
- 11% found regulation and ISM limited.
- 8% found simulator limited.

Superfluous:

- 32% found celestial navigation superfluous.
- 9% found old navigation systems superfluous.

All of the navigators on the HSCs had attended the mandatory five-day HSC course. They were therefore asked to comment on positive and negative sides of the course. This is important information to have when new regimes and simulator training shall be implemented. Most significant results of these comments are:

Positive:

- 16% found the physiological approach important.
- 16% found possibility for exchange of experience important.
- 12% find communication training important.

Negative:

- 28% found the lack of simulator use negative.
- 12% found lack of crises management negative.

On a question to comment on what kind of issues they would like to be updated on, the most significant feedbacks (from both groups) were:

- 23% would like electronic navigation systems.
- 23% would like IT and computers.
- 15% would like ship technology and cargo handling.
- 13% would like economical and administrative issues.

In addition to this, issues like law, ISM, and language, were mentioned.

To those with education level limited to upper secondary school (ref. tab. 5), the following question were asked: Have you considered additional and higher education to obtain a formal academic degree, giving better opportunities for other jobs and further studies. On this question 84% of the HSC navigators, and 72% of the other navigators, answered yes.

3.3 Navigators workload and incidents

An important safety factor is the workload of the navigator. Therefore the navigators were asked to consider their own situation. The response to some assertions is presented in table 8.

Table 8

Navigators opinion on their own work situation. Type 1 is HSC navigators and 2 are from the other ships ([1]) ref. tab. 3 and chapt.3.1).

Assertions	Type	Score [1]	SD	% agree [1]
I never feel stress while navigating the route	1	4.1	0.9	92
	2	4.1	0.9	91.7
Extensive regulations does not effect safe operation	1	2	1.5	32
	2	2.3	1.2	36.7
Non-conformities and hazardous occurrences are always reported	1	2.9	1.8	60
	2	2.1	1.6	37.5
I am rewarded for doing my work in a safe manner	1	1.9	1.1	8.4
	2	2.6	1.4	27.2
Operational considerations never hinder us to operate safely	1	2.2	1.2	25
	2	2.7	1.4	44.4
Work share between crewmember work perfectly	1	5.3	0.6	100
	2	4.5	1.1	81.3
I am not worried for more technology on the bridge	1	3.2	1.3	72
	2	3.5	1.1	79.6
We never deselect alarms to reduce disturbing noise	1	4.3	1.3	92
	2	3.9	1.3	81.6
It is enough crew to operate the vessel safely	1	3.1	1.5	64
	2	3.4	1.4	69.4
I thrive as navigator / captain on this ship	1	5.0	1.3	88
	2	5.2	1.1	89.4
If I was offered a new job ashore, I would not have taken it	1	3.2	1.7	64
	2	3.6	1.3	84

Table 9
Accidents experienced by navigators on their present vessel. Sum of accidents adds up to more than 100% because many had experienced more than one type of damage.

Ship type	Damage, % of vessels	Collision, terminal [%]	Grounding [%]	Collision, floating object [%]	Rough weather damage [%]	Collision, other ship [%]	Fire [%]
HSCs	76	45.8	37.5	29.2	20.8	4.2	4.2
Other	62.5	20.8	16.7	4.2	31.3	25	8.3

On a more general comment on stress factors, approximately 10% of the navigators responded that too much "paper work" is stressing. Approximately the same number reported telephones and communication as stressing. Since no HSC had installed Voyage Data Recorder (VDR) the navigators were asked if future implementation would contribute to more insecurity and poor working condition on the bridge. Here 32% of the HSC navigators, and 28% of the other answered confirming.

In addition to the assertions in table 8 all navigators were asked if they had experienced damages or accidents with their vessel. Response from the two groups is listed in table 9.

3.4 Navigation technique and route

For development of new training regime, as well as R&D on HSC navigation it is necessary to have a thorough understanding of what is the opinion of the navigators performing this type of job on an everyday basis. This includes both how they actually operate the ships and their views on the available infrastructure. It is therefore attached great importance to this issue in the survey, as well as spent a lot of time onboard different ships. Different regulatory bodies and shipyards have also been visited. In table 10 different aspects of this is listed. In an initial comment the navigators ranked the two most important navigational aids for coastal navigation as follows (all ships):

1) 98% regarded the radar as the most important instrument (2% ECS/GPS).
2) 67% regarded the ECS/GPS as the second most important system.

It can here be added that the HSC group weighted the ECS/GPS lower than the other ships. This had probably to do with the fact that a lower fraction of the HSCs had installed ECS (ref. tab. 2).

In addition to the assertions given in table 10 the navigators were asked to comment on the most frequent reason for speed reduction and what they assumed to be the most dangerous factor on the route (table 11). This gave the following results:

The most frequent reason for speed reduction:
HSCs:
64% rough weather
24% poor visibility
8% restricted waters
4% busy traffic

Other:
54% restricted waters
22% rough weather
15% poor visibility
9% busy traffic

A question was also related to the most frequent used mode on the radar during coastal navigation. Here 88% of the HSC navigators preferred "head up", while only 16% of the other group preferred this mode.

Further the navigators were asked to suggest actions that could improve the safety for coastal navigation. The most significant suggestions are presented in table 12.

3.5 Management and safety culture

During the last years there has been growing focus on the importance of proper ship management. This has lead IMO to launch the International Ship Management code (ISM-code) (IMO 2002). The ISM code has applied to all HSCs (SOLAS def.) and passenger vessels since July 1998, and to other cargo vessels since July 2002. Since other research is done in this field at AUC (Håvold 2002), this survey had no intension to go in depth in this matter. Some elements were nevertheless interesting as background for later discussions. The elements investigated are listed in table 13. From the HSC group 92% responded that the ISM code was implemented, while the figure from the other ships was 85%.

Table 10
Navigators response on issues regarding navigational technique and navigation infrastructure. Type 1 is HSC navigators and 2 are from the other ships([1] ref. tab. 3 and chapt.3.1).

Assertions	Type	Score [1]	SD	% agree [1]
The radar is of vital importance for coastal navigation	1	4.9	0.27	100
	2	5.0	0.14	100
The ECS is of vital importance for coastal navigation	1	4.6	0.65	100
	2	4.5	1.0	96
Visual bearings are often used during coastal navigation	1	1.8	1.3	8.6
	2	2.1	1.4	13.3
Speed is often reduced due to poor visibility and rough sea	1	4.8	0.8	96
	2	4.5	0.9	93.8
We seldom need to give way in the "last minute" to avoid collision	1	4.8	0.4	100
	2	4.5	0.7	97.9
We often use automatic plotting (ARPA) on the radar	1	1.7	1.9	39.1
	2	4.5	1.4	87.8
The route is very easy to navigate	1	1.6	1.3	20
	2	1.8	1.4	29.8
The route have good aids for navigation at night	1	3.0	1.6	40
	2	4.2	1.4	70.2
The route have good aids for navigation in poor visibility	1	2.7	1.4	24
	2	3.5	1.4	48.9
Required navigation corridors would improve safety on this route	1	3.5	1.8	54.2
	2	3.8	1.4	68.8
Track steering is used on most of the route	1	2.7	1.8	31.8
	2	3.7	2.0	53.3
Track steering and use of ECS / GPS is safer than manual steering and visual position fixing.	1	2.0	0.9	8.0
	2	2.9	1.5	36.2
When track steering is used, same waypoints are never used in both directions	1	1.9	2.0	35.3
	2	3.6	1.5	75.6
It never happen that we deviate from authorized route	1	2.1	1.4	28
	2	1.6	1.0	8.7
Radar surveillance of the route would improve safety	1	2.9	1.5	34.8
	2	2.5	1.5	18.7
We often meet ships that are not giving way	1	4.0	1.5	64.0
	2	4.3	1.0	85.7
VHF is often used to clarify give way manoeuvres	1	3.5	1.7	76
	2	3.9	1.3	85.7
Use of pilot would not improve safety on this ship	1	5.2	1.1	96.0
	2	4.1	1.7	65.4

Table11
The most dangerous conditions referred to the navigators comments in the questionnaire. The figures indicate percentage of navigators considering the related factor as the most dangerous. The categories are defined by the navigators.

Dangerous factor	HSC (n=26) [%]	Other (n=49) [%]
Harsh environment / rough weather	41.7	12.2
Restricted waters	25.0	9.8
Limited Navigation marks	16.7	2.4
Leisure crafts	8.3	4.9
Busy traffic	4.2	12.2
Drift ice	4.2	0
Technical errors	0	34.1
Workload	0	19.5
Inexperienced navigators	0	4.9

Table 12
Suggestions for improvement of navigational safety along the Norwegian coast. The percentage indicated all responding vessels, and the fraction consisting of HSCs. The categories are defined by the navigators.

Suggestion	HSC (n=26) [%]	All vessels (n=75) [%]
More racons (fixed and floating objects)	12.0	30.7
More quick / very quick flashing lights	20.0	20.0
Fixed indirect lighting	32.0	14.7
More floating marks	8.0	8.0
Better electronic charts (ENC)	5.3	5.3
More defined / required traffic systems and leads	5.3	5.3
Better colour on the green buoys	0.0	4.0
More VTS systems	4.0	4.0
Wave measurement systems	8.0	2.7
Reflex on more objects	8.0	2.7
More lead lines	0.0	2.7

Table 13
Navigators response on issues regarding safety culture and management. Type 1 is HSC navigators and 2 are from the other ships (1) ref. tab. 3 and chapt.3.1).

Assertions	Type	Score [1)]	SD	% agree [1)]
Procedures are always updated and evaluated when new equipment are installed	1	3.6	1.6	84
	2	3.6	1.3	76
We always use checklists before departure	1	4.1	1.9	64
	2	4.8	1.7	75.4
Implementing the ISM code has improved the safety	1	4.1	1.0	79.2
	2	4.1	1.4	75.6
The company provide us with important information for safety	1	3.9	1.4	68.0
	2	4.3	1.3	73.4
The safety culture in this company is significantly better than in my previous company	1	4.5	1.2	87.5
	2	4.6	1.3	77.3

In addition to the assertions in table 13 the navigators were asked to comment on if regulations some times could be a restrictive regarding the operation of the ship. The most significant response on this was:

- 17% too much paperwork taking attention from navigating the ship.
- 15% too much regulations.
- 9% ISM and quality control take focus away from navigating the ship.

The categories are made from the written comments from all the navigators.

4. DISCUSSION AND CONCLUSION

This chapter will discuss many of the results presented in chapter 3. The emphasis will be on HSCs, but with reference to other group of ships navigating the Norwegian coast on a regular basis.

4.1 Ship and instrument performance

As we can see from table 1 the average maximum speed of the HSCs are 31.8 knots. Sailing year round along the Norwegian coast with this speed will call for the best possible performance from vessel, instruments and crew. If we analyse the performance of different instruments it is important to have in mind the information given in chapters 3.1 and 3.4, where it is documented that electronic navigation with radar and ECS play a very important role. One should also bear in mind that the gyrocompass, which is an essential part of radar and ECS systems, will have significant problems to stabilize when a HSC is sailing with full speed on various courses on such high latitudes as Norway. As an example the latitudes of busy HSC ports, such as Tromsø, Hammerfest and Honningsvåg is located at respectively N69°40', N70°40' and N71°00'. Since the gyro is connected to radars and ECS, both this system will be affected by the gyro error.

In table 3 there are significant indications of limitations in most instruments and systems used during coastal navigation. With the important role that ECS is reported to have it is alarming to observe that as many as 24% of the HSC fleet don't have this tool for coastal navigation. It is also alarming to see that only 6% of the chart databases are authorized data. The situation on the other vessels seems to be better, but authorized maps are also there lacking. The limited use of authorized ENC is also reflected in table 3, where the score of the assertion related to database errors is extremely low. The gap between the fact that ECS is extensively used and no system has ECDIS standard is another important issue. IMO should absolutely notice this reality and speed up the work with alternative regimes in addition to the high-end ECDIS regime. A standard for performance and installation of ECS would regulate the wilderness of PC based systems on the market. The extensive use of electronic chart systems should also make requirement of "ECS/ECDIS" certificates a most relevant issue. This should definitely been included in the STCW-95 upgrade, but only Australia used this "golden opportunity" to have the navigators updated on ECS/ECDIS (Holden 2002).

Autopilot with track-steering mode is another frequently used instrument. This provides a convenient method of navigation, but it requires high knowledge of limitations in electronic charts, GPS and gyro. The risk involved with this mode of navigation is basically related to the performance of the GPS/DGPS and the electronic nautical chart (ENC). From table 2 we can se that the navigators report a relatively low performance on both GPS, DGPS and the ENC. Altogether this is a dangerous mix, and there will probably be many hazardous occurrences and accidents in the future because of this.

One should believe that a HSC cockpit could be designed in a more or less perfect and standardized layout, much like in the modern airplanes. This is unfortunately not the case, and a large proportion of the navigators complain on poor location of instruments and controls, as well as much disturbing lights during night operations. The problems related to lack of standardization has been highlighted in many accident reports, and are quite clear an area where research and new regulatory development should be done. At AUC work related to bridge design and human interface will have high priority in the future (Kjerstad 2002).

The survey indicates quite clearly the importance of the radar for coastal navigation. All navigators considered the radar as the most important navigation aid. In spite of this, almost 50% of the navigators report the radar to be too sensitive to sea and rain clutter. This has in fact been an issue in accident reports after HSC accidents. The Norwegian Maritime Directorate (NMD) have also discussed the issue in special groups, and many radar experts claim that most of this problem could be solved by means of new technology (Haugen & Eliassen 1997, Isaksen 2001). Other organizations such as RIN (Royal Institute of Navigation) and DGON (Deutsche Gesellschaft für Ortung und Navigation) have also this issue on their agenda for discussion (Williams 1998, Fisher 2002). One of the most difficult hinders in this discussion is the performance standard for maritime radars, issued by IMO. It is therefore reason to assert that the present standard is based on old technology and operational needs outside the needs of the navigators sailing in demanding coastal waters with high speed. Ships are forced to use pulsed radar with rigid standards to antenna performance and user interface. The only HSC approach done in the regulations is related to higher rotation speed of the antenna, but this don't solve the fundamental problem. Detection probability can in fact be reduced by this solution, as well as with modern filtering and correlation techniques (Kjerstad 2002b). Even if all navigators in the survey have had radar included in their education there is reason to believe that the intricate limitations mentioned here was not covered.

The consequences of regulatory limitations, and to some extent lack of insight, are that high performance radars would not apply to IMO standards. It will therefore be a significant extra cost for companies considering high performance radar technology. Another example of limitations is the display, which in IMO standards is very much related to old circular CRT presentations. Because of this we today see that most radars present the important radar data on just above 50% of the screen. The rest is used to present menus and loads of more or less redundant information. There has in fact been initiated a two year work in IMO for revising the performance standard. Summing up this I would conclude that the time has come to apply fundamental new radar technology for all vessels sailing under such demanding conditions as the HSC. To ensure compatibility to RACON and SART the "old" radar will be needed as well in the foreseeable future.

In the HSC-code IMO require use of night vision devices, but give the flag states an option to give exemption for this requirement. Since the survey indicates that only 4% of the HSCs have such devices, it can be concluded that exemption is the normal situation when the authorities approve new ships. In table 3 90% of the navigators indicate quite clearly that night camera would significantly improve the safety. Taken into consideration that many ships operate in almost 24-hour darkness and are vulnerable for collisions with floating objects as well as groundings, it is remarkable that the authorities still allow exemption to the IMO requirement. The lack of night vision cameras can together with lack of ECS also explain some of the low score on the assertion related to the company provision of needed equipment. This situation is most prominent for the HSC operators compared to other ships.

4.2 Navigators background and competence

As showed in chapter 3.2, more than 80% of the navigators have education only from upper secondary school. For most of them this have been a 1+1 year education after primary school and some seafaring experience. On the basis of their feedback we are tempted to say that the STCW convention and the practise of this from the various schools has failed dramatically for these navigators. This assertion can be supported by a few examples. If we focus on the fact that as 100% of the navigators consider ECS as important for coastal navigation (tab. 10), and compare their background in related subjects it is alarming. In the HSC group only 36% have had education in GPS, 12% in ECS, and as few as 4% in track steering. The lack of education in stability and floating characteristics for catamarans is another critical observation from table 6. Only 8% of the HSC navigators reported such education from school. From table 7 we can see that approximately 75% are not satisfied with their education regarding coastal navigation, and special subjects like electronic navigation, simulators, instruments, ISM and textbooks are found limited. Many comment also on issues like celestial navigation and old instruments as superfluous. The reasons to this can be different, but much of this is probably related the problems many schools (and teachers) have to be in phase with the industry, as well as their ability and competence to use the flexibility that actually is build into the STCW convention. A conservation factor is also the centralized assessment systems. This tends to motivate teachers and students to train on previous, and in many cases outdated tests, in the preparation to pass the next exam. With the high number of small upper secondary schools the Norwegian system will have very limited resources, both regarding experts and equipment like simulators and laboratories. Another indicator from the survey in this direction is that as many as 84% of the HSC navigators and 72% of the other navigators have considered to take a university college education to improve their knowledge. This raise the question if the education only applying to the minimum requirements in STCW are not good enough for operation and management of advanced vessels like HSCs and the coastal steamers. From tables 4 and 5 it is indicated that the HSC navigators in average have lower education and licences than other coastal navigators. As much as 36% of the HSC navigators don't have licence on management level (above Class 3) according to STCW. For the other group this figure was only 8%. Age and seafaring experience in the two groups are almost equal.

In 2002 all navigators should be updated to STCW-95. Many navigators found this update not very relevant to improved safety. The subjects that the navigator missed in their education (ref. chapt. 3.2) was only in very limited proportions given in the update program. In e.g. Australia a more up to date approach to the update have been done. There an ECDIS course, applying to IMO Model course 1.27, has been given. According to the survey and our experience from the industry this should be a requirement for all navigators sailing with ECS / ECDIS. As many as 92% of the navigators gave a quite clear indication that training ships are important for realistic coastal navigation training. It is therefore disappointing to note that the last of the Norwegian training ship was laid up 5 years ago. There is no present plan to introduce new ships for this purpose. Since the figure for visual simulators in this issue was around 70% it can be concluded that navigation simulators and such training still have potential for improvement.

Correlations

Based on the survey results a comprehensive correlation test was performed to test if there were significant differences in attitudes to the assertions given in the questionnaire. The results are based on a 2-tailed Pearson Correlation, and results only within significance level 0.01 are presented or mentioned. Since the scale used in the questionnaire was close to an ordinal level scale, also a Kendell Correlation was tested. The two different methods did not differ in correlating at the given level, even if the correlation factors could vary with minor values. We would like to test if there has been an improvement in the education, based on correlating age, experience, licence class and position with the different assertions in the questionnaire. In the assertions related to good and modern education there was no correlation, which means that the progress of the education system must have been limited. Educational background made no difference. Since the different educations are quite different it could be concluded that the main problem seem to be related to the content and the focus of the STCW convention. There was in fact a correlation that indicates that elder navigators meant that they were better first time navigators than young navigators of today. The importance of astronomical issues correlated also to age of the navigators. When test were done regarding modern equipment and electronic navigation there was correlation to younger navigators and navigators from the highest educational level (University College). It was also interesting to observe the correlation between what was considered as a good education and the comprehension of a well-equipped school. This must be a good argument to improve facilities for navigation training. When it comes to operational aspects the experienced navigators tends to represent an attitude that administrative and regulatory aspects don't reduce safety.

4.3 Navigators workload and incidents

In the survey we focus on the navigators opinion on own workload as a possibility for reduced safety. Here it is a pleasure to notice that all navigators thrive in their job. This is an important factor for the performance of a professional seamanship. This good working environment is also confirmed by the fact that only 64% of the HSC navigators would consider a new job ashore (tab. 8). The feedback also uncovers issues of a more critical character. On the HSCs there are reported that operational considerations often hinder the crew to operate safely. Both this and the increasing element of "paper work" are issues that have to be taken seriously by the companies. When it comes to the reports on non-conformities and hazardous occurrences, as required according to ISM, the practice seem to have a clear potential of improvement. The management of the HSC vessels should also pay attention to the fact that only 8% of the navigators feel they are rewarded if they do their work in a safe manner – that is absolutely not complying with safe management. The situation from other vessels is somewhat better, with positive response from 27% of the navigators. It is also alarming to notice that as many as 76% of the navigators have experienced different damages on the vessel. As indicated in table 9 as much as 37.5% was related to groundings. Such figures indicate a risk rate above what is average or acceptable for modern passenger transportation (Kristiansen 2001), and there should be a major potential of improvement. This is contradictionary to the navigators' opinion on own safety culture documented in table 13.

4.4 Future work and improvements

A natural approach to improve safety is to study and improve the factors pointed out as most dangerous by the navigators. This will include elements of education, management, infrastructure, navigation instruments and regulatory regimes. For the HSCs, rough weather and restricted waters is considered as the two most dangerous factors (tab. 11). There should therefore be improved routines to make better decisions on weather limitations. The work to improve navigation infrastructure such as racon and light systems must be continued. This is clearly indicated in table 12 where 30% of all navigators would like more racons on fixed and floating marks. As much as 60% of the HSC navigators find the route poorly marked for navigation during nighttime. This is confirmed in table 12 where 32% of the HSC navigators pointed on the need for more indirect illumination on prominent navigation objects. It is very interesting to notice that only 4% of the navigators considered VTS systems (radar surveillance) important for improved safety. This is a contradiction to public and media focus after many accidents. The education system seems to have quite a lot of challenges, but much has to be done in cooperation with the regulatory bodies. To comply with the most prominent lack of competence a special course in ECS / electronic navigation should be required. The HSC course should also be updated with simulator training. More than 90% of the navigators meant that a dedicated HSC simulator would be important for training in coastal navigation. The AUC has taken these challenges very serious and new training regime as well as a new HSC simulator is implemented. The simulator will also serve as the most important tool in R&D related to high-speed operation. The project "The virtual seaway" is also launched to build up geographical databases for simulator training, along the entire Norwegian coast.

4.5 International comparison

Since IMO has defined international education standards in their STCW-convention it is reason to believe that issues related to education and training described in this paper shouldn't differ much between various nations. Nevertheless, after analysing the Norwegian survey it was found interesting to compare the described Norwegian training regime and HSC operation with other countries with busy activity in this field. Because the high activity, as well as their recognized education system, Singapore was selected for this purpose. Surprisingly it was found very difficult to do a full comparability study. The main reason to this was that most HSCs in the region had Indonesian flag and crew. Because of language problems the English questionnaire therefore proved not be very useful. By interviewing different operators and experts, representing education, operation, shipbuilding and authorities, a quite good picture of the situation was gained. When it comes to education no special HSC requirement was imposed, excluding a short "Pilot course" to be permitted sailing to Singapore port without pilot. Most navigators had lower education than what was presented in the Norwegian survey, and many of the interviewed experts claimed that many of the navigators didn't have a valid licence. The same was the case with some of the ships. Even if the papers indicated compliance with the HSC code, clear lack in safety equipment, design, as well as fire protection was observed. Another difficulty was to find official damage statistics for this trafficked waters divided between Singapore, Indonesia and Malaysia. It was reported that no serious accident had happened, except of some groundings and minor collisions. This different approach to what is considered as a serious accident also make risk and safety comparison difficult. Even if the HSCs around Singapore have to encounter very busy traffic it is probably less demanding, as the navigators encounter on year round navigation in Norway. Most important in this case is that most routes are not operated during nighttime, and there are indeed no snowstorms.

During the interviews it was drawn a picture indicating that the regimes done most to improve HSC safety in the SE-Asian region was Hong Kong and Australia. Description of operation in these regions is outside the scope of this paper.

5. ACKNOWLEDGES

I am grateful to most of the individuals at the maritime departments at Ålesund University College (AUC) and Singapore Polytechnic (SP) that actively have been discussing coastal navigation safety issues with me. A special tanks to Associate Professor Ove Bjørneseth, which have been one of the driving forces during the construction of the HSC simulator at AUC. Also special thanks to Captain Gopala Krishnan for his professional support during may stay at SP. I would also thank AUC and the Norwegian Research Council for their financial support to my work.

6. REFERENCES

Converse,J.,Presser,S. *Survey Questions, Quantitative Applications in the Social Sciences Series*, Volume 63. Sage Publication, London, 1986.

Green, S., Salkind, N. *Using SPSS for Windows and Macintosh, Analysing and Understanding Data, Third Edition*. Prentice Hall, New Jersey, 2002.

Fisher, K. The Future of Maritime Radar. Navigation News, *The Magazine of the Royal Institute of Navigation*, July/August 2002. Royal Institute of Navigation, London, 2002.

Haugen & Eliassen Evaluering av radarteknologi for hurtiggående fartøyer, *Prosjektoppgave i Nautikk, NTNU*, Trondheim, 1997.

Holden, J. Opening address at the fifth international MARTECH conference. *MARTECH 2002*, Singapore.

Håvold, J.I. Development of a scale for measuring safety orientation. *19th International Systems Safety Conference*, Huntsville, Alabama, 2001.

Håvold, J.I. Some elements of safety orientation: Safety-culture factors and group differences between nationalities, occupation and vessels in a Norwegian shipping company, *Draft paper, Ålesund University College*, Ålesund, 2002.

IMO 2001a. *2000 HSC Code, International Code of Safety for High-Speed Craft, 2000.* Resolution MSC.97(73), adopted on 5 December 2000. International Maritime Organization, London, 2001.

IMO 2001b. *Seafarer's Training, Certification and Watch keeping Code (STCW Code),* including resolution 2 of the 1995 STCW Conference, as amended in 1997,1998, and 2000. International Maritime Organization , London, 2001.

IMO 2001c *SOLAS Consolidated Edition, 2001.* Consolidated text of the International Convention for the Safety of Life at Sea, 1974, and its Protocol of 1988: articles, annexes and certificates. International Maritime Organization , London, 2001.

IMO 2002 *ISM Code International Safety Management Code, and revised guidelines on implementation of the ISM Code, 2002 edition.* International Maritime Organization, London.

Isaksen, P. Sleipners forlis relatert til trygg og sikker navigasjon. *Skipsrevyen*, Bergen, 2001.

Kjerstad, N. Simulator for training and R&D in high-speed navigation, Proceedings of the *Fifth international MARTECH conference*, MARTECH 2002, Singapore.

Kjerstad, N. *Elektroniske og Akustiske navigasjonssystemer for Maritime studier.* Textbook, Tapir Forlag, Trondheim, 2002b.

Kristiansen, S. *Risk Analysis and Safety Management of Maritime Transport,* MARINTEK, Trondheim, 2001.

Kristiansen, S. Maritime Causality Analysis – Can simulation close information gaps. Proceedings of the *Fifth international MARTECH conference,* MARTECH 2002, Singapore.

NMD *Forskrift om bygging, utrustning og drift av hurtiggående fartøy som anvendes som passasjerskip eller lasteskip,* siste endring 04.06.2002. Norwegian Maritime Directorat, Oslo, 2002.

Ringdal, K. *Enhet og mangfold, samfunnsvitenskaplig forskning og kvantitativ metode.* Fagbokforlaget, Bergen, 2001.

SPSS *SPSS (Statistical Package for the Social Sciences) for Windows*, Release 10.0.05, 27 November 1999,Chicago.

Williams, P. Civil marine radar - a review and a way ahead. *The Journal of Navigation.* Vol.51, pages 394-403. Royal Institute of Navigation. Cambridge University Press, 1998.

Incorporating Model Uncertainty in Ship Reliability Analysis

Joško Parunov, Visitor, and **Ivo Senjanović**, Member, University of Zagreb, Croatia

ABSTRACT

Uncertainty analysis of calculated extreme vertical wave bending moment, which is the most important wave load component, is performed. The main sources of uncertainty that are analyzed are: choice of wave scatter table, inaccuracies in calculation of transfer functions, methods for prediction of long-term extreme values, human actions and non-linear effects. The aim of the study is to propose a rational statistical model describing the uncertainty of the extreme vertical wave bending moment for use in ship hull reliability analysis.

The work analyzes the wave scatter tables for the North Atlantic found in the Global Wave Statistics wave atlas. Extreme significant wave heights for different return periods, applicable in large ocean-going merchant ship safety assessment, are calculated. The results are compared with similar analyses from literature, as well as with measured and observed extreme significant wave heights. The wave steepness of extreme significant wave heights together with their influence on the extreme wave loads is studied. A method is proposed for the assessment of the uncertainty of extreme vertical wave bending moments that are calculated using linear strip theory. The method is applied to ships for which results of model experiments in towing tanks are available. Short-term and long-term methods for extreme wave load prediction are compared. Influence of human actions on extreme values is assessed. Approximate procedures for calculation of non-linear effects of vertical wave bending moments are analyzed.

Reliability methods are used to assess ultimate longitudinal strength for three different ship structures: an FPSO ship, a modern double-hull oil tanker and an old single-hull oil tanker. For the purpose of subsequent reliability analysis, uncertainty assessment of pertinent random variables influencing ship longitudinal strength is performed. It is demonstrated how the described procedures may be applied in practice for the calculation of section modulus of FPSO ship as well as for the safety assessment of corroded hulls of aged ship structures.

NOMENCLATURE

$\hat{\chi}_w$ - random variable representing uncertainty in calculation of wave loads under linear conditions

$\hat{\chi}_{nl}$ - random variable representing uncertainty in non-linear effects

$\hat{M}_w$ - random variable extreme vertical wave bending moment

COV - coefficient of variation

$S_\eta(\omega)$ - wave spectrum

ω - wave circular frequency

H_s - significant wave height

T_z - mean zero-up-crossing period

$f(H_s,T_z)$ - joint probability density function of H_s and T_z

$F(H_s), f(H_s)$ -marginal cumulative distribution function, probability density function of H_s

$\hat{H}_s$ - random variable significant wave height

N - number of data in sample of $\hat{H}_s$ used for calculation of its empirical probability distribution

m_i - number of observation higher or equal to certain value of H_s

$\varepsilon, \theta, \lambda$ -the location , the scale and the shape parameters of Weibull distribution

P - probability that certain value of H_s is not exceeded

Q - probability that certain value of H_s is exceeded

RP - return period in years

H_s^* - the most probable extreme significant wave height in given RP

T_{REG} - time interval in years of observation of the random variable $\hat{H}_s$

$F(T_z|H_s)$ - conditional distribution of mean zero-up-crossing periods

Φ - standard Gaussian cumulative distribution function

$\mu_{\ln T_Z}, \sigma_{\ln T_Z}$ - mean, standard deviation of logarithms of mean zero-up-crossing periods

μ_{T_Z}, σ_{T_Z} - mean, standard deviation of mean zero-up-crossing periods

s_{T_Z} - average wave steepness of random sea state

a^*, H^* - the most probable extreme wave amplitude, height

$\tilde{H}_x(\omega), H_x(\omega)$ - transfer functions of measured, calculated response variable X

$\tilde{n}, n$ - number of response cycles of measured, calculated response variable X

$\tilde{X}^*, X^*$ - the most probable extreme value of measured, calculated response variable X

$\tilde{\sigma}_x, \sigma_x$ - standard deviation of measured, calculated response variable X

$\tilde{X}_s, X_s$ - significant values of measured, calculated response variable X

ϕ - modelling error

μ_ϕ, σ_ϕ - mean value, standard deviation of modelling error

L_{pp}, B, D, T - ship length between perpendiculars, breadth, depth, draught

C_B - block coefficient

M_w^* - the most probable extreme vertical wave bending moment in n response cycles

$M_{w,s}$ - significant vertical wave bending moment

$F(x_p)$ - cumulative (Weibull) distribution of response peaks (amplitudes) x_p

$F(x_e)$ - cumulative (Gumbel) distribution of extreme values x_e of the response

α, x_e^* - parameters of Gumbel distribution (x_e^* is in the same time the most probable extreme value in n response cycles)

μ_e, σ_e - mean value, standard deviation of Gumbel distribution

M_H, M_S - non-linear vertical wave bending moments in hogging, sagging

Φ_H, Φ_S - correction factors for non-linearities in hogging, sagging

ψ - load combination factor

$\hat{\chi}_u$ - random variable representing uncertainty in prediction of ultimate strength

$\beta_{FL}, \beta_{PL}, \beta_{BL}$ - safety indices for full load, partial load, ballast condition

$P_{f,FL}, P_{f,PL}, P_{f,BL}$ - probability of failure for full load, partial load and ballast condition, respectively

β_t, P_{ft} - total annual safety index, probability of failure

$G(\cdot)$ - the failure hyperplane in reduced space of standard normal variates y_i

α_i - sensitivity factors

y_i^*, x_i^* - the most probable failure point in reduced, original space

μ_{sw}, σ_{sw} - mean value, standard deviation of still-water bending moment for one voyage (Gaussian distribution)

μ_{se}, σ_{se} - mean value, standard deviation of still-water bending moment for one year (Gumbel distribution)

n_{sw} - for FPSO, number of loading and offloading cycles per year; for oil tankers, number of voyages per year for each loading condition

F_{sw}, f_{sw} - Gaussian cumulative distribution and Gaussian probability density function with moments μ_{sw}, σ_{sw}

Special characters

$*$ - superscript, denotes the most probable extreme value of random variable

$_s$ - subscript, denotes significant value of random variable

$\wedge$ - superscript, used to distinguish random variable from actual value that random variable may take on

$\sim$ - superscript, used to distinguish measured from calculated value of response variable

$_p$ - subscript, denotes random variable representing peaks (amplitudes)

$_e$ - subscript, denotes random variable representing extreme values

FL, PL, BL - subscripts, denote loading conditions: full load, partial load, ballast condition.

1. INTRODUCTION

For the longitudinal ship strength, the most important global load component is the vertical wave bending moment amidships. The design value of that load component is given by Unified Requirement (UR) S11 of the International Association of Classification Societies (IACS) by explicit formulas for sagging and hogging. The theoretical basis of the requirement UR S11 is given by Nitta et al. (1992) where the rule vertical wave bending moment is defined as the bending moment with probability of exceedence of 10^{-8}. In other words, the rule value is the most probable extreme value for the return period of 20 years, which is the ordinary ship lifetime. The rule develops a design wave bending moment based on only the main dimensions of the ship: length, breadth, block coefficient and, in some of the rules, also on bow flare and forward speed. Important details, as operational profile, mass distribution and hull form are not taken into account. These parameters may be accounted for by direct hydrodynamic and statistical analysis.

It is the general opinion that the lifetime vertical wave bending moment given by UR S11 is conservative, i.e. that if one would apply direct calculation methods to evaluate extreme loads, the rule value would not be exceeded. However, in recent years, there have been more and more studies showing that this may not be true. Some very respectable researchers have shown in their studies that ship may be loaded by much more severe wave loads than those given by the rules (Faulkner 1997a, 1997b, 1998, Guedes Soares, 1996). Furthermore, some of them have expressed the opinion that this could be the cause of some unexplained ship losses (Faulkner 2002).

When classification societies and other research institutions apply complex direct hydrodynamic and statistical methods for calculation of lifetime vertical wave bending moments according to their own knowledge and experience, significantly different results are obtained. This is illustrated in Figure 1 which presents the results of a comparative study of long-term distributions of vertical wave bending moment performed by leading classification societies for the same ship and under the same conditions. As shown in Figure 1, the maximum and minimum lifetime vertical wave bending moment calculated for a probability of exceedence of 10^{-8} were different by up to 80% (Nitta et al. 1992). These differences could have a significant impact on ship safety.

It is clear from Figure 1, that direct calculation methods are not yet sufficiently standardized to assure a consistent level of structural integrity. This is unacceptable, since it could lead to frequent damage of ship structures. Namely, damage to ship structures occurs today despite the application of relatively conservative IACS rules. Therefore, one should be concerned about what could happen if potentially less reliable methods were to be allowed. Discrepancies in Figure 1 are the consequence of modeling uncertainty, since they reflect different mathematical models and assumptions used in calculations.

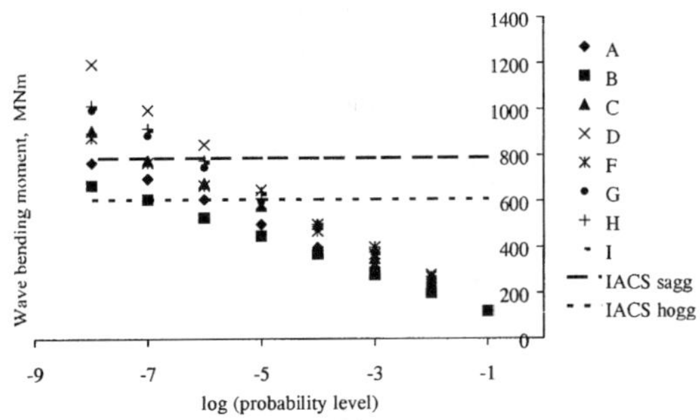

Figure 1: Comparative analysis of long-term distribution of vertical wave bending moments on containership (from Nitta et al. 1992).

Another important point that can be observed in Figure 1 is that the results of direct analyses have clear tendency to considerably overestimate the IACS rule vertical wave bending moments. IACS wave bending moments in sagging and hogging are shown in Figure 1 as horizontal lines. If the linear part of rule wave bending moment is assumed to be the mean value between sagging and hogging, then maximum overestimation of the rule value is 71%, while average overestimation is 33%. Other researches confirm such rather disturbing conclusion. Evidence has been presented that the bulk carrier "Derbyshire", lost in 1980, might, on her last voyage, have been loaded such that the wave bending moments were 80% and 40% higher than IACS values in sagging and hogging, respectively (Faulkner 1998). Guedes Soares (1996) obtained a 43% higher value than the IACS requirement for containerships, and 24% and 26% higher values for a large and a small tanker, respectively. Ayyub et al. (1998) determined that the difference between the lifetime vertical wave bending moments predicted by the US Navy and ABS for the same ship was 60%.

Significant efforts have been made by ISSC in recent years to explain modeling uncertainty in predicted extreme wave loads (Guedes Soares et al. 1997). As result of such studies, the main sources of modeling uncertainty of extreme vertical wave bending moments are now recognized to be as follows:

1) Uncertainty of wave loads calculated under linear assumptions
   a) Uncertainty of the shape of wave spectra
   b) The choice of wave scatter diagram

i. The choice of wave data source (Global Wave Statistics, instrumental buoys, satellites)

ii. The choice of representative ship route

c) Uncertainty of transfer functions

d) Uncertainty of the method for prediction of long-term extreme values

e) Uncertainty of human actions

i. Voluntary changes of shipping route to avoid heavy weathers

ii. Maneuvering in heavy weather (course and speed changes)

2) Uncertainty of non-linear effects

a) Different sagging and hogging bending moments mainly due to non-vertical ship sides

b) Influence of slamming and whipping on extreme vertical bending moments.

Each of these sources of uncertainty is a rather complex problem requiring specific study.

The uncertainty of extreme vertical wave bending moments is particularly important for the successful application of ship reliability analysis. Ship reliability methods have been under development for almost 30 years, but they have not yet become the standard engineering tool. This is because of serious problems related to the rational choice of statistical models of pertinent random variables. The main difficulty lies in the variability and unpredictability of wave loads, in particular the vertical wave bending moment. This is confirmed by numerous studies describing the state-of-the-art in application of ship reliability methods (e.g. Ayyub et al. 1998, Parunov & Senjanović 1998).

In applying ship hull-girder reliability analysis, the statistical model of the vertical wave bending moment is usually described as a product of three random variables $\hat{M}_w, \hat{\chi}_w$ and $\hat{\chi}_{nl}$. $\hat{M}_w$ is the extreme vertical wave bending moment, which is usually described using the Gumbel distribution. This distribution represents an inherent (physical, natural) uncertainty which can not be avoided or reduced. $\hat{\chi}_w$ and $\hat{\chi}_{nl}$ are modeling uncertainties due to the calculation of wave loads under linear assumptions and the uncertainty of non-linear effects, respectively. Modeling uncertainty $\hat{\chi}_w$ is the consequence of simplifications, assumptions and inaccuracies of the linear engineering models used to predict extreme loads on ship hulls. This type of uncertainty may potentially be reduced by using more accurate models to describe certain physical phenomena (Friis Hansen et al. 2000). Uncertainty $\hat{\chi}_{nl}$ is partially inherent in nature because of the non-linear behavior of structural response. The inherent part of the non-linearity is unavoidable. However, part of uncertainty $\hat{\chi}_{nl}$ is due to the method used to predict non-linear effects. This part of $\hat{\chi}_{nl}$ may be reduced by using more sophisticated non-linear hydrodynamic methods.

To estimate modeling uncertainty of wave loads in rational manner, large amount of collected full-scale measurement data from actual ships, would be required. Since this would be quite an expensive and time consuming request, researchers are faced with only a limited quantity of available data. Consequently, in ship reliability studies, they have used statistical models based on their subjective judgment. These have resulted in quite different calculated safety indices, for the same or similar ships, i.e. for the same physical reality. These discrepancies have slowed down practical implementation of reliability methods, a situation that has yet to be resolved.

To illustrate the problem, the researches done by Mansour et al. (1996) and Casella et al. (1996) are compared. In the former study, the target safety index for the ultimate strength of the tanker was β =4 for the return period of 20 years (Mansour et al. 1996). In the latter study, the target safety index for the same failure mode was β=3-3.7, but for a return period of 1 year. This is an absurd situation, since it follows that the safety index of the tanker hull should be higher for 20 years than for 1 year. This clearly shows that safety indices obtained in different studies must not be analyzed without considering the applied statistical models.

Section 2 of the paper addresses the uncertainty $\hat{\chi}_w$ of extreme vertical wave bending moment calculated under linear conditions. In the next section, uncertainty $\hat{\chi}_{nl}$ of non-linear effects is analysed. Section 4 is concerned with the application of reliability methods to ship structures. In Section 4 reliability assessments of the ultimate longitudinal strength of three different ship structures is presented: a Floating Production, Storage and Offloading (FPSO) ship, a modern double-hull oil tanker and a corroded single-hull oil tanker.

2. UNCERTAINTY ASSESSMENT OF WAVE LOADS CALCULATED UNDER LINEAR ASSUMPTIONS

This section deals with the modeling uncertainty $\hat{\chi}_w$ of wave loads calculated under linear assumptions. Uncertainty $\hat{\chi}_w$ accounts the following effects:

a) uncertainties in the sea model

b) uncertainties in the ship linear response

c) uncertainty of human actions.

Uncertainty in the sea model comprises uncertainties of the shape of wave spectra and the choice of wave scatter diagram. These uncertainties are addressed in Subsections 2.1 and 2.2, respectively. Uncertainty in the ship linear response may be divided into the uncertainty of transfer functions (Subsection 2.3) and the uncertainty of the prediction method for extreme values (Subsection 2.4). Uncertainty of human actions is analysed in Subsection 2.5. Each of subsections ends with a conclusion related to the mean value and *COV* (coefficient of variation) of the source of uncertainty under consideration.

In the last subsection, a summary of uncertainties analysed in this section, is given. The procedure and expressions necessary for calculating the inherent uncertainty $\hat{M}_w$, are presented. The application of the uncertainty assessment for the correction of long-term extreme vertical wave bending moments is also shown.

2.1 Uncertainty of the shape of wave spectra

In order to carry out stochastic response analysis for the design of marine systems, wave spectra representing the desired sea states are mandatory. It would be highly desirable to use wave spectral formulations that match the sea state where the marine system is to be operated. However, this is a rather difficult task for ships with an unrestricted operational area.

One of the first widely used spectral formulations for analysis of ocean-going ships was one parameter Pierson-Moskowitz (P-M) spectrum. This spectrum was developed from the analysis of the measured data obtained in the North Atlantic (Ochi, 1998). The one parameter P-M spectrum describes the fully developed sea states that appear when a moderate wind blows on open seas without the influence of the sea shore. In order to more realistically describe partially developed seas in addition to fully developed seas, two parameter modifications of the original P-M spectrum have been developed. The most widely used of these is the ITTC spectrum, described by the significant wave height Hs and mean zero-up-crossing period Tz:

$$S_\eta(\omega)=\frac{H_S^2T_z}{8\pi^2}\left(\frac{\omega T_z}{2\pi}\right)^{-5}e^{-\frac{1}{\pi}\left(\frac{\omega T_z}{2\pi}\right)^{-4}} \qquad (1)$$

The JONSWAP spectrum was developed in order to represent a steeper spectrum, for the same total energy as P-M spectrum, and to describe the shift of the peak frequency that occurs in such a case. That spectrum has found wide application in the design of fixed offshore structures in the North Sea. Very often, the parameters of the JONSWAP spectrum are adjusted to match as closely as possible the actual sea states in a particular region. This is possible, because very often the details of the spectra for different sea states are known from measurements at the fixed location.

There are other, more sophisticated spectral formulations, introduced in order to achieve better agreement with the measured spectra. Thus, spectra with 3, 4 and 6 free parameters have been developed. The shortcoming of such formulations is that it is difficult to estimate the additional parameters. This is especially true in the analysis of ship structures, where usually only two parameters are available – significant wave height and mean zero-up-crossing period.

The wave spectrum represents a simplified model of the long-crested waves, progressing in only one direction. Observations of ocean waves show, however, the existence of component waves progressing in various directions, composing the so called short-crested seas. This phenomenon is modeled by a wave energy spreading function that generally depends on wave direction and wave frequency.

It may be concluded that the mathematical description of sea states is a rather simplified representation of a complex physical reality. For the analysis of ship structures the most frequently used spectrum is the ITTC spectrum defined by (1) assuming long-crested waves, i.e. without energy spreading function. This is because the wave data format consists of pairs of significant wave heights and mean zero-up-crossing periods, so that the two-parameter spectral formulation becomes a natural choice.

Within the frame of the presented uncertainty assessment, the ITTC spectrum and long-crested seas are considered as “the best estimates”. Therefore, they do not modify the mean of calculated extreme values. Also, there is not enough data to rationally estimate *COV* of extreme values because of this uncertainty. For the purpose of the present study, a *COV* of 5% is assumed. It should be noted that this assumption is subjective, and should be verified by rational analysis.

2.2 The choice of wave scatter diagram

The choice of the wave scatter diagram suitable for ship strength and reliability analysis represents the most important source of uncertainty. To perform long-term prediction of extreme wave loads it is necessary to:

- choose the source of the wave data,
- within the chosen source of the wave data select a representative ship route.

For the analysis of ship structures, it is a usual practice to use wave data from the wave atlas Global Wave Statistics (GWS, Hogben et al. 1986). Data in GWS are obtained by observations of volunteers from merchant ships on usual shipping routes. Since the shipping route is often not known in the design phase, the usual practice is to assume navigation in the North Atlantic as the geographical area with the most severe weather conditions (Guedes Soares 1996, Folsø et al. 2002). It should be noted that some shipping routes in the North Pacific have even more severe weather conditions than the North Atlantic (Mansour and Preston 1994).

Comparative analysis of classification societies methods has shown that most differences in long-term distributions in Figure 1 originate from the choice of wave data (Faulkner 1995, Nitta 1995). The coefficient of variation of this type of uncertainty may be estimated to be 16%. It is interesting to note that all classification societies claim to use data for the North Atlantic with slight modifications according to their experience.

Another significant analysis of this type of uncertainty was performed by Guedes Soares (1996) for a containership and a small and a large tanker. He obtained a *COV* of the most probable extreme value for different wave scatter diagrams of 10-15%.

Closely related to the choice of wave scatter diagram is the question of the most severe sea states that ship may encounter in its lifetime (Faulkner 1997). To answer this question, one needs to perform analysis, to apply subjective engineering judgment as well as to study results of other researches and literature describing the observed and measured extreme sea states. The following paragraphs present a procedure for estimating the most probable extreme sea states applicable to ship reliability analysis.

Since most ships are designed for global, unrestricted service, the North Atlantic has been chosen as the reference operational area for calculating design wave loads. The GWS is considered a convenient source of wave data for analysis of ship structures, since these data implicitly account for the world-wide practices of shipmasters operating along the main shipping routes. The wave scatter diagram which is analyzed herein is the one proposed by some of the leading classification societies and ship research institutions within the European project SHIPREL (Reliability Methods for Ship Structural Design) (Guedes Soares et al. 1996). The SHIPREL scatter diagram is obtained by combining information from zones Nos. 8, 9, 11, 15, 16 and 17 assuming the probability of occurrence of each of the wave zones proportional to their relative size (Guedes Soares 1999). SHIPREL wave scatter diagram is presented by Guedes Soares (1996).

The proposed reference wave scatter diagram is not completely rational since wave zones are not of equal severity. Consequently, the ship speed would not be equal in each of them, influencing percentage of time spent in each of the wave zones (Mansour & Preston 1995). Despite this limitation, the SHIPREL scatter diagram may be considered as fairly accurate and may give a good indication of the extreme sea state that ships encounter.

The joint probability distribution of Hs and Tz is usually defined as product of the marginal distribution of Hs and the conditional distribution of Tz, i.e.:

$$f(H_s,T_z) = f(H_s)\cdot f(T_z\,|H_s) \tag{2}$$

Different theoretical models have been used to define the joint probability distribution given by (2) (e.g. DnV 1992, Ochi 1978, Cramer & Friis Hansen 1994). The best approximation of empirical data is achieved if the Weibull 3-parameter distribution and log-normal distribution are chosen as the marginal distribution of significant wave heights and the conditional distribution of mean zero-up-crossing wave periods, respectively.

2.2.1 Marginal distribution of significant wave heights

The extreme significant wave height is a parameter of primary importance in the analysis of marine structures. To calculate long-term extreme values of significant wave height, a three-step procedure is used. In the first step, the theoretical probability distribution of the random variable $\hat{H}_S$ on the given shipping route is defined. This is achieved by selecting a theoretical distribution to approximate the empirical distribution. In the second step, the probability of exceedence $Q(H_S^*)$ corresponding to the return period *RP* is calculated. In the third step of the procedure, the most probable extreme significant wave height H_S^* is calculated using the theoretical probability distribution function.

Empirical probabilities of exceedence

The empirical distribution is obtained by calculating empirical exceeding probabilities. Empirical probabilities of exceedence are calculated by one of a number of approximate expressions (Chow 1964). The most frequently used is the Weibull expression given as:

$$Q(H_{S,i}) = \frac{m_i}{N+1} \tag{3}$$

where m_i is the number of observations higher or equal to $H_{S,i}$, while N represents the total number of observations. Empirical probabilities of exceedence obtained by employing (3) on SHIPREL scatter diagram are shown in Figure 2.

Fitting of Weibull 3-parameter distribution

The 3-parameter Weibull distribution is given as:

$$F(H_S) = P(\hat{H}_S \le H_S) = 1 - e^{-(\frac{H_S-\varepsilon}{\theta})^\lambda}, \quad \hat{H}_S, H_S \ge \varepsilon \tag{4}$$

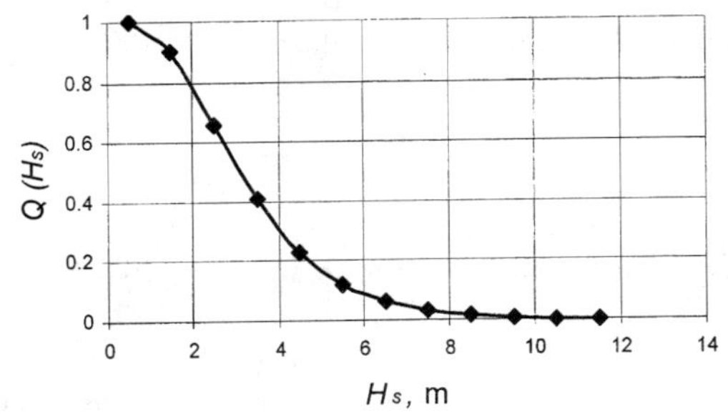

Figure 2: Empirical probabilities of exceedence according to SHIPREL scatter diagram

where $\hat{H}_S$ is the random variable significant wave height, while *Hs* represents actual value that random variable may take. In (4) ε, θ, and λ are the location parameter, the scale parameter and the shape parameter, respectively. The location parameter ε has an important physical meaning as the small significant wave height associated with permanent sea activity (Sarpkaya & Isaacson 1981). Since the location parameter ε is not calculated explicitly, it has to be assumed before starting the analysis. Generally, ε has to be determined by an iteration procedure to minimize approximation error (Cramer & Friis Hansen 1994, Mansour & Preston 1995). In the present paper, the location parameter ε is calculated by an efficient procedure presented by Rao (1992). The calculated parameters of Weibull distribution are $\lambda=1.475$, $\theta=2.802$ and $\varepsilon=0.71$. Comparison of empirical and theoretical probability density functions and cumulative distributions is shown in Figures 3 and 4, respectively. It may be concluded that Weibull 3-parameter distribution approximates quite well the empirical data.

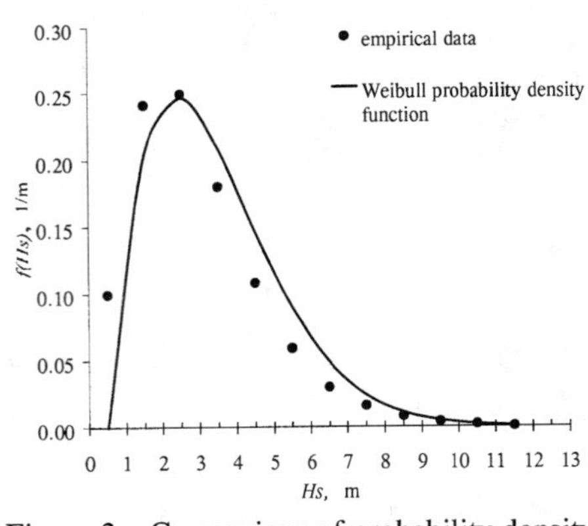

Figure 3: Comparison of probability density functions

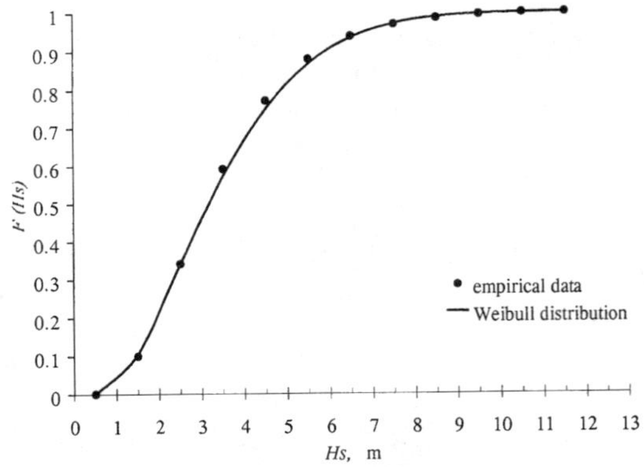

Figure 4: Comparison of cumulative distributions

<u>Return period</u>

The return period *RP* is defined as average number of years when sea state characterized by significant wave height H_S^* would be exceeded at least once (Sarpkaya & Isaacson 1981). The return period may be calculated for a given measured value based on the total sampling of the random variable, but more often it is used to define the predicted value of a random variable.

The probability of exceedence corresponding to a return period *RP* is given by the following expression (Sarpkaya & Isaacson 1981):

$$P\left(\hat{H}_S \geq H_S^*\right) = Q(H_S^*) = \frac{T_{REG}}{N} \cdot \frac{1}{RP} \tag{5}$$

where T_{REG} is the time interval in years of the observation (measurement) of the random variable $\hat{H}_S$. *N* in (5) is the sample size, i.e. quantity of data in the sample of the random variable $\hat{H}_S$.

<u>The most probable extreme significant wave heights for design and analysis of ship structures</u>

Loads used for design of ship structures are usually those with a return period of 20 years. In ship reliability analyses, however, very often a return period of 1 year is used. In that way, various problems disappear related to extrapolation of loads for larger return periods.

In order to calculate the most probable extreme significant wave height H_S^* for the given return period *RP*, the corresponding probability of exceedence $Q\left(H_S^*\right)$ needs to be determined by applying (5). For that purpose, it is assumed that $T_{REG}=1$ year, since data in GWS refer to average year, and that $N=1000$ (Mansour & Preston 1995). The most probable extreme significant wave heights for different return periods are presented in Table 1.

Table 1: The most probable extreme significant wave heights for different return periods *RP*

RP, years	$Q(H_S^*)$	H_S^*, m
1	$1 \cdot 10^{-3}$	11.1
5	$2 \cdot 10^{-4}$	12.7
10	$1 \cdot 10^{-4}$	13.3
20	$\mathbf{5 \cdot 10^{-5}}$	**14.0**
50	$2 \cdot 10^{-5}$	14.8
100	$1 \cdot 10^{-5}$	15.4

2.2.2 Conditional distribution of mean zero-up-crossing periods

The log-normal distribution is chosen as the conditional distribution of mean zero-up-crossing periods. The distribution function is given by the following expression:

$$F(T_Z|H_S) = \Phi\left(\frac{\ln T_Z - \mu_{\ln T_Z}}{\sigma_{\ln T_Z}}\right) \qquad (6)$$

where Φ is the standard Gaussian distribution with zero mean and unit variance, while $\mu_{\ln Tz}$ and $\sigma_{\ln Tz}$ represent mean and standard deviation of the logarithms of the mean zero-up-crossing periods, respectively. Details of fitting a log-normal distribution to empirical data employing the method of moments are given by Parunov & Senjanović (2003).

The comparison of the empirical and theoretical frequencies of occurrence for significant wave height of 6.5m, calculated for interval width of 1s, is presented in Figure 5. Figure 5 shows good agreement between theoretical and empirical frequencies of occurrence.

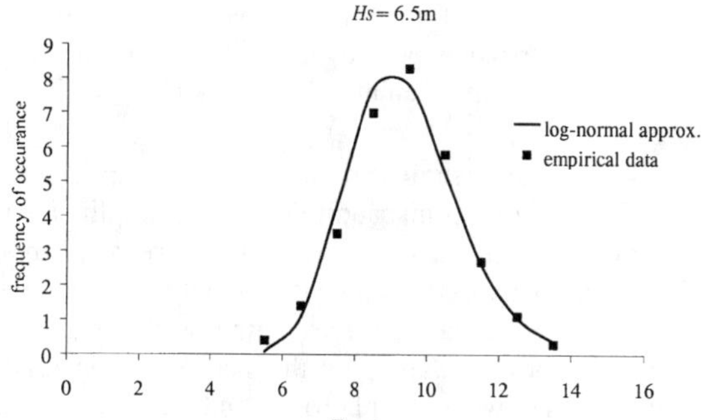

Figure 5: Comparison of the log-normal distribution and empirical data

The functional dependencies of mean values and standard deviations of the mean zero-up-crossing period on significant wave height are shown in Figure 6. Dependencies are presented in co-ordinate systems (ln*Hs*, μ_{Tz}) and (ln*Hs*, σ_{Tz}) where these can be reasonably approximated by straight lines. The straight lines from Figure 6 are applicable for estimating mean zero crossing periods of extreme sea states that are not covered by the wave scatter diagram. By extrapolating straight lines from Figure 6, the mean zero-up-crossing periods of very severe sea states with significant wave height greater than 12m, may be calculated. Thus, for significant wave height *Hs*=14m, the calculated mean zero-up-crossing period reads μ_{Tz}=10.24s, while the standard deviation reads σ_{Tz}=2.44s.

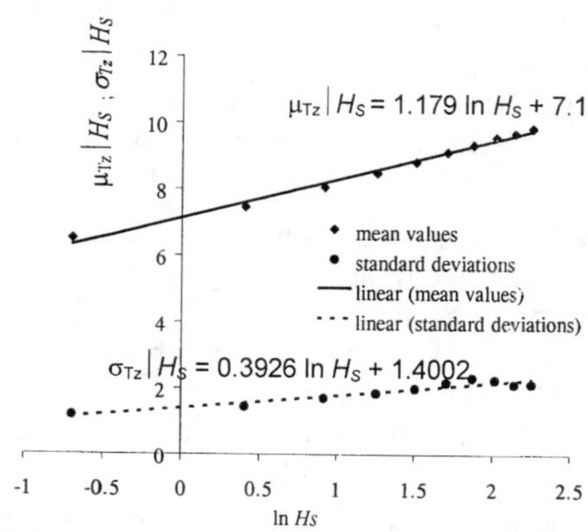

Figure 6: The mean values and standard deviations of mean zero-up-crossing periods as functions of significant wave heights

2.2.3 Analysis of wave steepness

Significant wave heights and mean zero-up-crossing periods are not independent, i.e. not all combinations are possible. The relationship between significant wave heights and mean zero-up-crossing periods is known as the average steepness of random sea states. The average wave steepness s_{T_Z} is given by the following expression (Bitner-Gregersen et al. 1995):

$$s_{T_Z} = \frac{H_S}{\lambda_z} = 2\pi \frac{H_S}{gT_z^2} \qquad (7)$$

where λz is the wave length associated with the mean zero-up-crossing period *Tz*, while *g* represents gravity constant. If the wave steepness is higher than the limiting wave steepness, waves become too steep and wave breaking occurs. Therefore, such sea states are not physically sustainable.

Based on the results presented, it may be concluded that the most probable extreme sea state with the return period of 20 years may be described by significant wave height *Hs*=14m and the mean zero-up-crossing period *Tz*=10.24s. The average wave steepness, according to (7), of this sea state is $s_{T_Z} = 1/11.7$.

The most severe sea states used for the design of marine structures are characterized by the wave steepness $s_{T_Z} = 1/10$ (Mitchel 1999, Faulkner 1998, Hogben 1998). This wave steepness may be generated by extremely strong wind suddenly developing on the open seas or by strong wind blowing in areas with limited fetch. As the wind duration and fetch increase, the wave steepness decreases and converges to the value appropriate for fully developed seas. The wave steepness of fully developed seas, described by the original one-parameter P-M spectrum, is $s_{T_z} = 1/19.6$. Bitner-Gregersen et al. (1995) have analyzed the limiting wave steepness of extreme sea states. They have recommended $s_{T_z} = 1/15$ as the limiting wave steepness appropriate for the analysis of ship structures. The mean zero-up-crossing wave period for the significant wave height *Hs*=14m and steepness $s_{T_z} = 1/15$, may be calculated from (7) as *Tz*= 11.6s. Bitner-Gregersen et al. (1995) have found that the data from GWS largely overestimate the recommended wave steepness. The results of the present study are in accordance with this reference, since the calculated expected wave steepness $s_{T_z} = 1/11.7$ largely overestimates the recommended value. The explanation for these discrepancies could be that extrapolation is not suitable for this particular problem and leads to unrealistic results.

Various assumptions about steepness of extreme sea states are summarized in Figure 7. Figure 7 represents the mean zero-up-crossing period as a function of the significant wave height using:

- one parameter P-M spectrum ($s_{T_z} = 1/19.6$),
- recommended wave steepness ($s_{T_z} = 1/15$),
- results from GWS (mean values from Figure 6),
- maximum sustainable wave steepness ($s_{T_z} = 1/10$).

From the analysis of the results presented in Figure 7, the following conclusions may be drawn:

- the data from GWS result in wave steepness of extreme sea states close to the maximum sustainable wave steepness,
- the difference between mean zero-up-crossing periods for the P-M spectrum and GWS with *Hs*=14m is about 3 seconds,
- the recommended wave steepness for analysis of ships $s_{T_z} = 1/15$ gives a wave period that is approximately the mean value between the GWS and the P-M spectrum for significant wave heights greater than 13m.

As an example, the influence of the mean zero-up-crossing period *Tz* on the vertical wave bending moment is shown in Figure 8, for a cruise ship (Senjanović et al. 2000). The curve in Figure 8 represents the significant value of the double amplitude vertical wave bending moment amidships as a function of *Tz* for unit significant wave height. It is obvious from Figure 8 that the significant vertical wave bending moment may differ up to 20%, for *Tz* between 10 and 12s.

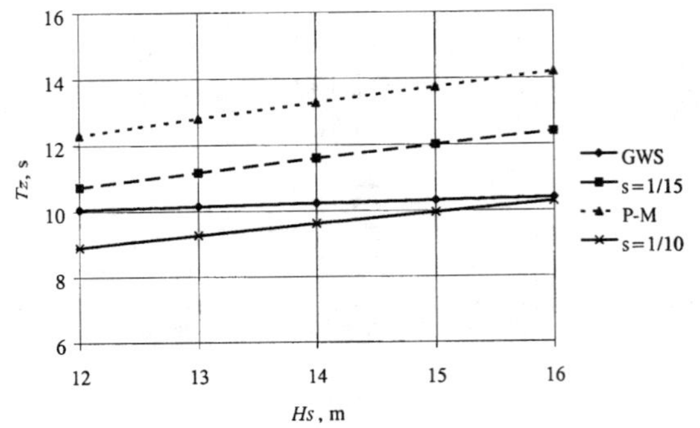

Figure 7: Dependence of mean zero crossing period *Tz* on significant wave height *Hs* for different assumptions about wave steepness

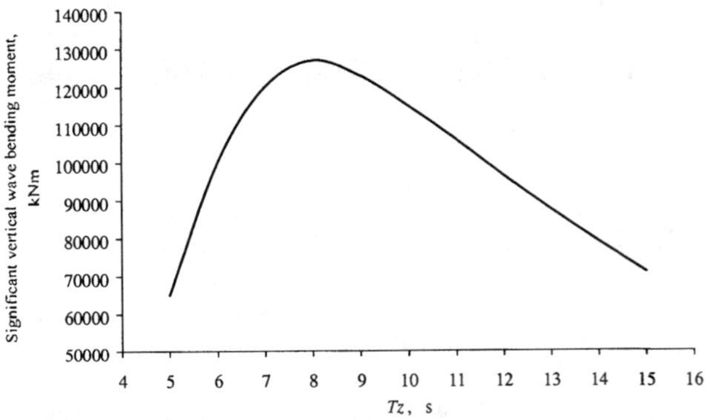

Figure 8: Significant value of the double amplitude vertical wave bending moment amidships as a function of the mean zero-up-crossing period for unit significant wave height (Senjanović et al. 2000)

2.2.4 Long-term distribution of wave amplitudes

The long-term distribution of wave amplitudes is important for the design of fixed offshore structures. The highest expected wave is not only used for the calculation of dynamic forces acting on the structure, but also for calculation of the necessary clearance between the working platform and the still water level. In the present paper, the long-term distribution of wave amplitude is used to indirectly estimate extreme sea states. The method for obtaining the long-term distribution of wave amplitudes is well known and described in numerous sources (e.g. Sarpkaya & Isaacson 1981, Isaacson & Foschi 1998, Guedes Soares 1998).

The calculation of the long-term distribution of wave amplitudes is performed using the program POSTRESP, that is a part of the package for structural analysis SESAM (DnV 1993b). Two wave scatter diagrams are analyzed: the SHIPREL wave scatter diagram based on data from GWS and the Nordenstrøm wave scatter diagram obtained by measurements from fixed weather stations in the North Atlantic (DnV 1993b). The calculated long-term distributions of wave amplitudes are shown in Figure 9. The long-term distribution for the SHIPREL scatter diagram is given by a dashed line, while long-term distribution based on Nordenstrøm data is presented by a solid line in Figure 9. The most probable extreme wave amplitudes calculated for different return periods are shown in Table 2.

Important differences between extreme wave amplitudes for two different wave scatter diagrams may be observed in Table 2, although both data sets refer to the same geographical area, i.e. the North Atlantic. The differences are the result of the way in which the data were collected, since Nordenstrøm data were obtained from fixed stations, while the data in GWS were collected by observations from merchant ships. In that way, the most severe sea states are omitted in GWS as the consequence of maneuvering to avoid heavy weather.

It is reasonable to assume that the most probable extreme wave amplitude is achieved within the most probable extreme sea state. Using basic expressions from the theory of extreme values, it is possible to relate significant wave height H_S^* and the most probable extreme wave height (double amplitude) H^* during the short-term sea state:

$$a^* = \frac{1}{2} H^* ; \qquad H^* = H_S^* \sqrt{\frac{\ln(n)}{2}} \tag{8}$$

In (8), a^* is the most probable extreme wave amplitude in n wave cycles during short-term sea state described by significant wave height H_S^*. The duration of the short-term sea state and the mean zero-up-crossing period may be assumed to be τ_S=3h and T_Z=10s, respectively. Therefore, the number of wave peaks is n=1080. The calculated significant wave heights using (8) with the most probable extreme wave amplitudes from Table 2, are presented in Table 3.

The results from Table 3, for the case of SHIPREL scatter diagram, may be compared with those from Table 1 that are obtained using a long-term distribution of significant wave heights. For the return period of 20 years, the most probable extreme significant wave heights were 14m and 14.1m, from Tables 1 and 3, respectively. This comparison validates the analysis that has been presented. There are discrepancies between results in Tables 1 and 3 for the shorter return periods of 1 and 5 years. For these return periods, the distribution of wave amplitudes leads to somewhat higher extreme significant wave heights. This may be explained by the difference in the assumed persistence of the sea states, being 8 hours and 3 hours for the long-term and the short-term prediction method, respectively. If the duration of short-term sea state is assumed to be 8 hours, both methods give almost identical results for shorter return periods. The physical explanation for this phenomenon may be that the less severe sea states, with shorter return periods, are more persistent (Ochi 1978).

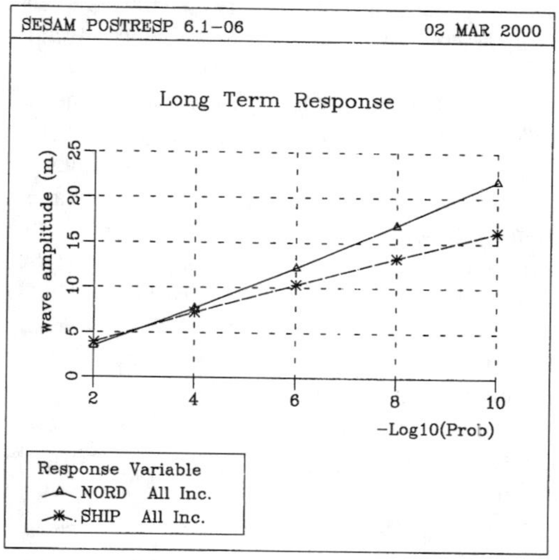

Figure 9: Long-term distributions of wave amplitudes

Table 2: The most probable extreme wave amplitudes a^* for different return periods

RP, years	a^*,*m* (SHIPREL)	a^*,*m* (Nordenstrøm)
1	11.2	13.8
5	12.3	15.4
10	12.7	16.1
20	**13.2**	**16.8**
50	13.8	17.8
100	14.2	18.6

Table 3: The most probable extreme significant wave heights H_S^* for different return periods

RP, years	H_S^*, *m* (SHIPREL)	H_S^*, *m* (Nordenstrøm)
1	12.0	14.8
5	13.2	16.5
10	13.6	17.2
20	**14.1**	**18.0**
50	14.8	19.0
100	15.2	19.9

2.2.5 A comparison with measured significant wave heights and other similar analyses

As may be seen from the analysis presented above, there are numerous uncertainties related to calculation of extreme sea states. In order to make the analysis more realistic, it is necessary to study the conclusions of other researchers as well as the literature describing the measured extreme sea states.

It is important to distinguish waves analyzed in the present study from episodic waves (Tedeschi et al. 2000). Episodic wave, also called freak wave, is abnormally high wave, not correlated with local weather conditions. Episodic wave can not be described by statistical methods as an extreme event which is the consequence of the severe sea condition. These waves are often met when wind waves propagate in opposite direction of the sea current, as Aghulas Current, Gulf Stream and Kuro Shio. Also, the shallow North Sea seems to provide suitable conditions for extremely high episodic waves (Duxbury 1991).

Extreme sea states have been often studied by Faulkner (1998). In a forensic analysis, he presented evidence to show that loss of the bulk carrier "Derbyshire" in 1980 was in a sea state with a significant wave height of 14m. In the same paper, the following expression was proposed, as general guidance, for the analysis of ships in extreme conditions:

$$H_S = 15 - (3 - L/100)^{2.5} \quad (9)$$

where L is the ship length. Thus, applying (9) for ship whose length is 200m results in a proposed significant wave height of 14m, in excellent agreement with results of the present study.

FPSO vessels operating in harsh environments are often designed for higher significant wave heights than 15m. For example, one FPSO unit operating west of Shetland is designed for a significant wave height of 18m (Faulkner 1998). One of the discussions for Faulkner's paper stated that the most severe measured sea state in the North Sea may be described with Hs=13.6m and Tz=10.5s, corresponding to wave steepness $s_{T_Z} = 1/12.66$ (Hogben 1998).

The waves with height of 34.2m and period 14.8s were reliably measured and documented in 1933 on the US Navy tanker "Ramapo" during a voyage from Manila to San Diego when the ship encountered a severe storm (Duxbury 1991).

There have been several reports of waves higher than 30m, although the waves were not accurately measured (Faulkner 1999).

A wave of 26m in height was accurately measured with a laser from the gas platform "Draupner", owned by "Statoil" in January 1995. This spectacular wave got its own name - "January Wave" (Lawton 2001).

The damage of the bow of FPSO "Schiehallion" in November 1998, was a recent example of structural damage due to extreme waves (*The Engineer* 1998, Faulkner 1999, Tedeschi et al. 2000). The side shell of the vessel in the bow area was seriously damaged by the impact of the 20-22m high wave.

When analyzing extreme sea states for design of marine structures, the standard reference is the paper by Ochi (1978). In that paper, the data collected from fixed weather stations in the North Atlantic are analyzed. Three measured significant wave heights were documented: 16.7m and 15.3m for return period of 8 years and 16.8m for return period of 14 years. For return period of 20 years, the calculated most probable extreme significant wave height was 20m. Ochi concluded that the measured extreme sea states are exceeded by the theoretical predictions by about 15%. These results are in good agreement with results obtained from Nordenstrøm data (Table 3). Ochi also gives an expression for calculating the most probable mean wave zero-crossing periods of sea states. Calculated wave steepness of severe sea states using expressions from that paper were approximately $s_{T_Z} = 1/15$.

Mansour et al. (1997) used a significant wave height of 13.7m for a ship reliability study. This significant wave height was calculated for a return period of 20 years, using a Weibull 3-parameter distribution for a route in the North Atlantic. The spatial statistical correlation between different wave zones in GWS was taken into account (Mansour & Preston 1995).

Bitner-Gregersen et al. (1995) analyzed extreme significant wave heights for all wave zones in GWS for a return period of 20 years. For zones Nos. 8, 9, 11, 15, 16 and 17, the data contained in the SHIPREL scatter diagram, the maximum and minimum calculated significant wave heights were 15.95m (zone 16) and 14.15m (zone 11), respectively. The average value for all zones was 15.16m, higher than the result from this study. This discrepancy is the consequence of differences in the assumptions regarding the recording interval. The measured extreme significant wave heights for three locations around the Norwegian coast are also included in the paper. For a return period of 20 years, the measured values of significant wave heights were 12.4m, 14.9m and 13.5m. The authors concluded that these values compared well with predictions using data from GWS. They compared measurements and theoretical estimates for the steepness of very severe sea states. $s_{T_Z} = 1/15$ was recommended as the limiting wave steepness appropriate for the analysis of ship structures. Bitner-Gregersen et al. (1995) found that the data from GWS largely overestimate the recommended and measured wave steepness.

Classification Note (DnV 1992), proposes a procedure for reliability assessment of FPSO ship operating in the North Sea. The joint probability distribution of significant wave heights and mean wave zero-up-crossing periods is defined. The theoretical model is similar to the one used in present paper. Using expressions given in that Note the expected wave steepness of the sea state with significant wave height of 14m would be 1/14.11.

ISSC 2000 paid special attention to the sea state steepness (Tedeschi et al. 2000). The general conclusion from the report is that wave steepnesses based on from GWS data are unrealistically high and that the accuracy of the data in GWS should be further investigated. The results from the present analysis are consistent with the conclusion of ISSC 2000.

2.2.6 Model uncertainty due to choice of wave scatter diagram

For the reliability analysis of merchant ships, the GWS wave atlas is the most reasonable source of wave data. The analysis presented above clearly shows that other sources of wave data may significantly overestimate extreme sea states that merchant ships actually encounter. The reason is that the effect of voluntary changes in shipping route to avoid extreme storms is implicitly included in GWS. Zones in the North Atlantic have to be chosen as operational area for calculation of design wave loads. Although this might be a conservative assumption for most ships, ship owners and classification societies would hardly accept any other assumption (Folso et al. 2002). Therefore, the SHIPREL wave scatter diagram, based on GWS data for the North Atlantic, is considered to be a reasonable reference wave scatter diagram. In addition, as shown above, the wave data in the SHIPREL scatter diagram result in extreme significant wave heights that have actually been encountered by numerous commercial ships. However, it is also shown that the application of this scatter diagram leads to an overestimation of the wave steepness of extreme sea states. Because, in most cases, this would lead to an overestimation of the lifetime vertical wave bending moments a mean value (bias) lower than one should be applied as the model uncertainty for lifetime values. Subsection 2.4 presents convenient method for estimating this bias. If yearly extreme values are analysed, this bias is not needed, i.e. its value may be set equal to one.

It should be noted that this uncertainty assessment is valid exclusively for the SHIPREL scatter diagram. In general, any other scatter diagram should be analysed using the method presented above. Without considering the extreme significant wave heights and wave steepness that are inherent in a particular scatter diagram, comparisons of the resulting extreme wave loads with those resulting from other wave scatter diagrams are not valid.

COV due to the choice of wave scatter diagram is within the range of 10-16%, according to results of various comparative studies published in literature. Considering that significant progress has been made in recent years in the field of uncertainty assessment, lower values for the *COV* due to the choice of wave scatter diagram might be appropriate. In this analysis, a *COV* equal to 7% is assumed for this uncertainty.

2.3 Modeling uncertainty of the extreme vertical wave bending moment due to calculation of transfer functions

Transfer functions of ship motion and load-effects are calculated using 3D panel and strip theory numerical hydrodynamics methods. For fixed offshore structures with large displacements, 3D panel methods are used. For ship structures, characterized by forward speed, two-dimensional strip theory is still the most popular method. This is confirmed by recent reports of ISSC'97 and ISSC 2000 (Guedes Soares et al. 1997, 2000). Therefore, the focus of this study is on linear strip theory.

Despite the fact that strip theory is considered to be a fairly accurate method, the calculated transfer functions would, in most cases, not perfectly match the transfer functions measured using ship models. These inaccuracies in transfer functions would lead to inaccuracies in the predicted extreme values. Discrepancies in extreme value predictions, which are the consequence of inaccurate calculations of transfer functions, represent a rather important source of uncertainty.

Guedes Soares (1990b, 1991) performed a comparison of measured and calculated transfer functions and the influence of the resulting uncertainty on the short-term ship responses. Schellin et al. (1996) performed a comparative study that included measured transfer functions for a containership model at various speeds and headings. These studies used the experimental transfer functions to develop a long-term distribution of vertical wave bending moments. This distribution was then compared with the long-term distribution based on transfer functions calculated by various hydrodynamic codes. A number of researchers have used the results of this study to assess model uncertainty for use in ship reliability studies (Casella et al. 1996, Guedes Soares & Teixeira 2000). It is interesting to note that Schellin et al. (1996) also included in their analysis transfer functions calculated by 3D panel method. However, they did not obtain any better correlation with the experimental results than with the strip theory.

One of the major recent concerns in structural analysis is due to a quite unexpected finding that there are significant discrepancies between transfer functions calculated by various linear strip theory codes (Nitta 1995, Guedes Soares 1996, 1999). These discrepancies are the consequence of differing details in the calculation of hydrodynamic coefficients. Results of comparative analyses indicate that the *COV* of predicted extreme values using different linear strip theory codes is about 10%.

In recent studies of ship hull-girder reliability, the most frequently used statistical model for uncertainties associated with the calculation of transfer functions is the one recommended by ISSC'91. The uncertainty is modeled using the normal distribution with a bias $\bar{\phi} = 0.9$ and a $COV_\phi = 0.15$. The bias of 0.9 means that the calculated value of the vertical wave bending moment amidships should be multiplied on average by 0.9 to obtain the actual value. It should be noted that, in this uncertainty model, the *COV* of 15% also includes uncertainty of the wave model and the correlation with full scale measurements under "linear conditions". This model was accepted by Mansour and Hovem (1994) and by Casella and Rizzuto (1998). However, it is questionable whether this statistical model is generally acceptable for all programs that use linear strip theory, since significant scatter between different programs exists. Therefore, it may be suspected that each of the computer codes has its own bias and *COV*, maybe even a different type of statistical distribution. This was recognized by Casella et al. (1996), where they used a bias of 0.845 based on a comparative study of Schellin et al. (1996). They decided to keep the *COV* of 15% as recommended by ISSC'91.

This paper presents a methodology for the rational assessment of the model uncertainty of the vertical wave bending moment due to the application of the linear strip theory. The WAVESHIP program, an element of the SESAM analysis system, is used for hydrodynamic analysis (DnV 1993a). The methodology uses the uncertainty of the extreme short-term ship response as a measure of uncertainty of the design values. This approach was originally proposed by Ando (1998). The POSTRESP program is used for the calculation of short-term responses. This method and the results are intended primarily for modeling the uncertainty of extreme ship responses. However, it could be extended for use in estimating the modeling uncertainty in calculating life cycle fatigue damage of a hull-girder.

The method presented in this section is described in more detail in Parunov & Senjanović (2000).

2.3.1 Modeling uncertainty of short-term ship response

The modeling error ϕ of the calculated response variable X is defined for the prescribed short-term sea state as:

$$\phi = \frac{\tilde{X}^*}{X^*} \tag{10}$$

where the most probable extreme value of the experimental response variable $\tilde{X}^*$ in a given short-term sea state is:

$$\tilde{X}^* = \tilde{\sigma}_x \sqrt{2 \ln \tilde{n}} \tag{11}$$

$\tilde{n}$ in (11) represents the mean number of zero-up-crossings in the short-term sea state of the given duration, while $\tilde{\sigma}_x$ is the standard deviation of the measured response variable. A similar expression for the calculated response variable X is:

$$X^* = \sigma_x \sqrt{2 \ln n} \tag{12}$$

where σ_x is the standard deviation of the calculated response variable, while n is the number of zero-up-crossings. The standard deviation of the measured response variable is obtained from the measured (experimental) transfer functions $\tilde{H}_x(\omega)$ and the given wave spectrum $S_\eta(\omega)$ as:

$$\tilde{\sigma}^2 = \int_0^\infty \tilde{H}_x^2(\omega) S_\eta(\omega) d\omega \tag{13}$$

The significant response (double amplitude) $\tilde{X}_S$ of experimentally obtained response variable is:

$$\tilde{X}_S = 4\tilde{\sigma} \tag{14}$$

Similar expressions are also valid for the standard deviation σ and the significant response (double amplitude) X_S of the calculated response variable, i.e.:

$$\sigma^2 = \int_0^\infty H_x^2(\omega) S_\eta(\omega) d\omega \tag{15}$$

$$X_S = 4\sigma \tag{16}$$

The mean numbers of zero-up-crossings $\tilde{n}$ and n appearing in (11) and (12) depend on the sea state duration and the mean zero-up-crossing period of the response variable. However, it is reasonable to assume that $\tilde{n} \cong n$ which enables (10) to be rewritten in a form convenient for practical application:

$$\phi = \frac{\tilde{X}_S}{X_S} \tag{17}$$

The wave spectrum used in this study is the two-parametric ITTC modification of Pierson-Moskowitz spectrum, as given by (1). By inserting (1) into (13) and (15) it becomes obvious that the modeling error ϕ varies only with *Tz*, since *Hs* is cancelled in the nominator and denominator of the expression (17).

Short-term response curves, representing the significant response (double amplitude) for unit wave height were calculated using the POSTRESP program, for a range of mean wave zero-up-crossing periods over a range of 5-15 sec, with an increment of 0.5 sec. Although the comparison may be made for an arbitrary value of the mean wave zero-up-crossing period, the wave periods associated with extreme sea states are the main concern of this analysis. Ship safety analysis is based on extreme values of the ship bending moment that occur in these sea states with probabilities of exceedence close to 10^{-8} based on long-term analysis. From the analysis of wave steepness presented in Subsection 2.2.3, it appears that a representative extreme sea state for ship safety analysis in extreme conditions may be described using a Hs=14m and a Tz=11.6s. Therefore, an 11.6s wave period is used for calculating modeling uncertainty.

Although the proposed approach is valid for estimating model uncertainty of extreme values in short-term sea conditions in head seas, it may be also considered as a very useful approximation for the modeling uncertainty of the long-term prediction of extreme vertical wave bending moments. This is because long-term extreme vertical wave bending moments are normally experienced in head seas in very severe storms because the shipmaster usually turns the ship's bow into the waves in order to avoid excessive rolling.

2.3.2 Hydrodynamic analysis

In order to assess the model uncertainty of linear strip theory, a literature survey was conducted to find papers with published test results on ship models including vertical wave bending moment amidships. The papers had to contain sufficient data to support hydrodynamic analyses that could reasonably be compared to the experiments. Experimental reports that met these criteria were found for five ships of different types, as described by Parunov & Senjanović (2000).

All ships were modeled with 20-22 strips. The analysis of each ship included calculation of motion and load components, and the comparison with measurements. Thus, all models were checked against gross errors.

One of the analyzed ships was a modern Canadian frigate whose model was tested in regular and irregular wave conditions (Mc Taggart et al. 1997). The transfer function of vertical wave bending moment amidships determined by WAVESHIP is shown in Figure 10 together with the experimentally obtained transfer function.

For the five ships analyzed, fifteen transfer functions of the vertical wave bending moment at amidships were calculated. Therefore, for some of the ships more than one transfer function was obtained. This is either because measurements were performed for different Froude numbers, or because were more than one measurement points in the midship area. For some of the ships, more transfer functions could have been obtained. These transfer functions were omitted intentionally in order to ensure that one or two ships would not dominate the statistical results.

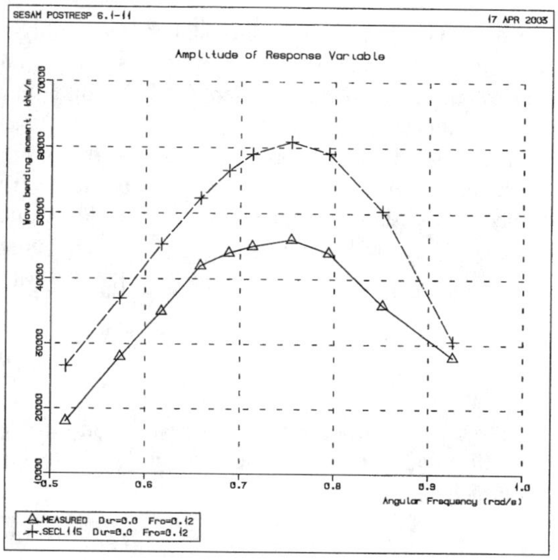

Figure 10: Transfer functions of vertical wave bending moment amidships for Canadian frigate

2.3.3 Short-term response

Short-term responses were calculated by the POSTRESP program, applying the Pierson-Moskowitz spectra with unit Hs. Double amplitudes of significant wave bending moments for transfer functions in Figure 10 are presented in Figure 11.

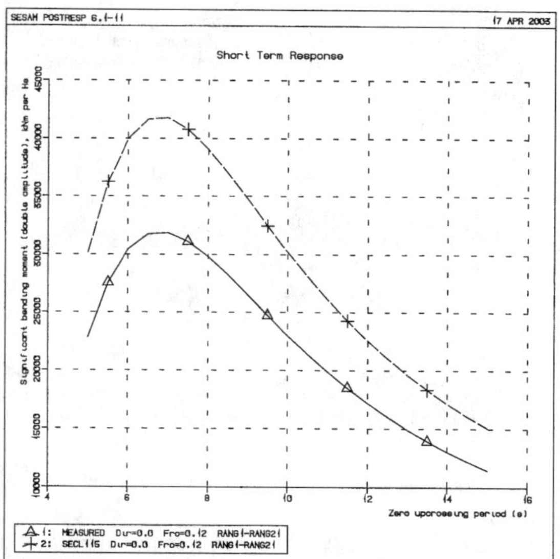

Figure 11: Short-term responses for transfer functions from Figure 10

The effect of differences between the calculated and experimental transfer functions on the design value of the vertical wave bending moment can be now seen more clearly than in Figure 10. At

Tz=11.6s, the ratio between the measured and calculated response is 0.76.

2.3.4 Statistical analysis of modeling uncertainty

The short-term response analysis was performed and the resulting fifteen calculated transfer functions were compared to the corresponding measured transfer functions. These results, representing modeling uncertainty of short-term response, were subjected to statistical analysis. The mean value μ_ϕ and standard deviation σ_ϕ of modeling uncertainty were calculated according to Rao (1992):

$$\mu_\phi = \frac{1}{n}\sum_{i=1}^{n}\phi_i; \qquad \sigma_\phi^2 = \frac{1}{n-1}\sum_{i=1}^{n}\left(\phi_i - \bar{\phi}\right)^2 \quad (18)$$

where n=15. The coefficient of variation of the modeling uncertainty as a measure of its dispersion is defined as:

$$COV_\phi = \frac{\sigma_\phi}{\mu_\phi} \quad (19)$$

Applying (18) to the set of fifteen calculated modeling errors, $\mu_\phi = 0.9$, $\sigma_\phi = 0.09$ and from (19) $COV_\phi = 0.1$, were obtained. The continuous normal distribution with these parameters compared to the discrete values of the modeling error is shown in Figure 12, where one can see that the normal distribution agrees quite well with this limited data set.

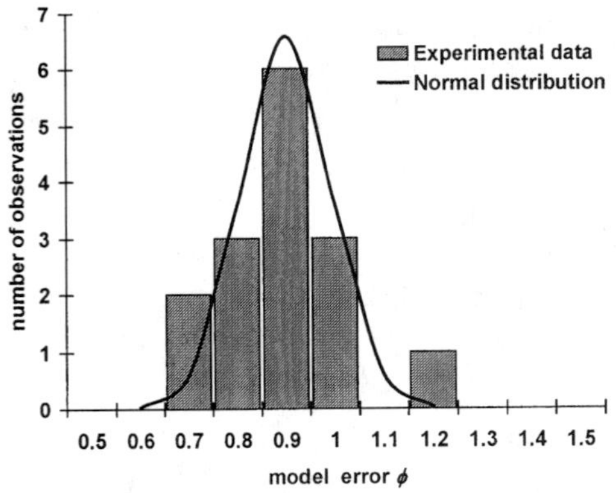

Figure 12: Variability of the modeling uncertainty

Since the analysis is based on a small number of discrete data points, it is useful to calculate confidence intervals. These intervals indicate the range of values within which mean and standard deviation would lie with certain probability if more data points were collected. Following the approach described by Rao (1992), it can be stated with 95% confidence that the mean value of model uncertainty would be in the range (0.85, 0.95). With the same confidence, standard deviation would be in the range (0.066, 0.142), if more data points were collected.

It should be mentioned that the calculated modeling uncertainty is valid only for particular linear strip theory code. To calculate the model uncertainty for other programs, an analysis similar to this one should be performed for each individual code.

The modeling uncertainty was evaluated only for Tz = 11.6s, an appropriate wave period for extreme sea states with return period of 20 years. This analysis assumes that the same model uncertainty would be valid for different wave periods, i.e. for extreme sea states with different return periods. This assumption should be further verified. Also, this method may be used to calculate the model uncertainty of service loads, applicable in fatigue reliability analysis. It should be noted that the model uncertainty might vary with other parameters, such as ship speed, ship type and ship main particulars, which have not been considered here.

2.4 Prediction methods for extreme wave loads

Generally, two methods are used to calculate extreme values of wave bending moments: the lifetime weighted sea method (LWS) and the design sea method (DS) (Hughes 1981, Mansour et al. 1996). The principle of the LWS method is to consider the total lifetime response history as a series of short-term episodes, and to combine them by a procedure that takes into account the relative amount of exposure to the various levels of sea severity (Hughes 1981). DS method is based on the assumption that the ship encounters an extreme sea state of specific duration. Short-term response analysis is then performed for such design sea state and extreme response is calculated. If these two types of analyses are performed consistently, they should lead to similar results. However, this is not always the case, since the choice of the short-term sea state is subjective to some extent. This is obvious from the results of Mansour et al. (1997), who obtained quite different extreme wave bending moments using the two approaches. This was also illustrated by an analysis of an FPSO vessel, where 10% lower extreme responses were obtained with DS method than with LWS method (DnV 1992).

The aim of this section is to compare these two methods in case of a large merchant ship. The comparison is based on a hydrodynamic and statistical analysis of a Flokstra containership (Flokstra 1974). The main scantlings of the containership are:

Length between perpendiculars	L_{PP}=270m
Breadth	B=32.2m
Draft	T=10.85m
Block coefficient	C_B=0.598

The hydrodynamic analysis was performed using the program WAVESHIP. The strip model of the vessel is shown in Figure 13. The analysis provided transfer functions of ship motion, wave bending moments and wave shear forces in different sections along the ship hull. Short-term and long-term statistical analyses were performed using the POSTRESP program.

2.4.1 Long-term analysis for head seas

The long-term analysis for head seas used the SHIPREL scatter diagram. Two ship speeds are considered, 24.5 knots and 4 knots, corresponding to the maximum service speed and the reduced speed in extreme conditions, respectively.

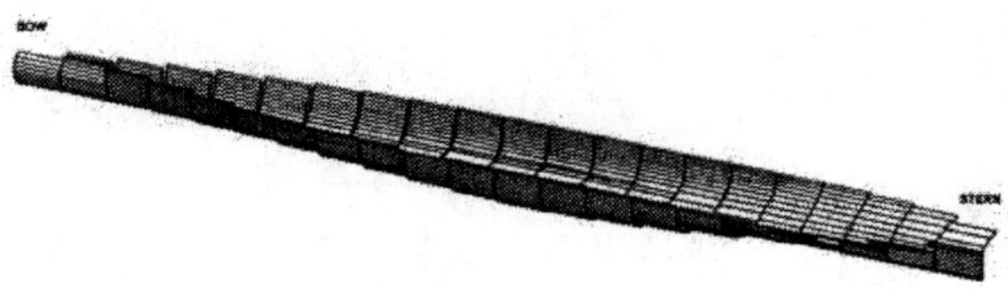

Figure 13: Hydrodynamic strip model of Flokstra containership

These two speeds correspond to Froude numbers 0.245 and 0.04. The long-term distributions of vertical wave bending moment amidships are shown in Figure 14. The most probable extreme vertical wave bending moments for return periods of 1 year and 20 years are presented in Table 4 for the two different ship speeds. For the purpose of comparison, the linear wave bending moment from IACS rules is 3186MNm.

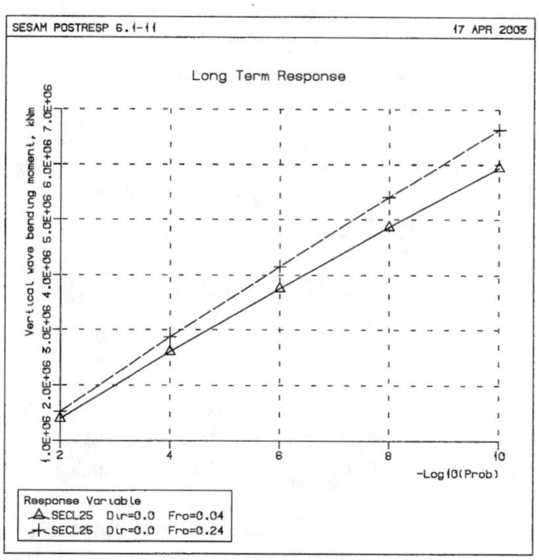

Figure 14: Long-term distributions of vertical wave bending moment for the head seas

Table 4: The most probable extreme vertical wave bending moments (long-term, head seas), MNm

Ship speed / Return period	4 knots	24.5 knots
1 year	4034	4602
20 years	4754	5411

2.4.2 Short-term analysis

It is interesting to compare the long-term results with those of the design sea method. For that purpose, the most probable extreme sea states with return periods of 1 and 20 years were used. From the analysis presented in Subsection 2.2 above, the most probable sea states are given as:

- Hs=11m, Tz=10s, duration T=8 hours for a return period of 1 year,
- Hs=14m, Tz=10.25s, duration T=3 hours for a return period of 20 years.

The results of the short-term statistical analysis, for these sea states and two different ship speeds are shown in Table 5.

Table 5: The most probable extreme vertical wave bending moments (short-term, head seas), MNm

Ship speed / Return period	4 knots	24.5 knots
1 year	3992	4551
20 years	4743	5438

By comparing the results from Tables 4 and 5 it may be concluded that the two methods result in quite similar extreme values. The maximum discrepancy between the two methods is less than 1%.

2.4.3 Short-term analysis applying spectral curves

The spectral curves, also called curves of short-term ship response, represent significant response (double amplitude) for unit significant wave height and for different mean zero-up-crossing periods. The curves may be used to estimate the short-term ship response, and at the same time provide insight into sensitivity of the response to changes of the mean zero-up-crossing period. Spectral curves of the vertical wave bending moment amidships, for the two different ship speeds are shown in Figure 15.

Because relationship between the significant wave bending moment and significant wave height is linear, the double amplitude of significant wave bending moment $2M_{w,s}$ is obtained by multiplying spectral curves by significant wave height. The most probable extreme vertical wave bending moment M_w^* in n response cycles is obtained as:

$$M_w^* = \frac{M_{w,s}}{\sqrt{2}} \sqrt{\ln n} \tag{20}$$

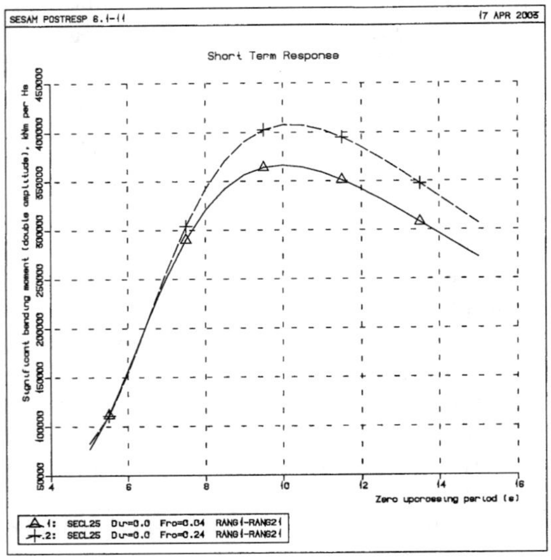

Figure 15: Spectral curves of significant vertical wave bending moment amidships for head seas

The number of cycles n, used in (20) depends on the duration of the sea state and the mean zero-up-crossing period of the response. It is reasonable to assume that the mean zero-up-crossing period of the response is approximately equal to the mean zero-up-crossing period of waves, i.e. $T_Z \cong 10\,\text{s}$ for both analyzed sea states. The most probable extreme vertical wave bending moments were calculated using these assumptions and shown in Table 6.

Table 6: The most probable extreme vertical wave bending moments calculated from spectral curves (short-term, head seas), MNm

Ship speed Return period	4 knots	24.5 knots
1 year	4029	4486
20 years	4782 (4474)	5335 (5041)

The maximum discrepancy between the values in Tables 6 and 5 is less than 2%. It is important to note that the wave steepness associated with the Tz assumed for the sea state with a return period of 20 years exceeds the recommended limits for the analysis of ship structures. This phenomenon was discussed in Subsection 2.2. A more reasonable mean zero-up-crossing period for Hs=14m is 11.6 s. It is obvious from Figure 15 that this would reduce the extreme values. The reduced values are shown in brackets in Table 6. In this particular case, the reduction is about 6%.

2.4.4 Long-term analysis for all headings

All the examples analyzed so far have been for head seas, being the most severe case for the vertical wave bending moments. For relatively small ships, less than about 200m in length, this is a reasonable assumption, since the shipmaster in extreme seas turns the ship into the seas in order to avoid excessive rolling. However, shipmasters of large ships rarely undertake such a maneuver (Guedes Soares 1990a). Therefore, it is more reasonable to consider all headings as equally probable. On the other hand, even very large ships experience speed reduction in heavy weather. The long-term distributions of vertical wave bending moment amplitudes for two different ship speeds with all heading angles equally probable are shown in Figure 16. The most probable extreme values are presented in Table 7.

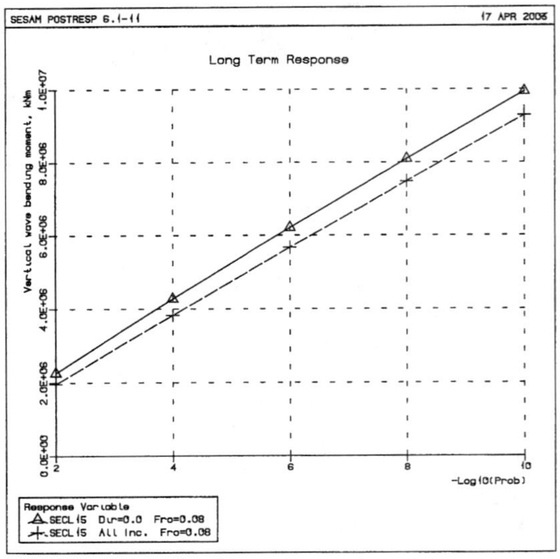

Figure 16: Long-term distributions of vertical wave bending moment for all headings equally probable

Table 7: The most probable extreme vertical wave bending moments (long-term, all headings), MNm

Ship speed Return period	4 knots	24.5 knots
1 year	3662	3993
20 years	4365	4777

Comparing the results from Table 7 with those in Table 4 which were obtained for head seas, a reduction of 10-15% in the extreme values may be noticed.

It is interesting to note that the probability of exceedence of the IACS vertical wave bending moment is higher than 10^{-6}, even for reduced speed. This corresponds to the return period shorter than 1 year. The same vessel was analyzed by Schellin et al. (1996), and their conclusions were similar. They determined that the probability of exceedence of the IACS wave bending moment was $10^{-3.5}$-$10^{-5.6}$, using the calculation methods of the leading classification societies.

2.4.5 Uncertainty assessment of the method for predicting long-term extreme values

Even when applying the same transfer functions and the same wave scatter diagram, classification societies still obtain discrepancies in extreme value predictions because of the different methods for numerical integration of long-term distributions. This problem was studied in a comparative analysis by Committee for Loads of ISSC'97 (Guedes Soares et al. 1997). They calculated a *COV* for this uncertainty of about 7%.

Quite often, uncertainty of long term extreme value prediction method has been included in analyses of other sources of uncertainties (scatter diagrams, transfer functions). This tends to inflate the resulting uncertainty values, as the uncertainty of the long-term prediction method is implicitly accounted for (Guedes Soares 1999).

The procedure for predicting extreme values proposed in this section included both long-term prediction and short-term prediction methods. As both methods lead to similar results, the procedure may be considered to be fairly accurate. It is reasonable to assume that the *COV* of the prediction uncertainty would be slightly lower than that determined in the ISSC study. Therefore, it is proposed to model the uncertainty of the prediction method as a normal variable with a mean equal to unity and *COV* equal to 5% .

2.5 Uncertainty of human actions

The uncertainty of human action may be divided into two parts:

- uncertainty of changes in shipping route in order to avoid heavy seas,
- uncertainty of master's actions in heavy seas.

The uncertainty of changes in shipping route is taken into account by accepting GWS as the source of the wave data. Data in GWS are obtained by observations from merchant ships that sail along standard trading routes. In that way, areas with the most severe weather conditions are implicitly avoided in the analysis.

When the ship is in heavy seas there are two well-known measures that ship's master can undertake to avoid excessive ship rolling and hull damage. These measures are:

- changing course from beam seas to head or following seas,
- voluntary speed reduction.

Course changes in heavy weather are mainly to avoid ship capsizing or excessive ship rolling amplitudes that may interfere with normal working activities on board (Guedes Soares 1990a). Consequently, the probability of head seas is much higher in heavy weather than in normal sea conditions. However, this is valid only for smaller ships (less than abt 200m in length). For large ships, course changes in heavy weather are not so frequent. The explanation for this finding could be that masters of large ships feel safe even in rather rough seas. However, the analysis is limited to observations of seas up to a significant wave height of 10m, lacking data for higher significant wave heights. Therefore, the likelihood of course changes of very large ships in extreme seas with significant wave heights greater than 10m is still an open issue.

Another important maneuver in rough seas is speed reduction. This action is not dependent on ship size (Guedes Soares 1990a). Reasons for speed reduction may generally be divided in two categories (Dubrovskiy 2000):

- technical reasons, such as the added resistance of wind and waves, change of submerged part of the hull, change of wake field and loss of thrust...
- reasons that are controlled by the shipmaster, such as very large motion amplitudes, velocities and accelerations, slamming, green seas, overload of the main engine.

Technical reasons influence ship speed at relatively low significant wave heights, while in heavy seas the shipmaster decides whether to reduce speed. There is no strict rule that determines under which conditions the shipmaster would reduce speed, so various authors have proposed different criteria (Dubrovskiy 2000).

Based on the arguments above, the following principles are proposed to take into account the effects of the heavy weather maneuvering:

- for the ships less than 200m in length, reduced speed and head seas should be taken into account for the return periods of 1 and 20 years,
- for ships larger than 200m in length speed reduction should be used. For a return period of 1 year, an equal probability of all headings should be used. For a return period of 20 years, there is uncertainty related to human actions in extreme seas. For the time being, taking the mean value of the results considering all headings equally probable and results considering head seas only, may be recommended.

Presently, there is not enough data to rationally describe a modeling uncertainty for the proposed procedure. For the purpose of this paper, result of the procedure is considered as the expected value. The coefficient of variation of the uncertainty is assumed to be 5%.

2.6 Summary of uncertainties of linear models

The two methods available for direct calculation of the linear part of the extreme vertical wave bending moment are the design sea method and the lifetime weighted sea method. Both methods are described and analyzed in Subsection 2.4 of the paper. From the comparative analysis, it was concluded that the result of both methods would be similar if they are applied consistently. It is recommended that both methods be used since the computational effort is negligible and the confidence in results is increased. The design sea states suitable for the analysis of ship structures for GWS wave zones 8,9,11,15,16 and 17 in the North Atlantic are:

- H_S=14m, T_Z=11.6s, duration 3 hours, for a return period of 20 years,
- H_S=11m, T_Z=10s, duration 8 hours, for a return period of 1 year.

It should be noted that for the return period of 20 years, the adjustment of the wave period is done in order to improve the accuracy of the results from GWS. This should be taken into account separately in the long-term analysis, since it has not been implicitly included in the wave scatter diagram.

2.6.1 Inherent uncertainty

The long-term probability distribution of wave loads is usually defined by a Weibull 2-parameter model:

$$F(x_p) = 1 - e^{\left(-\left(\frac{x_p}{\theta}\right)^{\lambda}\right)} \tag{21}$$

where θ and λ are the scale parameter and the shape parameter, respectively. It is well known that the Weibull distribution is an excellent approximation of the amplitude of ship response in waves. $F(x_p)$ in (21) represents the probability that the amplitude of the response variable is less than x_p in one, randomly chosen cycle. The probability that the response amplitude remains less than a given value x_e in longer time period, e.g. 1 voyage, 1 year or 20 years, is given by Gumbel law:

$$F(x_e) = e^{-e^{\left(-\frac{x_e - x_e^*}{\alpha}\right)}} \tag{22}$$

whose parameters x_e^* and α are derived from the parameters of the Weibull distribution (21) by the following relationships (e.g. Guedes Soares 1998):

$$\alpha = \frac{\theta}{\lambda}(\ln n)^{\frac{1-\lambda}{\lambda}} \tag{23}$$

$$x_e^* = \theta(\ln n)^{\frac{1}{\lambda}} \tag{24}$$

n in (23) and (24) represents the number of response cycles in given long-term return period, while x_e^* calculated by (24) represents the most probable extreme value in n cycles. The mean value and the standard deviation of the Gumbel distribution are given as:

$$\mu_e = x_e^* + \alpha \cdot 0.5772 \tag{25}$$

$$\sigma_e = \frac{\pi}{\sqrt{6}}\alpha \tag{26}$$

In the short-term analysis, the response amplitudes (peaks) follow a Rayleigh distribution (Mansour 1986). The Rayleigh distribution is a special case of Weibull distribution, (21), with $\lambda = 2$ and $\theta = \sqrt{2}\,\sigma_R$ where σ_R is the standard deviation of the short-term response. Expressions (23-26) for transformation of peak distribution into extreme value distribution are equally valid for the Rayleigh distribution, providing that n is the number of response cycles during a short-term sea condition of given duration.

The Gumbel distribution obtained by this procedure represents inherent uncertainty of extreme vertical wave bending moment. The coefficient of variation of this uncertainty is usually in the range 6-10%.

2.6.2 Model uncertainty

Based on the analysis presented in Subsections 2.1-2.5 above, the proposed statistical model of extreme vertical wave bending moment is shown in Table 8.

Table 8: Statistical model of extreme vertical wave bending moment

Source of uncertainty	mean	*COV*
Shape of wave spectra	1	0.05
Choice of wave scatter diagram	≤1[1]	0.07
Transfer functions	0.9	0.1
Method for prediction of long-term extreme values	1	0.05
Human actions	-[2]	0.05

Notes:

[1]-if a return period of 20 years is considered, the extreme wave bending moment should be reduced to take into account the limiting wave steepness.

[2]- the uncertainty of human actions on extreme wave bending moments should be assessed using procedure described in Subsection 2.5.

It is convenient to combine all of the uncertainties presented in Table 8 into a single, normally distributed random variable $\hat{\chi}_w$. This variable represents the total model uncertainty in the linear models. The mean value of the product of uncorrelated random variables is approximately the product of the mean values in Table 8. The combined coefficient of variation may be calculated using $COV = \sqrt{\sum_{i=1}^{5} COV_i^2} \cong 0.15$. This value of COV

has been used in other recent ship reliability studies (e.g. Casella & Rizzuto 1998, Teixeira & Guedes Soares 2002).

2.6.3 Effect of applying modeling uncertainty

The effect of applying the proposed modeling uncertainty can be demonstrated using the most probable extreme bending moments calculated for the Flokstra containership discussed above in Subsection 2.4. The most probable extreme values, corrected for modeling uncertainty are obtained by multiplying results of direct calculations with mean values from Table 8, and are presented in Table 9.

Table 9: The most probable extreme vertical wave bending moments, MNm

RP, years	head seas	all headings	corrected	IACS
1	4034	3662	3296	3185
20	4754	4365	3857	

From Table 9, the following interesting conclusions may be drawn:

- the most probable extreme values from direct analysis greatly exceed the IACS linear wave bending moment
- the corrected linear most probable extreme vertical wave bending moments exceed the IACS value by 3.5% and 21%, for return periods of 1 and 20 years, respectively.

Although the corrected bending moment for a return period of 20 years considerably exceeds the IACS vertical wave bending moment, it should be noted that other researchers have obtained even higher bending moments. This analysis supports opinion that IACS wave bending moment could lack conservatism in some cases.

3. APPROXIMATE PROCEDURES FOR NON-LINEAR CORRECTIONS

In the preceding sections, only linear hydrodynamic theory was considered. This theory includes the assumptions that hydrodynamic coefficients and hydrodynamic excitation forces are time invariant, being obtained by integration up to still water level. Variation of the wetted surface because of relative motion between ship and waves is thus neglected in linear hydrodynamic theory (Wu & Hermundstad 2002).

The most important source of non-linear behavior in wave loads is non-vertical ship sides. The main consequence of this non-linearity is the difference between sagging and hogging vertical wave bending moments. Using linear theory, it is not possible to distinguish sagging and hogging bending moments since only a single linear bending moment is obtained. Usually, the sagging bending moment is higher, while the hogging bending moment is lower than the linear prediction. Full scale measurements and the results of numerical calculations have shown that long-term distribution of sagging and hogging bending moments is not symmetrical with respect to distribution of linear moments (Guedes Soares 1993). Quadratic strip theory is capable of predicting differences in sagging and hogging bending moments using a perturbation approach (Jensen 2001).

Since the main source of non-linearity is non-vertical ship sides, ships with low block coefficients, such as containerships, experience rather significant non-linear effects. On the other hand, for ships with full forms, as large oil tankers, non-linear effects are not so important. However, recent researches have shown that in some cases even full form tankers of medium size may have significant non-linear contributions to the total vertical wave bending moments (Guedes Soares 1999). This is explained by the simultaneous effect of relatively short length and bulbous bow. In this case, both hogging and sagging vertical wave bending moments may be significantly higher than the linear prediction.

Despite quadratic strip theory and other non-linear methods, linear strip theory is the most popular tool for hydrodynamic computations. Non-linearity, especially the difference in sagging and hogging wave bending moments has been taken into account by approximate procedures. The most practical approach is to apply non-linear corrections for sagging Φ_S and hogging Φ_H to calculated linear extreme wave bending moments (Guedes Soares 1993, 1996, Guedes Soares & Schellin 1996). In this way, non-linear bending moments are obtained $M_S = \Phi_S M_L$ and $M_H = \Phi_H M_L$, for sagging and hogging, respectively. The simplest way to determine correction factors is to use IACS expressions for wave bending moments. According to IACS rules, the ratio between hogging and sagging bending moment is given as:

$$\left|\frac{M_H}{M_S}\right| = \frac{1.73\ C_B}{C_B + 0.7} \tag{27}$$

where the only parameter that influences the ratio of sagging and hogging bending moments is the block coefficient C_B.

A more advanced approach for determining correction factors is given by a regression analysis of data obtained by direct computations (Guedes Soares 1996). The advantage of this procedure is that the sag/hog ratio may be represented as a function of other parameters than C_B, such as ship length L. The linear regression analysis results in the following equations:

$$\Phi_S = 1.43241 + 0.00118L - 0.55032C_B$$
$$\Phi_H = 0.66981 - 0.00105L + 0.61029C_B \tag{28}$$

Regression equations are useful, since they show the trend of the correction factors with the independent variables, such as C_B and L. However, significant discrepancies may be observed in some cases between the results of direct calculations and (28). The most important discrepancy occurs in estimating the hogging moment of the medium-size tanker. The regression equation in this case underestimates the calculated value by 19%.

The drawback of the regression equations is that they are obtained for constant ship speed. Thus, heavy weather maneuvering is neglected, with potentially important consequences on the lifetime values.

Unlike the described regression approach, which is based on long-term analyses of non-linear wave bending moments, Mansour & Wasson (1995) has proposed an approach based on the short-term analyses of non-linear wave bending moments. According to this method, the correction factors for non-linear wave loads depend on the skewness and kurtosis of the response probability density function (Mansour & Jensen 1995). Both skewness and kurtosis are presented as functions of the significant wave height H_S, mean zero-up-crossing period T_Z, ship geometry, ship speed U and heading angle. The governing parameter for describing the effect of ship geometry is the flare coefficient C_f. The method requires results of linear hydrodynamic and short-term statistical analyses to calculate the response frequency. The complete procedure is presented in detail by Mansour & Wasson (1995), and is not repeated herein. It should be noted that the procedure has been recently extended to long-term non-linear prediction (Jensen & Mansour 2002).

To get a better insight into the accuracy of various approximate methods, a comparison against the results of direct non-linear hydrodynamic calculations is performed. The direct calculations are those published by Guedes Soares (1996). The mean values and standard deviations of ratios between direct calculations and correction factors calculated by various approximate methods are presented in Table 10. From Table 10, it appears that IACS sag/hog ratio (27) gives results that significantly deviate from direct analysis. Of particular concern is the large underestimation of the correction factor for sagging.

Table 10: Statistical moments of various approximate methods compared to direct non-linear hydrodynamic calculations

Method	sag		hog	
	mean	*COV*	mean	*COV*
IACS (27)	1.18	0.09	0.96	0.13
Equation (28)	1	0.07	1	0.08
Mansour & Wasson (1995)	1.03	0.08	0.94	0.09

3.1 Uncertainty modeling of non-linear effects

The unpredictability of non-linear effects may be seen in contemporary reliability analyses of ship hull. Authors have arrived at a variety of statistical models to describe uncertainty $\hat{\chi}_{nl}$. All of them have used the normal distribution, but with significantly different parameters. This is illustrated in Table 11, that shows the assumed parameters for tanker hulls. Although the data in Table 11 do not refer to the same ship, the shape of the hull is similar in all cases. Therefore, no significant differences in the predicted corrections for non-linear effects are expected.

Table 11: Parameters of normal distribution for modeling of non-linear effects

Reference	hog		sag	
	mean	*COV*	mean	*COV*
Dnv (FPSO) (1992)	-	-	0.92	0.09
Mansour (1994)	-	-	1.15	0.03
Casella et al. (1996)	0.97	0.15	1.03	0.15
Casella & Rizzuto (1998)	1.12	0.15	1.14	0.15
Guedes Soares (2000)	1.01	0.15	0.99	0.15

The discrepancies in Table 11 are the consequence of particular methods used in the non-linear analyses. The most sophisticated method is the time integration approach used for the analysis of an FPSO ship. Green water effects resulted in a mean value of less than one in sagging for the FPSO. Casella et al. (1996) used IACS rules, Casella & Rizzuto (1998) used quadratic strip theory, while Guedes Soares & Teixeira (2000) used regression equations similar to equations (28).

It may be concluded that a normal distribution is generally used to describe the uncertainty of non-linear effects $\hat{\chi}_{nl}$. The correction factors calculated by methods described in this section may be used to define the mean value of the normal distribution. One should be very careful in selecting non-linear corrections as significant differences may be obtained by various methods. The coefficient of variation of this uncertainty may be taken as 0.15, the highest value in Table 11. This rather high value of *COV* is due to the lack of confidence in non-linear corrections.

3.2 Influence of slamming and green water on extreme vertical wave bending moments

Bottom slamming occurs when forward part of the ship emerges from the water and the bottom portion subsequently experiences a severe hydrodynamic impact on re-entry. This impact may be particularly dangerous if it results in transient vibration, called whipping, considerably increasing the vertical bending moment amidships. The usual approach is to consider slamming and whipping loads separately from the wave-frequency loads (Ramos & Guedes Soares 1998, Senjanović et al. 2003). From such studies, it may be concluded that there are large uncertainties in the modeling of the intensity of the slamming impact and its time history. The statistical combination of the extreme high-frequency whipping load and wave-frequency non-linear load is also a rather difficult problem that has been the subject of numerous studies (Friis Hansen 1994, Jiao 1996).

The influence of green water on vertical wave bending moments is rarely taken into account (Jensen 1997). The usual opinion is that green water has no influence on the global ship loads. However, in the analysis of FPSO ships, significant reductions of wave bending moments in sagging due to green water have been calculated using time integration methods (Sogstad 1995).

In the present reliability formulation, the non-linear contributions of slamming and green water to total the wave load are not considered.

4. SHIP RELIABILITY ANALYSES

Reliability methods are intended as a tool that rationally takes into account uncertainties in loading, response and structural strength. The basic idea of reliability methods is to represent pertinent variables as random variables each with an associated distribution type and parameters. After that, by suitable procedures, the probability of structural failure (or its counterpart, the safety index) may be calculated. It should be noted that the failure probability determined by the reliability method is not a realistic value based on the tabulation of actual structural failures. It should be considered to be a nominal measure of safety, contingent upon the analysis model applied and the quality of the input data. Reliability methods for ship structural design are intended for two main purposes:

- direct application to individual designs when there is not enough accumulated experience,
- calibration of partial safety factors (PSF) in new ship rules for design of "ordinary" merchant ships.

The purpose of the direct application of reliability methods to certain ship structures is to establish rational safety margins between load and resistance in cases when accumulated past experience does not exist. FPSO ships, for which many recent designs exist, can serve as examples of this application (Lotsberg et al. 1991, DnV 1992, Wang et al. 1996, Maerli et al. 2000, Sun & Bai 2000). These are offshore structures that are similar to ships. However, they are at fixed locations, always turned to head seas, thus the application of the rules for ocean-going ships to these structures is not sufficient to ensure their structural adequacy. Therefore, reliability methods have been used to quantify the safety margin between load and resistance in order to determine the required section modulus.

PSFs can be calibrated against the structures and loads for standard ship types for which experience exists, in order to enable a more uniform safety level when a large number of ships is considered (Faulkner & Sadden 1978, Faulkner 1981, Guedes Soares & Moan 1985, Ostergaard 1991, Mansour et al. 1984, 1996, Madsen et al. 1986, Huther 2000). Rules based on the PSF concept would provide a prescribed safety level for each failure mode. That would provide significant economic benefits without any loss of safety. The rules for designing civil engineering and off-shore structures, where PSFs are successfully employed can serve as models for new reliability based ship rules (API 1989, Mansour et al. 1996).

The first studies of ship reliability were those of Mansour (1972, 1984, 1986), Mansour & Faulkner (1973), Faulkner & Sadden (1978), Stiansen et al. (1979), Faulkner (1981), Hart et al. (1985), Thayamballi et al. (1987). Recent papers by Beghin et al. (1992), Lee (1992), Mansour & Hovem (1994), Casella et al. (1996), Mansour & Wirsching (1997), Casella & Rizzuto (1998), Maerli et al. (2000), Teixeira & Guedes Soares (2002) may be considered as contemporary reliability analyses. From the beginning researchers have worked on two different problems. The first was to develop numerical methods for the calculation of failure probability, while the other problem was to collect and document data necessary to describe the statistical models of the basic random variables that influence ship safety. Numerical methods for calculating safety indices have now been sufficiently developed to be used for practical application, although methods for reliability assessment of complex systems are still under development (Mansour et al. 1996). Various computer codes for estimating the probability of failure are available today (Mansour & Wirsching 1995). These codes are usually based on an AFOSM (Advanced First Order Second Moment) reliability method that gives results of satisfactory engineering accuracy. On the other hand, there are serious problems related to the rational choice of statistical models of pertinent random variables.

Experience with new ships has shown that the structures designed according to UR S11 are fairly safe, since hull-girder damage in new ships is rare. However, the safety margin applied in new designs is not clearly defined (Faulkner 1998). This is the consequence of the way in which ship rules have been developed. They are basically of empirical nature relying strongly on past experience and good engineering judgment, partially supported by theoretical analyses and model tests. These types of structural design rules are called deterministic codes, where all uncertainties of the pertinent design parameters are compensated by a single safety factor, also known as "factor of ignorance" (Mansour et al. 1996, Huther et al. 2000). It is well known that such rules are hardly applicable for new designs that have features that deviate from normal practice. Of even greater concern is the rather difficult assessment of aged ships, subjected to unpredictable corrosion wastage. Since the safety margin is not explicitly accounted for in deterministic rules, it is not easy to make rational judgments about the adequacy of structures weakened by unevenly distributed corrosion. Unfortunately, this is confirmed in practice, since we witness frequent losses of old ships because of the hull-girder breakage. The losses of the oil tanker "Erika" in December 1999, the chemical carrier "Ievoli Sun" in October 2000 and the tanker "Prestige" in November 2002 could belong to this category. These tankers were all aged ships that suffered hull-girder damage in heavy weather. The application of direct calculation methods and ship reliability analysis, which is the subject of this paper, could help in the safety assessment of old ships (Friis Hansen et al. 2000).

4.1 Description of the example ships

In this study, three types of ship structure are analyzed: an FPSO ship, a large new double-hull tanker (D-H) and an aged single-hull tanker (S-H). All of them are frequently encountered today and may adequately represent the applicability and importance of reliability methods. The ship principal dimensions are presented in Table 12.

Table 12: Principal dimensions of example ships

	L_{pp}, m	B, m	D, m	T, m	C_B
FPSO	233	42	21.3	14.7	0.85
D-H	266.6	48.2	23	17.1	0.84
S-H	230.4	41.6	19.7	12.65	0.80

4.2 Reliability formulation

For all three structures, annual safety indices were calculated with respect to the ultimate failure of hull-girder. The analysis was performed for new ships as well as for corroded hulls assuming 20-year corrosion according to ship rules. In addition, the corroded single-hull tanker is analyzed for the condition following a regular survey, with measured thickness of structural elements.

The following limit-state equation is considered in this study:

$$\hat{\chi}_u \hat{M}_u - \hat{M}_{sw} - \psi \hat{\chi}_w \hat{\chi}_{nl} \hat{M}_w < 0 \qquad (29)$$

where:

M_u - deterministic ultimate longitudinal hull-girder strength

$\hat{M}_{sw}$ - random variable extreme vertical still-water bending moment

$\hat{M}_w$ - random variable extreme vertical wave bending moment

ψ - load combination factor between still water loads and wave loads

$\hat{\chi}_u, \hat{\chi}_w, \hat{\chi}_{nl}$ - random variables representing modeling uncertainty of ultimate strength, linear wave load and non-linearity of wave load.

Annual safety indices were calculated for two independent failure modes - sagging and hogging. The reliability assessment was performed separately for each of the elementary loading conditions - full load condition (FL), ship in ballast (BL) and partial loading conditions (PL), based on the assumed operating scenario. The assumed operating scenario for oil tankers is given by Casella et al. (1996) and Guedes Soares & Teixeira (2000), while the operating scenario of FPSO vessel is given by Maerli et al. (2000).

Safety indices $\beta_{FL}, \beta_{PL}, \beta_{BL}$, and associated failure probabilities $P_{f,FL}, P_{f,PL}, P_{f,BL}$ were calculated for each of the elementary loading conditions. The global annual safety index β_t is then determined as:

$$\beta_t = -\Phi^{-1}\left(P_{f,FL} + P_{f,PL} + P_{f,BL}\right) \qquad (30)$$

Failure probabilities were calculated using the EOSA program (Event Oriented System Analysis) that uses AFOSM and simulation procedures for reliability and sensitivity analysis of the components and the systems (Žiha 2000). In all cases, the reliability analysis was preceded by an exhaustive uncertainty assessment of pertinent random variables. The intention was to use, to the extent possible, assumptions that are similar to those generally used in contemporary reliability analyses. Only in that way can various analyses be compared.

Sensitivity analysis is an important part of reliability assessment since it enables identification of parameters that have the most important impact on safety indices (Mansour & Wirsching 1995). The sensitivity factors used in this study are those defined as:

$$\alpha_i = \frac{\left(\dfrac{\partial G}{\partial y_i}\right)_{y_i^*}}{\sqrt{\sum_i \left(\dfrac{\partial G}{\partial y_i}\right)^2_{y_i^*}}} \qquad (31)$$

where $G(.)$ is the failure hyperplane in the reduced space of the standard normal variates y_i (Madsen et al. 1986). y_i^* are the coordinates of the point in the reduced space closest to the origin. After transformation of y_i^* in original space, the most probable failure point x_i^* is obtained. This point, alternatively called the design point, is also included in the results of this analysis.

4.3 Uncertainty model of ultimate longitudinal strength of example ships

The ultimate longitudinal strengths of the example ships was calculated using an available computer code that has been appropriately calibrated and is in use by one of the leading classification societies. The results of the ultimate longitudinal strength calculations for new and corroded hulls are presented in Tables 13 and 14, respectively.

Table 13: Ultimate bending moments of new ships, MNm

	Sag	Hog
FPSO	9136	11575
D-H	12702	14832
S-H	7072	8020

Table 14: Ultimate bending moments of corroded ships, MNm

	Sag	Hog
FPSO	7977	10239
D-H	11150	13118
S-H	6162 (6381)	7173 (7597)

Values in brackets for the single-hull tanker in Table 14 represent ultimate bending moments calculated with corrosion deduction based on thickness measurements during a regular survey after 20 years of service.

The calculated moment was assumed to be the expected value of the ultimate bending moment. Therefore, the mean value of the random variable $\hat{\chi}_u$ was taken equal to unity. In most of the recent reliability studies, $\hat{\chi}_u$ is assumed to follow a log-normal distribution with COV=0.15 (e.g. Casella et al. 1996, Teixeira & Guedes Soares 2002).

4.4 Stochastic model of still-water bending moments

A Gaussian distribution was used for the stochastic model of still-water bending moment for one voyage. The distribution parameters were calculated from the loading manual, according to method proposed by Casella et al. (1996) and Guedes Soares & Dogliani (2000). For each of the elementary loading conditions (full load (FL), ballast (BL), and partial loading (PL)) mean values and standard deviations were calculated separately for departure and arrival conditions. Mean values and standard deviations of the resulting normal distributions were then obtained as average values for departure and arrival.

For the analysis of the single-hull oil tanker, the loading manual only included the design values for the still-water bending moments. In this case, the parameters for the normal distribution were developed according to Guedes Soares & Moan (1988). The method is based on statistical analysis of records from a large number of ships in departure.

When the mean value μ_{sw} and the standard deviation σ_{sw} of normal distribution are known, the extreme value distribution for a given time period Tc may be approximated using a Gumbel distribution (22) with following parameters:

$$x_e^* = F_{sw}^{-1}\left(1-\frac{1}{n_{sw}}\right), \quad \alpha = \frac{1-F_{sw}}{f_{sw}} \tag{32}$$

where n_{sw} is the number of occurrences of a particular load condition in the reference period Tc (1 year in our case). F_{sw} is the cumulative probability distribution, F_{sw}^{-1} its inverse while f_{sw} is probability density function of normal distribution with parameters μ_{sw} and σ_{sw}. The mean value μ_{se} and standard deviation σ_{se} of the Gumbel distribution are given by (25) and (26).

For the FPSO vessel, the number of loading and offloading cycles per year was n_{sw} =73, based on data from Maerli et al. (2000). For the oil tankers, n_{sw} represents number of voyages per year. According to the operating scenario proposed by Guedes Soares & Teixeira (2000), n_{sw} =5.4 for full load and ballast, while n_{sw} =27.4 for the partial loading condition.

Table 15 presents the parameters of the stochastic model of the still-water bending moment for one voyage and one year calculated according to described procedure.

Table 15: Parameters of still-water bending moment distributions, MNm

Ship	L.C.	One voyage[1]		One year	
		μ_{sw}	σ_{sw}	μ_{se}	σ_{se}
FPSO	FL	574	29	644	14
	PL	683	444	1763	223
	BL	2040	102	2286	51
D-H	FL	-1968	569	-2706	506
	PL	384	957	2352	560
	BL	3829	87	3943	78
S-H	FL	-396	452	-982	402
	PL	0	649	1335	380
	BL	549	674	1422	599

[1] – for FPSO, one loading and offloading cycle

4.5 Stochastic model of vertical wave bending moment

4.5.1 Inherent uncertainty

Hydrodynamic analysis of FPSO ship used the WADAM program that applies a linear 3D radiation-diffraction theory (DnV 1993). The double-hull oil tanker was analysed using WAVESHIP program that applies a linear strip theory. All subsequent statistical predictions were performed using POSTRESP program. The hydrodynamic analysis of single-hull oil tanker used the seakeeping tables of Loukakis & Chryssotomidis (1975), that were based on linear strip theory.

A short-term analysis was used to establish the stochastic model of vertical wave bending moment. The most probable extreme sea states with a return period of one year were calculated according to applicable wave scatter diagrams. The wave scatter diagram for the FPSO is obtained from measurements of sea states at a fixed location in the North Sea where the structure would operate. The scatter diagram used for the two sea-going oil tankers was the SHIPREL scatter diagram, discussed in Subsection 2.2. Details of the most probable sea states that were used for computation of extreme wave loads are presented in Table 16. In all cases, head seas were assumed. The FPSO ship has no forward speed, while the assumed speed of the oil tankers was 8 knots, corresponding to reduced speed in rough seas.

Table 16: Sea-states used for calculation of extreme vertical wave bending moments

Ship	Wave spectrum	Hs, m	Tz, s	Duration, hours
FPSO	JONSWAP	9.9	10.7	8
D-H	ITTC	11.1	10	8
S-H	ITTC	11.1	10	8

Based on short-term analysis, the standard deviation σ_R of short-term response and number of response cycles n, were obtained. Applying equations (23-26), mean value μ_e and standard deviation σ_e of the extreme (Gumbel) distribution were calculated. Table 17 presents calculated parameters of Gumbel low, which was used to model the inherent uncertainty of vertical wave bending moment.

The mean extreme values in short-term sea conditions presented in Table 17 are compared with long-term predictions for head seas for the double-hull oil tanker. The results of the long-term analysis of the double-hull oil tanker in head seas are presented in Table 18. The Gumbel moments were obtained from Weibull parameters using (23-26). Comparing the Gumbel moments from Table 18 with those from Table 17, maximum discrepancies of 1% and 4% in mean values and standard deviation, respectively, may be observed. These differences are rather low and could not have any practical implications. This example is further evidence of the equivalence of the short-term and long-term analyses, previously illustrated by the containership example in Subsection 2.4.

Table 17: Model of inherent uncertainty of extreme vertical wave bending moment

Ship	L.C.	σ_R, MNm	n	μ_e, MNm	σ_e, MNm	IACS, MNm
FPSO	FL	990	2148	4023	323	3860
	PL	915	2177	3721	299	
	BL	790	2227	3218	258	
D-H	FL	1472	2696	6064	475	5958
	PL	1682	2695	6930	543	
	BL	1481	2734	6105	477	
S-H	FL	945	2880	3906	304	3565
	PL					
	BL					

Table 18: Inherent uncertainty for double-hull oil tanker – long-term prediction for head seas

L.C.	Weibull param.			Gumbel moments (1 year)	
	θ, MNm	λ	n $(\cdot 10^6)$	μ_e, MNm	σ_e, MNm
FL	486	1.08	3.65	6130	465
PL	553	1.09	3.66	6981	527
BL	501	1.10	3.50	6115	458

Because the single-hull tanker was not analysed with the same strip-theory program as the double-hull tanker, the relative accuracy of these methods should be assessed. This was performed for the fully loaded double-hull tanker. The mean μ_e and the standard deviation σ_e of the Gumbel distribution were calculated using the seakeeping tables in Loukakis & Chryssotomidis (1975). Compared to the WAVESHIP calculation, the differences were only 2% for the mean values and 4% for the standard deviations. Therefore, further uncertainty assessment was performed using WAVESHIP for hydrodynamic analysis of both ships.

4.5.2 Model uncertainty

Model uncertainty of two ocean-going oil tankers was performed according to the methodology for linear models described in Subsection 2.5.2 and for non-linear effects described in Subsection 3.1.

According to Table 8 in Subsection 2.6, for return period of 1 year, the corrections that need to be applied to the mean value of $\hat{\chi}_w$ are the transfer function uncertainty and uncertainty of human actions. The mean value of the former uncertainty is fixed at 0.9, while the mean value of latter uncertainty needs to be calculated. By referencing the methodology proposed in Subsection 2.5, for ships larger than 200m and return period of 1 year, all heading angles should be considered as equally probable. Therefore, the long-term prediction for the double-hull oil tanker was performed assuming an equal probability for all heading angles. The reduction factors obtained were 0.96, 0.91 and 0.92 for full load, partial load and ballast condition, respectively. The same reduction factors were applied for single-hull tanker. Figure 17 presents the comparison of long-term distributions for head seas and assuming an equal probability for all headings for partial loading condition.

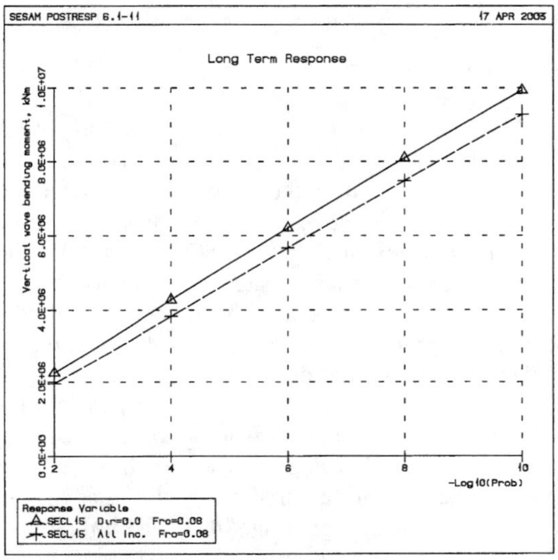

Figure 17: Comparison of long-term distributions for head seas (full line) and all headings equally probable (dashed line) for partial loading condition

The mean value of the random variable $\hat{\chi}_w$ was eventually obtained as the product of reduction factors for transfer function uncertainty and for heavy weather maneuvering. The coefficient of variation of model uncertainty of wave loads calculated using linear methods was 0.15, according to Table 8.

The model uncertainty of wave loads using linear methods for FPSO ship differed from that for the oil tankers. Transfer functions were calculated using the more accurate 3D panel hydrodynamic method. In addition, the wave scatter diagram and spectral shape were estimated more precisely based on measurements from the site where the FPSO would operate. Following the recommendations of DnV Classification Notes "Structural Reliability Analysis of Marine Structures" (1992) and analysis presented by Lotsberg et al. (1991), the model uncertainty of linear wave loads on FPSO calculated by 3D panel method was described using a normal distribution with the mean equal to 1 and a coefficient of variation of 0.1. This model is consistent with those developed in this paper for ocean-going ships. The mean value of 1 and the lower coefficient of variation for the FPSO are a consequence of the more accurate and reliable calculation of extreme wave loads. Also, it should be noted that FPSO vessel is not capable of heavy weather maneuvering and it is always turned bow to the seas.

The uncertainty of non-linear effects $\hat{\chi}_{nl}$ was estimated according to the analysis presented in Section 3. Since the FPSO and single-hull oil tanker have the hull shape similar to medium-size oil tankers, the mean values of this uncertainty were taken as 1.14 and 1.12 in sagging and hogging, respectively. These values, because they are specific to medium-size tankers (Guedes Soares 1999), were not calculated using the approximate methods described in Section 3. The non-linear corrections for large double-hull oil tanker were calculated according to method proposed by Mansour (Section 3). The normal distribution with a COV=0.15 was used to describe the uncertainty of non-linear effects.

A summary of the model uncertainties of wave loads used in these reliability calculations is presented in Table 19.

Table 19: Summary of model uncertainty of extreme wave loads

Ship	L.C.	$\hat{\chi}_w$		$\hat{\chi}_{nl}$	
		mean	*COV*	mean	*COV*
FPSO	FL	1	0.10	Sag 1.14 Hog 1.12	0.15
	PL				
	BL				
D-H	FL	0.86	0.15	Sag 1.03 Hog 0.97	0.15
	PL	0.82			
	BL	0.83			
S-H	FL	0.86	0.15	Sag 1.14 Hog 1.12	0.15
	PL	0.82			
	BL	0.83			

4.6 Load combination factors

The load combination factor ψ that appears in (29) is used to account for the fact that the extreme values of wave loads and still-water loads usually do not occur at the same time. Accordingly, the total extreme vertical moment must be reduced somewhat. Load combination factors are usually calculated using Ferry-Borges and Castanheta load combination method based on an assumed operating scenario (e.g. Guedes Soares 1992, Wang & Moan 1996). In the present analysis, the load combination factors used for the reliability assessment of the FPSO were those calculated by Maerli et al. (2000). For the analysis of the oil tankers, combination factors were calculated by Guedes Soares & Teixeira (2000).

Values of load combination factors adopted in the present analysis are given in Table 20.

Table 20: Load combination factors

Ship	FL	PL	BL
FPSO	0.91	0.84	0.78
D-H & S-H	0.92	0.91	0.80

4.7 Results of the reliability analysis

The annual safety indices were calculated by EOSA program using AFOSM method. Safety indices β and associated failure probabilities P_f used for new ship calculations are presented in Table 21 for two separate failure modes (hogging and sagging) and three different loading conditions. Safety indices and associated failure probabilities for corroded ship (20 years corrosion form ship rules) calculations are presented in Table 22. Safety indices for the corroded single-hull oil tanker based on actual corrosion as measured during its regular survey after 20 years are presented in Table 23.

Table 21: Safety indices in sag and hog for "as-built" ships

Ship	Cnd.	β_{FL}	β_{PL}	β_{BL}	P_{ft}	β_t
FPSO	Sag	3.69	4.94	6.16	1.14E-04	3.69
	Hog	4.00	3.92	4.36	8.41E-05	3.76
D-H	Sag	2.34	4.77	5.35	9.64E-03	2.34
	Hog	5.44	3.62	2.93	1.84E-03	2.90
S-H	Sag	1.87	4.30	3.67	3.09E-02	1.87
	Hog	3.83	2.91	2.25	1.41E-02	2.20

Table 22: Safety indices in sag and hog for corroded ships (20 years corrosion from ship rules)

Ship	Cnd.	β_{FL}	β_{PL}	β_{BL}	P_{ft}	β_t
FPSO	Sag	3.16	4.49	5.75	8.01E-04	3.16
	Hog	3.46	3.32	3.73	8.16E-04	3.15
D-H	Sag	1.72	4.34	4.98	4.27E-02	1.72
	Hog	5.06	3.06	2.30	1.18E-02	2.26
S-H	Sag	1.27	3.83	3.20	1.03E-01	1.27
	Hog	3.43	2.42	1.79	4.48E-02	1.70

Table 23: Safety indices in sag and hog for corroded single-hull tanker (measured thicknesses during regular survey after 20 years of service)

Ship	Cnd.	β_{FL}	β_{PL}	β_{BL}	P_{ft}	β_t
S-H	Sag	1.42	3.95	3.32	7.83E-02	1.42
	Hog	3.64	2.67	2.03	2.51E-02	1.96

Sensitivity analyses were performed for cases that resulted in the lowest safety indices, i.e. full load in sagging for corroded hull. The sensitivity factors α_i were normalized to sum to 100%. These normalized sensitivity factors are presented in Table 24. In the same table, coordinates of the design point x_i^* are included.

Table 24: Sensitivity factors and coordinates of design points

Ship		$\hat{\chi}_u$	$\hat{\chi}_w$	$\hat{\chi}_{nl}$	$\hat{M}_w$	$\hat{M}_{sw}$
FPSO	α_i, %	34.2	14.9	27.2	23.7	0
	x_i^*	0.75	1.12	1.42	4569	641
D-H	α_i, %	50.5	19.1	19.1	6.9	4.3
	x_i^*	0.82	0.96	1.15	6197	2941
S-H	α_i, %	42.1	22.1	22.1	7.1	6.7
	x_i^*	0.87	0.94	1.24	3954	1160

4.8 Analysis of results

4.8.1 FPSO ship

In order to use these results for the assessment of hull-girder safety, it is necessary to compare the calculated values to target safety indices. Maerli et al. (2000) have used 3.71 as a acceptable value of the safety index for new FPSO's. Although the calculated safety index in sagging (3.69) is slightly lower than the acceptable value (3.71), if the corrosion protection is efficient, the hull-girder reliability may be considered to be satisfactory.

It is interesting to note that the safety indices for the FPSO ship are approximately equal in sagging and hogging. The balanced safety indices are the consequence of careful structural and hydrodynamic analyses in the design stage that lead to rational distribution of the material.

It can be seen from the analysis of the sensitivity factors in Table 24 that the most important influence on safety index is the uncertainty in ultimate strength. Second place belongs to the uncertainty of non-linear effects. This variable could be calculated more accurately using non-linear hydrodynamic analysis thus reducing its coefficient of variation. As an illustrative example, the analysis was repeated for the case that resulted in the lowest safety index by taking *COV* of the $\hat{\chi}_{nl}$ as 0.1. The safety index increased from 3.16 to 3.44. The analysis was again repeated by reducing the mean value of this uncertainty to 1.03, which corresponds to the IACS rules, while *COV*=0.1. This increased the safety index significantly to 3.90. This analysis shows that advancing the state-of-the-art in non-linear hydrodynamics is an efficient way of increasing the safety of FPSO's.

4.8.2 Double-hull tanker

From the results presented in Table 21 for a new ship, the annual safety indices are 2.34 and 2.90 in sagging and hogging, respectively. These values are lower than 3.0, which is the proposed target safety index within the SHIPREL project for tankers with effective corrosion protection (Guedes Soares et al. 1996). The low safety index in sagging is influenced mainly by the full load condition. It is interesting to note that the mean still water bending moment in full load condition is much lower than its permissible value. However, the variability of the still water bending moment is rather high, which may be seen from Table 15. On the other hand, the still water bending moment in ballast is close to its permissible value, but the variability is rather low.

The safety indices for corroded hull are rather low (Table 22). The annual probability of failure in sagging is 0.0427, i.e. 4.27%. The importance of the corrosion protection on the reliability of the ship hull can be observed by comparing the results in Tables 21 and 22.

The sensitivity factors shown in Table 24 indicate that the most important variable is the modeling uncertainty of the ultimate longitudinal strength. Other important variables are the modeling uncertainty of the linear wave bending moment and non-linear effects.

The comparison of these results with those for the FPSO ship, shows that the FPSO has much higher annual safety indices. This is to be expected, since FPSO ships have to comply with more severe offshore rules. For new structure, without corrosion, the FPSO ship has a 53 times lower failure probability in sagging, and a 15 times lower failure probability in hogging.

Casella et al. (1996) have calculated the annual safety indices of a new double-hull oil tanker which is shorter than the present one (L=169m). The results of their study were similar to the present results, since they obtained safety indices 2.74 and 2.97 in sagging and hogging, respectively.

It should be noted that the calculated safety indices are dependent on the assumed shipping route. In the present analysis, it is assumed that a ship spends the whole time in the North Atlantic. For a similar tanker, but for the North Europe-Persian Gulf shipping route, significantly higher safety indices are obtained (Casella & Rizzuto 1998). They calculated annual safety indices of 4 and 3.4, in hogging and sagging, respectively. The main reason for higher safety indices compared to the present study are the wave bending moments, being much lower on the given ship route than in the North Atlantic.

4.8.3 Single-hull tanker

From Tables 21 and 22, it may be seen that the single-hull oil tanker has significantly lower reliability than the modern double-hull oil tanker. Particularly low reliability was calculated for the hull after it experienced corrosion according to the classification rules. The annual safety index in sagging was only 1.27, corresponding to a failure probability of 10.3%. This indicates that renewal of the material would be necessary before corrosion reached the permissible level. The safety index of the ship with measured thickness is 1.42 in sagging, lower than the value for the double-hull oil tanker with 20-year corrosion (1.72 from Table 22). Therefore, it would be necessary to undertake some action to improve it to an acceptable safety level. Measures to improve the safety of aged ships include:

- steel renewal to increase longitudinal strength,
- reduction of permissible still-water bending moments by appropriate ballasting,
- imposing operational restriction such as allowable sea zones where ship may operate safely.

Although rather low, the safety indices calculated in this study, are in good agreement with results of Guedes Soares & Teixeira (2000) for the case of a single-side bulk carrier. They obtained safety indices of 1.92 and 2.21 for a new ship in sagging and hogging, respectively. For the case of the corroded hull, safety indices calculated by Guedes Soares & Teixeira (2000) are also close to those from Table 22.

Although the calculated results are dependent on the assumptions used in the analysis, high values of calculated failure probabilities deserve attention and additional comment. The annual failure probability of the single-hull tanker with measured structural scantling thickness was 7.8% (Table 23). The physical interpretation of this value is the annual loss of 8 out of every 100 single-hull oil tankers may be expected. Fortunately, these figures have not been confirmed in practice, since the actual number of hull failures is much lower. This is mainly because the analysis was performed assuming that the ship spends all the time in the North Atlantic, where the weather conditions are the worst. Since ships sail on other shipping routes, the extreme sea states that they encounter are less severe than those used in this analysis.

This analysis demonstrates the applicability of reliability methods for the assessment of the structural adequacy of aged ships. The proposed approach is a compromise between the quality of the input data, the accuracy of the method for reliability calculation and the applicability of the obtained data. In order to make this method applicable in practice, it should be accepted by classification societies, which would, based on their own experience, propose an acceptable target safety index. This would contribute to an improvement in the rational safety assessment of corroded ship hulls.

5. CONCLUSION

Reliability methods are currently used to rationally take into account the uncertainties of the loads, response and structural strength. These methods are used either for direct application on individual structures when successful past experience is not available, or for calibration of partial safety factors in new rules for the design of ordinary ship types. Although reliability methods for ship structural design have been under development since early nineteen seventies, they have not yet become standard engineering tools. This is not because of the methods for calculating reliability, which are sufficiently developed for practical application, but because generally accepted statistical models for description of various uncertainties of pertinent random variables are lacking. Thus, various researchers have used different statistical models based on their subjective judgment, obtaining discrepancies in calculated reliability for the same physical reality. This has resulted in confusion, because only analyses based on the same or at least similar assumptions may be compared.

This paper considers longitudinal ship strength, as the most important strength component of the ship hull. The most important variables influencing ship longitudinal strength are the still-water bending moment, vertical wave bending moment and the ultimate longitudinal strength. All of these variables show significant scatter for actual structures and the analyses of associated uncertainties are currently the subject of much research. ISSC 2000 was a good demonstration of this: uncertainty analyses played a significant part in all ISSC committee discussions.

The modeling uncertainty of extreme vertical wave bending moments appears to be the most important problem and the subject of conflicting opinions of numerous researchers. This type of uncertainty is caused mainly by the choice of the wave data, inaccuracies in the calculation of transfer functions and non-linear effects. There are also other sources of modeling uncertainty, such as the uncertainty of human actions, the calculation method for estimating extreme values and the shape of wave spectrum. All of these uncertainties contribute to significant differences in the statistical models used by various researchers in their ship reliability studies.

In this paper, the choice of the wave scatter diagram was analyzed. Based on the data from GWS for the North Atlantic area, the joint probability distribution of significant wave heights and the mean zero-crossing periods was defined. Parameters of extreme sea states for different return periods, that are applicable in ship reliability analysis, were then calculated. The wave steepness of extreme sea states was analyzed. It was found that extrapolation of the data from GWS leads to a significant overestimation of the actual wave steepness that is appropriate for the analysis of ordinary ships. The results obtained in this analysis were then validated by comparison with measured extreme sea states and with recommendations of some relevant researchers. It should be noted that the present rules for the design of merchant ships do not contain any information about extreme sea states for which conventional ships are designed.

A method for estimating of the uncertainty of the extreme vertical wave bending moment due to calculation of transfer functions was proposed. The method was applied to ships for which well-documented model tests are available. The results were then subjected to statistical analysis. It was shown that this type of uncertainty can be described very well using a normal distribution with a mean value of 0.9 and a coefficient of variation of 0.1. It should be noted that this model is valid only for the strip-theory program used in this analysis. Other hydrodynamic codes may have different model uncertainty.

There are two well-known methods for calculating extreme values of wave bending moments: the design sea method and the lifetime weighted sea method. The methods were compared and it was shown that if the methods are applied in a consistent manner, both of them would lead to similar results. The influence of wave steepness on extreme vertical wave bending moments was analyzed, as well as the influence of heavy weather maneuvering.

Extreme linear vertical wave bending moments have to be corrected for non-linear effects. The most important sources of non-linearity of vertical wave bending moments are non-vertical ship sides, that result in different sagging and hogging vertical wave bending moments. Non-linear hydrodynamic methods are currently the subject of intensive development, as described by ISSC 2000. Since computer programs for direct calculation of non-linear hydrodynamic loads are not yet available for general use, approximate procedures were used for estimating non-linear corrections. It was shown, however, that such approximate procedures may lead to widely differing results.

These results were used as the basis for an approximate procedure for a proposed rational statistical model of extreme vertical wave bending moments.

The proposed uncertainty assessment was applied in three ship reliability analyses. Three different ship structures were analyzed: an FPSO ship, a large double-hull tanker and a aged single-hull tanker. Annual safety indices were calculated for two independent failure modes - sagging and hogging. The analysis was performed for new structure as well as for corroded hull with 20-year corrosion according to classification society rules. In addition, the single-hull oil tanker was analyzed based on regular survey results, i.e. with measured thickness of structural scantlings. In all cases, reliability analysis was preceded by a detailed uncertainty assessment. The intention was to use statistical models that are similar to those used by other researchers in contemporary ship reliability studies. Only in that way can various results be compared.

The FPSO ship was analyzed first. The analysis showed that the FPSO ship had a fairly high reliability, mainly because careful analyses had been performed in the design phase and because of the strict offshore regulation that was applied. It was shown that non-linear wave loads, that are not calculated rather accurately, may have a significant impact on the safety of the FPSO ship.

The purpose of the reliability analysis of the large double-hull oil tanker was to gain insight into the safety indices implied in current ship rules. The results of the analysis showed that the safety indices of the tanker were much lower than those of the FPSO ship. A rather low reliability of the corroded ship hull in sagging was observed.

The final example was the reliability analysis of an aged single-hull oil tanker. Since the "Erika" accident single-hull tankers are being analyzed more carefully. Furthermore, they should be gradually put out of service until 2015. The results in this paper show that safety indices of single-hull oil tankers are substantially lower than those of the double-hull oil tankers. Reliability assessments were recommended as additional means to assess aged ship structures.

ACKNOWLEDGMENT

This study was initiated while the first author was on special assignment in Bureau Veritas, Paris, thanks to a French Government Scholarship, which is gratefully acknowledged. The on-going cooperation between the Bureau Veritas and the University of Zagreb through the "Structural Analyses of Ships and Offshore Floating Units" project was also very useful to this paper.

The opinions presented herein are those of the authors and should not be construed as reflecting the views of any institution.

REFERENCES

ANDO, S., "Quantification of Correlation of Predicted and Measured Transfer Functions for Ship Motions and Wave Loads", *RINA, International Conference on Ship Motions and Maneuverability*, London, 1998.

API, *Draft recommended Practice for Planning, Designing and Constructing Fixed Offshore Platforms – Load and Resistance Factor Design*, API RP2A-LRFD, American Petroleum Institute, Dallas, TX, 1989.

AYYUB, B.M., *Reliability-Based Design of Ship Structures: Current Practice and Emerging Technologies*, T&R Report No. 53, SNAME, New Jersey, 2001.

BEGHIN, D.,HUTHER, M. and PARMENTIER, G., "The Probabilistic Approach – A Tool for Ship Rules Improvement", *Practical Design of Ships and Mobile Units (PRADS)*, 2.847-2.859, Newcastle, 1992.

BEGHIN, D. et al., "Design Principles and Criteria", 13th *International Ship and Offshore Structures Congress*, Vol.1, 351-405, Trondheim, Norway, August 1997.

BITNER-GREGERSEN, E.M., CRAMER, E.H.and LØSETH, R., "Uncertainties of Load Characteristics and Fatigue Damage of Ship Structures", *Marine Structures* 8, 97-117, 1995.

CASELLA, G., DOGLIANI, M. and GUEDES SOARES, C., "Reliability Based Design of the Primary Structure of Oil Tankers", *OMAE*, Vol.II, 217-224, 1996.

CASELLA, G. and RIZZUTO, E., "Second-level reliability analysis of a double-hull oil tanker", *Marine Structures*, 11, 373-399, 1998.

CHOW, V.T., *Handbook of Applied Hydrology*, McGraw-Hill, New York, 1964.

CRAMER, E.H., FRIIS HANSEN, P., "Stochastic Modeling of Long Term Wave Induced Responses of Ship Structures", *Marine Structures* 7, 537-566, 1994.

DET NORSKE VERITAS, Classification Notes No. 30.6., *Structural Reliability Analysis of Marine Structures*, July, 1992.

DET NORSKE VERITAS, *WAVESHIP – Wave Loads on Slender Vessels*, SESAM User's Manual, Høvik, 1993a.

DET NORSKE VERITAS, *POSTRESP – Interactive Postprocessor for General Response Analysis*, SESAM User's Manual, Høvik, 1993b.

DET NORSKE VERITAS, *WADAM – Wave Analysis by Diffraction and Morison Theory*, SESAM User's Manual, Det Norske Veritas Sesam AS, Høvik, 1994.

DUBROVSKIY, V.A., "Complex Comparison of Seakeeping: Method and Example", *Marine*

Technology, Vol. 37, No. 4, 223-229, October 2000.

DUXBURY, A.C. and DUXBURY, A.B., *Oceans*, WCB, Dubuque, 1991.

FAULKNER, D. and SADDEN, J.D., "Toward Unified Approach to Ship Structural Safety", *Transactions RINA*, paper No. 3, Spring Meeting 1978.

FAULKNER, D., "Semi-Probabilistic Approach to the Design of Marine Structures", *Transactions SNAME*, 213-230, 1981.

FAULKNER, D., On A. Nitta et al. "Basis of IACS Unified Longitudinal Strength Standard", *Marine Structures* 8, 335-336, 1995.

FAULKNER, D., Discussion on "Environment", 13th *International Ship and Offshore Structures Congress*, Vol. 3, 10-11, Trondheim, Norway, August 1997a.

FAULKNER, D., Discussion on "Design Principles and Criteria", 13th *International Ship and Offshore Structures Congress*, Vol. 3, 171-172, Trondheim, Norway, August 1997b.

FAULKNER, D., "An Independent Assessment of the Sinking of the MV DERBYSHIRE", *Transactions SNAME*, Vol. 106, 59-103, 1998.

FAULKNER, D., "Showing up the cracks", *Offshore Engineer*, January 1999.

FAULKNER, D., Letters to the editor: "Structural progressive collapse of bulk carriers", *The Naval Architect*, pp. 88, 2002.

FLOKSTRA, C., "Comparison of Ship Motion Theories with Experiments for a Container Ship", *International Shipbuilding Progress*, 21, 168-189, 1974.

FOLSØ, R., OTTO, S. and PARMENTIER, G., "Reliability-based calibration of fatigue design guidelines for ship structures", *Marine Structures* 15, pp. 627-651, 2002.

FRIIS HANSEN, P., "On Combination of Slamming- and Wave-Induced Responses", *Journal of Ship Research*, Vol. 38, No. 2, 104-114, June 1994.

FRIIS HANSEN, P. et al., "Design Principles and Criteria", 14th *International Ship and Offshore Structures Congress*, Vol. 1, 393-460, Nagasaki, Japan, October 2000.

GUEDES SOARES, C. and MOAN, T., "Uncertainty Analysis and Code Calibration of the Primary Load Effects in Ship Structures", 4th *International Conference on Structural Safety and Reliability (ICOSSAR)*, III501-III512, 1985.

GUEDES SOARES, C. and MOAN, T., "Statistical Analysis of Stillwater Load Effects in Ship Structures", *Transactions SNAME*, Vol. 96, 129-156, 1988.

GUEDES SOARES, C., "Effect of Heavy Weather Maneuvering on the Wave-Induced Vertical Bending Moments in Ship Structures", *Journal of Ship Research*, Vol. 34, No. 1, 60-68, March 1990a.

GUEDES SOARES, C., "Comparison of Measurements and Calculations of Wave Induced Vertical Bending Moments in Ship Models", *International Shipbuilding Progress*, 37, no. 412, 353-374, 1990b.

GUEDES SOARES, C, "Effect of Transfer Function Uncertainty on Short-term Ship Responses", *Ocean Engineering*, Vol. 18, No. 4, 329-362, 1991.

GUEDES SOARES, C., "Combination of Primary Load Effects in Ship Structures", *Probabilistic Engineering Mechanics* 7, 103-111, 1992.

GUEDES SOARES, C, "Long Term Distribution of Non-Linear Wave Induced Vertical Bending Moments", *Marine Structures* 6, 475-483, 1993.

GUEDES SOARES, C., "On the Definition of Rule Requirements for Wave Induced Vertical Bending Moments", *Marine Structures* 9, 409-425, 1996.

GUEDES SOARES, C. and SCHELLIN, T.E., "Long Term Distribution of Non-Linear Wave Induced Vertical Bending Moments on a Containership", *Marine Structures*, 9, 333-352, 1996.

GUEDES SOARES, C., DOGLIANI, M., OSTERGAARD, C., PARMENTIER, G. and PEDERSEN, P.T., "Reliability Based Ship Structural Design", *Transactions SNAME*, Vol. 104, 357-389, 1996.

GUEDES SOARES, C. et al., "Loads", 13th *International Ship and Offshore Structures Congress*, Vol. 1, 59-122, Trondheim, Norway, August 1997.

GUEDES SOARES, C., "Stochastic Modelling of Waves and Wave Induced Loads", *Risk and Reliability in Marine Technology*, 197-211, Balkema, Rotterdam, 1998.

GUEDES SOARES, C, "On the Uncertainty in Long-term Predictions of Wave Induced Loads on Ships", *Marine Structures* 12, 171-182, 1999.

GUEDES SOARES, C. et al., "Loads", 14th *International Ship and Offshore Structures Congress*, Vol.1, 63-132, Nagasaki, Japan, October 2000.

GUEDES SOARES, C. and DOGLIANI, M., "Probabilistic Modelling of Time-varying Still-water

Load Effects in Tankers", *Marine Structures* 13, 129-143, 2000.

GUEDES SOARES, C. and TEIXEIRA, A.P., "Structural Reliability of Two Bulk Carrier Designs", *Marine Structures* 13, 107-128, 2000.

HART, D.K., RUTHERFORD, S.E. and WICKHAM, A.H.S., "Structural Reliability Analysis of Stiffened Panels", *Transactions RINA*, 293-310, 1985.

HOGBEN, N., DACUNHA, N.M.C. and OLLIVER, G.F., *Global Wave Statistics,* British Maritime Technology Ltd., Feltham, 1986.

HOGBEN, N., Discussion on "An Independent Assessment of the Sinking of the MV DERBYSHIRE", *SNAME Transactions*, Vol. 106, 59-103, 1998.

HUGHES, O.F., *Ship Structural Design*, SNAME, New York, 1988.

HUTHER, M. et al., "The Use of Results of R&D in Probabilistic Approaches", *International Conference on Ship and Shipping Research (NAV)*, Vol I, 5.1.1-5.1.13, Venice, Italy, September 2000.

ISAACSON, M. and FOSCHI, R., "On the Selection of Design Wave Conditions", *International Offshore and Polar Engineering Conference (ISOPE)*, Vol. III, 111-118, Montreal, Canada, May 1998.

JIAO, G., "Probabilistic Prediction of Extreme Stress and Fatigue Damage for Ships in Slamming Conditions", *Marine Structures* 9, 759-785, 1996.

JENSEN, J.J. and PEDERSEN, P.T., "Wave-induced Bending Moments in Ships - a Quadratic Theory", *Transactions RINA*, Vol. 121, 151-165, 1979.

JENSEN, J.J, Discussion on "Structural Safety of Ships", *Transactions SNAME*, Vol. 105, 61-98, 1997.

JENSEN J.J. et al., "Extreme Hull Girder Loading", 14th *International Ship and Offshore Structures Congress*, Vol. 2, 263-320, Nagasaki, Japan, October 2000.

JENSEN J.J. and MANSOUR A.E. «Estimation of Ship Long-term Wave-induced Bending Moment Using Closed-form Expressions», *RINA Transactions*, 2002.

LAWTON G. «Monster of the Deep», *New Scientist*, 28-32, June 2002.

LEE, J.-S., "Reliability Based Limit State Design Format of Ship Structures", *Practical Design of Ships and Mobile Units (PRADS)*, 2.820-2.833, Newcastle, 1992.

LOTSBERG, I., BACK-GANSMO, O. and DALANE, J.I., "Probabilistic Design of a Ship Type Floating Production Vessel", *Offshore Mechanics and Arctic Engineering (OMAE)*, Volume II: Safety and Reliability, 1991.

LOUKAKIS, T.A. and CHRYSSOSTOMIDIS, C., "Seakeeping Standard Series for Cruiser-Stern Ships", *Transactions SNAME*, 67-127, 1975.

MAERLI, A., DAS, P.K. and SMITH, S.N., "A Rationalisation of Failure Surface Equation for the Reliability Analysis of FPSO Structures", *International Shipbuilding Progress*, 47, no. 450, 215-225, 2000.

MADSEN, H.O., KRENK, S. and LIND, N.C., *Methods of Structural Safety*, Prentice-Hall Inc., Englewood Cliffs, 1986.

MANSOUR, A.E., "Probability Design Concepts in Ship Structural Safety and Reliability", *Transactions SNAME*, 64-88, 1972.

MANSOUR, A.E. and FAULKNER, D., "On Applying The Statistical Approach to Extreme Sea Loads and Ship Hull Strength", *Transactions RINA*, 277-313, 1973.

MANSOUR, A.E. et al., "Implementation of Reliability Methods to Marine Structures", *Transactions SNAME*, Vol. 92, 353-382, 1984.

MANSOUR, A.E., "A Note on Extreme Wave Load and the Associated Probability of Failure", *Journal of Ship Research*, Vol.30, No.2, pp.123-126, 1986.

MANSOUR, A. and HØVEM, L., "Probability-Based Ship Structural Safety Analysis", *Journal of Ship Research*, Vol. 38, No. 4, 329-339, Dec. 1994.

MANSOUR, A.E. and JENSEN, J.J.,"Slightly Nonlinear Extreme Loads and Load Combinations", *Journal of Ship Research*, Vol. 39, No. 2, 139-149, 1995.

MANSOUR, A.E. and PRESTON, D.B., "Return Periods and Encounter Probabilities", *Applied Ocean Research* 17, 127-136, 1995.

MANSOUR, A.E. and WIRSCHING, P.H., "Sensitivity Factors and their Application to Marine Structures", *Marine Structures*, 8, 229-255, 1995.

MANSOUR, A.E. and WASSON, J.-P., "Charts for Estimating Nonlinear Hogging and Sagging Bending Moments", *Journal of Ship Research*, Vol. 39, No. 3, 240-249, Sept. 1995.

MANSOUR, A., WIRSCHING, P., WHITE, G., AYYUB, B., "Probability Based Ship Design : Implementation of Design Guidelines", *Ship Structure Committee*, SSC-392, 1996.

MANSOUR, A., WIRSCHING P. et al., "Structural Safety of Ships", *Transactions SNAME*, Vol. 105, 61-98, 1997.

McTAGGART, M. et al., "Motions and Loads of a Hydroelastic Frigate Model in Severe Seas", Transactions SNAME, Vol. 105, 1997.

MITCHEL, W., "Sea Spectra Revisited", *Marine Technology*, Vol. 36, No. 4, 211-227, Winter 1999.

NITTA, A. et al., "Basis of IACS Unified Longitudinal Strength Standard", *Marine Structures* 5, 1-21, 1992.

NITTA, A., On D. Faulkner Discussion of Paper by A. Nitta et al. "Basis of IACS Unified Longitudinal Strength Standard", *Marine Structures* 8, 337-339, 1995.

OCHI, M.K., "Wave Statistics for the Design of Ships and Ocean Structures", *SNAME Transactions*, Vol. 86, 47-76, 1978.

OCHI, M.K, *Applied Probability and Stochastic Processes,* John Wiley & Sons, New York, 1990.

OCHI, M.K, *Ocean Waves – The Stochastic Approach*, Cambridge University Press, Cambridge, 1998.

OSTERGAARD, C., "Partial Safety Factors for Vertical Bending Loads on Containerships", *OMAE*, Volume II, Safety and Reliability, 221-228, 1991.

PARUNOV, J., SENJANOVIĆ, I., "Review of Statistical Models for Ship Reliability Analysis", *Practical Design of Ships and Mobile Units (PRADS)*, 273-280, The Hague, September 1998.

PARUNOV, J., SENJANOVIĆ, I. and TOMAŠEVIĆ, S., "On the Modelling Uncertainty of the Linear Wave Bending Moment", *International Conference on Ship and Shipping Research (NAV)*, Vol. II, 7.8.1-7.8.10, Venice, Italy, September 2000.

PARUNOV, J. and SENJANOVIĆ, I., "Influence of Wave Steepness on Extreme Ship Hull Vertical Wave Bending Moments", *MARINE 2003*, Szczecin, Poland (accepted)

RAMOS, J. and GUEDES SOARES, C., "Vibratory Response of Ship Hulls to Wave Impact Loads", *International Shipbuilding Progress*, Vol. 45, No. 441, 71-87, 1998.

RAO, S.S., *Reliability-Based Design*, McGraw-Hill, New York, 1992.

SCHELLIN, T.E., OSTERGAARD and GUEDES SOARES, C., "Uncertainty Assessment of Low Frequency Load Effects for Containerships", *Marine Structures*, 9, 313-332, 1996.

SENJANOVIĆ, I., PARUNOV, J., TOMAŠEVIĆ, S., "Strength Analysis of a Cruise Ship in Early Design Stage", *IMAM*, Iscia, Italy, 2000.

SENJANOVIĆ, I., TOMAŠEVIĆ, S., PARUNOV, J., "Ship Slamming and Whipping Analysis in Rough Seas", *Shipbuilding*, Vol. 51, No. 1, 2003.

SOGSTAD, B.E., "A Sensitivity Study of Hull Girder Wave Loads for Ship Shaped Oil Production and Storage Vessels", *Offshore Technology Conference (OTC)*, Paper No. 7769, 1995.

STIANSEN, S.G. et al., "Reliability Methods in Ship Structures", *Transactions RINA*, 381-406, 1979.

SUN, H.-H. and BAI, Y., "Reliability Assessment of a FPSO Hull Girder Subjected to Degradation of Corrosion and Fatigue", *International Offshore and Polar Engineering Conference (ISOPE)*, Vol. IV, 355-363, Seattle, USA, May 2000.

TEDESCHI, R. et al., "Environment", 14th *International Ship and Offshore Structures Congress*, Vol. 1, 1-62, Nagasaki, Japan, October 2000.

TEIXEIRA, A.P. and GUEDES SOARES, C., "Assessment of Partial Safety Factors for Tankers", *JCSS Workshop on Reliability Based Code Calibration*, Zurich, Suisse, March 2002.

THAYAMBALLI, A.K., CHEN, Y.-K. and CHEN, H.-H., "Deterministic and Reliability Based Retrospective Strength Assessments of Oceangoing Vessels", *Transactions SNAME*, 1987.

THE ENGINEER, "BP hull breach still a mystery", December 1998.

WANG, X. and MOAN, T., "Stochastic and Deterministic Combinations of Still Water and Wave Bending Moments in Ships", *Marine Structures* 9, 787-810, 1996.

WANG, X., JIAO, G. and MOAN, T., "Analysis of Oil Production Ships Considering Load Combination, Ultimate Strength and Structural Reliability", *Transactions SNAME*, Vol. 104, 3-30, 1996.

WANG, X., "Ultimate Strength and Reliability Analysis of a Midship Transverse Frame in an Offshore Production Ship", *OMAE*, 1996.

WU M. and HERMUNDSTAD O.A. "Time-domain simulation of wave-induced nonlinear motions and loads and its applications in ship design", *Marine Structures* 15, 561-597, 2002.

ŽIHA, K., "Event Oriented System Analysis", *Probabilistic Engineering Mechanics*, Vol.15, 2000, 261-275.

Development of Reliability Based Classification Rules for Tankers

J. S. Spencer, Member, American Bureau of Shipping, **P. H. Wirsching,** Visitor, University of Arizona, **X. Wang,** Member, American Bureau of Shipping, and **A. E. Mansour,** Life Fellow, University of California

ABSTRACT

An approach has been developed for formulating a code for design of ship structures to achieve uniform structural safety (reliability index) independent of ship size or magnitude of the design variables. The approach is based on the partial safety factor format and a prescribed reliability level determined from calibration with existing classification society rules. Examples are presented for determining criteria for section modulus and initial scantlings of tankers. Formulations are based on material yield strength, buckling, ultimate strength, and fatigue reliability.

NOMENCLATURE

A	fatigue strength coefficient
ABS	American Bureau of Shipping
ASD	allowable stress design
B	random variable expressing uncertainty in Miner's stress
COV	coefficient of variation
EVD	extreme value distribution
D_0	fatigue damage
f_1	maximum allowable bending stress for load case 1
f_y	nominal yield strength
FS	factor of safety FS, where $FS \geq 1$
g, g(**x**)	limit state function
k	load combination factor
LN	lognormal distribution
LRFD	Load Resistance Factor Design
m	fatigue strength exponent
M_{SW}, M_W	still water and wave induced bending moments, respectively
MPP	most probable point
N	fatigue cycles to failure
N_Q	number of cycles at knee of two segment S-N curve
N_S	service life
P_0	allowable probability of failure
p_f	probability of failure
PSF	partial safety factor
r	ratio of design M_W / M_{SW}
R	strength
$\underline{R}$, Ratio	SM_{LRFD}/SM_{ABS}, i.e. the ratio of the section modulus
$\underline{\underline{R}}$	ratio of the minimum section modulus according to the reliability formulation and the minimum section modulus per the current ABS requirements
R_n	nominal strength
S	stress, stress range
S_0	design stress range for fatigue
S_e	Miner's equivalent stress
S_m	strength reduction factor
S_n	nominal value of stress
S_Q	stress at knee of two segment S-N curve
t_1	minimum required bottom plating thickness for loads in the longitudinal direction
WSD	working stress design
X, Y	normalized random variables
α	ratio of mean wave to stillwater bending moments
β	safety index
β_0	target safety index
Δ	fatigue damage at failure

Δ_0	target fatigue damage level
γ	exponent of Weibull distribution
γ_R	partial safety factor associated random variable R
γ_i	partial safety factor for load uncertainty i
Λ	fatigue bias factor
ϕ	partial safety factor for resistance uncertainty
σ_R	standard deviation of random variable R
μ_R	mean value of random variable R

INTRODUCTION

Probability based design has the promise of producing well designed systems, i.e., structures having optimized design and/or improved reliability relative to design based on conventional practices. The basic advantage of probability based design is the inclusion of more information into the design process that is not generally included in conventional design procedures. Not only are basic geometric characteristics of the structure considered, but also information about the uncertainties associated with the strength of various elements and the loading are also considered. Probability based design allows the engineer, through good judgment, to design on the basis of a target reliability for the structure thus designing a more efficient structure.

Partial safety factors (PSF) for both strength (resistance) and load account for the variability associated with these quantities. Generally, the higher the uncertainty, the greater the partial safety factor will be for loads. For strength variables, greater uncertainty yields smaller partial safety factors. A target probability of failure, or reliability index, must be established as a prerequisite to PSF selection.

While probability based design has the potential to benefit engineers in every field, this paper specifically focuses on the application of probability based design to establish a consistent code for tankers. The goal is to provide a more complete design guideline for shipbuilding not only to optimize weight in design, but also to provide a baseline reliability index that every ship will achieve regardless of size or operation.

PROBABILITY BASED DESIGN CRITERIA USING PARTIAL SAFETY FACTORS

Relative to a conventional factor of safety code, a probability based design code has the promise of producing better engineered structure. Specifically,

- there is a possibility of weight savings and/or improvement in reliability;
- uncertainties are treated more rigorously;
- Because of improved perspective, important advances in structural engineering can be stimulated;
- reliability based codes can be easily revised periodically to include new sources of information; and
- a framework is provided for extrapolating existing design practice to new ships where experience is limited.

A review is provided in the following section of the historical development of reliability based codes.

CODE FORMATS

Conventional Safety Check Using Factor Of Safety

While conventional (traditional) design code safety check expressions may vary in format, they are essentially of the same form,

$$S_n < \frac{R_n}{FS} \tag{1}$$

for a safe design. S_n is nominal value of stress. It is a value on the safe (or right) side of the distribution of S, although sometimes it is the best estimate of S. R_n is nominal strength, and is usually a value on the safe or left side of the distribution of R although it also could be a mean or median value. Additional uncertainty is accounted for by the factor of safety FS, where $FS > 1$. Historically, FS, S_n, and R_n are defined by committees charged with writing the code. These values are typically experience based.

In summary, structural integrity in the face of significant uncertainties is maintained by proper choice of FS, S_n, and R_n. Note that no meaningful index of quality or reliability can be explicitly implied from these values.

In traditional design practice, the factor of safety is generally applied to the strength, as shown in Eq. (1). The right-hand side can be interpreted as the maximum allowable stress, and the use of Eq. (1) is frequently referred to in the literature as allowable stress design (ASD) or working stress design (WSD). Usually the allowable stress is published in the design criteria document with no explicit reference to the factor of safety.

While traditional ASD and WSD have served the engineering profession well, it has been suggested that the design codes lack consistency. Some provisions of the codes produce over-designs, some produce under-designs. There has been a suspicion that this inconsistency results in a serious economic penalty. It has been argued in many papers that the use of probabilistic-statistical methods, giving explicit consideration of the statistical distribution of each, has the promise of producing better engineered designs. Indeed, many organizations and agencies worldwide have already implemented probability based design codes and have demonstrated significant cost savings.

Probability Based Safety Check Expressions

Options for the development of safety check expressions using probabilistic methods and reliability technology include the following:

- Specify the maximum allowable probability of failure, p_0. The designer would be required to perform a probabilistic analysis of the component or system to estimate the probability of failure, p_f. The design is considered safe if

$$p_f < p_0 \tag{2}$$

  Subjective judgments would have to be made on the quality of data sets and distributional models of the random design factors. Generally, this is not a practical option for general design.
- Specifying a minimum allowable (target) safety index, β_0. Again, the designer would perform a reliability analysis to estimate the safety index, β. The safety check expression would be,

$$\beta > \beta_0 \tag{3}$$

- Using a partial safety factor format. Among code writers, this seems to be the method of choice. A target safety index is translated into safety factors applied separately to the key stress and strength variables. For example, if R_n represents the nominal strength of a component in a ship, and S_{SW} and S_W represent nominal stillwater and wave induced stresses respectively, then the safety check expression would be of the form,

$$\gamma_R R_n > \gamma_{SW} S_{SW} + \gamma_W S_W \tag{4}$$

Upon review of the options, it is suggested that the partial safety factor format provides the best compromise. The partial safety factors are derived from full distributional information from all of the design factors, and the format of the safety check equations are similar to those used by designers in other accepted reliability based codes.

Appendix A gives examples of formats employing partial safety factors that have been considered by various agencies in implementing probability based code requirements.

APPROACH

The approach developed consists of a sequence of steps that begin with adopting a general partial safety factor format. First, the general format for the code is established as

$$\phi R \geq \sum_{i=1}^{n} \gamma_i k_i f_i \tag{5}$$

where

f_i = nominal value of load component i
k_i = load combination factor for load component i
R = nominal resistance or strength
γ_i = partial safety factor for load uncertainty i
ϕ = partial safety factor for resistance uncertainty

This general format is applicable to most requirements in the ABS Rules, although in some cases a slightly different format might be preferable. Next, a limit state function, $g(\underline{x})$, is developed from each code provision.

$g(\underline{x}) < 0$ represents failure state
$g(\underline{x}) > 0$ represents safe state
$g(\underline{x}) = 0$ represents the limit state equation (boundary between failure and safe states)
and $\underline{x}$ = main design variables represented as random variables

The nominal values (or characteristic values) are then defined in terms of average (or mean) values and the standard deviations of the design variables.

A target reliability is then assigned. The target reliability is determined using a calibration procedure based on current ABS Rules for at least one set of the design variables or a non-dimensional value of their combination. For example, in the case of the hull girder section modulus formulation, the calibration is made at a specific value of the ratio $r = M_W / M_{SW}$ where M_W is the design wave bending moment and M_{SW} is the design stillwater moment (i.e., the nominal values). At this specific ratio, r, the section modulus determined from the new code formulation is equated to the section modulus determined from the existing code formulation, which is based on ABS Rules. For the selected ratio, r, the implied safety index and the corresponding partial safety factors are determined using a first order reliability method and a simulation technique developed by Wu and Du (1998) for automating the calculation of partial safety factors. The computed safety index is taken as the target reliability and is kept constant in the remaining calculation of the partial safety factors in order to develop a consistent formulation.

By varying r and maintaining constant target reliability, a set of partial safety factors is determined for each new r. These partial safety factors can be plotted versus the ratio of the nominal wave to stillwater bending moment, r. The ratio $\underline{\underline{R}}$ of the minimum section modulus according to the new formulation and the minimum section modulus per the current ABS formulation can be calculated for the range of practical interest of r (1.2 to 2.1 for tankers). In the range of $\underline{\underline{R}} < 1$, weight savings can be achieved without falling below the minimum target reliability adopted in the calibration procedure. Other ship components have been analyzed in a similar manner and are discussed later in the paper.

ESTIMATION OF THE SAFETY INDEX

The first step of the calibration process is to estimate the safety index implied by the present ABS Rules. In general, when the reliability is evaluated for all provisions of a code, adjustments are made, e.g., to ensure that the reliability level is uniform throughout the code. The safety index β implied in a requirement in the ABS Rules is used as the target safety index β_0 to derive a probability based safety check expression using partial safety factors. The following is an illustration of how β implied by the present Rules is estimated.

An example is used to illustrate the process. However, the same basic process is used in all of the Rule scantling requirements.

The safety check expression for t_1 for bottom shell plating (Sec. 5/3A.4.3.2 of the ABS Rules) will be derived. A copy of the ABS Rules for this component is given in Appendix B. The symbols used in the following equations are based on the Rules in Appendix B and are defined there.

The goal of analysis is to estimate the safety index associated with the rule on t_1, the minimum required thickness associated with loads in the longitudinal direction .

Definition Of The Problem

The requirement in this example is that the permissible (maximum allowable) bending stress for load case 1 – "a" is

$$f_1 = 0.40 S_m f_y \tag{6}$$

where S_m is a strength reduction factor and f_y is the nominal yield strength. For development of a reliability based safety check expression, it is necessary to estimate the safety index implied by this requirement.

Following notation of conventional probabilistic design theory, formulate the problem in terms of stress (S) and strength (R) both considered to be random variables to account for uncertainties in the estimated loads and the yield strength. The event of failure is

$$R < S \tag{7}$$

and the limit state is

$$g = R - S \tag{8}$$

To normalize Eq. (6), let $f_y = 1.0$. The following assumptions are made:

- f_1 is the best estimate which implies that f_1 is the mean of S. From Eq. (6), the mean of S is
$$\mu_S = 0.40 S_m \tag{9}$$
- The coefficient of variation (COV) of S is 0.15 (Nikolaidis and Kaplan, 1991). The COV is defined as standard deviation divided by the mean.
- The distribution of S is the extreme value distribution (EVD). This is a reasonable distribution for the largest stress that would occur during the service life.
- f_y is the nominal value, which has been assumed to be 1.0.
- The mean value of R is 1.2 times f_y. The factor of 1.2 is based on a large volume of data on tests of structural steel. Thus,
$$\mu_R = 1.2 \tag{10}$$
- The COV of R is 0.08. This value is based on test data on structural steel.
- The distribution of R is lognormal (LN). This assumption is based on test data on structural steel.

Determining the Safety Index

A simple reliability program is used to compute the safety index, β, and the probability of failure, p_f. As an example, when $S_m = 1.0$,

$$\beta = 5.145$$

$$p_f = 1.338 \times 10^{-7}$$

Note that safety check expressions for the t_1 requirement of side shell and bottom shell plating provide a margin of safety which includes consideration of longitudinal bending stresses and the true factor of safety. The example given above was the requirement for t_1 and therefore the β does not reflect the true factor of safety. However, this value is appropriate for developing a safety check expression for t_1. Comments on the true factors of safety are provided later in the next section.

In the derivation of the partial safety factors for the safety check expressions, it is assumed that the target safety index is equal to the safety index as implied by the present rules.

$$\beta_0 = \beta \tag{11}$$

Thus, for the example of the t_1 requirement for bottom shell plating,

$$\beta_0 = 5.145 \tag{12}$$

True Factor Of Safety

Safety check expressions for the t_1 requirement for side shell and bottom shell plating provide a margin of safety that includes consideration of longitudinal bending stresses and the true factor of safety. In theory, the limit state function should contain an explicit term for the longitudinal bending stress, and the true factor of safety should be applied. (The ABS strategy has been to compensate for the longitudinal bending stress by increasing the factor of safety.)

To develop a probability based safety check expression that includes longitudinal bending, via the calibration process, it is necessary to determine the true safety factor of the present code. This can be done by considering the ABS side shell plating requirement. There are no longitudinal stresses at the neutral axis, so the margin of safety in the safety check expression at the neutral axis includes only the true factor of safety.

At the neutral axis, the permissible bending stress is

$$f_1 = 0.80 S_m f_y \tag{13}$$

Let $S_m = 1.0 f_y$ be the nominal yield strength. The mean yield strength is assumed to be 1.2 times the nominal. The coefficient of variation of yield strength and applied stress are assumed to be 0.08 and 0.15, respectively. Yield strength is assumed to be lognormal (LN), and applied stress is assumed to have an extreme value distribution (EVD).

The limit state can be written as

$$g = R - S \tag{14}$$

where, normalizing to $f_y = 1.0$,

$$R \sim \text{LN}(1.2, 0.096) \tag{15}$$
$$S \sim \text{EVD}(0.89, 0.12)$$

and where the numbers in parentheses are the mean and standard deviation, respectively.

The safety index implied by the requirement of Eq. (13) is obtained from a program that calculates reliabilities,

$$\beta = 2.24 \tag{16}$$

and the notional probability of failure is

$$p_f = 0.0126 \tag{17}$$

So the target safety index, for the case of $S_m = 1.0$, is assumed to be,

$$\beta_0 = 2.24 \tag{18}$$

A summary of the notional failure probabilities for each steel type is given in Table 1.

In theory, these values of β would be the target reliabilities for those limit state functions where the longitudinal bending stress is included.

Table 1 Notional failure probabilities for steel

	Ordinary Steel	H32	H36	H40
S_m	1.00	0.95	0.908	0.875
β	2.24	2.50	2.70	2.85
p_f	1.26E-2	6.21E-3	3.47E-3	2.19E-3

HOW THE PARTIAL SAFETY FACTORS ARE DERIVED

Illustrated in this section is the procedure for deriving partial safety factors (PSF) for scantlings. A specific example is provided, but the process is similar for all requirements in the ABS Rules.

Consider the safety check expression for t_1 for bottom shell plating (Sec. 5/3A.4.3.2) in the ABS Rules. This requirement is shown in Appendix B.

$$t_1 = 0.73 s \left(\frac{k_1 p}{f_1} \right)^{1/2} \tag{19}$$

For load case 1 – "a"

$$f_1 = 0.40 S_m f_y \tag{20}$$

Eq. (19) is the limit state where p and f_y are random variables. All terms are defined in Appendix B.

The goal of analysis is to derive PSF's for p and f_y, which correspond to a target safety index of β_0. The safety check expression will be

$$t_1 \geq 0.73 s \left(\frac{k_1 \gamma_p p_n}{\phi f_y} \right) \tag{21}$$

Analysis

It is assumed that p has an extreme value distribution (EVD) and that f_y has is lognormal (LN).

$$p \sim \text{EVD}(\mu_p, \sigma_p) \quad (22)$$

$$f_y \sim \text{LN}(\mu_f, \sigma_f)$$

The mean and standard deviation are respectively in the parentheses.

Eq. (19) can be reduced to

$$p = A \cdot f_y \quad (23)$$

where

$$A = \frac{1}{(0.73)^2 k_1}\left(\frac{t_1}{s}\right)^2 (0.40 S_m) \quad (24)$$

To simplify the notation, now considering f_y to be a random variable, which is denoted simply as f.

Define normalized random variables, X and Y, which have the following distributions,

$$X = \frac{p}{\mu_p} \qquad Y = \frac{f}{\mu_f} \quad (25)$$

The μ's are mean values. From basic properties of random variables,

$$X \sim \text{EVD}(1.0, C_p) \quad (26)$$

$$Y \sim \text{LN}(1, C_f)$$

where C_p and C_f are the standard deviations of the normalized random variables relating to load and strength, respectively.

The limit state can then be written,

$$X = \frac{A\mu_f}{\mu_p} Y \quad (27)$$

For a given target safety index, β_0, the computer program "PSF" (Wu and Du, 1998) automates the derivation of the partial safety factors (PSF). The most probable point (MPP) or design points (X^*, Y^*) corresponding to β_0 are identified, thus producing the safety check expression

$$X^* \ge \frac{A\mu_f}{\mu_f} Y^* \quad (28)$$

The PSF's for X and Y are defined as

$$\gamma_X = \frac{X^*}{X_n} \qquad \phi_Y = \frac{Y^*}{Y_n} \quad (29)$$

where the subscript n indicates the nominal value. Let the nominal values equal the means, i.e. $X_n = \mu_X = 1.0$ and $Y_n = \mu_Y = 1.0$. The safety check expression then becomes,

$$\gamma_X \ge \frac{A\mu_f}{\mu_f} \phi_Y \quad (30)$$

The nominal value of p, denoted as p_n, is defined to be the best estimate and equal to μ_p. The nominal value of f, denoted as f_y, is a value in the left tail of the distribution, defined by $\mu_f = 1.2 f_y$. The factor of 1.2 is based on extensive strength tests of structural steel.

Denoting $\gamma_X = \gamma_p$ and letting $\phi = 1.2\phi_Y$, the safety check expression of Eq. (27) can be written in terms of t_1 as

$$t_1 \ge 0.73 s\left(\frac{k_1 \gamma_p p_n}{\phi f_y}\right)^{1/2} \quad (31)$$

Example

X, normalized pressure p, is $X \sim \text{EVD}(1.0, 0.15)$.
Y, normalized yield strength f, is $Y \sim \text{LN}(1.0, 0.08)$.
The target safety index is $\beta_0 = 5.145$.

From the program "PSF", the resulting PSF's are

$$\gamma_p = 2.60 \qquad \phi_Y = 0.869$$

But

$$\phi = 1.2\phi_Y = 1.2(0.869) = 1.04$$

HULL GIRDER SECTION MODULUS FOR TANKERS

Yield Strength Formulation

Formulation of a consistent code based upon reliability requires a limit state function to be established. In accordance with the approach outlined above, the limit state equation for yield strength Rule formulation is:

$$g(\underline{x}) = f_y SM - M_T \quad (32)$$

where

f_y = the yield strength of the material considered as a random variable

$M_T = M_{SW} + M_W$ = the total bending moment (stillwater bending moment plus the wave bending moment, both considered as random variables)

SM = section modulus of the ship (considered deterministic)

ABS Rules gives the following yield strength formula for global total strength assessment (see ABS Rules):

$$SM = \frac{M_T}{f_b} = \frac{M_s + M_w}{f_b} \quad (33)$$

with variables defined by their nominal values, and f_b = permissible bending stress.

The reliability based formulation for yield strength based on the above $g(\underline{x})$ function gives:

$$SM \geq \frac{\gamma_1 M_s + \gamma_2 M_w}{\phi f_y} \tag{34}$$

where γ_1, γ_2, and ϕ are partial safety factors applied to the stillwater bending moment, wave bending moment, and yield strength, respectively

A target reliability is then assigned based on current ABS Rules. The target reliability is based on equating Eqs. (33) and (34) and rearranging the terms to include α, which is the ratio of the mean wave to stillwater bending moments. It is necessary to use the means of these values, because the partial safety analysis program gives PSF based upon mean values. Using means also allows the random variables (stillwater bending moment, wave bending moment, and yield strength) to be represented by normalized probability distributions, with mean values equal to unity. The section moduli resulting from Eqs. (33) and (34) can be made equal at either a midpoint value or at an extreme value of the ratio α. The range of α was determined from several tankers currently in operation (see Ship Characteristics - ABS Tankers, Appendix C). Eq. (35) shows that for a certain value of α, specific partial safety factors can be found.

$$\frac{f_y}{f_b} = \frac{\gamma_1 + 0.6\alpha\gamma_2}{(1+0.6\alpha)\phi} \tag{35}$$

with variables as defined above and,

$$\alpha = \frac{\mu_{M_W}}{\mu_{M_{SW}}} \tag{36}$$

In order to relate the nominal values to the mean values, the biases applied to $M_W, M_{SW},$ and f_y are 1.0, 0.6, and 1.2, respectively.

Equating the section moduli given by Eqs. (33) and (34) at the chosen value of α, a target reliability index is established. This target safety index will be applied for the remaining values of α in order to insure a consistent standard of reliability. Based on the established reliability index and the nominal bending moment ratio, r (r is based on the biases applied to α), a value representing the left hand side of Eq. (35) is found and compared with a value determined from the ABS Rules. It must be noted that the biases given above should be applied to the partial safety factors given by the analysis program, to allow the comparison to be made based on nominal values. R, which is the ratio of section modulus resulting from the reliability formulation (Eq. 34) to that of ABS requirements (Eq. 35), denotes this comparison in the analysis.

Ultimate Strength Formulation

The reliability based code formulation for a ship's ultimate strength is similar to the formulation for yield strength. Ultimate strength formulation is based upon a $g(\underline{x})$ function given by:

$$g(\underline{x}) = cf_y SM - \left[M_{SW} + k_W\left(M_W + k_d M_d\right)\right] \tag{37}$$

where

c = buckling knockdown factor; default values are 0.85 for mild steel, and 0.65 for high strength steel

f_y = yield strength of the material

k_W = load combination factor between stillwater bending moment and the combined wave and dynamic moments; default value is 1.0

k_d = load combination factor between the wave and dynamic bending moments; default value is 0.7

M_{SW}, M_W, M_d = stillwater bending moment, wave bending moment, and dynamic bending moment, respectively

SM = section modulus

ABS Rules provide no formulation for section modulus based on ultimate strength. However, for comparison purposes, Eq. (34) may be used.

Probability based partial safety factor design characterizes the ultimate strength requirement, based on the limit state equation above as:

$$SM \geq \frac{\gamma_1 M_{SW} + \gamma_2 M_W + k_d \gamma_3 M_d}{c\phi f_y} \tag{38}$$

with variables described above and γ_1, γ_2, γ_3, ϕ are partial safety factors applied to stillwater bending moment, wave bending moment, dynamic bending moment, and yield strength, respectively. The default value of $k_W = 1.0$ has been used.

Equate Eqs. (33) and (38) to calibrate the reliability analysis on the basis of ABS standards:

$$\frac{cf_y}{f_b} = \frac{\gamma_1 + (0.6\gamma_2 + 0.063\gamma_3)\alpha}{\phi(1+0.6\alpha)} \tag{39}$$

where α equals the ratio of mean wave to stillwater bending moments, and all other variables are as described above. The dynamic bending moment is related to the wave bending moment through the correlation factor given above.

A similar procedure is used to develop the relation between r and R as described earlier for the case of yield strength formulation.

Section Modulus Requirements Based On Yield Strength

Because the formulation is based on random variables that include stillwater bending moment, wave bending moment, and material yield strength, probability distributions must be established for each variable. This input, including mean, standard deviation, and distribution type are used in the "PSF" analysis program. All variables are normalized to give means equal to unity. The yield strength variable, f_y, is characterized by a lognormal distribution with a coefficient of variation (COV) of 0.08. The stillwater bending moment, M_{SW}, is represented by a normal distribution with a COV of 0.25 while the wave bending moment, M_W, is characterized by an extreme value distribution with COV of 0.1.

The yield strength formulation is calibrated at two points. The first point corresponds to $\alpha = 2.78$ equating to $r = 1.67$. This value is chosen because it represents the median nominal wave to stillwater bending ratio of the 13 case study ships used. The second calibration point is at a maximum value of $r = 2.00$ ($\alpha = 3.33$). From this, the fixed safety indices associated with each calibration point are 4.5 and 4.35 for calibration at 1.67 and 2.00, respectively.

In Tables 2 and 3 the variables are defined as follows:

β = safety index

γ_1 = stillwater bending moment partial safety factor

γ_2 = wave induced bending moment partial safety factor

ϕ = yield strength partial safety of factor

$\underline{R}$, Ratio = SM_{LRFD}/SM_{ABS}, i.e. the ratio of the section modulus based on Load Resistance Factor Design, which is the same as partial safety factor design, to the section modulus based on ABS Rules.

Partial safety factors are plotted in Figs. 1 and 2. They are plotted vs. r (defined as the ratio of nominal wave bending moment to stillwater bending moment). The partial safety factor applied to the yield strength is about 1.0, as expected. The partial safety factor applied to wave bending moment is above 1.0, which is also expected. However, the partial safety factor applied to the maximum nominal stillwater bending moment is less than 1.0. The safety index associated with calibration 1.67 is 4.5. When the moment ratio is raised to 2.0, the safety index is reduced to 4.35.

Fig. 3 plots R vs. r (the section modulus ratio vs. the moment ratio). Also included on this plot are the design moment ratios for 13 ships in both hogging and sagging conditions. The table from which the nominal wave to stillwater bending moment ratios were calculated is included in Appendix C. This plot describes the differences between probability based design and conventional design codes. Based on the calibration point, where R equals 1.0, the probability based design is equivalent to the ABS Rules. All other points on the calibration curves represent ratios of reliability based section modulus to ABS Rule required section modulus as calculated using constant values of the safety index β. The r represents different design conditions of wave to stillwater bending moments. It follows from the formulation of the $g(\underline{x})$ function that the curve will yield an acceptable design based on the calibrated probability of failure. The section modulus of the reliability based design can be inferred from R. In some cases, this section modulus is less than that specified by ABS ($R < 1.0$). In other cases, the section

Table 2 Results for Calibration at r = 1.67, $\beta = 4.5$, p_f =2.21E-5, Based on Yield Strength

α	1.67	2.00	2.33	2.67	3.00	3.33	3.67
$r = M_W / M_{SW}$	1.00	1.20	1.40	1.60	1.80	2.00	2.20
ϕ	0.993	1.001	1.006	1.011	1.015	1.017	1.019
γ_1	0.870	0.826	0.794	0.770	0.754	0.736	0.725
γ_2	1.580	1.638	1.675	1.702	1.720	1.734	1.744
A	0.202	0.179	0.160	0.135	0.135	0.135	0.135
β	4.5	4.5	4.5	4.5	4.5	4.5	4.5
$\underline{R}$, Ratio	0.920	0.946	0.970	0.992	1.011	1.028	1.044
ϕ mean	0.8279	0.8339	0.8386	0.8426	0.8456	0.8475	0.8495
γ_1 mean	1.45	1.376	1.324	1.284	1.256	1.227	1.209
γ_2 mean	1.58	1.638	1.675	1.702	1.72	1.734	1.744

Table 3 Results for Calibration at r = 2.0, $\beta = 4.35$, p_f =3.35E-5, Based on Yield Strength

α	1.67	2.00	2.33	2.67	3.00	3.33	3.67
$r = M_W / M_{SW}$	1.00	1.20	1.40	1.60	1.80	2.00	2.20
ϕ	0.997	1.005	1.011	1.015	1.018	1.021	1.023
γ_1	0.866	0.823	0.793	0.769	0.749	0.736	0.723
γ_2	1.537	1.595	1.630	1.656	1.675	1.687	1.689
A	0.207	0.184	0.164	0.148	0.135	0.124	0.115
β	4.35	4.35	4.35	4.35	4.35	4.35	4.25
$\underline{R}$, Ratio	0.900	0.924	0.946	0.967	0.985	1.00	1.01
ϕ mean	0.8307	0.8375	0.8422	0.8456	0.8487	0.8509	0.8526
γ_1 mean	1.443	1.371	1.322	1.281	1.249	1.226	1.205
γ_2 mean	1.537	1.595	1.63	1.656	1.675	1.687	1.689

modulus is greater than the ABS section modulus ($R > 1.0$). For any given ship moment ratio value, a comparison can be made between the two design methods.

When the probability based design code gives a section modulus less than that required by ABS Rules, weight reduction can be achieved while maintaining the design reliability level. For calibration at r equals 1.67, this means about half of the ship case studies can save weight while the other half would technically have to add weight to achieve the desired safety level. However, if the assertion is made that the ship with the highest nominal wave to stillwater bending moment ratio, r, is operating safely, then all case studies except that ship have the potential to save weight. If the ship at r = 2.02 (therefore a value of r at 2.02 is the new calibration point) is operating safely, and the probability of failure of that ship is acceptable, then all other ships could have a reduced section modulus and operate at the same acceptable level of safety (see lower line of Fig. 3).

Section Modulus Requirements Based On Ultimate Strength

The distributions used to characterize yield strength, stillwater bending moment, and wave bending moment are identical to those given in the yield strength results. Additionally, the dynamic bending moment is represented by an extreme value distribution with a COV equal to 0.3.

The ultimate strength formulation is first calibrated at $\alpha = 2.67$ or $r = 1.60$. This value is chosen because it approximates the median nominal wave to stillwater bending ratio of the 13 case study ships used. The second calibration point is at $r = 2.00$ or $\alpha = 3.33$. The fixed safety indices associated with each calibration point are 3.2 and 2.02 for calibration at 1.60 and 2.00, respectively.

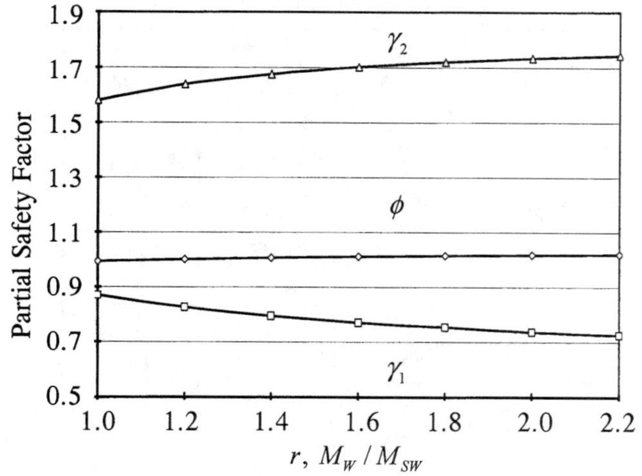

Figure 1 Partial Safety Factors for Calibration at 1.67, Based on Yield Strength

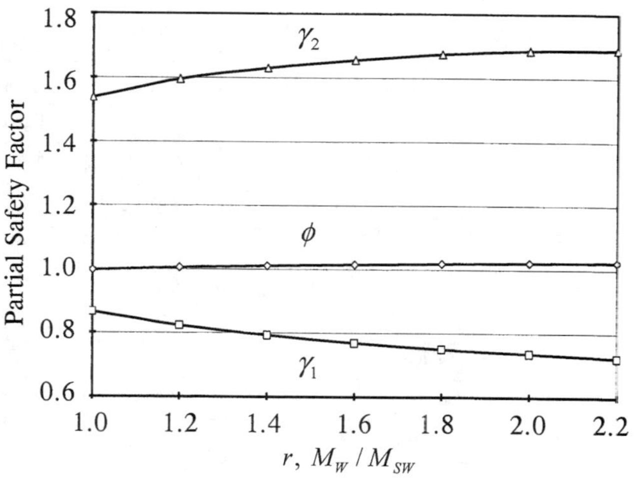

Figure 2 Partial Safety Factors for Calibration at 2.0, Based on Yield Strength

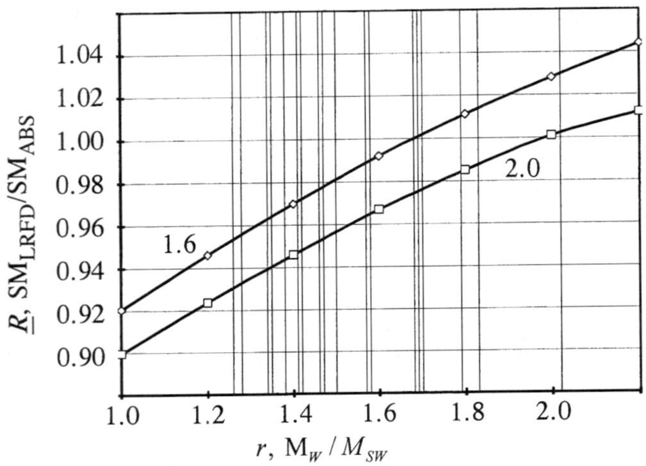

Figure 3 $\underline{R}$ vs. r for Calibrations 1.67 and 2.0, Yield Strength Formulation, Showing range of M_W / M_{SW} for Sample Ships

As can be seen in Tables 4 and 5, the results for ultimate strength conditions are similar to those for the yield strength in Tables 2 and 3. Figs. 4 and 5 are plots of the partial safety factors for ultimate strength analysis. The trends of these factors are identical to those in the yield case, with the addition of γ_3 reflecting the presence of dynamic bending moments. The dynamic bending moment safety factor is above 1.0 as expected.

For the ultimate limit state, the safety indices are 3.2 and 3.02 for calibrations 1.6 and 2.0, respectively, i.e., smaller than for the yield case. Again, this is as expected because of the inclusion of the dynamic moment, because ships, which have hulls that are comprised of thin plates and long members, are more likely to fail in buckling than yield.

Fig. 6, like Fig. 3, is based on the intersection of the calibration curves with R=1.0. The curves give a baseline probability of failure associated with a specific moment ratio, r. If a ship's r-value intersects with the calibration curve at an R-value less than 1.0, the ship can save weight while maintaining the prescribed safety index established during calibration. These observations are identical to those made in the yield strength formulation.

INITIAL SCANTLING CRITERIA FOR TANKERS

The following analysis describes the minimum strength requirements for determining initial scantlings, which include shell and bulkhead plating, longitudinals/stiffeners and main supporting members.

Plate Thickness Formulation

As outlined in the approach, a function, $g(\underline{x})$, must be established to determine the reliability implied by the ABS Rules. The limit state equation associated with plate thickness is

$$g(\underline{x}) = f_y - k\left(\frac{s}{t}\right)^2 p \tag{40}$$

where

f_y = the nominal yield strength of the material
k = load factor
p = nominal pressure
s = spacing of side longitudinals
t = plate thickness

Table 4 Results for Calibration at r = 1.6, $\beta=3.2$, p_f =6.87E-4, Ultimate Strength Formulation

α	1.67	2.00	2.33	2.67	3.00	3.33	3.67
$r = M_W / M_{SW}$	1.00	1.20	1.40	1.60	1.80	2.00	2.20
ϕ	1.029	1.033	1.036	1.041	1.044	1.048	1.049
γ_1	0.838	0.806	0.777	0.755	0.740	0.728	0.715
γ_2	1.251	1.290	1.321	1.340	1.354	1.368	1.375
γ_3	1.048	1.051	1.055	1.055	1.054	1.057	1.057
A	5.036	5.660	6.297	6.962	7.613	8.269	8.947
β	3.2	3.2	3.2	3.2	3.2	3.2	3.2
$\underline{R}$, Ratio	0.94	0.96	0.98	1.00	1.02	1.03	1.04
ϕ mean	0.8575	0.861	0.865	0.8673	0.8699	0.8733	0.8738
γ_1 mean	1.397	1.343	1.295	1.259	1.234	1.213	1.192
γ_2 mean	1.251	1.29	1.321	1.34	1.354	1.368	1.375
γ_3 mean	1.048	1.054	1.055	1.055	1.054	1.057	1.057

Table 5 Results for Calibration at r = 2.0, $\beta = 3.02$, $p_f = 1.26\text{E-}3$, Ultimate Strength Formulation

α	1.67	2.00	2.33	2.67	3.00	3.33	3.67
$r = M_W / M_{SW}$	1.00	1.20	1.40	1.60	1.80	2.00	2.20
ϕ	1.037	1.041	1.043	1.047	1.049	1.053	1.053
γ_1	0.832	0.800	0.776	0.757	0.741	0.726	0.715
γ_2	1.221	1.258	1.281	1.301	1.314	1.328	1.335
γ_3	1.049	1.047	1.055	1.050	1.049	1.055	1.051
A	4.915	5.520	6137	6.780	7.412	8.047	8.947
β	3.02	3.02	3.02	3.02	3.02	3.02	3.02
$\underline{R}$, Ratio	0.92	0.94	0.96	0.97	0.99	1.00	1.01
ϕ mean	0.864	0.867	0.869	0.873	0.874	0.877	0.878
γ_1 mean	1.387	1.333	1.293	1.262	1.235	1.210	1.191
γ_2 mean	1.221	1.258	1.281	1.301	1.314	1.328	1.335
γ_3 mean	1.049	1.047	1.055	1.050	1.049	1.055	1.051

If $g(\underline{x})$ is less than zero, the member is considered to be in failure state.

ABS Rules give equations for plate thickness based on location of the plate. In general, these guidelines can be described as:

$$t_i = cs\left(\frac{k_i p}{f_i}\right)^{\frac{1}{2}} \tag{41}$$

where

c = constant applied in ABS Rules format
f_i = permissible bending stress
k_i = load factor associated with index i
p = nominal pressure
s = spacing of longitudinals
t_i = net thickness of plating
i = 1, 2, or 3 for longitudinal and transverse stress formulation

The reliability based format for plate thickness based on the given limit state equation is:

$$t_i \geq cs\left(\frac{k_i \gamma p_n}{\phi f_y}\right)^{\frac{1}{2}} \tag{42}$$

with variables as described above including

γ, ϕ = partial safety factors applied to nominal pressure and material yield strength, respectively

The reliability based format, Eq. (42), is equated to ABS Rules, Eq. (41), for determination of the safety index and partial safety factors. This insures that all results given are a direct representation of the ABS Rules.

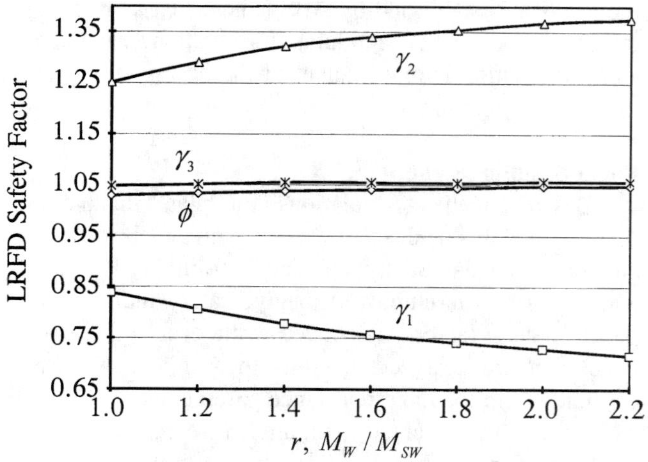

Figure 4 Partial Safety Factors for Calibration at 1.6, Based on Ultimate Strength Formulation

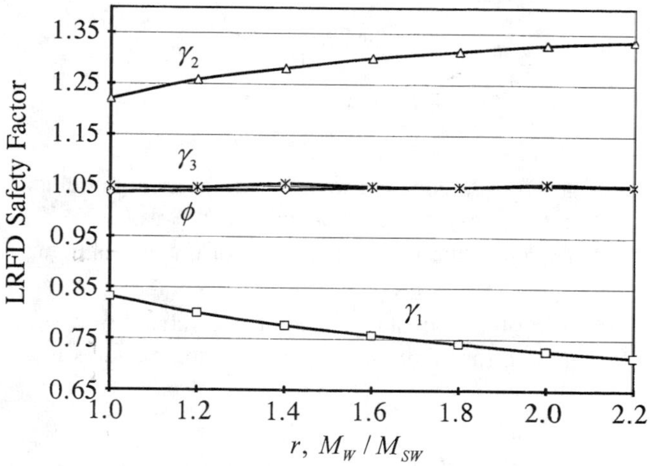

Figure 5 Partial Safety Factors for Calibration at 2.0, Based on Ultimate Strength Formulation

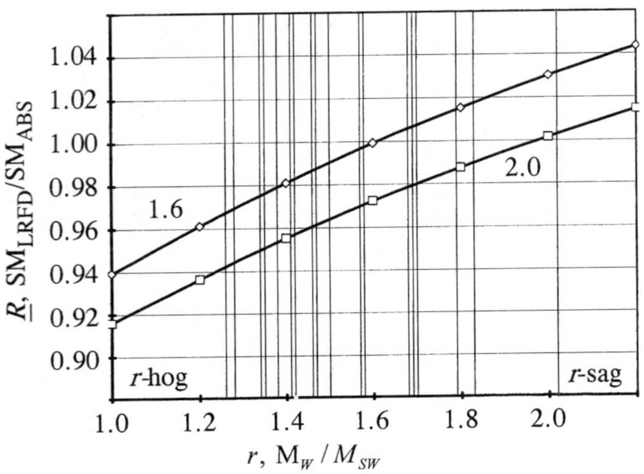

Figure 6 $\underline{R}$ vs. r for Calibrations 1.6 and 2.0, Ultimate Strength Formulation, Showing range of M_W / M_{SW} for Sample Ships

Stiffener Section Modulus Formulation

Section modulus requirements are also given in ABS Rules for initial scantling criteria. The limit state equation associated with these guidelines is similar to that which was given in the Hull Section Modulus for Tankers section of this paper. It is

$$g(\underline{x}) = f_y SM - M$$

where

f_y = nominal yield strength of material

M = bending moment associated with nominal pressure acting on analyzed region (deck, bottom, etc.)

SM = section modulus of analyzed region

ABS Rules give equations for the minimum section modulus required based upon the region being analyzed. In general, this equation is:

$$SM \geq \frac{M}{f_b} \quad (44)$$

with variables as described above and

f_b = permissible bending stress

In the Rule equation for required section modulus, different equations are given to find the values of M and f_b depending upon which area of the hull is being considered.

Using the limit state equation, the reliability based formulation for minimum required section modulus is

$$SM \geq \frac{\gamma M}{\phi f_y} \quad (45)$$

with variables described above and

f_y = yield strength of the material

γ, ϕ = partial safety factors applied to bending moment and yield strength, respectively

Eqs. (44) and (45) are then equated to give safety indices and partial safety factors directly related to the standards established by the ABS Rules.

Sectional Area Formulation

A probability based formulation for minimum sectional area requirements is also achieved using the following limit state equation:

$$g(\underline{x}) = f_y A - F \quad (46)$$

where

A = net sectional area of web portion at specific region

f_y = yield strength of material in shear

F = shear force applied to web portion

Generally, the ABS Rules' guideline for minimum sectional area required is:

$$A = \frac{F}{f_s} \quad (47)$$

with variables as described above including

f_s = permissible shear stress

ABS guidelines give different formulas for the shear force and permissible shear stress variables based on the part of the hull being analyzed.

The probability based design formula for minimum sectional area gives:

$$A \geq \frac{\gamma F}{\phi f_y} \quad (48)$$

with variables described above and

γ, ϕ = partial safety factors applied to shear force and yield strength, respectively

This gives reliability output based on the minimum requirements established by ABS Rules. Just as in plate thickness and section modulus formulation, sectional area formulation is independent of the design stillwater to wave bending moment ratio.

Initial Scantling Criteria

Based on the formulations for plate thickness, section modulus and sectional area given above, the automated Partial Safety Factor program "PSF" (Wu and Du, 1998) is employed to give safety indices and partial safety factors associated with each rule. The Rules that are analyzed are only those given by ABS explicitly with load and resistance parameters.

Some Rules include a parameter representing the distance of the component from the neutral axis. This parameter is included in the ABS formulation of permissible bending stresses. It is treated at regular intervals from the neutral axis (i.e., y/y_n as given in Rule 5-1-4/9.1, Side Shell Plating Thickness requirements). The interval is clearly defined in the results as Case I, Case II, Case III, etc. In addition, for all Rules the permissible bending stress quantities always include a strength reduction factor related to steel grade. This allows the reliability formulation of each Rule to be established based upon steel grade.

The strength reduction factors associated with each steel grade are given in the numeric table on each page. The values of the strength reduction factor (ordinary steel, $S_m = 1.0$) are taken from Rule 2-1-2/Table 2.

Finally, in Rules such as 5-1-4/13.1 Longitudinal Bulkhead Plate thickness requirements, a parameter of normalized distance from the centerline of the ship is used in the permissible bending stress definition. This is given as z/B where z is the transverse distance from the centerline of the section to the region under consideration and B is the vessel's breadth. This is treated in a manner similar to that of the y/y_n ratio, as Case I, Case II, Case III, etc. varied at regular intervals.

The results of the analysis of each ABS Rule requirement are given as partial safety factors and safety indices for each material. For each rule, two plots are created versus material yield strength. One shows the PSF's and the other gives safety indices. An example of bottom shell plating results is given in Figs. 7a and 7b.

The results are consistent with the ABS Rules. The first plot gives partial safety factors as a function of applied to the strength variable, and γ applied to the load variable. The second plot is of the safety index versus steel grade. These partial safety factors are labeled as ϕ versus steel grade. The probabilities of failure corresponding to each safety index are given in Table 6.

It is expected that as the yield strength of the material increases (i.e., the steel grade increases), the partial safety factor applied to the strength will decrease and the partial safety factor applied to the load will increase. This is generally shown in the results. The trends of the results generally also show a linear relationship of PSF's between steel grades. It is also expected that as steel grade increases, the probability of failure will decrease, and the safety index will increase. This is also represented in the results. Rule 5-1-4/11.5.2, Deck Girder Sectional Area, is the one rule for which distinct values could not be established for all steel grades. Both H36 and H40 steel have safety indices greater than 10, which the PSF program (Wu and Du, 1998) could not calculate. However, as given in the data chart, they must be greater than those for ordinary and H32 steels.

New criteria for most other structures have been developed in a similar manner as described above. These include scantlings of bottom, inner bottom, side, deck, and transverse and longitudinal bulkhead plating and stiffeners; bottom, side and deck transverse girders; and corrugated bulkheads.

FATIGUE STRENGTH ASSESSMENT

Fatigue Strength

Fatigue strength is defined herein by the characteristic S-N curve. A fundamental model is

$$NS^m = A \tag{49}$$

where N is cycles to failure, S is stress range (assumed

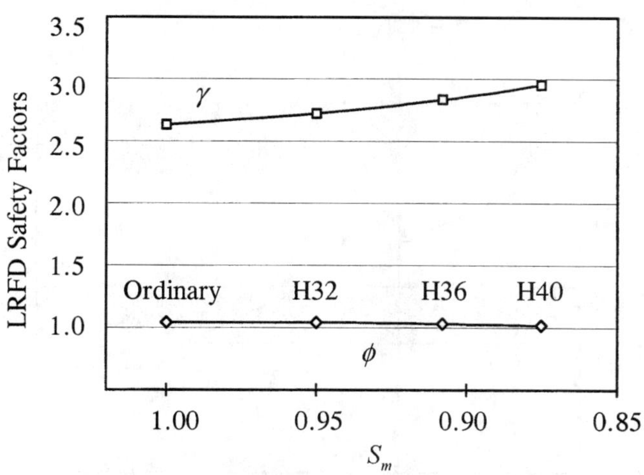

Figure 7a LRFD Safety Factors for Bottom Shell Plating Thickness, t_1, vs. Steel Grade

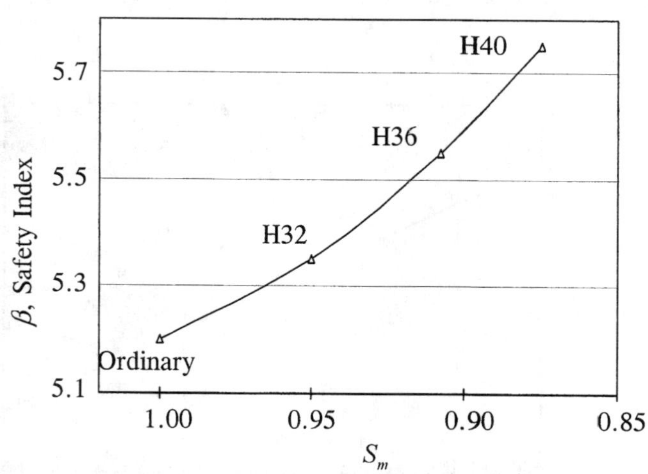

Figure 7b Safety Index for Bottom Shell Plating Thickness, t_1, vs. Steel Grade

Table 6 Failure Probabilities for Steel Grades

Steel Grade	Ordinary	H32	H36	H40
S_m	1.0	0.95	0.908	0.875
ϕ	1.04	1.04	1.03	1.02
γ_p	2.63	2.72	2.83	2.95
$\beta_{\text{calibrated}}$	5.2	5.35	5.55	5.75
p_f $(\times 10^{-8})$	9.98	4.41	1.43	0.448

to be constant amplitude), m is the fatigue strength exponent, and A is the fatigue strength coefficient. This S-N curve is illustrated in Fig. 8.

A more refined model is shown in Fig. 9. This model is a two segment curve that reflects data in the high cycle range. r and C are the fatigue strength exponent and coefficients, respectively, for the high cycle component. S_Q is the stress at the knee. The basic design S-N curves used in ABS design criteria are the two segment type and are based on the UK Department of Energy curves shown in Appendix D.

Fatigue Stresses

Stress at fatigue sensitive points is a random process, a sample function of which is shown in Fig. 10.

Over the long term, it is assumed that stress ranges, defined by S, have a Weibull distribution. The distribution function is,

$$F_S(s) = 1 - \exp\left[-\left(\frac{s}{\delta}\right)^{\gamma}\right] \qquad (50)$$

where γ and δ are the Weibull exponent and scale parameters, respectively.

For safety check purposes, it is useful to define a design stress range S_0 that characterizes the largest stress anticipated during the design service life. This design stress is defined by the probability statement,

$$P(S > S_O) = \frac{1}{N_S} \qquad (51)$$

where N_S is the number of cycles in the service life. From the definition of the distribution function, it follows that

$$S_o = \left[(\ell n\, N_S)^{1/\gamma}\right] \delta \qquad (52)$$

Fatigue Damage For Single Segment Model

Consider a variable amplitude stress process. Assume a linear damage accumulation rule (Miner's Rule). Referring to the one segment characteristic S-N curve (Fig. 8) and the strength model Eq. (49), damage at n stress cycles is

$$D_0 = \frac{n\,\mathrm{E}(S^m)}{A} \qquad (53)$$

where $\mathrm{E}(S^m)$ is the expected value of S^m, defined as

$$\mathrm{E}(S^m) = \int_O^{\infty} s^m\, f_S(s)\, ds \qquad (54)$$

and f_S is the probability density function of S.

If the stress process were constant amplitude of magnitude S, then $\mathrm{E}(S^m) = S$, and

$$D_0 = \frac{n\,S^m}{A} \qquad (55)$$

Equating Eqs. (53) and (55), define an equivalent constant amplitude stress which produces the same damage as the variable amplitude stress,

$$S_e = \left[\mathrm{E}(S^m)\right]^{1/m} \qquad (56)$$

S_e is called Miner's stress.

For the special case where S has a Weibull distribution,

$$S_e^m = S_o^m (\ell n\, N_S)^{-m/\gamma}\, \Gamma\!\left(\frac{m}{\gamma} + 1\right) \qquad (57)$$

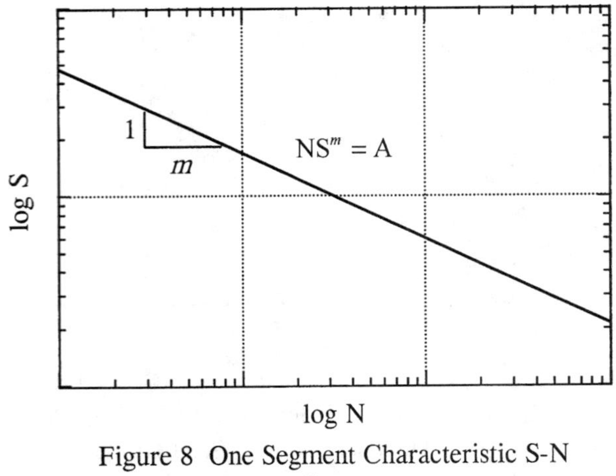

Figure 8 One Segment Characteristic S-N Curve

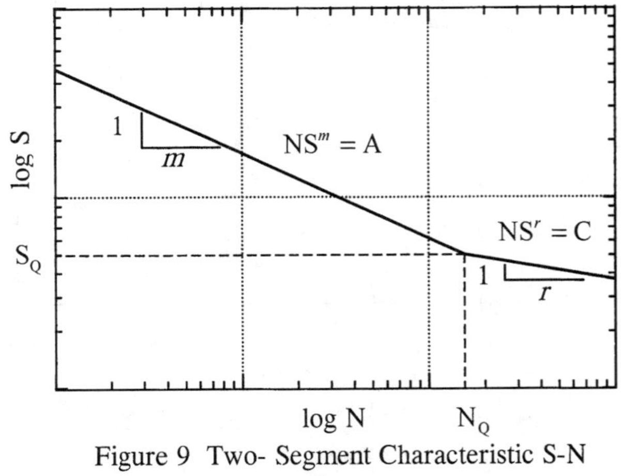

Figure 9 Two- Segment Characteristic S-N Curve

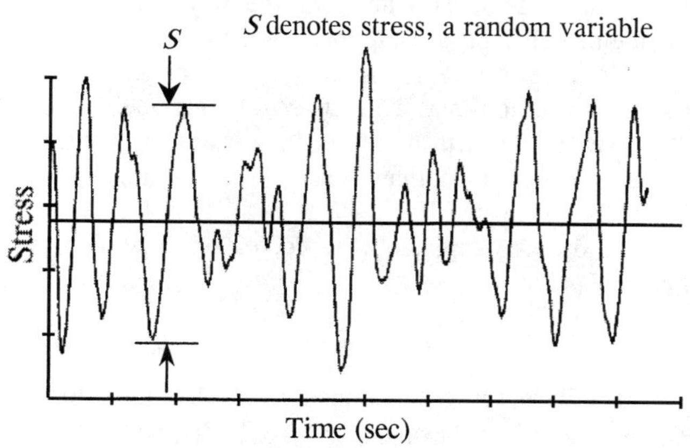

Figure 10 Realiziation of Random Process and Definition of Stress Range

where $\Gamma(x)$ is the gamma function, defined as,

$$\Gamma(x) = \int_O^\infty t^{x-1} e^{-t}\, dt \tag{58}$$

Fatigue Damage For Two Segment Model

Fatigue damage for the two-segment case of Fig. 9 is,

$$D = \Lambda \cdot D_0 \tag{59}$$

Where the bias factor, Λ is (Wirsching and Chen, 1988),

$$\Lambda = \frac{A}{C}\,\delta^{r-m}\,\frac{\Gamma_o(z,b)}{\Gamma(a)} + \frac{\Gamma(z,a)}{\Gamma(a)} \tag{60}$$

and where $\Gamma(z,a)$ and $\Gamma_o(z,a)$ are incomplete gamma functions (integrals z to ∞ and 0 to z, respectively)

$$\Gamma(z,a) = \int_z^\infty t^{a-1}\, e^{-t}\, dt \tag{61}$$

$$\Gamma_o(z,a) = \int_O^z t^{a-1}\, e^{-t}\, dt \tag{62}$$

$$a = \frac{m}{\gamma} + 1 \tag{63}$$

$$b = \frac{r}{\gamma} + 1 \tag{64}$$

$$z = \left(\frac{S_Q}{\delta}\right)^\gamma \tag{65}$$

Note that $\Lambda \le 1.0$ in general, and $\Lambda = 1.0$ when $S_Q = 0$.

The general expression for damage (two-segment case) can be written as

$$D = \frac{\Lambda\, n\, S_e^m}{A} \tag{66}$$

At failure, $D = 1.0$ and $n = N$, where N denotes cycles to failure. Thus,

$$S_e^m = \frac{A}{\Lambda N} \tag{67}$$

Note that $S_e^m = E(S^m)$ represents strength. S_e is the largest value of equivalent stress that can be tolerated without failure for a life of N stress cycles. The design stress range S_0, a more convenient parameter, can be computed from S_e^m using Eq. (57).

Definition of Fatigue Strength in Current ABS Rules

This example illustrates how the maximum allowable stress range, S_0, is computed. It is assumed that there is no factor of safety applied. Parameter values in this example are taken from the ABS Rules, Part 5 (1998-1999), pp. 96-97.

Consider the "D" S-N curve in Appendix D. Also, consider the case where the Weibull exponent $\gamma = 0.70$. Assume that the service life is $N_S = 5\times10^7$ cycles in 20 years.

In the following, all of the parameters required for calculating Λ are developed (also see Table D-1 in Appendix D). Parameters for the first segment of the design "D" S-N curves are,

$$m = 3.0 \qquad A = 1.52\text{E}12 \text{ (MPa units)}$$

Compute S_Q. Note that the knee is at $N = 10^7$ cycles.

$$NS^m = A \qquad \rightarrow (10^7)\, S_Q^3 = 1.52\text{E}12$$

$$S_Q = 53.4 \text{ MPa}$$

Obtain the parameters of the high cycle segment of the S-N curve. The equation is

$$NS^r = C \tag{68}$$

The fatigue model of ABS rules defines r,

$$r = m + 2 = 5 \tag{69}$$

Thus at the knee ($N = 10^7$ cycles),

$$C = 10(53.4)^5 = 4.33\text{E}15$$

Eqs. (63) and (64) are used to calculate other parameters.

$$a = \frac{m}{\gamma} + 1 = \frac{3}{0.7} + 1 = 5.29$$

$$b = \frac{r}{\gamma} + 1 = \frac{5}{0.7} + 1 = 8.14$$

Also, from Eq. (52),

$$\delta = S_O(\ell n\, N_S)^{-1/\gamma} \tag{70}$$

To compute the value of δ, it is necessary to know S_0. But this is the value that we seek in this analysis. From the ABS Rules, use

$$S_0 = 59.9 \text{ kg/mm}^2 = 58.7 \text{ N/mm}^2$$

Using this value enables the calculation of Λ and avoids the necessity, at least in this example, of solving a transcendental equation. This can be done because Λ is not highly sensitive to S_0. Thus,

$$\delta = 587(\ell n\, 5E7)^{-1/0.7} = 9.66 \text{ N/mm}^2$$

And from Eq. (65),

$$z = \left(\frac{S_Q}{\delta}\right)^{\gamma} = \left(\frac{53.4}{9.66}\right)^{0.7} = 3.31$$

The computation of Λ is done very effectively using a MatLab program having the gamma function and incomplete gamma functions in the system. (The incomplete gamma function *gammainc* is in the Stat toolbox.)

$$\Lambda = 0.905$$

Determine the design stress range, S_0. Substitute Eq. (57) into Eq. (67) and solve for S_0 setting the cycles to failure N equal to the service life, N_S. Note also that there is no factor of safety applied.

$$S_o = \left[\frac{A(\ell n\, N_S)^{m/\gamma}}{\Lambda\, N_S\, \Gamma\left(\dfrac{m}{\xi}+1\right)}\right]^{1/m} \tag{71}$$

The value of S_0 for this example is

$$S_0 = 59.9 \text{ kg/mm}^2 \quad (587 \text{ N/mm}^2)$$

This results in an exact solution to the transcendental equation and it verifies the value listed in ABS Rules, Part 5 (1998-1999), p. 97.

Strength values computed by the present analysis are listed in Table 8. These are identical to the ABS Rules values. The bias factor related to the knee of the S-N curve is also listed.

Assessment Of Reliability Implied By Current ABS Rules

The following reliability analyses are described in detail in the fatigue references, e.g., Mansour, Wirsching, et al (1997), The lognormal format is employed.

Uncertainty in the fatigue process is assumed to come from three sources:

- variability in fatigue strength as evidenced by scatter in S-N data;
- unpredictability of Miner's Rule; and
- uncertainties in the process of computing the fatigue stress at fatigue sensitive points.

First consider stress. The uncertainty in Miner's stress can be quantified by the random variable, B. Thus, in a limit state formulation, Miner's stress is written as, $(BS_e)^m$. The random variable B accounts for uncertainty associated with both the physical process and the model, e.g., assumptions made in the analysis.

With regard to fatigue strength, the exponent, m, is assumed to be constant. The fatigue strength coefficient, A, is now assumed to be a random variable accounting for the scatter in fatigue data. (See parameters in Appendix D.)

The damage expression of Eq. (66) is modified to include B.

$$D = \frac{n\Lambda\, B^m\, S_e^m}{A} \tag{72}$$

At failure, $n = N$, defined as cycles to failure. $D = \Delta$, defined as damage at failure, is assumed to be a random variable quantifying uncertainty in Miner's Rule.

Solving the damage expression for N,

$$N = \frac{\Delta A}{\Lambda\, B^m\, S_e^m} \tag{73}$$

Note that N is a random variable because Δ, A, and B are random variables.

The event of failure is the event that the cycles to failure (strength) is less than the service life (stress).

$$N < N_S \tag{74}$$

The probability of failure is

$$P_f = P(N < N_S) \tag{75}$$

Assume that the random variables, A, B, and Δ have lognormal (LN) distributions (median and coefficient of variation in parentheses).

$$A \,\square\, \text{LN}(\tilde{A}, C_A)$$

$$B \,\square\, \text{LN}(\tilde{B}, C_B)$$

$$C \,\square\, \text{LN}(\tilde{\Delta}, C_\Delta)$$

The safety index and probability of failure are respectively,

$$\beta = \frac{\ell n\,(\tilde{N}/N_S)}{\sigma_{\ell n\, N}} \tag{76}$$

$$p_f = \Phi(-\beta) \tag{77}$$

where Φ is the standard normal distribution function. The median of N is,

$$\tilde{N} = \frac{\tilde{\Delta}\,\tilde{A}}{\Lambda\,\tilde{B}^m\, S_e^m} \tag{78}$$

$$\sigma_{\ell n N} = \sqrt{\ell n\left[(1+C_\Delta^2)(1+C_A^2)(1+C_B^2)^{m^2}\right]} \tag{79}$$

As an example, consider the requirement for a component governed by the "D" S-N curve and for which the stress distribution is defined by $\gamma = 0.7$. ABS Rules specify that the permissible stress range is $S_0 = 59.9 \text{ kg/mm}^2$. All of the parameters for this example are summarized in Table 7.

The results using Eqs. (76) through (79) are

$$\beta = 1.34$$

$$p_f = 0.089$$

The estimated β and failure probability for each case of the ABS Rules are given in Table 8. Failure probabilities seem to be inappropriately large. This is due to the fact that no factor of safety is applied. Failure is defined as damage, $D = 1.0$. The sole implicit safety margin is the use of the lower bound S-N curve in computing *D*.

Table 7 Parameters Used in Example (N-mm units)

	Note	
$\tilde{A}$	(a)	3.988E12
C_A	(a)	0.51
m	(a)	3.0
N_S	(b)	5E7
$\tilde{\Delta}$	(c)	1.0
C_C	(c)	0.30
$\tilde{B}$	(c)	1.0
C_B	(c)	0.15
γ	(d)	0.70
S_0	(e)	59.9

Notes:

(a) See Table D-1.

(b) Defined in ABS Rules; corresponds to 20 years.

(c) See Mansour, Wirsching, et al. (1997) for discussion of choice of parameter values.

(d) Chosen for this example. ABS Rules correspond to the "D" S-N curve and $\gamma = 0.70$. Units are kg/mm^2, to covert to N/mm^2, multiply by 9.807.

INTERPRETATION OF THE RESULTS OF THE ASSESSMENT OF RELIABILITY IMPLIED BY CURRENT ABS RULES

Goals

The goals of the exercises leading to the results presented in Table 9 are to

- provide an estimate, in an absolute sense, of the level of reliability implied by the ABS Rules, and
- provide a basis for calibrating the code, i.e., adjusting the design requirements so that a uniform level of safety is applied.

Reliability in an Absolute Sense

A target safety index of $\beta_0 = 3$ is generally considered to be reasonable for general redundant structures for which the consequences of failure are not serious. Yet the permissible stress approach of the current ABS Rules implies a relatively low safety index (1.07 to 1.46) and relatively high notional failure probability (0.07 to 0.14). From a historical perspective, fatigue has not been a principal concern.

While fatigue is a major consideration during the design phase, for many parts of the structure, it is not the governing criteria. Where it may be a likely failure mode, such as in longitudinals near the water line, hard spots such as intersections with bulkheads, and some brackets, it drives the required scantlings.

Corrosion margins improve the fatigue strength. Yet the aging process (both corrosion and fatigue can be accelerated significantly by a breakdown of corrosion protection.

Cracks form at welded details and are usually localized. Crack growth is often arrested as the crack grows out of a region of high stress concentration. Also because of the high redundancy of ship structures, fatigue cracks do not grow to a critical size that leads to fracture before they can be repaired. Also because some cracks are detected because of leakage (water or liquid cargo) and ships undergo extensive inspection and gauging at 5-year intervals, cracks are usually detected and repaired before they become a serious safety problem.

There are regions of the ship that seem to be particularly fatigue sensitive. It may be good engineering practice to allow higher failure probabilities when there are a limited number of components involved. In general offshore practice most of the structure is designed to the ultimate load and as a result only a few joints are considered to be fatigue critical. Criteria having a relative high fatigue failure probability will result in only a few joint failures. The tendency with ships is to design the highest stressed component and then make the rest of the components identical to it, even though they will be over designed. The extra material cost more than offsets the additional cost of custom designing and building each component individually.

Calibration

The ultimate goal of design is to provide rational and consistent reliability throughout, e.g., if the consequences of failure of all components are the same, then the reliabilities should be identical. Results of the reliability analysis of Table 8 are used as a basis for selecting a target safety index to be applied to all

Table 8 Design Stress Range and Safety Index Implied by ABS Rules

S-N Curve	Distribution Parameter γ	ABS Permissible Stress Range (kg/mm^2)	Bias Factor Λ	Safety Index β	Probability of Failure
B	0.7	92.2	0.953	1.07	0.142
	0.8	75.9	0.937	1.07	0.142
	0.9	64.2	0.921	1.07	0.142
	1.0	55.6	0.906	1.07	0.142
C	0.7	79.2	0.933	1.23	0.109
	0.8	63.9	0.914	1.23	0.109
	0.9	53.3	0.896	1.23	0.109
	1.0	45.7	0.878	1.23	0.109
D	0.7	59.9	0.905	1.34	0.089
	0.8	47.3	0.882	1.34	0.089
	0.9	38.9	0.861	1.34	0.089
	1.0	32.9	0.842	1.34	0.089
E	0.7	52.8	0.905	1.46	0.072
	0.8	42.7	0.882	1.46	0.072
	0.9	34.2	0.861	1.46	0.072
	1.0	29.0	0.842	1.46	0.072
F	0.7	44.7	0.905	1.37	0.086
	0.8	35.3	0.882	1.37	0.086
	0.9	29.0	0.861	1.37	0.086
	1.0	24.5	0.842	1.37	0.086
F2	0.7	39.3	0.905	1.41	0.080
	0.8	31.1	0.882	1.41	0.080
	0.9	25.5	0.861	1.41	0.080
	1.0	21.6	0.842	1.41	0.080
G	0.7	32.8	0.905	1.21	0.113
	0.8	25.9	0.882	1.21	0.113
	0.9	21.3	0.861	1.21	0.113
	1.0	18.0	0.842	1.21	0.113
W	0.7	28.3	0.889	1.22	0.110
	0.8	22.3	0.862	1.22	0.110
	0.9	18.4	0.839	1.22	0.110
	1.0	15.5	0.813	1.22	0.110

Table 9 Probability Based Design Stress Ranges for S-N Curve "D"

S-N Curve	Distribution Parameter γ	Bias Factor Λ	ABS Permissible Stress Range (kg/mm^2)	$\beta_0 = 2.0$ Permissible Stress Range (kg/mm^2)	$\beta_0 = 3.0$ Permissible Stress Range (kg/mm^2)
D	0.7	0.905	59.9	51.2	40.3
	0.8	0.882	47.3	40.1	31.6
	0.9	0.861	38.9	32.7	25.7
	1.0	0.842	32.9	27.5	21.6

requirements. While these β values seem low relative to other applications, the present requirements do have a history of successful application.

For the ABS Rules demonstration project, a target safety index of $\beta_0 = 1.4$ was selected. This value was arbitrarily chosen as a value that is slightly on the conservative side of the list of Table 8.

PROBABILITY BASED FATIGUE DESIGN CRITERIA

The maximum allowable probability of failure, P_0, or the minimum allowable safety index, β_0, is specified as the basic design requirement. Using probabilistic analyses, it is customary to translate this requirement into one involving physical parameters. Any of the following options is possible:

Four formats for probability based fatigue design criteria are presented:

(1) The design stress approach. The maximum stress range during the service life is to be less than a specified maximum permissible stress, S_0. This is the approach of the ABS Rules.

(2) Target damage level, Δ_0. The estimated damage D over the service life is to be less than Δ_0.

(3) Factor of safety on life, FS_{life}. The design life, expressed as time to failure of the component, is defined as FS_{life} times the intended service life.

(4) A partial safety factor is applied to each of the random design factors.

The Limit State

As shown in Eq. (74) the cycles to failure (strength) is less than the service life (stress). The fatigue strength is given by Eq. (73). From those equations, the limit state function can be written as

$$g = \frac{\Delta A}{\Lambda B^m S_e^m} - N_S \tag{80}$$

The limit state function is defined so that $g < 0$ is the failure event. The random variables are Δ, A, and B. Using the procedures described in Mansour, Wirsching, et al (2000) partial safety factors can be derived for each random variable.

Design Stress, S_0 (Permissible Stress Range)

The design stress approach to fatigue requirements is used in ABS Rules. The lognormal format will be used to derive S_0. From Eq. (76) the safety index is

$$\beta = \frac{\ell n(\tilde{N}/N_S)}{\sigma_{\ell nN}} \tag{81}$$

where $\tilde{N}$ is the median cycles to failure.

$$\tilde{N} = \frac{\tilde{\Delta}\tilde{A}}{\Lambda \tilde{B}^m S_e^m} \tag{82}$$

Define the target safety index, β_0. The safety check requires that $\beta > \beta_0$, therefore

$$\beta_o < \frac{\ell n\left\{\dfrac{\tilde{\Delta}\tilde{A}}{\Lambda \tilde{B}^m S_e^m N_S}\right\}}{\sigma_{\ell nN}} \tag{83}$$

From Eq. (56)

$$S_e^m = S_0^m (\ell n\, N_S)^{-m/\gamma}\, \Gamma\left(\frac{m}{\gamma}+1\right) \tag{84}$$

Substitute Eq. (84) into Eq. (83) and solve for S_0.

$$S_0 < \left[\frac{\tilde{A}(\ell n\, N_S)^{m/\gamma}}{\Lambda N_S\, \Gamma\left(\dfrac{m}{\gamma}+1\right)\exp(\beta_o \sigma_{\ell nN})}\right]^{1/m} \tag{85}$$

S_0 defines the strength, i.e., the largest stress range that the joint can experience so that the safety index equals or exceeds β_0. In the ABS Rules, this term is called permissible stress range.

Example Of The Design Stress Approach

Consider the example defined in Table 7. Let the target safety index be $\beta_0 = 2.0$. The bias factor, Λ, is actually a function of S_0, which makes Eq. (85) transcendental. However, Λ is not a strong function of S_0. For all of the cases considered here, Λ is chosen as the computed value used to calibrate to present ABS Rules. This is not exact, but errors should be small. For this example $\Lambda = 0.905$ (see Table 8).

From Eq. (79), the standard deviation of the log of N is, for this example

$$\sigma_{\ell nN} = 0.7195$$

Then the permissible stress range (strength) from Eq. (85) becomes

$$S_0 = \left[\frac{3.988E12(\ell n 5E7)^{3/0.7}}{0.905(5E7)\,\Gamma\left(\dfrac{3}{0.7}+1\right)\exp(2\times 0.7195)}\right]^{1/3}$$

$$S_0 = 502 \text{ N/mm}^2 = 51.2 \text{ kg/mm}^2$$

Table 10 Probability Based Design Stress Ranges

S-N Curve	Distribution Parameter γ	Bias Factor Λ	ABS Permissible Stress Range (kg/mm^2)	$\beta_0 = 1.4$ Permissible Stress Range (kg/mm^2)
B	0.7	0.953	92.2	88.1
	0.8	0.937	75.9	72.3
	0.9	0.921	64.2	61.0
	1.0	0.906	55.6	52.7
C	0.7	0.933	79.2	77.4
	0.8	0.914	63.9	62.2
	0.9	0.896	53.3	51.6
	1.0	0.878	45.7	44.0
D	0.7	0.905	59.9	59.2
	0.8	0.882	47.3	46.4
	0.9	0.861	38.9	37.8
	1.0	0.842	32.9	31.7
E	0.7	0.905	52.8	53.7
	0.8	0.882	41.7	42.1
	0.9	0.861	34.2	34.3
	1.0	0.842	29.0	28.8
F	0.7	0.905	44.7	44.4
	0.8	0.882	35.3	34.8
	0.9	0.861	29.0	28.3
	1.0	0.842	24.5	23.8
F2	0.7	0.905	39.3	39.5
	0.8	0.882	31.1	30.9
	0.9	0.861	25.5	25.2
	1.0	0.842	21.6	21.1
G	0.7	0.905	32.8	31.5
	0.8	0.882	25.9	24.7
	0.9	0.861	21.3	20.1
	1.0	0.840	18.0	16.9
W	0.7	0.889	28.3	27.2
	0.8	0.862	22.3	21.3
	0.9	0.839	18.4	17.4
	1.0	0.813	15.5	14.6

Values of permissible stress ranges are presented in Table 9 for $\beta_0 = 2$ and 3 for the "D" S-N curve. These numbers are presented only to illustrate the penalty for using what might be considered more conventional reliability requirements.

The requirements are calibrated to $\beta_0 = 1.4$ (see calibration section above), and the results are shown in Table 10.

Target Damage Level, Δ_0

Using the design S-N curve defined by the fatigue strength coefficient, A, fatigue damage at the end of the service life is

$$D_0 = \frac{N_S \Lambda S_e^m}{A} \tag{86}$$

The design is considered safe if

$$D_0 \le \Delta_0 \tag{87}$$

The value of Δ_0 corresponding to the target safety

Table 11 Target Damage Levels and Factors of Safety on Life for Each of the S-N Curves

S-N Curve	$\beta_0 = 1.4$		$\beta_0 = 2.0$		$\beta_0 = 3.0$	
	Δ_0	FS_{life}	Δ_0	FS_{life}	Δ_0	FS_{life}
B	0.771	1.30	0.481	2.08	0.219	4.56
C	0.879	1.14	0.556	1.80	0.259	3.86
D	0.958	1.03	0.622	1.61	0.302	3.30
E	1.050	0.95	0.654	1.53	0.298	3.36
F	0.977	1.02	0.628	1.59	0.301	3.32
F2	1.005	1.00	0.642	1.56	0.304	3.29
G	0.88	1.14	0.587	1.71	0.299	3.35
W	0.887	1.13	0.589	1.70	0.299	3.35

index, β_0, is derived by solving Eq. (86) for S_e^m, substituting into Eq. (83), and solving the inequality for D_0. Comparing with Eq. (87), it follows that

$$\Delta_o = \frac{\lambda \tilde{\Delta}}{\tilde{B}^m \exp(\beta_o \sigma_{\ell n N})} \tag{88}$$

where

$$\lambda = \frac{\tilde{A}}{A} \tag{89}$$

Example Of The Target Damage Level Approach

Using the parameters of Table 7 and a target safety index of $\beta_0 = 2.0$, it follows from Eqs. (88) and (89) that

$$\lambda = \frac{\tilde{A}}{A} = \frac{3.988E12}{1.52E12} = 2.62$$

$$\Delta_0 = \frac{(2.62)(1)}{(1)^3 \exp(2 \times 0.7195)} = 0.622$$

Values of Δ_0 for each of the S-N curves are given in for the calibration level of $\beta_0 = 1.4$, and, for reference, $\beta_0 = 2.0$ and 3.0.

Factor Of Safety On Life

It is seen from Eq. (86) that a factor of Δ_0 on damage corresponds to a factor of FS = $1/\Delta_o$ on life. In the above example, $FS_{life} = 1.61$. Values of FS_{life} for each of the S-N curves are given in Table 11 for the calibration level of $\beta_0 = 1.4$, and, for reference, $\beta_0 = 2.0$ and 3.0.

Partial Safety Factor Format

The fatigue limit state is defined in Eq. (80). The three random design factors are, Δ, *A*, and *B*. A deterministic analog of the limit state that expresses a safe condition is

$$\frac{(\gamma_\Delta \Delta_n)(\gamma_A A_n)}{\Lambda(\gamma_B B_n)^m S_e^m} \geq N_S \tag{90}$$

where γ_Δ, γ_A, and γ_B are partial safety factors and Δ_n, A_n, and B_n are nominal (characteristic) values for Δ, *A*, and *B*, respectively. The goal is to derive γ_Δ, γ_A, and γ_B given the target safety index, β_0. Thus, when the safety check expression of Eq. (90) is satisfied, a reliability exceeding β_0 is ensured.

The partial safety factor concept is described in Mansour, Wirsching, et al. (1996). Application to the fatigue limit state is described in the following.

The limit state can be written as,

$$g = F_0 \frac{\Delta A}{B^m} - 1 \tag{91}$$

where

$$F_0 = \frac{1}{\Lambda S_e^m N_S} \tag{92}$$

The limit state is now rewritten in a form for automated analysis by the program "PSF" to derive the partial safety factors. Let X_1, X_2, and X_3 be random variables having mean values of 1.0 and standard deviations (and coefficient of variations since the mean is 1.0) equal to those of Δ, *A*, and *B*, respectively. Also, the *X*'s will have the same distribution families as Δ, *A*, and *B*.

$$X_1 \sim LN(1, C_\Delta)$$
$$X_2 \sim LN(1, C_A) \tag{93}$$
$$X_3 \sim LN(1, C_B)$$

Thus it follows that

$$\Delta = \mu_\Delta X_1$$
$$A = \mu_A X_2 \tag{94}$$
$$B = \mu_B X_3$$

where μ_Δ, μ_A, and μ_B are the mean values of Δ, *A*, and *B*. The limit state can now be written as

$$g = \frac{F X_1 X_2}{X_3^m} - 1 \quad (95)$$

where

$$F = \frac{\mu_\Delta \mu_A}{\mu_B^m} \cdot \frac{1}{\Lambda S_e^m N_S} \quad (96)$$

The limit state is now in a form ready for analysis for program "PSF", which derives the partial safety factors for the X's for a given β_0. The safety check analog of the limit state would be

$$\frac{F(\gamma_1 X_{1n})(\gamma_2 X_{2n})}{(\gamma_3 X_{3n})} > 1.0 \quad (97)$$

Program "PSF" computes partial safety factors for nominal values defined by the mean values of the random variables. Because those means are unity, it follows that the safety check expression is

$$\frac{F \gamma_1 \gamma_2}{\gamma_3^m} > 1.0 \quad (98)$$

This is translated into the appropriate form of Eq. (90) involving the basic design factors as illustrated in the following example.

Example of the Partial Safety Factor Format

In the example given previously (Table 7), it is assumed that the "D" S-N curve applies. The target safety index is $\beta_0 = 2.0$. It is assumed that Δ, A, and B have lognormal (LN) distributions (mean and standard deviation in parentheses). Thus,

$$\begin{aligned} X_1 &\sim LN(1, C_\Delta) \quad C_\Delta = 0.30 \\ X_2 &\sim LN(1, C_A) \quad C_A = 0.51 \\ X_3 &\sim LN(1, C_B) \quad C_B = 0.15 \end{aligned} \quad (99)$$

and,

$$\begin{aligned} \gamma_1 &= 0.756 \\ \gamma_2 &= 0.468 \\ \gamma_3 &= 1.19 \end{aligned} \quad (100)$$

Substituting the results into Eq. (98), and assuming $m = 3$, the resulting safety check expression would be

$$(\gamma_1 \mu_\Delta)(\gamma_2 \mu_A) > \Lambda(\gamma_B \mu_B)^3 S_e^3 N_S \quad (101)$$

All three variables have lognormal distributions. For analytical reasons it is more appropriate to use the median as a measure of central tendency and the coefficient of variation as a measure of uncertainty. For the lognormal, the mean values in terms of the medians (denoted with the tilde) are,

$$\begin{aligned} \mu_\Delta &= \tilde{\Delta}\sqrt{1 + C_\Delta^2} = 1.044\tilde{\Delta} \\ \mu_A &= \tilde{A}\sqrt{1 + C_A^2} = 1.12\tilde{A} \\ \mu_B &= \tilde{B}\sqrt{1 + C_B^2} = 1.01\tilde{B} \end{aligned} \quad (102)$$

The safety check expression should contain the design value of A. For the "D" S-N curve, the median in terms of the design value is,

$$\frac{\tilde{A}}{A} = \frac{3.988E12}{1.52E12} = 2.62$$

so that

$$\mu_A = 1.12(2.62A) = 2.94A$$

The safety check expression becomes

$$\left[(0.756)(1.044\tilde{\Delta})\right] \cdot \left[(0.468)(2.94A)\right]$$
$$= \Lambda(1.19 \times 1.01\tilde{B})^3 S_e^3 N_S$$

or

$$(0.789\tilde{\Delta})(1.376A) > \Lambda(1.202\tilde{B})^3 S_e^3 N_S \quad (103)$$

Thus the three partial safety factors corresponding to $\beta_0 = 2.0$ are defined.

$$\gamma_\Delta = 0.789$$
$$\mu_A = 1.376$$
$$\mu_B = 1.202$$

Using the ABS calibrated target safety index of $\beta_0 = 1.4$, partial safety factors (PSF) are computed. Just for reference, the PSF's for X_1, X_2, and X_3 are provided in Table 12.

It has generally been assumed that,

$$\tilde{\Delta} = \tilde{B} = 1.0$$

Thus the safety check expression becomes

$$1.086\,A > \Lambda\,(1.202\,S_e)^3\,N_S \quad (104)$$

It is seen that the strength term is on the left side of the inequality; and stress is on the right side.

A simple safety check expression involving the basic variables of fatigue stress (S) and fatigue strength (A) can be presented as the more general form of Eq. (104).

$$\gamma_A\,A > \Lambda\,(\gamma_S\,S_e)^3\,N_S \quad (105)$$

Values of γ_A and γ_S are provided in Table 12.

Taking this one step further, the partial safety factors can be combined into one factor.

Table 12 PSF's for X_1, X_2, and X_3 ($\beta_0 = 1.4$)

S-N Curve	γ_1 (Δ)	γ_2 (A)	γ_3 (B)	γ_A	γ_S
B	0.822	0.668	1.160	1.45	1.17
C	0.818	0.594	1.141	1.45	1.15
D	0.809	0.568	1.126	1.41	1.14
E	0.822	0.468	1.114	1.50	1.13
F	0.812	0.541	1.122	1.43	1.13
F2	0.813	0.523	1.120	1.46	1.13
G	0.801	0.645	1.135	1.33	1.15
W	0.802	0.635	1.134	1.34	1.15

$$A > \Lambda S_e^3 [1.60 N_S] \quad (106)$$

The requirement reduces to a factor of safety on life of 1.60. The results agree with the earlier analysis in which lognormal mathematics was employed to derive the factor of safety with respect to life.

SUMMARY AND CONCLUSIONS

The technical approach used for developing reliability based Rules is based on defining, for each rule requirement, a failure surface using a limit state equation. Target reliabilities implied in the current ABS Rules and a corresponding set of partial safety factors were then determined. The initial calibration process was intended to give scantlings comparable to those obtained from ABS Rules. However, in order to ensure uniform reliability and consistent designs, the LRFD formulation deviated in some cases from the current ABS Rules, such as for hull girder section modulus. This deviation was assessed for each provision and provided an opportunity for scantling reduction over a certain range of the design parameters that were found to be relatively over-designed. An additional opportunity for weight savings is provided by comparing the magnitudes of the safety indices calculated for similar structures located in the same loading environment, but which have different magnitudes.

Each requirement in the ABS Rules for tankers has been converted to LRFD format. These included LRFD rules for hull girder section modulus, shell bottom plating, shell and inner bottom longitudinals, side and deck plating and longitudinals, bulkheads, etc.

In the case of the hull girder section modulus, two reliability based formulations were developed. The first is an initial yield criterion similar to the current ABS Rules. The second is an ultimate strength criterion that explicitly includes a dynamic moment term associated with slamming. In both formulations, the partial safety factors that produce a uniform reliability level were found to be dependent on the ratio of the design wave bending moment to the stillwater bending moment. Such dependence is not accommodated in the current ABS Rules and leads to inconsistency in the safety levels associated with the section modulus of ships that have different design wave to stillwater bending moment ratios. When this inconsistency is removed, the new reliability based formulations offer a reduction in the section modulus of up to 7% for ships with low design wave to stillwater bending moment ratio (about 1.2).

Other requirements have also been converted to LRFD format. For example, in the reliability based standard for bottom shell plating, two formulations were developed similar to current ABS Rules (t_1 and t_2). In the first, the plate thickness is calculated from considerations of bending stresses in the longitudinal direction of the ship (t_1), while the second is based on transverse bending stresses (t_2). In both cases, the safety indices were calibrated to give the same plate thickness values as obtained from the current ABS standard. The corresponding values for the safety indices were found to be $\beta_1 = 5.2$ in the longitudinal direction and $\beta_2 = 2.2$ in the transverse direction for ordinary steel. The large difference in the safety indices is due to the presence of large longitudinal in plane stresses that are not explicitly accounted for in the bending formulation for t_1. In addition, the safety indices inherent in the Rules for higher strength steels (H32, H36 and H40) were found to be higher than those for ordinary steel, which seems reasonable to allow for susceptibility to buckling.

Since neither of the plate bending formulations explicitly takes into consideration in plane loads, a new reliability based formulation is proposed to account for the ultimate buckling strength of plates under in plane loads. This formulation has been also calibrated with current ABS Rules. However, for uniform reliability the resulting plate thickness depends on the ratio of the design wave to stillwater bending moments, i.e., on the two components that contribute to the in plane stresses. Because of this dependency, the plate thickness resulting from this formulation may deviate from the current ABS standard.

In cases where a stiffener's section modulus has to be determined, the safety index was calibrated to give the same section modulus as that obtained from current ABS Rules for all steel grades. The inherent safety index in the Rules was determined in each case and used to derive the partial safety factors.

As an example of a result that may lead to weight savings, a comparison between safety indices for the bottom shell plating and the bottom shell longitudinals revealed large differences, e.g., 5.20 versus 3.77, respectively, for ordinary steel. Since these two structural components are located in basically the same load environment, such a large difference in their safety indices are not rational, and thus provide an opportunity for adjustment.

In almost all ABS requirements, it was possible to convert the current Rules to LRFD format. However, in some instances, where the scantlings are determined in the current Rules through proportions (e.g., web thickness to web height), it was not possible to convert the provisions to LRFD format. Such rules of thumb could be kept in a new reliability based Guide.

Reliability based fatigue strength guidance has also been developed with target reliability equal to the implied reliability in the current ABS Fatigue Guide for tankers' structural components in Part 5, Section 2, Appendix 5/2AA of the ABS Rules. The requirements are based on characteristic S-N curves published by the UK Department of Energy and the assumption that the long term distribution of stress ranges is defined by the Weibull distribution. Permissible stress ranges are published for four values of the Weibull shape parameter for each of the eight design S-N curves. The design requirement is stated in terms of a permissible stress range relating to the largest expected stress during the service life.

Documentation on the derivation of permissible stress has been provided. An estimate was made of the reliability implied by the ABS Rules. For the combination of Weibull shape parameters and S-N curves, safety indices ranged from 1.07 to 1.46 and notional failure probabilities ranged from 0.072 to 0.142. A calibrated safety index of $\beta_0 = 1.4$ was suggested.

Four formats for probability based fatigue design criteria are presented:

(1) The design stress approach. The maximum stress range during the service life is to be less than a specified maximum permissible stress, S_0. This is the approach of the ABS Rules.

(2) Target damage level, Δ_0. The estimated damage D over the service life is to be less than Δ_0.

(3) Factor of safety on life, FS_{life}. The design life is defined as FS_{life} times the intended service life. The design life, or time to failure of the component is to be greater than the service life.

(4) Partial safety factors. A safety factor is applied to each of the random design factors.

Each of these methods were demonstrated using the calibrated target value of $\beta_0 = 1.4$.

In conclusion, reliability based standards demonstrate certain benefits over the traditional standards:

- Efficient structures of known reliability independent of ship size or design wave to stillwater bending moment ratio.
- More efficient structures providing more uniform safety over a range of parameters governing the design of a structural member.
- Direct comparison of the magnitudes of safety indices determined for structural members located in similar load environments.

It should be noted, however, that, in general, the proposed LRFD design equations are based on safety indices and safety levels implied in the current ABS Rules, and deviate only for the cases where inconsistencies were found.

REFERENCES

American Bureau of Shipping. *Rules for Building and Classing Steel Vessels*, Part 5, Specific Vessel Types, 2000.

Offshore Installations, Guidance on Design, Construction, and Certification (1990), Department of Energy, UK, 4th Edition, London, HMSO.

Mansour, A.E., Wirsching, P.H., White, G., and Ayyub, B. (1996), *Probability Based Ship Design Implementation of Design Guidelines for Ships: A Demonstration*, SSC-392, Ship Structure Committee, NTIS #PB97-109961.

Mansour, A.E., Wirsching, P.H., Luckett, M., and Plumbton, A. (1997), *Assessment of Reliability of Existing Ship Structure*, SSC-398, Ship Structure Committee, NTIS #PB97-141584.

Nikolaidis, E. and Kaplan, P. (1991), *Uncertainties in Stress Analysis on Marine Structures*, SSC-363, Ship Structure Committee.

Reliability Based Optimum Structural Fabrication tolerances (1998), final report to NAVSEA, Contract Number N00024-96-C-4123, by Mansour Engineering.

Wirsching, P.H. and Chen, Y.N. (1988), "Considerations of Probability Based Fatigue Design for Marine Structures," *Marine Structures*, 1, pp. 23-45.

Wirsching, P.H. (1998), "Fatigue Reliability," in *Progress in Structural Engineering and Materials*, CRC Ltd., London, Vol. 1, No. 2, pp. 200-206.

Wirsching, P.H., and Mansour, A.E. (1997), Reliability of Fatigue and Fracture Analysis of Ship Structure, *Prevention of Fracture in Ship Structure*, Ship Structure Committee, Report #PFSS95, NTIS #PB97-141568.

Wu, Justin Y-T. (Southwest Research Institute) and Xiaoping Du (The University of Iowa). Partial Safety Factor Analysis Program, version abs3b, 1998.

APPENDIX A
EXAMPLES OF PARTIAL SAFETY FACTOR DESIGN REQUIREMENTS

Following are examples of forms, employing partial safety factors, have been considered by various agencies in implementing probability based code requirements.

National Bureau of Standards (A58)

In general, the NBS format is:

$$\phi R_n \geq \sum_{i=1}^{n} \gamma_i Q_i$$

More specifically,

$$\phi R_n \geq \gamma_E \left(\gamma_D C_D D_m + \gamma_L C_L L_m + \gamma_W C_W W_m + ...\right)$$

where

ϕ = resistance factor

R_n = nominal resistance

γ_E = PSF representing uncertainty in structural analysis

$\gamma_{D,L,W}$ = deterministic factors that transform load intensities to load effects

$C_{D,L,W}$ = PSF's to account for the uncertainty in the determination of the loads

D_m, L_m, W_m = mean dead, live, wind loads, etc.

American Institute of Steel Construction

The format of AISC is identical to that recommended by NBS. Some details differ.

American Petroleum Institute

Original WSD format:

$$\frac{R}{SF} \geq D + L + W + ...$$

LRFD format:

$$\phi_{R_i} R_i > \gamma_D D + \gamma_L L + \gamma_W W + ...$$

where

L, D, W = nominal values for live, dead, and wind loads

ϕ_{R_i} = partial resistance factor for component i

= $\phi_i \phi_S$

ϕ_i = component strength uncertainty factor

ϕ_S = system consequence factor

and each of the load factors has the form,

γ = $\gamma_l \gamma_A$

γ_l = uncertainty in load intensity

γ_A = uncertainty in the analysis

Comite Euro-International Du Beton (CEB)

The general format is:

$$R_d \geq S_d$$

where

$$S_d = S\left\{\gamma_D D + \gamma_p P + \gamma_Q \left[Q_{1k} + \sum_{i>1}(\psi_{oi} Q_{ik})\right]\right\}$$

D = nominal value of dead load

P = nominal value of prestressing force

Q_{1k} = nominal value of principal variable load

Q_{ik} = nominal values of other less important variable loads

$\gamma_0, \gamma_P ...$ = partial safety factors

ψ_{0i} = load combination probability factor

f_k = characteristic or nominal strength

γ_m = factor to account for uncertainty in strength

National Building Code of Canada

NBC uses a split load factor format for load effects

$$\text{Load effects} = S\{\gamma_D D + \psi(\gamma_L L + \gamma_W W + \gamma_T T)\}$$

where

D,L,W,T = nominal loads (dead, live, etc.)

S{...} = load effects due to loads in the bracket

γ_D, γ_L ... = load factors

ψ = load combination probability factor

= 0.7 or 0.6 depending on whether one, two or three loads are included

APPENDIX B

5/3A.4.3.2 Bottom Shell and Inner Bottom Plating (1996)

The net thickness of the bottom shell and inner bottom plating, over the midship 0.4*L*, are to satisfy the hull-girder section modulus requirements in 5/3A.4.2.1. The buckling and ultimate strength are to be in accordance with the requirements in 5/3A.5.2.2. In addition, the net thickness of the bottom shell and inner bottom plating are to be not less than the following.

A Bottom Shell Plating (1997)

The net thickness of the bottom shell plating, t_n is to be not less than t_1, t_2, and t_3 specified as follows:

t_1 = $0.73s(k_1 p / f_1)^{1/2}$

t_2 = $0.73s(k_2 p / f_2)^{1/2}$

t_3 = $cs(S_m f_y / E)^{1/2}$

where

s = spacing of bottom longitudinals, in mm (in.)

k_1 = 0.342

k_2 = 0.500

p = nominal pressure in N/cm^2(kgf/cm^2, $lbf/in.^2$) as specified in Table 5/3A.3.2

f_1 = permissible bending stress, in the longitudinal direction in N/cm^2(kgf/cm^2,$lbf/in.^2$)

= $(0.95 - 0.67 \cdot SM_{RB} / S_{MB}) S_m f_y \leq k_3 S_m f_y$

k_3 = 0.40 for load case 1 – "*a*" in Table 5/3A.3.2

= 0.36 for load case 1 – "*a*" in Table 5/3A.3.2

SM_{RB} = reference net hull-girder section modules based on the material factor of the bottom flange of the hull girder in cm^2–m (in^2–ft)

= $0.9 \cdot SM$

SM_B = net design hull-girder section modules to the bottom in cm^2–m (in^2–ft)

SM = required hull-girder section modules in accordance with 3/6.3.4 and 3/6.5.3 based on the material factor of the bottom flange of the hull girder in cm^2-m (in^2-ft)

f_2 = permissible bending stress, in the transverse direction in N/cm^2 (kgf/cm^2, lbf/in^2)

= $0.80 \cdot S_m f_y$

S_m = strength reduction factor

= 1 for Ordinary Strength Steel as specified in Table 2/1.1-2

= 0.95 for Grade H32 as specified in Table 2/1.2-2

= 0.908 for Grade H36 as specified in Table 2/1.2-2

= 0.875 for Grade H40 as specified in Table 2/1.2-2

APPENDIX C
Ship Characteristics – ABS Tankers

Ship	Units	A	B	C	D	E	F	G	H	I	J	K	L	M
Deadweight	T	17.5K	35K	45K	65K	72K	105K	105K	150K	148K	301.5K	308K	300K	296K
Length between Perpendiculars	M	154.0	174.5	174.0	219.0	219.0	232.00	233.0	258.00	264.0	317.00	320.0	320.00	320.00
Breadth	M	26.80	27.40	32.20	32.20	32.24	42.00	42.00	46.00	47.80	58.00	58.00	60.00	60.00
Depth	M	11.50	17.60	18.20	17.85	20.20	21.20	21.30	23.90	22.80	31.40	31.00	29.30	29.55
Draft	M	7.00	11.00	12.20	12.95	14.00	14.90	14.75	16.84	16.10	22.00	22.70	20.85	21.10
C_b		0.807	0.828	0.800	0.845	0.862	0.826	0.833	0.847	0.831	0.836	0.820	0.848	0.825
M_{SW} sag	T-M	-60,000	-77,018	-88,900	-153,000	-130,000	-214,200	-232,000	-360,000	-347,000	-669,000	-633,500	-780,000	-651,809
M_{SW} hog	T-M	61,000	85,362	100,800	168,000	149,000	246,100	260,000	400,000	390,000	740,000	739,500	840,000	724,361
M_W sag	T-M	-94,173	-130,419	-150,739	-260,738	-262,871	-384,381	-389,884	-540,198	-587,317	-1,051,824	-1,074,077	-1,120,000	-1,103,573
M_W hog	T-M	87,000	122,007	138,953	246,190	250,441	359,190	365,742	510,602	550,345	988,314	1,000,328	1,059,035	1,030,676
SM req bottom	cm^2-M	67,408	116,276	134,391	181,262	182,745	267,217	273,395	398,133	411,138	694,641	760,687	830,718	767,488
SM offered bottom	cm^2-M	89,287	192,154	229,426	237,113	269,333	398,521	386,837	512,474	484,261	946,834	1,097,948	1,182,141	1,035,070
Excess SM bottom		1.32	1.65	1.71	1.31	1.47	1.49	1.41	1.29	1.18	1.36	1.44	1.42	1.35
Bottom Plate – s/t	mm	780/12.5	800/15	760/15.5	780/16	780/15	805/16.5	820/16.5	850/16.5	820/17.5	864/18.5	920/20	850/18	860/18
Bottom Plate – Pressure	kgf/cm^2	1.358	1.832	2.171	2.146	1.954	2.430	2.412	2.207	2.134	2.772	2.821	2.632	2.657
Bottom Plate – Material Yield	kgf/cm^2	3200	2400	2400	3200	3200	3200	3200	3200	3200	3600	3200	3200	3200
Bottom Long - Pressure	kgf/cm^2	1.358	1.832	2.171	2.146	1.954	2.430	2.412	2.207	2.134	2.772	2.821	2.632	2.657
Bottom Long – Material Yield	kgf/cm^2	3200	2400	3200	3200	3200	3600	3200	3200	3200	3600	3200	3200	3200
M_W / M_{SW} Sag		1.57	1.69	1.70	1.83	2.02	1.79	1.68	1.69	1.50	1.57	1.70	1.44	1.69
M_W / M_{SW} Hog		1.42	1.43	1.38	1.47	1.68	1.46	1.41	1.41	1.28	1.34	1.35	1.26	1.58
S/t		62.4	53.3	49.0	48.8	52.0	48.8	49.7	46.9	51.5	46.7	46.0	47.2	47.8
$B = \frac{S}{t}\sqrt{\frac{f_y}{E}}$		2.44	1.79	1.65	1.91	2.04	1.91	1.95	1.84	2.02	1.93	1.80	1.85	1.87

(*) Note : Side Ballast Tank Length is twice the Cargo Hold Length

Range of $r = M_W / M_{SW}$: 1.26 to 2.02

Range of S/t: 46.0 to 62.4

Range of B: 1.65 to 2.44

APPENDIX D
S-N Design Curves for Welded Joints

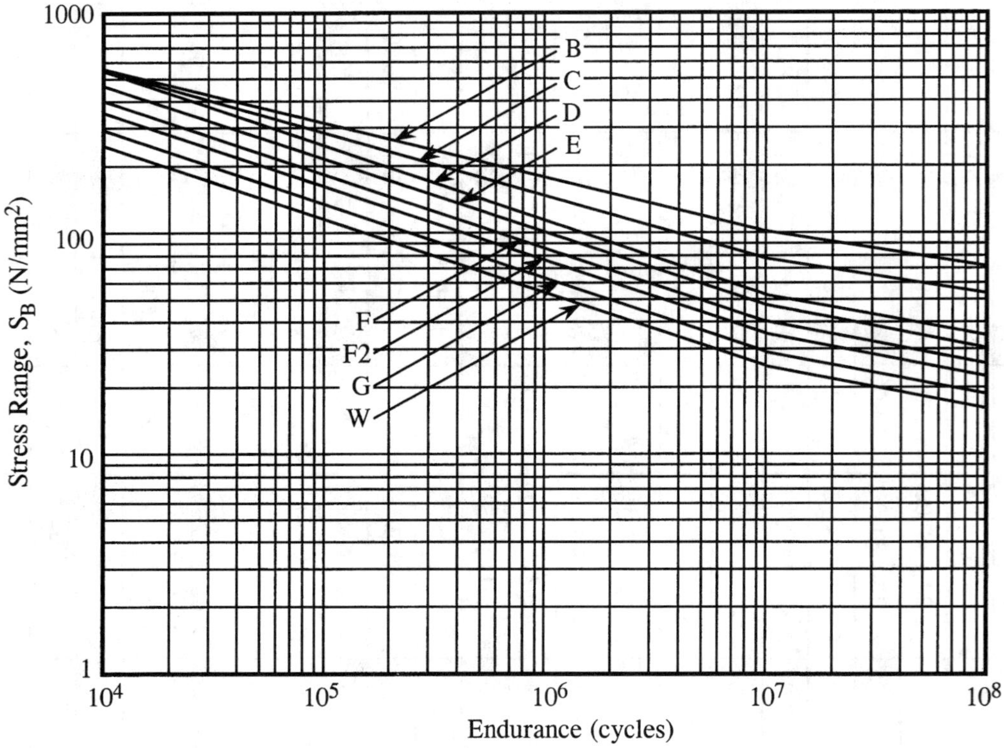

Figure D-1 Basic S-N Design Curves for Welded Joints ("Offshore Installations: Guidance on Design, Construction, and Certification," Department of Energy, U.K., Fourth Edition, 1990, London, HMSO)

Table D-1 Parameters of the S-N Curves (N/mm^2 units) as shown in Fig. D-1

S-N Curve	A Design Value*	m	C	r	Stress at knee S_Q (N/mm^2)	$\tilde{A}$ Median	COV of A
B	1.01E+15	4.0	1.02E+19	6.0	100.0	2.343E+15	0.44
C	4.23E+13	3.5	2.59E+17	5.5	78.2	1.082E+14	0.50
D	1.52E+12	3.0	4.33E+15	5.0	53.4	3.988E+12	0.51
E	1.04E+12	3.0	2.30E+15	5.0	47.0	3.289E+12	0.63
F	6.30E+11	3.0	9.97E+14	5.0	39.8	1.726E+12	0.54
F2	4.30E+11	3.0	5.28E+14	5.0	35.0	1.231E+12	0.56
G	2.50E+11	3.0	2.14E+14	5.0	29.2	5.660E+11	0.43
W	1.60E+11	3.0	1.20E+14	5.0	25.2	3.680E+11	0.44

Numerical and Experimental Investigation to Evaluate Wave-Induced Global Design Loads for Fast Ships

Thomas E. Schellin (M), **Christian Beiersdorf** (V), Germanischer Lloyd, Hamburg, Germany; **Xiao-Bo Chen** (V), Bureau Veritas, Paris, France; **Nuno Fonseca** (V), **Carlos Guedes Soares** (M), Instituto Superior Tecnico, Lisbon, Portugal; **Adolfo Maron Loureiro** (V), Canal de Experiencias Hidrodinamicas de El Pardo, Madrid, Spain; **Apostolos D. Papanikolaou** (M), National Technical University Athens, Athens, Greece; **Antonio Perez de Lucas** (V), IZAR, Madrid, Spain; **Juan M. Ponce Gomez** (V), Canal de Experiencias Hidrodinamicas de El Pardo, Madrid, Spain

The European cooperative research project WAVELOADS investigated use of a linear frequency-domain numerical technique and developed a nonlinear time-domain numerical technique to predict wave-induced global loads for a high-speed ferry and round-bilge fast monohull. For the frequency-domain technique, besides adding viscous effects, an advanced software module was developed to evaluate hydrodynamic influence coefficients, whereby forward speed effects were fully accounted for. The procedure relied on the Fourier-Kochin free-surface forward-speed Green function method that promised to provide a solid mathematical basis for obtaining robust and computationally efficient predictions. Incorporation of this module into two existing frequency-domain panel codes resulted in accurate predictions of wave-induced global response for a slender Wigley hull advancing in regular head waves at high forward speed. However, for the reference ships, predictions deviated excessively from experimental data because numerical inaccuracies could not be resolved. Therefore, subsequent frequency-domain computations were performed on the basis of the so-called encounter-frequency approach. To validate the numerical techniques and to specify limits of application, numerical predictions were compared to experimental data obtained from systematic model tests. For ship speeds corresponding to Froude numbers up to about 0.4, this encounter-frequency approach produced accurate predictions, albeit only in waves of small steepness. To obtain long-term design values in a realistic seaway consisting of finite amplitude waves, linearly computed loads were corrected for nonlinear effects. As a practical demonstration, a statistical analysis yielded long-term (design) values of wave-induced midship vertical bending moments for the high-speed ferry. In sagging, the resulting bending moment deviated by five percent from classification society rule values; in hogging, by thirteen percent.

INTRODUCTION

High-speed craft rules that include formulas for safe design loads are not always suitable for designing the optimum hull structure, because they mostly reflect service experience of existing ships. Current and future designs call for larger ships for which insufficient operating experience has been gathered to serve as a basis for establishing adequate rules. In addition, hull structures of economically competitive high-speed ships need to be optimized for least weight. Thus, shipyards as well as classification societies need practical methods to predict wave-induced global loads for large, modern fast ships.

In addressing this need, the European research project WAVELOADS (2001) investigated the use of a linear frequency-domain numerical technique and a nonlinear time-domain numerical technique to predict wave-induced global loads. Although in many ways these techniques represent standard methods currently used for conventional ships, the emphasis here was directed towards accounting for effects of high forward speed and significant nonlinearities. WAVELOADS also supplied a systematic set of experimental data suitable for calibration and validation of these techniques and for specifying their limits of application. Such data were not available at the start of the project. Two fast ships of modern design were tested, namely, a high-speed ferry and round-bilge fast monohull. Table 1 lists main particulars of the two reference ships; Fig. 1 shows body plans.

From a practical standpoint, it is generally advantageous to use a frequency-domain technique to obtain appropriate transfer functions that can be statistically analyzed to obtain long-term (design) loads. Such computations are efficiently accomplished within classical three-dimensional potential flow boundary element methods by solving a set of partial differential equations to model the wave diffraction and radiation fields around the advancing ship. Influence coefficients, the building blocks of this boundary element method, define the potential flow generated by a distribution of singularities. In solving this three-dimensional radiation/diffraction boundary value problem (BVP) for modern fast ships, the high forward speed must be adequately accounted for.

For the numerical frequency-domain technique, the approach followed consisted of developing an advanced software module to evaluate influence coefficients on the basis of the free-surface forward-speed Green function method of Noblesse and Chen (1995). This method differs from two previously tested Green function methods of Iwashita and Ohkusu (1992) and Ba and Guilbaud (1995) in that it relies on the so-called Fourier-Kochin formulation. The procedure promised to provide a solid mathematical basis for obtaining robust and computationally efficient numerical predictions of wave loads and ship motions within the framework of a three-dimensional frequency-domain analysis, whereby forward speed effects are fully accounted for.

Upon implementation of this module into frequency-domain (panel) codes of two of the WAVELOADS partners (Chen and Malenica, 1998; Papanikolaou and Schellin, 1992; Östergaard and Schellin, 1995), sample computations conducted for a slender Wigley hull demonstrated the effectiveness of this software module as its use led to favorable agreement between numerical predictions and available model test data. However, similar computations for the two reference ships were unsuccessful because, first, the required computer running times were too long for practical purposes and, second, the results were plagued by numerical inaccuracies that could not be resolved. Therefore, subsequent computations for both reference ships were conducted according to the so-called encounter-frequency approach, i.e., with panel codes relying on the zero-speed Green function.

For ship speeds corresponding to Froude numbers of up to about 0.4, the encounter-frequency approach produced accurate predictions of wave-induced global loads; however, only in waves of small steepness. Therefore, for a statistical analysis, subsequently performed to arrive at long-term (design) loads, the (linearly) computed loads had to be corrected for nonlinear effects, yielding so-called pseudo transfer functions that depended on wave amplitude. A procedure shown to be successful for many standard ships (Guedes Soares and Schellin, 1996; Rathje and Schellin, 1997; Rathje et al., 2000) was used here to obtain sample computations of midship vertical bending moment for the high-speed ferry. The resulting values deviated by five percent in sagging and thirteen percent in hogging from classification society rules.

Of course, the linear frequency-domain technique was not capable of accurately predicting ship responses under all conditions. Certain nonlinear behavior, such as large amplitude hull motions, was modeled more effectively in the time domain. For this reason, a time-domain method was developed within WAVELOADS. This technique was complementary to the frequency-domain technique as it covered conditions where the frequency-domain method was not valid. Large amplitude relative motions may cause strong nonlinear structural loads.

The time-domain method that was developed extended the formulation of Fonseca and Guedes Soares (1998 a,b), who also compared numerical results with published experimental data while emphasizing nonlinearities (Fonseca and Guedes Soares, 2002). Although forward speed effects were considered in a more simplistic way, the dominant nonlinear

Table 1 Main particulars of reference ships

	High-speed ferry	Fast monohull
Length between perpendiculars	110.0 m	122.0 m
Breadth max.	18.7 m	16.1 m
Height to main deck	11.2 m	9.2 m
Draft at service speed	2.8 m	4.66 m
Displacement at service condition	2230 t	4315 t
Service speed	39 knots	28 knots
Froude number	0.61	0.42

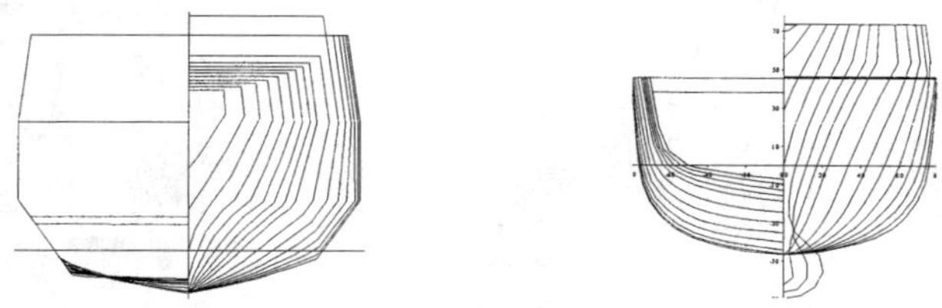

Fig. 1 Body plans of reference ships: high-speed ferry (left) and fast monohull (right)

contributions for the vertical motions and loads were accounted for by integrating the hydrostatic and Froude-Krylov pressures over the exact wetted hull at each time instant. Radiation and diffraction forces, treated as linear, were represented by a consistent time-domain model. Nonlinear vertical viscous forces were also included, using a simplified empirical model of Fonseca and Guedes Soares (2000).

FREQUENCY-DOMAIN TECHNIQUE

Influence coefficients for the frequency-domain technique were evaluated according to the Fourier-Kochin formulation. These coefficients, decomposed into a wave component and a local component, were evaluated directly by considering a distribution of singularities (sources or dipoles) on a hull panel or along a waterline segment. Unlike previous studies, the Green function associated with a point source was not explicitly evaluated, such that all difficulties related to the evaluation of classical Green functions were avoided.

The wave component defined by a single integral along all dispersion curves determined by the dispersion relation in the Fourier plane provides a relatively simple and elegant representation of time-harmonic ship flows, useful not only from an analytical point of view in that a direct relationship between the dispersion curve and the associated wave system was found, but also from the numerical standpoint in that an efficient method to evaluate the wave component was developed. The local component was defined by a double Fourier integral whose integrand is continuous everywhere in the Fourier plane. However, one major practical difficulty arose in the evaluation of the local component. For the numerical evaluation of the double Fourier integral, the computing time was too long. For example, to evaluate influence coefficients for one vertical panel and for 251 x 101 points at the free surface required nearly 48 *hours* of CPU time on an IBM workstation RS/6000. Although CPU time was less for other cases, such as two deeply submerged panels, the actual formulations for the local component were, unfortunately, not practical.

The most appropriate way to overcome this difficulty was to solve the problem directly after having identified its nature. Regarding time-harmonic ship flows, the critical problem involved accounting for special properties of potential flow along the track of the perturbation (a source or a distribution of sources) at the free surface. Indeed, recent work by Chen (1999) showed that the ship motion Green function is singular and highly oscillatory along the track of a source at the free surface. Based on a new analysis, Chen's work revealed for the first time the peculiar properties of time-harmonic ship waves some 40 years after Ursell (1960) applied the steepest descent technique on the special case of steady flow generated by a translating point pressure. This encouraging breakthrough in understanding free surface dispersive flows led to the development of the New Green Function (NGF) for ship flows (Chen and Noblesse, 1998).

This NGF can be thought of as representing a special class of influence coefficients. Regarding the ship motion problem, using the NGF offers several advantages. First, analytical expressions describe the wave component instead of a single integral; second, single integrals for the local component are used instead of a double integral; third, the evaluation of spectral functions is suppressed when solving for the influence coefficients. Most important, the NGF method provides a relatively efficient way to numerically evaluate the influence coefficients. In fact, from the local component different treatments are developed for the wave component. The integrand of the wave component integral can be highly oscillatory, making it necessary to develop a special algorithm based on analytical integration after numerical approximations of the amplitude and phase functions of the integrand. It was shown that the method developed is effective and accurate to evaluate highly oscillatory integrals. The integrand of the local component is not oscillatory, and the classical Gauss-Legendre quadrature is efficient. A logarithmic singularity appears in the integrand when both field point and source point approach or are located at the free surface. Techniques including integration by parts were used to eliminate this difficulty.

Surface elevations of the ring wave system and the V-wave system, caused by one pulsating and translating submerged source of unit strength, are shown in Figs. 2 and 3, respectively. The flow is characterized by Brard number $\tau = U\omega/g = 0.2$, where U is the translational speed of the source, ω is the circular frequency of pulsation, and g is the acceleration of gravity. The source is situated at $x/F_n^2 = 0$, $y/F_n^2 = 0$, and $z/F_n^2 = -0.1$, where the Froude number F_n is 0.15. The origin of the right handed x,y,z coordinate system is located on the surface above the source, with the x-coordinate pointing in the direction of U and the z-coordinate pointing positive upward. Free surface elevations are shown in Fig. 2 for an area defined by coordinates $-15.75 \leq x/F_n^2 \leq 6.75$ and $-8.775 \leq y/F_n^2 \leq 8.775$ and in Fig. 3 for an area defined by coordinates $-0.8165 \leq x/F_n^2 \leq 0.0430$ and $-0.3438 \leq y/F_n^2 \leq 0.3438$. The wave components, the local component, and the sum of both along a cut at the centerline ($y = 0$) are presented. Arrows indicate the direction of wave propagation (for the wave components) or surface elevation (for the local component).

Development of Software Module ICship.a

Software module ICship.a was developed to evaluate the influence coefficients in ship motion problems. This module, comprising nearly 10000 statements, is an archive library of 25 subroutines written in FORTRAN. Two principal subroutines, PrepIC.o and CalcIC.o, are called, one before the looping computation of influence coefficients and the other in the computational loop itself. No major difficulties were encountered in the implementation of both principal subroutines as the calling arguments are short, while a common block accomplishes the exchange of other data between different subroutines.

Software module ICship.a was tested by comparing numerical results with the analytical solutions for a submerged sphere obtained by Wu (1995). Wu used a multipole expansion to directly derive the velocity potential around a submerged sphere advancing in water waves. The analytical results are believed to be accurate up to four significant digits for wave resistance and lift forces, added masses and damping coefficients, as well as wave-induced loads. The favorable correlation between results based on the use of ICShip.a and results of the ship motion Green function found in previous studies and the analytical solution of global loads on an immerged sphere validated ICship.a.

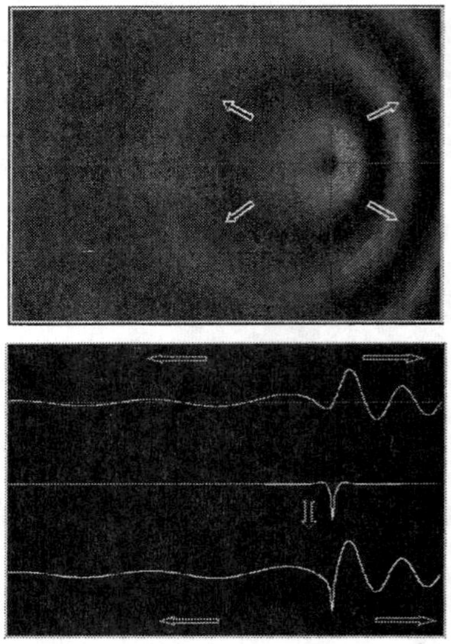

Fig. 2 Ring wave system. Upper graph: complete view; lower graph: decomposition of wave profile along centerline, showing wave component (upper trace), local component (middle trace), and total wave (bottom trace)

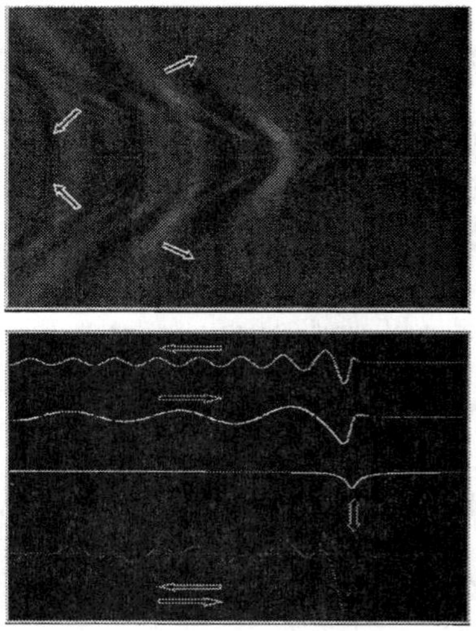

Fig. 3 V-wave system. Upper graph: complete view; lower graph: decomposition of wave profile along centerline, showing the inner-V wave component (upper trace), the outer V-wave component (upper middle trace), the local component (bottom middle trace), and the total wave (bottom trace)

Incorporation of ICship.a in Panel Codes

Incorporation of software module ICship.a in the panel code of Bureau Veritas as well as the panel code of Germanischer Lloyd involved exchanging existing routines that computed Green functions and their gradients and, in addition, solving for the line integral. Some work on optimization of the panel codes was necessary to improve numerical efficiency. The original version of these codes considers forward speed effects with some simplifications regarding the free surface boundary condition and the slenderness of the body (encounter-frequency approach). As a consequence, the line integral along the waterline does not need to be evaluated. The body boundary conditions for the radiation problem rely on the simplified interaction between the displacement of the body and the free stream flow. Furthermore, the convective pressure terms caused by velocity components in the vertical and lateral directions are neglected. Thus, to obtain a valid linear method for computing seakeeping problems for non-slender bodies, these simplifications had to be removed.

The closed (line) integral about the intersection of the body and the free surface was subdivided into a number of straight-line segments. For a simplified geometric treatment, these line segments coincided with panel edges at the free surface. The code relies on a so-called indirect boundary element method (BEM), where the density of the Green function is evenly distributed over the line segment or the panel area. Panels at the free surface share their upper edge with the length of a line segment. Therefore, the Green function density of the panel and the line segment is assumed to be equal, and no new unknown source densities have to be introduced to compute the potentials and their derivatives. The boundary conditions on the body surface are satisfied at panel centers, which are always situated under the free surface. Therefore, problems arising from the singularity of the Green function located at the free surface and the improperly defined boundary condition at the intersection of the body and the free surface are avoided.

To obtain the radiation potentials, the linearized boundary conditions on the body surface couple the steady flow with the time harmonic flow. This coupling, involving first and second order derivative of the steady flow potential, results in the so-called m_j-terms. The determination of the steady flow potential can be simplified by replacing the Neumann-Kelvin free surface flow with a double body flow. The panel code was modified in that the derivatives of the double body flow potential were calculated in a preprocess by applying a BEM. To obtain second order derivatives without executing numerical differentiation schemes, the method presented by Chen et al. (2000) was used. As the boundary conditions are satisfied at the panel centers, the m_j-terms are also evaluated at these panel centers.

Unlike the flow around a slender body, the flow around a real body contributes (convective) pressure terms caused by velocity components from all directions. To account for the time harmonic flow superimposed on a steady flow, the panel codes were modified in that the linearized Bernoulli equation was expanded by two terms. The first derivatives of the time harmonic and the steady flow potentials in the vertical as well as the lateral direction had to be evaluated.

The Green functions for the seakeeping problem were decomposed into two Rankine terms and one wave term. The two Rankine terms accounted for sources distributed over the submerged body surface and additional sources distributed over the image body surface, mirrored about the free surface. If flat panels and straight line segments idealize the hull, they have a constant source density. Therefore, the integration of the Rankine terms over a panel or a line segment was performed in closed form. The integration of the wave term over a panel or a line segment followed the midpoint rule. Thus, the wave term was evaluated only once, namely, when calculating the influence of a source panel on any field point. Depending on the distance between source point and field point, the number of source points for the integration of the wave term over a line segment ranged from one point (for a large distance) up to seven points (for a small distance). The numerical integration scheme was based on Simpson's rule.

The library module ICship.a computed not only the influence coefficient of the wave term, but also the Rankine term of the mirrored sources. To achieve an exact solution for the integral of the Rankine term, this term was subtracted from the influence coefficient and integrated separately.

Nonlinear Correction of Hydrodynamic Pressure

The application of linear panel methods is fundamentally restricted to small amplitude waves, resulting in equal magnitudes of vertical bending moments for hogging and sagging. To obtain realistic predictions in finite amplitude waves, the major nonlinear effect, caused by nonvertical sides of the hull at the waterline, was taken into account by implementing a procedure developed by Hachmann (1986) and applied by Rathje and Schellin (1997) to the panel code of Germanischer Lloyd.

This procedure uses a time-harmonic velocity potential to describe the two-dimensional flow field at the still waterline in a plane defined by the tangent to the ship's section contour and the still waterline. Considering only linear terms of a Taylor expansion of this flow potential, applying the Bernoulli equation along the section contour, and substituting the linear dynamic pressure at the still waterline yields the hydrodynamic pressure variation under the wave profile. This procedure also includes a criterion that limits the height of the wave at the ship's side. The resulting wave elevation always remains below the theoretical limit height of a breaking wave.

Viscous Effects

Viscous lift and drag forces acting on the ship's hull as well as on appendages, such as rudders, stabilizing fins and bilge keels, were computed in terms of relative fluid velocities and semi-empirical expressions that utilized experimental data from forced excitation tests of the two reference ships. This procedure, documented by Papanikolaou et al. (2000), accounted for the quadratic dependence of viscous forces on relative velocities. As it was impossible to introduce viscous lift and drag terms directly into the linearized motion equations, an harmonic energy balance, based on the principle of equivalent linearization, obtained linearized approximations.

TIME-DOMAIN TECHNIQUE

The nonlinear time-domain strip method combined linear and nonlinear terms – radiation and diffraction forces were kept linear. The dominant nonlinear contributions were assumed to arise from hydrostatic restoring and Froude-Krylov forces. Additionally, the method considers effects of green water on deck and viscosity in the vertical direction.

The time-domain seakeeping boundary-value problem was formulated using potential flow theory as it was assumed that the flow is inviscid and irrotational. The exact formulation of the BVP led to the nonlinear free surface problem with kinematic and dynamic conditions applied at the instantaneous unknown boundaries and with appropriate radiation conditions at infinity. The flow is governed by the Laplace equation. For practical applications, this complex problem had to be simplified to achieve numerically efficient solutions.

Regarding the BVP, simplifications consisted of removing nonlinearities from boundary conditions and simplifying three-dimensional aspects of the flow. Thus, it was assumed that amplitudes of incoming waves and oscillatory ship motions as well as the speed of advance are small and that the ship is slender. Although the resulting BVP is linear, it is fully three-dimensional in a linear sense. To remove most of the three-dimensional effects from the flow, it was also assumed that the frequency of oscillation is high. The resulting method computed hydrodynamic response and wave exciting forces in terms of two-dimensional coefficients associated with a finite number of two-dimensional cross sections representing the hull. A multipole method coupled with a multi-parameter conformal mapping technique was used to compute these coefficients. Forward speed effects are related only to the angle between the longitudinal hull centerline and the steady flow at infinity.

A fourth-order Runge-Kutta method advances the nonlinear equations of motions in the time domain. At each time step hydrostatic and Froude-Krylov pressures are integrated over the instantaneously wetted surface defined by the intersection of the incident wave elevation with the hull. Because radiation forces depend of past history of fluid motion, they are represented by convolution integrals of memory functions, radiation restoring coefficients, and infinite frequency added masses. Memory terms are a function of time, geometric configuration of hull cross sections, and ship speed. Radiation restoring coefficients depend on hull geometry and ship speed. Forces proportional to these coefficients are radiation restoring forces. They represent a correction to hydrodynamic forces acting on the ship caused by the steady flow. The infinite frequency added masses depend only on hull geometry.

When abrupt changes in curvature characterize the ship's cross sectional shape, flow separation is likely to occur during the hull's vertical motion. A semi-empirical procedure that relies on lift and drag coefficients and is based on the cross flow approach of aerodynamics (Thwaites, 1960) accounts for the associated viscous force at each cross section by assuming that it depends on the square of the hull section's vertical velocity relative to the fluid and appropriate lift and drag coefficients. Global viscous loads are obtained by integration along the ship's length.

Systematic computations indicated that viscous forces affect ship responses only near resonance frequencies (Fonseca and Guedes Soares, 2000). Forced excitation model tests showed that such viscous forces depend on the cross sectional shape of the hull, the frequency of oscillation, and the ship's speed of advance. A lift coefficient of 0.24 and a drag coefficient of 0.30, deduced from forced excitation tests, were used for subsequent computations. They closely agreed with the experimental data of Beukelman (1980, 1983).

In head and bow waves, large amplitude motions may occur, especially in the bow region, causing water to flow onto the deck. To predict these green water loads, the momentum of water on deck, the hydrostatic pressure, and the inertial forces caused by vertical deck acceleration are considered.

Four program blocks working together as a package compute nonlinear vertical ship motions and wave-induced loads. The first block performs frequency-domain computations, including frequency dependent hydrodynamic coefficients and diffraction forces as well as infinite frequency added masses. The second block computes memory functions and radiation restoring coefficients for motions and loads. The third block uses this information to perform time-domain computations by solving the nonlinear equations of motion to yield all forces at each instant in time. The results are time histories of absolute and relative motions, accelerations, and wave-induced loads. The last block analyzes these time histories, yielding mean values, harmonic components, and averages of maxima and minima of the responses.

MODEL TESTS

To validate the numerical techniques, appropriate comparative experimental data had to be provided. A series of benchmark model tests were performed for both reference ships to obtain experimental measurements suitable for comparison with computations. These experiments comprised forced excitation tests in calm water to yield experimental added masses and damping coefficients and seakeeping tests in regular head waves to obtain wave-induced loads. The models were allowed to execute only heave and pitch motions. The high-speed ferry was model tested at a scale of 1:25; the fast monohull, at a scale of 1:27.

Instrumentation

Both models were segmented at stations 5, 10 and 15 to obtain force measurements for each of the four segments as well as for the whole ship. Previous experience with segmented models was small, and difficulties were encountered with wooden models because of high humidity. Therefore, the models were made of GRP. The resulting weight reduction allowed for greater freedom to arrange instrumentation. To prevent water entering the models, the cuts between segments were sealed by means of a flexible strip. It was necessary to test several sealing systems. A thin piece of canvas impregnated with sealing rubber was found to be the best solution.

Model construction was complex and time consuming. First, it was necessary to construct a wax model. Then a GRP mould had to be made to laminate the final GRP ship model. The four segments were held in place by means of an instrumented longitudinal aluminum backbone of rectangular cross section. A full Wheatstone bridge installed at an angle of 45 deg to the neutral axis on this beam measured shear forces. To achieve sufficient output from the highly sensitive bridge, it was necessary to reduce the beam thickness. As it was practically impossible to machine the beam, a hole was drilled directly through the beam, and a thin steel sheet was bonded on the beam covering this hole. The strain gage bridge was then bonded on the thin metal sheet. Another full bridge with strain gages located on the upper and lower edges of the beam measured bending moments.

The reduced sensitivity of this arrangement required stable signal conditioners. It was convenient to use these conditioners together with other sensors, such as potentiometers, inductive transducers and wave probes. It was also important to reduce the influence of electric noise. Therefore, 4.8 kHz carrier frequency amplifiers were selected. To reduce length of wiring, these amplifiers were located close to the sensors.

Resistance wave probes measured wave heights. These devices consisted of two stainless steel bars connected to a constant amplitude 5 kHz AC source. The measured current in the circuit was proportional to the submerged sensor length. It was necessary to develop a sensor that incorporates a printed circuit in its design.

A special device that allowed only heave and pitch motion was developed for seakeeping tests. The device consisted of two vertical cylindrical bars running along the corresponding linear bearings supported by an aluminum structure. Ball joints connected the two bars to the model. Load cells measured towing forces, using a standard three-component dynamometer.

To perform forced motion excitation tests in calm water, a vertical planar motion system was developed. The equipment had two linear ball screw actuators, each one driven by an AC servomotor controlled by a corresponding electronic drive. Both actuators were connected to the model beam through joints and load cells. Since both actuators acted independently, it was possible to obtain the desired combination of heave and pitch motions. The actuator screws had maximum motion amplitude of 500 mm, and their step was 20 mm per revolution. The servomotors had maximum turning speed of 1500 RPM, providing a torque of 8 Nm. In accordance with these characteristics, the maximum linear speed was 500 mm/s and the maximum force 1500 N. (The frontispiece depicts the high-speed ferry undergoing a forced excitation test.)

The motion control system consisted of controllers, motors and encoders. The driver had to be correctly sized according to the servomotor power requirements. An external computer monitored and controlled the excitation system. Since the desired motion was sinusoidal, the computer evaluated the velocity reference signal for each driver in real time. Moving the actuators with a sinusoidal motion produced the desired excitation. For heave, the motions were of the same amplitude and in-phase for both actuators. For pitch, the motions of the two actuators were 180 deg out of phase.

The control application was first developed in Visual Basic with extensions for special instrumentation. However, the time base of the Windows operating system was not reliable. Therefore, an independent time base was used where the real time counter board had a clock precision of 1.0 μs. Initially, the application provided acceptable response times, but the first model tests showed the impossibility of working with this operating system. Because Visual Basic is an event-based environment that depends on system events and latencies, unexpected responses were registered. During one run the fast monohull model sank. However, no damage occurred. A new DOS based program was then developed. This robust and effective operating system provided measurements at small time steps (of the order of 0.001 s) and, by adjusting gain control parameters, yielded precise position measurements, accurate within 1.0 mm.

The Model Test Program

During the forced excitation model tests the linear actuators forced the models to float at their design draft. For this reason, the weight distribution was not adjusted and was left as it resulted from model fabrication. Of course, it was necessary to register the actual weight and mass distribution to account for inertia forces.

During seakeeping tests, wave height was measured forward and away from the models. The undisturbed wave elevation at the ship's center of gravity was obtained from the linear dispersion relationship. All signals were zeroed before each test run. Therefore, vertical forces and moments did not include calm water values. All signals were conditioned, low pass filtered, and sampled at a rate of 100 Hz model scale. Samples were continuously stored in memory and dumped on disk at the end of each run. Time series were converted to full scale before storing them on disk. Froude scaling was applied with no scale effect correction.

Model tests were performed at various ship speeds. For the high-speed ferry, forced excitation tests and seakeeping tests were carried out at five ship speeds: 0, 10, 20, 30 and 40 knots full scale, corresponding to Froude numbers of 0, 0.16, 0.31, 0.47 and 0.63, respectively. For the fast monohull, forced excitation tests and seakeeping tests were carried out at four ship speeds: 0, 13.45, 20.17 and 26.89 knots full scale, corresponding to Froude numbers of 0, 0.2, 0.3 and 0.4, respectively.

For seakeeping tests model displacement, longitudinal position of the ship's center of gravity, and longitudinal mass distribution were adjusted to given values. Total displacement was obtained by adding ballast. The longitudinal mass distribution was checked by means of an inertia table. The vertical center of gravity was measured by means of an inclining test and checked to assure it had a reasonable value. The transverse mass distribution was not considered, because it did not affect vertical motions in head seas.

It was impossible to attain a realistic distribution of mass for each model segment, because of the restrictions imposed by the backbone and crossbars. This left little ballast available for adjustment. However, because the weight and inertia of each segment were fundamental as an input to the numerical models, it was necessary to measure them. This was achieved by using the device for the forced excitation tests. The model was hung in air from the two actuators and forced to move harmonically in pure heave and pure pitch, while the forces and moments on each section and the forces exerted by the actuators were measured. From these measurements the mass and moment of inertia of each separate segment were calculated.

Seakeeping tests were carried out in four sets of regular head waves. For one set of test runs in waves of small wave steepness (ratio of wave length to wave height, H_w/L_w = 1/80) the corresponding wave periods covered the full range of interest for the ships. For the other three sets of test runs in waves of larger wave steepness (H_w/L_w = 1/50, 1/35 and 1/20) the corresponding wave periods covered a smaller range around pitch resonance. Initially, it was intended to run tests at three wave steepnesses (H_w/L_w = 1/80, 1/50 and 1/20). However, in some cases the largest wave steepness produced violent motions and excessive loads. Therefore, in those cases it was decided to substitute a smaller steepness for the larger one to avoid damaging the instrumentation and the model.

Processing of Forced Excitation Test Data

A hardware filter processed the recorded analogue forces exerted by the actuators, yielding time series of external forces and moments acting on the whole ship and time series of shear forces and bending moments acting at each model segment. A specially developed software filter was implemented to eliminate the model natural frequency signal at around 8.0 Hz (full scale) for the high-speed ferry. Therefore, a cut-off frequency of 6.0 Hz was chosen for the analysis of all the recorded data. The highest frequency was measured at about 1.0 Hz, corresponding to third harmonic oscillations with a fundamental period of 3.13 s.

Excitation tests in air were carried out to determine inertial loads that had to be known to obtain hydrodynamic response forces and moments. For this purpose, the models were suspended from the actuators and oscillated at different heave and pitch frequencies. As expected, the resulting inertial forces and moments were sinusoidal functions. Their phase relationship relative to heave and pitch motions made it possible to remove these forces and moments from the signals measured during tests in water, yielding time series of forces exerted by the actuators and vertical shear forces and bending moments acting at stations 5, 10 and 15. These forces were then processed to obtain time series of total fluid forces and moments acting on each segment and on the whole ship.

To determine their hydrodynamic components, these time series had to undergo two additional processing steps to account for, first, hydrostatic loads in undisturbed water caused by the continuous variation of sinkage and trim at zero speed and, second, the corresponding quasi-hydrostatic loads at different ship speeds. To estimate these loads, the models were oscillated in heave and pitch with a long period of 250 s, first at low forward speed and then at different forward speeds. Subtracting these forces and moments from total external forces and moments yielded the corresponding hydrodynamic components. Their time series were then analyzed, providing mean values, amplitudes and phases of the first harmonic and amplitudes of the second and third harmonics and, finally, added masses and inertias and damping coefficients for each segment as well as for the whole ship.

Processing of Seakeeping Test Data

The seakeeping analysis program of Canal de Experiencias Hidrodinamicas processed the resulting time series, yielding the mean offsets for heave, mean trim for pitch, and mean wave-induced forces and moments acting on the four segments of the models and on the ship models as a whole. A subsequent harmonic analysis produced transfer functions of input signals (waves) as well as output signals (ship motions and wave-induced loads). A nonlinear least squares procedure was used to determine the best fit for mean values, amplitudes and phase angles of the first three harmonics. The difference between the phase of the response signals and the phase of the undisturbed wave at the ship's center of gravity defined the phase angles. Positive phase angles represented a lead of the response signal relative to the wave.

Second and third harmonics were accurate only when the corresponding amplitudes were relatively large because only then was the signal to noise ratio high enough. When this was not the case, estimates of superharmonics were noise contaminated.

During tests in high amplitude waves, especially in steeper waves (H_w/L_w = 1/35 and 1/20), large amounts of water flowed onto the deck. This led to nonlinear motions of the models and made it difficult to obtain reliable response amplitudes for all periods.

Uncertainties of Experimental Data

Measured responses included higher order (nonlinear) portions. Since higher order responses were an order of magnitude less than first order responses, uncertainties of second and third order responses were relatively larger than uncertainties of first order responses. When model tests were planned, quantification of third order effects was not an objective. However, as third order harmonics turned out to be consistent functions of wave frequency as well as wave steepness with relatively low dispersion, it was decided to retain these results.

It was impossible to carry out an uncertainty analysis of experimental data according to the ITTC Quality Manual, because many of the error sources related to measuring devices and calibration procedures could not be quantified. However, for wave frequencies in the resonance range, some test runs were repeated.

The scatter of first order harmonics of measured response showed practically no deviation and was consistently around one to two percent. As expected, the scatter increased for second and third order harmonics for structural loads. This scatter was always small compared to the mean value of the corresponding response function. The scatter of sagging and hogging peaks was also small, except for hogging moments at the quarter stations where it was as large as 20 percent. The uncertainty of these measurements was larger because their magnitudes were small.

A significant source of uncertainty was introduced when analyzing the records, specifically, when obtaining harmonic contents and positive and negative peaks. This is because the measured time series were not strictly stationary, but varied slowly in time. Part of this variation was caused by the exciting wave itself. The advancing reference system measured a constant wave period, but the wave amplitude varied slowly. The reason could not be determined; it may have been related to the natural modes of oscillation of the water in the tank. Thus, the results of the signal analyzer depended on the time samples chosen for analysis. All records were relatively long, comprising more than 40 useful cycles, and the samples used for the analysis included at least 20 cycles where amplitudes were practically stationary.

VALIDATION OF FREQUENCY-DOMAIN TECHNIQUE

Slender Wigley Hull

Sample computations were first performed for a slender Wigley hull form with a midship section coefficient of 0.67 and a length-to-beam ratio of 10. This hull was chosen because reliable model test measurements in regular head waves were available for comparison (Journee, 1992). Surge motions were restricted during these tests as well as in the comparative computations. A radius of gyration equal to one quarter of the ship's length represented the longitudinal mass moment of inertia. The forward speed considered corresponded to a Froude number of $F_n = 0.3$.

In Figs. 4, 5 and 6 results are plotted together with corresponding experimental measurements. Shown are transfer functions for heave and pitch motions ς and ν, respectively, heave exciting force F_3, pitch exciting moment F_5, added masses a, and damping coefficients b. Motions and exciting forces are plotted against the ratio of wave length to ship length λ/L; added masses and damping coefficients, against the normalized wave frequency $\omega(L/g)^{1/2}$. For comparison, results obtained from the original version of one of the panel codes that uses the zero-speed Green function are shown as well. Heavy straight lines connect computed results based on the module ICship.a, incorporated into a panel code to fully account for forward speed effects (with FS); light lines connect computed results from a panel code based on the zero-speed Green function (w/o FS), i.e., the encounter-frequency approach. Circles, triangles and squares mark experimental data. These figures demonstrate that results obtained from a panel code using software module ICship.a leads to significantly closer agreement between numerical predictions and experimental test data.

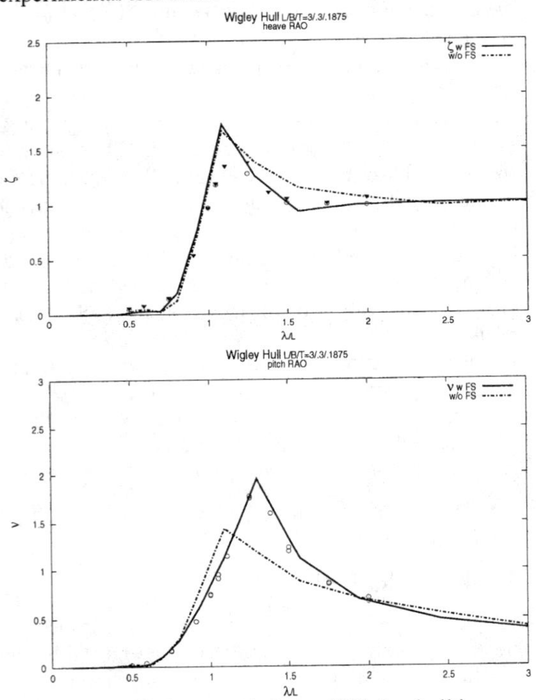

Fig. 4 Heave and pitch of Wigley hull in regular head waves

Fast Monohull

With the forward-speed Green function module ICship.a incorporated into the Bureau Veritas panel code, computations were performed for the fast monohull in regular head waves. A mesh of 134 bi-quadratic curved patches idealized the ship's hull. Selecting curved patches enabled a continuous representation of the velocity potential over the hull surface, and use of a Galerkin procedure enhanced numerical accuracy (Chen et al., 2000). Similar computations were also performed with software module ICship.a incorporated into the panel code of Germanischer Lloyd. For these computations, 952 flat panels modeled the submerged hull (Schellin et al., 2002). Four forward speeds were considered, corresponding to Froude numbers of 0, 0.2, 0.3 and 0.4. At zero forward speed, predictions generally compared favorably with model test data. However, at the low ship speed (F_n = 0.2) computed results already deviated by about 20 percent from measurements, and at higher ship speeds computed results showed excessive deviations from measurements over nearly the entire frequency range investigated.

Recently the inability to predict wave-induced loads using the new forward-speed Green function presented itself again. While analyzing non-wallsided ship hulls, Chen and Wu (2001) found that, for surface panels touching the waterline, numerical inaccuracies disturb the global solution and yield numerical errors that are of the same order of magnitude as the solution itself. Computational inaccuracies could not be resolved. Therefore, for the purpose of validation of the frequency-domain technique and for specifying limits of its application, all subsequent computations for both reference ships were conducted only on the basis of the zero-speed Green function.

Ship Motions of Reference Ships

For the two reference ships considered here, Figs. 7 to 20 show comparative heave and pitch motions in regular head waves of unit amplitude plotted against wave length to ship length, λ/L_{BP}. For the high-speed ferry, these response amplitude operators (RAOs) are shown in Figs. 7 to 14 for four ship speeds, corresponding to the Froude numbers 0, 0.31, 0.47 and 0.61. For the fast monohull the RAOs are shown in Figs. 15 to 20 for three ship speeds, corresponding to the Froude numbers 0, 0.3 and 0.4.

Viscous effects reduced peak motion response and generally improved correlation of numerical predictions with experimental measurements. Empirical viscous lift and cross-flow drag coefficients depended on amplitude and frequency of oscillation and on the ship's hull form. A systematic analysis of data from forced excitation tests of both reference ships was performed to find the best fit of measured hydrodynamic masses and damping coefficients on the basis of the zero-speed Green function formulation corrected for viscous effects. This led to the average value of 0.1 for the viscous lift coefficient A_0 and values ranging from 0.1 to 0.4 for the cross-flow drag coefficient C_d in Figs. 7 to 26, respectively (Papanikolaou et al., 2000). These coefficients were used in the numerical validation study athough effects of higher speeds could not properly be accounted for, because the application of the potential-theory approach, employed was limited to ship speeds corresponding to a Froude number of up to about 0.4.

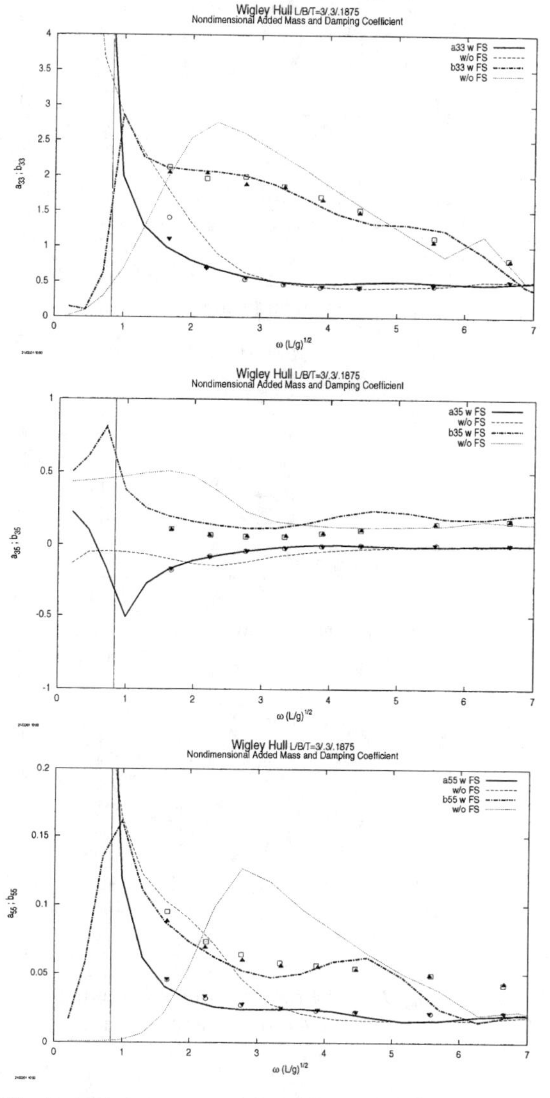

Fig. 5 Added masses and damping coefficients of Wigley hull in regular head waves

Wave-Induced Global Loads of Reference Ships

Sample comparisons of numerical and experimental wave-induced loads for the fast monohull at a section of the hull 10.4 m forward of amidships are presented in Figs. 21 to 26. Shown are vertical bending moments (VBMs) and vertical shear forces (VSFs) for the ship advancing in regular head waves at three speeds, corresponding to Froude numbers of 0, 0.2 and 0.4. These load samples are given as transfer functions, i.e., as ratios of amplitudes M_Y and F_Z to wave height H plotted against the non-dimensional circular wave frequency $\omega(L_{BP}/g)^{0.5}$.

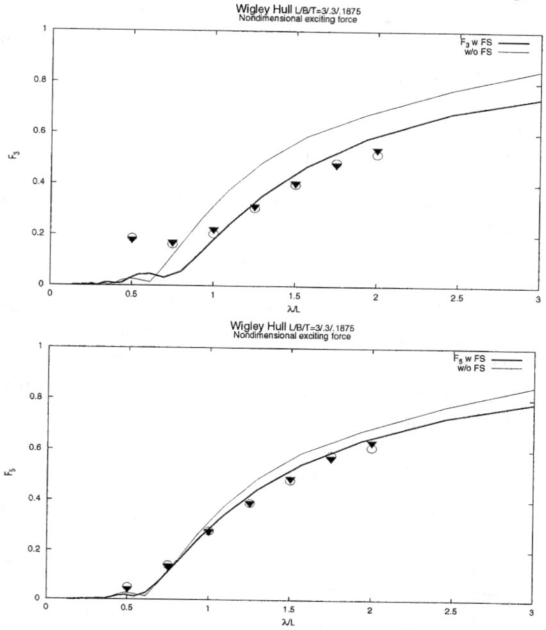

Fig. 6 Wave excitation in heave and pitch of Wigley hull in regular head waves

Sectional loads at stations near the ends of the monohull compared less favorably with measurements than sectional loads near amidships. This deteriorating quality of the results away from amidships was more pronounced for bending moments than for shear forces.

Overall agreement of computed sectional loads with measurements was most favorable for test runs in waves of small steepness. This was the case for both reference ships and for all speeds investigated. To draw attention to this finding, Fig. 27 shows, here for the monohull, plots of computed VBMs together with comparable measurements obtained only in waves of lowest steepness (H_w/L_w = 1/80). Agreement is seen to be generally favorable over the entire range of wave lengths tested, even for the high ship speed corresponding to a Froude number of 0.4. Crosses connected by solid lines identify computed (normalized) VBMs; unconnected crosses mark comparable measured data. Although not shown here, agreement was also favorable for loads at other ship sections (Schellin et al., 2000).

Although, for most cases investigated, effects of viscous corrections on wave-induced sectional loads were relatively weak, some results indicate that these effects were inconsistent, i.e., for some cases force and moment amplitudes that included viscosity were larger and for other cases smaller than (potential theory) amplitudes without these corrections. Especially at sections close to hull ends, some results corrected for viscous effects deviated close to 25 percent from potential-theory based results. However, in absolute terms, force and moment amplitudes near hull ends were generally small and, therefore, less significant from a design point of view.

Numerical Uncertainties

Concerning uncertainties of numerical predictions, the parameter that had the greatest influence on the accuracy of results is the size of surface panels used to idealize the wetted hull. Experience with standard ships (i.e., Östergaard and Schellin, 1995; Rathje et al., 2000) showed that choosing a large number of sufficiently small panels ensures adequate accuracy. For the investigated ships, it was felt to be adequate to choose the number of surface panels on the basis of previous experience.

Limits of Application

Generally, predicted ship motions as well as predicted wave-induced global loads compared favorably with measurements. An exception was some of the predictions for the high-speed ferry operating at its design speed corresponding to a Froude number (F_n) of about 0.63. These results were less satisfactory, indicating that the application of the frequency-domain technique for design practice should be restricted to moderately high-speeds corresponding to a Froude number up to about 0.4.

VALIDATION OF TIME-DOMAIN TECHNIQUE

To validate the time-domain technique against model test data and to establish limits of its applicability, systematic computations were performed for the two reference ships. This task was comprehensive not only because the extensive experimental program produced large amounts of measured data, but also because each of the analyzed conditions, taken one at a time, required separate simulations in regular waves. Additionally, numerous response components had to be reanalyzed to extract harmonic contents and to determine average positive and negative peaks. In general, the experimental data was of superior quality, with consistent results for different conditions and for all ship segments.

Hydrodynamic Response

Added mass and damping coefficients derived from forced excitation tests were largely independent of motion amplitude. As seen in Figs. 28 to 47, for three motion amplitudes, A, representative samples are plotted together with comparative numerical predictions against period of oscillation, T. Global coefficients are given in Figs. 28 to 39; coefficients for selected hull segments, in Figs. 40 to 47. Solid lines represent numerical predictions; symbols stand for results based on experimental measurements.

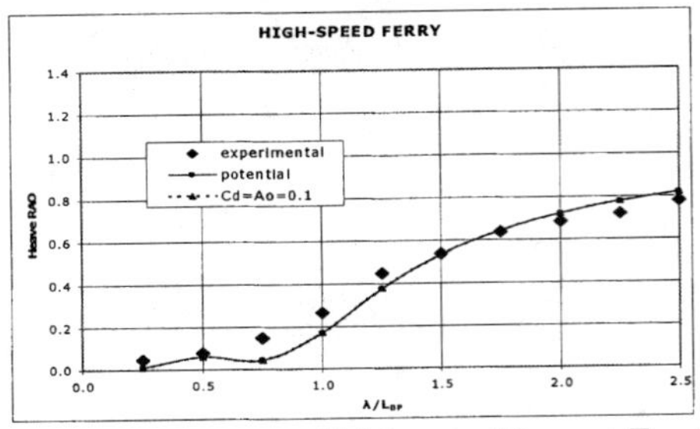

Fig. 7 Heave response of high-speed ferry at $F_n = 0$ in regular head waves

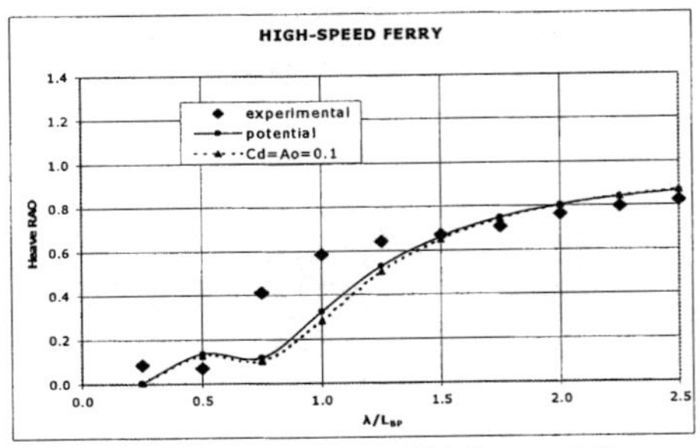

Fig. 8 Heave response of high-speed ferry at $F_n = 0.31$ in regular head waves

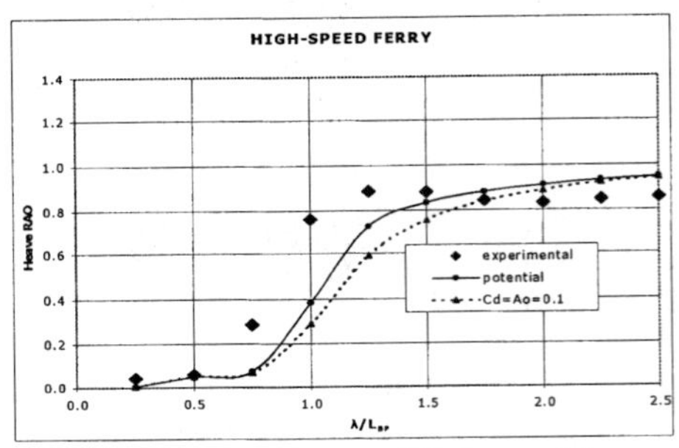

Fig. 9 Heave response of high-speed ferry at $F_n = 0.47$ in regular head waves

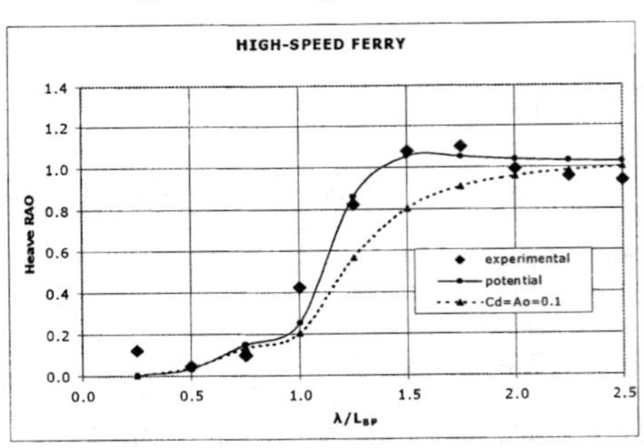

Fig. 10 Heave response of high-speed ferry at $F_n = 0.63$ in regular head waves

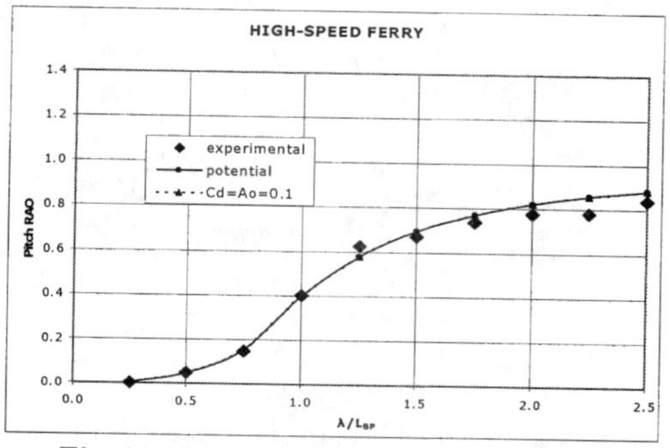

Fig. 11 Pitch response of high-speed ferry at $F_n = 0$ in regular head waves

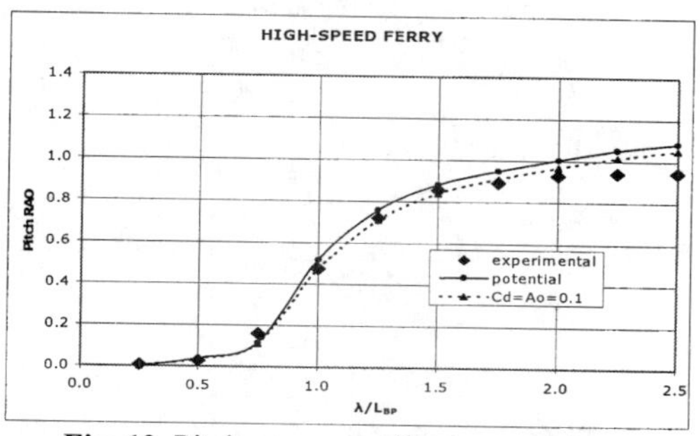

Fig. 12 Pitch response of high-speed ferry at $F_n = 0.31$ in regular head waves

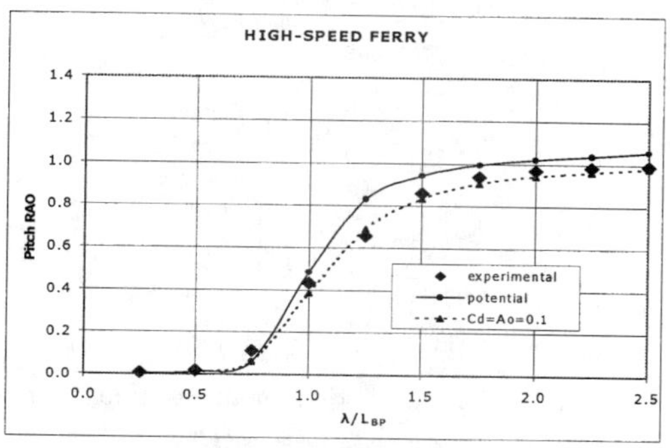

Fig. 13 Pitch response of high-speed ferry at $F_n = 0.47$ in regular head waves

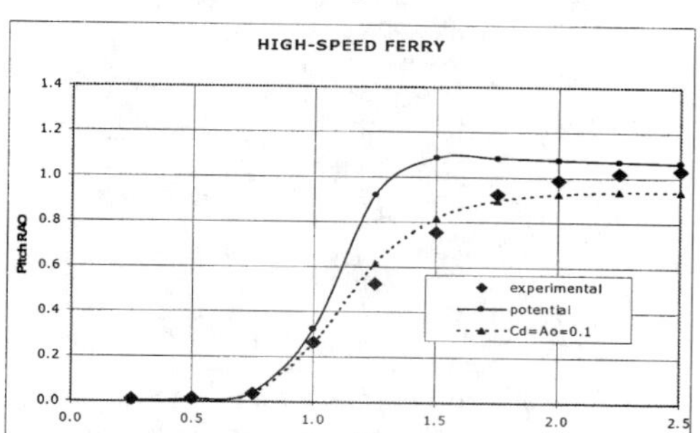

Fig. 14 Pitch response of high-speed ferry at $F_n = 0.63$ in regular head waves

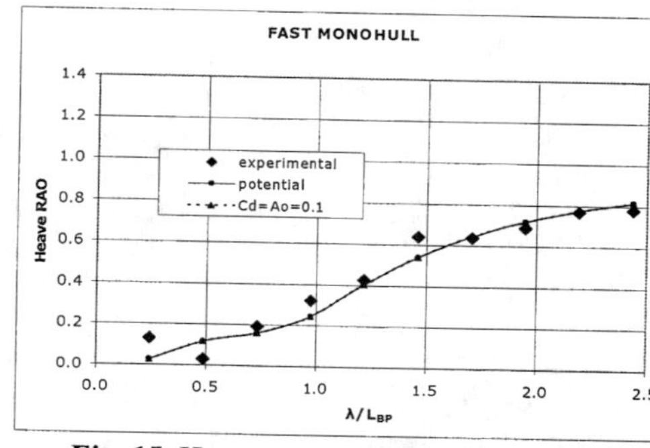

Fig. 15 Heave response of fast monohull at $F_n = 0$ in regular head waves

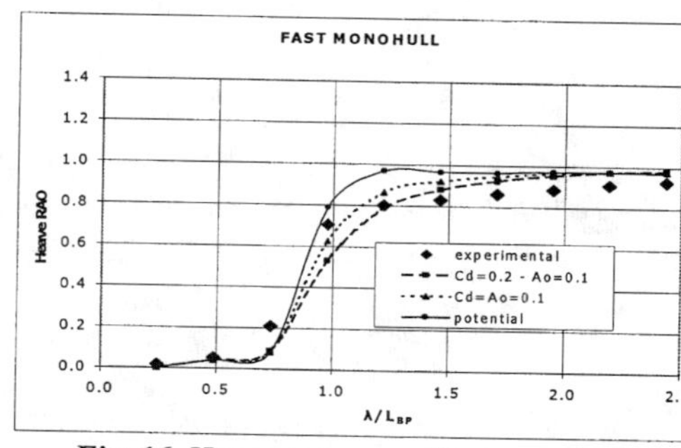

Fig. 16 Heave response of fast monohull at $F_n = 0.3$ in regular head waves

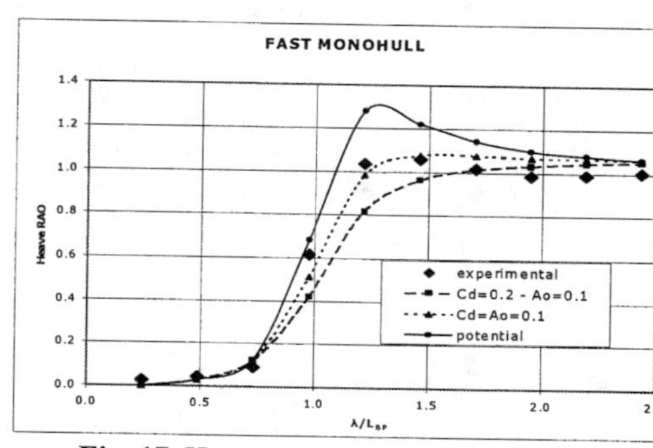

Fig. 17 Heave response of fast monohull at $F_n = 0.4$ in regular head waves

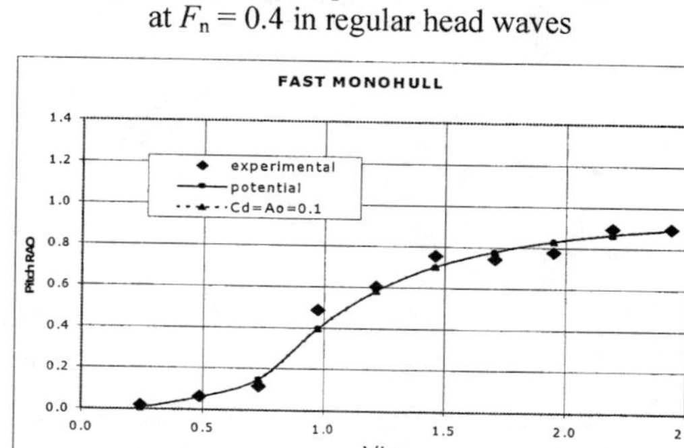

Fig. 18 Pitch response of fast monohull at $F_n = 0$ in regular head waves

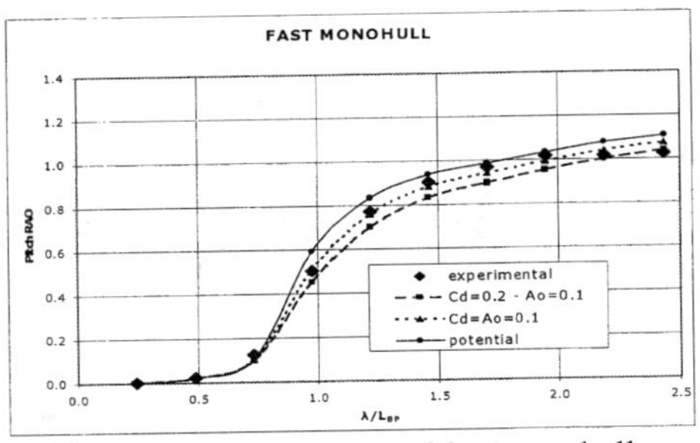

Fig. 19 Pitch response of fast monohull at $F_n = 0.3$ in regular head waves

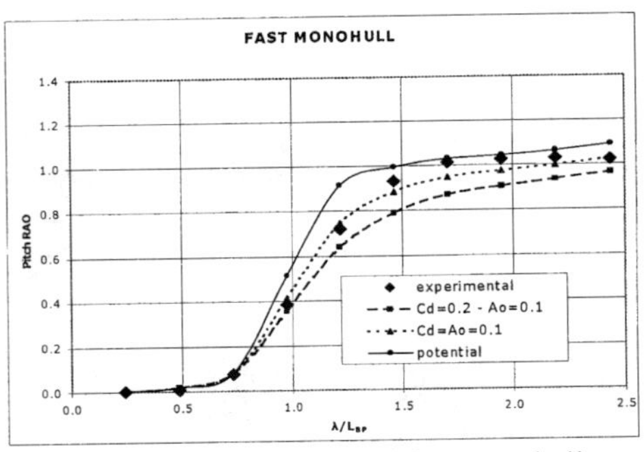

Fig. 20 Pitch response of fast monohull at $F_n = 0.4$ in regular head waves

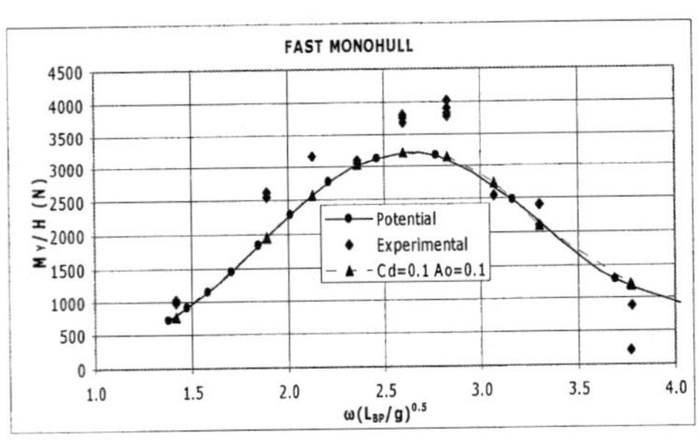

Fig. 21 Vertical bending moments of fast monohull at $F_n = 0$ in regular head waves

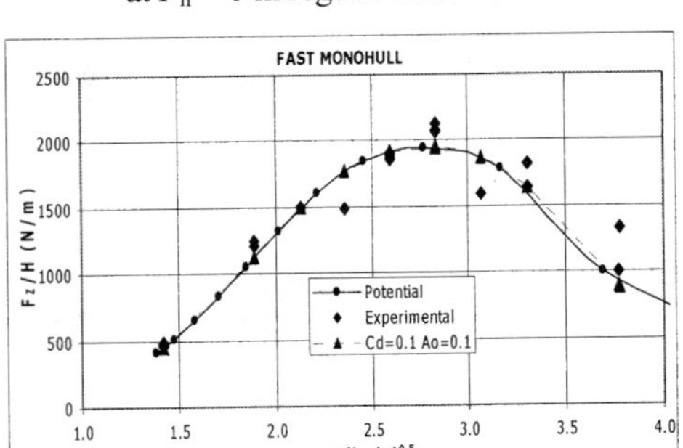

Fig. 22 Vertical shear forces of fast monohull at $F_n = 0$ in regular head waves

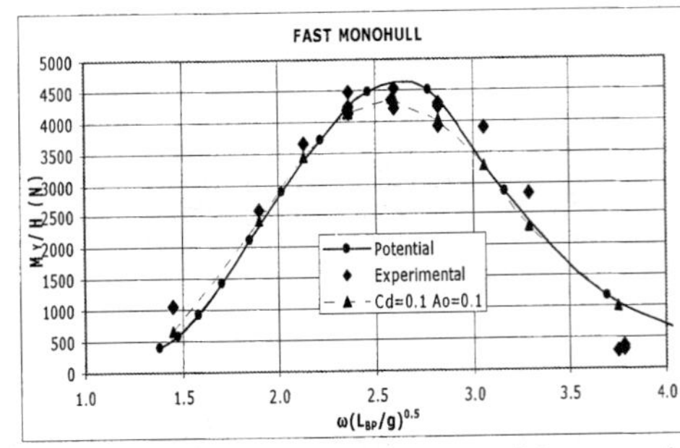

Fig. 23 Vertical bending moments of fast monohull at $F_n = 0.2$ in regular head waves

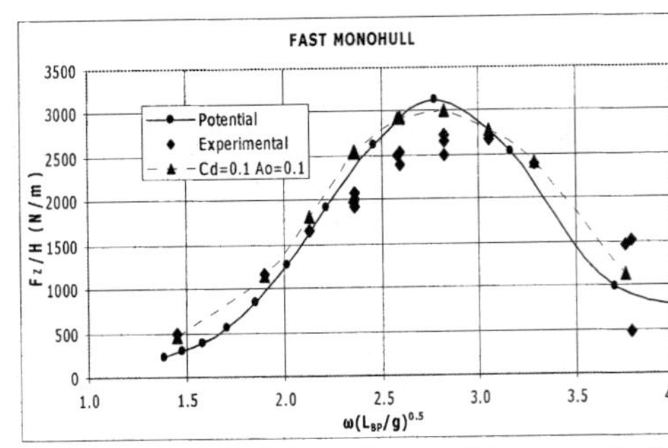

Fig. 24 Vertical shear forces of fast monohull at $F_n = 0.2$ in regular head waves

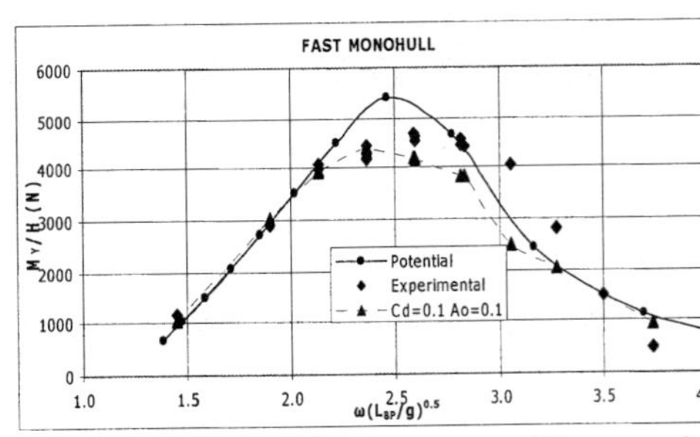

Fig. 25 Vertical bending moments of fast monohull at $F_n = 0.4$ in regular head waves

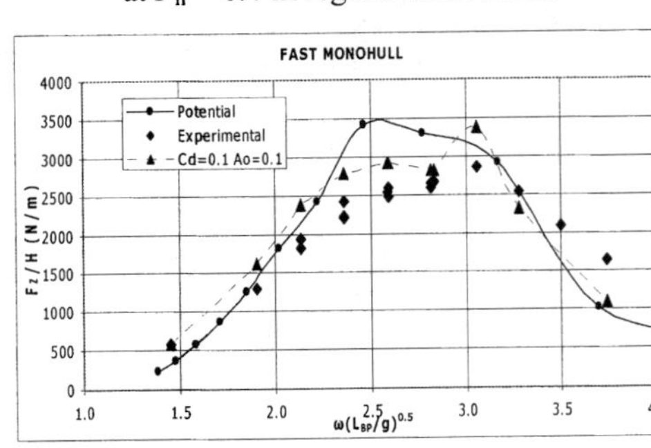

Fig. 26 Vertical shear forces of fast monohull at $F_n = 0.4$ in regular head waves

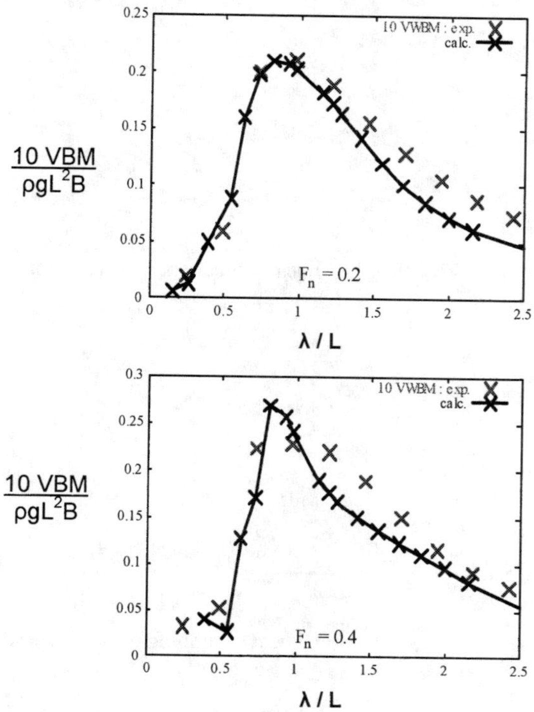

Fig. 27 Vertical bending moment of fast monohull at $F_n = 0.2$ (upper graph) and $F_n = 0.4$ (lower graph) in regular head waves of small steepness

This independence of motion amplitude was unexpected, considering the cross-sectional shape of the tested hulls and the relatively large amplitude range of the forced motions. These results justified one of the basic assumptions of the theoretical time-domain model, namely, that the major nonlinear contributions arise from hydrostatics and Froude-Krylov effects. As damping was relatively independent of the forced motion amplitudes, it may be concluded that, for the hulls tested, viscous effects are of secondary importance for the vertical motion problem.

The overall consistency of the experimental data together with the relatively favorable agreement with numerical results led to the conclusion that the experimental data is reliable and can be regarded as benchmark data. Additionally, since added mass and damping were basically linear, it was acceptable to apply a linear method to compute these response forces. Given these premises, comparison of experimental data and numerical results identified several trends. First, predictions of hydrodynamic response were generally satisfactory and compared more favorably for lower speeds and small periods, i.e., at high-speeds and long wave periods some pitch-related responses showed large deviations from experiments. Second, predictions for heave compared more favorably than predictions for pitch. Third, global response for the ship as a whole compared more favorably than response for individual ship segments, indicating that the computations were subject to inaccuracies that tended to cancel each other when integrated along the ship length to yield global loads.

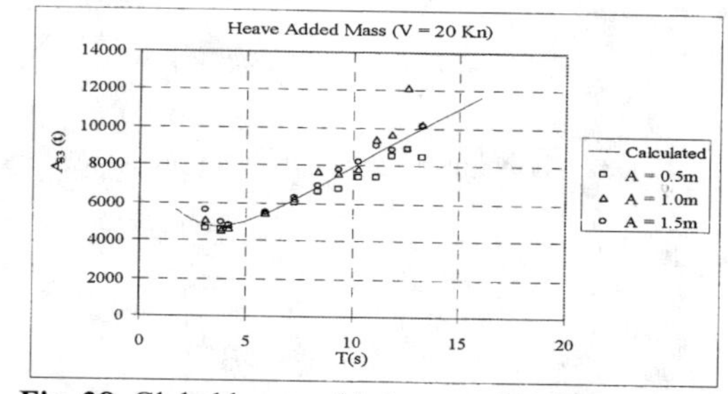

Fig. 28 Global heave added mass of high-speed ferry at $F_n = 0.31$

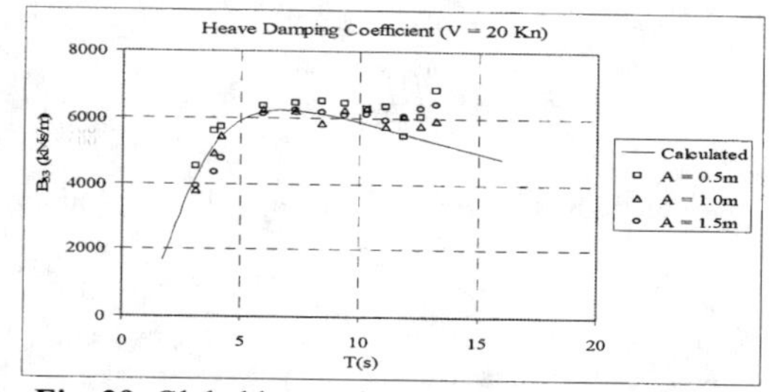

Fig. 29 Global heave damping of high-speed ferry at $F_n = 0.31$

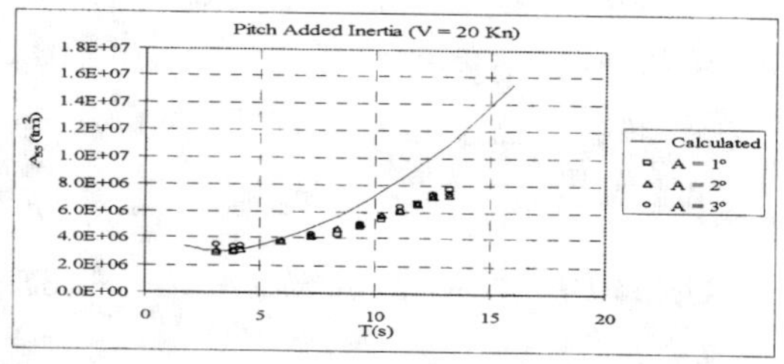

Fig. 30 Global pitch added mass of high-speed ferry at $F_n = 0.31$

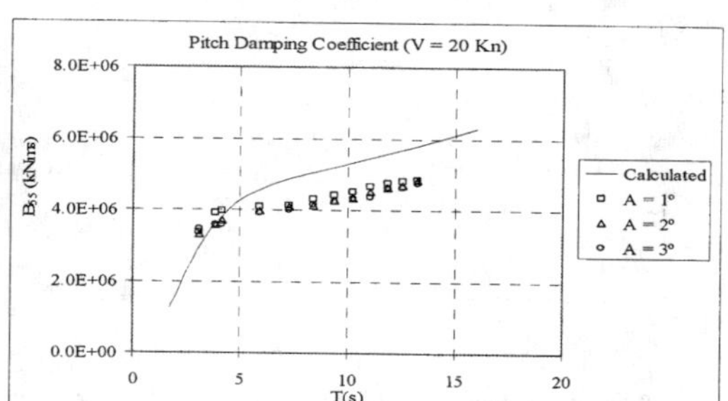

Fig. 31 Global pitch damping of high-speed ferry at $F_n = 0.31$

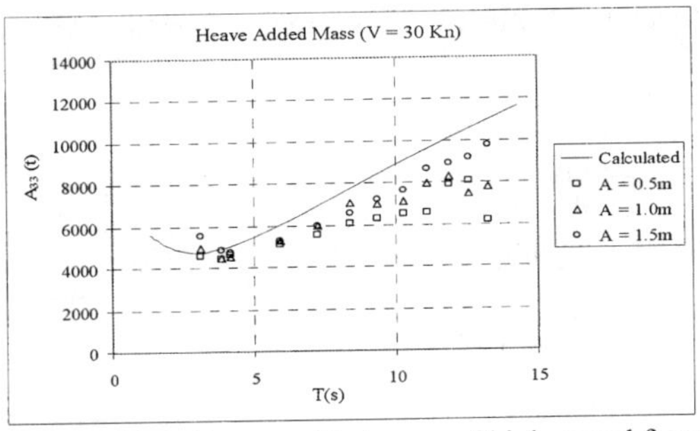

Fig. 32 Global heave added mass of high-speed ferry at $F_n = 0.47$

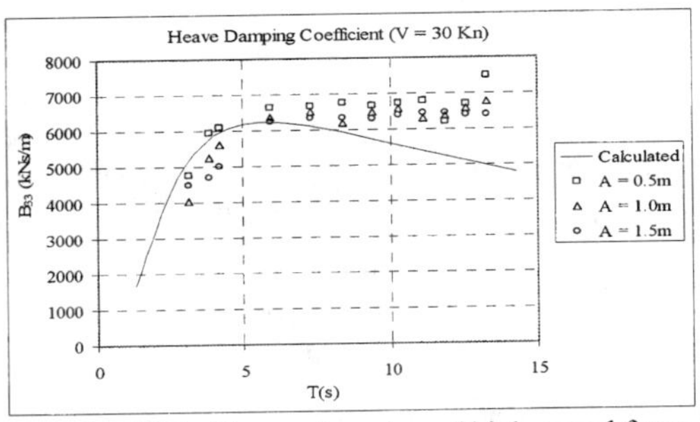

Fig. 33 Global heave damping of high-speed ferry at $F_n = 0.47$

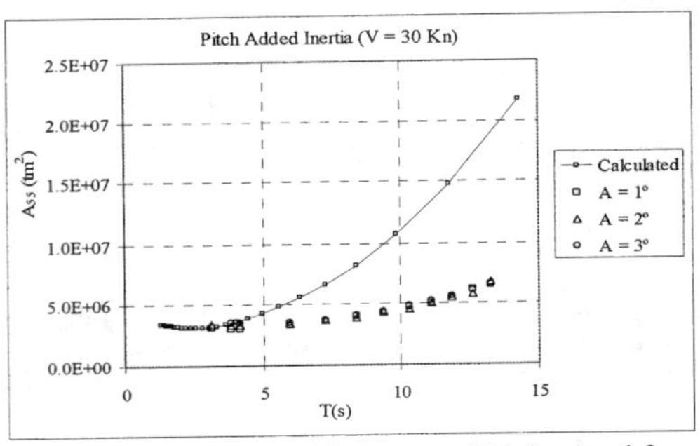

Fig. 34 Global pitch added mass of high-speed ferry at $F_n = 0.47$

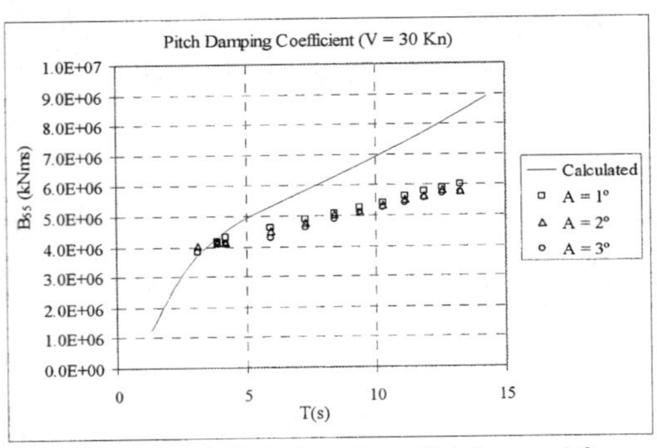

Fig. 35 Global pitch damping of high-speed ferry at $F_n = 0.47$

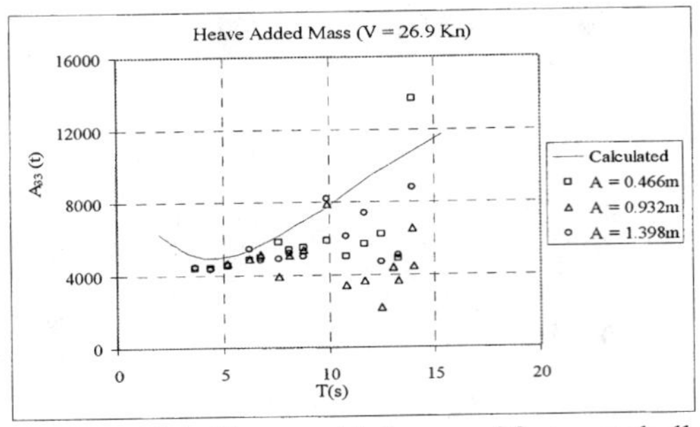

Fig. 36 Global heave added mass of fast monohull at $F_n = 0.4$

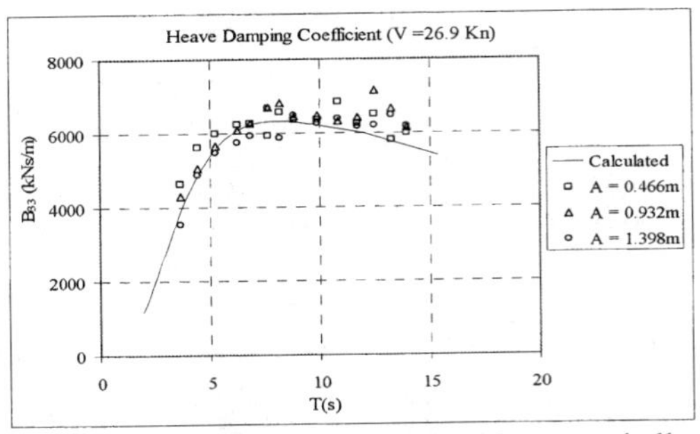

Fig. 37 Global heave damping of fast monohull at $F_n = 0.4$

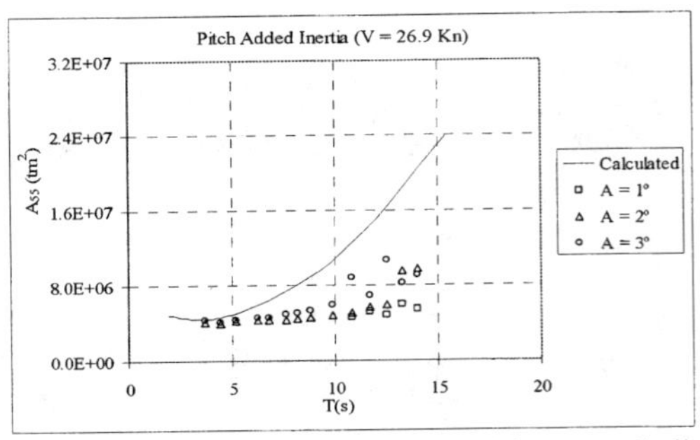

Fig. 38 Global pitch added mass of fast monohull at $F_n = 0.4$

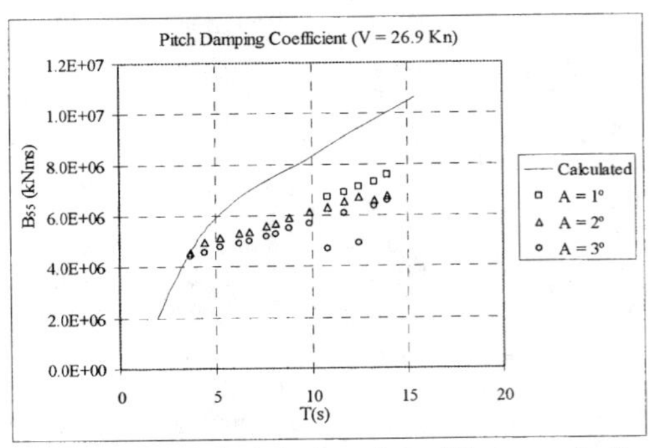

Fig. 39 Global pitch damping of fast monohull at $F_n = 0.4$

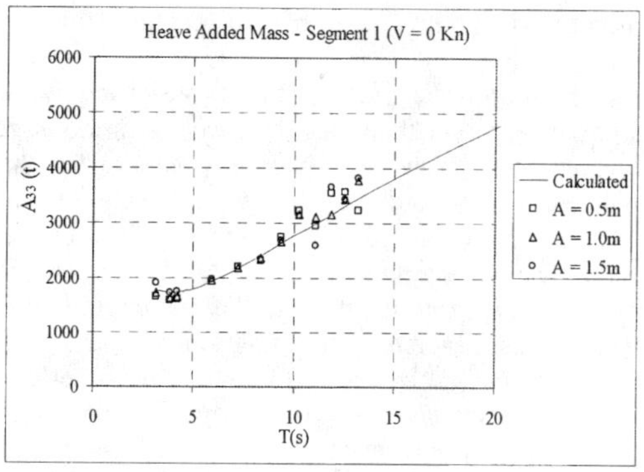

Fig. 40 Heave added masses for segment 1 of high-speed ferry at $F_n = 0$

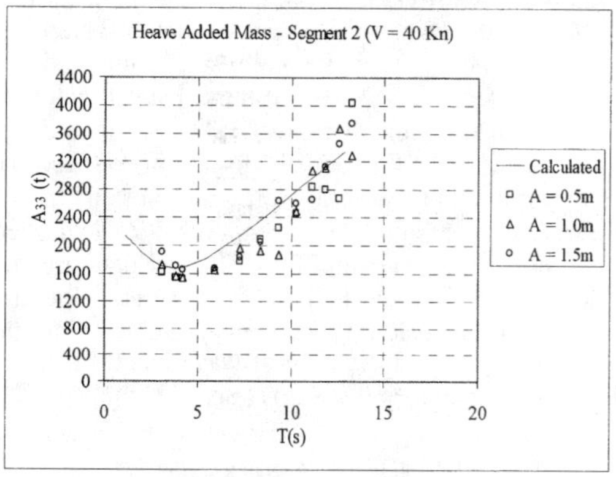

Fig. 41 Heave added mass for segment 2 of high-speed ferry at $F_n = 0.63$

Ship Motions

The time-domain technique predicted ship motions relatively well for the tested conditions although the predictions were less reliable as ship speed increased. For the high-speed ferry at zero speed and at speed of 40 knots ($F_n = 0.63$), first harmonics of heave were correctly predicted. At other speeds, predictions slightly underestimated experimental data, but only in the intermediate frequency range. High forward speed combined with low frequency of oscillation occasionally resulted in somewhat excessive pitch motions. For some cases such high pitch motions led to unstable solutions of the equations of motion for the high-speed ferry.

For the fast monohull, the time-domain technique correctly predicted first harmonics of heave and pitch motions, except around the resonance frequency, where predictions of heave slightly overestimated experimental data. Higher order effects for measured motions were small. The highest values of second harmonic motions extended up to approximately four to six percent of first harmonic motions.

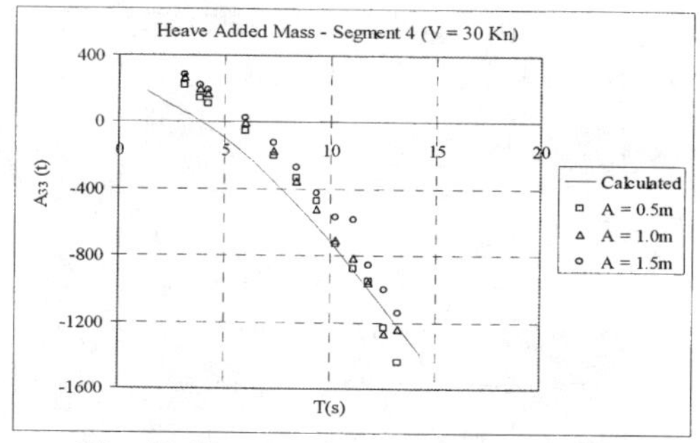

Fig. 42 Heave added mass for segment 4 of high-speed ferry at $F_n = 0.47$

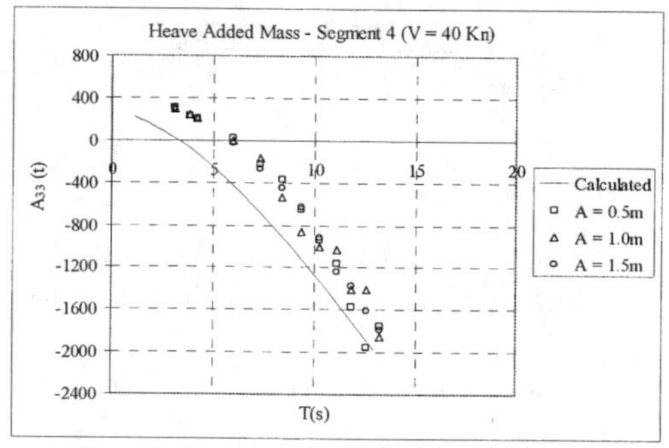

Fig. 43 Heave added mass for segment 4 of high-speed ferry at $F_n = 0.63$

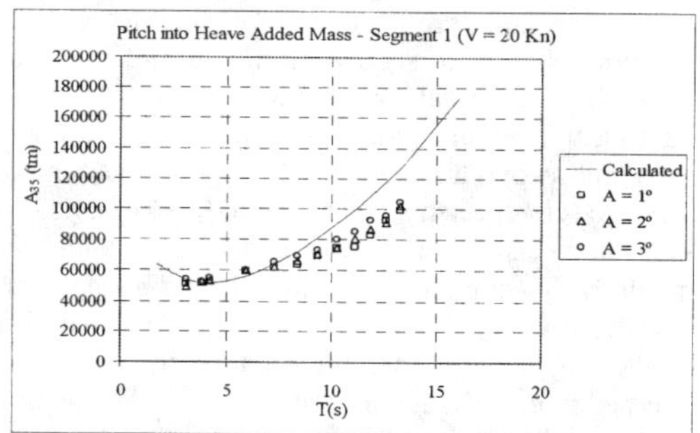

Fig. 44 Pitch into heave added mass for segment 1 of high-speed ferry at $F_n = 0.31$

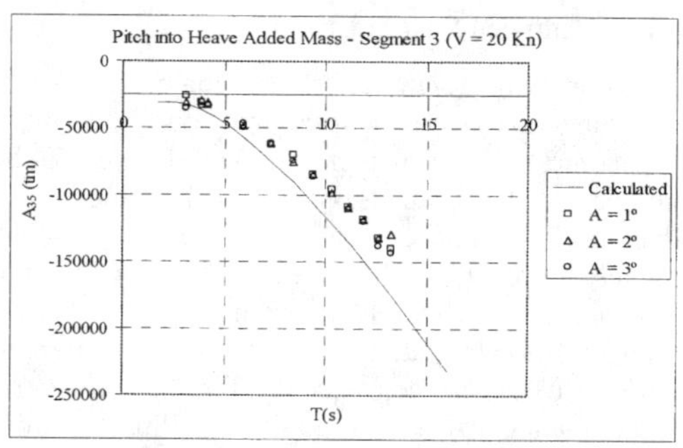

Fig. 45 Pitch into heave added mass for segment 3 of high-speed ferry at $F_n = 0.31$

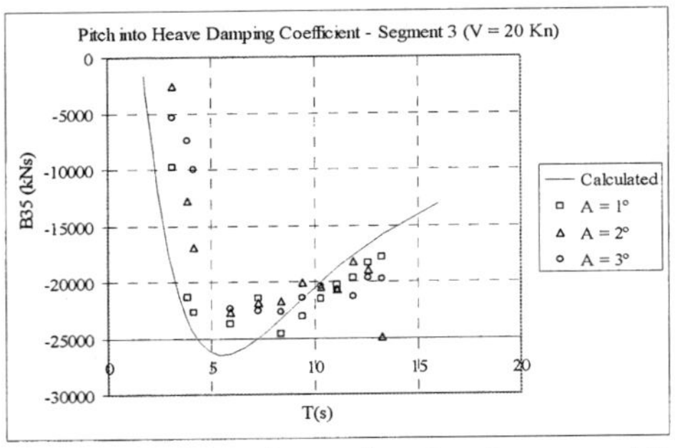

Fig. 46 Pitch into heave damping for segment 3 of high-speed ferry at $F_n = 0.31$

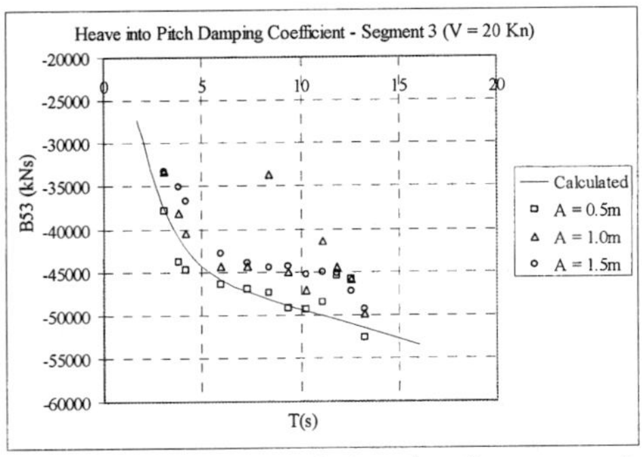

Fig. 47 Heave into pitch damping for segment 3 of high-speed ferry at $F_n = 0.31$

In general, predictions were able to represent the correct behavior of second harmonic motions, indicating that the numerical model was able to capture this nonlinear phenomenon. A selection of typical RAOs of first and second harmonics of heave and pitch motions for the two reference ships is shown in Figs. 48 to 53. A solid line connects linearly computed results; a dashed line, nonlinearly computed results. Experimental data are marked by dots. Values for the high-speed ferry are plotted against the nondimensional frequency; values for the fast monohull, against the ratio of wave length to ship length. All results shown refer to the smallest wave steepness tested ($H_w/L_w = 1/80$).

Wave-Induced Global Loads

For both reference ships, comparative numerical computations were carried out for the first three harmonics of vertical shear force (VSF), vertical bending moment (VBM), and sagging and hogging peaks. First harmonics of VSF compared relatively well with experiments; however, results were better for the fast monohull than for the high-speed ferry. In almost all cases, first harmonics of computed VBM exceeded measurements around the resonance frequency. For the high-speed ferry, experimental first harmonic amplitudes of VBM decreased with wave amplitude at larger wave slopes. This behavior was observed also for sagging peaks. The numerical model was able to capture this nonlinear characteristic.

Higher order harmonics of VSF and VBM were somewhat larger than was the case for motions, reaching values up to about ten to twenty percent of first harmonic values. The time-domain numerical technique only approximately captured this nonlinear effect. Experimental sagging and hogging peaks of VBM were characterized by a significant offset, and the magnitude of sagging peaks was larger then the magnitude of hogging peaks. This was caused by the combination of two effects. First, forward speed gave rise to a steady VBM that existed also when the ship advanced in calm water. Second, the non-vertical hull geometry (bow flare) caused nonlinear unsteady forces to act when the ship was underway. The mathematical model did not account for this first effect although the second effect was well represented.

In general, the numerical results compared well with sagging peaks of VBM, but they overestimated hogging peaks. A selection of typical RAOs of first and second harmonics of VSF and VBM for the high-speed ferry and the fast monohull are shown in Figs. 54 to 67. A solid line connects computed nonlinear results; a dashed line, computed linear results. Experimental data are shown as dots. Values for the high-speed ferry are plotted against the non-dimensional frequency; values for the fast monohull, against the ratio of wave length to ship length. The wave steepness (H_w/L_w) is noted in the legend.

Numerical Uncertainties

Uuncertainties of numerical predictions were primarily influenced by the parameters describing the hull geometry. These are the number of strips representing the hull and the number of segments for each cross section. For the investigated ships, it was felt to be adequate to choose these parameters on the basis of previous experience with similar ships. To verify the accuracy of hydrodynamic constants and functions incorporated in the time-domain method, linear time-domain solutions were systematically compared to corresponding frequency-domain solutions. Nonlinear computations were performed only after these convergence tests were satisfactorily completed.

Limits of Application and Suggested Improvements

The computational procedure used to solve the equations of motions combines linear contributions from radiation and diffraction effects and nonlinear contributions from hydrostatics and Froude-Krylov effects. The nonlinear effects resulted from the integration of pressures over the instantaneous wetted hull surface. Regarding damping, the inviscid option was used in that only potential damping was considered since the experiments demonstrated that viscous effects on vertical ship motions and associated wave-induced loads were small.

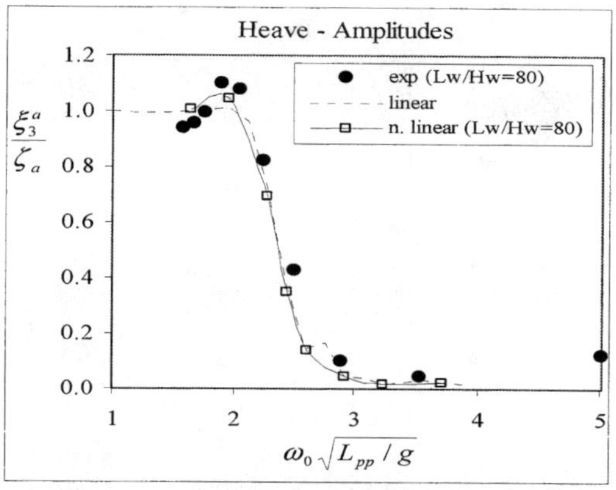

Fig. 48 First harmonic of heave for high-speed ferry at $F_n = 0.63$

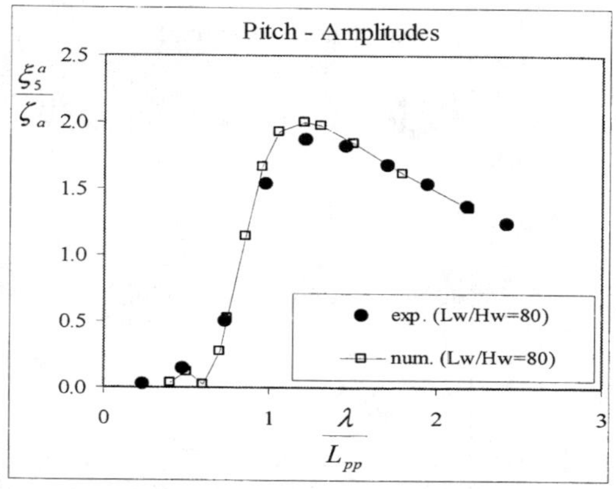

Fig. 51 First harmonic of pitch for fast monohull at $F_n = 0.3$

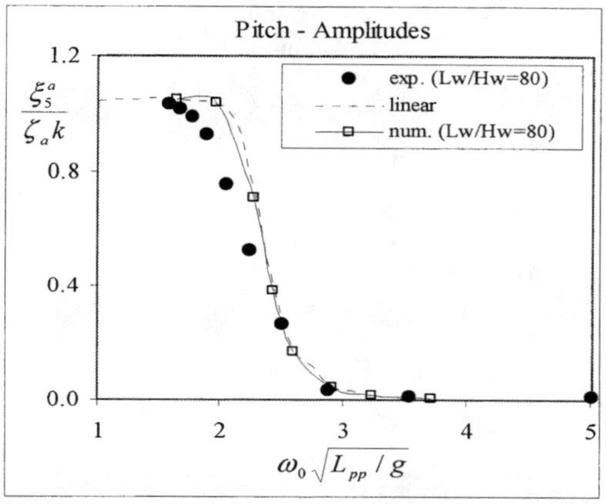

Fig. 49 First harmonic of pitch for high-speed ferry at $F_n = 0.63$

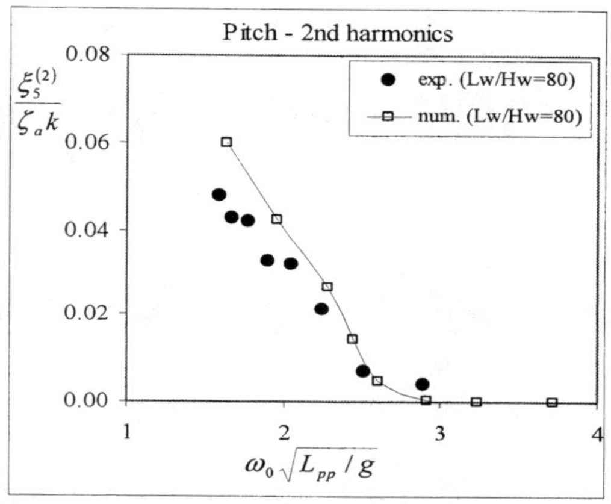

Fig. 52 Second harmonic of pitch for high-speed ferry at $F_n = 0.47$

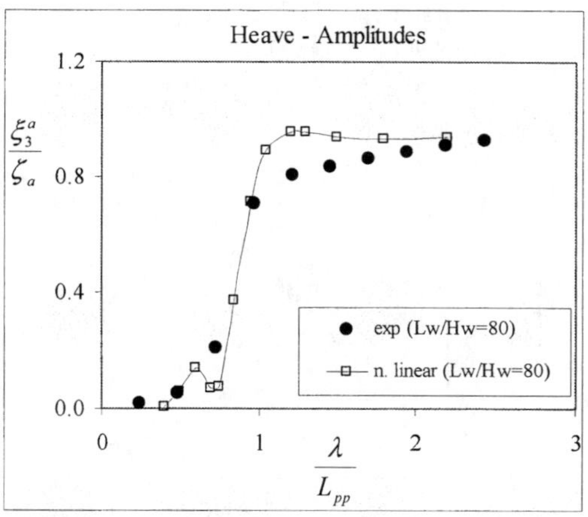

Fig. 50 First harmonic of heave for fast monohull at $F_n = 0.3$

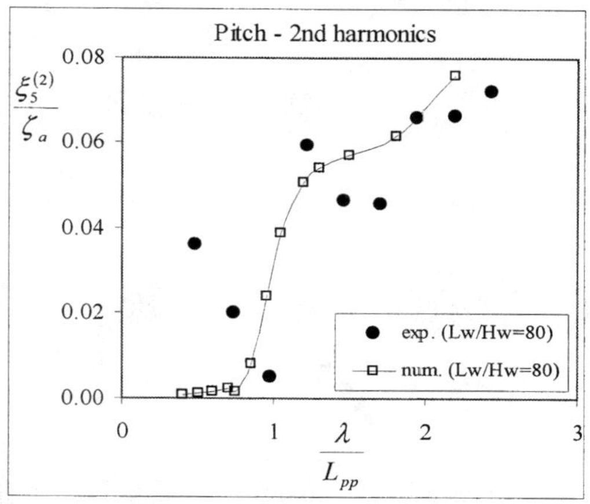

Fig. 53 Second harmonic of pitch for fast monohull at $F_n = 0.3$

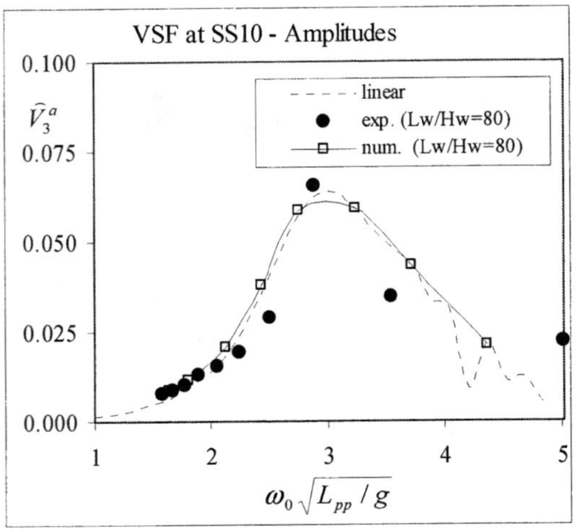

Fig. 54 First harmonic of midship VSF for high-speed ferry at $F_n = 0.47$

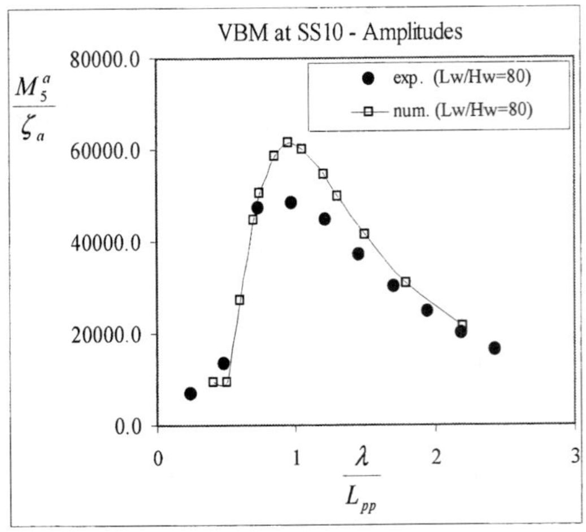

Fig. 57 First harmonic of midship VBM for fast monohull at $F_n = 0.3$

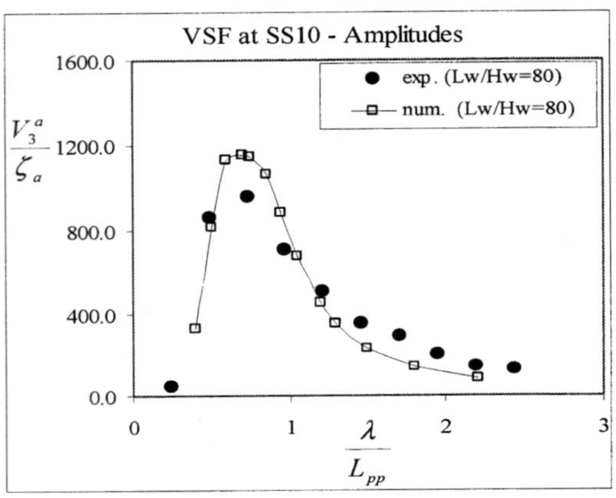

Fig. 55 First harmonic of midship VSF for fast monohull at $F_n = 0.3$

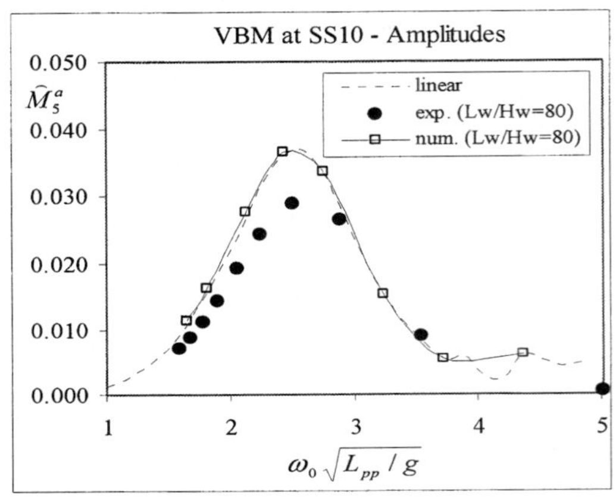

Fig. 58 First harmonic of midship VBM for high-speed ferry at $F_n = 0.47$ in waves of $H_w/L_w = 1/80$

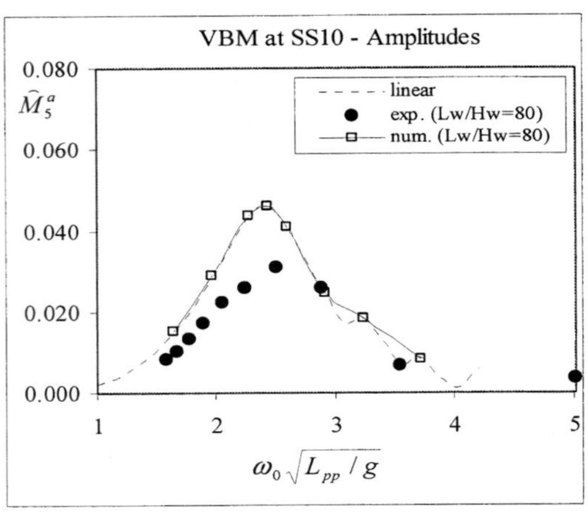

Fig. 56 First harmonic of midship VBM for high-speed ferry at $F_n = 0.47$

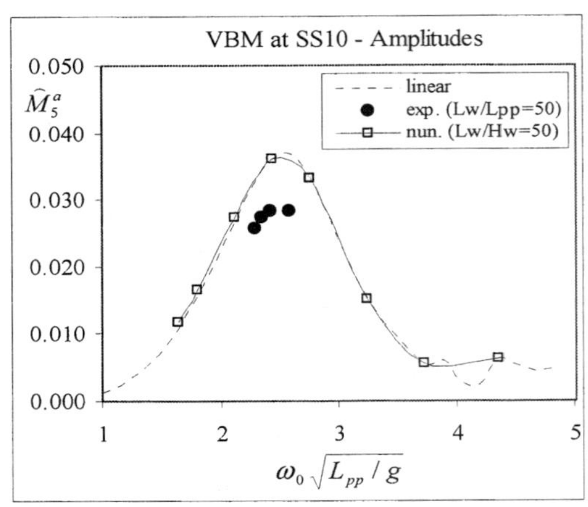

Fig. 59 First harmonic of midship VBM for high-speed ferry at $F_n = 0.47$ in waves of $H_w/L_w = 1/50$

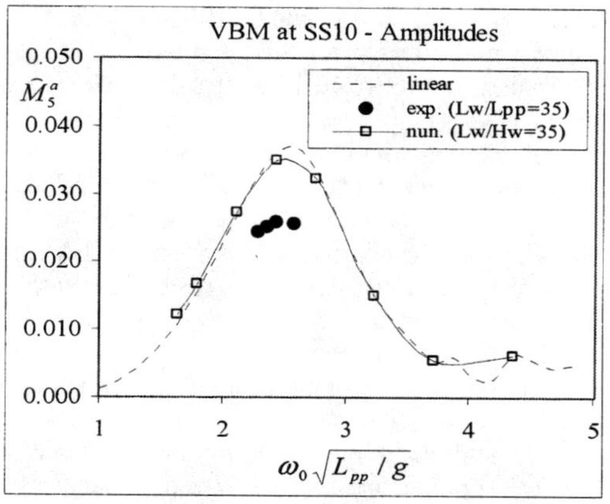

Fig. 60 First harmonic of midship VBM for high-speed ferry at $F_n = 0.47$ in waves of $H_w/L_w = 1/35$

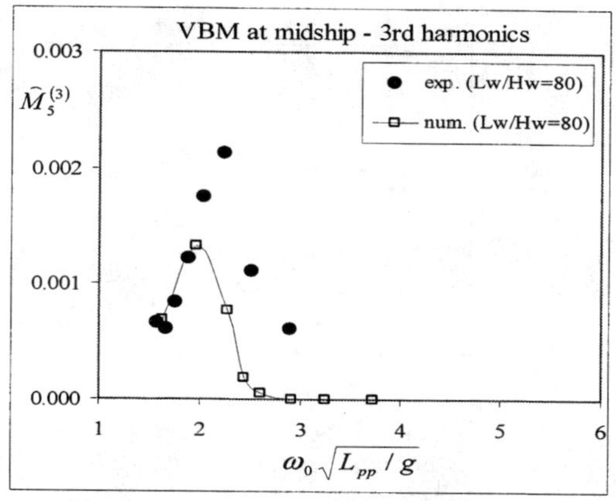

Fig. 63 Third harmonic of midship VBM for high-speed ferry at $F_n = 0.63$

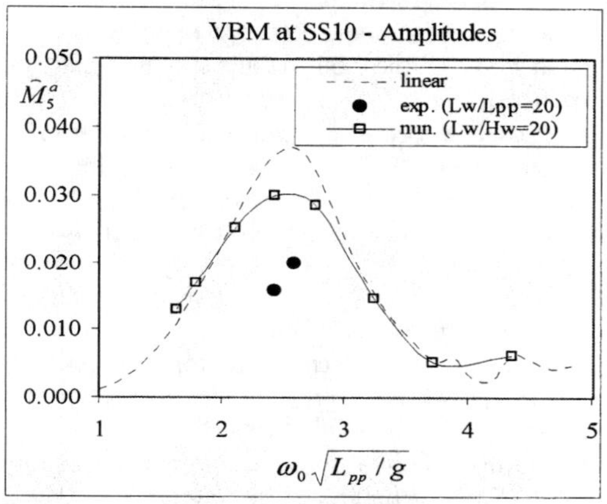

Fig. 61 First harmonic of midship VBM for high-speed ferry at $F_n = 0.47$ in waves of $H_w/L_w = 1/20$

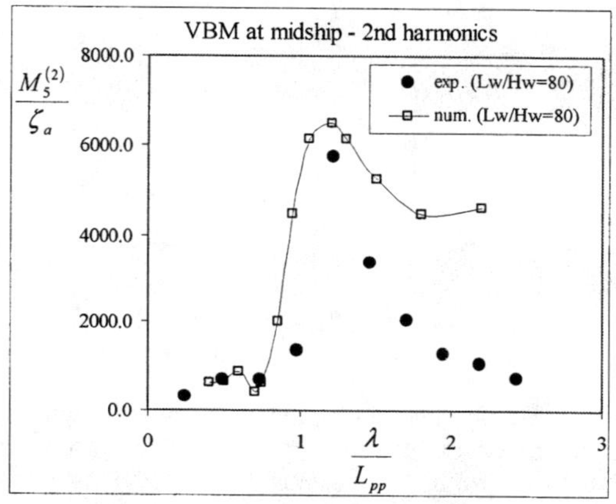

Fig. 64 Second harmonic of midship VBM for fast monohull at $F_n = 0.3$

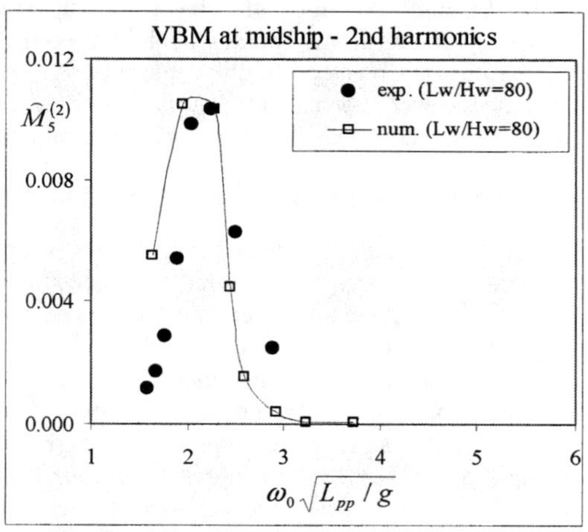

Fig. 62 Second harmonic of midship VBM for high-speed ferry at $F_n = 0.63$

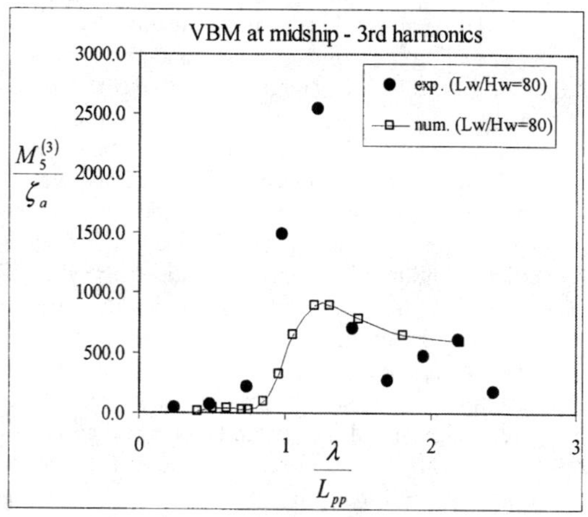

Fig. 65 Third harmonic of midship VBM for fast monohull at $F_n = 0.3$

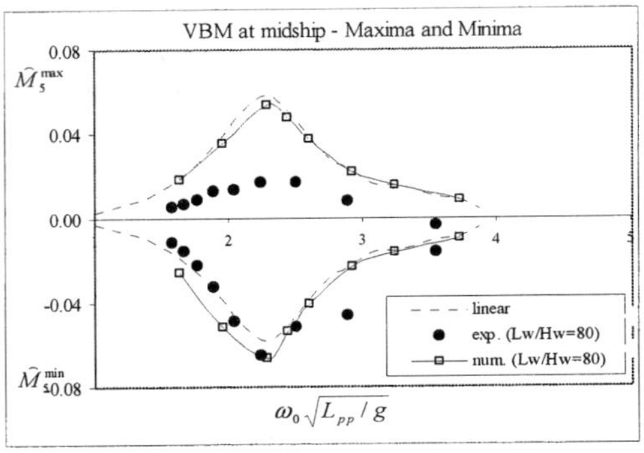

Fig. 66 Sagging and hogging peaks for high-speed ferry at $F_n = 0.63$

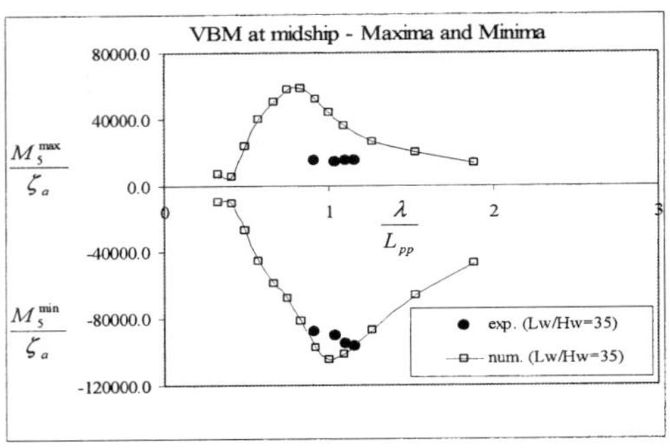

Fig. 67 Sagging and hogging peaks for fast monohull at $F_n = 0.4$

The formulation for radiation (and diffraction) forces was based on four assumptions that by themselves restrict the application of this numerical method. Accordingly, the ship must be sufficiently slender, the speed of advance should be relatively small, the frequency of oscillation must be high, and the amplitude of the motions must be small. To establish limits of application by varying the corresponding control parameters was not straightforward, because these parameters are interdependent. Thus, changing one parameter affected the range of applicability with respect to the others.

However, the following observations were made. First, the small amplitude of the forced motion assumption was not a limitation. Second, the high frequency of oscillation assumption was a limitation, especially for the computation of pitch hydrodynamic response, where predictions failed in the low frequency of oscillation range. Third, the small speed of advance assumption was a limitation because, in general, the predictions were worse as ship speed increased. High speed combined with low frequency of oscillation resulted in overestimated pitch response forces in some cases for the high-speed ferry, leading to unstable solutions of the equations of motion. Fourth, the requirement of slenderness of the hull can not be discussed alone, because its effects were also related to the frequency of oscillation and the forward speed. At the time it seemed justified to treat the ships as being slender and, with this in mind, to accept that the limitations were related to the low-speed and high-frequency assumptions.

Regarding motions and wave-induced loads, the time-domain technique predicted motions relatively well for the test conditions. However, experimental data showed significant offsets of VSF and VBM associated with forward speed effects. The mathematical model does not account for this effect and tended to overestimate vertical bending moments.

Based on the comparison of experimental and numerical results, it was possible to identify several aspects that should be improved to increase the accuracy of numerical predictions for high-speed monohull ships.

- First, forward speed effects should be more completely formulated to account for interactions between the steady flow associated with forward speed and the unsteady flow in waves. This should improve hydrodynamic response predictions in the high-frequency range and the high-speed range.
- Second, effects of dynamic sinkage and trim as well as the steady part of VSF and VBM at high speeds should be incorporated. Some of the observed discrepancies of numerical results would probably be eliminated and the accuracy improved.
- Third, exciting forces above the still waterline should not be represented solely by the Froude-Krylov components. Not accounting for diffraction/radiation effects possibly contributed to overestimated wave-induced loads in several cases.
- Forth, slamming forces are important for high-speed ships. They certainly cause relevant local loads and induce associated whipping vibration of the hull, thus contributing towards the degradation of fatigue life of the ship structure. Although the level of importance of slamming effects on maximum global loads on the ship is not clear, it is expected that the contribution is relatively small. However, not considering slamming effects is a limitation of the numerical time-domain technique.

DESIGN VBM FOR THE HIGH-SPEED FERRY

The VBM for the high-speed ferry was computed to demonstrate the application of the frequency-domain numerical technique for design purposes. As stated above, unresolved numerical inaccuracies did not allow performing the hydrodynamic analysis with a frequency-domain panel code using the newly developed software module ICship.a. However, investigations showed that computed wave-induced global loads from the frequency-domain technique using the zero-speed Green function (encounter-frequency approach) correlated favorably with model test data, albeit only for data obtained in low steepness waves and for ship speeds corresponding to a Froude number of less than about 0.4. Therefore, the resulting (linear) computations were subsequently corrected for nonlinear effects, thereby obtaining load predictions for finite amplitude waves. Also, to account for speed reduction in severe seastates, computations were performed for operation at two-third design speed, that is, at 26 knots ($F_n = 0.4$).

The panel method used 5650 flat triangular and quadrilateral surface elements to discretize the hull. The resulting RAOs of midship VBMs, after correction for nonlinear effects, consisted of two sets of pseudo RAOs. One set corresponded to the hogging condition; the other set, to the sagging condition. Each RAO depended on wave amplitude. As an example, Fig. 68 depicts such RAOs of wave-induced midship VBM for the high-speed ferry in 4.5 m amplitude head waves. Standard stochastic analysis methods, commonly used for linear systems, yielded long-term distributions of wave-induced midship VBMs. The wave directionality, averaged over all annual seasons, was assumed to be equally distributed, and a cosine squared distribution for the directions of wave energy specified the wave's short-crestedness. The Pierson-Moskowitz spectrum described the spectral density of all short-term seastates comprising the wave climate. The wave climate of the western part of the Mediterranean Sea was assumed to be valid, with the ship operating under an OC3 service restriction that limits the operational significant wave height to 4.0 m.

Wave-induced design VBMs were here defined by their long-term values at a probability of exceedence of 2.3×10^{-8}, a level corresponding to an estimated time at sea of about six years during the ship's life. The design midship VBMs turned out to be – 283 MNm in sagging and 268 MN in hogging. They deviated by five percent and thirteen percent from respective classification society rule values of – 296 MNm and 236 MNm according to EEIG UNITAS (1998).

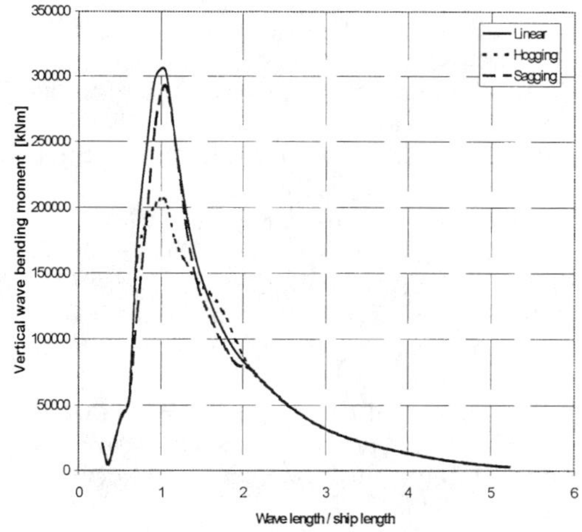

Fig. 68 Sample RAOs of midship VBM for high-speed ferry in 4.5 m head waves

CONCLUDING REMARKS

The use of a linear frequency-domain numerical technique and a nonlinear time-domain numerical technique was investigated for the purpose of predicting wave-induced global loads for modern, single hull fast ships. The emphasis was directed towards accounting for effects of high forward speed and significant nonlinearities. Systematic model tests were performed, yielding experimental data required to validate the numerical methods and to specify their limits of application. Two modern designs were tested, namely, a high-speed ferry and round-bilge fast monohull.

For the frequency-domain technique, an advanced software module was developed. This module, when incorporated into an appropriate frequency-domain computer code, evaluated influence coefficients according to the Fourier-Kochin formulation. This theory provided a solid mathematical basis for obtaining robust and accurate numerical predictions of wave loads and ship motions within the framework of a three-dimensional linear frequency-domain analysis while fully accounting for forward speed effects.

For a slender Wigley hull, implementation of this module led to favorable agreement between numerical predictions and model test data. However, this was impossible to achieve for the two reference ships, because numerical inaccuracies could not be resolved and computer running times were too long for practical design applications. Therefore, subsequent computations were performed according to the so-called encounter-frequency approach, using panel codes that relied on the zero-speed Green function. For ship speeds corresponding to Froude numbers of about 0.4, this produced accurate predictions, albeit only in waves of small steepness. Therefore, to obtain long-term design values from a statistical analysis, the (linearly) computed loads had to be corrected for nonlinear effects, yielding so-called pseudo transfer functions that depend on wave amplitude.

The developed time-domain technique accounted for high levels of nonlinearity, including effects of large body motions as well as viscosity. This technique was complimentary to the frequency-domain technique as it covered conditions where the frequency-domain technique was not valid. Results from the time-domain technique generally compared favorably with model test measurements. However, some uncertainties were encountered at high Froude numbers and in following seas.

The numerical procedure of the time-domain method was simpler and the computational effort smaller than the frequency-domain panel method. This can be advantageous for practical application although the method is restricted to slender ships. As modern fast ships are generally slender, this restriction is of minor practical relevance.

The large number of systematic model tests at various ship speeds in waves of different steepness comprised not only seakeeping tests, but also forced excitation tests. A systematic analysis of the test data yielded hydrodynamic response coefficients as well as viscous lift and cross-flow drag coefficients. As tests were carried out with segmented models, they provided experimental data not only for the ships as a whole, but also for individual ship sections.

To carry out the model tests, new instrumentation had to be developed, such as a six component dynamometer small enough for installation in ship models, an improved signal conditioning system, and a forced motion excitation device for added mass and damping tests. A shear force measuring method integrated into the thin walled aluminum backbone was also developed.

ACKNOWLEDGEMENTS

The presented work was performed within research project BE 97-4406, Advanced Method to Predict Wave Induced Loads for High-speed Ships (WAVELOADS). This project was partially funded by the Commission of the European Community under the BRITE/EURAM program, Contract BRPR-CT97-0580. The following institutions participated in the project: Germanischer Lloyd (project manager); Instituto Superior Tecnico, Technical University of Lisbon; Bureau Veritas; Canal de Experiencias Hidrodinamicas de El Pardo, Madrid; Ship Design Laboratory, National Technical University of Athens; and Direccion de Innovacion y Tecnologia, IZAR Construcciones Navales, S.A. Opinions expressed herein are those of the authors.

REFERENCES

BA, M. and GUILBAUD, M. (1995) "A Fast Method of Evaluation for the Translating and Pulsating Green's Function," *J. Ship Technology Research*, 42:60-80.

BEUKELMAN, W. (1980) "Added Resistance and Vertical Hydrodynamic Coefficients of Oscillating Cylinders at Speed," Technical University Delft, Report No. 510.

BEUKELMAN, W. (1983) "Vertical Motions and Added Resistance of a Rectangular and Triangular Cylinder in Waves," Technical University Delft, Report No. 594.

CHEN, X.-B. and NOBLESSE, F. (1998) "Super Green Functions," *Proc. 22nd Symp. on Naval Hydrodynamics*, Washington, D.C., National Academy Press, Washington, D.C., USA.

CHEN, X.-B. and MALENICA, S. (1998) "Interaction Effects of Local Steady Flow on Wave Diffraction-Radiation at Low Forward Speed." *J. Offshore and Polar Engineering*, 8:102-109.

CHEN, X.-B. (1999) "An Introductory Treatise on Ship Motion Green Functions." *Proc. 7th Int. Conf. on Numerical Ship Hydrodynamics*, Nantes, France.

CHEN, X.-B., DIEBOLD, L., and DOUTRELEAU, Y. (2000) "New Green Function Method to Predict Wave-Induced Ship Motions and Loads." *Proc. 23rd Symp. on Naval Hydrodynamics*, Val de Reuil, France, National Academy Press, Washington, D.C.

CHEN, X.-B. and WU, G.X. (2001) "On Singular Highly Oscillating Properties of the Green Function for Ship Motions." *J. Fluid Mechanics*, 445:77-91.

EEIG UNITAS. (1998) Rules for Classification and Construction, I – Ship Technology, Part 1 – Seagoing Ships, Chapter 5 – High-Speed Craft. Germanischer Lloyd, Hamburg.

FONSECA, N. and GUEDES SOARES, C. (1998a) "Time-Domain Analysis of Large-Amplitude Vertical Motions and Wave Loads." *J. Ship Research*, 42(2):100-113.

FONSECA, N. and GUEDES SOARES, C. (1998b) "Nonlinear Wave-Induced Responses of Ships in Irregular Seas." *Proc. 12th Int. Con. on Offshore Mechanics and Arctic Engineering (OMAE 1998)*, ASME, New York, Paper 98-0446.

FONSECA, N. and GUEDES SOARES, C. (2000) "Viscous Effects on the Vertical Motions of Ships in Waves." *Proc. 7th National Conf. on Naval Architecture and Marine Engineering*, Tech. Univ. Lisbon, Lisbon, 1-2 March 2000.

FONSECA, N. and GUEDES SOARES, C. (2002) "Comparison of Numerical and Experimental Results of Nonlinear Wave-Induced Vertical Ship Motions and Loads." *J. Marine Science and Technology*, 6:193-204.

GUEDES SOARES, C. and SCHELLIN, T.E. (1996) "Long-Term Distribution of Nonlinear Wave-Induced Vertical Bending Moments on a Containership." *J. Marine Structures*, 9:333-352.

HACHMANN, D. (1986) "Determination of Wave Elevation at Ship Sections Based on Pressure Variations at Design Waterline under the Influence of the Smith Effect." Report MTK 325 II, Germanischer Lloyd, Hamburg (in German).

IWASHITA, H. and OHKUSU, M. (1992) "The Green Function Method for Ship Motions at Forward Speed." *J. Ship Technology Research*, 39:3-21.

JOURNEE, J. M. J. (1992) "Experiments and Calculations on Four Wigley Hull Forms." Technical University Delft, Report No. 909, Delft.

NOBLESSE, F. and CHEN, X.-B. (1995) "Decomposition of Free Surface Effects into Wave and Near Field Components." *J. Ship Technology Research*, 42(2):167-186.

ÖSTERGAARD, C. and SCHELLIN, T.E. (1995) "Development of an Hydrodynamic Panel Method for Practical Analysis of Ships in a Seaway." *Trans.*

Schiffbautechnische Gesellschaft, 89:561-576 (in German).

PAPANIKOLAOU, A.D. and SCHELLIN, T.E. (1992) "A Three-Dimensional Panel Method for Motions and Loads of Ships with Forward Speed." *J. Ship Technology Research*, 39(4):147-156.

PAPANIKOLAOU, A.D., ZARAPHONITIS, G., MARON, A. and KARAYANNIS, T. (2000) "Viscous Damping on Vertical Plane Motions of Ships with Forward Speed in Waves." *Proc. 4th Int. Coll. on Seakeeping Performance of Ships*, Cosmosquare International Education and Training Center, Osaka, pp. 17-22.

RATHJE, H. and SCHELLIN, T.E. (1997) "Enhancement of Wave-Induced Load Effects for Ships in Abnormal Conditions." *Proc. Int. Conf. on Design and Operation for Abnormal Conditions*, Paper No. 18, RINA, Glasgow.

RATHJE, H., SCHELLIN, T.E., OTTO, S. and ÖSTERGAARD, C. (2000) "Predicting Nonlinear Wave-Induced Design Loads for Ships." *Proc. 19th Int. Symp. on Offshore Mechanics and Arctic Engineering (OMAE 2000)*, New Orleans, New York: ASME, 2000, Paper OMAE 00-6122.

SCHELLIN, T.E., BEIERSDORF, C., CHEN, X.-B. and MARON, A. (2002) "Comparative Frequency-Domain Seakeeping Analysis of a Fast Monohull in Regular Head Waves." *Proc. 21st Int. Conf. on Offshore Mechanics and Arctic Engineering (OMAE 2002)*, Oslo 2002, ASME, New York, Paper OMAE 02-28502.

THWAITES, B. (1960) *Incompressible Aerodynamics*. Oxford University Press.

URSELL, F. (1960) "On Kelvin's Ship-Wave Pattern." J. Fluid Mechanics, 8:418-431.

WATANABE I., KENO, M. and SAWADA, H. (1989) "Effect of Bow Flare on Shape to Wave Loads of a Containership." *J. Society of Naval Architects of Japan*, 166:259–266.

WAVELOADS (2001) "Advanced Method to Predict Wave Induced Loads for High-Speed Ships." Final Technical Report, July 2001, BRITE / EURAM III, Contract BRPR – CT 97 0580.

WU, G.X. (1995) "Radiation and Diffraction by a Submerged Sphere Advancing in Water Waves of Finite Depth." *Proc. Royal Society*, London, A. 448:29-45.

Time to Flood Analysis For the Fishing Vessel *ARCTIC ROSE*

Bruce Johnson, Fellow, U. S. Naval Academy and **LT George Borlase, Student Member,** U.S. Coast Guard Marine Safety Center

ABSTRACT

This paper outlines the hydrodynamic forensic analysis performed for the U.S. Coast Guard Marine Safety Center's analysis of the sinking of the F/V ARCTIC ROSE. An intact and dynamic stability analysis of the ARCTIC ROSE in the reported wave conditions in that area of the Bering Sea on April 2, 2001 demonstrate that the vessel most likely sank due to flooding through an open WT Door. A time-domain analysis of various possible progressive flooding scenarios was developed in spreadsheet format to determine how quickly the ARCTIC ROSE could capsize using the assumptions for various ship motions scenarios. Using an iterative time-stepping approach to calculate the static equilibrium righting arms for increasing flooding loads, the righting arms for each time step were calculated based on the vessel's position on the wave, flooding water height and location, and the free surface effect of the flooding water. The amount of water flooding from one space to another was calculated based on the difference in water height between spaces at open doors or hatches. The results show the ARCTIC ROSE most likely capsized between 1 minute 30 seconds and 2 minutes 40 seconds after progressive flooding began, due largely to the free surface effect of the water trapped inside the vessel. The fully pocketed free surface analysis methodology developed for this investigation is applicable for rectangular spaces in all vessels and has already been used to write an improved stability letter for a fishing vessel.

NOMENCLATURE (Mathematical Symbols are defined in the Appendix where they are used.)

FMBI	**Formal Marine Board of Investigation**
FS	**Free Surface Effect**
F/V	**Fishing Vessel**
GHS	**General Hydrostatics Program**
GM	**Metacentric Height**
IMO	**International Maritime Organization**
KG	**Distance from keel to center of gravity**
LT	**Long Ton**
MAIB	**Marine Accident Investigation Board (UK)**
MSC	**Marine Safety Center (USCG)**
NMFS	**National Maritime Fisheries Service**
RA	**Righting Arm**
SSPA	**Swedish Hydrodynamics Laboratory**
USCG	**U. S. Coast Guard**
WT	**Watertight**

Figure 1 Bow view of the ARCTIC ROSE

1. INTRODUCTION AND BACKGROUND

At approximately 3:30 a.m. on April 2, 2001, the 92 foot (waterline length) F/V *ARCTIC ROSE* (Figure 1) disappeared in the Bering Sea, approximately 200 miles west of St. Paul Island, killing all fifteen men onboard. No Mayday calls were heard, nor were any distress signals sighted. The U. S. Coast Guard convened a Formal Marine Board of Investigation (FMBI) to determine what happened to the *ARCTIC ROSE*, why it happened, and how casualties like this could be prevented in the future.

At the time of the loss of the F/V *ARCTIC ROSE*, a low pressure system was moving through the area, and an occluded front moved over the last known position of the vessel. The reported on-scene weather at the time of the casualty was winds at 20 knots from 090° true, seas 6-8 feet also from 090° true, and a swell likely from the southwest, which would result in confused seas. Based on a hindcast created by the National Weather Service Forecast Office in Anchorage, AK and provided to the FMBI, the maximum significant wave height of the seas was estimated at 24 feet and a wave period between 8 and 12 seconds. Additionally, the winds are estimated to have been from the southeast at a maximum of 45 knots.

Because of the location of the *ARCTIC ROSE* in the Bering Sea, the U.S. Coast Guard was unable to arrive on scene until over 3 hours after the vessel's Emergency Positioning Indicator Radio Beacon (EPIRB) began broadcasting its alarm. When the search aircraft and nearby fishing vessels approached the last-known location of the vessel, the vessel had already disappeared

Using side-scan sonar and remote-operated underwater vehicles, the Formal Marine Board of Investigation found the *ARCTIC ROSE* in 422 feet of water. The vessel was resting upright on her keel and starboard bilge keel. The vessel appeared to be intact, with no signs of external damage to the hull or superstructure, and the windows in the pilothouse were all intact. The video pictures indicated that the aft door entrance to the processing room appeared to be open as was frequently the case for normal operations. Estimates of how the vessel flooded helped pinpoint which openings in the hull and superstructure were critical in allowing flooding water into the hull, and helped determine if human error might have been responsible for the sinking. Conversely, if analysis showed flooding of certain compartment combinations would prevent the vessel from settling on her keel, then the investigators could at least determine how the vessel did not sink.

The MSC provided naval architecture support to the FMBI throughout their investigation. The MSC developed and evaluated 19 scenarios (see Appendix A) that could have potentially led to the loss of the *ARCTIC ROSE*, and used a variety of tools to evaluate the likelihood of each scenario, including static righting arm calculations, dynamic stability calculations to evaluate capsize resistance, and progressive flooding calculations to evaluate time to capsize and time to sink. The details of both the MSC analysis and the new analysis are shown in Appendices A-D. According to the analysis, the most likely reason the *ARCTIC ROSE* sank was progressive flooding from the aft weather deck to the processing space through an open door connecting the two areas, and the flooding then progressed to the mudroom/galley, fish hold, and engine room (See Figures 2 and 3) through non-watertight doors and hatches, all on the starboard side as shown in Figure 4. This sketch, showing the general arrangement of the main deck and below deck of the *ARCTIC ROSE*, was provided to the FMBI by Jensen Maritime Consultants, Inc. .

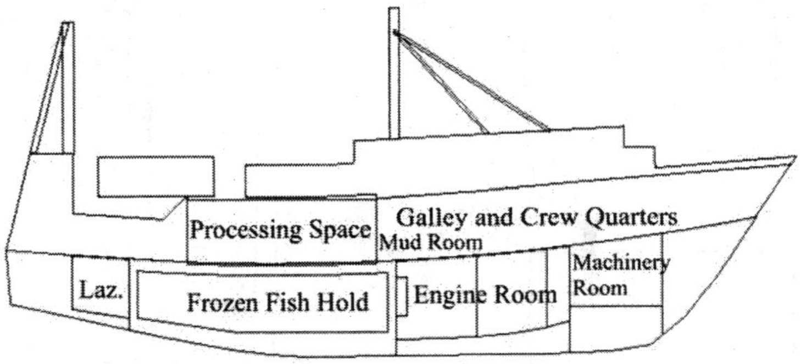

Figure 2 Interior and exterior profile showing floodable spaces (including engine room wing tanks, double bottom fuel tank, and superstructure for wind-heel calculations.) supplied by Jensen Maritime Consultants.

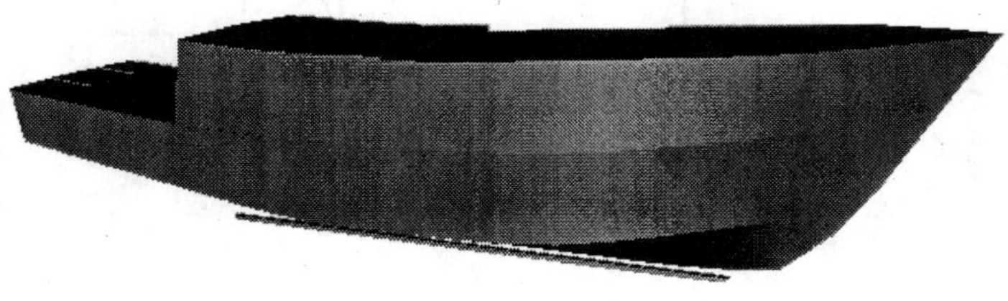

Figure 3 Exterior hull form of *ARCTIC ROSE* showing watertight volume and superstructure from GHS file supplied to the Marine Board by Jensen Maritime Consultants. Note the large watertight volume above the main deck which added considerably to the range of stability.

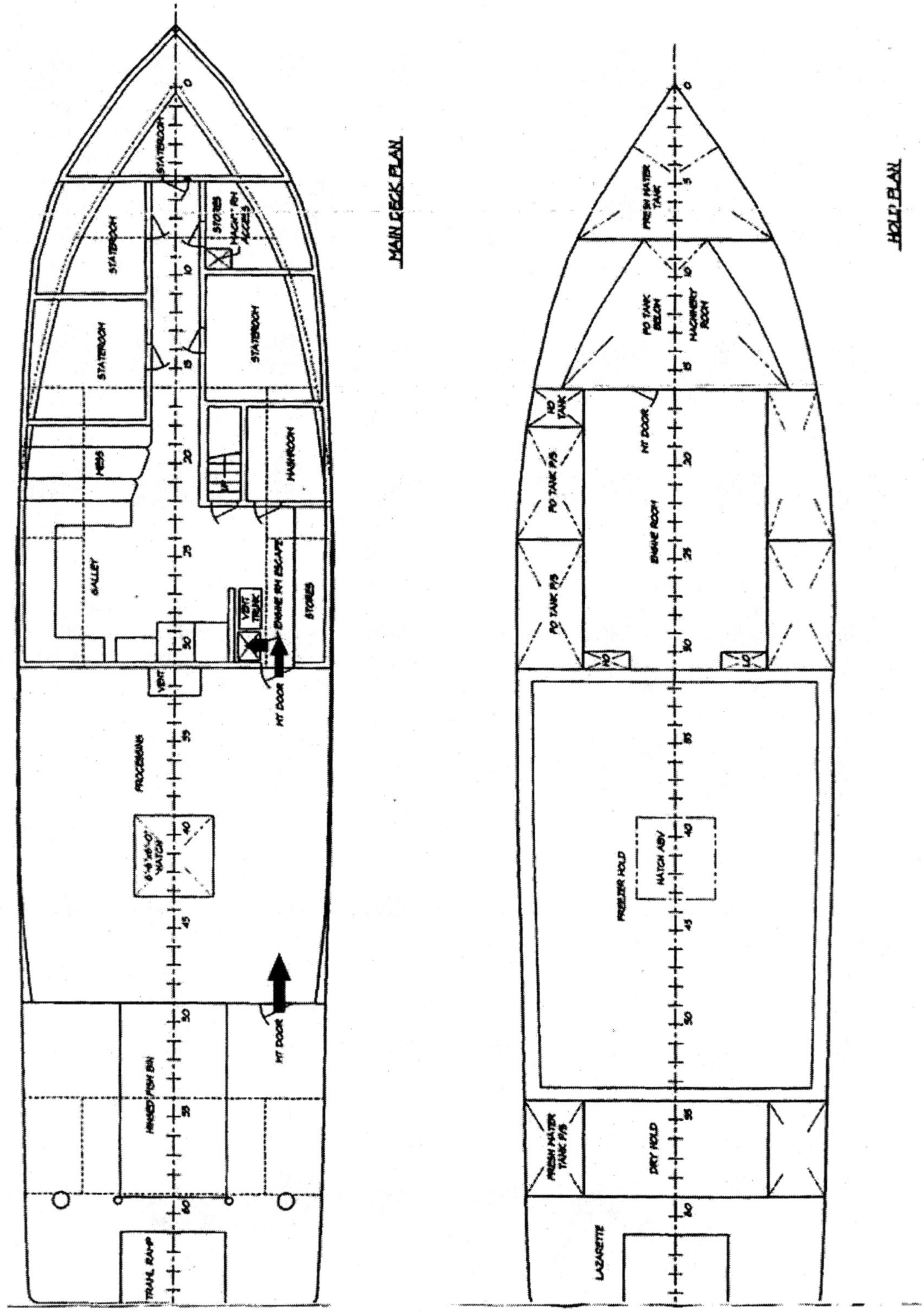

Figure 4. Deck Plan of the *ARCTIC ROSE* showing the direct path of water entering the open WT Door on the starboard side of the main deck to the interior starboard side door to the galley area and directly down to the engine room below.

2. INTACT STATIC AND DYNAMIC STABILITY ANALYSIS

As shown in Figure 9 in Appendix A, a number of scenarios involving an intact vessel were analyzed as part of the investigation. The *ARCTIC ROSE* very likely met the minimum intact stability including righting arm requirements in the estimated loading condition at the time of the casualty with the processing space watertight. It also very likely met the minimum Severe Wind and Roll requirements in the intact condition. (Cleary, 1993, Franscesutto, 2002, IMO 1995, Womack, 2002) The angle of equilibrium when the vessel encountered a 65 knot beam wind was 13.6 degrees, only 0.4 degrees less than the maximum allowable heel angle. This small margin is due to the small righting arm values at small heel angles. The vessel had 500% more area under the righting arm curve than required; because of the significant rise in righting arm out to 55 degrees (Figure 5) resulting from the large watertight deck (Figure 3). Thus, capsize in anything less than a 40-50 foot seaway was very unlikely and a direct hit by a rogue wave more than 1.5 times the beam of the vessel was also unlikely (Zseleczky 1988, 1989, Johnson 2000).

The MSC conducted a one-dimensional nonlinear safe-basin analysis to evaluate the *ARCTIC ROSE*'s capsize resistance, and to determine the vessel's time domain response to excitation by regular (sinusoidal) waves (Borlase 2002, Falzarano 1995).

Based on the apparent suddenness of the vessel's sinking, evaluating its intact dynamic roll response in a seaway was important for three reasons. First, the MSC could quantitatively evaluate the likelihood of a rogue wave capsizing the vessel in the intact condition. Second, the MSC could evaluate the wave height and period necessary to roll the vessel beyond the angle in which downflooding through open hatches and vents occurs. Finally, the calculations provided a baseline for further evaluation of the vessel's roll response in the damaged or flooding condition.

2.1 INTACT DYNAMIC STABILITY-GUIDANCE FROM MODEL TESTS

The intact stability guidance from model tests was based on previous fishing vessel model tests performed by Stefan Grochowalski at the SSPA model tank during the 1980s (Grochowalski, 1989, 1993). The fishing vessel used for these tests was based on a typical small Canadian, hard-chine stern trawler of 19.8 m length which in a model scale of 1/14 was 1.33 m on the waterline. It was tested using four loading conditions as shown in Table 1 which shows the results of scaling the model results back up to the original scale ratio of 14 and noting the wave group height for survivability.

Table 1

Scale Ratio =14 Length = 18.6m Beam = 6.1m

Matches Canadian FV @	Port Departure		Full Load	
Loading condition	IA	IB	IIA	IIB
Displacement, tons	149	149	185	185
Meets IMO Req.	yes	no	no	yes
Survived in Wave Hts. to	7.0m	5.3m	2.5m	9.1m

Excel spreadsheets which summarize the results of the SSPA tests in the model scale have been developed (Johnson and Grochowalski, 2002). The results of these model tests were used in the forensic analysis of the sinking of the F/V *ARCTIC ROSE*. By increasing the scale ratio to 17.3 in Tables 2 and 3, it was possible to match the beam, displacement and block coefficient of the *ARCTIC ROSE*.

Table 2

Scale Ratio = 17.3 Length = 23m Beam = 7.5m

Matches Canadian FV @	Port Departure		Full Load	
Loading condition	IA	IB	IIA	IIB
Displacement, tons	282	282	349	349
Meets IMO Req.	yes	no	no	yes
Survived in Wave Hts. to	8.7m	6.6m	3.1m	11.2m

Table 3 Comparison of vessel particulars at a scale ratio of 17.3

Vessel	Canadian FV		*ARCTIC ROSE*	
Vessel Particulars	m	ft	m	ft
Displacement, tons	350	350	350	350
LOA	24.44	80.2	31.1	102
L, LWL	23	75.4	29	95
B, Beam	7.5	24.7	7.5	24.5
T, Draft, Full Load	3.88	12.7	3.1	10.3
D, Depth	4.55	14.9	3.6	11.7
Bulwark height	1.12	3.69	0.9	3
Block coefficient	0.51	0.51	0.51	0.51
L/B	3.05	3.05	3.88	3.88
B/T Port Departure	2.19	2.19	2.39	2.39
B/T Full Load	1.94	1.94	2.42	2.42

Then the scaled righting arm curves of the model tests can be compared with those of the *ARCTIC ROSE* for intact stability as shown in Figure 5. Using this comparison, the Marine Investigation Board concluded it was unlikely that the *ARCTIC ROSE* capsized and sank in the intact stability condition. However, the observed vessel responses from the capsize test videos provided good insight in creating the motion sequence scenarios used in the time to flood analysis in various wave sequences. The amount of water trapped on deck contributes to intact capsize as observed in the

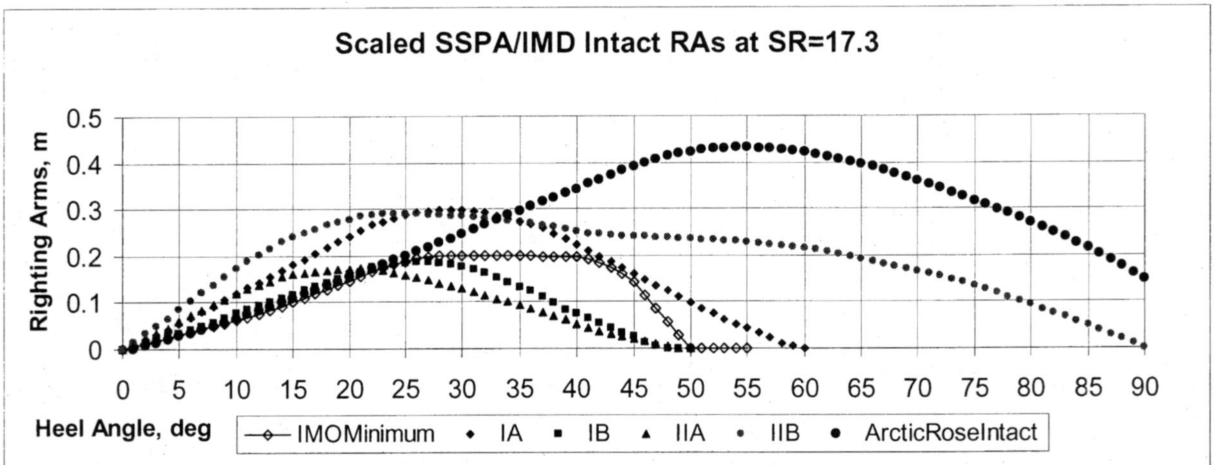

Figure 5 shows the intact righting arm curves for the four conditions compared to the *ARCTIC ROSE* and the IMO minimums, whose strange shape is the result of the inconsistency in the area requirements (Womack 2002).

Grochowalski videos and in other capsize studies (Francescutto, 2001, Umeda, 1999, 2002). In the case of the *ARCTIC ROSE*, the water on the small after deck (Figures 2 - 4) represents a survival risk only if the processing space door was open during the sinking.

The *ARCTIC ROSE* comparison is near the upper bound of scale ratios which give realistic bulwark heights. Something to remember when planning generic model tests for capsizing studies is to use several different adjustable bulwark heights to enable scaling to both larger and smaller scale ratios. For example, changing the scale ratio to 22.5 (a 30 m vessel) brings Condition IB up to the IMO criteria with a survival wave height of 8.6 meters and a bulwark height of 1.46 m, which could trap a lot of water on deck. On the other hand, changing the scale ratio to 11.3 (a 15 m vessel) lets a vessel loaded as condition IA barely meet IMO standards with a survival wave height of 5.7 m and a bulwark height of 0.71 m which wouldn't be high enough for crew safety. This approach is consistent with a new thrust towards the direct assessment of ship stability using model test results (Cramer 2002).

Note that the *ARCTIC ROSE* barely meets the IMO minimum intact stability standard at small angles but has a very large righting arm at 55 degrees and a range of stability beyond 90 degrees because of the watertight processing space above the main deck (See Figure 3). The area under the *ARCTIC ROSE* righting arm curve exceeds that of Condition IIB, which survived scaled model test conditions equivalent to a wave group height of 11.2 meters, well beyond the estimated wave conditions at the time of sinking.

3. DEVELOPMENT OF PROGRESSIVE FLOODING ANALYSIS SPREADSHEET

A progressive flooding analysis Excel spreadsheet was developed to assist in determining the effect progressive flooding would have on the stability of the *ARCTIC ROSE*, and to determine how long it would take the *ARCTIC ROSE* to capsize (heel to an angle from which it cannot return upright without external assistance) and sink due to progressive flooding. The quasi-equilibrium time stepping technique is consistent with the observation that "Dynamic behavior of a damaged ship in waves can be regarded as the sum of large-amplitude, slow motions and small-amplitude, fast motions." (Umeda 2002b) The present analysis neglects the fast motions, as was the case in the Umeda and Papanikolaou (2002) numerical investigations.

Creation of Spreadsheet Data Input Page

The first step was to create a basic intact stability analysis spreadsheet. This had been done previously for a variety of naval vessels using lookup tables for the hydrostatic characteristics and cross curves of stability based on output from the General Hydrostatics Program (GHS) by Creative Systems, Inc. A shipbjnh.xls spreadsheet, developed to make up test questions in basic naval architecture, calculated corrected stability parameters, starting with the initial drafts, KG, and weight shifts, additions and removals along with their locations within the vessel. The output included the corrected KG, GM, list, trim, and final drafts, fore and aft as well as a simple (non-pocketed) free surface and free communication correction. It displayed graphs of the corrected stability curves in one degree increments and a 3-D diagram of the stability surface.

The previous spreadsheet Data Input page was modified to handle the exact static equilibrium free surface correction for the processing space as described in Appendix B. Once this was validated using known results and a smooth curve between zones was obtained, the additional flooding spaces were placed on a second data analysis page which was linked to all the data inputs contained on the data input page.

The hydrostatic properties and cross curves of stability for the *ARCTIC ROSE* were calculated using GHS and imported into the spreadsheet as look-up tables. The hydrostatic properties and cross curves of stability were computed for displacements from 350 to 650 long tons and a vertical center of gravity of 11.2 ft, the assumed VCG of the vessel at the time of the casualty. The free-to-trim cross curves of stability from GHS were considered valid up to a heel angle of 90 degrees. Therefore, all righting arms calculated using the progressive flooding spreadsheet stop at 90 degrees. A hydrostatic analysis found the change in righting arm with upright trim angle was correctable by a nearly constant factor to 60 degrees of heel and thus could be represented by a look up table of correction factor vs. upright trim angle. Therefore, the righting arms could be automatically corrected for trim by applying a correction factor based on the calculated upright trim of the vessel as a function of flooding loads.

The processing space, fish hold, galley, engine room, machinery space, and lazarette of the *ARCTIC ROSE* were included in the spreadsheet as rectangular box compartments that may progressively flood as shown in Figures 2 and 4. The dimensions, permeability, and volume of each of the six compartments were taken from the GHS computer model and imported into the Excel spreadsheet. The amount of flooding water in any one compartment was limited by the rectangular box volume times the assumed permeability factor.

Basic Assumptions:

- **All calculations are quasi-equilibrium static calculations at relatively short time steps equal to the encounter period at the assumed vessel speed divided by 4.**
- **Hydrostatic parameters are functions of displacement and trim created as lookup tables using GHS.**
- **Free-to-trim righting arm tables are from GHS as a function of displacement with a correction for trim.**
- **All flow calculations are based on equilibrium head differences across hatches at the appropriate heel and trim angles.**
- **A permeability correction is applied to total space volume.**

The size and location of doors and hatches leading from one compartment to another on the *ARCTIC ROSE* were taken from the general arrangements drawings prepared by Jensen Maritime Consultants in 1999, and former crewmember testimony. The height, width, distance off centerline, height above the deck, and longitudinal location of each door and hatch were entered into Excel as variable cell values. The cross-sectional area of the door or hatch that allows flooding water to pass through was calculated as a function of heel angle and water depth above the outboard hatch corner. For flooding into the outside WT door, GHS was used to calculate the distance from the base of the door to the still water surface as a function of heel angle. Internally, the heads were calculated from the flooding water height above the hatch corner. Nested if statements were used to calculate the weir flow area as a function of hatch geometry and the extent of flooding water as a function heel angle.

The flooding rates into and out of each compartment were therefore calculated as a function of the difference in water height between the two compartments and as a function of the portion of the door or hatch submerged using a weir/orifice equation with a coefficient of 0.6. The details of this analysis are in Appendix C.

The user specifies which of the six compartments on the *ARCTIC ROSE* are allowed to progressively flood by specifying a "1" or "0" in the appropriate box on the Scenario Page (Figure 6). The free surface effect in each flooding compartment of the *ARCTIC ROSE* was calculated exactly by determining the transverse and vertical center of gravity (centroid) of the flooding water using a four zone methodology, depending on the average heel angle of the vessel and the amount of water in the space. (See Appendix B for details of the calculations and sketches and nomenclature.)

The scenarios assumed progressive flooding paths based on the general arrangements of the vessel, the results of the ROV dives, and testimony provided to the FMBI by former crewmembers, Coast Guard boarding officers, and NMFS observers. Initial flooding of compartments from through-hull fittings below the waterline was not considered in the scenarios for two reasons. First, the engine room, machinery space, and fish hold were fitted with bilge alarms that would sound if leaking occurred, and there is no indication from radio broadcasts that the alarms had sounded. Second, any flooding through leaking bearings, sea chests, keel coolers, or other through-hull fittings would be significantly slower than the above-deck progressive flooding into the processing space. While water entering the below decks compartments could have begun the above-deck progressive flooding, the MSC determined it was very unlikely that the loss of the *ARCTIC ROSE* was due solely to flooding from through-hull fittings.

The progressive flooding analysis on the *ARCTIC ROSE* was time-dependent, and accounted for the vessel's approximate position on the wave at each time step. Using the weather information provided to MSC

Table 4 Example scenario analysis spreadsheet page for the first eight time steps and flow into four compartments.

Time Step	Scenario Aw2deg for Arctic Rose Action	Heel Angle Deg	RA Factor deg	Excess Height Ft	Wave Slope deg	Duration Sec	Runn. Total Sec	Mean Square Error	Processing Space #1 Flow rate tons/sec	Processing Space #1 Water in tons	Processing Space #1 Total FS tons	Net in #1 tons	Galley, Space #2 Flow rate tons/sec	Galley, Space #2 Water in tons	Galley, Space #2 Total FS tons	Engine Room, Space # Flow Rate tons/sec	Engine Room, Space # Water in tons	Engine Room, Space # Total FS tons	Freezer Hold, Space Flow rate tons/sec	Freezer Hold, Space Water in tons	Freezer Hold, Space Total FS tons	Total Flooding Tons
0	Vessel in Trough	0	1	0	0	2.67	2.67	0	0	0	0	0	0	0	0	0	0	0	0	0	0	0
1	Wave Raises Stern	25	0.9	2	-2	2.67	5.34	0.0	1.19	3.18	3.18	2.77	0.15	0.41	0.41	0.00	0.00	0.0	0.00	0.00	0.00	3.18
2	Ride Wave Crest	45	0.7	2	0	2.67	8.01	0.0	4.30	11.48	14.25	12.19	0.77	2.06	2.47	0.00	0.00	0.00	0.00	0.00	0.00	14.66
3	Wave Lowers Stern	25	0.9	0	2	2.67	10.68	0.0	-0.29	-0.77	11.42	10.20	0.46	1.22	2.57	0.42	1.12	1.12	0.00	0.00	0.00	13.89
4	Vessel in Trough	10	1	0	0	2.67	13.35	0.0	-0.06	-0.16	10.04	8.94	0.41	1.10	2.32	0.51	1.35	2.48	0.00	0.00	0.00	13.73
5	Wave Raises Stern	40	0.9	2	-2	2.67	16.02	0.0	3.79	10.13	19.07	15.35	1.40	3.73	4.69	0.51	1.35	3.83	0.00	0.00	0.00	23.86
6	Ride Wave Crest	85	0.7	2	0	2.67	18.69	0.0	5.77	15.40	30.74	27.44	1.24	3.31	8.43	-0.16	-0.43	3.39	0.00	0.00	0.00	39.26
7	Wave Lowers Stern	45	0.9	0	2	2.67	21.36	0.0	0.99	2.65	30.08	27.79	0.86	2.29	9.21	0.57	1.52	4.91	0.00	0.00	0.00	41.91
8	Vessel in Trough	20	1	0	0	2.67	24.03	0.0	-0.98	-2.62	25.17	20.91	0.95	2.55	9.16	0.97	2.60	7.51	0.64	1.71	1.71	39.29
9	Wave Raises Stern	40	0.9	2	-2	2.67	26.70	0.0	3.40	9.08	29.99	25.74	1.59	4.25	10.77	0.99	2.63	10.14	0.00	0.00	1.71	48.36
10	Ride Wave Crest	55	0.7	2	0	2.67	29.37	0.0	3.90	10.43	36.16	32.30	1.44	3.86	12.49	0.80	2.14	12.28	0.00	0.00	1.71	58.79
11	Wave Lowers Stern	40	0.9	0	2	2.67	32.04	0.0	-0.85	-2.28	30.02	29.03	0.29	0.79	11.23	0.76	2.04	14.33	0.08	0.21	1.92	56.51
12	Vessel in Trough	30	1	0	0	2.67	34.71	0.0	-0.80	-2.14	26.89	23.95	0.94	2.52	10.91	1.06	2.84	17.16	0.18	0.42	2.34	54.36

by the Formal Marine Board of Investigation, a mean wave period of 10 seconds was assumed. As shown in Table 4, the vessel's movement through the waves was broken into four distinct motions based on equal time steps equal to one fourth of the encounter period, T_e.

During the "Vessel in Trough" time step, the vessel was assumed to be traveling in the trough of the wave, and accumulate water on the starboard side aft deck well if the heel angle was less than the angle of bulwark submergence. No righting arm correction was applied because no loss of waterplane area was assumed. The 10 second wave period gave a ship length to wave length ratio of 95/512.

During the "Wave Raises Stern" time step, the wave at the aft deck was assumed to be two feet above the normal waterline and a pitch down correction could be applied to the internal gravity head flooding calculations for sensitivity purposes as discussed in Appendix D. Additionally, a righting arm factor of 0.9 was applied. This factor accounts for the loss of waterplane area as the vessel travels up the face of the wave.

During the "Ride Wave Crest" time step, the water was assumed to have cleared off the deck through the freeing ports and aft stern ramp and a righting arm factor of between 0.7 and 0.9 was applied to reflect the loss of waterplane area at the bow and stern of the vessel at the crest of the wave. The wave was assumed to be up to 2 feet above the normal still waterline.

During the "Wave lowers Stern", the vessel would shed the water on deck, including that which flowed out of the processing space because of the attitude of the vessel on the wave and a pitch up correction could be applied to the internal gravity head flooding calculations. At each time step, the amount of flooding water that entered each of the spaces was calculated based on the amount of water already in the space, the average roll angle, the height of water at the door or hatch location, the amount of water flowing into the adjacent space and the time spent at the average roll angle

The righting arm for the *ARCTIC ROSE* was calculated from the cross-curves of stability at the estimated at-loss displacement and KG. This curve was used as the baseline for the righting arm calculations at each time step, and is shown in the righting arm graph for each time step as AZ0 in Figure 6. A second righting arm curve is calculated from the cross-curves of stability based on the displacement and trim of the vessel and the weight of the flooding water. Additionally, a constant free surface moment of 62.1 ft-LT was applied to the righting arm curve to reflect the free surface of the slack tanks onboard the vessel, and the trim scale factor and wave crest factor were applied. This second righting arm curve is shown in the righting arm graph for each time step as RA Corrected for KGF. The free surface correction was then calculated for each flooded compartment, and applied to the second righting arm curve. This third righting arm curve is shown in the righting arm graph for each time step as RA Corrected for FS, and shows the actual righting arm curve at each time step.

The time-stepping analysis required a time-domain description of the seaway at the time of the accident, which was impossible to accurately describe due to the absence of eyewitnesses or NOAA data buoy information. Therefore, two distinct time-domain wave patterns were used to predict the roll motion of the *ARCTIC ROSE* and bracket the likely wave patterns encountered by the vessel. The first wave pattern reflects the vessel encountering a constant-height wave train of 20 feet, with no larger or smaller waves assumed. This wave pattern, although unlikely out at sea, describes the minimum wave excitation the vessel likely encountered. The second wave pattern reflects two very large waves approximately 30-36 feet high striking the vessel, and then the vessel encountering a wave train of seas 20 feet high. This wave pattern describes the maximum wave excitation the vessel likely encountered, and simulates the vessel's response to very large waves.

Additionally, the user also specifies the average roll angle of the *ARCTIC ROSE* at each time step. The equilibrium angle was assumed to be zero before the first time step. A dynamic analysis was used to determine the angle of roll for the vessel in a 20 foot seaway, and calculated that the vessel would roll approximately 22 degrees. Therefore, for one scenario, a roll angle of up to 20 degrees past equilibrium was used for the "Wave Raises Stern" time steps. Videotapes of model tests for vessels of similar size and stability characteristics in irregular seas were reviewed to establish the roll behavior of a vessel on the crest of the wave. Each model rolled further on the crest of the wave then either on the front face or the trough of the wave, and therefore the vessel was allowed to roll 25 degrees past equilibrium for the "Ride Wave Crest" time step. The lolling angle from the previous "Ride Wave Crest" time step was halved to calculate the average equilibrium angle for the "Vessel in Trough" time step. If the righting arm during the "Ride Wave Crest" was mostly negative, then the average equilibrium angle was steadily increased from 35 degrees. The subsequent time steps, "Wave Raises Stern" and "Ride Wave Crest", were taken to be 20 to 25 degrees more than the "Vessel in Trough" time step equilibrium angle.

In the original 100 second scenario developed for the Marine Board Report (Table 5) and new scenarios A, B, and C developed in 2003 (Table 4, Table 6, Figure 6 and Appendix D), the vessel was assumed to be hit by two large steep waves in succession to simulate a wave group effect. Successive wave crest heights of up to 1.5 times the significant wave height can occur one percent of the time according to Markov Theory (Dawson 2002, Wist, et. al, 2002). There is the joint probability problem of being in the wrong place at the wrong time, but it is not uncommon given the population of fishing vessels in storm condition.

These scenarios assume the first of the successive waves heeled the vessel to around 45 degrees and then it recovered to 25 degrees. It was then hit again and rolled to 85 degrees. When it recovered during time step 8, it had already taken on 30 tons of water in the processing space and has started the progressive flooding as shown in Figure 7. The average heel angle during the remaining time steps was assumed to be within 10 degrees of the calculated lolling angle, arranged in an oscillating pattern. (See Figure 8a)

In all scenario steps the mean square error (blue column in Table 4) between the assumed amount of flooding water in each space and the calculated amount which satisfied equilibrium conditions was reduced as close to zero as computing time permitted. It turned out to be faster to use a trial and error solution combined with a solver solution, which had difficulty converging as the vessel neared capsize. The apparent reason for this is illustrated in Figure 20 of Appendix C which shows the problem in the lookup table for the Moment to Trim 1 degree.

Arctic Rose Scenario A Analysis				Time Step	12			Vessel in Trough	0
Do not enter 0's for flooding data			Flooded Depth		Percent	Count, 1	Lookup	Excess Wave Height	0
Space	56		@0 de	@90deg	Filled	Not Counted	Tons	RA Factor on Wave Crest	1
1	Process Space	23.96	1.58	4.75	19.8%	1	23.96	RA Scale Factor for Trim	1.04
2	Galley/Passageway	10.92	1.54	4.42	19.2%	1	10.92	Combined RA Scale Factor	1.04
3	Engine Room	17.17	2.12	2.72	19.4%	1	17.17	ArcticR Average Heel Angle	30
4	Freezer Hold	2.35	0.13	0.41	1.9%	1	2.35	ResRE/ Maximum RA, ft	0.63
5	Water on Deck	5.00	1.17	0.00	39.0%	0	Avg Floodin	IMO RE GMEFF	-1.60
6	Machinery	0.01	0.00	N/A	0.0%	0	Rate 1.57	1.3 FSCorrection, ft	3.08

Sig Wave Period		Ship Speed	
10	Seconds	2	Knots
Encounter Period		Ship Heading	
10.68	Seconds	15	deg
Each Time Step		Wave Length	
2.67	Seconds	512	ft
Wave Slope Correctio		Excess Wave I	
2.00	Degrees	2	ft
90	orrecteTrim in		

Total Flooding Weight, ton	54.4	Upright Freeboard at outside door to Proc. S	1.2
Vessel Displacement, tons	404	Heel angle for bullwark immersion, degrees	19
Displacement Reserve, tor	461	Residual Righting Energy, ft-tons	9,623
Reserve Buoyant Volume,	16,121	Input Remaining Avg. Flooding Rate	2.1
Elapsed Time	34.7	Remaing time to sink at avg FR, sec	219

Draft Fwd Ft	Draft Aft Ft	Upright Trim Ft	Upright Trim deg	Degree Trim Degrees	Pitch Angle deg	Proc. space Ft
12.95	9.71	-3.25	-1.96		-1.96	-0.82
	Scenario Total Time to Sink, sec					254

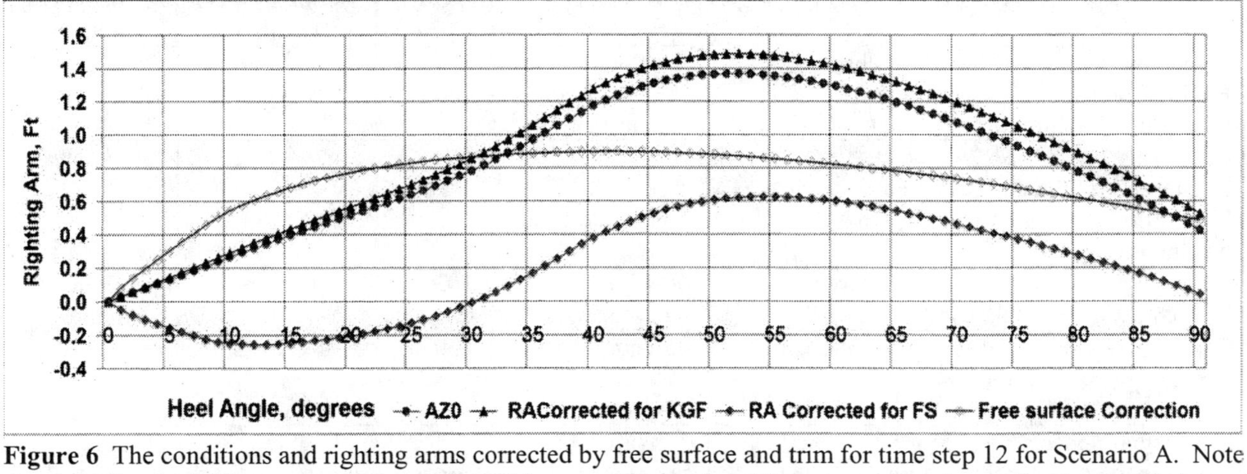

Figure 6 The conditions and righting arms corrected by free surface and trim for time step 12 for Scenario A. Note that vessel at the lolling angle of 30 degrees still has more residual righting energy than is required by the minimum IMO curve.

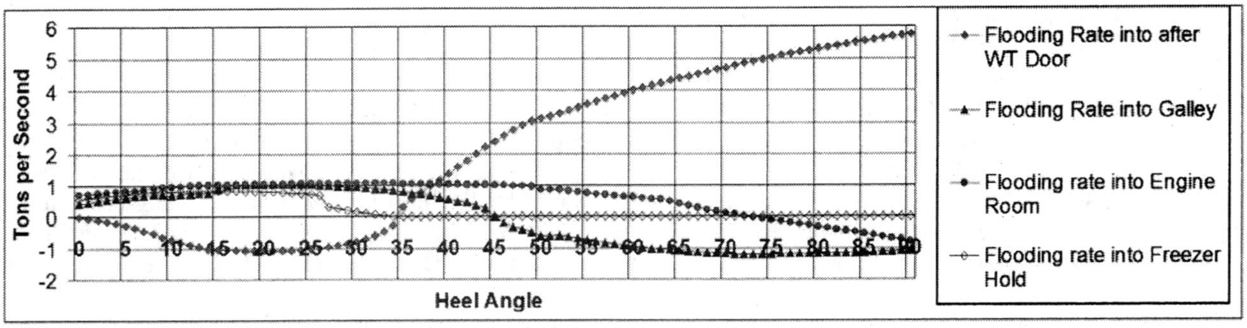

Figure 7a Flooding rates as a function of heel angle for time step 12 for Scenario A. Note that at about 20% filled in all but the freezer hold, the flow can be reversed into the galley and engine room at higher heel angles.

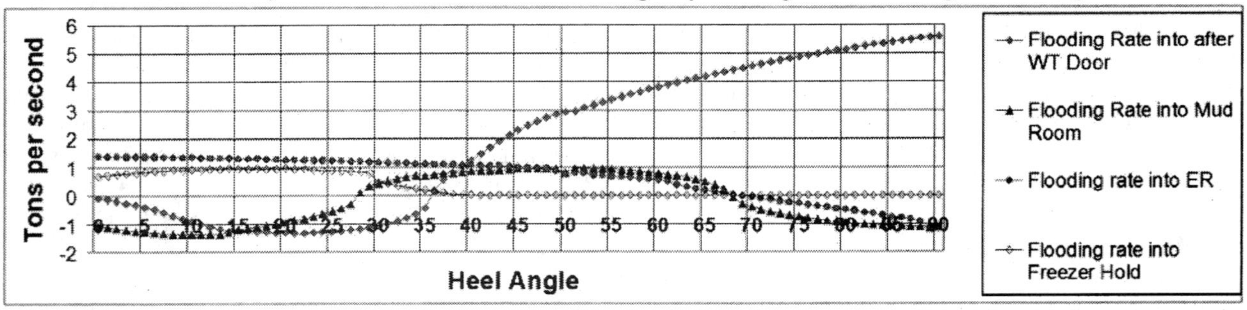

Figure 7b Flooding rates as a function of heel angle for time step 12 for Scenario C. Note that the flows can reverse at both high and low heel angles since the mudroom was about 70% filled.

Capsizing of the *ARCTIC ROSE* was considered to occur when the vessel's maximum righting arm during the "Vessel in Trough" time step were very nearly zero or negative. A negative righting arm meant the vessel could not recover without assistance from any heel angle. The elapsed time when the righting arm was negative was the time it would take for the vessel to capsize due to progressive flooding. Using the GHS computer model, it was calculated that the *ARCTIC ROSE* had approximately 865 long tons of total buoyancy. To calculate an approximate total time to sink, the vessel's displacement with flooding water was subtracted from the 865 long tons of buoyancy, and the resulting reserve buoyancy was divided by the average flooding rate. The time to sink would be affected by many factors that could not be incorporated into this analysis, including trapped air in the hull and superstructure, the actual angle of the vessel on the surface of the vessel, and wave action. (Deakin, 2000)

A summary of the different progressive flooding scenarios, the time to capsize, and the time to sink is shown in Table 5 and all scenarios are summarized in Appendix A

Table 5 Summary of Progressive Flooding Scenarios and Results submitted to the FMBI

Scenario	# Time Steps	Time to Capsize	Flooding Water at Capsize	Total Time to Sink
Processing Space, Fish Hold, Galley, and Engine Room Flood (Attachment 1 to Enclosure 5) assuming a set of 20 foot waves	48 with 2,3,5 second sequence	163 seconds	198 tons	7.9 minutes
Processing Space, Fish Hold, Galley, and Engine Room Flood (Attachment 2 to Enclosure 5) assuming a large, two wave group hits the vessel	21 with 2,3,5 second sequence 1	100 seconds	208 tons	4 minutes
Processing Space and Fish Hold Only Flood (Attachment 1 to Enclosure 6)	60 with 2,3,5 sec. sequence	203 seconds	66 tons	25.6 minutes
Lazarette Flooded, Processing Space, Fish Hold, Galley, and Engine Room Flood (Attachment 1 to Enclosure 7)	45 with 2,3,5 second sequence	150 seconds	193 tons	5.3 minutes
Lazarette Flooded, Processing Space and Fish Hold Flood (Attachment 1 to Enclosure 8)	51 with 2,3,5 sec. sequence	173 seconds	66 tons	13.2 minutes

Table 6 Summary of Progressive Flooding Scenarios and Results subsequent to the submission to the Coast Guard investigation

Scenario	# Time Steps	Time to Capsize	Flooding Water at Capsize	Total Time to Sink
A. Processing Space, Fish Hold, Galley, and Engine Room Flood with 2 deg pitch correction	32 2.67 sec. steps	88 seconds	197 tons	4 minutes
B. Processing Space, Fish Hold, Galley, and Engine Room Flood with 0 pitch correction	32 2.67 Sec. steps	88 seconds	187 tons	4 minutes
C. Processing Space, Fish Hold, Mud room and Engine Room Flood with 0 pitch correction	31 2.75 Sec. steps	88 seconds	197 tons	4 minutes

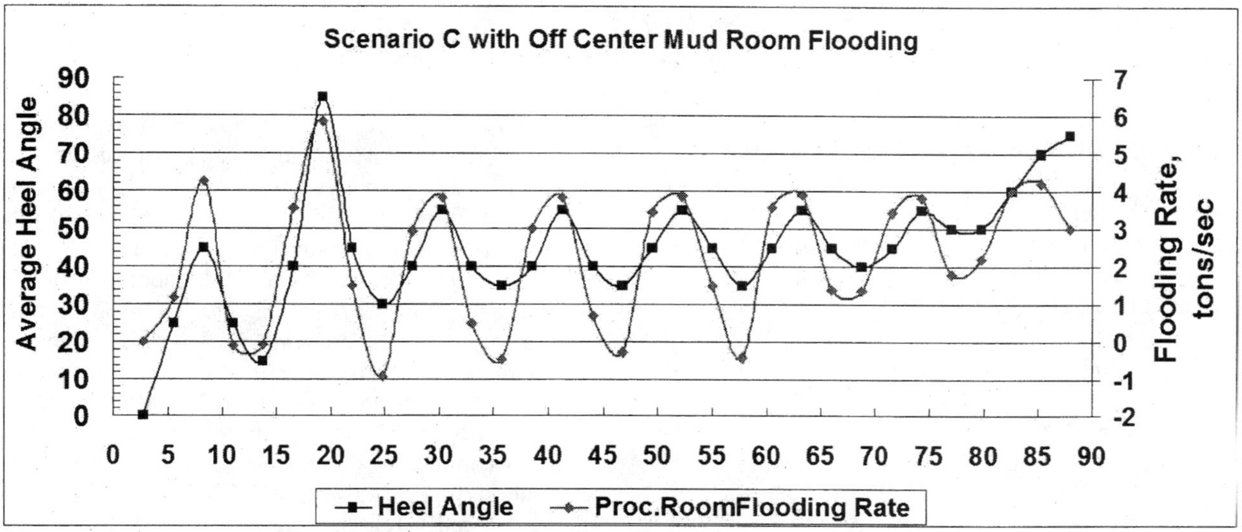

Figure 8a Assumed heel angles and resulting flooding rates as a function of time for Scenario C with galley flooding limited to the starboard side mud room

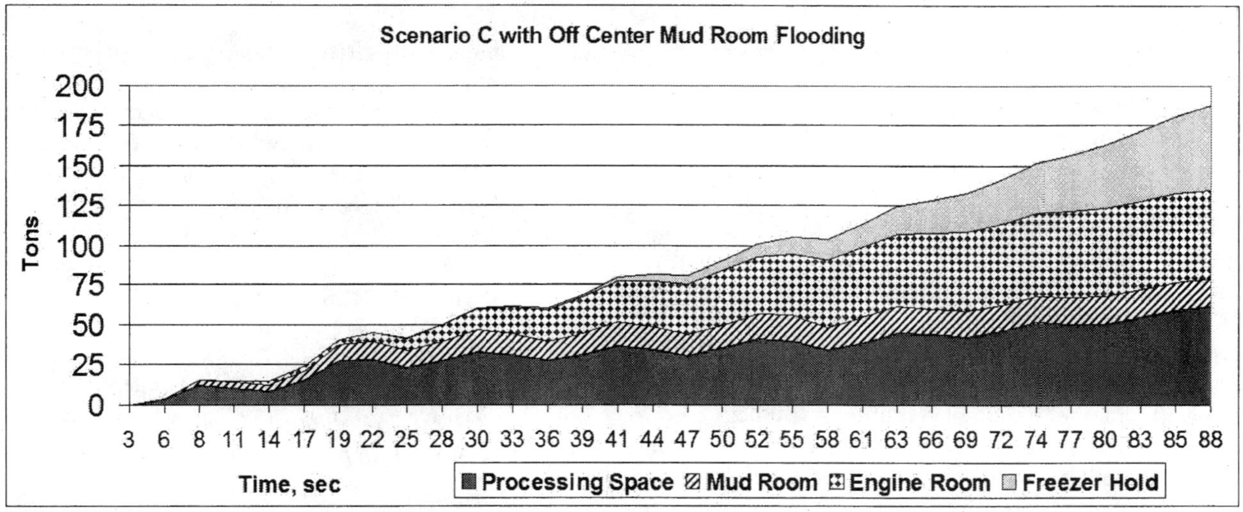

Figure 8b Flooding Diagram showing buildup of progressive flooding in the floodable spaces as a function of time and positive flooding rates in and out of the spaces. The mud room remains fully flooded beyond 80 seconds.

Scenario C represents the most likely scenario since the testimony of former crew members indicated the WT door to the processing space was always open, the door to the engine room hatch was probably propped open to give better ventilation to that space, and the "flimsy" wooden door to the galley from the mudroom was closed to reduce the engine room noise and the smell of the processing space in the galley/stateroom area. Being that far off center, the mudroom filled completely during the capsize phase at large angles of heel, so that further flooding depended on the head difference between the heeled and trimmed processing space and back flow head in the engine room now restricted by the size of the engine room hatch. All flows can go in and out of the spaces, depending on the relative heads as a function of heel as shown in Figure 7b.

5. CONCLUSIONS AND RECOMMENDATIONS

The scenario analysis methodology developed for this project is useful for all marine board investigations of sinking accidents where the internal details are available. The methodology for full pocketing free surface analysis described in detail in Appendix B is applicable to all rectangular spaces in all vessels. The method is currently being used to prepare stability letters for fishing vessels with large, wide tanks which are not correctly handled in a number of widely used computer based ship design software packages.

The U.S. Coast Guard Marine Safety Center utilized a number of different tools to evaluate the most likely causes of the disappearance of the *ARCTIC ROSE* and its fifteen man crew. However, a number of questions went unanswered because the analytical techniques needed to answer these questions were beyond the current state of research. More research is needed in the capsize resistance and dynamic stability of a vessel while progressive flooding is occurring. Additionally, a better understanding is needed of a vessel's behavior after capsizing, until the vessel either reaches an equilibrium position on the ocean surface or comes to rest on the ocean floor.

The progressive flooding scenarios that most likely sank the *ARCTIC ROSE* are similar to those assumed to have resulted in the loss of the GAUL, a Norwegian trawler that sank in February 1974, killing all 36 men onboard (Morrell 1980, MAIB 1999). Additionally, during the 1994-2000 time periods, the loss of 190 U.S. fishing vessels was attributed to capsizing, flooding, or severe weather. Continuing the extensive efforts to better understand the dynamics of fishing vessels, especially in the flooding condition, will help naval architects design safer vessels, assist fishermen in understanding how to prevent their vessels from sinking, and ultimately save lives.

6. DISCLAIMER

It must be emphasized that the opinions expressed in this paper are solely those of the authors and are not necessarily those of the Marine Safety Center or the United States Coast Guard.

Additionally, the authors stress, "The fundamental purpose of investigating an accident under these Regulations is to determine its circumstances and the cause with the aim of improving the safety of life at sea and the avoidance of accidents in the future. It is not the purpose to apportion liability, nor, except so far as is necessary to achieve the fundamental purpose, to apportion blame." The above extract is taken from Merchant Shipping (Accident Reporting and Investigation) Regulations 1999 published in the UK Marine Accident Investigation Board reports (MAIB 2003), and certainly applies to this paper.

7. ACKNOWLEDGEMENTS

LT Borlase would like to thank Dr. Armin Troesch at the University of Michigan for his assistance with the dynamic stability analysis, and CAPT Ron Morris, CDR John Bingaman, and LCDR Jim Robertson for the opportunity to assist in the FMBI investigation. The authors would like to thank the members of the SNAME Ad Hoc Panel on Fishing Vessel Operations and Safety, especially John Womack for his insights on vessel operations and for validating the free surface analysis as a marine survey tool. We also thank Dr. Stefan Grochowalski for providing the capsize videos which enabled the authors to estimate the time sequences in the loss of the *ARCTIC ROSE*. The authors thank Jensen Maritime Consultants and specifically Mr. David Green, Mr. Jonathan Parrot, and Mr. Eric Blumhagen for their extensive assistance with understanding the vessel's layout and operations. The authors thank Kirk Kristensen of the SNAME papers committee for his excellent suggestions for improving the paper. Finally, the authors thank Jennifer Eichelberger, a NMFS observer who departed the *ARCTIC ROSE* on March 21, 2001, for her assistance in validating the likelihood of the new mud room flooding Scenario C.

8. REFERENCES

Borlase, G., 2002, Research Opportunities Identified during the Casualty Analysis of the Fishing Vessel *ARCTIC ROSE*, *Proceedings of the 6th International Ship Stability Workshop*, Webb Institute, 14-16 October 2002

Cleary, W., 1993, The Regulation of Ships Stability Reserve, *Proc. of the U. S. coast Guard Vessel Stability Symposium*, New London, CT, March 15-17, 1993.

Comstock, J. P., ed., 1967 *Principles of Naval Architecture* (PNA 1967), SNAME 1967, pp78-88

Cramer, H., and Tellkamp, J., 2002. Towards the Direct Assessment of a Ship's Intact Stability, *Proceedings of the 6th International Ship Stability Workshop*, Webb Institute, 14-16 October 2002

Dawson, T. H., 2002, Markov Theory for Groups of Very High Waves, *International Journal of Offshore and Polar Engineering*, Vol. 12, No. 2, June 2002

Dahle, E. A., and Myrhaug, D., 1995. Risk Analysis Applied to Capsize of Fishing Vessels. *Marine Technology*, Vol. 32, No. 4, October 1995, pp. 245-257.

Deakin, B., 2000, Deployment of Liferafts From Capsized or Sinking Vessels, Sea Australia 2000, Sydney, February 2000

Falzarano, J.M., Esparza, I., and Taz Ul Mulk, M., 1995. A Combined Steady-State and Transient Approach to Study Large Amplitude Ship Rolling Motion and Capsizing, *Journal of Ship Research*, Vol. 39, No. 3, September 1995, pp. 213-224.

Francescutto, A., 2002 Intact Stability, The Way Ahead, *Proceedings of the 6th International Ship Stability Workshop*, Webb Institute, 14-16 October 2002

Francescutto, A., Russo Krauss, G., and Cardo, A., 2001 "Dynamic Stability and Effect of Water on Deck on Small Fishing Vessels", Paper n. 6, *Proceedings International Conference on "Small Craft Safety"*, The Royal Institution of Naval Architects, London, 22-23 May 2001.

Gillmer, T., and Johnson, B., 1882, *Introduction to Naval Architecture*, Naval Institute Press, Annapolis, MD. pp119-122.

Grochowalski, S., 1989, Investigation into the Physics of Ship Capsizing by Combined Captive and Free-Running Model Tests. *SNAME Transactions,* 1989 pp 169-212

Grochowalski, S., 1993. Effect of Bulwark and Deck Edge Submergence in Dynamics of Ship Capsizing. *Proceedings, US Coast Guard Vessel Stability Symposium*, New London, Connecticut, USA. March 1993.

IMO, 1995. 1993 Torremolinos Protocol and Torremolinos International convention for the Safety of Fishing Vessels. *Consolidated Edition*, 1995

Johnson, B. 2000, Capsize Resistance and Survivability When Smaller Vessels Encounter Extreme Waves, *Proceedings of the Rogue Wave 2000 Conference*, Brest France, 29-30 November 2000

Johnson, B., and Womack, J. 2001, On Developing a Rational and User-Friendly Approach to Fishing Vessel Stability and Operational Guidance, *Proceedings of the 5th International Workshop on Stability and Operational Safety of Ships,* Trieste, Italy, 12-13 September 2001.

Johnson, B., and Grochowalski, S., 2002. Development of a Performance Based Fishing Vessel Stability Criteria, *Proceedings of the 6th International Ship Stability Workshop*, Webb Institute, 14-16 October 2002

Lewis E. V. editor 1988, *Principles of Naval Architecture* (PNA 1988) Volumes 1 , SNAME, 1988, pp 87-93.

MAIB 1999,'Report on the Underwater Survey of the Stern Trawler GAUL H.423 and the Supporting Model Experiments, August 1998-January 1999,' MAIB Accident Report No. 4/99.

MAIB 2003, "Report of investigation into sinking of *fv Tullaghmurry Lass N246* with loss of three lives in the Irish Sea on 14 February 2002 Report No. 4/2003.

Morrall, A.,1980 'The GAUL Disaster: An Investigation into the Loss of a Large Stern Trawler,' *Naval Architect, Royal Institution of Naval Architects,* 1980

Papanikolaou, A, and Spanos, D.,2002, On the Modeling of Floodwater Dynamics and its Effects on Ship Motion. *Proceedings of the 6th International Ship Stability Workshop*, Webb Institute, 14-16 October 2002.

Umeda, N., Matsuda, A., Hamamoto, M, and Suzuki, S. 1999. Stability Assessment for Intact Ships in the Light of Model Experiments. *J. of Marine Science and Technology*, SNAJ, Japan, Vol. 4, pp 45-57, 1999.

Umeda, N., and Peters, A., 2002. Recent Research Progress on Intact Stability in Following/Quartering Seas. *Proceedings of the 6th International Ship Stability Workshop*, Webb Institute, 14-16 October 2002.

Umeda, N., Kamo, T and Ikeda, Y., 2002b. Some Remarks on theoretical Modeling of Damage Stability. *Proceedings of the 6th International Ship Stability Workshop*, Webb Institute, 14-16 October 2002

Wist, H. T., Myrhaug, D, and Rue, H. 2002 Joint Distributions of Successive Wave Crest Heights and Successive Wave Trough Depths for Second-Order Nonlinear Waves, *Journal of Ship Research,* Vol. 43, No. 3, Sept 2002, pp 175-185.

Womack, J., 2002, Small Commercial Fishing Vessel Stability Analysis, Where are We Now Where are We going, *Proceedings of the 6th International Ship Stability Workshop*, Webb Institute, 14-16 October 2002.

Zseleczky, J. J., 1988. Evolving Methods for Estimating Capsize Resistance in Breaking Waves. *Proceedings of the SNAME New England Sailing Yacht Symposium,* New England, March 1988.

Zseleczky, J. J. and Cohen, S. H. 1989, Model Tests to Evaluate the Capsize Resistance of a Motor Lifeboat in Breaking Waves, *Proceedings of the 22nd American Towing Tank Conference*, St. Johns, Newfoundland, 1989

APPENDICES

Appendix A Scenario Analysis Methodology (\
Appendix B Free Surface Analysis Methodology
Appendix C Flooding Analysis Methodology
Appendix D Sensitivity Analysis

Appendix A Scenario Analysis Methodology

With no mayday call identifying problems and no survivors to describe the final moments onboard the vessel, the Formal Marine Board of Investigation needed to evaluate a wide range of possible sinking scenarios. Therefore, the Coast Guard Marine Safety Center developed a flowchart to identify the different scenarios that could lead to the sinking of the *ARCTIC ROSE* (Figure 9). While the flowchart was kept as general as possible to ensure all legitimate scenarios were considered, the flooding scenarios were tailored to describe the specific general arrangements of the vessel.

The flowchart was developed working backwards from the time the vessel sinks because weight exceeded buoyancy. The sinking occurred with the vessel in one of two conditions: intact, with all external and internal doors closed and considered effective; or damaged, with either flooding from below decks or external and internal doors considered open and non-watertight. The intact capsizing conditions were developed to cover "pure" capsize scenarios, in which intact righting energy is unable to overcome an external heeling moment due to winds, waves or trawling (Dahle 1995). The damaged and progressive flooding scenarios were developed to illustrate sinking and capsizing scenarios where the residual static righting arm was reduced because of progressive flooding inside the vessel, whether the water enters the vessel from above or below the waterline. Both the intact and damaged or progressive flooding scenarios were identified from testimony given during the Formal Marine Board of Investigation hearings, or through published casualty investigations (Morrell 1980, MAIB 1999).

After developing the complete flowchart to identify the *ARCTIC ROSE* sinking scenarios, the Marine Safety Center used an extensive number of different techniques to evaluate the likelihood of occurrence of each scenario, the results of which are included as Table 5. To evaluate the static righting arm characteristics of the vessel and its compliance with the applicable federal stability requirements, Creative Systems' General Hydrostatics (GHS) V7.50 hydrostatic program was used. The digitized hull and tankage model was provided to the Formal Marine Board of Investigation by Jensen Maritime Consultants (JMC), along with the most current information on lightship and probable loading conditions.

Using the information taken from the Formal Marine Board of Investigation hearings and the vessel particulars provided by Jensen Maritime Consultants, a static righting arm analysis was performed to evaluate the following scenarios: loss of keel ballast, water trapped on after deck, severe wind capsizes vessel, and all three overloaded scenarios. To evaluate the effect of flooded below-deck compartments on the *ARCTIC ROSE*'s stability, MSC performed a static analysis with various compartments independently flooded. The damaged static righting arm analysis was performed to evaluate the likelihood of all through-hull fitting failures, and striking a submerged object. All scenarios were deemed either unlikely or very unlikely, based on the results of the static righting arm analysis.

To evaluate the probability of synchronous roll, the MSC enlisted the assistance of Dr. Armin Troesch, a Professor of Naval Architecture/Marine Engineering at the University of Michigan, to calculate the natural frequency of the vessel in the roll direction using the University of Michigan ship motion prediction program SHIPMO. The natural frequency calculation is based on the hull form of the vessel, and its displacement and centers of gravity. MSC then used the Nomograph for Determining Conditions for Synchronous Oscillation in Regular Waves, Figure 78 in Principles of Naval Architecture, Volume III, (Lewis, 1988) to establish the vessel heading and speed at which synchronous roll could have occurred. The MSC determined that it was very unlikely synchronous rolling caused the vessel to capsize (Borlase 2002).

The MSC used two separate methods to evaluate the likelihood of the *ARCTIC ROSE* capsizing after encountering an extremely large wave. To simulate the roll motions of the *ARCTIC ROSE* in a large seaway, MSC performed a one dimensional dynamic analysis on the vessel to determine the minimum wave height necessary to capsize the vessel within one wave period. With the assistance of Dr. Armin Troesch, MSC used the computer program SHIPMO to calculate the added mass, viscous damping forces, and wave forces that would act on the vessel in a seaway. Using the righting arm of the vessel at the time of the casualty as the stiffness term, MSC calculated the roll response of the vessel over time for increasing wave heights until the wave was sufficiently large to roll the vessel past its angle of vanishing stability.

Additionally, the MSC used the method described in Section 2.1 to parametrically evaluate the vessel's resistance to capsizing in a large seaway. The results of both analyses predicted a single 52 foot high wave would likely have to strike the vessel broadside to capsize it. Based on the hindcast weather information and the lack of structural damage of the vessel during Remote Operated Vehicle (ROV) dives, the MSC determined that it was very unlikely a rogue wave or extremely large wave capsized the *ARCTIC ROSE.*

SCENARIOS THAT COULD LEAD TO THE LOSS OF THE F/V ARCTIC ROSE

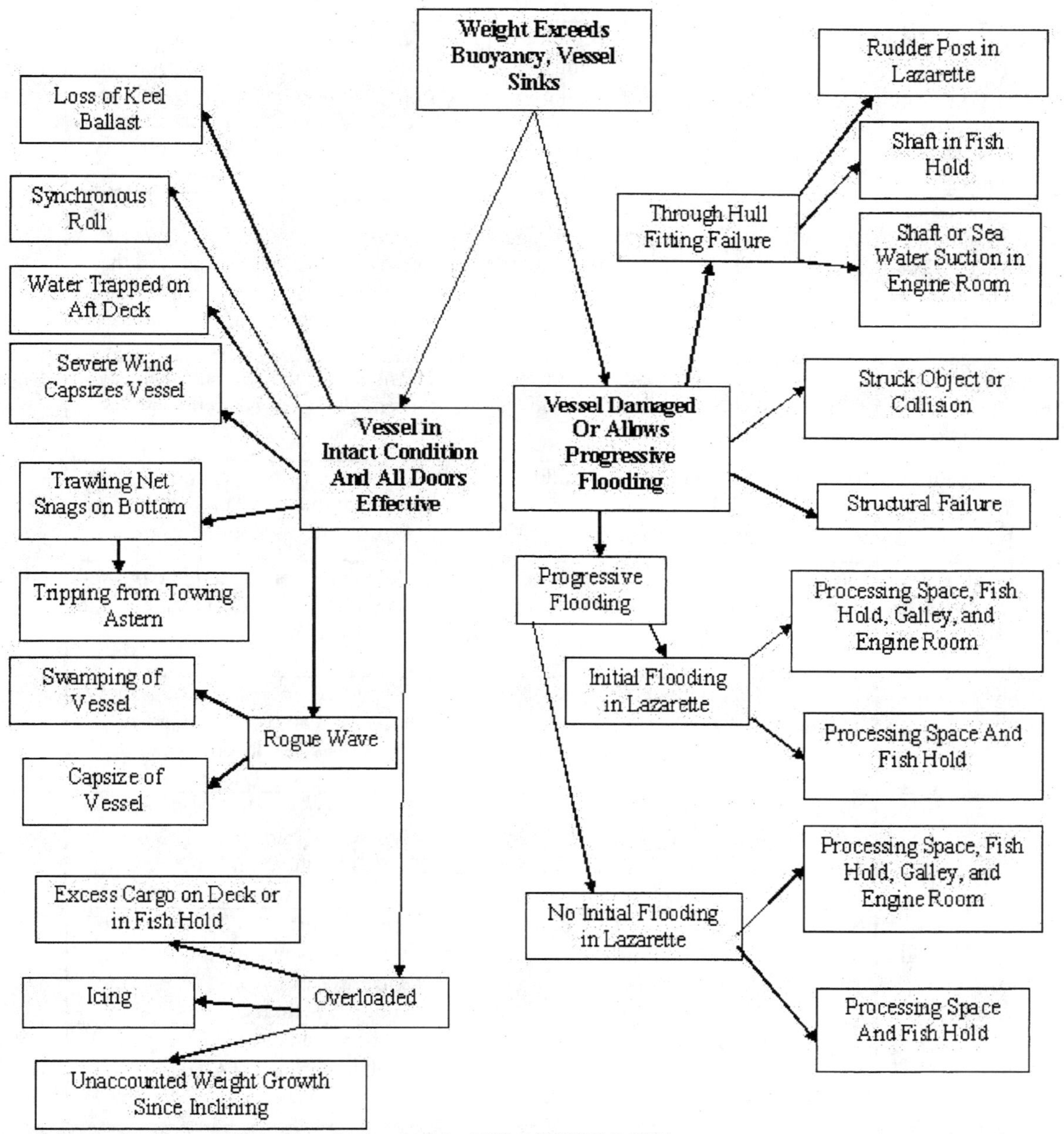

Figure 9 Scenarios that could lead to the loss of the F/V *ARCTIC ROSE*

Scenario That Could Lead to the Loss of the *ARCTIC ROSE*	**Assumptions**	**Findings**	**Likelihood**
Progressive Flooding - No Initial Flooding in Lazarette - Processing Space, Galley, Engine Room, and Fish Hold Flood	*ARCTIC ROSE* in damaged condition; Aft starboard door in processing space open and hatch to fish hold allows progressive flooding; Doors leading to engine room and galley allow progressive flooding	*ARCTIC ROSE* would capsize between 1 minute 40 seconds and 2 minutes 43 seconds after progressive flooding begins, and the *ARCTIC ROSE* would sink between four minutes and eight minutes after progressive flooding began	Most Likely
Progressive Flooding - No Initial Flooding in Lazarette - Processing Space and Fish Hold Flood	*ARCTIC ROSE* in damaged condition; Aft starboard door in processing space open and hatch to fish hold allows progressive flooding	*ARCTIC ROSE* would capsize approximately 3 minutes 23 seconds after progressive flooding begins, and the *ARCTIC ROSE* would sink approximately 25 minutes 35 seconds after progressive flooding began	Likely
Progressive Flooding - Initial Flooding in Lazarette - Processing Space, Galley, Engine Room, and Fish Hold Flood	*ARCTIC ROSE* in damaged condition; Aft starboard door in processing space open and hatch to fish hold allows progressive flooding; Doors leading to engine room and galley allow progressive flooding	*ARCTIC ROSE* would capsize approximately 2 minutes 30 seconds after progressive flooding begins, and the *ARCTIC ROSE* would sink approximately 5 minutes 19 seconds after progressive flooding began	Likely
Progressive Flooding - Initial Flooding in Lazarette - Processing Space and Fish Hold Flood	*ARCTIC ROSE* in damaged condition; Aft starboard door in processing space open and hatch to fish hold allows progressive flooding	*ARCTIC ROSE* would capsize approximately 2 minutes 53 seconds after progressive flooding begins, and the *ARCTIC ROSE* would sink approximately 13 minutes 11 seconds after progressive flooding began	Likely
Through Hull Fitting Failure - Rudder Post in Lazarette	*ARCTIC ROSE* in damaged condition; Lazarette and Dry Stores Flood	*ARCTIC ROSE* met damage stability requirements when lazarette and dry stores flooded.	Unlikely
Through Hull Fitting Failure - Shaft in Fish Hold	*ARCTIC ROSE* in damaged condition; Fish Hold Floods	*ARCTIC ROSE* met intact and damage stability requirements when lazarette and dry stores flooded.	Unlikely
Through Hull Fitting Failure - Shaft or Sea Water Suction in Engine Room	*ARCTIC ROSE* in damaged condition; Engine Room and Machinery Space Flood	*ARCTIC ROSE* met intact and damage stability requirements when lazarette and dry stores flooded.	Unlikely
Struck Object or Collision	*ARCTIC ROSE* in damaged condition; Forepeak damaged or Wing Fuel Oil Tanks Damaged	*ARCTIC ROSE* met almost all intact stability and all damage stability requirements when lazarette and dry stores flooded.	Unlikely
Water Trapped on Aft Deck	*ARCTIC ROSE* in intact condition; Water trapped on aft deck and cannot escape from freeing ports or stern ramp	*ARCTIC ROSE*'s stability would be severely reduced, but water would clear quickly out freeing ports and aft stern ramp	Unlikely
Overloaded - Unaccounted Weight Growth Since Inclining	*ARCTIC ROSE* in intact condition	Weight additions to *ARCTIC ROSE* did not significantly change the stability characteristics of the vessel since the 1999 stability report	Unlikely
Rogue Wave - Capsizing of *ARCTIC ROSE*	*ARCTIC ROSE* in intact condition; Very large wave rolls *ARCTIC ROSE* to angle of vanishing stability	Rogue wave would have to be at least 36 feet high to capsize *ARCTIC ROSE*	Very Unlikely

Scenario That Could Lead to the Loss of the *ARCTIC ROSE*	**Assumptions**	**Findings**	**Likelihood**
Loss of Keel Ballast	*ARCTIC ROSE* in intact condition; Keel ballast falls off *ARCTIC ROSE*	*ARCTIC ROSE*'s stability would not be significantly affected by loss of ballast, and no evidence of structural damage	Very Unlikely
Overloaded - Excess Cargo on Deck or in Fish Hold	*ARCTIC ROSE* in intact condition	*ARCTIC ROSE* was not overloaded at time of casualty	Very Unlikely
Structural Failure	*ARCTIC ROSE* in damaged condition	*ARCTIC ROSE* did not appear to suffer a structural failure	Very Unlikely
Synchronous Roll	*ARCTIC ROSE* in intact condition	*ARCTIC ROSE* was not in synchronous roll conditions	Very Unlikely
Severe Wind Capsizes *ARCTIC ROSE*	*ARCTIC ROSE* in intact condition; Strong wind heels ARCTIC ROSE to angle of vanishing stability	100 knot wind would be insufficient to capsize *ARCTIC ROSE*	Very Unlikely
Rogue Wave - Swamping of *ARCTIC ROSE*	*ARCTIC ROSE* in intact condition; Aft starboard door in processing space open and hatch to fish hold allows progressive flooding	*ARCTIC ROSE* would have to take on 500 tons of water in one wave period for vessel to be swamped	Very Unlikely
Trawling Net Snags on Bottom	*ARCTIC ROSE* in intact condition	*ARCTIC ROSE* was most likely not fishing at time of accident	Very Unlikely
Overloaded - Icing	*ARCTIC ROSE* in intact condition	*ARCTIC ROSE* was most likely not in icing conditions	Very Unlikely

Table 5 Summary of the Scenario Analyses Illustrated in Figure 9

Appendix B Free Surface Analysis Methodology

An exact static equilibrium free surface analysis for all pocketing conditions in rectangular spaces was developed for this project. It involves a four zone analysis for calculating center of gravity shifts of the flooding water as illustrated in Figures 10 and 11. which show the free surface zones for the 24 foot wide by 8 foot high by 25 foot long processing space of 3:1 aspect ratio.

Zone 1. For small angles of heel to starboard when the entire deck was covered with water, a trapezoidal solution was used until the port side deck is exposed.

Zone 2. For larger angles of heel at less than 50% flooded when a portion of the deck was exposed and water has collected against one side shell, a triangular solution was used until the water height against the side shell reaches the overhead.

Zone 3. For very large angles of heel when the water reaches from deck to overhead on the starboard side, a trapezoidal solution was used until the water completely covers the deck and side shell.

Zone 4. Finally, once the compartment was over half filled, a rectangular solution subtracting the remaining triangle of air was used to calculate the free surface effect.

NOMENCLATURE

The following is the nomenclature used for the rectangular flooding space. Subscript 0 refers to initial, upright conditions and Subscript 1 refers to a flooded condition at any heel angle

b - space breadth
h - space height
l - space length
μ - space permeability
b_1 – extent of water on deck away from heeled side
d_1 – depth of water on heeled side
o_1 – extent of water along overhead on heeled side
w_1 - flooding weight in tons in a given space
$d_{1\text{-}0}$ – flooded depth @ 0 degrees heel = $h\ (V_f / V_{max})$
$b_{1\text{-}90}$ – flooded depth at 90 degrees heel = $b\ (V_f / V_{max})$
V_{max} – floodable volume of given space = $b\ h\ l\ \mu$
w_{max} – flooding capacity in tons = $V_{max}/35$
V_f – flooded volume = $V_{max}\ (w_1/w_{max})$
Δ_0 - initial displacement of the vessel in tons
Δ_1 – flooded displacement of the vessel in tons
φ - vessel heel angle
φ_{bp} – pocketing angle on deck opposite heeled side
φ_{hp} – pocketing angle at overhead on heeled side
KG = Vertical location of the cg of the vessel
Kg – Vertical location of the cg of the flooding water
GG_1 - shift in the cg of the vessel
gg_1 – shift in the cg of the flooding water with heel
δ - specific volume of flooding water, vol/wt
V_s – Buoyant Volume of vessel = $\delta * \Delta_1$

Pocketing Angle Analysis

With the space less than or equal to 50% filled, the free surface will pocket first on the deck opposite the heeled side during the transition from Zone 1 to Zone 2 as illustrated in Figure 10 at a heel angle of

$$\phi_{bp} = \arctan(2d_{1\text{-}0} / b_0)$$

Pocketing will then next occur at the overhead on the heeled side during the transition from Zone 2 to Zone 3 at a heel angle of

$$\phi_{hp} = \arctan(h / 2b_{1\text{-}90})$$

With the space greater than 50% filled, the free surface will pocket first on the overhead on the heeled side during the transition from Zone 1 to Zone 4 as illustrated in Figure 11 at a heel angle of

$$\phi_{hp} = \arctan((h - d_{1\text{-}0}) / (b/2))$$

Pocketing will next occur on the deck opposite the heeled side during the transition from Zone 4 to Zone 3 at a heel angle of

$$\phi_{bp} = \arctan((h/2) / (b - b_{1\text{-}90}))$$

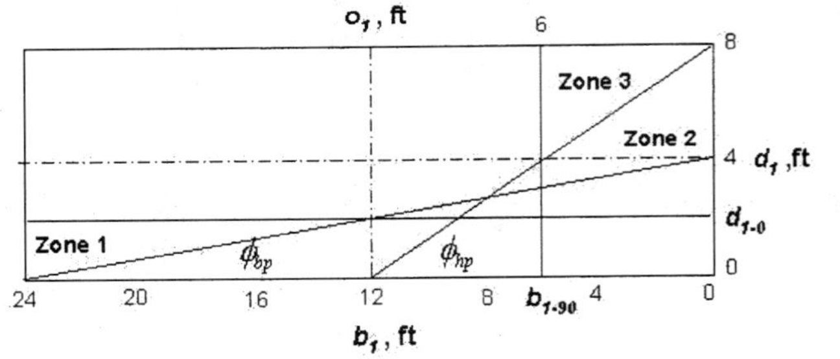

Figure 10 Free Surface Zones for Processing Space 25% filled

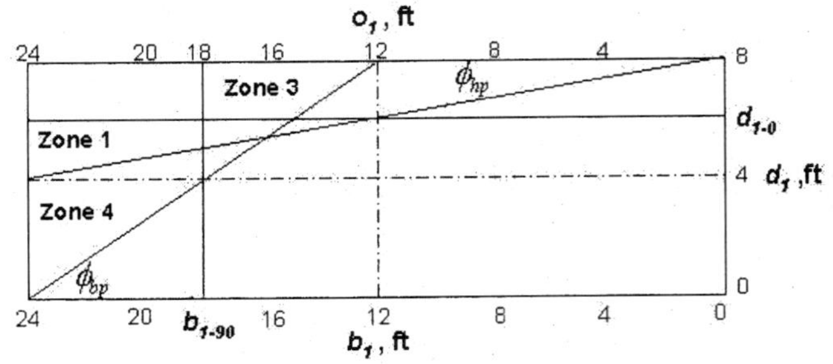

Figure 11 Free Surface Zones for Processing Space 75% filled

Shift in upright KG caused by flooding water

The intact KG is corrected for the weight and center of gravity of the flooding water using the traditional method (Gillmer and Johnson 1983)

$$KG_1 = \frac{\Delta_0 KG_0 + w_1 Kg_1}{\Delta_0 + w_1}$$

where the subscript 0 is used for the original intact conditions and the subscript 1 is for a given flooding weight at a vertical location of the keel-to-space deck baseline distance plus $d_{1\text{-}0} / 2$. This calculation is used to correct the intact stability righting arm curves for the vertical shift in the upright center of gravity.

Shift in transverse and vertical centers of gravity caused by free surface effect.

Heeling causes the centroid of the flooding water to shift both transversely and vertically. By calculating the transverse shift gg_{1T} and vertical shift gg_{1V} the shift in the vessel's center of gravity can be calculated from

$$GG_1 = \frac{w_1 gg_1}{\Delta_1}$$

With the critical heel angles defined one can now do a zone analysis on the free surface center of gravity shifts, both transversely and vertically. The transverse and vertical shifts in the centroid of the flooding water with heel angle can be analyzed as follows:

Zone 1: centroid of a trapezoid for $\varphi <= \varphi_{bp}$ and $\varphi <= \varphi_{hp}$

$b_1 = b$

$d_1 = d_{1\text{-}0} + b/2 * \tan\phi$

$o_1 = 0$

$$GG_{1T} = \frac{\mu b^3 l}{12\nabla_s} \tan(\varphi)$$

$$GG_{1V} = \frac{\mu b^3 l}{12\nabla_s} \frac{\tan(\varphi)^2}{2}$$

Zone 2: centroid of a triangle for $\varphi_{bp} <= \varphi <= \varphi_{hp}$

$$b_1 = \left(\frac{2V_f}{l \tan\phi} \right)^{0.5}$$

$d_1 = b_1 * \tan\phi$

$o_1 = 0$

$$GG_{1T} = (b/2 - b_1/3) \frac{w_1}{\Delta_1}$$

$$GG_{1V} = (-d_{1-0}/2 + d_1/3) \frac{w_1}{\Delta_1}$$

Zone 3 centroid of a trapezoid for $\varphi > \varphi_{bp}$ and $\varphi > \varphi_{hp}$

$$b_1 = b_{1-90} + \frac{h}{2}\tan(90 - \phi)$$

$d_1 = hb$

$$o_1 = b_{1-90} - \frac{h}{2}\tan(90 - \phi)$$

$$GG_{1T} = \left[\frac{b}{2} - \left(\frac{b_1^2 + b_1 o_1 + o_1^2}{3(b_1 + o_1)} \right) \right] \frac{w_1}{\Delta_1}$$

$$GG_{1V} = \left(-\frac{d_{1-0}}{2} + \frac{h(2o_1 + b_1)}{3(b_1 + o_1)} \right) \frac{w_1}{\Delta_1}$$

Zone 4 centroid of a rectangle minus a triangular void for $\varphi_{bp} > \varphi > \varphi_{hp}$

$b_1 = b$

$d_1 = h$

$$o_1 = b - \left(\frac{2(V_{\max} - V_f)}{l \tan(\phi)} \right)^{0.5}$$

$$GG_{1T} = \left[\frac{b}{2} - \left(\frac{\frac{b}{2}bh - (o_1 + \frac{2}{3}(b - o_1))(h - d_{1-0})b}{bd_{1-0}} \right) \right] \frac{w_1}{\Delta_1}$$

$$GG_{1V} = \left[-\frac{d_{1-0}}{2} + \left(\frac{\frac{h}{2}bh - (h - \frac{1}{3}(b - o_1)\tan\phi)(h - d_{1-0})b}{bd_{1-0}} \right) \right] \frac{w_1}{\Delta_1}$$

These formulas can be coded in spreadsheet columns using appropriate nested if statements as a function of the critical heel angles for pocketing (as shown in Figures 12 and 13) and heel angle as illustrated on Figures 14-16 for 10%, 20%, 30%, 40% 50% flooded. (The gg_1 corrections are a mirror image about the 50% flooded calculation (60% overlays 40%, and so forth), so higher flooding percentages differ only by the change in w_1 / Δ_1 in Figures 14 and 15.

The combined righting arm correction can now be made using:

$$RA_{cor} = GG_{1T}\cos\phi + GG_{1V}\sin\phi$$

And this curve for a single 3:1 aspect ratio flooded space is illustrated in Figure 16

This exact static solution provides a more rational and computer friendly analysis of the effect of free surface than the moment of transference method outlined in PNA (Comstock 1967, Lewis 1988). The moment of transference is only tabulated in tables for tanks 50%, 95% and 98% full (see Figure 17) which is not particularly useful in damaged stability analysis of flooded spaces. As formulated in PNA, the method does not account for permeability.

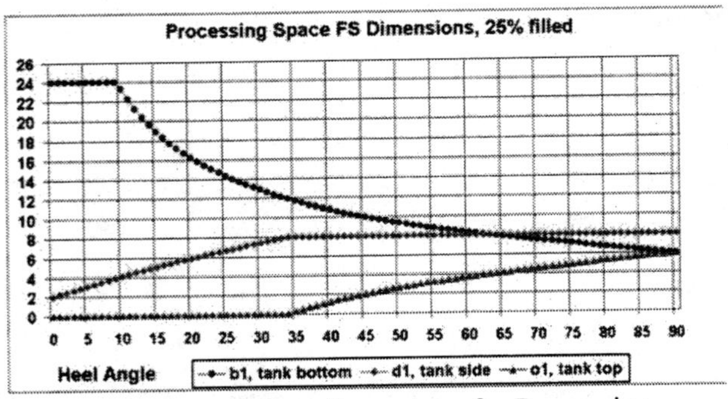

Figure 12. Free surface dimensions for Processing Space 25% flooded with a permeability of 0.92

Note that the free surface is trapezoidal from 0 to 9.5 degrees, triangular from 9.5 to 33.5 degrees, and then trapezoidal from 33.5 to 90 degrees when it becomes a 6' x 8' rectangle.

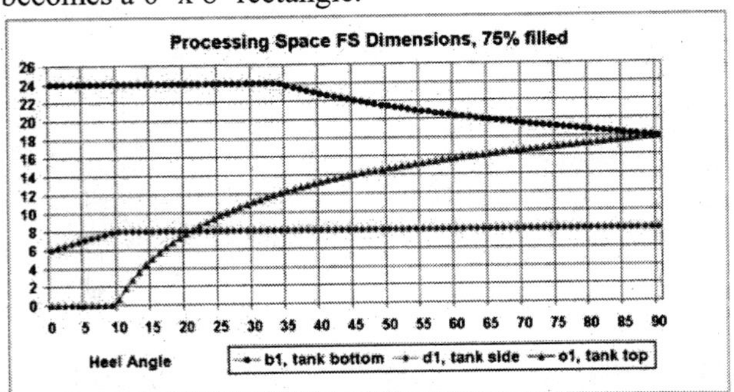

Figure 13 Free Surface dimensions for Processing Space 75% flooded

Note that the free surface is trapezoidal from 0 to 9.5 degrees, rectangular with a triangular air space from 9.5 to 33.5 degrees and trapezoidal from 33.5 to 90 degrees when it becomes an 18' x 8' rectangle

The using the zone analysis and the equations for the centroid of the free surface used in the GGT and GGV formulations enables one to calculate the exact static transverse and vertical weight shifts in a rectangular space as shown in Figures 14-15. Since the processing space was above the main deck, the flooding calculations did not exceed 50% before the vessel capsized.

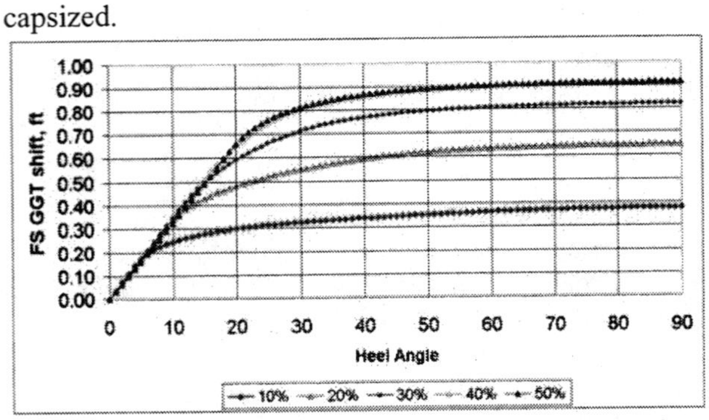

Figure 14 Transverse CG shift caused by free surface in Processing Space at initial displacement of 350 tons and a permeability of 0.92. Note that the 40% curve essentially overlays the 50% filled curve.

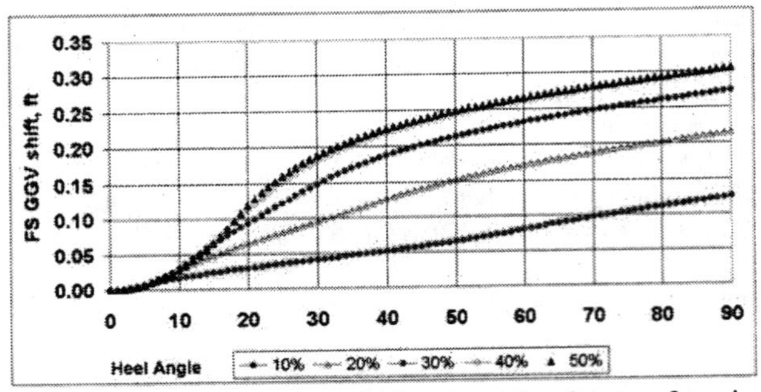

Figure 15 Vertical CG shift caused by free surface in processing space at an initial displacement of 350 tons and a permeability of 0.92. Note that the 40% curve essentially overlays the 50% filled curve.

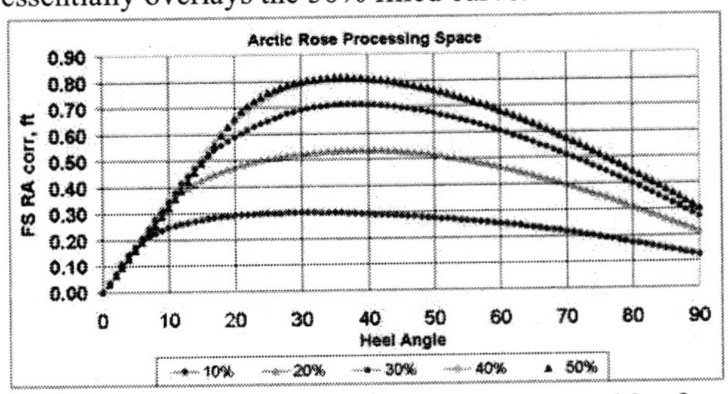

Figure 16 Net Righting Arm correction caused by free surface in processing space at an initial displacement of 350 tons and a permeability of 0.92. Note that the 40% curve essentially overlays the 50% filled curve

Moment of Transference Method

The righting arm corrections can be converted to a moment of transference method used in PNA (Comstock 1967 and Lewis 1988) using the expression

$$C = \frac{\nabla_s}{b^3 l/12} FSRA_{corr}$$

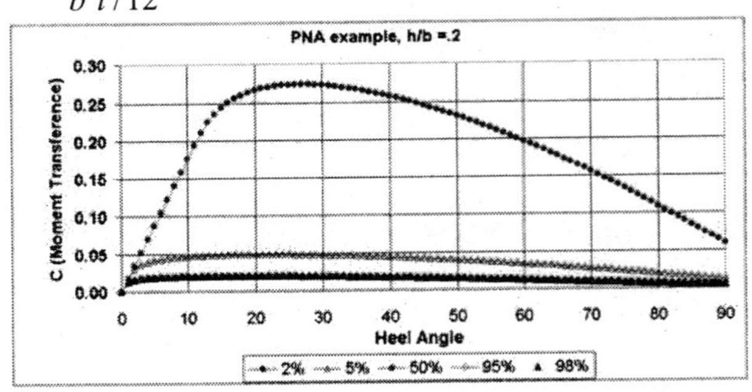

Figure 17 Non-Dimensional Moment of Transference for example in the Intact Stability chapter of PNA

Note the 50%, 95%, and 98% curves match the table values in PNA exactly, but also show that the moment of transference is mirror imaged about 50%. The new free surface analysis method requires no interpolation and works at any heel angle for any rectangular space.

The next figures illustrate the moment of transference for the *ARCTIC ROSE* processing space and the engine room space.

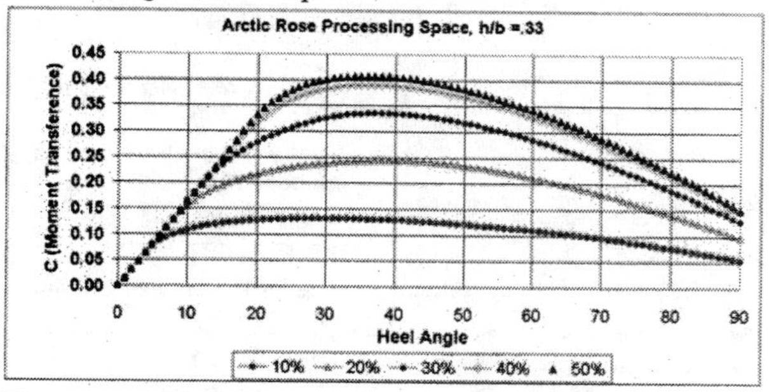

Figure 18 The moment of transference curves for the *ARCTIC ROSE* processing space which was 24 feet wide and 8 feet high and had an assumed permeability of 0.92.

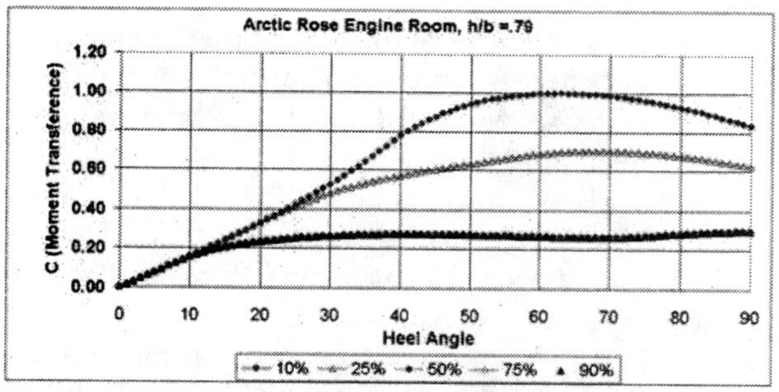

Figure 19 The moment of transference curves for the *ARCTIC ROSE* engine room which was 14 feet wide and 10.9 feet high and had an assumed permeability of 0.90. In the current analysis the engine room flooded to about 60% full, before the water flowed back into the galley area during the large heel angles and the vessel neared the capsize conditions.

The use of the four zone analysis correctly accounted for pocketing, and was therefore much more exact than merely using the traditional moment of inertia method to calculate free surface effect. The next spreadsheet development should consider correction factors for non rectangular spaces similar to the factor F developed in Sections 5.5 of PNA (Comstock 1967) or possibly using numerical integration as is done by some of the more sophisticated software packages

Appendix C Flooding Analysis Methodology

The flooding analysis for each scenario was based on the following sequence of actions:

1. Assemble vessel geometry data and hydrostatics data from a separate hydrostatics program.
2. Format as lookup tables for spreadsheet use the hydrostatic tables, righting arm tables and trim as a function of heel tables.
3. Identify the floodable spaces, the likely flooding paths, and the details on the space and hatch dimensions and locations.
4. Set up the space by space analysis to calculate the needed parameters to perform the 4 zone free surface calculations outlined in Appendix B. Check the charts of each space (Figures 12-16) for discontinuities and make corrections as needed.
5. Set up the hatch flow calculations as a function of heel and trim including the height of water over each hatch as a function of the flooding water level. Check for discontinuities by graphing the results. Compare the results with known values from sketches of the space and hatches with various water surfaces at know heel angles. (Figure 21)
6. Set up the Scenario analysis page as illustrated in Table 4 for sufficient steps to take the analysis to capsize.
7. Perform a step by step static equilibrium analysis by driving the mean square error between the assumed tonnage in each space and the equilibrium values to as close to zero as possible. Use the Excel solver for the early stages. The later stages may require a trial and error manual solution as the hydrostatic parameters start to diverge from well behaved. (see below)
8. Monitor the graphical output of all calculations to make sure the solution makes sense! (Figure 6)
9. Save the scenario steps as each is completed using the copy and paste values only method.

The hydrostatic outputs from GHS were graphed to get an idea of how each parameter behaved as the displacement of the vessel increased with time.

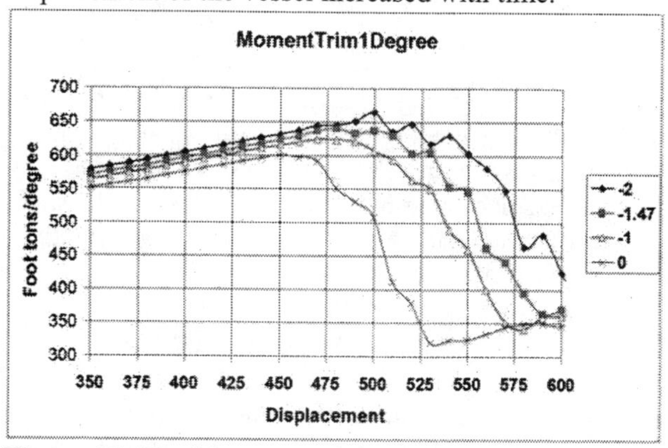

Figure 20 The Moment to Trim 1 degree from GHS as a function of upright trim angle.

At displacements above 475 tons, the solution oscillated between various final trim angles and equilibrium states until the LCF variation was locked at the original upright trim angle of -1.47 degrees for the

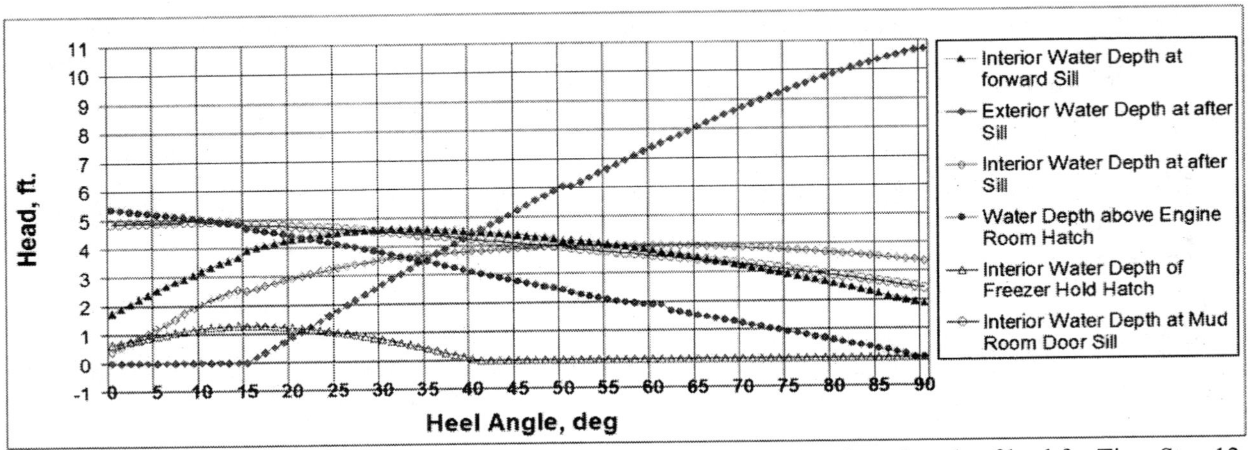

Figure 21 Variation of flooding water heads in each floodable space as a function of angle of heel for Time Step 12 in Scenario C.

lookup table. The LCF was chosen since it had the least variation in value with increasing displacement.

GHS can output the distance to the still water line from the forward and after hatch corners as a function of both heel angle and displacement. This data was added as another lookup table. A graph of the table values used to calculate the height of the water above the WT outside door sill (which eliminated the need for the roll axis assumption) and the static trim correction from the upright trim angle are shown in Figure 22.

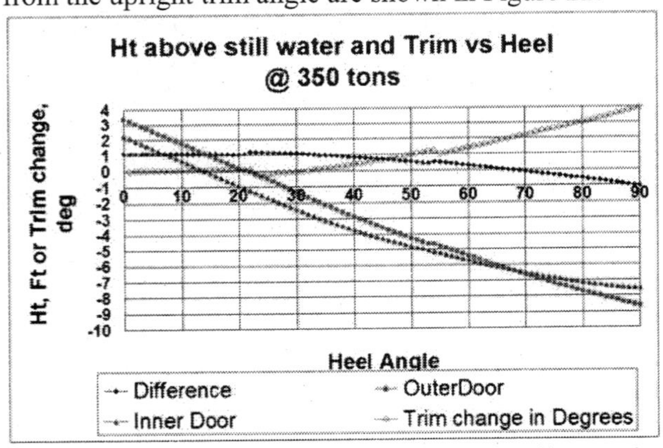

Figure 22 The height above the still water line at the outside door and the trim change as a function of heel angle at displacement increments of 10 tons. The small wiggles in the GHS generated curves carry through to the flooding calculations.

Appendix D Sensitivity Analysis

At the suggestion of the Papers Committee, the authors performed a sensitivity analysis concerning several of the assumptions used in this analysis.

This required a complete rewrite of the flooding analysis sections of the spreadsheet between February and May. The original assumptions for the Marine Board analysis included

- **All flow calculations were based on head differences at the static upright trim angle**
- **Roll axis was fixed for the external water head analysis**

These assumptions are no longer necessary since a methodology for calculating trim change as a function of both heel and displacement was developed for the project as explained in Appendix C.

The original two wave scenario performed for the Marine Board allowed the engine room to fill to capacity which took 100 seconds in several trials including changing the encountered wave period to 12 seconds, lengthening the time step to 3 seconds.

Adding a dynamic pitch angle correction to the "Wave Raises Stern" and "Wave Lowers Stern" time steps enabled the authors to check to see if the analysis was sensitive to whether or not dynamic pitch on the front and back of a wave would have any large change in the results. A comparison of a Scenario A with an average pitch angle correction of +- 2 degrees and Scenario B with no pitch corrections revealed a slight difference in amount of flood water taken on board during the equal capsize times of 88 seconds. Scenario A took on 197 tons of flooding water and Scenario B took on 187 tons of flooding water. All of the most recent scenarios allowed the engine room flooding water to flow back into the galley at high heel angles. (See Figure 7b).

This can be seen in the following comparison of flooding Scenarios A and B (Figures 23-26) which differ only by A having a 2 degree pitch correction. Note that Scenario A had more time steps with outflows from the processing space, especially during the "Wave Lowers Stern" step. Thus it capsized with a slightly lower amount of flooding water on board.

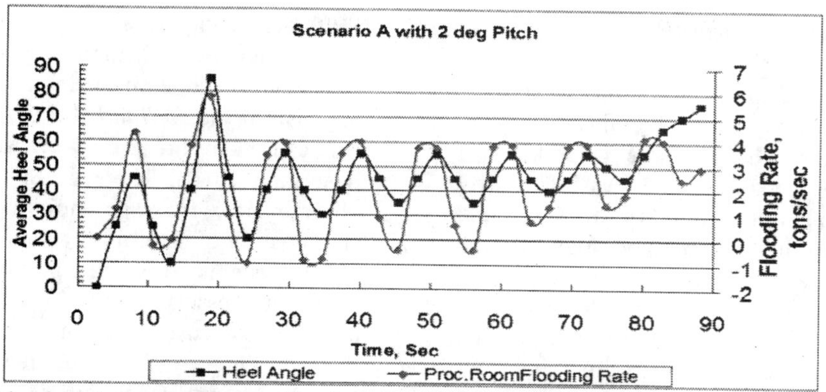

Figure 23 Assumed heel angles and resulting flooding rates as a function of time for Scenario A

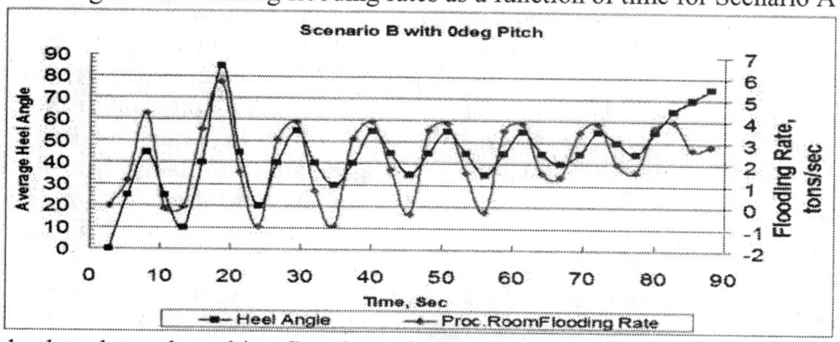

Figure 24 Assumed heel angles and resulting flooding rates as a function of time for Scenario B

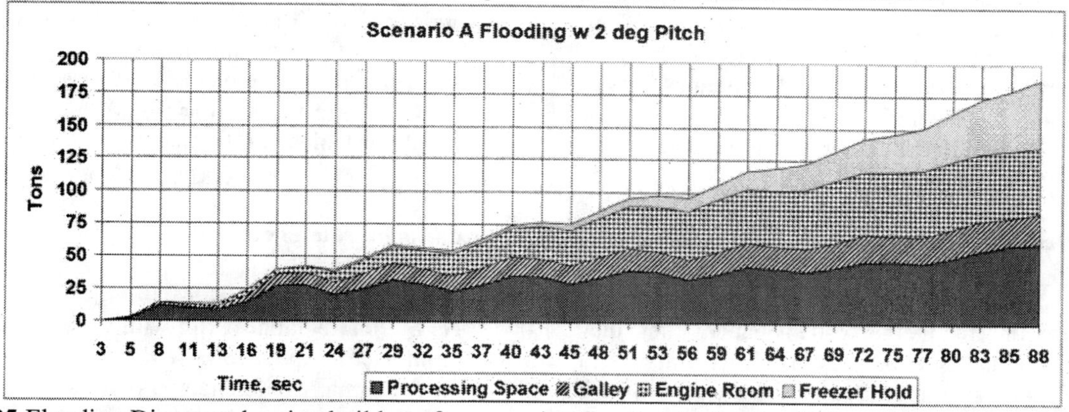

Figure 25 Flooding Diagram showing buildup of progressive flooding in the floodable spaces as a function of time and positive flooding rates in and out of the spaces for Scenario A

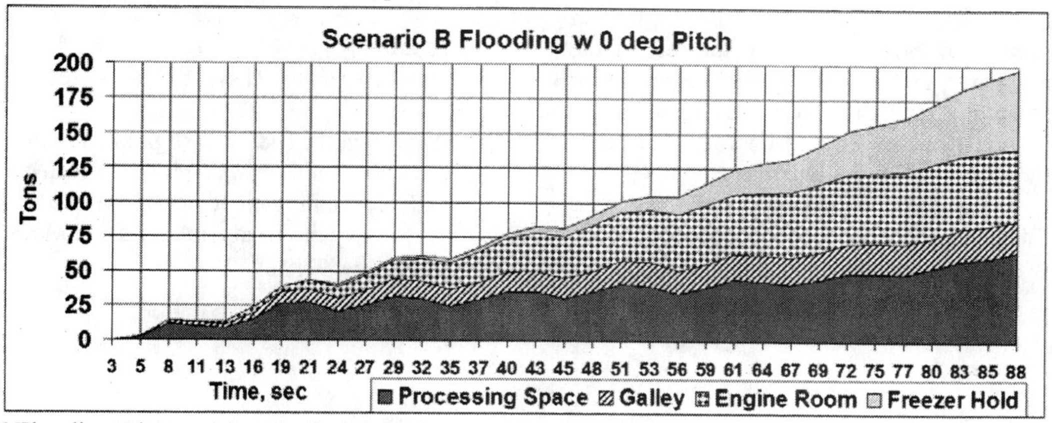

Figure 26 Flooding Diagram showing buildup of progressive flooding in the floodable spaces as a function of time and positive flooding rates in and out of the spaces for Scenario B.

DISCUSSION

William S. Peters, Member

I want to thank the authors for their very comprehensive treatment of this subject. They have shown by example what kind of analysis can be performed today using readily available computational tools. In this sense, the simplified time-to-flood methodology for the assessment of casualties caused by progressive flooding represents an important addition to the ways by which this type of casualty can be analyzed without specialized software.

The authors also should be recognized for their eagerness to share their work in progress with the SNAME Stability Panel SD-3 earlier this year.

Having been involved in time-to-flood analyses for ships much larger than fishing vessels and with considerably more internal compartments than the *Arctic Rose*, I would suggest that there is a practical limit to the number of flooding compartments that can be included in the flooding methodology using a spreadsheet program presented here. That said, I ask the authors to comment on what they think the limitations of the spreadsheet time-to-flood methodology are and the conditions under which the effect of trapped air and dynamics effects should be considered.

Again, the authors are to be thanked for this timely paper.

V. Belenky, Member

The authors have to be congratulated on two points for their excellent paper. First, they have addressed a very socially important problem: fishermen safety. The perils of the commercial fishing industry are not completely recognized by the public, and the risks to which fishermen are exposed are much higher in comparison to those existing in the rest of the food industry. Second, the paper is impressive because of the depth of analysis performed.

One of the peculiarities of this case is that the ship was lost while it was compliant with the current stability regulations (as far as it was possible to judge). This does not mean, however, that the current stability regulations are not strict enough. It might indicate that there are no regulations with 100% safety warranty. There are other examples of fishing vessels that capsized with evident violation of stability regulations: in the case of the *Helland Hansen*, the capsizing was caused by breaking wave action (Dahle & Kjaerland 1980). The way to achieve a safer commercial fishing industry may be a combination of stability regulations and onboard safety systems. However, putting such devices on fishing vessels that need it most might not be simple due to cost limitations.

The paper also demonstrated the importance of water on deck and flooding through the deck. This problem is not new: there is a significant amount of information available, including comprehensive model test data of SSPA tests (Grochowalski 1989). These test data might be useful for the development of a new generation of stability regulations for fishing vessels considering more realistic situations related to the dynamic influence of water on deck.

Recent numerical study has shown that such influence is significant for correct prediction of both ship behavior and mass of water on deck (Belenky et al 2003). Realistic modeling of the changing of amount of water on deck seems to be one of the keys to performing accurate time-to-flood analysis, a perfect example of which can be found in the presented paper.

Additional references

Belenky, V. L., Weems, K. M., Liut, D., and Shin, Y. S. 2003 Nonlinear roll with water-on-deck: numerical approach, *Proceedings*, 8th International Conference on Stability of Ships and Ocean Vehicles, Madrid, Spain.

Dahle, E. A., and Kjaerland, O. 1980 The capsizing of M/S HELLAND HANSEN, *Naval Architect*, RINA, **122**, 51–70.

Grochowalski, S. 1989 Investigation into the physics of ship capsizing by combined captive and free-running model tests, *Transactions*, the Society of Naval Architects and Marine Engineers, **97**, 169–212.

Maneuvering Simulations – An Application to Waterway Navigability

Bruce L. Hutchison, P.E., Life Fellow, **David L. Gray, P.E.,** Member and **Thomas Mathai, Ph.D.,** Visitor, The Glosten Associates, Inc.

ABSTRACT

Gray et al. (2003) have identified channel width as an emerging important feature of channel design. In this paper a powerful new procedure is advanced for the analysis of navigational aberrancy and channel width requirements. The procedure combines fast-time Monte Carlo maneuvering simulations with least squares regression of extreme value functions and the application of order statistics methods. Minimum required navigation channel width (between bridge piers) was estimated by the authors using the PIANC (Permanent International Association of Navigation Congresses) concept design method, fast-time Monte Carlo maneuvering simulations, and full-mission, real-time simulation. Excellent agreement between the fast-time Monte Carlo and full-mission, real-time simulation is reported.

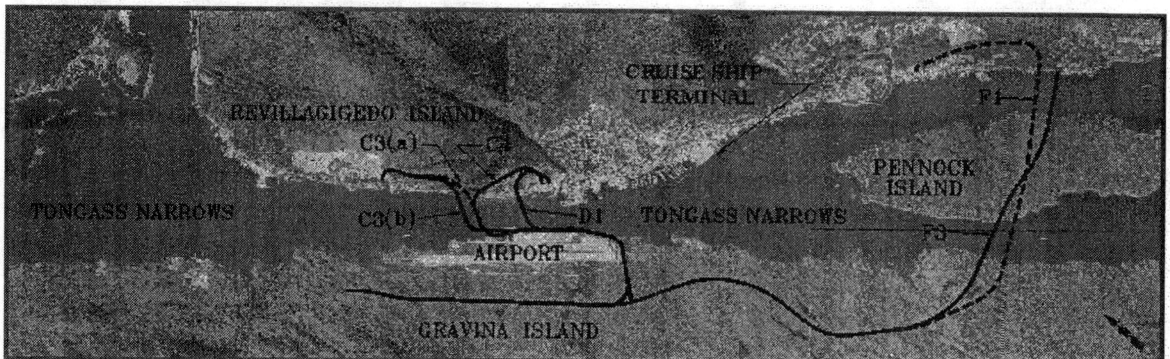

Fig. 1 Gravina access project alternatives

PART I

INTRODUCTION

Tongass Narrows is the most heavily trafficked waterway in Alaska. The U.S. Army Corps of Engineers Waterborne Commerce Statistics Center reports nearly 14,000 vessel trips in or through Tongass Narrows in 1999 (the latest year for which final statistics have been compiled). While total cargo tonnage of waterborne commerce through Tongass Narrows has seen a decreasing trend over the past decade the total number of vessel transits has seen an increasing trend and the character of the marine traffic has been shifting from commercial cargo towards cruise ships.

As indicated in Table 1 there were over 519 cruise ship calls at Ketchikan in 2001 and 385 of those calls were by large cruise ships. Overall, cruise ships calling at Ketchikan carried over 650,000 passengers in 2001.

Table 1
2001 Cruise Ship Traffic Calling at Ketchikan

Ship Size	Passengers	Ships	Port Calls
Small	11,962	24	134
Large	640,524	15	385
Total	652,486	39	519

The year-to-year increase in the annual cruise ship passengers has increased at a rate averaging 37,700 passengers each year over the past twelve years. Large cruise ship traffic is projected to increase to at least 450, and perhaps as many as 700, port calls by 2020.

Currently Ketchikan averages 3 large cruise ship calls per day during mid-summer and on peak days Ketchikan is visited by 5 large cruise ships. According to high estimate projections the average daily port calls by large cruise ships could grow to 5 by the year 2020.

In addition to cruise ships Ketchikan hosts about 1000 annual scheduled departures by vessels of the Alaska Marine Highway System (AMHS). During a recent August AMHS made 116 port calls at Ketchikan.

Tongass Narrows also serves as an important waterway for commercial barge traffic, major U.S. Coast Guard vessels, commercial fishing vessels, and countless private recreational watercraft.

GRAVINA ACCESS PROJECT

Tongass Narrows separates the city of Ketchikan, located on Revillagigedo Island, from The Ketchikan International Airport on Gravina Island. Access between Ketchikan and its airport is currently accomplished via a short ferry crossing at location C4 in Figure 1. The Alaska Department of Transportation and Public Facilities is investigating ways to improve access to Gravina Island, both to improve access to the airport and to gain access to developable land and resources located on Gravina Island. Several proposed bridge locations were studied, first with fast-time Monte Carlo maneuvering simulator techniques and then with full-mission, real-time simulations. This paper reviews these processes and compares their results.

Project Alternatives

The project alternatives crossing Tongass Narrows to provide access to Gravina Island are shown in Figure 1. The alignments of the four bridge alternatives used for the navigation simulations are shown. Alternatives C3(a) and C4 (C3(a)/C4) have identical bridge alignments across the Tongass Narrows marine navigation channel. The alignment of bridge alternatives F1 and F3 cross Pennock Island and both the East and West Channels of Tongass Narrows.

Table 2 presents the design vertical clearance and approximate natural channel width for all bridge alternatives, and Table 3 presents statistics of the effective navigational widths of the natural channels for large cruise ships. The effective natural channel widths are determined at the 5 fathom depth contour perpendicular to the nominal navigational track.

Alternatives C3(a)/C4. Alternatives C3(a)/C4 are the two bridge alternatives with vertical clearance suitable for large cruise ships.

Alternatives C3(b) and D. The 120-foot vertical clearance of alternatives C3(b) and D is suitable for such AMHS ferries and small cruise ships, but unsuitable for large cruise ships.

Alternative F1. Alternative F1 comprises a bridge with vertical clearance suitable for large cruise ships over East Channel, the channel currently favored by large cruise ships, and a bridge with 120-foot vertical clearance over west channel, affording passage for AMHS ferries.

Alternative F3. Alternative F3 comprises a bridge across East Channel with a vertical clearance of 60 feet and a bridge across West Channel, with a vertical clearance suitable for large cruise ships. The 60-foot bridge is not suitable for cruise ships, AMHS ferries, Inter-Island Ferry Authority ferries, or the larger U.S. Coast Guard cutters.

Table 2
Gravina Access Project Bridge Alternatives

Bridge Alternative	*Design Vertical Clearance (feet)*	*Class of Design Vessel*
C3(a)	185 or 200	Large Cruise Ships
C3(b)	120	AMHS Ferries
C4	185 or 200	Large Cruise Ships
D	120	AMHS Ferries
F1	185 or 200	Large Cruise Ships
F3	185 or 200	Large Cruise Ships

Table 3
Tongass Narrows Natural Channel Width Statistics by Channel Branch

Tongass Narrows Channel Branch	*Effective Navigational Minimum for Large Cruise Ships (feet)*	*Lowest 5% Channel Width (feet)*	*Median Channel Width (feet)*
North	673	950	1,690
East	477	1,220	1,680
West	476	580	1,930

The approximate natural channel width is between 5-fathom (30-foot) depth contours.
Channel widths are estimated perpendicular to vessel trackline.

PIANC RECOMMENDED PROCEDURES

The Permanent International Association of Navigation Congresses (PIANC) has headquarters in Brussels, Belgium. It is an organization concerned with technical aspects of navigation and port infrastructure, and with the associated safety, economic and environmental matters. PIANC was founded in 1885 and is sponsored by 40 national governments, including the United States, which joined in 1902.

The four technical committees of PIANC's U.S. section: Environment, Shallow-Draft Waterways and Ports, Deep-Draft Waterways and Ports, and Sport and Recreation Navigation complement the structure of the international organization.

Concept Design Methods

Chapter 5 of PIANC's "Approach Channels – A Guide for Design," PIANC (1997), outlines a concept design method for channels based on a design ship (or ships) and determines the minimum recommended channel width as a multiple of the design ship beam. In addition to the intrinsic maneuverability of the design ship(s), the considered factors are:

- Vessel speed (knots): fast, moderate, slow
- Prevailing cross wind (knots)
- Prevailing cross current (knots)
- Prevailing longitudinal current (knots)
- Significant wave height and wave length
- Depth of waterway relative to design ship draft
- Aids to navigation:
  Excellent with shore traffic control
  Good
  Moderate with infrequent poor visibility
  Moderate with frequent poor visibility
- Bottom surface:
  Smooth and soft
  Smooth or sloping and hard
  Rough and hard
- Cargo hazard level: low, medium or high
- Additional width for passing distance in two-way traffic
- Additional width for bank clearance

PIANC's concept design method was applied to Tongass Narrows using the *Carnival Conquest* class of cruise ship as the design ship. The principal dimensions of the ships are given in Table 4, and the results are summarized in Table 5.

Table 4
Dimensions of Cruise Ships in Design Ship Class

Parameter	*Dimension*
Length	894.0 feet
Beam	141.7 feet
Draft	27.2 feet
Speed *	7.0 knots

** Federal regulation restricts speed to 7 knots in Tongass Narrows.*

In the calculation of the estimated minimum channel widths shown in Table 5, the following assumptions were made in all cases:

- The cross current is 0.5 knots;
- The longitudinal current is 3.0 knots;
- The cargo hazard is low; and
- The bank configuration is steep and hard with a rough and hard bottom surface.

According to the PIANC concept design method, under the most favorable of assumptions and circumstances, the minimum recommended channel width could be as little as 326 feet, while under the least favorable of assumptions and circumstances, it might be as large as 609 feet.

Table 5
Minimum and Maximum Channel Widths of Tongass Narrows
Estimated Using PIANC Concept Design Method (One-Way, Light-Density Traffic Only)

Aids to Navigation	*Wind Speed (knots)*	*Water Depth (ft)*	*Channel Widths (in feet) at Various Intrinsic Vessel Maneuverability*		
			Good	*Moderate*	*Poor*
Good	10.15	40	411	439	482
		41	326	354	397
	33	40	482	510	553
Moderate with Frequent Poor Visibility	10.15	40	468	496	538
		41	383	411	453
	33	40	538	567	609
Minimum Channel Width			326	354	397
Maximum Channel Width			538	567	609

Design ship is Carnival Conquest *class cruise ship.*

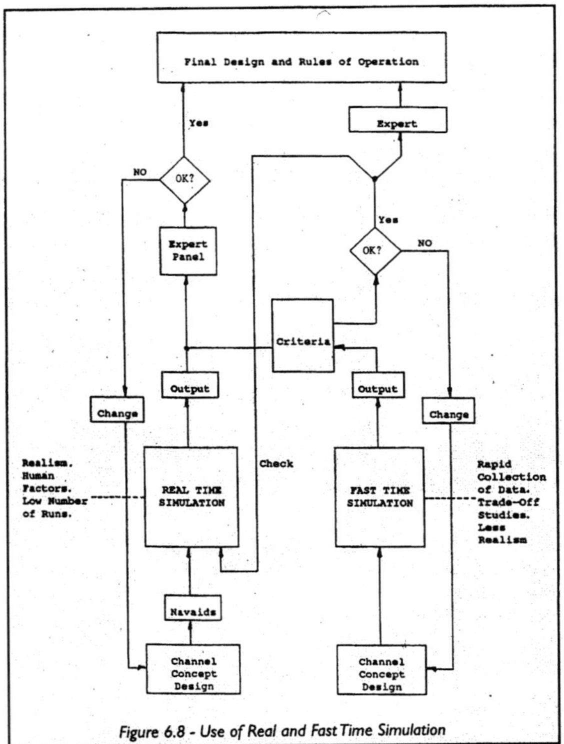

Fig. 2 Use of real- and fast-time simulation (Figure 6.8 from PIANC (1997))

Preliminary Design Method

PIANC does not specifically describe a method for the preliminary design stage, but Chapter 6 of PIANC (1997) describes detailed design methods for channels. Two methods are described, as depicted in Figure 2, the first using real-time simulation methods and the second using fast-time simulation methods. Further applications of simulators are described in Chapters 7 and 8 of PIANC (1997).

Fast-time simulation is appropriate to a preliminary design for which alternative bridge alignments must be evaluated. The advantage of fast-time simulation is the substantially lower cost and elapsed time per run compared with real-time simulation, which makes fast-time simulation more economical in the preliminary evaluation of multiple sites. The disadvantage is that the assessment with fast-time simulation is less realistic, particularly regarding human factors.

Monte Carlo fast-time simulations were used for the preliminary design of this study. Maneuvering models of suitable vessels are used in multiple fast-time simulator runs through the channel associated with a bridge alignment to be evaluated. The channel model include:

- Local current
- Water depth (affecting vessel squat and maneuvering characteristics)
- The location and nature of banks (giving rise to the "bank effects" that plague vessels in restricted channels)

The simulated maneuvers operate under the command of programmed autopilots. Conditions and parameters for each run were drawn from underlying probability distributions for direction of travel (northbound or southbound), tidal stage and associated current, unsteady wind, initial off-track error, initial heading error, and initial yaw rate).

A run begins some distance (e.g., one mile) before the passage under the bridge alignment and continues a short distance beyond it. When passing the plane of the bridge, the closest points of approach to the bank, both on the left and on the right, are recorded as state variables for the run. Runs are repeated hundreds of times to obtain sample probability distributions for the closest point of approach to each bank, as illustrated in Figure 3.

Final Design Method

Full-mission, real-time simulation methods are appropriate for final design once the potential bridge alignments have been winnowed down to one alternative. This is because they best evaluate human factors and also because such techniques are the only methods that can garner the confidence of marine pilots.

Full-mission simulators not only include the hydrographic, hydrodynamic, wind and current models common with the fast-time simulator, but also display visual projections of above-water features (topography, buildings, aids to navigation, etc.) that are viewed by the marine pilots operating the simulated vessel(s). Additionally, weather phenomena such as fog can be simulated, and the simulated time of day can be adjusted for daytime, dusk, or night-time. The marine pilots operate the simulated vessel from a realistic mockup of a navigation bridge. The mockup includes engine, rudder, and bow thruster controls; radar; fathometer; compasses; global positioning system (GPS), and radio. Visual and radar images of other ship traffic can also be generated. However, compared with fast-time Monte Carlo only a limited number of full-mission simulations can be accomplished as the simulations are conducted in real-time.

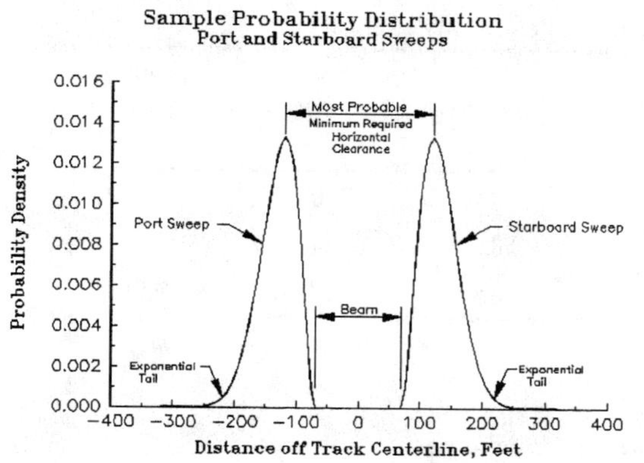

Fig. 3 Sample probability distributions for closest point of approach to each bank

PART II

MONTE CARLO FAST-TIME SIMULATIONS

Monte Carlo maneuvering simulations as set forth in Table 6 were carried out for large cruise ships for alternatives C3(a)/C4, and for AMHS ferries for alternatives C3(b) and D. An identical analysis was undertaken for bridge alternatives F1 and F3. Only the Monte Carlo simulation results from the C3(a)/C4 alternative are presented here.

Synopsis of Results

Marine vessel transits were simulated at the bridge alignments of four project alternatives: C3(a)/C4, C3(b) and D. The number of transits simulated for each bridge site alternative is summarized in Table 6. Cruise ships were used for the full-mission simulations, so the focus here is only on cruise ship transits. The greatest horizontal clearance directly sampled by the fast-time Monte Carlo simulator at each bridge alignment was 1,222 feet for alternatives C3(a)/C4, 502 feet for alternative C3(b), and 664 feet for alternative D.

The horizontal clearances required for a 50-year service life were estimated using the probability distributions obtained from these simulations,. The predications are from a sample of 26,639 transits by large cruise ships and 45,550 transits by AMHS ferries. Summarized statistics for horizontal clearances are given in Table 7.

Table 6
Summary of Monte Carlo Maneuvering Simulations

Class of Vessel and Bridge Alternatives	*Direction of Travel*	*Number of Distinct Vessels*	*Number of Transits Simulated*	*Number of Independent Climate Realizations*
Large Cruise Ships—Alternatives C3(a)/C4	Northbound	8	7,944	993
	Southbound	14	41,860	2,990
	Total		49,804	
AMHS Ferries—Alternatives C3(b) and D	Northbound	6	29,988	4,998
	Southbound	6	29,994	4,999
	Total		59,982	

Table 7
Maximum Horizontal Clearance Statistics for 50-Year Service Life

Bridge Alternative	*Vertical Clearance (feet)*	*Design Class of Vessel*	*Most Probable Clearance (feet)*	*Probability of Exceeding Most Probable Clearance*	*Horizontal Clearance with Exceedance Probability of 0.9999*	*50% Confidence Extreme Clearance (feet)*
C3(a)/C4	185 or 200	Large cruise ships	1,002	0.76	792	1,117
C3(b)	120	AMHS ferries	364	0.64	296	377
D	120	AMHS ferries	587	0.70	393	653

Table 8
Horizontal Clearance Statistics at Common Probability Levels

Bridge Alternative	*Vertical Clearance (feet)*	*Design Class of Vessel*	*0.9 (1 in 10)*	*0.99 (1 in 100)*	*0.999 (1 in 1000)*	*0.9999 (1 in 10,000)*
C3(a)/C4	185 or 200	Large Cruise Ships	298	503	710	918
C3(b)	120	AMHS Ferries	124	184	244	318
D	120	AMHS Ferries	124	185	287	451

These statistics represent horizontal clearances exceeded by one ship. Additional statistics for horizontal clearances are discussed later.

The autopilot parameters necessary to ensure good performance in Tongass Narrows confirm that the professional marine pilots are applying a high level of skill when piloting the large cruise ships through Tongass Narrows. It was observed that the gusting wind was most responsible for the rare large off-track performance. Once a vessel was forced off-track by the wind, the off-track course was further exacerbated by bank suction effects. Not included in these Monte Carlo simulations are emergency maneuvering actions to avoid an actual grounding or impact with bridge piers. Those pilot actions might include crash stop maneuvers, application of bow thrusters, and/or dropping of one or both anchors. It is assumed that the marine pilots and ship masters would have the bow thrusters on standby and the anchors rigged for emergency deployment before transiting Tongass Narrows, especially on a windy day. Thus, the rare off-track behavior that dominates the horizontal clearance requirements identified by this Monte Carlo study may not represent actual groundings or impacts with bridge piers, but merely the potentiality of such groundings or impacts.

The majority of ship transits are confined within a narrow corridor around the intended trackline. To illustrate this, Table 8 provides the horizontal clearances at common probability levels.

For alternatives C3(a)/C4 (large cruise ships), there are real, measurable, nonzero probabilities of allision with a bridge pier over the 50-year service life, unless the bridge piers are located outside the natural channel, i.e., located so that the ship will go aground first. The probability of a large cruise ship excursion to the limits of the natural channel (≈ 1,340 feet) over a 50-year period is approximately 36%. Consequently, unless the bridge piers are located outside the natural channel, pier protection is important.

Analysis Procedure

Tracklines

Cruise ship tracklines through Tongass Narrows were provided by Capt. Robert G. Winter of the Southeastern Alaska Pilots Association.

Autopilot

Two- and three-term trackline autopilots are alternatively applied in these Monte Carlo simulations to mimic the actions of a local marine pilot and quartermaster (helmsman) navigating Tongass Narrows. The autopilot attempts to follow a predetermined trackline that is defined by waypoint coordinates and headings between waypoints. The autopilot functions in two different modes.

Mode 1 is a steering mode that mimics a mechanical or human autopilot commanded to steer to maintain a course heading (e.g., "steer 285° true"). This is accomplished using a standard two-term autopilot function (i.e., heading error and yaw rate error):

$$\delta = c_1(\psi - \psi_o) + c_2(\dot{\psi}' - \dot{\psi}_o') \qquad (1)$$

where:

δ is the commanded rudder angle

ψ is the actual (instantaneous) vessel heading

ψ_o is the desired course heading

c_1 and c_2 are constants (i.e., "gains")

The dot indicates a temporal derivative. The prime denotes non-dimensionalization by

$\sqrt{g/L_{pp}}$ where g is gravitational acceleration and

L_{pp} is the length between perpendiculars.

Mode 2 is a mode that attempts not only to steer a particular course but also to simultaneously adhere to the predetermined trackline, minimizing the off-track distance. This is accomplished by using a three-term autopilot function. If the commanded rudder angle exceeds 10 degrees it is applied in steps of 5 degrees, thus imitating a human pilot:

$$\delta = c_1(\psi - \psi_0) + c_2(\dot{\psi}' - \dot{\psi}_o') + c_3 \frac{\varepsilon(s)}{L_{pp}} \qquad (2)$$

where:

s is a parameter (e.g., distance along track between waypoints)

$\varepsilon(s)$ is the cross-track position error, distance measured perpendicular to the trackline

c_3 is an additional constant (i.e., "gain")

Autopilot Gains

Gains were determined using trial-and-error experimentation. Gains used in these Monte Carlo simulations were (c1, c2, c3) = (3, 0.2, 2).

Looking Ahead and Overshoot at Waypoints

The heading and off-track error are computed by the autopilot based on an anticipated ship position half a ship length directly ahead of the bow. This creates some "look ahead" that anticipates future course changes and avoids a vessel's overshooting at waypoints. Human pilots do look ahead and anticipate changes.

Dead Zone to Sensitivity to Off-Track Error

Human pilots are insensitive to small off-track errors. When the off-track error is small, the human pilot is, in general, content to command the quartermaster to steer a course for heading alone, which results in a track that is essentially parallel to the intended track. However, when the off-track error exceeds some threshold, the human pilot will command measures to bring it back to within acceptable bounds.

Discussions with a marine pilot at the Southeastern Alaska Pilots Association indicated that, for passage under a fixed bridge, a pilot would correct for any detectable off-track distance. Fixed bridges are normally provided with range marks and range lights. If the range marks were 8 inches wide and separated by 60 feet, then a misalignment equal to the width of the range mark would correspond to an off-track error of 33.75 feet at a distance of one-half nautical mile (approximately 3½ ship lengths). The detectable off-track error would become progressively smaller as the ship approached the bridge.

The off-track dead zone modeled in these Monte Carlo simulations is half the beam of the ship on either side of the track line.

Dead Zone for Sensitivity to Heading Error

Similar to the dead zone for sensitivity to off-track error, the marine pilot and quartermaster will exhibit insensitivity to small errors in heading. The helmsman will not act to correct heading below this threshold. The heading error dead zone used in these Monte Carlo simulations was ±1 degree.

Initial Position and Heading

Initial position and heading at the start of each Monte Carlo trial are treated as random variables. Northbound vessels start from a random position within a 1,000-foot-diameter maneuvering area off the Ketchikan terminal, with a random heading that is between 15 degrees to either side of the trackline heading. Southbound vessels start from a random position within a circular area near Waypoint 5. The diameter of this circular area is equal to the beam of the ship. The initial ship heading is random between 5 degrees to either side of the trackline heading.

Speed

Federal regulations, *33 CFR § 162.240 (b),* limit to 7 knots the speed of vessels in the sections of interest in Tongass Narrows. These speed restrictions effectively limit or eliminate the likelihood of overtaking traffic in the most restricted areas of Tongass Narrows.

Northbound vessels start a simulation from within the maneuvering area off Ketchikan terminals with an initial speed of 1 knot and a commanded speed of 7 knots. The southbound vessels start from their initial position near Waypoint 5 with an initial speed of 8 knots and a commanded speed of 7 knots. The command speed is held steady at 7 knots throughout every passage.

Bank Effects

Ships transiting narrow channels are subject to hydrodynamic forces known as "bank effects." These are forces of attraction towards the bank brought about by the accelerated fluid flow between the vessel and the bank. For large R, these forces decay roughly at the rate of 1/R, where R is the distance from the ship to the bank. The attraction forces to opposing banks for a ship navigating down the center of a channel are equal and opposite, thereby canceling. The forces to the near bank become stronger as the vessel moves off centerline, and there is a net force towards the near bank.

Bank effects in these fast-time Monte Carlo maneuvering simulations were modeled using the method of images (Newman, 1965). The banks were located in the hydrodynamic model of Tongass Narrows at the mean lower low water (MLLW) contours on each side of the channel.

One-Way vs. Two-Way Traffic

While two-way traffic is legal in Tongass Narrows, the *de facto* practice of vessels requiring pilotage is to make courtesy passing arrangements outside the speed restricted region. As a consequence of these practices, two-way meetings between large vessels are in actuality rare or nonexistent.

Large Cruise Ship Principal Dimensions

Principal dimensions of the eight large cruise ships modeled for northbound transits and the fourteen vessels modeled for southbound transits are given in Tables 9 and 10. These sets of ships are augmented with respect to 2001 data to capture the continuing trend towards larger cruise ships. Some of the smaller ships in the 2001 Ketchikan cruise calendar have been replaced by larger ships of the same cruise line.

Table 9

Principal Dimensions of Cruise Ships Modeled Using the Fast-Time Monte Carlo Simulator - Northbound

Ship I.D.	*% of Northbound Transits*	*LWL (feet)*	*Water Line Beam (feet)*	*Maximum Beam (feet)*	*Draft (feet)*	*Displacement (L.T.S.W.)*	*Approximate GRT*	*Transverse Windage Area (sq. ft)*	*Lateral Windage Area (sq. ft)*
N-1	10.64%	858	118	158	26.25	49,050	109,000	20,262	114,322
N-2	3.19%	713	97	97	26.33	28,180	48,621	11,303	69,972
N-3	1.06%	870	105	121	26.00	45,578	91,000	12,918	99,038
N-4	9.57%	781	103	119	25.50	40,798	77,713	13,994	87,196
N-5	27.66%	773	106	132	26.58	39,372	77,441	12,918	98,499
N-6	9.57%	826	106	116	25.25	38,309	78,491	13,456	97,961
N-7	29.79%	648	101	118	25.26	31,450	55,451	9,958	69,972
N-8	8.51%	807	110	142	27.20	50,006	101,500	21,221	125,518

Table 10

Principal Dimensions of Cruise Ships Modeled Using the Fast-Time Monte Carlo Simulator - Southbound

Ship I.D.	*% of Southbound Transits*	*LWL (feet)*	*Water Line Beam (feet)*	*Maximum Beam (feet)*	*Draft (feet)*	*Displacement (L.T.S.W.)*	*Approximate GRT*	*Transverse Windage Area (sq. ft)*	*Lateral Windage Area (sq. ft)*
S-1	3.30%	858	118	158	26.25	49,050	109,000	20,262	114,322
S-2	1.83%	713	97	97	26.33	28,180	48,621	11,303	69,972
S-3	7.33%	870	105	121	26.00	45,578	91,000	12,918	99,038
S-4	4.40%	781	103	119	25.50	40,798	77,713	13,994	87,196
S-5	7.69%	680	93	105	23.00	31,103	50,760	11,845	83,808
S-6	10.26%	773	106	132	26.58	39,372	77,441	12,918	98,499
S-7	9.89%	826	106	116	25.25	38,309	78,491	13,456	97,961
S-8	8.06%	765	106	118	26.25	38,391	78,000	13,994	87,196
S-9	3.66%	557	88	88	27.30	22,528	23,500	8,237	54,142
S-10	11.72%	648	101	118	25.26	31,450	55,451	9,958	69,972
S-11	3.66%	807	110	142	27.20	50,006	101,500	21,221	125,518
S-12	5.86%	921	117	156	28.90	63,467	142,000	23,363	143,348
S-13	7.33%	722	96	106	23.60	32,566	53,900	11,845	83,808
S-14	15.02%	704	106	124	26.58	33,344	63,000	12,918	77,508

Ship Maneuvering Characteristics

The ease with which a vessel can be held to a desired course in open water or in a restricted channel is determined by the inherent controllability of the vessel and the actions of the helmsman/ pilot. There are a number of definitive maneuvers that serve to characterize the controllability of a ship. These include the direct or reversed spiral, zigzag, turning circle, etc. The results of the zigzag maneuver are indicative of the ability of a ship's rudder to control the ship. An example of the zigzag trials results for one of the ships in our study is shown in Figure 4. Results for ships in this study were compared with simulated results for the same ships to ensure that the models behave reasonably.

Zigzag, turning circle and spiral maneuver characteristics were collected for as many of the cruise ships currently calling at Ketchikan as possible. The zigzag maneuver characteristics for typical cruise ships are summarized in Table 11 and a typical zigzag maneuver trial record is provided as Figure 4.

Table 11

Zigzag Maneuver Characteristics

10°/10° Zigzag Maneuver				
Cruise Ship I.D.	**Speed, knots**	**Overshoot Angles (degrees)**		**Period (seconds)**
		1st Overshoot	**2nd Overshoot**	
N-1	21.3	-6.6	7.0	201
N-2	21.0	-5.0	8.8	195
N-5	21.4	-7.5	11.0	194
20°/20° Zigzag Maneuver				
Cruise Ship I.D.	**Speed, knots**	**Overshoot Angles (degrees)**		**Period (seconds)**
		1st Overshoot	**2nd Overshoot**	
N-1	20.7	-13.0	19.2	241
N-2	--	--	--	--
N-5	21.5	-18.2	21.4	222

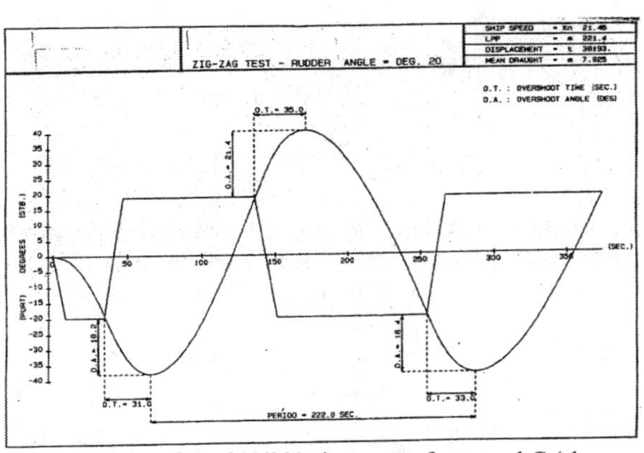

Fig. 4 Results of the 20°/20° zigzag test for vessel C (shown as example)

Wind

A Monte Carlo procedure was used to generate a sophisticated pseudo-time function for unsteady wind acting over Tongass Narrows. The unsteady wind process is divided into two steps, the first to determine the mean wind speed and direction and the second to determine the magnitude and direction of gust perturbations.

Average Wind Parameters

Hourly vector (magnitude and direction) mean wind (one-minute average) is available from the Ketchikan Airport, which is located parallel to and immediately adjacent Tongass Narrows. Monthly joint probabilities of wind speed and direction were developed from the Ketchikan Airport wind data. Monthly joint probabilities for May, June, July, August and September, corresponding to the cruise season in Alaska, were composited to obtain a joint probability distribution of wind speed and direction for the period May through September. This composite joint probability is for a 25-year period of record extending from 1974 through 1999.

Table 12 is the resulting cumulative probability distribution for one-minute average wind speed at the Ketchikan Airport during the period May through September, and Figure 5 depicts the joint probability of mean wind speed and direction *from*.

Table 12

Cumulative Probability, P(U), of Wind Speed at Ketchikan Airport for May through September (Period of Record, 1974-1999)

U (knots)	*P(U)*	*U (knots)*	*P(U)*
0	0	26	0.999799
2	0.000696	28	0.999907
4	0.093681	30	0.999907
6	0.272504	32	0.999923
8	0.566432	34	0.999923
10	0.710326	36	0.999923
12	0.886132	38	0.999938
14	0.955805	40	0.999938
16	0.976085	42	0.999938
18	0.989636	44	0.999954
20	0.997479	46	0.999985
22	0.997974	48	0.999985
24	0.999567	50	1.000000

The implemented Monte Carlo procedure generates a one-minute average wind speed, U (knots), and average wind direction, U_dir (degrees true).

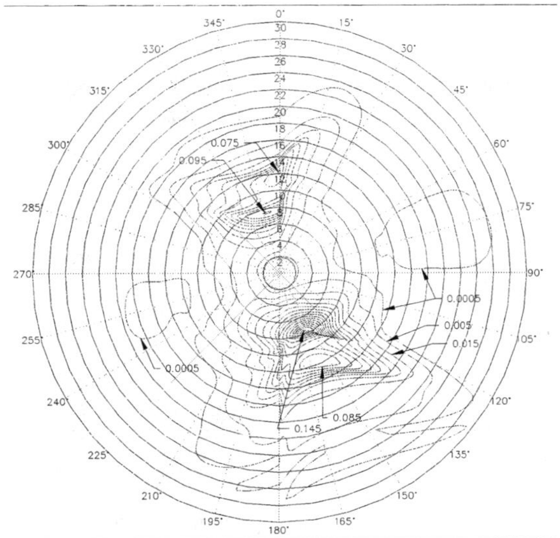

Fig. 5 Joint probability of wind speed and direction

Pseudo-Time Domain Gusting Wind

The average hourly maximum gusts at Ketchikan are 152% of the one-minute average wind speed. The standard error of the regression fit shown in Figure 6 is 4.6.

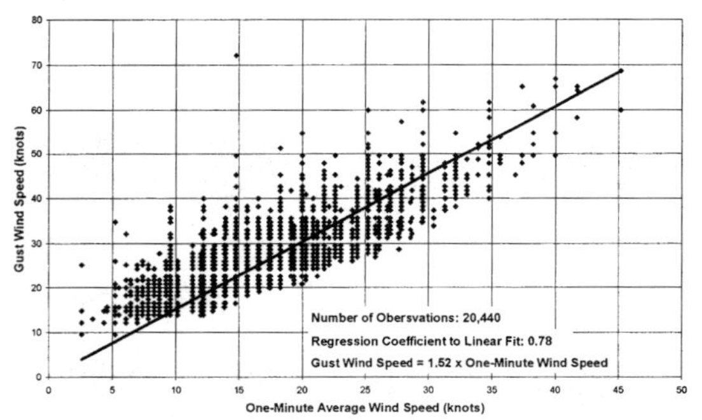

Fig. 6 Correlation between peak gusts and one-minute average wind speed at Ketchikan Airport where the gust amplitude is in knots

A Monte Carlo procedure was implemented that returns gust amplitude, direction and duration during each simulation, conditioned on the one-minute average wind speed. The instantaneous wind realization is the vector sum of the one-minute average wind vector and gust vector. Figure 7 is a sample of a typical simulated time history of wind.

Current

Current in Tongass Narrows is primarily tidal in origin. A Monte Carlo scheme was developed that exploits the cumulative distribution for current velocity to generate a current velocity and direction for each simulation.

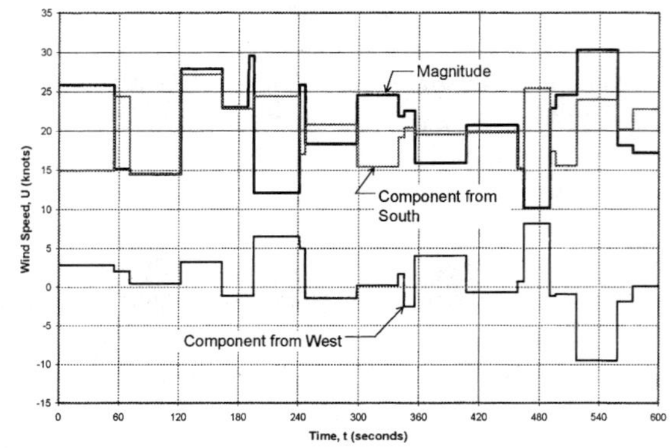

Fig. 7 Sample of typical simulated time history of gusting wind, U = 20 knots (one-minute average)

Discussion of Results for Alternatives C3(a)/C4

Figure 8 shows the cumulative probability distributions of east and west bank sweeps at the bridge alignment of alternatives C3(a)/C4 for combined north- and south-bound large cruise ships. A sweep is defined as the extreme extents of the intersection of the vessel footprint with the bridge alignment line. It is given in terms of the distance from the east end to the two extreme points measured along the bridge alignment line. At the 0.9 probability level, the width between east and west bank sweeps is approximately 298 feet; at the 0.99 level, the width is approximately 504 feet.

Figure 9 shows the probability density function for the horizontal clearance at bridge alternatives C3(a)/C4. The most frequently occurring clearance is on the order of 131 feet, approximating the weighted average of the maximum beams of the cruise ships.

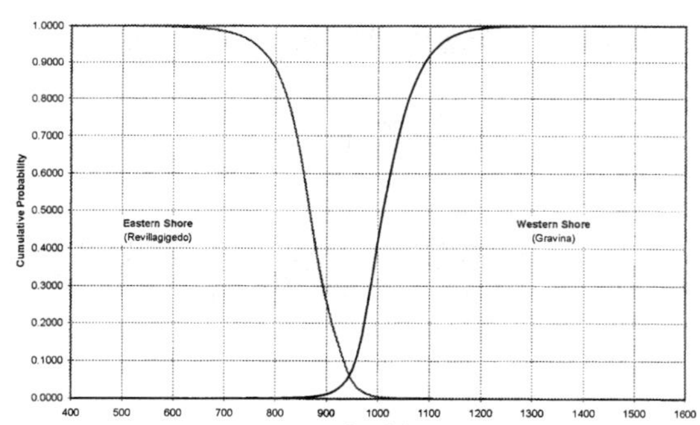

Fig. 8 Cumulative probability distributions of east and west bank sweeps for combined north- and southbound large cruise ships for alternatives C3(a)/C4

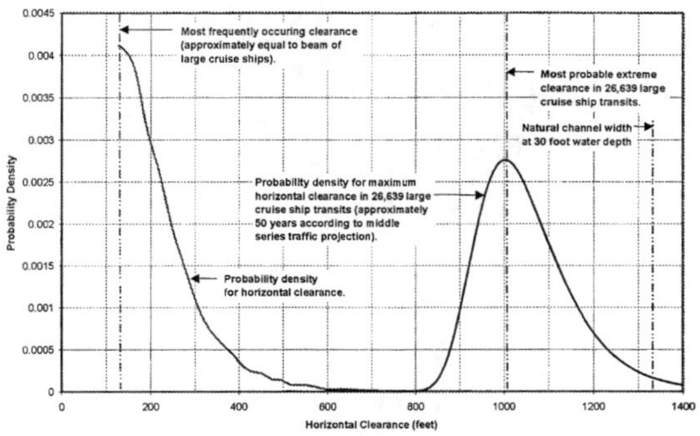

Fig. 9 Probability density distributions of horizontal clearance for combined north- and southbound large cruise ships for alternatives C3(a)/C4

The distribution to the right in Figure 9 was developed using order statistics procedures applied to the cumulative and density distributions at bridge alternatives C3(a)/C4. Order statistics provide a method whereby the cumulative density distributions of extreme values can be determined from the underlying distributions. If f(x) is the probability density and F(x) is the cumulative probability, then the cumulative probability distribution of the maximum in N independent trials is $[F(x)]^N$ and the probability density function (found by differentiation) is $N f(x) [F(x)]^{N-1}$.

The distribution given on the right in Figure 9 is the probability density function for the maximum horizontal clearance in 26,639 large cruise ship transits. That is the number of large cruise ship transits through Tongass Narrows projected to occur over the next 50 years, according to the middle projection in Glosten (2001). The most probable extreme value is approximately 1,002 feet. However, the probability of exceeding this most probable value (in 50 years) is nearly 76%. For 50% confidence in 50 years the horizontal clearance should be 1,117 feet. A natural channel width of 1,340 feet is given in Table 2. At that horizontal clearance, the confidence is approximately 64% in 50 years.

Table 13 provides order statistics for the horizontal clearance at bridge alternatives C3(a)/C4. Order statistics are provided for the horizontal clearance expected to be violated by one, two, three, four, and five ships over a 50-year period in which 26,639 large cruise ships transit.

The fourth column gives the horizontal clearance that is expected to be exceeded with near certainty (probability of 0.99999). Over a 50-year period it is nearly certain that at least one ship would violate a 792-foot horizontal clearance. Likewise, it is nearly certain that at least five large cruise ships would violate a 646-foot horizontal clearance over the same period.

The final column of Table 13 gives the average return period between large cruise ship allision events. If there are five large cruise ship allision events in fifty years the average return period is ten years.

Table 13
50-Year Order Statistics for Horizontal Clearance for Bridge Alternatives C3(a)/C4

Number of Large Cruise Ships Allisions	*Most Probable Horizontal Clearance*	*Probability of Exceeding Most Probable Horizontal Clearance*	*Horizontal Clearance with Exceedance Probability of 0.99999*	*Average Return Period (Years)*
1	1,002 ft	0.76	792 ft	50.0
2	942 ft	0.70	728 ft	25.0
3	906 ft	0.68	692 ft	16.7
4	881 ft	0.67	666 ft	12.5
5	862 ft	0.66	646 ft	10.0
6	846 ft	0.65	629 ft	8.3
7	832 ft	0.65	615 ft	7.1
8	820 ft	0.65	603 ft	6.3
9	810 ft	0.65	592 ft	5.6
10	801 ft	0.64	583 ft	5.0
11	792 ft	0.64	574 ft	4.5
12	785 ft	0.64	566 ft	4.2
13	778 ft	0.64	559 ft	3.8
14	771 ft	0.64	552 ft	3.6
15	765 ft	0.64	546 ft	3.3

Order statistics for 26,639 large cruise ship transits corresponding to 50 years, according to the middle series projections of Glosten (2001)

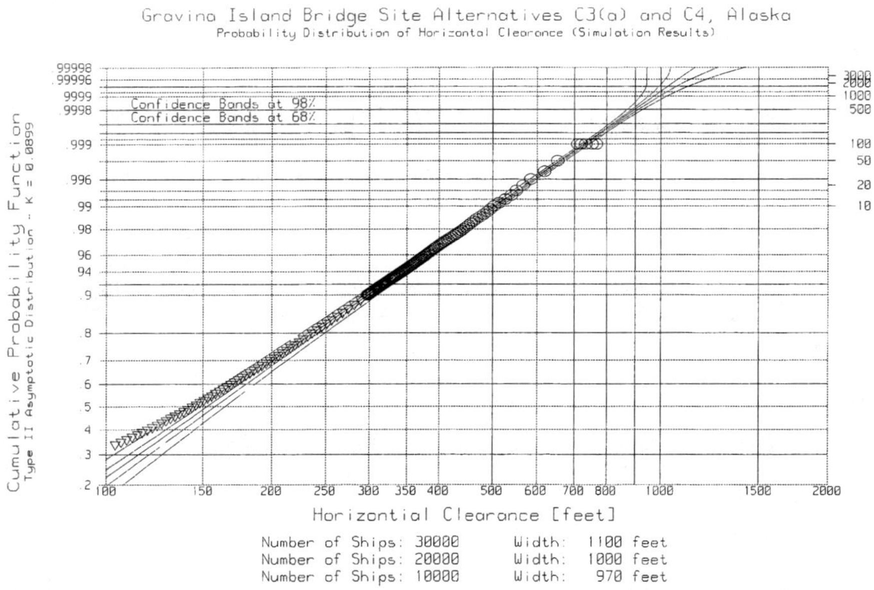

Fig. 10 Extrapolation of Type I extremal probability for horizontal clearance for combined north- and southbound large cruise ships for alternatives C3(a)/C4

Figure 10 presents a Type II extremal probability fit of the tail of the cumulative probability distribution for large cruise ships at bridge alternatives C3(a)/C4. Also shown are 68% and 98% confidence bands for the fit. The narrowness of the bands indicates that an excellent fit was obtained (especially in the region of horizontal clearances between 600 and 850 feet). Above about 940 feet, the confidence bands widen. The number of ship transits is indicated by the vertical scale on the right.

Monte Carlo Maneuvering Simulations – Summary and Conclusions

Monte Carlo maneuvering simulations have been carried out for the Gravina Access Project for 49,804 large cruise ship transits for bridge alternatives C3(a)/C4, and for 59,982 transits of large conventional AMHS ferries for bridge alternatives C3(b) and D. Gusting wind has been identified as the principal cause of the rare large off-track excursions; when such excursions occur, they are exacerbated by bank suction effects.

Probability distributions for common transit operations and for the maximum channel width requirements in 50 years have been developed from the Monte Carlo database. Statistics descriptive of common transit operations are given in Table 8, and for the maximums in 50 years in Table 7. The statistics in Table 8 are determined by interpolation within the Monte Carlo database while the maximums in 50 years given in Table 7 are determined using order statistic methods.

There are real, nonzero 50-year probabilities for large cruise ship excursions extending to the limits of the natural channel. If bridge piers are located within the limits of the natural channel, then pier protection is required.

This Monte Carlo maneuvering study did not consider the mitigating effects of emergency maneuvering actions to avoid grounding or allision with bridge piers. These actions, which may be anticipated in actual practice, include crash-stop maneuvering with the main propulsion, applying bow thrusters, and/or dropping one or both anchors. The ameliorating effects of these actions have not been simulated or otherwise measured in this study, but may be anticipated to reduce the probability of allisions and also to reduce the speed at which impacts (if any) occur.

PART III

REAL-TIME NAVIGATION SIMULATION

A full mission (360° bridge view) simulation of cruise ship transits of Tongass Narrows was conducted at STAR Center in Dania, Florida.

This study looked at navigation under a proposed high bridge at three sites: 1) over East Channel with bridge option F1; 2) over West Channel with bridge option F3; and 3) over 'North Channel' (near the airport) with bridge option C3(a)/C4 (see Figure 1). Options C3(a)/C4 present the same navigational opening for ships transiting North Channel. The navigation opening was set for each bridge at 550'. This is the distance between the east and west bridge pier fenders.

The simulations were conducted with four ships. They were: the *Carnival Destiny*, *Carnival Spirit*, *Voyager of the Seas* and *Golden Princess*. The *Voyager of the Seas* is the largest of the ships considered. It is an azipod ship, as is the *Carnival Spirit*, which was used for one day only. The *Carnival Destiny* and *Golden Princess* are conventionally steered ships (rudders behind twin open propellers) and are slightly smaller than the *Voyager of the Seas*.

Nine licensed pilots and one pilot in training from Alaska participated in this study. The group included pilots from the Southeast Alaska Pilots Association and Alaskan Coastwise Pilot Association (subsequently these two organizations have merged). The licensed pilots were selected by their respective pilot associations to represent a cross section of experience and familiarity with the Ketchikan waterway. An additional pilot was provided by STAR Center. This pilot was not experienced in Ketchikan waters, but was experienced in handling large cruise ships in the Caribbean and elsewhere. The participation and comments from a pilot who was not familiar with pilotage in Ketchikan was requested by the project sponsor. His runs and comments were evaluated separately from those of the Ketchikan pilots to see if there were prejudices about the bridge/channel options that were affecting the tests.

Variations of environmental conditions that were tested included wind speed and direction, currents, visibility (both day and night, and in fog) and run direction (northbound and southbound). Four wind speed conditions were simulated: calm, 15 knots, 20 knots and 30 knots. The 30 knot wind speed was proposed by the Ketchikan pilots as an extreme case, but one that they are faced with in piloting cruise ships into and out of Ketchikan in the summer months. It should be noted that winds of this velocity are rarely measured at Ketchikan airport (see Table 12). The wind was modeled as fluctuating in both magnitude and direction. The magnitude fluctuation was ±50% with direction fluctuating ±15°. The wind speed model was frequently criticized as not being realistic, and indeed it did not include all of the local variations that may be experienced due to local typography and thermal variations.

The tidal current model also did not accurately represent the real world condition. Again the model did not include all of the local variations that may be experienced due to local bathymetry and other effects. However, it is not expected that the lack of reality of the currents as modeled will affect the degree of difficulty of navigating either the East or North Channel bridge options. Based on the comments of the pilots, the West Channel navigation is expected to be more difficult than demonstrated herein, because of localized current effects.

The latitude, longitude, heading, COG, SOG, port and starboard propeller revolutions, rudder or thruster angles, and bow and stern thruster thrust settings were recorded at 5-second intervals for each simulated transit. The record was used to determine the swept path of the ship as it passed under each of the bridge options. Example tracks are show in Figs. 11 and 12.

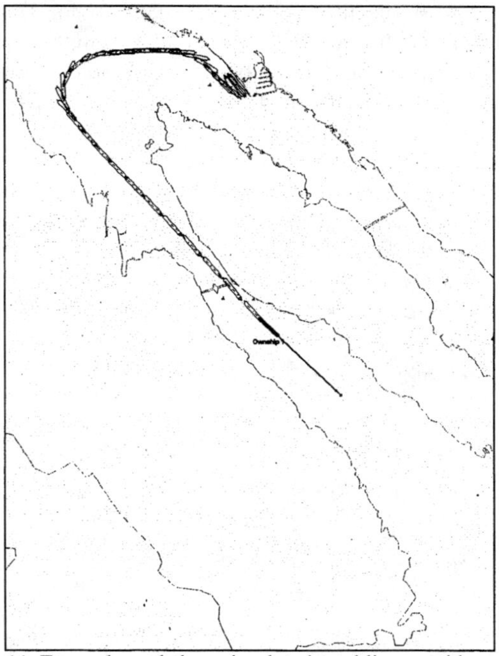

Fig. 11 Example real-time simulated trackline southbound in West Channel

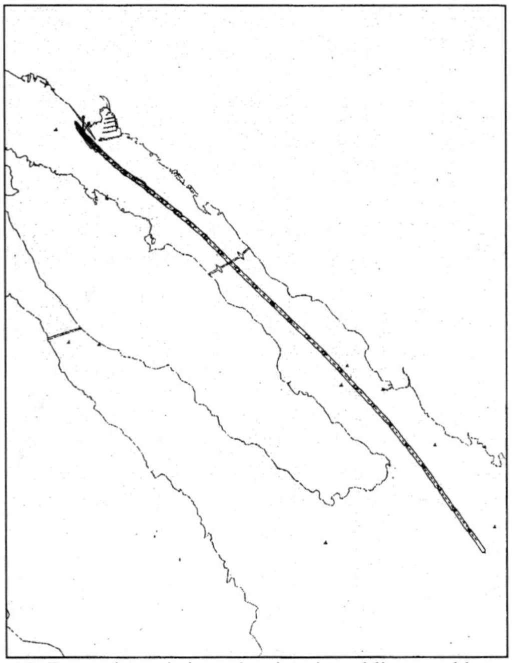

Fig. 12 Example real-time simulated trackline northbound in East Channel

Analysis of Navigation Data

The objective of the real-time analysis was to evaluate the probability of aberrancy (aka probability of allision) with bridge piers for each of the bridge site alternatives and to compare the results with the Monte Carlo simulation.

The data were also evaluated for other measures of the degree of difficulty in navigating the different channels with the bridge options simulated. This analysis looked at the number of adjustments the pilots made to the rudder (or thruster direction for azipods), the use of bow thrusters, and adjustments to RPM as they transited the final approach regions shown in Figures 13 and 14. These adjustments are generally accepted as one measure of the degree of apprehension when approaching a difficult maneuver.

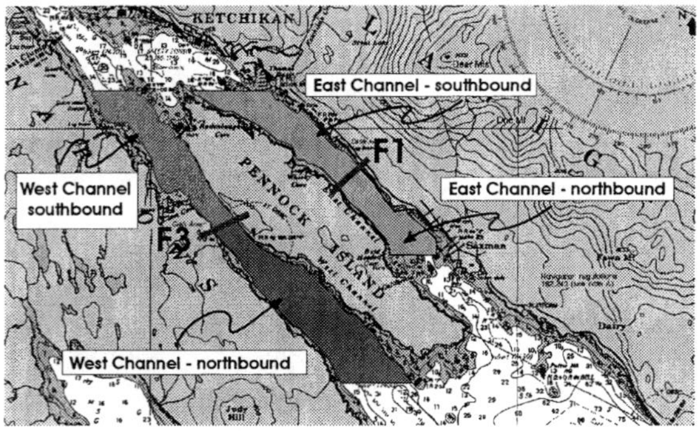

Fig. 13 Regions considered as final approach to bridges F1 and F3 adjustments in these regions are counted as a measure of degree of difficulty

Average speed in each of the channels for each direction was calculated. An additional measure of difficulty was the standard deviation of the rudder angle (or thruster direction for azipods). This measure looks at how much the rudder was varied during the approach to the bridge, not the absolute value of the rudder angle.

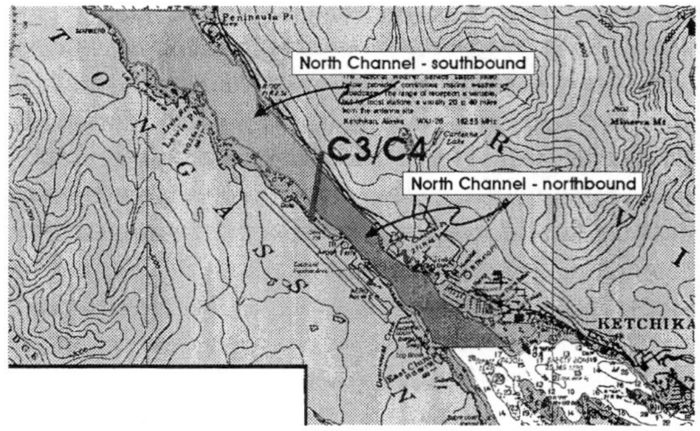

Fig. 14 Regions considered as final approach to bridge C3(a)/C4 adjustments in these regions are counted as a measure of degree of difficulty

The final navigation analysis measure was of the time it took for the simulated voyages to reach the F1 and F3 bridges when proceeding southbound from a standstill at the dock, and the times it took for northbound vessels to go from either the F1 or F3 bridges up to the dock. These times were used in a separate economic impact studies in which the additional time required to maneuver into and out-of West Channel was considered.

Probability of Aberrancy

Calculation of the Probability of Aberrancy

The probability of aberrancy was calculated using the same methodology as it was in the Monte Carlo analysis. For the purposes of this discussion aberrancy is defined as the deviation from the intended track including the swept path of the vessel. A sample of the results is shown in Figs. 15, 16 and 17.

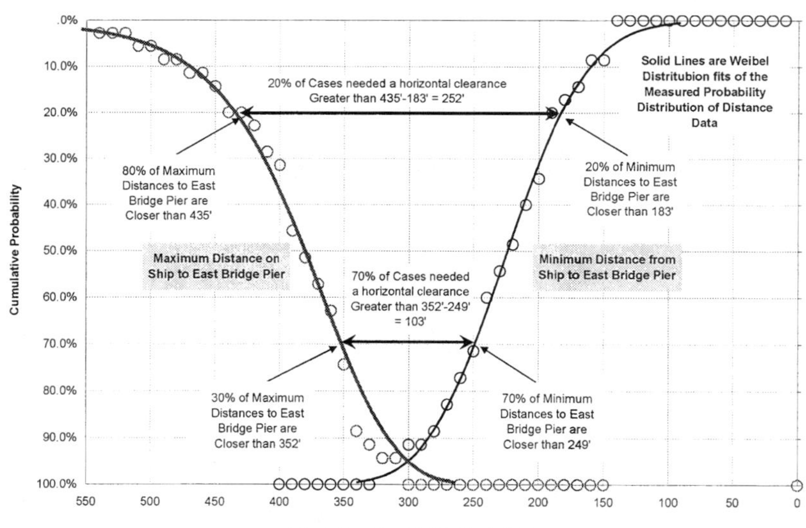

Fig. 15 Cumulative probability distribution (raw data & fit) for distances to east pier North Channel – bridge option C3(a)/C4

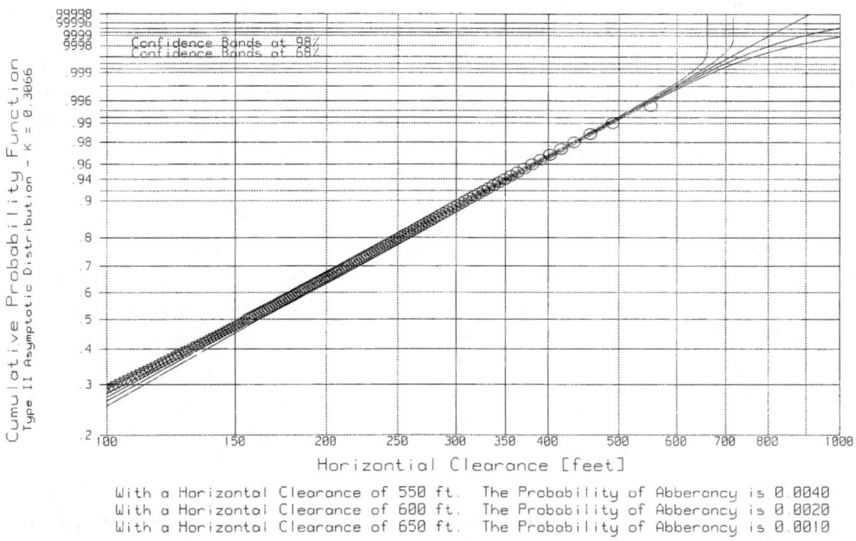

Fig. 16 Extrapolation of Type II extremal probability for horizontal clearance for combined north- and southbound transits of North Channel real-time (STAR Center) simulation results - bridge option C3(a)/C4

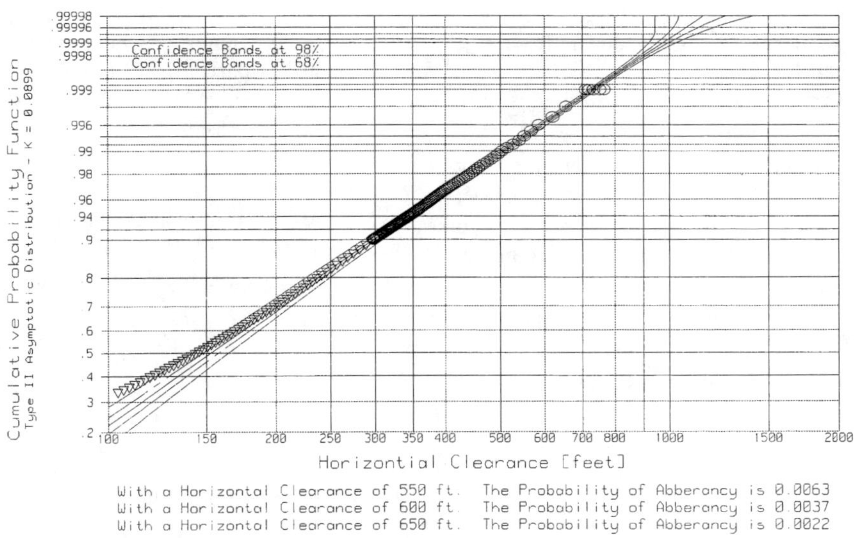

Fig. 17 Extrapolation of Type II extremal probability for horizontal clearance for combined north- and southbound transits of North Channel fast-time (Monte Carlo with autopilot) simulation results - bridge option C3(a)/C4

Degree of Difficulty

Statistical Measures of the Degree of Difficulty

The simulation data files contain continuous records of the rudder angle (or thruster angle for azipods), propeller RPM, bow and/or stern thruster power and speed through the water. It is generally accepted that the number of changes that the pilot makes to rudder angle, RPM and the thrusters is one measure of the apprehension that the pilot is experiencing when approaching a critical navigational point. In addition, the number of adjustments (changes) is dependent on the wind, current and bank conditions as well as the on the type and size of ship. This is not an exact measure and is only proposed as an additional way of looking at the data.

The method chosen and described here does not include the degree of difficulty of maneuvering into or out of each channel. It is proposed only as a measure of the degree of difficulty of passing through the bridge opening.

To calculate the number of adjustment made during the approach to the bridges, the following steps were taken:

1. A region was defined for each channel in each direction as the final approach to the bridge. The region is where the ship is on a direct alignment to the bridge opening. It was chosen to exclude the navigation required to enter the channel. The idea is to choose a region were the navigational challenge is directly comparable. The selected regions are shown in Figures 13 and 14.

   The location of the GPS antenna on the simulated ship is used as the reference location to determine whether the ship is in or out of the region of interest.

2. The length of each region is determined from the position data so that the number of adjustments can be normalized. The number of adjustments is divided by the length of the region to give the number of adjustments per nautical mile. Thus the different approaches with slightly different lengths can be compared on a common basis. This comparison of adjustments per nautical mile is only valid for approaches of approximately equal length.

3. The number of adjustment was determined by counting the number of times the rudder angle, propeller RPM and thruster power were changed (in 5 second intervals) while the ship was in the region of interest.

4. On the advice of the STAR Center staff, the number of adjustments in RPM and bow/stern thrusters was cut in half from the actual count. They reported that in their experience pilots do approximately twice as many adjustments in RPM and thrusters on a simulated ship as they do with a real ship. This is apparently due to the fact that in the simulated ship the pilot had direct control of RPM and thrust and will therefore adjust it more frequently than on a real ship. On a real ship a command must be given to a quartermaster (or ship's officer) to make the adjustment.

5. The standard deviation of the amount of rudder angle (thruster angle for azipods) is also calculated. This is a measure of the amount of movement of the rudder, not the actual angle of the rudder. This measure also captures the level of maneuvering (thus degree of difficulty) required to navigate through the channel and under the bridge.

6. The average speed through the water is calculated for each channel in each direction. This can also be related to the degree of difficult of using the waterway, as higher speeds are sometimes required to maintain heading and course in difficult wind, current or bank suction conditions. It should be noted that there is a 7 knot speed limit in each of the regions evaluated except south of F3 in West Channel. Therefore, the higher average speed recorded for northbound transits of West Channel might in part be attributed to the lack of this restriction and not just to the requirement for maintaining control.

Results

Table 14 contains results of the above described measures of degree of difficulty for navigation in each channel in each direction based on the real-time simulated transits.

Table 14

Summary of Results for Measures of Degree of Difficulty

		Average of All Runs in each Channel		
		East Channel	North Channel	West Channel
Adjustments per nautical mile	(both directions)	97	76	88
Standard Deviation of Rudder Angle (deg.)		11.5	10.0	11.9
Average Speed (knots)	(both directions)	8.0	8.2	8.6

Discussion

These measurements do not indicate that the West Channel bridge option (F3) is significantly more difficult to navigate than either the East Channel bridge option (F1) or the North Channel option (C3(a)/C4). In navigating the three channels the pilots required a comparable number of adjustments and rudder deflections as they approached the bridges. The slightly higher number for the southbound East Channel maneuvers may indicate that the ships were not quite up to speed in the early part of the analysis region and thus more rudder movements were required to counter wind and current effects. It is also possible that the extra width in East Channel allowed for more adjustments; i.e., the pilots had more lateral room and freedom to make adjustments to improve their alignment with the bridge opening.

There may be other psychological and physical factors at work that caused the pilots to hold commands longer in the C3(a)/C4 and F3 approaches than in the F1 approach; however, these reasons are not evident in the data or in the post-exercise pilot evaluations. The high levels of stress and task difficulty noted for West Channel in the post-exercise pilot evaluations are not evident in the way the ship was piloted.

Measurements of Time Delays for Ships Using West Channel

Each simulation run data file contains the times for vessel position in 5 second intervals. With this information it is possible to determine the time it took for the simulated ship to go from the dock to the bridge or from the bridge to the dock. The simulations departing from the dock started with zero forward speed, but with the ship several feet off the dock. The times do not include casting-off and any delays due to other traffic in the area. The time of arrival at the dock was chosen by the pilot as indicating that the ship had arrived. The speed was near stop at this point. Again the times do not include making up lines and any delays due to other traffic in the area. Only a subset of the total number of cases was simulated starting from the dock or was run all the way to arrival at the dock.

On average, West Channel required 22 more minutes when southbound and 24 more minutes when northbound compared with East Channel. Differences in run time due to maneuvers at the south end of either channel were not computed. This set of tests was not designed to evaluate maneuvers south of Pennock or Gravina Islands.

Analysis of Post-Exercise Pilot Evaluations

Description

After each exercise the pilot for that exercise was asked to fill out (without consultation with his associates) a questionnaire about the exercise. The questionnaire and all the original answers are contained in RTM STAR Center (2002). Several of the questions were in regard to the controllability of the simulated ship model. The answers to these questions are not evaluated herein. Of particular interest were the answers questions regarding vessel trackline (vessel position with regard to centerline; CPA to channel boundaries and/or buoys; transiting bridge span), overall safety, task difficulty and stress level.

The answers are given in Table 15.

Table 15
Results of Post-Exercise Pilot Evaluations

	Average of All Runs in each Channel		
	East Channel	**North Channel**	**West Channel**
All Pilot Evaluations (Scale 1 to 5)			
Vessel position with regard to centerline (1 = Not at all Satisfactory, 5 = Extremely Satisfactory)	4.1	3.9	3.3
CPA to channel boundaries and/or buoys (1 = Not at all Satisfactory, 5 = Extremely Satisfactory)	4.1	4.0	2.8
Transiting bridge span (1 = Not at all Satisfactory, 5 = Extremely Satisfactory)	4.0	3.8	2.7
Ketchikan Pilot[1] Evaluations (Scale 1 to 5)			
Overall Safety (1 = Not at all Safe, 5 = Absolutely Safe)	3.6	3.5	2.3
Task Difficulty Index (1 = Not at all Difficult, 5 = Extremely Difficult)	3.3	3.2	4.3
Stress Level Index (1 = Not at all, 5 = Extremely High)	3.1	3.1	4.1
STAR Center Pilot Evaluations (Scale 1 to 5)			
Overall Safety (1 = Not at all Safe, 5 = Absolutely Safe)	3.4	2.8	2.4
Task Difficulty Index (1 = Not at all Difficult, 5 = Extremely Difficult)	2.6	3.0	3.5
Stress Level Index (1 = Not at all, 5 = Extremely High)	2.8	2.5	3.1

[1] Responses include both licensed pilots and pilot-in-training

Results

An attempt was made to determine if the evaluations of the pilots for overall difficulty and stress level correlate with the degree of difficulty as measured by the number of adjustments per nautical mile. Some correlation was shown; however, a very large scatter in the adjustments data indicate that correlation is tenuous. The results are shown in Figs. 18 and 19.

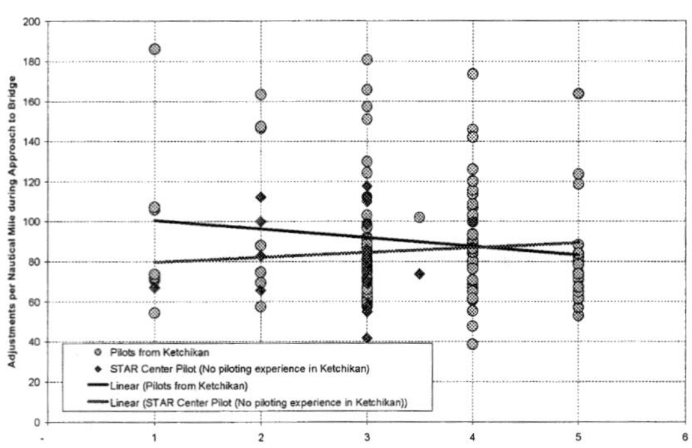

Fig. 18 Correlegram for adjustments per nautical mile and pilot indicated stress

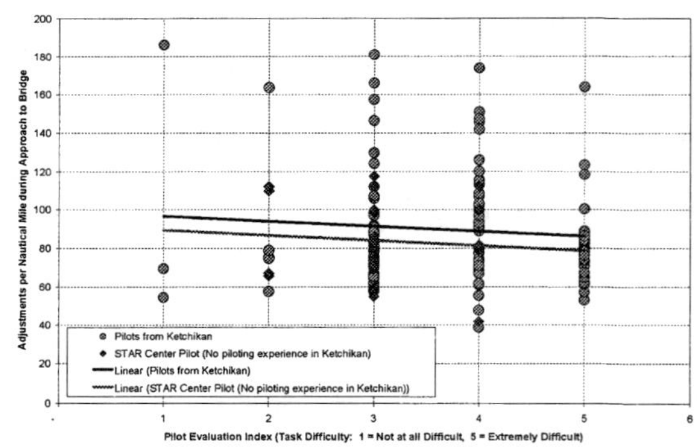

Fig. 19 Correlegram for adjustments per nautical mile and pilot indicated task difficulty

Comments on Post-Exercise Pilot Evaluations

The post-exercise pilot evaluations contain a perspective on navigation in East, West and North Channels that is not captured in the analysis of the ship data. It is apparent in the opinion of the pilots that navigation in West Channel is more stressful, more difficult and is less safe overall. Their evaluations of vessel position with respect to centerline, channel boundaries and when transiting the bridge also show dissatisfaction with their simulated transits of West Channel. There is almost no difference in their post-exercise evaluations of a bridge in North Channel or a bridge in East Channel.

It is noteworthy that the comments of the STAR Center pilot do show some small prejudice on the part of the Ketchikan pilots with respect to stress, difficulty and safety when comparing East and West Channels. The responses of the STAR Center pilot indicate that he felt the task of navigating West Channel was slightly less difficult and slightly less stressful than the other pilots. However, he rated the overall safety of both East and West Channels as being about the same as the Ketchikan pilots. Because the data sets are so small, we have not done tests to see if this difference of opinion is statistically significant. It is also possible that the Ketchikan pilots are commenting on real-world experience with respect to stress and difficulty in using West Channel and the STAR Center pilot is reacting only to the simulated world experience. It is intuitive that the simulated experience is neither as stressful nor as difficult as the real world experience. Supporting this observation is the fact that the STAR Center pilot on average found transiting all three channels to be less stressful and less difficult than did the Ketchikan pilots. Nonetheless, he thought that the C3(a)/C4 bridge option and navigating North Channel was less safe than did the Ketchikan pilots.

It is also of interest to find that there is no evidence of an increasing number of adjustments in rudder angle and thrust settings when compared with post-exercise evaluation of increased stress or increased task difficulty. Only the responses of the STAR Center pilot indicate that he made more adjustments in those scenarios he ranked as more stressful. On average the responses and behaviors of both sets of pilots showed a very sight decrease in the number of adjustments per nautical mile with an increase in their opinion as to the difficulty of the task. Again the scatter in the data is very high and correspondingly the confidence in the correlation is very low.

Real-Time Simulations – Summary and Conclusions

The STAR Center real-time simulation program was proposed as a verification and refinement procedure for a selected bridge site alternative. Real-time simulation is most useful in identifying navigation problem areas, refining aids to navigation, developing pilot confidence in the selected alternative, and for training pilots in techniques for navigating the selected channel. Also, since the cases in the real-time matrix tended toward severe and extreme weather conditions, these simulations demonstrate the possibility and difficulty of worst case conditions. The fact that the three bridge options can be navigated with similar ease in benign conditions does not tell the decision makers the whole story. The real-time simulations revealed varying navigating difficulties in worst case conditions. The real-time program is also of value in determining the navigation time differences between the options.

Using real-time simulation to prepare statistical evaluations of alternatives is limited in its usefulness. Statistical significance improves with experimental repeats. Unfortunately real-time simulation takes time. The 15 day program at STAR Center completed 144 transits averaging 9.4 cases per day. The cases were set up to include; 3 bridge sites, 2 transit directions, 2 ship types, 4 wind conditions, day, night and fog conditions and harbor modifications. Nominally the program could complete only 1 case for each combination of the above variables. Multiple repeats were possible only by doing a limited number of night and fog conditions. In the final analysis the real-time simulations taken alone provide only synoptic and anecdotal information about the alternatives.

PART IV

COMPARISON OF METHODS

Results

The probabilities of aberrancy (probability of allision) determined from the fast-time and real-time full-mission simulations are shown in Tables 16 through 19. Table 19 shows the probabilities of aberrancy for the 550' horizontal clearance for all three bridges.

The probabilities shown in these tables are the probabilities for each transit that a cruise ship (from the population simulated) will hit one or the other side of the navigational opening.

The real-time (STAR Center) simulations are for only four ships, all of which were over 100,000 GRT. Thus it can be assumed that the probabilities determined from the STAR Center tests are conservative because of the large ship sizes chosen (assuming smaller ships are easier to control). The estimates are also conservative in that the real-time simulations did not use a distribution of wind speeds based on statistics from the site. The set of cases simulated over-represented severe and extreme wind conditions.

The fast time simulations are for a population of cruise ships representative of those currently calling in Ketchikan, with some modifications to model expected growth in very large ship traffic. The fast-time simulations are also calculated for a distribution of wind speeds based on statistics from the site.

Table 16
Extrapolation of Type II Extremal Probability of Aberrancy for Combined North- and Southbound Transits of East Channel Fast-Time and Real-Time Simulation Results - Bridge Option F1

Horizontal clearance (feet)	Monte Carlo fast-time simulation estimate of aberrancy	STAR Center real-time simulation estimate of aberrancy	Ratio of real-time estimate to fast-time estimate
477	1.3519%	1.5307%	1.13
500	1.0662%	1.1146%	1.05
525	0.8203%	0.7852%	0.96
550	0.6286%	0.5503%	0.88
575	0.4798%	0.3838%	0.80
600	0.3649%	0.2664%	0.73
625	0.2765%	0.1840%	0.67
650	0.2088%	0.1266%	0.61

Table 17
Extrapolation of Type II Extremal Probability of Aberrancy for Combined North- and Southbound Transits of West Channel Fast-Time and Real-Time Simulation Results - Bridge Option F3

Horizontal clearance (feet)	Monte Carlo fast-time simulation estimate of aberrancy	STAR Center real-time simulation estimate of aberrancy	Ratio of real-time estimate to fast-time estimate
476'	1.6742%	1.1849%	0.71
500	1.3014%	0.8526%	0.66
525	1.0040%	0.6050%	0.60
550	0.7768%	0.4292%	0.55
575	0.6027%	0.3044%	0.51
590†	0.5183%	0.2478%	0.48

† The channel width (from 5 fathom to 5 fathom bathymetry contour) at the location of the F3 bridge is approximately 590' measured perpendicular to the trackline (course) of ships at this point. Thus the probability of aberrancy to 590' is the probability of grounding at this point. The probability of aberrancy for widths greater than 590' have no meaning, for the ship would be aground on either the west or east side of the channel.

Table 18
Extrapolation of Type II Extremal Probability of Aberrancy for Combined North- and Southbound Transits of North Channel Fast-Time and Real-Time Simulation Results - Bridge Option C3(a)/C4

Horizontal clearance (feet)	Monte Carlo fast-time simulation estimate of aberrancy	STAR Center real-time simulation estimate of aberrancy	Ratio of real-time estimate to fast-time estimate
500	1.0674%	0.8171%	0.77
525	0.8183%	0.5761%	0.70
550	0.6289%	0.4049%	0.64
575	0.4846%	0.2837%	0.59
600	0.3743%	0.1983%	0.53
625	0.2898%	0.1382%	0.48
650	0.2248%	0.0961%	0.43

Table 19
Extrapolation of Type II Extremal Probability of Aberrancy for Combined North- and Southbound Transits of all Channels Fast-Time and Real-Time Simulation Results for 550' Horizontal Clearance

Horizontal clearance (feet)	Monte Carlo fast-time simulation estimate of aberrancy	STAR Center real-time simulation estimate of aberrancy	Ratio of real-time estimate to fast-time estimate
E. Channel Option F1 550'	0.63%	0.55%	0.88
W. Channel Option F3 550'	0.78%	0.43%	0.55
N. Channel Option C3(a)/C4 550'	0.63%	0.40%	0.64

Discussion

There are two comments worth noting with respect to the probability of aberrancy.

The first is the demonstration that the fast-time Monte Carlo simulations are conservative estimates with respect to the piloted real-time simulations. This was expected. The autopilot of the fast-time simulator is not as good in decision making with respect to course keeping and way point arrivals as are experienced human pilots. It is interesting to note that the improvement with respect to the autopilot simulations was greatest for the West Channel cases. The hypothesis here is that the human pilots demonstrated a heightened level of effort and caution when using West Channel compared to the wider East Channel. The autopilot was programmed to treat each of the channels the same. The increased probability of allision with the F3 bridge calculated in the fast-time Monte Carlo simulations was due to the narrowness of the channel and resulting bank suction effects. The autopilot is not programmed to be prepared for bank suction, but the human pilots are.

The second observation is that the probability of allision for a 550' bridge opening is less in the West Channel with the F3 option than it is in either North Channel or East Channel. In addition, the North Channel bridge option is less likely to be struck than the East Channel bridge. Thus based on the data from the real-time simulations, a 550' horizontal opening bridge in the West Channel is less likely to be struck than either of the other two options.

This conclusion is most probably the result of the shape of the approach channel and does not indicate the level of safety of using the channel. The pilots were very skillful in navigating the cruise ships in the simulations through each of the channels. The narrowness of West Channel and also of North Channel forced the pilots into a narrower horizontal clearance when using these two bridges. In East Channel there is more room both to the north and to the south of the bridge (F1). The data indicate that it is not necessary to cross under the F1 bridge in as narrow a width as is necessary in the other two locations. The evidence suggests that the risk of using West Channel is not due to the probability of hitting the bridge.

An estimate of the probability of running aground can also be developed from the data. However, this estimate will be only for the probability of running aground at the site of the bridges. It is possible to develop probabilities of running aground at other sites along the length of the channel; however, the data required to make this determination were not retained from the simulations. However, looking at the probability of going aground at the bridge location can be used to develop a reasonable relative measure of the risks of grounding in the three waterways.

The minimum natural channel widths for each of the three channels as presented in the Monte Carlo simulation report (Glosten, 2002) are shown in Table 20 along with the estimates of the probability of exceedance (probability of potential grounding) from both the fast-time and real-time simulations. The distances are measured perpendicular to the nominal trackline (course) of the ships using the channel and measured from the 5 fathom contour on the east side to the 5 fathom contour on the west side. The minimum widths are measured as the projection of the natural opening perpendicular to the ship track. The probabilities shown are measured at the bridge site and assumed to be comparable to the probabilities of width required at other narrow sites in the waterway. Again it must be noted that these values are based on the population set of cruise ships used in simulations and for the distribution of wind conditions simulated. The fast-time simulated probabilities of grounding are assumed to be conservative with respect to real world probabilities because the ships were driven by an autopilot. The real-time simulated probabilities of grounding are assumed to be conservative with respect to real world probabilities because the experimental set used only large cruise ships and was overpopulated by severe and extreme wind conditions. However, in spite of these limitations, the relative risk of grounding in the three channels is instructive.

The results of the real-time simulation study at STAR Center as presented in Table 20 found that the risk for a potential grounding in East Channel is somewhat higher than was determined by the fast-time simulation program. The real-time study found that the risk for a potential grounding in West Channel is somewhat lower than was determined by the fast-time study.

Table 20

Extrapolation of Type II Extremal Probability of Grounding for Combined North- and Southbound Transits of all Channels Fast-Time and Real-Time Simulation Results

Minimum Horizontal Clearance (feet)	Estimate of the Probability of Grounding, including cases were avoidance maneuvers would prevent grounding		Ratio to North Channel Risk	
	Monte Carlo Fast-time simulation	STAR Center Real-time simulation	from Fast-time cases	from Real-time cases
East Channel at Idaho/Calif. Rocks 477'	1.3519%	1.5307%	**8.66**	**27.4**
West Channel north of G"5" 476'	1.6742%	1.1849%	**10.72**	**21.2**
North Channel at Charcoal Pt. 687'	0.1561%	0.0559%	**1**	**1**

Table 21

Fast-Time and Real-Time Simulation Results Extrapolation of Type II Extremal Probability 50-Year[1] Order Statistics for Potential Groundings[5] of Large Cruise Ships Operating in East and West Channels

Channel	*Most Probable Number of Potential Groundings*	*Probability of Exceeding Most Probable Number of Potential Groundings*	*Median Number of Potential Groundings*	*Expected Number of Potential Groundings*	*Average Number of Potential Groundings per Year*
		Fast-Time Simulation Results			
East[2]	181	0.865	523	1,353	27
West[3]	224	0.866	649	1,583	32
North[4]	20	0.873	60	243	5
		Real-Time Simulation Results			
East[2]	205	0.865	593	1,483	30
West[3]	158	0.866	459	1,227	25
North[4]	7	0.881	22	102	2

[1]Order statistics for 26,639 large cruise ship transits corresponding to 50 years, according to the middle series projections of Glosten (2001).

[2]Natural channel width of East Channel is approximately 477 feet between the 5 fathom (30 foot) depth contours.

[3]Natural channel width of West Channel is approximately 476 feet between the 5 fathom (30 foot) depth contours.

[4]Natural channel width of North Channel is approximately 687 feet between the 5 fathom (30 foot) depth contours.

[5]Potential groundings are events that would result in an actual grounding if extreme avoidance measures are not implemented in a timely manner.

The potential number of groundings given in Tables 21 and 22 are the result of a statistical extrapolation of the swept widths calculated in the fast-time and measured in the real-time simulation studies. However, it is important to understand that each potential grounding does not imply that a grounding occurs. A potential grounding is comparable to an aviation near-miss. Most aviation near-misses do not become mid-air collisions and most potential groundings do not become actual groundings. Most can be prevented by avoidance maneuvers.

The extrapolation gives a measure of the relative risk of using the two waterways, but is not a measure of absolute risk. An estimate of actual risk can be postulated by comparing the potential groundings with actual groundings. In the 10 year period from 1991 to 2000 there were 3,304 large cruise ship calls in Ketchikan. There were no groundings in East or North Channels. Thus the real world sample of the upper bound of the conditional probability of a grounding given a potential grounding is less than 1 grounding in 10 years or 1 in the number of potential groundings in 3,304 transits. The expected number of potential groundings determined by fast-time simulation for 10 years in North Channel is 50; the expected number in East Channel is 270. Thus the sample conditional probability for a grounding given a potential grounding is less than 1/(50+270) i.e. less than 0.31%. In other words for every 1,000 potential groundings, avoidance maneuvers by the pilots would prevent 997 of them from becoming actual groundings. For 50 years of transits of East Channel between California and Idaho Rocks with 1,353 predicted potential groundings only 4 are predicted to be actual groundings. The upper-bound risk of grounding per transit over a fifty year period assuming that the risk is the same in the next fifty years as it has been in the previous 10 years is 4/26,639, approximately 0.00016. Assuming this accident avoidance rate is the same for both East and West Channels, the risk of grounding per transit for West Channel is estimated to be (0.0031 x 1,583) / 26,639 = 0.00019. These calculations used the results of both fast-time and real-time results and are presented in Table 22.

Table 22
Estimate of the Upper-Bound Limit of the Risk of Grounding of Large Cruise Ships Operating in East and West Channels

Channel	*Predicted Number of Potential Groundings in 10 Years*	*Upper-bound Conditional Probability of Grounding given a Potential Grounding*	*Expected Number of Potential Groundings in 50 years*	*Expected Number of "Actual" Groundings in 50 years*	*Risk of Grounding per Transit*
Fast-Time Simulation Results					
East/North	270 + 50	1/(270+50)=0.003125	1,353	4	0.000159
West/North		*same as East Channel*	1,583	5	0.000186
Real-Time Simulation Results					
East/North	300 + 20	1/(300+20)=0.003125	1,483	5	0.000174
West/North		*same as East Channel*	1,227	4	0.000144

Order statistics for 26,639 large cruise ship transits corresponding to 50 years, according to the middle series projections of Glosten (2001).
Natural channel width of West Channel is approximately 476 feet between the 5 fathom (30 foot) depth contours.
Natural channel width of East Channel is approximately 477 feet between the 5 fathom (30 foot) depth contours.
Potential groundings are events that would result in an actual grounding if extreme avoidance measures are not implemented in a timely manner.

These estimates from the two simulation studies indicate a comparable risk of grounding in East Channel at Idaho/California Rocks and in West Channel north of buoy G"5". However it can be reasonably argued that the grounding potential at Idaho/California Rocks is less then it is in West Channel. The hazard in East Channel is a point hazard. It can be maneuvered through in a slight zigzag (thus widening the effective opening slightly). In addition, there is a fair amount of maneuvering room both north and south of the rocks. The minimum clearance in West Channel lasts for a significant distance (approximately 2,500 feet, equal to about 2-½ to 3 ship lengths), and thus does not allow for any maneuvers that might ease the passage. Also there is little additional room south of or north of the constriction. There were two groundings of interest in the real-time simulation cases. Both were in West Channel. One occurred during a northbound transit when the ship hit the shoal just north of the F3 site on the east side of the channel (Pennock Island side). The other also occurred on a northbound transit just north of the F3 site, however on the west side of the channel (Gravina Island side). One transit was handled by the STAR Center pilot on the conventionally propelled ship, the other was on the large azipod ship and was handled by an individual who is not yet qualified as a Ketchikan pilot. It should be noted again that the STAR Center pilot was not familiar with navigating in Tongass Narrows. Given the circumstances of these two cases, the authors feel that these groundings are not representative of the grounding rate to be expected from experienced Ketchikan pilots. However these cases support the generalization that West Channel is more difficult and risky than East Channel, but they are not adequate to develop a quantitative measure of the level of risk.

There were three near miss groundings in East Channel. Runs 12a, 12b and 301 were all within 75 feet of the 5 fathom shoal at Idaho Rock. All were northbound transits. Southbound runs were not continued down as far as Idaho/California Rocks.

There was also a grounding at Gravina Point during the turn into West Channel from Nichols Passage. However this difficult turn can be avoided by entering West Channel from Revillagigedo Channel.

The upper-bound limit on the grounding risk per transit predicted by both the fast-time and real-time simulations studies can be used in a risk factored cost analysis of the two bridge options. The predicted risks are very low and essentially identical given the differences in the uncertainties associated with the two studies. The real-time study with a small sample size has a lower statistical significance and higher uncertainty. Again is it to be understood that the real-time simulation program looked only at very large cruise ships with human pilots and the fast-time simulation program looked at a representative mixed of cruise ship sizes guided by an autopilot.

Comparison of Methods – Summary and Conclusions

The statistical information that was developed from the real-time simulations has been enormously useful as a confirmation of the larger and statistically more significant body of work prepared using Monte Carlo methods and the PC-based fast-time simulator. Despite the level of uncertainty given the limited of cases from the real-time simulations, the probability of aberrancy, relative risk, probability of potential grounding and risk of grounding are astonishing similar to the more statistically significant numbers from the fast-time program. Correlation diagrams for the probability of aberrancy between the real-time and fast-time simulations are presented in Figs. 20 and 21.

The real-time simulations can be seen as a verification of the models and conclusions of the fast-time program. However again it must be emphasized that there were significant differences in the two programs. The fast-time program depended on a computerized three-term autopilot to handle the ships and included a matrix of actual ships representative of the distribution of ships calling at Ketchikan and included a distribution of wind conditions representing the statistics of the site. Approximately 50,000 transits with large cruise ships were simulated. The real-time simulation program although guided by experienced human pilots, was limited to 2 ships and 4 wind conditions, two of which were a statistical rarity for summer months in Ketchikan. The program was skewed toward worst case scenarios. Yet despite the differences the results of both approaches are of the same order of magnitude.

The primary conclusion of the real-time simulation project is that there is a significant difference in the perception of risk of using West Channel and the statistics of risk based on pilot performance in West Channel. The pilots were exceedingly skillful in their ability to safely simulate the transit of very large cruise ships in severe and extreme wind conditions up West Channel and under the F3 bridge option. The confines of West Channel forced them to use a smaller amount of the navigational opening under the F3 bridge option than they did under the F1 bridge option. The measures of the difficulty of the task, the number of navigation adjustments as the bridge is approached, also demonstrate that the F1 and F3 bridge options are nominally similar with respect to navigation. However, in spite of their success, the pilots found the transit of West Channel to be more stressful, difficult and unsafe than transits of East Channel.

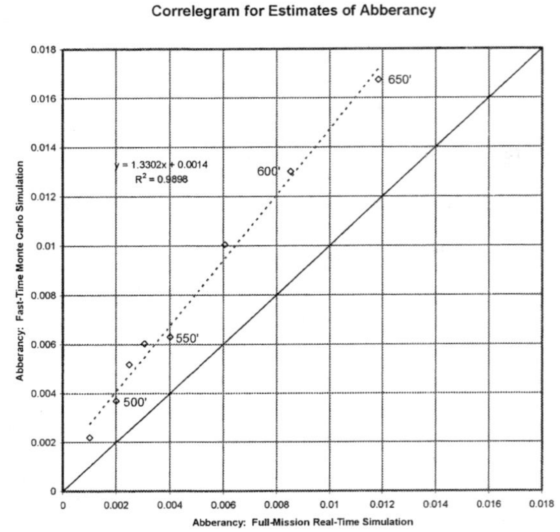

Fig. 20 Correlegram for estimates of abberancy

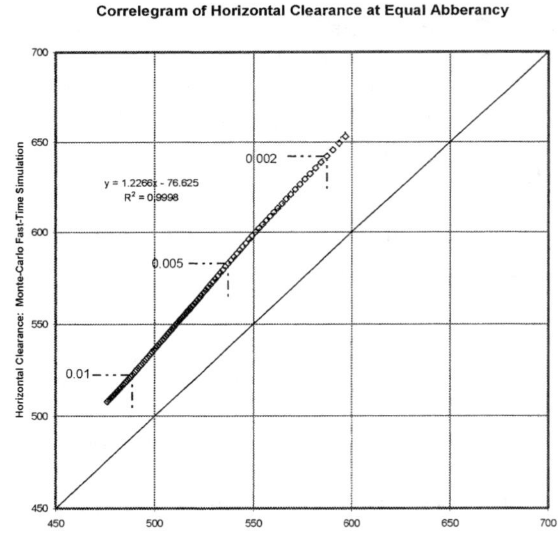

Fig. 21 Correlegram for estimates of horizontal clearance at equal abberancy

The other significant conclusion is that the measures of relative risk developed by fast-time simulation are upheld by real-time full-mission simulations. The comparative risk of potential groundings and allisions are as follows in Table 23:

Table 23
Normalized Risk Factors for Tongass Narrows Passages of Large Cruise Ships

Location	Normalized Risk Relative to the Natural Channel at Charcoal Point
Charcoal Point	1.0
North Channel with C3(a)/C4 550' bridge	4.0
East Channel at Idaho/California Rocks	8.7
West Channel north of G"5"	10.7

The risk of using West Channel at its narrowest point is estimated to be 24% greater than the risk of transiting East Channel at its narrowest point.

PART V

SUMMARY AND CONCLUSIONS

In a recent paper reporting on a 2001 international workshop on channel design Gray et al. (2003) have identified channel width as an emerging important feature of channel design. This paper has reported on the application of fast-time Monte Carlo and full-mission simulations to aid the assessment of navigational abberancy and potential impacts of new bridge alternatives on large cruise ship transits of a narrow natural waterway.

By combining fast-time Monte Carlo maneuvering simulations with least squares regression of extreme value functions and the application of order statistics methods, a powerful new procedure has been advanced for the analysis of navigational aberrancy and channel width requirements. Salient observations include:

- Fast-time Monte Carlo provides much larger sample sets than does full-mission simulation.
- Estimates of horizontal clearance and navigational abberancy from fast-time Monte Carlo are comparable, with estimates from fast-time Monte Carlo providing conservative estimates of the values obtained from full-mission simulations.
- The conservatism of the fast-time Monte Carlo estimates is most likely attributable to the skill of the marine pilots and the inability to fully represent all of the elements of that skill in a trackline autopilot function.
- In the case of large cruise ships operating in Tongass Narrows, the rare instance of extreme navigational abberancy was typically the result of perturbations initiated by gusting winds being exacerbated by bank suction.
- The channel widths obtained from the PIANC concept design method correspond to significant probabilities for large cruise ship allisions. It is the authors' opinion that the PIANC concept design method was not a conservative estimator in the example application.

The Gravina Access Project provides an example of proactive application of maneuvering simulation in advance of a civil infrastructure project with a potential impact to navigation. It is the understanding and opinion of the authors that the fast-time Monte Carlo and full-mission simulations were significant factors in the decision process leading to the recommendation of a preferred alternative in the Gravina Access draft environmental impact statement.

ACKNOWLEDGMENTS

The authors would like to thank The Alaska Department of Transportation and Public Facilities and HDR Alaska, Inc. for their funding support for these maneuvering simulation studies and for permission to present these investigations to The Society of Naval Architects and Marine Engineers.

REFERENCES

The Glosten Associates, Inc., "Cruise Ship Traffic Projections," Gravina Access Project Technical Memorandum, September 2001.

The Glosten Associates, Inc., "Monte Carlo Navigation Simulation, Technical Memorandum, Draft", prepared for the State of Alaska, Department of Transportation and Public Facilities, DOT&PF Project 67698, January 2002.

Gray, W.O., Waters, J.K., Blume, A.L. and Landsburg, A.C., "Channel Design and Vessel Maneuverability: Next Steps," *Marine Technology*, Vol. 40, No. 2, April 2003.

Permanent International Association of Navigation Congresses (PIANC). 1997. "Approach Channels – A Guide for Design" PTC II-30 Final Report, Supplement to Bulletin No. 95.

Newman, J.N. 1965. "The Force and Moment on a Slender Body of Revolution Moving Near a Wall," DTMB Report 2127.

RTM STAR Center, "Ketchikan Bridge Project, Summary Report," July 2002.

DISCUSSION

Roderick A. Barr, Member

This paper presents a comprehensive, concise, and generally clear discussion of a large and complex study, and the authors are to be congratulated for making a significant contribution to the subjects of waterway design and practical applications of maneuvering simulation techniques.

I have a number of questions and comments for the authors:

1. A brief description of what the authors mean by "Monte Carlo simulations" would be useful as this is not a widely used descriptor—does this relate only to the definition of a random population of initial conditions for the simulations or to something much broader?
2. In applying the PIANC method, I was surprised that cargo hazard was defined to be "low." I would have thought that thousands of passengers would have, from a safety standpoint, constituted a high level of potential hazard, and use of a higher hazard level would have brought the PIANC prediction into better agreement with the simulation results.
3. How was the autopilot "tuned"—by a straightforward method, such as minimizing RMS cross-track error, or by some rather more sophisticated method?
4. What was the source of the ship mathematical models and how well did the maneuvers simulated using these models compare with the trial data, and were the trial data used considered to be of high quality?
5. The wind time history shown in Fig. 7, which appears to be typical of those used, is made up of a series of discreet gusts, superimposed on the mean wind. Based on the airport data, was this considered to be a more realistic modeling of the wind velocity than using a continuously varying speed, pseudorandom wind velocity based on a spectral energy distribution?
6. Table 14 indicates mean speeds significantly greater than the legal 7 knots. What were the maximum ship speeds during the runs and why were the pilots not instructed to limit their speed to 7 knots? This could be important because of increasing difficulty imposed by wind of a given speed as ship speed decreases.
7. The comment that "real-time simulations taken alone provide only synoptic and anecdotal information about the alternatives" does not seem consistent with the author's general conclusions.
8. Is it valid to assume that the probability of grounding throughout the passage is comparable to that at the bridge location? Such a correlation should exist only when all conditions at the bridge are comparable to those throughout the passage?

Captain Henk Hensen, Member

Introduction

With pleasure I comply with the request to comment on the above-mentioned paper, which I found very interesting because it discusses very practical aspects, viz. channel width and in particular the required horizontal clearance of bridge passages. The latter is an issue of much attention nowadays.

The paper can be divided into two parts: the simulation and the analyses. The analysis method has been addressed in a scientific way. It is indeed a fact that appropriate analysis methods of simulator (or simulation) study results are often a point of discussion and still require much attention. However good an analysis method may become, a realistic input for the analyses remains crucial, viz. results based on simulations that reproduce the real-world situations and conditions carefully.

Both aspects, the simulations and the analyses, will be considered, with most attention to the degree of realism of the simulations because of its importance. It is in no way my intention to criticize, but to offer a well-meaning attempt to help to achieve the best possible results from future simulation studies, including the analysis method used.

Simulations

Fast-time simulations

It has been accepted by the authors that the outcome of the fast-time simulations is conservative in relation to the outcome of the real-time simulations. However, the opposite is normally the case, because the program can accurately calculate the rudder and engine actions needed much better than a human being. A reason for the conservative outcome might be the tuning of the looking-ahead distance, as this is very sensitive for obtaining realistic results. Has the looking-ahead distance as implemented been tested on its realism?

Other questions can be asked with respect to the conventional outcome and/or the realism of the simulations, for instance, regarding the tuning of the sensitivity for off-track and heading error. The degree of sensitivity for these aspects depends in the first place on channel/passage width and alignment, while environmental conditions may play a role as well; the smaller the available margins, the smaller off-track and heading errors are acceptable. Have such practical aspects been considered sufficiently?

With the maneuvering models, realistic models are assumed, but on what data are the mathematical models based and have they been tested, including the effects of wind and currents?

These are just a few questions, which all have to do with the degree of realism needed in the simulations.

Real-time simulations

An important question also here is whether a validation has taken place of all the relevant simulation aspects, prior to the actual simulation runs? Again, assuming realistic mathematical models, have these models been tested on the simulator by experienced pilots, particularly regarding slow speed maneuvering characteristics (only deep-water full-speed maneuvering characteristics are shown in the paper), as well as regarding bank, current, and wind effects on the ship? A crucial question is, why is not a realistic wind model implemented? Wind is such an important factor for high-sided cruise ships particularly in that specific area.

Real-time simulations introduce the human aspects into the simulations, such as maneuvering a ship from a far-forward and high-lying position on board a cruise ship that is moving on. A very human aspect is the handling of the controls. Handling controls by the pilots is not the way it is done on board ships (also a question of responsibility) and should not be allowed during important simulator studies. The reason to have the controls (rudder, engine, thrusters) operated by an "officer" as on board real ships is that the simulated cruise ships are then maneuvered as in reality, viz. based on a realistic number of orders, including the time delay for executing the orders, and on realistic fixed engine control settings. In addition, the possibilities of human failures can become apparent.

Summarizing, I would like to ask the following question: Has it sufficiently been realized how important the degree of realism of real-time simulations is? See, for example, the number of groundings and near-miss groundings during the real-time simulation, while such groundings and near-miss groundings do not occur in reality?

Analyses

Comparability

Even in this particular case where for both types of simulations size of ships differ, conditions differ, and ships are controlled in different ways, a comparison between the fast-time and real-time simulation results could be useful for certain aspects. However, taking into account the large differences, the question could be asked: Have the authors not overestimated the value of such a comparison?

Three additional remarks with respect to the analyses:

- During the real-time simulations, average speeds were 8 knots and more, so reducing wind effect and consequently ship's path width compared to the 7 knots speed required. Has it been realized that generally speeds on ship bridge simulator are higher than in reality?
- The rudder angle, as well as the time used, can be an important indication of how marginal a situation is.
- The risks of emergencies and human failures have not been taken into account. However, the possibility that an emergency or a human failure may lead to a grounding or a bridge allision should not be underestimated. It depends on ship's speed, the type and location of an emergency or human failure, as well as on the environmental conditions whether a grounding or bridge allision can be avoided. Risks

can be high, and would it not have been better to have such risks included in the analyses?

References

There are several publications dealing with the required or recommended horizontal clearance of bridge passages. Is there a reason why no reference is made to, for instance, publications by AASHTO (American Association of State Highway and Transport Officials; *Guide Specification and Commentary for Vessel Collision Design Specification of Highway Bridges*) or to the interesting book *Ship Collision Analysis* (Proceedings of the International Symposium on Advances in Ship Collision Analyses, May 10–13, 1998, Copenhagen, Denmark)?

Alexander C. Landsburg, Member

The authors have put together a very comprehensive and valuable paper describing an extensive approach to assessing the relative safety of several channel design alternatives. The paper, although very complete, has described a number of innovative approaches to analyzing risk in making critical alternative decisions and begs the reader to delve deeper into the various elements to better understand relationships and methodology. I look forward to reviewing the complete report to gain more insight into the approaches and studies undertaken.

The analysis of safety for ships carrying passengers is an area where we must be very proactive to prevent the disaster that can happen even in the safest of industries (cruise ship operations being near the very top). I believe that with the kind of tools the authors have pioneered for design alternative analysis we can move eventually to building a set of tools that can analyze the safety levels of the system and its components. Such tools would offer the ability to do an ongoing analysis of operations as ship sizes and features change along with other things, such as procedural approaches, improvements made in training, learning from experiences, instituting preventative emergency exercises, and ship operational team practices as examples. As learned in the chemical and other very dangerous industries, continual (1) analysis, (2) refocus, and (3) improvements are the keys to achieving a long-term record of safety. As an example of not continually improving, the supersonic transport (SST) was the safest plane flying until their unfortunate mishap (not that long ago), and of course, as a result, they are now out of business.

A key thought I would like the authors to address is the ship size question. When Bill Gray and others of us reported channel width as an emerging important issue of channel design (as referred to in the paper), there were a number of elements behind that thought. One of those was that when channels are designed, there is a "design" ship used in the evaluations of possible alternatives. Because channel improvements are often decades in coming, projections are made of what these future ships will be. Even with such future planning considerations, the result that almost always occurs is that the ships operating when the channels are finally built are much larger or quite different than contemplated, usually making operations more difficult. The authors have used the *Carnival Conquest* as their design vessel, and I am wondering what thought was given to effects on the study results if a much larger or quite different type of vessel becomes the standard.

As you can see, I am looking for some analysis schema and tools that can be used on a regular basis to recalculate levels of safety as designs and procedures change. I believe it is clear that eventually there will be a need to class waterways as suitable for particular vessels and operations—that is, to set limits. We must have a set of rational tools with which to work if we are to act proactively to prevent the disastrous accident that can lead to drastic changes in our transportation system.

I thank the authors for providing us with this very valuable paper and for their pioneering of important tools with which to make critical decisions for our future transportation infrastructure—tools that I believe will lead to continuous analysis of levels of safety for future operations.

Authors' Closure

The authors wish to thank all of the discussers for their contributions. The value of technical papers is greatly enhanced by thoughtful discussions such as we have received.

Dr. Barr has asked a number of questions that we will attempt to answer. First, he suggests that we elaborate on what we mean by Monte Carlo simulations. In the present instance, the randomness certainly includes initial conditions of position and velocity but also includes the probability weighted selection of cruise ships from a representative population, the direction of travel, and the random temporal process of gusting wind. The reader interested in learning more about Monte Carlo simulation might wish to consult a text on probabilistic safety analysis, such as *Probability Concepts in Engineering Planning and Design*, vol. 2, *Decision, Risk and Reliability*, by Alfred Ang and Wilson Tang, John Wiley & Sons, 1984.

Dr. Barr expresses surprise that we designated cargo hazard as "low" when we applied the PIANC concept design method. We understand his perspective but frankly do not regard the prospect of a cruise ship grounding or allision in Tongass Narrows as presenting a significant hazard to the safety of cruise ship passengers. Nonetheless, if one were to regard the cargo hazard as "high," it would increase the recommended channel widths by 113 ft and indeed bring the PIANC prediction into better agreement with the simulation results.

The gains in the autopilot equation were arrived at after experimenting with different combinations so as to minimize the off-track error. We did not apply any sophisticated optimization techniques. The look-ahead distance and dead zones were set after discussions with local marine pilots.

The maneuvering models used in this study were based on Goodman et al (1977) and Bertram (2000). At the beginning, an attempt was made to collect maneuvering trials data for all the vessels in the study. This yielded a set of 10/10 zig zags, 20/20 zig zags, turning circles, and spirals. These data were admittedly limited, and we did not have a way to assess their quality. These data were compared with corresponding simulations. In cases where such comparisons made a compelling case for model refinement, the maneuvering coefficients were tuned to improve the correlation.

Both Dr. Barr and *Captain Hensen* correctly observe that the mean speeds measured at the full-mission real-time simulator exceeded the legal 7 knots. We believe that the operations sampled during the full-mission simulations are in fact representative of the speeds that the pilots would employ under similar real-world situations, despite the statutory 7-knot speed limit. However, moderating this observation is our conviction that the scenarios run at the full-mission simulator overly represented conditions with high mean wind speeds and gusts, and thus the average ship speeds are not statistically representative of the population.

Dr. Barr asks if the discrete gust model is thought to be more realistic than a pseudorandom wind velocity based on a wind spectrum. Our answer is no, it is not more realistic, but we are of the opinion that it adequately captures the salient features of the unsteady wind that are of importance to cruise ship and ferry maneuvering.

Finally, Dr. Barr asks if it is valid to assume that the probability of grounding throughout the passage is comparable to that at the bridge location. This is an excellent question and one that we wish, in hindsight, we had endeavored to address when the original simulation work was undertaken. The simulation work was originally undertaken with the perspective that the simulations would be used to help determine the width of the horizontal clear navigation opening in the bridge at each site considered. Only later, after all of the simulation work had been performed, did it develop that alternative risk comparisons were desired between cruise ship passages of East and West Channels. To properly address the question of alternative channels, the aberrancy should be retained at locations stationed along the length of each channel. As this was not done, the aberrancy and risk at the bridge sites were used as the best available metric for the wider issue of channel passage. In defense of this approach, the narrowest confines of each channel extend only for a limited distance in a location nearly coincident with the contemplated bridge alignments.

Captain Hensen states that it is normally the case that the automatic pilot of a fast-time simulation program can maneuver a ship more precisely than a human pilot. He asks about our selection of look-ahead distance and the tuning thereof. Captain Hensen has insightfully seized on an important parameter, which frankly was set based only on discussions with local marine pilots. However, we think that the fault in the fast-time control algorithms (when compared to human pilots) may lie in not considering enough variables. We believe that real human pilots may mentally integrate weighted information from many look-ahead distances considered simultaneously (and perhaps unconsciously). Furthermore, while the control algorithm reflects only difference ordinates and rates, we believe that real human pilots may in fact be sensitive at least to the next derivative (acceleration). While in principle we agree with Captain Hensen that an automated pilot should be able to excel when compared with a human pilot, in practice this can be realized only if the automated pilot considers all of the sensory input considered by the human pilot. Frankly, the correct way to model the human pilot would be to implement a neural network model and tune that model to perform similarly to the human pilot.

Captain Hensen has also provided several very important comments regarding real-time full mission simulation to which we would like to respond. Captain Hensen asks if the real-time simulator was validated to all relevant aspects prior to the simulation runs. Specifically, were the simulator models tested by experienced pilots regarding slow-speed maneuvering characteristics, as well as bank suction, current, and wind effects. The answer is yes. A concerted effort had been done by STAR Center with their in-house cruise ship pilots to prepare and validate the models for slow-speed maneuvering prior to the start of the Ketchikan Gravina Bridge study. In addition, the pilots from Ketchikan prior to the data collection runs experimented with the models. There was considerable discussion as to the validity of the bank suction and the unsteady wind models, and opinions differed as to their reality. The differences were never fully resolved. However, the consensus was that the models were adequate for the project at hand and that the testing could proceed with reservations. It is the authors' observation that it is and probably will always be nearly impossible to get any diverse group of experienced pilots and ship captains to accept that a numerical simulation of the real-world of ship handling is anywhere near perfect.

The problem with the wind model is important to discuss further as it is an issue that should be carefully dealt with in other fast-time and real-time simulations programs. The fast-time simulations implemented a sophisticated pseudotime function for the unsteady wind based on wind data from Ketchikan Airport. The fast-time simulations included an implementation of a Monte Carlo procedure that sampled for gust amplitude, direction, and duration conditioned on the 1-minute average wind speed. The instantaneous wind realization was the vector sum of the 1-minute average wind vector and the gust vector, an example of which is given in Fig. 7 in the paper. The implementation at the real-time simulation program did not include the vector sum of the underlying average wind and a randomly selected gust vector. Instead, the total wind vector was randomly varied in magnitude and direction with a normal distribution about the mean position. A randomly selected gust duration was not included. Thus, a 20-knot 1-minute average wind speed with a 50% gust factor varied in magnitude ±10 knots and varied in direction ±30 deg. It was resampled at every time step. The result, at least in the opinion of the ship pilots, was a wind that felt as if it were averaging out and thus approximately constant in speed and direction. This did not seem realistic, because in part it could be ignored as its effects tended to average out.

Captain Hensen has also asked how the handling of the ship controls was carried out during the real-time simulations. Captain Hensen points out that "handling controls by the pilots is not the way it is done on board ships (also a question of responsibility) and should not be allowed during important simulator studies." The answer is the procedure during the real-time testing varied. However, it was primarily done as is done on board real ships with the pilots calling out the commands and a "simulated" quartermaster handling the controls. The exception to this procedure occurred mostly during docking with the azipod ships where the pilots often took direct control (however, this part of the simulation was not of direct interest with respect to the primary objective of comparing bridge sites and bridge span widths).

Finally, Captain Hensen asks if it has been sufficiently realized how important the degree of realism is to real-time simulations, especially considering the differences in the number of groundings and near misses that occurred in the simulations, whereas such groundings and near misses do not occur in reality. We would like to assure our readers that every attempt was made to incorporate a very high degree of realism into the real-time simulation program. However, what constitutes reality was subject to considerable debate, especially with respect to the chosen wind speeds and variation with respect to the geography. It should be pointed out that the conditions chosen for simulation primarily reflected the extremes of the environment and some of the average and benign conditions were passed over. The objective of the real-time simulation study was to compare the risks of the three bridge siting alternatives in a relative sense, not necessarily to determine the magnitude of the actual (real-world) risk. As a consequence, instances of groundings and allisions are expected to be higher in the simulated environment than in the real world.

Mr. Landsberg has asked what, if anything, was done to project the future sizes of ships (and for that matter, ship types) expected to operate in Tongass Narrows. Such projections are of course subject to substantial uncertainty and are, accordingly, only marginally reliable. Nonetheless, sound planning practices demand that an attempt be made, and the Gravina Access Project is no exception in this regard. Historical trends in vessel traffic (port calls, transits, ship types and sizes) were analyzed, using Army Corps of Engineers waterborne commerce data, Alaska Marine Highway System records, and cruise industry data. New buildings of cruise ships were assessed as were other infrastructure dimensional constraints along the inside passage (particularly, the Lions Gate Bridge at Vancouver, British Columbia, and the Seymour Narrows aerial cable crossing, also in British Columbia). Projections of cruise ship traffic volumes were developed using demographics and measures of market penetration.

Mr. Landsberg notes our use of the *Carnival Conquest* as a design vessel. More accurately, the *Carnival Conquest* was used as the design vessel in the PIANC concept design method, but *Carnival Conquest*–class vessels were only 1 of 22 distinct classes of cruise ship considered in the fast-time Monte Carlo simulations. Eight classes of northbound cruise ship were considered, and 14 classes of southbound cruise ship were considered. The classes and directions of travel reflected recent patterns of cruise ship operations at Ketchikan, excepting that older and smaller classes of vessels were replaced by newer, modern classes of vessels under construction for or recently delivered to the same traditional cruise operators in the Alaska trade.

Additional references

Goodman, A., Gertler, M., and Kohl, R. 1977 *Experimental Techniques and Methods of Analysis used at Hydronautics for Surface-ship Maneuvering Predictions*, IMechE.

Bertram, V. 2000 *Practical Ship Hydrodynamics*, Butterworth-Heinemann.

Assessment of Operational Risks of Parametric Roll

Marc Levadou, MARIN
Lionel Palazzi, MARIN

ABSTRACT

In October 1998, a post-Panamax, C11 class containership encountered extreme weather and sustained extensive loss and damage to deck stowed containers. The motions of the vessel during this storm event were investigated through a series of model tests and numerical analysis. This confirmed that parametric roll was the most likely cause of the accident. The results of this study were presented in a technical paper in 2003 (France et al 2003).

Developing operational guidance to avoid parametric roll requires study of the long-term probability (risk) of such an incident considering the key factors that must be satisfied for parametric roll to occur. These include the loading condition, speed, wave height and period.

This paper presents the results of an initial attempt to quantify the operational risk of parametric roll for a large containership. The operational risk is determined by analyzing the simulated motions of a ship steaming a given route over a five-year period. In the simulations a weather hind cast (waves and wind) is used to evaluate the sustained speed and ship behavior for each 2-hour time step. Both voluntary and involuntary speed loss are taken into account. The paper pays attention to the statistical aspects of the interpretation of the results.

NOMENCLATURE

k_{xx} – Transverse radius of gyration **(m)**
RMS – Root mean square
T – Period **(s)**
T_p – Peak period **(s)**
H_a – Wave amplitude **(m)**
H_s – Significant wave height **(m)**
V_{calm} – Calm water speed **(knots)**
μ – Relative wave heading **(deg)**, bow to the seas = 180 deg.

INTRODUCTION

In late October 1998 a laden, post-Panamax, C11 class containership, eastbound from Kaohsiung to Seattle, was overtaken by a violent storm in the North Pacific Ocean. The encounter with the storm continued for some 12 hours, mostly at night, during which the master reduced speed and attempted to steer into increasingly higher seas off the vessel's starboard bow. Ultimately, the seas became completely confused and violent.

Significant wave heights (H_s) steadily increased from 10.9 m with a peak period (T_P) of 13.5 s at the beginning of the storm to 13.4 m and 15.4 s at middle of the storm. The maximum H_s was 14.9 m with a T_p of 16.4 s.

More significant than the violence and magnitude of the seas, however, were reports by experienced engine and deck officers of unexpectedly extreme and violent ship motions during the worst of the storm. At times yaw angles of 20 deg port and starboard made course keeping almost impossible. Severe pitching resulted in main engine overspeed trips and shaft vibration together with pounding. Port and starboard rolls up to 35 deg to 40 deg were reported to have occurred simultaneously with the extreme pitching.

When the crew surveyed the vessel the following morning they found devastation of the cargo. Of the almost 1300 on-deck containers, one-third, with their contents, had been lost overboard. Another one-third, with their contents, were in various stages of damage and destruction. Containers and their contents hung over both sides of the vessel.

Figure 1 Representative Container Damage

The motions of the vessel during this storm event were investigated through a series of model tests and numerical analysis. This confirmed that parametric roll was the most likely cause of the accident. The results of this study were presented during SNAME's 2001 annual meeting and reported in a technical paper (France et al 2003).

In that paper several recommendations for operational guidance, design guidance and further research were made. Following that paper two major actions have been taken. First, SNAME established the ad hoc panel No. #13: "Investigation of Head-sea Parametric rolling and Resulting Cargo Security Loads". Second, the IMO subcommittee on "stability, loadlines and on fishing vessel safety" was asked during the July 2002 meeting to consider head-sea parametric roll.

One of the remaining questions is the long-term probability (risk) of a large amplitude parametric roll event in head or bow quartering seas. The occurrence of parametric roll behavior depends on several conditions, including loading condition, speed, heading with respect to the waves and wave conditions. It is likely that the probability of conditions conducive to parametric roll is small, and this may be among the reasons that ship designers and ship operators do not address the problem. However, parametric roll has caused significant damage on a container vessel, so the probability is not zero. But how high is it and can we expect a similar incident in the near future?

This paper presents an inital attempt to quantify operational risks of parametric roll for a large containership. The operational risk is determined by analyzing simulated motions of a ship steaming a given route over a five year period. The probability of parametric roll is evaluated for each two hour time step over the five year period.

The probability of parametric roll is taken from a database of simulation results generated for a range of speeds, wave headings and wave conditions. Numerical tools are used to generate this database.

In the simulations a weather hind cast (waves and wind) is used to evaluate the sustained speed and ship behavior for each two-hour time step. Both voluntary and involuntary speed loss are taken into account.

THEORY

The theory behind parametric roll has been studied and described by many investigators. The references list several papers and articles that develop the theory behind parametric roll (Kempf 1938, Graff and Heckscher 1941, Paulling 1959 and 1961, Oakley 1974, Dunwoody 1998, Dallinga et al 1998, Luth 1998). In this paper only the principles of parametric roll will be described. Some examples from model tests will be given.

In "normal" conditions the motions of a vessel are caused by direct wave excitation. Resonant conditions can occur when the combination of wave period, vessel speed and heading with respect to the waves leads to an encounter wave period close to the natural roll period of the vessel. These resonant conditions can lead to large amplitude motions.

Resonant roll motion can occur in beam waves and stern quartering waves. Roll motion caused by direct head sea wave excitation is not possible. Nevertheless, under certain conditions of encounter period, roll motion can be excited in head seas. The roll motion, once started, may grow to large amplitude, limited by roll damping and, in extreme conditions, may result in danger to the ship or its contents. This phenomenon is referred to as "auto parametrically excited motion" which is usually shortened to "parametric motion". The term describes a state of motion that results not from direct excitation by a time-varying external force or moment but from the periodic variation of certain numerical parameters of the oscillating system. For a ship in head or stern seas the uneven wave surface together with the pitch-heave motion of the ship results in a time-varying underwater hull geometry. This varying geometry, in turn, results in time-varying changes in the metacentric height, i.e., in the static roll stability. The variation of the static roll stability can cause instability if it occurs in the appropriate period. The instability can be triggered when a small excitation, such as rudder movement, initiates roll motion. The roll angles can become very large if the stability variation and the wave height are large.

From theory and as validated by model tests (Dallinga 1998, Luth 1998, France 2003), parametric roll occurs when the following requirements are satisfied:

1. The natural period of roll is equal to approximately twice the wave encounter period.
2. The wavelength is on the order of the ship length (between 0.8 and 2 times LBP).
3. The wave height exceeds a critical value.
4. The roll damping is low.

Typical Results from Model Tests

The figures below show model test time traces in regular wave head seas of roll motion, pitch motion and wave height.

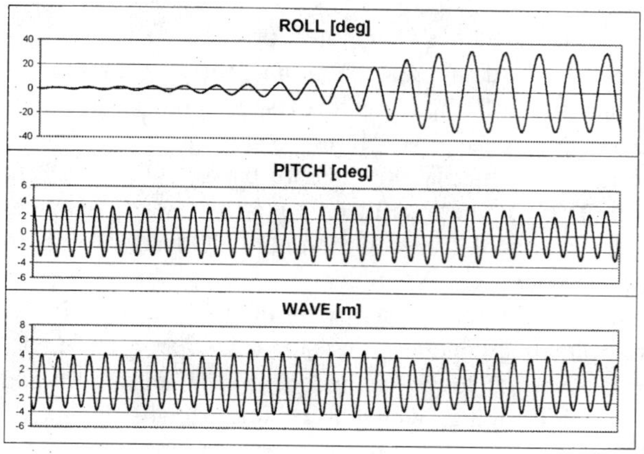

Figure 2 Time Trace of Roll, Pitch and Wave Height during Regular Wave Test

Figure 2 shows that at the start of the test the model was pitching to angles of about 4 deg, with negligible roll response. A small excitation, likely introduced by a rudder movement, caused the vessel to begin rolling. Quite unexpectedly, roll angles then increased from a few degrees to over 30 degrees in only five roll cycles. This behavior is parametric rolling. Once parametric roll was initiated, the model continued to roll violently. The pitch amplitude was not affected by the violent rolling.

Figure 3 shows an expanded view of the roll and pitch response in regular waves. Positive pitch values mean the vessel is pitched down by the bow. It can be seen from Figure 3 that there are two pitch cycles for each roll cycle, and that the model is always pitched down by the bow at maximum roll. That is, when the model is at maximum starboard roll it is pitched down by the bow, when upright at zero heel it is pitched down by the stern, and when at maximum port roll it is pitched down by the bow again. Throughout the test program, this relationship between pitch and roll motions existed whenever parametric roll was induced.

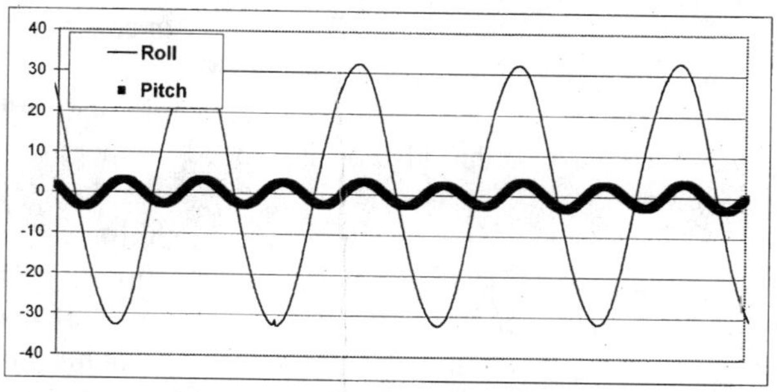

Figure 3 Expanded View of Roll and Pitch Motions during Regular Wave Test

Parametric roll was observed in most of the irregular wave and short-crested sea tests as well. In Figure 4 the roll and pitch motions and the wave height are shown for a test in irregular seas. The 2:1 ratio between roll period and pitch period is again apparent. When the model encounters a sequence of wave components of a certain period and height, parametric rolling is initiated if the conditions are right. As in regular waves, the roll quickly builds to large amplitudes. When the wave period changes or the wave height diminishes, the parametric roll response quickly dissipates.

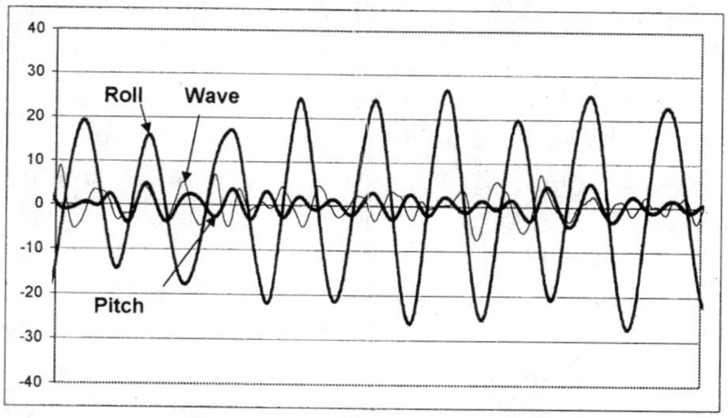

Figure 4 Expanded view of Roll and Pitch Motions at Time of Largest Roll in Short Crested Sea Test

Numerical Analysis

Nonlinear, time domain seakeeping computer codes based on potential theory are able to predict the phenomenon of parametric roll. During the investigation of the incident described above, (France et al 2003) two computer codes were successfully used to predict parametric roll. SAIC's LAMP (Large Amplitude Motion Program, Lin and Salvesen, 1998, Lin and Yeu, 1990, 1993, Treakle 2000) and MARIN's FREDYN computer code (Hooft, 1987, de Kat, 1989,

2002, McTaggart, 2000) were used to simulate ship motions for comparison with model test results.

FREDYN is a nonlinear, time domain ship motion simulation program developed by MARIN over the last 10 years for predicting motions of naval frigates as part of the Cooperative Research Navies Dynamic Stability project. However, it has also been utilized for merchant vessel motion predictions.

FREDYN takes into account the external forces on the ship due to wind and waves, rudders, bilge keels and active stabilizer fins and the reaction forces of the ship due to the motions. The simulation is nonlinear because the total Froude-Krylov exciting force is calculated by integrating the hydrostatic and dynamic wave pressures up to the instantaneous waterline,. Since FREDYN is a full 6 degrees of freedom model it includes the couplings between the individual modes of motion. Both nonlinearity of the excitation forces and coupling between the 6 motions is required to be able to predict parametric roll motion. The maneuvering model is based on frigate type ships; all other routines are independent of the ship type. Since maneuvering was not a major aspect in this investigation, the FREDYN model was applicable.

FREDYN models a ship as a free running vessel in waves, comparable to a free sailing model in a seakeeping basin. The heading of the vessel is controlled by an autopilot that reacts to the instantaneous motions of the ship. The RPM of the motor is set to propel the ship at the desired speed in calm water. Due to the waves, the speed and course of the ship change during the runs. FREDYN is capable of modeling uni-directional long-crested waves and bi-directional waves.

Several studies (France 2003, Luth and Dallinga, 2000) have shown that although FREDYN is capable of predicting the occurrence of parametric roll, the prediction of roll amplitude seems of limited accuracy. This has several reasons. First it is difficult to predict the non-linear (viscous) roll damping correctly. Second, the speed loss and speed variations in waves, which can be observed in model tests, are not modeled in FREDYN. These reasons likely account for the inaccuracy in roll amplitude.

SELECTION OF CASE

For the study presented in this paper the post-Panamax, C11 containership which encountered the storm, as described in the introduction, is used. It is a logical starting point because FREDYN simulations and model tests were previously performed on the same vessel at MARIN.

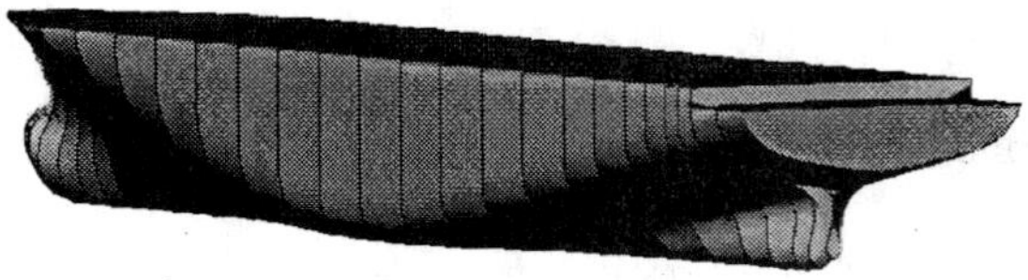

Figure 5 C11 Hull Form C11 type case

A rendering of the C11 hull form is presented in Figure 5. Typical of post-PANAMAX containerships, the C11 has extensive bow and stern flare. The ship's natural roll period as used in this study was 25.7 seconds. This corresponds to the estimated roll period of the ship at the time of the incident.

FREDYN SIMULATIONS

The FREDYN computer program was used to generate a database containing the probability of parametric roll for a range of speeds, wave headings and wave conditions. Before running the systematic series of simulations, representative data from FREDYN simulations of C11 motions were compared to the results of C11 ship motion model tests in order to validate our FREDYN model.

The simulation was performed for the ship in the same loading condition that was used during the model tests. Special focus was placed on the roll angles and the roll damping, which is calculated with the Ikeda Himeno (1978) method. In the following section the model test and FREDYN roll decay are compared for three different speeds.

Roll Decay Tests Comparison

In Figure 6 the roll decay results from model tests and FREDYN computations for a speed of zero knots are presented. As can be seen in the figure, the natural roll period as predicted by FREDYN is not exactly the same as from the model test. Also the roll damping is somewhat higher as predicted by FREDYN (faster decay).

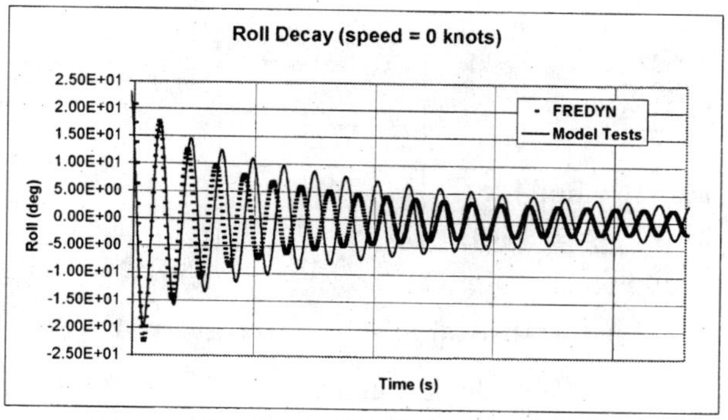

Figure 6 Roll Decay at Zero Knots

The FREDYN simulation can be improved by "tuning" the roll model by modifying the model radius of gyradius k_{xx} to improve correlation between FREDYN and model test roll period and roll damping. In Figure 7 the zero knot roll decay results from model tests and a FREDYN simulation with the tuned k_{xx} are presented. The figure shows now a very good comparison between FREDYN and the model tests.

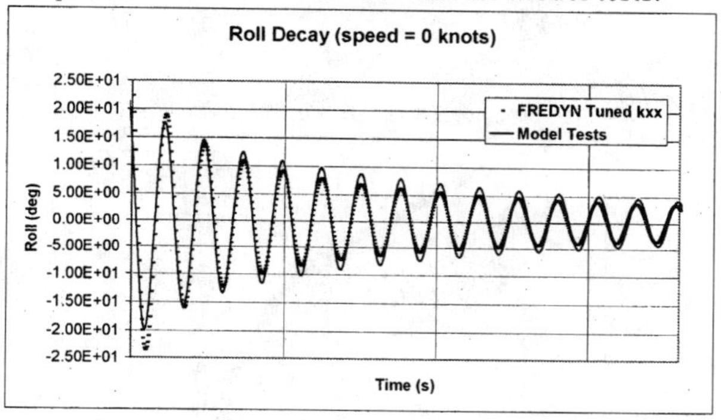

Figure 7 Roll Decay at Zero Knots

In Figure 8 and Figure 9 the same type of comparison is given for speeds of 5 knots and 10-knots for the k_{xx} tuned based on the results at zero knots. The results show that for these speeds the "k_{xx} tuned" FREDYN results are also very close to the model test results.

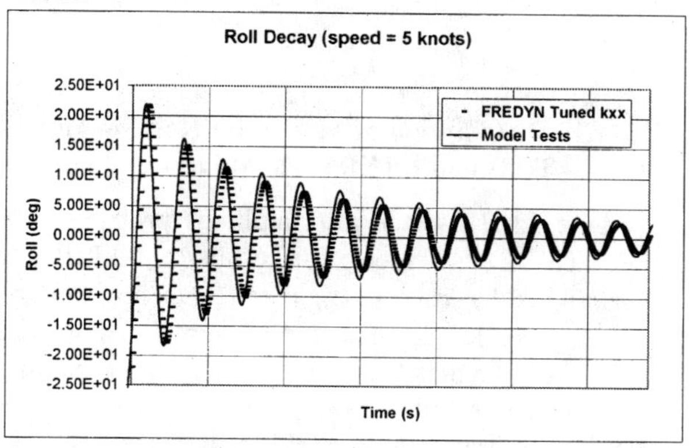

Figure 8 Roll Decay at Five Knots

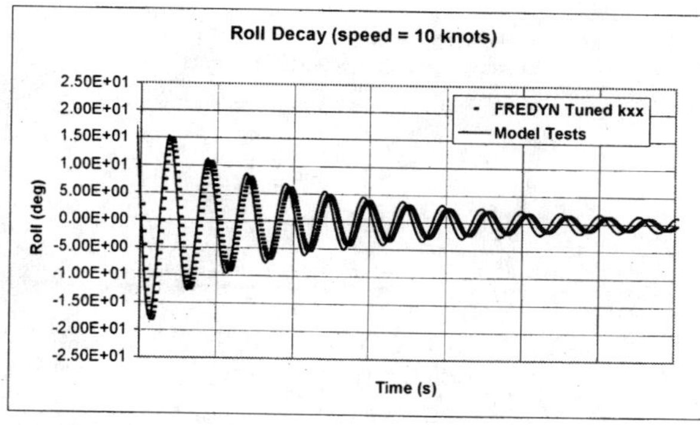

Figure 9 Roll Decay at Ten Knots

The rest of the FREDYN computations were performed with a tuned k_{xx} value. No other tuning parameters were used.

Regular Wave Test Comparison

In Figure 10 the results of FREDYN computations for regular waves are compared to model tests (H_a = 4 m, T = 14.1 s, μ = 180 deg and V_{calm} = 16 knot). The figure shows time traces of roll, pitch and wave amplitude. Examination of the time traces shows that the model test wave amplitudes and resulting pitch amplitudes vary slightly over time. By contrast, the corresponding FREDYN are constant. In spite of this slight inconsistency, the comparison shows a very good correlation between simulated and model test pitch motions. There is also excellent correlation for roll motion. In this condition, regular waves with the correct wave encounter period for parametric roll to occur; the vessel will be subject to continuous parametric roll. This is found from the model tests and predicted by FREDYN. The associated roll amplitudes are around 30 degrees.

In Figure 11 the time trace of roll motion as predicted by FREDYN and measured during model tests is given for another wave condition and heading (H_a = 5 m, T = 14.1 s, μ = 150 deg and V_{calm} = 16 knot). The figure also shows good correlation between FREDYN simulation and model test results and the continuous large (parametric) roll amplitudes.

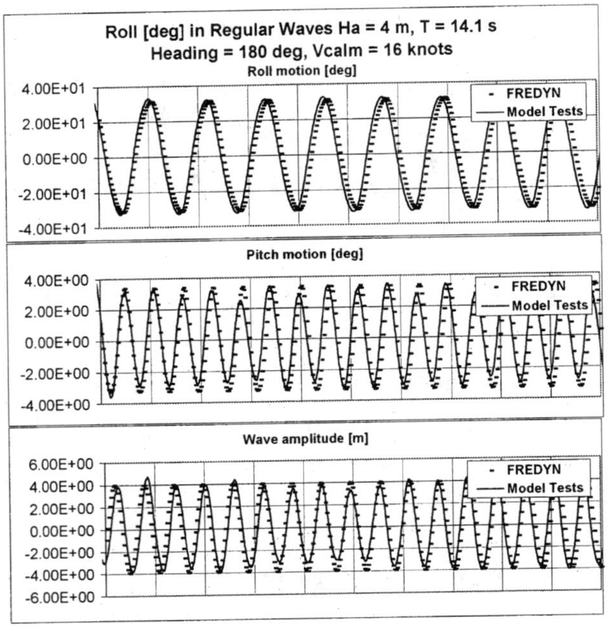

Figure 10 Time Trace of Roll, Pitch and Wave Height for Model Test and FREDYN

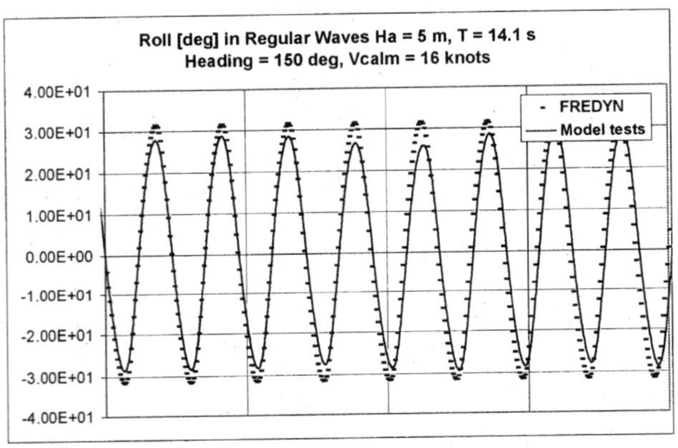

Figure 11 Time trace of Roll for Model Test and FREDYN

Irregular Wave Tests Comparison

The table below shows a comparison between the statistical results of model test measurements and FREDYN calculations in irregular waves (H_s = 13 m, T_p = 15.1 s, μ = 180 deg and V_{calm} = 16 knot).

The RMS value is the root mean square, the Max Ha+ roll value is the maximum amplitude to starboard and the Max Ha- roll value is the maximum amplitude to port.

	Model tests	FREDYN
H_s Wave [m]	13.2	12.96
RMS Roll [deg]	9.89	9.50
Max H_a+ Roll [deg]	34.74	30.57
Max H_a- Roll [deg]	-32.50	-31.61
RMS pitch [deg]	2.18	2.14

Table 1 Statistical Results in Irregular Waves

From the table it can be seen that in irregular waves the FREDYN results are also very close to the results of the model tests. The root mean square (RMS or standard deviation) of roll is within 4 % and the mean of the maximum and minimum values within 8%. Also the pitch motion is predicted very well by FREDYN.

Although the maximum values (positive and negative) are quite close, there are differences between the distribution of the maximum roll angles calculated by FREDYN and the model test results. This can be seen in Figure 12.

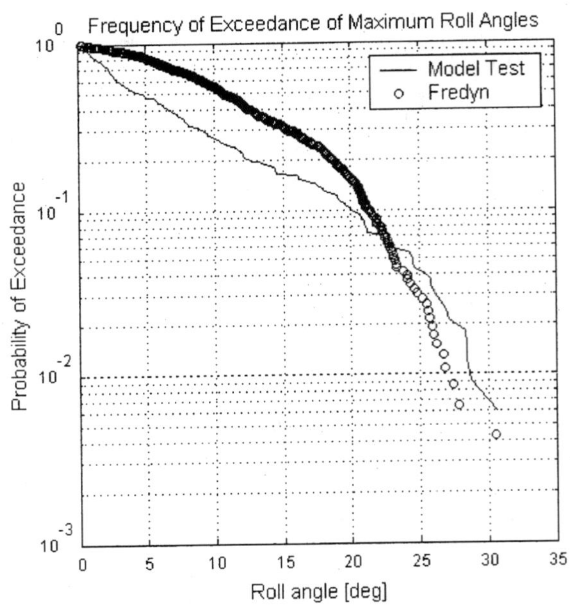

Figure 12 Distribution of Maximum Roll for Model Tests and FREDYN simulations

Figure 12 shows the roll amplitude probability of exceedance for the model test results and FREDYN simulation results (H_s = 13 m, T_p = 15.1 s, μ = 180 deg and V_{calm} = 16 knot). The figure shows that for the lower angles (smaller then 20 degrees) the probability of exceedance as found by FREDYN is higher. For high roll amplitudes the probability of exceedance as found by FREDYN is lower. This means that

FREDYN predicts extreme roll amplitudes that are lower than those found during the model tests. The differences in the extreme values are considered small and it is felt that FREDYN results, although slightly optimistic, can be used for this study.

Conclusion of Comparison FREDYN and Model Tests

Results of FREDYN calculations showed generally good correlation with the results of roll decay tests, regular wave tests and irregular wave model tests of the C11 hull form. After "tuning" k_{xx} to correct the natural roll period, the FREDYN results showed good to very good agreement with the model tests. It is assumed that also in other wave conditions the FREDYN predictions will be sufficient for the purposes of this study. Therefore it can be concluded that the FREDYN results can be used to predict the parametric roll behavior of the real vessel in arbitrary conditions.

PARAMETRIC ROLL PROBABILITY DATABASE

In order to assess the operational risk of parametric roll the probability of parametric roll must be known for all wave conditions (relative wave heading, wave height, wave period and vessel speed), that the vessel might encounter during her normal operation.

For this purpose a large number of FREDYN simulations were performed. The program was used to simulate motions for the C11 class containership with a natural roll period of 25.7s for a matrix of speeds, wave headings, wave periods and wave heights. The speed range was from 2.5 knots to 20 knots with a 2.5 knot increment. Three wave headings were used (two bow quartering waves 120 deg and 150 deg and head waves 180 deg). The sea states were modeled using periods (T_p) that varied from 3.5 s to 17.5 sec with one second increment and the wave heights that varied from H_s = 0.5 to H_s = 14.5 with one meter increment. By eliminating all unrealistic combinations of wave height and wave period the total amount of calculations was reduced to 3165 simulations. Each simulation was performed for three hours real time.

Identification of Parametric Rolling Events

The result of each individual calculation was a three hour time trace of the motions, position of the vessel, speed etcetera as discussed above. The program also gives statistical results (RMS, minimum, maximum etc.), which however cannot be used to determine if the vessel suffered from parametric roll in that particular condition. These statistical results will not be greatly affected by a short duration parametric roll event that produced large amplitude rolling with the potential for significant damage. We have developed an approach to analyze the time traces to identify parametric roll events.

For each time step the roll angle was compared to a threshold level. If two subsequent roll amplitudes (negative or positive) are higher than that threshold the group of roll angles is marked as parametric roll event (see Figure 13). For each simulation the number of these groups was determined as well as the total duration of all groups in one simulation. The total group time is a good indication of the probability and seriousness of parametric roll in a particular sea state. To identify parametric roll groups, a threshold of 10 degrees roll amplitude was used.

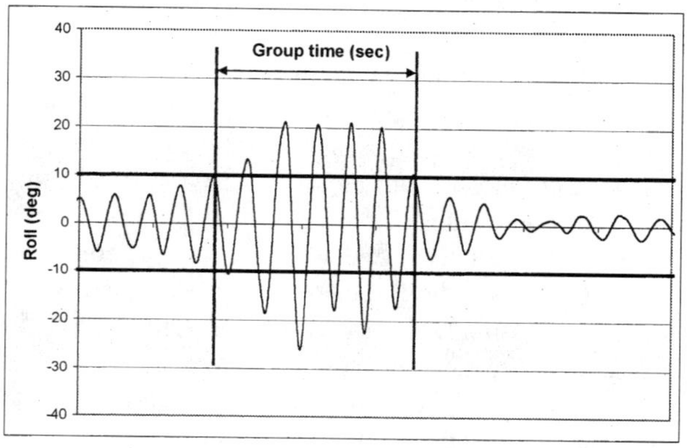

Figure 13 Determination of Group Time

Several different criteria were tried to determine if in a sea state the vessel suffers from parametric roll. The first was the number of parametric roll groups per 3 hour simulation. It was expected that the number of groups would increase with the wave height. It turned out that above a certain wave height the number of groups decreases because the vessel is continually subject to parametric roll. Finally, two criteria were used to determine if in a sea state the vessel suffered from parametric roll. First, a number of parametric roll groups larger then zero. Second, the average of the 1/10th highest roll amplitudes was used. If that value was above 10 degrees the sea state was marked as one in which the vessel would have parametric roll.

Results of FREDYN Calculations

In the table below a sample of the FREDYN results is given. The table shows the average of the 1/10th highest roll amplitudes for a speed of 5 knots in head waves for a range of wave heights and wave peak periods.

Hs (m) \ Tp(s)	8.5	9.5	10.5	11.5	12.5	13.5	14.5	15.5	16.5	17.5
14.5						35.7	30.2	27.3	24.7	17.7
13.5						33.5	28.7	26.6	23.7	8.3
12.5						31.6	28.1	25.8	22.6	3.4
11.5					29.6	29	27.5	24.7	4.4	4.4
10.5				3.3	26.5	24	26.5	23.5	1.6	3.7
9.5				5.1	17.5	22.8	25.3	21.6	0.7	5.6
8.5			0.2	8	14.1	22.2	24	1.3	0.4	5.8
7.5			0.1	8.1	15	20.8	22.2	0.4	0.3	5.2
6.5		0.1	0.1	0.6	13.8	19.2	19.6	0.3	0.2	0.1
5.5	0.1	0.1	0	0.2	4.3	16.6	1.9	0.2	0.1	0.1
4.5	0	0	0	0.1	0.3	1.3	0.4	0.1	0.1	0.1
3.5	0	0	0	0	0.1	0.3	0.2	0	0	0
2.5	0	0	0	0	0	0.1	0	0	0	0
1.5	0	0	0	0	0	0	0	0	0	0
0.5	0	0	0	0	0	0	0	0	0	0

Table 2 Mean of 1/10th highest Roll Amplitudes, Speed = 5 knots, Head Waves

The shadowed area shows the sea states where the average of the 1/10th highest roll amplitudes is higher than 10 degrees. These sea states are thus denoted as ones where the vessel will suffer from parametric roll. As can be seen the occurrence of parametric roll depends on wave height and wave period (at constant speed and wave heading). In the table below the total group time (in minutes) is given for the same speed (5 knots) and wave heading (head waves) as Table 2. Such tables where made for all wave headings and speeds used in the calculations.

Hs (m) \ Tp(s)	8.5	9.5	10.5	11.5	12.5	13.5	14.5	15.5	16.5	17.5
14.5						135	101	115	114	22.5
13.5						121	101	117	108	0
12.5						113	109	119	99.8	0
11.5					130	101	124	120	0	0
10.5				0	90.8	62	133	112	0	0
9.5				0	27.6	67.7	142	91	0	0
8.5			0	0	16.6	83.7	143	0	0	0
7.5			0	0	29.4	100	144	0	0	0
6.5		0	0	0	29.7	115	133	0	0	0
5.5	0	0	0	0	0	98.9	0	0	0	0
4.5	0	0	0	0	0	0	0	0	0	0
3.5	0	0	0	0	0	0	0	0	0	0
2.5	0	0	0	0	0	0	0	0	0	0
1.5	0	0	0	0	0	0	0	0	0	0
0.5	0	0	0	0	0	0	0	0	0	0

Table 3 Total time of Parametric Roll in 3 hour Simulation in minutes (Group Time), Speed = 5 knots, Head Waves

In the figures below the average of the 1/10th highest roll amplitudes is plotted against the significant wave height for a few different conditions.

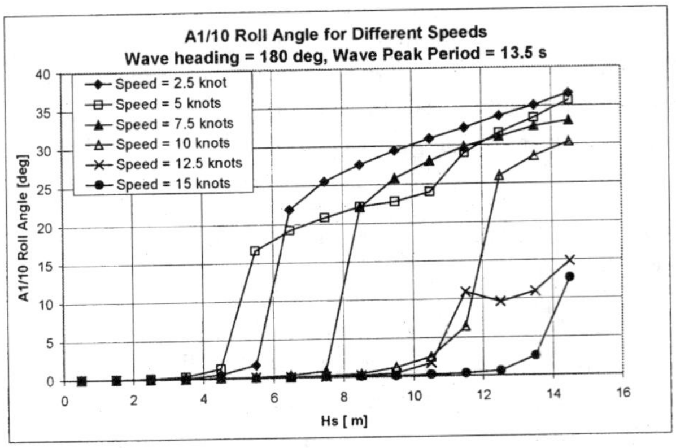

Figure 14 Mean of 1/10th highest roll amplitudes for different speeds, head waves, T_p = 13.5 s

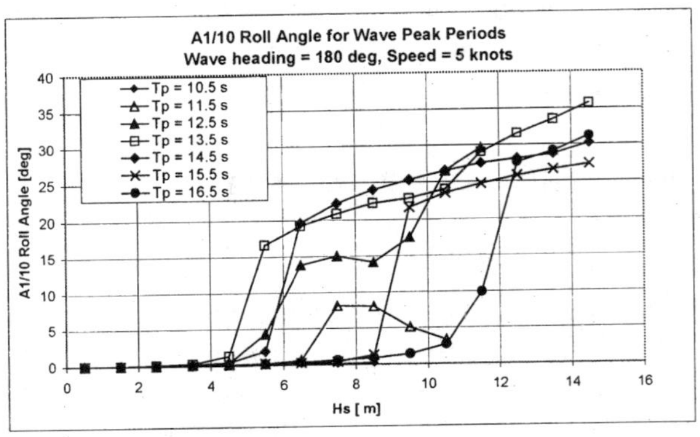

Figure 15 Mean of 1/10th highest roll amplitudes for different wave periods, head waves, Speed = 5 knots

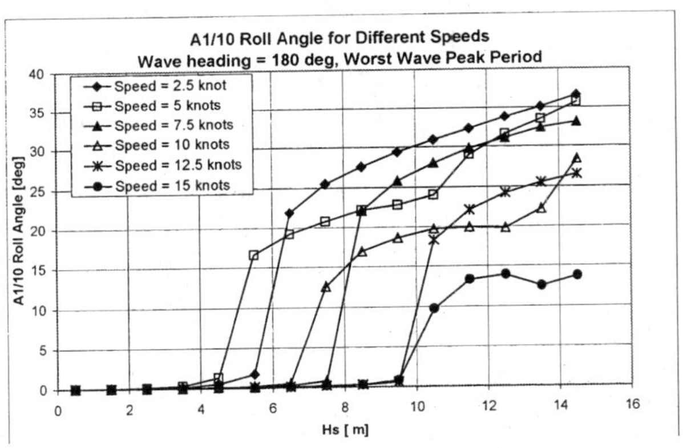

Figure 16 Mean of 1/10th highest roll amplitudes for different speeds, head waves, worst period

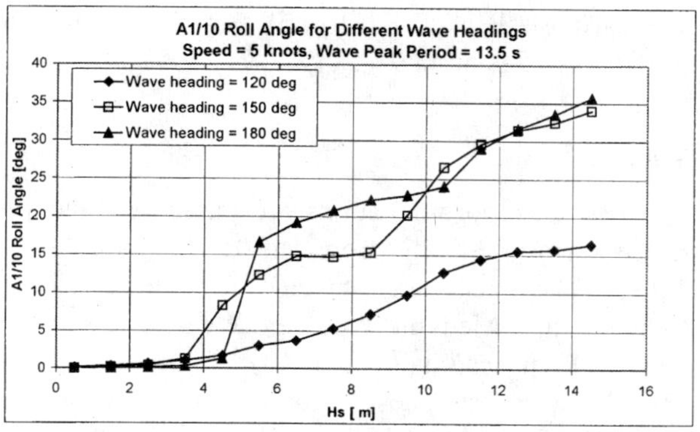

Figure 17 Mean of 1/10th Highest Roll Amplitudes for different Wave Headings, Speed = 5 knots, T_p = 13.5 s

First of all the figures clearly show that parametric roll is a phenomenon which is characterized by a threshold wave height. The sudden increase roll angle with increasing wave height is immediately apparent in all of the figures. This effect is less in bow quartering waves (120 deg) than in head waves, as seen in Figure 17. Figure 14 shows that at high speeds parametric roll will only occur in very high wave heights. This is due to two effects. First, the roll damping is higher at higher speeds. Second, the wave peak period is longer in high sea states, thus increased speed is necessary to produce a wave encounter period that is close to half the natural roll period, a condition for parametric roll.

ASSESSMENT OF OPERATIONAL RISKS OF PARAMETRIC ROLL

In the previous section a database containing information on parametric roll for a matrix of wave conditions, wave headings and speeds was described. By combining that information with information about the probability of these wave conditions and speeds on a particular route one can determine the operational risks of parametric roll on that route. In the following sections two approaches will be described, the wave scatter diagram approach and the scenario simulation approach.

It must be noted that the database as presented in the previous section was determined for one loading condition only. With different loading conditions, and associated different natural roll periods, the wave conditions in which the vessel is subject to parametric roll will be different. A fully developed method for assessing the operational risk of parametric roll would include the probability of operating in a range of loading conditions and consider the parametric roll behavior for the ship in each loading condition.

Wave Scatter Diagrams Approach

A wave scatter diagram gives the probability of a certain combination of wave height and wave period for a given sea area or route. The wave scatter diagram however doesn't give information on the relative wave direction and the speed that the vessel will sail. When combining the information of the wave scatter diagram with the parametric roll database a wave heading and vessel speed must be assumed. The probability of this wave heading and vessel speed is however not known.

To combine the parametric roll database with the wave scatter diagrams the results presented in Table 2 are used, that is, for each combination of H_s and T_p the probability of parametric roll is assumed to be either 1 or 0 regardless of the number of, or duration of, parametric roll events in each three hour simulation. The shaded areas in the table identify the combinations of wave height (H_s) and wave period (T_p) for which the probability of parametric roll is 1. The scatter diagram gives the probabilities of occurrence of each combination of wave height (H_s) and wave period (T_p). Thus, the probability of parametric roll is assumed to be identical to the probability of occurrence of each sea condition. To determine the probability of parametric roll for a given (annual and all direction) scatter diagram the probabilities of a each wave combination are summed. The method is visualized in Table 4.

Speed = 5 knots, head waves
Speed = 10 knots, head waves
[illegible]

Hs (m) \ Tp(s)	5.5	6.5	7.5	8.5	9.5	10.5	11.5	12.5	13.5	14.5	15.5	16.5	17.5
14.5													
13.5													
12.5													
11.5									1				
10.5								1	1	1			
9.5						1	1	2	2	1	1	1	
8.5					1	2	2	3	3	2	1	1	
7.5					1	3	5	6	5	3	2	1	
6.5				1	3	7	10	11	9	5	3	2	1
5.5				2	8	17	21	20	15	8	4	2	1
4.5			1	5	20	34	39	33	21	11	5	1	
3.5		1	4	15	42	60	59	44	22	10	4	1	
2.5		2	11	32	71	78	64	38	15	7	2	1	
1.5		5	21	50	71	56	34	14	4	2			
0.5	2	7	14	19	14	7	3	1					

Table 4 Probability of Occurrence of Parametric Roll for different Speeds in North Atlantic

The probability of each combination of H_s and T_p is given in thousands. The wave scatter diagram used was area 9 (North Atlantic) all directions from the Global Wave Statistics (Hogben 1986). The colors indicate the sea states for which parametric roll occurs for different speeds in head waves. From the table one can determine the probability of parametric roll by summing the probabilities within the colored fields. The shaded areas are overlapping. All shaded areas, which have parametric roll at 15 knots also have parametric roll at 10 and 5 knots. And all shaded areas, which have parametric roll at 10 knots also have parametric roll at 5 knots.

Thus, for the vessel in head seas with a speed of 5 knots, the probability of parametric roll is 7.3%. For 10 knots the probability is 0.7% and for 15 knots 0 %. The same analysis was performed for all speeds and wave headings from the parametric roll database. The results are presented in Table 5.

Speed [knots]	Wave Heading [deg]		
	120	150	180
2.5	1.4	9.4	8.9
5.0	0.6	6.5	7.3
7.5	0.2	1.9	2.1
10.0	0	0.5	0.7
12.5	0	0.1	0.2
15.0	0	0	0
17.5	0	0	0
20.0	0	0	0

Table 5 Probability of Parametric Roll Event (in % time) for North Atlantic area for Different Combinations of Speed and Wave Heading

From the table the influence of wave heading and speed on the occurrence of parametric roll can again be seen. At a wave heading of 120 deg the probability of parametric roll is very small even at very low speeds. In bow quartering waves of 150 deg and head waves the probability of parametric roll can be significant for speeds smaller than ten knots.

The table clearly shows that the vessel with this loading condition would have a non-zero probability of a parametric roll event. The probability of wave heading and speed is however not known. With the statistical method, using only wave scatter diagrams, the probabilities associated with speed and heading cannot be taken into account. In addition, the results of a statistical approach do not include the effect of persistence of storms or the number of affected trips. Also the impact that operational measures (like a temporary reduction of speed) have on the number of occurrences of parametric roll are not included. A scenario based simulation is more approprate for that purpose. But the main advantage is that at each time step of the simulation the wave and wind conditions and the sustained speed are known.

Scenario Based Simulation Approach

A practical way around the problems and limitations of a statistical approach is based on a deterministic step-wise simulation of a given "mission". The use of hind cast data (from wind-wave models) to model the sea conditions solves the problem of achieving the right coherence between wind and waves and the varying spectral characteristics of the waves. We used MARIN's GULLIVER scenario simulation program for this purpose.

Weather

The scenario simulations are based on hind cast data for North Atlantic from Jan 01, 1995 to Dec 31, 1999. The data were obtained from the European Center for Medium Range Weather Forecasts (ECMWF) in Reading in the UK.

The wind estimates are based on a mathematical model for the atmosphere; the data were correlated daily with meteorological station measurements. Wave spectrum data are then obtained after processing of the wind field with a wind-wave model.

The wind data is available in terms of speed and direction (see Figure 18 for example). The wave data consist of significant wave height, mean period and direction of wind-sea and swell. The data are available on a grid of 1.5 deg x 1.5 deg in time steps of 6 hours.

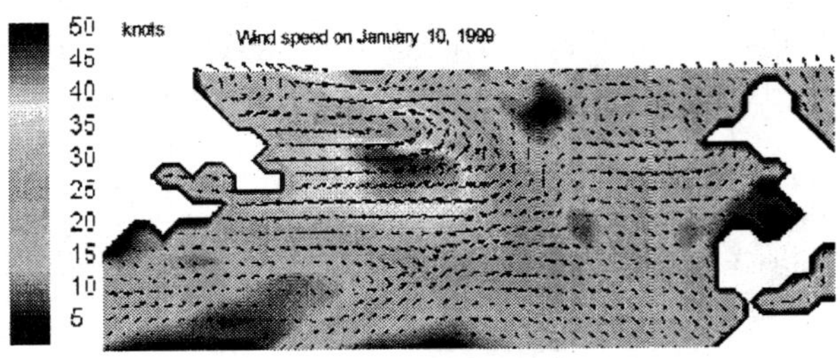

Figure 18 Example of ECMWF Data, Wind Speed on North Atlantic, January 10, 1999

For each time step an irregular short crested sea is reconstructed from the above global parameters. A JONSWAP wave spectrum with variable spectral shape and directional spread was used according to the Mitsuyatsu formulation.

Route

Calculations were performed on a great circle route between Southampton and New York. A northern trip was simulated in both westbound and eastbound directions. Because weather information is missing for about 60 miles from the British coast, trips do not exactly start at Southampton but at co-ordinate: 50.16N / 3.65 W. The route is 2937 nautical miles long.

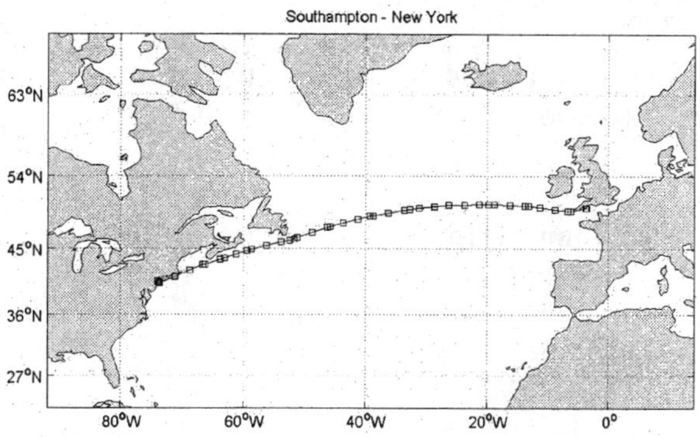

Figure 19 Route Used for Scenario Simulations

Calculation Procedure

An individual voyage consists of a fixed route, which consists of a number of waypoints. A voyage starts at the first waypoint at a pre-defined moment in time; it ends when the last waypoint has been reached. Time is increased with a fixed quantity (in this case 2 hours) at the end of each time step. At each time step the simulation:

- Interpolates the wind and waves at the ships position;
- Re-constructs the wave spectrum from the wind sea and swell parameters;
- Evaluates the involuntary speed loss due to wind drag and added resistance from waves;
- Calculates the progress at each time step.

To obtain reliable statistics a large number of trips were simulated. The present simulations account for a departure on every fourth day over a period of five years. A period of four days is regarded sufficient to obtain statistically independent results. It is assumed for these simulations that the vessel doesn't reroute or change course.

Analysis Procedure

The results of the simulations were time traces of encountered wave conditions (wave height, wave period and wave direction) and the sustained speed on the route. Unlike the wave scatter diagram approach, the probability of each wave heading and sustained speed combination is known. At each time step the wave condition and sustained speed were used to enter the parametric roll database (for example Table 2) to obtain parametric roll data. This simulation counted the total numbers of parametric roll events as well as the number of trips during which one or more events occurred.

Results of Scenario Simulations

A total number of 456 trips eastbound and 456 trips westbound were simulated. All simulations assumed that the vessel is sailing with the same loading condition.

In Figure 20 the time trace of the wave height for all simulations is given. The seasonal effects on the wave height can clearly be seen. The winter periods can be recognized as the periods with more severe wave conditions. It is interesting to note that the summer seasons are not without relatively high wave conditions.

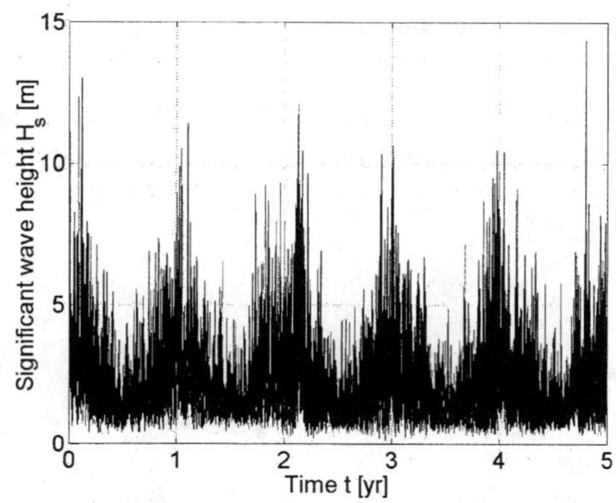

Figure 20 Time Trace of Wave Height for all Simulations West Bound

Figure 21 and Figure 22 show statistical results of the analyzed wave height time trace.

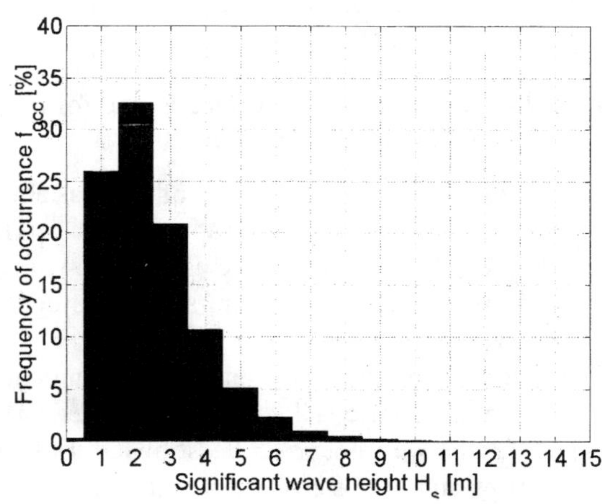

Figure 21 Frequency of Occurrences of Wave Height (All Trips West bound)

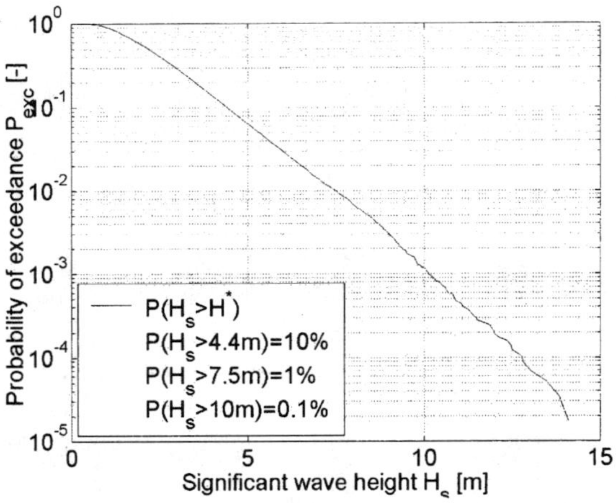

Figure 22 Probability of Exceedance of Wave Height (All Trips West bound)

Figure 21 and Figure 22 show that on this route high waves can occur. The probability of waves higher than 7.5 m is 1 % and the probability of waves higher than 10 m is 0.1 %.

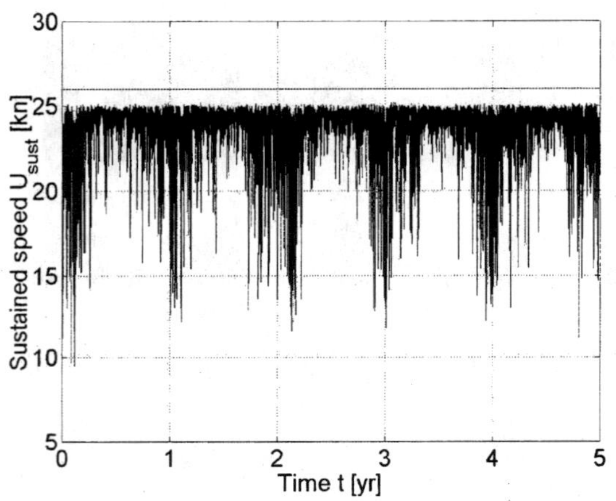

Figure 23 Time Trace of Sustained Speed for all Simulations West Bound

Figure 23 shows the time trace of the sustained speed. The seasonal effects are apparent. The maximum speed in calm water at the adopted maximum power is 25 knots and the maximum speed loss due to wind and waves is about ten knots.

In Table 6 the results of the analysis of simulation results and the parametric roll database are given. The table gives for the west bound and east bound trips the number of trips and time steps where, according to the parametric roll database, the vessel would have a parametric roll event. The criteria are that the vessel is sailing in head or bow quartering waves and that the average of the $1/10^{th}$ highest roll amplitude is higher than 10 degrees. A similar approach as previously described (see Table 2 and explaining text). In the table the percentage of trips and time steps with parametric roll events is also given. The table also gives the number of trips and events with two successive roll angles (due to parametric roll) above 20 degrees and 30 degrees.

	Trips	Steps	Trips	Steps
West Bound Trips	N	N	%	%
A1/10 >= 10 deg	3	6	0.66	0.02
Successive Angles >= 20 deg	2	5	0.44	0.02
Successive Angles >= 30 deg	0	0	0.00	0.00
East Bound trips				
A1/10 >= 10 deg	0	0	0.00	0.00
Successive Angles >= 20 deg	0	0	0.00	0.00
Successive Angles >= 30 deg	0	0	0.00	0.00

Table 6 Number of Trips and Time Steps with Parametric Roll Events. Max. Sustained Speed

From the table one can see that in approximately one percent of the trips west bound a parametric roll event occurs. The number of trips with parametric roll is three. In two of the trips the roll angles are higher than 20 degrees East bound no parametric roll events occur. The reason for the difference between the results for the west bound and the east bound route is the direction of the depressions on the North Atlantic. The depressions travel from west to east and the main direction of the waves along the selected route is in the same direction. A vessel going to the west will encounter more head and bow quartering waves than going to the east. Thus the probability of parametric roll is higher going from east to west.

In Figure 24 the maximum sustained speed for all the west bound trips is plotted against wave height. The scatter is due to the relative wave heading. In head seas the vessel will loose speed and in following waves and stern quartering waves the speed will not be greatly affected.

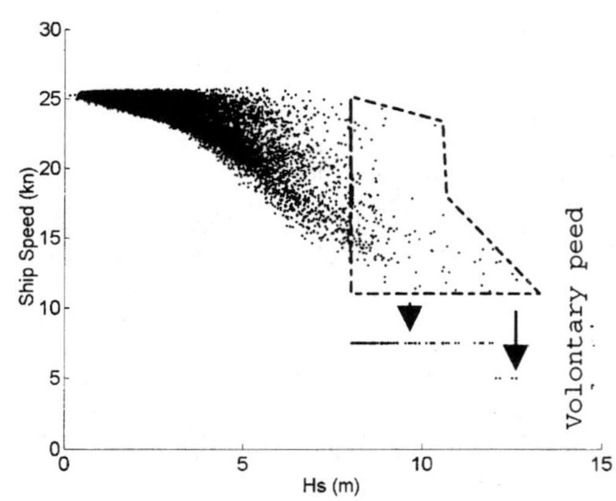

Figure 24 Maximum Sustained Speed versus Wave Height

From the figures it can be seen that if one only takes into account involuntary speed loss (due to wind and waves), the vessel would sail at very high speeds in high waves (between 12.5 and 15 knots in waves above ten meter). At these high speeds the probability that the vessel would suffer from bow flare slamming and green water is high. It is likely that a captain would reduce speed in these high sea states.

To illustrate the influence of voluntary speed loss on the probability of parametric roll a simple captain's mimic was assumed. The mimic (illustrated on Figure 24) assumes that the speed is reduced to 7.5 knots when the waves are higher than H_s = 8 m but lower than H_s = 12 m and further reduced to 5 knots in waves above H_s = 12 m. The results of using this scenario together with the results of the analysis of simulation and the parametric roll database are given in Table 7.

	Trips	Steps	Trips	Steps
West Bound Trips	N	N	%	%
A1/10 >= 10 deg	22	104	4.82	0.38
Successive Angles >= 20 deg	21	88	4.61	0.32
Successive Angles >= 30 deg	2	5	0.44	0.02
East Bound trips				
A1/10 >= 10 deg	3	5	0.66	0.02
Successive Angles >= 20 deg	3	5	0.66	0.02
Successive Angles >= 30 deg	0	0	0.00	0.00

Table 7 Number of Trips and Time Steps with Parametric Roll Events. Voluntary Speed Loss

From Table 7 one can see that when voluntary speed loss is taken into account, parametric roll occurs in nearly 5% of the trips, with nearly 5% events having roll angles higher than 20 degrees (west bound). The table shows the threshold behavior of parametric roll: the number of events with roll amplitudes less than 20 degrees is virtually identical to the number of events with roll amplitudes greater than 20 degrees with no gradual increase in amplitude. East bound the risk is less than 1%.

It is clear that voluntary speed loss and course changes will influence the risk of parametric roll events. The measures to reduce "normal" ship motions, vertical and transverse accelerations, slamming or green water can put the vessel into situations were it is subject to parametric roll.

The "theoretical" maximum number of parametric roll events can be determined by assuming that the vessel would sail at zero speed in all wave conditions where parametric roll could occur but where the sustained speed is too high for parametric roll. These results are presented in Table 8.

	Trips	Steps	Trips	Steps
West Bound Trips	N	N	%	%
A1/10 >= 10 deg	92	692	20.18	2.50
Successive Angles >= 20 deg	88	609	19.30	2.20
Successive Angles >= 30 deg	7	22	1.54	0.08
East Bound trips				
A1/10 >= 10 deg	11	35	2.41	0.13
Successive Angles >= 20 deg	10	34	2.19	0.12
Successive Angles >= 30 deg	0	0	0.00	0.00

Table 8 Maximum Number of Trips and Time Steps with Parametric Roll Events. Zero Speed

From the table one can see that the maximum percentage of trips where sea conditions are encountered in which the vessel would suffer from parametric roll is 20% (with 19% of these trips experiencing roll angles higher than 20 degrees). East bound the values are much lower (2%).

The results show that the potential risk of parametric roll is significant. But, if the vessel can maintain speed and avoid head seas and bow quartering seas when the waves are high the risk drops significantly. However, even than the risk is not zero. This means that if loading conditions with a high natural roll period cannot be avoided, the vessel should use weather routing to avoid wave heights above the threshold value determined for that ship. On the other hand, if the vessel cannot re-route, hazardous loading conditions, determined for that ship, should be avoided.

For ship designers this means that they should design vessels with sufficient power that high speeds can be maintained in rough weather. And use hull lines (bow configuration and bulwark heights) that don't force the master to reduce speed in order to avoid green water and bow flare slamming.

CONCLUSIONS

In this paper an initial attempt to quantify the operational risks of parametric roll for a large container vessel is presented. The vessel analyzed is a C11 class post-panamax container vessel. It can be assumed that vessels in the same class will behave in the same way.

For the study the information from weather statistics from wave scatter diagrams and scenario simulations is coupled with a parametric roll database. This database gives the probability of parametric roll for a given vessel speed, wave heading, wave height and wave period. The database was developed by conducting a very large number of non-linear seakeeping simulations using MARIN's FREDYN program.

Before developing the database, representative FREDYN simulations of C11 ship motions were compared to model tests results. The comparison against roll decay tests, regular waves tests and irregular wave tests showed that after the tuning of the transverse radii of gyration k_{xx} the results of the simulation were very close to the results of the model tests.

In this paper only one loading condition, that was known to be subject to parametric roll, was used. Normally a container vessel sails in a variety of different loading conditions, each of which may or may not be subject to parametric roll

An analysis based on a wave scatter diagram was used to determine the probability of parametric roll at given speeds and headings; however, the operational risk could not be determined by this method because it doesn't include the probability of the vessel sailing the given wave headings and speeds.

These conditions were accounted for in a voyage analysis of a container vessel on a route from Southampton to New York. The sea conditions along the route were modeled using a five-year hind cast database from ECMWF. The database contains information on the wind, wind driven seas and swell that was used for each voyage time step to calculate speed loss due to wind and waves and to enter the parametric roll database to identify parametric roll behavior. A total number of 456 trips west bound and 456 trips east bound were simulated.

Using voyage simulations it was determined that the maximum theoretical number of trips where wave conditions were encountered which could cause parametric roll was 20% when going west bound. East bound the probability was much lower due to the fact that the prevailing wave direction did not result in head or bow quartering waves.

Simulations assuming that maximum sustained speed would be ordered by the captain resulted in probabilities of encountering parametric roll of less than 1% going west bound and zero going east bound. These simulations did not assume that a captain would voluntarily reduce speed to avoid bow flare slamming, green water or high motions in certain conditions.

Simulations assuming a simple voluntary speed reduction scheme increased the probability of encountering parametric roll to 4.8% going west and less than 1% east bound. The results were consistent with the threshold nature of the parametric roll phenomenon.

For all the simulations it was assumed that the vessel would sail with the same loading condition and wouldn't reroute or change heading. Rerouting to avoid heavy weather would reduce, in most of the cases, the probability of parametric roll; however, change of heading could increase the probability of parametric roll if the captain decided to heave to into the waves.

The results showed that the occurrence of parametric roll results from an interaction among wave encounter period, natural roll period, and wave height and that paremattric roll experiences a threshold behavior with increasing wave height. In some cases this threshold wave height, and associated period, has a steepness that is outside the range of naturally occurring wave steepnesses. Thus, the probability of parametric roll might be expected to vary significantly with loading condition.

The results presented in this paper show that the current fleet of C11 type container vessels have a non negligible risk of encountering condition in which parametric roll can occur. The ship operators should be aware of this risk and take measures to reduce it. Avoid loading conditions with a high natural roll period or if that is not possible use weather routing to avoid high waves.

It was also shown that both involuntary and voluntary speed reductions can result in increased probability of experiencing parametric roll. For ship designers this means that they should design vessels with sufficient power and with bow configurations to maintain high speeds in rough weather. Measures to increase roll damping, such as (larger) bilge keels or (active) anti roll tanks could also reduce the risk of parametric roll.

RECOMMENDATIONS FOR FURTHER RESEARCH

Future work in developing a methodology for operational guidance to minimize the probability of parametric roll could address the following:

- Take into account the variations in loading condition.
- Evaluate the influence of design changes (stern, bow flare, bilge keel etc.) on the probability of parametric roll.
- Evaluate the influence of different routes (North Pacific route for example).
- Variation in used or installed power.

REFERENCES

Dallinga, R.P., Blok, J.J and Luth, H.R., Excessive rolling of cruise ships in head and following waves,

RINA International Conference on ship Motions & Maneuverability, London, February 1998.

De Kat, J.O., Pinkster, D.J. and McTaggart, K., Random Waves and Capsize Probability Based on Large Amplitude Motion Analysis, *Proceedings 21st International Conference on Offshore Mechanics and Arctic Engineering, OMAE 2002, J.V. Wehausen Symposium on Water Waves*, Oslo, June 2002.

De Kat, J. O. and Paulling, J. R., The Simulation of Ship Motions and Capsizing in Severe Seas, *Transactions*, The Society of Naval Architects and Marine Engineers, vol. 97, 1989.

Dunwoody, A.B., Roll of a ship in astern Seas – Metacentric height spectra, *Journal of Ship Research* vol 33, No. 3, September 1989, pp 221-228.

Dunwoody, A.B., Roll of a ship in astern Seas – Response to GM fluctuations, *Journal of Ship Research* vol 33, No. 4, December 1989, pp 284-290.

France, W. N., Levadou, M., Treakle, T. W., Paulling, J. R., Michel, R. K., and Moore, C., An Investigation of Head-Sea Parametric Rolling and its Influence on Container Lashing Systems, *Marine Technology vol 40*, No. 1, January 2003.

Graff, W. and E. Heckscher, Widerstands und Stabilitäts versuche mit drei Fischdampfermodellen, *Werft-Reederei-Hafen*, vol. 22, 1941 (also DTMB Translation No.75, June 1942).

Hogben, N., Dacuhna, N.M.C. and Olliver, G.F., *Global Wave Statistics*, BMT, London, 1986.

Hooft, J.P., Mathematical Description of the Maneuverability of High Speed Surface Ships, *MARIN report No. 47583-1-MOA*, 1987.

Ikeda, Y., Himeno, Y., Tanaka, Y., A prediction method for ship roll damping, *Technical Report 00405*, University of Osaka, 1978.

Kempf, G., Die Stabilitätsbeanspruchung der Schiffe durch Wellen und Schwingungen, *Werft-Reederei-Hafen,* vol. 19, 1938.

Lin, W.M., and Salvesen, N., Nine Years of Progress with LAMP – The Large Amplitude Motion Program, *SAIC Report No. 97/1079*, December 1998.

Lin, W.M., and Yue, D.K.P., Numerical Solutions for Large-Amplitude Ship Motions in the Time-Domain, *Proceedings of the Eighteenth Symposium of Naval Hydrodynamics*, The University of Michigan, U.S.A., 1990.

Lin, W.M., and Yue, D.K.P., Time-Domain Analysis for Floating Bodies in Mild-Slope Waves of Large Amplitude, *Proceedings of the Eighth International Workshop on Water Waves and Floating Bodies*, Newfoundland, Canada, 1993.

Luth, H.R. and Dallinga, R.P., Prediction of excessive rolling of cruise vessels in head waves and following waves, PRADS. The Hague Netherlands, 1998.

McTaggart, K., and DeKat, J.O., Capsize Risk of Intact Frigates in Irregular Seas, *Transactions*, The Society of Naval Architects and Marine Engineers, Vol. 108, 2000.

Oakley, O. H., Paulling, J. R. and Wood, P. D., Ship Motions and Capsizing in Astern Seas, *Proceedings,* Tenth ONR Symposium on Naval Hydrodynamics, ONR, ACR 204, 1974.

Paulling.J. R. and Rosenberg, R. M., On Unstable Ship Motions Resulting from Nonlinear Coupling, *Journal of Ship Research,* vol 3, no.1, 1959.

Paulling, J. R., The Transverse Stability of a Ship in a Longitudinal Seaway, *Journal of Ship Research,* vol. 4, no. 4, 1961.

Treakle, T.W., Nonlinear Active Control of Dynamical Systems (ONR Grant No. N000014-96-1-1123), *Proceedings of the Eighth Semi-Annual MURI*, Virginia Polytechnic Institute and State University, October 16-17, 2000.

DISCUSSION

V. Belenky, Member

The authors have to be congratulated for the fine paper on a very important subject. Parametric roll of containerships is a serious risk factor, which might cause damage to a ship and her cargo (France et al 2003).

The authors used voyage simulations to evaluate large roll motions. These simulations involved generation of long realizations of irregular waves. This is the understanding of the discusser that the authors used Fourier series to present the stochastic process of irregular wave elevation:

$$\zeta_W(t) = \sum_{i=1}^{N} c_i \cos(\omega_i t + \varepsilon_i) \qquad (1)$$

Theoretically, $\zeta_w(t)$ is a periodical function: its period equals the least common multiple of all components if frequencies are rational numbers. In fact, this period is large enough not to worry about possible exact repetition of wave pattern. However, the duration of time when $\zeta_w(t)$ still represents the adequate model of irregular wave elevation is limited by factors other than formal period of repetition. The autocorrelation function of the process presented in equation (1) is also a periodic function (see Fig. 1).

The second peak of the autocorrelation function indicates a very strong statistical relationship between time moments $t = 0$ and $t = T$. This second peak has no physical background. As a result, equation (1) represents an adequate model for stochastic process $\zeta_w(t)$ only for the time interval less than T.

As it is known, the period autocorrelation function is related to the frequency step:

$$T = \frac{2\pi}{\Delta\omega} \qquad (2)$$

This means that if we need to simulate a relatively long realization, several hours, for example, which is quite realistic for voyage simulation, the frequency step should be small and the number of frequencies necessary to cover the spectrum range might be large. (For example, for the spectrum from France et al [2003], with the range from 0.25 to 1.00 rad/sec, in order to simulate a long realization of 6 hours, it would be necessary to use 2,578 frequencies.)

Increasing the number of frequencies raises computation costs, which might make the development of the operational guidance expensive. However, the cost of cargo lost or damaged by parametric roll most probably would justify such computational expenses.

Additional references

France, W. M, Levadou, M., Treakle, T. W., Paulling, J. R., Michel, K., and Moore, C. 2003 An investigation of head-sea parametric rolling and its influence on container lashing systems, *Marine Technology,* **40**, 1, 1–19.

Belenky, V. L., and Sevastianov, N. B. 2003 *Stability and Safety of Ships*, vol. 2, "Risk of Capsizing", Elsevier, Amsterdam.

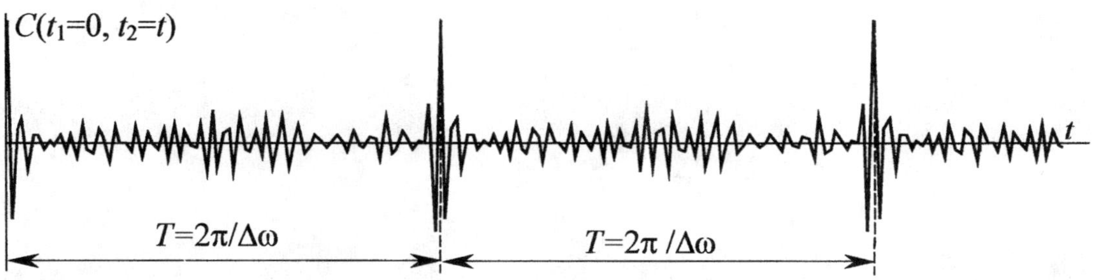

Fig. 1 Repeating pattern of the autocorrelation function calculated for the stochastic process presented with Fourier series (Belenky & Sevastianov 2003)

Authors' Closure

The authors would like to thank *Dr. V. Belenky* for his excellent comments and questions on our paper. As the discusser accurately states, the repetition of similar waves is of concern when performing time-domain seakeeping simulations with a long realization.

Before answering the question, it must be made clear that the voyage simulation does not use a Fourier series to present the stochastic process of the irregular wave elevation in a long term. The voyage simulation uses a 2-hour time step to evaluate if, for the encountered weather (defined as significant wave height and wave period) and sustained speed, parametric roll would occur. This is done by looking into a database, which for each encountered wave condition and speed predicts whether parametric roll would occur.

To generate this database, a program (FREDYN) was used, which uses a series of superimposed sine waves to present the stochastic process of the irregular wave elevation. However, the voyage simulation and the simulation time used in this program is independent from the voyage simulation time step.

For generating this database, the authors used a simulation time of 3 hours. If no repetition of waves occurs, 3 hours is enough simulation time in order to evaluate whether parametric roll would occur in that specific sea state. However, no long-term statistics can be derived from a 3-hour simulation time, as the occurrence of parametric roll is not Rayleigh distributed. But the intentions of the authors were to evaluate if there were parametric roll events and not the exact roll angles associated with the events. Even then, a 3-hour simulation time could give erroneous results if repetition of waves would occur. The simulation tool we used (FREDYN) uses nonequidistant frequency steps. By using smaller frequency steps at the peak of the wave spectrum, the number of frequency steps can be reduced. Besides, the theoretical frequency step following from a simulation time of T ($T = 2\pi / \Delta\omega$) is valid only for zero speed. Most of our calculations were conducted with forward speed, which reduces the risk of repetition of waves significantly.

The nonequidistant frequency step and the forward speed for a simulation time of 3 hours were sufficient to prevent wave repetition events from occurring during the FREDYN simulations.

A Route Or Mission-Dependent Approach For The Calculation Of Rational Structural Dynamic Loads For High-Speed Multihulls

Manish Gupta (AM), **Mark McCain** (AM), **Drew Eisele** (AM)
CDI Marine – Systems Development Division, Severna Park, MD

ABSTRACT

A two-phased integrated methodology to predict the route/mission-dependent rational structural dynamic loads for high-speed multihulls was developed. The project was funded by the Center for the Commercial Deployment of Transportation Technologies (CCDoTT). The specific program of work described in this paper represents the first phase of the two-phased effort to develop a ship motion and dynamic load calculation methodology that would be suitable for use by Designers, Classification Societies and the U.S. Navy. The method developed would be suitable for advanced hullforms, including slender monohulls, catamarans and trimarans, SES, and other advanced hullforms. In Phase I, an overall working model of a frequency-domain ship motion and loads model was developed, incorporating a route-based mission profile module to provide a description of the ships' overall lifetime loads. A commercial ship motions program that can accommodate multihulls, titled SHIPMO, was utilized to model the motion characteristics of advanced hullforms, including catamarans and trimarans. The results of the numerical predictions were compared with model and full-scale data, and analytical rationales were developed for the differences between the experimental and predicted data. Subsequently, modifications to the numerical modeling procedures were applied to overcome any such limitations to the greatest extent possible within the bounds of the Phase I effort, allowing the utilization of SHIPMO to compute short-term frequency-domain vessel motion and load statistics. A comparison of existing extreme value estimation methods was developed, in addition to a methodology to estimate extreme loads from the calculated frequency-domain statistics. Thereupon, a procedure and rationale was developed for determining the load for a probability-based structural design method. Phase I validation efforts were conducted using a notional east coast AutoFerry concept that was previously developed by the authors. This paper summarizes the contributions to the state-of-the-art on rational structural design methodology accomplished during Phase I of the project.

INTRODUCTION

Structural load determination is a critical part of the design spiral for high-performance vessels. Such vessels often feature high-speed capability and are, thus, inevitably weight sensitive. Therefore, it becomes essential that adequate structural strength be achieved without over-designing the structure. Such a structural design must possess a reasonable probability of surviving the environmental loads imposed on it during its lifetime.

The determination of the success of the structural design hinges on the determination of the loads to which it will be exposed and the specification of a reasonable probability level for its survival. The environmental loads imposed on the structure will have to be determined from an analysis of the modes of operation of the vehicle and its environment, and then by making use of this information to define a number of deterministic load conditions for analysis. Each of these load conditions is to be assigned a probability of occurrence and compounded to develop a single long-term probability distribution to each type of load. A failure in the accurate determination of structural loads can have significant impact on the design and operation of the vehicle. Underestimation of the loads can result in structural failure or limit operations. Overestimation of loads, on the other hand, will cause the vessel to carry excess structural weight, resulting in reduced operational range and payload. Hence, the need for a rational approach to loads determination is evident.

Key to the prediction of ship structural loads is the prediction of the response of the ship to the sea. Structural loads such as transverse bending and torsional moment are especially important for advanced hullforms. Therefore, there is a need for a ship motions model that can accurately model the motions of advanced hullforms. Such a motions model would have to be capable of handling high forward speed and multihull configurations such as catamarans, trimarans, SES and SWATH, and the potential impact of lifting bodies. The current state-of-the-art in frequency-domain models for ship motions is limited to low Froude numbers and has not been extensively validated for novel hullforms.

The specific objective of the tasks conducted in the first phase of a two-phased effort was to develop a ship motion and dynamic load calculation methodology that is suitable for use by designers, Classification Societies and the U.S. Navy. The resulting methodology will be suitable for advanced hullforms, including slender monohulls, catamarans and trimarans, SES, and other advanced hullforms. In Phase I, the development of an overall working model of a frequency-domain ship motion and loads model that incorporates a mission profile module to provide a description of the ships' overall lifetime loads is being developed.

The calculated values from the frequency-domain program, developed in Phase I, will be the statistical values of ship's wave-induced loads for various combinations of sea states, ship heading angle, ship displacement and forward speed. The probability of slamming, and the loads associated with slamming, will then be predicted using time-domain methods to be developed in Phase II.

The final deliverable from this two-phased effort will be a methodology to quickly calculate ship motions and dynamic loads, not only for monohulls, but also advanced multihulls, for operation within any specified speed and sea state lifetime scenario. The methodology will include a combination of frequency-domain simulations that utilize the same theoretical approach as the U.S. Navy Standard Ship Motion Program (SMP) and time-domain methods for predicting the slamming loads on high-speed ships. Figure 1 is a visual representation of the approach.

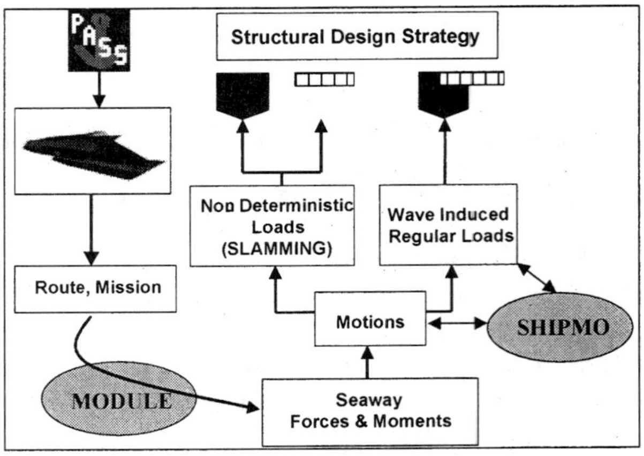

Fig. 1 The Overall Approach to the Rational Structural Loads Prediction

At the outset, a very detailed literature survey was performed to ensure that the most appropriate hydrodynamic, ship motion and statistics theories were chosen to describe multihull ship motions and loads. Data sources that characterize ship motion characteristics and measured structural loads on multihull vessels, including catamarans and trimarans, were identified and reviewed. Such data provided significant value in ascertaining the numerical correctness of the numerical modeling effort that was undertaken. Following these tasks, a more detailed program framework was developed and can be described as follows:

1. Modeled the motion characteristics of catamarans and trimarans utilizing SHIPMO. Compared the results of the numerical predictions with model and full-scale data. Developed an analytical rationale for any differences between the experimental and numerically-predicted data.
2. Investigated modifications to the numerical modeling approach in SHIPMO to overcome limitations in its applicability.
3. Used the SHIPMO model to predict motions and structural loads. Developed statistical descriptions of these variables.
4. Identified applicable extreme value estimation methods available to estimate extreme loads from the SHIPMO-calculated frequency-domain statistics. Developed a procedure and rationale for determining the load for a probability-based structure design method.

The tasks outlined above were initiated and performed. Several of the tasks that followed the validation effort of the motions model were performed using the example of an east coast AutoFerry concept that was previously developed by the authors.

ROUTE / MISSION BASED STRUCTURAL DESIGN METHODOLOGY

The loads experienced by an ocean-going ship depend upon a very large number of influences; they vary with time in a random manner and they never occur in exactly the same way twice. Some of the influences are vessel-related, such as speed, weight distribution and control settings, and some are dependent on the environment, such as wave height and wind speed. The structural designer has to convert this constantly changing scene to a limited number of design load cases that can be used to determine the vessel's structure.

The method adopted here consists basically of the following steps:

- Analysis of the operational environment.
- Selection of representative loading cases.
- Determination of the probability of occurrence of each of the loading cases.
- Calculation of the loads for each representative case in the form of an average or rms value together with a short-term probability distribution to describe the possible variations of loads under these conditions.
- Computation of long-term probability distributions by combining the short-term distributions.
- Selection of design loads.
- Structural design using rational design loads.

In order to define the problem, therefore, the first requirement is to define the operational environment. The steps addressing the selection of representative loadings and structural design were discussed in the previous section.

Operational Environment

Figure 2 shows a typical operational envelope for a high-performance vessel; in this case, an early version of a very high-speed amphibious assault SES. The vessel is expected to operate within the speed wave-height envelope shown. The proportion of time spent in each area of the envelope is indicated. The distribution on the wave-height axis is derived directly from ocean wave-height data, as suggested by Figure 3. The horizontal distribution along the speed scale was derived from a typical mission profile. The sloping lines dividing the speed zones represent the fact that the vessel's speed must fall off as sea state increases. This particular multihull vessel, for example, could reach 90 knots in calm water, but was not expected to exceed 40 knots in SS4. It was not expected to operate on-cushion in sea states higher than SS4. The loads experienced in this case in SS4, even at 40 knots, were so severe that consideration was given to reducing speed still more in the higher sea states. The two lines labeled "structural loads limitations" represent this consideration. One point of interest in this diagram is that the percentage of time spent in the most severe conditions is really quite small. 2.2% of the total time is spent in SS4 between wave heights ($H_{1/10}$) of 5.8 and 8.7 feet and between speeds of 25 and 48 knots. However, only a small proportion of this 2.2% is actually spent near the extreme corner of the diagram (at 40 knots in wave heights of 8.7 feet). If this distribution within each area is not taken into account, the results can be misleading. In the example shown in Figure 2, the load cases actually used for computation are shown by the small open circles.

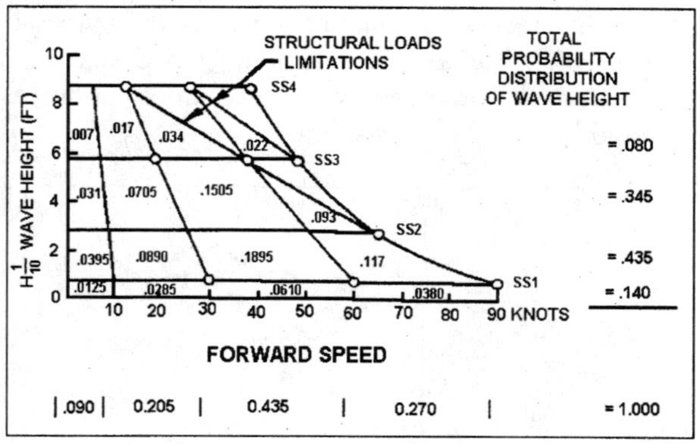

Fig. 2 Typical Operating Envelope for Advanced Marine Vehicles

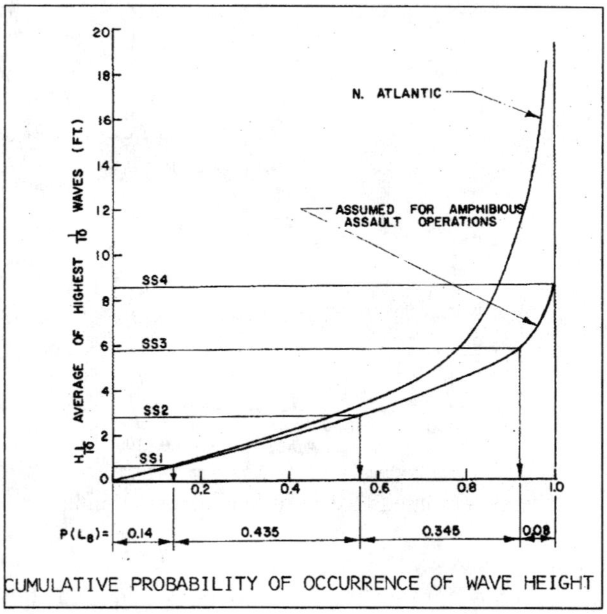

Fig. 3 Sea State Probabilities for Typical Amphibious Assault Mission

Speed and sea state are not the only quantities that define the operational environment. Equally important is the need to define the distribution of heading angle to the dominant wave direction and the distribution of gross weight. Typical distributions of these quantities are shown in Figures 4 and 5. All headings are normally assumed to be equally probable, and a gross weight distribution can usually be derived by a brief analysis of the typical mission profiles or the top-level requirements.

By means of the foregoing considerations, therefore, a formidable matrix of representative cases has been developed. For each case, a set of loads is required which can be generated either by computational methods or by making use of experimental results. In practice, a combination of both sources is often used. For each condition, the designer is interested in a wide variety of load information, as discussed in the previous section.

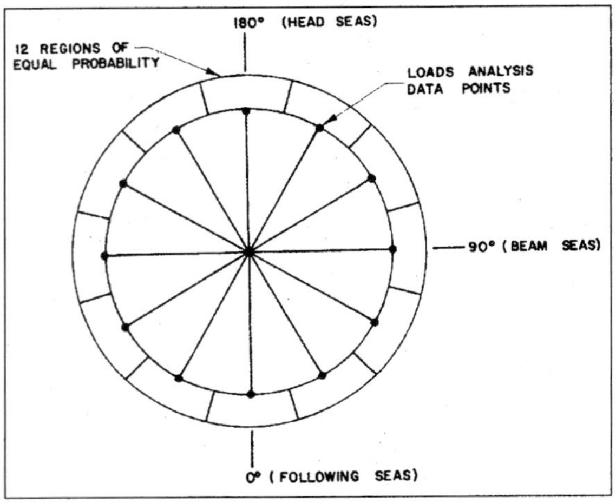

Fig. 4 Assumed Distribution of Ship Heading

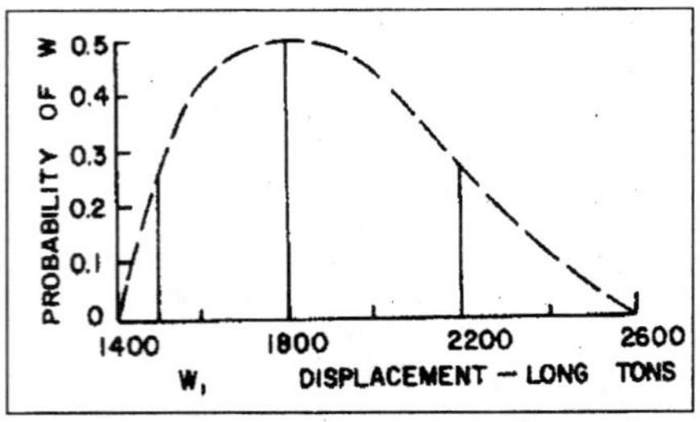

Fig. 5 Assumed Distribution of Gross Weight

Loads Prediction

Specific information on loads is required for structural design. This may include the largest midship bending moment, for example, which a ship will be subjected to within a twenty-year operational life. It is almost impossible to provide this deterministic figure. However, what can be provided is a specific bending moment value that has only a ten percent, or one percent, probability of being exceeded during the lifetime of the ship. Once the probabilistic criteria (in this case the ship's life and the acceptable probability of exceedance) have been specified, a single limit design bending moment can be calculated. The methods of calculation vary. In broad terms, there are two schools of thought:

(a) One method is to design for the "worst expected sea condition." This method proposes to design the ship by analyzing the ship's response to the "worst storm" that it is likely to encounter. It is well known that the waves encountered in severe storms are very dissimilar to those of lesser storms. The duration of extreme storms is normally so limited that the seas do not have time to become anywhere near fully developed. The time required to "fully develop" the seas corresponding to the higher sea states exceeds the probable time that the sea state exists in one year. In this method, therefore, the design loads are determined by exposing the ship, analytically or experimentally, to seas corresponding to the worst storm records available. The main drawback to this method is that the available records of such storms are very limited in number. The statistical sample used as a base is, thus, very limited and, in any case, recorded data shows that the maximum loads probably do not necessarily occur in the worst sea states.

(b) The second method is the statistical method in which extreme loads are postulated by extrapolating from a larger database using all available sea state data. The weakness of this method is that there is no guarantee that the extrapolations are valid. Historical evidence, however, tends to support this procedure. It may be argued that a conventional ship is basically a linear system, whereas most advanced ships are not, but it is suggested here that, provided suitably flexible probability distributions are used to describe the available data, the nonlinearities do not affect the validity of the method. This method, at least, allows prediction to be made from data that is generally available or obtainable, and is flexible enough to allow refinement as time goes on. It can also make use of the data provided by the extreme sea state cases and provides a method for properly ordering and weighting such contributions.

Short-Term Statistics

In order to predict extreme values of loads or motions, it is necessary, therefore, to be able to extrapolate from available samples of experimental or analytical evidence. In general, the experimental samples are very small compared with the expected life of the ship, so that the extrapolation process must be done with considerable care. The probability distribution used in general to approximate such experimental or analytical data is the Weibull distribution, which is a special two-parameter case of the three-parameter generalized gamma distribution. The cumulative Weibull distribution function is given by:

$$P(x) = \exp[-(\lambda x)^c]$$

where, λ is a constant and the index, c, is the gradient. If the index, c (gradient), is 1.0, then the distribution is exponential, and if it is 2.0, the distribution is Rayleigh, as both of these distributions are special cases of the Weibull distribution. The experimental data tend to lie consistently between these two extremes. The Weibull distribution is also acceptable from the point of view that it is closely related to the Rayleigh distribution that is generally accepted as representative of wave heights and wave-induced phenomena. The effect of the choice of distribution on the long-term loads, moments and stresses may be seen graphically in Figure 6, which shows the long-term distributions of x/x_{rms} for various c values. At the higher levels of probability, which are those of interest in establishing extreme values, the long-term cumulative distribution curves fan out over a wide range for different values of the Weibull parameter c.

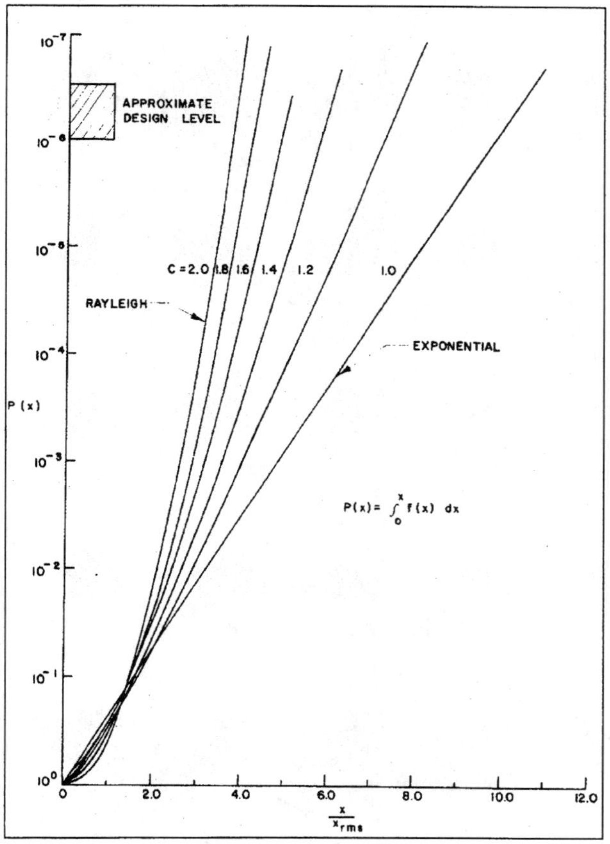

Fig. 6 Cumulative Weibull Probability Distributions for a Range of Values of the Index c

Long-Term Statistics

Each run in the test tank or of the full-scale ship is necessarily a short-term event. During each short-term event, it is assumed that the ship's speed and heading and the sea state remain constant. The operational life of a ship can be considered as a summation of a very large number of short-term events. If the behavior in each short-term event is known, and also if the manner of distribution of short-term events throughout the ship's life is known, then a long-term picture of the vehicle's life can be built up. This is the method that is used here to predict the long-term behavior of the ship so that, eventually, a single long-term distribution of each applied load or stress can be built up to represent the ship total experience in all speeds, sea states, headings and loading conditions. In order to accomplish this, use is made of the description of the operational environment given earlier in this section.

If the probability of exceeding a given wave-induced load x at a speed V in sea state S at heading H and gross weight W is $P_{VHSW}^{(x)}$, then the total probability P(x) of exceeding the load x is given by summing the component probabilities for all conditions.

$$P(x) = \sum_V \sum_H \sum_S \sum_W P_V P_H P_S P_W P_{VHSW}^{(x)}$$

where, P_V, P_H, P_S, P_W are the probabilities of occurrence of each velocity, heading, sea state and gross weight. Combined loads can be treated in the same way. Suppose, for example, that the probable occurrence of a still water load (or thermal load, or etc.) is $P(x_{swl})$, and that this distribution is independent of the wave-induced load, then the total combined probability $P_c(x)$ of exceeding the same load x is given by:

$$P_c(x) = \sum_i P(x_{SW_i}) P(x - x_{SW_i})$$

By basing the short-term extrapolations on experimentally measured or calculated stresses, the effects of the combined components of wave-induced loads are taken into account. The complex combinations of vertical and transverse bending moment and torsion can be represented in the model and full-scale tests in an adequate manner. The number of cases selected from the operational envelope shown in Figure 2 is limited by the test and computational time available. As mentioned before, it may be misleadingly conservative to use the points at the top right-hand corner of each section of the operational envelope to define the conditions pertaining within that segment. Instead, it is usually assumed that the distribution within each segment is uniform and a correction is applied based on this assumption. Figure 7, for example, shows the effect on the long-term probability of exceeding a value x_j of a uniform spread of short-term Weibull distributions with rms values ranging from (upper limit) to (lower limit). The larger the value of x, and the smaller the ratio, the greater the effect on the total probability. The ordinate $P'(x_j)_{AB}$ in Figure 7 is the ratio of the probability of exceeding x_j due to the uniformly distributed set of distributions to the probability of exceeding x_j due to a single distribution.

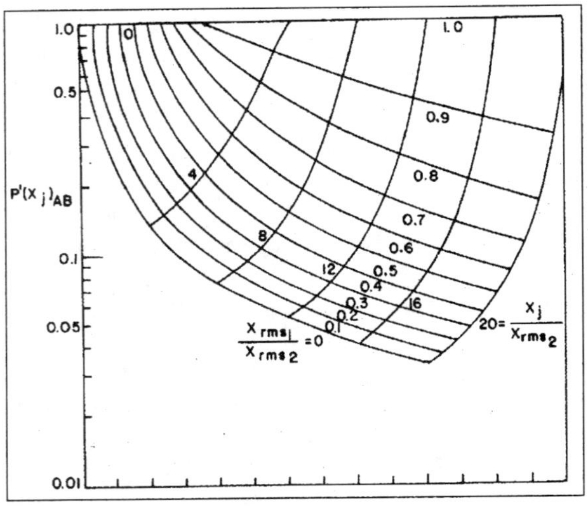

Fig. 7 Cumulative Distribution Characteristics of the Weibull Distribution (c = 1.0)

Selection of Design Limit Loads

By using the summation and integration procedures outlined in the previous sections, it is possible to generate a single, long-term probability distribution of each type of load or stress, such as the familiar curves illustrated in Figure 8. The abscissa is a logarithmic scale that may represent probability of exceedance in each stress cycle, or number of cycles per return period, which may be readily converted to ship life in years. The left-hand end of this curve is relevant to rarely occurring limit loads and the right-hand end to multiple-cycle fatigue loads.

The design limit load may be read off this curve at one, ten and one-hundred ship lives. It is simple to show that by selecting one-hundred ship lives as the return period, then each ship has a one-percent chance of exceeding this design load during its operational life. Fortunately, the increase in stress level between one, ten or one-hundred lives is, normally, not very large, so that a conservative choice may be made without too great a penalty. Sensitivity analysis of typical design bending moment to some of the principal parameters involved can be easily done to obtain the limit loads. One such analysis is shown in Figure 9, where the effect of operational life is seen to be rather small compared with the effect of the Weibull parameter c.

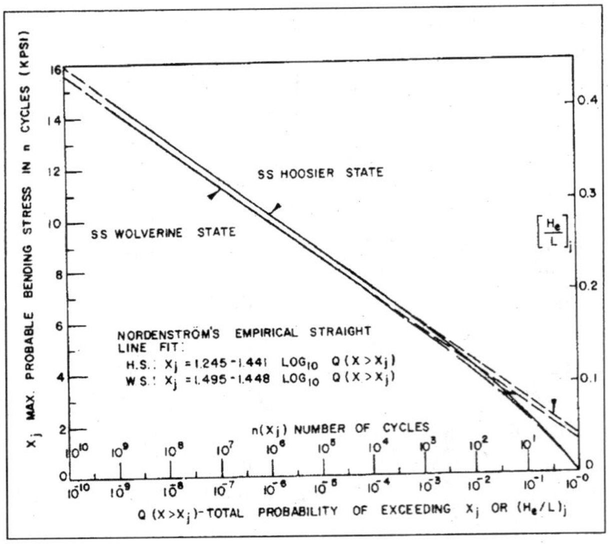

Fig. 8 Comparison of Long-Term Probability Curves for Actual Weather (Band 1966)

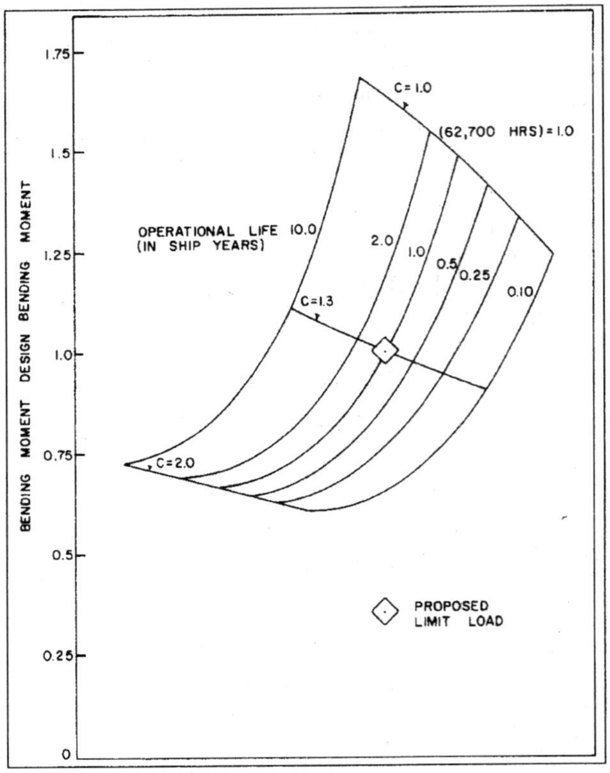

Fig. 9 Sensitivity of Design Bending Moment to Selected Return Period

SEAKEEPING & MOTIONS OF MULTIHULLS

The waterplane of each hull of a catamaran or trimaran usually incorporates a high length-to-beam (L/B) ratio and fine angles of entrance to minimize resistance in comparison to a monohull. Therefore, the longitudinal moment of inertia of the waterplane for the more slender individual hulls of a multihull is substantially smaller than that of a conventional monohull. In addition, having two or more widely separated hulls, the transverse moment of inertia can be higher than a similar-sized monohull. Therefore, multihull craft possess motion characteristics that are quite different from those of a monohull.

In practice, potential flow theory and solvers are used almost exclusively in seakeeping predictions. The most frequent application is the computation of the linear seakeeping properties of a ship in elementary waves. Potential flow solvers are usually based on boundary element methods and need only to discretize the boundaries of the domain, not the whole fluid space. This reduces the effort in grid generation (the main cost item in most analyses) considerably. Different numerical techniques are available to solve the potential flow theory and often account for the differences in the results. The most common of the linear techniques is the Strip Theory method.

Strip methods have emerged as the standard tools for ship seakeeping computations. As the name suggests, the method reduces the three-dimensional problem to a set (e.g. typically 10 to 30) of two-dimensional boundary value problems. They omit the potential of the steady flow disturbance completely and approximate the remaining unsteady potential in each strip independently of the other strips. The methods originated in the late 1950s with the work of Korvin-Kroukovsky and Jacobs. Most of today's strip methods are variations of the strip method proposed by Salvesen, Tuck and Faltinsen (1970). Details on the potential flow theory and strip methods can be obtained from the references.

Catamaran Motion Prediction Using SHIPMO

The primary tool utilized for the development of multihull motion and loads predictions in the first phase of this investigation was the University of Michigan seakeeping prediction program, SHIPMO.BM. SHIPMO employs a combination of strip theory, source distribution techniques, viscous roll damping calculations, and empirical corrections to predict the six degree-of-freedom linear motions and loads of displacement hulls in the frequency-domain. SHIPMO is unique in comparison to other commonly used motion prediction programs, such as the U.S. Navy Standard Ship Motion Program (SMP), in its ability to accommodate monohull *or* catamaran hullforms.

The first step in performing a validation of computer-generated catamaran and trimaran motion predictions involved the accumulation of a pool of controlled experimental data for which a series of SHIPMO match-runs could be developed. This proved to be a difficult task in the world of fast displacement catamarans; a highly competitive field in which access to design information and performance data is carefully controlled. However, several sources of model test results were identified in a wide-ranging search of the open literature, including the three primary papers utilized in this initial study; Molland, et.al. (2000), Hudson, et.al (2001), and Centeno, et. al. (2001).

The body of work reported in the above references consists of the experimental results of model tests on three common catamaran hullforms, including two round bilge hulls (NPL Round Bilge and Series 64) and a single hard chine variant (Vosper Hard Chine). Principal characteristics of the catamaran models are contained in Table 1, and body plans of each of the hullforms investigated are shown in Figure 10. The ratio S/L is the sidehull separation ratio, with S being the sidehull centerline separation, as shown in Figure 10. In summary, the salient research findings of these references are:

- For catamarans, sidehull separation is not a very significant influence on vertical plane motions. Sidehull separation is driven primarily by resistance considerations, and designing to minimize interference at design speed is, thus, more important.
- For higher Froude numbers, experimental investigations report response characteristics that indicate that catamaran motions in head seas are similar to those observed for monohulls.
- Catamaran motion characteristics are greatly improved when a pitch foil is incorporated between the hulls. The ideal location for the foil is forward of midship.

Experimental results for all three hullforms were presented in the form of response-amplitude-operators (RAOs). Data reported for each hullform largely consisted of pitch and heave RAOs at various Froude numbers for the head sea case, although limited data was included regarding response in oblique seas for the NPL Round Bilge and Series 64 models. The oblique sea data was limited to headings of 120 and 150 degrees, and roll RAOs were reported in each case.

Following the data collection phase, SHIPMO run files were developed to reproduce the conditions recorded in the model test reports. Offsets were digitized, based on the body plans illustrated in the data sources identified above, and verified for dimensional consistency with the principal characteristics reported in the literature. The resulting SHIPMO hullform geometry was also verified for hydrostatic similarity with the tested models through a comparison of the predicted displacement and longitudinal center of gravity (LCG).

Table 1. Principal Particulars of Evaluated Hullforms

Model	4b	5b	6b	5s	5b	5s	V40	V60
Hullform Type	Round Bilge	Round Bilge	Round Bilge	Series 64	Round Bilge	Series 64	Hard Chine	Hard Chine
Length, m	1.6	1.6	2.1	1.6	4.5	4.5	2.05	2.05
$L/\nabla^{1/3}$	7.4	8.5	9.5	8.5	8.3	8.3	6.07	6.07
L/B	9	11	13.1	12.8	11	12.8	12.97	12.97
S/L	0.2 / 0.4	0.2 / 0.4	0.2 / 0.4	0.2 / 0.4	0.2 / 0.4	0.2 / 0.4	0.195	0.3
B/T	2	2	2	2	1.9	1.9	1.86	1.86
C_B	0.397	0.397	0.397	0.537	0.4	0.54	0.695	0.695
Max. Froude No.	0.8	0.8	0.8	0.8	0.67	0.67	0.75	0.75
Tests	Head Seas	Head Seas	Head Seas	Head Seas	Oblique Seas	Oblique Seas	Head Seas	Head Seas

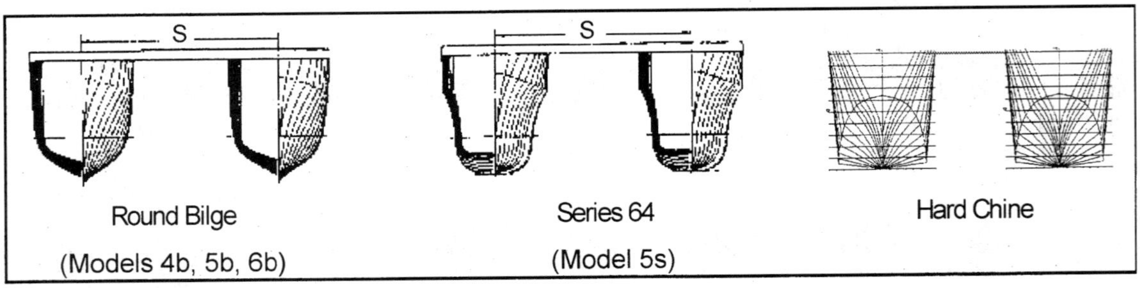

Fig. 10 Body Plans of Evaluated Hullforms

In addition to the development of catamaran run files consistent with the experimental data, numerical models were also generated for conditions in which a single demihull in each of the three hullform categories is treated individually to determine whether the hydrodynamic interaction generated in multihull vessels is significant from the standpoint of motion simulation, and to allow exploration of the hypothesis that vertical plane motions (pitch, heave, etc.) of catamarans are similar to the equivalent demihull (as a monohull) and can therefore be modeled as such for the typically limiting head sea case.

SHIPMO-generated RAOs for each match run were plotted for comparison against the actual test data as a function of non-dimensional encounter frequency. In keeping with the preceding paragraph, demihull-only data was also included in each comparison to determine whether multihull response in the vertical plane (pitch and heave) could be adequately reproduced, and therefore predicted, utilizing a simplified numerical geometry model (i.e., as a monohull).

Figure 11 shows a sample comparative plot; this particular case illustrating the heave RAO for the NPL Round Bilge hullform in head seas at a Froude Number of 0.5. This plot shows a trend that is fairly characteristic of the hullforms investigated, particularly at higher Froude Numbers (0.5 and above). The SHIPMO-predicted catamaran response at the resonant frequency is distinctly higher than the measured data, whereas the predicted demihull-only response is reasonably consistent with the experimental catamaran results over the full range of encounter frequencies. At lower Froude Numbers, the correlation of the SHIPMO-predicted catamaran response tended to improve, while the predicted demihull-only response remained generally consistent. A complete set of comparative plots (~75 conditions) was developed for all three hullforms, at different Froude Numbers, and the characteristic trend discussed above was consistent.

Following the one-to-one comparisons of the SHIPMO-predicted RAOs to the experimental data, attention was focused on capturing the large amount of information generated, in a condensed and collective manner, which would allow for the identification and analysis of trends in the data. With this in mind, Figures 12 and 13 were generated to show the absolute error in the prediction of RAOs at the peak response for pitch and heave in head seas. In this case, absolute error is the numerical value of the predicted RAO less the numerical value of the experimental RAO. This method of error investigation was selected because it inherently retains a consideration of magnitude in the analysis. Put to an example, an error of 50% or 100% in a very small magnitude RAO may produce an acceptable prediction of physical vessel motion, particularly in lower sea states. On the other hand, a small error (based on percentages) in a large magnitude RAO can produce greatly exaggerated motion predictions. This is important in analyzing speed-dependent trends, since the average magnitude of RAOs at slower speeds is generally much lower than the corresponding data at high speed.

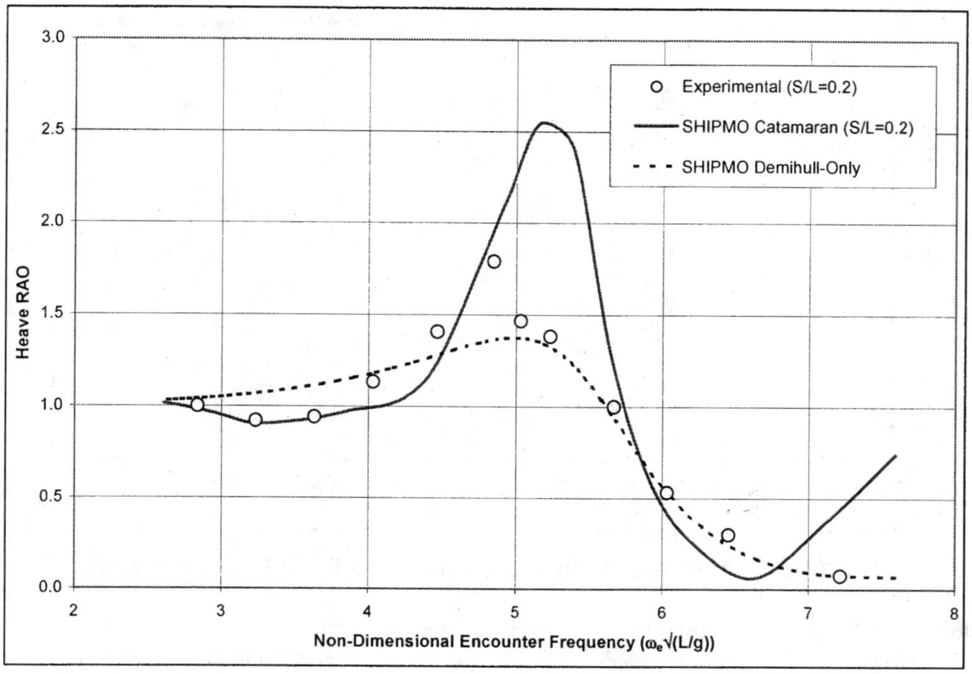

Fig. 11 Typical RAO Comparison (NPL Round Bilge in Head Seas at FN=0.5)

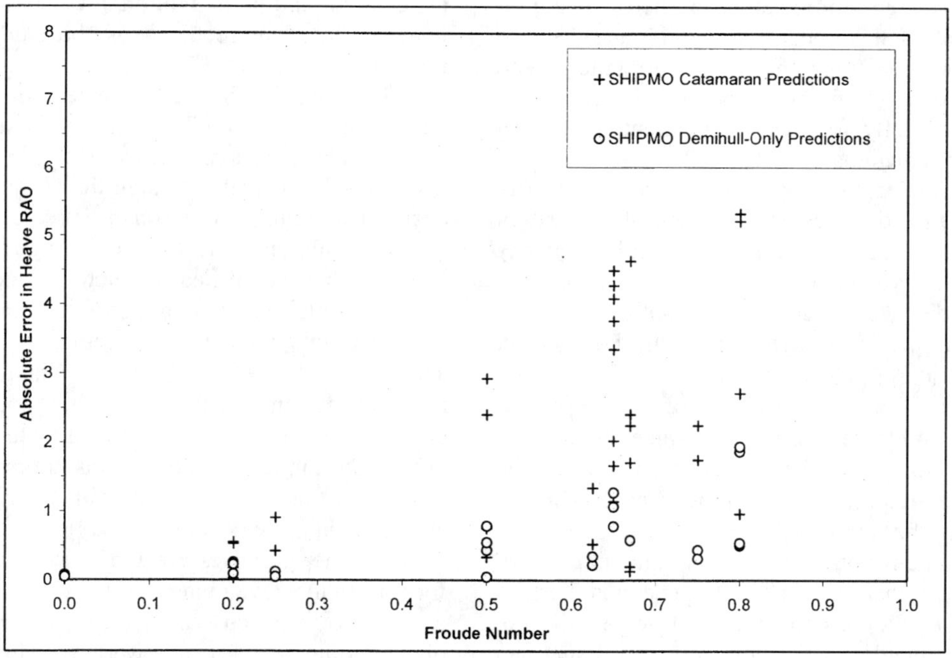

Fig. 12 Absolute Error in Heave RAO Predictions for SHIPMO Monohulls

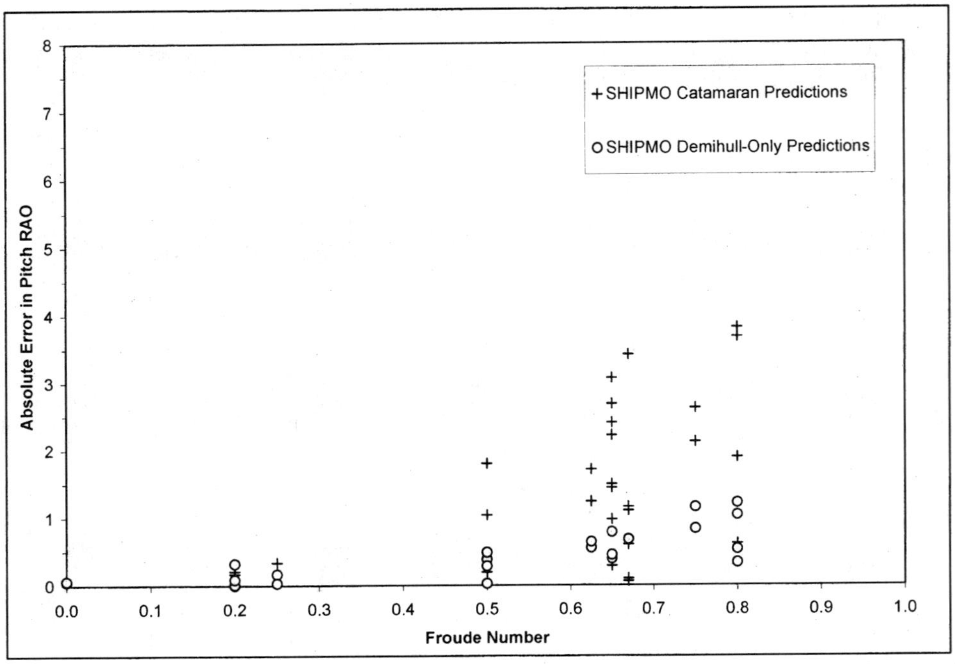

Fig. 13 Absolute Error in Pitch RAO Predictions for SHIPMO Monohulls

In fact, the degradation of the correlation between predicted and experimental data with increasing vessel speed can be observed easily in Figures 12 and 13, in keeping with a key hypothesis of this analysis, as the assumptions inherent to linear strip theory begin to break down at high speed. Figures 12 and 13 also clearly demonstrate the improved correlation with experimental data when the individual demihulls are treated independently for the development of vertical plane motion predictions.

It is recognized that the comparison of the peak response RAO is not necessarily an absolute indication of the accuracy of the prediction over the full frequency range, particularly since the experimental data is collected at discrete speeds and smoothed in post-processing (i.e., not necessarily capturing the resonant response accurately). However, this simplified method of comparison allows the assessment of a large pool of data in a very condensed format, and since the ultimate objective of this two-phased effort involves the prediction of extremes, this was deemed acceptable for the purposes of this study.

Based on the results of the initial validation effort, it was apparent that the SHIPMO-generated vertical plane motion predictions for multihull vessels did not display an acceptable level of reliability. As such, an investigation into the root of the computational discrepancies was undertaken, focusing on the development of the added mass and damping coefficients for the numerical predictions in comparison to the coefficients derived based on the equivalent Lewis-form approximation. This side study, while not treated here due to length restrictions, demonstrated instability and oscillatory behavior in the SHIPMO-generated added mass and damping coefficients that may contribute to the poor correlation with experimental data for multihull motion prediction, and will be investigated in more depth as the study moves into Phase II.

However, the finding that a reasonable correlation with experimental multihull data can be achieved in the vertical plane (pitch, heave, etc.) by treating the demihulls individually is significant, as it allows the designer to achieve a reasonable prediction of vertical multihull motions, and therefore longitudinal bending loads, with a simplified geometry model. However, similar simplification is not possible in the lateral plane for the development of roll predictions, lateral loads, and torsional loads, which are equally critical to multihull design. Unfortunately, the pool of available experimental data available to validate the SHIPMO predictions in the lateral plane was limited, as discussed earlier in this section, and will be required to be addressed in more detail in Phase II.

For the purposes of the initial effort, moving forward to the development of the east coast AutoFerry example case, discussed in later sections, all vertical plane motions and loads were developed based on numerical models of an individual demihull treated independently, whereas lateral plane motions and loads were based on the full catamaran geometry.

STRUCTURAL LOADS AND DESIGN OF MULTIHULLS

From a structural design standpoint, the dominant global load on a ship moving through waves is vertical bending along the ship's longitudinal axis. Transverse bending and torsional loads are also important design considerations, especially for multihull vessels. When a ship is subjected to longitudinal vertical bending, it tends to respond around two distinct frequencies. The lower of the two frequencies is a narrowband spectrum of frequencies that closely matches the frequency spectrum of the waves encountered. The higher of the two frequencies is the first natural frequency of the ship's hull-girder. The low frequency component of the response is the forced vibration response of the ship to the waves, while the high frequency response is a damped free-vibration due to wave impact called whipping, or a resonant phenomenon known as springing. For multihull vessels, similar whipping response can be observed in the transverse bending direction, depending upon the structural stiffness characteristic of the vessel in the lateral direction. Along with the global bending due to seaway, still water bending loads also need to be assessed to be added to the wave-induced bending. In addition to the wave-induced vertical and lateral global bending and whipping, pressures from hull slamming and pitching also govern the design of the bottom and side hull structures. The slamming-induced pressures can be significant on the cross-structure of multihull vessels and will dominate the global loads for small vessels.

An easy to use procedure is required to consider each of these loads in turn, estimate their magnitudes and probability distributions, and to determine to what extent they should be superimposed on each other. The method described here attempts to carry this out. It is inevitable that some of the steps in the method are considerably weaker than others, but the important thing is that it does produce usable numbers for the designer. As time goes on, each step can be re-examined and the validity of the total method can be steadily improved. In order to determine these loads, however, hard data is required for validation. The sources of data have to come from experimental and/or full-scale trials. As more and more research and development work is done on an ever-increasing number and variety of advanced hullforms, a larger and larger database will become available.

Hull-Girder Bending Load

For a route and mission-dependent approach for load assessment, a characterization of the most probable sea states and ship operating scenarios needs to be established first. Thereupon, using a two-dimensional strip theory-based tool, such as SHIPMO, a frequency-domain assessment of the motions of the vessel can be developed for the predetermined sets of sea states and operating conditions. The input to this effort will be information such as the wave spectra, vessel operating speeds and probable headings, and the output will be the vessel's motions response spectra and associated statistical values, for example root-mean-square (rms) values. Along with the motion responses, this frequency-domain assessment will also provide the vessel's low frequency ordinary wave-induced bending load spectra. The ordinary wave-induced vertical bending in both longitudinal and transverse directions and torsional bending can then be obtained. It should be noted that the other part of the global load, wave impact-induced whipping, cannot be estimated via this approach since the vessel's structural stiffness and its viscous damping characteristics are required as priors.

Extensive studies have shown that wave elevation time histories tend to have a Gaussian or normal distribution. Provided that the response of the vessel to a seaway can be assumed to be linear, the ordinary wave-induced component of the total bending moment has a similar distribution. Since the ordinary wave bending moment has a narrow band response, then the extreme values of a normal random process are close to Rayleigh distributed. However, since the responses are nonlinear, especially in high sea states, extreme total ship load peak amplitudes tend to follow somewhere between a Rayleigh and exponential distribution (determined by model tests), represented by a general Weibull distribution or generalized gamma distribution, as discussed later. Using these extreme value distributions, predictions of extreme values for the route and mission-dependent operational conditions can be made. The great advantage of a frequency-domain tool, such as SHIPMO, is that it is simple, flexible and inexpensive, and can generate usable values of rms response, providing a valuable design tool.

Slam Pressure Load

Slamming-induced pressure loads are generated from hull slamming resulting from pitching due to wave-induced motions. The slam pressures are considered local loads and govern the design of the bottom and side structures of the hull. For multihulls, slam pressures also dictate the design of the cross-structure, mainly in the forward sections. Several analytical and experimental studies are published explaining the nature and occurrence of slamming-induced peak pressures. The important factors concerning slam pressures are, first, the vertical acceleration at the structural location of interest; second, the relative entry velocity of the structure with respect to the water; third, the angle of entry of the structure, i.e. the shape of the structure experiencing slamming; and last, the structural unsupported panel size subjected to slam. By virtue of the third variable governing the slam pressure, it is apparent why the slam pressure on a wedge entering the water is lower than that of a flat plate. This is precisely the reason for concern for the slamming-induced pressures on the cross-structures of multihulls, since they are usually flat. Recent designs of several high-speed catamarans have modified their forward cross-structure wet-deck section to a shallow third bow in order to reduce the slam pressures. A typical slam pressure profile along a vessel's length is shown in Figure 14. In this figure, areas I and II denote the wet-deck and internal deck of the cross-structure, whereas areas III and IV denote the same with a shallow third bow.

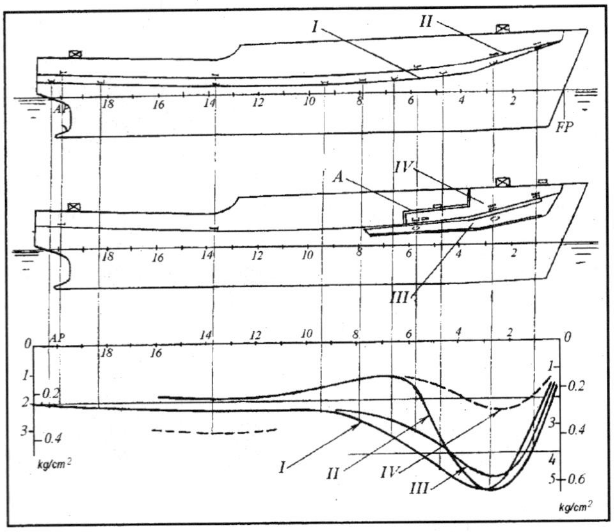

Fig. 14 Slam Pressure Profile along Vessel's Length (Dubrovsky, et.al, 2001)

Since slamming-induced loads are characterized as non-deterministic and more closely influence the design of the local structures than the hull-girder loads, it is imperative to accurately assess the slamming-induced loads. Designs of small high-speed vessels have shown that the local loads essentially control the total structural design, and slam pressures are an essential part of the local loads for the bottom, side and cross-structures. Traditionally, rule-based approaches have provided guidance for predicting slamming-induced pressures, which are typically based on empirical and experimental data and, therefore, provide reasonable accuracy.

For a route and mission-based load prediction approach, a frequency-domain tool to predict the vessel's motions response is not enough to estimate the slamming loads. A time-domain motions response tool is required, which can simulate the motion response time series for the desired operational conditions. Since slamming is a complex nonlinear phenomenon, it is common practice to analyze the response in random waves through time-domain simulations. In order to adequately estimate extreme responses, typically 50-70 one-hour simulations are required. A major problem is that the simulation costs grow with nonlinearity. Quadratic drag effects lead to a roughly four-fold increase in the required simulation length. Hence, several factors work against the use of time-domain simulations: (1) The number of simulations required to adequately estimate the extreme responses, and (2) the need for response analysis over multiple sea states. However, with a route and mission-based response prediction approach, the number of simulations required can be significantly reduced by focusing on the operational parameters and sea states of interest.

In parallel, a slam event will need to be described in terms of a threshold relative velocity that will be attributed to the vessel's characteristics. Determining the number and characteristics of slam occurrences will help in statistically defining a slam prediction model using an extreme value distribution such as a Weibull distribution. Using such a time-domain based statistical prediction model, slamming induced loads can be estimated for various route/mission scenarios.

Structural Design Of Multihulls

Multihulls, in common with other high-speed craft, are designed to withstand the static and dynamic loads that can act under specific operating conditions. The size and mission of the craft often determines the design driving criteria. For multihulls, in general, the moments and forces resulting from the vertical bending moment and shear force on the cross-structure act together with the pitch connecting moment where one hull is rotating relative to the other, and the associated torsional moment has to be considered. All of the major Classification Societies (ABS, BV, DnV, LR) have developed rules specifically for high-speed and lightweight craft. A generalized approach to structural design followed by the major Classification Societies is highlighted in the next section.

Rulebook-Based Design

Ship structures are today designed using rules and guidance given by the Classification Societies and the International Maritime Authorities. It is important that these design rules ensure that the ships are built with an acceptable level of safety. The rules from Classification Societies are based on analytical models for response and strength that are empirically modified to obtain agreement with observations. The safety level of the existing ship structures is, hence, to a large extent, defined through registered failure statistics. Consequently, it is not straightforward to compare alternative designs if these designs are not closely related to traditional design. Recent trends in ship design are towards the use of more innovative materials and the development of new hullforms. In these cases, it has become very important to estimate the actual safety level of ship structures, making it possible to compare traditional designs to new ones, aiming at the same safety level.

In general, local pressures generated through slamming, wave slap, and hydrostatics govern Classification Society rulebook structural design. The overall procedure for a typical rulebook design is as follows:

- The craft geometry is defined and rulebook values for individual parameters are developed. Simple parameters, such as the craft length, are uniquely defined based on the particular set of rules employed (ABS, DNV, Lloyd's, etc.).
- The design accelerations due to ship motions are developed through empirical relations. The vertical acceleration at the craft LCG plays a significant role in the development of local slamming pressures.
- Upon determining the local loads, the thickness of plate elements is established. Similarly, loads on stiffening elements (girders, webs, stiffeners) are determined and the stiffeners are sized accordingly.
- Once the local design of the hull structure (and cross-structure, if applicable) is complete, the section modulus at the craft midship section (and centerline of the cross-structure) is determined and compared to a minimum value that is provided by the rules. If the section established through local design is sufficient, the design is considered complete.

Generally, the rulebook-based approaches require an additional check of the local buckling strength of individual members due to global loads. The global loads are those induced through ship's longitudinal bending and they induce a compressive stress in local members. By and large, the buckling check is a formality, as the local design due to slamming and hydrostatic pressures govern the strength of these members. This procedure is the backbone of each of the rulebook-based design spirals. Variations exist in individual steps, and those differences make the rules for each Classification Society unique.

The development and application of global loads on the ship structure varies based on the Classification Society rules being used. Even if the methodologies employed by each of the Classification Societies differ in their approach, there is a common thread between the approaches in that the governing global loads (longitudinal bending moment (BM), transverse BM, transverse torsion) are developed through empirical relations. However, DnV is the only Society to provide a longitudinal torsion moment. In a later section, a complete structural design that follows two major Classification Society's rules is accomplished to provide a background for the first-principles-based structural design methodology that is being developed as part of this project.

Route or Mission-Based Determination of Design Criteria

Since rule-based designs can only be accomplished for similar ship types and hullforms, the need for first-principles-based rational design methodology is apparent. The success of the structural design depends on the determination of the loads to which the vessel will be exposed and the estimation of a reasonable probability level for its survival. The environmental loads imposed on the structure will have to be determined from an analysis of the modes of operation of the vessel and its environment, and then, making use of this information, by defining a number of deterministic load conditions for analysis. Each of these load conditions is to be assigned a probability of occurrence and compounded to develop a single long-term probability distribution for each type of load. A failure in the accurate determination of structural load can have significant impact on the design and operation of the vehicle. Loads, if underestimated, can result in structural failure or can limit operations. Overestimation of loads, on the other hand, will make the vessel carry excess structural weight, resulting in reduced operational range and payload. Hence, a rational approach to loads determination is paramount.

EAST COAST AUTOFERRY (EXAMPLE)

For this example exercise, the operational characteristics of a proposed AutoFerry service were chosen to represent a vessel that can operate successfully in competition with existing modes of transportation. A preliminary estimate of the transport capabilities of the proposed service, such as speed, range and payload capacity, was developed based on the capabilities of the AutoTrain. The speed requirement of the AutoFerry was based on the transit time of the AutoTrain between Sanford and Lorton. This required that the ferry cover the 800 NM distance between Port Canaveral and Norfolk at an average speed of 40 knots to be comparable with the AutoTrain. The AutoTrain's maximum payload capacity of 200 cars and 450 passengers was used as the payload capacity for the proposed service. A ferry route from Port Canaveral to Norfolk to New York City was deemed ideal in that service would be provided beyond that offered by the AutoTrain.

PASS™, a proprietary ship design synthesis program developed by Band, Lavis & Associates, was used to evaluate the design space and determine an optimal design for a catamaran AutoFerry. There are several catamarans in operation with load-carrying capabilities similar to those described in Table 2, and the principal characteristics of these vessels were used as a guide to assess available design space for the AutoFerry. However, it must be stated that most of the fast catamarans in service do not possess ranges in excess of 500 NM, and a majority of the catamarans operate in coastal ferry routes and kind sea conditions. For the AutoFerry, craft length was varied from 225 ft to a maximum of 350 ft, with demihull length-to-beam ratios varying from 15 to 20. The selection of a design point for the AutoFerry is made by determining a choice of principal parameters that provides for efficient mission performance, i.e., load, carry and unload 200 cars for the lowest cost while providing the requisite operational speed and passenger comfort. Table 2 identifies the principal parameters for the chosen design for the catamaran AutoFerry.

Table 2. Catamaran AutoFerry Principal Characteristics

Length Overall (ft)	291.6
Waterline Length (ft)	270.0
Beam Overall (ft)	72.13
Draft (ft)	13.74
Depth (ft)	33.60
Full-Load Displacement (LT)	1,666
Passengers/Cars	450/200
Prime Movers (Gas Turbines) (HP)	4 @ 14,563
Propulsors Dia. (Waterjets) (ft)	4.30

Route & Environmental Description

Typically, the environmental description of the route is presented in terms of significant wave properties over the route, typically divided into sections, where environmental data is available from historical measurements or compiled weather almanacs. Along the Atlantic seaboard, the Gulf Stream current dominates the wave climate. The AutoFerry route between Port Canaveral, Norfolk and New York is depicted in Figure 15 for both the northbound and southbound segments. In Figure 15, markers representing buoys on the route are highlighted. Data from these buoys, in terms of the statistical descriptions of wave heights and modal periods for over 20 years, were obtained and evaluated for environmental assessment. Data from the buoys along the AutoFerry route have been summarized in Table 3. Included are data on both the average values of significant wave height and period, as well as the extreme measured values of the same. The data from Table 3 have been used to develop a multi-segmented route for the AutoFerry. The route has been discretized into 15 voyage segments for both the northbound and southbound routes. Table 4 represents data for the northbound route and highlights the segment distance, vessel speed in the segment, the time spent in each segment as a fraction of the total voyage time, and identifies the applicable buoy for each segment. The data shown in Tables 3 and 4 form the basis of the route description module that is being developed.

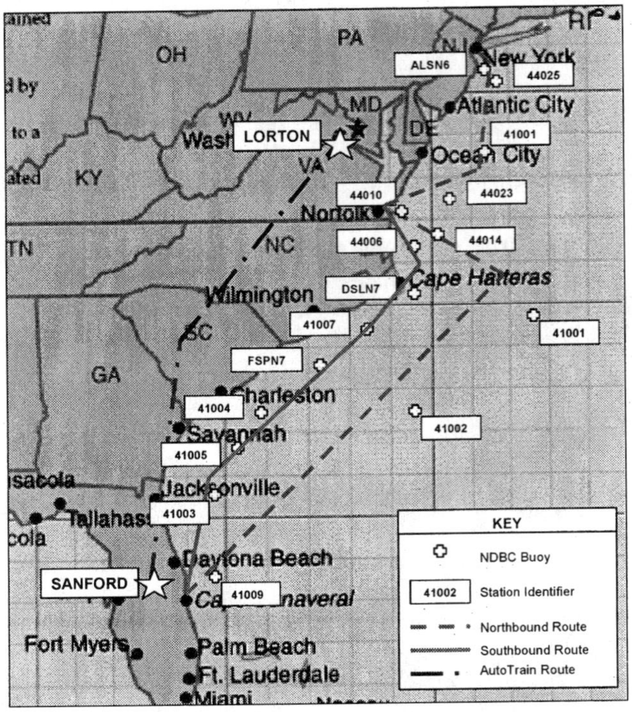

Fig. 15 Atlantic AutoFerry Route and Locations of Wave Buoys

Numerical Modeling of AutoFerry Motions & Loads Determination

The utilization of frequency-domain seakeeping prediction software, such as SHIPMO, allows the rapid generation of rational first-principles motions and wave-induced loads based on actual route-specific operational data. Frequency-domain modeling results in motions and loads data that are represented as short-term single-amplitude statistics based on a user-supplied wave energy spectrum. Thus, this approach is fundamentally linked to a robust analysis of the environmental conditions inherent to the proposed route. Short-term statistical descriptions of the frequency-domain data can then be extrapolated into expected lifetime structural loads using the procedures outlined in previous sections. Time-domain modeling, as proposed for the Phase II effort, will allow the investigation of specific non-linear response occurrences over a specified discrete time series.

Table 3. Environmental Data Compilation for Atlantic Route AutoFerry

ROUTE BASED ENVIRONMENTAL DATA FOR NOTIONAL EAST-COAST CATAMARAN PASSENGER / CAR FERRY "AUTOCAT"															
NBDC Buoy or C-MAN Station	Location			Sample Duration	Historical Significant Wave Height Data					Historical Dominant Wave Period Data					
	General Description	Latitude	Longitude		Avg Yearly	One Sigma		Extreme		Avg Yearly	One Sigma		Extreme		
						High	Low	Max	Min		High	Low	Max	Min	
		(deg)	(deg)	(years)	(m)	(m)	(m)	(m)	(m)	(sec)	(sec)	(sec)	(sec)	(sec)	
41009	CANAVERAL - 20M East of Cape Canaveral FL	28.50 N	80.18 W	5.25	1.2	1.7	0.6	6.2	0.2	8.3	10.0	6.5	20.0	2.5	
41002	S. HATTERAS - 250M East of Charleston SC	32.27 N	75.42 W	18.50	1.9	2.7	1.0	16.0	0.0	8.1	10.0	6.1	20.0	2.5	
41001	HATTERAS - 150M East of Cape Hatteras NC	34.68 N	72.23 W	17.50	2.0	3.0	1.0	10.4	0.0	8.0	9.9	6.1	20.0	3.0	
44014	VIRGINIA BEACH - 64M East of Va Beach VA	36.58 N	74.84 W	3.17	1.4	2.1	0.7	8.2	0.5	7.8	9.9	5.6	20.0	2.5	
44010	Chesapeake Bay Entrance	36.90 N	75.70 W	1.00	0.9	1.3	0.6	3.6	0.1	6.8	8.8	4.7	17.0	2.5	
44023	100M NE of Va Beach VA	37.50 N	74.40 W	1.00	1.5	2.2	0.8	5.0	0.1	7.8	9.8	5.9	14.0	2.8	
44001	100M NE of Ocean City MD	38.70 N	73.60 W	15.50	1.4	2.0	0.7	7.8	0.0	7.5	9.6	5.3	12.5	0.0	
44025	LONG ISLAND - 33M South of Islip NY	40.25 N	73.17 W	2.67	1.2	1.9	0.5	9.3	0.0	7.0	9.2	4.7	20.0	0.0	
ALSN6	Ambrose Light NY	40.46 N	73.83 W	4.00	0.9	1.4	0.5	7.4	0.0	7.7	10.6	4.9	20.0	0.0	
44006	Northern Outer Banks - 60M SE of Va Beach VA	36.30 N	75.40 W	7.08	1.2	1.7	0.6	4.3	0.3	8.2	10.7	5.7	17.0	2.5	
DSLN7	Diamond Shoals Light NC	35.15 N	75.30 W	5.08	1.5	2.2	0.8	10.8	0.3	7.8	9.8	5.9	20.0	2.8	
41007	75M East of Wilmington NC	34.20 N	76.50 W	1.00	1.7	2.4	1.1	4.4	0.6	8.2	9.7	6.6	12.5	3.0	
FPSN7	Frying Pan Shoals NC	33.49 N	77.59 W	1.00	1.6	2.3	0.9	4.8	0.4	7.8	9.8	5.8	14.0	2.8	
41004	EDISTO - 41M SE of Charleston SC	32.50 N	79.10 W	13.58	1.3	1.9	0.7	6.6	0.2	7.2	9.1	5.3	16.5	2.3	
41005	100M SE of Savannah GA	31.70 N	79.70 W	1.25	1.4	2.0	0.7	5.2	0.3	7.2	9.5	5.0	17.0	2.5	
41003	80M East of Jacksonville FL	30.40 N	80.10 W	2.25	1.2	1.7	0.7	4.4	0.2	7.1	9.4	4.8	17.0	2.5	

Table 4. Operational Profile for the Northbound Leg of the AutoFerry Route

OPERATIONAL PROFILE OF NOTIONAL EAST-COAST CATAMARN PASSENGER / CAR FERRY "AUTOCAT"									
Voyage Segment		Description	Approx Segment Distance	Vessel Speed	Time on Segment	% Total Voyage Time	NBDC Buoy or C-MAN Station	Avg Significant Wave Height	Avg Dominant Wave Period
			(nm)	(knots)	(hours)			(m)	(sec)
NORTHBOUND LEG	1 N	Port Canaveral Terminal (Onload / Offload)	0.0	0.0	18.00	17.0	N/A	0.0	0.0
	2 N	Transit Port Canaveral Harbor	3.0	6.0	0.50	0.5	N/A	0.0	0.0
	3 N	Port Canveral to GS 110M NE of Daytona Beach	120.0	40.0	3.00	2.8	**41009**	1.2	8.3
	4 N	Daytona Beach to GS 180M SE of Wilmingrton NC	400.0	40.0	10.00	9.4	**41002**	1.9	8.1
	5 N	Wilmington to GS 100M East of Cape Hatteras	130.0	40.0	3.25	3.1	**41001**	2.0	8.0
	6 N	Cape Hatteras to Virginia Beach (50M Offshore)	100.0	40.0	2.50	2.4	**44014**	1.4	7.8
	7 N	Virginia Beach Offshore to Chesapeake Bay Entrance	50.0	40.0	1.25	1.2	**44010**	0.9	6.8
	8 N	Transit Norfolk Harbor	3.0	6.0	0.50	0.5	N/A	0.0	0.0
	9 N	Norfolk Terminal	0.0	0.0	6.00	5.7	N/A	0.0	0.0
	10 N	Transit Norfolk Harbor	3.0	6.0	0.50	0.5	N/A	0.0	0.0
	11 N	Chesapeake Bay Entrance to 50M NE Virginia Beach	50.0	40.0	1.25	1.2	**44010**	0.9	6.8
	12 N	NE Virginia Beach to 75M SE of Ocean City MD	90.0	40.0	2.25	2.1	**44023**	1.5	7.8
	13 N	Ocean City to 50M East of Atlantic City NJ	90.0	40.0	2.25	2.1	**44001**	1.4	7.5
	14 N	Atlantic City to Outer New York Harbor	90.0	40.0	2.25	2.1	**44025**	1.2	7.0
	15 N	Transit New York Harbor	3.0	6.0	0.50	0.5	**ALSN6**	0.9	7.7

Through time-domain simulation, data can be generated to support the development of wave impact loads and secondary hull-girder response loads, such as springing and whipping, by generating relative velocities at various discrete points on the vessel's demihulls and cross-structure. Relative velocities refer to the difference in vertical velocity between a specified point on the craft and the surface of the seaway, and are therefore a key component in the dynamic pressures acting on the vessel's structure. Typically, a slam occurs when the relative velocity exceeds 20 ft/sec.

The AutoFerry motions model was developed in SHIPMO, incorporating the lessons learned from the program validation performed during this project, utilizing a simplified demihull-only model for the development of vertical plane data and a full catamaran section for the development of lateral plane motions and loads.

One of the key elements for the motions model is the longitudinal weight distribution, which is critical to the development of realistic environmentally-induced structural loads. The weight and buoyancy distribution developed for the notional AutoFerry is shown in Figure 16 in addition to the still water (static) bending moment. As the intent of this analysis is largely to illustrate procedure rather than perform a fully detailed design of the AutoFerry, only the full-load weight distribution has been investigated to date.

All motions modeling performed utilizing SHIPMO was based on an ITTC two-parameter wave spectrum (modified Bretschneider). Characteristics of the idealized spectrum are developed through a user-supplied significant wave height and modal period.

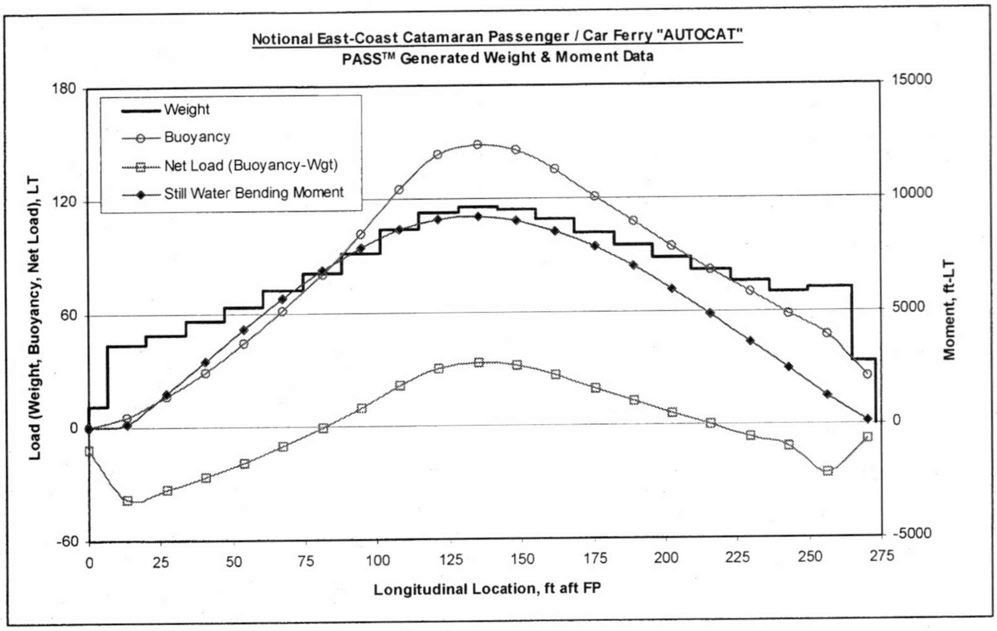

Fig. 16 Weight, Buoyancy and Static Moment at Full-Load

This is the same basic data compiled in the historical database of NDBC buoys along the proposed AutoFerry route off the eastern coast of the United States. For the initial study phase, all motions modeling has been performed for the buoys on the AutoFerry route that were deemed to be representative of limiting operational criteria. NDBC stations 41001 and 41002 off the coast of Cape Hatteras were thus chosen. These buoys are located in relatively robust seaway conditions, including larger waves, stronger winds, and a characteristic fog as compared to the more protected and benign seas along the remaining sections of the route. In addition to displaying very similar sea conditions, these buoys encompass a substantial portion of the total distance covered by the AutoFerry in typical operation (~530 miles or 25% of the total transit distance).

While Phase II of this project entails a more detailed statistical description of loads along the entire AutoFerry route, as described in previous sections, for the time being motions have been modeled for the mean (M) and mean-plus-one standard deviation (M+1σ) yearly significant wave height at the extreme buoys to provide an initial benchmark against the parallel 'rulebook' designs. In considering the M+1σ wave height case (3m), approximately 84% of all wave encounters are expected to be less than or equal to the design values at these extreme buoys. A simple compounding of probabilities demonstrates that utilization of the M+1σ wave represents an approximate 0.005 total probability of exceedance over the lifetime of the AutoFerry. This is due to various factors, including the fact that 75% of the AutoFerry route is in sea conditions less severe than those at the limiting buoys, in addition to the fact that the design loads are developed at the worst vessel heading, while it is presumably fair at this point in the analysis to assume an equal probability of all headings. It should also be noted that currently reported loads represent single significant amplitude statistics, or the average of the highest 1/3 encounters (twice the rms response). The preliminary design wave conditions utilized for this study will most assuredly be updated with the expanded statistical treatment slated for Phase II of this study.

Vessel motions and design loads have been compiled for both longcrested and shortcrested irregular seas. Longcrested seas imply that all waves propagate from the same direction, a relatively idealized assumption that allows the user to identify particular headings and encounter frequencies that may result in excessive motions and structural loads, generally in resonant conditions. The shortcrested seas case is more representative of physical reality, as a directional spectrum is applied to the seaway, resulting in wave directions that are varied around a single principal or dominant direction. Directionality is applied to the spectrum utilizing a cosine squared spreading function and is accepted procedure. The utilization of shortcrested seas tends to filter the resonant responses, as the likelihood of multiple resonant oscillations is low given the varying wave direction and irregular wave frequency associated with this assumption. In this analysis, the longcrested sea data was generally utilized in an investigation of the resonance characteristics of the proposed AutoFerry. Resonance occurs when the forcing frequency of the encountered seaway matches the natural frequency of the hullform, a condition resulting in potentially large amplitude motions and thus loads. An investigation to identify particular headings at the design speed of 40 knots in which resonance was likely in heave, pitch or roll was performed. These headings could then be singled out for further analysis as the dominant wave direction in a more realistic shortcrested sea.

The natural periods of the proposed AutoFerry were calculated, converted to frequencies, and subsequently plotted along with the encounter frequency of the AutoFerry at design speed in headings covering the full range from head to following seas. Encounter frequencies were calculated for the dominant wave frequency at the limiting buoys on the proposed route, in addition to the high and low ends of a one-standard-deviation range. This data is shown in Figure 17. Note that a negative encounter frequency simply implies that the forward speed of the vessel results in a situation where the AutoFerry is overtaking following waves (i.e., following seas at 40 knots). The shaded vertical bands overlaying Figure 17 indicate the headings in which the maximum load responses (in both the vertical and lateral planes) were calculated based on the idealized longcrested sea assumption. As expected, these extreme loads correspond with encounter frequencies in which resonant conditions are likely. As a result of this investigation, headings of 105 degrees (region of maximum lateral loads) and 150 degrees (region of maximum vertical loads) were singled out for analysis in more realistic shortcrested seas, along with a typical investigation of all headings at 45-degree intervals. As alluded to earlier in this section, the transition to shortcrested seas tends to filter the resonant responses, and the response maxima shifted to more intuitive headings in which the limiting vertical loads occur in head and following seas and the maximum lateral loads occur in beam seas. All design loads reported for this analysis are based on data from simulations using a shortcrested sea.

Design loads for analyses completed in Phase I were compiled for both the average (mean) yearly significant wave height at the limiting buoys and the mean-plus-one standard deviation case ($M+1\sigma$) to provide a range of comparison against the 'rulebook' designs. The applicable design loads are shown in Table 5. The loads displayed represent both the significant single amplitude (average of the highest 1/3), as well as the average of the 1/10 highest responses at the worst heading. The complete set of compiled loads in longcrested and shortcrested seas at all headings, from which the design values were extracted, is shown in Tables 6 and 7.

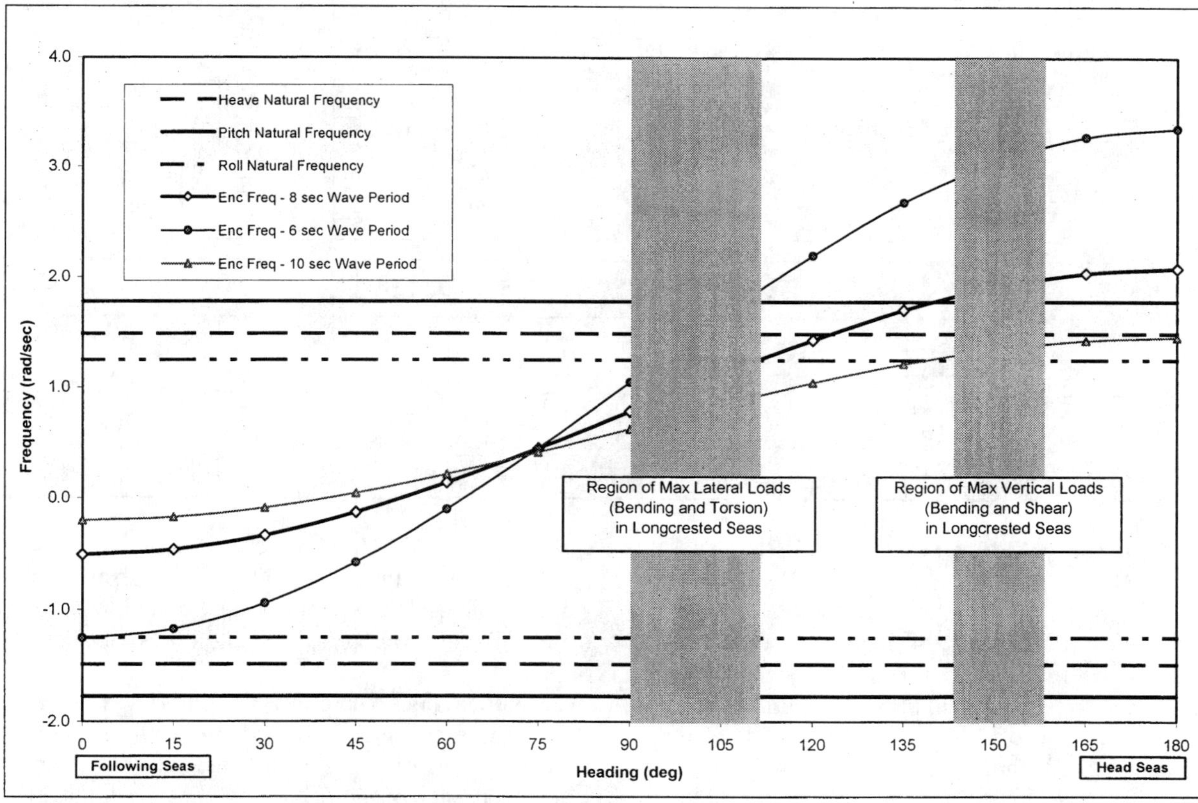

Fig. 17 Identification of Resonant Conditions

Table 5. AutoFerry Design Loads at 40-knot Design Speed

STATIC LOAD		
Still-Water Bending Moment (Midship)	LT-ft	8564.29

WAVE-INDUCED LOADS		Mean Signf. Wave Height - 2m -		Mean + 1σ Signf. Wave Height - 3m -	
		Avg. 1/3	Avg. 1/10	Avg. 1/3	Avg. 1/10
Vertical Bending Moment (Midship)	LT-ft	10446.42	13319.19	31696.42	40412.94
Lateral Bending Moment (Midship)	LT-ft	5883.92	7502.00	7178.58	9152.58
Torsional Bending Moment (Midship)	LT-ft	1517.86	1935.27	1794.64	2288.17
Vertical Shear (Midship)	LT	56.16	71.60	83.22	106.11
Lateral Shear (Midship)	LT	31.70	40.42	37.32	47.58

Design Values Developed Assuming Shortcrested Irregular Seas (ITTC Two-Parameter Wave Spectrum).
Pertinent Parameters (Wave Height and Modal Period) Consistent With Historical Data for NDBC Buoys 41001 & 41002.

Table 6. Compiled Loads Data for Notional AutoFerry in 2m Seas

Still-Water Bending Moment (Midship)	8564.29	LT-ft

Significant Wave Height	2.0 m	Yearly Averages for NDBC Buoys 41001 & 41002
Dominant Wave Period	8.0 sec	
Average Wave Period	5.83 sec	

LONGCRESTED SEAS

- ITTC Two-Parameter Wave Spectrum (Longcrested) -
Short Term Statistics Based on Avg Yearly Wave Height & Period Data for Buoys 41001 & 41002

Heading (0 - Following Seas, 180 - Head Seas)	Vertical Bending Moment (LT-ft)				Lateral Bending Moment (LT-ft)				Torsion (LT-ft)				Vertical Shear (LT)				Lateral Shear (LT)			
(deg)	RMS	Avg	Signf	Avg 1/10	RMS	Avg	Signf	Avg 1/10	RMS	Avg	Signf	Avg 1/10	RMS	Avg	Signf	Avg 1/10	RMS	Avg	Signf	Avg 1/10
0	2178.57	2723.21	4357.14	5555.35	0.00	0.00	0.00	0.00	0.00	0.00	0.00	0.00	27.32	34.15	54.64	69.67	0.00	0.00	0.00	0.00
15	2290.18	2862.73	4580.36	5839.96	741.07	926.34	1482.14	1889.73	207.14	258.93	414.28	528.21	27.32	34.15	54.64	69.67	8.93	11.16	17.86	22.77
30	2066.96	2583.70	4133.92	5270.75	1508.93	1886.16	3017.86	3847.77	415.18	518.98	830.36	1058.71	26.43	33.04	52.86	67.40	16.29	20.36	32.58	41.54
45	2803.57	3504.46	5607.14	7149.10	2482.14	3102.68	4964.28	6329.46	566.96	708.70	1133.92	1445.75	27.23	34.04	54.46	69.44	18.48	23.10	36.96	47.12
60	4464.29	5580.36	8928.58	11383.94	2593.75	3242.19	5187.50	6614.06	633.93	792.41	1267.86	1616.52	29.78	37.23	59.56	75.94	16.12	20.15	32.24	41.11
75	2308.04	2885.05	4616.08	5885.50	1500.00	1875.00	3000.00	3825.00	347.76	434.70	695.52	886.79	35.67	44.59	71.34	90.96	15.00	18.75	30.00	38.25
90	843.75	1054.69	1687.50	2151.56	4410.71	5513.39	8821.42	11247.31	651.79	814.74	1303.58	1662.06	8.35	10.44	16.70	21.29	7.10	8.87	14.20	18.10
105	1258.93	1573.66	2517.86	3210.27	4120.54	5150.68	8241.08	10507.38	1258.93	1573.66	2517.86	3210.27	10.36	12.95	20.72	26.42	16.70	20.88	33.40	42.59
120	1754.46	2193.08	3508.92	4473.87	2584.82	3231.03	5169.64	6591.29	955.36	1194.20	1910.72	2436.17	13.30	16.63	26.60	33.92	13.79	17.24	27.58	35.16
135	2272.32	2840.40	4544.64	5794.42	1549.11	1936.39	3098.22	3950.23	750.00	937.50	1500.00	1912.50	16.96	21.20	33.92	43.25	11.88	14.85	23.76	30.29
150	5089.29	6361.61	10178.58	12977.69	901.78	1127.23	1803.56	2299.54	491.07	613.84	982.14	1252.23	38.13	47.66	76.26	97.23	14.38	17.98	28.76	36.67
165	3093.75	3867.19	6187.50	7889.06	504.46	630.58	1008.92	1286.37	233.93	292.41	467.86	596.52	25.09	31.36	50.18	63.98	5.80	7.25	11.60	14.79
180	2901.79	3627.24	5803.58	7399.56	0.00	0.00	0.00	0.00	0.00	0.00	0.00	0.00	23.04	28.80	46.08	58.75	0.00	0.00	0.00	0.00

SHORTCRESTED SEAS

- ITTC Two-Parameter Wave Spectrum (Shortcrested w/ $\cos^2$ Spreading Function) -
Short Term Statistics Based on Avg Yearly Wave Height & Period Data for Buoys 41001 & 41002 (Heading corresponds to Principal Wave Angle)

Principal Wave Direction (0 - Following Seas, 180 - Head Seas)	Vertical Bending Moment (LT-ft)				Lateral Bending Moment (LT-ft)				Torsion (LT-ft)				Vertical Shear (LT)				Lateral Shear (LT)			
(deg)	RMS	Avg	Signf	Avg 1/10	RMS	Avg	Signf	Avg 1/10	RMS	Avg	Signf	Avg 1/10	RMS	Avg	Signf	Avg 1/10	RMS	Avg	Signf	Avg 1/10
0	5223.21	6529.01	10446.42	13319.19	1544.64	1930.80	3089.28	3938.83	383.04	478.80	766.08	976.75	28.08	35.10	56.16	71.60	15.85	19.81	31.70	40.42
45	4089.29	5111.61	8178.58	10427.69	2241.07	2801.34	4482.14	5714.73	531.25	664.06	1062.50	1354.69	27.05	33.81	54.10	68.98	15.49	19.36	30.98	39.50
90	3616.07	4520.09	7232.14	9220.98	2941.96	3677.45	5883.92	7502.00	758.93	948.66	1517.86	1935.27	20.63	25.79	41.26	52.61	14.43	18.04	28.86	36.80
135	3008.93	3761.16	6017.86	7672.77	2406.25	3007.81	4812.50	6135.94	669.64	837.05	1339.28	1707.58	23.21	29.01	46.42	59.19	13.66	17.08	27.32	34.83
180	3348.21	4185.26	6696.42	8537.94	1276.79	1595.99	2553.58	3255.81	468.75	585.94	937.50	1195.31	25.89	32.36	51.78	66.02	10.09	12.61	20.18	25.73

STRUCTURAL DESIGN OF AUTOFERRY

As a point of comparison to the rational structural design paradigm developed through this project, two rulebook structural designs were investigated for the catamaran AutoFerry. The American Bureau of Shipping (ABS) Guide for Building and Classing High-Speed Craft (HSC), issued February 1997, and Det Norske Veritas (DNV) Rules of Classification of High-Speed, Light Craft and Naval Surface Craft, issued January 2001, were employed. The AutoFerry structural design was assessed using the ABS guide, and a cursory look at the required plating thickness and global loads based on DNV rules was completed and is described.

First-Principles Structural Design of AutoFerry

The first-principles-based approach relies on the development of realistic design loads based on the intended route of the AutoFerry. The design loads as developed and used are already discussed and are shown in Table 5. Based on this data and an allowable stress of 80% of the welded strength of aluminum, the minimum section modulus requirements for the AutoFerry were determined and are shown in Table 8. It should be noted that the first-principles structural design was only performed for the global loads since local loads, such as slam pressures, governing the structural scantling design are ear-marked under the Phase II effort.

Structural Design of AutoFerry Using ABS Rules

Using the ABS HSC guide and following the general procedure, as discussed earlier under the section on rulebook based design, the local structures were designed first. Upon completion of the local structure design, the global hull-girder section properties of the midship section were determined to compare with the requirements imposed by ABS. Table 9 shows the global loads and minimum required midship section properties as determined using the ABS HSC guide. The section modulus and moment of inertia of the midship section based on the local structure design were found to be 99.0 ft^3 and 1832.0 ft^4, respectively. The values dictated by the local design present global factors of safety of 3.1 and 5.4 for the section modulus and hull-girder moment of inertia.

In addition to the longitudinal bending check, the transverse strength of the cross-structure was checked to verify that it would withstand the global loads imposed on it. The maximum bending and shear stress induced in the cross-structure due to the global loads were 8,132 psi and 1,209 psi, respectively. Based on the allowable stresses provided by ABS, these represent safety factors of 2.1 and 8.2 in bending and shear. Figure 18 illustrates the AutoFerry midship section design using the ABS HSC guide.

Table 7. Compiled Loads Data for Notional AutoFerry in 3m Seas

Still-Water Bending Moment (Midship)	8564.29	LT-ft

Significant Wave Heigh	3.0 m	+1σ Data for NDBC Buoys 41001 & 41002
Dominant Wave Period	10.0 sec	
Average Wave Period		

LONGCRESTED SEAS — ITTC Two-Parameter Wave Spectrum (Longcrested) - Short Term Statistics Based on Yearly +1σ Wave Height & Period Data for Buoys 41001 & 41002

Heading (0 - Following Seas, 180 - Head Seas)	Vertical Bending Moment (LT-ft)				Lateral Bending Moment (LT-ft)			
(deg)	RMS	Avg	Signf	Avg 1/10	RMS	Avg	Signf	Avg 1/10
0	3941.96	4927.45	7883.92	10052.00	0.00	0.00	0.00	0.00
15	6785.71	8482.14	13571.42	17303.56	1138.39	1422.99	2276.78	2902.89
30	5133.93	6417.41	10267.86	13091.52	2276.79	2845.99	4553.58	5805.81
45	5443.43	6804.29	10886.86	13880.75	3419.64	4274.55	6839.28	8720.08
60	5491.07	6863.84	10982.14	14002.23	3334.82	4168.53	6669.64	8503.79
75	2308.04	2885.05	4616.08	5885.50	1674.11	2092.64	3348.22	4268.98
90	870.54	1088.18	1741.08	2219.88	4553.57	5691.96	9107.14	11611.60
105	1397.32	1746.65	2794.64	3563.17	5089.29	6361.61	10178.58	12977.69
120	2129.46	2661.83	4258.92	5430.12	3781.25	4726.56	7562.50	9642.19
135	2941.96	3677.45	5883.92	7502.00	2540.18	3175.23	5080.36	6477.46
150	8169.64	10212.05	16339.28	20832.58	1513.39	1891.74	3026.78	3859.14
165	4732.14	5915.18	9464.28	12066.96	852.68	1065.85	1705.36	2174.33
180	4343.75	5429.69	8687.50	11076.56	0.00	0.00	0.00	0.00

Heading (deg)	Torsion (LT-ft)				Vertical Shear (LT)				Lateral Shear (LT)			
(deg)	RMS	Avg	Signf	Avg 1/10	RMS	Avg	Signf	Avg 1/10	RMS	Avg	Signf	Avg 1/10
0	0.00	0.00	0.00	0.00	37.81	47.26	75.62	96.42	0.00	0.00	0.00	0.00
15	254.91	318.64	509.82	650.02	37.41	46.76	74.82	95.40	11.12	13.90	22.24	28.36
30	500.00	625.00	1000.00	1275.00	37.81	47.26	75.62	96.42	20.13	25.16	40.26	51.33
45	678.57	848.21	1357.14	1730.35	37.19	46.49	74.38	94.83	23.21	29.01	46.42	59.19
60	745.54	931.93	1491.08	1901.13	34.96	43.70	69.92	89.15	19.73	24.66	39.46	50.31
75	415.18	518.98	830.36	1058.71	34.87	43.59	69.74	88.92	15.67	19.59	31.34	39.96
90	696.43	870.54	1392.86	1775.90	8.66	10.83	17.32	22.08	7.23	9.04	14.46	18.44
105	1428.57	1785.71	2857.14	3642.85	11.61	14.51	23.22	29.61	17.32	21.65	34.64	44.17
120	1245.54	1556.93	2491.08	3176.13	16.43	20.54	32.86	41.90	16.34	20.43	32.68	41.67
135	1017.86	1272.33	2035.72	2595.54	21.70	27.13	43.40	55.34	14.82	18.53	29.64	37.79
150	669.64	837.05	1339.28	1707.58	60.71	75.89	121.42	154.81	21.43	26.79	42.86	54.65
165	342.86	428.58	685.72	874.29	37.23	46.54	74.46	94.94	6.56	8.20	13.12	16.73
180	0.00	0.00	0.00	0.00	33.13	41.41	66.26	84.48	0.00	0.00	0.00	0.00

SHORTCRESTED SEAS — ITTC Two-Parameter Wave Spectrum (Shortcrested) - Short Term Statistics Based on Yearly +1σ Wave Height & Period Data for Buoys 41001 & 41002

Principal Wave Direction (0 - Following Seas)	Vertical Bending Moment (LT-ft)				Lateral Bending Moment (LT-ft)			
(deg)	RMS	Avg	Signf	Avg 1/10	RMS	Avg	Signf	Avg 1/10
0	15848.21	19810.26	31696.42	40412.94	2156.25	2695.31	4312.50	5498.44
45	7544.64	9430.80	15089.28	19238.83	2816.96	3521.20	5633.92	7183.25
90	7901.79	9877.24	15803.58	20149.56	3589.29	4486.61	7178.58	9152.69
105	4598.21	5747.76	9196.42	11725.44	3558.04	4447.55	7116.08	9073.00
135	4250.00	5312.50	8500.00	10837.50	3062.50	3828.13	6125.00	7809.38
150	4642.86	5803.58	9285.72	11839.29	2584.82	3231.03	5169.64	6591.29
180	5267.86	6584.83	10535.72	13433.04	1919.64	2399.55	3839.28	4895.08

Principal Wave Direction (deg)	Torsion (LT-ft)				Vertical Shear (LT)				Lateral Shear (LT)			
(deg)	RMS	Avg	Signf	Avg 1/10	RMS	Avg	Signf	Avg 1/10	RMS	Avg	Signf	Avg 1/10
0	459.82	574.78	919.64	1172.54	41.61	52.01	83.22	106.11	18.66	23.33	37.32	47.58
45	616.07	770.09	1232.14	1570.98	35.31	44.14	70.62	90.04	18.21	22.76	36.42	46.44
90	897.32	1121.65	1794.64	2288.17	27.99	34.99	55.98	71.37	16.65	20.81	33.30	42.46
105	897.32	1121.65	1794.64	2288.17	29.11	36.38	58.21	74.22	18.04	22.55	36.08	46.00
135	825.89	1032.36	1651.78	2106.02	32.28	40.35	64.56	82.31	16.61	20.76	33.22	42.36
150	763.39	954.24	1526.78	1946.64	35.58	44.48	71.16	90.73	13.71	17.14	27.42	34.96
180	629.46	786.83	1258.92	1605.12	38.84	48.55	77.68	99.04	13.21	16.51	26.42	33.69

Table 8. First-Principles Global Sectional Property Requirements

Section Modulus	2 m Wave Height		3 m Wave Height	
	Avg 1/3	Avg 1/10	Avg 1/3	Avg 1/10
Min. Longitudinal Bending SM at Midship (ft^3)	19.15	22.04	40.55	49.33
Min. Transverse Bending SM along CL (ft^3)	5.93	7.56	7.23	9.22
Min. Transverse Torsional SM (ft^3)	1.53	1.95	1.81	2.30

Table 9. ABS Global Loads and Sectional Property Requirements

Longitudinal Bending Moment (LT-ft)	31,721
Vertical Shear due to Longitudinal BM (LT)	210
Transverse Bending Moment (LT-ft)	3,415
Transverse Torsional Moment (LT-ft)	7,805
Vertical Shear due to Transverse Loads (LT)	60
Min. Longitudinal Section Modulus at Midship (ft^3)	31.84
Min. Hull-girder Moment of Inertia at Midship (ft^4)	340.7
Min. Transverse Bending Section Modulus along CL (ft^3)	4.17
Min. Transverse Torsional Section Modulus (ft^3)	8.39

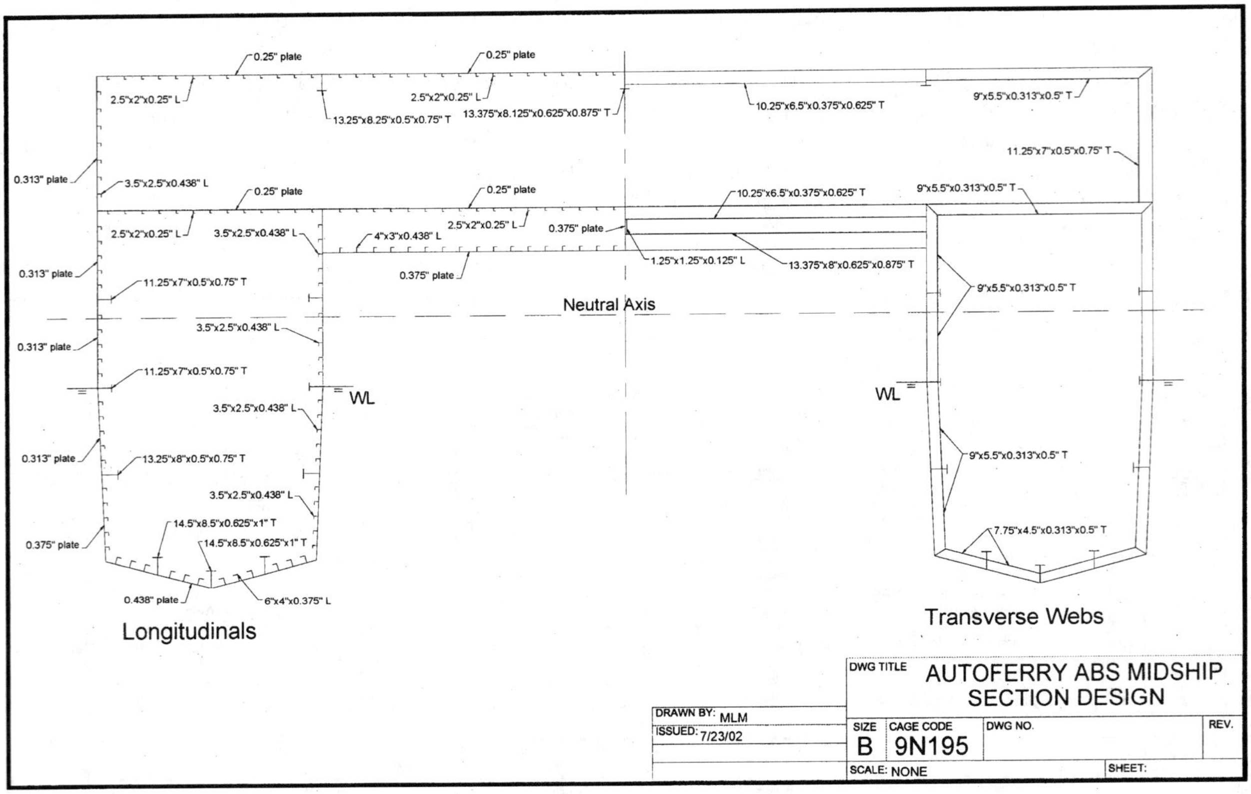

Fig. 18 AutoFerry Midship Section Structural Design

Structural Design of AutoFerry Using DNV Rules

Following a similar rulebook approach, the local structures for the AutoFerry were redesigned, this time using the DNV rules. In order to make a fair comparison, the structural arrangement (stiffener spacing and location) employed in the ABS design was maintained for the DNV design. For the bottom hull plating and cargo deck plating, the DNV requirements result in a thicker plate than those using ABS. The opposite is true for the side hull and wet deck cross-structure. Based on these observations, it appears that the total weight of the plates comprising the structure analyzed will be approximately equal. Finally, the global loads were determined using the DNV requirements and are summarized in Table 10. It is important to mention here that unlike ABS, DNV does not provide explicit guidance for minimum required global section properties, but recommends direct calculation methods to verify the global structural adequacy under the prescribed loads and allowable stresses. By glancing at the data in Table 10, it may appear that DNV required loads are much higher than ABS and even those obtained through the first-principles route/mission approach. However, DNV's approach to global load application is through load combination with a certain percentage of each load type and not the entire calculated load. In addition, the design allowable stresses are much higher than those proposed by ABS. Thus, by putting these three factors in perspective, the DNV design would not be radically different than the design using ABS or the route/mission approach.

Table 10. DNV Global Loads Requirements

Longitudinal Bending Moment (LT-ft)	86,087
Vertical Shear due to Longitudinal BM (LT)	1,275
Longitudinal Torsional Moment (LT-ft)	23,627
Transverse Bending Moment (LT-ft)	22,681
Transverse Torsional Moment (LT-ft)	56,252
Vertical Shear due to Transverse Loads (LT)	417

CONCLUSIONS & RECOMMENDATIONS

The specific objective of the study reported herein was to develop a route or mission-dependent approach to predict structural loads for high-speed multihulls. This is being performed in a two-phased program of work to produce an easy to use ship motion and dynamic load calculation program and an overall statistical methodology that is suitable for use by Designers, Classification Societies, and the U.S. Navy. The specific work discussed here is for Phase I of the program.

Under the Phase I effort, a 2D hydrodynamics and strip theory approach was used to develop an easy to use but reliable frequency-domain ship motion and dynamic loads prediction method for high-speed multihulls. As part of this process, the strip theory-based ship motion program, SHIPMO, was validated against several experimental investigations on the motions of high-speed catamarans. Overall, the strip theory frequency-domain model was found to provide an acceptable level of fidelity for motions data when separate numerical models are utilized to develop data in the vertical and horizontal planes, respectively. The model performance tends to degrade with increasing Froude number as a result of the limitations of the underlying two-dimensional hydrodynamics theory. This degradation is less extreme when the catamaran demihulls are treated independently for vertical plane motions, but this remains a concern moving forward into Phase II.

The program had been successfully combined with a route/mission-based statistical module, utilized to combine short-term wave load predictions into a long-term (ship lifetime) load prediction, and to identify segments of the mission that are likely to be of concern from a ship slamming standpoint. Probabilistic theories of short-term events, and culmination of these short-term statistics into a long-term probabilistic model to predict extreme load statistics, were also briefly discussed. The example of an AutoFerry operating between Port Canaveral, FL and New York has been utilized to develop a notional route/mission module as well as the determination of preliminary statistical descriptions of seaway loads on a notional design of a catamaran operating on this route. The resultant statistical description of the seaway loads (wave-induced loads) has been determined and compared to rulebook-based design criteria for a multihull craft.

The route/mission approach-based global loads for the preliminary AutoFerry example were close to those obtained using the ABS HSC guide. A safety margin of 20% on allowable stress was used for the route/mission approach under the assumption that the loads obtained from motions modeling represent a realistic assessment of the loads that are expected along the intended route of the AutoFerry. If a greater level of confidence can be achieved with the expanded statistical treatment planned for Phase II of this study, the margin could be reduced accordingly to further optimize the structural design.

For smaller craft, local loads due to slamming generally drive the structural design. In order for global loads to take precedence, the craft must be large or slender (high L/B). Although the AutoFerry was a relatively large craft by rulebook definition, the structural design according to ABS showed that the local slamming and hydrostatic pressures on individual structural elements were the governing factors. Thus, the global hull-girder requirements were inherently met and exceeded by the local design, with reasonable safety factors. Therefore, the need for a first-principles-based local slamming model to accurately determine the slamming pressures is further emphasized.

Based on these conclusions, and the potential for a very significant contribution to the state-of-the-art in the development of a rational approach to structural design for high-performance vessels, it is highly recommended for this effort to continue into the second phase, as discussed earlier. This will provide the complete rational prediction process needed to calculate the load statistics and estimated extreme loads for practical ship structure design.

ACKNOWLEDGEMENT

The authors would like to thank the Center for Commercial Deployment of Transportation Technologies (CCDoTT) at California State University, Long Beach, CA, for funding this work, and especially Mr. Stan Wheatley of CCDoTT for his continuous support throughout this project and for the much needed Phase-

II effort. Additionally, the authors would also like to thank Dr. Sathish Balasubramanian for his excellent contribution that laid the groundwork for this paper.

REFERENCES

BALASUBRAMANIAN, S., CULLINA, J.C. & MISH, W.H. (2001). "Performance and Cost-Based Design Discrimination for Fast-Ferry Hullform Selection," Proceedings, 17th Fast Ferry Conference, New Orleans.

BAND, E.G.U. (1966). "Analysis of Ship Data to Predict Long-Term Trends of Hull Bending Moments," American Bureau of Shipping, New York.

BECK, R.F. AND TROESCH, A. W. (1990). "Documentation and User's Manual for the Computer Program SHIPMO.BM," Department of Naval Architecture and Marine Engineering Report No. 89, University of Michigan.

BERTRAM, V. (2000). "Practical Ship Hydrodynamics," Butterworth Heinemann, Oxford, UK.

BHATTACHARYYA, R. (1978). "Dynamics of Marine Vehicles. A Volume in Ocean Engineering," A Wiley.

Series Edited by Michael E. McCormick. John Wiley & Sons, Inc., New York, 1978.

CENTENO, R. VARYANI, K.S., and GUEDES SOARES, C. (2001). "Experimental Study on the Influence of Hull Spacing on Hard-Chine Catamaran Motions," Journal of Ship Research, Vol. 45, No. 3, pp. 216-227.

CHAPMAN, R.B. (1975). "Free Surface Effects for Yawed Surface Piercing Plates," Journal of Ship Research, 20, pp. 125-132.

DUBROVSKY, V. and LYAKHOVITSKY, A. (2001). "Multihull Ships," Backbone Publishing Company, New Jersey.

HUDSON, D., MOLLAND, A., PRICE, W. and TEMAREL, P. (2001). "Seakeeping Performance of High Speed Catamaran Vessels in Head Seas and Oblique Waves," Transactions of the Royal Institution of Naval Architects, pp. 247- 257.

LEWIS, E.V., HOFFMAN, D., MACLEAN, W.M., VAN HOOF, R., and ZUBALY, R.B. (1973). "Load Criteria for Ship Structural Design," SSC-240, Ship Structure Committee, Washington, DC.

MOLLAND, A.F., WELLICOME, J.F., TEMAREL, P., CIC, J., and TAUNTON, D.J. (2000). "Experimental Investigation of the Seakeeping Characteristics of Fast Displacement Catamarans in Head and Oblique Seas," Transactions of the Royal Institution of Naval Architects, pp. 78-97.

SALVESEN, N., TUCK, E.O., AND FALTINSEN, O. (1970). "Ship Motions and Sea Loads," Transactions, SNAME 78, pp.250-287.

Nonlinear Time Domain Simulation Technology for Seakeeping and Wave-Load Analysis for Modern Ship Design

Y.S. Shin, Associate Member, American Bureau of Shipping, **V.L. Belenky**, Member, American Bureau of Shipping, **W.M. Lin**, Member, Science Applications International Corporation, **K.M. Weems**, Member, Science Applications International Corporation, **A.H. Engle**, Associate Member, Carderock Division, Naval Surface Warfare Center.

ABSTRACT

This paper describes recent developments and new applications in the field of numerical simulation for nonlinear ship motions using the Large Amplitude Motion Program (LAMP), following up on our 1997 paper (Shin, et al. 1997). The objective of LAMP's development is to develop an analysis tool for highly realistic prediction of wave loads and behavior for a ship in severe seas. This approach is based on physics and does not rely on statistical information gathered from model tests or existing ships, so it is expected to be especially useful for new ship types. The kernel of LAMP is panel-based potential flow solution of the ship-wave hydrodynamic problem with many numerical options, including a recently-added Rankine singularity model with a damping beach and an option to rapidly compute the perturbation potential using pre-computed Impulse Response Function (IRF) potentials. The paper describes important implementation details for these and other computational options added since 1997. The LAMP System is structured so that numerical or empirical models of other systems or effects can be directly incorporated into the time domain ship motions and load calculations using a series of optional features. One such feature is a multi-level green-water-on-deck model including a finite-volume solution of 3-D shallow water flow over the deck. This model can be used for analyzing ship behavior with water on deck and loads caused by green-water-on-deck. Numerical results illustrate the effect of green-water-on-deck on the pitch motion and vertical bending moment of a cruiser in head seas and on the roll behavior of a fishing vessel. Another feature involves anti-roll fin and tank systems, including an integrated anti-roll tank model that solves for fluid motion in a U-tube tank concurrently with the wave-body hydrodynamics. The paper evaluates the use of passive anti-roll tanks to mitigate parametric roll resonance. Recent development has focused on applying LAMP to unconventional ships, multi-hull high-speed displacement ships, and non-ship-like configurations. Results are presented from studies for an advanced Naval hull form, a trimaran, and a semi-submersible platform.

INTRODUCTION

New ship designs could be characterized by large variation of lines and speed range. New container carriers have not only become larger, but their hull shapes have also experienced a significant transformation as large bow flares and stern overhangs become standard features. Bulk carriers have also become larger and wider, but relatively more shallow (Shin, 2001). Increasing demand for high speed generates more multi- and mono-hull high-speed designs both for commercial and naval applications.

Such a variety of new and unusual hull forms poses a new challenge for the maritime industry as design methods based on semi-empirical processes become suspect. Even direct calculation methods with proven track records for accuracy and reliability may lose their applicability, as their assumptions are no longer met.

A typical example for such a situation is a recent case of parametric rolling in head seas (France *et al.*, 2003). The physical phenomenon of parametrically excited roll motions in following seas has been known since the 1950s (Paulling and Rosenberg, 1959), but the appearance of parametric roll in head seas conditions was unexpected. The cause was the large bow flare and stern overhang that created strong coupling between longitudinal and transverse ship motions, a phenomena that is not observed for more traditional hull shapes. This resulted in unusually large roll motions for which the cargo securing system was not designed. Another example of new problems generated by modern hull shapes is bow flare slamming in oblique seas (Finn, *et al.*, 2001). Problems associated with the influence of green-water-on-deck on ship motions and loads are also becoming serious issues for new hull shapes where operational experience has not yet been gathered.

The natural reply to this challenge is to develop calculation methods suitable for these new designs. Such new methods were thoroughly reviewed by Beck and Reed (2001), who pointed out that time domain simulations of ship motions based on potential flow formulations is becoming a more practical way to achieve engineering results for new ship designs.

BRIEF DESCRIPTION OF LAMP

The LAMP System is a time-domain simulation model specifically developed for computing the motions and loads of a ship operating in extreme sea conditions. LAMP System development began with a 1988 DARPA project for advanced nonlinear ship motion simulation, and has continued under the sponsorship of the U.S. Navy, U. S. Coast Guard, the American Bureau of Shipping (ABS), and Science Applications International Corporation's (SAIC) IR&D program.

LAMP uses a time stepping approach in which all forces and moments acting on the ship, including those due to the wave-body interaction, appendages, control systems, and green-water-on-deck, are computed at each time step and the 6-DOF equations of motions are integrated in the time-domain using a 4th-order Runge-Kutta algorithm. In addition to motions, LAMP also computes main hull-girder loads using a rigid or elastic beam model and includes an interface for developing Finite-Element load data sets from the 3-D pressure distribution (Shin *et al.*, 1997; Weems *et al.*, 1998).

The central part of the LAMP System is the 3-D solution of wave-body interaction problem in the time-domain (Lin and Yue, 1990, 1993). A 3-D perturbation velocity potential is computed by solving an initial boundary value problem using a potential flow boundary element or "panel" method. A combined body boundary condition is imposed that incorporates the effects of forward speed, the ship motion (radiation), and the scattering of the incident wave (diffraction). The potential is computed using a hybrid singularity model that uses both transient Green functions and Rankine sources (Lin *et al.*, 1999). Once the velocity potential is computed, Bernoulli's equation is then used to compute the hull pressure distribution including the second-order velocity terms.

The perturbation velocity potential can be solved over either the mean wetted surface (the "body linear" solution) or over the instantaneously wetted portion of the hull surface beneath the incident wave (the "body nonlinear" approach). In either case, it is assumed that both the radiation and diffraction waves are small compared to the incident wave and the incident wave slope is small so that the free-surface boundary conditions can be linearized with respect to the incident-wave surface. Similarly, the incident wave forcing (Froude-Krylov) and hydrostatic restoring force can also be computed either on the mean wetted surface or on the wetted hull up to the incident wave.

The combinations of the body linear and body nonlinear solutions of the perturbation potential and the hydrostatic/Froude-Krylov forces provide multiple solution "levels" for the ship-wave interaction problem. These levels are:

- LAMP-1 (Body linear solution): Both perturbation potential and hydrostatic/Froude-Krylov forces solved over the mean wetted hull surface
- LAMP-2 (Approximate body nonlinear solution): The perturbation potential is solved over mean wetted hull surface while the hydrostatic/Froude-Krylov forces are solved over the instantaneous wetted hull surface
- LAMP-4 (Body nonlinear solution): Both the perturbation potential and the hydrostatic/Froude-Krylov forces are solved over the instantaneous wetted hull surface

For most problems, the most practical level is the "approximate body-nonlinear" (LAMP-2) solution, which combines the body-linear solution of the perturbation potential with body-nonlinear hydrostatic-restoring and Froude-Krylov wave forces. This latter approach captures a significant portion of nonlinear effects in most ship-wave problems at a fraction of the computation effort for the general body-nonlinear formulation. However, body-nonlinear hydrodynamics and nonlinear incident wave effects can be important depending on ship geometry and operating conditions.

Mixed Source Formulation

In the context of time-domain potential-flow boundary-element methods, the most commonly used approaches fall in two categories: (1) methods that use transient Green functions and (2) methods that use Rankine sources (Weems, *et al.,* 2000).

For the methods in the first category (*e.g.* Lin *et al.*, 1994), the transient Green function satisfies the linearized free surface boundary conditions and radiation conditions in the far field, so that the singularities need to be distributed only on the wetted portion of the body surface. For ships with non-wall-sided geometry, numerical difficulties may arise in the area where the intersection angles between the body surface and the free surface become small, as with wave-piercer ships. This is mainly due to the highly oscillatory nature of the transient Green function adjacent to the free surface.

For the methods in the second category (e.g. Nakos *et al.,* 1993), the Rankine source is used as a kernel in the boundary integral equation. Rankine sources are fairly robust for modeling either wall-sided or non-wall-sided geometries. To satisfy the free surface boundary condition, Rankine sources must be distributed not only on the body surface but also on the free surface. In order to limit the size of the computation domain, the free surface region is typically truncated at several ship lengths away from the ship, and a numerical damping zone is employed to absorb wave energy.

In view of the pros and cons of these two approaches, a hybrid numerical approach was developed to use both transient Green functions and Rankine sources (Lin *et al.*, 1999). This approach is implemented in the LAMP System as the "mixed source formulation." In the mixed source formulation, the fluid domain is split into two regions as shown in Figure 1. The outer domain is solved with transient Green functions distributed over an arbitrarily shaped matching surface, while the inner domain is solved using Rankine sources. The advantage of this formulation is that Rankine sources behave much better than the transient Green function near the body and free surface juncture, and the matching surface can be selected to guarantee good numerical behavior of the transient Green functions. The transient Green functions satisfy both the linearized free surface boundary condition and the radiation condition, allowing the matching surface to be placed fairly close to the body. This numerical scheme has resulted in robust motion and load predictions for hull forms with non-wall-sided geometries.

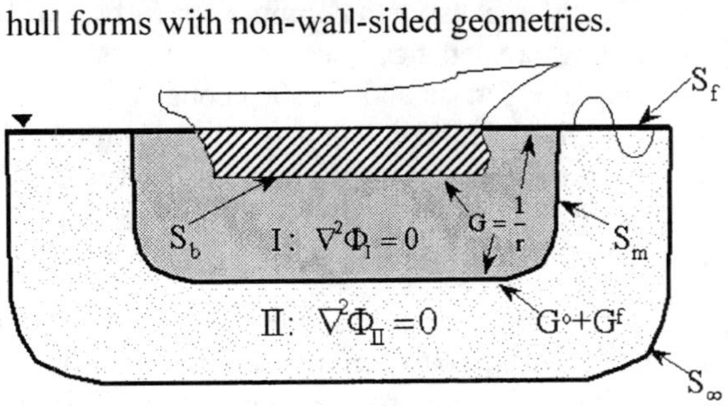

Figure 1. Mixed source formulation

Another advantage of the mixed formulation is that the local free surface elevation is part of the solution, and no additional evaluation is needed as in the case of the transient Green function approach. In addition, a nonlinear free surface boundary condition can be implemented at modest computational cost. In the LAMP System, a 2nd-order free surface boundary condition can be applied on the local portion of the free surface; see more details in Weems *et al.* (2000). However, in the case of nonlinear free-surface boundary condition in the local portion of the free surface, the matching surface has to be placed further away from the body to minimize errors caused by a mismatch of the free surface condition.

Rankine Singularity with Damping Beach

While the mixed-source singularity model works very well for low to modest speeds ($Fr \leq 0.5$), it can be difficult to obtain a stable solution at higher speeds. For this reason, an alternative singularity model has been implemented that replaces the external domain and the matching surface with a numerical damping region on the outer edge of the inner region's free surface. The body and free surface boundary conditions are otherwise identical to the mixed-source singularity model.

While this singularity model typically requires a considerably larger free surface grid then the mixed-source model, it has been successfully applied at $Fr=0.85$. The model also allows shallow water to be modeled in the hydrodynamic problem by panelizing the bottom or using image sources.

IRF-Based Formulation

A drawback to time domain hydrodynamics is the computational cost. To mitigate this, an Impulse Response Function (IRF) based hydrodynamic formulation (Liapis 1986, King *et al.*, 1988, Bingham *et al.* 1993) was integrated into the LAMP System to complement the mixed source formulation. In the IRF formulation, velocity potentials are pre-computed for steady forward speed, impulsive motion in up to six modes, and impulsive incident waves for each speed and heading angle. The hydrodynamic problem is reduced to a convolution of the IRF potentials with the actual ship motions and incident wave elevation, thereby allowing numerical simulations to be performed faster than real time using modest computational resources and without compromising the accuracy of the hydrodynamic calculation.

The IRF potentials are convoluted and summed on a panel-by-panel basis, so that the complete potential distribution of the hull can be computed in the time domain. This allows the panel pressure to be computed directly, including the nonlinear terms in Bernoulli's equation, in the same fashion as in the mixed source formulation. The only restriction is that the IRF formulation can only be used with body linear (LAMP-1) and approximate body nonlinear (LAMP-2) hydrodynamics.

In the IRF-based formulation, the perturbation velocity potential on each body panel is decomposed as

$$\Phi_k(\vec{x},t) = \sum_{k=1}^{6} \Phi_k(\vec{x},t) + \Phi_7(\vec{x},t) + \Phi_8(\vec{x},t)$$

where the Φ_k, k=1..6 are the radiation potentials for the six rigid-body motions, Φ_7 is the diffraction potential related to the incident wave potential Φ_0, and Φ_8 is the steady state potential related to the constant forward speed U. To solve for the six radiation potentials Φ_k, six corresponding impulse response functions φ_k are introduced via the convolution integral

$$\Phi_k(\vec{x},t) = \int_0^t \varphi_k(\vec{x},t-\tau)\dot{X}_k(\tau)d\tau$$

where X_k is the ship motion in mode k and the dot signifies the derivative with respect to time. Each of the radiation IRF potentials is the solution for an impulsive velocity in that mode and is obtained by solving the initial boundary value problem:

$$\nabla^2\varphi_k = 0 \qquad \text{in } V$$

$$\frac{\partial\varphi_k}{\partial n} = n_k\delta(t) + m_k h(t) \qquad \text{on } S_b$$

$$\frac{\partial^2\varphi_k}{\partial t^2} + g\frac{\partial\varphi_k}{\partial z} = 0 \qquad \text{on } S_f$$

$$\varphi_k = \frac{\partial\varphi_k}{\partial t} = 0 \quad \text{at } t=0 \qquad \text{on } S_f$$

where $\delta(t)$ and $h(t)$ are the Dirac delta and Heavyside step functions with respect to time t, respectively, and k runs from 1 to 6.

To solve for the diffraction potential Φ_7, the diffraction IRF φ_7 is introduced via the convolution integral

$$\Phi_7 = \int_{-\infty}^{\infty} \varphi_7\zeta_0(\tau)d\tau$$

where ζ_0 is the incident wave elevation at the origin of the ship-fixed frame. The diffraction IRF φ_7 is the diffraction potential due to an impulsive unit incident wave and is obtained by solving the initial boundary value problem:

$$\nabla^2\varphi_7 = 0 \text{ in} V$$

$$\frac{\partial\varphi_7}{\partial n} = -\frac{\partial\varphi_0}{\partial n} \text{ on} S_b$$

$$\frac{\partial^2\varphi_7}{\partial t^2} + g\frac{\partial\varphi_{k7}}{\partial z} = 0 \text{ on} S_f$$

$$\varphi_7 = \frac{\partial\varphi_7}{\partial t} = 0 \text{ at} t = 0 \text{ on} S_f$$

where φ_0 is the potential due to an impulsive unit incident wave at the origin of the ship-fixed frame:

$$\varphi_0 = \text{Re}\frac{g}{2\pi}\int_{-\infty}^{\infty}\frac{i}{\omega}e^{kz-ik(x\cos\beta+y\sin\beta)+i\omega t}\,d\omega_e$$

where β is the wave heading angle relative to the ship, ω is the wave frequency, and ω_e is the frequency of encounter.

For the head seas ($\pi/2 \le \beta \le 3\pi/2$) and zero speed cases, there is single valued relationship between ω and ω_e so there is a single impulsive wave, and hence a single diffraction IRF, for each wave heading angle. However, for the following seas cases, there are up to three wave frequencies for each frequency of encounter, so there are three impulsive incident waves, and hence three diffraction IRFs, for each wave heading angle.

The final component of the decomposed wave potential, the steady forward potential, can be determined by simply solving a forced steady forward speed problem with no incident waves until a converged solution is reached. All of the IRF potentials can be computed using the mixed-source formulation's numerical scheme and saved in a series of files. Each of the IRFs depends only upon the ship geometry, speed, and incident wave heading angle.

In principle, the convolution of the radiation IRFs must start at the beginning of the calculation and the convolution of the diffraction IRFs must run from $\tau=-\infty$ to $\tau=\infty$. In practice, however, the radiation convolutions need only include the most recent 400 or so time steps (for a typical time step of 0.15 seconds) while the diffraction convolutions need only include ±400 steps for head seas and ±800 steps for following seas. As a result, the IRFs are not excessively expensive to compute – a full set of IRFs for a long crested head sea case requires the same calculation effort as about 200 seconds of simulation using the mixed-source formulation. The computational cost of IRFs for short-crested and/or following sea cases will be higher as additional diffraction IRFs will be required.

Once computed, however, the resulting IRF-based LAMP calculation can be run 40-100 times faster than the mixed source formulation. This allows faster-than-real-time calculations for most configurations using very modest computational resources. Furthermore, the IRFs are independent of the incident waves' frequency and amplitude, so an entire series of regular or irregular wave conditions can be run with the same IRFs (Weems *et al.*, 2000).

As mentioned earlier, a restriction of the IRF-based formulation is that it can only be used with the body-linear solution for the perturbation potential – where the perturbation potential is solved over the mean wetted surface of the hull. However, this does not imply that the IRF formulation is restricted to a classical "linear" seakeeping solution. Classical "linear" seakeeping linearizes Bernoulli's equation, hydrostatic restoring forces, and the equations of motion and assumes a direct linear relation between the incident wave and the response. In contrast, the IRF-based velocity potentials due to ship motion (radiation) are linear in the *instantaneous* ship velocity as computed by the general solution of the equations of motion. Also, the IRFs are used to compute a distribution of velocity potential, so that the second order terms in velocity in Bernoulli's equation are preserved. In the approximate body-nonlinear calculation (LAMP-2), the incident wave forcing and hydrostatic restoring forces are computed using a nonlinear approach. Additional forces such as those due to viscosity, control surfaces, green-water-on-deck, etc. can also be treated using this nonlinear approach. As a result, LAMP's approximate body-nonlinear solution is capable of capturing significant large amplitude wave phenomena that are missed in classical linear seakeeping, such as parametric rolling and the dependence of motions and loads on wave slope. Using the IRF-formulation with the approximate body-nonlinear calculation can provide faster-than-real-time calculation of large amplitude responses, thereby facilitating the large number of relatively lengthy simulations required to properly characterize a ship's response in large amplitude seas.

LARGE AMPLITUDE WAVE LOADS

Comparisons Between the Mixed and Rankine Source Formulations

In an effort to assess possible differences in results between the Mixed and Rankine Source formulations in LAMP, several comparisons were made against available model test data. The seakeeping and loads model data used for these comparisons were compiled during an advanced hull form development study at Carderock Division, Naval Surface Warfare Center (NSWC/CD).

To support this analysis, a free running 1/20-scale segmented model was built, with a continuous longitudinal backspline beam and instrumented to provide structural loads data (see Figure 2). The test program consisted of calm water, regular wave, and irregular wave measurements over a range of ship speeds and wave headings.

Figure 2. Advanced monohull structural loads model

Subsequent to testing, the LAMP System was used to compare motions and loads predictions with model test results. The visualization of the LAMP simulation for an extreme wave case is shown in Figure 3.

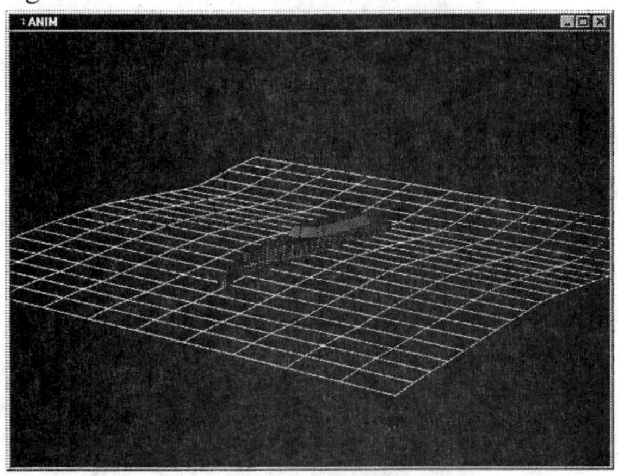

Figure 3. LAMP simulation for advanced Monohull form

A sample of the body nonlinear formulation (LAMP-4) results obtained using both the Rankine and the mixed formulations is shown in Figures 4 and 5. The first series of figures compare the two formulations for a regular wave run with a wave steepness of 1/30 and a ship speed of 15 knots. Although there are some difference in phasing between the model test results and the predictions, both formulations show fairly good agreement.

The next set of comparisons shows the model in a Hurricane Camille seaway. As can be seen from Figure 5, predicted motions and loads using either formulation are quite good, the main differences between the predicted and measured loads being that slam-induced whipping predictions were not performed. Hence the high frequency component to the load time history is not accounted for.

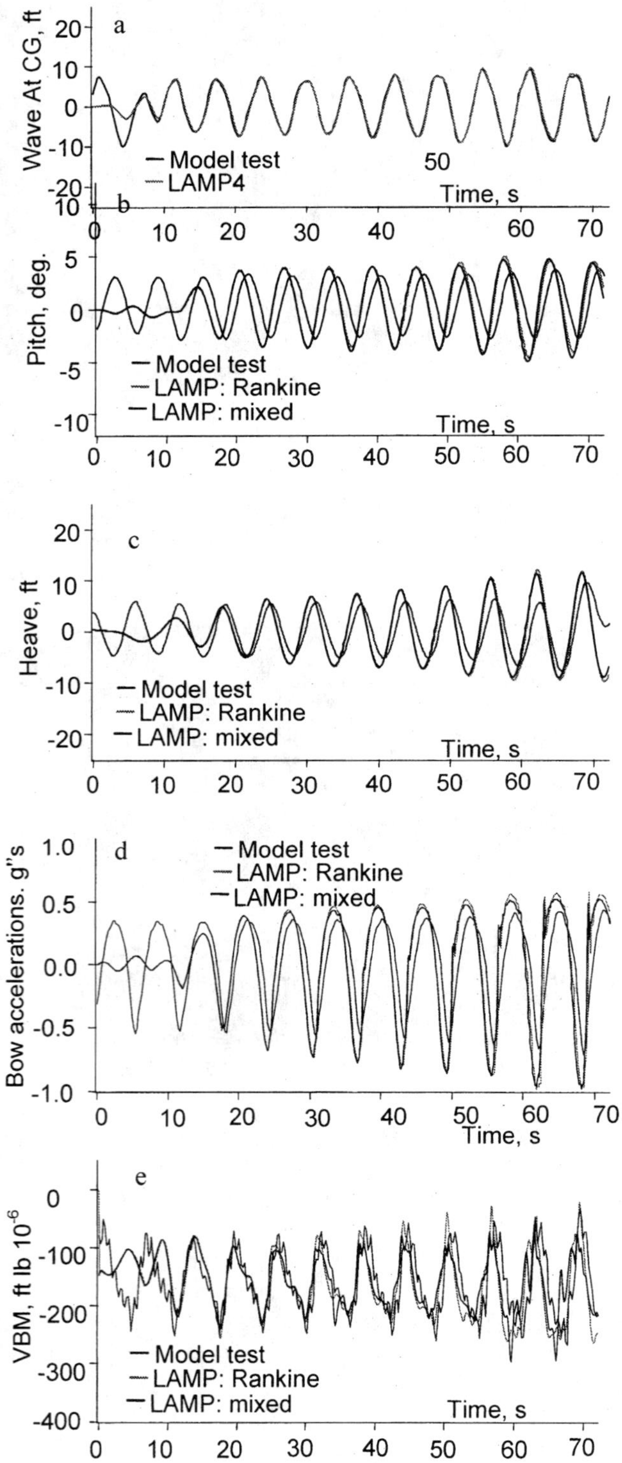

Figure 4. Run 727: λ/L=1.0, H/λ=1/30, 15kts. (a) Wave at CG. (b) pitch, (c) heave, (d) bow acceleration, and (e) vertical bending moment

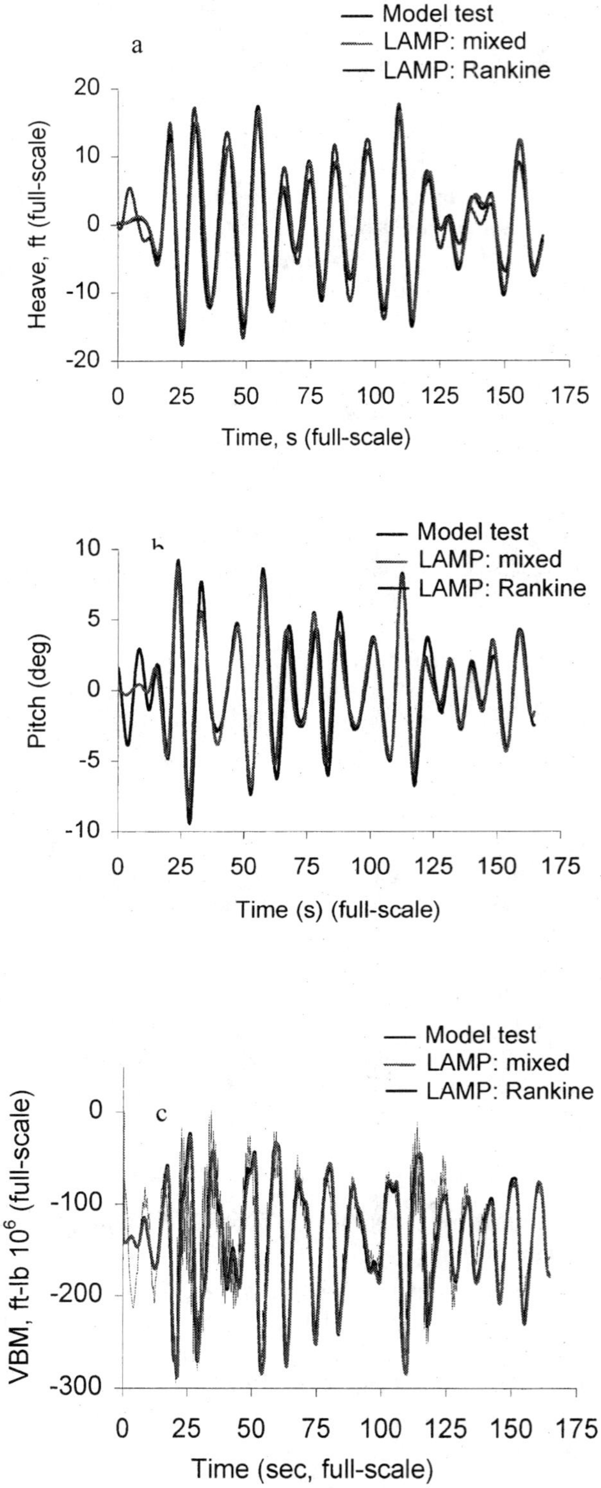

Figure 5. Run 127: (a) heave, (b) pitch, (c) vertical bending moment

Application of LAMP-FEM Interface for Computation of Wave Loads for Large Container Carrier

Shin *et al.* (1997) contains detailed information on the interface between LAMP and finite element method (FEM) analysis software. Another sample application is shown in Figure 6. This time, a large container ship was the subject of the case study. Since no experience existed in designing a container ship of such a size, the designer recognized the simulation results as an important source of information for making technical decisions.

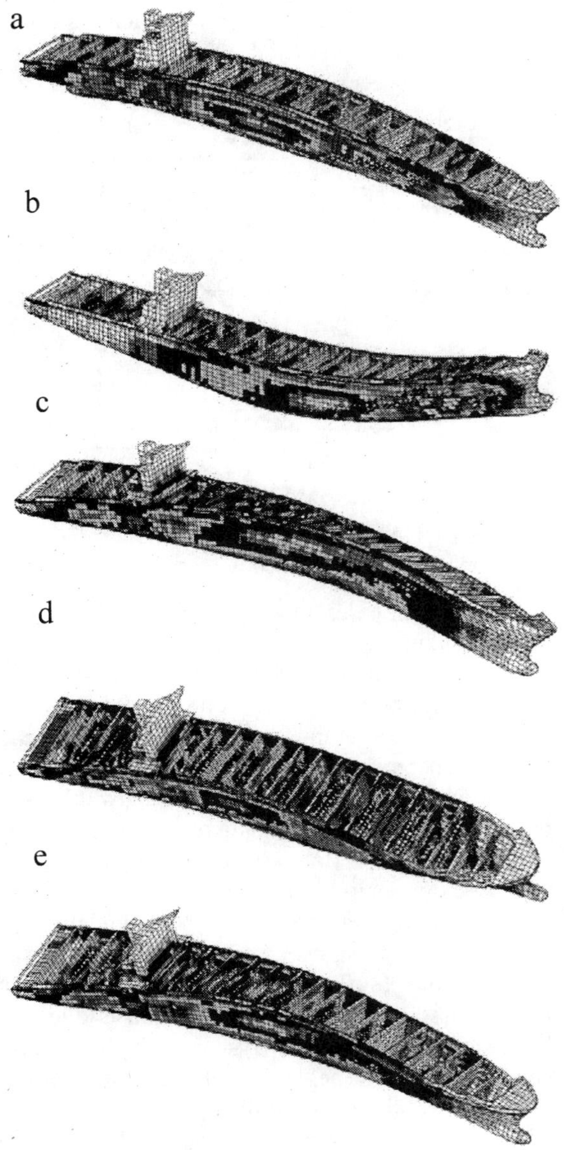

Figure 6. Structure responses of large container ship for different loading cases (a) hogging; (b) sagging; (c-e) extreme torsion for different location of wave crest along the ship

LARGE MOTION PHENOMENA AND ASSOCIATED LOADS

Parametric Roll and its Stabilization

Parametric excitation in head seas is known to be a factor in the very high amplitude roll motion that caused lost cargo (France *et al.*, 2003). One possible way to mitigate this kind of motion is a roll stabilization system.

A method for approximating a U-tube type passive anti-roll tank system has been implemented as an optional "plug-in" module in LAMP. The system calculates fluid motion in the U-tube tank and determines the coupled nonlinear forces acting on a floating body in full six degrees of freedom. These forces are added to LAMP's calculation of motion and loads as an external (non-pressure) force. A schematic of the anti-roll tank is shown in Figure 7.

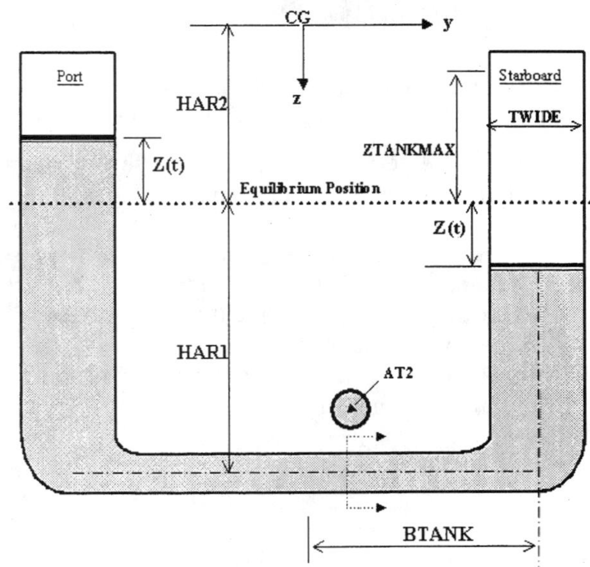

Figure 7. Schematic of anti-roll tank system

The LAMP anti-roll tank formulation is based on the improved passive anti-roll tank model developed by Youssef, *et al.* (2002). It has been successfully applied in the nonlinear motion range (including parametric roll suppression) as shown in the above reference.

Several assumptions are used in the anti-roll tank formulation including:

- Vertical tank columns and connecting tubes are assumed to be circular.
- Shear stress at the wall is related to a friction factor that is dependant on Reynolds number and based on either the Colebrook formula or the Swamee and Jain (1976) formula.

- Energy loss in the elbows connecting the vertical columns to the cross pipe is expressed in terms of a loss coefficient.
- Tank free surface correction does not account for sloshing effects and assumes a flat calm free surface that remains parallel to the global horizontal plane.

An initial investigation into the implementation and effectiveness of the anti-roll tank module was performed for a regular synchronous roll case. In this example, the sample container ship shown in Figure 8 was simulated at 15 knots in regular beam sea waves.

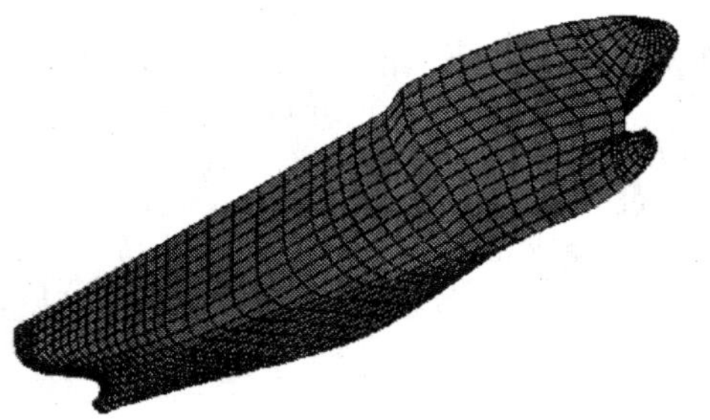

Figure 8. Panel representation of a sample large modern container ship

The wave period was equal to the ship's natural roll period of 25 seconds, and the wave height was 12 meters. Figure 9 compares roll motions for a case with no anti-roll tanks and a case with two identical anti-roll tanks. The case with no anti-roll tanks has a steady-state roll angle of approximately 22 degrees while the case with the anti-roll tanks has a maximum roll angle of 3.5 degrees. This investigation clearly shows the ability of an appropriately tuned anti-roll tank system to reduce resonant roll motions.

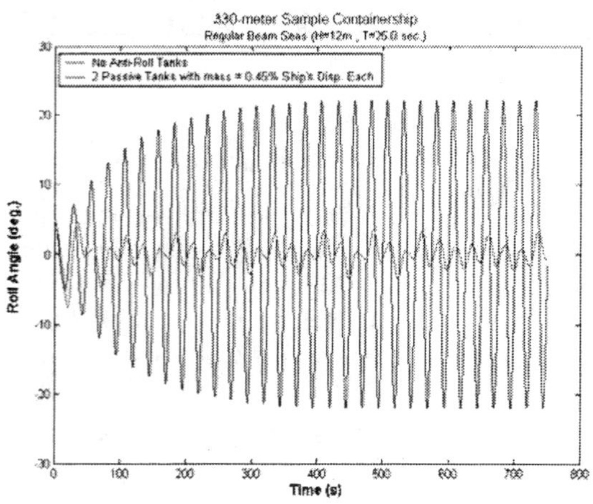

Figure 9. Roll motion in regular beam seas (wave period = ship's natural roll period)

Several calculations were performed on a sample container ship to determine the effectiveness of the passive anti-roll tank system in reducing the ship's susceptibility to parametric roll. Figure 10 shows the predicted maximum roll angles for this container ship in regular head seas as a function of encounter frequency. A range of encounter frequencies in which this particular ship might be susceptible to parametric roll can be clearly seen in the curve with no anti-roll tanks. The roll natural frequency for this ship is 0.251 rad/sec, which corresponds to a 25 second period, so parametric rolling would be expected near an encounter frequency of 0.502 rad/sec, which is clearly shown in the results. The nonlinear dependence of parametric roll on the incident wave amplitude is also clearly illustrated. This ship had a large range of encounter frequencies where parametric roll is predicted, which indicates that the ship is very susceptible to parametric roll.

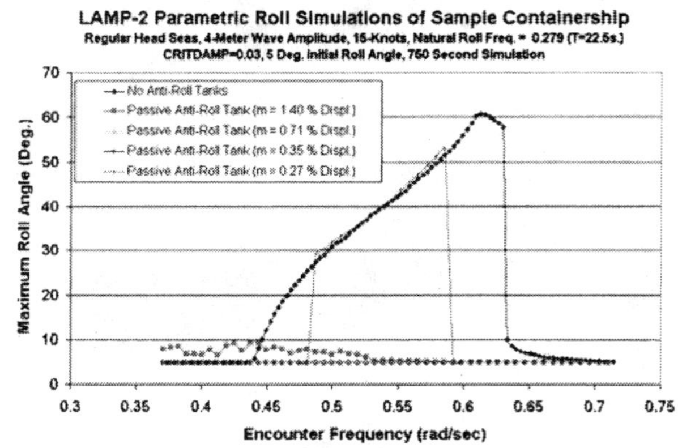

Figure 10. LAMP computed parametric roll domain for sample container ship

The other cases plotted in Figure 10 show the same container ship with a single passive anti-roll tank where the tank mass is equivalent to 0.27, 0.35, 0.71, and 1.4 percent of the ship's displacement, with a tank system natural period equal to the ship's natural roll period of 25 seconds. The 0.27 percent case shows a reduction in the bandwidth where parametric roll occurs, but there is still a region of large roll angles. Both the 0.35 percent and 0.71 percent cases show virtually no roll past the initial 5-degree roll angle at the start of the simulation. The 1.4 percent case shows a region of slightly elevated roll motion (<10 degrees) for low encounter frequencies. The increased roll motion is due to the large mass in the tank system, which is not in equilibrium at the beginning of the simulation. The tank fluid continues to oscillate for a short period of time until it eventually damps out as the simulation progresses.

Analogous principles and procedures have been used to couple a numerical solution of sloshing in cargo and anti-rolling tanks with time-domain ship motions (Kim, 2002).

Influence of Green-Water-on-Deck on Motions and Loads

Recent efforts have developed a sophisticated green-water-on-deck model that can be integrated directly into LAMP, allowing water-on-deck effects to be included in time domain calculations of nonlinear ship motions and loads. In the approach that was selected, the ship motions and green-water-on-deck calculations run concurrently in the time domain. As illustrated in Figure 11, this approach subdivides the computational domain into an "outer" problem where the ship-wave interaction problem is solved and an "inner" problem where the green-water-on-deck problem is solved (Belenky, *et al.*, 2002).

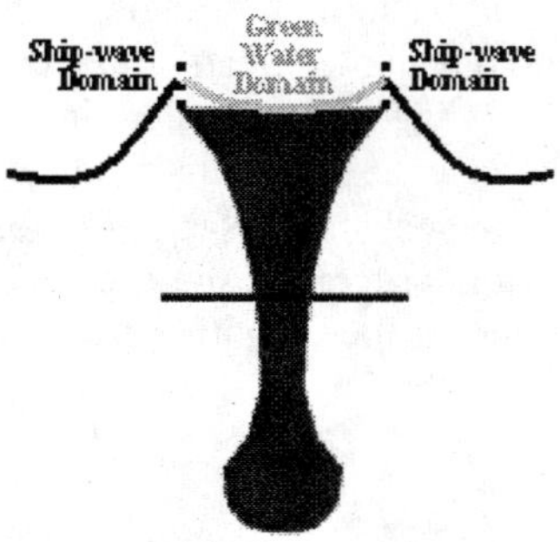

Figure 11. Decomposition of the computation domain

At each time step of the LAMP calculation, the ship motion and wave definition are used to compute the relative motion at the deck edge (often called the deck edge exceedance), the deck tilt, and the deck acceleration. These values are in turn passed to the green-water-on-deck calculation. The green-water-on-deck calculation is subsequently updated, after which the green-water-on-deck pressure and forces are passed back to the ship motions calculation to be used in solving the equations of motions and computing sectional loads. The structure of this computational approach is illustrated in Figure 12.

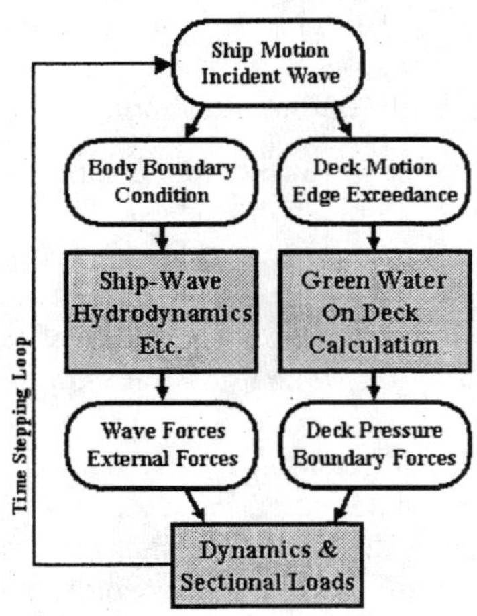

Figure 12. Structure of the LAMP System with green-water-on-deck

The green-water-on-deck calculation in the LAMP System has been integrated in a "modular" fashion so that different "levels" of calculations are available.

Hydrostatic and Froude-Krylov pressure: Deck pressure is computed from the submergence of each deck element and the linear incident wave pressure ($\partial\Phi/\partial t$). This is a straightforward modeling of the deck as a "body-nonlinear" hydrostatic/Froude-Krylov surface and represents the basic deck-in-water calculation.

Semi-empirical model: Water-on-deck is calculated directly from deck edge exceedance using an empirical expression derived from Zhou *et al.* (1999). The resulting deck pressure, including the effects of deck motion, can then be evaluated following the approach proposed by Buchner (1995). This model is intended to provide a quick estimate of the effect of foredeck green-water-on-deck on ship motions and main girder loads.

Shallow water flow calculation: Computes the longitudinal and transverse flow over the deck with the assumption of shallow water using a finite-volume technique. This is the principal green-water-on-deck calculation used in the present study and is described in detail below.

Fully 3-D flow calculation: Computes the flow over the deck, including vertical gradients, using a 3-D volume grid and a finite-volume technique. This new technique is under development and is a likely future path for green-water-on-deck flow calculation methods.

The most sophisticated of LAMP's current green-water-on-deck calculation methods is the one based on the solution of the shallow water flow equations using a novel finite volume strategy. In this method, the equations for conservation of mass and momentum are solved in the time domain, shallow-water flow is assumed, and viscous effects are ignored except for a relatively simple deck friction term. The shallow water assumptions are that the fluid acceleration normal to the deck can be ignored while the tangential fluid velocity and pressure are constant across the depth of water-on-deck (Stoker, 1957). These assumptions reduce the 3-D fluid domain to a 2-D computational domain. This method can handle a variety of boundary and initial conditions, and it is capable of supporting arbitrary motions and general ship-deck geometries, including partial height walls (*e.g.* bulwarks), "infinite" walls (*e.g.* deckhouses), and stepped or raised sections (*e.g.* hatch covers). This approach has been validated with available experimental data and has been successfully integrated with the LAMP System (Liut, *et al.,* 2002).

For this approach to be useful as an integrated part of the LAMP simulation, the green-water-on-deck calculation must be reasonably fast, robust, and capable of calculating the flow on a deck that is moving with large-amplitude six-degrees-of-freedom motions and also exchanging water with the environment. Because of these requirements, a considerable portion of the effort in developing and implementing the numerical method has been spent in ensuring a stable and reasonable solution even when the assumptions of shallow water flow are stretched to their limits.

In order to test and validate the numerical solution of the shallow-water calculation, computational studies have been made for several configurations for which detailed theoretical and experimental data are available. One such study examined the "dam-breaking" problem, studying the level of water behind a dam after the dam is suddenly removed. A second study examined the outflow from a box whose end is suddenly removed. Both studies show good agreement between the present method and published data (Liut *et al.,* 2002).

In order to test the green-water-on-deck implementation in the LAMP System, a series of calculations were made for a US Navy CG-47 Class cruiser, a U.S. Coast Guard Cutter, and a large modern container ship. The qualitative validation continues with the fishing boat study described below.

The fishing vessel analyzed in the present study is the 21-meter stern dragger *Italian Gold*, which sank in heavy seas off Massachusetts in 1994. All calculations were made at a heavily loaded condition with a displacement of 194 long tons and the center of gravity 0.08 meters above the mean waterline. The resulting transverse metacentric height (GM_t) is 0.22 meters. The basic geometry model used in the hydrodynamic and hydrostatic calculations is shown in Figure 13.

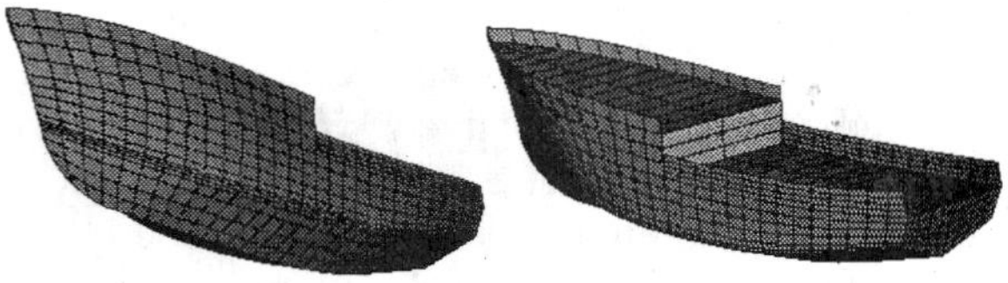

Figure 13. Fishing boat geometry

In the simulations, the skeg and rudder were modeled using "non-pressure" force models, so their effect was included in the equations of motions but not in the wave-body interaction potential. Green-water-on-deck was calculated over the afterdeck, with inflow/outflow allowed over both sides and the stern of the ship.

The first and, to this point, most extensive portion of the study involved a series of roll decay calculations using different computational models and options. For clarity, these calculations were made in calm water at zero speed with the ship free only to roll. The initial roll angle was 30 degrees, at which the deck edge and bulwark are reasonably deeply submerged. For the case involving the shallow water green-water-on-deck calculation, the deck elevations were initialized with stationary water-on-deck up to the calm waterplane $Z = 0$.

An initial calculation was made for the boat with no water-on-deck calculation of any kind on the afterdeck. A time history of the roll angle, in degrees, is shown in red in Figure 14. The blue "square-tooth" at the bottom of the graph indicates the relative motion of the "minimum freeboard point" of the deck edge; a high value indicates that the deck edge is submerged while a low value indicates that it is above the wave surface. This point, which is located just aft of the deckhouse, submerges at about 10° roll angle in calm water. To some extent, this indicates the beginning of the transition from "water-on-deck" to "deck-in-water."

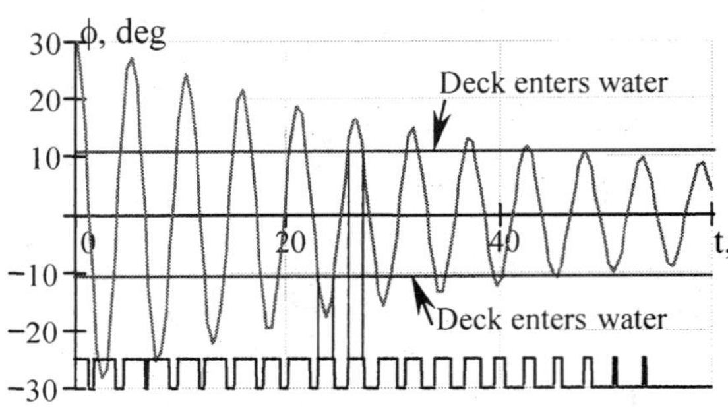

Figure 14. Roll decay angle and deck edge submergence without effect of water-on-deck

The initial roll period is 5.17 seconds and increases slightly to 5.47 seconds as the roll angle decreases. This change in roll period is due primarily to the change in hydrostatic restoring moment of the "body-nonlinear" formulation.

A second calculation computed water-on-deck pressure from hydrostatics. The results for this calculation are shown in Figure 15. In addition to the roll angle and deck edge submergence shown in the bottom graph, the top graph shows a red line indicating the volume of water-on-deck in m^3 and a blue line shows the resulting heeling moment in Newton-meters. Since the model contains no "lag" for water entering or leaving the deck, the volume of water and the heeling moment are exactly in phase with the roll angle. In other words, only the "deck-in-water" situation is modeled. The initial roll period increases to 5.96 seconds, an effect caused by the water-on-deck heeling moment, which in this case effectively decreases the restoring moment when the deck is submerged. As the roll angle decreases, the roll period decreases to 5.47 seconds, matching the period from the previous calculation. A small but noticable decrease in the roll damping can be observed over the early portion of the run, where the longer roll period has decreased the roll velocity and reduced the damping due to skin friction and appendages.

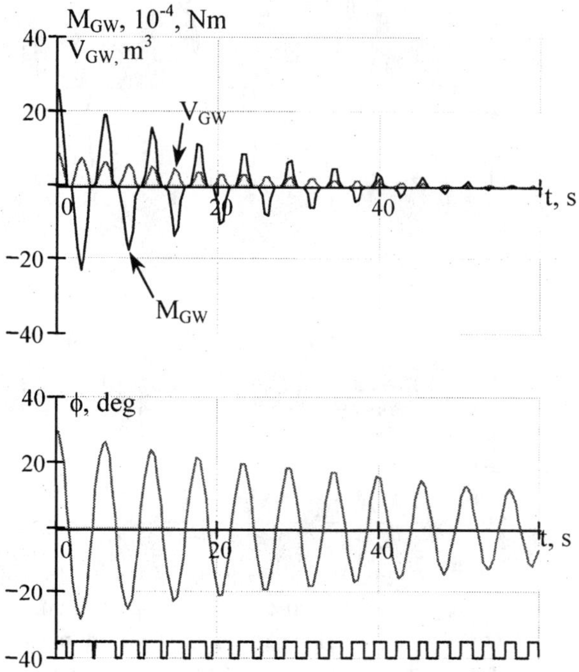

Figure 15. Roll decay results including hydrostatic forces on deck: water-on-deck volume (V_{GW}, top, red) and roll moment (M_{GW}, top, blue), roll angle (ϕ, bottom, red), and deck edge submergence (bottom, blue)

The next calculation uses the finite-volume solution of the shallow water flow over a deck with bare deck edge (*i.e.* no bulwark). The volume of water-on-deck starts out large (as noted above, the finite-volume model is initially "full" to mean Z = 0) and generally decreases in volume as the water flows off the deck, although there are small periods of inflow during subsequent deck edge submergences. The initial roll period is similar to the hydrostatic pressure case, which is reasonable as the water-on-deck heeling moment is similar, while subsequent roll periods match both of the previous calculations at similar roll angles. The influence of water-on-deck and deck-in-water is realized mainly through a significant increase of roll damping (see Figure 16).

Since the problem is otherwise nearly identical to that of the previous calculation, the damping influence must be credited to the water-on-deck. The most likely explanation is that the "lag" of the water flowing off the deck or to the other side of the boat when the boat rights itself results in a significant volume of water on the "rising" side of the deck, where the resulting deck pressure produces a significant moment in phase with but opposed to the roll velocity. Evidence for this interpretation is offered in Figure 17, which shows a snapshot of the boat, which was initially healed to starboard, with the computed water-on-deck surface as it rolls through 0-degree heel in the first cycle of the calculation.

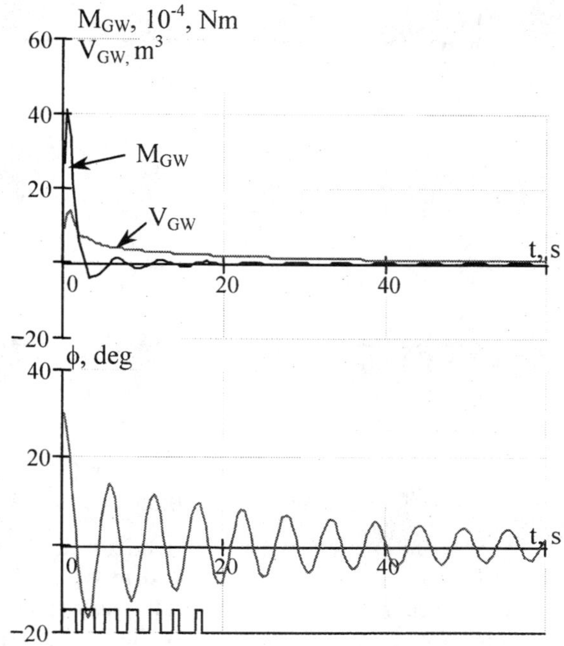

Figure 16. Roll decay results including finite-volume water-on-deck effect without bulwarks

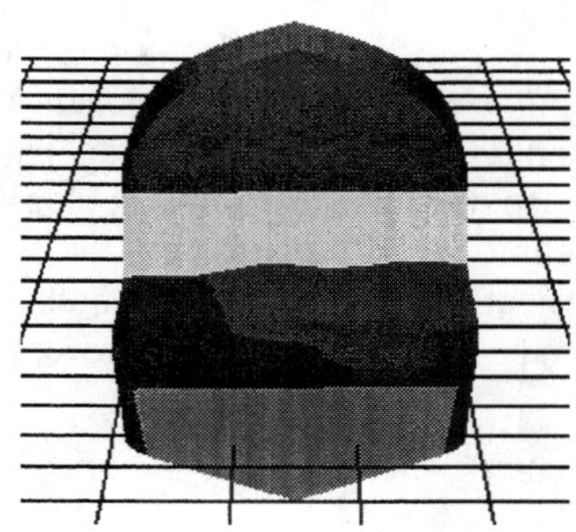

Figure 17. Computed water-on-deck surface for the fishing boat as it rolls though 0° heel

The next calculation, whose results are shown in Figure 18, introduces a solid (no freeing ports) bulwark with height 0.853 meters to the finite-volume calculation. The bulwark, which traps the water-on-deck and raises the angle at which water can enter the deck in calm water to over 20 degrees, changes the whole picture of motions. Damping is such that the second peak of roll angle is too small to immerse the bulwark top or to allow significant inflow. The amount of green-water-on-deck stays constant and the ship quickly assumes a static heel angle of about 7 degrees.

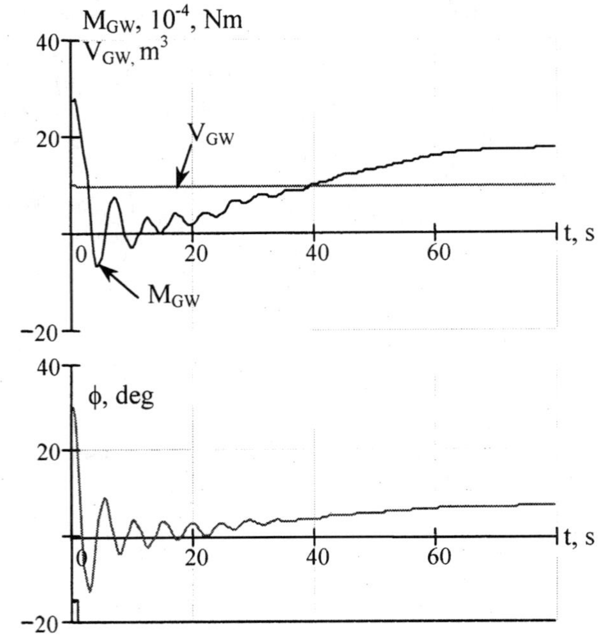

Figure 18. Roll decay results including finite-volume water-on-deck calculation with 0.853 m bulwarks

The second portion of the fishing boat study involved calculations for the boat at zero speed in regular (single frequency) beam waves. As in the roll decay calculation, the boat was only free to roll. While not a realistic situation, it simplifies the comparison of the different calculations and reduces the risk of capsize, especially during the initial transient period. The wave height was 3.05 meters with the period of 6.28 sec, which makes the wave forcing close to the roll resonance regime. We consider only steady state rolling in the present discussion, so the results shown in Figures 19-22 are plotted beginning 520 seconds into the simulations. The initial transient behavior, which is both interesting and very important, is simply beyond the scope of the present discussion.

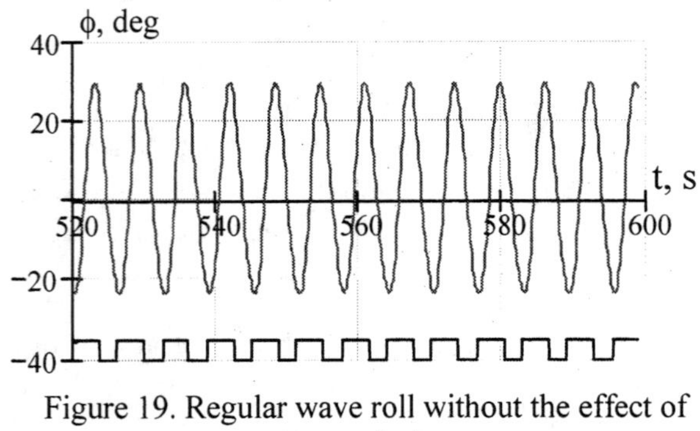

Figure 19. Regular wave roll without the effect of water-on-deck

Figure 19 shows the results of the regular wave calculation for the fishing boat with no water-on-deck model at all. As for the roll decay results, the blue "square-tooth" indicates the submergence of the deck's minimum point, which now is a function of the roll angle and the incident wave elevation.

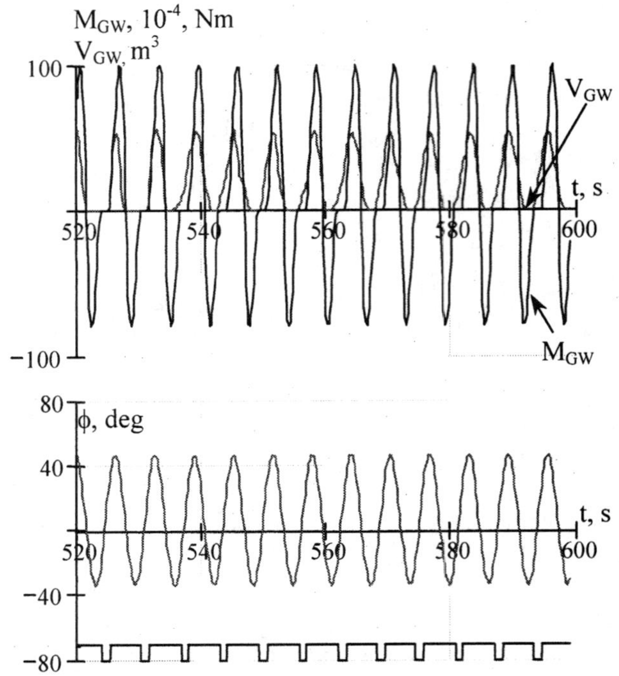

Figure 20. Regular wave results including hydrostatic forces on deck: water-on-deck (v_{gw}, top, red) and roll moment (m_{gw}, top, blue), roll angle (ϕ, bottom, red), and deck edge submergence (bottom, blue)

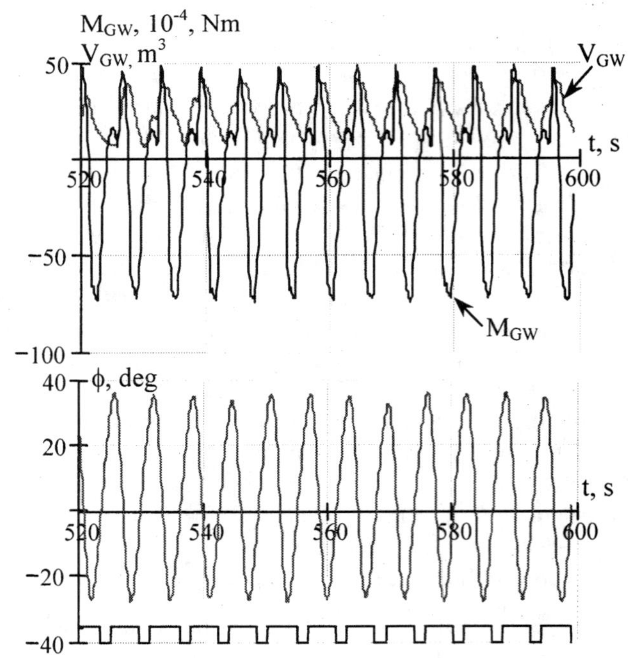

Figure 21. Regular wave results including finite-volume water-on-deck calculation but with no bulwark

Figure 20 shows the results of the regular wave calculation with the deck pressure calculated as the hydrostatic and linear incident wave (Froude-Krylov) pressure only. The response is very regular and primarily periodic at the forcing frequency. An increase in roll amplitude, attributed mostly to the decrease in restoring moment and increase in wave forcing when the deck is submerged, leads to an increasing amount of time "deck-in-water."

The next regular wave calculation used the finite-volume calculation of the shallow water flow over the deck with bare deck edges (no bulwark). The volume of water-on-deck, plotted in red on the top graph, changes by approximately 80 percent over the wave cycle. The roll response is smaller than the hydrostatic and linear incident wave pressure calculation, probably because of the water-on-deck damping seen in the roll decay study, and shows a bias of about 5° toward the "up-wave" direction. A weak subharmonic roll response can be observed.

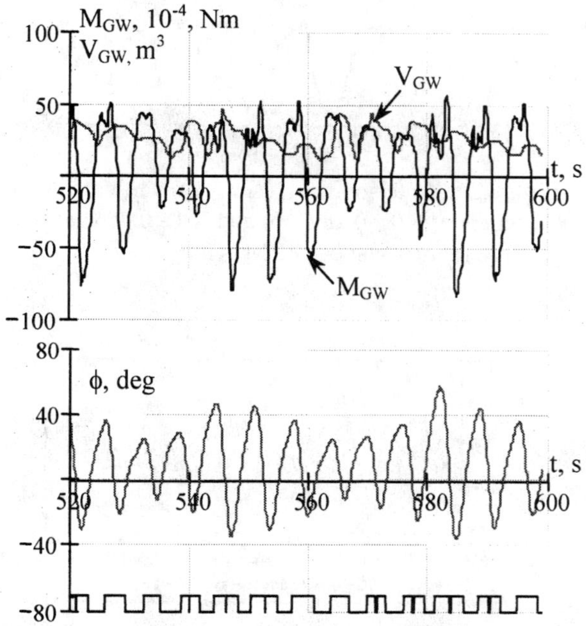

Figure 22. Regular wave results including finite-volume water-on-deck calculation with 0.853 m bulwarks

Figure 22 shows the regular wave results with a 0.843 m bulwark added to the finite-volume calculation. The maximum volume of water-on-deck is slightly higher than in the no-bulwark calculation, but the mean value of water-on-deck has increased substantially. Subharmonic roll response and increased roll amplitudes are observed.

Comparing Figures 21 and 22, it can be concluded that accumulation of water-on-deck has a major impact on ship dynamics: the subharmonic character of the response might be attributed to water-on-deck influence. Garkavy (1991) observed subharmonic rolling of a ship with small freeboard (and as a result, one with water-on-deck) during model tests.

Observations of subharmonic roll could be a result of period-doubling bifurcation, which is known to happen for nonlinear rolling in beam seas; see, for example, Nayfeh and Sanchez (1990). Accumulation of water-on-deck decreases the instantaneous restoring moment; as a result, ship roll could be shifted close to the nonlinear region in the vicinity of the GZ curve's peak, which might facilitate the bifurcation of the periodic solution. However, in order to state that the observed behavior is the manifestation of period-doubling bifurcation, it is necessary to analyze the stability of steady state motion, which is beyond of the scope of this study. At the same time, the period-doubling bifurcation could be seen as a possible physical explanation for the observed results and as an indicator of validity for the numerical model of water-on-deck.

To check influence of the green-water-on-deck on longitudinal motions and global loads another set of calculations have been made for the US Navy CG-47 Class cruiser in two sea conditions:

- Head seas, steep regular waves, $\lambda = 810$ feet and $H = 48$ feet (ship length is 528 feet)
- Head seas, storm sea condition, $H_{1/3} = 37$ feet

The latter case matches an experimental recreation of the waves measured during a hurricane and would be classified as Sea State 8.

For the green-water-on-deck calculations, a CG-47 input geometry was prepared with an additional component representing the deck forward of the superstructure. This input geometry is presented in Figure 23.

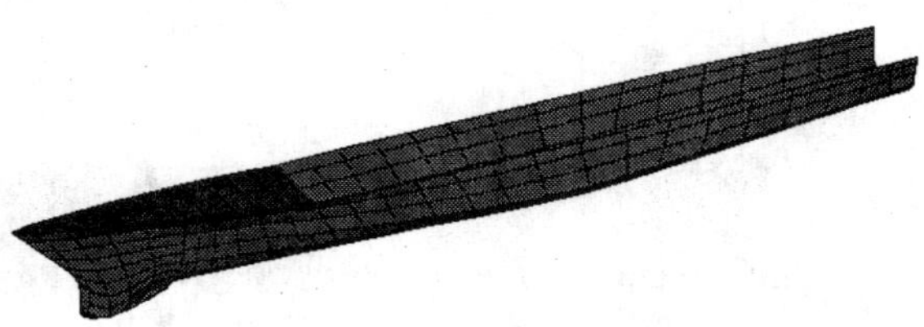

Figure 23. LAMP geometry for CG-47 including forward deck

The first of the two test cases is somewhat unrealistic but was intended to test the robustness of the calculation and exaggerate the green-water-on-deck effects for evaluation purposes. The heave and pitch motion histories are shown in Figure 24. Note that the heave is normalized by ship length, the pitch is in radians, and a positive pitch angle is bow down in this and all subsequent motion plots. The horizontal axis is nondimensional time, in which unit time would correspond to about 4.0 seconds for the full-scale ship. As seen in these plots, water-on-deck does not significantly change the motion response.

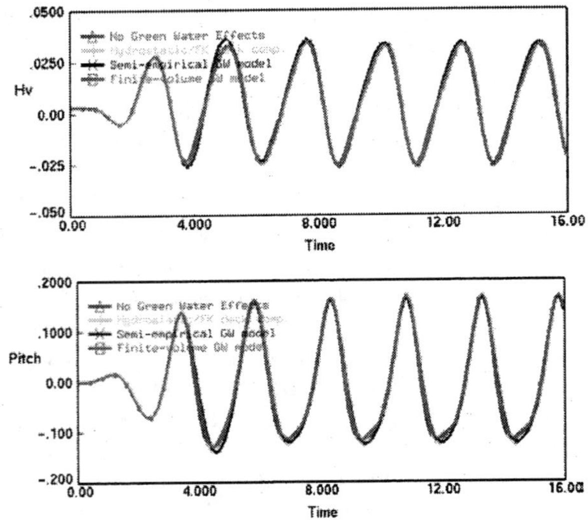

Figure 24. Water-on-deck effects on heave (top) and pitch (bottom) motion for CG-47 in large regular waves

Figure 25 shows a detail of the pitch response for two cycles of the calculation. All four of the calculations agree at the maximum bow down position (+ pitch) but three calculations that include water-on-deck effects all predict a lag in the bow's recovery and some differences in the bow up (– pitch) peak.

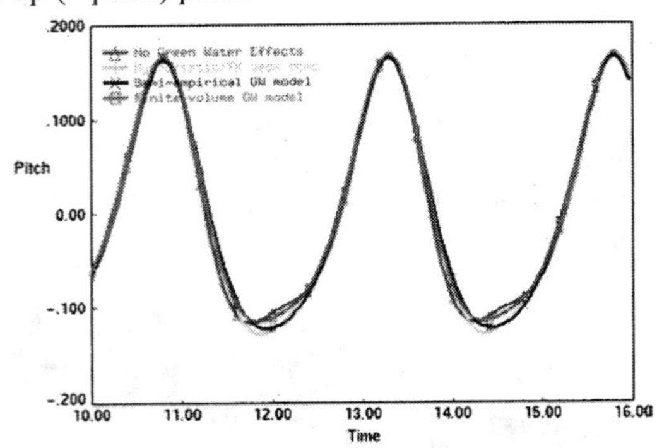

Figure 25. Detail of water-on-deck effects on pitch motion of CG-47 in large regular waves

The green-water-on-deck effect is far more noticeable in the load calculations shown in Figure 26. In all of the calculations, the water-on-deck reduces the peak sagging (+ VBM) moment. The magnitude of this reduction is similar for the three methods but the details of the peak differ. As might be expected, the water-on-deck effect is greatest for the loads near the bow and smallest for those near the stern.

The second of the two test cases duplicates a model test that recreated the sea spectrum derived from wave probe data taken during a major hurricane. This case is thought to represent physically realistic extreme sea conditions that a ship might encounter. As for the regular wave case, the various green-water-on-deck models do not have a significant apparent effect on the motions, shown in Figure 27, although the bow emerges from the water somewhat more slowly when green-water-on-deck effects are present.

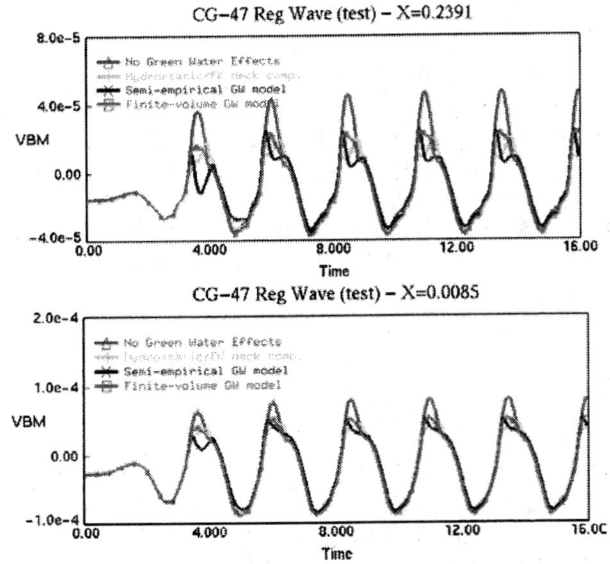

Figure 26. Water-on-deck effects on vertical bending moment at X/L=0.26 (top) and X/L=0.49 (bottom) in large regular waves (VBM is normalized by $\rho g L^4$)

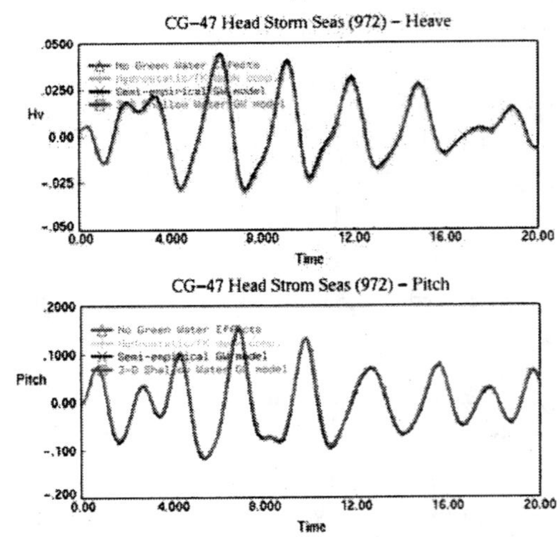

Figure 27. Water-on-deck effects on heave (top) and pitch (bottom) motions for CG-47 in irregular head seas, $H_{1/3}$ = 37 feet.

Figure 28 shows the total green-water-on-deck force and moment for the semi-empirical and 3-D shallow water models. The results for the two models are similar, although the forces from the 3-D model appear to be larger and slightly delayed. The force results also show how the 3-D model allows water on deck to "linger" somewhat after the free surface falls below the deck edge.

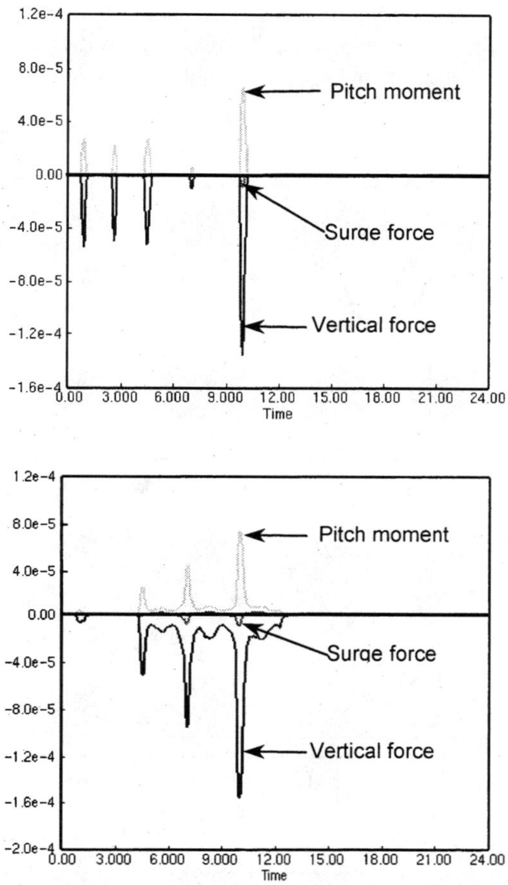

Figure 28. Water-on-deck force (normalized by $\rho g L^3$) and moment (normalized by $\rho g L^4$) for CG-47 in irregular head seas by semi-empirical (upper) and 3-D shallow water (lower) models for water-on-deck.

Figure 29 shows the predicted vertical bending moment near midships for the storm sea case. As with the regular wave case, the green-water-on-deck models all predict a reduction in the peak sagging bending moment.

For the storm sea run, pressure output files were written that include the deck geometry and computed pressure for any green-water-on-deck calculations. LAMP's pressure data post-processor can use the computed deck pressure, along with the computed hull pressure distribution, to calculate nodal pressure and FE load data sets. Figure 30 shows a plot of the deck and hull pressure distributions at T=10.04 of the storm sea run.

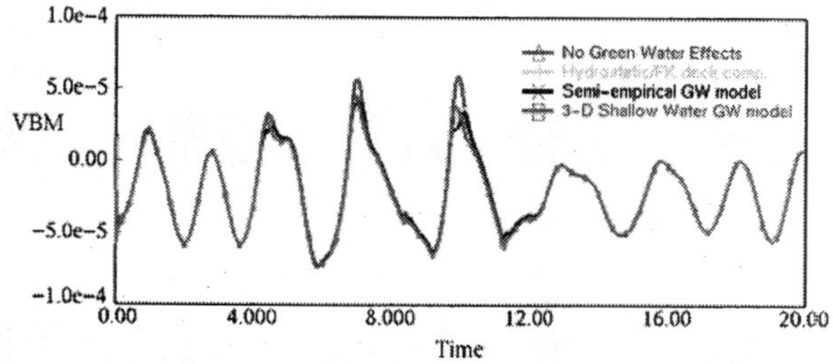

Figure. 29 Green-water-on-deck effects on vertical bending moment at X/L=0.49 for CG-47 in head storm seas

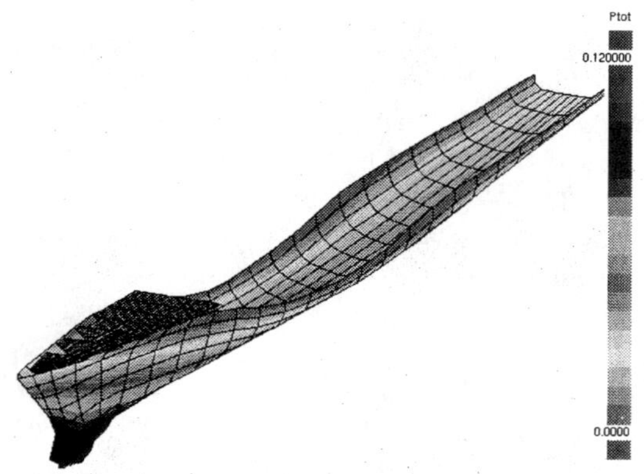

Figure 30. Plot of hull and deck pressure (normalized by $\rho g L$) for CG-47 in storm seas at T=10.05

Multi-Hull Applications

With its 3-D formulation and general geometry model, the LAMP System is well suited for analyzing multi-hull ships such as catamarans and trimarans. Figure 31 shows the geometry of a notional trimaran design for which some LAMP validation has been performed.

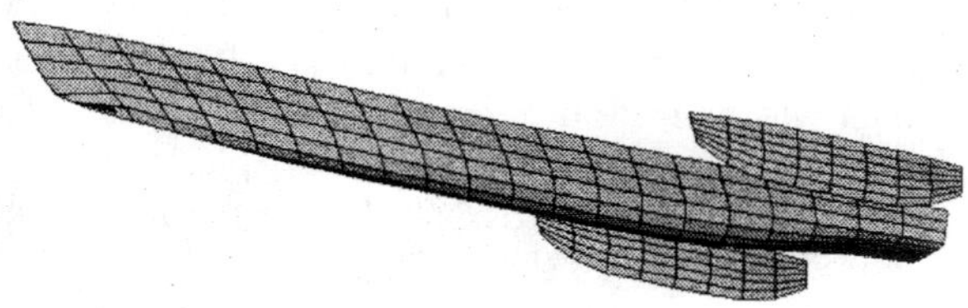

Figure 31. Geometry of sample trimaran

This trimaran was designed by a team of graduate students and tested at the U.S. Naval Academy (Harris, 1999). The tests analyzed several longitudinal and transverse locations of the outer hulls and included seakeeping tests in regular and irregular head seas. Figure 31 compares the predicted and measured heave and pitch motions for the regular wave runs at a Froude number of 0.15. Like the experiments, the LAMP predictions were computed as a series of regular wave runs that used the approximate body-nonlinear (LAMP-2) model and matched the wave slope of the experimental runs.

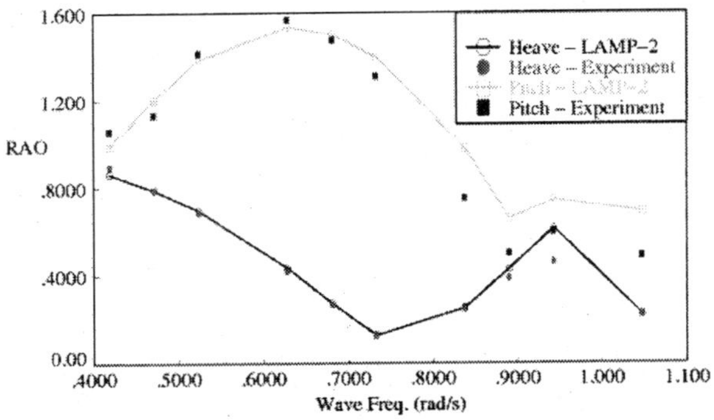

Figure 32. RAO for heave (m/m) and pitch (deg/m) motion in head seas for trimaran at Fr=0.15

Figure 33 offers a similar comparison at a Froude number of 0.30. The comparison for both cases is good. LAMP calculations have been made for this configuration for speeds up to a Froude number of 0.74, but no experimental data is available at that higher speed.

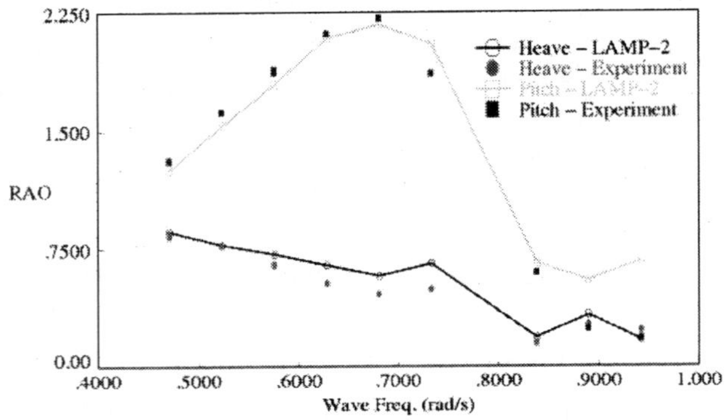

Figure 33. RAO for heave (m/m) and pitch (deg/m) motion in head seas for trimaran at Fr=0.15

The LAMP System has also been used to analyze large offshore structures, including a detailed evaluation of the disturbed wave field beneath a large semi-submersible platform (Weems *et al.*, 1999). The study, which was part of an effort to examine the risk of wave impact on the underside of the main platform structure, used both the linear and 2nd-order free surface boundary conditions. Figure 34 shows the predicted local wave field at one time instant of a calculation for a regular wave propagating down the long axis of the platform.

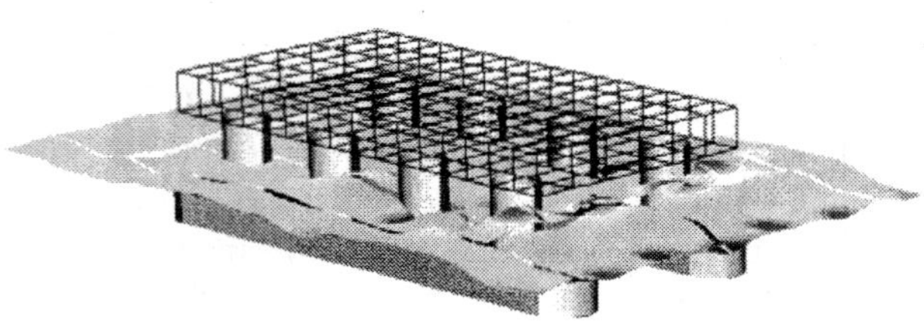

Figure 34. Predicted local wave elevations for semi-submersible platform

LARGE AMPLITUDE MOTIONS IN FREQUENCY DOMAIN

One of the main advantages of the time-domain simulation approach is its more precise description of the forces on a ship, compared with frequency domain calculations. Once the amplitude of response grows large, its nonlinear nature becomes evident. Figure 35 illustrates this through motion and load predictions for a 195 m container ship.

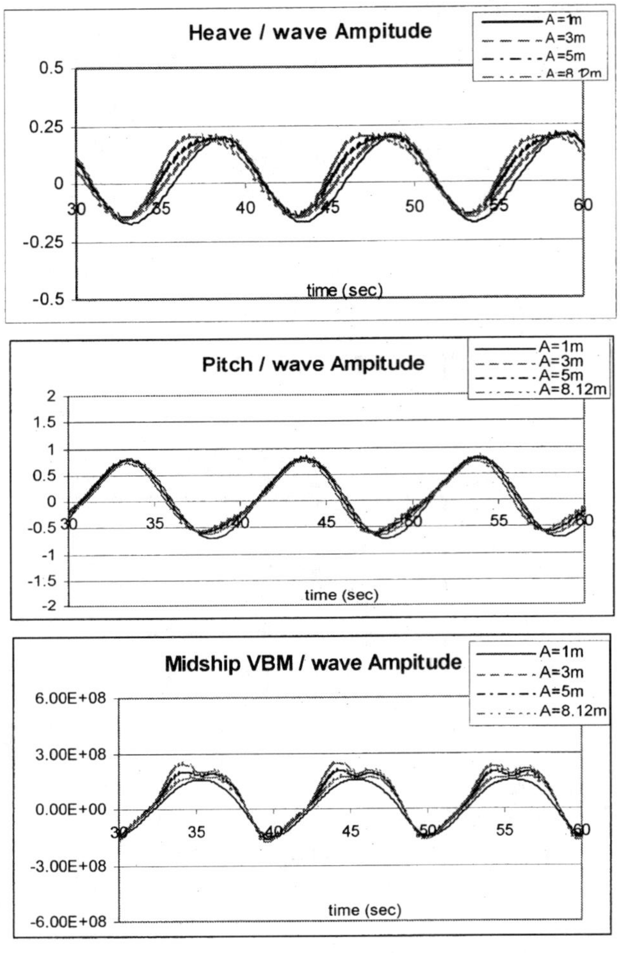

Figure 35. Dimensionless responses of 195 m container ship

The heave, pitch and VBM response is shown in dimensionless form in Figure 35, so it is very clear how the responses lose their sinusoidal shape with increasing wave amplitude. It is noticeable that the VBM reacts more to increasing wave amplitude than do heave and pitch. However, the heave and pitch responses are also quite nonlinear for the 8.12 m wave amplitude.

Increased fidelity of ship motion and load predictions comes with a price. Results are nonlinear and they no longer can be presented in a form of an

RAO, because the amplitude of response (and shape, too) now depends on the amplitude of wave excitation. Moreover, the amplitude of a nonlinear response may depend on initial conditions, so several responses may be possible at the same frequency.

At the same time, amplitude vs. frequency is a conventional way to present a design's ship motion. This makes it important to find the appropriate method of frequency domain presentation for nonlinear simulation results. The natural way to do it is to apply methods of nonlinear dynamics. However, these methods were developed for objects, behavior of which can be described by a system of ordinary differential equations. This is not the case for a ship in waves. Radiated and diffracted waves interfere with incident waves, which causes excitation to depend on motion history beyond the previous time step. As a result, initial conditions (including the excitation phase) are no longer enough to determine the status of a dynamical system. There are additional issues that make the application of nonlinear dynamics non-trivial.

One of the standard ways to present nonlinear responses in a frequency domain is the response curve. The response curve shows dependence of amplitude on frequency. Its main difference from an RAO is that each response curve corresponds to specific excitation amplitude.

Sample calculations to illustrate nonlinear dynamics were performed for the rolling fishing vessel used in the green-water-on-deck simulation (shown in Figure 13). The only difference was that for these calculations, the forecastle was removed to smooth the GZ curve and to simplify the corresponding nonlinear term. The righting arm (GZ) is plotted in Figure 36.

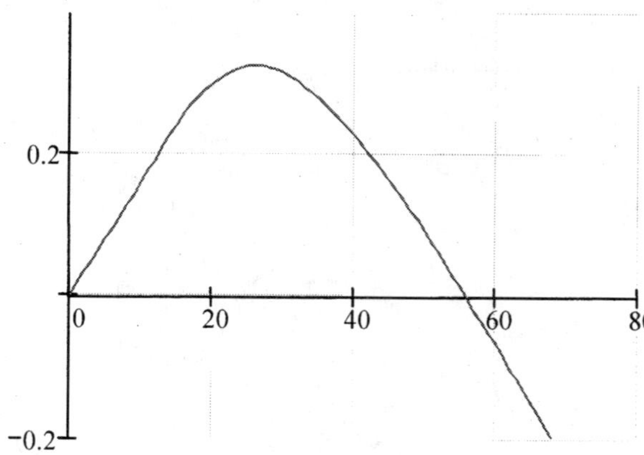

Figure 36. Righting arm (GZ) curve in meters of the fishing vessel for response curve calculation

Response curves were calculated in the following manner. The calculation started from a wave frequency that was far from resonance, so that roll response was almost linear. This calculation was used to determine initial conditions corresponding to steady state roll motions. However, these initial conditions could only be considered as an approximation, since they correspond to certain state of wave field that might not be reproduced in the next run. Nevertheless, the transition process initiated with approximately steady state conditions was much faster than calculations initiated from rest, and the resulting response could be considered as "almost steady state." This can be seen in Figure 37, where process (a) started from approximate steady state initial conditions and process (b) from zero initial conditions.

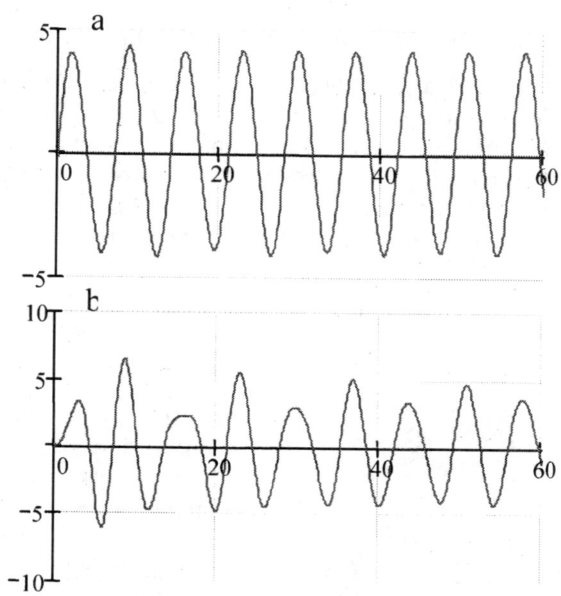

Figure 37. Transition in roll response started with approximate steady state initial conditions (a) and zero initial condition (b). Wave height 0.4 m, circular frequency 0.9 1/s

Once approximate steady state initial conditions were found, then the response curve could be calculated from two series of LAMP runs. Each series contains runs with small changes of wave frequency. Approximate steady state initial conditions were determined using the previous run. As a result, the frequency step had to be small enough for the initial conditions to be close. The first series of predictions was with the wave frequency increasing, the second series with the wave frequency decreasing.

The results are presented in Figure 38. As would be expected, there is an area where two different responses exist for the same wave frequency (points A and B).

LAMP calculations were supplemented with direct numerical integration of the approximate differential equation for rolling using the actual GZ curve and Froude-Krylov wave excitation.

$$\ddot{\phi} + 2\delta\dot{\phi} + \omega_{\phi}^2 f(\phi) = \alpha_W \omega_{\phi}^2 \sin \omega t$$

with ω_ϕ natural roll frequency, α_W amplitude of wave slope, and

$$f(\phi) = \frac{W \cdot GZ(\phi)}{A_{44} + I_x}$$

Here W is weight displacement and I_x is the transverse moment of inertia of ship mass. Damping was assumed to be linear and the damping coefficient δ was determined by means of a special roll decay run using LAMP.

Figure 38 also shows the theoretical response curve developed using equivalent linearization. In the equivalent linearization, the nonlinear differential equation for roll was replaced with a linear one, providing equality of roll natural periods for large amplitude roll using both the nonlinear and linearized equations.

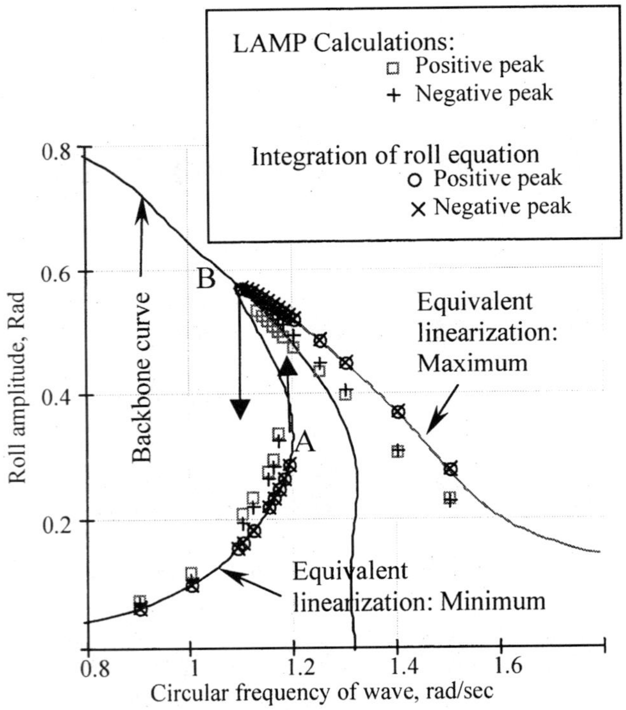

Figure 38. Response curve of roll based on LAMP calculation, numerical integration of nonlinear roll equation and equivalent linearisation.

It is known from nonlinear dynamics (see, for example Thompson and Stewart, 1986), that the structure in Figure 38 with two steady state responses might constitute fold bifurcation, existence of which has been confirmed experimentally (Francescutto *et al.*, 1994).

The conventional way to confirm a bifurcation is by a motion stability analysis. The behavior of the eigenvalues of a Jacobean identifies the type of bifurcation. A technique similar to those employed by Virgin (1987) and Belenky (2000) was used here. Three short LAMP runs were used for the motion stability analysis. The first run was carried out using approximate steady state conditions to produce values of the state variables (roll angle and roll velocity) after exactly one wave period. For the second run, the initial roll value was perturbed (0.5–1 percent perturbation typically gives reasonable results) and the run was repeated. The third run used approximate steady state initial roll angle and perturbed the roll velocity. The eigenvalues of the Jacobean matrix and further analysis indicated no difference from conventional analysis (Virgin, 1987).

Figure 39(a) shows the behavior of the eigenvalues in the circular frequency range 0.9–1.18 rad/sec calculated as described above, part (b) of this figure shows the eigenvalues calculated using the equivalent linearized differential equation for roll over the same frequency range, and part (c) shows the theoretical behavior before and during fold bifurcation. The similarity is quite evident; moreover, the numerical values of the eigenvalues are close, taking into account the approximate method used to determine the steady state conditions with LAMP. However, loss of accuracy was observed for the same frequency range for large amplitude and for the circular frequency range 1.18–1.55 rad/sec.

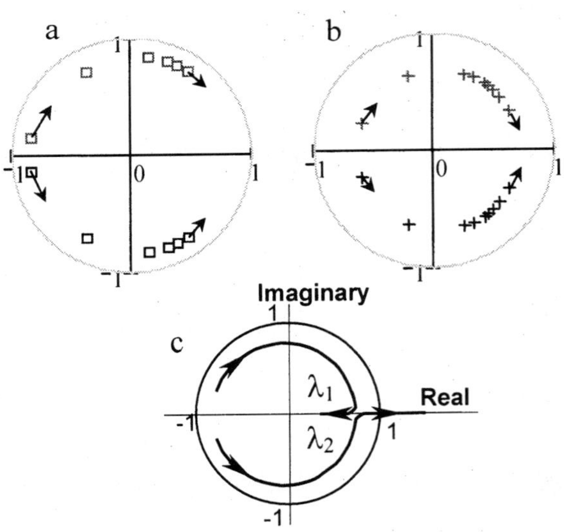

Figure 39. Eigenvalues of Jacobean matrix for (a) LAMP calculation, (b) nonlinear roll equation (c) theoretical prediction; circular frequency range 09-1.18 rad/sec: low response, wave amplitude 0.4 m

The above study was the first attempt to apply nonlinear dynamics methods to the results obtained with LAMP simulations. Despite the difficulties in defining steady state initial conditions, the outcome looks promising. Accuracy problems with steady state initial conditions still must be addressed but, in general, nonlinear dynamics gives a tool to interpret results of time domain simulation and helps to present them in the frequency domain.

The next problem will be to determine how to use the response curves for further design analysis, rather than an RAO. It will also be important to apply the prescribed methodology to other ship responses in order to ascertain the influence that nonlinearities have on motions and loads.

CONCLUDING REMARKS

This paper is a sequel to Shin, *et al.* (1997). As such, we have described the latest developments in nonlinear simulation technology and its applications for new ship designs that took place since 1997. Therefore, this paper is a review rather than a focused study.

We do not consider state-of-the-art nonlinear simulation technology as a "ready" replacement for linear theory (in both 2-D and 3-D formulations); it is a complement to rather than a replacement for linear theory.

However, the progress achieved during the last five years allows us to state that the next steps toward a more sophisticated seakeeping analysis tool, indeed, have been made. In particular:

- New numerical procedures based on impulse response functions (IRF) allow calculation efficiency comparable with linear theory. Limitations of this procedure do not prevent adequate modeling of new hull shapes, since the hydrostatic and Froude-Krylov forces are calculated on the instantaneous wetted surface, radiation forces are computed with the instantaneous ship velocity, and there is no need to linearize the pressure evaluation or the equations of motion.
- The approximate body-nonlinear approach (LAMP-2), using the direct or IRF-based approach, is capable of reproducing known nonlinear phenomena like fold bifurcation. Since it has already been shown that LAMP-2 could predict the correct amplitude of parametric rolling (France, *et al.*, 2003), we can conclude that it provides an adequate description of the nonlinear dynamics of a ship.
- Comparisons between model tests and simulation results have shown that reasonable agreement between both measured and predicted motions and loads in extreme sea conditions is possible. However, a rigorous validation effort is still required.
- The time-domain solution of the equations of motion and the modular structure of the code makes it possible to incorporate numerical or empirical models of green-water-on-deck, roll control, maneuvering systems, tanks, and other effects directly into the ship motion simulation. This provides an opportunity to improve both the completeness and fidelity of the ship motion simulation, and provids a simulation platform for designing and evaluating such systems. As an illustration of this, it was shown how LAMP's anti-roll tank module could be used to evaluate different engineering solutions for suppressing parametric roll. The green-water-on-deck module allows not only observations of changes in ship dynamics due to the influence of water-on-deck, but also the evaluation of additional loads caused by green-water-on-deck.

Despite the progress reported above, many issues remain for making time domain simulation more practical. Issues to be addressed include:

- Current versions of the simulation codes need reliable information on roll damping for accurate prediction of extreme roll motion. Reliable methods for the *a priori* prediction of viscous and vortex components of roll damping will be needed in the future.
- Ship motion simulations in large, oblique seas can be sensitive to rudder and hull side forces, so sophisticated models of the autopilot and of maneuvering forces will be required to accurately model them. Like roll damping, reliable methods for developing control and maneuvering force models will be needed.
- Consideration of nonlinear effects greatly expands the amount and complexity of seakeeping data required to "characterize" a ship, both in terms of analyzing simulations and developing long-term extreme statistics. The methods of nonlinear dynamics could be a great help in interpreting simulation results.

Finally, consideration of nonlinear incident waves seems to be a logical continuation of the present effort: extreme motions and loads have to be modeled in large waves that may not be well predicted by linear wave theory. The construction of nonlinear regular and irregular wave fields, and the modeling of these fields in LAMP, is part of an ongoing development effort.

ACKNOWLEDGEMENT

The authors wish to express their appreciation and gratitude to the management of the American Bureau of Shipping, Science Application International Corporation, and Carderock Division, Naval Surface Warfare Center. The authors also wish to thank W. F. Belknap (NSWC/CD), Y. Kim (currently MIT), D. Liut (SAIC), T. Treakle (SAIC), G. Wang (ABS), and S. Zhang (SAIC) for their help in developing the LAMP System and numerical results presented in this paper.

The development of the LAMP System has been supported by the U.S. Navy, the Defense Advanced Research Projects Agency (DARPA), the U.S. Coast Guard, ABS, and SAIC. The green-water-on-deck development has been supported by the Office of Naval Research (ONR) under Dr. Patrick Purtell, by the Ship Structure Committee under Mr. Peter Minnick, and by ABS.

REFERENCES

Beck, R.F. and Reed, A.M. (2001). "Modern Computational Methods for Ships in Seaway", *Trans. SNAME,* **109**:1–48.

Belenky, V. L. (2000). "Piecewise linear approach to nonlinear ship dynamics", in *Contemporary Ideas of Ship Stability*, D. Vassalos, M. Hamamoto, A. Papanikolaou and D. Moulyneux (Editors), Elsevier Science, pp. 149–60.

Belenky, V.L., Liut, D., Weems, K. and Shin Y.S., (2002). "Nonlinear Ship Roll Simulation With Water-On-Deck," *Proc. 6th International Ship Stability Workshop*, Webb Institute, New York.

Bingham, B., Korsmeyer, F., Newman, J., and Osborne, G. (1993), "The Simulation of Ship Motions," *Proc. 6th International Conference on Numerical Ship Hydrodynamics*, Iowa City, Iowa, U.S.A., pp. 27–47.

Buchner, B. (1995). "On the Impact of Green Water Loading on Ship and Offshore Unit Design," *Proc.6th Symposium on Practical Design of Ships and Mobile Units*, pp.1.430–1.443.

Finn, P. Beck, R.F., Troesch, A.W. and Shin, Y.S. (2001). "Nonlinear Impact Loading in an Oblique Seaway", *Proc.* of *20th International Conference on Offshore Mechanics and Arctic Engineering*, Rio de Janeiro, 11 p.

France, W.G., Levandou, M. Treakle, T.W., Paulling, J.R., Michel, R.K. and Moore, C. (2003). "An investigation of head seas parametric rolling and its influence on Container Lashing Systems", *Marine Technology,* **40**(1):1–19.

Francescutto, A., Contento, G. and Penna, R. (1994). "Experimental Evidence of Strong Nonlinear Effects in the Rolling Motion of the Destroyer in Beam Seas," *Proc. of STAB'94: 5th International Conference on Stability of Ships and Ocean Vehicles*, vol. 1, Melbourne, Florida.

Garkavy, V.V. (1991). "Deterministic chaos in the task of the roll motion ship with a small freeboard," *Proc. of HADMAR'91: International Symposium On Hydro- And Aerodynamics In Marine Engineering*, Paper 49, Vol. 2, Bulgarian Ship Hydrodynamics Centre, Varna.

Harris, N. (1999). "Effective horsepower and seakeeping tests on a trimaran model," Report EW-01-99, United States Naval Academy.

Kim, Y. (2002). "A Numerical Study on Sloshing Flows Coupled with the Ship Motions," *J. Ship Research*, **46**(1):1–11.

King, B., Beck, R., and Magee, A. (1988). "Seakeeping Calculations with Forward Speed Using Time Domain Analysis," *Proc. 17th Symposium on Naval Hydrodynamics*, The Hague, The Netherlands.

Liapis, S. (1986). "Time-Domain Analysis of Ship Motions," Report 302, Department of Naval Architecture and Marine Engineering, University of Michigan, U.S.A.

Lin, W.M., and Yue, D.K.P. (1990). "Numerical Solutions for Large-Amplitude Ship Motions in the Time-Domain," *Proc. 18thSymposium on Naval Hydrodynamics*, The University of Michigan, U.S.A., pp. 41–66.

Lin, W.M., and Yue, D.K.P. (1993). "Time-Domain Analysis for Floating Bodies in Mild-Slope Waves of Large Amplitude," *Proc. 8th International Workshop on Water Waves and Floating Bodies*, Newfoundland, Canada.

Lin, W.M., Meinhold, M., Salvesen, N., and Yue, D.K.P. (1994). "Large-Amplitude Ship Motions and Wave Loads for Ship Design," *Proc. 20th Symposium on Naval Hydrodynamics*, The University of California, U.S.A

Lin, W.M., Zhang, S., Weems, K., and Yue, D.K.P. (1999). "A Mixed Source Formulation for Nonlinear Ship-Motion and Wave-Load Simulations," *Proc. 7th International Conference on Numerical Ship Hydrodynamics*, Nantes, France, pp. 1.3.1–12.

Liut, D., Weems, K., and Lin, W.M. (2002). "Nonlinear Green Water Effects On Ship Motions and Structural Loads," *Proc. of the 24th Symposium on Naval Hydrodynamics*, Fukuoka, Japan.

Nakos, D., Kring, D., Sclavounos, P. (1993), "Rankine Panel Methods for Transient Free-Surface Flows", *Proc. 16th Symposium on Naval Hydrodynamics*, Iowa, U.S.A

Nayfeh, A.H., and Sanchez, N.E. (1990). "Stability and complicated rolling response of ships in regular beam seas," *International Shipbuilding Progress*, **37**(412):331–52

Paulling J.R. and Rosenberg R.M. (1959). "On Unstable Ship Motions Resulting from Nonlinear Coupling", *J. Ship Research*, **3**(1):36–46.

Shin, Y.S., Chung, J.S., Lin, W.M., Zhang, S. and Engle, A. (1997). "Dynamic Loadings for Structural Analysis of Fine Form Ships Based on Non-linear Large Amplitude Motions and Loads Method," *Trans. SNAME*, **105**:127–54.

Shin, Y.S. (2001). "Official Discussion to Report of Special Tack Committee on Extreme Hull Girder," *Proc. of 14th International Ship and Offshore Structure Congress*, Elsevier, **3**:297–303.

Stoker, J.J. (1957). *Water Waves* The Mathematical Theory with Applications, Interscience Publishers, Inc., New York, U.S.A., xxviii+567 p.

Swamee, P.K. and Jain, A.K. (1976). "Explicit equations for pipe-flow problems," *Journal of Hydraulic Division Proceedings*, ASCE, pp. 657–64.

Thompson, J.M.T. and Stewart, H.B. (1986). *Nonlinear dynamics and chaos*, John Wiley, Chichester.

Virgin, L.N. (1987). "The nonlinear rolling response of a vessel including chaotic motions leading to capsize in regular seas", *Applied Ocean Research*, **9**(2):89–95.

Weems, K., Zhang, S., Lin, W.M., Shin, Y.S, and Bennett, J. (1998). "Structural Dynamic Loadings Due to Impact and Whipping," *Proc. 7th International Symposium on Practical Design of Ships and Mobile Units*, The Hague, The Netherlands, 7 p.

Weems, K., Lin, W.M, Zhang, S., and Treakle, T. (1999). "Application of the Large Amplitude Motion Program (LAMP) for Design and Operation of A Mobile Offshore Base (MOB)," *Proc. of 3rd International Workshop on Very Large Floating Structures (VLFS '99)*, Honolulu, Hawaii.

Weems, K., Lin, W.M Zhang, S, and Treakle, T. (2000). "Time Domain Prediction for Motions and Loads of Ships and Marine Structures in Large Seas Using a Mixed-Singularity Formulation", *Proc. OC2000*, Osaka, Japan.

Youssef, K.S., Mook, D.T., Nayfeh, A.H., and Ragab, S. A. (2002). "Roll Stabilization by Passive Anti-Roll Tanks Using an Improved Model of the Tank-Liquid Motion," submitted for publication, *J. Vibration and Control*.

Zhou, Z.Q., De Kat, J.O., and Buchner, B. (1999), "A Nonlinear 3-D Approach to Simulate Green Water Dynamics on Deck," *Proc. 7th International Conference on Numerical Ship Hydrodynamics*, Nantes, France.

DISCUSSION

Kevin McTaggart, Member

I had the pleasure of seeing the related 1997 paper on LAMP presented in Ottawa. Although I must pass on my regrets for the current meeting, I feel privileged to be a discussor.

The authors note that this paper is a "sequel" to the 1997 paper. I saw a television program in which a learned film critic explained the difference between a sequel and a serial. A sequel tends to be more of the same, with no significant character development from the original. In contrast, a serial contains significant character development in progressive installments. The Harry Potter books are a good example of a serial. I'm of the opinion that the current paper is an installment of a serial rather than a sequel.

The body nonlinear solution of LAMP-4 is certainly among the most complex approaches available for prediction of seakeeping and sea loads. The approximate body nonlinear solution of LAMP-2 is more widely used. Although potentially more accurate, the body nonlinear solution of LAMP-4 has high costs in terms of code development, code maintenance, and computational time. Based on comparisons with model tests and sea trials, are there cases for which LAMP-4 gives significantly better accuracy than LAMP-2? Furthermore, could you give representative ratios of CPU time to real time using LAMP-2 and LAMP-4.

At DRDC Atlantic we continue to devote significant effort to comparison of ship motion and sea load predictions with experimental data from model tests and sea trials. We have observed significant differences among roll predictions from various codes due to varying treatments of viscous and lift forces. Could you comment on the treatment of these forces within LAMP?

We all look forward to the next installment of the serial. In your conclusions you note the requirement for rigorous validation, which I think would be of particular interest to readers.

Jeffrey M. Falzarano, Member

The authors are to be congratulated on a comprehensive and complete paper. I am mainly interested in the section "Large amplitude motions in the frequency domain." The frequency domain is a well-known and useful tool to analyze small-amplitude linear ship motions motion, but its use for large-amplitude nonlinear motions is generally not as widely understood.

The authors' Fig. 38 is particularly interesting in showing the close comparison between the simple equivalent linearization results and the time domain simulations. From this figure one can see the existence of multiple solutions over a narrow frequency range, which is common in nonlinear systems. It is satisfying to see this nonlinear behavior verified using a complicated time domain simulation computer program that includes memory functions. Previously, Professor Francescutto in Italy was able to verify this nonlinear type of behavior experimentally, as referenced in the paper. However, I would like to ask the authors if they found that the solution to the approximate differential equation always gave such good comparisons and if so why?

An alternative approach to studying this problem is described in a paper that we at UNO prepared in 1995 (Falzarano et al 1995). In this paper we look at how magnification curves change with increasing amplitude and how initial conditions and basins of attraction determined which of the multiple steady state solution solutions were eventually achieved.

Over the last several years much progress has been made in applications of nonlinear dynamic systems theory to analyze large-amplitude ship rolling motions. Unfortunately, to date much of this analysis has been done using simplified and approximate analysis using single degree of freedom (uncoupled) and periodic excitation. The future lies with more sophisticated dynamical systems analysis considering multiple degrees of freedom (Falzarano & Zhang 1993) and random excitation (Vishnubhotla et al 1998), and more accurate hydrodynamic models (Falzarano et al 2002). Finally, I would like to thank the authors for the opportunity to comment on their paper.

Additional references

Falzarano, J., Cheng, J., and Rodrigues, W. 2002 Transit draft heave and pitch motion analysis of the mobile offshore base (MOB) using reverse MI/SO techniques, *Proceedings,* 21st Offshore Mechanics and Arctic Engineering Conference, June, Oslo, Norway.

Falzarano, J. M., Esparza, I., and Taz Ul Mulk, M. 1995 A combined steady-state and transient approach to study large amplitude ship rolling motion and capsizing, *Journal of Ship Research,* **39,** 3.

Falzarano, J., and Zhang, F. 1993 Multiple degree of freedom global analysis of transient ship rolling motion, *Proceedings*, ASME Winter Annual Meeting, Symposium on Nonlinear Dynamics of Marine Vehicles, November.

Vishnubhotla, S., Falzarano, J., and Vakakis, A. 1998 Nonlinear ship rolling motion in a random seaway, *Proceedings*, Conference on Computational Stochastic Mechanics, Santorini, Greece.

Bruce L. Hutchison, Life Fellow

The authors are to be congratulated both on this paper and on its underlying accomplishments. In this, the 50-year anniversary of the legendary St. Denis-Pierson paper, the authors present us with very practical advances beyond the frontier of the linear seakeeping world. It would seem, at last, that practical nonlinear seakeeping studies may be carried out using time domain simulation as well as by physical model testing.

This rich paper provides important extensions to those previously reported to the Society in 1997. There are too many advances to address within the limitations of a formal discussion, so I will limit my comments to the impulse response functions, the modeling of antiroll tanks, and nonlinear rolling.

The origins of this paper's impulse response functions (IRFs) may be traced to papers by Tick (1959) and Cummins (1962), but the implementation described by the authors has original elements that extend the usefulness of the method. Earlier papers applied the convolution integral to the simulation of motions and thus retained a restriction to the linear domain similar to that of conventional strip theory. In this paper the convolution integrals are applied to the components of the perturbation velocity potential. The perturbation velocity potentials for the radiation processes are linear in the (convoluted) ship velocity. Furthermore, the methods described permit retention of the quadratic terms in Bernoulli's equation. The instantaneous body wetted surface may be estimated consistent with the assumptions that the authors identify as approximate body nonlinear, and the Froude-Krylov forces can be computed on that instantaneous wetted surface. Because the IRFs are available for each body panel, there is an immediate connection to loads suitable for dynamic loads analysis.

A valuable observation is made in the references cited by the authors for the IRF method. That observation is that the diffraction potential is derived as the solution to a canonical diffraction problem forced by the incident wave potential, given by the authors in the present paper, where the incident wave potential contains all frequencies with equal weight and describes a Dirac function at time zero.

The reference for the U-tube antiroll tank model implemented by the authors has only recently been published (Youssef et al 2003). As noted in the present paper, the U-tube antiroll tank was implemented as a complete 6-DOF system suitable for large angle rotational motions. It is the experience of this commenter that the 3-DOF local transverse plane motions (sway, heave, and roll) are particularly important to the modeling of U-tube antiroll tank performance. Within the context of ship coordinates, local sway and heave depend also on pitch and yaw, resulting in 5 deg of freedom being important. Youssef et al provide a comparison between a 1-DOF (roll only) model and a full 6-DOF tank model, where the resulting roll motions differ by a factor of two. Using the 6-DOF model, they also demonstrate that longitudinal location of U-tube antiroll tanks within the ship can influence performance, a result that cannot be revealed by a 1-DOF model.

As regards the application of LAMP time domain simulations to nonlinear rolling, and associated frequency domain and phase plane representations of those nonlinear roll responses, the authors present a righting arm curve as their Fig. 36. A comparison is shown in the authors' Fig. 38 between LAMP simulations and direct integrations of an approximate 1 deg of freedom nonlinear differential equation for roll. As a point of clarification, it is this discusser's understanding that the righting arm shown in Fig. 38 was used in the approximate 1-DOF roll equation but that all of the restoring forces in LAMP are implicit consequences of instantaneous pressure integrations over the body wetted surface inclusive of hydrostatic pressure terms. Furthermore, it is understood that the LAMP simulations were for 6-DOF motions with the possibility of water over the deck edge (but presumably without any bulwarks). If this understanding is correct, then the agreement shown in Fig. 38 between the LAMP simulations and the direct integration of the approximate 1 deg of freedom nonlinear differential equation for roll is quite remarkable and the small differences shown between the two solutions are entirely understandable.

Additional references

Cummins, W. E. 1962 The impulse response function and ship motions, Symposium on Ship Theory, Institute for Schiffbau, Hamburg, January; *Schiffstechnik*, **9**, 47.

Tick, L. J. 1959 Differential equations with frequency-dependent coefficients, *Journal of Ship Research*, **3**, 2, 45–46.

Youssef, K. S., Mook, D. T., Nayfeh, A. H., and Ragab, S. A. 2003 Roll stabilization by passive anti-roll tanks using an improved model of the tank-liquid motion, *Journal of Vibration and Control*, **9**, 7.

Miroslaw Gerigk, Visitor

This paper addresses the latest developments in nonlinear simulation technology and its applications for the new ship designs. It has been proven in the paper that it is possible to develop more sophisticated seakeeping analysis tools.

The general comments may concern as follows:

1. The new numerical nonlinear procedures for the seakeeping calculations based on the impulse response function (IRF) enable the quantitative, efficient, and risk-based analysis of hydrodynamic characteristics of a ship with accuracy higher than in the case of linear procedures.
2. The LAMP-2 system using the approximate body-nonlinear approach is able to analyze the nonlinear phenomena, including the large motion phenomena and associated loads for all the modern designs including the multihulls. The time domain simulation enables one to analyze the green water on deck, roll control, tanks, and other effects. It has been shown in the paper that the LAMP-2 system enables one to identify the fold bifurcation, which is a distinguished achievement.
3. The LAMP-2 system guarantees a very high accuracy of results using both well-known and novel methods and techniques. The accuracy has been confirmed by validation using the experimental data.
4. A great achievement shown in the paper is that by applying the nonlinear dynamics methods to the results obtained with the LAMP-2 simulations, it was possible to prove that the results of time domain simulation may be presented in the frequency domain.

A few more comments regarding the problems presented in the paper are as follows:

Parametric roll and its stabilization: A big advantage of the LAMP system is to have the possibility of considering the parametric roll and its stabilization using a U-tube–type passive antiroll tank system. It has been shown that the passive antirolling tanks may substantially reduce the resonant roll motions. The paper presented how the parametric rolling may be limited using the single passive antiroll tank with the different values of the tank mass. It could be very interesting to see how the relative longitudinal position of two passive tanks may affect the coupled nonlinear forces acting on a floating body in full 6 deg of freedom and how it may finally affect the roll motions and parametric rolling.

Influence of green water on deck on motions and loads: It certainly is a great achievement to develop the proposed green-water-on-deck model, which allows consideration of the water-on-deck effects during the time domain calculations of the nonlinear ship motions and loads. Having the different "levels" of calculations of the green water on deck, the LAMP modular system enables one to consider the water-on-deck effects for different designs. including the partial height walls, "infinite" walls, and stepped or raised sections. This is a big advantage to have the possibility of considering the water-on-deck problem for the different types of ships, including navy ships and fishing boats as examples. The LAMP system has become more powerful in this respect after applying the sophisticated method based on the solution of the shallow water flow equations using the finite volume strategy. This is a big step forward from the hydrodynamic development point of view.

From the scientific, research, and design point of view, the value of the LAMP system is very high as it enables one to consider the water-on-deck effects using the hydrostatic and Froude-Krylov pressure, semiempirical model, shallow-water-flow calculation, fully three-dimensional flow calculation, and shallow water flow using a novel finite volume strategy model, which can be used independently. It enables one to validate the results obtained applying the different codes.

Using the different "levels" of calculations, the LAMP system enables one to take into account the stages from "water on deck" to "deck in water," including the precise calculation of the volume of water on deck and heeling moment. Using the system, the dumping process may be controlled at the design stage and the roll amplitudes may be predicted for different designs using the bulwarks or with no bulwarks. The subharmonic rolling, resonance conditions, and how much the restoring moment is decreased due to water on deck may be identified using the LAMP system as well. A big step forward from the design analysis point of view is to implement the LAMP system to consider the influence of the green water on deck on the longitudinal motions and global loads. It is interesting that there is no influence of the water on deck on the heave and pitch values for the ship considered, although there is such an influence regarding the loads. It may be that such a phenomena is typical for the naval ship under consideration. Possibly the new investigations are necessary.

Multihull applications: The three-dimensional formulation and general geometry model enables the LAMP system to be applied easily for analyzing the multihull ships, including high-speed craft and offshore structures. Further investigations regarding the mentioned designs are needed.

Large-amplitude motions in frequency domain: The time domain simulation approach applied within the LAMP system enables one to find how the nonlinear nature becomes evident for large amplitudes of response. It has been presented in the paper that for the nonlinear responses the results cannot be presented in the form of a response amplitude operator. It follows from the fact that the amplitude of response depends on the amplitude of wave excitation and may depend on initial conditions. This is a good example of progress in modern ship hydrodynamics. A big achievement presented in the paper is the proposal of how the nonlinear responses can be presented in a frequency domain. This is distinguished work to show how, after finding the steady-state initial conditions, two different responses for the same frequency have been obtained. Thus, two steady-state responses might constitute the fold bifurcation confirmed experimentally. The work presents new information regarding how to consider ship safety in the case of large-amplitude motions in frequency domain at the design stage. By applying the nonlinear dynamics methods to the results obtained with the LAMP simulations, it was possible to prove that the results of time domain simulation may be presented in the frequency domain. This is a great idea, and naval architects look forward to seeing it developed.

Stefan Grochowalski, Member

The authors are to be commended for continuing the development of the LAMP program. The time-domain simulation approach seems to be, at present, the only way to represent realistically ship behavior and wave loads in severe wave conditions. It provides tools for analysis and prediction of ship dynamics in the range of amplitudes of motions where the strong nonlinearities preclude the use of traditional frequency domain methods.

In order to be a reliable tool, the time-domain numerical model has to represent the hydrodynamic forces for the full range of motion amplitudes, including ship positions at which the deck gets immersed in water.

The deck–green-water interaction may dramatically change ship behavior and load distribution in extreme waves. It consists of two different physical phenomena:

_ Flow of water trapped on deck, called in the paper "water-on-deck"

_ Additional nonlinear hydrodynamic forces on the deeply immersed part of the deck, induced if there is a significant transverse relative motion of the immersed deck with respect to the surrounding water.

The latter was presented by Grochowalski (1989, 1993) and is called "deck-in-water effects." It has to be emphasized that *water on deck* and *deck in water* symbolize two different physical phenomena and require two different theoretical modeling (Grochowalski et al 1998).

The paper presents a significant improvement of the LAMP program by incorporating calculation of green-water flow on deck. Of the free approaches to the water-on-deck problem presented, the shallow-water-flow model seems to be the best one.

The flow in restricted space in the form of waves induced by deck motion can be well modeled, and the resulting loads and exciting forces subsequently calculated (Huang & Hsiung 1996, Grochowalski et al 1998).

The essential element in the water-on-deck problem is calculation of the instantaneous volume of water in the deck well. The paper indicates that in the calculation of water flow on deck, the water inflow and outflow is computed in order to find the real volume of sloshing water. However, the methods used are not given. Could the authors provide some details on how the mass of water shipping on deck and escaping off the deck is calculated?

The examples presented in the paper show the advantage of the modular structure of the simulation program. This is a very welcome and very important feature. It provides possibility of further development of the program by incorporation of various numerical or empirical models of individual physical phenomena, which otherwise could not be considered.

In order to achieve completeness in the modeling of deck–water interaction, the missing element, the *deck-in-water forces* (Grochowalski 1989, 1993) could be incorporated as a separate module. A preliminary mathematical model of the forces has been developed and was used in the simulations presented in Grochowalski et al (1998). This model has yet to be validated experimentally, but the qualitative results are good.

The other element that is missing in the development of LAMP, so far, is validation of numerical simulations in severe, oblique waves when the deck is frequently, deeply immersed, against appropriate model test results, including capsize events. Such experimental data for ship behavior in extreme, oblique waves exist and are well documented (Grochowalski 1989). It could provide an invaluable basis for the ultimate validation of the LAMP code.

The authors indicate that they are aware of the problem of simulating the maneuvering and course control in large, oblique seas. From my own experience, I can say that this is one of the most challenging tasks, and the relevant simulation of course keeping defines the accuracy of the final prediction of ship behavior in extreme waves.

Once again, I wish to congratulate the authors for the excellent paper, and I am pleased to learn that the development of the LAMP simulation program will be continued.

Additional references

Grochowalski, S. 1989 Investigation into the physics of ship capsizing by combined captive and free running model tests, *Transactons of the Society of Naval Architects and Marine Engineers,* **97**.

Grochowalski, S. 1993 Effect of bulwark and deck edge submergence in dynamics of ship capsizing, *Proceedings,* US Coast Guard Vessel Stability Symposium, New London, CT.

Huang, Z. J., and Hsiung, C. C. 1996 Nonlinear shallow-water flow on deck, *Journal of Ship Research,* **40,** 4.

Grochowalski, S., Hsiung, C. C., Huang, Z. J., and Cong, L. Z. 1998 Theoretical modeling of ship motions and capsizing in large and steep waves, *Transactions of the Society of Naval Architects and Marine Engineers,* **106**.

Authors' Closure

The authors appreciate the opportunity provided by SNAME to present the paper on this important topic. We also would like to thank each of the discussers for their interest in the paper, questions, and comments.

As brought up by *Dr. McTaggart*, the general body-nonlinear (LAMP-4) approach is potentially more accurate than the approximate body-nonlinear (LAMP-2) approach but at a not-inconsiderable cost: the LAMP-4 calculation will typically take 5 to 10 times more computational time than the corresponding LAMP-2 run using the conventional, direct calculation technique. In the validation studies done to date, significant differences between these two motions and load calculations have primarily been seen only in very large, very steep regular wave cases, which is why we feel comfortable in using LAMP-2 for general engineering calculations. On the other hand, LAMP-4 does provide a more consistent hull pressure distribution for detailed pressure postprocessing, such as generating load data sets for finite-element structural analysis.

In terms of computational time versus real time, the comparison will depend greatly on the size and type of problem being computed, but a typical LAMP-2 ship calculation may run at a computational to real-time ratio of between 1:1 and 4:1 on a modern workstation or high-end PC, whereas LAMP-4 will typically run a ratio of 10:1 to 40:1. However, a major advantage of LAMP-2 is that it can use the IRF-based approach, which precalculates portions of the hydrodynamic problem. With the IRF-based technique, LAMP-2 response calculations can run significantly faster than real time, with ratios from 1:5 up to 1:100.

Another question raised by Dr. McTaggart concerned viscous and lifting forces. In the present calculation, the potential flow computation of the wave-body hydrodynamic forces is complemented by a series of models for the forces due to viscous roll damping, bilge keels, rudders, internal tanks, and so forth, some of which include active control systems, such as the course-keeping autopilot. Most of these models are based on empirical formulas, simple analytic expressions, or even externally specified derived coefficients, but we are also working on more sophisticated models, including a fully integrated vortex lattice calculation for unsteady appendage forces.

Dr. Grochowalski raises the question of how the inflow and outflow are computed in order to determine the mass of water on deck. At each time step, the amount of water shipped out (or escaping) through each deck edge element is calculated from the current deck edge water height and velocity of the internal (water-on-deck) domain and from the relative motion of the deck to the incident wave surface in the external (wave-body) domain. The details of this calculation can be found in Liut et al (2002) and Belenky et al (2002).

Dr. Gerigk comments on the lack of influence that the green water on deck has on the heave and pitch of the CG-47 cruiser in head seas. In these test cases, the mass of green water is relatively small in comparison with the ship's mass and is on the deck for a relatively short time. During that time, however, the green water on the foredeck has significant influence on loads, especially on local loads on foredeck structure. The result is not dissimilar to the comparison calculations for this ship that was presented in Weems et al (1996): the large bow flare has small effect on the motions in extreme waves and a very large effect on the sectional loads.

Mr. Hutchison is correct in his comment on the frequency domain roll analysis that the restoring forces and moments are calculated by integrating hydrostatic pressure over the instantaneous submerged hull surface.

Mr. Hutchison and *Professor Falzarano* discuss a number of issues related to the nonlinear dynamics part of the paper. The focus of this part was to check how numerical simulations (LAMP, in this case) are able to reproduce nonlinear phenomena, known both from nonlinear dynamics and model tests by Francescutto et al (1994), as referenced in the paper. To make such comparison comprehensible and vivid, only roll calculations were performed for this analysis.

Nonlinear dynamics in general is a very powerful tool to interpret the results of numerical simulation. What we seek first is a qualitative similarity between approximate ordinary differential equation and simulation model that includes a more accurate description of hydrodynamics. In other words, it is not necessary to get good quantitative comparison between an approximate model and the simulation (numerical simulation would be close to reality anyway); it is much more important to get good qualitative comparison, which would enable us to distinguish between physical phenomena and numerical artifact.

As noted in the very beginning, this paper summarizes developments, experience, and results gathered since the previous LAMP presentation at an annual SNAME meeting in 1997. As a result, the paper was composed from many different researches, which sometimes aimed at different objectives and applications. These sources, however, are united in their approach to the problems they pursue: time-domain numerical simulation.

Professor Yeung in an oral comment addressed his earlier work in the application of Rankin source approach in time domain as described in the additional references below.

The authors are especially grateful to the principal reviewer of this paper, *Dr. Arthur Reed*. He spotted and corrected many hard-to-find inconsistencies, small technical errors, and typos, and his comments on the paper's contents and presentation made the paper better and significantly improved its readability.

Additional references

Yeung, R. W. 1973 *A Singularity-Distribution Method for Free-Surface Flow Problems with an Oscillating Body*, Report NA73-6, College of Engineering, University of California, Berkeley, CA.

Bai, K. J., and Yeung, R. W. 1974 Numerical solutions to free-surface flow problems, with K. J. Bai, *Proceedings,* 10th Symposium on Naval Hydrodynamics, Cambridge, MA, 609–646.

Yeung, R. W. 1985 A comparative evaluation of numerical methods in free-surface flows, invited review, *Proceedings*, IUTAM Symposium on Hydrodynamics of Utilization of Ocean-Wave Energy, July, Lisbon, Portugal, 325–356.

Yeung, R. W., and Cermelli, C. 1993 Shell functions: a global method for computing time-dependent free-surface flows, *Proceedings*, 8th International Workshop on Water Waves and Floating Bodies, May, St. Johns, Newfoundland, Canada.

Crude Oil Under Negative Pressures and Hydrocarbons Emission Containment

M. Husain, (M) President, MH Systems, Inc., San Diego, CA
H. Hunter, MH Systems, Inc. San Diego, CA
D. Altshuller, MH Systems, Inc., San Diego, CA
E. Shtepani, Hycal Energy Research Labs Ltd., Calgary, AB, Canada
S. Luckhardt, Energy Research Center, University of California, San Diego, CA

ABSTRACT

This paper presents the results of experimental and theoretical analysis of crude oil under negative (sub-atmospheric) pressures. The use of negative pressure in the ullage space of tankers has been proposed for the mitigation of cargo losses resulting from accidental hull rupture. This potential application resulted in detailed laboratory tests of various crudes over a range of temperatures and pressures. The findings were then applied to prevent or minimize VOC (Volatile Organic Compound) emissions occurring during transport of crude oil.

Equilibrium between liquid and gas/vapor components is reached at moderate hydrocarbon partial pressures. Predictions of liquid-vapor transformation can be made very accurately from known relationships, in the present case, the Peng-Robinson equation of state. The amount of material vaporized in reaching equilibrium is small, and the composition of the source liquid is changed only slightly. It can be shown that the partial pressures of hydrocarbon components are low, and can be accommodated by a closed system.

The adverse environmental and economic losses caused by VOC emissions from crude oil tankers are significant. A brief discussion on the design of a closed, negative pressure system for VOC emission minimization is discussed. The closed loop system, along with a seawater heat exchanger, mitigates the effects of temperature and pressures as well and will significantly reduce VOC venting to the atmosphere.

NOMENCLATURE

a, b	constants in cubic equation of state
A, B, C	constants in Antoine equation
C	ullage capacity coefficient
C_p	specific heat
D	duct diameter
h_D	turbulent heat transfer coefficient 5.4 BTU/hr ft^2 °F for a flow speed of 30 ft/sec.
H	enthalpy
K	control loop gain
L	duct length
$\dot{m}$	gas mass flow
M_g	molecular weight of gas
M_i	molecular weight of ith component
M_o	molecular weight of liquid
Ma	Mach number
P	Pressure
Q	heat load
$Q(s)$	volume flow rate in the Laplace domain
$\dot{Q}$	heat flow rate BTU/sec
R	gas constant
s	Laplace-domain variable
S	entropy
t	time
T	temperature
$T_{inlet}(t)$	inlet gas temperature
T_w	wall temperature
T_x	gas temperature at distance x

ΔT	temperature rise °F.
V	volume
x_i	molar fraction of a component in liquid phase
y_i	molar fraction of a component in gas phase
Z	thermodynamic compression factor

Greek Letters

γ	specific heat ratio
ε	emissivity
κ	thermodynamic constant
ρ	density
σ_{SB}	Stefan-Boltzman constant
ω	acentric factor
Ω	thermodynamic constant

INTRODUCTION

The transport of crude oil by sea represents a continuing source of pollution by cargo loss due to collision and grounding and by venting of volatile organic compounds (VOC) evolved from the cargo. Cargo losses due to collision and grounding have been estimated at about 25,000 tons per year, with occasional catastrophic losses. INTERTANKO (Gunner 1999) estimates losses of VOC to the atmosphere at about 1.6 million tons per year. Other estimates put this number at 4 million tons or more per year.

Over the past several years, the authors have been conducting research, testing and conceptual design of ullage gas/vapor management systems. Primarily, these studies have focused on containment of oil spills by maintaining subatmospheric ullage pressure.

However, these studies have also demonstrated a collateral benefit that can be achieved by maintaining constant ullage pressure - a practical approach to the reduction of VOC emissions.

It is currently required that the ullage space be (i) suitably inerted, normally using oxygen-depleted flue gas, and (ii) that the pressure be maintained at a positive value, typically about 1.5 psig. At the same time, it is also required that ullage pressure may not exceed a given level, typically about 2 psig.

These pressure bounds are close together, requiring that a nearly constant pressure be maintained between them. Pressure excursions beyond these limits require either venting or the addition of inert gas. This need can be reduced or eliminated if subatmospheric pressure is maintained in the ullage, because then the margin between the pressure bounds increases.

The ullage mixture is generally not in equilibrium with the liquid cargo at the completion of cargo loading. Therefore, VOC evaporation continues during transit. A closed-system equilibrium (at saturation) may not be reached because of frequent venting and inerting.

Diurnal heating and cooling of the cargo tanks increases the variation in ullage pressure necessitating additional venting and inerting. Under current management practice, periodic venting can continue during an entire voyage.

The concept to be discussed in this paper is to contain in-transit VOC emissions by maintaining a near constant ullage pressure. Ullage gases are circulated in a closed-loop arrangement at sub-atmospheric pressure via a blower and seawater heat exchanger. It is the circulation that makes it possible to maintain constant ullage pressure under perturbations caused by mass and heat transfer during the voyage. The system can be implemented at moderate cost and with the use of essentially standard marine components.

Part 1 presents a summary of the laboratory tests conducted in order to define crude oil vaporization processes under moderately reduced pressures. The laboratory test data formed the basis for the development of an equation of state model. This model allows the test data to be extrapolated to conditions that were not explicitly present in the tests.

A brief reference is made to a full-scale tanker test of a reduced ullage pressure system. The test (Husain et al 2001) demonstrated that the proposed pressure control system could be implemented and managed without undue cost or operational difficulty.

The final section presents a brief description and proof-of-concept evaluation of an operating system for crude oil tankers satisfying these requirements:

- Contain VOC emissions by eliminating the need for venting or topping, as currently practiced in tanker transport.
- Sustain saturated conditions in the ullage gases throughout the voyage.
- Provide for a wider pressure excursion band by operating at moderate sub-atmospheric pressures.
- Minimize pressure fluctuations resulting from mass and heat transfer in the ullage spaces.

It is recognized that the reduction of the lower pressure bound from 1.5 psig to subatmospheric level would conflict with current regulations for tankers, and would thus require careful development, qualification testing and regulatory approval.

PART I – ANALYSIS OF CRUDE OIL UNDER NEGATIVE PRESSURES

Overview

An experimental study, together with a theoretical analysis, was conducted to determine the vaporization properties of representative crude oil samples at pressures below atmospheric and at the moderately elevated temperatures typical of crude oil transport at sea. This section of the report summarizes these investigations.

In brief, the experimental phase placed a sample of the as-received crude in a closed container under an inert gas atmosphere. Temperature was maintained at a predetermined level, and pressure was set to a specified value by adding inert gas or by removing the mixture. When equilibrium had been obtained, pressures, temperatures and liquid and vapor compositions were measured.

In the experiments the ullage volume was 25% of the total volume. By applying the equations of state (see Appendix 2) the results for any volume of the ullage space can be predicted.

These data were fitted to appropriate theoretical relationships in order to verify that vapor composition was related to liquid composition in accordance with theory. The correspondence was excellent. The observed test points were also compared to predictions made using a computer-based equation of state (EOS) model. In this case, too, the observations corresponded closely with predictions. These operations are described in the following sections.

The conclusions derived from the laboratory tests and analyses are as follows:

- Steady-state equilibrium between the liquid and gas/vapor components is reached at very moderate partial pressures of hydrocarbon mixture. The amount vaporized increases as the temperature increases and as ambient pressure decreases.
- The amount vaporized will adjust in either direction as temperature or pressure change.
- The composition of the vapor mixture is essentially as predicted analytically, based on the liquid composition.
- This prediction can be made very accurately from known relationships using a computer model to carry out the calculations. In the present case, the Peng-Robinson equation of state, a relation frequently used in the petroleum industry, correlated very well with test data.
- The amount of material vaporized in reaching equilibrium is very small, and the composition of the source liquid is changed only slightly. In relation to the transport of crude, if pressure can be held constant this vaporized material is a one-time loss of cargo, not a continuing one.

Experimental Work

To measure gas evolution from the crude oil at reduced pressures and at given temperature, the following procedure was performed:

- While purging the autoclave mixer with nitrogen, pour in 75 % in volume the crude oil to be tested.
- Seal the mixer with appropriate valves and gauges.
- At zero psig and lab temperature, after stabilization, measure a small gas analysis (about 50cc is required), being careful not to pull a negative pressure.
- To achieve the reduced pressures, a sampling syringe is used. The syringe is pre-purged with helium and evacuated.
- The volume of gas taken is measured in the gasometer and analyzed for composition using Gas Chromatograph.

This procedure is repeated for every pressure stage and at given temperature. Always the system is isolated and the temperature is stabilized, then the syringe is used to obtain the reduced pressure. The schematic of equipment used is shown in Figure 1.

Three crude oils were selected, which cover the range from 12 API to 37 API. The compositional analyses of crude oils were performed. For each of them a series of gas evolution tests was performed at pressures of 0 psig, -2 psig, -3 psig and –5 psig at temperatures 67 F, 80 F and 110 F and the liberated gas compositions were measured.

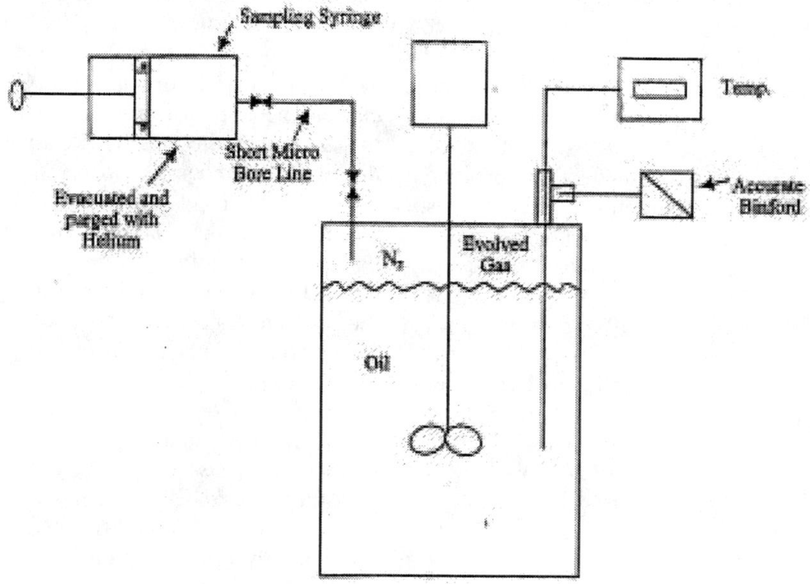

Figure 1.

As a general conclusion the gas evolution study indicated that the vaporization of light components from crude oil depends on:

- API gravity of crude oil -- For heavy oils, the vaporization can be considered insignificant.
- Temperature -- A temperature increase favors the vaporization of intermediate components.
- Pressure -- As pressure decreases below atmospheric pressure, the gas composition becomes richer with intermediate components.

The Table 1 summarizes the mole fractions of crude oil vaporized at a given pressure and temperature.

Table 1. Mole fraction of Vaporized Crude Oil

CRUDE OIL	Temperature \ Pressure	0 psig	-2 psig	-3 psig	-5 psig
	F	Mole Fraction			
12 API	67	0.0013	0.0500	0.0744	0.1234
	80	0.0092	0.0568	0.0807	0.1288
	110	0.0245	0.0703	0.0933	0.1398
30 API	67	0.0009	0.0327	0.0507	0.0956
	80	0.0167	0.0535	0.0756	0.1371
	110	0.0868	0.1626	0.2240	0.5298
37 API	67	0.0007	0.0300	0.0470	0.0906
	80	0.0139	0.0479	0.0687	0.1284
	110	0.0683	0.1329	0.1837	0.4093

The vaporization described above is in accordance with expectations. In terms of quantity, even in the worst-case scenario of highest temperature and lowest pressure, the amount of vaporised components was within 0.55% in mole, as shown in Table 1. That means that the change in the crude oil composition is practically insignificant. A detailed explanation of mechanism and compositional description of oils exposed to volatility and extraction processes is described in the technical paper of F.B. Thomas et al.

A summary of each crude oil compositions tested at stock tank conditions and after the highest vaporization (highest temperature and lowest pressure) has occurred is given in Appendix 1.

The differences in the composition before and after the process of vaporization are practically within the error of chromatography measurements. This is a good indicator that the properties of crude oil are not affected by this degree of vaporization. On the other side, because the gas vaporised remains in equilibrium with the crude oil, depending on conditions the evaporants may again go into solution either partially or fully.

In addition to laboratory tests, the compositions of liberated gases were calculated using the Peng-Robinson equation of state (PREOS). The PREOS was chosen because of its superiority in fluid phase behaviour predictions as compared to equations, developed for the prediction of ideal gas behaviour. The EOS parameters were tuned to match the data such as the density of the crude oil, the bubble point pressure and gas composition associated measured at 0 psig and 67 F. The tuned parameters were then used to predict the gas compositions at reduced pressures and at given temperatures. This procedure was followed for each crude oil.

The EOS results are in good agreement with the measured ones, making the EOS an accurate tool to predict the gas-oil phase equilibrium at reduced pressures and at given temperature. Comparisons of measured and calculated results are given in Figures 1 through 3 for the 30 API oil.

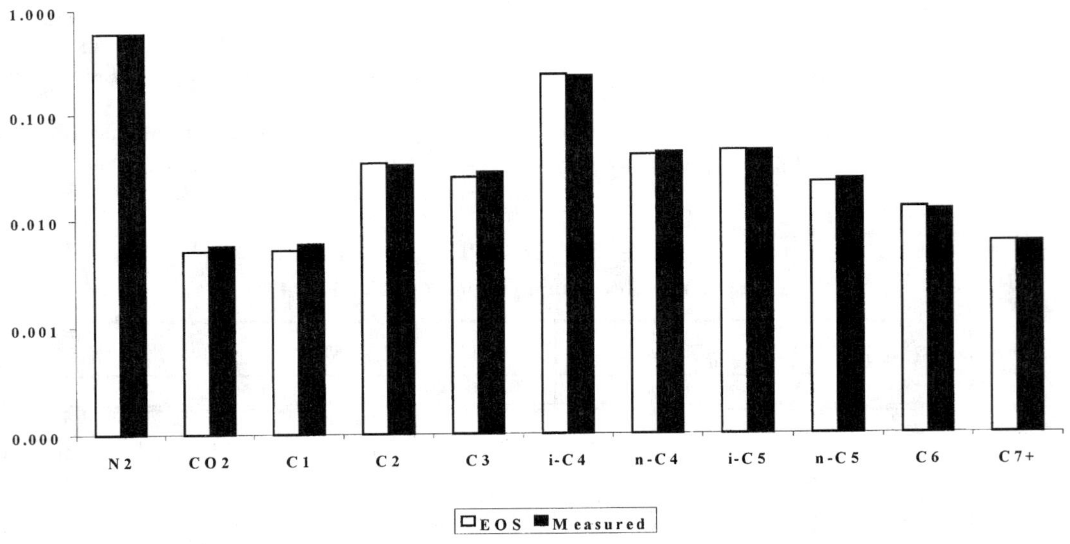

Figure 2 30 API CRUDE OIL LIBERATED GAS COMPOSITION @ -5 PSIG & 67 F

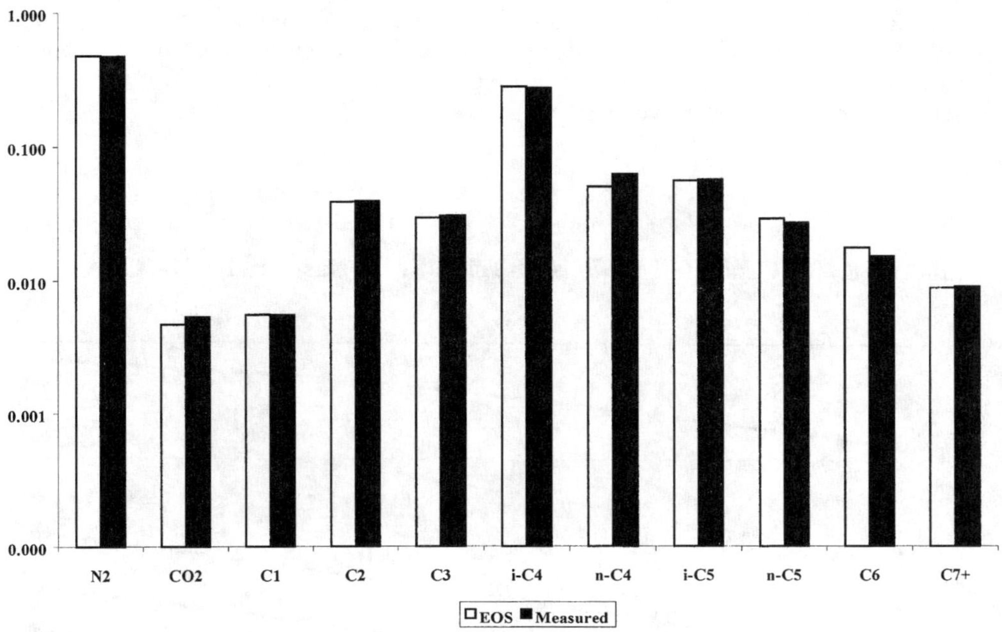

Figure 3. 30 API CRUDE OIL LIBERATED GAS COMPOSITION @ -5 PSIG & 80 F

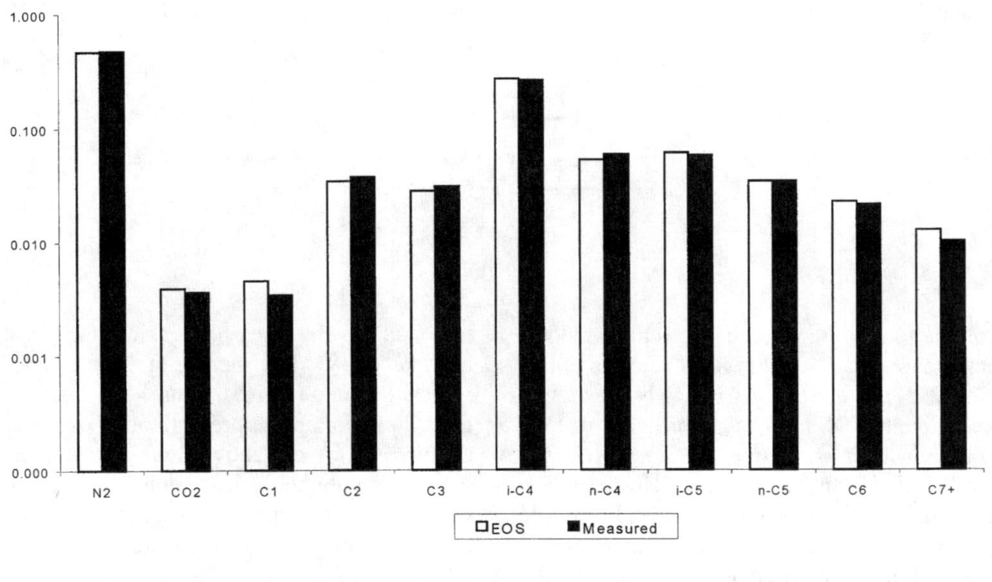

Figure 4. 30 API CRUDE OIL LIBERATED GAS COMPOSITION @ 0 PSIG & 110 F

A summarized description of EOS characterization and calculation techniques used is given in Appendix 2.

The average emission value of 0.25 % in weight is derived from the approach based on statistical data. In order to quantify the ultimate vaporization that can occur in the inert gas-crude oil system, a series of more sophisticated calculations by using the equations of state on systems with different N_2 gas-crude oil mixture ratios has been performed. The objective of these calculations was to investigate the relationship of N_2 +hydrocarbon gas liberated during the equilibration process with initial N_2 gas, with temperature and pressure.

The relationship of liberated gas in equilibrium for a given crude oil for two different temperatures and two pressures with respect to initial N_2 gas/crude oil ratio is given in the figure below.

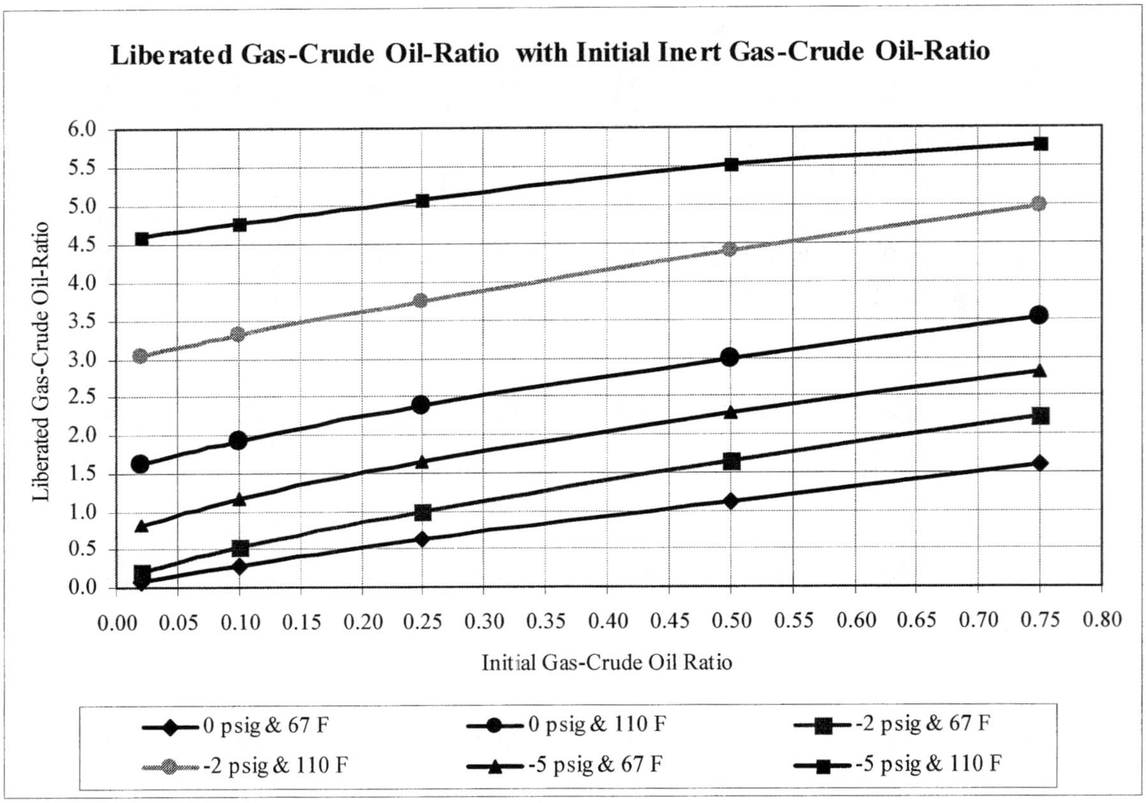

Figure 5.

In the above figure, the volume of gas liberated has been corrected at standard conditions. The calculations indicated that with the increase of initial N_2 gas/crude oil ratio the amount of N_2 +hydrocarbon gas/crude oil ratio increases, which subsequently decreases the percentage of N_2 gas in the liberated N_2 +hydrocarbon gas mixture. This process depends on the API of crude oil, temperature and pressure. Thus, e.g., for an initial N_2 gas/crude oil ratio in volume equal to 0.1, the liberated N_2 +hydrocarbon gas mixture /crude oil ratio becomes 0.285, that means, that the excessive amount of gas liberated (hydrocarbon + N_2) has to be vented in order to maintain the pressure at stock tank conditions. Converting for the above initial gas/crude oil ratio of 0.1 from volumes into weight %, after using some simple calculations, the results are shown in the following tables (all numbers are weight percentages):

Table 2. For Initial V/L = 0.02 (By Weight)

	Initial N2/Oil Ratio	After Equilibration Liberated Gas/Crude Oil Ratio					
	0 psig & 67 F	0 psig & 67 F	0 psig & 110 F	-2 psig & 67 F	-2 psig & 110 F	-5 psig & 67 F	-5 psig & 110 F
	Wt (%)	Wt (%)	Wt (%)	Wt (%)	Wt (%)	Wt (%)	Wt (%)
Gas	0.003	0.023	0.616	0.060	1.013	0.292	1.754
Oil	99.997	99.977	99.384	99.940	98.987	99.708	98.246
Hydrocarbons in Gas (%)		0.014	0.596	0.046	0.993	0.273	1.734

Table 3. For Initial V/L = 0.1(By Weight)

	Initial N2/Oil Ratio	After Equilibration Liberated Gas/Crude Oil Ratio					
	0 psig & 67 F	0 psig & 67 F	0 psig & 110 F	-2 psig & 67 F	-2 psig & 110 F	-5 psig & 67 F	-5 psig & 110 F
	Wt (%)	Wt (%)	Wt (%)	Wt (%)	Wt (%)	Wt (%)	Wt (%)
Gas	0.015	0.084	0.719	0.144	1.100	0.408	1.806
Oil	99.985	99.916	99.281	99.856	98.900	99.592	98.194
Hydrocarbons in Gas (%)		0.052	0.678	0.106	1.060	0.365	1.764

Thus, an initial N_2 gas/crude oil ratio in volume equal to 0.1 corresponds to 0.015 % in weight of N_2 gas in the total system of N_2 gas and crude oil. The liberated N_2 +hydrocarbon gas mixture /crude oil ratio in volume equal to 0.285 corresponds to 0.084 % in weight of liberated N_2 +hydrocarbon gas mixture in the total system at atmospheric pressure and 67 F, while the amount of gas liberated at atmospheric pressure and 110 F is 0.719 % in weight.

Tables 2 and 3 show the effect of V/L ratio on equilibrium gas mixture/crude oil for a range of pressures and temperatures. As expected, the weight fractions of the liberated gases increase with the available gas volume. It also appears that the gas mixture composition is not significantly affected by the change in the initial V/L ratio.

PART 2 - EMISSION MINIMIZATION BY NEGATIVE PRESSURES

Overview

As stated earlier in this paper, the original objective of the underpressure system is to prevent or reduce oil spillage--, by maintaining sub-atmospheric pressure in the ullage space of the crude oil tanker. The crude oil was tested under negative pressures. Test results and analysis demonstrate that maintaining constant negative pressure will eliminate or materially reduce venting.

A circulating closed-loop arrangement includes an in-line blower, a seawater heat exchanger and pressure modulating valves. This system will provide VOC emission containment and will maintain constant ullage pressure under perturbations caused by in-transit heat and mass transfer.

This section of the paper was organized as follows. First, the closed-loop system is defined. Second, the operational modes of the closed-loop system are identified and the modes relevant to VOC emission or its mitigation are discussed briefly. Third, the critical issues of the closed-loop system including the effect of diurnal heating on ullage pressure are discussed. Fourth, the control system is discussed. Because of page limitations, the discussion of the important issues had to be very brief, hopefully without sacrificing either the accuracy or understanding of the paper.

What Is A Closed-Loop System?

Definition

The closed-loop arrangement prevents VOC emissions from tankers in transit, with the added capability of containing spills in the event of accidental rupture of the hull. Dedicated piping recirculates the ullage gas using an in-line blower and a seawater heat exchanger. The transport of gases regulated by the control system effectively normalizes all perturbations in-transit that could change underpressure or inertness of the ullage gases.

It is important to distinguish the term "closed-loop" used to describe the underpressure system arrangement from the one commonly used in control engineering. The former refers to the fact that the ullage gases are recirculated through the system as opposed to venting to the atmosphere or to the vapor recovery facility (open-loop arrangement). The loop regulating the ullage and the header pressure (described later in this paper) is always closed in the sense of control engineering.

In the closed-loop underpressure system, negative pressure must be maintained at a predetermined level to prevent cargo loss in the event of accident, while oxygen content must be held within specified limits to eliminate fire danger. Inert ullage gases are circulated through the ullage spaces, primarily to dilute air that could leak into the reduced-pressure ullage spaces, and the gases are returned to the tanks via a blower and seawater heat exchanger. The blower compensates for the pressure loss during transfer and the heat exchanger keeps the gas/vapor temperature low.

All piping, components and controls are independent of the tanker supporting services. There are other operational modes employed in contingencies and side-breaches that employ an open-loop configuration with an inert gas supply. These operational modes still contain spills, but not VOC emissions.

The ullage pressure is maintained between –2 and –3 psi (set pressure). This is accomplished by controlling the flow rate through the blower and by restricting the flow into each tank through the inlet valve. The ullage pressure is continuously monitored.

The controller (including both hardware and software) is designed to respond to pressure perturbations resulting from mass (from leakage or evaporation) and heat transfer in a circulating closed-loop system.

Operational Modes

The underpressure system is designed to operate in the following modes:

- Mode 0. Cargo loading
- Mode 1. Initial reduction of ullage pressure -2 to -3 psi
- Mode 2. Normal transit
- Mode 3. Contingency when oxygen concentration exceeds 5% but not 8%
- Mode 4. Contingency when oxygen concentration exceeds 8%
- Mode 5. Grounding
- Mode 6. Grounding with strong tidal variations
- Mode 7. Side damage of a cargo tank
- Mode 8. Routine cargo unloading at destination
- Mode 9. Cargo unloading in the event of a tank damage.

In the operational mode 0 (cargo loading) the underpressure system is off line. The ship's supply of inert gas is used to maintain the inert atmosphere in the cargo tank. Vapors are collected and transferred to a vapor recovery facility.

In the operational mode 1 (See Figure 6a) the underpressure system is turned on to bring ullage pressure down to –2 to -3 psi. The exhaust main and the intermediate head blower are used to pump ullage gas out of the tank through valve *1*. The inlet valve *2* is closed in all tanks. The heat exchanger is bypassed. The inlet main is connected to the vapor recovery facility through valves *3* and *4*. This is an example of an open-loop operation.

The normal operational mode for the underpressure system is mode 2 shown in Figure 6b. The inert gas enters each cargo tank via an inlet valve *2*. The flow rate through the valve is controlled in order to maintain the ullage pressure at –2 to 3 psi. The inert gas is circulated through the inlet and exhaust mains by the low head blower. The flow rate through the blower can be adjusted as deck temperature varies during the day. The heat exchanger is used in this operational mode to keep the inert gas temperature low. The valves *3* and *4* are closed. This is an example of the closed-loop operation.

Operational mode 2 is very similar to the operational mode 5 (grounding without strong tidal changes). The only difference is that in mode 5, the intermediate head blower is used because a more rapid response to the change in the ullage pressure is needed.

In the operational modes just described, the VOC emissions are completely contained. The other operational modes (except unloading modes 8 and 9) represent contingencies. Spill can be contained in operational modes 3, 6, and 7, but emission containment is no longer possible. In the operational mode 4 (high concentration of oxygen) it becomes necessary to vent into the atmosphere. Neither spill nor emission containment is possible. Detailed discussion of these operational modes is outside of the scope of this paper.

Closed-Loop System Critical Issues

The underpressure system design provides for spill and emission containment on cargo tankers when employing a closed-loop arrangement. Spill containment requires the ullage gases to be maintained at a slight negative pressure (between -2 and -3 pounds per square inch) and emission containment requires no

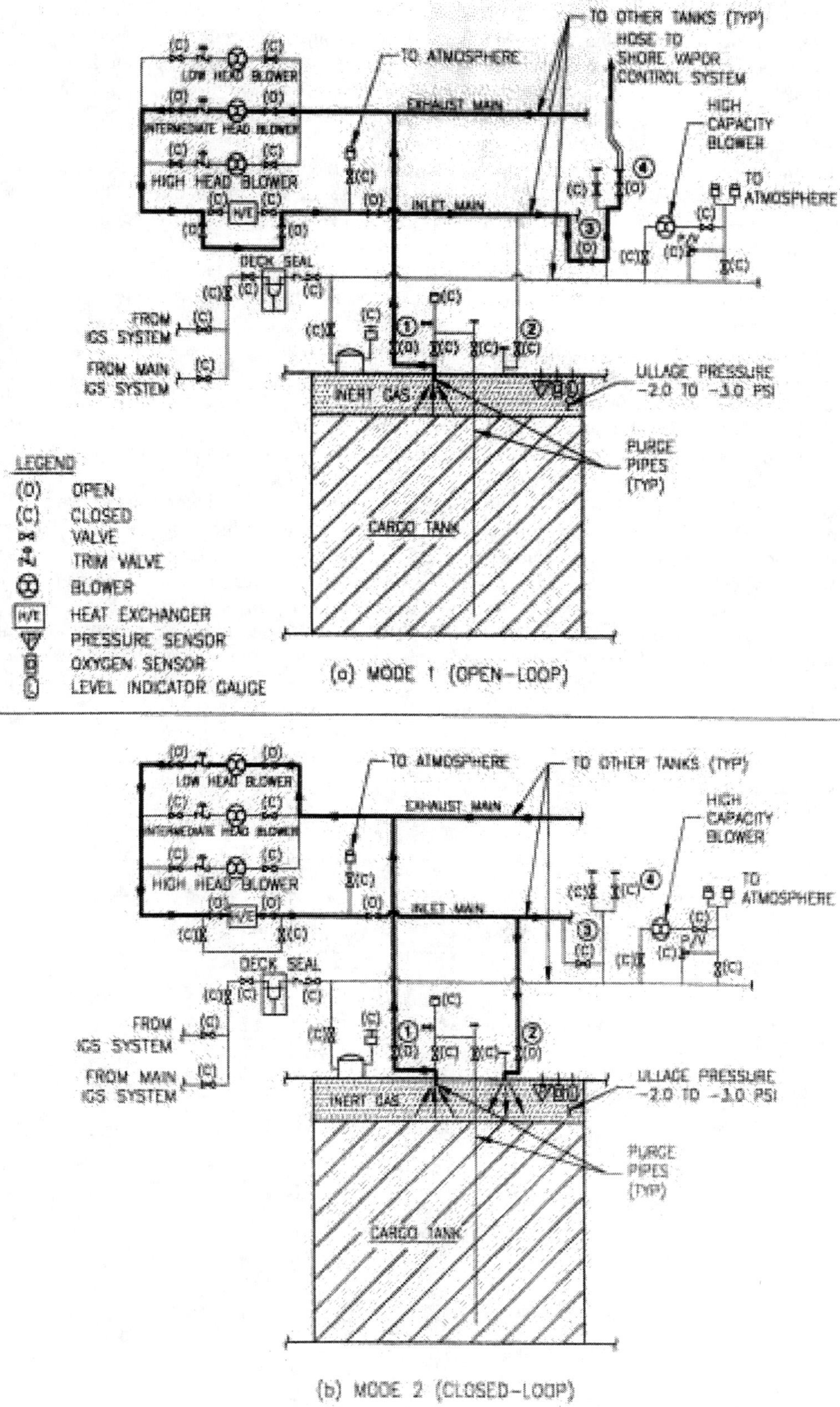

TO OTHER TANKS (TYP)
HOSE TO SHORE VAPOR CONTROL SYSTEM
TO ATMOSPHERE
LOW HEAD BLOWER
INTERMEDIATE HEAD BLOWER
HIGH HEAD BLOWER
EXHAUST MAIN
INLET MAIN
HIGH CAPACITY BLOWER
TO ATMOSPHERE
DECK SEAL
FROM IGS SYSTEM
FROM MAIN IGS SYSTEM
ULLAGE PRESSURE -2.0 TO -3.0 PSI
INERT GAS
PURGE PIPES (TYP)
CARGO TANK
LEGEND
(O) OPEN
(C) CLOSED
VALVE
TRIM VALVE
BLOWER
HEAT EXCHANGER
PRESSURE SENSOR
OXYGEN SENSOR
LEVEL INDICATOR GAUGE
(a) MODE 1 (OPEN-LOOP)
(b) MODE 2 (CLOSED-LOOP)

Figure 6

venting of the ullage gases during in-transit routine operations and during a grounding accident. The closed-loop arrangement essentially re-circulates the ullage gases at near seawater temperature while maintaining inertness by keeping oxygen content low.

A number of "proof-of-concept" evaluations have been made in order to validate the key issues associated with the use of a closed-loop system. These are summarized below together with the improvements in cargo loss that are achieved:

Potential Air Leakage and Its Contribution to Pressure and Oxygen Level Increases

Leakage estimates at the operational underpressure system was determined experimentally during full-scale testing aboard USNS Shoshone, a US Navy reserve fleet tanker. The effective leakage area derived from the test data at two pounds per square inch underpressure was less than 2×10^{-6} square feet.

Leakage is most likely caused by the penetrations of ducting or instruments and is essentially independent of tank size. The observed leak, extrapolated for an ullage volume of 10^4 cubic feet, shows an increase of 1 psi over a 15-day voyage, for a closed non-circulating ullage volume. However, the appropriate controls of the circulating stream can still maintain the desired underpressure by adjusting the blower flow rate.

Air leakage, based on field measurements, has a negligible effect on oxygen build-up over a 15-day voyage. However, the underpressure system provides a contingency mode, of an open-loop circulating stream, that replenishes the ullage with fresh inert gas from a combustion gas generator to an oxygen level <5% by volume. This contingency mode is enabled when O_2 levels are >5% by vol., through faulty assemblies or ducting stressed in high sea states.

Negative Pressure Precludes the Need for Vapor Release or inert gas addition

Typically, tankers frequently add or release ullage gas/vapor. Venting occurs when the tank pressures approach a limit setting, which is 0.6 psi above the pressure lock-up at the completion of cargo loading (Figure 7). The frequency can reach every four hours. One plausible explanation for the pressure rise is the unsaturated condition of the ullage gases at the completion of cargo loading, with dissolved gases in the oil continuing to evolve to reach equilibrium (saturation), at a pressure in excess of the set venting pressure (16.2 psia).

This process of periodic venting will continue as long as the equilibrium pressure is in excess of the venting pressure. Diurnal heating further exacerbates these conditions. The underpressure system, on the other hand, draws down on the loading pressure and allows sufficient time for the gas to equilibrate, before setting the required underpressure. This setting is secured at -2 to –3 psi below the venting pressure so an ample excursion is available to the more volatile fluids if required. Venting will not be required with an enabled underpressure system.

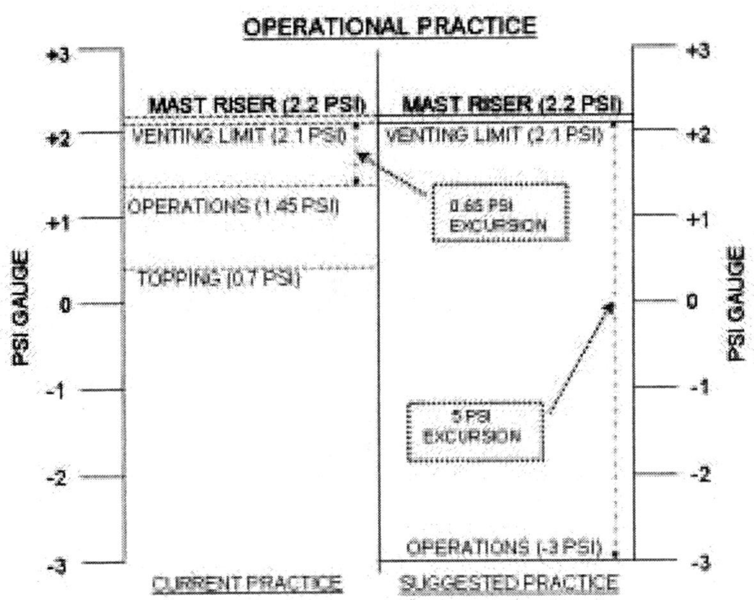

Figure 7

Topping under current practice is required by the loss of positive ullage pressure through leakage to the atmosphere mainly due to poor fitting in the mast riser valving. If leakage occurs with the underpressure system the reverse happens, ullage pressure increases and the consequence is addressed. In any case, topping is not a required procedure with the underpressure system installation.

In summary, both venting and topping will not be necessary with the underpressure system.

Ullage Pressure Reduction due to Cargo Oil Cooling & Air Leakage

Typically, cargo oil temperature decreases in transit due to heat transfer to the ocean. The hydrocarbon vapor pressure will accordingly decrease and reduce the total pressure of the gas mixture. The control's modulating valves at the inlet to each tank effectively alters the flow rate into the ullage as required to maintain the underpressure.

Diurnal Heating of Ullage Gas and its Effect on Ullage and Header Gas Pressure, Velocity, and Density

There is a concern that heat input due to the sun's radiant heat load could significantly raise the gas temperature and pressure adversely affecting the underpressure set for hydrostatic balance for spill control. A heat transfer analysis on the closed-loop underpressure system configuration installed on a large tanker operating on a mid-summer day in the Persian Gulf is addressed follows.

The effect of temperature change on the pressure, velocity, and density can be calculated by using the following equations of gas dynamics (Zucrow and Hoffman 1976):

$$\frac{dP}{P} = \gamma \mathrm{Ma}^2 f(\mathrm{Ma}) \frac{dT}{T} \tag{1}$$

$$\frac{dV}{V} = f(\mathrm{Ma}) \frac{dT}{T} \tag{2}$$

$$\frac{d\rho}{\rho} = -f(\mathrm{Ma}) \frac{dT}{T} \tag{3}$$

In these equations *Ma* is the Mach number (the ratio of the gas flow speed to the speed of sound). The function $f(\mathrm{Ma})$ has the property $f(0)=1$. Since the gas flows at a speed much slower than the speed of sound, it is reasonable to set $Ma=0$. Therefore, the right-hand side of the Equation (1) vanishes, and the conclusion is that temperature has no effect on gas pressure.

From integration of the Equation (2) it follows that the speed of gas flow is directly proportional to the temperature. Gas density in the ullage can be determined from the equations of state. Gas density in the header is determined by integrating Equation (3), which yields the conclusion that it is inversely proportional to temperature. The overall conclusion is that gas compressibility can be neglected for flow that is much slower than the speed of sound.

Ullage gases exposed to the deck can reach as high as 165° F. However, this temperature change is not proportional to the pressure, as would occur in a closed non-circulating ullage space.

A detailed analysis of the thermal load on the gas in both ullage and header is given in Appendix 3. The conclusion is that diurnal heating can be adequately addressed by a heat exchanger.

The control system includes a variable speed blower which can speed up to offset these losses, at no change in the circulating flow rate or ullage pressure. The in-line seawater heat exchanger removes the accumulation of heat deposited in the gas stream from the sun's radiant heat and blower compression heat. Its size and pressure loss is moderate, and has a minimal effect on the design.

VOC Emission Due to Loading

Loading cargo in a tanker can be a primary source of emissions. These losses can be prevented by a vapor balance service or a shore vapor recovery system. These services would also reduce emissions occasioned by the initial depressurization in preparation for the voyage.

At the completion of loading at positive pressure, the hydrocarbon (HC) content is usually 30-40 percent by volume, which is approaching saturation. The ullage is now subject to depressurization to below atmospheric pressure, which causes the gas composition to become richer with additional outgassing to reach a saturation some 5% higher. This means a requirement for the vapor recovery facility to remain operable until the final adjustment to the preselected pressure is achieved and stabilized.

In-transit, the saturated gases are circulated in a closed-loop arrangement consisting of an in-line blower, seawater heat exchanger and individual inlet valves to each tank that control flow to maintain tank pressure. The automated controls include the hardware, multiple sensors and software to ensure the underpressure is maintained through all operational perturbations.

Tankers Structural Capability to Withstand Negative Pressures

Extensive theoretical analysis has been performed to determine the effects of negative pressures in ullage spaces of tanks, and strain gage tests were performed on the USNS Shoshone (a reserve fleet tanker) as well. Analysis (Mansour 1996) of 245,000 DWT VLCC shows that even at –6 psig the increase in stresses was 9.4%. An analysis submitted by Japan (IMO 1993) for a 233,000 DWT tanker, shows that the negative pressure required to cause structural failure of the deck was found to be 19.8 psi (0.137 MPa). Strain gauge tests performed on the structures of the *USNS Shoshone* in 2001 show good correlation with mathematical models (Husain et al. 2001).

Risk assessment

A preliminary risk assessment was performed for the underpressure system in all of its operational phases. This analysis identified potential adverse events, frequency of occurrence, consequences if failure and possible mitigating measures.

As a result, the related design features include:

- An autonomous control system that ensures functional units and shipboard supporting services are operable, especially during a collision when the man/machine interface may not be fully available.

- Functional units are independent and separate from ship's systems, and are located where they are least susceptible to damage from a collision. The key units are the standby inert gas supply module, the electric/pneumatic power supply system, the gas distribution

- system and the controls and data transmission system.

- Design featuring redundancies and degraded modes of operation built into the hardware and software systems, including dedicated inert gas generator.

CONTROL SYSTEM

Closed-loop control architecture is shown schematically in Figure 8. The Control System provides for a fully automated operation of the underpressure system in both routine and casualty modes. Personnel in the control room will be provided with consoles that display the critical parameters for monitoring and start/stop plant operations. There is minimal interface with ship services with the underpressure system containing its own electric power and inert gas supply.

The principal hardware are blowers to circulate and meet the pressure loss demands, modulating inlet valves to each tank to maintain ullage pressure, and the instruments that measure the pressure, oil and O_2 level in the ullage spaces and blower performance. The software architecture will control the functions under dynamic conditions in response to sensor data with minimal human intervention.

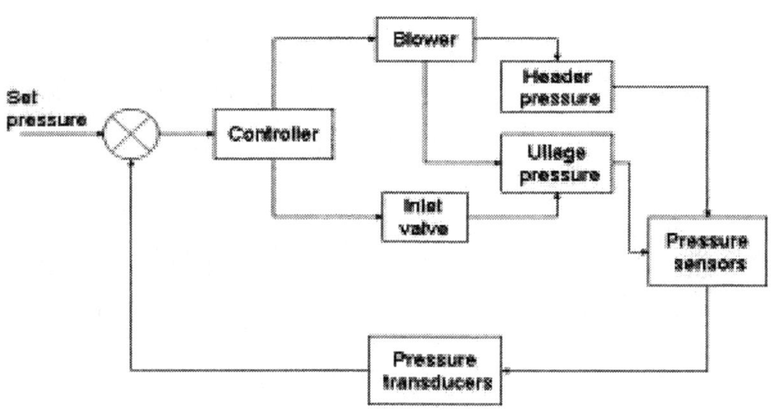

Figure 8

For the sake of simplicity, consider the dynamics of a single cargo tank. Following the commonly accepted control engineering practice, the ullage space in the tank is modeled as a simple integrator. This gives the following equation in the Laplace domain:

$$P(s) = \frac{C}{s} Q(s) \tag{4}$$

The following diagram shows the loop, including the proportional control action:

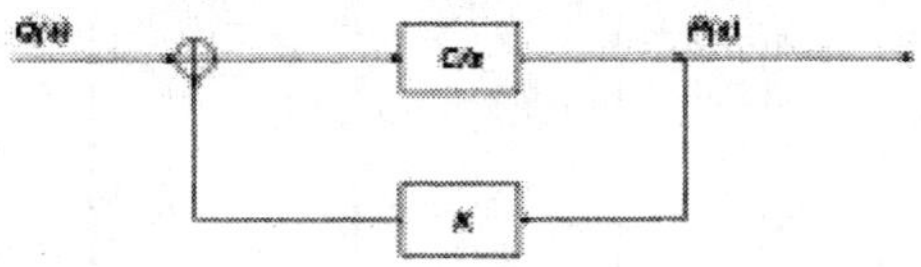

Figure 9

Therefore, the pressure responds to the change in flow rate according to the following Laplace-domain equation:

$$P(s) = \frac{C}{s + KC} Q(s) \tag{5}$$

It is well known (Dorf 1989) that a simple integrator is stable for any value of the proportional controller gain. Thus, the conclusion is that the pressure can be kept at the desired level by proportional control.

CONCLUSIONS

The effects of physical behavior of crude oil under negative pressures were investigated by tests and analysis. The derived results were then applied to crude oil tankers to minimize or prevent VOC emission during transport and the following conclusions were made:

- VOC emission during transport of crude oil by tankers can be eliminated or greatly reduced by maintaining a sub-atmospheric (negative) pressure in the ullage space.
- Maintenance of slight sub-atmospheric (negative) pressures in the ullage space of tanker crude oil during its transport is practical and safe. This method of maintaining sub-atmospheric pressures eliminates the need for repeated venting. Therefore, the VOC emissions are eliminated or reduced.
- Maintenance of negative pressures will not result in evaporation of the lighter elements once the ullage gas is saturated with hydrocarbons. There will be little effect on sludge formation.
- Effects of diurnal heating are minimized and manageable by employing a heat exchanger in a closed-loop and by control system.
- Control of VOC emission can be achieved by automated closed-loop control system.

APPENDIX 1.

Table 4. Compositions of Crude Oil Before and After Vaporisation

Component Name	Chemical Symbol	12 API Crude Oil		30 API Crude Oil		37 API Crude Oil	
		Before	After	Before	After	Before	After
		Mole Fraction					
Nitrogen	N_2	0.0031	0.0017	0.0017	0.0008	0.0009	0.0002
Carbon Dioxide	CO_2	0.0000	0.0000	0.0000	0.0000	0.0000	0.0000
Hydrogen Sulphide	H_2S	0.0000	0.0000	0.0000	0.0000	0.0000	0.0000
Methane	C_1	0.0000	0.0000	0.0000	0.0000	0.0000	0.0000
Ethane	C_2	0.0001	0.0001	0.0003	0.0000	0.0003	0.0000
Propane	C_3	0.0004	0.0004	0.0024	0.0022	0.0071	0.0060
i-Butane	i-C_4	0.0002	0.0002	0.0045	0.0022	0.0079	0.0074
n-Butane	n-C_4	0.0004	0.0004	0.0606	0.0605	0.0310	0.0302
i-Pentane	i-C_5	0.0004	0.0004	0.0280	0.0276	0.0220	0.0219
n-Pentane	n-C_5	0.0002	0.0002	0.0415	0.0414	0.0310	0.0310
Hexanes	C_6	0.0004	0.0004	0.0579	0.0581	0.0418	0.0418
Heptanes	C_7	0.0017	0.0017	0.0738	0.0741	0.1055	0.1059
Octanes	C_8	0.0045	0.0045	0.0787	0.0791	0.0582	0.0585
Nonanes	C_9	0.0087	0.0087	0.0546	0.0549	0.0456	0.0458
Decanes	C_{10}	0.0163	0.0163	0.0420	0.0422	0.0467	0.0469
Undecanes	C_{11}	0.0389	0.0390	0.0408	0.0410	0.0442	0.0444
Dodecanes	C_{12}	0.0535	0.0536	0.0365	0.0367	0.0370	0.0372
Tridecanes	C_{13}	0.0677	0.0678	0.0361	0.0363	0.0349	0.0351
Tetradecanes	C_{14}	0.0705	0.0706	0.0338	0.0340	0.0304	0.0305
Pentadecanes	C_{15}	0.0640	0.0641	0.0271	0.0272	0.0275	0.0276
Hexadecanes	C_{16}	0.0598	0.0599	0.0272	0.0274	0.0244	0.0245
Heptadecanes	C_{17}	0.0569	0.0570	0.0269	0.0271	0.0257	0.0258
Octadecanes	C_{18}	0.0467	0.0467	0.0242	0.0244	0.0212	0.0213
Nonadecanes	C_{19}	0.0488	0.0488	0.0198	0.0199	0.0178	0.0178
Eicosanes	C_{20}	0.0481	0.0482	0.0195	0.0196	0.0164	0.0165
Heneicosanes	C_{21}	0.0391	0.0392	0.0173	0.0174	0.0148	0.0148
Docosanes	C_{22}	0.0379	0.0379	0.0157	0.0158	0.0138	0.0139
Tricosanes	C_{23}	0.0319	0.0319	0.0141	0.0142	0.0126	0.0127
Tetracosanes	C_{24}	0.0278	0.0278	0.0125	0.0126	0.0112	0.0112
Pentacosanes	C_{25}	0.0268	0.0268	0.0120	0.0121	0.0109	0.0110
Hexacosanes	C_{26}	0.0220	0.0220	0.0104	0.0105	0.0097	0.0097
Heptacosanes	C_{27}	0.0224	0.0225	0.0095	0.0096	0.0086	0.0086
Octacosanes	C_{28}	0.0218	0.0218	0.0092	0.0093	0.0083	0.0083
Nonacosanes	C_{29}	0.0186	0.0187	0.0082	0.0082	0.0077	0.0078
Tricontanes Plus	C_{30+}	0.1547	0.1549	0.0639	0.0642	0.1708	0.1715
NAPHTHENES							
Cyclopentane	C_5H_{10}	0.0000	0.0000	0.0007	0.0007	0.0010	0.0010
Methylcyclopentane	C_6H_{12}	0.0000	0.0000	0.0185	0.0186	0.0152	0.0152
Cyclohexane	C_6H_{12}	0.0000	0.0000	0.0103	0.0103	0.0003	0.0003
Methylcyclohexane	C_7H_{14}	0.0000	0.0000	0.0306	0.0308	0.0058	0.0058
AROMATICS							
Benzene	C_6H_6	0.0000	0.0000	0.0012	0.0012	0.0017	0.0017
Toluene	C_7H_8	0.0000	0.0000	0.0013	0.0013	0.0021	0.0021
Ethylbenzene & p,m-Xylene	C_8H_{10}	0.0012	0.0012	0.0133	0.0134	0.0126	0.0127
o-Xylene	C_8H_{10}	0.0007	0.0007	0.0051	0.0052	0.0057	0.0057
1, 2, 4-Trimethylbenzene	C_9H_{12}	0.0038	0.0038	0.0082	0.0082	0.0096	0.0097
Total		**1.0000**	**1.0000**	**1.0000**	**1.0000**	**1.0000**	**1.0000**

APPENDIX 2. BACKGROUND FROM THERMODYNAMICS

Phase Equilibrium

The thermodynamic behavior of a multicomponent mixture depends strongly on composition, pressure and temperature. At given reservoir temperature and different pressures, the components may be distributed between phases.

Phase and volumetric behavior of real reservoir system is most conveniently described by the respective *phase diagrams*, which are often called pressure-temperature diagrams.

The relationship between the temperature of a liquid and its vapor pressure for a pure substance is given by Clausius equation.

$$\frac{dP}{dT} = \frac{S^V - S^L}{V^V - V^L} = \frac{\Delta H}{T\Delta V}$$

It is an exact relationship between the slope of the saturated vapor pressure curve and two easily measured physical properties of the system, viz, the latent heat of vaporization (ΔH) and the volume change on vaporization (ΔV).

There is a commonly used approximation to this equation for vapor-liquid systems, due to Clausius, and so called the *Clausius-Clapeyron* equation. This makes three assumptions, which are generally reasonable for low-pressure systems for pure substances:

1. The latent heat of vaporization does not change over the temperature range being considered
2. The vapor is an ideal gas
3. Vapor volume is much larger than liquid volume

With those assumptions, the following approximate equation can be derived for saturated vapor pressure:

$$\ln(P^{sat}) = D - \frac{\Delta H}{RT}$$

The integration constant D, can be deduced from vapor pressure data at one temperature. This equation is similar in form to the empirically based Antoine equation; however, the Antoine equation has three empirical constants.

$$\ln(P^{sat}) = A - \frac{B}{T+C}$$

where

P^{sat} is the vapor pressure
T is the temperature in K

The behavior of a mixture of multi-components is not as simple as the behavior of a pure substance. Instead of a single line representing the vapor-pressure curve, there is a broad region in which two phases coexist. This region is called phase envelope. The phase envelope of multi-components systems can be predicted by equations of state.

Cubic Equations of State

Cubic equations of state are simple equations relating pressure, volume and temperature. Based on the first cubic equation of state that was proposed by van der Waals in 1873, an evolution of cubic EOS' took place in focusing the interest on the application to petroleum engineering, as described above in the *Path 1*. The equations of state of this path may be written in the following generalized form:

$$P = \frac{RT}{V-b} - \frac{a(T)}{V^2 + ubV + wb^2}. \tag{6}$$

where P is the pressure, T the temperature, V the molar volume, R the gas constant and a and b are equation of state parameters which for a pure component are determined by imposing the critical conditions

$$\left(\frac{\partial P}{\partial V}\right)\Bigg|_{Crit.Point} = \left(\frac{\partial^2 P}{\partial V^2}\right)\Bigg|_{Crit.Point} = 0. \tag{7}$$

$$a(T) = a_c \alpha(T) \tag{8}$$

$$a_c = \Omega_a \frac{R^2 T_c^2}{P_c} \tag{9}$$

$$\sqrt{\alpha} = 1 + \kappa\left(1 - \sqrt{T_r}\right) \tag{10}$$

$$b = \Omega_b \frac{RT_c}{P_c} \tag{11}$$

The values of constants u, w, Ω_a, Ω_b, and κ are shown in **Table 5**.

Table 5

Cubic EOS	u	w	Ω_a	Ω_b	κ
Van der Waals	0	0	0.421875	0.125000	
Redlich-Kwong	1	0	0.427480	0.086640	
Soave-Redlich-Kwong	1	0	0.427480	0.086640	$0.480 + 1.574\omega - 0.176\omega^2$
Peng-Robinson	2	-1	0.457236	0.077796	$0.37464+1.54226\omega - 0.2699\omega^2$ for $\omega \leq 0.4$ $0.3796+1.485\omega - 0.1644\omega^2 + 0.01667\omega^3$ for $\omega > 0.4$

Most of these modifications have been empirical and arbitrary, with parameters that are adjustable to fit certain kinds of experimental data such as vapor pressures, densities, or enthalpies.

In review, the PREOS and SRKEOS are the two most widely used cubic equations of state. The PREOS is comparable to SRK equation in simplicity and form. Peng and Robinson report that their equation predicts liquid densities better. They provide the same accuracy for VLE predictions and satisfactory volumetric predictions for vapor and liquid phases when used with volume translation.

The method of volume translation as proposed by Peneloux *et al.*, modifies a two-constant cubic equation by introducing a third EOS constant c, but without changing the equilibrium calculations of the original, two-constant equation. The volume translation constant c eliminates the inherent volumetric deficiency suffered by all two constant equations. A simple correction term is applied to the EOS calculated molar volume:

$$V = V^{EOS} - c \qquad (12)$$

where V is the corrected volume, V^{EOS} is the calculated volume using EOS, and c is component specific constant.

In general, an equation of state is developed first for pure substances, and then extended to mixtures using mixing rules for combining the pure component parameters. The conflict between accuracy and simplicity is a big dilemma in developing an equation of state. Despite the wide use of high-speed computers, simplicity is still highly desired for easy and unequivocal applications of the equations to complex problems.

Mixing Rules for Cubic Equations of State

As discussed earlier in this section, the equations of state are generally developed for pure fluids first and then extended to mixtures. The mixture extension requires so-called mixing rules, which are simple means of calculating mixture parameters equivalent to those of pure substances. Most of the simple equations of state evolved from the van der Waals' equation use van der Waals' mixing rules with or without modifications.

Several authors have introduced in the mixing rules for *the binary interaction coefficients* k_{ij}, which are empirically to be determined from the experimental vapour-liquid equilibrium data, for each binary present in the mixture. As pointed out by several authors, each k_{ij} factor can be considered independent of system temperature, pressure and composition. Some others have proved the strong dependence of k_{ij} from the temperature.

Heavy Fraction Characterization

The complete description of the performance of a petroleum reservoir is a complex calculation. Naturally occurring reservoir fluids typically consist of pure well-defined components including:

$$CO_2,\ N_2,\ H_2S,\ C_1,\ C_2,\ C_3,\ i\text{-}C_4,\ n\text{-}C_4,\ i\text{-}C_5,\ n\text{-}C_5,\ C_6,$$

and hundreds of heavy-end components (Heptanes and heavier, C_{7+}). It is not possible to isolate all C_{7+} compounds or to assign accurate physical properties to each of them. Therefore, an approximate characterization of C_{7+} fractions using either experimental methods or mathematical methods has to be carried out. These methods include:

- Dividing C_{7+} fraction into a number of sub-fractions with known composition
- Defining the molecular weight, specific gravity, and boiling point of each C_{7+} fraction.
- Estimation of critical properties, acentric factor and binary interaction coefficients according to the cubic EOS used.

There are some experimental methods for quantifying the C_{7+} into discrete fractions. True boiling point (TBP) distillation provides the necessary data for a complete C_{7+} characterization, including specific gravity, molecular weight, and boiling point for each fraction, while gas chromatography (GC) analysis can only quantify the mass of C_{7+} fractions.

The choice of the characterization method certainly affects the outcome of reservoir fluid PVT prediction. However, the most reliable basis for C_{7+} characterization is the experimental data obtained by TBP distillation or chromatography.

Molecular Weight

The *molecular weight* of the gas and liquid phase, respectively, is given by

$$M_g = \sum_i y_i M_i \tag{13}$$

$$M_o = \sum_i x_i M_i \tag{14}$$

Density

Phase density is given by

$$\rho_p = \frac{PM_p}{ZRT} \tag{15}$$

Because the Z-factor is calculated as part of the EOS equilibrium solution, densities can be directly evaluated using the above formula.

Component Properties and Correlations

A simple two-(or three) parameter equation of state requires the thermodynamic properties for each component in the mixture to be modeled. The *critical pressure* and *critical temperature* are needed. The *acentric factor* is also required. The *binary interaction coefficients* and volume translation parameters may be another input to the equation of state. This subset of input variables is used in the equation of state to predict thermodynamic properties of the mixture, such as saturation pressure, liquid volume in the two-phase region, composition of equilibrium phases, etc. Finally, when the phase density and viscosity are to be predicted by the equation of state the value of *molecular weight*, *volume translation* parameter, and the *critical volume* for each component has to be known.

The thermodynamic properties for the defined reservoir hydrocarbons and non-hydrocarbons and injectants are taken from the literature. It includes Molecular weight MW, Critical Pressure P_c, Critical Temperature T_c, Critical Volume V_c, Acentric Factor ω, Boiling Temperature T_b and Specific Gravity γ.

For C_{7+} fractions the thermodynamic properties are found from empirical correlations in absence of experimental *TBP* data. Paraffin molecular weights are usually assumed when converting simulated distillation results to a mole-fraction basis. The best estimates of γ and T_b would come from a *TBP* analysis of a sample from the same field, and the next-best source would be the TBP data from a field producing from the same geological formation.

APPENDIX 3. THERMAL LOAD ANALYSIS

In this section the thermal properties of the inert gas flow system used to control the pressure in the ullage space of the seaborne oil tanks are analyzed. In this system there are a number of modular oil tanks through which an inert gas is pumped at a regulated pressure. The primary heat load on this system is solar radiation.

The model configuration, Figure 10, consists of a tank containing oil and an inert gas layer in the ullage space above the oil surface. The top of the tank is illuminated by solar radiation and the bottom of the tank is in contact with ocean water heat sink at temperature $T = T_{water} = 85^o\ F$. The surface of the oil is subject to heat loads from thermal radiation coming from the undersurface of the ship's steel deck. The flowing gas is heated by conduction from these bounding surfaces.

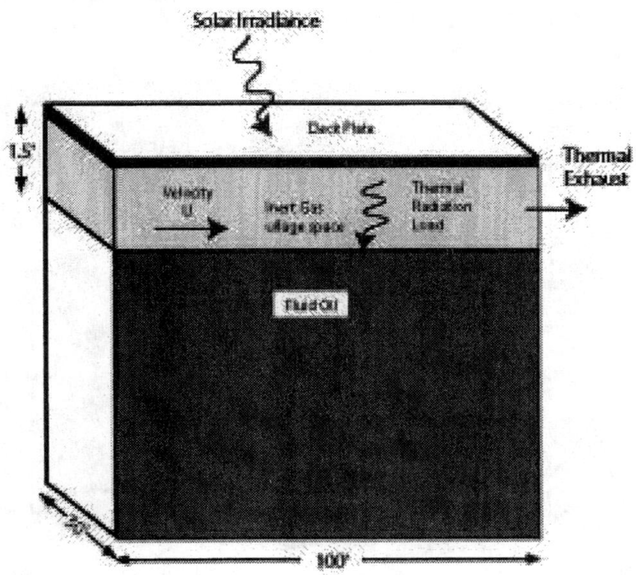

Figure 10. Inert gas flows with velocity U through the ullage space of the oil tank module. Thermal radiation from the underdeck heats the oil surface and the flowing gas is heated by thermal conduction from the surfaces.

The dimensions of the oil tank are shown in Figure 10, L_x=*30 ft*, L_y = *100 ft* and the depth of the gas layer is L_g=*1.5 ft*. The deck material is taken to be steel of L_d=*1 inch* thickness. The inert gas flows over the oil surface at mean velocity *U*.

Some of the characteristic time scales and heat loads can be estimated as follows. The thermal diffusion time constant of the steel deck is τ-*deck* = d^2/α_{steel} ~ *1 minute*, so the top and bottom surfaces of the steel deck are equilibrated on the timescales of interest. The thermal time constant of the oil mass is longer, τ-*oil* = $(L_{oil}/2)^2\alpha_{oil}$ ~ $(15ft)^2/(3x10^{-3}\ ft^2/hr)$ ~ *8.5 year*. Hence, neglecting convection and mixing, the oil temperature distribution will not equilibrate during the voyage. Further, under conditions of surface heat load, the oil surface temperature can deviate from its initial temperature. The diurnal variations in heat load to the oil surface will change the temperature in a thin layer of oil δ ~*sqrt(24hrs x* α_{oil}*)* ~ *3.3 in*. Such heating can lead to increase evaporation of volatile hydrocarbons.

The heat load Q_{rad} on the oil surface from radiative transfer depends on the difference between the deck temperature and the oil surface temperature and can be estimated as

$$Q_{rad} = \frac{4\sigma_{SB} T_{deck}^3}{\frac{1}{\varepsilon_1} + \frac{1}{\varepsilon_2} - 1} \left(T_{deck} - T_{oil}\right)$$

where ε_1=*0.95 and* ε_2 = *0.90* are the emissivities of the deck and oil surfaces respectively. These emissivity values may vary from these numbers in practice because of surface composition, roughness, and other factors. Taking T_{oil} = *90 °F.* and T_{deck} = *165* °F. to approximate conditions of a summer day at noon, gives an estimate of Q_{rad} ~ *60 BTU/ft²/hr*. This heat flux will heat the oil surface significantly. In the following calculations the bulk oil temperature below this surface layer is assumed to be T_{bulk} = *90 °F.*

The other heat load on the oil surface is thermal conduction through the gas layer; this turns out to be much smaller than the typical radiative heat load. Estimating the conductive heat load to the oil surface as

$$Q_{gc} = \frac{K_{gas}}{L_{gas}}(T_{deck} - T_{oil})$$

where K_{gas} = *0.0262 W/m-K, 0.015 BTU/hr/ft/F* and T_{deck} is the deck temperature, and T_{oil} is the oil surface temperature. Taking T_{deck} = *165 °F.* gives an estimate for the conducted heat flux of Q_{gc} ~ *0.4 BTU/hr/ft²*. Hence, the radiation heat load will be the dominant factor in heating of the oil surface layer.

Next, consider heating the inert gas (nitrogen) flowing in the ullage space. The gas flow rate is taken to be *F=50 cfm*, cross-sectional area of the ullage space is *45ft²*, and the flow speed is *U=1.1 ft/m*. The Reynolds number for this flow is *Re* = *11,000*, hence the non-turbulent thermal diffusion model to calculate the temperature distribution of the flowing gas within the ullage space can be used. The model of the diurnal temperature variation of the deck underside is

$$T_{deck} = T_{\min} + \frac{\Delta T}{16}\left(1 - \cos(\theta(t) - \theta_0)\right)^4$$

where t is the time elapsed in hours from midnight, $\theta(t)=2\pi t/24$ and θ_0 is the time delay factor for thermal penetration of the deck plate θ_0 = $4x10^{-3}$ for the 1 inch deck. For this calculation take the minimum daily temperature as T_{min} = 85 °F. and the temperature rise ΔT = *80 °F*. The maximum deck undersurface temperature is then 165 °F with diurnal variation shown in Figure 11.

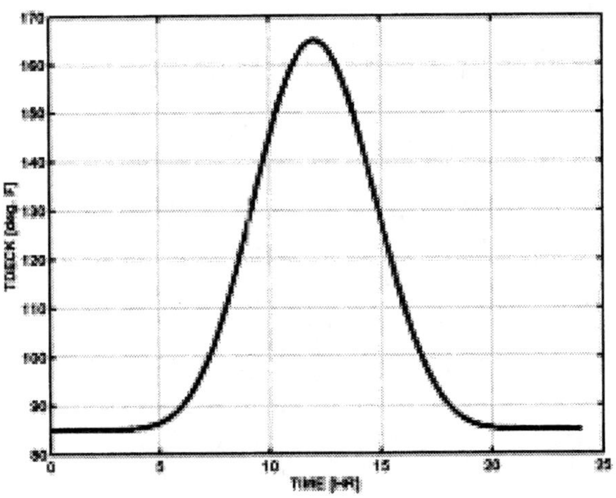

Figure 11. Model for deck temperature diurnal variation on hot day.

Since the gas transit time through the tank is much less than 24 hrs, the time dependence of deck temperature for this problem can be neglected. The gas temperature in the ullage spaces is modeled by the convection-diffusion equation.

$$\vec{v}\cdot\vec{\nabla}T(x,y)-\alpha\nabla^2T(x,y)=0\,,$$

where v is the gas velocity and α is the gas thermal diffusivity. The boundary conditions are

$$\begin{cases} T(0,x)=T_{deck}=165^o\,F. \\ T(-1.5,x)=T_{oil}=90^o\,F. \\ T(y,0)=85^o F.\ \text{Gas inlet} \\ \dfrac{\partial T(y,100)}{\partial x}=0\ \text{Gas outlet} \end{cases}$$

The equation was solved numerically for the two cases: low and high flow speed. The results showed that the gas thermal boundary layer expands to fill the ullage space at the outlet. The average gas temperature at the outlet is then T_{out} = *(165 + 85)/2=125F*. In case of 400 cfm flow the gas thermal boundary layer expands to approximately half the ullage space and the average temperature of the outlet gas is T_{out} = *105F*. Finally, for flow rates in the range of 50 cfm, it can be concluded that the outlet average temperature of the gas will be

$$T_{out}(t)=\frac{T_{deck}(t)+T_{oil-surface}(t)}{2}$$

The time dependence indicates the diurnal variation. The rate of heat transfer by the gas flow is then

$$Q(t)=\dot{m}C_p\left[\frac{T_{deck}(t)+T_{oil-surface}(t)}{2}-T_{inlet}(t)\right]$$

Header Heat Load

Finally, the heating of the gas manifold or header duct that carries the working gas to and from the oil tank array is addressed. The header assembly is located above deck consisting of cylindrical ducts having reflective paint surfaces. In the case without reflective paint, the duct wall temperature will be comparable to the deck temperature, having a maximum of $T_w = T_{deck}$ = *165 °F*. Coating the duct with reflective aluminum paint having emissivity of ε = *0.26* can lower the duct temperature to the range of T_w = *123 °F*. under optimum conditions. Generally, the duct wall temperature and resulting gas stream are significantly affected by the surface coating. The header duct consists of a *D=10inch* diameter pipe of length *L= 800ft*, a flow rate of 1000cfm, and average flow velocity of *U=30 ft/sec*. The Reynolds number for the flow $Re = UL/n = 1.5 \times 10^5$ is in the turbulent flow regime, hence the turbulent convective heat transport coefficient used.

$$h_D^{turb}=\frac{k_{gas}}{D}Nu$$

The turbulent Nusselt number is evaluated using the Petukhov equation (Pitts and Sissom 1998) with parameters for nitrogen gas as Nu = 298. The convective heat transfer coefficient is approximately h_D = 5.4 BTU/hr ft^2 °F

The gas temperature distribution along the header is then given by

$$\frac{T_w-T_x}{T_w-T_1}=e^{-\kappa x},$$

where

$$\kappa=\pi Dh_D\Big/\dot{m}C_p$$

The gas temperature variation along the header duct is plotted in Figure 12 for the case of 30ft/sec flow rate and a wall temperature of 165 °F. It appears that the critical length for the working gas to reach the wall temperature is approximately 25% of the duct length.

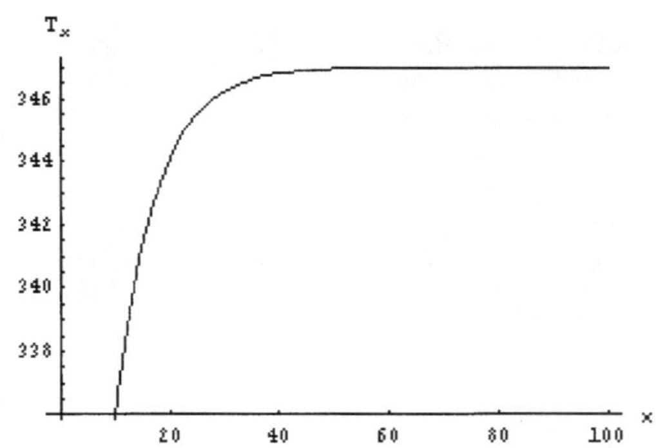

Figure 12

The corresponding heat absorbed is

$$\dot{Q}=\dot{m}C_p\Delta T$$

For the candidate tanker configuration the heat load generated by the header leg can be sizably reduced. For the same inlet temperature (95° F) comparison is:

Unpainted wall	=	65×10^3 BTU/hr
Painted wall	=	29×10^3 BTU/hr

This is a consideration in the design. The conditions for the heat exchanger are based on a painted inlet header and a blower temp rise of 20° F

Inlet Temp. to Heat Exchanger	= 123 + 20 = 143° F
Weight Flow	= 1 LB/sec
Outlet Temp.	= 95° F
Heat Exchanger Load	= 82×10^3 BTU/hr

This is a relatively small heat-changer with a minimal pressure drop.

REFERENCES:

BIRD, R.B., STEWART, W.E, and LIGHTFOOT, N.,Transport Phenomena, John Wiley & Sons, NY (2002).

CRABBE, D. and MCBRIDE, R., The World Energy Book, Nichols Publishing Co. NY (1978).

DORF, R.C., *Modern Control Systems.* 5th Ed., Addison-Wesley (1989).

GUNNER, T.J., *CRUCOGSA - The Physical Behaviour of Crude Oil Influencing its Carriage by Sea,* An INTERTANKO Publication (1999).GUNNER, T.J., "Physical Behavior of Crude Oil During Transportation and Its Impact on the Carriage of Crude Oil by Sea". *Marine Technology* (2002) vol.39, No.4, pp.256-265.

HUSAIN, M., APPLE, R., THOMPSON, G., and SHARPE, R., "Full-Scale Test of the American Underpressure System (AUPS) on Board the USNS *Shoshone.*" SNAME Northern California Section Meeting. (Sept. 2001).

IMO (International Maritime Organization) Subcommittee On Ship Design And Equipment, 37th Session, Agenda Item 9.2 – DE 37/Inf .8 - 23rd December, 1993: "Alternative arrangements under Marpol 73/78 regulation I/13G - Ultimate Strength Analysis of Deck Structure of a crude oil tanker with an application of an Underpressure method."

LIDE, D.R. and FREDERIKSE, H.P.R., Eds. CRC Handbook of Chemistry and Physics, CRC Press (1994).

MANSOUR, A.E. "Effect of Negative Pressure on Tanker Structure". In: MH SYSTEMS. *Concept Definition Documentation.* Submitted to US Department of Transportation, Maritime Administration, Washington, D.C (1996).

PENG, D.Y. and ROBINSON, D.B.: "A New Two Constant Equation of State", *Ind. Ing. Fund.* (1976) 15, 59.

PITTS, D. and SISSOM, L., Theory and Problems of Heat Transfer, 2nd Ed., McGraw-Hill NY (1998).

REDLICH, O. and KWONG, J.N.S.: "On the Thermodynamic of Solutions - An Equation of State. Fugacities of Gaseous Solutions", *Chem. Reviews* (1949) 44, 233.

SHTEPANI, E., WEINHARDT, B.E., and POTSCH, K.T.: "A New Modification of Cubic EOS Improves Prediction of Gas-Condensate Phase Behaviour"; 1996, SPE 36924, Europec '96, Milan, Italy, Vol. 2, p. 459 - 464.

SOAVE, G.: "Equilibrium Constants from a Modified Redlich-Kwong Equation of State", *Chem. Eng. Sci.* (1972) 27, 1197.

THOMAS, F.B., SHTEPANI, E., IMER, D., and BENNION, D.B.: "How Many Pseudo-Components Are Needed to Model Phase Behavior", *Journal of Canadian Petroleum Technology*, (Jan. 2002), Vol. 41, No. 1, pp. 48-54.

WALAS, S.M.: *Phase Equilibria in Chemical Engineering*, Butterworth Publishers (1985), 3.

ZUCROW, M.J. and HOFFMAN, J.D., *Gas Dynamics, V.1.* John Wiley and Sons (1976).

DISCUSSION

Norman W. Lemley, Member

The authors are to be commended for a number reasons:

- For a thorough and interesting presentation of their research results
- For addressing a means for reducing the adverse environmental and economic losses caused by VOC emissions. Reducing tanker VOC emissions is important environmentally and economically.
- For pursuing additional ways to mitigate oil spills resulting from tanker accidents.

A couple of questions come to mind. I look forward to the answers and hope that these issues can be addressed during the paper presentation.

It seems that negative pressures on crude oil may cause the cargo to evaporate and/or boil at a lower temperature. Will this disturb the equilibrium in the ullage space of the tank?

Could you please elaborate further on your control system? How does your system control the evaporation of the lighter element of the crude oil subject to negative pressures?

Continuing to improve ship and system design and to advance technology is key and an important goal for SNAME and its members. The well-being of the profession and the Society is dependent on members pushing the envelope in exploring the design options.

I applaud the authors for their inquisitiveness and for their perseverance.

John C. Daidola, Member

I would first like to commend Mr. Husain and his team for their persistence and commitment in the work to develop underpressure systems. Their research has provided substantial contributions to the technical literature increasing our understanding of a number of technical considerations, which have been identified in regard to such systems.

It appears that the identification of the emission containment potential of subatmospheric pressure in the ullage space of tankers has been a direct result of work carried out addressing mitigation of outflow during collision and grounding. The application would appear plausible with the current data presented by the authors. The authors methodically moved to full-scale testing while investigating cargo containment using underpressure, and this would appear to be a step to take in this current case somewhere in the future.

In fact, it would seem that as the underpressure approach, regardless of application, has not been implemented in operation, the most significant potential for its becoming a reality would be an application. Considering the number of years that underpressure has been investigated, what do the authors see as the requirements for moving ahead? Is there a plan in place for such activity?

In closing, I wish to thank the authors for their contribution and wish them success in the development of their systems.

Kjeld Aabo, Visitor

Due to limited time, it has not been possible to make a detailed evaluation of the thermal dynamic used, but the idea to limit the VOC emission with this technique seems very promising. The report convincingly describes the technical possibilities, and it must then be up to other reports to evaluate the economic and environmental impact of using this system.

David Hirst, Visitor

Thank you for the opportunity to review this paper, which I have read with great interest. I am not qualified to comment on the technical elements, but I welcome technology that enables reduction of VOC emissions. At present, this is not an area SEAaT is pursuing, although I suspect there are opportunities for emissions trading of VOCs (and thus lower cost reduction and simpler implementation) within the tanker industry.

Does the technology change the demand for inert gas? It does not look as though this demand is reduced much during loading and unloading, but primarily during the voyage. Are there possible developments that could minimize inert gas demand when loading or unloading and so reduce the need for scrubbing flue gases in port?

Authors' Closure

The authors thank the discussers for their comments and contribution to this paper.

Mr. Lemley:

1. Regarding outgassing or evaporation of crude oil under negative pressures, extensive experiments conducted at HYCAL laboratories of Calgary, Canada, prove this not to be the case. Once the equilibrium is attained after the loading of the cargo

and initial pull-down of the ullage pressure, it is maintained by keeping temperature and pressure in the ullage space as close to constant as possible. Under such conditions no additional evaporation or outgassing takes place.

2. The control system is designed with an objective to keep the ullage pressure constant. As stated in the answer to the previous question, this controls the evaporation. For the purpose of design of the control system, each tank is treated as a feedback loop. The flow of inert gas in and out of the ullage space is modeled as a simple first-order system (integrator). Under purely deterministic conditions, this system is stable for all control gains. Any mass transfer processes that may take place between liquid oil and vapors in the ullage space are treated as unmodeled dynamics. In our future research we plan to use methods of robust control theory to minimize the effects of unmodeled dynamics and maintain stable equilibrium in the system.

The authors certainly agree with *Dr. Daidola* that an application would be the most significant potential for the underpressure system becoming a reality. However, in order for this to take place, there is a need for rule-making change by the United States Coast Guard (USCG).

The USCG provided MH Systems a set of criteria, which must be met for such rule-making change. They will be addressed in the upcoming Technical Report to be issued by MH Systems under the conditions of a contract from the Office of Naval Research (ONR). This report will also incorporate a document, which MH Systems will submit to USCG.

Once USCG agrees that all its requirements are met and issues the required rule-making change, MH Systems, Inc., will proceed with marketing and implementation of the underpressure system.

The authors emphasize the full-scale test of the underpressure system under the aegis of the ONR and associated laboratory tests and analysis. This made it possible to substantiate the method of crude oil emission control from tankers by the underpressure method.

The authors appreciate *Mr. Aabo*'s comments.

Emissions trading of VOCs is an intriguing concept, and relative to this *Mr. Hirst* brings up an interesting point of cost reduction in the tanker industry vis-à-vis the reduction of VOC from the crude-oil-carrying tankers. Another point Mr. Hirst brings up is the issue of reduction of inert gas usage in our technology. The use of inert gas is primarily a safety issue and always has been. In the technology we described in our paper, we have attempted to optimize the volume of inert gas utilized without compromising safety.

A Quantitative Risk Assessment Model for Oil Tankers

Robert B. Cross (V), ABSG Consulting and **Jorge E. Ballesio** (V), Principal Engineer, American Bureau of Shipping

ABSTRACT

The use of risk based techniques applied to the classification of marine vessels and offshore structures is becoming more commonplace. Classification Societies are publishing Guides that the industry would follow to make use of risk assessment techniques to support the request for classification approval. The risk-based classification process can involve any of a wide range of risk assessment approaches, from simple qualitative tools to full quantitative risk models, depending on the extent and complexity of the new or alternative design to be classed. The use of tools such as quantitative models will significantly assist the industry in taking advantage of this innovative risk-based classification scheme. One of the developed tools is a comprehensive Quantitative Risk Assessment (QRA) model for an oil tanker. This model may be used to evaluate competing designs, the relative benefits of redundancy, and the impact of equipment unavailability during operations. This paper will present the basic design and methodology used to build the model, along with samples of results from the applications performed.

NOMENCLATURE

ABS – American Bureau of Shipping
ESD – Event Sequence Diagram
FSA – Formal Safety Assessment
HAZID – Hazard Identification
IACS – International Association of Classification Societies
IMO – International Maritime Organization
MGL – Multiple Greek Letter
NOAA - National Oceanic and Atmospheric Administration
POB – Personnel on board
QRA – Quantitative Risk Assessment
RAC – Risk Assessment Criteria
TAPS – Trans-Alaskan Pipeline Service
USCG – United States Coast Guard
USNRC – United States Nuclear Regulatory Commission

INTRODUCTION

Classification societies have been investigating the use of risk based techniques applied to the classification of marine vessels and offshore structures. Such an approach is particularly well suited in cases where designers are proposing new designs for which no applicable classification rules are published or designs that would benefit from alternatives to more prescriptive rules. A range of risk assessment techniques can be used to support classification requests by demonstrating that the proposed designs offer an equivalent level of safety to that of traditional designs or related industry standards.

This risk-based classification process can involve any of a wide range of risk assessment approaches, from simple qualitative tools to full quantitative risk models, depending on the extent and complexity of the new or alternative design to be classed. New tools are being developed that the industry will be able to use when making risk-based classification submittals. The use of these tools will significantly assist the industry in taking advantage of this innovative risk-based classification scheme.

One of the tools which has been developed is a comprehensive Quantitative Risk Assessment (QRA) model for an oil tanker. While this model has been developed as a generic tanker, it is expected that with modifications to reflect specific features, the model can be used by the industry to represent specific tanker vessels. Comparisons may then be made between competing designs and with alternate approaches to existing class rules, as described in this paper.

This paper contains background information on risk assessment and the history of applications in the maritime industry, describes the development of the tanker quantitative risk assessment model, gives a sample of the results obtained, and describes an application of the model.

RISK ASSESSMENT IN THE MARITIME INDUSTRY

Background

Though some isolated risk assessments were performed in the maritime industry, it wasn't until 1997 when a major move occurred towards such applications. It was in this year when the Maritime Safety Committee and the Marine Environmental Protection Committee of the International Maritime Organization (IMO) adopted the Interim Guidelines for the Application of Formal Safety Assessment (FSA) to the IMO Rule-Making Process (IMO 1997). In 2002, the guidelines were revised (IMO 2002). IMO developed these Guidelines because "FSA is a rational and systematic process for assessing risks associated with shipping activity and for evaluating the costs and benefits of IMO's options for reducing these risks". The application of FSA should lead to the development of "transparent" models to be reviewed by any number of relevant parties (i.e. regulatory bodies, operators, class societies, etc.) that lead to objective decisions on how to manage the risks. However, the lack of consistent modeling techniques and FSA applications can lead to a significant variance in results and create confusion. Inconsistencies may occur in proper and consistent hazard definition and identification, accurate determination of the variables affecting risk, determining when quantitative measures are sufficient, and obtaining consistency in risk measurement and defining criteria for acceptable risk.

FSA has varied applications, depending on the interests of the organization performing the assessment. Example applications include crew training, availability-reliability monitoring, evaluation of proposed changes, prioritization and allocation of manpower and resources, assessment of regulatory/class requirements. IMO has defined FSA as a five-step process as follows:

(1) identification of hazards;
(2) risk assessment;
(3) risk control options;
(4) cost benefit assessment; and
(5) recommendations for decision making.

Classification societies have been active participants in the IMO committees during the development of the FSA methodology. Currently, major societies are embracing the FSA approach in order to enhance the classification services to the maritime industry.

The development of the FSA for Bulk Carriers, carried out by an international team using the IMO as a technical forum, is the latest example of an application of risk assessment to the IMO regulatory process, with its repercussions in classification rules.

Basic Concepts in Risk Assessment

This section provides basic concepts on risk and reliability, and how they relate to risk assessment (IACS 2000).

What is Risk?

Risk is a measure of the likelihood that an undesirable scenario will occur together with a measure of the resulting consequences.

What is Reliability?

Reliability is a statistical prediction of desired performance over time. Reliability calculations may involve the development of analytical models combined with reliability data, primarily derived from maintenance and performance records.

What is the Relationship between Risk and Reliability?

Risk was defined as the product of frequency and consequence of undesirable events. If the event involves the unsuccessful performance of parts of the installation, then one of the contributors to the "frequency" of the event is the unreliability of such "parts". These "parts" may be:

- Machinery or systems (mechanical reliability)
- Structures (structural reliability), and/or
- People (human reliability).

What is Safety?

Safety is related to the degree of absence of risk. Because no activity is free of risk, an activity is considered safe when the level of risk is within acceptable limits.

What is Risk Assessment?

Risk assessment is an analytical technique that can successfully integrate diverse aspects of design and operation in order to assess risk. Risk assessment tools comprise an integrated array of techniques, including reliability, availability and maintainability engineering, statistics, decision theory, systems engineering and human behavior. In simple terms, risk assessment involves three key questions:

- What can go wrong? ⇨Scenario
- How likely is it? ⇨Likelihood
- What are the consequences? ⇨Consequence

Key aspects of a risk assessment

A risk assessment can be a complex project with special requirements of time and manpower. Depending on the objectives, they may range in scope from an analysis of a system, to a full risk assessment. The following aspects are key for the successful completion of a risk assessment:

- Completeness/Comprehensiveness
- Consistency
- Tractability/Documentation
- Credibility/Realism
- Efficiency

The following paragraphs describe each one of these key aspects.

Completeness/Comprehensiveness:

If a risk assessment is not comprehensive or complete, then the validity of the results are questionable. However, a risk assessment does not, and cannot, identify all of the potential events that might constitute risk contributors. But, through disciplined analysis and thorough review, it can be reasonably assured that a risk assessment identifies all of the probabilistically significant risk contributors.

Consistency:

Risk assessments involve considerable judgment, which depend on the objectives of the study and on the areas that need to be emphasized. Consistency in all assumptions, approximations and methodologies are essential for a credible risk assessment. It is worth stressing that one of the main values of risk assessment is the ranking or comparative values of risk contributors. If different parts of the risk assessment are developed inconsistently, such comparisons would be invalid.

Tractability/Documentation:

Each study should clearly indicate what is included and what is not, and any screening, assumptions, and approximations performed should be properly documented. The ability to retrace the steps taken is very important, not only for the review and use of the study, but also for any future updates.

Credibility/Realism:

It is very important that all assumptions made during the course of the study accurately reflect the design and characteristics of the vessel under study to ensure that the analysis is realistic. Assumptions or methodologies that are too conservative should be avoided if possible because conservatism may mask the real risks.

Efficiency:

Being that a risk assessment can be a complex activity, making adequate assumptions, simplifications and screening of low probability, low consequence scenarios will assure an efficient use of resources, concentrating in those areas that contribute the most to the risk, and which the analysis wants to emphasize.

It is also important to keep in mind what are the strengths and limitations of quantitative risk assessments. Generically, a risk assessment has the following main strengths:

- Provides a rigorous, systematic analysis tool
- Allows for information integration
- Allows for the consideration of complex interactions
- Supplies qualitative design insights
- Develops quantitative measures for decision making
- Provides structure for sensitivity studies

On the other hand, some of the most common limitations found with quantitative risk assessments are:

- Uncertainties in failure data
- Level of understanding of physical processes
- Analytical assumptions
- Modeling constraints
- Uncertainty in consequence assessment

It should be noted that a risk assessment would not reach the desirable results unless it is properly scoped and formulated at the beginning of the study.

THE CLASSIFICATION PROCESS AND RISK-BASED TOOLS

The classification process can be divided into three parts (Ballesio 2002):

- The classification rule development process
- The class assignment to vessels and offshore structures
- The maintenance of class during the life of the vessel or structure.

ABS is developing tools to apply risk assessment techniques to all three parts of the process. This paper however will concentrate in risk assessment applications to class assignment.

All major Classification Societies, are starting to offer means for owners to request specific alternatives to applicable prescriptive classification rules. The key factor for approval of such alternatives is to prove that the proposed designs are at least as safe as the equivalent "rule-compliant" design. Depending on the nature and extent of the alternative arrangement, this "proof of equivalency" can be of varying degrees of effort, and involving a wide range of risk and reliability techniques. While in most cases, it is expected that the owner's proposed alternatives to the rules will be limited to a few items, the risk-based classification path is open for any extent of proposed alternative, and two distinctive cases will be encountered as shown in Figure 1:

- Approval of Alternatives

This is the case when the proposed alternatives to the applicable prescriptive rules are isolated to a limited number of issues. In this case, the first step is to identify what is the intent of each of the specific rule requirements that are being complied with through an alternative arrangement. After such intents are identified, the risk-based classification procedure consists of proving that the proposed design complies with such intents, in some equivalent way. The nature and extent of the alternative arrangements, and the corresponding identified intents, will drive the type of analysis needed to achieve class assignment. The type of analysis needed for proving equivalency include hazard identification, change analysis, failure modes and effects analysis, fault trees, and event trees.

- Novel Designs

The proposed design does not comply with most of the applicable prescriptive rules, for example in the case of revolutionary or novel designs, where existing classification rules may just not be applicable. The classification approach may be based on an integrated risk assessment model of the design.

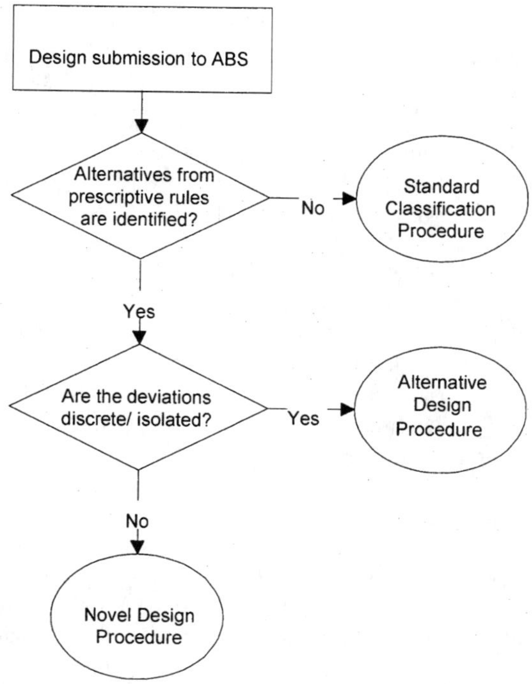

Figure 1. Risk-based Decision-Making for Classification

DEVELOPMENT OF THE TANKER RISK ASSESSMENT MODEL

Selection of Design and Mission

The development of the model started with defining the tanker overall design parameters, and the location(s) where the tanker was assumed to operate. The selection of these parameters is important because of the potentially significant influence on the results and they also impact how widely applicable the "generic" model is for use in applications.

Overall Design Parameters

The general philosophy used in model design parameter selection was to utilize a broadly applicable design based on the current and projected future tanker fleet. Based on industry trends and regulations, size and hull design were chosen as:

- Size – 100,000 – 150,000 DWT
- Hull – Double Hull

Many newbuild tankers fall within these parameters, and were therefore thought to be the most applicable looking forward to designs being considered by owners. A new trend in engine room design for some newbuild tankers is to build redundant engine rooms. This design feature is beyond current regulatory requirements, but was noted as an emerging trend in some areas. In order to be able to provide the maximum analytical capability for the model, it was decided to model the engine room using a redundant design, with the capability to "turn off" one of the engine rooms and therefore simulate a single engine room. One caveat used in this decision was that both cases (single and redundant engine rooms) must meet Classification requirements for their respective designs.

Assumed Mission

With the overall design selected, the next step was to select a mission for the tanker. The geographic area where a tanker operates influences risk through weather, duration and complexity of course in restricted waters, etc. Therefore, in selecting a mission, such factors need to be defined. A Trans-Alaskan Pipeline Service (TAPS) mission was chosen because operations in this area have a relatively high amount of data available. Specifically, the mission chosen for the model assumed is the following:

- Loading in Valdez
- Fully loaded transit to Washington (state) with a partial offload
- Transit to Long Beach to complete offload
- Transit to Valdez empty

In addition to the availability of data, this mission allows the analysis to examine all nominal loading conditions (fully loaded, partially loaded, empty) a tanker would experience.

Scope of Analysis – Hazards and Consequences

Consequence Selection

Hazards (through accident scenarios) and consequences are interrelated in that the potential consequence depends on the type of hazard. Therefore, a selection of consequences of interest to Class Societies and owners was performed in order to identify the appropriate hazards for the scope of the model. The consequences of interest were found to be:

- Personnel safety
- Environmental Damage
- Downtime
- Costs

Many hazards potentially lead to multiple consequences. For instance, an explosion could result in all of the above consequences. From the above potential consequences, personnel safety and environment were selected as the risk metrics to be analyzed. Tanker hazards vary significantly depending on the mode of operation and potential consequences. For instance, a contaminated bilge water discharge is a hazard since environmental damage could result, however, a drift grounding with subsequent severe weather would have much higher potential for environmental damage. For the scope of the model, the events considered of most interest were those with significant potential consequences.

Hazard Selection

In order to assess the hazards associated with the consequences selected, a Hazard Identification (HazId) exercise was performed with tanker and risk assessment experts. Using a top down approach, hazard categories were identified that could result in fatalities or environmental damage. For tankers, four critical functions were identified that are necessary for tankers to be "fit for purpose." Failure of any one of these functions was considered as a hazard category as described below:

- Buoyancy Failure – Any hazard that threatens the buoyancy or stability of the vessel. Examples include structural failure and internal flooding.
- Steering Failure – Any hazard or event that can lead to a steering failure such as loss of electric or hydraulic power.
- Propulsion Failure – Any hazard that threatens the propulsion capabilities of the vessel such as loss of electric power.
- Stationkeeping Failure – Any hazard such as mooring or anchor failure that impacts the ability to effectively keep stationed.

Categories of events that could lead to fatalities and/or possible environmental damage, without necessarily involving loss of one of the critical functions above were then identified. This review resulted in the following five additional hazard categories described below.

- External Hazards – Any hazard that is initiated outside the boundary of the vessel. Examples include severe weather, collision, etc.
- Personnel Hazards – Any hazard that impacts personnel safety, but is limited in potential threat to overall crew safety. Examples include maintenance incidents and dropped objects.

- Explosion – Any hazard or event that has the potential to lead to an explosive atmosphere, a release of stored kinetic energy, or disassembly of high energy equipment. Examples of these are failure of the inert gas system, overpressure of a hydraulic accumulator, or a diesel engine crankcase explosion.

- Fire – Any hazard or event that has the potential to lead to a fire such as fuel oil or other hydrocarbon release.
- Environmental Damage – Any hazard that leads to an environmental impact and does not threaten personnel safety or the overall functionality of the ship. An example would be a contaminated bilge water discharge.

The HazId then proceeded and identified causes for each of these hazard categories. For instance, loss of electrical power, fuel system failure, lube oil failure, etc. were identified as causes for loss of propulsion. A risk matrix was developed to qualitatively rank the causes in order that they could be prioritized for modeling. The matrix used was a 3x5, with three consequence levels and five frequency levels as shown in Figure 2.

Increasing Frequency ↑ \ Increasing Consequences →	A	B	C
	1	2	3
	1	2	3
	1	2	3
	1	2	3
	1	1	2

Figure 2. Risk Matrix Used to Rank Tanker Hazards

The matrix is consequence oriented in that an objective of the model was to focus on severe consequences rather than more frequent less severe consequences. Note that the first four frequency categories had the same consequence definitions associated with them. They could have been lumped together in one frequency category, however, more detailed frequencies were used to facilitate the possibility of comparing the QRA results with the original qualitative estimates. Each hazard cause was assigned a frequency (1-5) and consequence (A-C). The combination of frequency and consequence resulted in a Risk Assessment Criteria (RAC) from which the modeling was defined. The RAC values (1, 2, 3) are shown in the matrix in Figure 2. The frequency and consequence definitions in the matrix were as follows:

Frequency

1 – Event occurs at least once per year (approximately 20 round trips)
2 – Event occurs at least once in tanker life (approximately 500 round trips)
3 – Event occurs at least once in fleet of 10 tankers (approximately 250 tanker years)
4 – Event occurs at least once in life of current ABS classed tankers (approximately 15,000 tanker years)
5 – Event not expected to occur in current ABS tanker fleet

Consequence

A – No fatalities / minor injuries. Minor oil release (< 100 bbls)
B – Localized fatality / injury potential (not a threat to the crew as a whole). Moderate oil release (> 100 bbls but < 1 cargo tank)
C – Potential to cause numerous fatalities / injuries – evacuation may be required. Major oil release (> 1 cargo tank).

The modeling priorities were assigned after each cause was ranked based on frequency and worst case consequence in the risk matrix. Hazards with a RAC of 3 were given a priority for modeling because of the potential severity of their consequences. A RAC of 2 were considered a lower priority and were modeled as appropriate to be consistent with the objective of modeling the events with the most severe potential consequences. Events with an RAC of 1 were not considered for inclusion because of their low severity.

Development of Event Sequences

Event Sequence Diagrams

Development of the model proceeded by detailing the possible event sequences for each hazard category considered (e.g. loss of propulsion). In order to do this, Event Sequence Diagrams (ESDs) were employed. ESDs inductively consider "what happens next" in a series of possible events. By defining the critical events that can occur and change the course of the consequences, ESDs identify critical systems and events to model with fault trees, as well as the logic required to be used in event tree development. The following sections detail the development of the ESDs for the hazard categories evaluated and the hazard causes that were included in the model.

Loss of Propulsion and Steering

Events related to loss of propulsion and/or steering were the first to be addressed. Events considered under this category are given in Tables 1 and 2 from the HazId exercise.

Table 1. Loss of Propulsion Hazard Causes Considered for Tanker Model

Hazard Cause	Risk Assessment Criteria (RAC)	Modeled (Y/N)
Loss of electric power	2	Y
Loss of hydraulic power	2	Y
Fuel system failure	2	Y
Lube oil failure	2	Y
Control system failure	2	Y - Implicitly
Air system failure	2	Y
Mechanical failure	2	Y
Cooling system failure	2	Y
Human Error	2	Y

Table 2. Loss of Steering Hazard Causes Considered for Tanker Model

Hazard Cause	Risk Assessment Criteria (RAC)	Modeled (Y/N)
Loss of electric power	2	Y
Loss of hydraulic power	2	Y
Autopilot failure	1	N
Control system failure	2	Y – Implicitly
Steering gear failure	2	Y
Human error	3	Y

The ESD for a loss of propulsion and/or steering was constructed with the events in the above tables in mind. Other events, such as fires, may lead to a loss of propulsion and/or steering (e.g. a fire in the fuel oil system fails the main engine and results in a loss of propulsion) so the applicability of this event sequence was developed in a global sense. The event sequences for propulsion and steering considered the following questions:

- Is the grounding likely to be powered or drift?
- How much time is available before grounding?
- Is the failure able to be repaired before grounding?
- Are tugs available to prevent grounding?
- Is anchorage an option to prevent grounding?
- What type of bottom is experienced when grounding?
- Is evacuation required?
- Are lifeboats available?
- Is helicopter rescue available?

A simplified version of the loss of propulsion and steering ESD is shown in Figure 3.

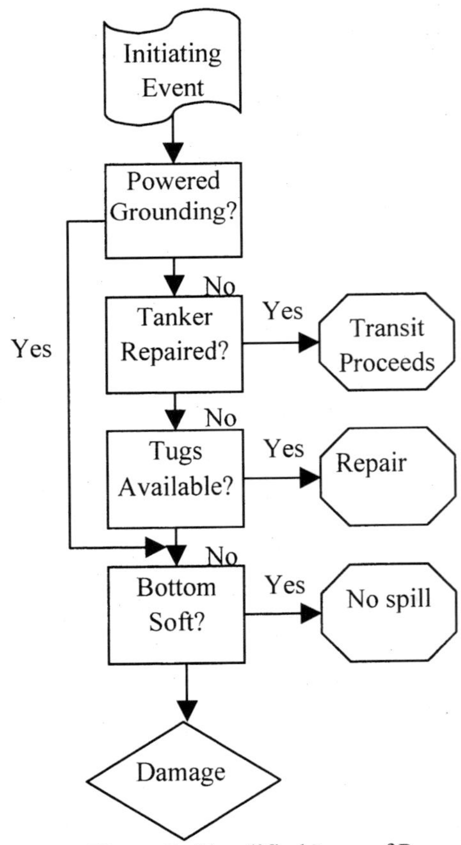

Figure 3. Simplified Loss of Propulsion and/or Steering Event Sequence Diagram

Fire

The fire events from the HazId exercise are shown in Table 3.

Table 3. Fire Hazard Causes Considered for Tanker Model

Hazard Cause	Risk Assessment Criteria (RAC)	Modeled (Y/N)
Accommodations	3	Y
Engine room	3	Y
Pump room	2	Y
Steering gear room	2	Y
Paint locker	2	N
Inert gas room	2	N
Bow thruster room	1	N
Battery room	2	Y
Deck (cargo block)	2	Y
Incinerator room	2	Y
Emergency diesel generator room	2	Y
Forecastle	1	N
Purifier room	3	Y

A "generic" ESD for fire events in Table 3 was developed to be able to evaluate fires in the various locations on the tanker. Questions considered in the development of the fire event sequences are:

- What is the source of the fire?
- Does the fire have the potential to escalate?
- Is the fire detected?
- Are there personnel in the area of the fire?
- Can portable suppression be used?
- Is fixed suppression available?
- Does the fire escalate?

The outcomes from these questions along with the specific location of the fire were used to determine what equipment or functions might be affected by the fire, if any fatalities may be expected, and if an evacuation was required. In some cases where loss of propulsion and/or steering was a result of a fire, the event sequences then consider the questions asked in the previous section.

Explosion

Explosions considered in the HazID and the modeling decision for each is shown in Table 4.

Table 4. Explosion Hazard Causes Considered for Tanker Model

Hazard Cause	Risk Assessment Criteria (RAC)	Modeled (Y/N)
Explosive atmosphere in cargo tanks	3	Y
Explosive atmosphere in ballast/void tanks	3	Y
Explosive atmosphere in non-classified areas	2	N
Explosive atmosphere on deck	2	N
Explosive atmosphere in pump room	3	Y
Explosive atmosphere in fuel tanks	2	N
Explosive atmosphere in battery room	1	N
Explosive atmosphere in oxygen / acetylene stores	2	N
Engine explosion	2	Y
Transformer / switchboard explosion	2	Y
Boiler explosion	2	N

A "generic" ESD was developed to define potential scenarios for cargo tank, ballast tank, and pump room explosions. The initiating event considered for these types of scenarios was assumed to be that an explosive atmosphere was present. The development of the ESD then proceeded by considering the following questions:

- Are there personnel in the area of the explosion?
- Is an ignition source present?
- Does structural damage occur?
- Is there a leak of hydrocarbons?
- Does the explosion cause flooding?

Like the fire event sequences, explosions were modeled to result in secondary effects such as fire, loss of propulsion, etc. In these instances, the scenarios considered the appropriate questions for the secondary events (e.g. fire).

Collision

Collisions as a result of human error were considered as initiating events by themselves with no detailed "causes" as with the previous events. The ESD development considered the following questions:

- Is the ship docking or underway?
- What part of the tanker is affected?
- Does the collision penetrate the double hull?
- Is there a leak of hydrocarbons?
- Is there seawater flooding?

The distinction was made between docking and underway collisions, since docking collisions are generally lower in energy and are less likely to penetrate a double hull. The part of the tanker affected was assumed to be random and is based on geometric considerations. A high-energy collision in the cargo block will cause a leakage of hydrocarbons and potential fire threat, while an impact in the engine room area may result in flooding and a loss of propulsion.

Structural Failure

A relatively simple model was used to estimate structural failure since machinery events were the primary focus of the model. HECSALV^{TM} (HEC 2000) calculations were performed at beginning of life and end of life of the tanker to estimate bending moments on the tanker. The end of life analyses considered the effects of maximum corrosion rates and very limited replenishment of steel. The conservative analysis showed that beginning of life of the tanker was very robust structurally. The calculated end of life bending moment was beyond yield strength of the deck plating. This result was true, however, only for a severe wave environment. The structural failure probability was then estimated using historical structural failure data, based on the North Pacific area.

Consequence Assignment

General Consequence Approach

As previously mentioned, the consequences of interest were selected as personnel safety and environmental damage. At the start of the model development, it was also noted that a second "level" of consequences existed that can occur before the ultimate consequences are realized. These "intermediate" consequences (e.g. loss of propulsion, loss of steering, etc.) are precursors to potentially significant events, and are generally based on design characteristics not environmental factors. Since, these consequences were considered when developing the ultimate consequence scenarios, the decision was made to include the ability to obtain results on both levels.

Intermediate Consequences

Intermediate consequences for the model are based on the design of the ship and not on external influences such as helicopter rescue, human navigation errors, etc. It is recognized, however, particularly for structure, that environmental conditions can have a large impact on reliability. The sets of intermediate consequences modeled by this study are:

- Total loss of propulsion and steering – This consequence is defined as the annual frequency of a concurrent loss of all propulsion and steering which includes a loss of normal shipboard power from the high voltage switchboards through the emergency switchboard.
- Total loss of the propulsion system – This consequence is defined as the failure of the propeller to turn. This event only involves loss of propulsion and not steering. For loss of propulsion cases with steering available, additional recovery time is assumed available relative to a total loss of propulsion and steering.
- Total loss of the steering system – This consequence is defined as a loss of the ability to control the rudder(s).
- Fires – This consequence is defined as the initiation of a fire in one of the areas identified in the HazId. Fire may have an immediate impact on personnel and could lead to loss of the critical functions above.
- Explosions – This consequence is defined as the initiation of an explosion. Like fires, explosions may have an immediate impact on personnel and could lead directly to environmental damage or to loss of the critical functions.
- Structural Failure – Structural failure as an intermediate consequence is defined as one that can put the vessel in an unsafe condition to operate or a through wall crack that can lead to cargo leakage.

Environmental Damage

The risk metric used to characterize environmental damage was barrels of oil spilled. A combination of factors was used to estimate the amount of oil spilled for each scenario. The software HECSALV^{TM} was used to calculate oil outflow from cargo tanks for various conditions. Key assumption used in the HECSALV^{TM} calculations were that a grounding event would result in a loss of containment midway between the waterline and the tanker bottom, or on the bottom with an equal likelihood. For collisions, it was assumed that the loss of containment would manifest itself at the waterline. The program accounts for hull configuration, location of break, static head of water, etc. to determine the expected oil outflow. Weather conditions were then accounted for in terms of escalating the expected damage. For example, Table 5 shows how some of these factors combine to result in different consequence severity.

Table 5. Estimated Average Environmental Consequences for Selected Grounding Scenarios

Scenario	Weather	Bottom	Spill Size (Barrels of Oil)
Drift Grounding, Fully Loaded	Calm	Mud	0
Drift Grounding, Fully Loaded	Calm	Rock	0
Drift Grounding, Fully Loaded	Moderate wave	Rock	30,000
Drift Grounding, Partially Loaded	Moderate wave	Rock	18,000

Fatalities

In terms of personnel safety, fatalities were chosen as the risk metric. Fatalities were estimated by considering two possibilities:

- Some events may result in immediate fatalities
- Evacuation (if required) may result in fatalities

In determining immediate fatalities, the type of event, location, and personnel manning was considered. To do this a Personnel On-Board (POB) distribution as shown in Table 6 was assumed.

Table 6. Assumed Personnel On-Board Distribution

	Cargo Deck	Accommodation (Including Bridge)	Engine Room Areas
Days	4	10	6
Nights	0	20	0
Average	2	15	3

Historical events were reviewed in order to determine the likelihood of experiencing immediate fatalities for events such as fire and explosion that are somewhat likely to lead to immediate fatalities. The historical data along with the assumed POB distribution were used to estimate the average number of immediate fatalities for specific events. Evacuation was assumed to be required for two conditions: fire escalation, and severe structural failure. For evacuation fatalities, a simple model based on Safety Case work done in the North Sea was used. This simple model considered lifeboat availability, liferafts availability, direct entry fatalities and helicopter rescue.

A simple event tree was made to estimate evacuation fatalities, which were then combined with immediate fatalities for each event. An example of this is shown in Table 7.

Table 7. Estimated Average Total Fatalities from Fires in the Accommodations Area

Immediate Fatalities	Evacuation Conditions	Evacuation Fatalities	Total Fatalities
0.75	Evacuation not required	0	0.75
0.75	Evacuation required, rescue successful	1.7	2.45
0.75	Evacuation Required, rescue fails	3.7	4.45

Analyses of Tanker Systems

The system analyses proceeded on a system by system basis. Systems required to be evaluated because of their critical functions (e.g. steering gear, firewater, etc.) were initially identified during the ESD development. Other systems that support the critical systems (e.g. electric power, seawater cooling, etc.) were also identified for evaluation as supporting systems. These supporting systems may be important, since often their failure may cause multiple other failures. For each system analysis that required a fault tree, several steps were involved:

Define the system boundary:

This step provides the definition of what equipment or components are assumed to be within the system analysis. Typically, systems were reviewed in terms of their drawings, manuals, etc. to find out what unique functions or impacts exist. In many cases, a component or components within a system defined on an engineering drawing may have unique impacts if they fail. For instance, the freshwater cooling loop consists of pumps, heat exchangers, and piping that if failed, would cause loss of the entire system and all downstream dependent functions. There are other components in the system like isolation valves for the ship's service diesel generator, for which failure would only cause loss of the diesel generator. In cases like this, it is generally required to model only the fresh water system components that would cause a loss of the entire function. The unique impacts such as the ship's service diesel generator isolation valve would be included in the system analysis where the impact is felt (i.e. the ship service diesel generator system).

Identify the system success criteria:

The QRA model event sequences are developed mainly based on successes and failures of events including the functionality of various systems as previously described. In many circumstances, a system can tolerate some failures and still perform its mission defined in the risk model. For instance, if a system has redundancy within it, part of the system may fail provided the redundant components work. Another example would be if the system was overdesigned to begin with, failures may leave the system functional enough to prevent a severe accident. In many cases, however, the whole system is required in order to meet its functional requirement. This step defines those requirements.

Identify the system interfaces:

System functionality generally depends not only on the system components within its boundary, but on other systems for support. The supporting functions required for functionality may be electric power, control, cooling, etc. This step identifies those interfaces, and they are used to develop conditional probabilities used in the event tree models. Another type of dependency evaluated at this step is whether any of the components are subject to common cause failures. A common cause failure is defined as a failure of two or more like components from the same cause (e.g. maintenance error, environmental factors, design flaw, etc.).

Develop the fault tree top event(s):

This step involves development of the fault tree logic. One or more top events may be required if the system is found to have unique impacts or is separated into trains.

Develop system conditional probabilities of failure:

Each system may have different probabilities of failure depending on the conditions of supporting systems. This step identifies the different conditions and develops the unique failure probabilities for each. For instance, the steering gear system has a different probability of failure with all switchboards available, than if the emergency switchboard was failed. The availability of the emergency switchboard alone does not fail the steering gear, but does affect the probability it will fail.

Data

Data Requirements

In order to populate the model for quantification, necessary data was obtained from a variety of sources. In general, the types of data required fell into three categories: equipment failure rates, external event data, and initiating event data.

Equipment Failure Rates

Equipment failure rates were obtained from several commercially available sources. In many instances, data was available for a specific piece of equipment in multiple sources. In those cases, often there was good relative agreement. In cases with large differences, knowledge of the sources and analyst experience were used to determine the most appropriate value. Common cause factors were also used extensively in the analysis. The Multiple Greek Letter (MGL) method was used (USNRC 1998) with the MGL data parameters obtained from existing commercial nuclear power data.

External Event Data

External event data used in the model consists of items primarily related to estimating the consequences of events such as weather, tug availability, helicopter rescue, etc. A variety of means were used to develop data for these types of events. For weather (wind and wave), National Oceanic and Atmospheric Administration (NOAA) data was used from buoys in locations relevant to the analyses. At each location wind and wave were correlated to capture the dependence of wave height on wind speed. Tug availability was estimated by reviewing factors such as tug locations versus time to grounding, weather, tug engine reliability, etc. These factors were then combined, and conditional probabilities developed as necessary for each specific location analyzed in the model. Some factors, such as bottom geology (sand or rock) where a tanker might ground was based on opinion after discussions with personnel familiar with the mission locations.

Initiating Event Data

Initiating event data was derived in two fashions. For machinery failure type events, fault trees were used with equipment failure. For human error initiators leading to grounding and collision, structural failure, and fire events, data was obtained from United States Coast Guard (USCG 2000) databases. For the human error initiators, only data from the ports assumed in the mission were used because it was felt that the human error rate would be heavily dependent on location. Specific initiating event frequencies were developed for each port and sea passage when required. Only historical data for tankers was used to develop the model data.

TANKER MODEL RESULTS

Results from the model may be obtained at various levels including:

- Overall consequence frequencies (loss of propulsion, drift grounding, fire, etc.)
- Scenario rankings
- System level importance
- Component level importance

Additionally, all of these levels of results may be applied to a particular phase of the mission or consequence. Included in this paper, are only samples of each of the types of results in order that the reader may understand the detail and complexity involved in the model. Presented here are the base case results for a single engine room tanker.

Single Engine Room Tanker Results

Results for the single engine room tanker were obtained using the starboard side engine room systems including steering gear. Some systems such as firewater, inert gas, etc. were assumed to be the same for each case.

Intermediate Consequence Results

The high-level intermediate consequence results from the baseline model are shown in Table 8.

These results show by far the most likely intermediate consequence is a loss of propulsion. The result is expected since the majority of equipment normally in use during transit is there to support propulsion. Explosions, a major concern for tankers, were shown to be the least frequent. This is because, in large part, to the fact that during transit tank operations are only infrequently performed. Each of the items in Table 8 may be broken down into individual contributors. Because machinery events have similar consequences, the events experiencing loss of propulsion and/or steering can be evaluated as a group. Table 9 lists the top scenarios for these events.

Table 8. Single Engine Tanker Estimated Frequencies of Intermediate Consequences

Consequence	Frequency (events per year)	One Event per (years)
Loss of propulsion	5.04E-1	1.98
Loss of steering	3.36E-2	29.8
Fire initiation	1.25E-2	80
Structural Failure	6.33E-3	158
Loss of propulsion and steering	5.28E-3	189
Explosion	9.37E-6	107,000

As expected, loss of propulsion is the consequences in all but one of the top scenarios. An examination of the results also shows that the major contribution to loss of propulsion comes from a variety of mechanical systems. If more scenarios were included, it would show that the loss of steering event is essentially all within the steering system itself. Losses of steering from supporting systems such as electrical power generally result in a concurrent loss of propulsion.

Ultimate Consequence Results – Environment

Ultimate consequence results related to the environment are shown in Tables 10 and 11. These two tables are included here to show that the frequencies and consequences are not necessarily related. Collisions were found to be the most frequent event that could lead to environmental damage during transit, however, they are only a small contributor to the expected risk. Sinking as a result of structural failure is a very low likelihood event, but because of the severity of the consequences, has a large impact to expected risk.

Table 9. Top 10 Intermediate Consequence Scenarios for Machinery Events

Rank	Scenario	Frequency (per yr)	Consequence
1	Main Engine Fails	1.19E-01	Loss of Propulsion
2	Freshwater Cooling Fails	2.54E-02	Loss of Propulsion
3	Lube Oil System Fails	1.70E-02	Loss of Propulsion
4	Fuel Oil System Fails	1.57E-02	Loss of Propulsion
5	Seawater Cooling System Fails	1.43E-02	Loss of Propulsion
6	Steering Gear System Fails	1.35E-02	Loss of Steering
7	Power Take Off Generator and Ship Service Diesel Generator Fail	9.92E-03	Loss of Propulsion
8	Main Engine and Ship Service Diesel Generator Fail, Ship Service Diesel Failure Has No Additional Effect	4.39E-03	Loss of Propulsion
9	Main Engine and Emergency Diesel Generator Fail, Emergency Diesel Failure Has No Additional Effect	4.36E-03	Loss of Propulsion
10	Failure of Both Sea Chest Intakes	1.65E-03	Loss of Propulsion

Table 10. Estimated Frequencies of Events Potentially Resulting in Environmental Damage

Event Type	Frequency (events per year)
Collision	3.15E-02
Powered Grounding	6.56E-03
Drift Grounding	4.45E-04
Structural Failure	1.76E-04
Explosion	9.32E-06
Total	3.87E-02

Table 11. Expected Risk for Environmental Damage by Event Type

Event Type	Expected Risk (barrels per year)
Structural Failure	6.96E+01
Powered Grounding	5.97E+01
Collision	5.87E+00
Drift Grounding	2.08E+00
Explosion	3.45E-01
Total	1.38E+02

Ultimate Consequence Results – Fatalities

Ultimate consequence results related to personnel safety are shown in Tables 12 and 13. For fatalities, it is seen that the expected risk again does not mirror event frequency for those events that results in fatalities. Fire is estimated to be the leading contributor to expected risk for fatalities, while explosions during transit are estimated to be very low contributors for the same reasons described in the previous section for environmental results.

Table 12. Estimated Frequencies of Events Potentially Resulting in Fatalities

Event Type	Frequency (events per year)
Collision	3.13E-02
Fire	1.22E-02
Structural Failure	1.76E-04
Explosion	9.33E-06
Total	4.37E-02

Table 13. Expected Risk of Fatalities by Event Type

Event Type	Frequency (fatalities per year)
Fire	2.29E-3
Structural Failure	8.78E-4
Collision	1.46E-4
Explosion	1.88E-5
Total	3.33E-3

APPLICATIONS

As previously mentioned, several applications are contemplated for the QRA models. Presented here is a high level comparison between the single engine room tanker and the twin engine room tanker. For the purposes of this comparison, only machinery failures are compared. The reason for this is that human error leading to collision/grounding are assumed to have the same contribution for each model. Also the ship structure is assumed to be the same for each. These events are dominant contributors to risk, and by excluding them a better comparison may be performed of the machinery contributions. The basic design features of each relevant to this comparison are shown in Table 14 below.

Table 14. Comparison of Single and Twin Engine Room Design Assumptions

Equipment	Single Engine Room Design	Twin Engine Room Design
Main engine	Single slow speed diesel	Twin slow speed diesel separated by bulkhead
Cooling / air auxiliaries	Single system with redundancy within system	Single system per engine room with redundancy within system
Lube / fuel oil	Single system with redundancy within system	Single system per engine room with redundancy within system
Generators	Power take off, ship service diesel in standby	Power take off, ship service diesel in standby for both engine rooms
Emergency power	Emergency switchboard with standby emergency diesel generator	Same as with single engine room, shared system
Steering gear	Single system with hydraulic and electric redundancy within system	Single system per engine room with hydraulic and electric redundancy within each system
Hull	Double hull	Double hull

Comparison of Intermediate Consequences

The major difference between the single and twin engine room designs is the redundancy of propulsion and steering equipment. The natural expectation then is that the likelihood of a loss of propulsion or steering is significantly reduced. Table 15 shows a comparison between the intermediate level of results for the single and twin engine room cases.

Table 15. Comparison of Intermediate Consequence Estimated Frequencies for Single Engine Room and Twin Engine Room Designs

Consequence	Single Engine Room Frequency (events per year)	Twin Engine Room Frequency (events per year)
Loss of propulsion	5.04E-1	1.12E-1
Loss of steering	3.36E-2	2.78E-3
Fire initiation	1.25E-2	1.91E-2
Loss of propulsion and steering	5.28E-3	2.13E-3
Explosion	9.37E-6	9.37E-6

From Table 15, it is seen that all intermediate consequence frequencies are reduced except fire. Fires are estimated to increase approximately 50 percent because of the redundancy of the engines. The amount of piping, flanges, tanks, etc. is increased with the redundant systems, and therefore there is a higher likelihood of a leak. Fires other than in the engine room (battery room, incinerator room, etc.) remain the same in both cases. The explosion frequency is the same since there is no redundancy as in the engine room. For loss of propulsion and/or steering a decrease in frequency of approximately 2.5 to 12 is estimated for the various conditions. The concurrent loss of propulsion and steering is reduced the least by the redundant engine room design. This result is because of the common emergency power design in the twin engine room case.

Using the model, more detailed comparisons may also be performed. Table 9 showed the top 10 scenarios for loss of propulsion and/or steering events for the single engine tanker case. Table 16 shows a similar list for the twin engine room case.

A comparison of the scenarios in Tables 9 and 16 shows that, in general, many of the same systems are in both the single engine room and twin engine room top ten scenarios, though at a reduced frequency for the twin engine room. A close look at the scenarios for the twin engine room case shows that all of the top scenarios contain redundant systems failing (e.g.

seawater cooling) and not diverse systems failing (e.g. port seawater cooling and starboard lube oil). This is a result of common cause failures being the dominant contributor to the scenarios. Common cause failures would include things like a common maintenance error on the port and starboard side for the seawater system, or both sea chests being clogged. Loss of propulsion and steering, however shows up in the twin engine room design top 10 scenarios as a result of the common reliance between the engine rooms on a single emergency power bus and diesel generator.

Comparison of Ultimate Consequences

The increased reliability shown in the comparison of intermediate consequences, also affects the ultimate consequences. Comparisons may also be done at this level. For simplicity, this paper compares the machinery events leading to a loss of propulsion and/or steering as discussed under the intermediate consequence results comparison. Loss of propulsion and/or steering was assumed to result in grounding scenarios for the model. Table 17 shows a comparison of estimated grounding frequencies based on machinery failures for the two cases.

From Table 17, it should first be noted that there is a significant difference between the powered grounding frequency for the single engine case with Table 10. Human error was found to be the biggest contributor to powered grounding and this cause was not included in Table 17. The drift grounding frequency in Table 17 is very close to that of Table 10 because events not included in the Table 17 results (e.g. fire, explosion, etc.) do not contribute significantly to the grounding frequency.

Also, in both cases shown in Table 17, the grounding frequency is significantly less than the intermediate event frequencies (e.g. loss of propulsion) that may lead to grounding. This difference is because the model allows for repair of equipment based on available time before grounding occurs, and external factors such as location, and whether a tug save is likely.

Table 16. Top 10 Twin Engine Room Intermediate Consequence Scenarios for Machinery Events

Rank	Scenario	Frequency (per yr)	Consequence
1	Total Loss Of Seawater Cooling	1.34E-2	Loss of Propulsion
2	Total Loss Of Freshwater Cooling	9.70E-3	Loss of Propulsion
3	Both Main Engines Fail	7.93E-3	Loss of Propulsion
4	Total Loss Of The Fuel Oil System	6.36E-3	Loss of Propulsion
5	Total Loss Of The Lube Oil System	6.20E-3	Loss of Propulsion
6	Failure Of Both Sea Chest Intakes	1.64E-3	Loss of Propulsion
7	Total Loss of Steering Gear	1.08E-3	Loss of Steering
8	Fire Initiation in the Accommodations	5.08E-4	Fire Initiation
9	Total Loss Of Seawater Cooling And The Emergency Diesel Generator Fails	4.93E-4	Loss of Propulsion and Steering
10	Total Loss Of Freshwater Cooling And The Emergency Diesel Generator Fails	3.56E-4	Loss of Propulsion and Steering

Table 17. A Comparison of Estimated Grounding Frequencies for Machinery Related Events

Event Type	Single Engine Room Frequency (per year)	Twin Engine Room Frequency (per year)
Powered Grounding	4.31E-5	5.64E-6
Drift Grounding	4.23E-4	9.49E-5
Total	4.66E-4	1.01E-4

A comparison between the single and twin engine room cases in Table 17 shows similar relative results to those of the intermediate consequences. For powered grounding, there is a reduction of approximately eight in terms of event frequency. This result is similar to that of the loss of steering case in the intermediate results. The drift grounding case shows a reduction factor of approximately 4.5 which is similar to the reduction in frequency of a loss of propulsion for the twin engine room case.

CONCLUSIONS

The oil tanker QRA model has been developed for use in the marine industry. This comprehensive QRA model is the first of its kind for an oil tanker, and will yield benefits for both the Class Society and tanker owners as a tool for evaluating risk tradeoffs, new designs, etc.

The results and comparisons in this paper are a small, high-level sampling of the results able to be obtained from the model. Results may also be obtained down to the component level (e.g. seawater pump, emergency diesel feeder breaker, etc.). It should be noted, however, that the baseline results presented are based on a specific set of design and mission assumptions used to develop the model, however when compared against industry statistics they show good overall agreement. Analyses for specific ships may be tailored by revising those assumptions. Ongoing evaluations include:

- Evaluation of Class Society Rules
- Comparison of single and redundant engine designs
- Specific analyses such as evaluating the change in risk from removing the emergency diesel generator from the redundant engine design.

This tool and the associated analyses are an integral part of the overall plan to move toward integrating risk processes into the maritime and offshore industries.

REFERENCES

BALLESIO, J. and DIETTRICH, D., Risk and reliability applications to marine classification, *Proc. European Safety and Reliability Conference, ESREL 2002, Lyon, March 2002.*

HEC, HECSALV 7.4.3, Herbert Engineering Corp., 2000.

IACS, International Association of Classification Societies, *FSA Training Course, Module 5*, August 2000.

IMO, International Maritime Organization, *Interim Guidelines for the Application of Formal Safety Assessment (FSA) to the IMO Rule-Making Process*, London, 1997.

IMO, International Maritime Organization, *Guidelines for Formal Safety Assessment (FSA) for use in the IMO Rule-Making Process*, MSC/Circ. 1023, MEPC/Circ. 392, London, 2002.

USCG, US Coast Guard Marine Safety Management System (marine casualty and pollution database), 2000.

USNRC, US Nuclear Regulatory Commission, *NUREG/CR-5485: Guidelines for Modeling Common Cause Failures in Probabilistic Risk Assessments, 1998.*

Obituaries

Herbert T. Anderson
Life Associate Member

Herbert T. Anderson, of Houston, Texas, passed away in 2003. He was born in Lincoln, Nebraska, on July 6, 1914, and attended the University of Nebraska.

He joined the Enterprise Engine Company, Inc., as a draftsman in 1942. The following year, he became a project engineer and applications engineer for Enterprise at New York and San Francisco. In 1946, he returned to New York as district manager and engineer.

He was later associated with marine engineering interests in Houston, Texas, and remained at Houston after his retirement.

Mr. Anderson joined the Society in 1947 as an Associate Member. He became a Life Associate Member in 1981.

Theodore W. Anderson
Affiliate

Captain Theodore W. Anderson, of Redmond, Washington, passed away in 2003. He was born in Alaska on November 6, 1920, and attended the University of Washington, where he studied engineering. He was employed by Standard Oil Company of California as an able seaman from 1939 and soon earned a license, first as a third officer and, by 1943, as a second officer. He also held an Alaskan pilot license.

From 1943 to 1947, he sailed for Pacific Tankers, Inc., as chief officer and master. Captain Anderson came ashore in 1947 as port captain, becoming general superintendent the following year. He later held various executive positions for Pacific Tankers and for affiliated companies, Joshua Hendy Corporation, and Hendy International Corporation.

He was vice president and director of Joshua Hendy Corporation from 1958 to 1969, and president and director of Hendy International Corporation from 1964 to November 30, 1977, at which time the Hendy interests were succeeded by West Coast Shipping Company. Captain Anderson then retired and acted as a marine and shipping consultant.

He was a Member of the Society's Council, representing Affiliates from 1978 to 1980.

Captain Anderson joined the Society as an Affiliate in 1972.

Samon Ando
Member

Mr. Samon Ando, of Dartmouth, Nova Scotia, passed away in 2003. He was born on July 26, 1947. He received a Bachelor of Science degree in Applied Mathematics and, in 1980, a Master of Applied Science degree from the University of Toronto. He then studied naval architecture at the Technical University of Nova Scotia and in 1985 received a Master of Science degree in that field from the University of California, Berkeley.

Since 1980, Mr. Ando had been employed in the Hydronautics Section of the Defense Research Establishment Atlantic in Nova Scotia.

Mr. Ando joined the Society as a Member in 1989.

Gunnar C. F. Asker
Member

Gunnar C. F. Asker, of New York, New York, died on April 28, 2003, at the age of 90. Born in Sweden on February 3, 1913, he came to the United States in 1937 on a Swedish Navy Scholarship after graduating from the Royal Institute of Technology in Stockholm as a naval architect and naval constructor. He was employed in the drafting room of the De Laval Turbine Company in Trenton, New Jersey.

During the World War II years, he returned to Sweden and served the Swedish Navy as a Chief Engineer of the flagship *Sverige*, engaged in neutrality patrols and convoy duty for shipping in Swedish territorial waters. He was also a staff engineer for the Stockholm Defense Flotilla. Later he was assigned to the Bureau of Ships of the Swedish Navy Department, where he served as a section chief of several Bureau of Ships sections concurrently and also taught Marine Engineering at the Swedish Naval Academy. He was also liaison engineer between the Swedish Navy Department and the War Production Board.

After the war, he was appointed assistant naval attaché for engineering duty at the Swedish Legation in Washington, DC, after having lectured at naval facilities in Sweden. He resigned his naval commission as commander and he became a naturalized citizen in 1950.

From 1949 to 1962, he worked as an engineer with Domestic Products, Inc., and became chairman and president of the corporation in 1959. He later was director and executive vice president of Cargocaire Engineering Corporation in New York. During this time, he introduced the Carl Munters driers and heat exchangers to the industry. He formed Air Economy Corporation with Erling Berner of Gothenburg and served as its chairman and president from 1964 to 1975, when he started Asker Enterprises, a consulting and research and development firm.

Mr. Asker held over 20 patents in such fields as driers, air conditioning systems, and heat exchangers, and he lectured around the world on the Jet-Sail wind assist systems for ship propulsion. He was the director and secretary of the American Society of Swedish Engineers.

Gunnar Asker became a Member of the Society in 1979.

Captain Jack E. Balkeman
Associate Member

Date of death unknown.

Robert C. Ballinger
Life Member

Robert C. Ballinger, of Annapolis, Maryland, and known to many as "Battleship Bob," passed away on May 18, 2003. He was born in Bowling Green, Kentucky, on January 18, 1921, and attended schools in Nashville, Tennessee, later moving to Memphis. He attended Vanderbilt University and the University of New Mexico, but his education was cut short when he enlisted in the U.S. Navy in 1942.

During his 27-year naval career, he rose from fireman to chief warrant officer, serving in World War II and the Korean and Vietnam Wars. He retired in 1969. A thoroughly practical, hands-on artisan, among his accomplishments in keeping the Navy's steam propulsion plants running was the development in 1965 of a method of early detection of pin-hole leaks in 1,200-psi boiler tubes. His naval activities were a reflection of his boyhood in Memphis, where he assisted his father, a steamfitter, in the maintenance of river steamboats.

After retirement, he was employed by George G. Sharp, Inc., for a number of years and later returned to the Navy attached to the Naval Sea Systems Command, retiring once again in 1992.

Mr. Ballinger joined the Society in 1973 as a Member. He became a Life Member in 1991.

Edgar H. Batcheller
Member

Rear Admiral Edgar H. Batcheller, USN (retired), of McLean, Virginia, died on May 3, 2003, the day after his 93rd birthday. He was born in Mattapoisett, Massachusetts, on May 2, 1910, graduated from the U.S. Naval Academy in the Class of 1934, and earned a Master of Science degree in Naval Architecture and Marine Engineering from Massachusetts Institute of Technology in 1939.

He experienced sea duty aboard USS *New Mexico* from 1934 to 1936. After the completion of his postgraduate work at MIT, his career as an engineering duty officer brought him to a wide variety of shoreside assignments of increasing responsibility in the field of naval vessel construction and repair, along with promotions in rank. He retired from the Navy in 1969 with the rank of rear admiral.

Admiral Batcheller met life head on, without the benefit of someone breaking trail or providing flank security. While on active duty, he dealt with the Great Depression of the 1930s, three wars, and tremendous technological change, all the while demonstrating discipline, thrift, imagination, wisdom, enthusiasm, and unfailing good humor.

Admiral Batcheller became a Member of the Society in 1947.

Lawrence J. Bates
Life Member

Lawrence J. Bates, of Huntington, New York, passed away at his home on February 11, 2003. He was born in Cleveland, Ohio, on October 18, 1921, and grew up on the shores of the Great Lakes, moving from Lake Erie to Lake Michigan. His schooling was in Chicago, and he later attended the University of Chicago, from which he received a Bachelor of Science degree in Mathematics in 1943. He then enlisted in the Navy, attended Columbia University, received his commission, and served aboard a minesweeper in the Pacific. After World War II, he received a Bachelor of Science degree in Naval Architecture and Marine Engineering from the University of Michigan, from which he graduated in 1949. His first employment was as Assistant Naval Architect at Calumet Shipyard and Dry Dock Company in South Chicago.

Mr. Bates joined the American Bureau of Shipping in 1951. After a decade at the ABS headquarters in New York, he was assigned to London, where he remained until 1975. During the latter years in London, he was principal surveyor for Western Europe. Returning to New York, he assumed the position of vice president of research and development and was later successively senior vice president and, in 1979, executive vice president. Retirement came in 1986 after a 35-year career with ABS.

He was a passionate small boat sailor, commencing in 1954 with the Thistle class. When he moved to the United Kingdom, he sailed the International 14-foot dinghies, then moved on to the Laser, Firefly, and Ensign classes. He was a member of several prestigious yacht clubs.

Mr. Bates joined the Society in 1948 as a Junior Member. He was later an Associate Member and Member. He became a Life Member in 1987.

Peter B. Bell
Associate Member

Date of death unknown.

Kai Hugh Bøggild
Member

Captain Kai Hugh Bøggild, of Halifax, Nova Scotia, died on November 28, 2002, at the age of 82. He was born of Danish parents in Haddiscoe (near Yarmouth), Norfolk, England, on November 3, 1920. He was raised on Nova Scotia's south shore, where he grew up loving the sea. As a youth, he would spend weeks at a time on fishing dories, where he learned to sail and navigate, sometimes receiving only cod livers as payment.

Captain Bøggild entered the Royal Canadian Navy in 1940. During the World War II, he served as the commander of the First Canadian Mine Sweeping Squadron, which worked off the Nova Scotia coast. He also served on a number of ships that escorted merchant shipping in the Atlantic.

In 1947 the Navy sent him to survey Canada's Western Arctic and correct existing admiralty charts. He was an officer aboard HMCS *Crescent*, a destroyer that rescued British diplomats along the Yangtze River during a period of civil unrest in China. During the Korean War, he was executive officer on the HMCS *Iroquois*, a Canadian destroyer. Other Navy appointments included the command of HMCS *St. Croix* and HMCS *Provider*. Later in his military career he came ashore and served as director of maritime operations at the National Defense Headquarters.

Captain Bøggild left the Navy in 1968 and began a career as a marine consultant, often acting as an expert witness in cases involving accidents. At this time he also worked as mooring master at Canaport, based in St. John, berthing VLCCs that called at that terminal.

Captain Bøggild became a Member in the Society in 1973.

Thomas Jay Breiwick
Member

Thomas Breiwick, of Edmonds, Washington, died tragically in December 2, 2003. While attending an industry exhibition in New Orleans, Mr. Breiwick was attacked near his hotel. Robbery was the motive for the attack.

He was born in Tacoma, Washington, on August 14, 1947. He received a Bachelor of Science degree in Mechanical Engineering from the University of Washington in 1969 and a Master of Science degree from the same institution in 1978.

Mr. Breiwick began his career working in the family business at Kolstrand Marine Supply Company in Ballard, while still in junior high school. His first professional employment after college was as a design engineer at the Mare Island Naval Shipyard, followed by four years with L. R. Glosten and Associates, Inc. He was later employed as a design engineer by Stacey Controls, Inc., and as vice president of engineering by B. F. Jensen and Associates, Inc. Still later, he was associated with Marine Efficiency Engineering and Polar Offshore Design.

In the late 1980s, Mr. Breiwick founded and was the president of Seattle-based Pacific Marine Equipment, Inc., a position he held until his passing.

Mr. Breiwick joined the Society as a Student Member in 1973. He became an Associate Member in 1974 and a Member in 1982.

Kwok Fai Chan
Member

Kwok Fai Chan, of Hong Kong, died during the summer of 2003. He was born on April 23, 1948, in China and attended the Hong Kong Technical College. In 1962, Mr. Chan apprenticed as an engineer at the Hong Kong and Whampoa Dock Company Ltd. and later sailed as an engineer for Worldwide Shipping Ltd. He later worked for San San Construction Company, and as a surveyor for Germanischer Lloyd.

Mr. Chan joined the Society as an Associate Member in 1975.

Poul A. Christensen
Member

Poul A. Christensen, of Mystic, Connecticut, died on July 13, 2003. He was born in Copenhagen, Denmark, on October 19, 1911. He was a 1937 graduate of the Royal Technical University, at Copenhagen, where he earned a Master of Science degree in Engineering. After a short tour in the Danish Navy, he was employed by Burmeister & Wain, serving in the engineering, testing and sales departments.

Mr. Christensen came to the United States about 1941, and was employed successively by Nordberg Manufacturing Company, F. L. Smith & Company and the American Locomotive Company, before returning to Burmeister & Wain in August 1945 as a consultant in New York. Burmeister & Wain formed its American subsidiary, Burmeister & Wain American Corporation, in late 1947, and Mr. Christensen became its president. He was later a consultant to Avco Lycoming, Inc., and president of Techtrade, Inc., before he retired in the mid-1980s.

He was honored in 1961 by the King of Denmark when he was presented the Knight Cross of the Order of the Dannebrog, a decoration for meritorious service to that nation first awarded in 1219. (The Dannebrog is the Danish national flag.)

Poul Christensen joined the Society as a Member in 1946.

Sabin Crocker, Jr.
Life Member

Sabin Crocker, Jr., was born in Philadelphia, Pennsylvania, on May 30, 1918. He died in 2003. Mr. Crocker attended the University of Michigan, where he earned his Bachelor of Science in Marine Engineering. While at the University, Sabin was employed at the Trenton Channel Power Plant of the Detroit Edison Company.

Amelio M. D'Arcangelo
Life Fellow

Professor Amelio M. D'Arcangelo, of Duluth, Georgia, passed away on March 2, 2003. He was born in Tres Arroyos, Argentina, on February 13, 1914. He graduated as an Ensign from the Argentine Naval Academy in 1937, earning a Bachelor of Science degree in Marine Engineering. He later attended Massachusetts Institute of Technology, receiving a Master of Science degree in Naval Architecture and Marine Engineering in 1941, and subsequently returned to Argentina.

While in the Argentine Navy, he held various positions, including three years as chief of the Hull Department of the naval shipyard at Puerto Belgrano, after which he was transferred to the Argentine Bureau of Ships. He served three years as chief of the Ships Technical Division and two years as chief of the Planning Division. Concurrently, he was professor of Naval Construction at the School of Engineering of the University of Buenos Aires. In 1951, he was assigned as chief naval architect of the Argentine Naval Mission to the Philadelphia Naval Shipyard, and a year later assumed the post of chief administrator of the Argentine Naval Shipyards. He retired from the Navy in 1953 as a commander.

In July 1953 he joined the faculty of Massachusetts Institute of Technology as assistant professor of Naval Architecture and Marine Engineering, remaining there until 1958. He then moved to Virginia Polytechnic Institute as professor of Mechanical Engineering. In 1964, another move brought him to Ann Arbor, Michigan, as a professor in the department of Naval Architecture and Marine Engineering of the University of Michigan.

It was at this time, after years of voluntary work with the National Academy of Sciences-National Research Council, that his definitive *A Guide to Sound Ship Structures* was published under the sponsorship of the Ship Structure Committee.

In 1983, he assumed the post of professor of Naval Architecture at the U.S. Merchant Marine Academy at Kings Point, where he remained until his retirement in 1991.

His activities in Society affairs included membership in the Committee on Scholarships (1969-1983), the Education Committee (1975-1980), the Hull Structure Committee (1978-1988), and Panel HS-4, Design Procedures and Philosophies (1978-1991). He also served on the Marine Technology Committee and as an Ad Hoc Visitor to the Accreditation Board for Engineering and Technology.

In 1991, Professor D'Arcangelo was awarded the William H. Webb Medal for his outstanding contributions to education in naval architecture. He was the second recipient of this prestigious award.

He had a passionate lifelong dedication to the education of future naval architects. In the introductory remarks made at the time of his acceptance of the Webb Medal, it was stated:

Amelio has had an impact on almost every education institution in our field in this country. He was at MIT on the faculty in the mid '50s, at VPI in the late '50s and early '60s, at Michigan for two decades, and after he retired from the University of Michigan he taught [at] the U. S. Merchant Marine Academy for another seven years. One job wasn't enough—therewas even one term at the time when he was also teaching at Webb Institute.

Professor Amelio D'Arcangelo joined the Society as a Member in 1953 at the start of his tenure at MIT, and later became a Life Member. He was elected a Fellow in 1977.

Conway Douglas Davis
Member

Conway Douglas Davis of St. Johns, New Brunswick, Canada, died on November 1, 2003.

Born in 1938 Mr. Davis received his BBME from the University of Illinois in Urbana. He worked as an engineering manager for several firms before going to work for Ingalls Shipbuilding Division of Litton Industries, where he served in several capacities, including vice president of engineering and director of advanced technology. In the early 1980s Mr. Davis was made vice president and manager of field offices of Designers and Planners.

John G. Davis
Member

John Gary Davis of North Bend, Oregon, died on February 13, 2003.

Born in 1931 in Crescent City, California, Captain Davis served in the Korean War. He was a commercial diver for Sause Bros. Ocean Towing and Coos Bay Pilots. He was a member of the Coast Guard Auxiliary, a project officer for District 13, a Flotilla 51 aid to the navigation officer, a Division 5 operations officer, and a Division 5 aviation officer.

Joseph A. Drelitcharz
Associate Member

Joseph A. Drelitcharz of West Springfield, Virginia, died in 2002.

Mr. Drelitcharz attended Michigan Technological University. He joined the Society in 1975 as a student.

Stanley G. Gawlik
Life Member

Stanley Gawlik died in 2003. Born in Nowy-Sacz, Poland, in 1924, he attended junior college in France and the naval academy in England.

Mr. Gawlik served in the Polish Navy in England from 1943 to 1947 and then in the British Navy. In 1957 he went to work at the David Taylor Model Basin.

Sal Giannoccolo
Member

Born in Maglis, Italy, in 1912, Sal Giannoccolo passed away in 2001.

Dr. Giannoccolo received his doctorate in naval architecture and marine engineering in 1937 from the Royal University in Genoa, Italy. He received his naval architecture license from the Italian Merchant Marine in 1946. Dr. Giannoccolo served in the Italian Royal Navy from 1938 to 1943. After the armistice, he fought with the Italian freedom movement and was entrusted with the running and operation of the Bagglietto Shipyard until 1946, when he went to work at the Costaguta Shipyard in Genoa. In 1947 Giannoccolo went to Venezuela, where he worked in the National Shipyards as chief of the Ship Designs and Projects Office.

William L. Giordan
Member

William L. Giordan, of Ashtabula, Ohio, passed away on January 3, 2002, at the age of 88. He was born in Ashtabula on December 10, 1913, and as an Ashtabula Harbor High School graduate in 1932, he began working for the Interstate Steamship Company (Jones & Laughlin) on the Great Lakes. After receiving his license as a marine engineer, he was employed in the fleet of the Wilson Transit Company.

During World War II, Mr. Giordan was commissioned as an officer in the U.S. Coast Guard, after which he returned to sailing in the Great Lakes fleet.

In 1968 he become a fleet engineer for Litton Industries, Kinsman Transit Company, and S.& E. Shipping Corporation, retiring in 1975 to enjoy a long period of life ashore.

Mr. Giordan became a Member of the Society in 1968.

Chester Gary Higgins
Member

Chester Gary Higgins, of Sturgeon Bay, Wisconsin, passed away on April 8, 2003. He was born in San Jose, California, on September 28, 1936, attended San Diego State University and San Diego City College, and served in the U.S. Marine Corps from 1953 to 1956.

He spent 30 years in shipbuilding management, working his way up to production control manager at National Steel & Shipbuilding Company, in San Diego. After a period as general manager at San Diego Marine Construction Corporation, he was vice president of the Marine Division of Campbell Industries, also located in San Diego. In 1977 he moved to Sturgeon Bay, where he was employed in various managerial positions at Peterson Builders, Inc., until his retirement in 1991.

Away from the shipyard, Higgins' real passion was classical music. Among his favorites was the beautiful, haunting aria "Mon coeur s'ouvre à ta voix" from Camille Saint-Saen's *Samson et Dalila*.

Gary Higgins became a Member of the Society in 1975.

Torgir E. Hinna
Life Member

Torger Hinna graduated from the Norwegian Technical High School. He worked at Kaldnes Mek Verksted as a designer, a chief draftsman, and a chief engineer. Mr. Hinna died in 2002.

Kazuo Hirata
Member

Mr. Kazuo Hirata, of São Paolo, Brazil, passed away on November 23, 2002. Born in São Paolo on October 10, 1951, Mr. Hirata graduated from Escola Politecnia of São Paolo University with a degree in Naval Architecture and Marine Engineering in 1975. He joined the Instituto de Pesquisas Tecnologicas (Institute for Technological Research) as naval architect in 1976. He received a Master of Science degree in Naval Engineering in 1984.

From 1989 he was the head of the hydrodynamics group, where he taught hydrodynamics at FATEC, the Brazilian faculty for inland construction and navigation.

Mr. Hirata became a Student Member of the Society in 1975 and advanced to Associate Member the following year.

George Hockley, Sr.
Member

George Hockley, Sr., died on January 10, 2003. Mr. Hockley studied at La Universidad De Las Americas in Mexico, Seattle Central Community College in Seattle, and Westlawn School of Yacht Design.

Mr. Hockley was employed as a designer and naval architect at various firms from the time of his graduation from Westlawn until the mid-1980s, when he went to work for Westinghouse Electric Corporation's Oceanic Division.

Arthur Louis Holderman
Life Member

Born in Lake Charles, Louisiana, in 1917, Mr. Holderman attended various navigation and cargo handling schools. The date of Mr. Holderman's death is unknown.

Mr. Holderman worked as a ship's master for Pope and Talbot and Alcoa from 1943 to 1947 and as a chief mate, relief mate, discharging mate, and ship's master. He was responsible for cargo loading and unloading and various safety initiatives. Besides his life membership with the Society of Naval Architects and Marine Engineers, Mr. Holderman was also a member of the Merchant Marine Officers Guild.

Thomas M. Hopkins
Member

Rear Admiral Thomas M. Hopkins, USN (retired), of McLean, Virginia, passed away on May 18, 2003. A native of Laurel, Maryland, he was born on February 3, 1927. He entered the Navy in 1944 as an apprentice seaman in the V-12 program at Cornell University and later transferred to the NROTC. He graduated from Cornell in 1948, as an ensign, with a Bachelor of Science degree in Mechanical Engineering.

From 1952 to 1955, he attended the Naval Construction and Engineering course at Massachusetts Institute of Technology, graduating with a Master of Science degree in Naval Architecture and Marine Engineering. His first assignment was as a ship superintendent and docking officer at Charleston Naval Shipyard. Promotions and advances in his naval career, which was devoted primarily to submarines, came regularly. In 1958, he was assigned to SUPSHIP Groton, where his duties included that of design project officer for *Triton* and the *Ethan Allen* class. Then came a tour at the Bureau of Ships Submarine Type Desk, followed by other billets culminating in a period as deputy commander for Ship Systems at the Naval Sea Systems Command. He retired from the Navy in 1982 with the rank of rear admiral. His decorations included the Legion of Merit and two Meritorious Service Medals.

After his retirement, Admiral Hopkins was a consultant in and around Washington, DC, specializing in ship survivability for the Navy, Maritime Administration, and commercial consulting firms. He was an adjunct research staff member of the Institute for Defense Analyses in support of the Department of Defense and held positions in several other similar boards and committees. In 1990 he was elected a member of the Board of Governors of the American Society for Testing and Materials (ASTM).

Admiral Hopkins joined the Society as an Associate Member in 1953 and became a Member in 1982. He served as a member of the Centennial Committee from 1988 to 1990.

Martin C. Kelsey, Jr.
Member

Martin C. Kelsey, Jr., was born in Dallas, Texas. He died in 2002 (date unknown). Martin Kelsey studied at Southern Methodist University in Dallas, where he received his Bachelor of Science degree in Geology. He then attended Miami University in Ohio, earning his MS in Geology and Geophyscis.

After graduation, Martin went to work for Atlantic Refining Company in Texas before moving on to Palmer Boats, Inc., where he began as vice president in charge of Sales and Services before being elected president.

Joseph F. Kemp
Associate Member

Joseph Francis Kemp, of Bel Air, Maryland, passed away on April 23, 2003. He was born on November 5, 1949. Mr. Kemp graduated from Hoboken (New Jersey) High School in 1968 and received a Bachelor of Science degree in Architectural Design and Landscape Architecture from Oklahoma State University in 1974.

After graduation, he went to work for Bethlehem Steel Corporation's Hoboken Yard. From 1985, he was employed by Tampa Shipyards, Inc., until the mid-1990s and later lived in Maryland.

Mr. Kemp became an Associate Member of the Society in 1975.

Robert N. Kershaw
Life Member

Robert N. Kershaw of Weymouth, Massachusetts, died January 3, 2003.

Born in Melrose, Massachusetts, Robert Kershaw graduated from Taber Academy in Marion and Bryant and Stratton Business School in Boston. Mr. Kershaw served in World War II and was a firefighter in Braintree. He was the owner of Robert N. Kershaw Marine Surveyors and Adjusters from 1955 until his retirement in 2001.

Bernard C. Lennon
Member

Bernard C. Lennon, Commander U.S. Navy, Ret. died September 13. 2003 at the age of 82.

Born in San Francisco, California, Mr. Lennon graduated from St. Ignatius High School in San Francisco and attended the University of California for 1 year before receiving his appointment to the U.S. Naval Academy, Annapolis. There he received his commission as an ensign in 1942 and went immediately into submarine service during World War II. After the war, the Navy sent Commander Lennon to Massachusetts Institute of Technology, where he received his degree in P.G. Engineering in 1949.

Mr. Lennon spent 20 years serving the Navy as planning officer supervisor of shipbuilding, in Portsmouth, New Hampshire, the Philadelphia Naval Shipyard, the New York Shipyard, and at the American Bureau of Shipbuilding offices in Washington, DC. After retirement, he went to work in the private sector of the New York Shipyard and J.J. Henry Shipbuilding Company for several years before going into private consulting for various shipbuilders.

Captain Walter A. Marshall USN (Retired)
Life Member

Walter Marshall died January 1, 2003. He was born in Barry, South Wales, in 1918. He attended Ile College, where he received a degree in Mathematics. He then attended the U.S. Naval Post Graduate School, where he received his diploma in Naval Architecture, a Bachelor of Science degree in Marine Engineering, and a Master's degree in Mechanical Engineering.

After a career in the Navy, which saw him stationed in the Philadelphia Naval Yard, Hunters Point, Montauk, San Francisco, San Diego, and Hawaii, he retired and went to work for Campbell Industries as the general manager of ship repair.

Archibald H. McComb, Jr.
Life Member

Archibald H. "Bud" McComb, Jr., of Winchester, Virginia, passed away on February 8, 2003. Born in New York, New York, on March 27, 1913, he received a Bachelor of Science degree in Naval Architecture and Marine Engineering from Webb Institute of Naval Architecture in 1937.

His first employment was with the American Bureau of Shipping, followed by a short time at the Navy's Experimental Model Basin as a junior naval architect.

In late 1938, Mr. McComb joined the Bureau of Marine Inspection and Navigation, later under the U.S. Coast Guard as merchant marine technical. He was commissioned as a lieutenant in the Coast Guard at the start of World War II, and remained in MMT until he retired as a captain in the late 1960s.

He then was employed by the Transportation Division of the American Petroleum Institute, from which he retired in 1975.

Mr. McComb's foremost contribution to the Society, and to maritime safety, was a paper presented at the 1950 Annual Meeting with coauthor Edwin E. Benzenberg. Although the subject of fires aboard ship had been a well-discussed topic since the *Morro Castle* disaster in 1933, the 1950 paper, "Stateroom Fire Test," provided valuable information about fire intensity and spread in a simulated passenger ship stateroom.

Mr. McComb joined the Society in 1938 as an Associate Member. He advanced to Member in 1944 and became a Life Member in the 1970s.

Joseph A. Melanson
Member

Joseph Melanson graduated from New York University with a Bachelor of Science degree in Mechanical Engineering followed by a degree in Naval Architecture from Stevens Institute Graduate School. His date of death is unknown.

His career encompassed journeyman for Consolidated Shipbuilding, naval architect for the U.S. Navy Department Bureau of Shipping, two years in the U.S. Army, marine engineer for Gibbs and Cox, naval architect for M. Rosenblatt & Son, nuclear engineer for Todd Shipyard, and naval architect for Sparkman and Stevens. In 1963 Melanson worked for the U.S. Navy as a naval architect from 1963 to 1997. Joseph Melanson died in 2003.

Kjell Melkvik
Associate Member

Mr. K. J. Melkvik of St. Thomas, Virgin Island, died July 16, 2002.

Mr. Melkvik joined the Society in 1998 as an Associate Member. Mr. Meklvik worked for the Caribbean Connections Company. His wife Jean writes that he was proud to be a member of SNAME and especially liked the sponsorship of young people that SNAME fosters so well.

George John Mercia USN (Retired)
Member

Captain George Mercia USN (retired) was born in Brooklyn, New York, in 1913 and died in 2003.

He attended Stevens Institute, where he studied mechanical engineering; New York University, where he studied thermodynamics; and Stevens again, to study ship construction. From 1941 to 1942 he received instruction from the U.S. Merchant Marine Academy in Marine Engineering before going to work for the U.S. Navy as a chief engineer. George was a marine surveyor for the U.S. Maritime Commission.

Robert Carter Morrell
Member

Robert Carter Morrell died in 2003.

Mr. Morrell was a graduate of the University of Michigan, where he received his BSE in Naval Architecture and Marine Engineering. After graduation, he went to work for Newport News Shipbuilding and Drydock Company until beginning his own naval architecture consulting firm, Robert W. and Robert C. Morrell in New York.

Hans A. Ording
Member

Hans A. Ording of Darien, Connecticut, died March 7, 2003.

Hans A. Ording was born in 1943 in Norway. He attended the Technical University of Norway, where he studied Naval Engineering. Afterward, he did research for Knut Langseth, the Royal Norwegian Navy, and the Technical University of Norway, where he also taught. He then moved on to a sales/application engineering position for the Gas Turbine Division of Kongsberg Vaapenfabrikk.

Sivert Överaas
Life Member

Silvert Överaas of Stabekk, Norway, died September 13, 2003.

Born in Alesund, Norway, in 1922, Sivert Överaas graduated from the Technical University of Norway at Trondheim with a degree in Naval Architecture and Marine Engineering. After a few years at AM Liaasen, he went to work for Det Norske Veritas.

Willard J. Pierson, Jr.
Member

Professor Willard J. Pierson, Jr., of West Hempstead, New York, passed away on June 7, 2003. He was born in New York City on July 7, 1922, attended Brothers College of Drew University, and was an aviation cadet (Meteorology) at the University of Chicago, receiving a Bachelor of Science degree in 1944. After wartime service as a commissioned radar meteorologist in the China-Burma-India Theatre, he entered New York University and received a PhD degree from its Department of Meteorology in 1949.

In 1949, he became an instructor in the NYU Department of Meteorology (later the Department of Meteorology and Oceanography) and achieved the position of professor in 1961. When the University closed its School of Engineering and Science in 1973, Professor Pierson moved to the University Institute of Oceanography at City College of New York (later City University of New York). The following year he became director of the Institute, and in 1976 became a professor at the CUNY Institute of Marine and Atmospheric Sciences. He retired from this position in 1992 and remained a research professor at CUNY's Remote Sensing Laboratory.

A widely published authority on oceanography, his Society activities included the presentation of two important annual meeting papers and membership in Panel H-7 (Seakeeping Characteristics).

Professor Pierson was the coauthor, with Manley St. Denis, of the landmark paper "On the Motions of Ships in Confused Seas," presented in 1953. An interesting conjunction of hydrodynamics, mathematics, oceanography, and philosophy, the paper was summarized by the authors in this way:

The representation of the oscillatory motions of a ship in confused seas by a response spectrum based on a statistical definition of the seaway is sufficient to give a characterization of these motions of such completeness that the naval architect may obtain practical solutions to many problems in this field which he now deems important.

It was a tall order, but the 1953 paper was the touchstone to the realization that naval architecture and oceanography are inseparable.

In 1968, Professor Pierson was one of the authors of the other significant annual meeting paper, "An Automated System for Optimum Ship Routing."

Professor Pierson became a Member of the Society in 1965.

Charles S. L. Robinson
50-Year Member

Charles S. L. Robinson of Hampton Falls, New Hampshire, died on November 21, 2002.

Mr. Robinson joined the Society in 1942 and was awarded the 50-Year Golden Member Certificate award in 1992.

Lester Rosenblatt
Past President, Honorary Member, Fellow

Lester Rosenblatt passed away peacefully on June 15, 2003, at his home in New York City at age 83. In 1947, with his father, Mandell, he founded M. Rosenblatt & Son, Inc., where he served as chairman, CEO, and naval architect. Through his perseverance, the company grew to be one of the world's largest and most prominent private engineering firms specializing in naval architecture and marine engineering.

He provided management and leadership to his company in the design and development of all types of naval and commercial ships, extending from aircraft carriers and submarines to tankers, tugs, and high-performance craft. Designs accomplished by his company included the *Spruance* Class Destroyers, *L.Y. Spear* Class Submarine Tenders, *Shimrit* Class Hydrofoils, T-AO187 Class Fleet Oilers, and T-AGS 39 Class Oceanographic Survey Ships, to name a few. Through its nationwide network of offices, M. Rosenblatt & Son, Inc., successfully completed over 27,000 assignments on such tasks as management and technical aspects of ship design, complete ship designs, including engineering analyses, conceptual, preliminary, contract, and detail designs, and installation and support services. Through his company, he successfully helped to establish standards in the field of naval architecture and marine engineering.

Through his personal efforts, M. Rosenblatt & Son, Inc., became a significant part of the U.S. naval and merchant ship design capability. With the strong belief that a strong navy, coast guard, and merchant marine are vital to the well-being of this nation, Lester Rosenblatt dedicated his career of over 59 years to excellence, integrity, technical proficiency, and innovation in ship design. The entire marine industry has been positively affected by him.

Mr. Rosenblatt earned undergraduate degrees from the University of Michigan and the City College of New York, and an honorary doctorate degree from Webb Institute. He took a personal leadership role in the industry, serving the Society of Naval Architects and Marine Engineers for many years. He was a member of the American Society of Naval Engineers (ASNE), an honorary trustee of Webb Institute, a member of the American Bureau of Shipping and Bureau Veritas, and honorary member of the Marine Society of the City of New York. Mr. Rosenblatt was an active participant on the National Academy of Sciences' Maritime Research Board's Committee on Innovation in the Maritime Industry.

In recognition of his accomplishments in ship design and marine engineering, he was honored in 1967 by his alma mater, receiving the University of Michigan's Sesquicentennial Award. In 1987 he was awarded ASNE's Harold E. Saunders Award for outstanding accomplishments in the field of naval engineering. In 1995 he was inducted into the International Maritime Hall of Fame and presented with the coveted Ship's Lookout Award for his dedication and leadership in the advancement of the marine industry.

He was known for his active support of our industry's professional societies. His lively discussions and comments on technical presentations at section meetings, symposiums, and other colloquia were highly respected. He was an ardent and strong advocate of the generation and dissemination of state-of-the-art technical information, as evidenced by his support of hundreds of employees' written papers on the subject of naval architecture and marine engineering.

Lester Rosenblatt's contributions to the Society of Naval Architects and Marine Engineers were many and varied. He was the author of a landmark paper at the 1960 Spring Meeting. *The Design of Modern Oceanographic Research Ships* was the textbook that turned the attention of the maritime world to these increasingly important vessels. In addition, the author's appreciation for the aesthetics of ship design was evident in the yachtlike profiles that illustrated the paper. This admirable trait of character was evident throughout his career.

With coauthors F. K. Serim and R. W. Goode, he displayed his continuing interest in oceanographic research a decade later, with a paper entitled "Ten Years of Progress in Research Ship Design: WHEO 701 Class Oceanographic Cutter," presented at the 1970 Spring Meeting.

In the New York Metropolitan Section, he was successively the membership chairman, secretary-treasurer, vice chairman, and chairman.

He was a member of the national Membership Committee from 1962 onward and served as the chairman of this important committee from 1966 to 1978. His other committee appointments included the Papers Committee in the early 1960s, the Long Range Planning Committee from 1983 to 1987, the Fellows Committee from 1984 to 2003, the Nominating Committee from 1985 to 1988, as chairman of the Audit Committee from 1986 to 1988, the Investments Committee from 1991 to 2003, and the Centennial Campaign Committee in 1993.

He was Member of Council, representing members and associate members from 1969 to 1974, and a member of the Executive Committee from 1972 to 1974. He was a vice president of the Society from 1974 to 1976, and served as its president in 1979 and 1980.

In 1984 he was awarded the Society's prestigious Vice Admiral "Jerry" Land Medal for his many outstanding accomplishments in the marine field.

Lester Rosenblatt joined the Society as a Junior Member in 1942. He transferred to Associate Member in 1948, became a Life Associate Member in 1960, and advanced to Life Member in 1962. He was elected Honorary Vice President for Life in 1975 and a Fellow in 1977, and received his 50-Year Certificate in 1992.

Many hundreds of people from near and far, including numerous dignitaries from government, industry, educational, and society organizations, paid their respects at his funeral services. This giant of our industry, who left his mark on the hearts of so many individuals with his professional expertise, his love of sailing and of his beautiful sailing yacht *Rosa II*, and his idiosyncratic sense of humor, will truly be missed.

CDR Robert A. Rourke USN (Retired)
Life Member

Robert A. Rourke of Fairfield, California, died October 3, 2004.

CDR Robert A. Rourke, USN (Retired) attended the University of California, where he earned his degree in Mechanical Engineering and Massachusetts Institute of Technology, where he earned his Master of Science degree in Naval Constructions and Engineering. He served in the U.S. Navy in such stations as Norfolk, San Francisco, Japan, Venezuela, and Long Beach.

Robert S. Sanford
Life Associate Member

Robert S. Sanford died May 31, 2003 at the age of 85.

Mr. Sanford was a native of Pittsburgh, Pennsylvania, and a U.S. Navy veteran who served in World War II. After working as vice president for Washington Aluminum Company in Maryland, he moved to Enterprise, Alabama, in 1970 and founded AlFab, Inc., a company specializing in metal fabrication for the marine industry.

Manley St. Denis
Honorary Member, Fellow

Professor Emeritus Dr. Manley St. Denis, an internationally renowned naval architect and ocean engineer, passed away in Honolulu, Hawaii, on October 28, 2003. He was born in Trieste, then in Austria-Hungary, on May 19, 1910.

His extensive education commenced with a Bachelor of Science degree in Naval Architecture and Marine Engineering in 1932. A two-year postgraduate course in mathematics followed at the University of Pennsylvania, after which he studied structures at Drexel Institute. In 1939, he received a Master of Science degree in Mathematics from the University of Hawaii and in 1956 was awarded a Doctorate in Aeronautical Engineering by Catholic University of America.

His experience in the nonacademic world began in 1932 with Victor Kondriaeff, Inc., in structural design. The following year he moved to the New York Shipbuilding Corporation, in Camden, New Jersey, where he was involved in the design of cruisers and destroyers. In 1936, he joined the Pearl Harbor Naval Shipyard as associate naval architect. (Dr. St. Denis used to enjoy telling the story of how he designed the only naval vessel ever built at that yard—the 1,000-ton tank vessel YO 43.)

He then left Hawaii and moved to Washington to become the first head of the Hull Scientific Section of the Bureau of Ships. A decade later he joined the David W. Taylor Model Basin. While at DTMB, his attention was drawn to seakeeping, and his work in this field led to much wider understanding of the dynamics of ship motions. His work included many assignments in government bodies, such as the Institute of Defense Analyses in 1959, the Center for Naval Analyses in 1961, and, from 1964, the National Engineering Science

Company, where he worked on Project Mohole. During this period, he became interested in seakeeping projects related to the offshore petroleum industry.

Academic positions during this long career included periods as lecturer at Catholic University in vibration (1950-1954) and aircraft stability and control (1954-1956), as adjunct professor of naval systems at Webb Institute of Naval Architecture, and as senior lecturer in naval systems (1968) at Massachusetts Institute of Technology. From 1968 to 1977, he was professor of ocean engineering at the University of Hawaii, becoming professor emeritus upon his retirement in 1977. He was NavSea Professor of Naval Systems Engineering at the U.S. Naval Academy from 1977 to 1978.

His participation in Society affairs included membership on the Hydrodynamics Committee, the Hull Structures Panel (S-3), the Ad Hoc Committee on Ship Slamming, and the Panel on Ocean Platforms and the Seakeeping Panel (H-7). He was chairman of Panel H-7 from 1978 to 1997. He also served on the Control Committees for Professor Korvin-Krakovsky's "Theory of Seakeeping" and for "Maneuvering and Seakeeping," by Captain H. E. Saunders.

Dr. St. Denis was a prolific author of notable SNAME annual meeting papers. His first, in 1948, with Richard B. Couch as coauthor, was "Comparison of Power Performances of Ten 600-Foot Tanker Hulls as Predicted from Model Tests." No less than 24 discussors commented on this important paper, which analyzed the comparative performance of 10 vessels designed at the very outset of the "supertanker" era. Rarely is the opportunity encountered in which 10 competitive designs can be so analyzed.

His next paper, written with Dr. G. Weinblum in 1950, was "On the Motion of Ships at Sea," a landmark work in which the introduction pointed out the "two great difficulties" in determining ship motion: First, finding simple analytic expressions for the surface of the sea, and second, "concluding how a vessel will respond to the struggle of the elements." Up until that time, little was available to the naval architect in that quest.

A complementary subject was covered by Dr. St. Denis in 1951, when he presented "On Sustained Sea Speed." He included a quotation from Joseph Conrad as a foreword to the paper: "The sea never changes, and its works, for all the talk of men, are wrapped in mystery." (That Conrad's words still apply—even with the great strides made by Manley St. Denis in understanding the mysteries of the sea—should never be forgotten.)

In 1953 came an even more significant landmark paper, when Dr. St. Denis and Professor Willard. J. Pierson, Jr., presented "On the Motion of Ships in Confused Seas." The 1953 paper provided further insight into how a vessel reacts in response to "the struggle of the elements." It may be said that in 1953 our understanding of Conrad's concerns was enhanced further. The presentation of the St. Denis-Pierson paper remains one of the major turning points in our understanding of the relationship between the sea and the ship. It was eminently outstanding during a period of impressive technological accomplishments—truly a golden age in naval architecture.

His last annual meeting paper came in 1966. "The Strike Aircraft Carrier: Considerations in the Selection of Her Size and Principal Design Characteristics" represented a change of pace, but was still presented with the same academic fidelity by which Dr. St. Denis was known.

In addition to these annual meeting papers, he contributed many other works to the Society in spring meetings, symposia, and T&R bulletins. His contributions to other organizations and technical meetings numbered several score. Nearly all related to the problems of seakeeping and allied topics, but one late paper (1979), presented at the Fifth Congress on Port and Ocean Engineering Under Arctic Conditions, was entitled "On the Origin of the Vikings."

In 1980, Dr. St. Denis was awarded the Davidson Medal, presented for outstanding accomplishment in ship research. It is of interest to note that nearly all of the previous 10 recipients of the Davidson Medal (including Dr. Davidson himself) were discussors of his notable annual meeting papers in the 1948-1953 period.

Dr. St. Denis joined the Society in 1936 as a Junior Member. He transferred to Associate Member in 1939, and to Member in 1950. He was elected an Honorary Member in 1977 and a Fellow in 1980, and received his 50-Year Certificate in 1986.

He was, in essence, much more than an eminent naval architect, hydrodynamicist, and ocean engineer. An accomplished linguist (fluent in four languages), he was not afraid to interject poetic or philosophic allusions into his engineering discussions. He once dreamed of being one of the great soccer players in Europe, was a foil fencing champion in New England, and after retirement earned a Master of Fine Arts degree from the University of Hawaii. Manley St. Denis was indeed a man of many parts.

Harvard P. Stewart
Life Member

Harvard "Bud" Stewart graduated with honors from the University of California, Berkeley, with a degree in Naval Architecture and Marine Engineering. His date of death is unknown at the time of this publication. Upon graduation he was commissioned an ensign in the U.S. naval reserve.

Mr. Stewart worked for 42 years for Bethlehem Steel Corporation in several divisions of the shipbuilding and repair departments. He began in the yard learning various crafts firsthand and later moved on to the administrative-management side, where he worked in the estimating and sales department.

During the war years, he served as a contract liaison officer between Bethlehem Steel and the U.S. Navy, Maritime Commission. After the war, he was head of Bethlehem's Ship Repair and Conversion Sales Engineering Department. Later he was assigned to be the yard's superintendent of planning and scheduling and chief of cost control. His time at Bethlehem Steel included many overseas assignments, including Iran, Chile, and Singapore. After retirement from Bethlehem, Mr. Stewart was a frequent consultant for Chevron and American President Lines.

Mr. Stewart was active in the Northern California Section where he was chair for a time. He contributed many technical papers to the Society, was awarded the distinguished service award, was elected a fellow in 1999, and was awarded the 50-year Golden Membership Certificate.

Joan Laurel Watts
Associate Member

Joan Laurel Watts of Churchtown, Maryland, died the week of June 9, 2003. Ms. Watts attended the University of Michigan and was award a Bachelor of Science degree in Engineering in December 1979.

Ms. Watts served on the SC-2 Sailing Committee from 1984 to 1986. Glen Ashe of the Chesapeake Section commented that Ms. Watts will be missed tremendously for her many talents and contributions to the fields of naval and marine engineering and that her coworkers will miss her talents and generosity exceedingly.

Ms. Watts joined the Society as a student in 1977.

William P. Welch
Member

William P. Welch, III, of Santa Cruz, California, died on September 16, 2003. It was his ninetieth birthday. Born in Selma, Alabama, he loved inventing as a boy, and as a mechanically inclined Eagle Scout, he was named "The Smartest Boy in Alabama." Young William won a trip to Thomas A. Edison's laboratory, where he met the inventor as well as Henry Ford and Harvey S. Firestone. This meeting left an indelible mark on this future engineer.

He received a Bachelor of Science degree in Mechanical Engineering from Georgia Institute of Technology in 1935 and a Master of Science degree in Engineering Mechanics from the University of Wisconsin in 1936.

After graduation, Mr. Welch entered the employ of Westinghouse Electric Corporation, where he stayed until 1941. During World War II, he worked in Washington, DC, as manager of the shock section of the electrical section of the Bureau of Ships.

He returned to Westinghouse in 1946 and devoted his entire career to the design of gears and turbines for large naval vessels, nuclear submarines, nuclear power plants, and large Japanese container vessels. He was a member of the design team for the aircraft carrier USS *Enterprise*, the nuclear submarines USS *Skipjack*, and USS *Nautilus*. His career as an engineer with Westinghouse, which took him from East Pittsburgh to Sunnyvale, spanned nearly half a century. He enjoyed a long retirement, which began on August 1, 1978.

At the age of 86, Mr. Welch became the oldest person on record to have hiked the 14-mile, 10,000-foot High Sierra Loop to Vogelsang in Yosemite National Park. He loved the outdoors and enjoyed swimming, canoeing, sailing, and many other activities.

Mr. Welch joined the Society in 1957 as a Member.

Raymond A. Yagle
Life Fellow

Professor Raymond A. Yagle, of Ann Arbor, Michigan, passed away on August 17, 2003. He was born in Aspinwall, Pennsylvania, on December 29, 1923.

He completed his undergraduate degree in naval architecture and marine engineering at the University of Michigan in 1944. After serving as an officer in the U.S. Navy, he returned to the University and earned a Masters of Science and Engineering degree in 1947. His initial appointment at the University came in 1949 as a research ;Associate in the Engineering Research Institute. He became a research engineer and lecturer in 1953 and an assistant professor in the Department of Engineering Mechanics in 1954. He taught for 35 years, progressing from associate to full professor in the Department of Naval Architecture and Marine Engineering. During that time, he taught a wide range of courses, some 17 in all, within the College of Engineering. Among those were six courses that he originated in computer applications or ocean engineering. He also served as the undergraduate program advisor for many years.

He was recognized as a leading authority on undergraduate education for the marine industry. Professor Yagle retired in 1994 but continued to contribute to the University's department in many ways.

He was the author of a 1966 Annual Meeting paper entitled "Naval Architectural Education Today." He was at that time correct, and in no small measure prescient, in concluding "that a field generally called ocean engineering or oceanographical engineering is evolving—the engineering problems associated with subterranean mining, undersea communication, the harvest of living resources or the gathering of knowledge concerning the oceans, about as many enterprises as might be included in 'land engineering'—and that many aspects of it are really naval architecture and marine engineering, as some, including the writer, view this engineering discipline today."

His other Society activities included longtime (1984-1993) membership on the Scholarships Committee. He was also a member of the Ship Design Committee and, from 1960, various panels and task groups. In 1961, he founded the Computer Panel under the Ship Technical Operations Committee.

Professor Yagle was elected a Junior Member of the Society in 1944, and later became an Associate Member. In 1959, he upgraded his membership to that of Member. He became a Life Member in 1989 and was elected a Life Fellow in 1994.

Memorial Resolutions

AMELIO M. D'ARCANGELO

The Society of Naval Architects and Marine Engineers unanimously adopted, at a meeting of its Executive Committee, the following memorial resolutions:

WHEREAS, The Society of Naval Architects and Marine Engineers and the maritime industry have suffered a great loss in the death of AMELIO M. D'ARCANGELO on March 2, 2003, a greatly respected, highly honored and deeply beloved Life Fellow of the Society, it desires to record these memorial resolutions.

Amelio M. D'Arcangelo first joined the Society as a Member in 1953 and later became a Life Member. He was elected to Fellow in 1977. His activities in Society affairs included membership on the Scholarships, Education, and Marine Technology Committees, as well as serving as an Ad Hoc Visitor to the Accreditation Board for Engineering and Technology. He was active in the Society's Technical and Research program serving on the Hull Structure Committee and on Panel HS-4, Design Procedures and Philosophies. In 1991, Professor D'Arcangelo was awarded the William H. Webb Medal, for his outstanding contributions to education in naval architecture. He was the second recipient of this prestigious award.

Born in 1914, in Tres Arroyos, Argentina, Professor D'Arcangelo graduated as an Ensign from the Argentine Naval Academy in 1937, earning a Bachelor of Science degree in Marine Engineering. He later attended Massachusetts Institute of Technology, receiving a Master of Science degree in Naval Architecture and Marine Engineering in 1941, and subsequently returned to Argentina.

While in the Argentine Navy, he held various positions, including three years as Chief of the Hull Department of the naval shipyard at Puerto Belgrano, after which he was transferred to the Argentine Bureau of Ships. He served three years as Chief of the Ships Technical Division and two years as Chief of the Planning Division, and concurrently was Professor of Naval Construction at the School of Engineering of the University of Buenos Aires. In 1951, he was assigned as Chief Naval Architect of the Argentine Naval Mission to the Philadelphia Naval Shipyard, and a year later assumed the post of Chief Administrator Argentine Naval Shipyards. He retired from the Navy in 1953 as a Commander.

In July 1953 he joined the faculty of Massachusetts Institute of Technology as Assistant Professor of Naval Architecture and Marine Engineering, remaining there until 1958. He then moved to Virginia Polytechnic Institute as Professor of Mechanical Engineering. In 1964, another move brought him to Ann Arbor, Michigan, as a Professor in the department of Naval Architecture and Marine Engineering of the University of Michigan. It was at this time, following years of voluntary work with the National Academy of Sciences-National Research Council, that his definitive "A Guide to Sound Ship Structures" was published under the sponsorship of the Ship Structure Committee. In 1983, he assumed the post of Professor of Naval Architecture at the U. S. Merchant Marine Academy at Kings Point, where he remained until his retirement in 1991.

In tribute to his long, distinguished, and invaluable service to the Society, the maritime industry and his fellow man, hereby be it

RESOLVED, That the Society of Naval Architects and Marine Engineers, and the maritime industry, have lost, in the death of AMELIO M. D'ARCANGELO, a beloved member of the Society, a distinguished, capable and fine gentleman and a true friend.

Further, be it

RESOLVED, That these resolutions be recorded in the TRANSACTIONS of the Society and that a copy be sent to his family as a token of the Society's deep regard for AMELIO M. D'ARCANGELO.

LESTER ROSENBLATT

The Society of Naval Architects and Marine Engineers unanimously adopted, at a meeting of its Executive Committee, the following memorial resolutions:

WHEREAS, The Society of Naval Architects and Marine Engineers and the maritime industry have suffered a great loss in the death of LESTER ROSENBLATT on June 15, 2003, a greatly respected, highly honored and deeply beloved Past President, Honorary Member and Fellow of the Society, it desires to record these memorial resolutions.

Lester Rosenblatt first joined the Society as a Junior Member in 1942 and transferred to Associate Member in 1948, having became a Life Associate Member in 1960 and advanced to Life Member in 1962. He was elected an Honorary Vice President for Life in 1975, elected to Fellow in 1977, elected President of the Society 1979-1980, elected an Honorary Member in 1981, and received his Fifty-Year Certificate in 1992. He served on the Society's Council, Executive Committee, the New York Metropolitan Section Offices of Secretary-Treasurer, Vice Chairman and Chairman, the national Membership, Papers, Long Range Planning, Fellows, Nominating, Audit, Investments Committees and the Centennial Campaign Committee. In 1984, recognizing his outstanding accomplishment in the marine field, he was awarded The Vice Admiral "Jerry" Land Medal.

Lester Rosenblatt's contributions to the Society were many and varied. He was the author of a landmark paper at the 1960 Spring Meeting."The Design of Modern Oceanographic Research Ships" was the text book that turned the attention of the maritime world to these increasingly important vessels. In addition, the author's appreciation for the aesthetics of ship design was evident in the yacht-like profiles that illustrated the paper. This admirable trait of character was evident throughout his career. This giant of our industry, who left his mark on the hearts of so many individuals with his professional expertise, his love of sailing and of his beautiful sailing yacht *Rosa II*, and his idiosyncratic sense of humor, will truly be missed.

Born in 1920, in New York City, Mr. Rosenblatt earned undergraduate degrees from the University of Michigan and the City College of New York, and an honorary doctorate degree from Webb Institute. In 1947, with his father, Mandell, he founded M. Rosenblatt & Son, Inc., where he served as Chairman, CEO and Naval Architect. Through his perseverance, the company grew to be one of the world's largest private engineering firms specializing in naval architecture and marine engineering. He provided management and leadership to his company in the design and development of all types of naval and commercial ships, extending from aircraft carriers and submarines to tankers, tugs and high performance craft. Designs accomplished by his company included the Spruance Class Destroyers, L.Y. Spear Class Submarine Tenders, Shimrit Class Hydrofoils, T-AO 187 Class Fleet Oilers, and T-AGS 39 Class Oceanographic Survey Ships, to name a few. M. Rosenblatt & Son, Inc., through its nationwide network of offices successfully completed over 27,000 assignments on such tasks as management and technical aspects of ship design, complete ship designs, including engineering analyses, conceptual, preliminary, contract and detail designs, and installation and support services. Through his personal efforts, M. Rosenblatt & Son, Inc., became a significant part of the United States of America's naval and merchant ship design capability. With the strong belief that a strong navy, coast guard and merchant marine are vital to the well being of this nation, Lester Rosenblatt dedicated his over fifty-nine year career to excellence, integrity, technical proficiency, and innovation in ship design.

Lester Rosenblatt's broad experience with the design, administrative, and financial activities of the marine industry was gained through leading one of the nation's most prominent naval architecture and marine engineering firms, and the entire marine industry has been positively affected by him. Through his company, he successfully helped to establish standards in the field of naval architecture and marine engineering.

In addition to his SNAME activities, Mr. Rosenblatt was a member of the American Society of Naval Engineers (ASNE); an Honorary Trustee of Webb Institute, a member of the American Bureau of Shipping and Bureau Veritas, and an Honorary Member of The Marine Society of the City of New York. Mr. Rosenblatt was an active participant on the National Academy of Sciences' Maritime Research Board's Committee on Innovation in the Maritime Industry.

In recognition of his accomplishments in ship design and marine engineering, he was honored in 1967 by his alma mater, receiving the University of Michigan's Sesquicentennial Award. In 1987 he was awarded ASNE's Harold E. Saunders Award for outstanding accomplishments in the field of naval engineering. In 1995 he was inducted into the International Maritime Hall of Fame and presented with the coveted Ship's Lookout Award for his dedication and leadership in the advancement of the marine industry.

In tribute to his long, distinguished, and invaluable service to the Society, the maritime industry and his fellow man, hereby be it

RESOLVED, That the Society of Naval Architects and Marine Engineers, and the maritime industry, have lost, in the death of LESTER ROSENBLATT, a beloved member of the Society, a distinguished, capable and fine gentleman and a true friend.

Further, be it

RESOLVED, That these resolutions be recorded in the TRANSACTIONS of the Society and that a copy, suitable engrossed, be sent to his family as a token of the Society's deep regard for LESTER ROSENBLATT.

MANLEY ST. DENIS

The Society of Naval Architects and Marine Engineers unanimously adopted, at a meeting of its Executive Committee, the following memorial resolutions:

WHEREAS, The Society of Naval Architects and Marine Engineers and the maritime industry have suffered a great loss in the death of MANLEY ST. DENIS on October 28, 2003, an Honorary Member and Life Fellow of the Society, it desires to record these memorial resolutions.

Dr. St. Denis first joined the Society in 1936 as a Junior Member. He transferred to Associate Member in 1939, and to Member in 1950. He was elected an Honorary Member in 1977, elected a Fellow in 1980, and received his Fifty-Year Certificate in 1986. His participation in Society affairs included membership on the Hydrodynamics Committee, the Hull Structures Panel (S-3), the Ad-Hoc Committee on Ship Slamming, the Panel on Ocean Platforms and the Seakeeping Panel (H-7) and was appointed Chairman of Panel H-7 in 1978. He also served on the Control Committees for Professor Korvin-Krakovsky's "Theory of Seakeeping" and for "Maneuvering and Seakeeping," by Captain H. E. Saunders. In 1980, recognizing his outstanding accomplishment in ship research, he was awarded the Davidson Medal.

Dr. St. Denis was a prolific author of notable SNAME Annual Meeting papers. His first, in 1948, with Richard. B. Couch as co-author, was "Comparison of Power Performances of Ten 600-Foot Tanker Hulls as Predicted from Model Tests." No less than twenty-four discussers commented upon this important paper, which analyzed the comparative performance of ten vessels designed at the very outset of the "supertanker" era. Rarely is the opportunity encountered in which ten competitive designs can be so analyzed. His next paper, written with Dr. G. Weinblum in 1950, was "On the Motion of Ships at Sea"—a landmark work in which the introduction pointed out the "two great difficulties" in determining ship motion: First, finding simple analytic expressions for the surface of the sea, and second, "concluding how a vessel will respond to the struggle of the elements." Up until that time, little was available to the naval architect in that quest.

In 1953 came an even more significant landmark paper, when Dr. St. Denis and Professor Willard. J. Pierson, Jr., presented "On the Motion of Ships in Confused Seas." The 1953 paper provided further insight into how a vessel reacts in response to "the struggle of the elements." It may be said that in 1953 our understanding of Conrad's concerns was enhanced further. The presentation of the St. Denis-Pierson paper remains one of the major turning points in our understanding of the relationship between the sea and the ship. It was eminently outstanding during a period of impressive technological accomplishments—truly a golden age in naval architecture.

Dr. St. Denis' last Annual Meeting paper came in 1966. "The Strike Aircraft Carrier: Considerations in the Selection of Her Size and Principal Design Characteristics" represented a change of pace, but still presented with the same academic fidelity by which Dr. St. Denis was known.

In addition to these Annual Meeting papers, he contributed many other works to the Society in spring meetings, symposia and T&R bulletins. His contributions to other organizations and technical meetings numbered several score. Nearly all related to the problems of seakeeping and allied topics, but one late paper (1979), presented at the Fifth Congress on Port and Ocean Engineering Under Arctic Conditions, was entitled "On the Origin of the Vikings."

Born in 1910 in Trieste, his extensive education commenced with a Bachelor of Science degree in Naval Architecture and Marine Engineering in 1932. A two-year post-graduate course in mathematics followed at the University of Pennsylvania, after which he studied structures at Drexel Institute. In 1939, he received a Master of Science degree in Mathematics from the University of Hawaii, and in 1956 was awarded a Doctorate in Aeronautical Engineering by Catholic University of America. His experience in the non-academic world began in 1932 with Victor Kondriaeff, Inc. in structural design. The following year he moved to the New York Shipbuilding Corporation, in Camden, New Jersey, where he was involved in the design of cruisers and destroyers. In 1936, he joined the Pearl Harbor Naval Shipyard as associate naval architect. He then left Hawaii and moved to Washington to become the first head of the Hull Scientific Section of the Bureau of Ships. A decade later he joined the David W. Taylor Model Basin. While at DTMB, his attention was drawn to seakeeping, and his work in this field led to much wider understanding of the dynamics of ship motions. Many assignments in government bodies—such as the Institute of Defense Analyses in 1959, the Center for Naval Analyses in 1961, and, from 1964, the National Engineering Science Company, where he worked on Project Mohole. During this period he became interested in seakeeping projects related to the offshore petroleum industry.

Academic positions during this long career included periods as Lecturer at Catholic University in vibration, 1950-1954, and aircraft stability and control, 1954-1956, as Adjunct Professor of Naval Systems at Webb Institute of Naval Architecture, and as Senior Lecturer in Naval Systems,1968, at M. I. T. From 1968 to 1977, he was Professor of Ocean Engineering at the University of Hawaii, becoming Professor Emeritus upon his retirement in 1977. He was NavSea Professor of Naval Systems Engineering, and the U. S. Naval Academy in 1977-1978.

In tribute to his long, distinguished, and invaluable service to the Society, the maritime industry and his fellow man, hereby be it

RESOLVED, That the Society of Naval Architects and Marine Engineers, and the maritime industry, have lost, in the death of MANLEY ST. DENIS, a beloved member of the Society, a distinguished, capable and fine gentleman and a true friend.

Further, be it

RESOLVED, That these resolutions be recorded in the TRANSACTIONS of the Society and that a copy be sent to his family as a token of the Society's deep regard for MANLEY ST. DENIS.

ISBN: 0-939773-45-7